TECTONOSTRATIGRAPHIC TERRANES OF THE CIRCUM-PACIFIC REGION

CONTENTS

PRINCIPLES AND APPLICATIONS OF TERRANE ANALYSIS

TECTONOSTRATIGRAPHIC TERRANES, PACIFIC NORTHEAST QUADRANT

TECTONOSTRATIGRAPHIC TERRANES, PACIFIC NORTHWEST QUADRANT

TECTONOSTRATIGRAPHIC TERRANES, PACIFIC SOUTHWEST QUADRANT

TECTONOSTRATIGRAPHIC TERRANES, PACIFIC SOUTHEAST QUADRANT

PREFACE

A topic that stimulates lively discussion among geologists is the concept that the circum-Pacific region, besides being a "ring of fire," is a rim of accretion. The fundamental geologic elements in this region of accretion are fault-bounded crustal bodies called terranes, sometimes referred to as tectonostratigraphic terranes. Besides carrying on the tradition of polysyllabic terminology, championed in the era of geosynclines, the modifier "tectonostratigraphic" specifies that a terrane is defined on the basis of stratigraphy and that its position and dimensions result from tectonic processes involving dislocations of tens to thousands of kilometers.

This volume is dedicated to the theme of terrane accretion throughout the circum-Pacific region. With the startling revelation that the present Pacific Ocean is younger than 200 m.y., one must try to understand the destruction of Panthalassa, the global ocean that surrounded the supercontinent of Pangea. The paradigm of plate tectonics provides subduction and midocean ridge spreading as complimentary processes in the destruction and creation of oceanic strata. A by-product of subduction is plutonism and the formation of volcanic arcs. These sialic massifs rarely remain autochthonous owing to backarc spreading or transcurrent slip along continental margins. The notion of terrane accretion, however, goes beyond plate tectonics; particularly involving collision and crustal buildup of allochthonous bodies. The accretion process seemingly controls mountain building and the growth of continents; we now believe terranes compose all Proterozoic and Phanerozoic foldbelts.

In September 1981, the Oji International Seminar on Accretion Tectonics was held in Tomakomai, Hokkaido, Japan. It was sponsored by the Japan Society for the Promotion of Science and the Fujihara Foundation of Science. The highlights of this important meeting are published in the 1983 proceedings volume: *Accretion Tectonics in the Circum-Pacific Regions*. At the conclusion of the "Oji I" meeting it was agreed that a follow-up conference was warranted. Thus, a second meeting for terrane aficionados was held at Stanford University during the week of August 28 to September 2, 1983. The conference was convened by D. G. Howell, D. L. Jones, Allan Cox, and Amos Nur, and was sponsored by the U.S. Geological Survey, Stanford University, the U. S. Department of Energy, and the Circum-Pacific Council for Energy and Mineral Resources. Following this meeting, a collection of 71 expanded abstracts was published as *Proceedings of the Circum-Pacific Terrane Conference*, edited by the above conveners.

The contents of this volume develop further the ideas presented in the aforementioned two volumes. The accompanying articles and the circum-Pacific terrane map demonstrate the extent and, in some instances, the possible significance vis a vis resource exploration of terranes in the circum-Pacific region. As we learn more about the distribution and composition of terranes as well as the process of accretion and dispersion, we should experience a parallel increase in our understanding of the resource potential of the Pacific region.

David G. Howell
U.S. Geological Survey

Principles and Applications of Terrane Analysis

Tectonostratigraphic Terranes of the Circum-Pacific Region

David G. Howell
David L. Jones
Elizabeth R. Schermer
U.S. Geological Survey
Menlo Park, California

Recent geological and geophysical studies have shown that much of the crust of the North American Cordillera has grown through the accretion of discrete tectonostratigraphic terranes. Further study suggests that terrane accretion has also occurred along the margin of most of the Pacific basin. We have explored the hypothesis of circum-Pacific terrane accretion by compiling a preliminary tectonostratigraphic terrane map of this vast region at a scale of 1:20,000,000. We present here a smaller scale version of the map, review the principles of terrane analysis, and discuss the characteristics of circum-Pacific terranes.

Tectonostratigraphic terranes are fault-bounded geologic entities of regional extent, each characterized by a geologic history distinct from that of neighboring terranes. Individual terranes can be classified into three types: (1) stratigraphic terranes, composed of coherent sequences that represent depositional environments of continental fragments, ocean or continental margin basins, and/or volcanic arcs; (2) disrupted terranes, characterized by blocks of heterogeneous lithology and age set in a matrix of foliated sandstone or serpentinite; and (3) metamorphic terranes, represented by structural blocks with a regional penetrative metamorphic fabric that obscures and is more distinctive than original lithotypes. Composite terranes consist of two or more terranes that amalgamated together prior to accretion onto a continental margin. Most of the terranes depicted on the map are "suspect" in that their origin elsewhere (allochthoneity) is surmised but not yet proved.

Terranes of the circum-Pacific vary enormously in size, from those too small to depict at 1:20,000,000, to large continental blocks. Many terranes are now disjunct, having been dismembered by postaccretionary strike-slip faulting, but unless a genetic linkage is unequivocal we take a conservative stance and treat the disjunct pieces as separate terranes.

Terrane boundaries are by definition faults or complex fault zones. Boundaries are commonly, but not necessarily, marked by blueschist or ophiolites. Details of the processes of accretion and amalgamation are still largely unknown. Thrust faulting seems to play an important role in accretion, but many thrust faults are later modified by folding and high-angle faulting. At many places around the Pacific, low-temperature metamorphism and melange formation accompany accretion.

Displacement between terranes need not be large but must be sufficient to juxtapose dissimilar rocks and disrupt original facies trends. Determining offset along faults has established displacements for some terranes on the order of a few hundred kilometers. Paleobiogeographic and paleomagnetic study of other terranes has shown more than 6,000 km (3,700 mi) of displacement.

The distribution and nature of oceanic plateaus in the modern Pacific basin depicted on the terrane map give insight into the accretion process. Oceanic plateaus comprise fragments of continents, oceanic islands, hotspot tracks, remnant arcs, and anomalously thick volcano-plutonic piles of unknown origin. The presence of these same lithotectonic elements in onshore accreted terranes, often dismembered and detached from their basement, suggests that continental growth and orogeny occur as a result of collision of oceanic plateaus at continental margins. Postaccretionary consolidation of terranes may be even more important than initial accretionary events in effecting deformation in a cordillera and its foreland.

The growth and shaping of continents may be viewed as a result of both terrane accretion and terrane dispersion. While the accretion of terranes results in continental growth or outbuilding, the dispersion of terranes, either rifting or sliding, results in the diminution of continents.

We discuss the stratigraphic and tectonic history of some of the relatively well-studied terranes from around the Pacific, including terranes in Alaska, British Columbia, the western conterminous United States, South America, Siberia, China, Australia, and New Zealand.

INTRODUCTION

In recent years, the concept that much of the crust of the North American Cordillera has formed through accretion of separate tectonostratigraphic terranes has gained support as new geologic, geophysical, and paleontologic data have become available (Jones et al, 1981; Jones et al, 1983; Coney et al, 1980; Davis et al, 1978; Blake et al, 1982a). This same process of accretionary growth seems to have affected much of the circum-Pacific margin. To explore this hypothesis, we have compiled a preliminary circum-Pacific terrane map. The purposes of this paper are to review the concepts of terrane analysis as stated in Coney et al (1980) and Jones et al (1983) and to present some examples of different types of terranes from the margins of the Pacific basin. This report presents an experimental interpretation that is designed specifically to elicit a dialogue concerning the tectonic evolution of this complex region. We have compiled the terrane map, 1:20,000,000 scale, on the AAPG Circum-Pacific Map Series base (Jones et al, 1982a; Howell et al, 1983). The map (Fig. 1) is folded into a pocket located on the inside back cover of this book. Terranes not discussed in detail within this can be located on the map and are described in the Appendix section that concludes this paper.

PRINCIPLES OF TERRANE ANALYSIS

Terrane Definition

A tectonostratigraphic terrane is a fault-bounded package of rocks of regional extent characterized by a geologic history which differs from that of neighboring terranes. Terranes may be characterized internally by a distinctive stratigraphy, but in some cases a metamorphic or tectonic overprint is the most distinctive characteristic. In cases where juxtaposed terranes possess coeval strata, one must demonstrate different and unrelated geologic histories as well as the absence of intermediate lithofacies that might link the two terranes. In general, the basic characteristic of terranes is that the present spatial relations are not compatible with the inferred geologic histories. As additional geologic, paleontologic, and geophysical data are accumulated, terrane boundaries and classifications can be modified and their nomenclature revised. Terranes can be grouped into three categories: stratigraphic, disrupted, and metamorphic; examples of each are discussed below. Subdivisions of these categories are differentiated by color on the map (Fig. 1).

(1) *Stratigraphic terranes.* These terranes are characterized by coherent sequences of strata in which depositional relations between successive lithologic units can be demonstrated. Basement rocks may or may not be preserved. Sequences within stratigraphic terranes may be subdivided into three broad categories; the complex geologic histories of some terranes comprise more than one of these categories:

a. Fragments of continents: These terranes are characterized by the presence of a generally Precambrian continental basement with an overlying sequence of shallow-water sedimentary rocks of Paleozoic and Mesozoic ages. Also included are sedimentary rocks of continental derivation that are detached from their basement substratum. Examples include the Nixon Fork terrane (NXF) of Alaska, the Tujunga terrane (TUJ) of southern California, the Omolon terrane (OMO) of Siberia, and the Tuhua terrane (TUH) of New Zealand.

b. Fragments of ocean basins: These terranes are characterized by sequences of mafic and ultramafic rocks characteristic of oceanic crust with overlying deep-sea sedimentary deposits, for example the Del Puerto terrane (DPO) of California, the Chulitna terrane (CHU) of Alaska, the South Anyui terrane (SAY) of Siberia, and the Kunlun (KUN) terrane of China. Each has ophiolitic basement and deep-sea sediments, yet younger strata in each indicate continental-margin depositional environments. These progradational sequences reflect translational mobility characteristic of many terranes. Also included are deep-sea deposits that are detached from their basement substratum, such as the Pingston terrane (PMW) of Alaska, and fragments of sea mounts and oceanic plateaus.

c. Fragments of volcanic arcs: These terranes are composed dominantly of volcanic rocks, or the plutonic roots of arcs, and sedimentary debris derived from volcanoes that are similar in composition to rocks of presently active volcanic arcs such as the Aleutians. Examples include the Stikine terrane (STK) of British Columbia, the Peninsular terrane (PEN) of Alaska, the Salinia terrane (SAL) of California, and the Hokonui terrane (HOK) of New Zealand. In many areas these terranes may be difficult to distinguish from oceanic terranes without detailed geochemical and geological data, for example in the Philippines and in the Qinling (QIN) terrane of China.

d. Fragments of continental margin basins: These terranes are composed of shallow to deep marine graywacke and are dominantly submarine fan lithofacies consisting of quartzofeldspathic debris shed from a continent. The strata composing these terranes are generally coherent but are often associated with disrupted terranes that commonly include pillow basalt, chert, and pelagic strata. Examples include the Torlesse terrane (TOR) of New Zealand, the Shimanto terrane (SHM) of Japan, and the Chugach terrane (CGH) of southern Alaska.

(2) *Disrupted terranes.* These terranes are characterized by blocks of heterogeneous lithology and age, usually set in a matrix of foliated shale, sandstone, or serpentinite. Most of these terranes contain fragments of ophiolitic rocks, blocks of shallow-water limestone, deep-water chert, and packages of graywacke with lenses of conglomerate; in addition, many disrupted terranes contain blueschist facies rocks both as exotic blocks or as a regional metamorphic overprint. Some disrupted terranes have been interpreted as fragments of subduction complexes, though alternative models are possible. Many disrupted terranes are intimately associated with continental margin strata terranes. Examples include parts of the Chugach terrane of Alaska, the Central terrane (CEN) of the Franciscan assemblage in northern California, and parts of the Cache Creek terrane (CCK) of British Columbia.

(3) *Metamorphic terranes.* These terranes are characterized by a regional, terrane-wide penetrative metamorphic fabric and development of metamorphic minerals to such a degree that original stratigraphic

features and relations are obscured. In addition to metamorphic differences, protolithic contrasts with adjoining terranes also must be demonstrable. Examples include the Yukon–Tanana terrane (YKT) of Alaska and the Badly terrane (BDY) of California.

Terrane Agglomeration

Two similar but unrelated events in the history of terranes should be distinguished. An early event, referred to as *amalgamation*, is the joining together of separate terranes to form a composite terrane prior to the addition of the amalgamated terrane to a continental margin. The later event, termed *accretion*, is the collision and welding of a terrane (either composite or individual) to the continent. These events may be widely separated in time, or may closely follow one another. Timing of amalgamation and accretion can be established by three main criteria: (1) overlap assemblages that depositionally overlie two distinctive, juxtaposed stratigraphic sequences (e.g., the Gravina–Nutzotin belt of southeastern Alaska that links Wrangellia [WRN] to the Alexander composite terrane [ALX], and the Nohi Rhyolite of Japan that links the Hida [HID] and Mino [MIN] terranes); (2) sudden appearance of detritus in one terrane derived from a dissimilar neighbor (e.g., in California, granite boulders in paralic facies of the Stanley Mountain terrane [SIM] indicate a Late Cretaceous suturing to the Salinia terrane, Cache Creek terrane debris in the Bowser Basin deposits of British Columbia demonstrate linkage to the Stikine terrane, and in southwest China Late Devonian clastic rocks in the Hunan–Jianxi terrane [HNJ] show provenance linking to the early Paleozoic volcanic island arc system of the Cathaysian foldbelt terrane [CTY]); (3) welding together of unlike sequences by intrusions (e.g., 55- to 60-m.y.-old granitic plutons that stitch all the terranes between the Denali and Border Ranges faults of southern Alaska).

Terranes composed of two or more distinct parts that became amalgamated and subsequently shared a common geologic history prior to their accretion are called *composite terranes*. Examples of these amalgamations include arc-arc amalgams (Alexander–Wrangellia), and arc-continental-oceanic-disrupted amalgams (Salinia–Tujunga–Stanley Mountain–San Simeon terrane amalgam of southern California, also called the Santa Lucia–Orocopia allochthon). The Tujunga terrane by itself is a composite with basement rocks comprising at least three distinct Precambrian terranes, and the Alexander terrane comprises the pre-Triassic Craig, Admiralty, and Annette terranes. Many of the terranes in the western Pacific are probably also composite, but we cannot distinguish the individual component terranes because of the paucity of literature in English translation describing the detailed geology of those areas.

By means of a flow chart we can schematically depict successive amalgamation stages of a variety of terranes. Figures 2 and 3 demonstrate the suturing sequence for some of the terranes in Alaska and in southern California, and Figure 4 depicts suturing events in Siberia. In all instances episodes of accretion and amalgamation are inferred from the geologic relations of overlap sequences, plutonic intrusion, or debris from a distinctive provenance.

Size of Terranes

The terranes of the circum-Pacific vary enormously in size. Some are of subcontinental dimensions, whereas others cover a few hundred square kilometers or less. The large foldbelts of Mongolia, China, and Australia are probably composed of many smaller terranes and terrane fragments that cannot be distinguished with the data presently available. The size of the terranes shown in Figure 1 is in part a reflection of the amount of study that has been done in various areas; in addition, many terranes have been combined as composite terranes because of the scale of the map. A few terranes are not now continuous bodies but consist of separate, disjunct patches that can be unequivocally correlated. The best example of a disjunct terrane is Wrangellia, which presently is distributed as isolated bodies from the state of Oregon, through British Columbia, to southern Alaska, a latitudinal spread of nearly 24°. However, paleomagnetic data indicate that the original latitudinal spread was likely less than 4°. The Santa Lucia–Orocopia composite terrane is currently being dispersed into northwest-oriented slivers owing to Neogene right-slip along the San Andreas fault system.

Terrane Boundaries

By definition, all terranes must be separated from adjoining terranes by major faults or complex fault zones. These suture zones are commonly characterized by a belt of melange, blueschist, and/or ophiolite, but in many instances terrane boundaries are cryptic or unimpressive fault zones. Where necessary, a fault may be inferred between areas with different stratal units if the boundary is not exposed and the units on either side cannot be linked together by normal facies variations.

Confusion may arise in discriminating between fault-bounded terranes and successive tectonic elements within a particular terrane. For example, in our definition, the development of a volcanic arc assemblage on top of an earlier formed sedimentary or igneous package does not constitute formation of a new terrane. This definition is crucial to terrane analysis, and failure to discriminate between terranes and contrasting lithogenetic elements within a single terrane is a potential source of confusion.

Accretion of Terranes

The major tectonic event affecting most terranes is their accretion to a continental margin. In many cases, this collisional event has produced intense folding, thrust faulting, penetrative deformation, and recrystallization to blueschist, greenschist, and amphibolite grade. The resulting tectonite fabrics in some instances are regional in scale and in others are thin or narrow tectonized zones. In other cases, emplacement by strike-slip faulting may leave little evidence of the accretion event. Igneous activity directly related to accretion seems to be rare, but low-temperature alteration is widespread, and local instances of high-temperature alteration, including anatexis, have been documented (Hudson et al, 1979; Monger et al, 1982).

Structural styles in accreted terranes vary widely; isoclinal folding is common, but vergence of folds is rarely consistent even within a single terrane. Thrust faults are common, although most fault surfaces have themselves been folded. Identification of initial accretionary structures

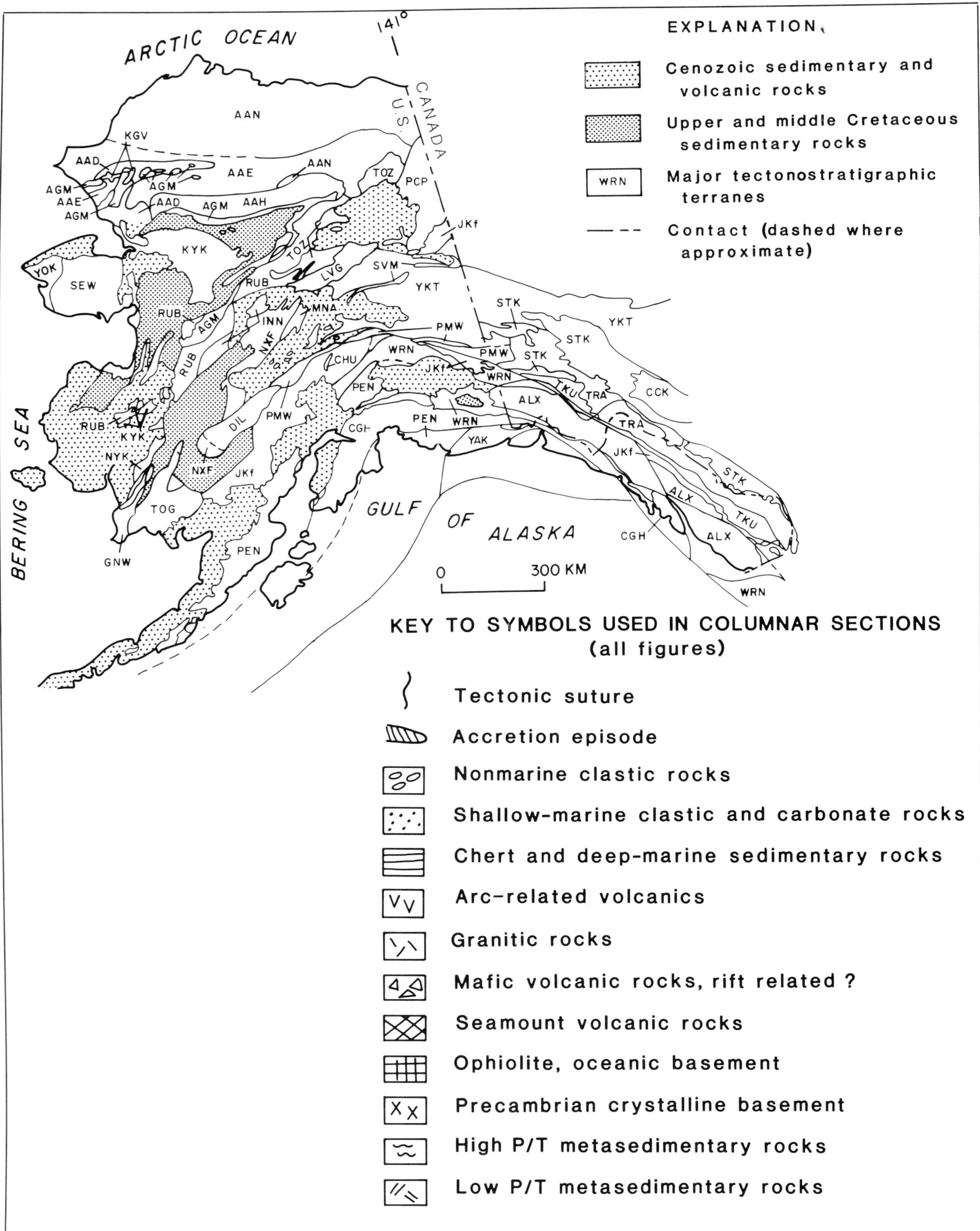

Figure 2—(a) Inset map for terranes of Alaska and explanation of symbols used in columnar sections, Figures 2–5.

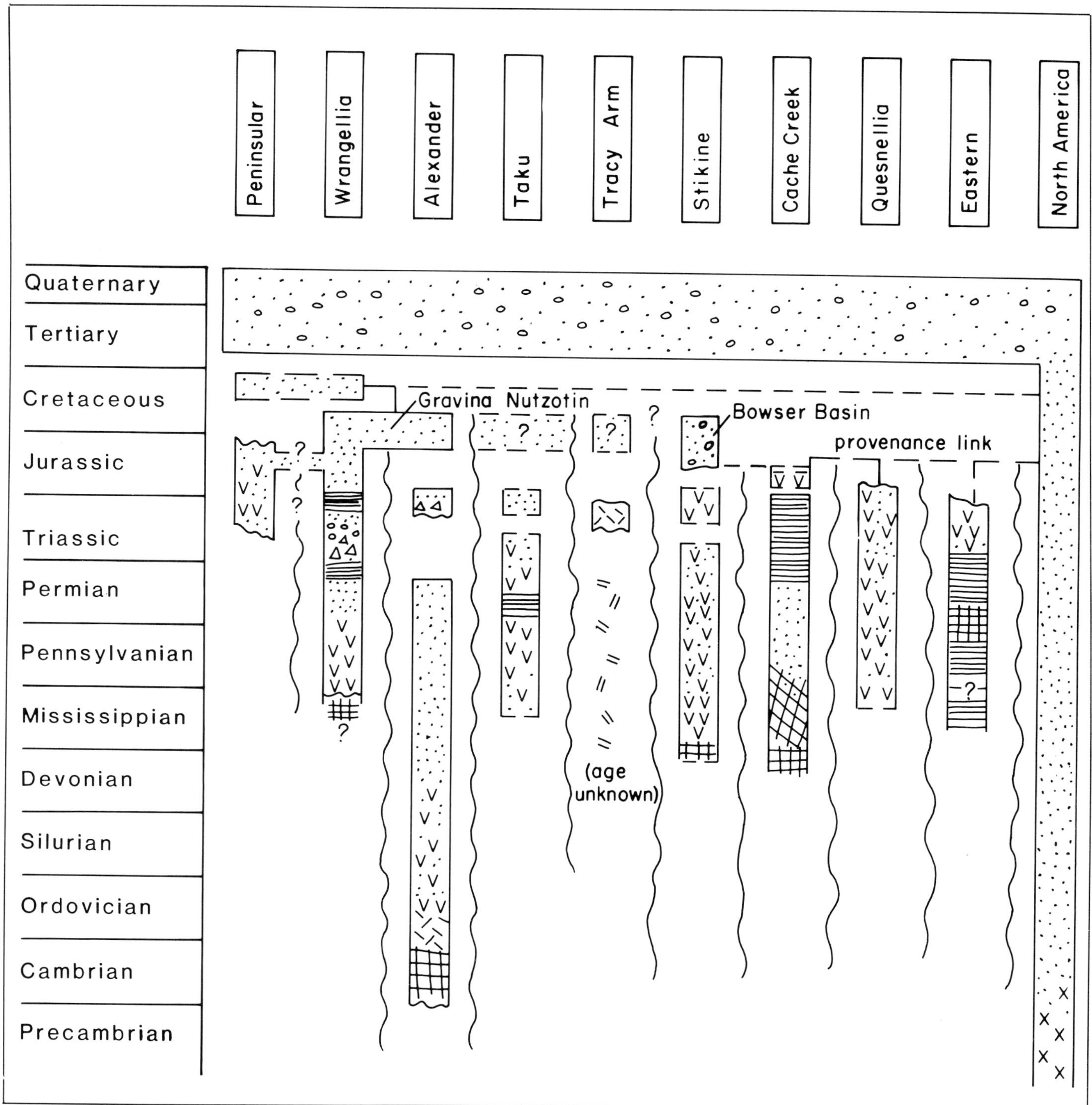

Figure 2—(b) Columnar sections and flow diagram showing accretionary history of Alaska.

that have not been altered by later movements is extremely rare.

The mechanisms of terrane accretion and amalgamation remain obscure; nevertheless, these phenomena are not to be confused with the concept of subduction accretion, which relates to off-scraping of unconsolidated pelagic or trench deposits during subduction (Scholl et al, 1980). Many terranes of the North American Cordillera and the western Pacific are composed of sedimentary units that indicate past involvement in subduction accretion, yet their present tectonic position reflects emplacement mainly as coherent, strongly lithified masses, either above the continental margin or on earlier accreted terranes. This implies that many terranes are obducted flakes that are mere remnants of once much larger plates that have now mostly been subducted.

Sediment subduction involves oceanic crust with a cover of pelagic and clastic deposits passing beneath an adjoining plate with or without the formation of a growing accretionary wedge; the formation of the latter may be

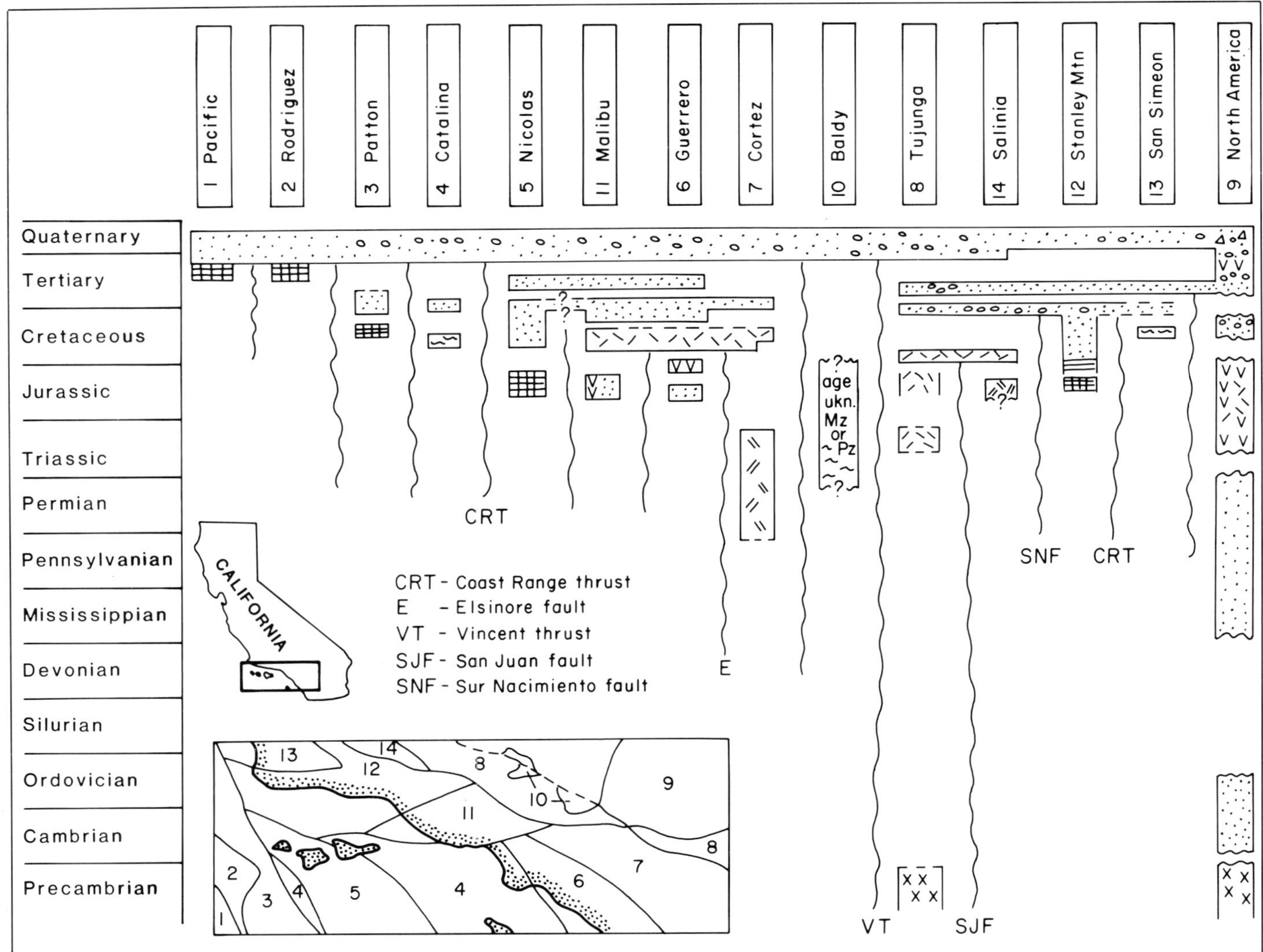

Figure 3—Columnar sections and flow diagram showing accretionary history of southern California. Symbols as in Figure 2.

dependent on the presence of a particularly thick component of clastic deposits above a zone of overpressured pore water (Scholl et al, 1980). Thick crustal bodies, such as sea mounts, oceanic plateaus and ridges, and continental fragments appear to be more difficult to subduct and are instead accreted as relatively intact blocks though commonly severed from their basement rocks (Ben Avraham et al, 1981). Complete kinematic and dynamic explanations of terrane accretion processes, however, remain to be elucidated.

Postaccretionary Dispersion

The main period of accretionary activity ended by Early Tertiary time in the Cordillera and northeastern Siberia but is still continuing in much of the circum-Pacific. These accretionary episodes have been followed by (and are often concurrent with) a long and still poorly known history of complex strike-slip faulting, folding, and thrust faulting resulting in the breakup of some terranes. This crustal rearranging is still locally active and represents continuing interaction between oceanic plates of the Pacific and the continental plates around the basin margin. The result of this neotectonic activity is to disperse the terranes and to further complicate the structures formed during accretion. In North America, large-scale right-slip faults such as the San Andreas, Fairweather, Denali, Fraser River, and Tintina all have minimum displacements of a few hundred kilometers, and some may have much more. The cumulative relative movement on all of these, plus innumerable subsidiary faults, must amount to several thousand kilometers, and some may have much more. In Japan, left-slip faults (e.g., the Median tectonic line) are smearing out and dispersing the terranes while accretion is still occuring, and in eastern China, east-west-trending left-slip faults resulting from the northeastward movement of India are fragmenting the collection of terranes in that area. In North America, Late Cretaceous to early Cenozoic structures of the Laramide orogeny have been interpreted as a final tightening of a poorly consolidated crust composed of previously accreted allochthonous terranes (Coney, 1981).

Measuring Terrane Displacement

The fact that fault-bounded terranes in the circum-Pacific have different stratigraphic sequences implies that they have moved relative to each other and to the craton. The amount of movement may not be large but must be sufficient to juxtapose dissimilar rocks and to disrupt original facies trends completely.

Much effort is now being expended in order to establish the kinematic history of the various terranes in the North American Cordillera; this sort of work is also needed for the rest of the circum-Pacific. The principal methods of determining relative movement are: (1) measuring the offset of linear geologic elements such as shorelines, dike swarms, or fold hinges; (2) matching offset similar stratigraphic sequences or distinctive rock types; (3) matching displaced biogeographic provinces; (4) matching climatically controlled lithologic features (e.g., red beds, sabkhas, etc.) with their world-wide regional extent; and (5) determining paleolatitude by means of paleomagnetic investigation.

Studies of offset permits precise analysis of slip, but offsets greater than 500 km (310 mi) are rarely determined, whereas paleomagnetic studies potentially can give quantitative measurements for displacements that have a very large latitudinal component. Determining offset in a longitudinal sense is not possible with paleomagnetic data unless detailed polar-wander paths are available. The other methods are mainly qualitative for large displacements, although direct matching of offset stratigraphic sequences can be quite precise for displacement ranges of a few tens to a few hundreds of kilometers. Paleomagnetic measurements now available from several terranes in the western Cordillera clearly substantiate large-scale northward displacement, thousands of kilometers, in some instances. Coupled with these translational movements are crustal rotations commonly of 60° or more (Beck and Cox, 1979).

Thorough terrane analysis involves all of the above parameters to assess relative differential movement of terranes. If two or more methods of estimating original paleolatitude are in essential agreement, much more confidence can be placed on palinspastic and paleogeographic reconstructions. A good example of concordance of geophysical and geologic data is afforded by Wrangellia, where paleomagnetic measurements from Triassic basaltic rocks show equatorial paleolatitudes (Hillhouse, 1977) and Triassic carbonates overlying the basalt are sabkha deposits characteristic of shallow tropical supratidal conditions (Armstrong and MacKevett, in press). These rocks are now at latitudes as high as 62°N and are completely out of place with respect to cratonal Triassic sequences.

Paleogeographic Reconstruction

The aims of terrane analysis are to: (1) identify, characterize, and portray terranes on terrane maps; (2) relate their faunal and floral characteristics through time to major paleobiogeographic provinces; (3) establish paleolatitudes through time; and finally (4) in the case of the circum-Pacific region, attempt paleogeographic reconstruction of the paleo-Pacific Ocean (Panthalassa) and surrounding cratonal regions. Deep-sea drilling has established that no part of the present Pacific Ocean floor is older than Middle Jurassic—thus, the pre-Jurassic history of the Pacific basin can only be gleaned from scraps and fragments plastered around the Pacific margin in the form of accreted allochthonous terranes.

CIRCUM-PACIFIC TERRANES

An extensive study of the geological and geophysical literature from around the Pacific basin indicates that terranes similar to those in western North America can be distinguished throughout the circum-Pacific region (Fig. 1). From these relations it is clear that a mosaic of terranes characterizes all regions of the Pacific. We hope that the generalized distribution of terranes depicted on Figure 1 will stimulate research to characterize more accurately the nature and extent of all terranes throughout this vast region. A brief discussion of a few of the relatively well-studied groups of terranes is given below, including North and South America, northeast Siberia, China, Australia, and New Zealand.

North America

Tectonostratigraphic terranes have been better studied in North America than any other part of the circum-Pacific. Terrane maps of Alaska and western Canada (Berg et al, 1978; Jones and Silberling, 1979; Jones et al, 1981; Irving et al, 1980); California (Blake et al, 1982a; Irwin, 1972, 1977); and Mexico (Campa and Coney, 1983) are available and show terranes in more detail than the scale of Figure 1 allows. The timing of accretion within the Cordillera is becoming reasonably well known (e.g., Monger et al, 1982; Coney et al, 1980; Saleeby, 1983; Churkin and Eberlein, 1977; Davis et al, 1978); we can provide only a brief summary of a few areas here.

Alaska and British Columbia

Alaska is composed of terranes accreted during the Mesozoic and Cenozoic (Fig. 2a). Active terrane accretion is now underway in southern Alaska, where the Yakutat block (YAK) has entered the Aleutian trench (Plafker et al, 1980). While details of timing of deformation, amalgamation, and accretion are uncertain in many cases, a preliminary tectonic history can be constructed. Much work currently in progress is attempting to unravel the complexities of terrane history in Alaska.

The Arctic Alaska composite terrane (AAC, AAD, AAE, AAH, AAN) has been a major "backstop" affecting terrane accretion in Alaska (Churkin et al, 1980). This terrane, composed of Paleozoic and Mesozoic continental-shelf sequences that are probably underlain by Precambrian basement, may have been rotated counterclockwise or translated from an original position in Northern Canada or the Arctic ocean. Several continental, oceanic, and arc terranes were accreted to the southern margin of the Arctic Alaska terrane in late Mesozoic time, in part contemporaneously with movement of the Arctic Alaska terrane itself. The main accretionary event in the Brooks

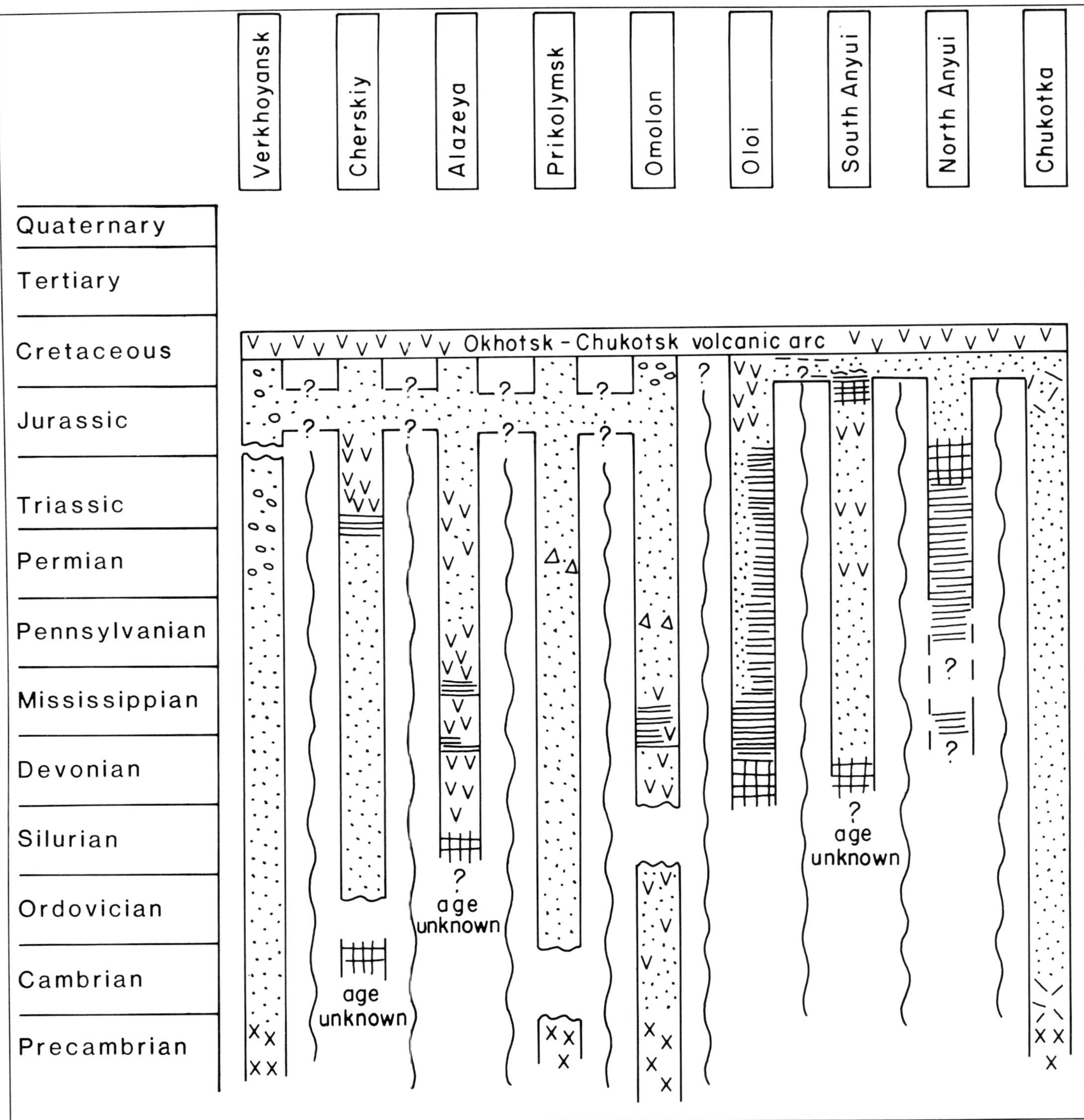

Figure 4—(a) Columnar sections and flow diagram showing accretionary history of northeast Siberia. Symbols as in Figure 2. Adapted from Churkin and Trexler, Earth and Planetary Science Letters, v. 48, p. 356–362, copyright © 1980 by Elsevier Science Publishers and Fujita and Newberry, Tectonophysics, v. 89, p. 337–357, copyright © 1982 by Elsevier Science Publishers. Used with permission.

Range is Early Cretaceous (Valanginian to Aptian?). The Angayucham (AGM) terrane, composed of upper Paleozoic to lower Mesozoic oceanic basalt, chert, limestone, and ophiolitic fragments, may represent oceanic islands and sea mounts that were emplaced at this time as enormous obducted thrust sheets which may have covered much of the area that is now the Brooks Range. Structurally complex deep-marine sedimentary and volcanic strata of the Innoko terrane (INN) may also have been emplaced at this time. During emplacement of these oceanic assemblages, other terranes of continental affinity were moved northward as discrete allochthons (Roeder and

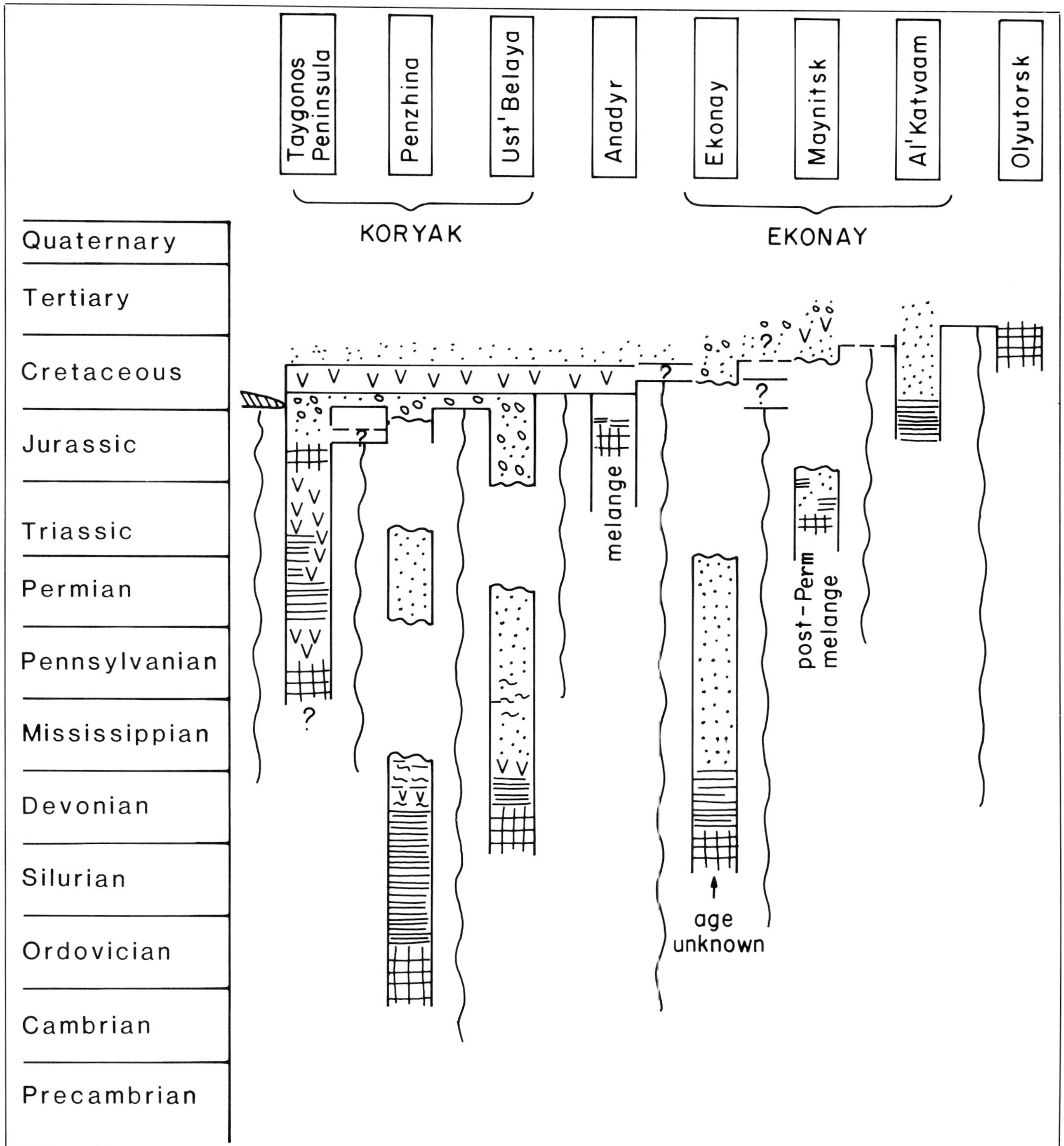

Figure 4—(b) Columnar sections and flow diagram showing accretionary history of the Koryak Highlands region of northeast Siberia. Symbols as in Figure 2.

Mull, 1978); these include the Endicott terrane (AAE) and other terranes too small to portray on Figure 1. To the south, other continental blocks, such as the Ruby (RUB) and Nixon Fork (NXF) terranes, were emplaced and covered by a Cretaceous (Albian and younger) overlap assemblage. The arrival of the Yukon-Tanana terrane (YKT), composed of metamorphosed continentally derived sedimentary, volcanic, and granitic rocks, appears to have deformed an Upper Jurassic to Lower Cretaceous flysch sequence in mid-Cretaceous time. The youngest metamorphism of the Yukon–Tanana rocks has been dated (K–Ar) as mid-Cretaceous (Bundtzen and Turner, 1979), but the relation of this metamorphism to tectonic events in the region is uncertain. Yukon–Tanana was probably accreted hundreds of kilometers farther south during the Early Cretaceous and later moved northward along the Tintina fault. Overlap assemblages and paleomagnetic data from volcanic rocks that stitch the terranes together north of the Denali fault indicate that these terranes were in place by Paleocene time (Jones et al, 1982b; Hillhouse and Grommé, 1982).

Within a broad suture zone in the central Alaska Range (north of the Peninsular terrane), at least 11 small terranes are enveloped by deformed Cretaceous flysch (JKf of Fig. 2a). Most of these terranes are too small to show on Figure 1, yet their stratigraphic sequences show distinct contrasts. These so-called mini-terranes are clearly fragments or flakes of much larger crustal bodies. An example of one of these terranes is Chulitna (CHU), a rootless nappe composed of Devonian ophiolite overlain by Paleozoic chert and shallow marine strata and Triassic red beds whose fauna indicate an equatorial paleolatitude (Jones et al, 1982b; Nichols and Silberling, 1979). The Triassic strata show no relation to the coeval and nearby Triassic rocks of Wrangellia. Postaccretionary transcurrent faulting along the Denali fault system has contributed to the dispersion of this group of small crustal fragments.

The earliest Mesozoic accretion in southeastern Alaska and British Columbia occurred with the Middle to Late Jurassic collision of a composite terrane, Stikinia, an amalgam of the Stikine (STK), Cache Creek (CCK), Quesnellia (QNL), and Eastern (EAS) terranes (Fig. 2). The Eastern terrane is composed of upper Paleozoic and lower Mesozoic ophiolitic and deep-marine strata, while Quesnellia is composed of upper Paleozoic to Lower Triassic volcanic, volcaniclastic, and carbonate rocks of volcanic arc complexes. The Cache Creek terrane is an assemblage of Mississippian to Upper Triassic chert, argillite, basalt, ultramafic, and carbonate rocks that are locally metamorphosed to blueschist grade. This terrane has been interpreted in part as a sea mount built on oceanic crust (Monger, 1977a; Ben-Avraham et al, 1981; Souther, 1977). The carbonate reefal strata in the Cache Creek terrane contain Permian Tethyan fusulinids, in contrast to those in the neighboring Stikine terrane, a volcanic island-arc complex that contains North American fusulinid assemblages. The Paleozoic and Triassic rocks of Stikine are overlain by Upper Triassic to Middle Jurassic volcanic-arc rocks of the Takla and Hezelton assemblages; these volcanics have been interpreted as overlapping Stikine, Cache Creek, and Quesnellia (Saleeby, 1983), but field evidence for such a connection is lacking. In addition, radiolarian cherts in the Cache Creek terrane indicate that deep-marine sedimentation continued until latest Triassic time.

The most definitive connection between Stikinia and the Cache Creek–Quesnellia–Eastern amalgam is shown by the Middle to Upper Jurassic Bowser basin. Strata of this basin lie entirely on the Stikine terrane, but the composition of the debris suggests provenance linkage to Cache Creek. Strata as young as Early to Middle Jurassic are deformed and imbricated, suggesting Late Jurassic accretion to the craton (Monger et al, 1982). However, Lower(?) Jurassic melange of the Stikine terrane containing detritus possibly of the Cache Creek terrane and the uppermost portions of the Eastern–Quesnellia assemblage seemingly indicates provenancial linkage and therefore suggests an earlier amalgamation of these three terranes (Davis et al, 1978; Monger et al, 1982). Paleomagnetic results from the postamalgamation arc rocks in southern Canada show paleolatitudes near 35°N in the early Mesozoic (Monger and Irving, 1980) and indicate significant northward displacement after amalgamation. Metamorphism and deformation of the continental margin and Precambrian basement rocks in the Omineca crystalline belt apparently began in mid-Jurassic time in possible response to collision of the Stikine composite (Monger et al, 1982). Lower Cretaceous strata overlying the Bowser basin contain debris derived from the Omineca crystalline belt, substantiating that accretion had occurred by that time. Postaccretion dispersion northward along the Tintina and related faults of at least 450 km is inferred based on geologic relations (see Tipper et al, 1981).

The next major accretionary event in the British Columbia segment of the Cordillera involves the Alexander–Wrangellia–Peninsular composite terrane and is shown schematically in Figure 2. Alexander (ALX) is a composite of three terranes comprising Ordovician and younger volcanic island-arc, clastic, and carbonate strata; Wrangellia (WRN) is a late Paleozoic andesitic volcanic arc overlain by thick Triassic submarine and subaerial flood basalts and plateau carbonate rocks. The Peninsular (PEN) terrane is characterized by an Early Jurassic andesitic arc that contrasts strongly with coeval nonvolcanic strata found nearby in Wrangellia. Upper Jurassic and Cretaceous stratigraphic units of the Wrangellia and Peninsular terranes are essentially identical, indicating amalgamation by that time. The amalgamation of the Wrangellia and Alexander terranes by Middle Jurassic time is documented by the presence of the Upper Jurassic to Middle Cretaceous Gravina–Nutzotin belt of flysch, deep-marine sedimentary, and volcanic rocks that overlaps the two terranes (Fig. 2). Thrust faulting and folding in Wrangellia is dated as post-Barremian and pre-Albian. This tectonism is thought to mark the initial collision between Wrangellia and North America. Final accretion of the large amalgam in Late Cretaceous (post-Cenomanian) time is indicated by regional penetrative deformation of the youngest basinal strata and the age of post-collision intrusives (Coney, 1981; Jones et al, 1982b; Saleeby, 1983). Late Cretaceous metamorphism and magmatic activity in the Coast Plutonic Complex in western British Columbia is thought to have occurred in response to this collision (Monger et al, 1982).

Western United States

A passive margin existed along western North America from approximately 800 to 400–350 m.y. ago. The first major Phanerozoic accretion occurred during the Mississippian Antler orogeny with the emplacement of the Roberts Mountains allochthon composed of ocean-margin strata. Other collisions at about this same time may explain the occurrence of such chert-pebble rich sequences as are found in the Canadian Rocky Mountains (Monger and Price, 1979) and the Copper Basin Formation of Idaho, where paleocurrent data indicate a western source terrane (Nilsen, 1977).

The Sonomia composite terrane (including GLC, WKL, EKL, NSI) in west-central Nevada contains late Paleozoic clastic strata of the Golconda allochthon (GLC) and the Walker Lake terrane (WKL), which Speed (1979) suggested was emplaced during the Sonoma orogeny in Permian and Triassic time. Evidence for other late Paleozoic to Triassic accretions is not found to the north or south of Nevada.

Several terranes comprising Paleozoic and Mesozoic deep-marine strata and ophiolitic material were accreted during the late Paleozoic and early Mesozoic, but the timing of these events is highly uncertain. These include the pre-Ordovician ophiolite and volcanic arc strata of the Eastern Klamath terrane (EKL) and a composite of lower Paleozoic ophiolitic and distal continental margin strata in the northern Sierra Nevada (NSI); these accretions may have been coincident with the Sonoma orogeny, so Speed (1979) included EKL and NSI in Sonomia. Other terranes contain Tethyan Permian fusulinids and upper Paleozoic to Triassic ophiolitic and deep-marine assemblages, including the western Paleozoic and Triassic composite terrane (TRP) of the Klamath Mountains and the Calaveras composite terrane (CLV) of the Sierra Nevada. A Middle Jurassic truncation of the continental margin, possibly by strike-slip faulting, is indicated by southward displacement of Precambrian belts and miogeoclinal strata (Silver and Anderson, 1974) and by truncation of the Paleozoic and early Mesozoic accreted terranes and tectonic trends (Davis et al, 1978).

During the Mesozoic and Cenozoic, terranes were accreted to Washington, Oregon, and northern California in several discrete events (Blake et al, 1982b). The first well-dated event, the Middle to Late Jurassic Nevadan orogeny, may be related to accretion of Jurassic ophiolite and arc terranes such as the Western Klamath (WKM) and Coast Range Ophiolite–Great Valley terranes (DPO, ELC, STM). These accretions were accompanied by blueschist metamorphism and melange formation. During the mid-Cretaceous (ca. 90 m.y. ago) Coast Range orogeny, sea mounts within the Franciscan assemblage (FRA) were accreted and some of the melange of the Central terrane (CEN) was formed. At approximately 55 m.y. ago, accretion of terranes such as the Santa Lucia–Orocopia allochthon (discussed below), and Siletzia (SLZ), composed of sea-mount volcanic and sedimentary rocks, was accompanied by renewed westward thrusting. Terranes accreted later in the Tertiary at approximately 38 m.y. include the Coastal (COS) and King Range (KIN) terranes of the Franciscan assemblage. Parts of these two composite terranes (COS, KIN) may have been amalgamated before accretion while others were accreted directly to the margin of California, but more work is needed to define such relations better. In the Peninsular Ranges and the California Continental Borderland of southern California, terranes were not emplaced until Miocene time (see below).

Southern California

An enlarged map of terranes in southern California and a flow diagram of the stratigraphic and accretion history of these terranes is shown in Figure 3. The first major Phanerozoic accretionary event affecting the cratonal margin in the southern California region was the accretion of the Santa Lucia–Orocopia allochthon approximately 55 m.y. ago (Vedder et al, 1983). Santa Lucia–Orocopia is a late-stage amalgam of composite terranes (SAL, SSM, STM, TUJ) in which earlier unrelated suturing events had occurred. The Salinia terrane (SAL) consists of a mid-Cretaceous volcanic arc that is built upon a heterogeneous prebatholithic assemblage. Further work on these metamorphic rocks may reveal the presence of one or more additional terranes. Tujunga (TUJ), a continental fragment, is composed of at least three Precambrian terranes (Powell, 1982); the Stanley Mountain terrane (STM) consists of Middle Jurassic ophiolite and overlying forearc sediments structurally underlain by the San Simeon terrane (SSM), a disrupted terrane of Franciscan melange. Pluton stitching and overlap sequences are inferred to have linked Salinia and Tujunga by Middle to Late Cretaceous time; provenance linkage and overlap assemblages indicate that the Stanley Mountain and Salinian terranes were amalgamated in Campanian time (Vedder et al, 1983). The Baldy terrane (BDY) is composed of Mesozoic(?) sedimentary strata metamorphosed under high P/T conditions in the Late Cretaceous to Early Tertiary and may have been accreted to Santa Lucia–Orocopia or to North America during the metamorphic event. Eocene strata overlap all parts of the Santa Lucia–Orocopia composite as well as the margin of North America, indicating that accretion had occurred by this time.

Paleomagnetic data document the far-traveled and composite nature of Santa Lucia–Orocopia (Champion et al, 1980; McWilliams and Howell, 1982). At 160 m.y., the Stanley Mountain terrane was in low northerly or southerly latitudes (14°), and in Cenomanian–Santonian time it was at 6°N or S. Paleomagnetic measurements from the overlap strata indicate that the Stanley Mountain–Salinian composite terrane then traveled northward, from 21°N in Campanian–Maestrichtian to 25°N in early Paleocene. The paleolatitude of the California margin during this time was approximately 45°N; this fact supports the geologic evidence that amalgamation occurred prior to the accretion in southern California.

The other important group of terranes in southern California amalgamated throughout Jurassic and Cretaceous time and were accreted during the early to middle Miocene, approximately 16–17 m.y. ago. Plutons 125–105 m.y. old stitch the Cortez terrane (CRZ, a continental fragment), the Guerrero terrane (GUE, an Upper Jurassic to Cretaceous volcanic island arc), and the Malibu terrane (a metamorphosed Jurassic arc; see Fig. 3). The Nicolas terrane (SNC), composed of Jurassic ophiolite and overlying Cretaceous strata, was accreted to this amalgam by the middle of the Cretaceous (90 m.y.), as evidenced by forearc basin and submarine fan strata that overlap all of these terranes. Overlap sequences that link the Catalina (CAT) metamorphic terrane and the Patton

Ridge (PTR) graywacke terrane to the Cortez-Guerrero-Malibu-Nicolas composite are no older than Middle Tertiary. The youngest apparently allochthonous rocks are middle Miocene volcanic rocks of the Nicolas and Malibu terranes; the oldest strata that can be shown unequivocally to overlap all these terranes and the California continental margin are upper Miocene.

These stratigraphic relations seem to indicate that most of coastal southern California is allochthonous and that accretion had not occurred until the late Miocene. This is substantiated by paleomagnetic data from Eocene rocks on the Nicolas terrane that show 19° of northward movement and from lower and middle Miocene volcanic rocks on the Nicolas and Malibu terranes that indicate 10–15° of northward translation (Luyendyk et al, 1982; Champion et al, 1981).

South America

Terrane analysis in South America is still preliminary. The Jurassic and younger volcanic cover of the Andean arc obscures terranes that may be present. However, the presence of terranes in southern Chile can be inferred from work by Dalziel et al (1974) and in the Caribbean region from papers by Maresch (1974) and Case et al (in press), who provide more complete references for terranes in the Caribbean area.

In the Caribbean mountains of Venezuela, the Cordillera de la Costa terrane (CDC) is composed of pre-Mesozoic gneissic basement overlain by Jurassic to Cretaceous quartzose clastics derived from a continental source and metamorphosed to amphibolite grade. This terrane also locally contains eclogite bodies. The Caucagua–El Tinaco terrane of the Caucagua–El Tinaco–Paracotos composite terrane (CET) also contains a pre-Mesozoic gneissic basement and Cretaceous quartzose clastics, but its unmetamorphosed to greenschist-grade metamorphic history is significantly different from that of the CDC terrane. The Paracotos portion of this composite terrane is composed of Upper Cretaceous prehnite-pumpellyite grade metagraywacke that may be a fragment of a marginal basin which collapsed during the collision of other terranes in the Caribbean mountains. Upper Middle Cretaceous island-arc volcanic and volcaniclastic strata compose the Villa de Cura terrane (VDC); these rocks overlie volcanic strata over 100 m.y. old metamorphosed to blueschist and greenschist facies (Maresch, 1984). Volcanic strata of the Villa de Cura terrane occur as klippen that have been correlated with the coeval Aruba–Blanquilla islands volcanic arc of the Curacao Ridge (Maresch, 1984), but the metavolcanic rocks of Villa de Cura may be significantly older than the Aruba–Blanquilla volcanics, so more work is required before such a correlation can be proved. In the southern part of the Caribbean mountains, the Guarico-Roblecito terrane (GRB) is composed of Paleogene flysch and melange containing exotic blocks from nearby terranes; this flysch basin may have been deformed during accretion of the terranes to the north. The northernmost edge of Colombia and Venezuela consists of a composite of several terranes, including the South Caribbean deformed belt terrane (SCB) and the Curacao Ridge whose Mesozoic and Paleogene rocks have paleomagnetic poles that show large translations and/or rotations (Hargraves and Skerlec, 1980; Skerlec and Hargraves, 1980). These terranes are dominantly composed of oceanic basement and disrupted pelagic and submarine-fan strata.

Case et al (in press) have distinguished several fault-bounded fragments of the Guyana shield that may have been rifted away from the craton. These include the composite Meridia (MER) and Cordillera Central-Sierra Perija (CPJ) stratigraphic terranes, the Santa Marta (SMA), Guajira (GUA), and Paraguana (PAG) terranes. Paleomagnetic data for MER and CPJ show stable Mesozoic pole positions. The Santa Marta, Guajira, and Paraguana terranes show extensive Mesozoic deformation and metamorphism, and Mesozoic to Paleogene paleomagnetic poles show significant translation and rotation (Case et al, in press; MacDonald and Opdyke, 1972; Skerlec and Hargraves, 1980).

In Colombia, the Cordillera Occidental terrane (COL) represents a fragment of Cretaceous ophiolite and deep-marine strata (Case et al, 1971; Goosens et al, 1977, Bourgois et al, 1982). This terrane is part of a larger belt composing several Mesozoic and Cenozoic oceanic and island-arc terranes called the Basic Igneous Complex (Goossens and Rose, 1973), which extends from Colombia to Ecuador and includes the Coastal and Western Cordilleras. Initial Sr 87/86 ratios of .704 and less in this area indicate an oceanic nature (Zeil, 1979). Within this belt is a late Paleozoic or early Mesozoic volcanic island-arc complex (Herbert, 1977; Goossens et al, 1977; see Zeil, 1979, p. 193). The Choco terrane (CHC) consists of a younger (Late Cretaceous to Paleogene) accreted ophiolite complex in the Coast Range of Colombia; these rocks may be correlative with similar age oceanic rocks in the Ecuador Coast Range terrane (ECR) (Goossens et al, 1977).

The presence and possible allochthonous character of Precambrian crystalline rocks seaward of the belt of suspect terranes (the eugeosynclinal complexes of others) have been a subject of recent debate (e.g., see Nur and Ben-Avraham, 1981). Several "suspect" fragments of Precambrian and Paleozoic rocks, including the Piura, Marañon, and Arequipa terranes, are shown in Figure 1. Piura (PIU) is composed of Paleozoic metamorphic rocks overlying Precambrian basement. The Marañon terrane (MAR) has been interpreted as a topographic high during the Mesozoic and contains graywacke with lithofacies that are in distinct contrast to those of neighboring terranes. The basic question with regard to the Marañon terrane (and other Precambrian blocks) is whether it is an uplifted welt of autochthonous Precambrian shield rocks or an exotic block of sialic crust.

Arequipa (ARQ) is composed of Precambrian (1.2 b.y.) basement overlain by Devonian clastic rocks. Paleomagnetic studies of the Devonian rocks of Arequipa are in progress to determine whether Arequipa is exotic with respect to previously correlated basement and Devonian sediments of the Eastern Cordillera on the other side of the Andes (McWilliams, 1982, personal communication). Previous paleomagnetic studies on Jurassic rocks of this terrane indicate no poleward movement of Arequipa with respect to the Brazilian shield since the Jurassic (Shackleton et al, 1979). Devonian faunas in the fine-grained strata of Arequipa are of warm-water,

Eastern Americas Realm affinity, while similar age fauna in the Eastern Cordillera are of cold-water, Malvinokaffric Realm affinity (Boucot et al, 1980). This faunal discrepancy has been attributed to a warm-water current linking the Middle America Realm to the Arequipa region (Boucot et al, 1980), but at present, the data are insufficient to rule out the allochthoneity of this terrane.

The Chile Coast Range terrane (CCR) is a composite of Permian to early Mesozoic blueschist-metamorphosed ophiolitic melange, and Carboniferous high P/T metagraywacke (Munizaga et al, 1973). Farther north, Carboniferous blueschist facies metagraywacke intruded by Paleozoic plutons composes the Curepto terrane (CUR) (Gonzalez-Bonorino, 1971).

Northeast Siberia

In northeast Siberia, accreted terranes consist of oceanic islands, carbonate platforms, volcanic island arcs, continental blocks, and ophiolitic suits (Fig. 4a). The character and distribution of terranes in this region as shown on Figures 1 and 4 are taken from studies by Fujita (1978), Fujita and Newberry (1982, 1983), Kosygin and Parfenov (1981), Shilo and Til'man (1981), and Natal'in and Parfenov (1983). The Pacific margin of Siberia has undergone accretion throughout the Mesozoic, similar to western North America. The Verkhoyansk terrane (VKH) is composed largely of eastward thickening continental margin clastic and carbonate strata that may not be exotic but rather represent a deformed part of the Siberian craton. The Cherskiy terrane (CSK) is an oceanic assemblage composed of lower to upper Paleozoic volcanic and carbonate strata and chert upon which a volcanic island arc was built in the Late Triassic to Jurassic. The Alazeya terrane (ALZ) is a late Paleozoic(?) to Mesozoic volcanic arc underlain by ophiolite. Both Cherskiy and Alazeya were probably accreted during the middle Mesozoic; these collisions may be responsible for deformation in the Verkhoyansk foldbelt.

Seaward of the oceanic terranes lie the Prikolymsk (PRK) and Omolon (OMO) terranes, each bearing continental affinities and Precambrian basement. These terranes are separated by a major fault and each contains Paleozoic marine sedimentary rocks with distinctly different lithofacies; the lower Paleozoic strata of Omolon are primarily volcanic, while those of Prikolymsk are dominantly coarse clastics. The Paleozoic paleogeographic relation between these two terranes is uncertain, although they were accreted during the Middle to Late Jurassic, as indicated by the similar stratigraphies of Cherskiy, Prikolymsk, and Omolon after that time. The eastern and southeastern margins of Omolon were sites of subduction and arc amalgamation since at least Permian time (Fujita and Newberry, 1982). Paleomagnetic data from the Omolon terrane indicate that in the Late Permian it lay at 33°N latitude in contrast to the much more northerly latitude of Siberia, 58°N for the central portion (McElhinny et al, 1981). The Albian–Aptian volcanic and plutonic rocks of the Okhotsk–Chukotsk belt overlap and stitch most of the terranes in this region as well as to the east and extend south to the northern margin of the Okhotsk Sea.

The Koryak highlands region of Siberia is composed of a complex assemblage of terranes, including the Koryak (KOY), Anadyr (ANY), Ekonay (EKO), and Olyutorsk–Kamchatka (OLY) terranes. These terranes represent accreted arcs, ophiolite, and deep-marine sediments that lie outboard of the Jurassic accreted belt of continental fragments. A simplified flow diagram (Fig. 4b) shows the amalgamation and accretion history of this region. Much of the Penzhina foldbelt of the Koryak terrane consists of Ordovician to Upper Cretaceous forearc basin strata (Natal'in and Parfenov,1983), metamorphosed to blueschist facies 320–350 m.y. (Churkin and Trexler, 1981), and deformed in numerous collisions from Late Jurassic to Tertiary. Ophiolitic material is dominant in the four Koryak terranes. Late Paleozoic, middle Mesozoic, and Cretaceous ocean crust, and a Cretaceous volcanic island arc were accreted to the eastern margin of Siberia in four episodes: Late Jurassic, Hauterivian, early Senonian, and Eocene (Fujita and Newberry, 1983). The Koryak terrane consists of Paleozoic ocean floor thrust over Jurassic to Lower Cretaceous sedimentary rocks; in the southwest, the Taygonos Peninsula terrane (included with the Koryak terrane because of the scale of Fig. 1) consists of a Permian and Triassic volcanic island arc accreted in the Middle Jurassic, and a 185-m.y.-old ophiolite accreted in Early to Middle Cretaceous time. In the Hauterivian, Chukotka (CHK), which may be equivalent to Arctic Alaska, collided with Siberia, resulting in the accretion of the North Anyui (NAY) and South Anyui terranes (SAY) composed of eugeosynclinal strata and oceanic crust and (possibly) the obduction of the early to middle Mesozoic Anadyr ophiolite terrane. Upper Jurassic to Lower Cretaceous island-arc volcanic rocks and coastal marine strata of the Oloi terrane (OLO) may also have been accreted at this time. Ekonay is a composite terrane composed of middle Paleozoic ophiolite obducted over mid-Cretaceous sediments and tuffs and Triassic(?) ophiolitic melange overlain by Campanian to Paleogene volcanics and sediments. The youngest of the terranes, Olyutorsk–Kamchatka, consists of latest Cretaceous to Tertiary oceanic crust (30–65 m.y.) emplaced during the Paleocene or Eocene, overlain by Oligocene deposits containing ultramafic clasts recycled from the ophiolite.

Accretion in the Koryak highlands ceased after the Eocene collision of the Shirshov ridge, which possibly resulted in the accretion of the Olyutorsk terrane, and was followed by a change in the locus of subduction from Koryak–Kamchatka to the Aleutian–Kurile trend.

China

China exhibits an extremely complex history of accretion, which began in the Late Precambrian in Mongolia and continued up to the collision of India in the middle Cenozoic. In the simplest sense, China is made up of three cratons separated by large foldbelts (from less than 100 km [60 mi] to more than 500 km [300 mi] wide) composed of accreted terranes. The terranes shown in Figure 1 were distinguished using data given in the Bally et al (1980) extensive summary of the geology of Tibet and adjacent regions, as well as maps and articles published by Terman (1974) and the Chinese Academy of Geological Sciences (1975, 1976, 1979). The amalgamation and accretion history of these large blocks, Tarim (TAR), Sino-Korea (SKE), and Yangtze (YGT), is shown in Figure 5a.

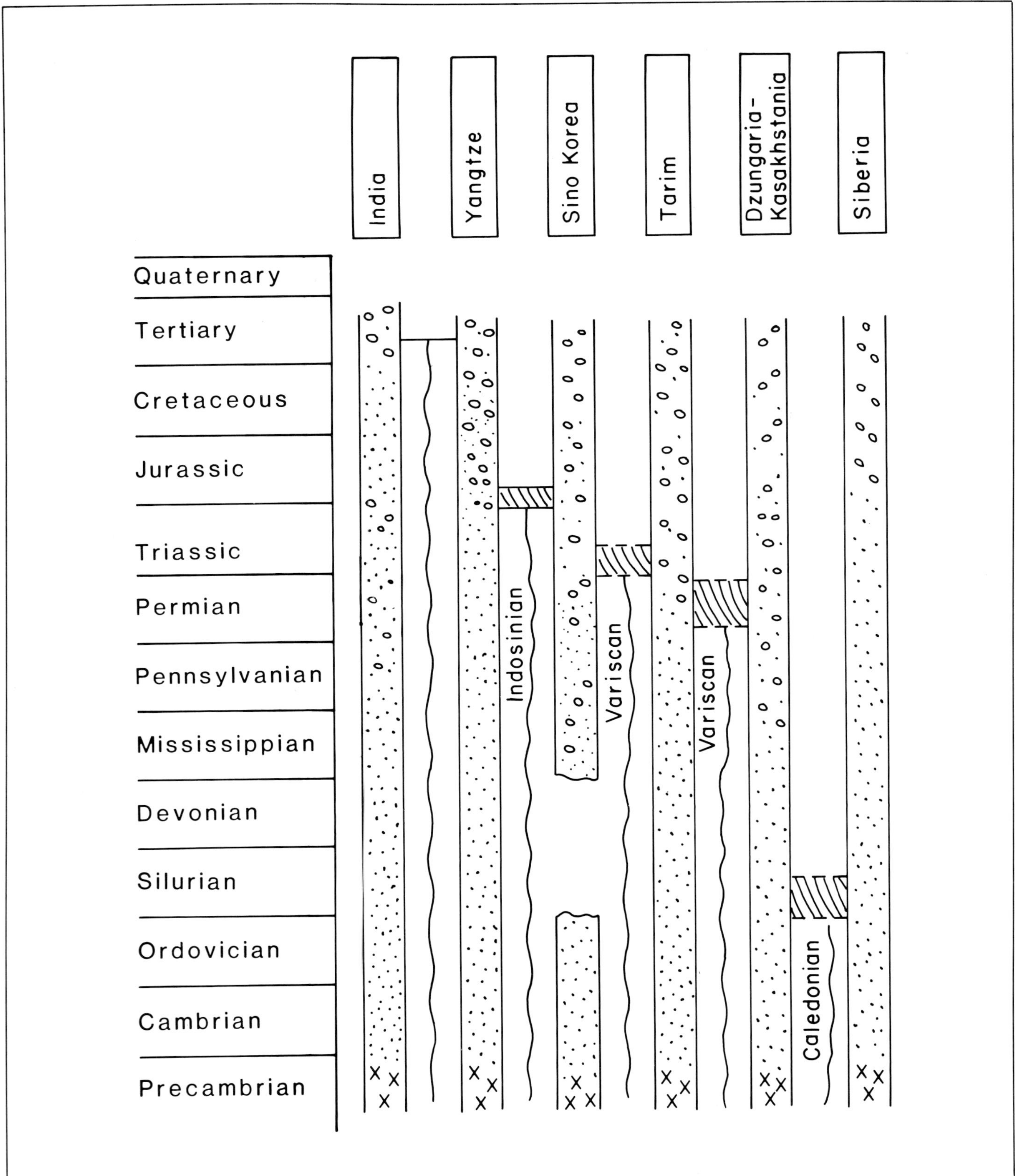

Figure 5—(a) Columnar sections and flow diagram showing general accretionary history of large cratonal blocks of China and Mongolia. Symbols as in Figure 2.

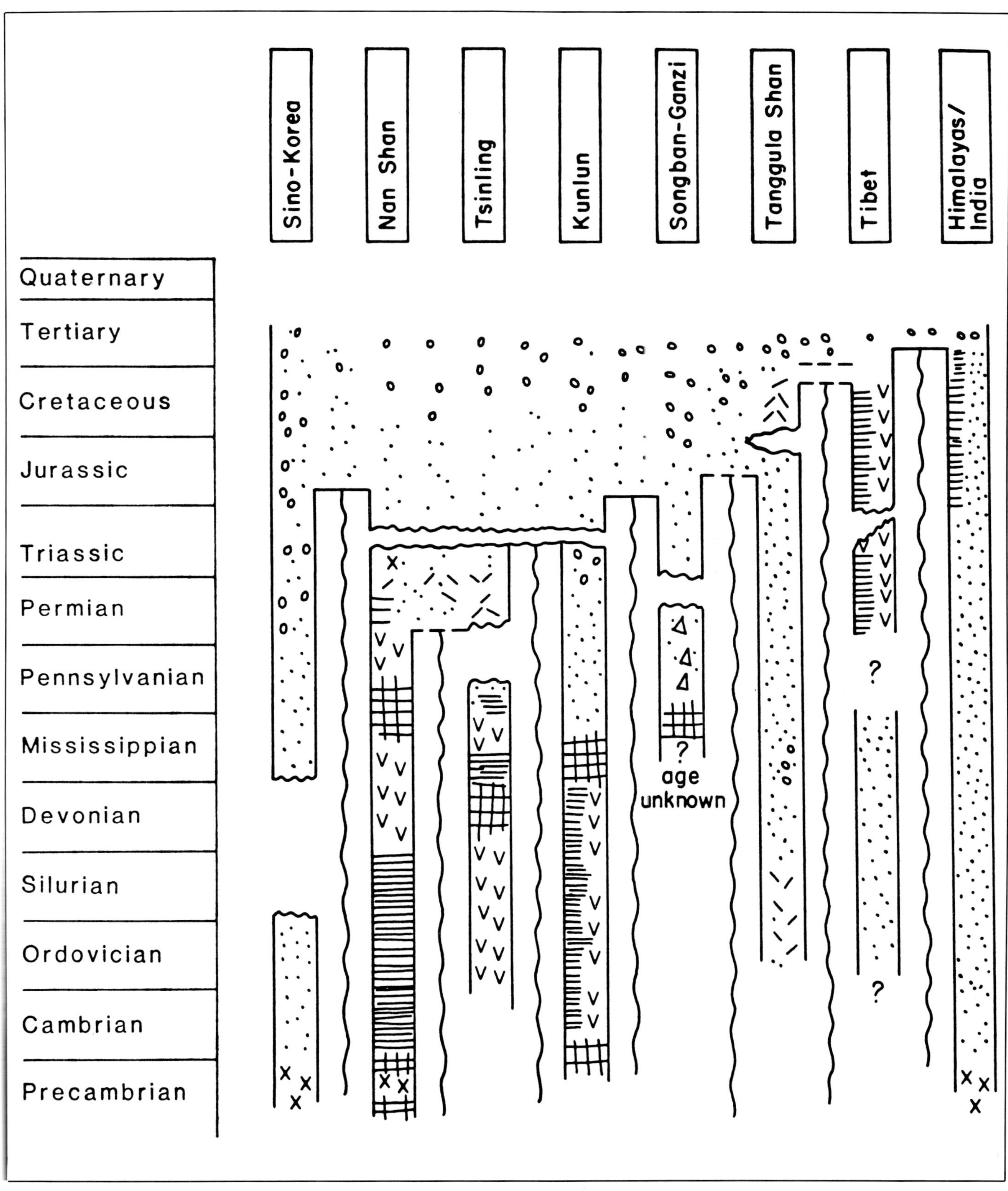

Figure 5—(b) Columnar sections and flow diagram showing accretionary history of Paleozoic and Mesozoic foldbelts of southern and central China. Symbols as in Figure 2.

Paleomagnetic studies show that these cratons were widely separated as recently as the Late Permian (McElhinny et al, 1981). Tarim had probably already been accreted to the southern edge of Siberia, a margin composed of many terranes accreted during the Late Precambrian Baikalian and early Paleozoic Caledonian orogenies. The paleolatitudes of Sino-Korea and Yangtze during this time were 10°N and 2°N, respectively, far south of the 40°N position of Tarim. In addition, Sino-Korea and Yangtze have rotated approximately 120° with respect to one another since the Late Permian.

Recent publications have proposed that a wedge-shaped ocean opening to the east existed between Siberia and Tarim–Sino-Korea during the Paleozoic (e.g., Li et al, 1980). Double subduction to the north under Siberia and to the south under Sino-Korea supposedly closed this ocean earlier in the west than in the east. However, evidence is insufficent to infer that Tarim, Ala Shan (ALS), and Sino-Korea were connected during most of this time; instead, several microplates probably moved northward toward Siberia (McElhinny et al, 1981). The composite Inner Mongolian (INM) and Gobi (GBI) terranes have several ophiolite belts that become younger toward the south, from Silurian to Permian; these belts may be interpreted as a series of volcanic island arcs and oceanic terranes accreted during the Paleozoic to the southern edge of Siberia. In contrast, the Yin Shan composite terrane (YNS) comprises several ophiolite belts of Late Devonian to Permian age that are youngest in the north. The Sung Liao terrane (SLO) is a continental fragment that may have been caught between these two oceanic belts. The Late Permian to Triassic Variscan orogeny can be interpreted as the final closure of the ocean between these two blocks and a consolidation of the arc and oceanic terranes between them. A Sino-Korean amalgam, comprising the Sino-Korea, Yin Shan, Sung Liao, and Inner Mongolia terranes, was accreted to a Siberian amalgam comprising the Gobi, Onon–Argun (ONA), Hangeyn–Hanteyn (HHC), Altay (ALT), and Sayan Tuva (SYT) terranes and the Siberian craton. This collision formed the Solon Mountains–Hegen Mountains suture (Li et al, 1980), which represents the boundary between the Cathaysian floral province to the south and the Angarian floral province to the north. Possibly the Siberian and Tarim blocks collided just prior to the collision of Yin Shan and Inner Mongolia, but the details of this accretion are unknown.

The timing of the accretion of the Yangtze block to Sino-Korea is poorly constrained, but the blocks seem to have been distinct through the Triassic (Wang et al, 1981). The collision appears to have formed the Yu Hwai (YHW) and Lower Yangtze (LYZ) foldbelt terranes. The Dabie Shan terrane (DAB), also called the Huaiyang massif, may be a microcontinent caught in the collision zone. At least part of the boundary in the Lower Yangtze foldbelt, the Tangcheng–Lujiang deep fracture of Huang (1978), shows evidence that accretion occurred during the Jurassic to Cretaceous. Evidence includes dynamic metamorphism from 230 to 190 m.y. ago near the fault zone in Sino-Korea, Late Jurassic volcanic rocks along the fault zone, and Cretaceous sedimentary rocks overlapping the fault (Weng and Kong, 1981). However, this deformation event has also been interpreted as the accretion of the Dongbei terrane (Xiong and Coney, this volume), equivalent to the Wan-ta (WTA) terrane of Figure 1.

During the closure of the Tethys Ocean that culminated during the Eocene with the collision of India against Asia, several terranes within the Tethys sea were accreted to the southern margin of the Sino-Korean–Tarim block. The Tethyan terranes are included in the Nan Shan–Qilian (NNS, NQL), Qinling (QIN), Kunlun (KUN), Songban–Ganzi (SBG), and Tanggula Shan (TGS) foldbelts. Several continental crustal blocks that appear to be caught in these foldbelts include Songpan (SGP), Yidun (YID), Oulungbruk (OUL), Central Qilian (CQL), and south-central Tibet (which is itself probably an amalgam of at least two and possibly three terranes, THM, CNT, and NNT, according to stratigraphic columns of the Tibet Bureau of Geology, summarized in Bally et al, 1980).

At the west and southwest edge of the Yangtze block, late Paleozoic to Jurassic subduction and the collision of Indochina (an amalgamation of the Shan–West Malay [SWM], Tak [TAK], Uttaradit [UTD], and Khorat–Kontum [KKM] terranes) during the Jurassic resulted in the accretion of several volcanic island-arc fragments, continental fragments, and oceanic terranes, including the Yushu Muli arc (YMU), the Salween foldbelt (SLW), and the Mekong foldbelt (MEK). The suture zone formed during this collision and marked by ophiolite and blueschist is called the Kexelili–Jinsha–Ailao Shan suture by Li et al (1980) and corresponds to the Red River depth fracture of Huang (1978).

The details of stratigraphy, amalgamation, and accretion, insofar as they are known (dates are very approximate), are shown in Figure 5b. Clearly, this whole region is a complex composite of volcanic island arcs, oceanic islands, deep-sea sediments and ophiolites of various ages, and small continental fragments. However, the southern margin of Sino-Korea–Tarim did not experience sequentially southward-stepping subduction on a small scale, as intermediate stages of amalgamation did occur, for example the amalgamation of Tsinling and Kunlun, which later accreted to Nan Shan and Sino-Korea–Tarim. Continued northward motion of India under China is resulting in dispersive effects throughout eastern China by left-slip faults and easterly extension.

Australia and New Zealand

In Australia, numerous Precambrian blocks can be distinguished that apparently have distinctly different histories prior to Phanerozoic time (Fig. 1). These Archean and Proterozoic massifs (such as the Yilgarn [EYA, NOW, WYA, YIL], Kimberly [KIM], and Pilbara [PIL] terranes) are surrounded by Proterozoic foldbelts that most probably represent early formed suture zones. Examples of these Proterozoic foldbelts include the Albany-Fraser (ALF), Ashburton (ASB), Gawler (GAW), and King Leopold–Halls Creek (KLH) terranes (Scheibner, 1983, written communication). The eastern margin of this Precambrian amalgam is the Cambrian rift zone. This rifting was followed by Paleozoic accretion of volcanic arcs and ophiolites represented by the Kanmantoo, Lachlan, and Thomson foldbelts, and includes terranes such as Anakie–

Nebine (ANN), Kanmantoo–Glenelg (KMG), Molong–Monaro (MMO), and Wagga–Omeo (WAO). Further to the east, the New England foldbelt is a composite of middle to late Paleozoic and early Mesozoic oceanic and volcanic-arc terranes, including Wandilla (WAN), Gympie (GYM), and Texas–Woolomin (TEW) (Scheibner, 1983, written communication); these terranes are overlapped to the south by Triassic strata, but to the north overlap assemblages are younger. In northwestern Australia, terrane accretion is still younger, characterized by the collision of Timor and the Banda arc with Australia during the Late Tertiary, forming the Timor–Seram terrane (TMS). In the Middle Cretaceous, the rifting away of New Zealand and the Lord Howe rise and the opening of the Tasman Sea during the Late Cretaceous to Paleocene caused a passive margin to develop in eastern Australia.

Terrane designations in New Zealand follow the scheme of Howell (1980). The Tuhua terrane (TUH) in western New Zealand represents the Paleozoic margin of Gondwana built on continental crust; it is a composite of the Karamea, Golden Bay, and Drumduan terranes (Bishop et al, this volume). The Hokonui terrane (HOK), east of the Tuhua terrane and separated by the Median tectonic line, is composed of upper Paleozoic to Mesozoic volcanic arc and ophiolitic assemblages that were accreted to the Tuhua terrane during the Middle Cretaceous. The Caples terrane (CAP) represents a margin of a volcanic arc; Caples strata pass eastward into a zone where the Haast Schist metamorphic overprint has obliterated stratigraphic relations. The eastern facies of the Haast Schist is composed of quartzofeldspathic rocks of the Torlesse terrane (TOR). The Haast Schist thus represents a cryptic suture along which two dissimilar but coeval stratigraphic assemblages were welded together. The Torlesse graywacke terrane contains several thrust sheets of Permian to Triassic submarine fan strata mixed with Permian and younger pelagic and volcanic oceanic rocks that are sequentially younger to the east. All these terranes were deformed during the Rangitata orogeny and assembled by the middle of the Cretaceous. They then rifted away from the cratonal Antarctica–Australia parts of Gondwana during the opening of the Tasman Sea. Late Cenozoic right-slip along the Alpine fault zone has slivered and dispersed the terranes.

Oceanic Plateaus

Oceanic plateaus are anomalously high parts of the sea floor that are not at present parts of continents, active volcanic arcs, or spreading ridges (Ben-Avraham et al, 1981; Nur and Ben-Avraham, 1982). As depicted in Figure 1, oceanic plateaus comprise fragments of continents, oceanic islands, hotspot tracks, remnant arcs, and other anomalously thick volcano-plutonic piles. Continental fragments such as the Chatham Rise, Lord Howe Rise, and Exmouth Plateau seem to occur relatively close to cratonal masses from which they have been rifted. Oceanic islands such as the Mid-Pacific Mountains and hotspot tracks such as the Cocos, Carnegie, and Hawaiian-Emperor Ridges are abundant throughout the Pacific basin. Examples of remnant arcs include the Bowers, Aves, and Palau-Kyushu Ridges. Several plateaus such as Ontong–Java and Hess Rise are composed of oceanic island basalt and sediments in their uppermost portions, but seismic-refraction data indicate that they have thick, continental-like structure at depth (Nur and Ben-Avraham, 1982; Vallier et al, 1981). Oceanic plateaus are not tectonostratigraphic terranes as defined above but may become accreted terranes in the future. When these anomalously thick and light bodies encounter a subduction zone, they are more likely to be obducted than subducted. For example, recent seismic data indicate that the southern edge of the Ontong–Java Plateau may have been obducted onto several of the Solomon Islands (Kroenke, 1972; Coleman and Kroenke, 1981). Other plateaus that have collided with fossil subduction zones include the Umnak Plateau, the Shirshov Ridge, and the Sea of Okhotsk Plateau; those encountering modern trenches include the Nazca and Carnegie Ridges and the Ogasawara Plateau.

THE GROWTH AND SHAPING OF CONTINENTS

From the perspective of tectonostratigraphic terrane analysis, one visualizes the shape of continents to be the result of both terrane accretion and terrane dispersion. The accretion of terranes results in continental growth or outbuilding, while the dispersion of terranes, by either rifting or sliding, results in the diminution of continents. As an example, the North American continent is composed of Archean massifs that are surrounded by several Proterozoic foldbelts that suggest successive terrane accretions between 2 and 1 b.y. ago. Upper Proterozoic rifting sequences outboard of upper Proterozoic glacial deposits suggest that North America was probably the interior of a large continental agglomeration at that time. During most of the Paleozoic, the Cordilleran and Ouachitan regions were passive margins while the Franklinian and Appalachian margins experienced growth owing to terrane accretion. By the late Paleozoic, the east and southeast margins were again interior regions of a supercontinent but were reshaped by rifting during the Mesozoic. Beginning in the late Paleozoic, and still active today, the Cordillera experienced growth owing to numerous accretionary events; these combined with dispersive processes to distend terrane slivers northward along the margin.

CONCLUSIONS

An extensive study of the geology of the circum-Pacific has revealed that this vast region is composed of a complex mosaic of tectonostratigraphic terranes. Stratigraphic, metamorphic, and disrupted terranes can be identified; these comprise fragments of continents, volcanic arcs, oceanic basement and cover, and continental margin sedimentary packages. For a few terranes, large translation and/or rotation has been documented; but most terranes remain "suspect" until further geologic and geophysical data can be collected.

Accretion of terranes seems to have occurred at least throughout Proterozoic and Phanerozoic time. In a general way, terranes have been accreted in discrete pulses that correspond with periods of orogenic activity in the hinterland and coeval deformation in the foreland. The presence of the same lithotectonic elements in oceanic plateaus as onshore accreted terranes, often dismembered and detached from their basement, suggests that continental growth and orogeny occurs as a result of collision of oceanic plateaus at continental margins. The present assemblage of volcanic arcs, continental fragments, and small ocean basins in the southwest Pacific provides a modern analogue for processes of terrane accretion. But further research detailing the nature, extent, and geologic history of individual terranes is needed to formulate a more precise explanation for the tectonic history of the Pacific Basin.

ACKNOWLEDGMENTS

The data and concepts developed in this manuscript reflect distillation of ideas discussed by many of our colleagues during the past several years. To write this paper we depended upon the expertise of others for many regions of the circum-Pacific, in particular, Erwin Scheibner for Australia, Edmund Zhang for China, Kazuya Fujita for Siberia, J. W. H. Monger for the Canadian Cordillera, N. J. Silberling for western North America, J. E. Case for the Caribbean region, Jose Corvalan for South America, and Zvi Ben-Avraham for the classification of oceanic plateaus. Early versions of the manuscript benefited from reviews by G. W. Moore and D. K. Larue.

REFERENCES

Armstrong, A. K., and E. M. MacKevett, Jr., in press, Geologic relations of Kennecott-type copper deposits, Wrangell Mountains, Alaska: U.S. Geological Survey Professional Paper 1212-A, 26 p.

Bally, A. W., et al, 1980, Notes on the geology of Tibet and adjacent areas—Report of the American Plate Tectonics Delegation to the People's Republic of China: U.S. Geological Survey Open-File Report 80-501, 100 p.

Beck, Jr., M. E., and A. Cox, 1979, Paleomagnetic evidence for large-scale tectonic rotations and translations along the western edge of North America, *in* J. M. Armentrout, et al, eds., Cenozoic paleogeography of the western United States: Society of Economic Paleontologists and Mineralogists, Pacific Section, Pacific Coast Paleogeography Symposium 3, p. 325.

Ben-Avraham, Z., et al, 1981, Continental accretion and orogeny: From oceanic plateaus to allochthonous terranes: Science, v. 213, p. 47–54.

Berg, H. C., et al, 1978, Map showing pre-Cenozoic tectonostratigraphic terranes of southeastern Alaska and adjacent areas: U.S. Geological Survey Open-File Report 78-1085, scale 1:1,000,000.

Blake, Jr., M. C., et al, 1982a, Preliminary tectonostratigraphic terrane map of California: U.S. Geological Survey Open-File Report 82-593, scale 1:500,000.

______, 1982b, Relation between orogeny, subduction, and changing-plate motion models for the Pacific coast region of the United States: EOS, American Geophysical Union Transactions, v. 63, p. 911.

Boucot, A. J., et al, 1980, An Early Devonian, Eastern Americas Realm faunule from the coast of southern Peru: Journal of Paleontology, v. 54, p. 359–365.

Bourgois, J., et al, 1982, The Andean ophiolitic megastructures on the Buga-Buena Ventura transverse (Western Cordillera-Valle Columbia): Tectonophysics, v. 82, p. 207–229.

Bundtzen, T. K., and D. L. Turner, 1979, Geochronology of metamorphic and igneous rocks in the Kantishna Hills, Mount McKinley quadrangle, Alaska: Alaska Division of Geology and Geophysical Surveys, Geologic Report 61, p. 25–30.

Burke, K., et al, 1977, World distribution of sutures—the sites of former oceans: Tectonophysics, v. 40, p. 69–99.

Campa, M. F., and P. J. Coney, 1983, Tectonostratigraphic terranes and mineral resource distributions in Mexico: Canadian Journal of Earth Sciences, v. 20, p. 1040–1051.

Case, J. E., et al, 1971, Tectonic investigations in western Colombia and eastern Panama: Geological Society of American Bulletin, v. 82, p. 2685–2712.

______, in press, Map of geologic terranes in the Caribbean region: Geological Society of America Memoir.

Champion, D. E., et al, 1980, Paleomagnetism of the Cretaceous Pigeon Point Formation and inferred northward displacement of 2500 km for the Salinian Block, California: EOS, American Geophysical Union Transactions, v. 61, p. 948.

______, 1981, Paleomagnetic evidence for 3800 km of northwestward translation of San Miguel Island, Southern California Borderland: EOS, American Geophysical Union Transactions, v. 62, p. 855.

______, in press, Paleomagnetic and geologic data indicating 2500 km of northward displacement for the Salinian, Sur Obispo, Tujunga, and Baldy(?) terranes, California: Journal of Geophysical Research.

Chinese Academy of Geological Sciences, 1975, Geological map of Asia, scale 1:5,000,000.

______, Compilation Group of the Geological Map of China, 1976, An outline of the geology of China: Peking, 22 p.

______, Institute of Geology, Section of Structural Geology, 1979, Tectonic map of China, scale 1:4,000,000.

______, 1979, Stratigraphy of China: (Abs.), 49 p.

Churkin, Jr., M., 1981, Continental plates and accreted oceanic terranes in the Arctic, *in* A. E. M. Nairn, et al, eds., The ocean basins and margins, v. 5, The Arctic Ocean: Plenum Press, p. 1–20.

Churkin, Jr., M., and G. D. Eberlein, 1977, Ancient borderland terranes of the North American Cordillera: correlation and microplate tectonics: Geological Society of America Bulletin, v. 88, p. 769–786.

Churkin,Jr., M., and J. H. Trexler, Jr. 1980, Circum-Arctic plate accretion—isolating part of a Pacific plate to form the nucleus of the Arctic basin: Earth and Planetary Science Letters, v. 48, p. 356–362.

Churkin, Jr., M., et al, 1980, Collision-deformed Paleozoic continental margin of Alaska: foundation for microplate accretion: Geological Society of America

Bulletin, v. 91, pt. 1, p. 648-654.

Coleman, P. J., and L. W. Kroenke, 1981, Subduction without volcanism in the Solomon Islands arc: Geo-Marine Letters, v. 1, p. 129-134.

Coney, P. J., 1981, Accretionary tectonics in western North America: Arizona Geological Society Digest, v. 14, p. 23-37.

Coney, P. J., D. L. Jones, and J. W. H. Monger, 1980, Cordilleran suspect terranes: Nature, v. 288, p. 329-333.

Dalziel, I. W. D., et al, 1974, Fossil marginal basin in the southern Andes: Nature, v. 250, p. 291-294.

Davis, G. A., et al, 1978, Mesozoic construction of the Cordilleran "collage," central British Columbia to central California, *in* D. G. Howell and K. A. McDougall, eds., Mesozoic paleogeography of the western United States: Society of Economic Paleontologists and Mineralogists, Pacific Section, Pacific Coast Paleogeography Symposium 2, p. 1-32.

de Almeida, F. F. M., 1978, Tectonic map of South America: Geological Society of America Map and Chart series, MC-32, scale 1:5,000,000.

Fujita, K., 1978, Pre-Cenozoic tectonic evolution of northeast Siberia: Journal of Geology, v. 86, p. 159-172.

______, 1983, Accretionary terranes and tectonic evolution of northeast Siberia, *in* M. Hashimoto and S. Uyeda, eds., Accretion tectonics in the circum-Pacific regions: Boston, D. Reidel Publishing Co., p. 43-58.

Fujita, K., and J. T. Newberry, 1982, Tectonic evolution of northeastern Siberia and adjacent regions: Tectonophysics, v. 89, p. 337-357.

Gonzalez-Bonorino, F., 1971, Metamorphism of the crystalline basement of Central Chile: Journal of Petrology, v. 12, p. 149-175.

Goossens, P. J., and W. I. Rose, Jr., 1973, Chemical composition and age determination of tholeiitic rocks in the basic igneous complex, Ecuador: Geological Society of America Bulletin, v. 84, p. 1043-1052.

Goossens, P. J., et al, 1977, Geochemistry of tholeiites of the Basic Igneous Complex of northwestern South America: Geological Society of America Bulletin, v. 88, p. 1711-1720.

Hargraves, R. B., and Skerlec, G. M., 1980, Paleomagnetism of some Cretaceous-Tertiary igneous rocks on Venezuelan offshore islands, Netherlands Antilles, Trinidad and Tobago: 9th Caribbean Geological Conference, Santo Domingo, Dominican Republic, p. 509-515.

Herbert, H. J., 1977, Die Grunschiefer der Ost-kordillere Ecuadors und ihr metamorpher Rahmen: University of Tubingen Dissertation, 183 p.

Hillhouse, J. W., 1977, Paleomagnetism of the Triassic Nikolai Greenstone, McCarthy Quadrangle, Alaska: Canadian Journal of Earth Sciences, v. 14, p. 2578-2592.

Hillhouse, J. W., and C. S. Grommé, 1982, Limits to northward drift of the Paleocene Cantwell Formation, central Alaska: Geology, v. 10, p. 552-556.

Howell, D. G., 1980, Mesozoic accretion of exotic terranes along the New Zealand segment of Gondwanaland: Geology, v. 8, p. 487-491.

Howell, D. G., et al, 1983, Tectonostratigraphic terranes of the frontier circum-Pacific region: Bulletin of the American Association of Petroleum Geologists, v. 67, p. 485-486.

Huang, T. K., 1978, An outline of the tectonic characteristics of China: Eclogae Geologicae Helvetiae, v. 71, p. 611-635.

Hudson, T., et al, 1979, Paleogene anatexis along Gulf of Alaska margin: Geology, v. 7, p. 573-577.

Irving, E., et al, 1980, New paleomagnetic evidence for displaced terranes in British Columbia, *in* D.W. Strangway, ed., The continental crust and its mineral deposits: Geological Association of Canada Special Paper 20, p. 441-456.

Irwin, W. P., 1972, Terranes of the western Paleozoic and Triassic belt in the southern Klamath Mountains, California: U.S. Geological Survey Professional Paper 800-C, p. 103-111.

______, 1977, Review of Paleozoic rocks of the Klamath Mountains, *in* J. H. Stewart, et al, eds., Paleozoic paleogeography of the western United States: Society of Economic Paleontologists and Mineralogists, Pacific Section, Pacific Coast Paleogeography Symposium 1, p. 441-454.

Jones, D. L., and N. J. Silberling, 1979, Mesozoic stratigraphy—the key to tectonic analysis of southern and central Alaska: U.S. Geological Survey Open-File Report 79-1200, 37 p.

Jones, D. L., et al, 1977, Wrangellia: a displaced terrane in north-western North America: Canadian Journal of Earth Sciences, v. 14, 2565-2577.

______, 1981, Tectonostratigraphic terrane map of Alaska: U.S. Geological Survey Open-File Report 81-792, scale 1:2,500,000.

______, 1982a, Preliminary tectonostratigraphic terrane map of the circum-Pacific region: Bulletin of the American Association of Petroleum Geologists, v. 66, p.972.

______, 1982b, Timing of major accretionary events in Alaska: EOS, American Geophysical Union Transactions, v. 63, p. 913-914.

______, 1983, Recognition, character, and analysis of tectonostratigraphic terranes in western North America, *in* M. Hashimoto and S. Uyeda, eds., Accretion tectonics in the circum-Pacific regions: Boston, D. Reidel Publishing Co., p. 21-36.

Kosygin, Y. A., and L. M. Parfenov, 1981, Tectonics of the Soviet Far East, *in* A. E. M. Nairn, et al, eds., The ocean basins and margins, v. 5, The Arctic Ocean: Plenum Press, p. 377-412.

Kroenke, L. W., 1972, Geology of the Ontong-Java Plateau: Hawaii Institute of Geophysical Research Report HIG-72-S, 119 p.

Li, C. (C. Y. Lee), et al, 1980, A preliminary study of plate tectonics of China: Bulletin of the Chinese Academy of Geological Sciences, Series I, v. 2, no. 1, p. 11-22.

Luyendyk, B. P., et al, 1982, Simple shear of southern California during the Neogene: paleomagnetic evidence: EOS, American Geophysical Union Transactions, v. 63, p. 914.

Macdonald, W. D., and N. D. Opdyke, 1972, Tectonic rotations suggested by paleomagnetic results from northern Colombia, South America: Journal of Geophysical Research, v. 77, p. 5720-5730.

Maresch, W. V., 1974, Plate-tectonics orgin of the Caribbean mountain system of northern South America: Geological Society of America Bulletin, v. 85, p. 669-682.

McElhinny, M. W., et al, 1981, Fragmentation of Asia in the Permian: Nature, v. 293, p. 212-216.

McWilliams, M. O., and D. G. Howell, 1982, Exotic terranes of western California: Nature, v. 297, p. 215-217.

Ministry of Geology of USSR, 1966, Geological map of USSR, scale 1:5,000,000.

Monger, J. W. H., 1977a, Upper Paleozoic rocks of the western Canadian Cordillera and their bearing on Cordilleran evolution: Canadian Journal of Earth Sciences, v. 14, p. 832-859.

______, 1977b, The Triassic Takla Group in McConnell Creek map-area, north-central British Columbia: Geological Survey of Canada Paper 76-29, 45 p.

Monger, J. W. H., and E. Irving, 1980, Northward displacement of north central British Columbia: Nature, v. 285, p. 289-294.

Monger, J. W. H., and R. A. Price, 1979, Geodynamic evolution of the Canadian Cordillera—progress and problems: Canadian Journal of Earth Sciences, v. 16, p. 770-791.

Monger, J. W. H., et al, 1982, Tectonic accretion and the origin of the two major metamorphic and plutonic welts in the Canadian Cordillera: Geology, v. 10, p. 70-75.

Munizaga, F., et al, 1973, Rb/Sr ages of rocks from the Chilean metamorphic basement: Earth and Planetary Science Letters, v. 18, p. 87-92.

Natal'in, B. A., and L. M. Parfenov, 1983, Accretion and collision eugeosyncline folded systems in the northwest of the Pacific framing, *in* M. Hashimoto and S. Uyeda, eds., Accretion tectonics in the circum-Pacific regions; Boston, D. Reidel Publishing Co., p. 59-68.

Nichols, K. M., and N. J. Silberling, 1979, Early Triassic (Smithian) ammonites of paleoequatorial affinity from the Chulitna terrane, south-central Alaska: U.S. Geological Survey Professional Paper 1121-B, 5 p.

Nilsen, T. H., 1977, Paleogeography of Mississippian turbidites in south-central Idaho, *in* J. H. Stewart, et al, eds., Paleozoic paleogeograpy of the western United States: Society of Economic Paleontologists and Mineralogists, Pacific Section, Pacific Coast Paleogeography Symposium 1, p. 275-299.

Nur, A., and Z. Ben-Avraham, 1981, Volcanic gaps and the consumption of aseismic ridges in South America, *in* L. D. Kulm, ed., Nazca Plate, crustal formation and Andean convergence: Geological Society of America Memoir 154, p. 729-740.

______, 1982, Oceanic plateaus, the fragmentation of continents, and mountain building: Journal of Geophysical Research, v. 87, p. 3644-3661.

Plafker, G., et al, 1980, Preliminary report on the geology of the continental slope adjacent to OCS lease sale 55, eastern Gulf of Alaska: U.S. Geological Survey Open-File Report 80-1089.

Powell, R. E., 1982, Crystalline basement terranes in the southern eastern Transverse Ranges, California, *in* J. D. Cooper, ed., Geologic excursions in the Transverse Ranges, southern California: Geological Society of America Guidebook, Cordilleran Section, 78th Annual Meeting, p. 109-136.

Roeder, D., and C. G. Mull, 1978, Tectonics of the Brooks Range ophiolites, Alaska: Bulletin of the American Association of Petroleum Geologists, v. 62, p. 1696-1702.

Saleeby, J. B., 1983, Accretionary tectonics of the North American Cordillera: Annual Review of Earth and Planetary Sciences, v. 15, p. 45-73.

Scholl, D. W., et al, 1980, Sedimentary masses and concepts about tectonic processes and underthrust ocean margins: Geology, v. 8, p. 564-568.

Shackleton, R. M., et al, 1979, Structure, metamorphism and geochronology of the Arequipa Massif of coastal Peru: Quarterly Journal of the Geological Society of London, v. 136, p. 195-214.

Shilo, N. A., and S. M. Til'man, 1981, The tectonic zones of northeastern USSR and the formation of its continental crust, *in* A. E. M. Nairn, et al, eds., The ocean basins and margins, v. 5, The Arctic Ocean: Plenum Press, p. 413-438

Silver, L. T., and T. H. Anderson, 1974, Possible left-lateral early to middle Mesozoic disruption of the southwestern North American craton margin: Geological Society of America Abstract with Programs, v. 6, p. 955-956.

Skerlec, G. M., and R. B. Hargraves, 1980, Tectonic significance of paleomagnetic data from northern Venezuela: Journal of Geophysical Research, v. 85, p. 5305-5315.

Souther, J. G., 1977, Volcanism and tectonic environments in the Canadian Cordillera—a second look: Geological Association of Canada Special Paper 16, p. 3-24.

Speed, R. C., 1979, Collided Paleozoic microplate in the western United States: Journal of Geology, v. 87, p. 279-292.

Terman, M. J., 1974, Tectonic map of China and Mongolia: Geological Society of America, scale 1:5,000,000.

Tipper, H. W., et al, 1981, Tectonic assemblage map of the Canadian Cordillera and adjacent parts of the United States of America: Geological Survey of Canada Map 1505A, scale 1:2,000,000.

Vallier, T. L., et al, 1981, The geology of Hess Rise, central north Pacific Ocean, *in* J. Thiede, et al, Initial Reports of the Deep Sea Drilling Project: Washington, DC, U.S. Government Printing Office, v. 62, p. 1031-1072.

Vedder, J. G., et al, 1983, Stratigraphy, sedimentation, and tectonic accretion of exotic terranes, southern Coast Ranges, California, *in* J. S. Watkins and C. L. Drake, eds., Studies in continental margin geology: American

Association of Petroleum Geologists Memoir 34, pp. 471–498.
Wang, Y.-G., 1981, An outline of the marine Triassic in China: International Union of Geological Sciences Publication 7, 21 p.
Weng, S., and Q. Kong, 1981, Mesozoic tectonism and magmatism of the lower Yangtze Valley: Bulletin of the Chinese Academy of Geological Sciences, Series I, v. 3, no. 1, p. 1–19.
Williams, H., and R. D. Hatcher, Jr., 1982, Suspect terranes and accretionary history of the Appalachian orogen: Geology, v. 10, p. 530–536.
Zeil, W., 1979, The Andes: A geological review: Berlin, Gebrüder Borntraeger, 260 p.
Zhang, Z. M., et al, 1984, An outline of the plate tectonics of China: GSA Bulletin, v. 95, p. 295–312.

APPENDIX

Key to Terrane Map Symbols

Plio = Pliocene	K = Cretaceous	Pz = Paleozoic	Dev = Devonian
Mio = Miocene	Jr = Jurassic	Ptz = Proterozoic	Sil = Silurian
Olig = Oligocene	Tr = Triassic	Perm = Permian	Ord = Ordovician
Eo = Eocene	Cz = Cenozoic	Carb = Carboniferous	Camb = Cambrian
Q = Quarternary	Mz = Mesozoic		
T = Tertiary (Ng, Neogene; Pg, Paleogene)			

SIBERIA

ALU Aluchin (South Anyui subterrane?): pre-Tr ophiolitic rocks

ALZ Alazeya: ophiolite overlain by late Pz–early Mz island arc; local blueschist metamorphism of chert, basalt, keratophyre

ANY Anadyr: lower-middle Mz ocean floor obducted in Late Jr–K, overlain by Cenozoic sed basin strata

BKJ Bureya–Khanka–Jiamusi massif: continental massif with Precambrian basement, Ptz to Pz sed and volc strata

CHK Chukotsk: continental block, probably part of North Slope Alaska (AAN)

CKC Cape Kamchatka: Upper K gabbro, ultramafic overlain by Upper K–Eo island arc volcs, basalt, chert, tuffs, clastics

CKM Central Kamchatka: greenschist, quartzite, metamorphosed mafic and ultramafic rocks of Jr or Early K age in Khavyen uplift; most of terrane is Olig–Qt volcaniclastic basin fill

CPM Central Primor'e: late Pz foldbelt; Carb–Early Perm basalt, chert, carbonate, and clastics, Early Perm ultramafics and gabbro, overlain by Late Perm island arc volcs, may be accreted by latest Perm; overlapped by Perm–Mz molasse

CSK Cherskiy: lower and upper Pz carbonate, volc, ultramafic rocks and Tr–Jr island arc volc rocks; basement possibly oceanic

EKM East Kamchatka: composite Pz(?) or Mz(K) ultramafic, mafic schist metamorphosed in Late K, thrust over Upper K (Maestrichtian) felsic and mafic schists of island arc origin, blueschist at contact, Olig–Mio volc overlap sequence

EKO Ekonay (Koryak subterrane): Sil–Carb ocean crust obducted in Middle K (pre-Coniacian)

EPM East Primor'e: Mz foldbelt; Tr–K basalt, chert, clastic, and carbonate strata, K clastic and carbonate rocks, may have Jr–Early K ophiolite at base, exposed on western border; Middle–Late K plutons stitch suture

ERP Eropol (Omolon subterrane?): possible island arc sequence within OMO; has deeper water facies, more volc rocks, Tr–Jr unconformity

KEP Keperveyem massif (North Anyui subterrane): Precambrian crust, Pz strata

KHR Khroma massif: possible continental microplate—geophysical data only

KOT Koryak–Taygonos Peninsula region: composite island arcs, subduction complex recording subduction migrating NW–SE from Carb–Jr. Arc accretion in Late Jr, 183 m.y. ophiolite obducted in Early K

KOY Koryak: 380 m.y. ocean crust, 320–350 m.y. blueschists thrust over Upper Jr–Hauterivian sed rocks; ophiolite may be obducted onto an island arc, which was later accreted to Siberia

KRO Kronitsky: sea-mount terrane; ultramafics at base, submarine volcs, chert, overlain by alkalic basalts; Late K–Olig age; accreted mid-Mio

KUH Kuhktuy (Okhotsk massif): pre-Pz continental massif, possibly collided with Siberia in Perm

KVK Kvakhon: Upper Jr–Lower K blueschist facies island arc metavolcanics, overlain by Lower–Upper K coarse clastics with blueschist pebbles, thrust over ophiolitic(?) melange and breccia; thrust over MAL in mid-K; composite with Omgon terrane of Fujita (this vol.): gabbro, chert, volcs overlain by mid-K turbidites, coarse clastics, Pg strata; Olig–Mio volcanic overlap sequence.

MAL Malkinsky: polymetamorphosed crystalline schist, gneiss, amphibolite, pelites intruded by granites, overlain by phyllites, metasandstones, metavolcanics, and Paleocene strata; basement may be Mz but has Pz and older(?) material included in it; metamorphosed at 100 m.y. and 50 m.y.

MGO Mongolo-Okhotsk: "eugeosynclinal" sed rocks, may be margin of Siberia from pre-Pz to late Pz; may be composite and contain Late Jr island arc rocks; Mz rocks intensely deformed in Late Jr to Middle K collision between SOK and BUR

NAY North Anyui: volcanic uplifts separated by fault-bound slate-filled basins; might be island arc(s), probably late Mz

NOV Novosibirsk: early Pz carbonate platform, possibly part of Siberian margin or Cherskiy terrane.

OLO Oloi: Upper Jr–Lower K island arc volc rocks, oceanic basement, deformed coastal marine rocks; accreted Early K(?)

OLY Olyutorsk: Upper K (Maestrichtian) oceanic crust obducted in Paleocene to Eo

OMO Omolon: continental basement; lower Pz volc rocks, Pz–Mz shallow-marine seds and volc rocks, accreted in Middle–Late Jr

PRK Prikolymsk: lower Pz coarse clastic, Pz–Mz marine sed

rocks over continental basement; accreted in Middle Jr(?)
SAY South Anyui: oceanic crust and deep-sea sed rocks—eugeosyncline; closed in Hauterivian–Valanginian; contains Upper Jr ophiolite, Jr–K arc volc and sed rocks
SKA Sikhote Alin: may have continental basement; Pz to Jr basalt, graywacke, shale, and siliceous rocks with Middle K–Aptian arc developed on top. Perm paleolat. = 33.9°N
SOK South Kamchatka: probable oceanic basement, (?) age
SPM South Primor'e: middle Pz foldbelt; Camb to Early Dev ultramafic, gabbro, basalt, chert, carbonate, and clastic rocks, Dev volc and granitic rocks; folded in Late Dev; overlain by Permo–Tr clastic rocks and basalt
SRE Sredinny Ranges: Cenomanian mafic volc and siliceous rocks, tuffs, overlain by Paleocene calc-alkaline volcs, cherty and terrigenous strata; ultramafics along contact with MAL; Mio volc, plutonic rocks overlap and stitch this terrane to others in Kamchatka
UML Uda–Murgal: accreted arc with ophiolite and blueschist border at south; age estimates of Jr and 100–130 m.y.
VKH Verkhoyansk: margin of Siberian platform; thick platform sed rocks in eastward thickening wedge
YRK Yarakvaam (South Anyui sub terrane?): deep-marine sed, arc volc rocks of ? age overlain by coastal-terrigenous sed, volc rocks; uplifted in Tr, accreted pre-Aptian

---

CHINA AND MONGOLIA

(*indicates subdivisions of the marine Triassic in China of Wang et al, 1981; many of the district names have been changed. All have Tethyan faunal affinities except NDH.)

ALS Ala Shan: probably a continental block connected to SKE, TAR(?)
ALT Altay fold belt: composite lower–upper Pz (mostly Dev) arc rocks, Ord–Sil metamorphic rocks; ultramafics; Late Perm island arc collision in south
CHA Chaidam basin: probable continental block, accreted to TAR and Qilian composite (NNS, NQL, QIN) in early Pz(?) or Dev with ophiolitic suture of Astin-Tagh Mtns; pre-K basement rocks of unknown type
CNT* Central Nianqing–Tangla: (equivalent to part of Gandesi–Nianqing terrane of Xiong (this vol) and Zhang et al (1984); possible pre-Pz basement overlain by Ord–Perm marine sed rocks; Dev–Carb submarine mafic volc rocks and turbidites may indicate opening of Neo-Tethys; Late Perm–Late Jr unknown, possible hiatus; Lower–Middle K volcs and marine deposits change to continental facies in Late K post accretion of this part of Tibet
CQL Central Qilian: microcontinent with Precambrian basement, early Pz marine clastic rocks, Dev–Carb red beds and nonmarine clastic rocks, Perm–Tr marine clastic rocks
CTS Central Tienshan (formerly Peishan): probably a continental block accreted pre-Tr
CTY* Cathaysian fold system: (= Huanan terrane of Xiong); pre-Dev arc, includes pre-Pz to Sil metamorphic and volc rocks, late Pz shallow-marine, Tr continental strata, amalgamated with YGT by Sil(?) or Carb; Mz accretion likely, stitched by Tr, Jr, K granites
DAB Dabie Shan: (= Huaiyang terrane of Zhang et al, 1984); may be a microcontinent between Yangtze and Sino-Korean blocks; has Precambrian crust, Pz clastic strata, upper Pz–Mz granitic rocks, and is bounded by ophiolite belts
GBI Gobi foldbelt: composite, contains Late Dev–Carb island arc, upper Pz ophiolite and geosynclinal sed rocks; probably connected to ALT
GYJ* Garze–Yajiang area: (= part of Songpan terrane of Xiong) Tr foldbelt; shallow-marine sed strata, flysch, minor volc rocks
HHC Hangayn–Hanteyn crystalline belt: middle–upper Pz geosynclinal sed rocks, lots of granitic plutons, possibly with lower Pz ophiolitic boundary to North
HNJ Hunan–Jiangxi: probably has pre-Pz basement, marine clastic strata deformed in early Pz, overprinted in late Pz; Tr shallow marine, Jr–K nonmarine
INM Inner Mongolia foldbelt: composite; apparently a series of accreted arcs, Sil–Perm, some built on lower Pz and older metamorphic rocks; several ophiolite suites of different ages and geosynclinal sed strata
IYS Indus–Yaluzangbu (Tsangpo) suture zone: Jr–K ophiolite thrust over Middle K–Eo volcanogenic sed rocks; Upper K–Paleocene outer arc sed rocks crushed as suture formed when Himalayas/India collided with Tibet
KCH Kachin: high-grade metamorphic basement may be pre-Pz; this continental block could either be a part of the Shan massif (SWM) or a part of Tibet (CNT) faulted to the south
KUN* Kunlun foldbelt: composite Ord–Dev island arc, subduction complex, Late Dev–Carb ophiolite emplacement and arc accretion, Permo–Carb geosynclinal sed and granitic rocks, Mz marine sed strata; accreted in late Pz (?)
LYZ* Lower Yangtze: may be margin of Yangtze block, with Precambrian basement, Dev–upper Pz geosynclinal strata deformed in Permo–Tr
MEK* Mekong foldbelt: (= Dianxinan terrane of Xiong); subduction zone, suture between Indochina and Yangtze; Pz–Mz deep-marine sed rocks, flysch, Tr–Jr(?) ophiolite and blueschist formed during Late Tr–Jr collision; may include Tr volc island arc, possible fragments of pre-Pz basement
MOT Motienling: early Pz ophiolite obducted during collision of SBP and YGT
NDH* Nadan hada foldbelt: upper Pz geosynclinal strata, Mz deep-marine-bathyal strata with Boreal fauna, tuffs, volc rocks
NNS* Nan shan: (= part of Nanqilianshan terrane of Xiong); composite early Pz(?) fold system; lower Pz deep-marine rocks, Upper Dev–Carb island arc volc rocks, upper Pz ophiolite, geosynclinal strata deformed in Permo–Tr collision of ALS and TAR and later during collision of SKE; NNS, HQL, QIN may be amalgamated by early–mid-Pz
NNT* North Nianqing–Tangla: (= part of Nianqing–Gandesi terrane of Xiong); Upper Tr–Upper K deep-marine chert, volcanogenic ss shale are intensely deformed and overlie Ord–Sil marble and shale (late Pz unknown); Upper K–T shallow marine to continental clastic rocks are undeformed; may indicate Late K accretion along ophiolitic suture to north with TGS
NQL North Qilian: (= Beiqilianshan and Zhongqilianshan terranes of Xiong); composite early–middle Pz foldbelt; early–middle Pz ophiolite, island arc volcanic, geosynclinal strata, Carb blueschist
NTH* Northern Tethyan Himalaya: pre-Pz gneissic basement overlain by Pz marine clastic strata, very thick Tr deep-sea flysch, chert, mafic volc rocks, overlain by Jr quartzofeldspathic ss shale, ls; possibly deep-marine northern margin of India during Mz, telescoped during collision of India with Tibet
NTS Nantienshan: late Pz shallow-marine deposits with volc detritus; probably accreted Perm–Tr
ONA Onon–Argun foldbelt: early Pz fold system; middle–upper Pz ophiolitic, ultramafic, granitic, and geosynclinal rocks
OUL Oulungbruk: microcontinent; Precambrian basement, Sinian–Pz clastic, carbonate strata, tillite, Middle–Upper Jr nonmarine rocks
QAM* Qamdo area: (= Sanjian terrane of Xiong, this vol.); possible Perm island arc built on Ord–upper Pz ocean crust, collides during northward movement of Tibet; marine deposition

until Middle Jr, followed by deposition of terrigenous coarse clastic rocks

QIN Qinling fold belt: (in part = Nanqilianshan terrane of Xiong); composite of upper Pz(?) to Mz ophiolite belts, possible lower-middle Pz arc rocks, Perm-Carb unstable continental shelf sequences, Tr graywacke, intermediate-silicic volc and intrusive rocks

SBG* Songpan-Ganzi fold system: mostly Tr feldspathic graywacke, Perm intermediate volc rocks; no pre- to late Pz rocks known; ophiolite emplaced in south during collision of Tibet in Tr

SEM Southeast Maritime: (= Zheminyanhai terrane of Xiong); late Pz foldbelt; Pz geosynclinal strata, deformed in late Pz; overlain and intruded by extensive Jr and K volc and plutonic rocks, deformed again in Mz; may have Mz ophiolite at western border suture; Late Tr accretion

SGP Songpan: possible continental block with Archean basement rifted from SCH block; bounded by deep fractures and thrusts

SKE Sino-Korea: (= Huabei + Yishan terranes of Xiong); microcontinent, Precambrian (1.7 b.y.) basement; Pz platform, shallow-marine deposits; shows characteristic, ubiquitous Sil-Dev hiatus; paleomag shows Perm paleolat. 11°N for central portion

SLO Songliao basin: may have pre-Pz basement and may be a microcontinent; thick K sed rocks may be overlap on previously accreted INM

SLW Salween fold belt: upper Pz platform deposits, Tr marine, geosynclinal strata; 217-197 m.y.-old blueschist; Jr ophiolite belt

STH* Southern Tethyan Himalaya: pre-Pz gneissic basement overlain by Pz marine clastic rocks, thick Tr fossiliferous miogeoclinal facies, marine sedimentation through Middle T; may be northern margin of India/Gondwana during Mz; has shallower water facies than NTH

SYT Sayan-Tuva foldbelt: composite early Pz fold system; ophiolitic and granitic rocks (early Pz?), upper Pz continental clastic and basin rocks; in north, late Precambrian and early Pz ophiolitic and deep-marine rocks thrust over Precambrian and Pz sialic rocks

TAR Tarim: microcontinent with complete Pz carbonate sequence, Mz platform deposits; ophiolite belt borders; accreted in Perm

TCR Taiwan coast ranges: Neogene volcanoclastic sed rocks; Plio-Pleistocene ophiolitic melange border with WTW.

TGS* Tangla shan: (= Zangbei terrane of Xiong); metamorphic basement age unknown, overlain by Pz (Dev?) to Tr marine platform facies; 99-20 m.y. intrusives in south, folding in Jr-K

THM* Transhimalaya range: (= part of Gandesi-Nianqing terrane of Xiong, and Zhang et al, 1984); possible Tr island arc built on Pz oceanic crust and sed rocks; overlain by Late K-Eo arc complex and intruded by widespread granitics averaging 40-70m.y.; paleomag shows more than 2,500 km (1,550 mi) northward movement since 50 m.y.B.P.

TNS Tien Shan: (= Beitienshan terrane of Xiong); composite of island arcs, flysch basins, ocean floor; ophiolite and arc rocks as old as Ord-Sil, Dev-Carb; may be part of early-late Pz paleo-Tethys; accretion in Perm, change to continental facies sedimentation

WTA Wan-ta foldbelt: (= part of Dongbei terane of Xiong); may have Precambrian basement, overlain by Jr and K nonmarine strata and mafic volc rocks, accreted Late Tr (?)

WTW Western Taiwan: Paleogene argillite-slate series and Neogene clastic rocks on pre-T metamorphic complex; Pleistocene andesites in north

YGT Yangtze: late Ptz basement, Sinian-mid-Tr marine strata; Perm basalts, Late Tr terrestrial deposits; Perm paleolat. 2°N; accreted Late Tr-Early Jr

YHW Yu-hwai: (= part of Qinba terrane of Xiong); may have pre-Pz basement; Pz and early Mz limestone; seems to be Late Tr (?) suture zone between SKE and Yangtze continents; Jr-K pluton stitching

YID* Yidun area: (= Baiyu-Yidun terrane of Xiong, Zhongza terrane of Zhang et al, 1984); may be a microcontinent caught in subduction complex of MEK; probable Precambrian basement overlain by Sinian-Lower Carb carbonate and marine clastic strata

YMU Yushu-Muli: Tr island arc volcs, Upper Tr melange; ophiolite and blueschist belt is eastern boundary

YNS Yin Shan: may be margin of SKE block; seems to have Precambrian basement; bordered on all sides by ophiolites (350-331 m.y. ultramafics), with northern border ophiolites becoming younger to north toward INM

ZFB Zunggarian foldbelt: may connect with ALT, GBI; Ord-Dev, some upper Pz ophiolitic and island arc rocks, Carb-Perm deep-marine clastic, volc rocks, and chert; ophiolite possibly obducted in Late Perm during collision of island arc between ALT and DZS

ZSB Zunggar stable block: continental block, probably accreted pre- to late Pz, may connect to ALS, INM

SOUTHEAST ASIA AND INDONESIA

AYM Arakan-Yoma: composite; Tr-K shallow-deep-marine strata, highly deformed Lower T ultramafic rocks, flysch in suture between AYM and CBB—may be equivalent to IYS

BAG Banggai-Sula: continental fragment from New Guinea; Pz and early Mz granitic and metamorphic rocks, Mz stable shelf, shallow-water strata, Lower T ss and ls

BAT Batjan: probably a fragment of a T island arc; undated granitic and metamorphic rocks, QT arc volc rocks

BEN Bentong: ophiolitic melange, suture of late Pz (?) age between continental blocks KKM, SWM, and possible intervening late Pz arcs; overlapped by Tr granitic batholithic rocks; may be equivalent of UTD

BUT Buton: T melange and dismembered ophiolite; may include Mz rocks

CCB Central Burma Basin: upper M melange; thick Upper T shallow-marine clastic and carbonate strata

CSE Central-SE Sulawesi: Tr, Jr, K, and T(?) low-grade metamorphosed pelagic and terrigenous-clastic, chert, and ls strata; structurally complex

CSW Central Sulawesi: composite; Upper K, Pg blueschist and melange suture zone between WSW, BAG, TBP

EHM East Halmahera: T and K(?) ophiolite, melange

ESE East, southeast arms of Sulawesi: (composite); K-T ophiolite and Tr(?)-T deep-marine sed strata obducted in Mio; Mio melange overlain by BAG

KKM Khorat-Kontum: composite; possible pre-Pz basement, middle Pz shallow marine, eugeosynclinal strata, upper Pz arc rocks; nonmarine sed rocks in Indochina, in Malaysia Carb-Tr deep-marine and volcaniclastic strata; Tr tin granites; very thick Tr-K molasse; much of post-Tr is nonmarine; Perm-Carb flora in East Sumatra is Chinese affinity, Tethyan fusilinids present in East Sumatra

KNJ Central Kalimantan-NE Java: Mz "melange" with K ophiolitic rocks, K-T foreland basin strata

SBA Sabah: meta-ophiolite; may be Jr or older; overlain(?) by Upper K to Eo chert; locally thrust over Olig-Mio strata

SBS Sarawak-Brunei-Sabah: Eo melange

SUM Sumba: continental fragment from Java Sea-Sulawesi shelf; Pg and older (Jr?) highly deformed graywacke, carbonate, volc rocks overlain by Upper T-Holo shallow-water strata

SWK Western Sarawak: Jr-Lower K oceanic melange; may also be composite with Tr island arc volc rocks (correlative with

UTD?); overlain by latest K or Early T strata, intruded by Late K granitic plutons
SWM Shan-West Malaysia-Sumatra: (includes Baoshan terrane of Xiong); continental fragment; pre-Pz basement, Camb-Ord platform marine seds, thick Perm-Carb graywacke, coarse clastic rocks with western source; Perm "Gondwana tillites"; Mz shallow- to nonmarine strata may be correlative with Tibet, west border with CBB is ophiolite belt suture
TAK Chiang Mai-Tak: lower-middle Pz marine strata, upper Pz-Tr strongly deformed arc volc, sed rocks and tholeiitic intrusive and volc rocks; thrusting, metamorphism, and change to continental facies in Early to Middle Jr after SWM and KKM collision
TBP Tukang-Besi Platform: continental fragment, probably from New Guinea; Tr continentally derived strata tectonically overlain by Jr-Mio imbricately faulted marine strata, ophiolitic slices on Butung Island
UTD Uttaradit: Sil or older ophiolite obducted in Perm-Tr or Middle Tr; upper Pz island arc volc rocks, graywacke, Tr-Jr arc volc rocks; collision in Jr accompanied by thrusting, metamorphism, nonmarine sedimentation; in SW Kampuchea, mafic-ultramafic rocks intruded by undeformed 180 m.y.-old granitic rocks; may extend south to central Malaysia (BEN).
WSW West Sulawesi: K ultramafics overlain by K-T clastic and blueschist rocks, overlain by Ng island arc

PHILLIPINES

CEB Cebu-Bohol: Lower K diorite (105 m.y. Rb-Sr) and Upper K volc, marine sed strata
COT Cotobato: island arc active K-QT; collided with LSM in Mio
CPH Central Phillipines: Eo-Holo island arc rocks
LSM East Luzon-Samar-Mindanao: island arc active from K to Olig; ophiolite at base obducted during collision of COT
PAL Palawan: microcontinent, possibly from southern China; Carb-Jr slightly metamorphosed continental shelf strata overlain by Cz marine sed, arc volc rocks
PMD West Panay-Mindoro: Jr-K ophiolite and Mio-Recent ls and clastic rocks occur as blocks in melange
SPA South Palawan: Paleocene-Eo ophiolite and metamorphic rocks overlain by upper Olig and younger marine and continentally derived sed rocks
ZAM Zambales-Bayo: middle Eo ophiolite overlain by marine sed rocks
ZBS Zamboanga-Sulu: Upper K ophiolite, Paleocene metased and metavolc rocks overlain by Mio and younger arc rocks

JAPAN

ABU Abukuma: Pz sed rocks, metamorphic rocks and granitics
ASH Ashio: Jr graywacke
CHB Chichibu, Kurosegawa, and Sambosan combined: middle Pz continental knockers, Tr chert, volcaniclastic, volc rocks, local blueschist metamorphism
HDK Hidaka: overturned meta-ophiolite, high-T metamorphic rocks overlain by flysch, ocean basin strata
HID Hida: Precambrian metamorphic basement, Pz shelf strata, Tr-Jr metamorphic and granitic rocks, Jr-K thick molasse, extensive Upper K granites and rhyolites; amalgamated to MIN by Upper Jr (?) or Upper K
ISH Ishikari: Pz (?) to Lower K ophiolite overlain by deep-water seds, volcs, K turbidites; correlative rocks on north Sakhalin give K-Ar age of 87 m.y. for ophiolite dikes
IWZ Iwazumi: Upper Pz-Mz graywacke, chert, ls, basalt
KIT North Kitakami: Jr(?) metagraywacke
KMK Kamuikotan: high-grade blueschist rocks
MIN Mino: Permo-Carb-Tr sea-mount and deep-sea rocks, imbricated and high-T metamorphosed in Lower K; overlapped by Upper K rhyolites
NEM Nemuro: K-T green tuff and granitic rocks overlain by shallow-marine to nonmarine sed strata
RYK Ryoke: pre-upper Pz banded gneiss, upper Pz-Jr marine sed and mafic volc rocks metamorphosed to greenschist, amphibolite and migmatite in Late Jr-K (?)
SAB Sanbagawa: upper Pz-K (?) deep-marine turbidites, basalts, metamorphosed to blueschist, structurally interleaved with eclogite, serpentinite, granulite, amphibolite, overlain by unmetamorphosed, deformed Upper K sed rocks and flat-lying Eo strata
SAG Sangun: Carb, Perm, and Tr(?) mafic, ultramafic igneous rocks, marine sed rocks metamorphosed to blueschist (in Permo-Tr[?]), unconformably overlain by Upper Tr-T deep to shallow to coastal marine strata, Upper K silicic volc rocks
SHM Shimanto: Upper Jr-Lower K ocean floor assemblage, including chert, basalt, ultramafic rocks, distal turbidites
SKI South Kitakami: Pz-Mz neritic-littoral clastic, ls strata
TOK Tokuro: ophiolite, Jr subduction complex(?)

AUSTRALIA-NEW ZEALAND REGION, NEW GUINEA, AND ANTARCTICA

ALF Albany-Frazer: Ptz mobile and metamorphic belt, metamorphic and magmatic events at 1.69-1.56 and 1.3-1.25 b.y.
ANN Anakie-Nebine: metamorphosed Late Ptz to Camb turbidites, basic volcs, ophiolite(?); intruded by Ord granite; Camb arc and forearc rocks, lower Pz shallow-marine to continental sed rocks; Dev granites, sed rocks
ARA Arunta: Ptz mobile and foldbelt, old rocks correlate with KLH(?), 1.8 b.y. sed, volc rocks, 1.8-1.7 b.y. metamorphism, magmatism; younger high-grade metamorphism
ASB Ashburton: Ptz mobile and foldbelt; sed, volc rocks metamorphosed, deformed 1.9 b.y.; 1.9-1.6 b.y. granitic rocks
BOW Bowers: Camb-Ord island arc volc and sed rocks; cut by Sil-Dev intrusives, Dev volcs overlap
COC Middle Pz assemblage of chert, siltstone, and ss including mafic ign rocks, probable ocean floor deposit
CAP Caples: greenstones, serpentinite melange (? age), Perm volc, volcanogenic sed rocks
CAU Central Australia: Composite (ARA, MUS, PAT, and unknown); Ptz mobile and foldbelt, 1.4-1.3 b.y. deformation, metamorphism
CGE Coen-Georgetown: Ptz mobile and foldbelt, marine sed rocks, mafic volcs, metamorphism, granites 1,570 m.y.; 1,470, 1,400-1,300 m.y. granites, volcs, metamorphism
COL Cooper-Lolworth-Ravenswood: Composite; late Ptz basement, early Pz sed, volc rocks (marginal basin and volc arc), lower Pz granites, metamorphics, mid-Pz shallow-marine to continental sed rocks
CRW Curnamona-Willyama: Ptz mobile and foldbelt, B.I.F., volcs 1.8 b.y.; metamorphism 1.7 b.y., younger Ptz granite
CYC Cyclops: Pg island arc rocks and ophiolite
DTN Dundas-Tyennan: Late Ptz-late Pz foldbelt; Eocamb-Camb sed rocks, dismembered ophiolite, Camb arc

volcs, early Pz shallow-marine to continental sed rocks, mid-Pz granites, Perm cover
EAN East Antarctica: Archean craton, includes Camb–Ord continental arc rocks near boundary with BOW; rifted from Australia mid-K
ENG East New Guinea: Pg or Upper K ophiolite, (116 m.y. age on basalts, Upper K fauna in inter-pillow marls); mid-Mio melange, volc arc rocks; emplaced in early Eo over Upper K blueschist and greenschist metased rocks
ETM East Tasmania: Ord to Early Dev micaceous qtzose turbidites, intruded by Dev–Early Carb granites; Perm cover
EYA East Yilgarn: Archean craton; granitoid-greenstone belt, 2.7 b.y.
GAS Gascoyne: Ptz mobile and foldbelt, sed and volc rocks metamorphosed, deformed 1.9 b.y.; granitics 1.9–1.6 b.y.
GAW Gawler: Ptz mobile and foldbelt, B.I.F., basic volcs, 1.6 b.y.; metamorphism, granites 2.5–2.3; 1.8, 1.7; 1.65 b.y. granites and felsic volcs
GIR Girilambone: ?Late Ptz to Camb qtzose turbidites, chert, basic volcs, ultrabasics (ophiolite?); possible basement to early Pz arc; Sil–Dev granitics, Dev shallow-marine sed and volc rocks, mid-Pz and younger sed rocks, cover
GYM Gympie: late Pz foldbelt sed and volc rocks, ophiolitic rocks
HAS Hastings: similar to YST, but could be independent terrane; foldbelt, Dev–Carb marine seds, volcanogenic turbidites, mass flow conglomerate with granite clasts of unknown provenance; Perm shallow-marine, Late Perm to Jr granitoids, Tr cover
HBY Hodgkinson–Broken River: composite; early-mid-Pz marine seds, volcanogenic turbidites; volc arc, mid- to late Pz shallow marine-continental strata, granitics; may have pre-Pz basement in part
HJB Hunstein–Jimi–Bismarck: Jr–K granitic and metamorphic rocks, with possible continental crust affinity, ophiolite borders to south
HOK Hokunui: Perm ophiolite, melange overlain by volcanogenic sed rocks, turbidites, Mz calc-alkaline volc, sed strata
HTS Howqua–Tabberabbera: Pz fold and thrust belt; early Pz ophiolite, island arc volcs, turbidites, mid-Pz shallow-marine strata, granites, volcs; Perm cover
JAM Jubilee–Adamsfield: early Pz seds, ophiolitic rocks (?); bounded by Precambrian basement blocks
KAI Kaikoura: early Mio–Recent turbidites, accretionary prism
KEB Kepala–Burung: Pz foldbelt involved in Mz–Cz plate-margin tectonics of Irian Jaya
KIM Kimberly: Archean craton; cover rocks 1,815–1,760 m.y.
KLH King Leopold–Halls Creek: Ptz mobile and foldbelt, sed, ign rocks 2.8–2.2 b.y.; deformation, metamorphism 1.96 b.y.; volcs, granitics 1.9 m.y.; final deformation >1.75 m.y.
KMG Kanmantoo–Glenelg: metamorphosed Late Ptz(?) to Camb turbidites, volcs, ophiolites(?), Ord granites, Sil volcs, continental cover strata
MEL Melbourne: Ord–mid-Dev turbidites; Late Dev granites, volcs
MIL Millen: isoclinally folded metasedimentary and metavolcanic rock, 500+ million-year-old K/Ar ages
MMO Molong–Monaro: early Pz volc arc, marine strata; qtzose turbidites in fore-arc basin; mid-Pz granites, seds, back-arc felsic volcs; mid- to late Pz shallow-marine, Late Dev–Carb continental strata
MTI Mt. Isa: Ptz mobile and foldbelt, 1.86 b.y. felsic ign rocks; 1.8–1.76 sed rocks, felsic, mafic volcs; granites, metamorphism 1.74–1.7 b.y.; younger Ptz seds, metamorphism
MUS Musgrave: Ptz mobile and foldbelt, 1.56 b.y. seds, 1.33 b.y. ign rocks; high-grade metamorphism, granites 1.2–1.1; volcs, granites, basic-ultrabasic dikes 1,050–900 m.y.
NAB Nabberu: cratonic cover of B.I.F., marine seds, basic volcs over Archean block and Ptz mobile and foldbelt
NAR Narooma: Ord qtzose turbidites, chert, volcs; accretionary prism?; Dev–Carb bimodal volcs, sed strata; late Pz granite
NAU North Australia: composite; Ptz mobile and foldbelt, sed, volc rocks metamorphosed 1.95–1.81 b.y.; granite, sed, volc rocks 1.77–1.65 b.y.
NBR New Britain: Eo–Holo composite volc island arc
NBU Nambucca: late Pz metaseds, some volcs; Perm–Jr granites
NNC North New Caledonia: K–Eo blueschist, melange; metamorphosed in late Eo–Olig
NOR Northland: Jr–K, T(?) oceanic crust; late K–Olig sed rocks emplaced 22 m.y.B.P.
NOW Norseman–Wiluna: Archean craton; greenstone belt
NUM New Caledonia ultramafic massif: Late K–T(?) serpentinized peridotite thrust over K–Eo sed rocks
NVI North Victoria: late Ptz–Ord qtzose flysch and schistose rocks; stitched to BOW by Sil and Dev plutons
OWS Owen Stanley: Pg–mid-T melange, K volcanogenic sed rocks from continental arc source; blueschist and greenschist with K(?) oceanic protolith, metamorphosed in Eo
PIL Pilbara: >3.5–2.5 b.y. granitoid-greenstone terrane
PTM Port Macquarie: similar to TEW, dismembered ophiolites with blueschist in melange zones
RUM Rum Jungle–Nanambu: Archean granitoid and gneiss dome terranes >2.5–1.8 b.y.; Ptz orogenic belt 2.4–1.8 m.y.
SBO Sydney–Bowen: Paleozoic foredeep; its basement may be composed of previously accreted terranes
STB Stavell–Bendigo: Paleozoic fold and thrust belt; Camb–Ord turbidites thrust eastward during Sil, Camb volcs and ophiolite at thrust sole, Dev granites, felsic volc rocks, Perm cover
TCM Tennant Creek–Murphy: composite; Ptz mobile and foldbelt, marine seds, felsic, mafic volcs; 1.9–1.8 b.y. metamorphism; ign rocks 1.8–1.66 b.y.
TEW Texas–Woolomin–Coffs Harbor: Dev–Carb qtzose to volcanogenic graywacke, chert, jasper, over basalt, ophiolite, Late Carb granites, Perm–Tr Andean arc volcs; Tr–Recent cover
TMS Timor–Seram: QT flysch(?) melange obducted onto Australia during collision of Banda arc
TOR Torlesse: imbricately faulted Perm to K quartzo-feldspathic graywacke turbidites with minor volc rocks, chert
TUH Tuhua: (composite of Karamea, Golden Bay, and Drumduan terranes of Bishop et al, this vol); pre-Pz–Dev metamorphic basement, Camb volc arc, unconformably overlain by Perm–Tr fluvio-deltaic qtzose conglomerate, intruded by K granitic rocks
WAN Wandilla: Dev–Carb qtzose to volcanogenic graywacke, chert, jasper, over basalt, ophiolite
WAO Wagga–Omeo: Ord qtz wacke turbidites, chert, basic volcs; metamorphosed and intruded by Sil granites
WAT West Antarctica: Mz foldbelt, probably contains Mz island arc and oceanic rocks in Antarctic Peninsula
WTM West Tasmania: Ptz metamorphic basement, Late Ptz sed strata, Camb sed and volc rocks; mid-Pz granites
WYA West Yilgarn: Archean craton; high-grade gneiss>3 b.y.
YIL Yilgarn: Archean craton; composite; high-grade gneiss>3 b.y.
YST Yarrol–Silverwood–Tamworth: foldbelt; Sil–Dev to Carb marine seds, volcanogenic turbidites-forearc basin; with Yarrol included Connors–Auburn volc arc rocks; Perm shallow-marine and bidmodal volcs

SOUTH AMERICA

ATP Altiplano: Precambrian metamorphic massif; may be part of Guyana shield; accreted by Sil

ARQ Arequipa: Precambrian massif and Pz sed strata; Dev faunal affinity possibly not native to South America; accreted in mid-Pz, overlapped by Tr and Jr strata, but ARQ rethrust over these cover rocks

BEL Bellon schist; 1.0 b.y. metamorphic rocks

BRZ Brazilianes: 900–550 m.y.-old orogenic belt; local ophiolitic rocks at western border

CCR Chile Coast Ranges: composite; pre-Jr and Pz (342 m.y.) ophiolitic melange and blueschist metamorphosed flysch thrust over continental margin strata; includes sea-mount terrane with Penn–Perm ls at 50°S; accreted by Early Tr; overlapped and intruded by Jr and K volc and plutonic rocks

CDC Cordillera de la Costa: composite; pre-Mz gneissic basement, Jr–K high-grade qtzose metased rocks and local eclogite; K medium- to high-grade metased and metavolc rocks thrust over Pg flysch

CET Caucagua–El Tinaco–Paracotos: composite; pre-Mz gneissic basement in part; K low-grade metamorphosed qtzose sed rocks; Upper K(?) mafic metased and metaign rocks (in Paracotos); ultramafic bodies occur along fault slices

CHC Choco: Late K–T ophiolitic and primitive magmatic arc rocks; emplaced in Mio

COL Cordillera Occidental: post-Barremian ophiolite in complex structural relation with metagraywacke, chert, tuff, and turbidites

CPJ Cordillera Central–Sierra Perija: composite, Precambrian(?) basement, Pz metamorphic rocks, Pz and Mz shallow- and nonmarine strata; complexly metamorphosed and deformed; Pz and Mz volcs and plutonics in Perija

CPU Central Peru: Mz mio- and eugeosynclinal strata; includes island arc volc rocks, Middle K–T plutons

CUR Curepto: Carb blueschist metamorphosed eugeosynclinal strata, intruded by upper Pz granites, metamorphism older than in PIC

EBR East Brazilian shield: Archean craton separated from west Brazilian shield by Ptz orogenic belt

ECR Ecuador Coast Range: may correlate with CHC; Late K–T ophiolitic rocks

GRB Guarico–Roblecito: Pg flysch and melange

GUA Guajira: Precambrian basement, Pz metamorphic rocks, thick Mz clastic and carbonate strata; deformed and metamorphosed in Mz

GUY Guyana and West Brazilian shields: Archean craton

IMA Imataca: Precambrian metamorphic terrane; may be portion of Guyana shield; cut by Tr dikes whose paleomag shows no movement since Tr

MAR Maranon: Precambrian basement(?), Mz graywacke of distinctly contrasting lithofacies to neighbors

MEJ Mejillo: fault-bounded lower Pz geosynclinal rocks and Jr eugeosynclinal rocks

MER Cordillera Meridia: Precambrian(?) basement, Pz metamorphic rocks, Pz and Mz clastic strata

NIR Nirivio: Carb high T–P metamorphosed continental margin strata, intruded by upper Pz granite—Pz arc rocks

NPU North Peru: pre-Ord schist

PAG Paraguana: Pz metaign basement; Mz metavolc and metased rocks; overlain by thick T strata; includes Mz(?) ultramafic complex

PAM Pampean ranges: early Pz metamorphic basement; middle Pz cover strata

PIC Pichilemu: Carb medium-grade metamorphosed eugeosynclinal strata; intruded by upper Pz granitic rocks

PIU Piura: Precambrian and lower Pz geosynclinal rocks

PTM Patagonian massif: Precambrian(?) basement, metachert, argillite, and marble

SAM Siquisque–Aroa–Mision: composite; melange of Jr–K ophiolitic, volc and sed rocks emplaced in Pg flysch; melange may contain Precambrian metamorphic blocks

SCB South Caribbean deformed belt: composite; K oceanic basement and Mz–Early T pelagic and submarine fan strata; paleomag shows large translation and/or rotation

SMA Santa Marta: Precambrian basement; Pz metamorphic and ign rocks, local Pz sed rocks; thick Mz volc and volcaniclastic rocks; extensive Mz metamorphism and deformation

VDC Villa de Cura; K island arc volc and volcaniclastic rocks, over basement of 120 m.y.-old blueschist and greenschist metavolc rocks

---

MEXICO, CARIBBEAN, AND CENTRAL AMERICA

CAB Caborca: 1.7–1.8 b.y. Precambrian crust and Pz continental margin strata; may be para-autochthonous North American craton

CHI Central Hispaniola: pre-K ultramafic rocks, blueschist, greenschist metavolcs overlain by K–T basalt, ls, clastic strata; may include Paleocene accretionary prism

CHO Chortis: continental crust, Pz metamorphic and ign rocks, Mz nonmarine and shallow-marine strata, volc rocks

COA Coahuila: deformed upper Pz flysch and andesite, Camb–upper Pz deformed sed rocks; probably accreted in late Pz

CYJ Cayo Coco–Yaguajay–Jatibonico–Las Villas: obducted Jr–K carbonate bank rocks of Bahamas; contains Eo deep-water carbonate slope-scarp strata; deformed, intruded latest K–T

DCA Domingo–Cabaiguan: Jr ophiolite belt emplaced lower-mid-Eo; also includes K diorite, granite, volcs; mica-garnet schist

ENH Enriquillo–Massif de la Hotte: Early–Late K basalt, ls, T transgressive coarse clastic rocks and ls; deformed by wrench tectonics; probably not emplaced before 10 m.y.

GUE Guerrero: composite; mostly Upper Jr–K volc arc rocks

HIS Hispanola: Jr metamorphic rocks, K arc, forearc, and accretionary strata

JUA Juarez: severely deformed Upper Jr calcareous clastic rocks, Neocomian cherty ls and arc rocks, ultramafic rocks; thrust over MAY

LAP La Paz: K batholithic rocks and pre-batholithic rocks of unknown age

MAY Maya: Precambrian(?) rocks, highly deformed and metamorphosed Pz flysch, Perm–Tr meta-plutonic complex; overlapped by Mz red beds, volc and marine sed rocks, accreted late Pz

MIX Mixteca: Pz metamorphic rocks, upper Pz terrigenous strata, Mz marine strata

NIC Nicoya: Late Jr–T ophiolitic rocks, turbidites, and limestone; emplaced in latest K or Pg

NPR North Puerto Rico: K volc, volcaniclastic rocks, chert; Upper K pillow basalts, Upper K–Eo intrusive rocks, Eo ls, mid-T carbonate cap

OAX Oaxaca: 1.1 b.y. gneisses and Pz sed strata with South American faunal affinities, Mz red beds and ls

PLC Placetas–Cifuentes: Lower K micaceous ss with continental source faulted against Eo clastic rocks; meta-arc basement?

RUS Rusias: Carb ls, clastic rocks

SBI Sonnabari: metaplutonic arc(?) rocks

SEP Cordillera Septentrional: composite; (1) blueschist

carbonate bank rocks from collision with Bahamas at 53 m.y.; (2) Early K ophiolite, tuffs, greenschist metavolcs overlain by Late K–T volcs, Eo olistostrome, subaerial clastic strata, Olig–Mio carbonate, clastic rocks
SMD Sierra Madre: upper Mz sed strata with Pz and Precambrian metamorphic basement
SPR South Puerto Rico: structurally complex >110 m.y. peridotite, pillow basalts, chert, overlain by Late K carbonate rocks; overthrust in later K by pelagic ls, unconformably overlain mid-T carbonate cap
TRI Trinidad Mtns: Jr passive margin clastic, carbonate rocks blueschist metamorphosed in K; K–Ar age 73–78; K diorite
VIS Viscaino: Tr and Jr ophiolites
XOL Xolapa: Jr metamorphic rocks
YOL Yolaina: Mz continental and marine strata, Cz sed and volc rocks; may be built on oceanic crust

ALASKA AND CANADA

AAC Arctic Alaska—Coldfoot subterrane: metagraywacke, phyllite, and quartz mica schist, polymetamorphosed in late Mz; age of protolith may be Miss
AAD AA—Delong Mountains subterrane: complex assemblage of Dev, Miss carbonates, younger chert, argillite
AAE AA—Endicott Mountains subterrane: Dev clastic rocks, Miss shale and carbonate, younger chert argillite
AAH AA—Hammond subterrane: structurally complex, polymetamorphosed assemblage of middle Pz carbonate, schist, quartzite, metavolc rocks; may have Precambrian basement
AAN AA—North Slope subterrane: Precambrian continental crust, Precambrian, Pz and Mz clastic and carbonate rocks; probably a displaced portion of North American craton
AGM Angayucham: tectonically complex assemblage of pillow basalt, Tr radiolarian chert, and underlying Miss chert; contains blocks of Pz ls, mafic and ultramafic rocks (ophiolitic)
ALX Alexander: Precambrian(?), Pz, and Mz volc, clastic, and carbonate rocks
CCK Cache Creek: Miss to Upper Tr melange and deformed chert, argillite, basalt, ultramafic rocks, carbonate, local blueschist; contains Perm Tethyan fauna
CGH Chugach: deformed upper Mz flysch and melange, including Mz chert, gabbro, ultramafic, and volc rocks, and ls
CHU Chulitna: Dev ophiolite overlain by Pz chert, conglomerate, ls, flysch; Mz ls, red beds, flysch, and chert
DIL Dillinger: complexly folded assemblage of lower and middle Pz graptolitic shale, micaceous, qtzose calcareous sed rocks, turbidites, ls; overlain unconformably by Lower Jr clastic rocks
EAS Eastern composite (Kootenay, Slide Mtn, McLeod, Cassiar, Monashee): upper Pz and lower Mz (?) basalt, ultramafic rocks, deep-marine strata
END Endicott: metamorphosed lower to upper Pz clastic and carbonate strata intruded by Pz granitic rocks
GNW Goodnews: structurally complex assemblage of lower Pz to Lower K pillow basalt, chert, ls, blueschist, ultramafic rocks, and graywacke
INN Innoko: deformed upper Pz to lower Mz chert, argillite, minor graywacke, volcanogenic clastic rocks and tuff, ls
KAM Kaminak: Archean craton
KGV Kagvik: disrupted and deformed assemblage of Miss–Tr radiolarian chert, argillite, shale, and minor volc rocks
KYK Koyukuk: Andesitic volc, volcaniclastic strata, and local Lower K ls; depositional base unknown
LVG Livengood: composite; (1) deformed, metamorphosed Ord–Dev chert, dolomite, volc rocks, serpentinite, clastic strata; (2) Ord volc, volcaniclastic rocks overlain by Sil and Dev carbonate; (3) Camb(?) qtz-rich clastic rocks, shale, and slate
MNA Minchumina: complex assemblage of chert, argillite, and quartzite; Ord–Dev(?) ages
MTB Methow–Tyaughton and Bridge River combined: Perm–Middle Jr melange, deformed chert, argillite, basalt, ultramafic rocks
NUT Nutak: Archean craton
NXF Nixon Fork: Precambrian metamorphic rocks overlain by Pz and Mz carbonate, clastic rocks, and chert
NYK Nyack: arc-related assemblage of volc and sed rocks with Middle Jr fossils
PCP Porcupine: Precambrian(?) phyllite, slate, quartzite, and carbonate rocks overlain by thick, structurally complex Camb to Upper Dev carbonates and shale; upper Pz clastic and carbonate rocks; local Jr strata
PEN Peninsular: rare Pz ls, Tr basalt, argillite, and ls, Lower Jr andesitic arc volc and volcaniclastic rocks, younger clastic rocks; may have originally formed on continental basement, though basement is unexposed
PMW Pingston, McKinley, and Windy combined: (1) upper Pz phyllite and Tr ls and shale; (2) Perm flysch, Tr chert, pillow basalt, and upper Mz flysch and conglomerate; (3) disrupted serpentinite, basalt, metachert, Dev shale and ls, in a matrix of upper Mz flysch
PRW Prince William: deformed lower Cz flysch and volc rocks
QNL Quesnellia: upper Pz and Lower Tr volc, volcaniclastic, and carbonate rocks; Upper Tr–Lower Jr volc, clastic strata, argillite also deposited on EAS, possibly on CCK; may include upper Precambrian to lower Pz continental crustal basement
RCU Reindeer–Circum–Ungava: 1.9–1.6 b.y.-old orogenic belt
RUB Ruby: middle–upper Pz metavolc and metased rocks
SEW Seward Peninsula: structurally complex assemblage of Precambrian to Pz metased, metavolc, carbonate strata
SHU Shukson: blueschist metamorphic rocks
SJN San Juan and Lopez combined: deformed Mz chert, argillite, graywacke, and volc rocks partly in melanges, with blocks of lower Pz plutonic rocks, Pz chert, carbonate, and volc rocks; Perm ls blocks contain Tethyan fusulinids
SKG Skagit: metamorphic terrane
SLV Slave: Archean craton
STK Stikinia: may have Upper Precambrian basement; Carb–Perm volcaniclastic, mafic to silicic volc rocks, carbonate; deformed intruded in Middle–Late Tr; overlapped by Upper Tr–Middle Jr andesites of (?) CCK
SUP Superior: Archean craton
SVM Seventy-mile: upper Pz (Perm?) disrupted ophiolite; tectonically overlies YKT
TKU Taku: structurally complex assemblages of upper Pz volcaniclastic rocks, ls, flysch(?), and lower Mz basalt, ls, and flysch
TOG Togiak: structurally complex, thick Jr to Lower K basaltic to andesitic flows, breccia, tuff, volc graywacke, and argillite, and chert, minor Lower K ls
TOZ Tozitna: structurally complex assemblage of Pz(?)–Mz gabbro, basalt, diabase, argillite, chert, tuff, graywacke, conglomerate, and Perm(?) ls; Miss–Tr cherts; Late Tr gabbro
TRA Tracy Arm: structurally complex assemblage of marble, pelitic gneiss, and schist of unknown ages
WOP Wopmay: 1.9–1.6 b.y.-old orogenic belt
WRN Wrangellia: Pz island arc overlain by ls, clastic rocks, and chert, Upper Tr pillow and subaerial basalts, Tr and Jr ls, clastic, volc rocks; amalgamated to PEN by Middle Jr, but paleomag shows low paleolatitude in Tr
YAK Yakutat: upper Mz graywacke and shale, with structurally interleaved chert, argillite, volc rocks, Eo basalt
YKT Yukon–Tanana: Precambrian and Pz polymetamorphosed continentally derived qtzose sed, volc, and granitic rocks; youngest metamorphism dated as mid-K
YOK York: Tectonic assemblage of lower Pz shelf carbonate thrust over Precambrian(?) to lower Pz slate; Ord rocks contain trilobites of non-North American affinity

WESTERN CONTERMINOUS UNITED STATES

BDY Baldy: high P–T metamorphosed graywacke, basalt, and chert, age unknown
BLM Blue Mtns: melange with blocks of Pz ophiolite, ls, chert, and Mz chert and ss structurally overlain by Tr and Jr volcaniclastic rocks
CAT Catalina: K blueschist, graywacke, basalt, and chert
CEN Central: K melange
CLV Calaveras: composite; western belt of melange with ophiolite and Mz chert, and eastern belt of qtzose clastic rocks argillite and minor Perm ls
CMM Central Metamorphic: Dev or older metamorphic complex composed of mica schist, amphibolite schist, and gneiss
COS Coastal: Upper K graywacke
CRE Crescent: T basalt sea-mounts
CRZ Cortez: pre-Pz sialic basement, Pz continental margin strata
DPO Del Puerto: Jr ophiolitic arc basement, Upper Jr and K volc arc sed strata
EKL Eastern Klamath: middle–upper Pz clastic, volc rocks, carbonate overlain by Tr, Jr volc rocks, and minor ls
ELC Elder Creek: upper Mz Great Valley turbidite strata over ophiolite
FRA Franciscan undifferentiated (includes CEN and YBP)
FUL Fulmer: Eo nonvolc ss and shale basin
GLC Golconda: deformed assemblage of chert, argillite, minor ls, and volc rocks of Miss–Perm age
GRE Grenville: 1.1 b.y.-old orogenic belt
HMA Huntington and Malheur combined: Mz volc arc and forearc strata
HOH Hoh: Olig(?) sed rocks
HSG High Sierra–Goddard: continental margin strata
KIN King Range: Upper K ophiolite overlain by Upper K–T sed rocks
MOJ Mojave desert: composite; includes Pz deep-marine strata, Mz volc, sed and granite rocks
NSI Northern Sierra: lower Pz clastic rocks, upper Pz and lower Mz volc and associated sed rocks
OLM Olympic core: lower Cz volc and deep- and shallow-water sed rocks; presumed to have oceanic basement
OPT Otter Point: Jr volc arc
OZT Ozette: middle Eo–middle Mio deep-marine turbidites, sed strata; olistostromal basalt blocks in Eo strata
PAR Pacific Rim: Upper Jr, Lower K flych and melange
PTR Patton: Upper K to T graywacke
RBM Roberts Mtn: structurally complex assemblage of chert, argillite, ss, basalt, and minor ls; Camb–Late Dev or Miss age
ROD Rodriguez: T sea mounts
SAL Salinia: meta-pelites, marble, and graywacke of unknown age intruded by K granite; continental crustal affinity
SCM Snow Camp Mountain: Middle Jr ophiolite overlain by Late Jr–Early K sed rocks
SFH Sierra foothills: Upper Jr andesite, volcaniclastic rocks, phyllite, slate, and graywacke, and Upper Jr ophiolite
SLZ Siletzia: lower Cz volc and sed rocks; paleomag data show post-Eo 70° clockwise rotation
SNC Nicolas: Jr ophiolite (?) and K graywacke
SSM San Simeon: melange; graywacke, argillite, chert, and serpentinite
STM Stanley Mtn: Jr ophiolite, Upper Jr–K graywacke
TRP Tr-Pz belt of Klamath Mtns: structurally complex assemblage of Lower Mz ophiolite, chert, basalt, Jr andesite and associated sed rocks
TUJ Tujunga: Precambian metamorphic rocks, Tr and K plutons, K graywacke
WKL Walker Lake: upper Pz and lower Mz arc volc rocks, Mz clastic rocks
WKM Western Klamath composite—Dry Butte, Briggs Crk, Rogue River, Smith River: Upper Jr island arc rocks
WYO Wyoming: Archean craton
YBP Yolla Bolly and Pickett Peak: blueschist metagraywacke, metachert, and metabasalt

---

APPALACHIANS AND SOUTHERN UNITED STATES

(Appalachian terranes from Williams and Hatcher, 1982)

AVA Avalon: upper Precambrian sed and volc rocks relatively unmetamorphosed, undeformed; local Precambrian intrusions, Camb shales with Atlantic realm trilobites; probably composed of several late Precambrian terranes, including arc, tholeiitic, ophiolitic sequences linked by Camb time. Linked to GAN by Dev plutons; Sil–Dev sed strata
BRU Brunswick: delineated based on magnetic, gravity data; low frequency, symmetrical, long-wavelength anomalies
DUN Dunnage: early Pz mafic island arc volcs, marine sed rocks, melanges over ophiolite, mildly deformed; structurally complex ophiolitic boundaries; overlapped to W and E by Ord cgl, olistostromes
GAN Gander: pre to mid-Ord clastic sequence in N, volcs, shale over gneiss domes to S; continental basement, mid-Pz intrusions common. Intense metamorphism and deformation in mid-Pz
HUA Humber Arm: a Taconic allochthon; complex thrust sheets of ign, ophiolitic, metamorphic rocks over early Pz sed rocks; mid-Ord overlap link with miogeocline
MEG Meguma: Camb-Lower Ord graywacke, shale with continental source to SE; overlain by Ord tillite to Dev terrestrial sed sequence; Dev plutons unlike AVA; Carb overlap with AVA
MRN Marathon: Pz eugeosynclinal deposits, flysch, deformed in Penn–Perm
OUA Ouachita: Pz eugeosynclinal deposits, flysch, deformed in Penn–Perm
PIE Piedmont: late Precambrian–early Pz metaclastics on Grenvillian basement; metamorphosed to upper greenschist-amphibolite; intensely deformed; ophiolitic mafic, ultramafic bodies throughout terrane; plutonism and metamorphism indicate thrusting over miogeocline during Perm–Carb orogeny
STL St. Lawrence: a Taconic allochthon; lower Pz sed rocks overthrust by volc, ign, ophiolitic rocks of a variety of environs; overlap by Sil, Dev rocks to E
TAC Taconic Ranges: composite, structurally complex; lower Pz sed rocks overthrust by volc, ign, and ophiolitic rocks that may represent sea mounts and oceanic crusts; overlap by mid-Ord strata
TAL Talladega: lower Pz or older sed strata; overlain by Dev chert, island arc mafic volcs
TAS Tallahassee–Suwannee: delineated by magnetic and gravity data; high-frequency magnetic pattern and positive gravity; Pz strata with European faunas from drillhole data

Natural Gas Generation in Sediments of the Convergent Margin of the Eastern Aleutian Trench Area

Keith A. Kvenvolden
Roland von Huene
U.S. Geological Survey
Menlo Park, California

Sediment being subducted in the eastern part of the convergent margin of the Aleutian Trench has a potential to generate large volumes of natural gas, perhaps as much as 2.8×10^6 cu m of methane per cu km (4.2×10^8 cu ft/cu mi) of sediment, even though the content of organic carbon in the sediment is very low, averaging about 0.4%. This high potential for gas generation results primarily from the enormous volume of sediment undergoing subduction. Along the eastern Aleutian Arc-Trench system a 3 km (1.9 mi) thick sheet of sediment is being subducted at a rate of about 60 km (37 mi) per million years. We estimate, based on considerations of the stability requirements for gas hydrates observed as anomolous reflectors in some of our seismic records, and on one measurement in a deep well, that the geothermal gradient in this region is about 30°C/km (87°F/mi). Such a gradient suggests a temperature regime in which the maximum gas generation in the subducting sediment occurs beneath the upper slope. Thus, the sediment of the upper slope, as opposed to that of the shelf and lower slope, could be the best prospect for gas accumulation if suitable reservoirs are present. This observation suggests that, throughout the circum-Pacific margin, potential gas accumulation may be present whenever significant volumes of sediment are being subducted.

INTRODUCTION

Plate tectonics and tectonostratigraphic terrane analyses provide conceptual models that can be powerful tools in the exploration for oil and gas resources of continental margins of the world's oceans. Continental margins have been broadly classified based on relative movements along crustal plates into passive (trailing) and active (sliding or convergent) types. Trailing margins generally consist of broad shelves, slopes, and rises and are characterized by lack of seismicity. Sliding margins are commonly characterized by a basin and ridge borderland morphology, whereas convergent margins are associated with deep trenches, volcanoes, and earthquakes. Passive or trailing margins have the requisites for the generation and accumulation of significant quantities of oil and gas (Thompson, 1976). Basins associated with sliding margins may contain large quantities of petroleum (e.g., the Neogene basins of southern California), but these settings often are vulnerable to destruction by transpressional tectonics. The requisites for significant hydrocarbon resources in sediments of convergent margins are poorly understood, but new information from studies during the last 10 years indicates that convergent margin settings could contain sites of important gas and possibly oil accumulations (Thompson, 1976; Roberts, 1981; Gwilliam, 1982). The peculiar nature of convergent margins, consisting of tectonically complicated subduction complexes, does not lend itself to "conventional" petroleum exploration strategies, and many of the basic tectonic mechanisms operating in subduction zones are only partially understood.

Studies of seismic records and samples from the Deep Sea Drilling Project (DSDP) have established the reality of extensive subduction of sediment along at least six modern convergent margins:

1. Japan Trench (Scientific Party, 1980)
2. Nankai Trough (Scientific Party, 1980; Karig et al, in press)
3. Mariana Trench (Hussong and Uyeda, 1981)
4. Middle America Trench (Watkins et al, 1981; Aubouin et al, 1982, 1983)
5. Barbados Ridge (Moore and Bijou-Duval, 1981)
6. Aleutian Trench (von Huene, 1979; von Huene et al, 1983)

At these margins, large volumes of oceanic sediment are being subducted. Organic matter in the subducting sediment undergoes alterations as this sediment plunges deeper and subsurface temperatures increase. Hydrocarbon gases, particularly methane, are expected to result from these thermogenic alteration processes. Although there are few direct indications of thermal gradients within subduction zones, model studies show that temperatures will eventually reach sufficient levels for the generation of gas (Delong and Fox, 1977). The upward migration of gas from the subducted sediment to traps within the front of the continental margins seems likely considering the Paleogene age of the sediment and the surprising stability of the tectonic environment in many modern convergent margins, i.e., the Aleutian, Middle America, and Japan Trenches.

Our studies were designed to test the idea that convergent margins can be the site of significant hydrocarbon gas generation and accumulation. We chose the convergent margin of the Aleutian Trench, at the northernmost edge of the Pacific Plate (Fig. 1), as an appropriate area for investigating concepts of gas generation and migration in a subduction zone. Previous DSDP drilling (Kulm et al, 1973; Creager et al, 1973) and extensive geophysical and geological studies (von Huene, 1979; Plafker et al, 1982; von Huene et al, 1983) provide basic information relevant to this problem. Along the Aleutian Trench, structural styles associated with subduction are diverse, ranging from subduction complexes with no accretion (Plafker et al, 1982) to complexes with extensive accretion as well as sediment subduction (von Huene, 1979; von Huene et al, 1983). In addition to this basic geological and geophysical information, some geochemical data in the form of organic carbon determinations on sediments of the Aleutian Trench system were available from DSDP Legs 18 and 19 (Bode, 1973a, 1973b). For our study we have:

1. Reprocessed multichannel seismic data to clarify the tectonic framework of the Aleutian subduction complex, to image deeper parts of the Aleutian subduction zone, and to estimate volumes of sediment involved in subduction.

2. Estimated the thermal gradient from the base of gas hydrate reflector on seismic records, from available drill data, and from generalized information along other convergent margins.

3. Compiled information on organic matter in sediments sampled during DSDP Legs 18 and 19 and in selected dredge samples from U.S. Geological Survey cruise S-79-WG.

4. Assessed the potential for gas occurrence based on kinds and volumes of sediments, types and amounts of organic matter, geothermal gradients, and the structural framework of the Aleutian subduction complex.

SEISMIC SURVEYS

Seismic records were taken from two grids (A and B) of seismic lines and two single seismic lines (C and D) across the Aleutian convergent margin near Kodiak Island (Fig. 2). Many records from these grids and lines have been processed to clarify and deepen the seismic information well beyond that of the original, more routinely processed records. Much structural variation is evident in the accreted sediment. Although this portion of the Aleutian convergent margin is truly accretionary at the present time, considerable sediment is being subducted as well.

Field Work

All of the records were obtained with the seismic system aboard the U.S. Geological Survey's (USGS) research vessel *S.P. Lee* during the field seasons of 1976, 1977, and 1981. The seismic system was a tuned array of five air-guns totaling 21.7 l (0.76 cu ft), a 2,400 m-24 group streamer, and a digital recording instrument. In all surveys, shotpoints were located by satellite navigation supplemented by doppler-sonar and Loran-C fixes.

Data Processing

The data were processed in two phases. Initial, routine processing was done to general standards for common depth point seismic data in records where the structure is relatively simple. In the second phase, the data were reprocessed and enhanced by migration and depth conversions to restore true structural relations. The processing sequence consisted of demultiplexing, editing, velocity analysis, normal-movement correction and stacking, deconvolution, filtering, automatic gain control, and display. These data were analyzed and reported in preliminary form in various publications (von Huene, 1979). Many important structural features were masked by diffractions, and more detailed reprocessing, including migration, was required before these features could be properly resolved.

Reprocessing began with a more detailed analysis of the stacking velocities. The improved velocities were then used to restack the data prior to migration. Each record was migrated at four or five constant velocities ranging from 1,450 to 1,850 m/sec (4,757-6,070 ft/sec). This sequence of migrated records was then inspected to determine the best migration velocity of each portion of the record. Migration velocities differed markedly from stacking velocities in areas of complex structure where large changes in dip occur over small distances. Often the best migration velocity for each small area could not be applied to produce a single record because inversion of interval velocity values caused the migration algorithm to fail. Thus, each final migrated record was a compromise and was the best record that could be assembled in one section, but local complex structures were clearer on some of the constant velocity displays. Where structure was depicted most clearly, tracings were made that comprise the final composite line drawing. Examples of reprocessed time sections are shown in Figure 3.

INTERPRETATION OF SEISMIC RECORDS

The seismic records were interpreted simultaneously with the processing of each record. The stacked records were studied to identify areas where the clarification of structure and elimination of diffractions would improve the imaging and thereby enhance the information for interpretation of geologic structure. In this manner, each change in velocity and scaling parameters was monitored with respect to the improvement of specific structural features, and a series of interpretations were made on successively improved records.

Generally, migration clarified structures in the first 2 or 3 sec of the record but was not as effective in deeper parts of the record. Folds at the beginning of the subduction zone were clearly imaged when masking diffractions were collapsed. The migration velocity that best collapsed diffraction in areas of folding was commonly different from the velocity of the rock measured by seismic refraction or from velocity measurements on DSDP drill cores. Faults were interpreted based on bed truncations, changes in the dips of reflections, and occasionally, shallow fault plane reflections.

The igneous oceanic crust, or basement of the sub-

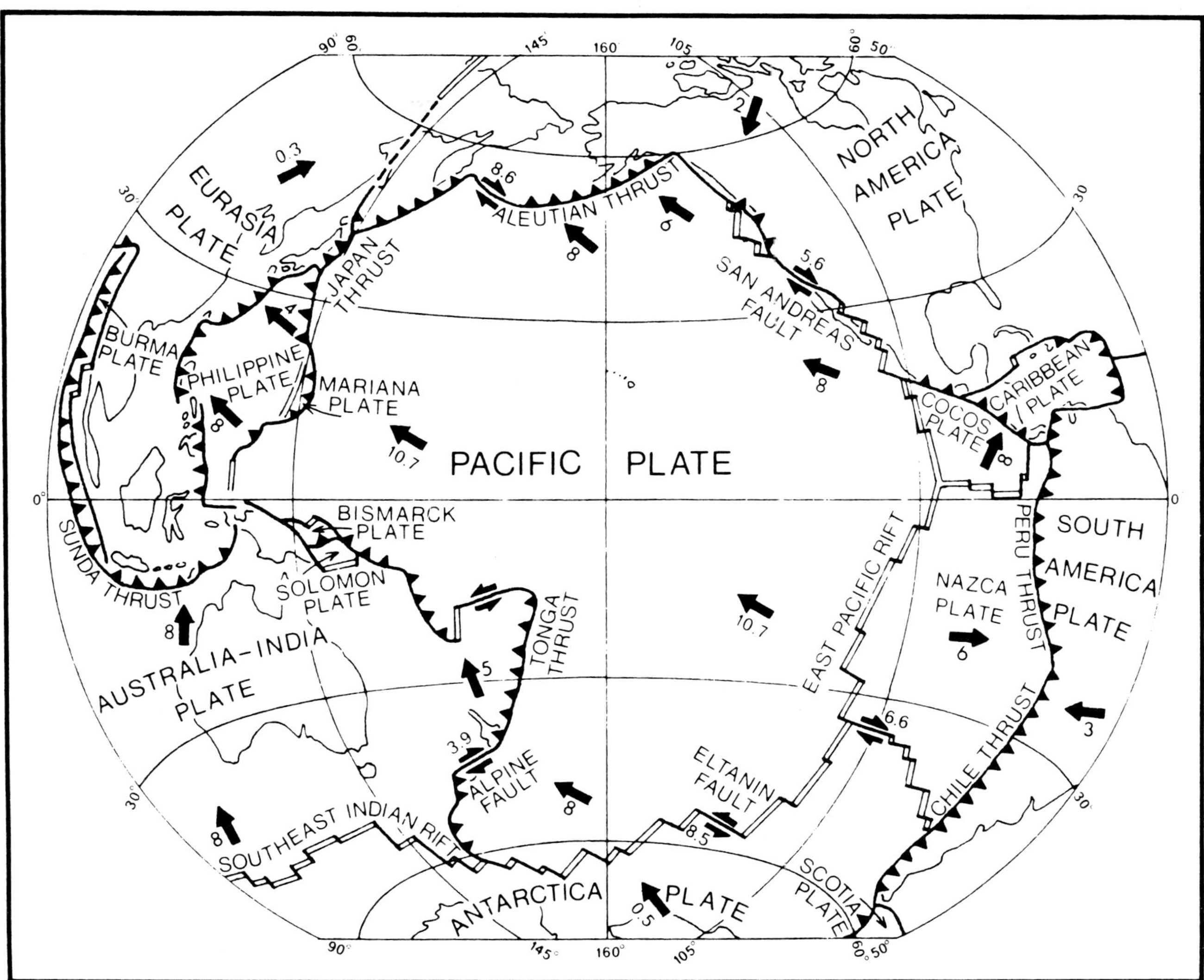

Figure 1—Plate tectonic map of the circum-Pacific region showing major plate boundaries including the Aleutian Thrust, the site of the Aleutian Trench subduction zone. Numbers and arrows indicate rates (cm/year) and directions of plate movements. The Cocos Plate is the site of the Middle America Trench; the Japan Thrust is the site of the Japan Trench. (Modified after Moore, 1982.)

ducting plate, is defined acoustically by high-amplitude reflections, low-frequency signal returns, and an irregular, diffraction-producing surface. These features give the basement reflection a very distinctive character. However, the basement is generally buried by more than 2 km (1.2 mi) and as much as 4 km (2.5 mi) of sediment before entering the subduction zone. Thus, down the subduction zone as the overlying sediment rapidly thickens, the basement reflection becomes difficult to follow, especially where acoustic dispersal in a complexly deformed overlying sediment sequence has scattered the reflective energy. The maximum depths to which we were able to image the basement are about 7 km (4.3 mi), and generally a reasonable record clarity was obtainable at 4 to 5 km (2.5–3.1 mi) depths; however, relative structural simplicity highly influenced depth and clarity of imaging.

Examples of the main structural features in the records used for this study are shown in Figures 3 through 6. There are differences in the reflections shown between depth and time presentations because each has a different character even though the original data are the same. This difference is primarily caused by the compression of scale required in the upper, low-velocity part of the sediment sequence during the conversion from time to depth and by the stretching seen in the lower, high-velocity sediment. It was often difficult to trace exactly the same reflections in both the time and depth presentations (compare time section of line 111 on Fig. 3 with depth section on Fig. 4).

The seismic records of the eastern Aleutian Trench show that large volumes of sediment are being subducted. The thickness of sediment in the trench is from a little more than 2 km (1.2 mi) (von Huene, 1972) to about 5 km (3.1 mi) (line 120, Fig. 3) prior to entering the subduction zone. The subduction zone shows a variety of structural styles, and even records only 10 to 20 km (6.2–12.4 mi) apart are significantly different. Therefore, each grid of lines is discussed separately.

The southwesternmost grid (grid A, and Figs. 2, 3) depicts a part of the trench where sediment is presently from 2.5 to 5 km (1.6–3.1 mi) thick. About 1.5 km (0.9 mi)

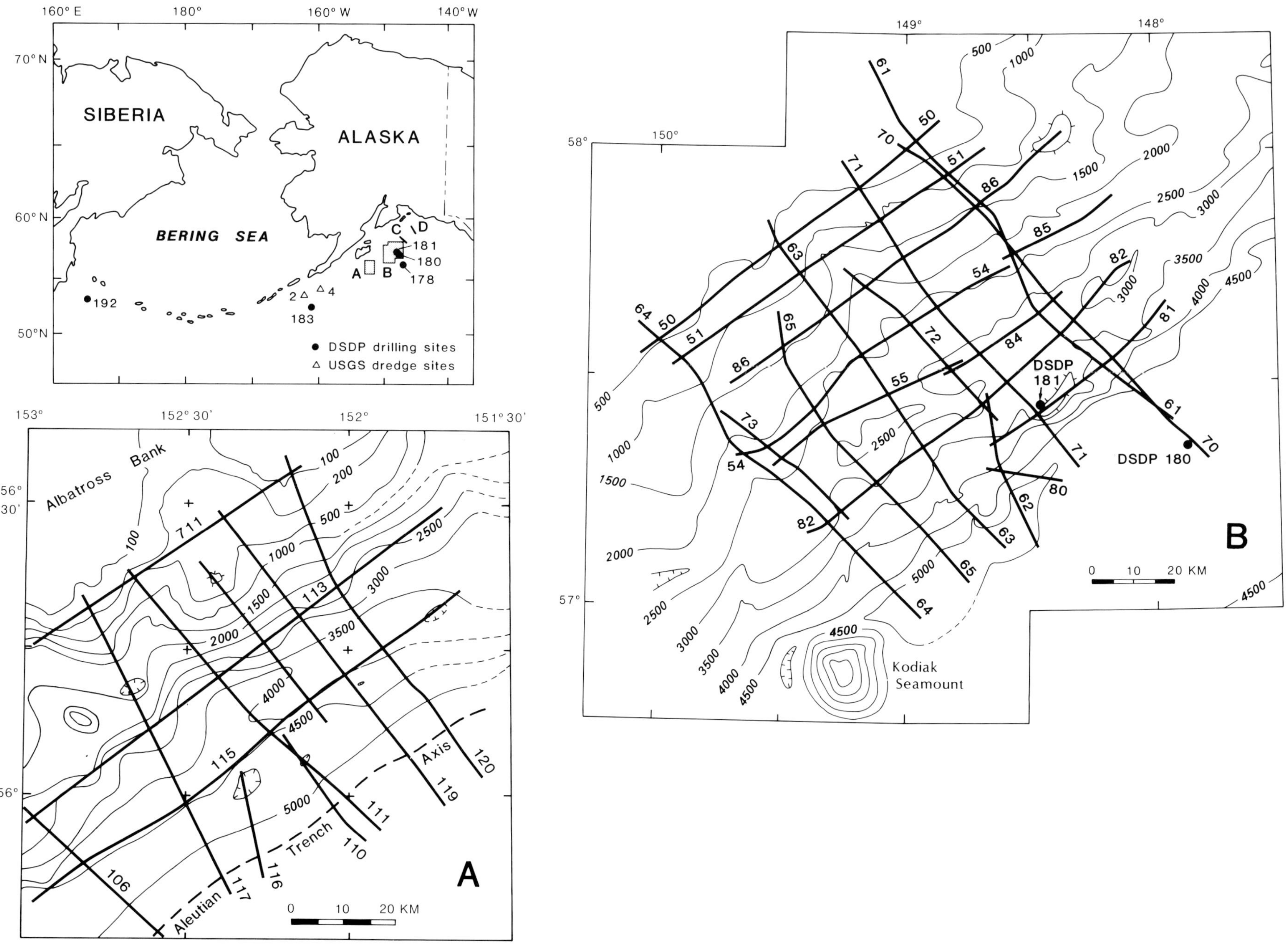

Figure 2—Locations of seismic-survey grids A and B with positions of multichannel seismic lines in the grids shown at expanded scale. Locations of single seismic lines C and D are also shown. In addition, DSDP sample sites 178, 180, 181, 183, and 192 (●) and USGS dredge sites 2 and 4 (Δ) are indicated.

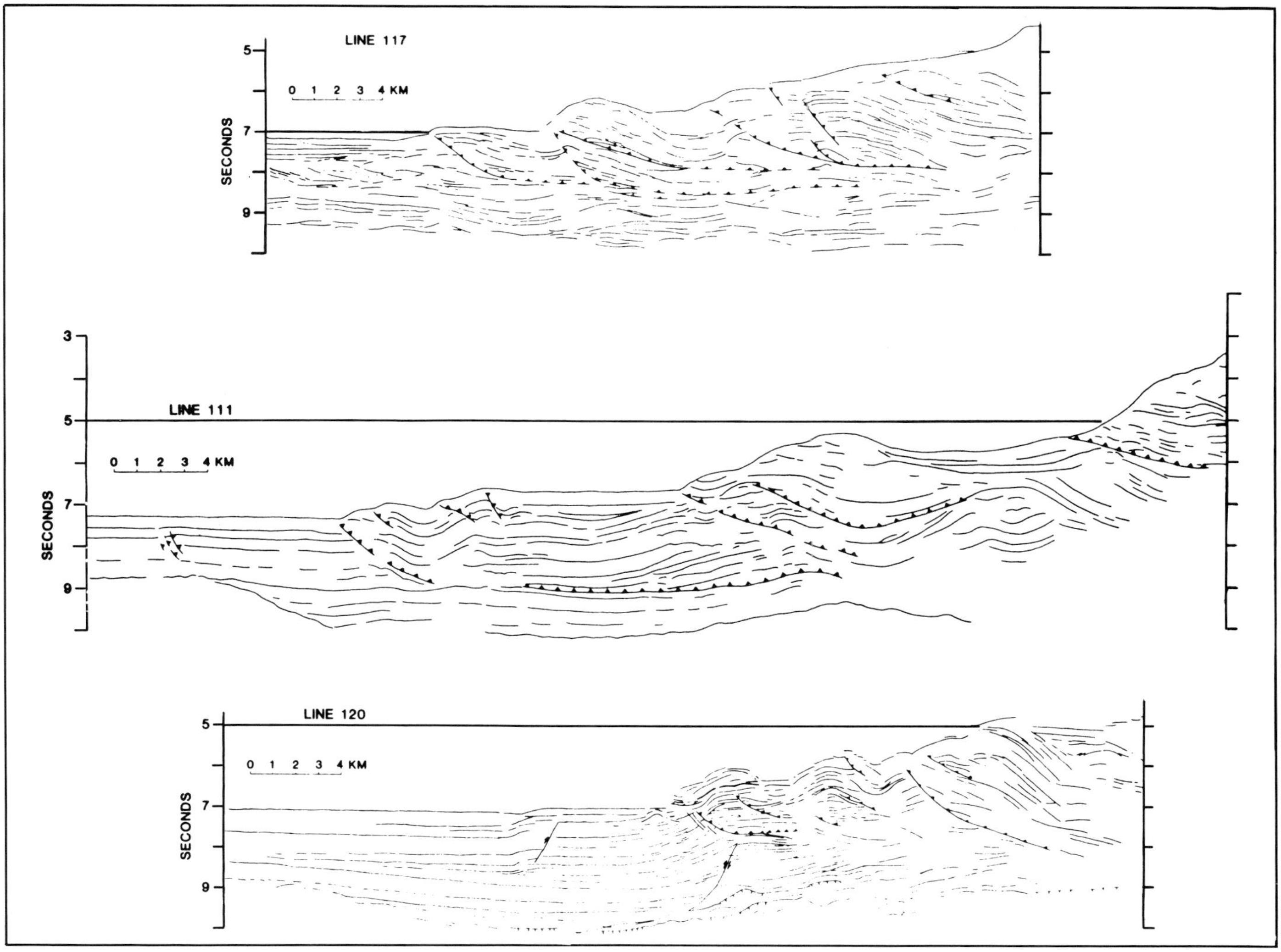

Figure 3—Time sections of lines 111, 117, and 120 from seismic grid A (Fig. 2).

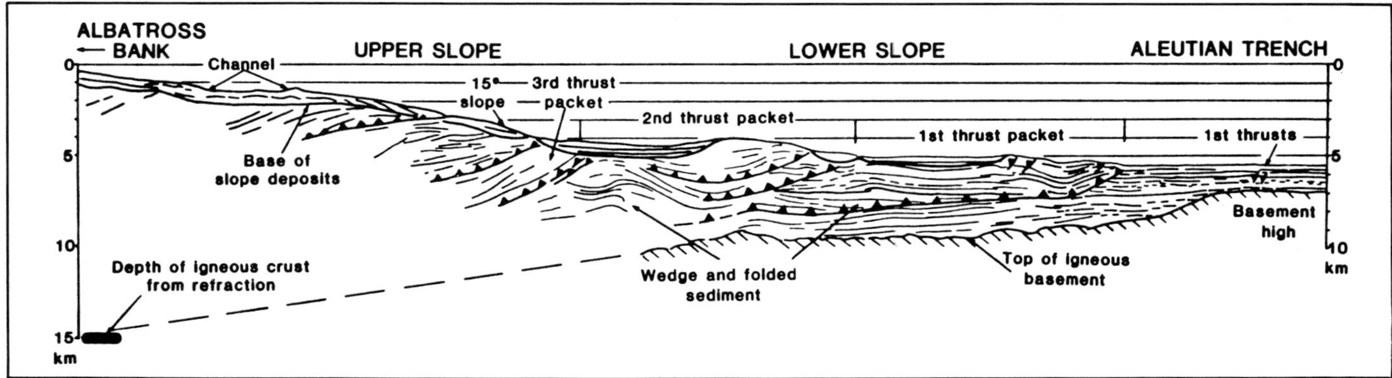

Figure 4—Depth section of seismic line 111 from Grid A (Fig. 2). (After von Huene et al, 1983.)

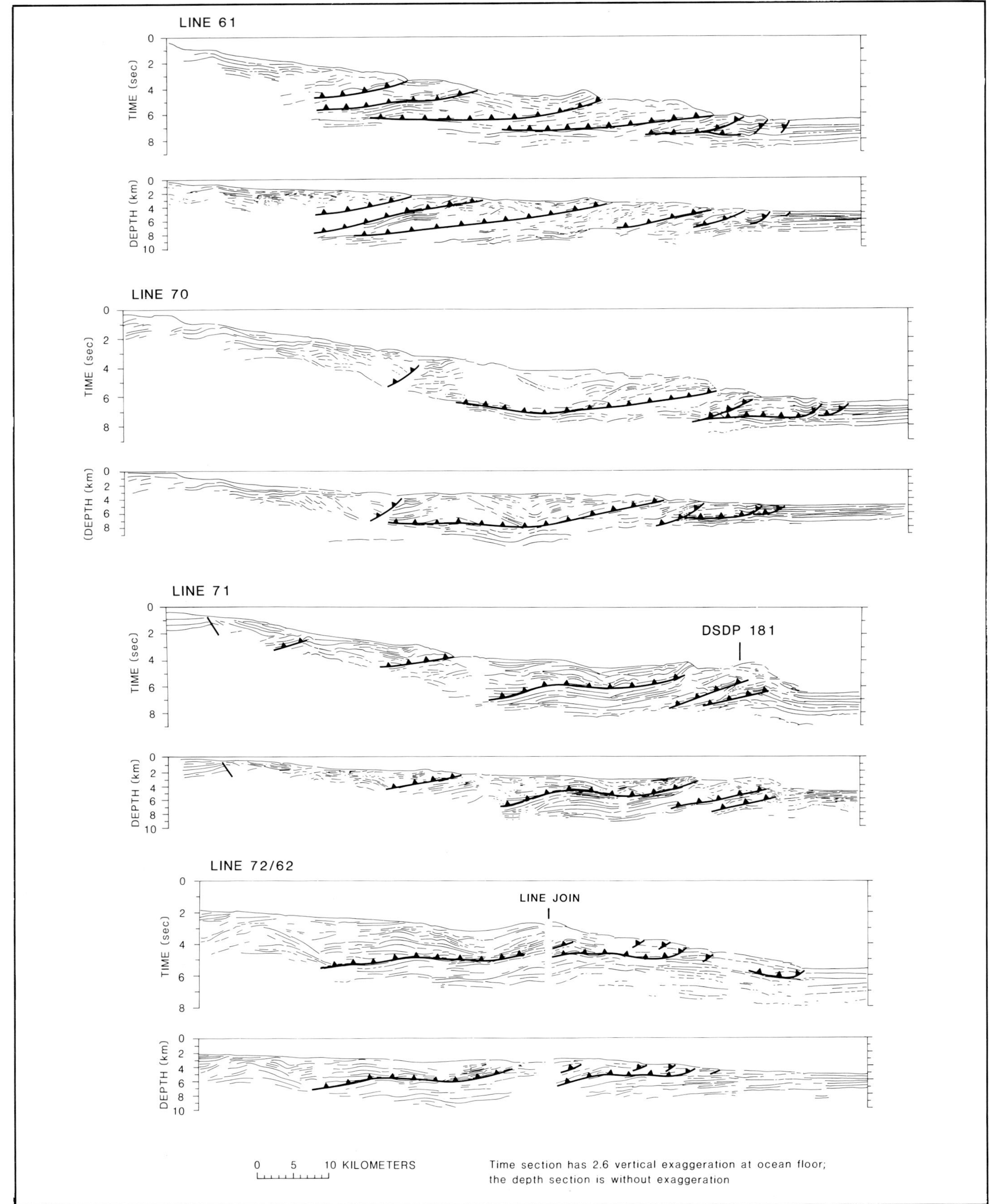

Figure 5—Time and depth sections of seismic lines 61, 70, 71, and 72/62 from seismic grid B (Fig. 2).

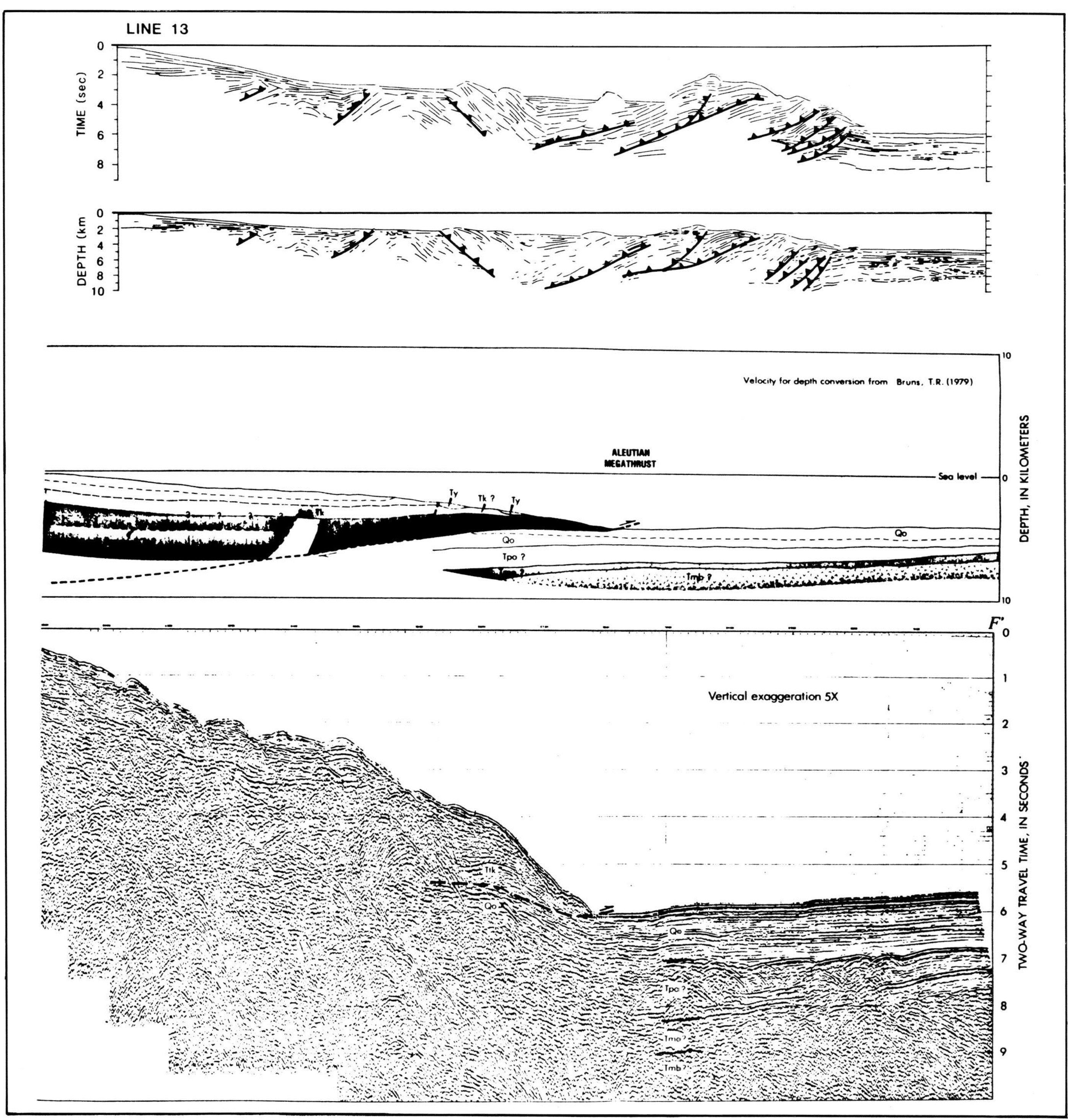

Figure 6—(Top) Time and depth sections of seismic line 13, which is the same as line C (Fig. 2). (Bottom) Depth and time sections of seismic line D (Fig. 2) from Plafker et al (1982).

of the sediment in the trench is turbidite and hemipelagic fill that is 0.6 m.y. old at DSDP site 180, just about 200 km (124 mi) northeast in the trench (von Huene and Kulm, 1973). As the sediment section passes into the trench, the first deformation is marked by small proto-reverse faults that do not reach the surface (Fig. 3). Here, the section is divided into two parts; the upper part is scraped off at the deformation front and attached to the front of the margin, whereas the lower part is subducted. The depth of this division can vary, and the division is found both above and below the base of the hemipelagic and turbidite sediment filling the trench axis. In record 111 (Fig. 4), the subducted sediment is 1.2 sec (2.0 km [1.2 mi]) thick; in records 117 and 120 (Fig. 3), the subducted sediment is 2.5 and 2.6 sec thick; using the same velocity as in record 111, the thickness along records 117 and 120 is 4.2 and 4.3 km (2.6

and 2.7 mi), respectively. An average thickness of subducting sediment at the front of the margin in this grid is 2 sec or about 3.7 km (2.3 mi). In record 111 (Figs. 3, 4) the subducted sediment thins slightly with distance landward from the deformation front. Landward of the midslope area the slope of the trench steepens, and the structure is mainly monoclinal (Fig. 4). The midslope area marks a change in tectonic regime where at least the upper 6 km (3.7 mi) of deformed sediment are no longer being extensively tectonized by horizontal compression but are rather being uplifted as a rigid body. The change in tectonic regime may mark a transition from largely ductile deformation of weak, offscraped sediment with elevated pore-fluid pressure at the front of the margin to a more rigid body being underplated by subducting sediment (von Huene, 1979; von Huene et al, 1983). Most pertinent to our work is that within the southern grid, which covers a 65-km (40.3 mi) stretch along the slope of the trench, the thickness of subducted sediment varies between 2.0 and 4.3 km (1.2 and 2.7 mi) and averages 3.7 km (2.3 mi). The subducted sediment here is mainly from a sequence of ocean basin turbidites. At DSDP site 178, this sediment sequence is of Neogene age and contains massive hemipelagic mudstones and some silt and sand turbidites (von Huene and Kulm, 1973).

The segment of the trench surveyed in grid B (Figs. 2, 5) is also characterized by subduction of large volumes of sediment; however, because the structural style differs from grid A, the nature of the sediment being subducted may also differ. In fact, it has been suggested that considerable trench fill and even slope sediment may be subducted along this segment of the trench (von Huene, 1979). This previous structural interpretation has been modified, however, based on the more recent reprocessing results. Records studied from grid B (Fig. 5) show a variable structural style, but each demonstrates subduction of a sediment section that varies in thickness from 1.2 to 3.3 km (0.7–2.1 mi) at the front of the margin. In records 71 and 72/62, the subduction involves both the trench fill and the underlying ocean basin sediment sequence; the zone of sediment subduction appears to break through the lower slope and could involve some slope sediment. Landward of the deformation front the maximum thickness on a subducting sediment section is about 4 km (2.5 mi) in record 71. An average thickness of all lines in the grid is 2.75 km (1.7 mi) and can vary from 1.2 km to 4.0 km (0.7–2.5 mi).

One record at line C (Fig. 2), equivalent to record 13 (Fig. 6), shows a very thick offscraped section, and no subducted sediment is obvious; however, the basement surface was not imaged. Thus, we cannot determine how much sediment is subducted along record 13 and whether record 13 is representative of the area.

In contrast to line C, nearby line D (Figs. 2, 6) shows subduction of the complete sediment section (Plafker et al, 1982). Although the seismic record does not show a well-developed decollement, the age relation required by the stratigraphy of a well on the shelf compels such an interpretation. This well penetrated a sediment sequence of at least lowest Eocene age (Rau et al, 1977; Keller et al, in press). The Oligocene–Miocene time horizon can be clearly followed from the drill hole to the base of the trench slope; the position of the base of the Eocene is inferred from its depth in the drill hole. From DSDP hole 180 the Quaternary section is traced to the base of the slope and must pass beneath the front of the margin (Plafker et al, 1982). If this record is representative of a segment of the trench, there is a 2.7 km (1.7 mi) thick sediment section being subducted that includes the uppermost trench sediment. This seismic record shows that the subduction of sediment may involve the total section in the trench, as has also been demonstrated in the Middle America Trench off Guatemala (Aubouin et al, 1982, 1983). The proximity of line D to line C (Fig. 2), which failed to record sediment subduction, emphasizes the variability possible from total sediment subduction to total sediment addition by offscraping.

GEOCHEMICAL ANALYSES AND RESULTS

The organic matter in sediment associated with the area of the Aleutian Trench can provide a guide to the organic matter that is being subducted beneath the present convergent margin. Forty samples have been studied in order to ascertain the content and properties of the organic matter in sediment postulated to be involved in one way or the other in the subduction process. Thirty-four of these samples came from five DSDP sites (178, 180, 181, 183, and 192), and six samples came from two USGS dredge sites (2 and 4). Figure 2 shows the location of these sampling sites relative to the seismic grids that were discussed previously. The basinal sediments at sites 178, 180, and 183 are believed to be equivalent to sediments currently undergoing subduction at this active margin. Sediments sampled at site 181 are lower slope deposits that may have been accreted during the subduction. Site 192 is on a sea mount where the sediment cover may be representative of basinal sediments involved in subduction at the western end of the Aleutian Trench. Dredge samples 2 and 4 came from outcrops on the upper slope. These samples may be lithified equivalents of older sediments undergoing subduction.

DSDP Samples

Selection of 34 samples from the DSDP repository was guided by the following considerations: Samples from DSDP Leg 18 were chosen from sites 178, 180, and 181; samples from DSDP Leg 19 were chosen from sites 183 and 192. These sites were selected because the sediments at these locations appear to be equivalent to sediments currently involved in some stage of subduction. Individual samples were selected on the basis of position in the core, organic carbon content, and on availability, with the samples containing the highest amount of organic carbon being preferentially collected. Approximately 40 cu cm (2.4 cu in) of sediment were removed for each sample. Weights of samples ranged from 62 to 114 g (2.2–4.0 oz). The samples are listed in Table 1.

USGS Samples

Six dredge samples from USGS cruise S-79-WG were analyzed (Table 1). Five samples, including three mudstone

Table I

Sample no.	DSDP Leg/ Cruise	Site	Depth of Water (m)	Depth of Sediment (m)	Lithology	Age	OC (%)	Carbonate Carbon (%)	Total Hydrocarbon Yield (%)	"Live Carbon" (%)	Volatile Hydrocarbon (ppm)	T_{max} °C
1	18	178	4,218	34.8	Silty clay, md dk gy	Pleistocene	0.62	0.09	0.05	8	37	475
2	18	178	"	108.1	Silty clay, md dk gy	Pleistocene	0.50	0.00	0.03	6	36	455
3	18	178	"	217.3	Silty clay, md dk gy	Late Pliocene	0.35	0.04	0.03	9	53	494
4	18	178	"	319.3	Silty clay, olive gy	Late Pliocene	0.53	0.03	0.06	11	47	522
5	18	178	"	327.2	Silty clay, gn-gy, w/diatoms	Late Pliocene	0.23	0.03	0.05	22	95	422
6	18	178	"	395.2	Fine silt and silty clay, md dk gy	Late Pliocene	0.35	0.14	0.05	14	32	523
7	18	178	"	459.8	Silty clay, olive gy	Late Pliocene	0.50	0.02	0.06	12	47	422
8	18	178	"	507.6	Silt and silty clay, dk gn-gy	Pliocene	0.33	0.04	0.04	12	43	437
9	18	178	"	591.5	Silt and silty clay, dk gn-gy	Pliocene	0.33	0.01	0.05	15	31	511
10	18	178	"	631.8	Diatomite, gn-gy	Early Pliocene	0.14	0.03	0.05	36	86	430
11	18	178	"	718.6	Silty clay, olive gy	Middle Pliocene?	0.33	0.01	0.05	15	38	433
12	18	178	"	747.7	Claystone, lt olive gy	Early Miocene	0.06	0.04	0.03	50	41	442
13	18	180	4,923	71.4	Clayey silt and silty clay, gy	Late Pleistocene–Holocene	0.67	0.07	0.07	10	79	448
14	18	180	"	246.0	Silty clay and clayey silt, gy	Late Pleistocene–Holocene	0.66	0.07	0.05	8	76	496
15	18	180	"	272.7	Silty clay and clayey silt, gy	Late Pleistocene–Holocene	0.50	0.09	0.04	8	67	481
16	18	180	"	347.3	Silty clay and clayey silt, gy	Late Pleistocene–Holocene	0.59	0.07	0.07	12	164	418
17	18	180	"	454.8	Silty clay and clayey silt, gy	Late Pleistocene–Holocene	0.70	0.13	0.06	9	60	465
18	18	181	3,086	15.2	Silty clay, olive gy, w/diatoms	Late Pleistocene–Holocene	0.59	0.15	0.06	10	75	464
19	18	181	"	71.3	Silty clay, med gy	Late Pleistocene–Holocene	0.63	0.07	0.05	8	81	502
20	18	181	"	137.9	Silty clay, med dk gy, w/diatoms	Late Pleistocene–Holocene	0.69	0.04	0.05	7	90	450
21	18	181	"	157.7	Silty clay, med dk gy, w/diatoms	Late Pleistocene–Holocene	0.87	0.03	0.05	6	73	466
22	18	181	"	177.4	Silty clay, gy-bk and gn-bk	Pleistocene	0.53	0.01	0.04	8	56	447
23	18	181	"	339.5	Silty clay, med dk gy to olive bk	Early Pleistocene?	0.35	0.01	0.03	8	43	459
24	19	183	4,708	35.3	Silty clay, olive gy, w/diatoms	Middle Pleistocene	0.36	0.06	0.08	22	165	452
25	19	183	"	101.6	Silty clay, olive gy, w/diatoms	Late Pliocene	0.29	0.00	0.05	17	34	458
26	19	183	"	167.5	Diatom ooze, lt yel-bn	Late Miocene	0.19	0.01	0.10	52	197	455
27	19	183	"	220.3	Clayey silt and clay olive gy and gn-gy	Miocene?	0.92	0.05	0.07	8	158	428
28	19	183	"	296.7	Clay, dk gy	Early Oligocene	0.64	0.07	0.05	8	93	443
29	19	183	"	476.2	Clay olive gy to dk gn-gy	Middle Eocene?	0.34	0.01	0.04	12	655	455
30	19	192	3,014	25.9	Silty clay, dk gy, w/diatoms	Middle Pleistocene	0.28	0.37	0.04	14	80	408
31	19	192	"	271.5	Clayey diatom ooze, olive gy	Early Pliocene	0.37	0.02	0.06	16	135	410
32	19	192	"	571.0	Silty clay w/diatoms, dk gy	Late Miocene	0.20	0.00	0.04	20	106	398
33	19	192	"	907.0	Claystone, dk gn-gy	Miocene?	0.13	0.02	0.03	23	37	505
34	19	192A	3,014	1019.5	Claystone, dk gy to gy-bk	Middle Eocene	0.05	0.25	0.04	80	99	473
35	S-79-WG	2	1,800	Outcrop	Mudstone, volcanic, w/tr ss, pumice	Early–Middle Miocene	0.45	0.00	0.03	7	37	440
36		2	"	"	Dolomite, massive, indurated, pale bn	Middle–Late Eocene	0.33	9.1	0.05	15	46	443
37		2	"	"	Dolomite, clastic, rexl	Early–Middle Miocene	0.24	8.8	0.03	13	20	456
38		2	"	"	Mudstone, mud, indurated, lt gy	Eocene?	0.66	0.01	0.05	8	61	437
39		2	"	"	Mudstone, lt gy	Early–Middle Miocene?	0.80	0.00	0.06	8	94	429
40	S-79-WG	4	2,200	"	Limestone, massive, gy, nodular	Early Pleistocene?	0.30	6.3	0.03	10	76	433

Table 1—Listing of geochemical results.

and two dolomites, were obtained from site 2, and one sample, a massive limestone, was used from site 4. These upper-slope outcrop samples ranged in age from middle Eocene to early Pleistocene. The water depths at sites 2 and 4 are 1,800 and 2,200 m (5,910 and 7,220 ft), respectively.

Organic Carbon

For the 34 DSDP samples used in this study, an estimate of the organic content (Table 1) was available through compilations by Bode (1973a, 1973b), which contain organic carbon values for equivalent core samples. Our determinations by high-temperature oxidation techniques, utilizing a LECO analyzer, are listed in Table 1 and are in remarkable agreement with the values obtained by Bode (1973a, 1973b). This close agreement establishes a high level of confidence in the reported organic carbon values.

Total Carbon

Total carbon was determined for 40 samples (Table 1). The results show that all but three samples contain only small (less than 0.37%) carbonate carbon. Three dredge samples contain 6.3 to 9.1% total carbon, and these samples are classified as dolomite or limestone.

Pyrolysis

All 40 samples were analyzed by a temperature programmed pyrolysis technique (Thermal Evolution Analysis, or TEA) in which the products of pyrolysis are measured by a flame ionization detector (FID), and the temperature of maximum pyrolysis yield is determined (Claypool and Reed, 1976). This pyrolysis method provides information on total hydrocarbon yield (measured in percent of organic carbon), volatile hydrocarbons (measured in parts per million of the total hydrocarbon yield), and T_{max} in degrees C representing the temperature at which the maximum amount of organic matter is thermally decomposed. "Live carbon" is a measure of the content of "hydrocarbon-prone" organic matter and is obtained by dividing the total hydrocarbon yield by the amount of organic carbon. "Live carbon" is considered to be the carbon that will, upon thermal evolution, still yield additional hydrocarbon products such as methane gas.

Discussion

The results of our organic geochemical analyses are listed in Table 1. These results show that all of the samples analyzed contain low amounts (less than 1%) of organic carbon (OC). The total hydrocarbon yield is also low and ranges between 0.03 and 0.10%. With the exception of four samples, the amounts of "live carbon" are less than 25%. Those four samples with "live carbon" exceeding 25% are samples with the least amount of organic carbon. The volatile hydrocarbon content is always less than 200 ppm. T_{max} ranges from 398 to 523°C (748–973°F), but the pyrograms from which these data are taken are so indistinctive that the results are generally considered unreliable for interpretive purposes. Profiles with depth of our geochemical results for the five DSDP sites are shown in Figure 7. These profiles emphasize the low amounts of the various geochemical parameters. Organic carbon decreases irregularly with depth at site 178, 181, and 192. Significant trends of other parameters are not obvious. The highest amount of organic carbon (0.9%) was found at a 220 m (720 ft) subbottom depth at site 183. This amount of organic carbon is interesting from the point of view of source-rock evaluation, but the low total hydrocarbon yield makes the amount of "live carbon" very low (8%), thus decreasing the potential of this sediment for hydrocarbon generation.

Table 2 summarizes our results. The low amounts of organic carbon, total hydrocarbon yield, "live carbon," and volatile hydrocarbons suggest that the sediments we have sampled and analyzed are poor potential source sediments for petroleum, both oil and gas. The average amount of organic carbon is equal to or less than 0.6%, a value considered near the lower limit for potential sources of hydrocarbons (Hunt, 1979). The amount of "live carbon," i.e., the carbon available for future hydrocarbon generation, is less than 30%, which indicates, according to work by Magoon and Claypool (1981), that the organic matter is prone to gas generation rather than oil generation.

Nine samples were selected for detailed examination of organic matter type (Table 3). These samples were chosen because they appeared to be richest in organic carbon, and there was sufficient sample on which to carry out the analyses. These samples were subjected to a specialized pyrolytic technique called Rock-Eval (Tissot and Welte, 1978). In addition, our analyses consisted of vitrinite reflectance measurements, visual kerogen analyses by both transmitted and incident light, and evaluation of the thermal alteration index (TAI).

Rock-Eval analyses yield hydrogen indices (HI) and oxygen indices (OI), listed on Table 3, from which a van Krevelen-type diagram (Tissot and Welte, 1978) can be constructed (Fig. 8). On this diagram the thermal evolution paths of different types of kerogen are indicated. For the most part all of our samples lie along the Type III pathway of this diagram. Type III organic matter is generally considered to be of terrestrial origin and gas prone (Tissot and Welte, 1978).

Vitrinite reflectance (R_0) values of primary vitrinite range from 0.36 to 0.50% (Table 3). Recycled vitrinite of greater R_0 values is also present in all samples. TAI evaluations range from 1.5 to 2.6 (Table 3). Both vitrinite reflectance and TAI values indicate that the organic matter of these samples is immature with respect to oil or gas generation and shows strong indications of reworking. Visual kerogen analyses (Table 3) indicate that, in addition to vitrinite and recycled vitrinite, the samples also contain exinite, inertinite, recycled sporinite, and amorphous material. The distribution of these kerogen macerals suggests that most of the organic matter is terrestrial in origin, and the position of these samples on the van Krevelen diagram (Fig. 8) indicates immaturity.

Our geochemical analyses show the following: The amount of organic carbon in the samples is small and approaches the lower limit as a potential source of hydrocarbons. The presence of inertinite further reduces the hydrocarbon potential of these samples. The organic matter is mainly from terrestrial sources, is gas prone, and is immature with respect to gas generation. Upon thermal

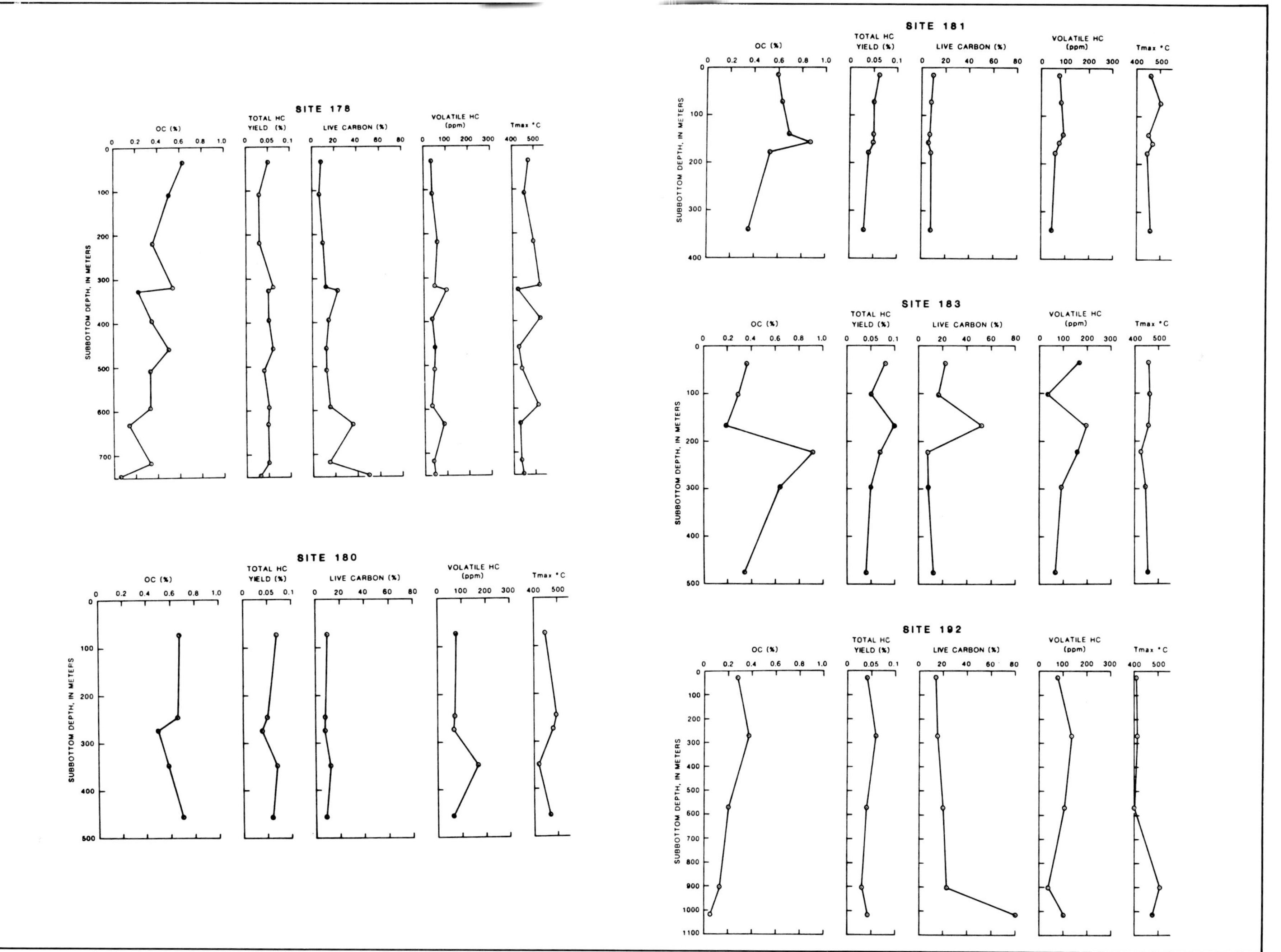

Figure 7—Profiles with depth of organic geochemical results [organic carbon (OC); total hydrocarbon (HC) yield; live carbon; volatile hydrocarbons; temperature at maximum yield of pyrolysis products (T_{max})] for DSDP sites 178, 180, 181, 183, and 192.

Table 2

Site	Number of Samples	OC (%)	Hydrocarbon Yield (%)	"Live Carbon" (%)	Volatile Hydrocarbons (ppm)	T_{max} (°C)
178	12	0.36 ± 0.16	0.05 ± 0.01	18 ± 13	49 ± 21	464 ± 39
180	5	0.62 ± 0.08	0.06 ± 0.01	9 ± 2	88 ± 43	462 ± 30
181	6	0.61 ± 0.17	0.05 ± 0.01	8 ± 1	70 ± 17	464 ± 20
183	6	0.46 ± 0.27	0.07 ± 0.02	20 ± 17	119 ± 64	448 ± 11
192	5	0.21 ± 0.13	0.04 ± 0.01	31 ± 28	91 ± 36	439 ± 47
2	5	0.50 ± 0.23	0.04 ± 0.01	10 ± 4	52 ± 28	441 ± 9
4	1	0.30	0.03	10	76	443

Table 2—Summary of geochemical results by sample sites.

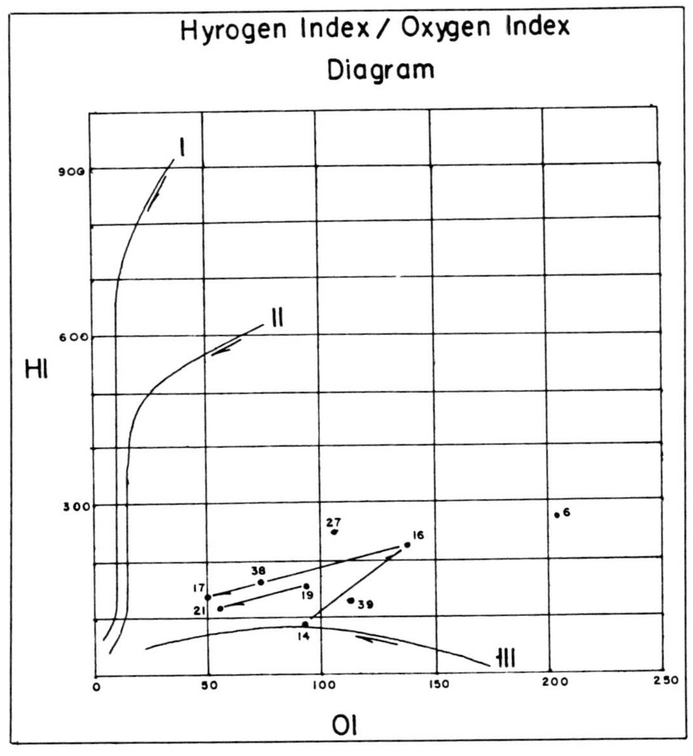

Figure 8—van Krevalen diagram showing the hydrogen index (HI) in mg HC/g OC and the oxygen index (OI) in mg CO_2/g OC for nine selected samples (Table 3). Results fall in the region of type III organic matter.

Table 3

Sample No.	HI (mg HC/ g OC)	OI (mg CO_2/ g OC)	R_o (%)	TAI	Exinite (%)	Vitrinite (%)	Inertinite (%)	Recycled Vitrinite (%)	Recycled Sporinite (%)	Other Amorphous (%)
6	270	202	0.40	—	4	32	26	24	4	10
14	95	92	0.40	2.0	1	13	7	68	7	6
16	223	136	0.45	1.5	—	12	6	62	—	20
17	130	50	0.36	2.0	4	38	13	40	5	—
19	156	91	0.47	2.0–2.3	3	39	4	32	18	4
21	119	53	0.46	2.4–2.6	—	13	—	18	7	2
27	245	105	0.45	2.0–2.3	2	14	10	28	—	48
38	160	71	0.50	2.0–2.4	5	20	5	30	10	30
39	122	111	0.38	2.3–2.5	7	15	—	24	16	40

Table 3—Results of analyses of kerogen.

evolution, the samples are expected to generate only small amounts of gas.

ESTIMATION OF GEOTHERMAL GRADIENT

To determine the region where gas is generated in the subduction complex of the Aleutian Trench requires some knowledge of the geothermal regime. Information regarding this regime in the Aleutian Trench is minimal, and direct readings of geothermal temperatures have been made at only one place. A temperature log from a well drilled just offshore of Middleton Island on the edge of the shelf showed an average temperature gradient of 28°C/km (81°F/mi).

An innovative approach to the determination of regional geothermal gradients has been described by Yamano et al (1982) for areas where gas hydrates are present and are manifest on seismic records as an anomalous bottom-simulating reflector or BSR. From the depth of the BSR, the geothermal gradients are estimated using the phase relations of the gas hydrate system. This method has been successfully applied to data from the Nankai Trough offshore Japan, around Central America including the Middle America Trench, and along the Blake Outer Ridge. Geothermal gradients in the Nankai Trough ranged from 41.5 to 65.8°C/km (121–191°F/mi), and along the Middle America Trench they ranged from 28.2 to 35.1°C/km (82–102°F/mi). Both of these areas are convergent margins, as is the Aleutian Trench. Macleod (1982), using the same methods, estimated the average geothermal gradient in sediment of the Gulf of Alaska to be 27.8 ± 3.5°C/km (81 ± 10°F/mi).

BSRs can be seen on some seismic records from the upper slope of the Aleutian Trench (Fig. 9). Depths to the BSR were obtained using a seismic velocity of 2.0 km/sec (1.2 mi/sec). Five depths to BSRs were calculated at five positions with different water depths. This information was applied to the phase diagram for gas hydrates given by Kvenvolden and McMenamin (1980) for five different bottom-water temperatures to obtain estimates of the geothermal gradient, listed on Table 4 along with geothermal gradients estimated from hydrate stability curves given by Macleod (1982). There are several reasons for the difference in gradients. Kvenvolden and McMenamin assume a pure methane–pure water, gas-hydrate system whereas Macleod uses a pure methane–Arctic seawater, gas-hydrate system. Also, the latter system assumes bottom-water temperatures that vary with depth of water and are all less than 1°C (34°F).

Our estimates for the geothermal gradient on the upper slope of the Aleutian Trench area, based on the depth of the BSR and bottom-water temperatures of 1° to 3°C (34–37°F), are 32 ± 4°C/km (93 ± 12°F/mi) (Table 4). This range includes the geothermal gradient from the Middleton Island well of 28°C/km (81°F/mi) and the estimates of Macleod (1982) of 27.8 ± 3.5°C/km (81 ± 10°F/mi) in sediment of the Gulf of Alaska. For simplicity in our study we use an average geothermal gradient of 30°C/km (87°F/mi).

A geothermal gradient of about 30°C/km (87°F/mi) for the upper slope of the Aleutian Trench is significantly lower than the average gradient of 53°C/km (154°F/mi) in the Nankai Trough but about the same as the average gradient of 32°C/km (93°F/mi) along the Middle America Trench and the 24 to 32°C/km (70–93°F/mi) gradient for the Japan Trench (Langseth and Burch, 1980). All of these areas are convergent margins at the edge of the Pacific Plate (Fig. 1). The Nankai Trough is exceptional in that it is a convergent margin where back-arc crust of the Philippine Plate is being subducted, and thus its thermal structure is unusual. Because of the uncertainties in the use of gas-hydrate BSRs to estimate geothermal gradients, the accuracy of the estimates is not high, but the values obtained are consistent with those measured by conventional means and sufficient for our purposes.

Temperature distribution across active margins is characteristically complex (cf. Watanabe et al, 1977). The uncertainties derive not only from the subduction of cool oceanic crust but also from the fluid flux and differential rates of tectonic transport as well. Most investigators model a zone of temperature reversal along the principal slip plane of a subduction zone (Anderson et al, 1977). Our temperature information is so rudimentary that any attempts to account for such variations are overshadowed by the uncertainties of sparse observations. Thus, for our estimates of thermal structure, we simply assume a linear vertical gradient but are aware of the imprecision this assumption incorporates.

ASSESSMENT OF POTENTIAL FOR NATURAL GAS GENERATION

An assessment of the potential of the Aleutian Trench subduction zone for natural gas generation requires basic geologic information. We now have preliminary ideas about the amount and nature of carbon in some of the sediments involved in the subduction process. Our seismic surveys show the tectonic framework of the area from which we can estimate subducted sediment thicknesses and assume sediment volumes. Finally, we have an estimate of the geothermal gradient based on considerations of gas hydrates and a measurement in a well as discussed above. We can now put this geochemical, geophysical, and geothermal information together to estimate the amount of natural gas generated by thermogenic processes. For our subduction model we have chosen to use average values of our data and to extrapolate these averages for the eastern Aleutian Trench subduction complex.

Our geochemical studies show that the average organic carbon content of the samples analyzed is 0.44% and that an average of 15% of this carbon is "live," i.e., able to react thermally further to produce hydrocarbons. These average values were calculated from Table 2. Because the kerogen in these samples is Type III, we assume a hydrogen to carbon (H/C) ratio of about 1 knowing that the H/C ratio of methane is 4. Therefore, we assume that only about one-fourth of this potential hydrocarbon could be methane gas, because of the gas-prone nature of the analyzed organic matter. The remaining material after gas generation would be elemental carbon if all hydrogen for methane formation comes from the organic matter. This assumption leads to maximum values of potential generation of natural gas. The

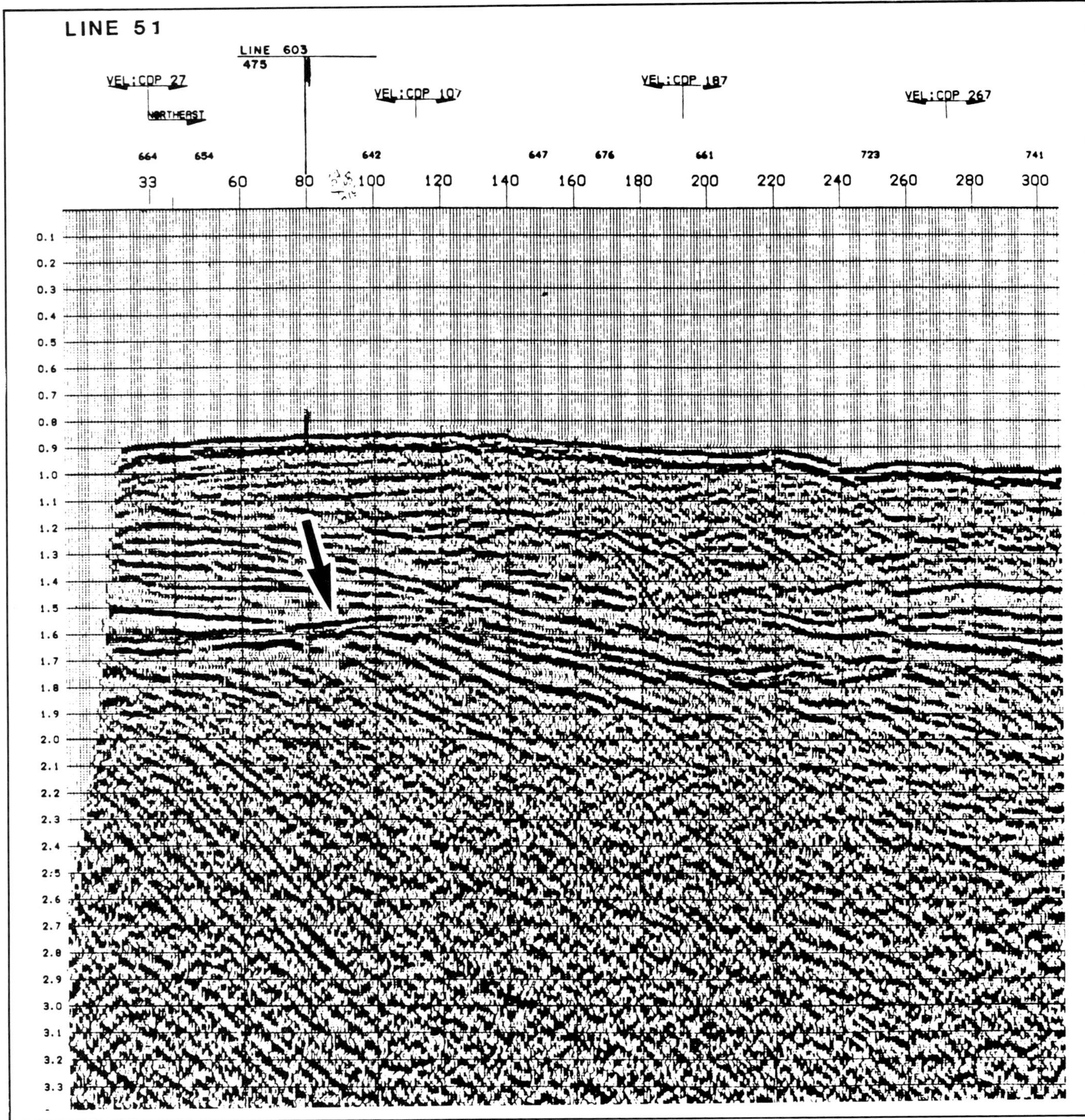

Figure 9—Example of BSR (arrow) on seismic record 51 from grid B (Fig. 2).

actual amount of methane possible could be two or three orders of magnitude lower. These average values coupled with the following assumptions led to a calculated potential methane content of these sediments of about 5.5×10^8 cu m of methane per cu km (8.2×10^{10} cu ft/cu mi) of sediment: (1) 12 g of "live carbon" equals 1 mole of "live carbon"; (2) 1 mole of "live carbon" equals 22.4 l (0.78 cu ft) of methane; and (3) the density of sediments is 1.8 g/cu cm. A value of 5.5×10^8 cu m of methane/cu km (8.2×10^{10} cu ft/cu mi) of sediment represents the maximum amount of methane dispersed in the sediment. Estimates of the ratio of dispersed and dissolved gas to reservoir gas in sedimentary basins in general range from 10–200 to 1 (Hunt, 1979). Because we saw little evidence for reservoir-type sediment, that is, sediment with good porosity and permeability, during our sampling, we selected the larger value of this ratio to apply in our estimate. Thus the amount of entrapped methane in each cu km of sediment is estimated to be about 2.8×10^6 cu m (4.2×10^8 cu ft/cu mi).

Table 4

Water Depth (m)	Sediment Depth (m)	Total Depth (m)	Bottom Water Temperatures (°C)					Temperature Gradient from Macleod (1982)
			1.0	1.5	2.0	2.5	3.0	
2085	648	2733	32.5	31.8	31.1	30.3	29.6	29
1928	567	2495	35.9	35.0	34.0	33.1	32.2	32
2325	648	2973	33.9	33.2	32.4	31.7	30.9	30
1830	666	2496	30.8	30.1	29.3	28.6	27.8	28
1575	612	2187	31.5	30.7	30.0	29.2	28.4	28

Table 4—Geothermal gradients in °C/km determined from base of gas-hydrate reflector (BSR) on marine seismic records.

Thicknesses of sediment involved in the subduction process of the Aleutian Trench area can be estimated from the seismic sections of the two grids and two lines described previously.

Grid/Line	Thickness (km/mi)	Comment
A	3.7/2.3	lower section only
B	2.75/1.71	including trench and slope(?) sediment
C	0	accretionary sediment only
D	2.7/1.7	whole trench section

The average thickness of subducting sediment is about 3 km (1.9 mi). In our calculation we use a 3-cu km (.72 cu mi) element of sediment measuring 1 km by 1 km of areal extent and 3 km thick and estimate the rate of subduction at 60 km/m.y. (37 mi/m.y.) (Fig. 1). Thus, in 1 m.y. 180 cu km (43 cu mi) of sediment would be subducted. The amount of reservoired methane to be expected from this volume of sediment is therefore $180 \times 2.8 \times 10^6 = 5.0 \times 10^8$ cu m (1.8×10^{10} cu ft). If this number is extrapolated over the 600 km (373 mi) length of the eastern Aleutian Trench, then the amount of thermally generated, entrapped methane per million years in this province is about 0.3×10^{12} cu m (11×10^{12} cu ft = 11 trillion cu ft [11 Tcf]). In 20 million years, about 6×10^{12} cu m (210 Tcf) of methane would be present and this number is about 9% of the current world reserves (66.4×10^{12} cu m [2,343 Tcf]) of natural gas (Rice and Claypool, 1981).

Other studies support the idea that thermally produced gases are present in sediments of the Aleutian Trench area, but none of these studies provide quantitative assessments of the amount of gas. For example, Claypool et al (1973) suggested that active thermal generation of gases was taking place at depths as shallow as 250 m (820 ft) in this area. Some of the gas represents early thermal generation as opposed to peak generation. On the basis of spore-coloration data, Grayson and LaPlante (1973) indicated that the level of maturation of a sample from DSDP site 181 from a subbottom depth of about 340 m (1,120 ft) was equivalent to an overburden depth in the Gulf Coast of about 2,600 m (8,530 ft). Dow (1978) calculated an equivalent $R_0 = 0.3\%$ for this sample. Our measured R_0 values for shallower sediments at the same site equal 0.46 and 0.47% (Table 3). The calculated and measured R_0 values indicate that the sampled sediments currently are at lower temperatures than required for gas generation; however, as we have shown, these sediments may eventually be included in the subduction complex and be exposed to temperatures at which the organic matter will be transformed to methane.

There are large uncertainties in our estimate of 0.3×10^{12} cu m (11 Tcf) of potential, reservoired methane generated each million years of subduction. Our calculated value results from extensive extrapolations and assumes that the methane becomes reservoired and is not lost from the system either downward into the subduction zone or upward into the atmosphere. Our model requires that the generated methane somehow migrates from the subducting sediments where it formed into the overlying, nonsubducted sediments where it becomes trapped. We have no direct evidence relating to conduits for gas migration or for traps of suitable size for exploration. Also, our estimates represent a maximum and could easily be two orders of magnitude smaller if, for example, only one percent of the "live carbon" is transformed thermally into methane. Our large number estimate results not from the richness of organic matter in these sediments but rather from the enormous volumes of sediment involved in the subduction process. In fact, the amount of organic matter is very small and, by itself, would indicate only a poor potential for significant gas generation.

Our calculations have not taken into account biogenically produced methane that was present in the modern sediments associated with the Aleutian Trench. Kulm et al (1973) and Creager et al (1973) refer to the gassy nature of sediment cores recovered during DSDP Legs 18 and 19, although no quantitative measure of the total amount of gas was made. Claypool et al (1973) showed that the carbon isotopic composition of methane from DSDP Site 180 (one of the sample sites used for this paper) ranged from −72.6 to −80.8 per mil relative to the PBD standard. This range of values is well within the range

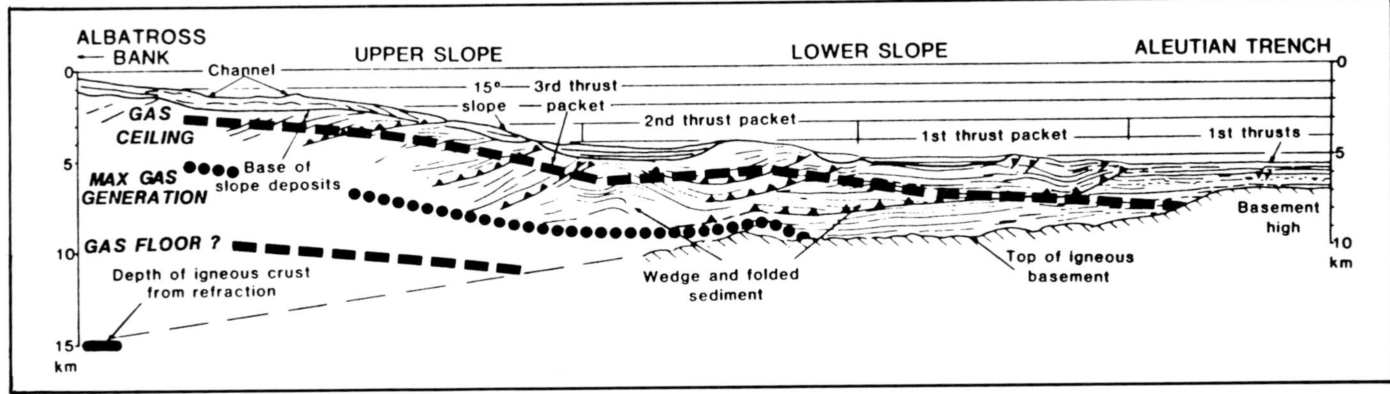

Figure 10—Isotherms at temperatures of 50°C (122°F) (gas ceiling, i.e., the lowest temperature for gas generation), 150°C (302°F) (maximum gas generation), and 250°C (482°F) (gas floor, i.e., the highest temperature for gas generation) superimposed on seismic line 111 (Fig. 4) based on a constant geothermal gradient of 30°C/km (87°F/mi).

diagnostic of biogenic gas (Fuex, 1977). Also we have not considered the gas volume consequences of the presence of gas hydrates observed in some of our seismic records. In spite of the low measured concentrations of organic matter in the sediments of the Aleutian Trench area, our study suggests that very large concentrations of both biogenic and thermogenic methane can be formed during sedimentation and subsequent subduction. A major unknown is where the gas is now. We still lack sufficient data on preservation, migration, and trapping to draw firm conclusions at this time. Clues from onshore geology (Fisher, 1980) indicate both poor reservoir and source potential at least for the area of the Kodiak shelf.

LOCATION OF GAS RESERVOIRS

If we accept the idea that significant quantities of methane can be thermally generated in the subduction zone of the Aleutian Trench, we can determine, based on the geothermal gradient, in what region of the subduction complex the gas might be present. Figure 10 shows the geologic section for seismic line 111 of Grid A. Superimposed on this section are isothermal lines for 50, 150, and 250°C (122, 302, and 482°F) based on our estimated constant vertical geothermal gradient of 30°C/km (87°F/mi). The three isothermal temperature lines represent the temperatures for the gas ceiling, maximum gas generation, and the gas floor, respectively. Of these three temperatures, the gas floor is the most uncertain and is likely a higher temperature than 250°C (482°F) (Barker, 1982). A 3 km (1.9 mi) thick section of marine sediment involved in subduction would pass through the temperature of maximum gas generation somewhere beneath upper-slope sediment. Thus, if our ideas are correct, the region below upper-slope sediment would be most prospective for future gas exploration if reservoirs and traps are present. This is the same region that was predicted by Thompson (1976), based on a thrust-fault-controlled model, for petroleum accumulation in the Aleutian Trench area. A diagram illustrating our subduction gas generation model is shown in Figure 11.

Besides the possible existence of gas in deep, upper-slope sediments, gas may also be present in the deep basins associated with the Aleutian convergent margin. Except for the great depth of water, these basins are similar to basins of the shelves and require conventional strategies for evaluation of petroleum prospects. Hedberg et al (1979) concluded that deep ocean basins, including those associated with convergent margins, are potential sites for petroleum accumulation. We have not considered these kinds of basins in our assessment. If in the Aleutian Trench area both the basins, as suggested by Hedberg et al (1979), and the upper-slope sediments, as suggested by Thompson (1976) and ourselves, are prospective for petroleum gas, then this vast area may become an important exploration target in the future.

SUMMARY

This paper provides preliminary data addressed to the question of the potential of subducted sediment within the Aleutian Trench convergent margin to generate significant quantities of natural gas. Our results suggest that even though the content of organic carbon in sediments of the ocean basin is low, averaging about 0.4%, the volume of sediment involved in subduction is so large, having an average thickness of 3 km (1.9 mi) that, indeed, significant quantities of gas can be thermally generated. We calculate that as much as 6×10^{12} cu m (210 Tcf) of natural gas could have been generated and stored during the last 20 million years of subduction. This number is equal to 9% of the currently known world reserves of natural gas. To this number can also be added an as yet unknown but probably large quantity of biogenically derived methane that is present in the sediments even before they undergo subduction. In addition, there is an unknown amount of natural gas in basins associated with this convergent margin. Our estimate of potential gas generation during subduction has great uncertainties, but because of the large volumes of sediment involved, a significant amount of generated gas is possible even though the content of organic matter in the sediment is small. A large unknown is where the gas migrates and is currently stored. Our model assumes that the gas is not lost through subduction or

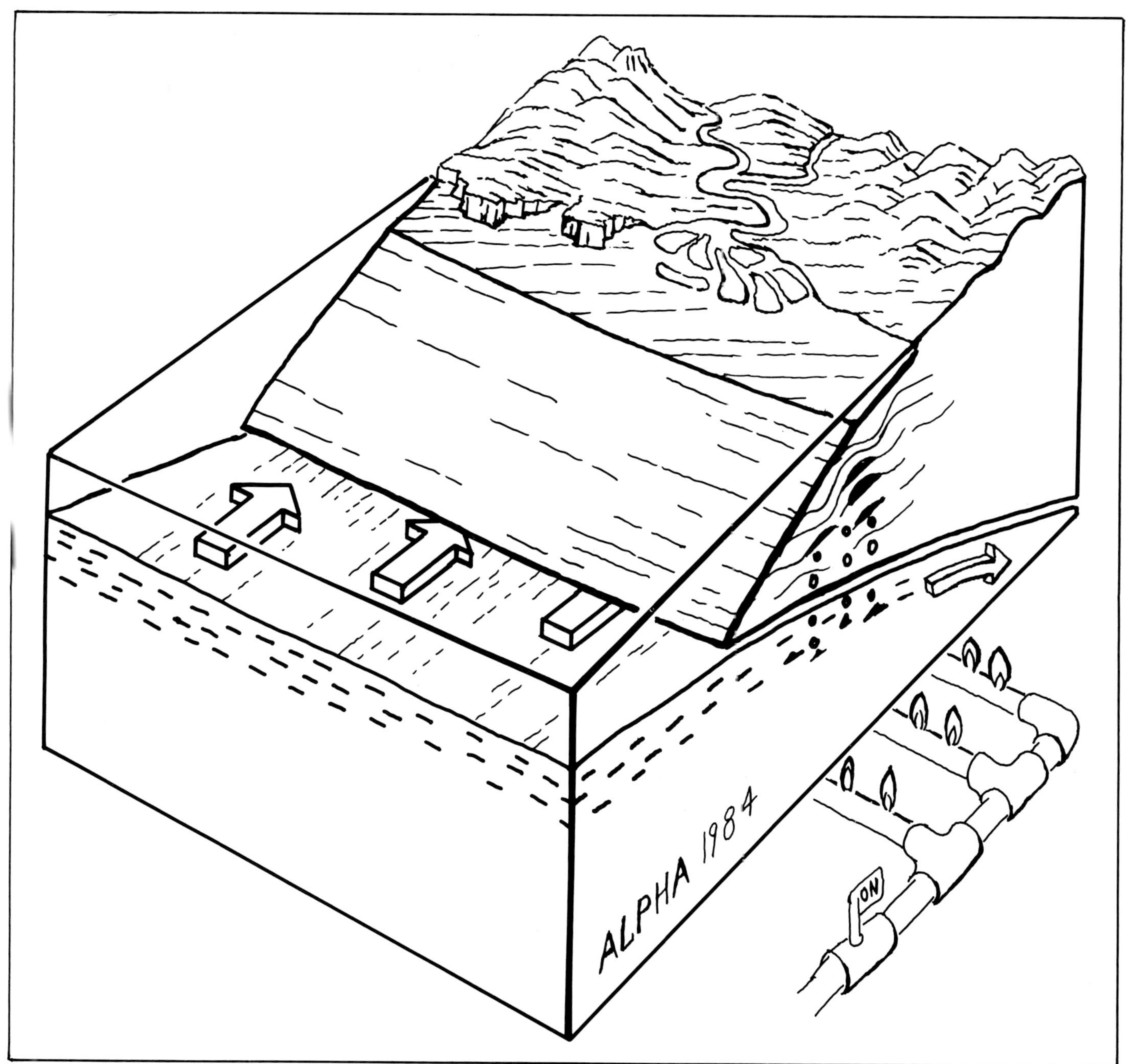

Figure 11—Diagram showing subduction gas generation model (drawn by T. R. Alpha).

through diffusion to the atmosphere. We have no direct evidence for migration pathways or for traps of suitable size for exploration. Our results suggest that, if significant quantities of natural gas can migrate upward above the area of maximum gas generation, then most of this gas would likely be found in sediment of the upper slope of the Aleutian Trench and not in shelf or lower-slope deposits. If we extrapolate our subduction gas generation model to the entire circum-Pacific margin, we conclude that potential gas accumulations may be present in this margin wherever large volumes of sediment are being subducted.

ACKNOWLEDGMENTS

Samples from DSDP Logs 18 and 19 were supplied by the Deep Sea Drilling Project through the assistance of the U.S. National Science Foundation. We thank GeoChem Research Laboratories of Houston, Texas, for geochemical analyses and Clark Geological Services of Fremont, California, for detailed analyses of kerogen. We appreciate the help of John Miller in processing the seismic data, Jen Blank for organizing the figures, and Kay McDaniel for typing and processing the manuscript. This work was partially funded by the Morgantown Energy Technology Center under U.S. Geological Survey-Department of Energy Interagency Agreement No. DE-AI21-83MC20422.

REFERENCES

Anderson, R. N., et al, 1977, Thermal model for subduction with dehydration in the downgoing slab: Journal of Geology, v. 86, p. 731–739.

Aubouin, J., et al, 1982, Initial Reports of the Deep Sea Drilling Project: Washington, DC, U.S. Gov't. Printing Office, v. 67, 799 p.

______, 1983, Leg 84 of the Deep Sea Drilling Project, Subduction without accretion: Middle American Trench off Guatemala: Nature, v. 297, p. 458–460.

Barker, C., 1982, Methane generation and survival in the deep subsurface, *in* W. J. Gwilliam, ed., Deep Source Gas Workshop Technical Proceedings: U.S. Department of Energy, Morgantown Energy Technology Center, 82-50, UC-92a, p. 86–107.

Bode, G. W., 1973a, Carbon and carbonate analyses, Leg 18; *in* L. D. Kulm, et al, Initial Reports of the Deep Sea Drilling Project: Washington, DC, U.S. Gov't. Printing Office, v. 18, p. 1069–1076.

______, 1973b, Carbon-carbonate, *in* J. S. Creager, et al, Initial Reports of the Deep Sea Drilling Project: Washington, DC, U.S. Gov't. Printing Office, v. 19, p. 663–665.

Claypool, G. E., and P. R. Reed, 1976, Thermal analysis technique for source-rock evaluation: quantitative estimate of organic richness and effects of lithologic variation: Bulletin of the American Association of Petroleum Geologists, v. 60, p. 608–626.

______, et al, 1973, Gas analyses in sediment samples from Legs 10, 11, 13, 14, 15, 18, and 19, *in* J. S. Creager, et al, Initial Reports of the Deep Sea Drilling Project: Washington, DC, U.S. Gov't. Printing Office, v. 19, p. 879–884.

Creager, J. S., et al, 1973, Initial reports of the Deep Sea Drilling Project: Washington, DC, U.S. Gov't. Printing Office, v. 19, 913 p.

Delong, S. E., and P. J. Fox, 1977, Geological consequences of ridge subduction: *in* M. Talwani and W. C. Pittman, eds., Island Arc, Deep Sea Trenches, and Back-Arc Basins: Washington, DC, Maurice Ewing Series 1, American Geophysical Union, p. 221–228.

Dow, W. D., 1978, Petroleum source beds on continental slopes and rises: Bulletin of the American Association of Petroleum Geologists, v. 62, p. 1584–1606.

Fisher, M. A., 1980, Petroleum geology of Kodiak shelf, Alaska: Bulletin of the American Association of Petroleum Geologists, v. 64, p. 1140–1157.

Fuex, N., 1977, The use of stable isotopes in hydrocarbon exploration: Journal of Geochemical Exploration, v. 7, p. 155–188.

Grayson, J., and R. E. LaPlante, 1973, Estimated temperature history in the lower part of hole 181 from carbonization measurements, *in* L. D. Kulm, et al, Initial Reports of the Deep Sea Drilling Project: Washington, DC, U.S. Gov't. Printing Office, v. 18, p. 1077.

Gwilliam, W. J., 1982, Subducted organic origin gas hypothesis for deep source methane, *in* W. J. Gwilliam, ed., Deep Source Gas Workshop Technical Proceedings: U.S. Department of Energy, Morgantown Energy Technology Center, 82-50, UC-92a, p. 140–161.

Hedberg, H. D., et al, 1979, Petroleum prospects of the deep offshore: Bulletin of the American Association of Petroleum Geologists, v. 63, p. 286–300.

Hunt, J. M., 1979, Petroleum geochemistry and geology: San Francisco, W. H. Freeman and Co., 617 p.

Hussong, D. M., and S. Uyeda, 1981, Tectonics in the Mariana Arc: results of recent studies including DSDP Leg 60: Oceanologica Acta, v. 4, no. SP, p. 203–212.

Karig, D. E., et al, in press, Initial Reports of the Deep Sea Drilling Project: Washington, DC, U.S. Gov't. Printing Office, v. 87.

Keller, G., et al, in press, Paleoclimatic evidence for Cenozoic migration of Alaskan terranes: Tectonics.

Kulm, L. D., et al, 1973, Initial Reports of the Deep Sea Drilling Project: Washington, DC, U.S. Gov't. Printing Office, v. 18, 1077 p.

Kvenvolden, K. A., and M. A. McMenamin, 1980, Hydrates of natural gas: a review of their geologic occurrence: U.S. Geological Survey Circular 825, 11 p.

Langseth, M., and T. Burch, 1980, Geothermal observations of the Japan Trench transect, *in* Scientific Party, Initial Reports of the Deep Sea Drilling Project: Washington, DC, U.S. Gov't. Printing Office, v. 56, 57, pt. 2, p. 1207–1210.

Macleod, M. K., 1982, Gas hydrates in ocean bottom sediments: Bulletin of the American Association of Petroleum Geologists, v. 66, p. 2649–2662.

Magoon, L. B., and G. E. Claypool, 1981, Petroleum geology of Cook Inlet Basin, Alaska—An exploration model: Bulletin of the American Association of Petroleum Geologists, v. 65, p. 355–374.

Moore, G. W., 1982, Plate-tectonic map of the circum-Pacific region—explanatory notes: Tulsa, OK, American Association of Petroleum Geologists, 14 p.

Moore, J. C., B. Bijou-Duval, 1981, Near Barbados Ridge scraping off subduction scrutinized: Geotimes, v. 26, n. 10, p. 24–26.

Plafker, G., et al, 1982, Cross-section of the eastern Aleutian arc, from Mount Spurr to the Aleutian Trench near Middleton Island, Alaska: Geological Society of America, Map and Chart Series MC-28-P.

Rau, W. W., et al, 1977, Preliminary foraminiferal biostratigraphy and correlation of selected stratigraphic sections and wells in the Gulf of Alaska Tertiary Province: U.S. Geological Survey Open-File Report 77-747, 54 p.

Rice, D. D., G. W. Claypool, 1981, Generation, accumulation, and resource potential of biogenic gas: Bulletin of the American Association of Petroleum Geologists, v. 65, p. 5–25.

Roberts, D. G., 1981, Geological issues in offshore hydrocarbon exploration, *in* The future of offshore

petroleum: New York, McGraw-Hill, Inc., p. 27–29.
Scientific Party, 1980, Initial Reports of the Deep Sea Drilling Project: Washington, DC, U.S. Gov't. Printing Office, v. 56 and 57, 1417 p.
Thompson, T. L., 1976, Plate tectonics in oil and gas exploration of continental margins: Bulletin of the American Association of Petroleum Geologists, v. 60, p. 1463–1501.
Tissot, B., and D. Welte, 1978, Petroleum formation and occurrence: New York, Springer-Verlag, 538 p.
von Huene, R., 1972, Structure of the continental margin and tectonism at the eastern Aleutian Trench: Geological Society of America Bulletin, v. 83, p. 3613–3626.
———, 1979, Structure of the outer continental margin off Kodiak Island, Alaska, from multichannel seismic records, *in* J. Watkins and L. Montadert, eds., Geological investigations of continental margins: American Association of Petroleum Geologists Memoir 29, p. 261–272.
———, and L. D. Kulm, 1973, Tectonic summary of Leg 18, *in* L. D. Kulm, et al, Initial Reports of the Deep Sea Drilling Project: Washington, DC, U.S. Gov't. Printing Office, v. 18, p. 1069–1076.
———, et al, 1983, An eastern Aleutian Trench seismic record, *in* A. W. Bally, ed., Seismic expression of structural styles—A picture and work atlas: American Association of Petroleum Geologists Studies in Geology 15, v. 3.
Watanabe, T., et al, 1977, Heat flow in back-arc basins of the western Pacific, *in* M. Talwani, and W. C. Pittman, eds., Island Arcs, Deep Sea Trenches, and Back-Arc Basins: Washington, DC, Maurice Ewing Series 1, American Geophysical Union, p. 137–161.
Watkins, J. S., et al, 1981, Initial Reports of the Deep Sea Drilling Project: Washington, DC, U.S. Gov't. Printing Office, v. 66, 864 p.
Yamano, M., et al, 1982, Estimates of heat flow derived from gas hydrates: Geology, v. 10, p. 339–343.

The Recognition of Transform Terrane Dispersion within Mobile Belts

John C. Crowell
University of California
Santa Barbara, California

Tectonostratigraphic terranes are sliced from margins of lithospheric plates by faulting and then carried far afield and dispersed, to perhaps dock long afterwards in accreted belts. The San Andreas transform belt in California, at present and during the past several millions of years, provides an example where fracturing and dispersion processes are operating. Here transform dispersion takes place along several long faults demarcating crustal blocks. The mosaic of slices is moving irregularly northwest relative to the North American lithospheric plate. As the mosaic of blocks moves, some slices are squeezed and rise while others stretch and sag. Through time, single blocks may alternately move from a compressive regime to one of extension. Some are broken from their neighbors and rotated clockwise.

A lateral movement picture is also displayed at places by the simple-shear pattern of folds and faults where terranes are caught within a couple. The shape of rhombic pull-apart basins may also be controlled by the simple-shear scheme. In addition, long and narrow pull-apart basins containing thick overlapping packages of strata that change facies into coarse conglomerates along fault-scarp margins are also characteristic of transform dispersion. In general, however, transform dispersion is satisfactorily documented only where the regional tectonic history is adequately deciphered.

INTRODUCTION

California is undergoing active tectonic deformation as the result of displacements along a broad transform boundary between lithospheric plates. Through the study of these active processes and those that have taken place during the last few millions of years much can be learned concerning ways that tectonic terranes are broken loose from the borders of adjacent lithospheric plates. In this paper we will look briefly at the style of tectonic splintering and the nature of deformation within such a broad transform belt, with the San Andreas transform belt as an example.

After the slices are separated from their sources, they are carried laterally and in time may dock at great distance within an accreted terrane. For example, bits and pieces of California now caught within the San Andreas transform system will in time drift far northwestward. Several millions of years in the future some of these slices can be expected to crowd in against the continental margin. They will then be parts of future accreted terranes of Canada or Alaska. Some concepts involved in working with such great crustal mobility are different from those trusted by geologists a few decades ago, and now require reappraisal.

A terrane is defined as a "fault-bounded geological entity of regional extent, characterized by a geologic history different from the histories of contiguous terranes" (Blake et al, 1982). During the past several decades much geological, geophysical, paleontological, and geochemical data have come in from crustal regions. Differences between crustal blocks show that contiguous terranes are indeed significantly different in their recorded geologic history. In fact, these differences are now only easily understood by conceiving of great displacements.

Throughout the history of tectonic studies, however, it is these very differences in geology across faults that have led to the discoveries of major thrusting, major strike-slip faulting, and even the hypothesis of continental drift. The giant tectonicists of the last century, such as Suess, Bertrand, Escher von der Linth, Logan, Heim, Tornebohm, Peach, Horne, and many others all established major thrust-faulting by recognizing that stratal sequences in crossing faults were so different that great displacements in the crust were required (Bailey, 1935). Wegener (1915, but refer to 1929 reprinted in 1966) drew upon the similarities in recorded geologic history of terranes now far apart as documentation of the hypothesis of continental drift. Du Toit (1937) assembled data from the various Gondwanan cratons and showed that oceans in the Southern Hemisphere now separated these Gondwanan terranes. The concept of displaced terranes is therefore an old one in the history of our science. The concept has been vital as the hypothesis of continental drift has evolved into the theory of plate tectonics during the middle part of our century. In mobile tectonic belts characterized by major strike-slip faults, however, the documentation of huge lateral displacements has grown only since mid-century. Only slight reorientation in tectonic thinking is now needed, however, as scientists grow to appreciate the mobility of lithospheric units, including tectono-stratigraphic terranes.

In mobile belts, tectonic terranes may be broken loose from contiguous terranes along three types of plate boundaries: transform, convergent, and divergent, as well as in transition zones between them. In this discussion, the following questions will be addressed: In complex regions, how do we recognize that initial disruption and dispersion took place within a transform belt rather than within another type? Is the tectonic style of transform displacement distinctive? Are there sedimentary facies that are characteristic?

THE TECTONIC STYLE OF THE SAN ANDREAS TRANSFORM BELT

Our understanding of the tectonic history of the San Andreas transform belt since within Miocene time is unusually complete, primarily because datable volcanic rocks and strata of late Cenozoic age are widespread. Moreover, at many places, facies associations with major faults document the times of fault movements. At other places, unconformities overlap faults and give a minimum age of displacement upon them. Although the mobile belt is complex and is still inadequately investigated, this very complexity helps in revealing the deformational history and style.

The San Andreas transform belt extends northwestward from the Gulf of California, the site of the divergent boundary between the Pacific and North American lithospheric plates (Fig. 1). At the northern head of the gulf, within the Salton Trough region, several major faults, including the San Jacinto and Elsinore, extend on northwestward with a braided pattern and mark the broad boundary between the two lithospheric plates (Crowell, 1981a, 1981b). This is a transitional region between the divergent and transform plate boundaries. Widening of the gulf, which began about 4 m.y. ago, takes place along a system of oblique transform faults separated by pull-apart basins, many of which are floored by young basalt covered with sediment. The sea-floor spreading mechanism is viewed as responsible for this widening, which, in turn, converts displacements northwestward into the San Andreas transform system. In northern Baja California and western California, major high-angle and extensive faults are slicing the continental terrane across a broad belt extending from the deep Pacific Ocean on the southwest to within the Mojave Desert and Basin and Range Province on the northeast (Ernst, 1981).

Blocks constituting tectonostratigraphic terranes between these faults related to the San Andreas system are at places squeezed so that they rise to make mountains. Elsewhere blocks are stretched and sag to make receptacles for sediments. Pull-apart basins of a variety of shapes and sizes, and blocks between them, are part of this mobile scheme. The total of the moving slices, rising and falling as they migrate northwestward with respect to the main North American continent on the east, results in "porpoise structure." These processes can be documented in coastal California as far northwest as Cape Mendocino, to the vicinity of the Mendocino triple junction. They are operating at present as shown by geomorphic, geodetic, seismic, and other lines of evidence and have operated back in time into the Miocene epoch.

During this northwestward migration of tectonic blocks and slices, some are rotated and translated as shown by paleomagnetic data (Luyendyk et al, 1980; Cox, 1980). At places, the interpretation of sedimentary facies belts and inferences concerning the direction of sediment flow from source areas, as in the Santa Monica Mountains (Jones et al, 1976), support these translations and rotations. The rotation of blocks is most easily visualized as occurring during crustal stretching within a simple-shear stress regime, including the isolation of discrete blocks within extensional domains and pull-apart basins (Fig. 2). Some sutures between different terranes are interpreted as marking where blocks have been caught between converging major faults and then squeezed and locked into place as the blocks have been pushed into each other (Vedder et al, 1983). This concept involving great mobility of slices helps to visualize how complex tectonic belts in California originated, but the investigations to document the pattern in detail are just beginning. Because some faults and sutures have acted as tectonic junctures between different blocks or terranes in different ways through geologic time, it may be difficult to find out when and within what type of tectonic environment the faults originated, especially when older rocks are involved that have been tectonically overprinted many times. Younger rocks have obviously been around only long enough to record events since their origin.

Within transform belts some features display characteristics suggesting that lateral movements predominate. Major strike-slip faults have long and straight traces across the topography and usually dip steeply. Fault zones are braided and complexly deformed slices are mangled within them. Locally rocks are squeezed upward and outward from the fault zone to form palm-tree or flower structures (Sylvester and Smith, 1976). The width of the fault zone ranges along the length of major faults from only a meter or so at the narrowest to several kilometers. Broader parts are commonly lenticular in map view where the lenses contain highly deformed slices. At places, these slices include exotic phacoids of rocks that are not present in either of the adjacent walls. Many fault zones constitute the fault system (Crowell, 1962), which may be several hundred kilometers wide (Atwater, 1970).

Some faults reveal a complicated history (Page, 1982a; Crowell, 1981a; Dickinson, 1983). The San Gabriel fault, for example, acquired low-angle oblique-slip displacement between about 11 and 5 m.y. ago, after which its major reach on the northwest was abandoned (Crowell, 1982). Locally along this reach it has been reactivated within the last million years. To the southeast, within the San Gabriel Mountains, strands of the same fault system are both young and old, and some are now abandoned and are cross-cut and displaced by younger faults. One strand now merges with the active Sierra Madre fault along the base of the range that in turn is aligned with the active Raymond fault to the west and the active Cucamonga fault to the east. But studies have not advanced far enough in this region to find out when major basement terranes were juxtaposed. These may well have been stitched together in Mesozoic time, for example, and not by late Cenozoic displacements. The tectonic gears along the margin of California have changed through time, from divergent to convergent to transform styles. Part of the investigator's

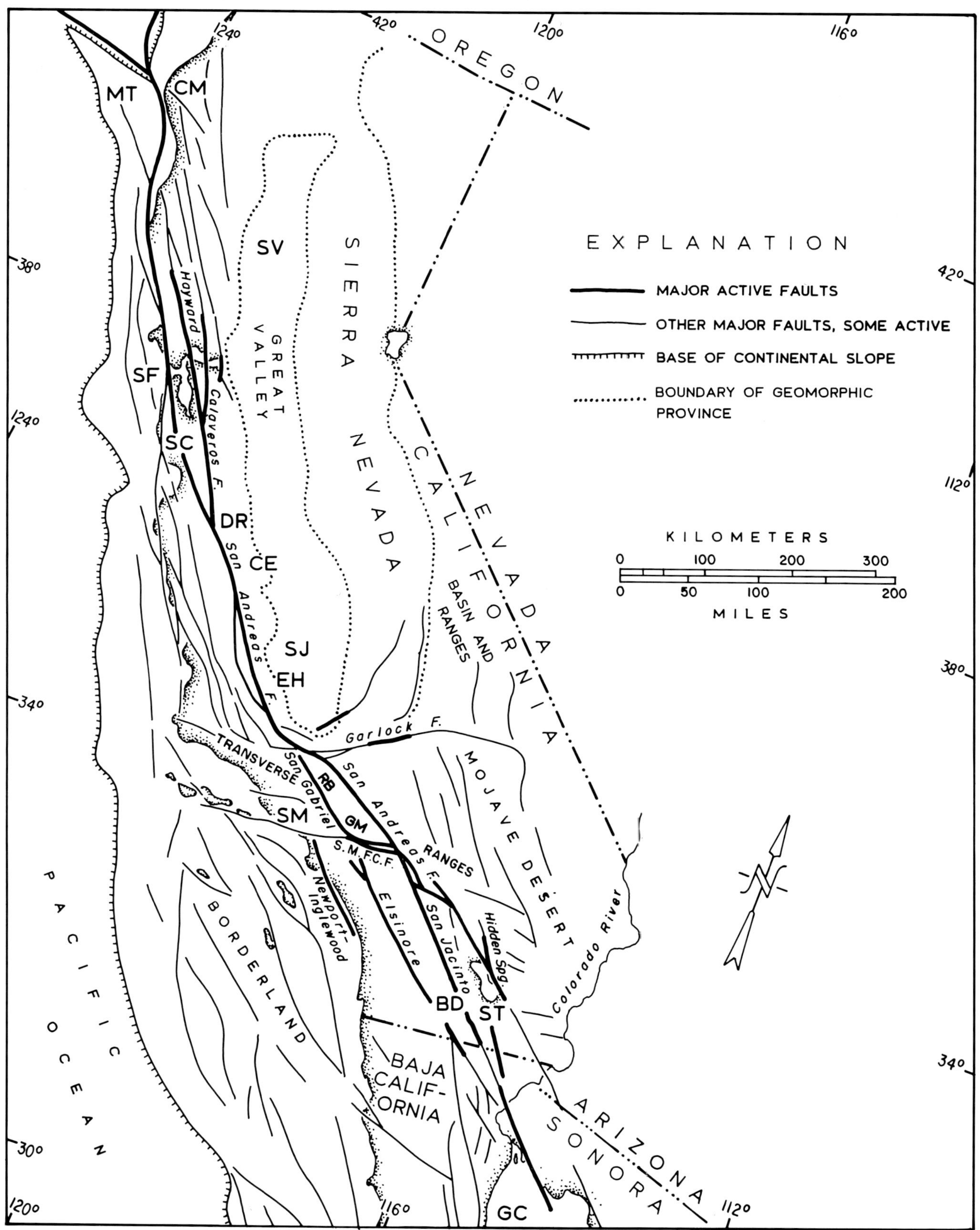

Figure 1—Fault map of California and environs. BD = Borrego Desert; CE = Coalinga Earthquake; CM = Cape Mendocino; DR = Diablo Range; EH = Elk Hills; GC = Gulf of California; GM = San Gabriel Mountains; MT = Mendocino Triple Junction; RB = Ridge Basin; SC = Santa Cruz Mountains; SF = San Francisco; SJ = San Joaquin Valley; SM = Santa Monica Mountains; SMFCF = Santa Monica fault–Cucamonga fault trend, including the Raymond and Sierra Madre faults; ST = Salton Trough; SV = Sacramento Valley.

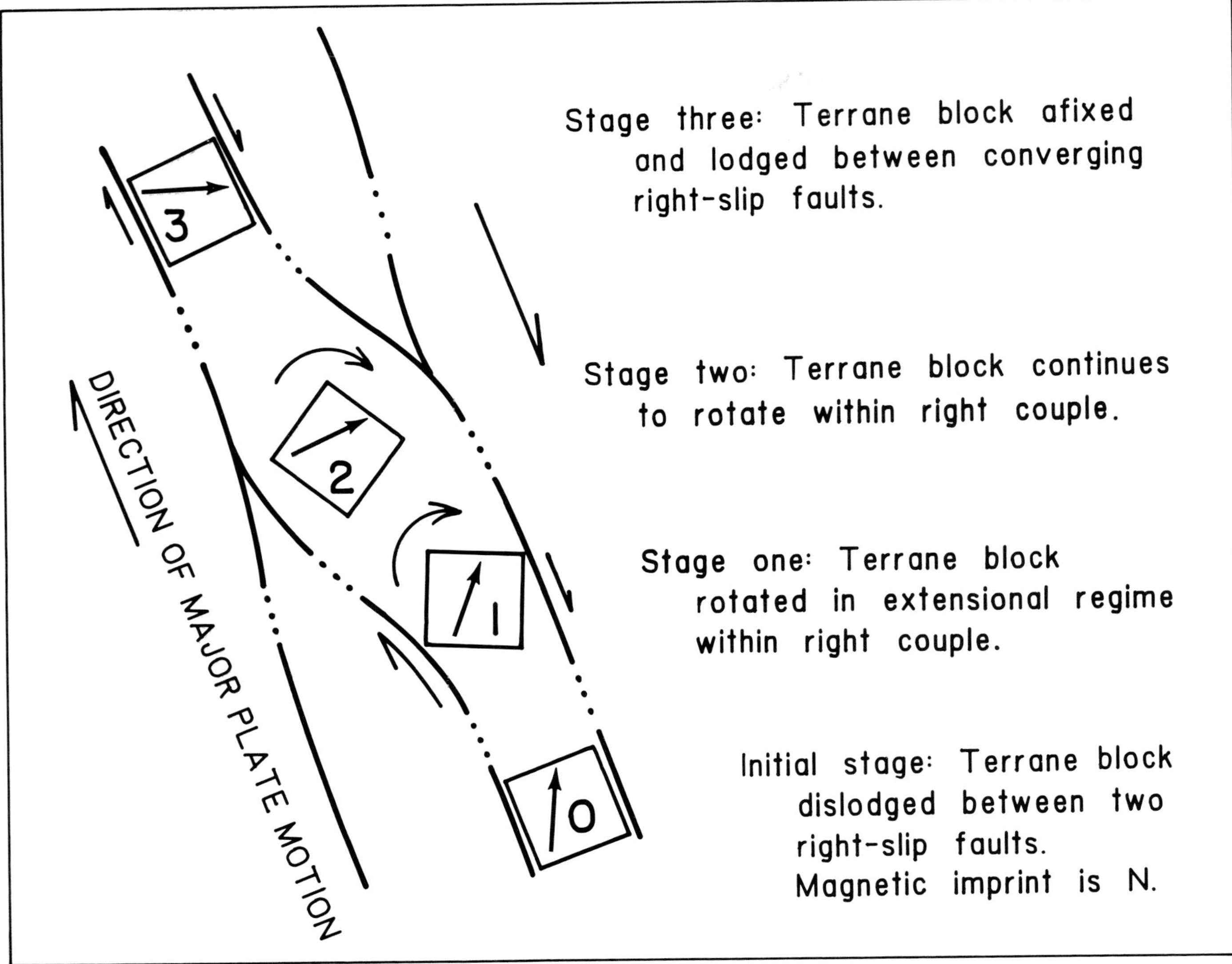

Figure 2—Diagram suggesting how a block may be dislodged within a transform belt and rotated as it moves into an extensional (pull-apart) regime and then affixed as it moves in turn into a region where the strike-slip faults converge.

problem is that different terranes made of ancient rocks may have been juxtaposed by young movements as well as ancient ones. For this reason research focused on documenting younger displacements will be especially helpful so that they can be distinguished from older displacements.

In addition, some major faults are sutures where terranes have been juxtaposed with a component of displacement normal to the fault zone itself, so that crustal material has been lost. It has been either squeezed upward and eroded away or depressed into depths. Others are major strike-slip faults with huge lateral displacements, and most have repeated histories that are different along different reaches. Displacements on major strands are at times abandoned and are taken up on splays. Major movements between plates switch from one fault to another within the wide transform belt. Investigations of these complexities are just beginning; they are most successfully approached by working back in time from the present into the past. The subdisciplines of tectonic geomorphology and those studies undertaken for the siting of critical engineering facilities are adding much detail to the tectonic history.

Terranes of different rock types and sequences implying very different geologic histories are juxtaposed across some of the major faults of the San Andreas system. In fact, it is this mismatch in inferred histories as well as the differences in rock types that drew attention to the concept of major strike slip on some of California's great faults (Crowell, 1952; Hill and Dibblee, 1953). Much of the work during the past several decades has concentrated on trying to resolve these mismatches in inferred history by finding offset counterparts and resolving mismatches into quantitative displacements. On only a few faults of the fault system, such as along the San Andreas fault itself and some of its abandoned strands such as the San Gabriel fault, has this approach so far been successful. Displaced counterpart terranes have usually been lost through erosion or through burial beneath young sediments or have not yet been recognized because of insufficient work. In fact, we are now learning that lateral displacements along them may be so great that geologists may not have gone far enough away from their study area along the fault zone in their search. Data from paleomagnetism and paleontology, showing latitudinal translations of many hundreds of kilometers, strengthen this likelihood (Coney et al, 1980; Dickinson, 1981; Jones et al, 1982). Moreover, major faults may be both sutures and strike-slip faults; the marked differences in terranes across them may be due to great crustal migration at one stage in the fault's history, and large strike-slip may have come along later. Modern research is always revealing that complexity in history is far more likely than simplicity; we should grow to expect this complexity.

In map view the braided pattern of faults within the San Andreas transform belt is noticeable at several scales, and even at 1:750,000 scale (Jennings, 1977). Only locally are fault traces parallel to the averaged trend of the boundary between the Pacific and North American lithospheric plates. As a rule the traces converge or diverge, an arrangement that is in part responsible for the rising and falling of slices and blocks associated with their stretching and extension, and constituting porpoise structure. As the slices migrate within the transform belt over great distances, they may alternately move up as they are squeezed, sag as they are stretched, and also broken apart and rotated. When high they are sediment source areas; when low, perhaps the floors of sedimentary basins.

The map pattern of faults near San Francisco provides an example. Here the San Andreas, Hayward, and Calaveras faults diverge northwestward and demarcate boundaries between blocks that are now being squeezed or stretched (Crowell, 1976; Prescott et al, 1981; Page, 1982b). The Santa Cruz Mountains and Diablo Range are underlain by terranes being compressed and uplifted, and the San Francisco Bay area, largely lying between the San Andreas and Hayward faults, is being stretched and is therefore sagging. In fact, these active tectonic processes may be viewed as largely responsible for the scenic attractiveness of the San Francisco Bay region. Fault zones such as the Calaveras have a similar braided design that shows up well in large-scale mapping (Page, 1982c). The braided pattern appears to be characteristic of transform belts elsewhere over the world, as in Venezuela (Smith, 1962), Spitzbergen (Lowell, 1972), the North Sea (Illing and Hopson, 1981), and New Zealand (New Zealand Geological Survey, 1972).

THE APPLICABILITY OF THE SIMPLE-SHEAR STRAIN PATTERN

The arrangement of structures along some of California's major faults, and in the terranes adjacent to them, at places follows the simple-shear scheme (Cloos, 1928; Riedel, 1929; Wilcox et al, 1973) (Fig. 3). Simple-shear strain results from the internal rotation of structural fabric elements within upper crustal layers operated upon by an external couple; in this case, the movements of the lithospheric plates. In this simple-shear scheme, crustal blocks are deformed within a couple operating parallel to the transform plate motion, or parallel to local lateral strain on a segment of a fault (Fig. 3b). Folds trend obliquely to a deeply buried movement zone or to a through-going fault in map view, and associated faults at many places can be classified into those of different types: synthetic (or R) antithetic (or R′), extension, thrust, or "P-shear" faults. Where nearly flat-lying strata are young adjacent to some faults, or where new breaks in alluvium or young strata overlie a buried fault, these classifications are made quite satisfactorily. Many of the younger structures in the Salton Trough region seem to fit this pattern (Fig. 4). Deformed Quaternary beds along the Newport–Inglewood zone, the Garlock fault along its central reaches, and locally along alluviated stretches of the San Andreas, are additional examples. Where the strata are older and have been deformed repeatedly for longer times, the pattern may not be easily discernible. Overprinting of early patterns as deformation has continued has modified and rotated and even obliterated earlier formed patterns at many places.

Despite overprinting of earlier structures by younger within continually deforming belts, the simple-shear scheme may be recognizable. If so, and if occurring on a regional scale, it is characteristic of a transform belt, or at least of a mobile belt with a large strike-slip component. Pull-apart basins, for example, commonly have a rhombic shape and have been termed rhomb-grabens (Fig. 4, 5). These pull-apart basins have two sets of margins: the borders parallel to the transform faulting and those parallel to the stretched or pull-apart borders. The orientation of the pull-apart margins, oblique to the trend of the major transform fault, may be controlled by the orientation of extension faults in the simple-shear system that are roughly perpendicular to the fold axes (Figs. 3a, 5). Where we have data on the local shapes of pull-apart basins, as in the Salton Trough region (Johnson et al, 1982; Fuis et al, 1984), they are roughly rhombic in shape but with pinched ends that bend into the transform borders (Fig. 4). On regional maps it is probably more realistic to depict pull-apart basins with a rhombic shape and oblique orientations rather than with pull-apart margins perpendicular to transform margins. An attractive speculation is that the shape of the rhombic pull aparts is controlled by the simple-shear pattern in a yieldable upper crust. When the pull aparts grow to a critical size, mantle material rises into the pull-apart holes from depth. Regional maps of the floor of the Gulf of California show such an arrangement of parallel transform faults separated by rhombic spreading centers with an orientation fitting the concept that they are indeed controlled by the extension direction of the simple-shear pattern. In addition, many of the marked east-west faults lying northeast of the Salton Trough follow the antithetic (R′) direction, and splays from the San Andreas fault in this region, such as the Hidden Springs fault, follow the synthetic (R) orientation. It is not yet known whether these orientations are coincidental or whether they are the response of broad sheets of the upper crust to simple-shear strain. Perhaps there is a rapid transition from a relatively brittle upper crust to ductile layers at relatively shallow depth across a structural discontinuity or zone of decollement.

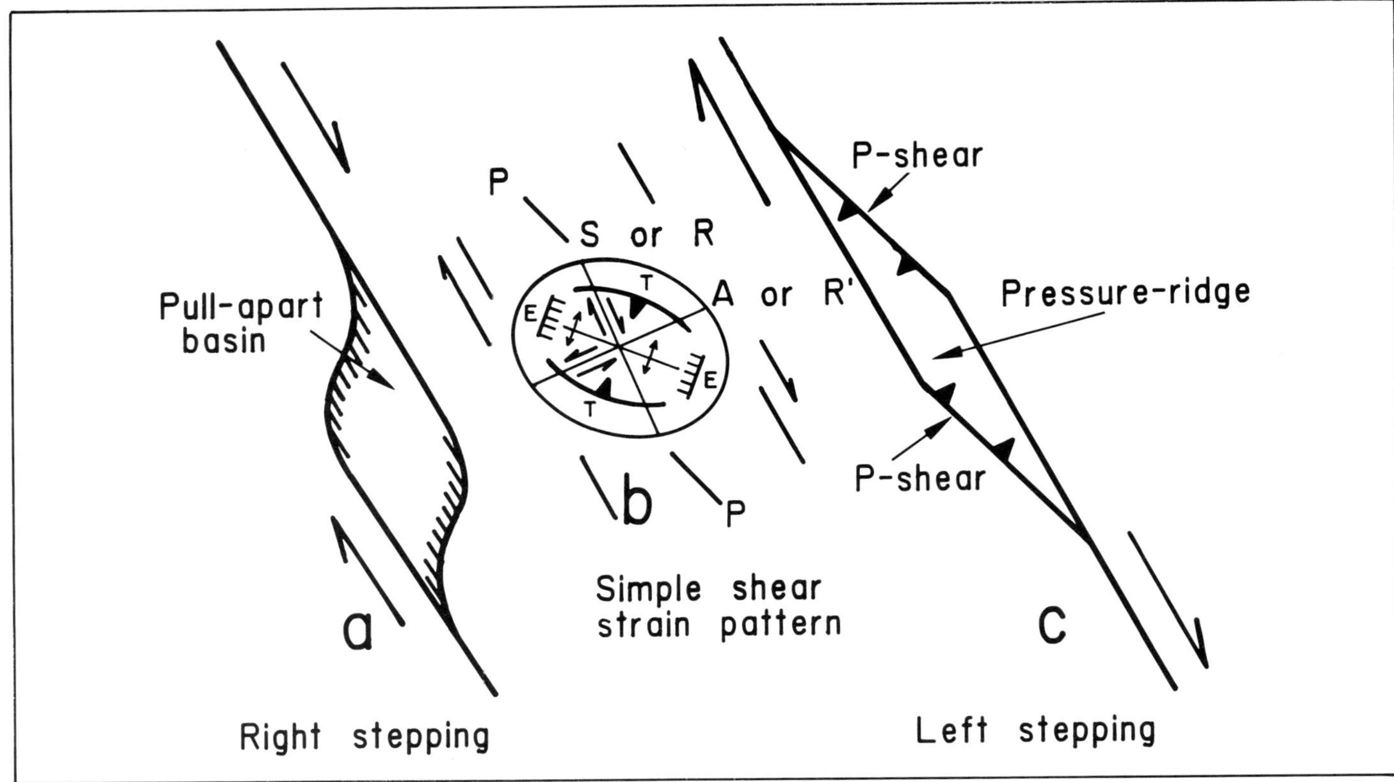

Figure 3—(a) A pull-apart basin along a right-stepping right-slip fault. **(b)** The simple-shear strain pattern showing oblique elongation of a reference circle into an ellipse when deformed by a right couple. Anticline symbols show orientation of oblique fold axes. E = normal-slip faults resulting from extension; T = thrust-slip faults; S or R = synthetic right-slip faults; A or R′ = antithetic left-slip faults; and P-shears. **(c)** One type of pressure ridge formed along a right-slip fault zone where en-echelon left-stepping P-shears meet.

The orientation of major folds along the western margin of the San Joaquin Valley are in accord with the simple-shear pattern (Harding, 1976). Pleistocene and younger beds are folded by these structures and indicate that they are being deformed under the present strain regime. The crests and troughs of the anticlines and synclines display an angle of about 40° to the strike of the San Andreas fault in the vicinity of the Elk Hills Oil Field (Jennings, 1977). The site is along the straight stretch of the fault, the stretch that is parallel to the movement direction between the Pacific and North American lithospheric plates. These fold-trends curve as they are traced northwestward to the San Andreas fault until they are subparallel to it. This swinging is noted also in clay-cake models where the displacement on a through-going basement couple is large (Wilcox et al, 1973). Farther north, in the vicinity of the epicenter of the Coalinga Earthquake of May 1983, the map pattern of the modern folds has swung clockwise so that they are nearly parallel to the master fault, about 30 km (18.6 mi) to the southwest. The compression direction and orientation of structures at depth that deformed during the earthquake (Wentworth et al, 1983; Eaton et al, 1983; Namson et al, 1983) are in accord with the simple-shear pattern. The speculation is attractive that the earthquake responded to strain in a pure-shear domain associated with a regional simple-shear scheme. In pure shear, irrotational strain prevails so that in map view a region is elongated in one direction and shortened in the other. Note that in Figures 3b and 5, pure shear is implied for a region set within an external couple, and that at a local scale, there is no distinction between rocks responding to either simple or pure shear. In addition, the more rigid crustal layers near Coalinga are deforming differently from those at depth and are apparently structurally detached.

Much smaller geomorphic forms within and alongside of major fault zones follow the simple-shear scheme (Davis and Duebendorfer, 1982). For example, some pressure ridges appear to originate between enechelon P-shears (Fig. 3c). Many of these landforms disrupt drainage; most ponds along the San Andreas fault are accordingly not true "sag ponds" if the term means a relative sagging of the ground surface as in the formation of a pull apart.

In summary, the simple-shear pattern of deformation appears to be applicable to many parts of the San Andreas transform belt. Where there are Quaternary deposits, the existence of the pattern can be established, but where there are only older rocks, its existence usually cannot be proved by examining these older rocks. There has been too much structural overprinting so that young strains cannot be untangled from older. Future geodetic, earthquake, and certain types of geomorphic studies, however, may establish whether the concept is applicable. If geologists find simple-shear arrangements in other mobile belts, they can suspect transform dispersion.

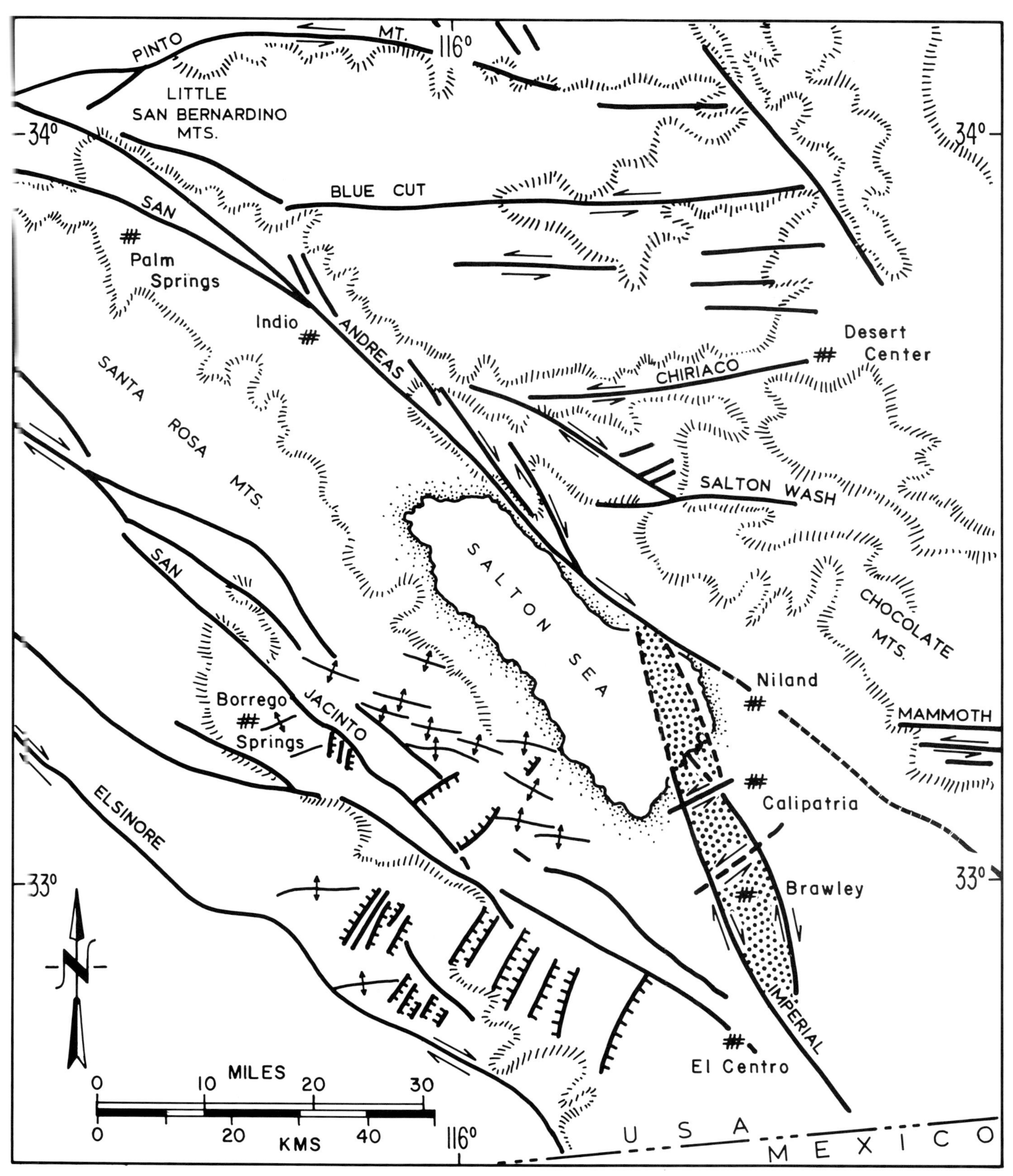

Figure 4—Sketch map of Salton Trough region showing the orientation of faults and folds (selected from Jennings, 1977, with the permission of the California Division of Mines and Geology) and the inferred shape of the pull-apart basin(s) near Brawley (simplified from Fuis et al, 1984, Journal of Geophysical Research, v. 89, p. 1165–1189, published by the American Geophysical Union).

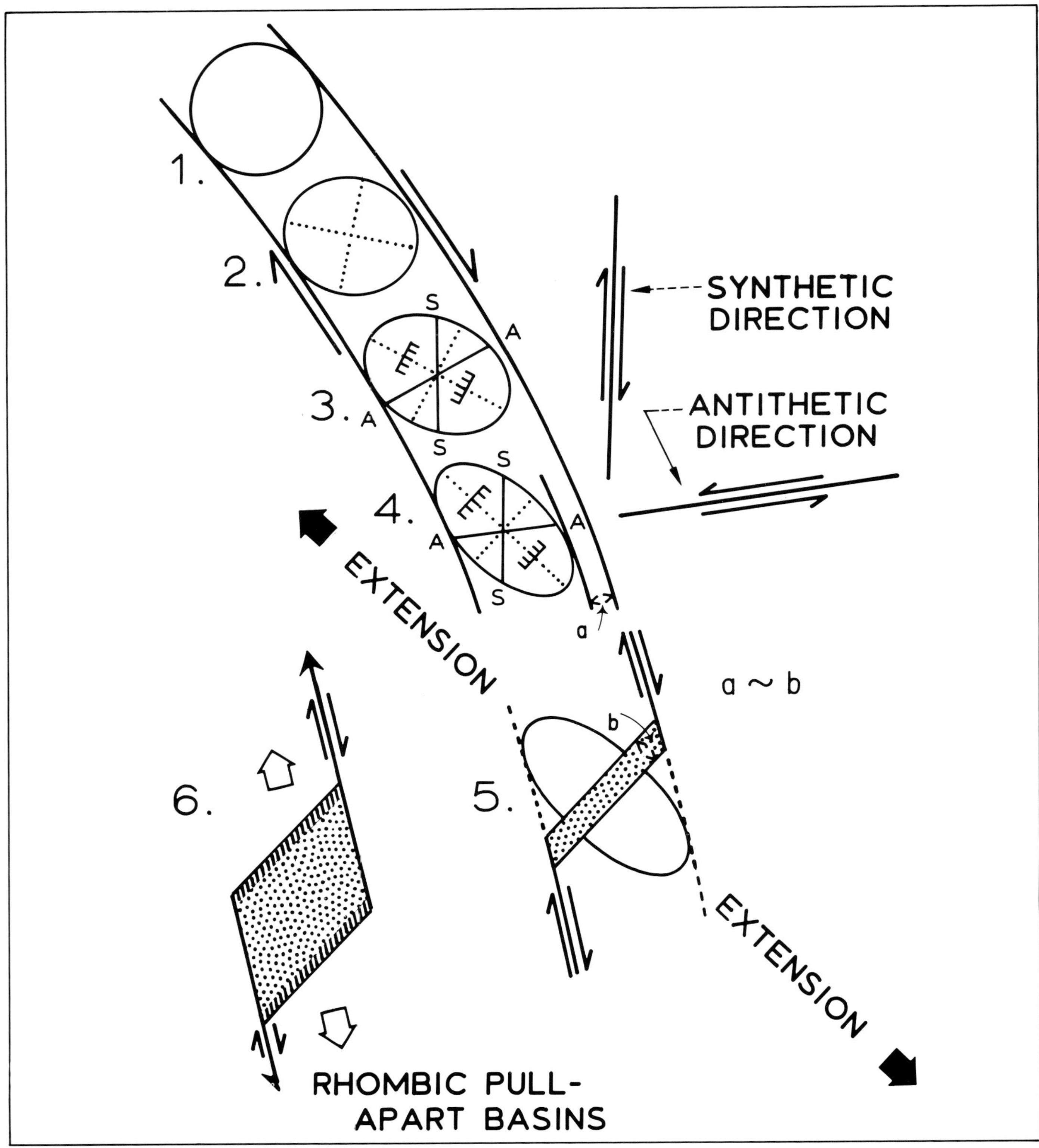

Figure 5—Diagram to explain speculation that rhombic shape of pull-apart basins at the surface may be expected in a simple-shear system where crustal plates at depth are moving laterally. The pull-apart margins are perpendicular to the extension directions and these operate obliquely to movements at depth. It is inferred that most of the stretching required (as at a) approximately balances that at b, the stretch-direction of the pull-apart basin. The areas of ellipses (2, 3, 4, and 5) are the same as that of the initial reference circle (1). When simple-shear deformation has been sufficient (by stage 3), a conjugate system of vertical faults results (the synthetic [s] and antithetic [a] faults) and extension faults (with normal-slip and dipping either way) begin to form. All of these faults rotate as displacement accumulates. By stage 5 side-stepping right-slip faults have broken through to the surface, and the rhombic shape of the basin is defined. Sketch 6 shows the map pattern after the rhombic pull-apart basin has enlarged. The basin continues to grow as the sectors move apart in the directions shown by the open arrows.

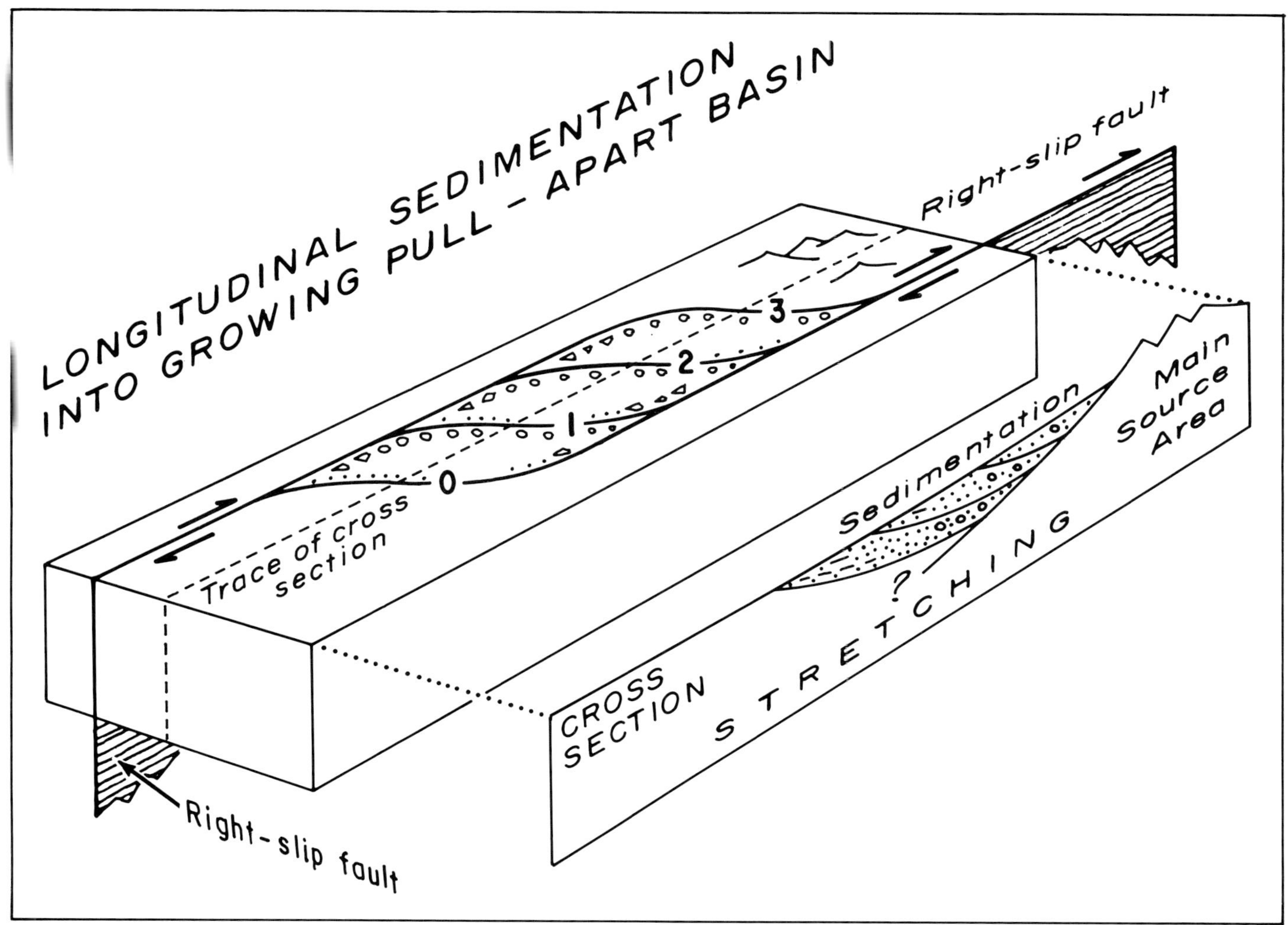

Figure 6—Block diagram and cross section illustrating a growing pull-apart basin. Sediments arrive into the opening basin from source areas largely at one end and are deposited as overlapping and shingling stratal sequences.

SEDIMENTATION PATTERNS ALONG THE SAN ANDREAS TRANSFORM BELT

The questions arise: Are there sedimentation patterns along the active San Andreas transform belt that are characteristic of transform belts in general? In tectonically complex regions consisting of accreted terranes is it possible to recognize from sedimentary facies that dispersion took place by transform displacements?

Long, deep, and narrow pull-apart basins (Fig. 6) are characteristic of transform belts (Crowell, 1974a, 1974b; Reading, 1980; Howell et al, 1980). Some contain huge thicknesses of overlapping and shingling strata that were laid down as the basin opened and as the depositional center migrated to keep pace with the opening (Crowell and Link, 1982; Steel and Gloppen, 1980). In Ridge Basin, within the central Transvere Ranges, for example, a narrow half-graben opened in Miocene time next to the then active strand of the San Andreas fault: the San Gabriel fault. About 13,500 m (44,000 ft) of both marine and nonmarine beds were laid down within it as the depositional center migrated relatively northwestward parallel to the basin axis. At any one place, no more than a third of this thickness was deposited in a vertical sequence. Other basins of this sort are documented in California (e.g., McLaughlin and Nilsen, 1982) and elsewhere.

These narrow basins are also characterized by marked lateral facies changes. Sedimentary breccias and conglomerates accumulated along marginal fault scarps and give way rapidly to finer grained facies within the basins. But the facies within the basins may be either marine or nonmarine, and shallow or deepwater.

Unfortunately from the viewpoint of the questions posed above, both great thicknesses of sediments laid down in shingling packages as well as very rapid facies changes accumulate in other tectonic environments and are not limited to transform basins. For example, in the forearc basins of the California Great Valley, Cretaceous beds reaching an aggregate thickness of well over 20,000 m (65,000 ft) were laid down in an overlapping stack, and there are even very rapid lateral facies changing within the stratigraphic section (Ingersoll, 1979). Only if facies changes can be related confidently to a long linear fault scarp, which in turn can be related tectonically to a transform belt, does this type of sedimentation document deposition within a transform belt. The matching of clast types in conglomerates to offset sources, however, may show that lateral displacements have taken place after the basinal sediments were deposited. Perhaps the stratigraphic record may document the rising and sinking of basin floors and source areas associated with porpoise structure. In short, studies of sedimentary units dissociated from regional relations will reveal little concerning the tectonic environment of their original deposition.

SUMMARY

Transform-belt dispersion is characterized by a braided system of long strike-slip faults, but the pattern is not unique. Structures arranged in the simple-shear pattern, if of the same age as the dispersion process, are helpful. Thick packages of sedimentary rocks, deposited in long and narrow pull-apart basins, are strongly suggestive. But only with reasonably full knowledge of the regional tectonics and sedimentary history will there be confidence that terrane dispersion has taken place within a transform belt. Mobile belts are complex because they are the consequence of interplate movements of great magnitude through long spans of geologic time with changing tectonic styles. Investigations of these belts should expect complexity rather than simplicity.

REFERENCES

Atwater, T., 1970, Implications of plate tectonics for the Cenozoic tectonic evolution of western North America: Geological Society of America Bulletin, v. 81, p. 3513–3536.

Bailey, E. B., 1935, Tectonic essays, mainly alpine: Oxford, Clarendon Press, 200 p.

Blake, Jr., M. C., et al, 1982, Preliminary tectonostratigraphic map of California: United States Geological Survey Open-File Report 82-593, 10 p.

Cloos, H., 1928, Experimente zur inneren Tektonik: Centralblatt fur Mineralogie, Geologie, und Palaontologie, Abteilung B, p. 609–621.

Coney, P. J., et al, 1980, Cordilleran suspect terranes: Nature, v. 288, p. 329–333.

Cox, A., 1980, Rotation of microplates in western North America, *in* D. E. Strangway, ed., The continental crust and its mineral deposits: Geological Association of Canada Special Paper 20, p. 305–321.

Crowell, J. C., 1952, Probable large lateral displacement on San Gabriel fault, southern California: Bulletin of the American Association of Petroleum Geologists, v. 36, p. 2026–2035.

______, 1962, Displacement along the San Andreas fault, California: Geological Society of America Special Paper 71, 61 p.

______, 1974a, Sedimentation along the San Andreas fault, California, *in* R. H. Dott., Jr., and R. H. Shaver, eds., Modern and ancient geosynclinal sedimentation: Society of Economic Paleontologists and Mineralogists Special Publication 19, p. 292–303.

______, 1974b, Origin of late Cenozoic basins in southern California, *in* W. R. Dickinson, ed., Tectonics and sedimentation: Society of Economic Paleontologists and Mineralogists Special Publication 22, p. 190–204.

______, 1976, Implications of crustal stretching and shortening of coastal Ventura basin, California, *in* D. G. Howell, ed., Aspects of the geological history of the California continental borderland: American Association of Petroleum Geologists, Pacific Section, Miscellaneous Publication 24, p. 365–382.

______, 1981a, An outline of the tectonic history of southeastern California, *in* W. G. Ernst, ed., The geotectonic development of California: Englewood Cliffs, NJ, Prentice-Hall, Inc., p. 584–600.

______, 1981b, Juncture of San Andreas transform system and Gulf of California rift: Oceanologica Acta, v. 4 (Supplement), p. 137–141.

______, 1982, The tectonics of Ridge basin, southern California, *in* J. C. Crowell and M. H. Link, eds., Geologic history of Ridge Basin, Southern California: Los Angeles, Society of Economic Paleontologists and Mineralogists, Pacific Section, p. 25–41.

______, and M. H. Link, (eds.), 1982, Geologic history of Ridge Basin, Southern California: Los Angeles, Society of Economic Paleontologists and Mineralogists, Pacific Section, 304 p.

Davis, T., and E. Duebendorfer, 1982, Surficial structure and geomorphology of the San Andreas fault, western portion of the Big Bend, *in* J. D. Cooper, ed., Neotectonics in Southern California: Anaheim, CA, Geological Society of America, Cordilleran Section, Volume and Guidebook, p. 77–106.

Dickinson, W. R., 1981, Plate tectonic evolution of the southern Cordillera, *in* W. R. Dickinson and W. D. Payne, eds., Relations of tectonics to ore deposits in the Southern Cordillera: Tucson, Arizona Geological Society Digest, v. 14, p. 113–135.

______, 1983, Cretaceous sinistral strike slip along Nacimiento fault in coastal California: Bulletin of the American Association of Petroleum Geologists, v. 67, p. 624–645.

Du Toit, A. L., 1937, Our wandering continents: Edinburgh, Oliver and Boyd, (3rd edition), 366 p.

Eaton, J., et al, 1983, Setting, distribution, and focal mechanisms of the 1983 Coalinga earthquake and its aftershocks (Abs.): EOS, American Geophysical Union Transactions, v. 64, p. 749.

Ernst, W. G., (ed.), 1981, The geotectonic development of California: Englewood Cliffs, NJ, Prentice-Hall, Inc., 706 p.

Fuis, G. S., et al, 1984, A seismic refraction survey of the Imperial Valley Region, California: Journal of Geophysical Research, v. 89, p. 1165–1189.

Harding, T. P., 1976, Tectonic significance and hydrocarbon trapping consequences of sequential folding synchronous with San Andreas faulting, San Joaquin Valley, California: Bulletin of the American Association of Petroleum Geologists, v. 60, p. 356–378.

Hill, M. L., and T. W. Dibblee, Jr., 1953, San Andreas, Garlock and Big Pine faults, California: Geological Society of America Bulletin, v. 64, p. 443–458.

Howell, D. G., et al, 1980, Basin development along the late Mesozoic and Cainozoic California margin: a plate tectonic margin of subduction, oblique subduction, and transform tectonics, *in* P. F. Ballance and H. G. Reading, eds., Sedimentation in oblique-slip mobile zones: International Association of Sedimentologists Special Publication 4, p. 43–62.

Illing, L. V., and G. D. Hopson, (eds.), 1981, Petroleum geology of the continental shelf of north-west Europe: London, Heydon and Son, 521 p.

Ingersoll, R. V., 1979, Evolution of the Late Cretaceous forearc basin, northern and central California: Geological Society of America Bulletin, pt. I, v. 90, p. 813–826.

Jennings, C. W., (ed.), 1977, Geologic map of California: Sacramento, California Division of Mines and Geology.

Johnson, C. E., et al, 1982, The Imperial Valley, California, Earthquake of October 15, 1979: United States Geological Survey Professional Paper 1254, 451 p.

Jones, D. L., et al, 1976, The four Jurassic belts of northern California and their significance to the geology of the southern California borderland, *in* D. G. Howell, ed., Aspects of the geologic history of the Southern California borderland: Los Angeles, American Association of Petroleum Geologists, Pacific Section, Miscellaneous Publication 24, p. 343–362.

———, 1982, The growth of western North America: Scientific American, v. 247, p. 70–84.

Lowell, J. D., 1972, Spitzbergen Tertiary orogenic belt and the Spitzbergen fracture zone: Geological Society of America Bulletin, v. 83, p. 3091–3102.

Luyendyk, B. P., et al, 1980, Geometric model for Neogene crustal rotations in southern California: Geological Society of America Bulletin, v. 91, p. 211–217.

McLaughlin, R. J., and T. H. Nilsen, 1982, Neogene non-marine sedimentation and tectonics in small pull-apart basins of the San Andreas fault system, Sonoma County, California: Sedimentology, v. 29, p. 865–876.

Namson, J., et al, 1983, Thrust-fold deformation style of seismically active structures near Coalinga, California (Abs.): EOS, American Geophysical Union Transactions, v. 64, p. 749–750.

New Zealand Geological Survey, 1972, Geological map of New Zealand, 1:1,000,000 scale (two sheets: North and South Island).

Page, B. M., 1982a, Migration of Salinian composite block, California, and disappearance of fragments: American Journal of Science, v. 282, p. 1694–1734.

———, 1982b, Modes of Quaternary tectonic movement in the San Francisco Bay Region, California: California Division of Mines and Geology Special Publication 62, p. 1–10.

———, 1982c, The Calaveras fault zone of California, an active plate boundary element: California Division of Mines and Geology Special Publication 62, p. 175–184.

Prescott, W. H., et al, 1981, Geodetic measurement of crustal deformation on the San Andreas, Hayward, and Calaveras faults near San Francisco, California: Journal of Geophysical Research, v. 86, p. 10853–10869.

Reading, H. G., 1980, Characteristics and recognition of strike-slip fault systems, *in* P. F. Ballance and H. G. Reading, eds., Sedimentation in oblique-slip mobile zones: International Association of Sedimentologists Special Publication 4, p. 7–26.

Riedel, W., 1929, Zur Mechanik geologischer Brucher-scheinungen: Centralblatt fur Mineralogie, Geologie, und Palaontologie, Abteilung B, p. 354–368.

Smith, Jr., F. D., (ed.), 1962, Mapa geologico-tectonico del norte de Venezuela: Primer Congresso Venezolano de Petroleo, scale, 1:1,000,000.

Steel, R. J., and T. G. Gloppen, 1980, Late Caledonian (Devonian) basin formation, western Norway: signs of strike-slip tectonics during infillings, *in* P. F. Ballance and H. G. Reading, eds., Sedimentation in oblique-slip mobile zones: International Association of Sedimentologists Special Publication 4, p. 79–104.

Sylvester, A. G., and R. R. Smith, 1976, Tectonic transpression and basement-controlled deformation in San Andreas fault zone, Salton Trough, California: Bulletin of the American Association of Petroleum Geologists, v. 60, p. 2081–2102.

Vedder, J. G., et al, 1983, Stratigraphy, sedimentation, and tectonic accretion of exotic terranes, southern Coast Ranges, California, *in* J. S. Watkins and C. L. Drake, eds., Studies in continental margin geology: American Association of Petroleum Geologists Memoir 34, p. 471–496.

Wegener, A., 1929, The origin of continents and oceans, (translated from the 4th revised German edition by John Biram): New York, Dover Publications, Inc., 246 p.

Wentworth, C. M., et al, 1983, Thrust and reverse faults beneath the Kettleman Hills anticlinal trend, Coalinga earthquake region, California, inferred from deep seismic-reflection data (Abs.): EOS, American Geophysical Union Transactions, v. 64, p. 747.

Wilcox, R. E., et al, 1973, Basic wrench tectonics: Bulletin of the American Association of Petroleum Geologists, v. 57, p. 74–96.

Biogeographic Significance of the Upper Triassic Bivalve *Monotis* in Circum-Pacific Accreted Terranes

N. J. Silberling
U.S. Geological Survey
Denver, Colorado

Various species of the pectinacid bivalve *Monotis* occur in marine Upper Triassic strata around the margins of the Pacific and in the Alpine-Himalayan suture belt, mainly in allochthonous accretionary terranes. These thin-shelled bivalves are generally believed to have been pseudoplanktonic in life habit, living attached to floating marine algae in surface waters. Different *Monotis* faunas can be recognized by the presence or absence of various common species, and these have discrete geographic distributions. Different *Monotis* faunas characterize (1) Cordilleran South America (and possibly the Alexander terrane in southeast Alaska), (2) the Pacific margin of Cordilleran North America, (3) Arctic Canada and possibly parts of northeastern U.S.S.R., (4) the Alpine-Himalayan suture belt, and (5) an elongate belt extending all the way from cratonic western Canada through northern Alaska, northeastern U.S.S.R., Japan, west Borneo, and New Zealand. Three of these faunas occur in strata depositionally or sedimentologically related to cratonic North America and apparently represent, respectively, the low, middle, and high northern paleolatitudes.

Further study is needed to better characterize some *Monotis* occurrences and the different *Monotis* faunas themselves, which as recognized are perhaps at too high a taxonomic level to show important provincialism. Nevertheless, large-scale displacement of accretionary terranes is evidently required to explain the present distribution of *Monotis* faunas, and biogeographic studies will clearly be of great importance in reconstructing pre- to mid-Jurassic intraoceanic plate boundaries and motions.

INTRODUCTION

Thin-shelled pectinacid bivalves of the genus *Monotis*, sensu lato, are widely distributed in Upper Triassic marine strata around the margins of the Pacific basin and in the Alpine-Himalayan suture belt. These strata relate to different parts of the single ocean—Panthalassa—that existed in Triassic time, and most of these deposits are parts of allochthonous (or "suspect") accretionary terranes (Coney et al, 1980; Howell et al, 1983). Because different *Monotis* faunas are geographically restricted, their biogeographic interpretation bears on the displacement histories of circum-Pacific accreted terranes and hence on plate motions in general within Panthalassa.

BIOGEOGRAPHY OF *MONOTIS*

To utilize *Monotis* faunas biogeographically, two prerequisites must first be met insofar as presently possible. First, a taxonomic scheme must be adopted, even if somewhat prematurely, in order to simplify the nomenclatorial maze that embraces these fossils, and second, the paleoecologic constraints on the original distribution of *Monotis* must be considered.

Modern authorities on the taxonomy of *Monotis* recognize several tens of different species (Kiparisova et al, 1966; Westermann, 1973; Tozer, 1979; Grant-Mackie, 1981; and other references to these same authors cited in these publications). As might be expected, different opinions exist on the usage of some specific names except for those applied to kinds of *Monotis* that are particularly well represented at one place or another. Several groupings of *Monotis* species, expressed as separate genera, subgenera, or simply as species groups under the Genus *Monotis*, have been proposed by Westermann (1973), Grant-Mackie (1978), and Tozer (1979). These groupings are generally similar but do embody some important differences among them. The different "*Monotis* faunas" adopted here are characterized by one or more species groups, generally of subgeneric rank, that express the predominant character of the fauna in any one region. The possible biogeographic significance of problematic or local species is thus mainly ignored, and, as discussed herein, this admittedly may cause some problems with interpretation. Even though the different faunas adopted here are characterized by kinds of *Monotis* at a relatively high taxonomic level, some faunas have several different kinds of *Monotis* because of (1) a real, but as yet poorly understood, evolutionary succession of different species and subgenera within the late middle Norian through early late Norian span of the genus, and (2) some degree of specific and subgeneric diversity. The four species groups used to characterize the different *Monotis* faunas herein are most similar in content to those of Westermann (1973) but embrace the subgeneric terminology of Grant-Mackie (1978). Representative examples of different upper Norian species groups are illustrated on Figure 1.

Figure 1—Representative examples of upper Norian species groups of *Monotis*. (**a–b**) *M. salinaria*, left and right valves, Salzkammergut, Austria. (**c–d**) *M. salinaria*, left and right valves, Wrangellia terrane, Wrangell Mountains, Alaska. (**e**) *M. (Entomonotis) ochotica*, near Sakawa, Shikoku, Japan. (**f**) *M. (Entomonotis) ochotica*, North Slope subterrane of Arctic Alaska terrane, Fire Creek, Shublik Mountains, northeast Alaska. (**g**) *M. (Entomonotis) ochotica*, Murihiku terrane, Nelson, New Zealand. (**h**) *M. subcircularis*, Cachapoyas, Peru. (**i**) *M. subcircularis*, Endicott Mountains subterrane of Arctic Alaska terrane, Killik River, Brooks Range, Alaska. (**j–k**) *M. subcircularis*, right and left valves, Wrangellia terrane, Wrangell Mountains, Alaska.

Five distinct *Monotis* faunas, here designated faunas A through E, are recognized on the basis of the mutual occurrence or exclusion of certain of the species groups. The world-wide distribution of the *Monotis* faunas is plotted on Figure 2, and their composition in terms of subgeneric-level species is shown in Table 1. Further morphologic and taxonomic characterization of these species groups is given in Appendix A. The resulting biogeographic pattern in terms of world distribution and kind is similar to that described for *Monotis*, sensu lato, by Westermann (1973) and Tozer (1979), and for North America it is essentially the same as that described by Tozer (1982).

The known *Monotis* faunas of Cordilleran South America consist of a single species (Westermann, 1970) that is designated fauna A. No other reasonably well-sampled occurrences are like this except possibly for those in the Alexander terrane of southeast Alaska (loc. 3, Fig. 2) as discussed later. Fauna B is restricted to the part of the North American Cordillera bordering the Pacific Ocean. Fauna C is typically developed in northeast British Columbia, Canada (loc. 2, Fig. 2), and is represented in northern Alaska, northeastern U.S.S.R. (e.g., loc. 4, Fig. 2) in the South Anyui fold belt, Sikhote Alin–Nadanhada Ling (loc. 6, Fig. 2), Japan, west Borneo (loc. 7, Fig. 1), New Caledonia, and New Zealand (Fig. 2). Fauna D is typical of Arctic Canada and can be differentiated from C only on the negative evidence of being less diverse. Fauna D may be represented in northeastern U.S.S.R. as shown on Figure 2, but this probably cannot be demonstrated on the basis of the existing sampling. Fauna E is unique to the Tethyan rocks of the Alpine–Himalayan belt and the Banda Sea region where such rocks occur as reworked blocks in Cenozoic melanges (Hamilton, 1979).

Shells of *Monotis* occur in a wide variety of open-marine deposits, including deep-marine strata in which known bottom-dwelling kinds of shelled invertebrates are totally absent. Because of this manner of occurrence and the morphology of the *Monotis* shell, which is remarkably thin (on the order of hundreds of microns) in comparison with its diameter (mainly in the 2 to 8 cm [.8–3 in.] range) and bears evidence of a feebly byssate, attached mode of life, the view is generally, but not universally, accepted that most *Monotis* lived attached to floating objects such as seaweeds (Tozer 1982, p. 1098; Gall, 1983, p. 150). This inferred pseudo-planktonic, near-surface mode of existence dictates that surface temperature, and thus latitude, should be among the primary controls on the original north–south distribution of most of the common kinds of *Monotis*. Thus, the kind of *Monotis* faunas in accreted terranes should provide a crude measure of their latitudinal displacement, but of course this says little about the possible amount of their longitudinal displacement.

The original distribution of different *Monotis* faunas would also have been affected by oceanic currents, as argued by Kristan-Tollmann and Tollmann (1983). However, as discussed below, the meager amount of paleomagnetic control on the paleolatitudes of *Monotis*-bearing accreted terranes of the north-Pacific margin generally corroborates the assumption that latitude was the primary biogeographic control. *Monotis* faunas inferred to be of low paleolatitude kind but occurring at high northern latitudes are in accreted terranes having paleomagnetic inclination anomalies of similar trend, and vice versa. The present-day circum-Pacific distribution of *Monotis* faunas shown on Figure 2 is not readily attributable to the reconstruction of Pacific ocean currents depicted by Kristan-Tollmann and Tollman (1983, fig. 9).

Establishing control on the original paleolatitude of different *Monotis* faunas is difficult because most known occurrences of *Monotis*, as shown on Figure 2, are in accreted terranes, and the early Mesozoic paleolatitude has been adequately established paleomagnetically for only a very few of these. Probable rates of Mesozoic plate motion and duration of travel times allow the possibility of trans-Pacific displacement of terranes, and for some, for example, Wrangellia on the Pacific margin of North America (Jones et al, 1977), paleomagnetic evidence actually documents post-Triassic displacements on the order of tens of degrees of latitude.

Only in Cordilleran and Arctic North America is more than one kind of *Monotis* fauna apparently distributed according to original paleolatitude in rocks depositionally related to a single cratonic block. Faunas B in northwestern Nevada (loc. 1, Fig. 2), C in the foothills of the Canadian northern Rocky Mountains (loc. 2, Fig. 2), and D in the Canadian Arctic correspond respectively to the low-, mid-, and high-paleolatitude "plate-bound" marine Triassic faunas recognized by Tozer (1982). These occurrences of faunas C and D are continentward of the limit of Mesozoic and younger terrane accretion and are thus bound to the North American plate. The occurrence of fauna B in northwest Nevada is in rocks, such as those in parts of the Auld Lang Syne Group, that can be regarded as being part of an accreted terrane. Nevertheless, these rocks certainly are not exotic with respect to North America and evidently are depositionally related to the North American craton at about their present latitude (Lupe and Silberling, this volume). As pointed out by Tozer (1982), an original paleoequatorial distribution of the Tethyan *Monotis* fauna E is generally accepted and is corroborated by some internal paleomagnetic control and by the paleolatitudinal constraints provided by the continental blocks to the north and south of the Alpine–Himalayan suture belt in which most of these faunas are found. The presence of the same species making up fauna E in *Monotis* fauna B tends to substantiate the paleoequatorial interpretation for the original setting of fauna B.

INTERPRETATION

If the evidence for an originally low paleolatitude of fauna B is accepted, the distribution of other occurrences of B in the accreted terranes of the North American Cordillera, as far north as the Denali fault (Fig. 2) in Alaska, expresses the northward tectonic displacement of these terranes with respect to the nonaccretionary occurrences of fauna C in western Canada, as previously pointed out by Tozer (1982). This displacement pattern generally confirms those based on the biogeography of Lower Triassic (Nichols and Silberling, 1979) and Lower Jurassic (Tipper, 1981; Smith, 1983) ammonites and most

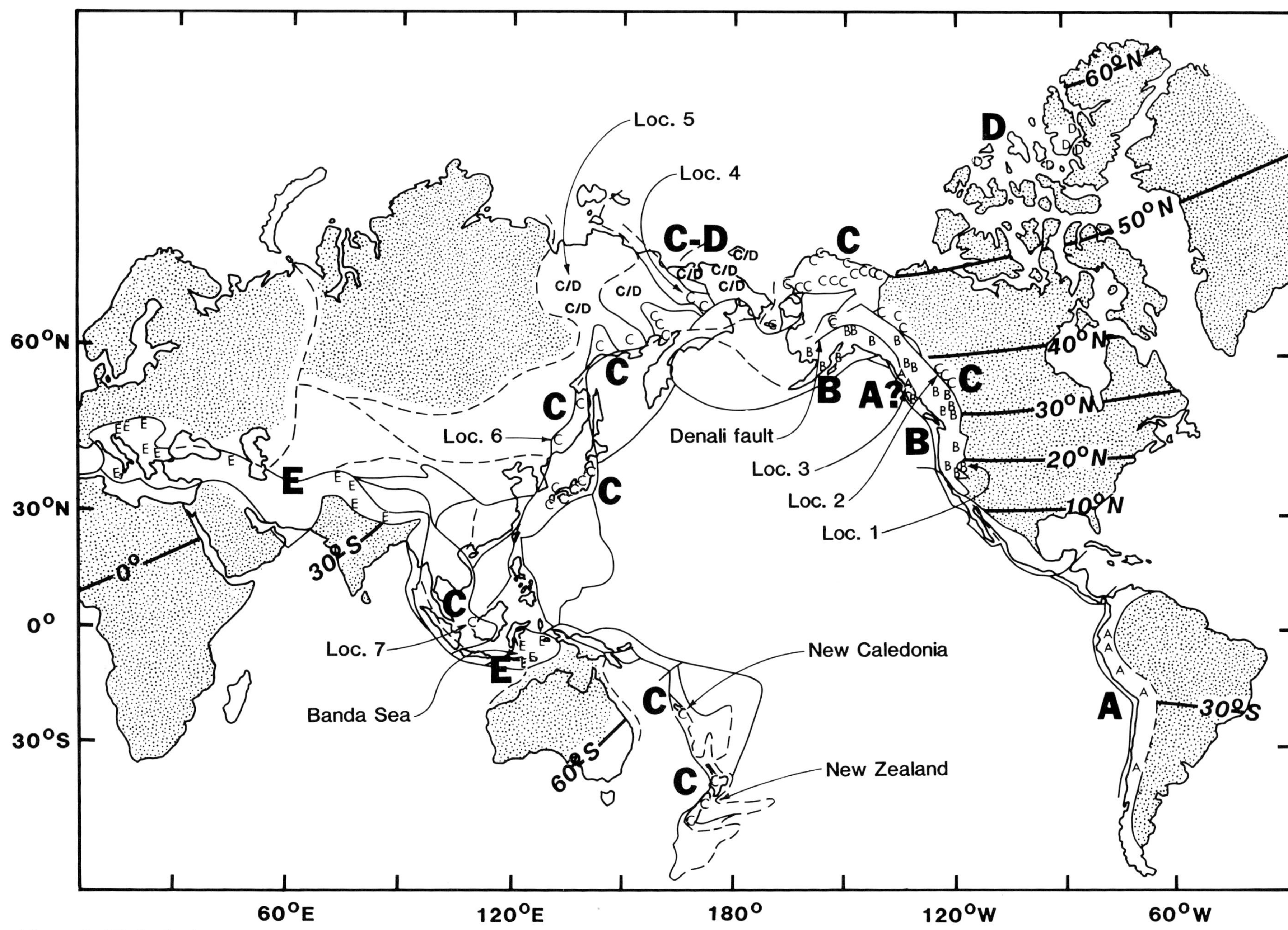

Figure 2—World distribution (Mercator projection) of *Monotis* faunas A through E, as defined in the text. Large, bold letters show the regional occurrences of the different faunas; small letters represent actual localities. Pattern of accreted terranes, which lie between the larger continental blocks (stippled) and the present-day major ocean basins, is highly generalized from Howell et al (1983), Fujita and Newberry (1982), and Wang et al (1981). Upper Triassic paleolatitudes, where shown for the continental blocks, are from Irving (1979).

Table 1

		Monotis Faunas				
		A	B	C	D	E
Species Groups or Subgenera	*Monotis (Monotis) subcircularis* group	X	X	X		
	Monotis (Monotis) salinaria group		X			X
	Monotis (Entomonotis) group			X	X	
	Monotis (Eomonotis) group			X	X	

Table 1—Composition of *Monotis* faunas plotted on Figure 1 in terms of subgeneric-level species groups.

paleomagnetic studies (e.g., Packer and Stone, 1974; Hillhouse and Grommé, in press; Monger and Irving, 1980). A more complete documentation of the *Monotis* faunas of Alaska, based on the collections of the U.S. Geological Survey, and their distribution among the various Alaskan accreted terranes is currently in preparation by the author.

Within the broad, tectonically complex belt characterized by *Monotis* fauna B, a special case has to be made for the Alexander terrane in southeast Alaska (loc. 3, Fig. 2) where paleomagnetic studies of Upper Triassic basalt by Hillhouse and Grommé (1980) indicate no significant difference in paleolatitude from that expected from continental North America. The *Monotis* fauna of these rocks is fairly well sampled from strata that in part are intercalated with the paleomagnetically sampled Triassic basalt. It differs in a negative way from fauna B in that *Monotis (Monotis) subcircularis* is virtually the only species of *Monotis* represented [with the exception of one middle Norian *Halobia* collection that possibly contains a specimen of *Monotis (Eomonotis)* sp.]. In its preponderance of *M. subcircularis*, the fauna of the Alexander terrane is like that designated fauna A and found along the Pacific margin of South America. Thus, on the basis of *Monotis* faunas, derivation of the Alexander terrane from a southern rather than a northern latitude is permissible and would agree with the moderately high paleomagnetically determined paleolatitude. However, among other uncertainties is the degree to which fauna A occurrences in South America are within accreted terranes themselves and therefore have uncertain displacement histories.

Fauna C is the most diverse taxonomically and the most widespread geographically of the five different *Monotis* faunas considered here. Its seemingly broad geographic range can be viewed as a result of terrane displacement, a manifestation of some peculiar paleoecologic distribution, an artifact of a superficially high taxonomic level for this analysis, or some combination of all of these possibilities. Along with fauna D, fauna C differs from the other *Monotis* faunas in two special ways. First, it has a longer stratigraphic range in that species of *Monotis (Eomonotis)* are abundantly represented in rocks of late middle Norian age in the areas characterized by fauna C. In contrast, *Monotis* is completely unknown in middle Norian rocks immediately underlying the many upper Norian occurrences of faunas A, B, and E. Second, the upper Norian species of fauna C include some that have morphologic extravagancies such as a pronounced inequivalved shape owing to relative inflation of the left valve of the shell, coarse ribbing, or loss of ribbing. These morphologic features could be manifestations of ecologic adaptation to, for example, relatively cold water such as that occurring at high latitudes or at depths greater than surface waters. A benthonic habitat has in fact been hypothesized by Westermann (1973) for some inequivalved species. At the subgeneric level of faunal characterization adopted here, the distribution of fauna C might be construed as bipolar by a tectonic stabilist. At a more refined taxonomic level, however, the middle or high northern and southern hemisphere parts of its range might actually be characterized by different species having only generally similar morphologies. Further detailed descriptive studies, of the kind being undertaken by Grant-Mackie in New Zealand, and comparisons of geographically disparate *Monotis* faunas are needed.

The occurrence of fauna C in nonaccretionary parts of interior western Canada is well documented (e.g., Westermann, 1962; Tozer, 1967, 1982) and is thus well controlled in this one region at mid-northern paleolatitudes. The abundant *Monotis* faunas of northern Alaska are assigned to fauna C rather than D because of the diversity of different species of *Monotis (Eomonotis)* and the occurrence of *Monotis subcircularis* in them. Because faunas C and D differ only in total diversity, this interpretation needs to be confirmed on other biogeographic evidence. If confirmed, it contradicts the popular notion of post-Triassic, counterclockwise tectonic rotation of northern Alaska away from the northwestern continental margin of Canada. The Mercator projection of Figure 1 distorts this, but restoration of northern Alaska to its Triassic position according to this hypothesis would nearly juxtapose faunas C and D.

Differentiation of *Monotis* fauna C from D is again difficult in northeastern U.S.S.R. because all of the few species characteristic of D also occur in the more diverse C. The westernmost of these faunas are in rocks of the Verkhoyansk fold belt (loc. 5, Fig. 2) that may be related depositionally to the Siberian continental block (Fujita and Newberry, 1982) for which a relatively high Permian paleolatitude, about the same as the modern latitude, is reported by McElhinny et al (1981). These Verkhoyansk *Monotis* occurrences and perhaps others in the terranes of northeastern U.S.S.R. could represent fauna D. Nevertheless, from descriptions by Kiparisova et al (1966), fauna C seems to be well represented at least in the South Anyui terrane or suture belt at locality 4 (Fig. 2).

Farther south along the western Pacific margin and in Indonesia, fauna C occurs in Sikhote Alin–Nadanhada Ling (Wang et al, 1981) (where again its separation from fauna D is uncertain), in many places in Japan, and apparently in west Borneo. Fauna C is particularly well represented in several of the accreted terranes of Japan. As described by Tamura (1965), for example, the Japanese occurrences can be nearly identical in both species composition and stratigraphic succession to those in northern Alaska. In collections from west Borneo, *Monotis (Entomonotis) ochotica* has been recognized by Westermann (1970). "*Monotis subcircularis*," described from another locality in west Borneo by Tamura and Hon (1977), is more likely assignable to a species of *Monotis (Eomonotis)*, but either way it could be in the same stratigraphic succession as *M. ochotica* and agrees with the character of fauna C (rather than, for example, the Tethyan fauna E that occurs nearby in terranes of the Banda Sea region). Providing some credence to the mid-northern paleolatitude interpretation for fauna C derived from its North American occurrences are the Permian northern hemisphere paleolatitudes of about 34° and 31° reported by McElhinny et al (1981), respectively, for central locations within the "Sikhote Alin block" and the "Southeast Asia block" to which the west Borneo occurrence of fauna C might be related paleogeographically. Because of its seemingly anomalous location, however, the relatively little-known *Monotis* fauna of west Borneo needs further characterization before its assignment to fauna C can be fully accepted.

Monotis faunas of New Caledonia and New Zealand are entirely within the Torlesse and Murihiku (or Hokonui) accreted terranes and have been described in ever increasing detail, especially by Grant-Mackie (1981, and references cited therein). They belong to fauna C at the relatively high taxonomic level of the present analysis, although appreciable differences between the Murihiku and Torlesse *Monotis* faunas are recognized by Grant-Mackie (1984, written communication). On this basis, derivation of at least the *Monotis*-bearing parts of the Murihiku and Torlesse terranes from mid-northern latitudes is a possibility. Tozer (1982) already has argued for this on the basis of both *Monotis* and Triassic cephalopod faunas of New Zealand. Support for this seemingly bizarre interpretation is provided by Howell (1980), who on sedimentologic and other geologic grounds has viewed the accreted terranes of New Zealand as being exotic with respect to the parts of New Zealand that can be related to Gondwanaland. Moreover, an appreciable southward component of plate motion, as recorded in the Torlesse terrane, is consistent with the occurrence there of tectonic blocks of upper Paleozoic limestone containing Tethyan fusulinid faunas (MacKinnon, 1983, p. 981).

Some evidence, on the other hand, favors original high southern latitudes for the accreted lower Mesozoic rocks of New Zealand. A paleomagnetically determined paleolatitude of about 66° and agreement with a magnetic direction expected for a site that was originally at the margin of Gondwanaland is reported by Grindley et al (1981) for Upper Triassic to Lower Jurassic andesitic rocks of the Murihiku terrane. Furthermore, at the specific or subgeneric level, some of the Murihiku *Monotis* fauna could be unique, such as *M. (Eomonotis) murihikuensis* and *M. (Maorimonotis)* as defined by Grant-Mackie (1978). As stated earlier, an argument can thus be mounted that fauna C, as defined here at generally subgeneric level, had an originally bipolar distribution.

Whatever their original paleogeographic site might have been, the Murihiku and Torlesse terranes both have *Monotis* faunas distinct from those of the accreted terranes, such as Wrangellia, that rim the northeastern Pacific margin of North America. This constrains speculation about originating allochthonous terranes of the Pacific by the break-up and plate-tectonic dispersal of a single hypothetical land mass such as "Pacifica" of Nur and Ben-Avraham (1977). If Pacifica were located along the eastern Australia–western Antarctic Gondwana margin, and if it contributed pieces now accreted to North America (Nur and Ben-Avraham, 1982) and also served as a source for clastic rocks of the Torlesse terrane in New Zealand (Kamp, 1980), it would have had to extend over a large span of latitudes in order to have been the habitat for markedly different, evidently paleolatitudinally distinct, *Monotis* faunas.

A bipolar distribution of *Monotis* fauna C is preferable to the assumption that it was originally distributed in deep, cold water because of the absence of fauna C in regions characterized by either faunas B or E. The great many occurrences of faunas B and E in North America and the Tethyan region, respectively, represent a wide range in water depths and both oceanic and miogeoclinal sedimentary environments. It therefore seems unlikely that fauna C ever ranged through the paleoequatorial seas populated by faunas B and E.

Another kind of biogeographic dilemma is exemplified by the various terranes (or "geotectonic divisions," Saito, 1984) in Japan that contain *Monotis* occurrences exclusively of fauna C. Although an original, either northern or southern hemisphere, mid- to high paleolatitude seems indicated by these occurrences, nearly all of them are in terranes, as presently recognized, for which preliminary paleomagnetic data (Hirooka et al, 1984) suggest low paleolatitudes during early Mesozoic time and in which limestone containing some of the classic occurrences of Permian Tethyan fusulinid faunas occur (Ozawa and Kanmera, 1984). In the terranes of Japan (and for the Torlesse terrane of New Zealand as well) incorporation of strata bearing *Monotis* fauna C in the same terranes containing indices of low paleolatitude might be explained by viewing these terranes as subduction-related accretionary composites that incorporate progressive records of fairly long-term episodes of large-scale oceanic plate motion.

In conclusion, with regard to *Monotis* fauna C, more questions probably remain than answers, but nonetheless, the problems posed by its widespread distribution seem difficult to reconcile without calling upon large-scale plate-tectonic displacement of some of the terranes, or the parts of terranes, characterized by this fauna.

IMPLICATIONS

The concept of accretionary tectonostratigraphic terranes and the possibility that some terranes have been greatly displaced as a consequence of intraoceanic plate motion that is partly independent of the drift of the larger continental blocks add a new dimension to biogeographic studies. The need to consider the possible consequences of accretionary tectonics is especially critical when the period of geologic time involved in one such as the Triassic when the extent of epicontinental seas was minimal, causing the present distribution of marine Triassic rocks and fossils to be largely restricted to the accretionary continental margins. Although a full understanding of the biogeographic patterns of the *Monotis* faunas discussed herein is as yet incomplete, no longer must the explanation of peculiar patterns such as this one be constrained to drifting continents, hypothetical oceanic currents, or the assumption of virtually unlimited latitudinal or paleoecologic constraints. Accretionary tectonics provides an additional important variable for the interpretation of biogeographic patterns, and, inversely, more detailed studies of the biogeography of accreted terranes can help reconstruct pre- to mid-Mesozoic intraoceanic plate boundaries and motions.

APPENDIX A

The "groups" of *Monotis* species (Table 1) whose presence or absence defines the "*Monotis* faunas" whose geographic distribution is the subject of this paper require some further definition. Although similar to previously published groupings, they differ in some respects. In terms of only the more abundant and well-defined species, the groupings used herein can be diagnosed as follows (mainly adapted from Grant-Mackie, 1978).

Monotis (Monotis) subcircularis group—Large *Monotis*; subequivalved (when fully grown); shape obliquely suboval; posterior ear large, smooth, well differentiated; ribbing regular, in two distinct orders of strength, concentric striae conspicuous. Example: *M. (M.) subcircularis*.

Monotis (Monotis) salinaria group—Small to medium *Monotis*; subequivalved; shape oblique; posterior ear smooth, well differentiated; ribbing fine to medium in strength, regular, commonly wrinkled or wavy posterodorsally. Examples: *M. (M.) haueri, salinaria*.

Monotis (Entomonotis) group—Small to large *Monotis*; inequivalved, inflation of left valve pronounced [even in medium to large species of *M. (Entomonotis)*]; shape subcircular to obliquely suboval; posterior ear smooth, generally well differentiated; ribbing mainly coarse, of one [e.g., in *M. (Inflatomonotis)*] or more orders of strength, tending to become obsolete in *M. (Maorimonotis)*. Examples: *M. (Entomonotis) densistriata, jakutica, ochotica, richmondiana*; *M. (Inflatomonotis) hemispherica, pachypleura*; *M. (Maorimonotis) calvata, routhieri, zaibaikalica*.

Monotis (Eomonotis) group—Small to medium *Monotis*; subequivalved; shape oblique to obliquely suboval; posterior ear poorly differentiated, obliquely truncated posteriorly, commonly ornamented by ribs; ribbing fine to medium in strength, regular. Examples: *M. (Eomonotis) anjuensis, obtusicostata, pinensis, typica, scutiformis*.

REFERENCES

Coney, P. J., et al, 1980, Cordilleran suspect terranes: Nature, v. 288, p. 329–333.

Fujita, K., and J. T. Newberry, 1982, Tectonic evolution of northeastern Siberia and adjacent regions: Tectonophysics, v. 89, p. 337–357.

Gall, J. C., 1983, Ancient sedimentary environments and habitats of living organisms: Berlin, Springer-Verlag, 219 p.

Grant-Mackie, J. A., 1978, Subgenera of the Upper Triassic bivalve *Monotis*: New Zealand Journal of Geology and Geophysics, v. 21, p. 97–111.

______, 1981, New Zealand Warepan (Upper Triassic) sequences: Murihiku Supergroup of the North Island: Journal of the Royal Society of New Zealand, v. 11, p. 231–256.

Grindley, G. W., et al, 1981, Lower Mesozoic position of southern New Zealand determined from paleomagnetism of the Glenham Porphyry, Murihiku Terrane, Eastern Southland, *in* M. M. Cresswell and P. Vella, eds., Gondwana five: Proceedings of the Fifth International Gondwana Symposium, Wellington, New Zealand, 11–16 February 1980, p. 319–326.

Hamilton, W., 1979, Tectonics of the Indonesian region: U.S. Geological Survey Professional Paper 1078, 345 p.

Hillhouse, J. W., and C. S. Grommé, 1980, Paleomagnetism of the Triassic Hound Island Volcanics, Alexander Terrane, southeastern Alaska: Journal of Geophysical Research, v. 85, p. 2594–2602.

______, in press, Northward displacement of Wrangellia: paleomagnetic evidence from Alaska, British Columbia, Oregon, and Idaho: Journal of Geophysical Research.

Hirooka, et al, 1984, Paleomagnetic evidence of accretion and tectonism of the Hida and the circum-Hida terranes, central Japan, *in* D. G. Howell, et al, eds., Proceedings of the Circum-Pacific Terrane Conference: Stanford University Publications, Geological Sciences, v. 18, p. 115–118.

Howell, D. G., 1980, Mesozoic accretion of exotic terranes along the New Zealand segment of Gondwanaland: Geology, v. 8, p. 487–491.

______, et al, 1983, Tectonostratigraphic terrane map of the circum-Pacific region: U.S. Geological Survey Open-File Report 83-716.

Irving, E., 1979, Pole positions and continental drift since the Devonian, *in* M. W. McElhinny, ed., The earth: its origin, structure, and evolution: London, Academic Press, p. 567–593.

Jones, D. L., et al, 1977, Wrangellia—A displaced terrane in northwestern North America: Canadian Journal of Earth Sciences, v. 14, p. 2565–2577.

Kamp, P. J. J., 1980, Pacifica and New Zealand: proposed

eastern elements in Gondwanaland's history: Nature, v. 288, p. 659–664.

Kiparisova, L. D., et al, 1966, Upper Triassic bivalved mollusks from northeast U.S.S.R. (in Russian): Vsesoyuznyi Nauchno-Issledovatel' skii Geologicheskii Institut, Magadan, 312 p.

Kristan-Tollmann, E., and A. Tollman, 1983, Ueberregionalle Züge der Tethys in Schichtfolge und Fauna am Beispiel der Trias zwischen Europa und Fernost, speziell China: Oesterreichische Akademie der Wissenschaften, Schriftenreihe der Erdwissenschaftlichen Kommissionen, v. 5, p. 177–230.

MacKinnon, T. C., 1983, Origin of the Torlesse terrane and coeval rocks, South Island, New Zealand: Geological Society of America Bulletin, v. 94, p. 967–985.

McElhinny, M. W., et al, 1981, Fragmentation of Asia in the Permian: Nature, v. 293, p. 212–216.

Monger, J. W. H., and E. Irving, 1980, Northward displacement of north-central British Columbia: Nature, v. 285, p. 289–294.

Nichols, K. M., and N. J. Silberling, 1979, Early Triassic (Smithian) ammonites of paleoequatorial affinity from the Chulitna terrane, south-central Alaska: U.S. Geological Survey Professional Paper 1121-B, p. B1–B5, 3 plates.

Nur, A., and Z. Ben-Avraham, 1977, Lost Pacifica continent: Nature, v. 270, p. 41–43.

______, 1982, Oceanic plateaus, the fragmentation of continents, and mountain building: Journal of Geophysical Research, v. 87, p. 3644–3661.

Ozawa, T., and K. Kanmera, 1984, Tectonic terranes of late Paleozoic rocks and their accretionary history in the circum-Pacific region viewed from fusulinacean paleobiogeography, *in* D. G. Howell, et al, eds., Proceedings of the Circum-Pacific Terrane Conference: Stanford University Publications, Geological Sciences, v. 18, p. 158–160.

Packer, D. R., and D. B. Stone, 1974, Paleomagnetism of Jurassic rocks from southern Alaska, and their tectonic implications: Canadian Journal of Earth Sciences, v. 11, p. 976–997.

Saito, Y., 1984, Pre-Tertiary geotectonic units in Japan, *in* D. G. Howell, et al, eds., Proceedings of the Circum-Pacific Terrane Conference: Stanford University Publications, Geological Sciences, v. 18, p. 164–166.

Smith, P. L., 1983, the Pliensbachian ammonite *Dayiceras dayiceroides* and Early Jurassic paleogeography: Canadian Journal of Earth Sciences, v. 20, p. 86–91.

Tamura, M., 1965, *Monotis (Entomonotis)* from Kyushu, Japan: Memoirs of the Faculty of Education, Kumamoto University, no. 13, sec. 1, p. 42–59.

______, and V. Hon, 1977, *Monotis subcircularis* Gabb from Sarawak, East Malaysia—Contributions to the geology and paleontology of southeast Asia, CLXXVII: Geology and Palaeontology of Southeast Asia, v. 18, p. 29–31, pl. 4.

Tipper, H. W., 1981, Offset of an upper Pliensbachian geographic zonation in the North American Cordillera by transcurrent movement: Canadian Journal of Earth Sciences, v. 18, p. 1788–1792.

Tozer, E. T., 1967, A standard for Triassic time: Geological Survey of Canada Bulletin 156, 103 p., 10 plates.

______, 1979, Latest Triassic (upper Norian) ammonoid and *Monotis* faunas and correlations: Rivista Italiana di Paleontologia e Stratigrafia, v. 85, p. 843–876.

______, 1982, Marine Triassic faunas of North America: their significance for assessing plate and terrane movements: Geolgische Rundschau, v. 71, p. 1077–1104.

Wang, Y.-G., et al, 1981, An outline of the marine Triassic in China: International Union of Geological Sciences Publication 7, 21 p., 2 tables.

Westermann, G. E. G., 1962, Succession and variation of *Monotis* and the associated fauna in the Norian Pine River Bridge section, British Columbia (Triassic, Pelecypoda): Journal of Paleontology, v. 36, p. 112–118.

______, 1970, Occurrence of *Monotis subcircularis* Gabb in central Chile and the dispersal of *Monotis* (Triassic Bivalvia): Pacific Geology, v. 2, p. 35–40.

______, 1973, The Late Triassic bivalve *Monotis*, *in* A. Hallam, ed., Atlas of palaeobiogeography: Amsterdam, Elsevier, p. 251–258.

Paleozoic Miogeoclines and Suspect Terranes of the North Atlantic Region: Cordilleran Comparisons

Harold Williams
Memorial University of Newfoundland
St. John's, Newfoundland, Canada

The Caledonian-Appalachian orogen comprises the Appalachian miogeocline of eastern North America, the Caledonides miogeoclines of the northern British Isles and east Greenland, the Mauritanides miogeocline of northwest Africa, and the Scandinavian miogeocline of Norway and Sweden.

Suspect terranes make up the axial and eastern parts of the Appalachians, the central and southern parts of the British Caledonides, higher nappes of the Scandinavian Caledonides, most of the Iberian Peninsula, Brittany, and Morocco, and the Caledonian inliers of France. Some suspect elements are continuous and recognized in eastern North America, the British Isles, and Brittany; others are of only local significance and extent.

Ophiolitic terranes were accreted during early Paleozoic time. Terranes more outboard from the miogeoclines were accreted in middle Paleozoic time. The earliest accretionary boundaries in the Appalachian orogen are marked by ophiolites and melanges. Later accretionary boundaries between more outboard terranes are steep mylonite zones or high-angle brittle faults. The first accreted terranes were cut from their roots and emplaced above the miogeoclines. The last accreted terranes have thick crusts and are probably microplates embedded in the mantle. Evidence for closure of Paleozoic seaways in the stratigraphic record of the Caledonian-Appalachian orogen agrees with the accretionary analysis.

The Caledonian-Appalachian orogen offers some interesting comparisons with the North American Cordilleran and other orogens of the circum-Pacific, in terms of accretionary history, orogenic times, and structural patterns.

The Appalachians of eastern North America, the Caledonides of the British Isles, Scandinavia, and east Greenland, the Mauritanides of Senegal and Mauritania, and the Variscides of Morocco, the Iberian Peninsula, and Brittany are all disjunct parts of a Paleozoic orogen that was dispersed by the opening of the North Atlantic Ocean. Restoration of the North Atlantic borderlands by closing the present ocean facilitates analysis of the Paleozoic miogeoclines and suspect terranes of the region (Fig. 1).

The western cratonic flank of the orogen is found in the Appalachian miogeocline of eastern North America and the Caledonides miogeoclines of the northern British Isles and east Greenland. The eastern flank of the orogen is represented by the Mauritanides miogeocline of northwest Africa and the Scandinavian miogeocline of Norway and Sweden. Whereas the western miogeocline was continuous, the opposing miogeoclines in Africa and Scandinavia appear to have been separated by a seaway along the Tornquist line. Thus, the orogen was not bilaterally symmetrical, although everywhere two-sided.

Suspect terranes make up the axial and eastern parts of the Appalachians, the central and southern parts of the British Caledonides, higher nappes of the Scandinavian Caledonides, most of the Iberian Peninsula, Brittany, and Morocco, and the Caledonian inliers of France. Some suspect elements are continuous and recognized in eastern North America, the British Isles, and Brittany; for example, the Dunnage, Gander, and Avalon terranes of the northern Appalachians are equated with the Dundee, Greenore, and Anglesey terranes of the British Isles (Fig. 1). Others are of only local significance and extent, for example, volcanic and ophiolitic terranes above the Appalachian and Scandinavian miogeoclines. Terranes that are suspect with respect to one margin of the orogen can in some cases be linked to another margin; thus, the Avalon terrane of North America is linked to the west African craton through correlation between the Avalon and Pan-African belts. Other terranes are suspect with respect to both margins; for example, African correlatives of the North American Meguma terrane are just as suspect with respect to the Moroccan miogeocline as is the Meguma with respect to the Appalachian miogeocline.

Ophiolitic terranes are almost everywhere coeval and of Late Cambrian to Early Ordovician age. Their emplacement represents the first accretionary event at the Scandinavian margin and at the opposing margin of the system in North America and the British Isles. This earliest phase of accretion (Early to Middle Ordovician) corresponds with Finnmarkian, Taconian, and Grampian orogeneses, respectively. Ophiolitic terranes were cut from

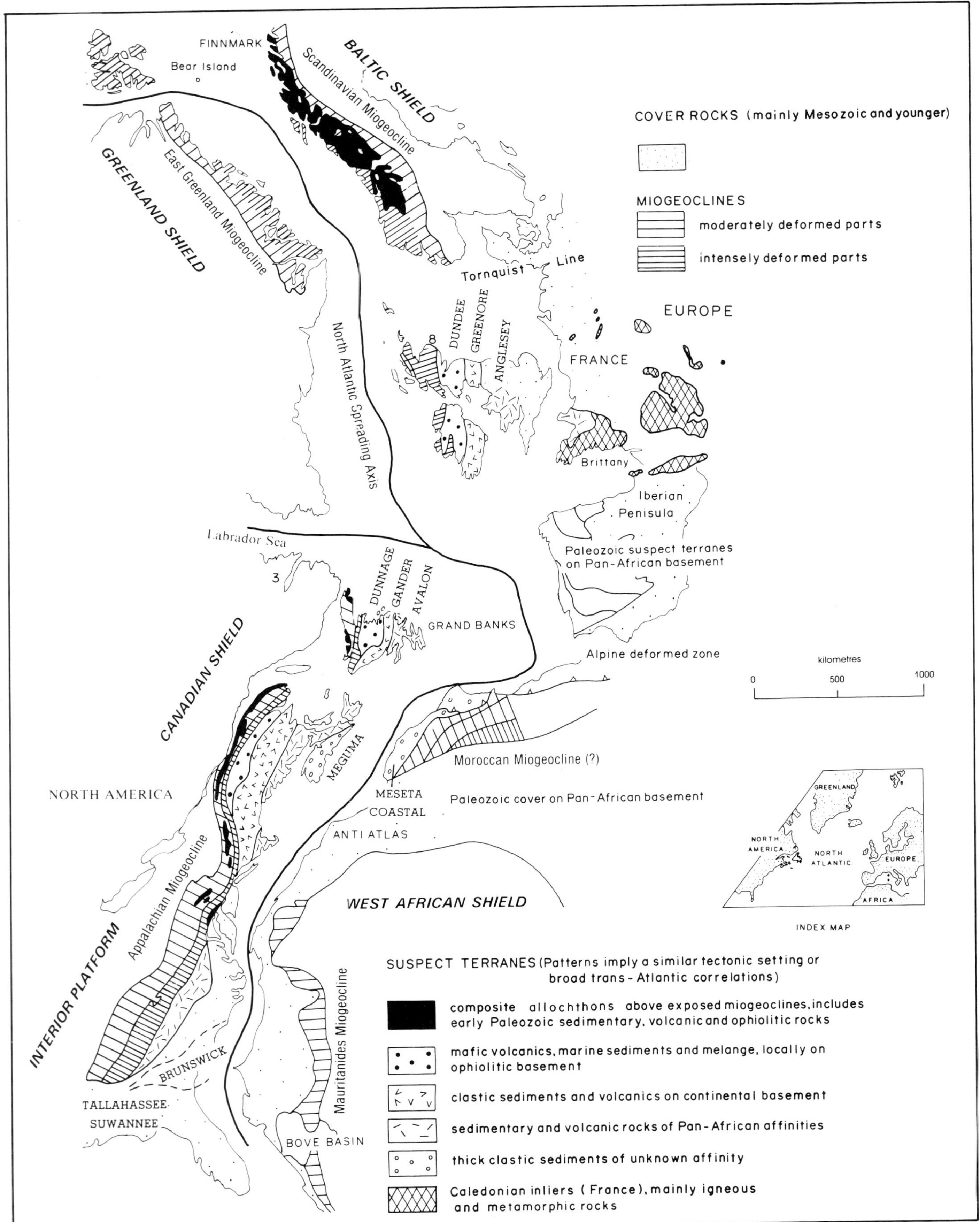

Figure 1—Miogeoclines and suspect terranes of the restored North Atlantic Region.

their roots at approximately 10 km (6.2 mi) beneath the Moho, and they moved first as thick slabs that acquired thin metamorphic soles of supracrustal rocks. Subsequent transport was accompanied by melange formation, and movement was along tectonic surfaces that truncate the ophiolite stratigraphic sequences and their metamorphic soles. In Scandinavia, final middle Paleozoic emplacement was achieved by hard thrusting and mylonite formation. Areas affected by early Paleozoic orogenesis, and by inference earliest accretion, are outlined in Figure 2. Whereas ophiolite accretion was in places simultaneous on opposite sides of the orogen in the Scandinavian Caledonides and North American Appalachians, other areas experienced rifting indicated by contemporaneous volcanism and intrusion, for example, France and parts of Morocco.

Terranes most outboard from the miogeoclines were accreted in middle Paleozoic time, for example, the Avalon and Meguma terranes of the Canadian Appalachians, and in some cases during late Paleozoic, for example, Brunswick and Tallahassee–Suwannee terranes of the U.S. Southern Appalachians. Areas of the Caledonian–Appalachian orogen affected by middle and late Paleozoic deformation are depicted in Figures 3 and 4, respectively.

The earliest accretionary boundaries in the Appalachian orogen are marked by ophiolites and melanges, and they parallel lithofacies belts of the miogeocline and adjoining terranes. Subhorizontal structures are dominant with polarity at right angles to the terrane boundaries. Thus, the earliest accretionary boundaries are soft structural zones interpreted to involve head-on obduction and/or subduction. Seismic studies indicate major decollement surfaces at shallow to intermediate crustal depths beneath the earliest accreted terranes. Later accretionary boundaries between more outboard terranes are steep mylonite zones or high-angle brittle faults that truncate lithofacies belts, implying oblique convergence or transcurrent movements. The last accreted outboard terranes have thick crusts, and they are probably microplates embedded within the mantle. A virtual absence of early Paleozoic deformation in eastern Appalachian terranes fits the pattern of a lack of early Paleozoic deformation in their African and French connections. Exactly how and where the accreted terranes were uncoupled from their place of origin remain problematic.

Evidence for closure of Paleozoic seaways in the stratigraphic record of the Caledonian–Appalachian orogen agrees with the accretionary analysis. Thus, the appearance of cosmopolitan Middle Ordovician faunas on opposite sides of the orogen supports Ordovician accretion and the Ordovician destruction of some miogeoclines. The appearance of Late Silurian and Devonian terrestrial rocks throughout the orogen, coupled with the wide extent of middle Paleozoic orogenesis, corresponds to an important middle Paleozoic accretionary event. Finally, a preponderance of late Paleozoic terrestrial rocks and the limited extent of late Paleozoic deformation signify final accretion, the closure of existing narrow seaways, and structural tightening of weak crustal areas.

The Caledonian–Appalachian orogen offers some interesting comparisons with the North American

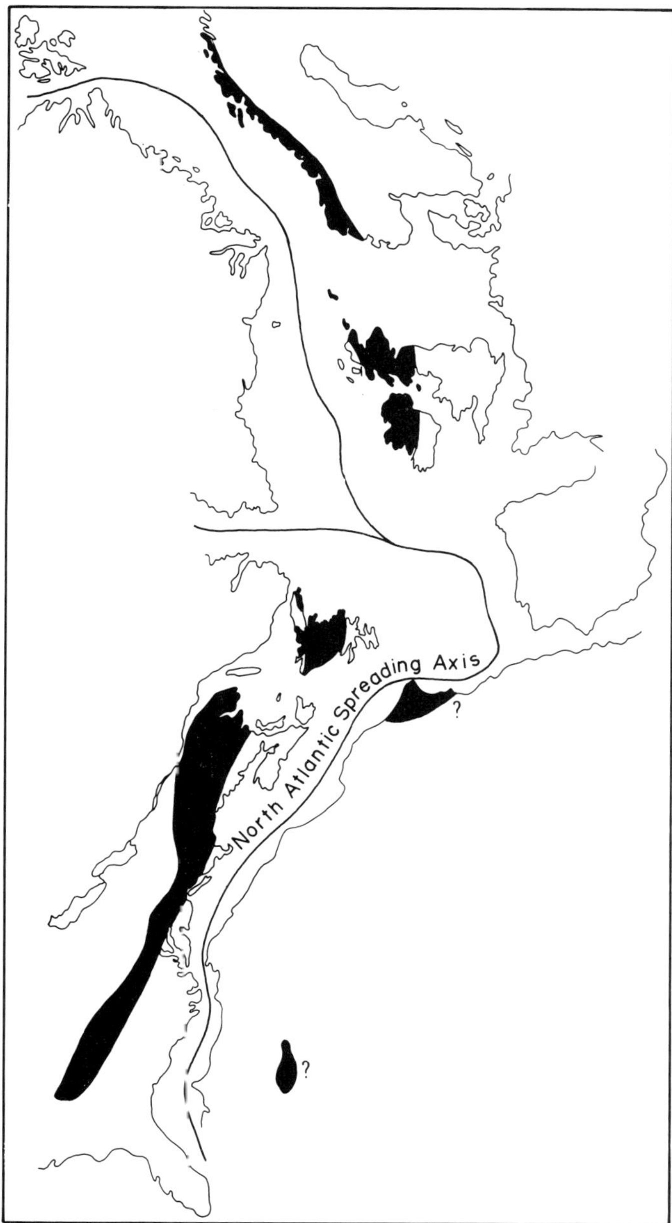

Figure 2—Areas of the North Atlantic Region affected by early Paleozoic orogenesis.

Cordilleran and other orogens of the circum-Pacific.

Caledonian–Appalachian accretion began in the Early Ordovician and continued throughout the Paleozoic for approximately 200 m.y., culminating with the complete destruction of Paleozoic seaways. Cordilleran and most circum-Pacific accretion began in the late Paleozoic and continues now. Circum-Pacific accretion began when Caledonian–Appalachian accretion ceased, and its initiation coincides with rifting and the commencement of Caledonian–Appalachian dispersion through opening of the North Atlantic. Thus, circum-Pacific accretion was balanced by Atlantic spreading, and a lack of Paleozoic circum-Pacific accretion reflects passive margins and ocean spreading that balances Paleozoic convergence evident in the Caledonian–Appalachian orogen.

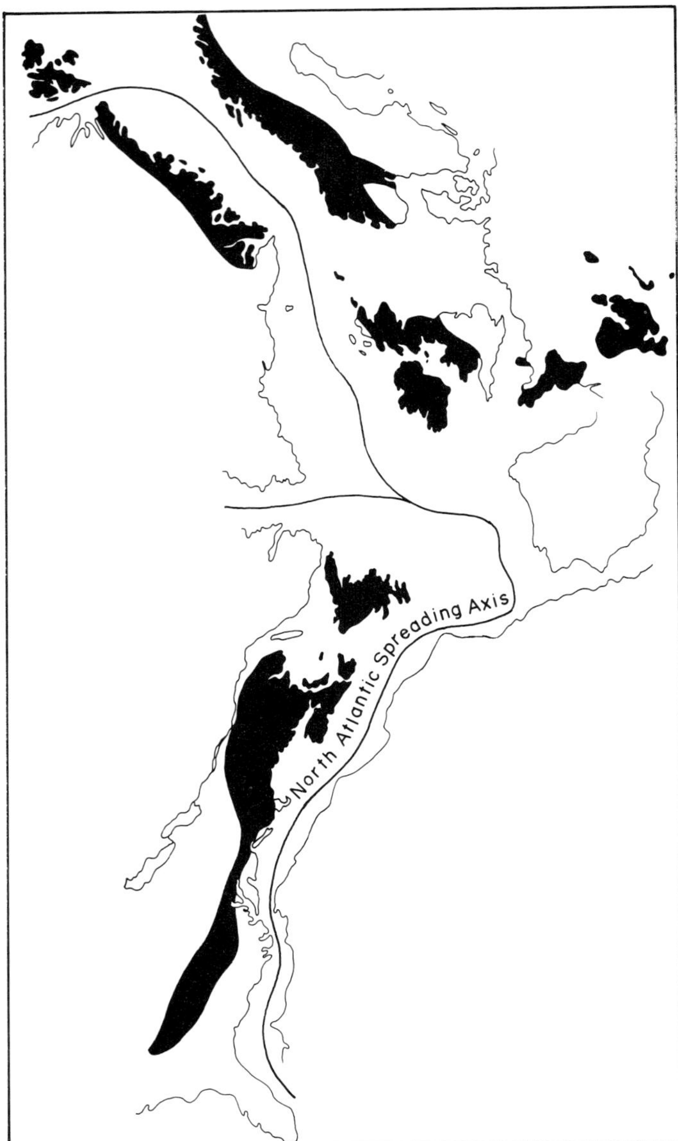

Figure 3—Areas of the North Atlantic Region affected by middle Paleozoic orogenesis.

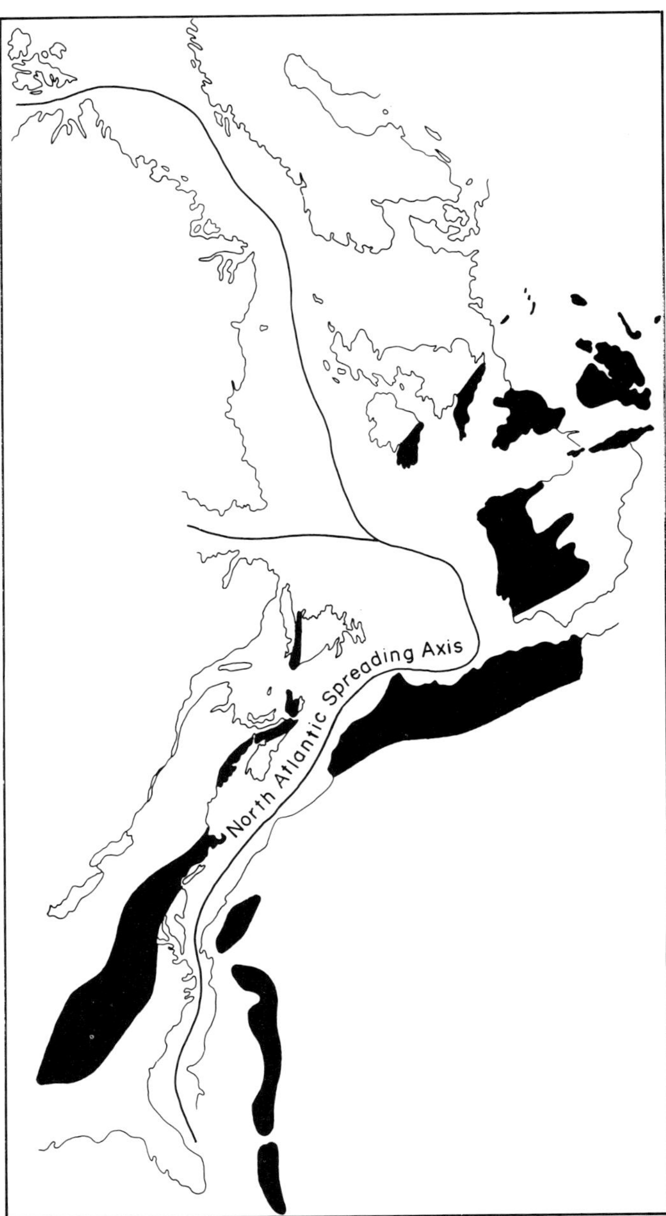

Figure 4—Areas of the North Atlantic Region affected by late Paleozoic orogenesis.

Accretion progressed outward from the miogeoclines along the Appalachian and Cordilleran margins of North America. In the Appalachians, early Paleozoic accretion was diachronous, earlier and more profound in the north than in the south. Middle Paleozoic accretion was more important in the north than the south, and late Paleozoic accretion was more important in the south. Thus, the accretion of the Appalachians progressed from north to south. In the Cordilleran, accretion progressed in the opposite sense, from south to north, with late Paleozoic events in the United States segment, early and late Mesozoic events in the Canadian segment, and late Mesozoic and Tertiary events in the Alaskan segment.

Orogeneses, or the combined effects of deformation, metamorphism, and plutonism, are equated with accretionary events in both the Appalachian and Cordilleran orogens. However, in both mountain belts the orogenic culminations followed accretion as established by stratigraphic and sedimentologic analyses.

Finally, whereas circum-Pacific orogens are one-sided and asymmetric because of the persistence of an open Pacific Ocean, the Caledonian-Appalachian orogen culminated as a two-sided symmetrical system with the complete destruction of Paleozoic seaways. The axis of opening of the North Atlantic was centered mainly in accreted parts of the Caledonian-Appalachian orogen and along internal zones of late Paleozoic deformation. Modern geometry of the North Atlantic mimics Paleozoic geometry, and Paleozoic geometry in some cases reflects earlier Grenvillian structural trends. Thus, the North Atlantic is patterned and exhibits ancestral controls to a far greater degree than is apparent in the Pacific Ocean and its bordering orogens.

ACKNOWLEDGMENTS

The foregoing is an expanded abstract based mainly on a review by Williams (1984) and proceedings of the Circum-Pacific Terrane Conference (Howell et al, 1984). Thanks are extended to members of the Caledonian–Appalachian working group of the International Geological Correlation Program and to the participants of the Circum-Pacific Terrane Conference.

REFERENCES

Howell, D. G., et al, (eds.), 1984, Proceedings of the Circum-Pacific Terrane Conference: Stanford University Publications, Geological Sciences, v. 18, 248 p.

Williams, H., 1984, Miogeoclines and suspect terranes of the Caledonian–Appalachian Orogen: tectonic patterns in the North Atlantic Region: Canadian Journal of Earth Sciences, v. 21, p. 887–901.

Tectonostratigraphic Terranes, Pacific Northeast Quadrant

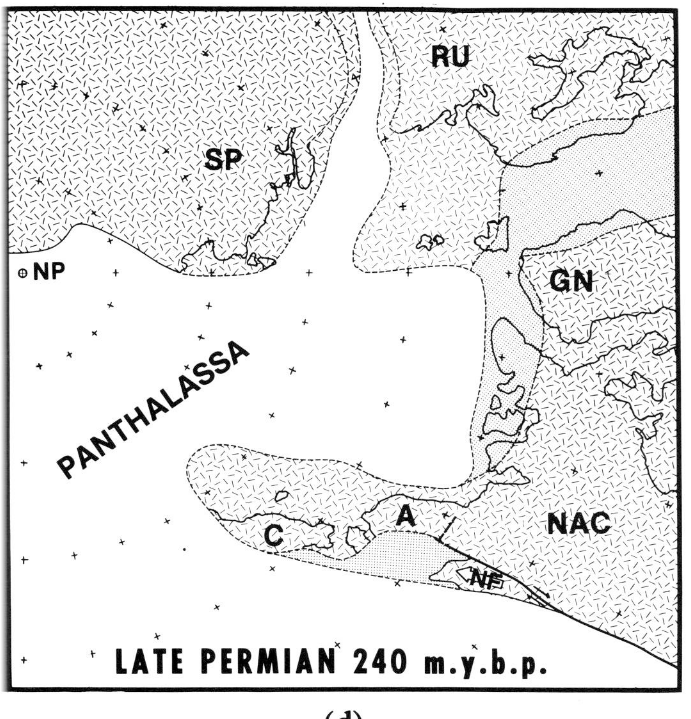

(d)

Jurassic time (Fig. 3c). The treatment of Arctic Alaska-Chukotka as a stationary peninsular extension of North America in Early Jurassic (Fig. 3c) or later Permian (Fig. 3d) is a conservative reconstruction based on available data. Mobilistic reconstructions of Arctic Alaska-Chukotka that either reconstruct Arctic Alaska-Chukotka against the Canadian Arctic Islands (Grantz et al, 1981) or from more southerly latitudes (Jones, 1982) are insufficiently documented. The land bridge joining the Siberian platform and Arctic Alaska-Chukotka formed by accretion of exotic terranes from the south, including the continental Omolon and Prikolymsk terranes. A Late Permian paleogeographic reconstruction (Fig. 3d) suggests that a wide Panthalassa oceanic basin separated North America from the eastern margin of the Siberian platform. Instead of an isolated Permian ocean in the Arctic, circum-Arctic stratigraphic, tectonic, and paleomagnetic data support an extension of Panthalassa across the northern margin of Pangea, which included NAC and SP.

IMPLICATIONS FOR PETROLEUM AND MINERAL EXPLORATION

The North American-Siberian connection and the terrane concept of accretionary growth of the Cordilleran continental margin have implications for petroleum and mineral exploration strategy in the region. Simply put, terranes that have different geologic histories have different resource potential. Coeval strata, on opposite sides of major terrane boundaries, are likely to be quite different, and one may be prospective for hydrocarbons while its neighbor is prospective for base metal deposits. Regions of Alaska currently targeted for petroleum exploration exemplify these differences.

The extensive oil and gas deposits of Alaska's North Slope occur in a tectonic environment that is epicontinental, cratonic, and, in part, passive-margin. Prudhoe Bay and Kuparuk fields are currently in production. Lisburne and Milne Point are planned for development, and West Sak, Ugnu, Endicott, Gwydr Bay, and Point Thompson are all awaiting improved economic conditions. Continued exploration in the Prudhoe area will exploit those stratigraphic and tectonic associations that have proven successful. Exploration in the overthrust belt north of the Brooks Range will use strategies similar to those being used in the overthrust belt of the Rocky Mountains.

In contrast, much of the remainder of Alaska and the adjoining continental shelf is underlain by accreted oceanic terranes (Fig. 1). Alaska's first major oil and gas province, Cook Inlet, is situated within the Peninsular terrane that began as an island arc and was later accreted to Alaska. The complex set of circumstances responsible for oil and gas occurrence there starts with an early oceanic setting and ends only after accretion and establishment of a successor basin in a postaccretionary setting. Middle Jurassic source beds were deposited in the island arc setting when the Peninsular terrane was possibly at a paleolatitude of about 20°S (Stone and Panuska, 1982). Mid-Cenozoic reservoir beds were deposited after the terrane accreted to Alaska. Current exploration south of Arctic Alaska-Chukotka is focused on Norton, Gulf of Anadyr, Navarin, St. George, and North Aleutian parts of the Bering Sea, as well as some of Alaska's interior basins and the Gulf of Alaska. Exploration strategy for these areas must account for diverse and exotic early-stage oceanic histories, followed by accretionary phases, that are ultimately modified by development of postaccretionary successor basins.

The mining industry in its search for minerals can also benefit from terrane analysis. In Alaska, numerous mining districts can be related to the geologic events that characterize individual accreted terranes. These include the old Juneau, Fairbanks, Forty-mile, Nixon Fork, and Wiseman gold mining districts and the Kennecott and Cassiar peninsula copper mines. Today much attention for development is proceeding on the molybdenum deposits of Quartz Hill in the Tracy Arm terrane, the polymetallic deposits of Greens Creek in the Alexander terrane, Johnson Creek in the Peninsular terrane, and Ambler mining district in Hammond terrane. All these deposits, including the lead-zinc Red Dog deposit in the Kagvik terrane, have been described as parts of accreted terranes.

REFERENCES

Churkin, Jr., M., et al, 1982, Terranes and suture zones in east-central Alaska: Journal of Geophysical Research, v. 87, p. 3718–3730.

———, 1984, Nixon Fork-Dillinger terranes—a dismembered Paleozoic craton margin in Alaska displaced from Yukon Territory (Abs.): Geological Society of America, v. 16, n. 5, p. 275.

Filatova, N. I., 1979, Cretaceous-Paleogene volcanism of the transition zone between the Verkhoyansk-Chuckchi and Koryak-Kamchatka regions: Geotectonics, v. 13, n. 5, p. 402–412.

Fujita, K., and J. T. Newberry, Accretionary terranes and tectonic evolution of northeast Siberia, *in* M. Hashimoto and S. Uyeda, eds., Accretion tectonics in the circum-Pacific regions: Tokyo, Terra Scientific Publishing Co., p. 43–57.

Grantz, A., et al, 1981, Geology and physiography of the continental margin north of Alaska and implications for the origin of the Canada basin, *in* A. E. M. Nairn, et al, eds., The ocean basins and margin, v. 5, The Arctic Ocean: New York and London, Plenum Press, p. 439–492.

Jones, P. B., 1982, Mesozoic rifting in the western Arctic ocean basin and its relationship to Pacific seafloor spreading, *in* A. F. Embry and H. R. Balkwill, eds., Arctic geology and geophysics: Canadian Society of Petroleum Geologists Memoir 8, p. 83–99.

Patton, Jr., W. W., and I. L. Tailleur, 1977, Evidence in the Bering Strait region for differential movement between North America and Eurasia: Geological Society of America Bulletin, v. 88, p. 1298–1304.

Seslavinskiy, K. B., 1970, Structure and development of the South Anyui fault trough, West Chukotka: Geotectonics, v. 4, p. 311–317.

Stone, D. B., and B. C. Panuska, 1982, Paleolatitudes versus time for southern Alaska: Journal of Geophysical Research, v. 87, n. B5, p. 3691–3707.

Templeman-Kluit, D., 1984, Counterparts of Alaska's terranes in Yukon (Abs.), *in* Cordilleran geology and mineral exploration: status and future trends: Vancouver, Geological Association of Canada, Cordilleran Section Symposium, p. 41–44.

Wallace, W. K., and D. C. Engebretson, 1984, Relationship between plate motions and Late Cretaceous to Paleogene magmatism in southwestern Alaska: Tectonics, v. 3, n. 2, p. 295–315.

Paleomagnetic Results from Alaska and Their Tectonic Implications

Robert S. Coe
Brian R. Globerman
Peter W. Plumley
Gordon A. Thrupp
University of California
Santa Cruz, California

We present the results from seven of our paleomagnetic studies in Alaska and collect and evaluate the results of 21 others from the literature. The results from lava flows are generally more consistent and reliable than those from sedimentary rocks. The discrepancies may be explained by depositionally induced inclination error and by undetected secondary overprinting in sedimentary rocks. Using the most reliable results and other selected data, we propose the following tectonic scenario for terrane accretion in southern Alaska. Terranes north of the Peninsular terrane were essentially in place by latest Cretaceous time. The paleomagnetic data suggest, however, that much of western and central Alaska has rotated about 40° counterclockwise about a vertical axis since then, in agreement with an earlier model of Grantz (1966) based on geological grounds. A major cause of the rotation was convergence between Eurasia and North America. The Peninsular terrane and Wrangellia collided with North America in the vicinity of present-day Cape Mendocino in mid- to Late Cretaceous time, about 20° south of their present position with respect to cratonic North America. Together with the associated Jura-Cretaceous flysch belt formed during initial convergence with North America, they were moved northwestward along strike-slip faults parallel to the ancient margin by oblique subduction of the Kula plate, arriving at interior Alaska some time before 52 Ma. The Prince William and Chugach terranes lay about 25° to the south of their present positions of 62 Ma ago and were also driven north by oblique subduction of the Kula plate and, after 42 Ma, the Pacific plate. Slivering by inter- and intraterrane dextral transcurrent faulting accommodated the post-52-Ma relative motion required by the paleomagnetic results between the most outboard part of the Prince William terrane and the most inboard part of Wrangellia.

INTRODUCTION

During the past dozen years, investigation and reinterpretation of the geology of Alaska have shown that the state consists of a large number of terranes that are characterized by disparate geological records (Jones et al, 1972, 1981, 1983; Coney, 1980). Natural questions to ask about each of these tectonostratigraphic terranes are where it came from and when it arrived at its present position relative to other terranes. Although tectonostratigraphic terranes by definition are fault bounded and have geological histories distinctly different from those of their neighbors prior to juxtaposition, they may or may not have undergone large relative displacements (Jones et al, 1983). Paleomagnetism is most helpful in answering these questions. With enough reliable paleomagnetic data one can make quantitative estimates of paleolatitude and paleogeographic orientation, and thus one can estimate as well north-south displacement and azimuthal rotation relative to continents or other terranes. In this paper we deal with the paleomagnetic evidence concerning the last stages of assembly of the collage of Alaskan tectonostratigraphic terranes—that is, with central, south-central, and southwestern Alaska during Late Cretaceous and Early Tertiary time.

Wrangellia (Fig. 1) is the Alaskan terrane that has been shown most clearly to be highly allochthonous. Both paleontological (Jones et al, 1977; Newton, 1983) and paleomagnetic (Hillhouse, 1977) evidence from Wrangellia point to a lower paleolatitude during Late Triassic time, implying that it traveled northward at least several thousand kilometers relative to the North American craton before it was emplaced in its present location. Jurassic and Cretaceous paleomagnetic data obtained even earlier by Stone and co-workers (Packer and Stone, 1972, 1974; Stone and Packer, 1977, 1979) suggest that the Peninsular terrane (Fig. 2) also was far to the south. This is consistent with a recent paleontological interpretation (Newton, 1983) that Wrangellia and the Peninsular terrane were in the same general region in the Late Triassic and with stratigraphic evidence that they were amalgamated by Late Jurassic time (Jones and Silberling, 1979).

Neither the paleomagnetic nor the geological evidence alluded to so far could distinguish between northern and

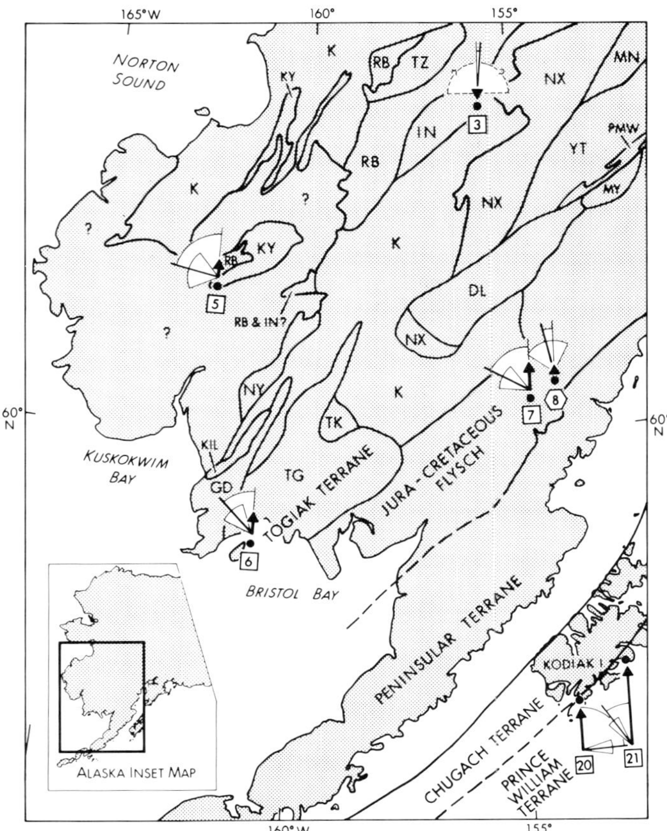

Figure 1—Latitudinal displacement and azimuthal rotation inferred from paleomagnetic results of our studies listed in Table 1. See legend and caption of Figure 2 for explanation of symbols.

southern hemisphere paleolatitudes. More recent paleomagnetic data for Wrangellia, obtained by Panuska and Stone (1981, and this volume) from rocks whose ages lie within the Permo-Carboniferous reversed interval, are capable of resolving this ambiguity because of their known magnetic polarity. These imply a northern hemisphere position for Wrangellia in the Late Triassic period and, assuming Wrangellia and the Peninsular terrane were close to each other as suggested above, a southern hemisphere position for both terranes in the Jurassic Period.

Surprisingly, the question of when these terranes accreted to interior Alaska and arrived at essentially their present position with respect to the North American craton is much more controversial than their earlier histories. Paleomagnetic results from rocks of Late Cretaceous and Early Tertiary age prompted Stone and Packer (1977, 1979) to suggest that the Alaska Peninsula still lay considerably south of its present position during that time. On the basis of these and additional data, Stone et al (1982) concluded that the Peninsular, Wrangellia, Chugach, and "Kuskokwim" terranes seem to have moved systematically northward at an average latitudinal velocity of 6 cm/year (2.4 in/year) from Middle Jurassic to the present. While the scatter certainly allows a range of interpretations, acceptance of those data makes it difficult to conclude that the Peninsular terrane and Wrangellia reached their present latitude relative to North America any earlier than 20 m.y. ago. On the other hand, the geological evidence appears to require that they accreted to interior Alaska no later than Late Cretaceous (Jones and Silberling, 1979; Silberman et al, 1981; Csejtey et al, 1982). The reason for this interpretation is that no Tertiary marine sediments have been found in the highly deformed Jura-Cretaceous flysch basin (Fig. 1) that borders the Peninsular and Wrangellia terranes to the northwest; the youngest fossil yet identified is Cenomanian in age (Jones et al, 1982). This basin is thought to record the closing of an intervening ocean basin and the subsequent collision of the terranes with interior Alaska. Radiometric dates on undeformed, "stitching" plutons in this region and thermally reset early Late Cretaceous ages of Nikolai Greenstone samples from Wrangellia are also consistent with accretion in the Late Cretaceous. The apparent lack of a Tertiary suture zone anywhere inboard of this flysch belt has led geologists to conclude that large-scale northward displacement was complete by Late Cretaceous time. This conclusion is in direct conflict with the paleomagnetic interpretation of Stone et al (1982).

The contradiction has been sharpened by several recent paleomagnetic results from Upper Cretaceous and Lower Tertiary volcanic rocks inboard of Wrangellia and the Peninsular terrane that imply little or no latitudinal displacement relative to North America. One of these studies (Hillhouse and Grommé, 1983a) even laps onto northernmost Wrangellia. However, the paleomagnetic results indicating large poleward displacements for the southern terranes (Wrangellia, Peninsular, Chugach, and Prince William) are so numerous that they demand serious consideration.

In this paper we briefly present our own data that bear on this problem. We then summarize and evaluate the considerable body of paleomagnetic results from Late Cretaceous onward for all but the southeastern part of Alaska. Finally, based on a selection of these data, we give our own interpretation (and speculation) concerning the latitudinal displacement and azimuthal rotation of the Alaskan terranes.

PALEOMAGNETIC RESULTS FROM UCSC LAB

During the past 5 years we have put considerable effort into the study of Alaskan terranes, with particular emphasis on the questions and problems raised above. In this section we describe the results of a number of our studies that have important bearing on these problems. Two of these studies are published in recent articles, and the others are in press or in preparation at the time of writing. Readers wishing more details of these studies should consult those references. The purpose here is to gather the main results together and to present them in a coherent framework.

Ghost Rocks Lava Flows, Kodiak Island

On Kodiak Island we conducted a paleomagnetic study of lava flows in the Paleocene Ghost Rocks Formation at two localities that are separated along strike by about 80 km (50 mi) (Plumley et al, 1982, 1983). These andesitic and basaltic flows are interbedded with sandstone and argillite, which make up the bulk of the formation, and the whole mass is intruded locally by distinctive

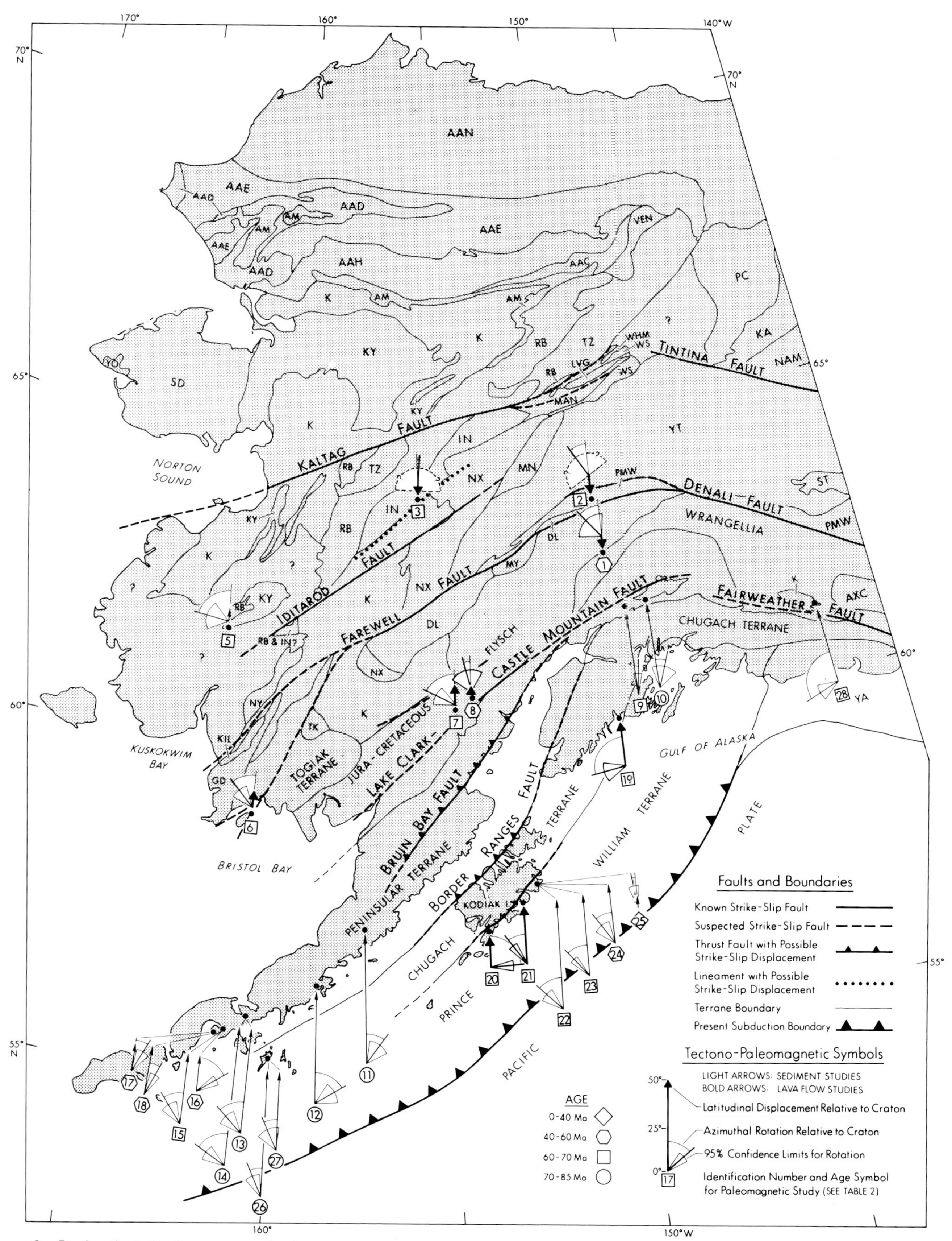

Figure 2—Latitudinal displacement and azimuthal rotation inferred from paleomagnetic studies listed in Table 2 (no. 4 omitted, see text). Bold accented symbols indicate results derived from studies of lava flows. Base is Alaska terrane map of Jones et al (1984), but incorporating some simplifications found in the circum-Pacific terrane map of Howell et al (1983). See those references for abbreviations of terrane names.

granodiorite plutons (Moore, 1969). The geology and tectonic significance of the Ghost Rocks Formation have been reexamined recently by Moore et al (1983), who explain this unusual lithological association as the result of interaction between a subduction complex and a spreading ridge. Ages as young as Paleocene from pelagic foraminifera at two localities and K–Ar dates as old as 62 Ma from the intrusive rocks appear to constrain the actual age of the Ghost Rocks Formation very tightly. A figure very close to 62 Ma is favored.

The Ghost Rocks Formation has been complexly deformed and dismembered, but coherent units are well preserved in places (Byrne, 1982). Dips are generally steep to the northwest, but with enough variation to enable definitive fold tests to be made. The Ghost Rocks Formation was metamorphosed to prehnite-pumpellyite facies, and both vitrinite reflectance and the maximum unblocking temperature of the secondary component of remanent magnetization suggest maximum temperatures reached 225 to 250° C (437–482°F). The fact that the secondary component failed the fold test indicates that the metamorphism was syn- or postfolding. Overlying Eocene and younger rocks are practically unmetamorphosed, so that metamorphism probably accompanied the Paleocene intrusive event. Structural relations show that igneous activity and regional deformation were contemporaneous (Byrne, 1982). Thus, deposition and extrusion, intrusion and metamorphism, and initial severe deformation occurred within a brief interval (probably less than 5 Ma).

Stepwise thermal demagnetization was performed on 186 cores from 28 lava flows and one tuffaceous unit. These results were analyzed by means of vector diagrams (Zijderveld, 1967) to determine whether magnetic overprints could be eliminated and an underlying characteristic component (defined by unidirectional decay toward the origin on a vector diagram) isolated. Except for the tuffaceous unit, thermal demagnetization was successful in the great majority of cases, whereas alternating field demagnetization was ineffective. Characteristic directions so obtained for each lava flow were then averaged to yield a flow or site mean direction, and these site means were again averaged for each of the two areas and examined for indications of consistency and reliability.

The two areas sampled, near Kiliuda Bay and Alitak Bay, are shown in Figure 1, and the corresponding paleomagnetic results are given in Table 1 (nos. 20 and 21). In both areas, the site mean directions pass the generalized fold test at a very high level of confidence, as shown by the much tighter grouping (smaller α_{95} and larger k values in Table 1) in stratigraphic coordinates than in geographic coordinates. Thus, the characteristic component was acquired before folding. Moreover, in the Kiliuda Bay area both normal and reversed polarity are found, and the mean normal and reversed directions are closely antiparallel. Because of these convincingly positive fold and reversal tests, and because good stable endpoints are commonly reached on vector diagrams during thermal demagnetization, we believe that the characteristic component is primary—that is, an original thermoremanent magnetization (TRM) acquired during primary cooling of each lava flow.

The mean primary directions of the flows at Kiliuda and Alitak Bays after structural restoration to paleohorizontal (see Table 1, stratigraphic coordinates) differ by 122° in declination. This discrepancy suggests that a large tectonic rotation between the two areas has occurred, a frequent occurrence near both convergent and strike-slip boundaries (e.g., Beck, 1976, 1980). Indeed, the strikes of Alitak Bay are rotated in the same sense as the declinations relative to those at Kiliuda Bay, though not enough to explain the entire discrepancy by simple rotation about a vertical axis. The mean inclinations differ by 12°, a much smaller but still significant discrepancy. The most likely explanation is a combination of (1) incorrect structural correction resulting from initial dips and errors in measuring the bedding attitudes, and (2) incomplete averaging of geomagnetic secular variation by the lava flows sampled.

The most important finding of this study is that the mean inclination of the primary remanence is significantly shallower than that which would be expected if the lava flows of the Ghost Rocks Formation had been in their present position relative to cratonic North America when they recorded the geomagnetic field direction about 62 m.y. ago. This result strongly suggests that these rocks were considerably further south with respect to North America in Early Paleocene time. Taking the paleomagnetic result of each area at face value, the discrepancy in paleolatitude is 31 ± 9° for Kiliuda Bay and 16 ± 9° for Alitak Bay. (See Appendix A for method and assumptions used in calculating latitudinal displacements, azimuthal rotations, and their uncertainties.) Since the two areas are composed of very similar rocks of the same formation and are presently separated by only 80 km (50 mi) along strike, it is likely that the differences in inclination found do not stem from actual differences in paleolatitude, but rather from reasons such as those mentioned in the paragraph above. Therefore, the results have been combined, using colatitude statistics (McFadden and Reid, 1982), to obtain an overall estimate for the paleolatitude of the Ghost Rocks Formation of 40 ± 6° N. The expected paleolatitude, calculated from the North American reference paleomagnetic pole for 60 Ma or 65 Ma (Table 4), is 65 ± 3 or 68 ± 5°, respectively. Using the former value, we conclude that the Ghost Rocks Formation has moved northward relative to North America by 25 ± 7°. (Any longitudinal movement is, of course, indeterminate by conventional paleomagnetic methods.) This conclusion—large northward displacement since Early Paleocene time—is of critical importance in later discussions.

Nowitna Lava Flows, McGrath Region

Almost 800 km (500 mi) north of the Kodiak Island sites (Fig. 1, no. 3), we conducted a pilot study on andesitic lava flows of the Paleocene Nowitna Volcanics exposed in the Nixon Fork–Innoko terranes (Plumley, 1984). In this region, the Nowitna Volcanics give K–Ar ages of about 64 Ma (Moll et al, 1981) and overlap both the Innoko and Nixon Fork terranes (Fig. 1). In a broader sense they belong to the Kuskokwim Mountains magmatic belt of Late Cretaceous to Early Tertiary age, which extends from

Table 1

ID No	Study Area	Rocks (Age)	N/N_0	POL N/R	COORD	Directions D	I	α_{95}	k	Poles ϕ	λ	A_{95}	κ
21	Kiliuda Bay, Kodiak (207.1E,57.3N)	Ghost Rocks lavas ($\lesssim$ 62 Ma)	11/12[a]	4/7	Geog	165.5	61.9	25.9	4.1				
					Strat	320.1	52.6	8.6	29.4	92.5	54.9	10.5	19.8
20	Alitak Bay, Kodiak (206.1E,56.9N)	Ghost Rocks lavas ($\lesssim$ 62 Ma)	16/16	0/16	Geog	169.9	44.9	20.5	4.2				
					Strat	82.1	65.0	7.4	25.8	269.4	43.1	10.6	13.0
3	McGrath Region (204.7E,63.8N)	Nowitna lava flows (~ 64 Ma)	6/7	0/6	Geog	315.4	65.8	13.8	24.5				
					Strat	350.2	82.9	13.5	25.6	198.1	76.9	25.3	8.0
6	N. Bristol Bay (199.4E,58.7N)	Unnamed volcanic series (~ 68 Ma)	74/85[b]	74/0	Geog	317.0	30.7	7.8	5.5				
					Strat	306.8	76.2	3.0	31.4	147.1	65.2	5.0	11.7
5	Lower Yukon River (198.1E,61.6N)	Unnamed tuffaceous rocks (overprinted ~ 70 Ma?)	27/31	27/0	Geog	269.4	73.9	6.4	20.1	159.5	56.6	11.0	7.3
					Strat	23.6	71.0	7.1	16.5				
7	Lake Clark Region (205.3E,60.3N)	Unnamed lava flows (~ 66 Ma[c])	30/30	25/5	Geog	337.9	75.4	4.8	31.6				
					Strat	298.6	74.4	4.7	32.8	146.8	60.7	7.9	12.0
8	Lake Clark Region (205.7E,60.6N)	Unnamed lava flows (~ 44 Ma)	6/7	0/6	Geog	288.2	59.8	13.9	24.1				
					Strat	341.6	74.8	13.9	24.1	147.4	79.6	20.0	12.2

ID No: Same numbers as for Table 2 and Figures 1 and 2; AGE: K-Ar dates on flows or associated igneous rocks, except for no. 5 (see text); N/N_0: Number of sites used in analysis/total number of sites; POL: Polarity (Normal sites/Reversed sites); COORD: Geographic (in situ) or Stratigraphic (structurally restored) coordinates; Directions: Mean of characteristic (most stable) site mean directions, with reversed polarity inverted to normal—D = declination, I = inclination, α_{95} = radius of 95% confidence cirlce, k = precision parameter; POLES: Paleomagnetic pole obtained by averaging VGPs derived from such site mean directions (see Appendix A)—ϕ = E. long., λ = N. lat., A_{95} = radius of 95% confidence cirlce, K = precision parameter.

[a]One tuff unit excluded from N_0;

[b]Six tuff units and one sedimentary unit excluded from N_0;

[c]Single date with unusually large error bars (± 13 Ma).

Table 1—Summary of latest Cretaceous and Early Tertiary paleomagnetic results from UCSC lab.

the Bering Sea in the southwest to at least as far as the Kaltag fault in the northeast (Moll and Patton, 1982; Wallace and Engebretson, 1984). In the area that we were able to reach (Fig. 1, no. 3) the exposures were not extensive, despite the considerable thickness of the section, with the result that only seven flows could be sampled. For the most part, attitudes had to be estimated from hillside benches that formed on top of the more erosion-resistant flows. Fortunately, dips were gentle, and the flows were essentially unmetamorphosed.

Large scatter in the direction of natural remanent magnetization (NRM) and occasional exceptionally strong intensities indicated overprinting induced by lightning. Demagnetization in alternating fields ranging from 200 to 500 Oe effectively removed this random overprint well enough to estimate a characteristic component in all but a few of the most intensely magnetized samples. This characteristic component is reversed in every case. Thermal demagnetization, which was employed in some samples, showed that it is carried by both magnetite and hematite. Thus, even though no fold or reversal test is available, the characteristic remanence is very probably primary.

Six of the flows give mean directions (Table 1, no. 3) that cluster relatively near the expected Paleocene field direction for their location. The flow mean for the seventh was not included with the others because it is transitional between reversed and normal polarity, about 90° away from the mean of the other six. Because no other reason for this discrepancy is apparent, we think it was erupted during a geomagnetic excursion (see, for example, Doell and Dalrymple, 1973). Such large excursions of field direction are thought to require at least a few hundred to a thousand years to happen. Moreover, the directions of the other flows that lie under and over it do not show significant serial correlation. Therefore, the flows probably span at least 2,000 years, enough to provide a fair sampling of secular variation (Champion, 1980). In contrast to the results from Kodiak Island, the mean direction is consistent with little or no latitudinal displacement and rotation relative to North America (Table 2). The 95%

Table 2

Region	ID No	Locality	Formation	Age	Ref	LF	N	K/K_{sv}	AF	TH	VC	RT	FT	PLAT	DISPL (°N)	ROT (°CW)
						Reliability Factors			Demag			Tests				
North of Peninsular Terrane	1	Talkeetna Mtns	Unnamed lava flows	~52 Ma	*A*,B	+	26*	0.4	+	+[a]	+	+	+	76±10	−8±10	−37±48
	2	Cantwell Basin	Cantwell	≲60 Ma	*C*	+	18*	0.8	+	+[a]	+		+	81±8	−9±8	−29±?
	3	McGrath Region	Nowitna Volcanics	~64 Ma	*D*	+	6*	0.5	+	+	+			75±20	−1±21	5±?
	4[b]	McGrath Region (Kus. 1.7)	Kuskokwim	L. Cret. (85 Ma?)	*E*	−	33	0.5	+					15±5[b]	67±7[b]	−7±51[b]
	5	Lower Yukon River	Unnamed tuffaceous rocks	E. Cret., RM is L. Cret. overprint (70 Ma?)	*F*	−	27*	0.4	+	+	+		−	70±9	7±10	−77±37
	6	N. Bristol Bay	Unnamed volcanic series	~68 Ma	*G*	+	74*	0.7	+	+	+		+	65±4	9±7	−43±23
	7	Lake Clark Region	Unnamed lava flows	~66 Ma	*H*	+	30*	0.7	+	+	+	+		62±6	10±8	−54±22
	8	Lake Clark Region	Unnamed lava flows	~44 Ma	*H*	+	6*	0.7	+	+	+			65±16	1±16	−12±42
Peninsular Terrane	9	Matanuska Valley	Chickaloon	Pal. (60 Ma?)	*E*,I	−	36	0.5	+		+	+	+	22±5	48±6	7±9
	10	Sheep Mtn. (SHP.1–2)	?	L. Cret. (80 Ma?)	*E*,J	−	22	0.8	+					31±6	48±8	20±32
	11	Painter Creek (PNC.1)	Chignik	L. Cret. (85 Ma?)	*E*,K,L	−	8	1.0	+					7±7	71±9	36±28
	12	Chignik LG. (CHG.1–2)	Chignik	L. Cret. (80 Ma?)	*E*,K,L	−	15	0.4	+					16±8	62±10	59±28

(continued)

Table 2 (continued)

Region	ID No	Locality	Formation	Age	Ref	LF	N	K/K_{sv}	Reliability Factors: Demag AF	Demag TH	Demag VC	Tests RT	Tests FT	PLAT	DISPL (°N)	ROT (°CW)
Peninsular Terrane	13	Herendeen Bay (HND.7)	Chignik	L. Cret. (75 Ma?)	*E*,L	−	7	0.5	+					18±14	56±15	−46±26
	14	Herendeen Bay (HND.6)	Chignik	L. Cret. (75 Ma?)	*E*,L	−	8	0.3	+					0±10	74±12	−59±23
	15	Canoe Bay (CNB.1)	Hoodoo	L. Cret. (70 Ma?)	*E*,K,L	−	22	0.2	+			+		33±11	39±12	−37±22
	16	Pavlov Bay (PVL.1)	Tolstoi	Pal.-Eoc. (55 Ma?)	*E*,K,L	−	23	0.4	+			+		43±9	20±9	55±13
	17	Pavlov Bay (PVL.2)	Tolstoi	Pal.-Eoc. (55 Ma?)	*E*,K,L	−	17	0.4	+					48±11	15±11	22±17
	18	Canoe Bay (CNB.4)	Tolstoi	Pal.-Eoc. (55 Ma?)	*E*,K,L	−	18	0.4	+			−[c]	−[c]	36±10[c]	27±10[c]	20±13[c]
Wrang-ellia	28	McCarthy Region	Maccoll Ridge	L. Cret. (70 Ma?)	*Q*	−	14	0.7	+	−	+		+	32±8	42±9	−107±22
Prince William–Chugach Terrane	19	Resur-rection Peninsula, G. Alaska	Valdez Group	L. Cret. (70 Ma?)	*M*,N	+	81[d]	[d]	+		+		+	51±10	24±12	−98±26
	20	Alitak Bay, Kodiak Is.	Ghost Rocks	≳ 62 Ma	*O*,P	+	16*	0.6	−	+	+		+	49±8	16±9	86±14
	21	Kiliuda Bay, Kodiak Is.	Ghost Rocks	≳ 62 Ma	*O*,P	+	11*	0.8	−	+	+	+	+	34±8	31±9	−34±12
	22	Kodiak Is. (KOD.2)	Kodiak	L. Cret. (70 Ma?)	*E*	−	12	4.4	+					12±3	60±6	−46±19
	23	Kodiak Is. (KOD.3)	Ghost Rocks	≳ 62 Ma	*E*	−	8	0.7	+					21±9	45±9	−39±11
	24	Kodiak Is. (KOD.1)	Sitkalidak	Eoc.-Oli. (40 Ma?)	*E*	−	8	0.2	+					30±18	32±19	−32±22

(continued)

Table 2 (continued)

Region	ID No	Locality	Formation	Age	Ref	Reliability Factors								PLAT	DISPL (°N)	ROT (°CW)
									Demag			Tests				
						LF	N	K/K_{sv}	AF	TH	VC	RT	FT			
Prince William-Chugach Terrane	25	Kodiak Is. (NRC)	Narrow Cape	L. Mio. (10 Ma?)	*E*	–	73	0.2	+					51±8	6±8	−6±13
	26	Shumagin Is. (SHM.1)	Shumagin	L. Cret. (75 Ma?)	*E*,L	–	19	1.5	+					6±4	68±17	−24±20
	27	Shumagin Is. (SHM.2)	Shumagin	L. Cret. (75 Ma?)	*E*,L	–	8	0.4	+					32±15	42±16	−26±27

ID NO: Study identification number, same as used in Figures 1 and 2 and Table 1.
Age: Approximate age of rocks studied, determined by K-Ar method or (when in parentheses) assigned on basis of paleontological and geological evidence.
Ref: A, Hillhouse and Grommé (1983a); B, Hillhouse et al (1983); C, Hillhouse and Grommé (1982); D, Plumley (1984); E, Stone et al (1982); F, Globerman et al (1983); G, Globerman and Coe (1983 and in preparation); H, Thrupp and Coe (1983, 1984); I, Stone (1983); J, Packer (1972); K, Stone and Packer (1977); L, Stone and Packer (1979); M, Hillhouse and Grommé (1977); N, Grommé and Hillhouse (1981); O, Plumley et al (1983); P, Plumley et al (1982); Q, Panuska (1983). Italics indicates source for quantitative paleomagnetic data actually used in table.
Reliability Factors: + or − indicates factor is affirmative (positive) or negative, respectively; a blank indicates no information regarding the factor; LF, lava flows sampled; N, number of samples used or, when asterisked, number of sites used (average usually 3 to 7 samples per site); K/K_{sv}, precision parameter for data/precision parameter for secular variation appropriate to the paleolatitude; AF, alternate field demagnetization used; TH, thermal demagnetization used; VC, some form of vector component analysis used; RT, reversal test passed; FT, fold test passed.
PLAT: Paleolatitude and 95% confidence limit inferred from mean paleomagnetic direction.

DISPL and ROT: Displacement northward (+) or southward (−) and azimuthal rotation clockwise (+) or counterclockwise (−) of sampling area with respect to cratonic North America, inferred from the difference between the mean paleomagnetic pole and the coeval North American reference pole. Uncertainties are 95% confidence limits estimated by the method of Demarest (1983) outlined in the Appendix.
[a]Stepwise thermal demagnetization conducted on pilot samples from each flow, confirming that the characteristic direction is essentially the same as that obtained by AF demagnetization (J. W. Hillhouse, 1984, personal communication).
[b]This result is now considered invalid because of previously unrecognized overprinting (D. B. Stone, 1983, personal communication).
[c]Examination of original data listing (ref. K, Table 2, p. 192) shows that the directions fail the fold and reversal tests. Hence the results given in ref. E and this table are not valid.
[d]Two-level analysis conducted on samples from pillow basalt and dikes in one area and dikes in another area. *N* and *K* not comparable with those of usual single-level analysis (J. W. Hillhouse, 1984, personal communication).

Table 2—Tectonopaleomagnetic data summary for southern, southwestern and central Alaska from Late Cretaceous to Present.

confidence limits, however, are large because of the small number of flows sampled.

Unnamed Lava Flows, North Bristol Bay

In the Togiak terrane (Fig. 1, no. 6), more than 600 km (373 mi) to the southwest of the Nowitna Volcanics sampling area, at the southwesternmost extremity of this same Kuskokwim Mountains magmatic belt, we collected about 700 samples from 91 subaerial volcanic flows and pyroclastic deposits on Hagemeister, Crooked, and Summit Islands in northern Bristol Bay for a detailed paleomagnetic study (Globerman and Coe, 1983, and in preparation). This volcanic sequence, which is well exposed along the shorelines, comprises mainly basaltic-andesite flows interbedded with airfall tuff, fine-grained sediments, and volcanic breccia. The sequence is homoclinally dipping 65 to 75° toward the southeast on Hagemeister Island, but on Crooked and Summit Islands the dips are shallow. Their chemistry is consistent with a volcanic arc setting (Globerman et al, 1983). K–Ar dates on three of the flows give a mean age of 68 ± 3 Ma.

These samples were magnetically straightforward. Both alternating field (AF) and thermal stepwise demagnetization experiments were analyzed with vector diagrams for about half of the samples to isolate a characteristic component. A stable endpoint, that is, univectorial decay toward the origin, was achieved in almost every case by 200 Oe or 400° C (752°F). Blanket treatment at one or two alternating field values or temperatures was employed on the other half of the samples. AF and thermal methods were equally successful, except for a small fraction of samples in which thermal demagnetization was necessary to remove a secondary overprint carried by hematite. Again, the characteristic component carried by magnetite and hematite was the same, which is consistent with our hypothesis that the hematite originated by auto-oxidation processes (Wilson and Haggerty, 1966; Grommé et al, 1969) during primary cooling.

The mean directions of six coarse-grained tuffaceous deposits and one sedimentary interbed were considered unreliable and were discarded. In addition, the results from ten lava flows with strongly discrepant directions (VGPs more than 50° from the mean) were eliminated. These discrepant directions, which occur in a few flows from two sections within the volcanic sequence on Hagemeister Island, show strong serial correlation and are statistically distinct from the near-Fisherian distribution defined by the directions of the remaining 74 flows. They have almost certainly recorded two geomagnetic excursions and should be eliminated in order to obtain the best estimates of paleolatitude, poleward displacement, azimuthal rotation, and their uncertainties (e.g., see Harrison's [1980] careful analysis of results from more than 1,000 lava flows in Iceland).

The remaining 74 sites all have unambiguous normal polarity. Thick, well-laminated sedimentary interbeds indicate, however, that the flows span a period sufficiently long to obtain a representative time average of the ancient field. Since there were several periods of normal polarity in Maestrichtian and early Paleocene time lasting 0.5 to 1.0 m.y. (Harland et al, 1982), it is not unlikely that the entire sequence accumulated within one of them. The interbeds also give us confidence in the structural attitudes that we determined. Rotation of the flow mean directions into stratigraphic coordinates using these attitudes yields a strongly positive fold test (McFadden and Jones, 1981), significant at the 95% confidence. Despite the lack of a reversal test, we are confident that the characteristic component is primary.

The inclination of the mean characteristic direction for the 74 flows is close to the expected value for latest Cretaceous time, but the declination is significantly rotated in a counterclockwise sense. Thus, little northward displacement (9 ± 7°) with respect to North America is indicated, but there is a strong suggestion of counterclockwise rotation (43 ± 23°). Note that if the ten directions discarded because they were thought to reveal field excursions had been retained, these mean values would have been little changed: 6 ± 7° and 47 ± 25°. Because the remanent magnetization is simple and stable, the structural corrections are unusually well determined, the number of flows is large, and the period of time they span appears to be relatively long, we believe the results of this study are particularly reliable.

Unnamed Tuffaceous Rocks, Lower Yukon River

Globerman et al (1983) came to a similar but much less well-constrained conclusion in their paleomagnetic study of beds of tuff and tuffaceous sediment exposed along the Yukon River about 300 km (186 mi) north of the Bristol Bay area (Fig. 1, no. 5). These rocks (Hoare and Coonrad, 1959; Hoare, 1961), which are part of an uppermost Jurassic to Lower Cretaceous belt of andesitic flows and volcaniclastic rocks that characterize the Yukon–Koyukuk province in the northeast (Patton, 1973), belong to the Koyukuk terrane of Jones et al (1984). In the study area, bedding attitudes vary by about 30°, enough to allow application of a weak fold test. Of 109 samples collected at 31 sites spanning about 1.5 km (0.9 mi) stratigraphic thickness, 87 (27 sites) yielded useable estimates of characteristic direction in thermal and AF demagnetization experiments (Table 1). All sites have normal polarity, and the scatter in directions increases slightly when the structural correction is applied. A few samples, however, revealed a reversed characteristic direction, usually after removal of a heavy normal component at lower temperatures or alternating field strengths. For all these reasons, we interpret the characteristic remanence to be a secondary overprint.

The age of the overprint is problematical. We have used 70 Ma because peak activity in the Kuskokwim Mountains magmatic belt appears to have occurred then (Wallace and Engebretson, 1984, Fig. 4). Assuming the beds have not been tilted since overprinting, the paleomagnetic results in geographic coordinates imply little or no northward displacement (7 ± 10°) but strong counterclockwise rotation (77 ± 37°) relative to the North American craton (Table 2). These conclusions, however, are sensitive to the exact assumptions. For instance, if the age of remagnetization was older, say between 90 and 120 Ma, the calculated displacement increases to about 12° and the rotation decreases to about 50°. If 30% of the average tilt of the beds occurred after the postulated 70 Ma remagnetization event, the values of displacement and rotation would both be cut roughly in half.

Unnamed 66 Ma Lava Flows, Lake Clark Area

About 400 km (249 mi) to the southeast, just northwest of Lake Clark, we sampled 30 andesite flows (Fig. 1, no. 7). According to the latest version of the Alaska terrane map (Jones et al, 1984), these lie on the Jura-Cretaceous flysch terrane that separates the southern terranes from the rest of Alaska. One flow has a rather poor K-Ar age determination of 66 ± 14 Ma, which suggests that these flows belong to the Alaska Range magmatic belt (Hudson, 1979; Wallace and Engebretson, 1984). Dips are shallow (less than 20°) to the southwest and poorly defined, so a conclusive fold test could not be obtained.

Preliminary results of this study have been given by Thrupp and Coe (1983), and a more detailed account is now in preparation. For the most part, isolation of a characteristic component by either thermal or alternating field demagnetization was straightforward. The less stable component that was removed appeared to be recent field viscous remanent magnetization (VRM) in some cases and lightning-induced isothermal remanent magnetization (IRM) in others. Five of the flows are reversed, with directions nearly antipodal to those of the other 25 flows. Thus, it is very likely that the characteristic component is primary and that enough time is spanned by these samples to provide a good time average of the ancient field. The mean direction in stratigraphic coordinates (Table 1) has inclination fairly close to that expected using the 65 Ma North American reference pole (Table 4), but the declination is again rotated counterclockwise. This corresponds to a modest northward displacement (11 ± 9°) and a sizeable counterclockwise rotation (52 ± 22°) relative to cratonic North America. Using either the 55 Ma or the 75 Ma reference pole does not significantly affect these conclusions.

Unnamed 44 Ma Lava Flows, Lake Clark Area

Some 40 km (25 mi) to the northeast of the 66 Ma sequence (Fig. 1, no. 8), we sampled 7 lava flows spaced through a series of 15 to 20 flows that dip about 25° to the east (Thrupp and Coe, in preparation). These have a K-Ar age of 44 Ma, suggesting that they represent early activity of the onland extension of the Aleutian arc (Wallace and Engebretson, 1984). All but one of these flows were also easily cleaned of minor secondary components of VRM and IRM. The six remaining flows are reversed with a mean direction close to that expected for their age (Table 1). Little or no northward displacement and counterclockwise rotation is indicated (Table 2), but the uncertainty is large because the number of flows is few and there is no independent assurance that a good time average of the ancient field has been obtained.

Summary

The poleward displacements and rotations inferred from our work are depicted graphically in Figure 1. Five results concur that the terranes north of the Peninsular terrane have undergone at most a modest latitudinal displacement during Tertiary time. They also suggest the possibility of systematic counterclockwise rotation. In contrast, the Kodiak Island study suggests that the Prince William terrane has undergone a substantial northward displacement of 25 ± 7° and is characterized by inconsistent azimuthal rotations. No direct information was obtained from the Chugach and Peninsular terranes, which lie between our northern and southern paleomagnetic data sets. However, the distinctive plutons of the Gulf of Alaska magmatic belt (Hudson, 1979; Hill et al, 1981; Moore et al, 1983; Wallace and Engebretson, 1984) provide a loose tie between the Prince William and Chugach terranes, and apparently to the southernmost Peninsular terrane as well (Davies and Moore, 1984). Taken at face value, these ties suggest that the southern part of the Peninsular terrane has also moved north with the Prince William terrane. We will return to this subject in the section entitled Proposed Scenario.

PALEOMAGNETIC DATA SUMMARY

In this section we examine these tentative conclusions in the context of other paleomagnetic data available in the literature. We convert these data (including our own) to the tectonopaleomagnetic parameters of interest—paleolatitude, latitudinal displacement, and azimuthal rotation—and summarize them in Table 2. Because the intrinsic quality can vary, we discuss below factors that affect the reliability of paleomagnetic data. Much of this can be skimmed by practitioners of paleomagnetism. The eight columns in Table 2 under the heading Reliability Factors, as well as the 95% confidence limits, provide the basis for a rudimentary evaluation that is key to the discussion that follows.

Reliability Factors

We take up these reliability factors under the general categories of confidence limits, rock types, sampling, demagnetization, and stability tests. We include a selective review of basic material that, in our opinion, is essential for a critical evaluation of the discrepancies in the paleomagnetic data from Alaska.

Confidence Limits

Random errors accumulate during all phases of collecting, measuring, and magnetic cleaning. In addition, natural causes of error usually have a random component. Such errors contribute to the scatter of sample directions and so are naturally accounted for in the estimate of confidence limits. Systematic errors, however, are not. Thus the confidence limit that is listed with each value of paleolatitude, latitudinal displacement, and azimuthal rotation in Table 2 is only one of several factors that contribute to the reliability of a result.

Rock Types

The reliability of paleomagnetic results depends directly on the fidelity with which rocks record and preserve the ancient field direction and on the accuracy with which structural tilting can be identified and corrected for. It is not surprising that rocks differ in these regards. Lava flows and near-shore clastic sediments are the rock types from which almost all the paleomagnetic results listed in Table 2 have been derived.

Lava flows are probably the most faithful class of paleomagnetic recorders commonly available in nature. Experiments have repeatedly demonstrated that most lava samples acquire a TRM during cooling that is parallel within 1 to 2° to the ambient magnetic field. Such TRM is usually strong and stable (especially those that have undergone auto-oxidation), and the mechanism by which it is acquired is relatively well understood (e.g., Nagata, 1961). For these reasons, we have indicated in the column headed LF (lava flow) in Table 2 whether or not the paleomagnetic result was derived from lava flows.

Sediments, on the other hand, may acquire their primary (or quasi-primary) remanence by a variety of mechanisms (e.g., Verosub, 1977; Tucker, 1983a). The fidelity with which they record the ambient field direction will depend on the mechanism or combination of mechanisms involved. Most of these are poorly understood, and it is very difficult, especially with shallow-water clastic sediments, to establish which mechanisms have operated in any given case. Depositional remanent magnetization (DRM) produced by redepositing varved lake sediments in the laboratory suffers from a systematic inclination error (King, 1955; Griffiths et al, 1960). The inclination of DRM in the redeposited sediments is too shallow by an amount that depends on the inclination of the field, ranging from 0 to values typically as high as 25° (Fig. 3). Paleolatitudes derived from such DRM would be systematically low, the error reaching a maximum of 25° at a true latitude of 58°. In naturally deposited varved lake sediments, evidence has been found also for inclination errors that vary from one-quarter to two times the experimental values depicted in Figure 3 (Ising, 1942; Johnson et al, 1948; Griffiths, 1955; Granar, 1958). The mechanism responsible appears to be mechanical rotation of the magnetic grains upon impact with the water-sediment interface (King, 1955; Griffiths et al, 1960).

Fortunately, in many types of sediments the remanence is not locked in until some time after deposition. Such post-depositional remanent magnetization (PDRM) is not susceptible to the inclination error that may result from the process of deposition (Irving and Major, 1964). The lack of systematic inclination error has been particularly well documented for deep-sea sediment cores (Opdyke and Henry, 1969). In such sediments low deposition rates, fine grain sizes, and intense bioturbation may promote the acquisition of PDRM. Alternatively, if the remanence resides in biogenic magnetite derived from organisms that lived in the bottom mud (Kirschvink, 1983), it would be postdepositional by definition. The precise factors are probably complex and are not well understood. For example, Tucker (1983b) has described a case in which the remanence of carbonate ooze was reset by slumping while the remanence of nearby clay-rich sediment involved in the same slump was unaffected.

In addition, the large uniaxial compaction that is experienced by some sediments, to the extent that it occurs after PDRM lock-in, would be expected to cause inclination error by rotating the magnetic grains (Blow and Hamilton, 1978). The effects of this mechanism would be more pronounced in dry outcrop samples than in saturated, incompletely compacted core samples. This mechanism has been suggested in explanation of average inclination shallower than that of the axial geocentric dipole field that has been observed in outcrop samples of Mono Lake sediments (Liddicoat and Coe, 1979). The shallowing of inclination by compaction should exhibit the same sort of dependence on field inclination as the depositional inclination error. In the extreme case, where all magnetic grains are so inequant that they rotate as homogeneous strain markers, a 60% compaction after lock-in of PDRM would produce the same inclination error as that illustrated in Figure 3. Compaction error, however, probably does not affect sediments that are cemented before the accumulation of a significant overburden, such as many limestones.

At the present time there are no widely accepted criteria for predicting inclination error, especially in coastal and marginal sea sediments. Simpson and Cox (1977) obtained essentially the same average direction both from fine-grained layers of the turbiditic Tyee Formation in western Oregon and from sequences of lava flows that immediately overlie and underlie the Tyee, thereby demonstrating the absence of significant inclination error. Thompson and Kelts (1974), however, found significantly shallower inclinations in the coarser parts of turbidite layers than in the finer grained, laminated parts. Nonetheless, fine grain size is not a guarantee against inclination error. Blow and Hamilton (1975) showed that the inclinations of DSDP sediment cores from the Arabian Sea were much more consistent with those expected from the northward movement of India if a correction for inclination error was applied to those that were deposited more rapidly than 700 m/Ma (2,297 ft/Ma). A similar example can be cited of probable inclination error in the fine-grained, laminated lutite core studied by Creer (1974). The average inclination of sediment spanning approximately 5,000 to 25,000 years B.P. is about 15° shallower than that of an axial, geocentric dipole field, whereas the average inclination during the same period found in a core from the Aegean Sea less than 1,500 km (932 mi) away shows no significant discrepancy (Opdyke et al, 1972). Finally, Alvarez and Lowrie (1984) demonstrated the absence of significant inclination error in white, calcarenitic turbidites from Italy. This result held true for both coarse- and fine-grained portions of the turbidite units. The authors point out that this might be expected for calcarenitic turbidites because the magnetite grains are exceedingly small and thus easily rotated within larger, water-filled pores of the matrix by Brownian movement after deposition.

In summary, although little is known of a systematic nature about the frequency and magnitude of inclination error in near-shore environments, we consider it to be a real danger. Sandstones and coarse siltstones are probably most prone to deposition-induced inclination error. Clay-rich sediments are probably most susceptible to compaction-induced inclination error. In the absence of deeper understanding, we regard tectonic conclusions derived from the inclination of such sediments with caution. On the other hand, we generally expect the mean declination to be recorded more faithfully, except in regions of strong,

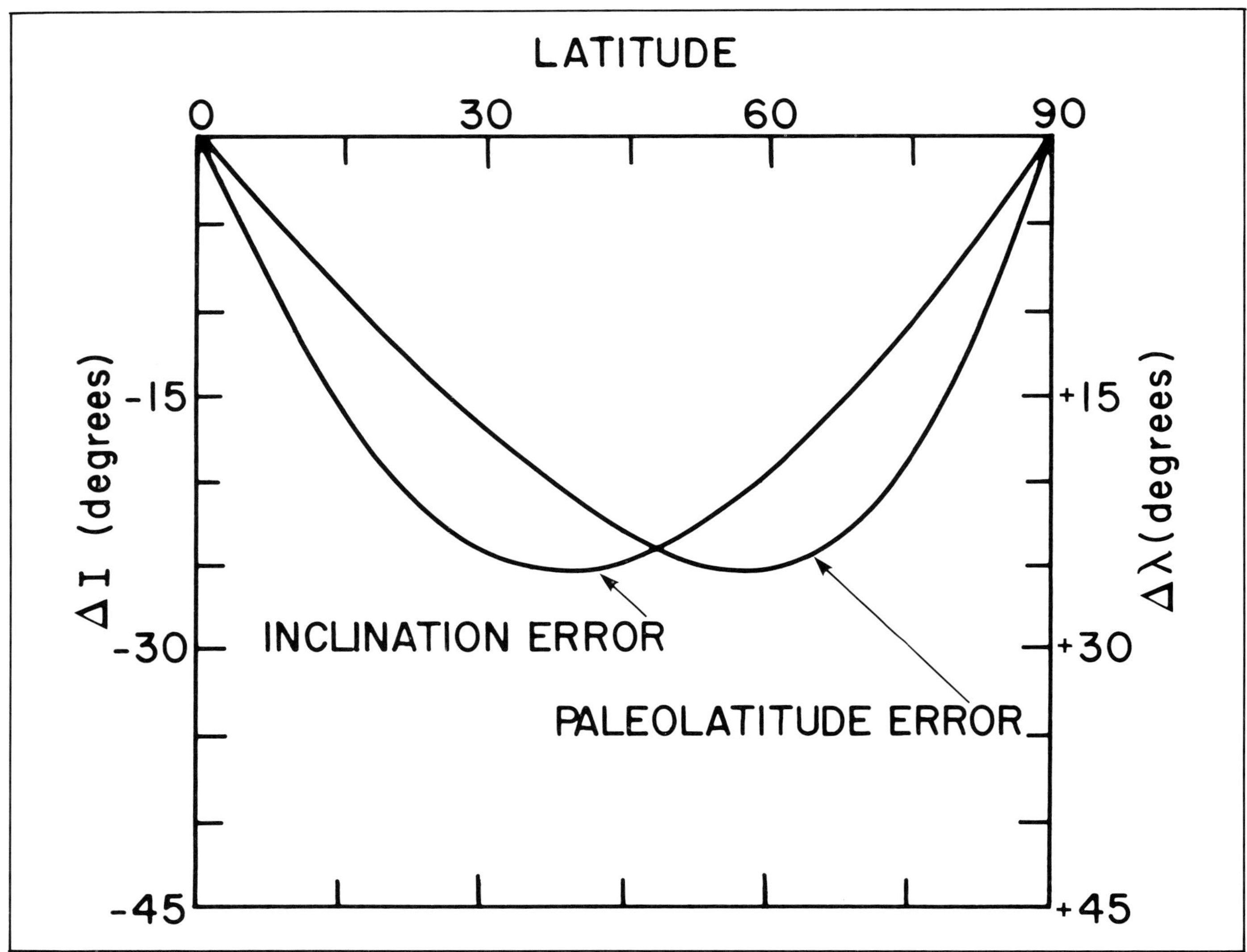

Figure 3—Inclination error and corresponding error in paleolatitude consistent with King's (1955) experiments redepositing varved sediments. Follows from relation between inclination of DRM in sediments I' and inclination of field I: $\tan I' = f \tan I$; and the dipole field relationship between inclination and paleolatitude λ: $\tan I = 2 \tan\lambda$. Combining these yields the inclination error

$$\Delta I = \tan^{-1}(2f \tan \lambda) - \tan^{-1}(2 \tan \lambda)$$

and the paleolatitude error

$$\Delta\lambda = \lambda - \tan^{-1}(f \tan \lambda).$$

as a function of paleolatitude.
The value of f consistent with the experiments is 0.4. The same rotations and value of f would describe the error resulting from passive rotation of elongated or flattened magnetic grains by an amount consistent with a 60% compaction after lock-in of remanence.

systematically directed bottom currents (King, 1955).

Conversely, more reliable structural corrections can generally be made on sediments than on lavas, for two reasons. First, the orientation of sedimentary bedding can usually be measured more precisely than the orientation of the top or bottom of a lava flow. Second, the risk of substantial initial dip is greater for lava flows than for most sedimentary rocks. Thus, structural corrections of lava flows to paleohorizontal are likely to be more reliable when attitudes are obtained from sedimentary interbeds than from the flows themselves, as is the case for several of the studies listed in Table 1. In the absence of sedimentary interbeds, a typical value for the error in structural correction expected for a single flow is around 5°. These errors, however, will tend to average out when there are a number of sites distributed over a large area.

Sampling

Reliable tectonic application of paleomagnetic data also hinges on the assumption that the geomagnetic field closely approximates the field of an axial geocentric dipole when averaged over a suitably long period of time. Except during episodes of unusual field behavior such as reversals and excursions, this assumption appears to be generally true within a few degrees (e.g., McElhinny and Merrill, 1975). Analysis of archeomagnetic data suggests that a period of 2,000 years or so may be a sufficiently long time average (Champion, 1980), although one would feel more comfortable with an averaging time at least several times this figure. If the ancient field is sampled over too short a period, the mean paleomagnetic pole position will not usually provide a good approximation of the paleogeographic pole because the effects of nondipole and nonaxial dipole fields will not have been averaged out.

Often it is difficult to know whether a representative time average has been obtained, especially for a result in the literature about which little detailed information is given. To answer this question rigorously one would need a realistic model of paleosecular variation and knowledge of the number of samples, the length of time they span, and their distribution within this time span. If, however, the precision parameter (Fisher, 1953), K, of a set of paleomagnetic data is significantly greater than that expected at that paleolatitude because of secular variation alone, K_{SV}, then one would strongly suspect that the sampling interval was too short (e.g., Cox and Gordon, 1984). This test is shown in Table 2 in the column headed K/K_{SV}, where K_{SV} has been taken from Harrison's (1980) model of secular variation. The great majority of the studies have K/K_{SV} values that range between 0.4 and 0.8, leaving a comfortable margin for the inevitable scatter resulting from experimental and natural causes of error. Only three studies have values greater than 1.0. One of these (no. 22) stands out from the rest with a value of 4.4, indicating that the remanence of all these samples was probably acquired within an interval of time that is short compared to the dominant periods of secular variation. This might be due to a chemical remanent magnetization (CRM) overprint or to sampling within a very restricted stratigraphic interval.

The number of sites (i.e., distinct time horizons) that constitute a paleomagnetic study, given in the column headed N in Table 2, can also give some indication of the overall reliability of the result. If N is too small, then, no matter whether the sites are well distributed throughout a long interval of time, the confidence limits will be unacceptably large. For example, to achieve a definitive result we would like to keep the contribution of secular variation to the 95% confidence circle about the mean pole to 5° or less. Using Harrison's (1980) model for secular variation, we find that a minimum of 22 sites would be required if the paleolatitude were 0°, 40 sites if the paleolatitude were 45°, and 55 sites if the paleolatitude were 90°. If a 10° confidence circle were acceptable, the minimum number of sites needed ranges from 7 to 15. In practice, particularly in the case of lava flows, sites may be bunched temporally, so that more than the minimum number would be needed (Cox and Gordon, 1984).

It should be pointed out that the number of sites, N, in Table 2, is equal to the number of samples for the paleomagnetic studies of sedimentary rocks, whereas for the lava flows an average of at least six samples per flow were employed to determine each site mean direction. The only exception is study no. 5, on tuffaceous rocks, in which an average of three samples per site was used. The statistical effect of taking the mean of multiple samples per site is to reduce scatter arising at the within-site level, thereby improving the overall precision. Comparing results for more than one sample from the same site also helps catch orientation mistakes that might otherwise propagate into the site mean. In addition, knowledge of the scatter about each site mean provides an objective basis for eliminating aberrant site directions from the overall mean.

Demagnetization

Demagnetization experiments are essential in paleomagnetic studies. This is especially true in tectonopaleomagnetic investigation of terranes, where the variety of rock types available to work with is often severely limited. A well-designed demagnetization procedure has a hierarchy of goals: (1) to establish the presence or absence of more than one component of remanent magnetization, (2) to determine the direction of each component, and (3) to provide information that aids in identifying the age and probable origin of each component. The characteristic (most stable) component is generally of greatest importance, but the properties of the less stable components are often useful in deciphering the history of the rock and in deciding whether the characteristic component is primary.

The columns headed AF and TH in Table 2 show whether alternating field and thermal demagnetization experiments were used successfully (+), were tried but were not successful (−), or were either not tried or not reported (blank). AF demagnetization is optimal for removing IRM overprints, for example, as caused by lightning strikes, and it is often satisfactory for dealing with the VRM induced at room temperature in magnetite, also. Thermal demagnetization is optimal for removing all types of VRM and partial TRM overprints, which are collectively referred to as VPTRM (the partial TRM acquired by cooling from some temperature in a magnetic field plus the VRM acquired as a result of being held at elevated temperature for an extended period of time). Thermal demagnetization probably holds the edge for removing most types of CRM, whereas AF demagnetization has the advantage that it can be used on materials that cannot stand heating. Thus, it is much more likely that overprints will be detected and completely removed if both types of demagnetization have been tried, and the best one (or combination) selected empirically. For example, in studies 20 and 21 in Table 2, a large secondary component, presumably VPTRM, is effectively removed from the characteristic component by thermal demagnetization, whereas with AF demagnetization it is not even detectable (Plumley et al, 1983). Conversely, we generally find that thermal demagnetization does a poor job of removing IRM and sometimes fails even to detect it.

Except for straightforward samples with minor overprints, stepwise demagnetization that removes remanence of progressively higher stability combined with some form of vector component analysis is generally more effective than demagnetization at one or a few blanket values and plotting the directions on an equal area net.

Many variations of such vector component methods have been presented and used to very good effect (Zijderveld, 1967; Halls, 1978; Hoffman and Day, 1978; Kirschvink, 1980). These methods allow one to discriminate better whether or not two or more components of remanent magnetization have been cleanly separated and to determine their directions more precisely. Thus, in Table 2 studies that are known to have employed vector component analysis are marked with a + in the column headed VC.

Stability Tests

One of the greatest pitfalls in the application of paleomagnetism to the tectonics of terranes is mistaking secondary characteristic remanence for primary. Demagnetization experiments provide no clues for recognizing when the primary remanence has been completely replaced by a secondary overprint, unless the overprint is carried by a mineral that is known to be secondary and that has magnetic properties that make it easily identifiable (e.g., goethite). If the overprinting postdates significant latitudinal displacement, azimuthal rotation, or simple tilting of the rocks, misinterpretation of the characteristic remanence as primary will result in serious errors in tectonic reconstructions. Fortunately, a number of stability tests exist that can guard against such errors (e.g., McElhinny, 1973). Two of these, the reversal and fold tests, have been applied successfully in some of the studies listed in Table 2.

If paleomagnetic study of a suite of samples from a formation reveals characteristic remanence of reversed polarity in some time-stratigraphic zones and normal polarity in others, it is likely that the characteristic component is primary. Otherwise, quite special circumstances would have to be invoked to explain this occurrence by secondary overprinting. If analysis shows that the angle between the mean normal and mean reversed directions is not significantly different from 180°, then it is even more probable that the characteristic remanence represents a primary recording of the geomagnetic field. Moreover, since the geomagnetic field takes a few thousand years to reverse, a positive reversal test demonstrates that the samples span an interval at least that long, a point that is often not easy to establish by other means if the units sampled are lava flows. Finally, when the normal and reversed components can be shown to be closely antiparallel, it provides assurance that the characteristic direction determined in that study is not significanly contaminated by residual secondary components. In Table 2, a + in the column headed RT denotes a positive reversal test, a − denotes a negative reversal test—i.e., normal and reversed directions significantly different from antiparallel—and a blank indicates that no reversal test was available or reported.

The other very useful test is the fold test, of which several versions exist (Graham, 1949; McElhinny, 1964; McFadden and Jones, 1981). All depend on the same principle that, if secondary remagnetization postdates significant folding, applying the various structural corrections to characteristic directions of the samples should increase their scatter. In such case the directions are said to fail the fold test, a result which is indicated in Table 2 by a − in the column headed FT. Conversely, if the directions converge convincingly on structural correction, the characteristic remanence probably predates all or most of the deformation and the fold test is termed positive (denoted by a + in Table 2). When enough variability in structural attitudes is available to provide a good test, a positive fold test is therefore a necessary but not a sufficient condition for the characteristic remanence to be primary. The fold test can also be inconclusive if the various samples were remagnetized at different times throughout the period of folding. A blank in the appropriate column of Table 2 indicates that the fold test was inconclusive, not available, or not reported.

Despite the ambiguities that sometimes arise, the fold test is extremely valuable in many instances. A convincingly negative fold test is virtually conclusive proof that the characteristic remanence is not primary. When the age of folding and other aspects of the geological history of the rocks are known, a convincingly positive fold test often constitutes strong evidence that the characteristic remanence is primary.

Evaluation

All the paleomagnetically inferred latitudinal displacements and azimuthal rotations contained in Table 2 are depicted graphically in Figure 2, with the exception of one result (no. 4, Table 2) that has been omitted for reasons given below. The general pattern of displacements is qualitatively similar to those shown by our own data in Figure 1: large northward displacements of the southern terranes and no significant displacements of the northern terranes. In detail, however, the picture is badly blurred by systematic inconsistencies between the displacements determined from lava flows and sediments. This can be seen at a glance in Figure 2, where displacements obtained from lava flows are much less than those obtained from sediments of comparable age. Table 3 shows a quantitative comparison, mainly for rocks ranging in age from 60 to 75 Ma. For the Prince William and Chugach terranes considered as a unit (Moore et al, 1983), the average displacement inferred from lavas is less than half that inferred from sediments (24 ± 4 versus 54 ± 6°, respectively). In the "northern" terranes (north of the Peninsular terrane), displacements derived from six different studies of lava flows, including four with dates from 60 to 68 Ma, all indicate at most a rather modest latitudinal displacement, whereas a surprisingly large displacement of 67 ± 7° was inferred from a study of clastic sedimentary rocks that are only 20 m.y. older (no. 4, Table 2). Further analysis of results from rocks in this latter area, however, has shown that the characteristic remanence, once thought to be primary, is probably a secondary overprint (D. B. Stone, 1983, personal communication), thereby invalidating the large displacement inferred above.

It is worth inquiring, therefore, whether any other of the studies with anomalously large displacements listed in Table 2 might also suffer from the problem of secondary remanence that has been incorrectly interpreted as primary. Reanalysis of the data given in the original publication for study no. 18 (Table 2) reveals that they fail the fold test. Moreover, even though both polarities are present, the normal and reversed directions are much more nearly

Table 3

Terranes	Lava Flow Data			Sediment Data		
	N. Displ.	Age	*N*	N. Displ.	Age	*N*
Prince William-Chugach	24 ± 4°	62–70 Ma	3	54 ± 6°	62–75 Ma	4
Peninsular-Wrangellia	—	—	—	52 ± 6°	60–75 Ma	5
"Northern"	3 ± 5°	60–68 Ma	4	(67 ± 7°)	(85 Ma)	(1)

Table 3—Comparison of average northward displacement of terranes and standard error inferred from lava flows and sediments.

antiparallel in geographic than in stratigraphic coordinates. This is strong evidence that the remanence is secondary, despite the fact that a reversal was recorded. There are 16 additional results listed in Table 2 for which no fold test or reversal test is available. These include all but one of the studies with anomalously large northward displacements. As already mentioned, one of these (no. 22, inferred northward displacement of 60°) has a very large value of K, 4.4 times that expected for secular variation at the paleolatitude calculated from its mean direction. Overprinting could easily account for the anomalously large values of both K and displacement. There are other possible victims of overprinting: for instance, no. 26 (Table 2, apparent poleward displacement 68°). This study has a value of $K/K_{SV} = 1.5$ and a structural correction of more than 50° that drastically shallows the inclination. These examples underscore the need for using stability tests to check for remagnetization.

It is most improbable, however, that undetected secondary characteristic remanence can explain all the instances in Table 2 of surprisingly large northward displacements. Among those for which no fold or reversal test is available there are several (nos. 11–15, 23, and 24) for which the direction of in situ remanence is so shallow or the structural correction is so small that post-tilting acquisition of a steeply dipping overprint cannot possibly account for the anomalously low paleolatitude. Moreover, there is one (no. 9, Chickaloon Formation, Matanuska Valley) that passes both local and regional fold tests and the reversal test (Stone, 1983). The inferred northward displacement is 48° since Paleocene time, which is 23° greater than that deduced from lava flows and dikes of similar age from the Chugach and Prince William terranes (no. 19, 20, and 21, Table 2). This large displacement appears excessive, particularly because these latter terranes lie outboard of the Matanuska Valley. A possible explanation of this and other systematically low paleolatitudes in Table 2 inferred from paleomagnetic studies of sediments is inclination error, as discussed earlier. Using the typical values shown in Figure 3 of inclination error found in experiments and in natural settings for clastic sedimentary rocks, we find that the apparent paleolatitude of 22°N deduced from the observed paleomagnetic inclination of the rocks would correspond to an actual paleolatitude of 45°. This adjusted value is in excellent agreement with the mean for the three studies to the south just mentioned. For the same reasons, we would guess that the paleolatitude of 32° deduced from sandstone of latest Cretaceous age on Wrangellia (Table 2, no. 28) is somewhat too low. Using the simple adjustment for inclination error, a value of 57° is found, which is only 6° higher than the paleolatitude of coeval igneous rocks immediately outboard in the Chugach terrane. Similarly, the anomalously low paleolatitude inferred from the Ghost Rocks Formation sediments (no. 23, Table 2) adjusts to 43°, only 3° more than the mean derived from the Ghost Rocks volcanics. It is also interesting and perhaps significant that the declinations of the sediments and the nearest lava flows (Kiliuda Bay, Table 1) agree within 4°. This agreement is consistent with the hypothesis of depositional or compactional inclination error. Finally, the adjusted paleolatitudes for no. 24 and for the mean of nos. 10 to 15 (70 to 85 Ma sediments of the Peninsular terrane) are also quite reasonable, though individually quite scattered in the latter case.

In summary, the paleomagnetic results in Table 2 derived from lava flows appear much more reliable than those derived from sediments. Not only are lava flows generally better recorders than clastic sedimentary rocks, but the lava flow studies were much more thorough than all the sediment studies with the exception of nos. 5 and 9. In contrast to the studies of sediments, all but two of the studies of lava flows employed at least some stepwise thermal cleaning in addition to AF cleaning and used some form of vector component analysis to extract the characteristic remanence. All but two were backed up by a positive fold test, reversal test, or both. Thus, the results derived from lava flows are highlighted in bold face on the map in Figure 2.

The results from the studies of sedimentary rocks are systematically biased toward shallow inclinations and, therefore, low paleolatitudes and large northward displacement. Undetected secondary overprints and inclination errors appear capable of explaining perhaps all of these discrepancies. Again, however, it should be noted that several paleomagnetic studies from fine-grained turbidites suggest that these rocks have faithfully recorded the geomagnetic field direction and are therefore free of inclination error. Declination is not affected by the processes responsible for inclination error, so that these results may be more useful for indicating azimuthal rotations than they appear to be for determining northward displacements.

Discussion

Northern Terranes

The paleomagnetically inferred latitudinal displacements for the terranes to the north of the Peninsular terrane are modest in magnitude and divided almost equally between positive (northward) and negative (southward). These include six results from lava flows and one from tuffaceous sediments. The eighth result indicating a large displacement (no. 4, Table 2) is not considered valid, as discussed previously. The apparent displacements range from −9 to +11°, with the negative values lying to the east of the positive ones (Fig. 2). We are uncertain, however, whether any of these values are significantly different from zero.

Three of the displacements appear to be statistically significant at marginally greater than 95% confidence (nos. 2, 6, and 7, Table 2). It is important to note, however, that the formal confidence limits are always minimum estimates for two reasons. First, they are often based on overestimates of the number of independent samplings of the geomagnetic field, especially in the case where a sequence of lava flows is sampled (e.g., Cox and Gordon, 1984). Preliminary analyses of the sequences of directions suggest that the overestimate is roughly a factor of two, which implies that the confidence limits given for them in Table 2 are too small by a factor of about 1.4. Second, as mentioned earlier, the formal confidence limits do not take account of possible sources of systematic geological errors. The most serious of these for lava flows is usually uncertainty in the structural correction. For instance, typical initial dips for lava flows on the flanks of shield volcanoes are 5 to 7°, and they may be considerably steeper than this. Such initial dips are difficult to distinguish in ancient environments from tectonic dip and thus undoubtedly lead to spurious estimates of latitudinal displacement. Since 5° error in inclination corresponds to 8 or 9° of apparent latitudinal displacement at the high paleolatitudes of these studies, it is entirely possible that any or all of the paleomagnetically inferred displacements that appear statistically significant are artifacts of the initial dip.

Averaging the results of various studies spread over a sufficiently wide area should lessen the effects of initial dip. To avoid the possibility of subconsciously biasing the results, we have treated the question of initial dip uniformly by structurally correcting beds to the horizontal in all studies regardless of whether their dips were steep or shallow. In other words, we take the intitial dip to be zero in all cases and assume that the errors involved will tend to average out. The average displacement for the four studies of lava flows with ages between 60 and 68 Ma (Table 2, nos. 2, 3, 6, and 7) is 2° northward, with a standard error of 5°. Inclusion of two results from younger flows (nos. 1 and 8), with ages of 52 and 44 Ma, respectively, changes this figure to 0 ± 3° northward. Finally, if we include the direction derived from remagnetized tuffaceous sediments (Table 2, no. 5), we still get only 1 ± 3° of northward displacement.

We conclude that the mean displacement of the terranes north of the Peninsular terrane since latest Cretaceous time is not significantly different from zero and that therefore the modest displacements found by us and others in individual studies might not be real. The fact that the northward and southward apparent displacements found so far group to the west and east, respectively, could be used as an argument against this latter interpretation. Further studies are needed to resolve this second-order problem.

The azimuthal rotations for the northern terranes (Fig. 2) show a distinct counterclockwise bias. The only exception is the small clockwise rotation of 5° indicated by the Nowitna Volcanics (no. 3, Tables 1 and 2). This determination has the largest error of all the lava flow studies because the number of flows is few and their structural attitudes were difficult to measure accurately. The other six results are rotated counterclockwise in azimuth by amounts ranging from 12 to 77°. Because the inclinations are so steep, relatively small errors in the paleomagnetic directions and North American reference poles can cause much larger errors in azimuthal rotations. Hence only three of the rotations are larger than the formal 95% confidence limits, which are themselves minimum estimates. These three rotations, however, do not exceed the confidence limits by a large margin (i.e., a factor of two). Systematic errors such as initial dip also would have large effects on the estimates of rotation. If such errors in structural correction were optimally oriented, dip errors from 10 to 20° would be required to account entirely for the three largest and most significantly counterclockwise rotations (nos. 5, 6, and 7, Table 2). Even greater errors would be necessary for nonoptimal orientations.

Even though the validity of any one rotation may be easily doubted, the probability of obtaining so many sizeable counterclockwise rotations over such a large region by coincidence alone is small. Furthermore, it is unlikely that the individual rotations are the entirely fortuitous result of independent rotations of small blocks, especially since the dextral slip that characterizes the major fault systems in Alaska would be expected to favor rotations of the opposite sense from that which is observed. For these reasons, we take the average of the results of various studies to get an estimate for possible regional rotation. The mean and standard error for the four studies of lava flows with ages between 60 and 68 Ma (nos. 2, 3, 6, and 7, Table 2) is −30 ± 13°. Each study was weighted equally, despite the obvious differences in quality, in order to take maximum account of possible regional variations in rotation. This may have led to an underestimation of the average rotation, however, because the relatively uncertain Nowitna Volcanics result (no. 3) with its 5° clockwise rotation is given too much influence. Taking only lava flow results that are each formally significant at greater than 95% confidence (nos. 6 and 7), the mean rotation is −48 ± 7°. If all seven results, with ages from 44 to 70 Ma, are averaged indiscriminately, a value of −35 ± 10° is obtained.

Thus, it appears probable that a large portion of central and southwestern Alaska has undergone counterclockwise rotation of about 30 to 50° during latest Cretaceous and Early Tertiary time. Various hypotheses involving rotation of large portions of Alaska have been suggested previously by several authors. Carey (1955) applied his orocline hypothesis to Alaska, postulating a 28° counterclockwise rotation of most of Alaska about a pivot point north of the Gulf of Alaska in the Alaska Range at an unspecified time after the Triassic period. Carey called upon this rotation to open the Arctic Basin "sphenochasm" and bend the mountain ranges of Alaska into their present configuration. Subsequent rotation models can be divided into two categories. (1) Northern Alaska undergoes Late Jurassic to mid-Cretaceous counterclockwise rotation of approximately 70° about a more northerly pivot point, thereby opening the Canada Basin and causing the Brooks Range thrusting (Rickwood, 1970; Tailleur, 1973; Sweeney et al, 1978; Mayfield et al, 1983). Early paleomagnetic results (Newman et al, 1979) appeared to support this model, but subsequent analysis showed these data to be inapplicable because the rocks were remagnetized (Hillhouse and Grommé, 1983b). A pervasive Cretaceous thermal event is thought to be the cause and appears thus far to have thwarted further paleomagnetic efforts in northern Alaska to test the hypothesis of counterclockwise rotation. (2) Central and southern Alaska rotate 40 to 55° counterclockwise during latest Cretaceous to Early Tertiary time, not as one block but as several narrower blocks separated by faults, very much like a huge chevron fold or kink band (Grantz, 1966). Simultaneously, northwestern Alaska and eastern Siberia buckle and shorten under east-west compression, and the part of Alaska in compression is largely decoupled by the Kaltag fault from the part lying to the south (Patton and Tailleur, 1977). The two categories of models are not mutually exclusive, because the rotations occur at different times and involve different parts of Alaska. In both, Eurasia–North America convergence is regarded as the motive force.

The paleomagnetic data under discussion here are clearly pertinent to the second type of model in which Late Cretaceous to Early Tertiary rotation is expected. The paleomagnetically inferred rotations are consistent in both sense and magnitude with those that Grantz (1966) predicted by restoring right-laterally offset Late Cretaceous features on the Kaltag, Iditarod, and Farewell faults (Fig. 2). Grantz also pointed out that these same rotations very nearly straighten out the great bends in the Denali–Farewell and Tintina–Kaltag fault systems. Recent plate-motion models based on marine magnetic anomalies in the North Atlantic (Engebretson, 1982) indicate that the period of convergence between North America and Eurasia was between about 95 and 53 Ma. The amount of convergence implied by Engebretson's (1982) stage poles and angular rotations is sufficient to have rotated Alaska south of the Kaltag fault by a maximum of 68° (less than that if Siberia underwent complementary clockwise rotation during collision). Using (1) the finite difference poles originally found by Srivastava (1978) from his analysis of magnetic anomalies associated with the opening of the North Atlantic, and (2) the anomaly time scale of Harland et al (1982), we calculate that 63% of the total convergence occurred between 67 and 53 Ma. This corresponds to counterclockwise rotation of 43°. The paleomagnetic results in Table 2 that are the most reliable and statistically significant (nos. 6 and 7; both from lava flows, and with mean azimuthal rotation approximately twice the 95% confidence limit) indicate a mean counterclockwise rotation of 48° since 67 Ma, in good agreement with the predicted value.

The counterclockwise rotation of the 52 Ma lava flows in the Talkeetna Mountains (no. 1) postdates this period of convergence between Eurasia and North America. This rotation ($-38 \pm 46°$), although not significant at 95% confidence, may still be real; if so, it clearly requires a different explanation if the analysis and dating of the marine magnetic anomalies are accurate. Assuming locally rigid plates, Stout and Chase (1980) have made a detailed analysis of the sharply curved segment of the Denali fault system 70 km (43 mi) northwest of the Talkeetna locality. They concluded that displacement around the bend has occurred in Early Tertiary time and that this would entail significant counterclockwise rotation. To explain the rotations observed in Bristol Bay by this mechanism, however, would require that region to also have been displaced around the bend. This would necessitate about 900 km (559 mi) of dextral displacement on the Denali fault since 68 Ma, which is more than twice the maximum estimate that has been proposed (Lanphere, 1978).

Southern Terranes

The $25 \pm 7°$ mean northward displacement of the Prince William terrane inferred from the lava flows of the lower Paleocene Ghost Rocks Formation is the most comprehensive and probably the most reliable paleomagnetic result from the southern terranes (nos. 20 and 21). Nonetheless, it is not unimpeachable, especially in isolation. Fortunately, it is reinforced by the $24 \pm 12°$ displacement determined for slightly older pillow lava and dikes (no. 19) of the Chugach terrane. Although the contact between the Chugach and Prince William terranes is a fault with evidence for right-lateral slip (J. Sample, 1984, personal communication), there is some reason to believe that the offset has not been great in paleomagnetic terms (Moore et al, 1983; Davies and Moore, 1984). As discussed above, the paleomagnetic determinations from sediments in these terranes are equivocal. Thus, there is a critical need for additional careful studies of paleomagnetically suitable material to test the reliability of these two results.

There are no results of comparable age from the Peninsular terrane or Wrangellia that we think are reliable. Insignificant, or even slightly southward displacement, however, is inferred from lava flows of early Eocene age (no. 1, Table 2) that lap from the Jura–Cretaceous flysch basin onto the northernmost part of Wrangellia. In addition, we have nearly completed studies of more than 50 lava flows with dates between 30 and 40 Ma at seven localities on the Peninsular terrane that are spread between the Lake Clark and Bruin Bay faults in the Lake Iliamna area (Thrupp and Coe, 1984). Taken together, these have a 95% confidence limit of around 5° and indicate no significant latitudinal movement since eruption. These two results are not compatible with the hypothesis of northward displacement at an average rate of 6 cm/year (2.4 in./year) throughout the Tertiary (Stone et al, 1982). The possibility remains, however, that more outboard portions of the Peninsular terrane could have somewhat greater northward displacements and thus have continued significant motion until later.

Eight of nine results in Table 2 and Figure 2 show counterclockwise azimuthal rotation of Prince William–Chugach terrane rocks. The one that does not (no. 20) appears to have been affected by structural complications not typical of most of the Ghost Rocks Formation. In addition, we made the case earlier that nos. 22 and 26 may well have been overprinted, and so we shall not consider them further. This still leaves six studies rotated counterclockwise at three widely separated places: the Shumagin Islands, Kodiak Island, and the Gulf of Alaska. As discussed earlier, the declinations from sediments might be reliable even when the inclinations are not. The mean counterclockwise rotation since about 70 Ma for the three places is $53 \pm 23°$, comparable with the paleomagnetically inferred counterclockwise rotation of the inboard portion of Alaska previously discussed. Rotation of the displaced southern Alaska terranes may have been caused by a combination of oroclinal bending and by jamming around a preexisting bend in the margin. Moreover, there appears to be a large decrease in amount of rotation from east to west in the Prince William–Chugach terranes. If this is a real trend and not just a coincidence, it would suggest the hypothesis of substantial indentation and shortening concentrated in the Gulf of Alaska area that has recently been proposed by Moore et al (1983). Such a process (Perez and Jacob, 1980; Plafker, 1983; von Huene et al, this volume) is apparently occurring today with the ongoing collision of the Yakutat block (YA on Fig. 2).

The azimuthal rotations indicated by the studies on the Peninsular terrane are inconsistent in both sense and magnitude. Although there might be a sensible pattern in time and space that will become apparent with the addition of more data, our impression at this time is that the variation is quite capricious. This may reflect the competing effects of different mechanisms of rotation: regional counterclockwise rotation caused by the same mechanisms as discussed above for the Prince William and Chugach terranes with superimposed local clockwise rotations of blocks in response to dextral displacement along strike-slip faults.

PROPOSED SCENARIO

The ultimate goal is to integrate the paleomagnetic, plate tectonic, and geologic data into a single coherent story. This would be a formidable task even without the large gaps and inconsistencies that are present in the data. What we will attempt to do instead is outline some of the more likely possibilities for the kinematic history, discuss some of the issues, and point out various problems.

The terranes north of the Peninsular terrane appear to have been, from the standpoint of the broad-brush characterization of paleomagnetism, essentially in place by latest Cretaceous time. The same paleomagnetic data suggest, however, that much of western and central Alaska rotated counterclockwise approximately 40° since latest Cretaceous, probably in a manner first described by Grantz (1966). A major cause of such an event could have been North America–Eurasia convergence between 67 and 53 Ma, the direction and magnitude of which appears able to account for much of the rotation.

For a variety of reasons mentioned previously, accretion of the Wrangellia–Peninsular composite terrane to North America probably occurred no later than the Late Cretaceous. Analysis of plate motions suggests it moved northeastward on the Kula plate until it collided with North America (Engebretson, 1982). The question is, Where did this initial collision occur? Although there are no paleomagnetic displacement data that we believe are quantitatively dependable, the Late Cretaceous and early Paleocene studies certainly convey the impression that these terranes lay to the south relative to cratonic North America (Fig. 2), as does the occurrence of pieces of Wrangellia along the North American margin. Moreover, paleomagnetic data from Cretaceous intrusive rocks in the Stikinia terrane and the Coast Plutonic Complex, both in the Canadian Cordillera, suggest 10 to 20° of northward displacement of inboard terranes (Irving et al, 1980, 1983), probably along the Tintina fault system, since mid- to Late Cretaceous time. According to Irving (1984, personal communication), the upper limit of 20° northward displacement of the Coast Plutonic Complex is presently favored, although that figure is subject to the uncertain error of ambiguous tilt inherent in intrusive rocks. Geological evidence has been cited as well for 1,400 km (870 mi) of cumulative dextral slip: 1,000 km (621 mi) along the Tintina fault system sometime after mid-Cretaceous and before late Eocene time (Gabrielse, in press), and 400 km (249 mi) of displacement, probably post-Early Tertiary, on the Denali fault (Lanphere, 1978; Nokleberg et al, in press). Since geological determinations of slip on faults tend to be minimum estimates of terrane displacements, the paleomagnetic estimate of 20° northward displacement is not unreasonable.

This would place the collision of the Alaskan part of Wrangellia at a present-day latitude of 40° relative to North America (just north of Cape Mendocino). Displacement of Wrangellia, the Peninsular terrane, and at least part of the Jura–Cretaceous flysch belt northwestward along the continental margin was probably driven by the oblique subduction of the Kula and perhaps the Izanagi plates (Engebretson, 1982). The important paleomagnetic result of Hillhouse and Grommé (1983a) from the Talkeetna Mountains places the arrival of at least the northern part of the Wrangellia–Peninsular composite terrane to interior Alaska earlier than about 52 Ma.

A problem arises with this and any other scenario that involves large terrane displacements on the Tintina or Denali faults. How was this displacement accommodated in interior Alaska? Only about 130 km (81 mi) of right-lateral offset is inferred for the Kaltag fault from Late Cretaceous to present (Patton and Tailleur, 1977), so that large Tintina offsets during that time would have had to be accommodated elsewhere. Perhaps the dextral motion used to continue northwestward along other faults such as the Kobuk on the south side of the Brooks Range, a possibility mentioned by Grantz in 1966. Yet another possibility is that Tintina motion was redistributed on a series of faults, both known and unknown.

Relative to North America, the mean paleomagnetic inclination found for the Ghost Rocks volcanics would put the Prince William terrane of Kodiak Island $25 \pm 7°$ further south in early Paleocene time. This result is reasonably consistent with the 20° of post-mid-Cretaceous displacement found for the Coast Plutonic Complex. Since the Coast Plutonic Complex is linked to the Vancouver Island segment of Wrangellia by Late Cretaceous time (Monger et al, 1982) and is separated from the Alaskan segment of Wrangellia by the Denali fault system, its displacement is presumably somewhat less than that of the Wrangellia and Peninsular terranes as well. Moore et al (1983) discussed the geological constraints for Kodiak Island and concluded that most likely Kodiak collided with the margin of proto-Alaska in the Eocene. One reason is that the expected uplift would account nicely for the formation of the Zodiak fan just to the south of Kodiak in Eocene–Oligocene time. Paleontological evidence from Middleton Island also suggests that Kodiak Island was at least as far north as 50 ± 5° paleolatitude by 42 Ma (Keller et al, 1984; von Huene et al, this volume). This is consistent with existing plate models. Assuming translation beginning at 62 Ma along a simplified, restored North American continental margin at a rate equal to the component of Kula plate velocity resolved parallel to that margin, a paleolatitude of 55° is reached by 42 Ma (Engebretson, 1982, and personal communication). There would still remain 30% of the total latitudinal displacement relative to North America to be accommodated. This residual motion would have to be driven by the slower moving Pacific plate, with attendant shortening of proto-Alaska, and continue until about 10 Ma to emplace Kodiak in its final position. However, the uncertainty of the paleomagnetically inferred displacement permits considerable range in the details of such kinematic modeling.

In general terms, the scenario described above appears quite consistent and reasonable for the most outboard, Prince William terrane. Problems emerge if one applies ties suggested by the presence of distinctive 60 Ma intrusive rocks in the Prince William, Chugach, and Peninsular terranes (Moore et al, 1983; Davies and Moore, 1984). These plutons occur in a long belt within the Prince William and Chugach terranes, but only one location is known thus far in the Peninsular terrane, just northwest of the Border Ranges fault on Kodiak Island (Davies and Moore, 1984). If these intrusive ties were tight, the Peninsular terrane and presumably Wrangellia would have had to move with the Prince William terrane and consequently would not have arrived at proto-Alaska by 52 Ma. In fact, they would only have completed about two-thirds of their total inferred displacement by then. However, arrival of the Wrangellia–Peninsular composite terrane by this time is required by the concordant paleolatitude determined from 52 Ma lavas of the Talkeetna Mountains, which overlie parts of Wrangellia and the flysch basin to the north.

It may be more realistic to regard these intrusive ties as loose, that is, to allow for the possibility of hundreds of kilometers of interterrane displacement while still maintaining juxtaposition of distinctive suites of igneous rocks. For example, it is possible to accommodate up to 500 km (311 mi) of slip on the contact fault between the Prince William and Chugach terranes and about 600 km (373 mi) of slip on the Border Ranges fault between the Chugach and Peninsular terranes, without completely offsetting the Gulf of Alaska magmatic belt. In addition, there are numerous intraterrane faults that could quite conceivably have accommodated significant transcurrent motion not attributable to interterrane displacements. Some possible examples are the Uganik thrust within the Chugach terrane and the Bruin Bay and Castle Mountain–Lake Clark faults within the Peninsular terrane (Fig. 2).

Further evidence in support of the notion of substantial dextral slivering within the southern terranes is provided by similarities in the geologic histories of the Leech River Complex of southern Vancouver Island and schists on Baranof Island, which were recognized by Cowan (1982). The interpretation that he favors involves a minimum of 1,000 km (621 mi) of dextral slip between the Chugach terrane and the Peninsular–Wrangellia composite terrane and requires that this major displacement occurred entirely *after* 40 Ma.

In view of this interpretation and other arguments presented above, there is scope for considerable relative displacement between the Peninsular–Wrangellia terranes on the one hand and the Prince William and Chugach terranes on the other. Thus, both the geologic evidence and the paleomagnetic constraints can be incorporated into a reasonably consistent kinematic history involving coastwise displacement along a series of strike-slip faults. In this scenario Wrangellia and the Peninsular terrane move approximately 20° northward relative to cratonic North America between mid-Cretaceous time and 52 Ma, while the Prince William and Chugach terranes travel about 25° northward throughout Paleogene and perhaps even during Neogene time.

ACKNOWLEDGMENTS

Financial support for this research was provided by National Science Foundation grants EAR-8008213 and EAR-8206918 (to Coe), and EAR-8115528 (to Coe and J. C. Moore); the U.S. Geological Survey (to Plumley and Coe); Sigma Xi Grant-in-Aid of Research and The Geological Society of America Harold T. Stearns Fellowship (to Globerman); Amoco Production Co., Arco Oil and Gas, Exxon USA, Mobil Exploration and Producing Services, Inc., Phillips Petroleum Co., Sohio Petroleum Co., and Union Oil Co. (to S. Box and J. C. Moore); and Arco (to Thrupp and Coe). We gratefully acknowledge the Alaska Maritime National Wildlife Refuge, the Bristol Bay Native Corporation, and Togiak Natives, Ltd., for permitting us to conduct field studies in the north Bristol Bay area; W. K. Wallace of Arco Alaska, Inc., and W. W. Patton, Jr., J. M. Hoare, and W. L. Coonrad of the U.S. Geological Survey for invaluable helicopter and logistical support and geological guidance; S. E. Box, D. G. Howell, and E. Irving for insightful reviews of the manuscript; and J. W. Hillhouse and D. B. Stone for unpublished details regarding their paleomagnetic results.

APPENDIX A

Analyses of geomagnetic and paleomagnetic field data (Cox, 1970; Harrison, 1980) have shown repeatedly that the distribution of directions recorded at a site over a period of time is significantly less Fisherian than that of the corresponding virtual geomagnetic poles (VGPs—Cox and

Doell, 1960). Thus, when using Fisher (1953) statistics to analyze paleomagnetic data, it is preferable to work with poles rather than directions. Sample directions are averaged to find the mean direction for a site, which is then transformed to the corresponding VGP using a standard algorithm (e.g., McElhinny, 1973, p. 25). Next, the N site-mean VPGs are averaged to yield the mean paleomagnetic pole and the associated 95% confidence limit A_{95} and precision parameter K are found using Fisher's formulas (e.g., McElhinny, 1973, p. 79).

Given the latitude and longitude of the sampling area (λ_s, ϕ_s) and of the experimentally determined paleomagnetic pole (λ_p, ϕ_p), the observed paleolatitude λ_0 can be found using

$$\mathrm{PLAT} = \lambda_0 = \sin^{-1}[\sin\lambda_p \sin\lambda_s + \cos\lambda_p \cos\lambda_s \cos(\phi_p - \phi_s)] \quad \text{(A1)}$$

The uncertainty in the observed paleolatitude (95% confidence limit) is

$$\lambda_0 = C\,A_{95} \quad \text{(A2)}$$

where C can be calculated by methods given by Demarest (1983). We have used $C = 0.78$ consistently, though the actual value deviates considerably when inclinations are steep and A_{95} is large.

Often, however, only the mean paleomagnetic direction and associated confidence limit α_{95} and precision parameter k are presented in the literature. Unless the individual site-mean directions are available, the analysis cannot be redone in terms of poles because the individual VGPs are not known. However, a reasonable estimate can be computed for the precision parameter of the distribution of poles from the precision parameter given for the distribution of directions and the paleolatitude λ_0 (Cox, 1970):

$$K = k(5 + 3\sin^2\lambda_0)/2(1 + 3\sin^2\lambda_0)^2 \quad \text{(A3)}$$

where paleolatitude is now found from the inclination I using the dipole formula

$$\lambda_0 = \tan^{-1}(0.5\tan I) \quad \text{(A4)}$$

The confidence limit A_{95} for the distribution of poles is then obtained from the relations of Fisher (1953) statistics:

$$A_{95} = \cos^{-1}[1-(N-R)(20^{1/(N-1)}-1)/R] \quad \text{(A5)}$$

$$R = N - (N-1)/K \quad \text{(A6)}$$

where N is the number of samples.

To find the latitudinal displacement with respect to the North American craton, the expected paleolatitude λ_e must be found by substituting the latitude and longitude of the appropriate North American reference pole (Table 4) for λ_p and ϕ_p in (A1). Then the latitudinal displacement is

$$\mathrm{DISPL} = d = \lambda_e - \lambda_0 \quad \text{(A7)}$$

and the 95% confidence limit on the latitudinal displacement follows from (A2):

$$\Delta d = (\Delta\lambda_0^2 + \Delta\lambda_e^2)^{1/2} \quad \text{(A8)}$$

To find the azimuthal rotation with respect to cratonic North America, the angle between the great circle passing through the site and the mean observed pole and the great circle passing through the site and the mean expected pole must be found. The azimuth of the former great circle is just the mean observed declination D_0 corresponding to the mean observed pole. It can be calculated by reversing the standard algorithm relating paleomagnetic direction at a site to its corresponding pole (e.g., McElhinny, 1973, p. 25). Likewise, the azimuth of the latter great circle is just the expected declination D_e corresponding to the reference pole, which can be calculated from the same algorithm.

The azimuthal rotation is therefore

$$\mathrm{ROT} = r = D_0 - D_e \quad \text{(A9)}$$

The 95% confidence limit on ROT is

$$\Delta r = (\Delta D_0^2 + \Delta D_e^2)^{1/2} \quad \text{(A10)}$$

where the individual confidence limit on D_0 or D_e is given by

$$\Delta D = C\sin^{-1}(\sin A_{95}/\cos\lambda) \quad \text{(A11)}$$

REFERENCES

Alvarez, W., and W. Lowrie, 1984, Magnetic stratigraphy applied to synsedimentary slumps, turbidites, and basin analysis: The Scaglia Limestone at Furlo (Italy): Geological Society of America Bulletin, v. 95, p. 324–336.

Beck, Jr., M. E., 1976, Discordant paleomagnetic pole positions as evidence of regional shear in the western Cordillera of North America: American Journal of Science, v. 276, p. 694–712.

______, 1980, Paleomagnetic record of plate-margin tectonic processes along the western edge of North America: Journal of Geophysical Research, v. 85, p. 7115–7131.

Blow, R. A., and N. Hamilton, 1975, Paleomagnetic evidence from DSDP cores of northward drift of India: Nature, v. 257, p. 570–572.

______ and ______, 1978, Effect of compaction on the acquisition of a detrital remanent magnetization in fine-grained sediments: Geophysical Journal of the Royal Astronomical Society, v. 52, p. 13–23.

Byrne, T., 1982, Structural geology of coherent terranes in the Ghost Rocks Formation, Kodiak Islands, Alaska, *in* J. K. Leggett, ed., Trench and forearc sedimentation and tectonics: Special Publications of the Geological Society of London, v. 10, p. 229–242.

Carey, S. W., 1955, The orocline hypothesis in geotectonics: Proceedings of the Royal Society of Tasmania, v. 89, p. 255–288.

Table 4

Ref	Age	Data Span	Long (°E)	Lat (°N)	A_{95}	N	Δ
A	10 Ma	0-20 Ma	101	88	3	18*	—
A	40 Ma	25-55 Ma	168	84	4	11*	1.3
A+B	45 Ma	(interpolated) 40-50 Ma	169.3	83.4	4	—	—
B	50 Ma	44-54 Ma	170.4	82.8	3.0	147	2.6
B	55 Ma	(interpolated) 50-60 Ma	182.4	82.3	3.1	—	1.1
B	60 Ma	55-67 Ma	192.6	81.5	3.2	68	0.9
A+B	65 Ma	(interpolated) 60-70 Ma	190.9	77.8	7	—	—
A	70 Ma	55-85 Ma	190	74	7	9*	—
A	75 Ma	(interpolated) 70-80 Ma	190.6	71.0	7	—	—
A	80 Ma	65-95 Ma	191	68	7	4*	—
A	85 Ma	interpolated 80-90 Ma	188.4	67.5	7	—	—
A	90 Ma	75-105 Ma	186	67	4	5*	—

Ref: A, Irving and Irving, 1982; B, Diehl et al, 1983.
N: Number of paleomagnetic sites or (*) studies (each with many sites).
Δ:Angular discrepancy between poles from references A and B.

Table 4—Paleomagnetic reference poles used for North American Craton.

Champion, D. E., 1980, Holocene geomagnetic secular variation in the western United States: implications for the global geomagnetic field: U.S. Geological Survey Open-File Report 80-824, 326 p.

Coney, P. J., et al, 1980, Cordilleran suspect terranes: Nature, v. 288, p. 329-333.

Cowan, D. S., 1982, Geological evidence for post-40 m.y. B.P. large-scale northwestward displacement of part of southeastern Alaska: Geology, v. 10, p. 70-75.

Cox, A., 1970, Latitude dependence of the angular dispersion of the geomagnetic field: Geophysical Journal of the Royal Astronomical Society, v. 20, p. 253-269.

———, and R. R. Doell, 1960, Review of paleomagnetism: Geological Society of America Bulletin, v. 71, p. 647-768.

———, and R. G. Gordon, 1984, Paleolatitudes determined from paleomagnetic data from vertical cores: Review of Geophysics and Space Physics, v. 22, p. 47-72.

Creer, K. M., 1974, Geomagnetic variations for the interval 7,000-25,000 yr BP as recorded in a core of sediment from station 1474 of the Black Sea Cruise of 'Atlantis II': Earth and Planetary Science Letters, v. 23, p. 34-42.

Csejtey, Jr., B., et al, 1982, The Cenozoic Denali fault system and the Cretaceous accretionary development of southern Alaska: Journal of Geophysical Research, v. 87, p. 3741-3754.

Davies, D. L., and J. C. Moore, 1984, 60 M.Y. intrusive rocks from the Kodiak Islands link the Peninsular, Chugach, and Prince William terranes (Abs.): Geological Society of America Abstracts with Programs, v. 16, p. 277.

Demarest, H. H., Jr., 1983, Error analysis for the determination of tectonic rotation from paleomagnetic data: Journal of Geophysical Research, v. 88, p. 4321-4328.

Diehl, J. F., et al, 1983, Paleomagnetism of the Late Cretaceous-Early Tertiary Central Montana alkalic province: Journal of Geophysical Research, v. 88, p. 10593-10609.

Doell, R. R., and G. B. Dalrymple, 1973, Potassium-argon ages and paleomagnetism of the Waianae and Koolau volcanic series, Oahu, Hawaii: Geological Society of America Bulletin, v. 84, p. 1217-1242.

Engebretson, D. C., 1982, Relative motions between oceanic and continental plates in the Pacific Basin: PhD Dissertation, Stanford University, 211 p.

Fisher, R. A., 1953, Dispersion on a sphere: Proceedings of the Royal Society of London, Series A, v. 217, p. 295-305.

Gabrielse, H., in press, Major dextral transcurrent displacements along the northern Rocky Mountain trench and related lineaments in north-central British Columbia: Geological Society of America Bulletin.

Globerman, B. R., and R. S. Coe, 1983, Paleomagnetic results from Upper Cretaceous volcanic rocks in northern Bristol Bay, SW Alaska, and tectonic implications, *in* D. G. Howell, et al, eds., Proceedings of the Circum-Pacific Terrane Conference: Stanford University Publications, Geological Sciences, v. 18, p. 98–102.

______, et al, 1983, Paleomagnetism of Lower Cretaceous tuffs from Yukon-Kuskokwim delta region, western Alaska: Nature, v. 305, p. 516–520.

______, et al, 1984, Petrochemical data from Upper Cretaceous volcanic rocks—Hagemeister, Crooked, and Summit Islands, SW Alaska (Abs.): Geological Society of America Abstracts with Programs, v. 16, p. 286.

Graham, J. W., 1949, The stability and significance of magnetism in sedimentary rocks: Journal of Geophysical Research, v. 54, p. 131–167.

Granar, L., 1958, Magnetic measurements on Swedish varved sediments: Arkiv för Geofysik, v. 3, p. 1–40.

Grantz, A., 1966, Strike-slip faults in Alaska: U.S. Geological Survey Open-File Report 267, 82 p.

Griffiths, D. H., 1955, Remanent magnetism of varved clays from Sweden: Monthly Notice of the Royal Astronomical Society, Geophysical Supplement, v. 7, p. 103–114.

______, et al, 1960, Remanent magnetism of some recent varved sediments: Proceedings of the Royal Society of London, Series A, v. 256, p. 359–383.

Grommé, C. S., and J. W. Hillhouse, 1981, Paleomagnetic evidence for northward movement of the Chugach terrane, southern and southeastern Alaska: U.S. Geological Survey Circular 823-B, p. 161–164.

______, et al, 1969, Magnetic properties and oxidation of iron-titanium oxide materials in Alae and Makaopuhi lava lakes, Hawaii: Journal of Geophysical Research, v. 74, p. 5277–5293.

Halls, H. C., 1978, The use of converging remagnetization circles in paleomagnetism: Physics of the Earth and Planetary Interiors, v. 16, p. 1–11.

Harland, W. B., et al, 1982, A geologic time scale: Cambridge, Cambridge University Press, 128 p.

Harrison, C. G. A., 1980, Secular variation and excursions of the earth's magnetic field: Journal of Geophysical Research, v. 85, p. 3511–3522.

Hill, M. D., et al, 1981, Hybrid granodiorites intruding the accretionary prism, Kodiak, Shumagin, and Sanak Islands, southwest Alaska: Journal of Geophysical Research, v. 86, p. 10569–10590.

Hillhouse, J. W., 1977, Paleomagnetism of the Triassic Nikolai Greenstone, McCarthy quadrangle, Alaska: Canadian Journal of Earth Sciences, v. 14, p. 2578–2592.

______, and C. S. Grommé, 1977, Paleomagnetic poles from sheeted dikes and pillow basalt of the Valdez(?) and Orca Groups, southern Alaska (Abs.): EOS, American Geophysical Union Transactions, v. 58, p. 1127.

______ and ______, 1982, Limits to northward drift of the Paleocene Cantwell Formation, central Alaska: Geology, v. 10, p. 552–556.

______ and ______, 1983a, Wrangellia in southern Alaska 50 my ago (Abs.): EOS American Geophysical Union Transactions, v. 64, p. 687.

______ and ______, 1983b, Paleomagnetic studies and the hypothetical rotation of Arctic Alaska: Journal of the Alaskan Geological Society, v. 2, p. 27–39.

______, et al, 1983, Paleomagnetism of early Tertiary volcanic rocks in the Northern Talkeetna Mountains, *in* D. G. Howell, et al, eds., Proceedings of the Circum-Pacific Terrane Conference: Stanford University Publications, Geological Sciences, v. 18, p. 111–114.

Hoare, J. M., 1961, Geology and tectonic setting of the Lower Kuskokwim-Bristol Bay region, Alaska: Bulletin of the American Association of Petroleum Geologists, v. 45, p. 594–611.

______, and W. L. Coonrad, 1959, Geology of the Russian Mission quadrangle, Alaska: U.S. Geological Survey Miscellaneous Geological Investigations Map I-292.

______ and ______, 1978, Geologic map of the Goodnews and Hagemeister Island quadrangles region, southwestern Alaska: U.S. Geological Survey Open-File Report 78-9-B.

Hoffman, K. A., and R. Day, 1978, Separation of multi-component NRM: a general method: Earth and Planetary Science Letters, v. 40, p. 433–438.

Howell, D. G., et al, 1983, Preliminary tectonostratigraphic terrane map of the Circum-Pacific region, *in* D. G. Howell, et al, eds., Proceedings of the Circum-Pacific Terrane Conference: Stanford University Publications, Geological Sciences, v. 18, p. 227–242.

Hudson, T., 1979, Mesozoic plutonic belts of southern Alaska: Geology, v. 7, p. 230–234.

Irving, E., and A. Major, 1964, Post-depositional detrital remanent magnetization in a synthetic sediment: Sedimentology, v. 3, p. 135–143.

______, et al, 1980, New paleomagnetic evidence for displaced terranes in British Columbia, *in* D. W. Strangway, ed., The continental crust and its mineral deposits: Geological Association of Canada Special Paper 20, p. 441–456.

______, et al, 1983, Paleomagnetic results from the southern Coast Plutonic Complex indicate northward transport and clockwise rotation since mid-Cretaceous (Abs.): Geological Association of Canada Program with Abstracts, v. 8, p. A35.

Ising, G., 1942, On the magnetic properties of varved clay: Arkiv för Matematik, Astronomi och Fysik, v. 29, p. 1–37.

Johnson, E. A., et al, 1948, Pre-history of the earth's magnetic field: Terrestrial Magnetism and Atmospheric Electricity, v. 53, p. 349–372.

Jones, D. L, and N. J. Silberling, 1979, Mesozoic stratigraphy—the key to tectonic analysis of southern and central Alaska: U.S. Geological Survey Open-File Report 79-1200, 40 p.

______, et al, 1972, Southeastern Alaska—A displaced continental fragment?: U.S. Geological Survey Professional Paper 800-B, p. B211–B217.

———, et al, 1977, Wrangellia—A displaced terrane in northwestern North America: Canadian Journal of Earth Sciences, v. 14, p. 2565–2577.

———, et al, 1981, Map showing tectonostratigraphic terranes of Alaska, columnar sections, and summary description of terranes: U.S. Geological Survey Open-File Report 81-792, 20 p.

———, et al, 1982, Character, distribution, and tectonic significance of accretionary terranes in the central Alaska Range: Journal of Geophysical Research, v. 87, p. 3709–3717.

———, et al, 1983, Recognition, character, and analysis of tectonostratigraphic terranes in western North America, *in* M. Hashimoto and S. Uyeda, eds., Accretion tectonics in the circum-Pacific regions: Tokyo, Terra Scientific Publishing Company, p. 21–35.

———, et al, 1984, Part A: Lithotectonic terrane map of Alaska west of the 141st meridian, *in* N. J. Silberling and D. L. Jones, eds., Lithotectonic terrane maps of the North American cordillera: U.S. Geological Survey Open-File Report 84-523, 12 p.

Keller, G., 1984, Paleoclimatic evidence for Cenozoic migration of Alaskan terranes: Tectonics, v. 3, p. 473–495.

King, R. F., 1955, Remanent magnetism of artificially deposited sediments: Monthly Notices of the Royal Astronomical Society, Geophysical Supplement, v. 7, p. 115–134.

Kirschvink, J. L., 1980, The least-squares line and plane and the analysis of paleomagnetic data: Geophysical Journal of the Royal Astronomical Society, v. 62, p. 699–718.

———, 1983, Biomagnetic geomagnetism: Review of Geophysics and Space Physics, v. 21, p. 672–675.

Lanphere, M. A., 1978, Displacement history of the Denali fault system, Alaska and Canada: Canadian Journal of Earth Sciences, v. 15, 817–822.

Liddicoat, J. C., and R. S. Coe, 1979, Mono Lake geomagnetic excursion: Journal of Geophysical Research, v. 84, p. 261–271.

McElhinny, M. W., 1964, Statistical significance of the fold test in paleomagnetism: Geophysical Journal of the Royal Astronomical Society, v. 8, p. 338–340.

———, 1973, Paleomagnetism and plate tectonics: Cambridge, Cambridge University Press, 358 p.

———, and R. T. Merrill, 1975, Geomagnetic secular variation over the past 5 m.y.: Reviews of Geophysics and Space Physics, v. 13, p. 687–708.

McFadden, P. L., and D. L. Jones, 1981, The fold test in paleomagnetism: Geophysical Journal of the Royal Astronomical Society, v. 67, p. 53–58.

———, and A. B. Reid, 1982, Analysis of paleomagnetic inclination data: Geophysical Journal of the Royal Astronomical Society, v. 69, p. 307–319.

Mayfield, C. F., et al, 1983, Stratigraphy, structure, and palinspastic synthesis of the western Brooks Range, northwestern Alaska: U.S. Geological Survey Open-File Report 83-779, 58 p.

Moll, E. J., and W. W. Patton, Jr., 1982, Preliminary report on the Late Cretaceous and Early Tertiary volcanic and related plutonic rocks in western Alaska: U.S. Geological Survey Circular 844, p. 73–75.

———, et al, 1981, Chemistry, mineralogy, and K-Ar ages of igneous and metamorphic rocks of the Medfra quadrangle, Alaska: U.S. Geological Survey Open-File Report 80-811C, 19 p.

Monger, J. W. H., et al, 1982, Tectonic accretion and the origin of the two major metamorphic and plutonic belts in the Canadian Cordillera: Geology, v. 10, p. 70–75.

Moore, G. W., 1969, New formations on Kodiak and adjacent islands, Alaska: U.S. Geological Survey Bulletin 1274-A, p. A27–A35.

Moore, J. C., et al, 1983, Paleogene evolution of the Kodiak Islands, Alaska: consequences of ridge-trench interaction in a more southerly latitude: Tectonics, v. 2, p. 265–293.

Nagata, T., 1961, Rock magnetism: Tokyo, Maruzen, 225 p.

Newman, G. W., et al, 1979, Northern Alaska paleomagnetism, plate rotations, and tectonics, *in* A. Sisson, ed., The relationship of plate tectonics to Alaskan geology and resources: Anchorage, Alaska Geological Society, p. 1–6.

Newton, C. R., 1983, Paleozoogeographic affinities of Norian bivalves from the Wrangellian, Peninsular, and Alexander terranes, western North America, *in* C. H. Stevens, ed., Pre-Jurassic rocks in western North American suspect terranes: Society of Economic Paleontologists and Mineralogists, Pacific Section, p. 37–48.

Nokleberg, W. J., et al, in press, Origin, migration, and accretion of the Maclaren and Wrangellia terranes, eastern Alaska Range, Alaska: Geological Society of America Bulletin.

Opdyke, N. D., and K. W. Henry, 1969, A test of the dipole hypothesis: Earth and Planetary Science Letters, v. 6, p. 138–151.

———, et al, 1972, The paleomagnetism of two Aegean deep-sea cores: Earth and Planetary Science Letters, v. 14, p. 145-159.

Packer, D. R., 1972, Paleomagnetism of the Mesozoic in Alaska: PhD Dissertation, University of Alaska, Fairbanks, 160 p.

———, and D. B. Stone, 1972, An Alaskan Jurassic paleomagnetic pole and the Alaskan orocline: Nature, v. 237, p. 25–26.

——— and ———, 1974, Paleomagnetism of Jurassic rocks from southern Alaska, and the tectonic implications: Canadian Journal of Earth Sciences, v. 11, p. 976–997.

Panuska, B. C., 1983, Paleomagnetic data from the Cretaceous MacColl Ridge Formation and tectonic implications for the Wrangellia terrane (Abs.): EOS, American Geophysical Union Transactions, v. 64, p. 688.

———, and D. B. Stone, 1981, Late Paleozoic paleomagnetic data for Wrangellia—prospects for a polar wander path: Nature, v. 293, p. 561–563.

Patton, Jr., W. W., 1973, Reconnaissance geology of the northern Yukon-Koyukuk province, Alaska: U.S. Geological Professional Paper 774-A, 17 p.

______, and I. L. Tailleur, 1977, Evidence in the Bering Strait region for differential movement between North America and Eurasia: Geological Society of America Bulletin, v. 88, p. 1298–1304.

Perez, O. J., and K. H. Jacob, 1980, Tectonic model and seismic potential of the eastern Gulf of Alaska and Yakataga seismic gap: Journal of Geophysical Research, v. 85, p. 7132–7150.

Plafker, G., 1983, The Yakutat Block: an active tectonostratigraphic terrane in southern Alaska (Abs.): Geological Society of America Abstracts with Programs, v. 15, p. 406.

Plumley, P. W., 1984, A paleomagnetic study of the Prince William terrane and Nixon Fork terrane, Alaska: PhD Dissertation, University of California, Santa Cruz, 190 p.

______, et al, 1982, Paleomagnetism of volcanic rocks of the Kodiak Islands indicates northward latitudinal displacement: Nature, v. 300, p. 50–52.

______, et al, 1983, Paleomagnetism of the Paleocene Ghost Rocks Formation, Prince William Terrane, Alaska: Tectonics, v. 2, p. 295–314.

Rickwood, F. K., 1970, the Prudhoe Bay Field, *in* W. L. Adkison and M. M. Brosgé, eds., Proceedings of the geological seminar on the North Slope of Alaska: American Association of Petroleum Geologists, Pacific Section, p. L1–L11.

Silberman, M. L., et al, 1981, Metallogenetic and tectonic significance of whole rock K-Ar ages of the Nikolai Greenstone, McCarthy quadrangle, Alaska: U.S. Geological Survey Open-File Report 81-355, p. 52–73.

Simpson, R. W., and A. Cox, 1977, Paleomagnetic evidence for the tectonic rotation of the Oregon Coast Range: Geology, v. 5, p. 585–589.

Srivastava, S. P., 1978, Evolution of the Labrador Sea and its bearing on the early evolution of North America: Geophysical Journal of the Royal Astronomical Society, v. 52, p. 313–357.

Stone, D. B., 1983, Concordance and conflicts in Alaskan paleomagnetic data, *in* D. G. Howell, et al, eds., Proceedings of the Circum-Pacific Terrane Conference: Stanford University Publications, Geological Sciences, v. 18, p. 186–189.

______, and D. R. Packer, 1977, Tectonic implications of Alaskan Peninsula paleomagnetic data: Tectonophysics, v. 37, p. 183–201.

______ and ______, 1979, Paleomagnetic data from the Alaskan Peninsula: Geological Society of America Bulletin, v. 90, p. 545–560.

______, et al, 1982, Paleolatitudes versus time for southern Alaska: Journal of Geophysical Research, v. 87, p. 3697–3707.

Stout, J. H., and C. G. Chase, 1980, Plate kinematics of the Denali fault system: Canadian Journal of Earth Sciences, v. 17, p. 1527–1537.

Sweeney, J. F., et al, 1978, Evolution of the Arctic Basin, *in* J. F. Sweeney, ed., Arctic geophysical review: Energy, Mines, and Resources Canada, Publications of the Earth Physics Branch, v. 45, p. 91–100.

Tailleur, I. L., 1973, Probable rift origin of Canada Basin, Arctic Ocean, *in* M. G. Pitcher, ed., Arctic geology: American Association of Petroleum Geologists Memoir 19, p. 526–535.

Thompson, R., and K. Kelts, 1974, Holocene sediments and magnetic stratigraphy from Lakes Zug and Zurich, Switzerland: Sedimentology, v. 21, p. 577–596.

Thrupp, G. A., and R. S. Coe, 1983, Preliminary results of a paleomagnetic study of Alaska Peninsula rocks (Abs.): EOS, American Geophysical Union Transactions, v. 64, p. 688.

______, 1984, Paleomagnetism of Tertiary volcanic rocks from the Alaska Peninsula: is the Peninsular terrane a coherent tectonic entity? (Abs.): Geological Society of America Abstracts with Program, v. 16, p. 337.

Tucker, P., 1983a, Magnetization of unconsolidated sediments and theories of DRM, *in* K. M. Creer, et al, eds., Geomagnetism of baked clays and recent sediments: Amsterdam, Elsevier, 324 p.

______, 1983b, Magnetic remanence acquisition in IPOD Leg 73 sediments, *in* Initial Reports of the Deep-Sea Drilling Project: Washington, DC, U.S. Gov't. Printing Office, v. 73, p. 663–672.

Verosub, K. L., 1977, Depositional and postdepositional processes in the magnetization of sediments: Review of Geophysics and Space Physics, v. 15, p. 129–143.

Wallace, W. K., and D. C. Engebretson, 1984, Relationships between plate motions and Late Cretaceous to Paleocene magmatism in southwestern Alaska: Tectonics, v. 3, p. 295–315.

Wilson, R. L., and S. E. Haggerty, 1966, Reversals of the earth's magnetic field: Endeavour, v. 25, p. 104–109.

Zijderveld, J. D. A., 1967, AC demagnetization of rocks: analysis of results, *in* D. W. Collinson, et al, eds., Methods in paleomagnetism: Amsterdam, Elsevier, p. 254–286.

Latitudinal Motion of the Wrangellia and Alexander Terranes and the Southern Alaska Superterrane

B. C. Panuska
Vanderbilt University
Nashville, Tennessee

D. B. Stone
University of Alaska
Fairbanks, Alaska

Paleomagnetic data for the terranes of southern Alaska are helping to define the terrane locations through time. Some of these data have an ambiguity of magnetic polarity that can be problematic in deciding between a northern and a southern hemisphere location. Interpreting these data on the basis of the simplest polar-wander paths and the least number of changes of direction of motion of the terranes leads to a scenario where the Alexander and Wrangellia terranes were spatially separate and at low northern hemisphere paleolatitudes in the Paleozoic. Both terranes moved to the southern hemisphere, the Alexander terrane in the late Paleozoic and Wrangellia in early Mesozoic time. These terranes, along with the Peninsular terrane, amalgamated in the Jurassic and together moved northward to their present locations as the Southern Alaska Superterrane (Alexander, Wrangellia, and Peninsula terranes together with the outboard terranes and the Gravina-Nutzotin terrane). Though paleolongitude is not determinable by our data, paleontologic evidence indicates locations in the eastern Pacific basin during these travels.

INTRODUCTION

Two key questions in any discussion of tectono-stratigraphic terranes are, Are they allochthonous, and if so where did they come from? To attempt to answer these questions at least partially for southern Alaska we have constructed paleolatitude versus time curves. To construct such a curve it is necessary to know the polarity of the magnetic field recorded in the rocks. The paleomagnetic data allow the paleolatitude to be determined, but without knowing the polarity it is not possible to use these data to distinguish between northern and southern hemisphere locations. To deal with the polarity ambiguity inherent in much of the available paleomagnetic data, paleomagnetic, geologic, and paleontologic arguments have all been brought to bear on the problem. Some of the available evidence is circumstantial, and the preferred model of latitudinal motion presented is by no means proven, but it does provide a framework on which to build, speculate, and plan future studies.

To determine a paleolatitude from paleomagnetic data, only the inclination of the ancient magnetic field relative to ancient horizontal needs to be known. The inclination (I) of the field is related to the magnetic latitude (λ) by the relationship:

$$\tan I = 2 \tan \lambda$$

The determination of paleolongitude is more difficult. In general, the relative paleolongitude of two areas can only be determined paleomagnetically when good apparent polar-wander (APW) paths are available for both areas for periods of time during which no relative motion occurred. Without relative motion the two paths should be identical and can thus be overlain to get the relative longitude. The classic example of matching APW curves to obtain both relative latitude and longitude is the matching of South America and Africa. In this case, the polar-wander paths for both continents were identical while they were part of Gondwanaland. With the opening of the South Atlantic, the APW curves began to diverge. None of the Alaska terranes have sufficiently well-defined APW curves to apply this technique at present; however, paleontologic data can provide circumstantial evidence of relative paleolongitude by recognizing the affinities of the various fossils from what are now widely separated geographic areas.

Useful information can also be deduced from the groupings of the virtual geomagnetic poles (VGP) for the various localities sampled, though in this case care has to be taken to demonstrate whether or not apparently discordant VGPs are the result of local block rotations and/or large-scale tectonic deformation (e.g., local block rotations versus regional oroclinal bending). Determining VGPs also gives the local declination of the paleofield, but again interpreting these declinations requires a detailed understanding of the structural history of the area sample, particularly with regard to the possibility of local small-scale rotations.

THE PROBLEM OF PALEOMAGNETIC POLARITY

One of the serious problems that has hampered the construction of satisfactory paleomagnetic paleolatitude versus time trajectories for allochthonous terranes in general is the problem of determining the original polarity of the rocks being measured. For young rocks at high latitudes, assigning an original polarity is relatively simple, since paleolatitudes in the far hemisphere would require prohibitive rates of translation. For older rocks, or low latitude sites, it is not commonly possible to assign an unambiguous polarity, and therefore the hemisphere of a paleomagnetic latitude determination, unless known polarities or complete APW paths are available. To know the polarity requires that the ages of the rocks sampled are known with sufficient precision to be able to place them in one of the periods of known polarity of the geomagnetic field. In practice this means that the ages have to fall in either the relatively well-known Cretaceous and younger polarity intervals or in the Pennsylvanian–Permian reversed interval previously known as the Kiaman event (Irving and Pullaiah, 1976).

Various criteria have been used to infer the polarity, including the least latitudinal translation required (Hillhouse, 1977; Hillhouse and Grommé, 1980; Stone and Packer, 1979; Van der Voo et al, 1980); the least amount of tectonic rotation required to account for the discordance of VGPs (Hillhouse, 1977; Van der Voo et al, 1980; Panuska and Stone, 1981); and the least number of required changes in the sense of latitudinal translation (Stone et al, 1982). The least total tectonic transport criterion is not by itself a very strong argument, especially if excessively rapid plate motions are not required. The least angular VGP discordance is a reasonable criterion; however, given the allochthonous nature of the terranes and the lack of information on their paleogeographic affinities, selection of a reference pole (e.g., North American, Asian, Pacific, etc.) is often debatable. Moreover, the effects of local tectonic rotation, arc segmentation, and oroclinal bending may be difficult to separate, especially with a limited data base, and can easily lead to erroneous conclusions. The least number of changes in latitudinal motion or APW paths is perhaps one of the best indirect criteria for polarity assignment. Averaged APW paths for continental blocks tend to show an "orderly" progression of pole positions with relatively few abrupt changes of trend. When sudden changes are observed, they can often be attributed to major tectonic events such as orogenies or opening of ocean basins (e.g., Irving, 1979). On the basis of the limited number of rifting and orogenic events preserved in the rock record of a tectonostratigraphic terrane, the smoothest paleolatitude versus time or APW curve is, thus, a rational guide to polarity interpretation.

WRANGELLIA PALEOMAGNETIC DATA

Figure 1 is a representation of virtual geomagnetic pole positions for existing paleomagnetic data from the Wrangell Mountains, Nutzotin Mountains, and the Alaska Range portion of Wrangellia (see Table 1 for summary data). The Triassic paleomagnetic data were taken from Packer (1972), Hillhouse (1977) and Stone (1982). The late Paleozoic and Triassic VGPs plotted in Figure 1 are considered to be good estimates of the north geographic pole positions with respect to the Wrangellia terrane (i.e., the pole that would lie on a continuous APW path connecting through to the present north pole). The selection of polarity is based on the close proximity of most of the VGPs to the VGP of known polarity from the Pennsylvanian and Permian aged rocks, since the later were deposited during the well-defined Permo-Carboniferous reversed polarity interval (PRC) (Irving and Pullaiah, 1976). Thus, the paleomagnetic data strongly suggest that Wrangellia occupied a northern hemisphere position (10 to 20°N latitude) during the late Paleozoic and early Mesozoic as proposed by Panuska and Stone (1981).

Paleolongitude control is not possible using the available paleomagnetic data; however, available biostratigraphic data provide some paleogeographic constraints. Newton (1983) has demonstrated a faunal similarity between Norian (Late Triassic) bivalves from Wrangellia and from locations in North America. Three species of *Septocardia*, which are currently considered endemic to the Americas and indicate tropical paleolatitudes, were identified in Wrangellia, although a confident biostratigraphic hemisphere determination is precluded by insufficient faunal data from the autochthonous rocks of both North and South America (Newton, 1983).

Silberling and Jones (1983) have identified three Late Triassic latitudinal zones in western North America based on species of the bivalve genus *Monotis*. The Wrangellian *Monotis* species are characteristic of low paleolatitudes similar to the species found in Nevada (Silberling and Jones, 1983). The *Monotis* fauna is apparently distinct from southern hemisphere varieties (Silberling and Jones, 1983).

The distribution of *Septocardia* and *Monotis* suggests that Wrangellia occupied a low-latitude position in the eastern paleo-Pacific Ocean. Wrangellia must have been close enough to North America to allow faunal exchange, although Wrangellia and North America need not have been in contact (Newton, 1983). The biostratigraphic data from Late Triassic bivalves tend to corrobate the low paleolatitude estimate based on paleomagnetism. Moreover, the *Monotis* fauna of Wrangellia suggests that the northern hemisphere interpretation based on paleomagnetic data is correct.

The geographic locations of the north VGPs for the south-central Alaska portion of Wrangellia yield a relatively tight distribution within the southwest Pacific Ocean, with two notable exceptions. VGPs from localities NBS.5+6 and NBS.2 from the Triassic Nikolai Greenstone (Stone, 1982) are significantly displaced from the southwest Pacific cluster. It is possible that these aberrant directions are due

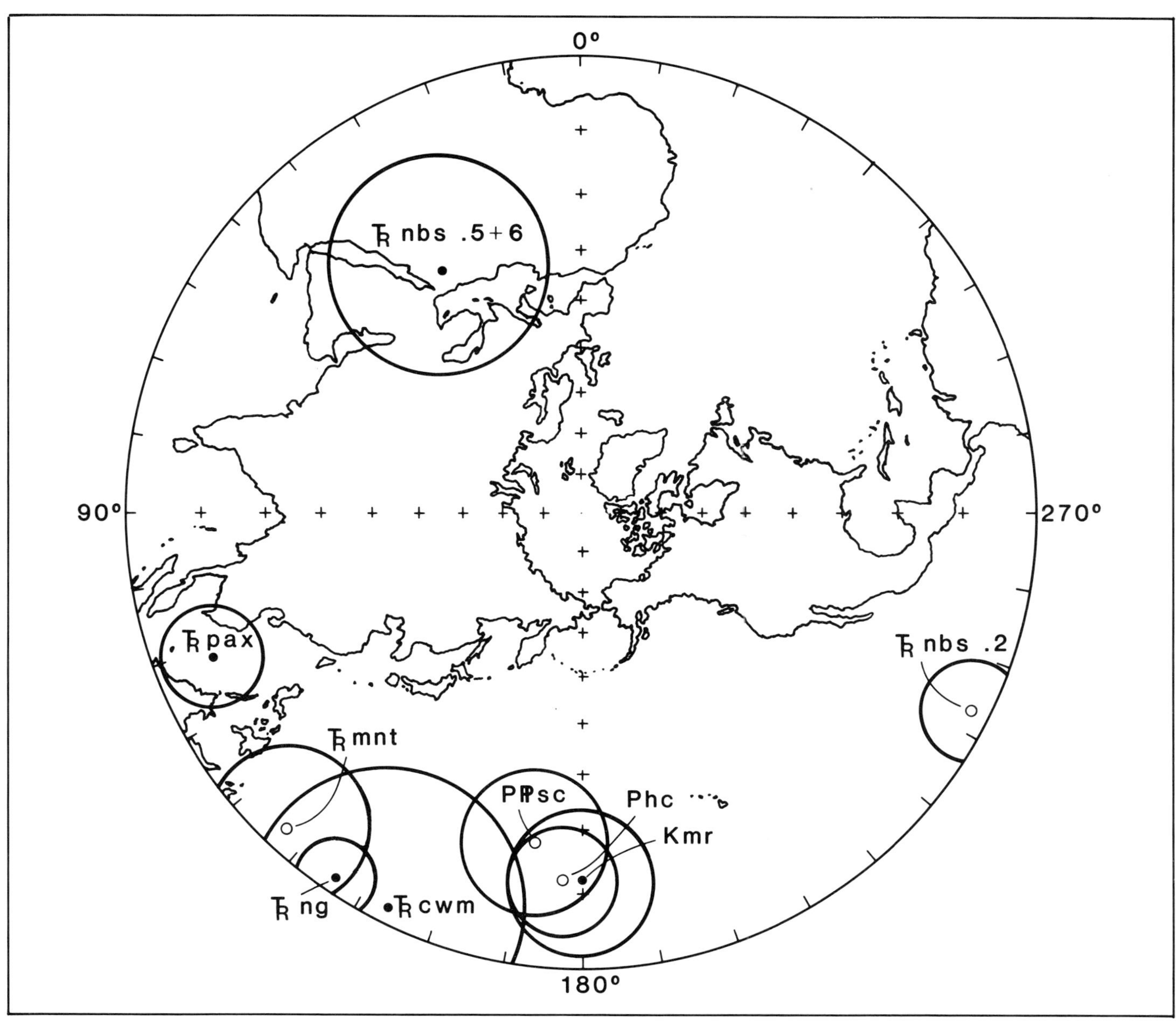

Figure 1—VGPs (virtual geomagnetic poles) for the Wrangell Mountains and nearby portions of Wrangellia. PIPsc = Permian/Pennsylvanian Station Creek Formation, Phc = Permian Hasen Creek Formation, from Panuska and Stone (1981). TR ng = Triassic Nikolai Greenstone, Hillhouse (1977). TR nbs.5+6, TR mnt, TR cwm and TR nbs. 2 = Triassic Nikolai Greenstone, Stone (1982). Kmr = Cretaceous MacColl Ridge Formation, Panuska (1984).

to remagnetization; however, the existence of reversals in locality NBS.5+6 (Stone, 1982) suggests that remagnetization has not occurred at this site. The fact that NBS.2 gives a paleolatitude similar to localities with demonstrably stable primary remanence directions is circumstantial evidence, albeit weak, that the NBS.2 direction is primary. The discordance of the NBS.5+6 and NBS.2 VGPs can be explained in terms of local tectonic rotations generated by a strike-slip fault system. This interpretation is reasonable in view of the proximity of these localities to the Totschunda fault (Stone, 1982).

The VGP for the Late Cretaceous MacColl Ridge Formation (Panuska, 1983, 1984) is also considered to represent the north VGP (Kmr in Fig. 1). The resultant magnetic vector used to determine the VGP for the MacColl Ridge is considered primary, because the data pass a fold test and contain reversals. The corresponding paleolatitude is 32°N, the northern hemisphere option being preferred because it requires only 30° of latitudinal motion instead of 90° since latest Cretaceous time. In addition, the late Campanian to early Maestrichtian (the age of the MacColl Ridge Formation) geomagnetic field is strongly biased toward normal polarity (Lowrie and Alvarez, 1981; Irving and Pullaiah, 1976), which is also the polarity bias observed in the MacColl Ridge Formation (Panuska, 1984).

The observation that the late Paleozoic and Mesozoic Wrangellia VGPs are grouped together suggests that the

Table 1

Formation	Age	N	Stratigraphic Dec	Inc	λ	k	α_{95}	VGP Long	Lat	kK	α_{95}
					Wrangellia Terrane						
MacColl Ridge	K	14/20	226	51	32	20.3	9.1	180	12	14.8	10.7 (1)[a]
Nikolai	TR	50/91	256	28	15	21.5	16.9	146	2	18.3	4.8 (3)
Hasen Creek	P	10	220	19[b]	10	24.1	10.0	177	−12	33.0	8.0 (1,2)
Station Creek	IP,P	10	223	4[b]	2	9.2	16.8	172	−18	16.6	12.2 (1,2)
					Alexander Terrane						
Hound Island	TR	12/139	235	63	44	32.7	7.7	189	23	17	11.0 (4)
Pybus	P	12/24	332	−18[b]	−9	11.9	11.8	75	19	20.1	9.0 (1)
Ladrones/Klawak	IP	33	101	−15	8	5.6	11.6	132	12		12.0 (5)
Peratovich	M	28	95	−27	14	5.1	13.3	141	14		14.9 (5)
Port Refugio	D	30	98	−37	21	10.1	8.7	143	21		10.9 (5)
Wadleigh }											
Coranados }	D	31	115	−20	10	8.6	9.4	122	23		10.0 (5)
St. Joseph }											
Descon	O	14	116	−14	7	7.4	15.7	119	20		16.2 (5)
					Peninsular Terrane						
Tolstoi	Te	36			49			305	68	38	13 (6)
Chignik/Hoodoo	K	25			19			337	40	16	20 (6)
Chignik/Hoodoo	K	64			16			85	39	14.0	17 (6)
Naknek/Chinitna	J	107			31			281	37	21	10 (6)
					Gravina-Nutzotin Terrane						
"Marsh Island"	K	4/22	56	20	10	44.9	10.5	342	28	86.5	7.5 (7)
Brothers	JK	7/21	87	16	8	71.5	7.2	314	9	123.7	5.4 (1)
					Prince William Terrane						
Ghost Rocks	Tp	11/73	320	53		29.4	8.6	93	55	19.8	10.5 (8)
Ghost Rocks	Tp	16/83	82	65		25.8	7.4	269	43	13.0	10.6 (8)
Ghost Rocks (both localities)					40.3	19.3	6.2				(8)

[a]Data references:
1. Panuska (1984)
2. Panuska and Stone (1981)
3. Hillhouse (1977)
4. Hillhouse and Grommé (1980)
5. Van der Voo et al (1980)
6. Stone and Packer (1979)
7. Panuska et al (1984)
8. Plumley et al (1983)

[b]Directions were inverted from those measured because these units were deposited during the Permo-Carboniferous reversed magnetic interval. Thus, the reported directions and paleolatitudes are considered to be of known polarity (hemisphere).

Note: Stratigraphic directions for Peninsular terrane data were not reported in the original reference; paleolatitudes were determined from the VGP data. Stratigraphic direction is corrected for tilt. N = number of beds or flows/number of specimens. λ = paleolatitude, K = Fisher precision parameter, α_{95} = circle of 95% confidence.

Table 1—Paleomagnetic summary data.

south-central Alaska portion of the terrane behaved, for the most part, as a coherent tectonic block. If this interpretation is correct, then the Wrangell Mountains portion of Wrangellia has undergone approximately 130° counterclockwise rotation as a unit (or 230° of clockwise rotation) since latest Cretaceous. If other parts of the terrane of significant size are similarly intact, then VGPs from these "stable" portions may eventually be used to test oroclinal bending or tectonic rotation models.

VGP positions for the Middle and/or Upper Triassic Nikolai Greenstone in the Alaska Range portion of Wrangellia (Hillhouse and Grommé, 1984) tend to support the concept of little internal deformation away from the major bounding faults. Although these VGPs show an elongate distribution, they plot in the general vicinity of the other VGPs shown in Figure 1. The three VGPs most displaced from the main cluster (Hillhouse and Grommé's localities 2, 6, and 7) can be explained as localized tectonic rotations because of their proximity to the Totschunda and Broxson Gulch faults. The other three VGPs that form part of the cluster offer support for rotation of a coherent block.

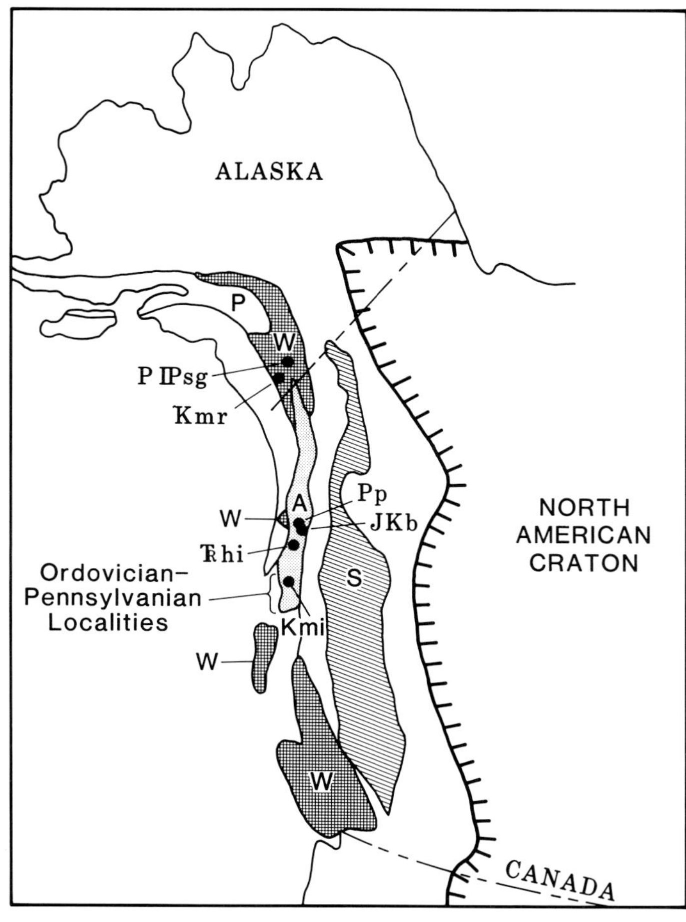

Figure 2—Map showing selected terranes in northwest North America and location of new (and critical) paleomagnetic sampling sites. A = Alexander terrane, W = Wrangellia, P = Peninsular terrane, S = Stikine terrane. PIPsg = Skolai Group, Kmr = McColl Ridge, Pp = Pybus, JKb = Brothers Volcanics, Ŧ hi = Hound Island, Kmi = Marsh Island.

ALEXANDER TERRANE PALEOMAGNETIC DATA

The first APW path segment for the Alexander terrane was constructed by Van der Voo et al (1980). Remanence directions were acquired from Middle Ordovician- through Middle Pennsylvanian-aged rocks from western Prince of Wales Island. The primary nature of the remanent magnetization is indicated by fold and reversal tests. The similarity of the North American and Alexander VGP positions and generalized APW paths, combined with a minimum latitudinal motion argument, led to a preferred northern hemisphere location for the Alexander terrane, adjacent to North America (Van der Voo et al, 1980). The best match of the paleolatitude versus time plots for the Alexander terrane and cratonic North America is obtained for a paleolatitude of the Alexander terrane equivalent to that of western Nevada (Van der Voo et al, 1980). This is strong evidence in support of an original location of the Alexander terrane near present-day Nevada, a concept originally proposed by Jones et al (1972). Based on the fit of the Alexander terrane paleolatitudes for Ordovician through Middle Pennsylvanian time with a western Nevada origin, Van der Voo et al (1980) concluded that the Alexander terrane was fixed with respect to ancient Nevada through Pennsylvanian time and subsequently was displaced northwards to its present position in southeast Alaska.

Hillhouse and Grommé (1980) have reported paleomagnetic data from the Upper Triassic Hound Island Volcanics in Keku Straight, part of the Alexander terrane, and determined a paleolatitude of 44°. This determination is in good agreement with the predicted paleolatitude, assuming that the Alexander terrane was in its present position with respect to North America by the Triassic. On this basis, Hillhouse and Grommé (1980) and Van der Voo et al (1980) suggested that the Alexander terrane rifted away from Nevada in post-Pennsylvanian time, moved approximately 15° northward, and accreted to North America by Late Triassic time. Such an interpretation, however, is difficult to reconcile with the accretion history of inboard terranes.

Based on paleomagnetic paleolatitude data, Monger and Irving (1980) suggest a Jurassic or Cretaceous age of accretion for one of these inboard terranes, the Stikine terrane (Fig. 2). Tempelman-Kluit (1979) proposed that the Late Jurassic deformation in the Stikine terrane is the manifestation of emplacement of the terrane during an arc-continent collision. If the Triassic accretion age for the Alexander terrane is correct, then the Stikine terrane would have to be "wedged behind" the earlier accreted terrane, a possibility put forward by Monger and Irving (1980), or thrust over the Alexander terrane. Neither scenario is appealing.

In addition to problems with the timing of Stikine terrane accretion, a Triassic Alexander terrane accretion is apparently precluded by paleomagnetic data from other southern Alaska terranes. Since the Wrangellia, Alexander, and the Gravina–Nutzotin terranes were apparently linked by Late Jurassic time (Berg et al, 1972; Panuska, 1984), paleomagnetic paleolatitude data postdating this amalgamation from any one of these terranes should apply to all. The very low paleolatitude of Wrangellia found in the MacColl Ridge Formation (Panuska 1983, 1984) and similarly low paleolatitudes found in the Brothers Volcanics of the Gravina–Nutzotin Belt, both of Cretaceous age (Panuska, 1984) imply that the Alexander terrane was also far south of its present location in Cretaceous time. These data do not rule out the possibility that the Alexander terrane accreted to North America in Triassic time, rifted away and amalgamated with Wrangellia, then reaccreted to North America in post-Cretaceous time; however, this scenario seems unduly complicated.

The Triassic Hound Island data can also be interpreted in terms of a southern hemisphere paleolatitude to circumvent the problems of the timing of accretion. The paleomagnetic data for the Alexander terrane taken as a whole tend to support such an inference. According to the data presented by Van der Voo et al (1980), the Alexander terrane was moving south in Devonian to Pennsylvanian time. During this time, the Alexander terrane moved from 21 to 8°N latitude, although it should be noted that this latitudinal motion is just barely within the limits of detectability. Data from the Permian Pybus Dolomite of the Alexander terrane are of known polarity and correspond to a 9°S paleolatitude (Panuska, 1984). Taken at face value, the Pybus Dolomite data, together with the data of Van der Voo et al (1980), suggest that there was net southerly motion of the Alexander terrane from Devonian through Permian time (Fig. 3). In addition, the Pybus Dolomite VGP plots relatively close to the earlier Paleozoic VGPs, which lends credence to the polarity preference of Van der Voo et al (1980) and also suggests that those parts of the Alexander terrane sampled have undergone little or no differential rotation. If the premise of little differential rotation is correct, then a preliminary terrane-wide APW path can be constructed.

Because of the importance of the Pybus Dolomite in our plate tectonic interpretations, a brief discussion of the data is in order. The magnetic intensity of the Pybus samples is unusually weak, which lowers the precision of directional measurements. Special care was taken to ensure reproducibility by making multiple repeat measurements at every level to try to average out random errors. A fold test was applied to the Pybus data and the difference in the directions of the two limbs reduced substantially after correcting for bedding tilt, suggesting a positive fold test. However, because of overprinting problems on many samples, the mean direction from one limb of the fold is based on only three specimens and a statistically rigorous fold test is not possible. Since the possibility of remagnetization cannot be unequivocally ruled out, the Pybus data can only be tentatively accepted as representing the original magnetization.

A limited paleomagnetic study on sediments of Cretaceous age from Marsh Island (from Panuska et al, 1984) and data from the roughly contemporaneous Brothers Volcanics (Panuska, 1984), both within the Alexander terrane, define a Cretaceous VGP of known polarity when considered together. The polarity call is based on the known polarity for the Marsh Island samples, since they were deposited during the Cretaceous long normal polarity interval, the polarity being extrapolated to the less well-dated Brothers Volcanics. Hillhouse (1983, personel communication) has expressed concern that the Marsh Island rocks are remagnetized. He found in situ (geographic) vector directions from apparently remagnetized Paleozoic and Triassic rocks that are similar to the Marsh Island in situ direction, suggesting remagnetization of the latter. These directions are close, but the Marsh Island direction is statistically distinct in all but one case. Unfortunately, the available data from Hillhouse represent the mean NRM directions, although AF and thermal demagnetization of some pilot specimens did not change the direction substantially (Hillhouse, 1983, personal communication). A recent K–Ar age date on a lahar deposited within the Marsh Island section gives a 100 m.y. age, which is in excellent agreement with the Albian paleontological age determination (Decker, 1983, personal communication). Since the K–Ar age of the rock has not been reset, it puts a constraint on whether or not the magnetic directions of the rocks can have been reset by heating above their blocking temperatures. The age quoted was determined for a hornblende separate (Blum, 1983, personal communication). The minimum argon blocking temperature of hornblende is approximately 460°C (860°F) (Odin, 1982). While the argon blocking temperature is below the magnetic blocking temperature of these rocks (Panuska et al, 1984), the possibility of high-temperature viscous thermoremanent remagnetization cannot be completely ruled out. The remagnetization curves of Pullaiah et al (1975) would allow the Marsh Island rocks to be remagnetized if they were heated to just below the argon blocking temperature and maintained at that temperature for about 50 m.y., a rather long but not impossible length of time.

Perhaps the strongest test for magnetic stability of the Marsh Island rocks is the general agreement of direction with that of the Brothers Volcanics. The stability of the Brothers Volcanics is suggested by intraformational conglomerate test, which showed that fragments of the volcanics included in a volcanic breccia were magnetized in random directions and thus had not been reset. Though the two lines of evidence for the stability of Marsh Island rocks are only circumstantial, namely the similarity to the Brothers Volcanics direction and the evidence against reheating, the Marsh Island data are tentatively accepted as representing a primary direction.

VGP data from the Alexander terrane are shown in Figure 4, along with two possible generalized APW paths. The Paleozoic VGP are considered north poles because of the known polarity of the Pybus pole and the arguments of Van der Voo et al (1980) previously discussed. A major concern in constructing an APW path is the polarity assignment for the Triassic Hound Island VGP. The southern hemisphere Triassic VGP is preferable for a variety of reasons. The southern pole lies on a more or less continuous great circle APW path from Pennsylvanian to

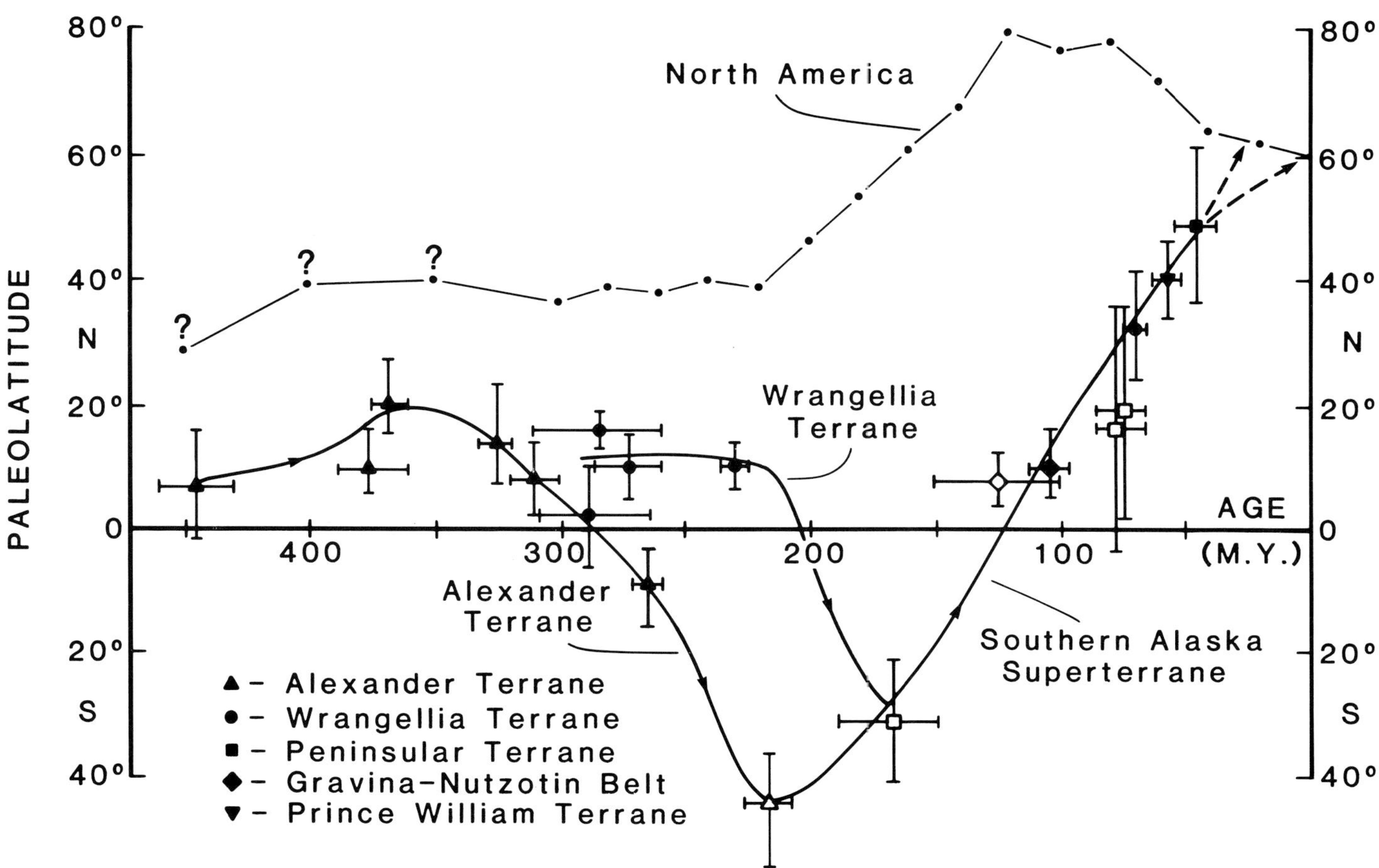

Figure 3—Paleolatitude data for Wrangellia, the Alexander terrane, and the Southern Alaska Superterrane. The Southern Alaska Superterrane comprises the Alexander terrane, Wrangellia, the Peninsular terrane, and the Gravina-Nutzotin belt, which amalgamated in Late Jurassic. The Chugach and Prince William terranes may also be part of this superterrane. Solid symbols indicate that polarity (thus hemisphere) is known with some confidence. Open symbols indicate equivocal polarity. The uppermost curve is a summary of paleolatitudes that the Southern Alaska Superterrane would have recorded if fixed to North America since the Paleozoic. Data are taken from the following sources: pre-Permian Alexander terrane data from Van der Voo et al (1980); Permian Alexander terrane data from Panuska (1984), Triassic Alexander terrane data from Hillhouse and Grommé (1980); Pennsylvanian-Permian Wrangellia data from Panuska and Stone (1981) and Panuska (1984); Triassic Wrangellia data from Hillhouse (1977); Jurassic, Cretaceous, and Tertiary Peninsular terrane data from Stone and Packer (1979); Jurassic and Cretaceous Gravina-Nutzotin belt data from Panuska (1984) and Panuska et al (1983); Cretaceous Wrangellia data from Panuska (1984); and Tertiary Prince William terrane data from Plumley et al (1983).

Triassic time. It requires only 76° APW motion and 35° of latitudinal motion since Early Permian time for the parts of the Alexander terrane that were sampled. In addition, the southern hemisphere pole is only 62° from the probable Brothers north VGP and only 53° from the Marsh Island north VGP. The alternative northern hemisphere VGP for the Hound Island rocks would require 102° of APW motion with 53° of latitudinal motion and would produce several pronounced kinks in the APW path. Thus, on the basis of simplest APW path and least latitudinal motion, the Hound Island VGP located in the southern hemisphere is considered the most likely estimate of the geographic north pole.

There is also some paleontological support for the Triassic southern hemisphere location for the Alexander terrane. Silberling and Jones (1983) report a *Monotis* fauna from the Alexander terrane that is unlike the North America cratonic *Monotis* fauna and may represent southern hemisphere middle paleolatitudes. In addition, Newton (1983) has identified *Septocardia* cf. S. *Pascoensis* in Alexander terrane Triassic rocks from Keku Straight. This species is markedly different from the North America *Septocardia*, and the sole previously reported occurrence is in Peru (Newton, 1983).

Although the Triassic biostratigraphic data ostensibly support a southerly-before-northerly tectonic motion for the Alexander terrane, the data are too sparse to demonstrate conclusive ties. Similarly, the paleomagnetic data are too limited to assess the likelihood of local tectonic rotations, and clearcut stability tests are lacking for some localities. The APW path presented in Figure 4 is offered as a working hypothesis and will no doubt be modified by future studies. Nevertheless, tectonic analyses based on this curve are useful in pointing out correlations and conflicts and for generating testable tectonic models.

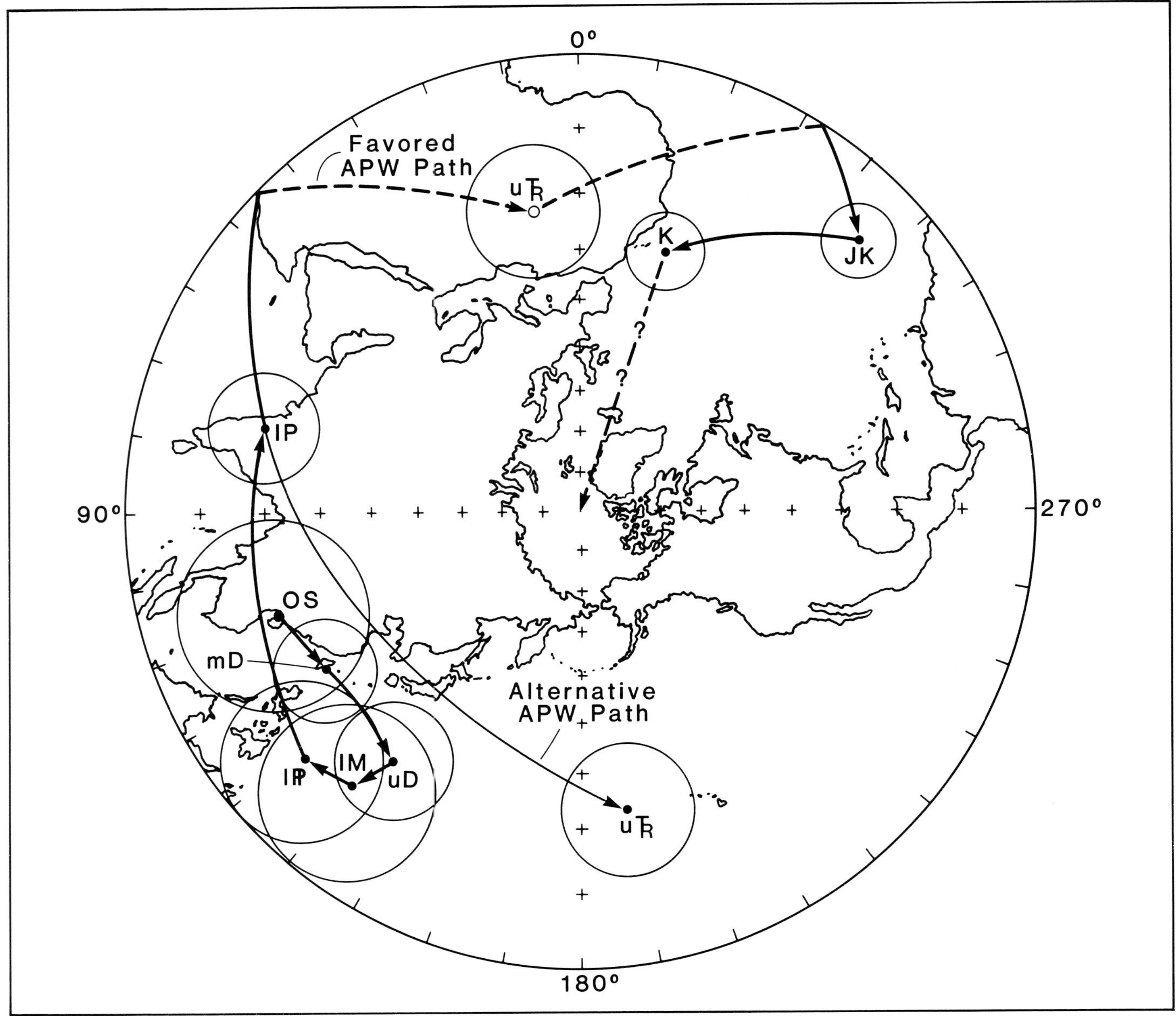

Figure 4—Alexander terrane VGPs and possible APW paths. Solid (open) dots are northern (southern) hemisphere. The northern hemisphere Late Triassic VGP and "alternative" APW path are not favored because of the abrupt change in APW trend and paleolatitude arguments (see text). Ordovician through Pennsylvanian VGPs are from Van der Voo et al (1980). Permian VGP is from Panuska (1984). Triassic VGP is from Hillhouse and Grommé (1980). Jura-Cretaceous VGP is from Panuska (1984). Cretaceous VGP is from Panuska et al (1983).

SOUTHERN ALASKA SUPERTERRANE

An interpretive paleolatitude versus time trajectory for the Wrangellia and Alexander terranes, as well as the other terranes comprising the Southern Alaska Superterrane as they joined the amalgam, is shown in Figure 3. [The name Southern Alaska Superterrane is preferred over Talkeetna Superterrane (Csejtey et al, 1982) because the Alexander terrane is not included in the original definition.] Several constraints and guidelines have been employed to assign polarity. As explained previously, the Paleozoic Alexander terrane data were assigned polarities on the basis of the general similarity to the North American APW path, the known polarity of the Pybus Formation data, and the gross stratigraphic similarity to the western U.S. (Jones et al, 1972). A southern hemisphere Triassic paleolatitude for the Alexander terrane is inferred on the basis of the smoothest APW path, the least tectonic translation from Permian to Triassic time, and on paleobiogeographic considerations.

The Paleozoic polarity for the Wrangellia terrane is known to be northern hemisphere because most of the units sampled were deposited during the Permo-Carboniferous reversed polarity interval. Triassic paleolatitudes for Wrangellia are almost certainly northern hemisphere by correlation with the known polarity late Paleozoic VGPs together with supportive paleobiogeographic evidence. Since the Southern Alaska Superterrane was probably assembled before the end of the

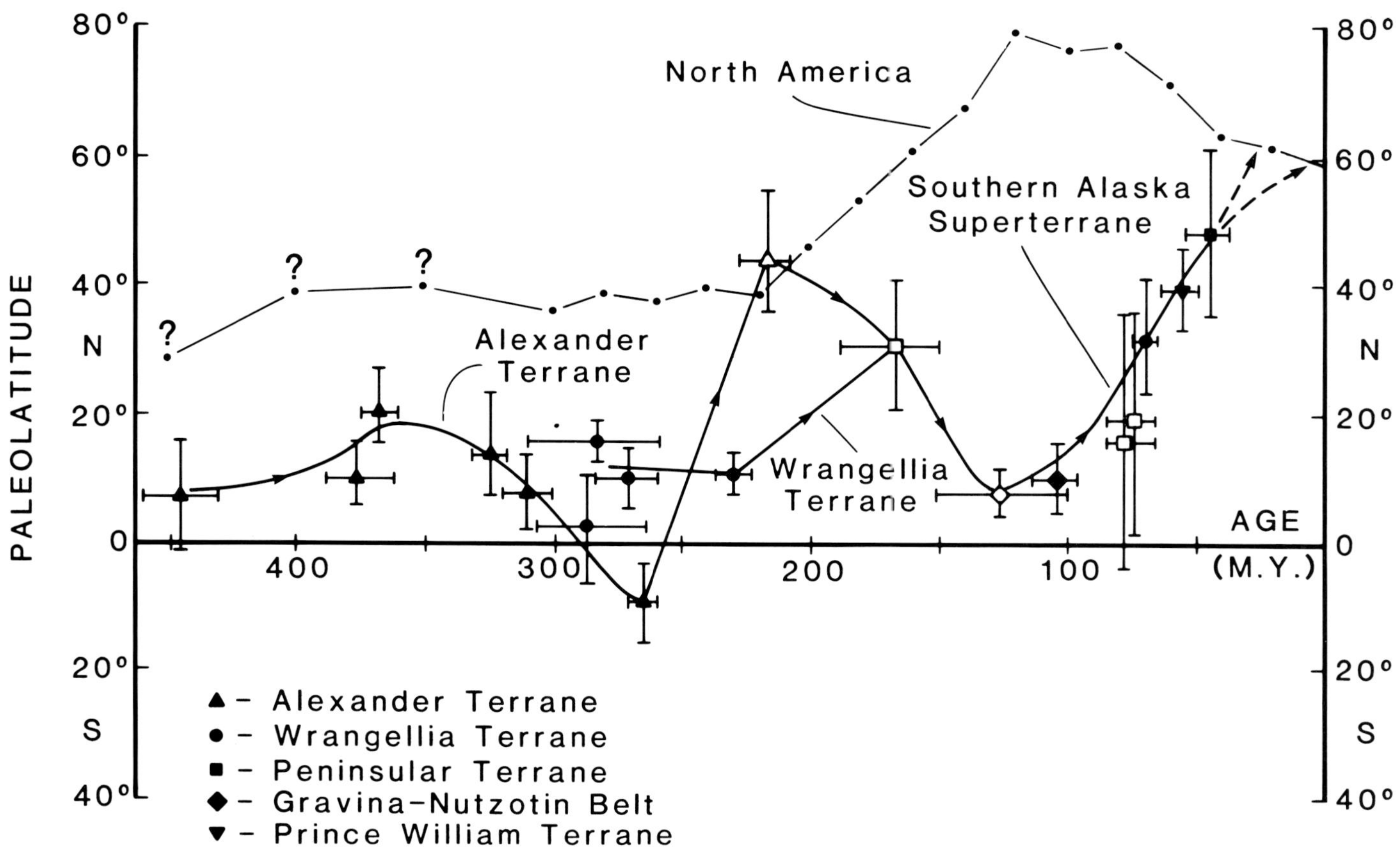

Figure 5—Paleolatitude data for Wrangellia, the Alexander terrane, and the Southern Alaska Superterrane showing the northern hemisphere polarity preference for Triassic Alexander datum and the Jurassic Peninsular datum. This interpretation is not favored because of the "jerky" paleolatitude trends and the relatively frequent changes in sense of paleolatitude motion (i.e., northward or southward). Data sources, symbols, and reference curve are described in Figure 3.

Jurassic (Berg et al, 1972) and probably during Middle to Late Jurassic time (Panuska, 1984), the paleomagnetic data from Middle to Upper Jurassic and younger rocks from the Peninsular terrane (Stone and Packer, 1979) can be applied to the whole superterrane. These data give a southern hemisphere position for the superterrane based on minimum motion considerations. The data also show that the Wrangellia and Alexander terranes are clearly latitudinally separated in the Triassic. Geological evidence requires that the terranes be joined by Late Jurassic, thus the Jurassic paleolatitudes for the Peninsular and Wrangellia terranes should have the same polarity as the Triassic Hound Island data from the Alexander terrane. The alternative to this (i.e., a southern polarity for the Alexander terrane in Triassic time and northern polarity for all terranes in the Jurassic) would require plate motions of at least 16 cm/year (6.3 in./year), which although possible are rather high. However, if the Triassic paleolatitude of the Alexander terrane is changed and a northern hemisphere location is preferred, then the least required motion argument would constrain the Jurassic data to be northern hemisphere polarity as well. The result of interpreting all Triassic and Jurassic data to be northern hemisphere is to generate a very "jerky" terrane motion (Fig. 5; see Fig. 6 for predicted paleolatitudes if southern Alaska was fixed with respect to North America). Such a latitude versus time curve would require the Alexander terrane to reverse its latitudinal direction in the Permian (from southward to northward), again in Triassic (from northward to southward), and finally in Late Jurassic–Early Cretaceous (from southward to northward). Similarly, Wrangellia would have an additional reversal in motion, namely a change in Jurassic time (from northward to southward). The latitude versus time plot shown in Figure 3 is thus favored because it requires only two changes in the sense of latitudinal motion (once for the Alexander terrane and once for Wrangellia as opposed to the four changes required by the polarity calls illustrated in Figure 5.

On the basis of the latitudinal motion changes just outlined, the APW path arguments, and the biogeographic arguments outlined earlier, the general latitude versus time trajectory in Figure 3 is considered to be the most likely for the Southern Alaska terranes. Based on this curve, the following tectonic model is put forward as a working hypothesis.

The oldest known rocks (early Paleozoic) of the Southern Alaska Superterrane occur within the Alexander terrane. The Alexander terrane probably originated as a volcanic arc adjacent to what is now western Nevada (Jones et al, 1972; Van der Voo et al, 1980). It is worth noting that before the southerly excursion, the Alexander terrane was

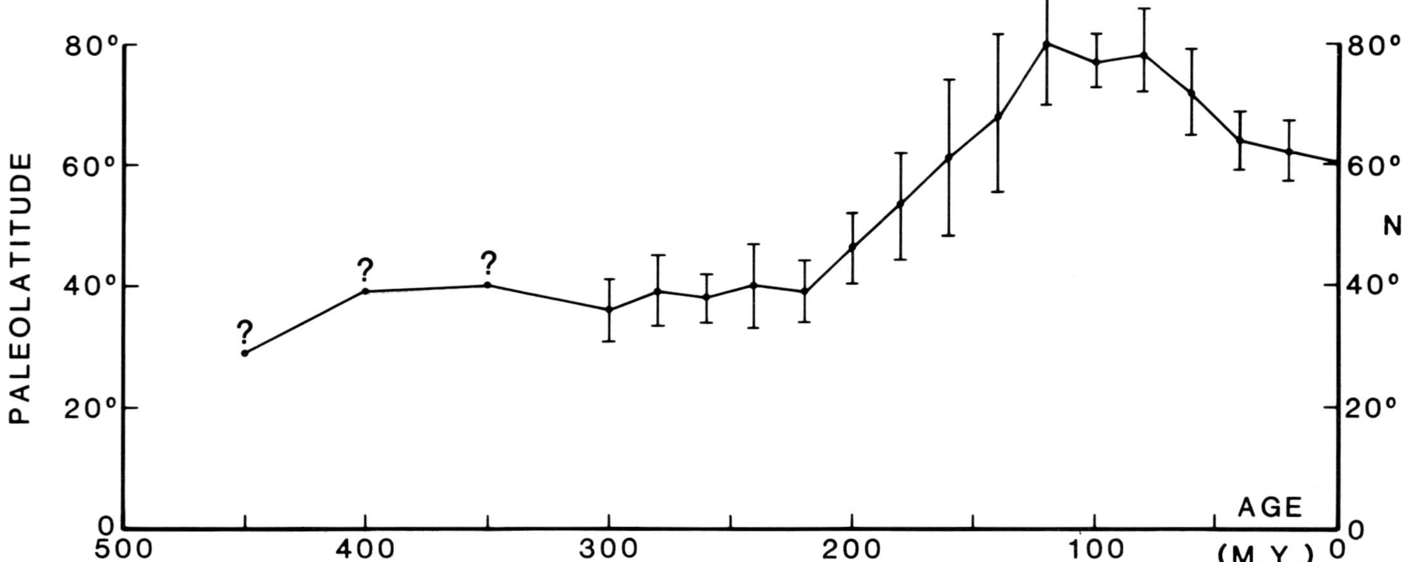

Figure 6—Paleolatitude versus time reference curve for a point on the Kenai Peninsula (150°W, 60°N), which would approximate the paleolatitude history of the Southern Alaska Superterrane if it were fixed to North America since the early Paleozoic. Paleolatitudes and error bars are based on Irving's (1979) compilation of North American poles.

at a paleolatitude that would allow it to participate in the Antler orogeny of western North America. The concept that the collision of this terrane with North America was related to the orogeny needs closer scrutiny.

Based on paleomagnetic evidence, the Alexander terrane began to move in a southerly direction beginning in Devonian to Mississippian time. It continued to move southward until it arrived at a latitude of 44°S in the Late Triassic. This scenario allows an explanation of the South American affinities of Alexander terrane *Septocardia* (Newton, 1983) and *Monotis* (Silberling and Jones, 1983), by suggesting close Triassic geographic proximity of the Alexander terrane to South America.

During Pennsylvanian through Triassic time, Wrangellia remained at an approximately constant paleolatitude implying at least one active plate boundary between the Alexander and Wrangellia terranes. By comparison with North American reference poles (Irving, 1979), Wrangellia's paleolatitude with respect to North America was equivalent to the paleolatitude of Las Vegas throughout the late Paleozoic and early Mesozoic, although the error limits on the paleomagnetic data permit a paleolatitudinal position between present-day Oregon and central Mexico. Faunal evidence (Silberling and Jones, 1983; Newton, 1983) suggests an eastern Pacific paleogeography because of similarities with fauna from Triassic cratonic rocks in Nevada. Jones et al (1977) suggested that the voluminous Nikolai Greenstone and equivalent rocks represented a Late Triassic rifting event. This interpretation is supported by trace element data from the Goondip Greenstone (Nikolai equivalent) reported by Decker (1982). Accordingly, the Nikolai Greenstone could be related to a postcollisional (Sonoma orogeny?) rifting event that severed Wrangellia from the North American margin. Following rifting, Wrangellia moved southward into the southern paleo-Pacific Ocean.

There is some support for this hypothesized southward plate motion from areas outside southern Alaska. Gose et al (1982) report paleomagnetic data from Mexico that indicate 130° counterclockwise rotation ending in Middle Jurassic. These data could be explained as a tectonic rotation generated in a left-lateral strike-slip fault system. It is possible that the rotation observed by Gose et al (1982) is related to the left-lateral Mojave–Sonora Megashear (Silver and Anderson, 1983). This timing and sense of faulting is consistent with the southward motion of an oceanic plate carrying Wrangellia.

By Late Jurassic, the Peninsular Wrangellia and the Alexander terranes had amalgamated, while located in low to middle latitudes of the southern hemisphere. The amalgamation event was probably an arc-arc collision corresponding to the Late Jurassic orogeny in the Wrangell Mountains recognized by MacKevett (1978). The newly formed Southern Alaska Superterrane then moved northward throughout the remainder of the Mesozoic. By latest Cretaceous time, the superterrane was at middle northern hemisphere paleolatitudes, and by Paleocene the superterrane was located at approximately 40°N latitude (Plumley et al, 1983). The timing of the accretion of the Southern Alaska Superterrane to North America is not known; it could have achieved its present position by the closing of a Tertiary ocean basin lying between it and the continent, by intracontinental shortening, by transcurrent faulting, or by a combination of these scenarios, as outlined by Moore et al (1983).

It will be recognized that the conclusions relating to the Alexander terrane rely heavily on the data from the Pybus Dolomite and from the Marsh Island and Brothers Volcanics. These data only pass stability tests based on circumstantial evidence and thus need to be substantiated. Translation history hypotheses based on these data are nonetheless useful for visualizing the options for interpretation and identifying the critical areas for future work.

ACKNOWLEDGMENTS

The authors would like to acknowledge the help and support of John Decker and the Alaska Division of Geological and Geophysical Surveys, Z-Axis Exploration, Arlen Ehm Consulting, Cities Service Oil Company, and National Science Foundation grants EAR-8213005, EAR-8121424, EAR-7800817.

REFERENCES

Berg, H. C., et al, 1972, Gravina-Nutzotin Belt—tectonic significance of an upper Mesozoic sedimentary and volcanic sequence in southern and southeastern Alaska: U.S. Geological Survey Professional Paper 800-D, p. D1-D24.

Csejtey, B., et al, 1982, The Cenozoic Denali fault system and the Cretaceous accretionary development of southern alaska: Journal of Geophysical Research, v. 87, p. 3741–3754.

Decker, J., 1982, Geochemical signature of the Goon Dip Greenstone on Chichasof Island, southeastern Alaska, *in* Short Notes on Alaskan Geology 1981: Alaska Division of Geological and Geophysical Surveys Geologic Report 73, p. 29–35.

Fisher, R. A., 1953, Dispersion on a sphere: Proceedings of the Royal Society of London, Series A, v. 217, p. 295–305.

Gose, W. A., et al, 1982, Paleomagnetic results from northeastern Mexico: Evidence for large Mesozoic rotations: Geology, v. 10, n. 1, p. 50–54.

Hillhouse, J. W., 1977, Paleomagnetism of the Triassic Nikolai Greenstone, McCarthy Quadrangle, Alaska: Canadian Journal of Earth Sciences v. 14, n. 11, p. 2578–2592.

———, and C. S. Grommé, 1980, Paleomagnetism of the Triassic Hound Island Volcanics, Alexander Terrane, southeastern Alaska: Journal of Geophysical Research, v. 85, p. 2594–2602.

——— and ———, 1984, Northward displacement and accretion of Wrangellia: New paleomagnetic evidence from Alaska: Journal of Geophysical Research, v. 89, p. 4461–4477.

Irving, E., 1979, Paleopoles and paleolatitudes of North America and speculations about displaced terrains, Canadian Journal of Earth Sciences, v. 16, p. 669–694.

———, and G. Pullaiah, 1976, Reversals of the geomagnetic field, magnetostratigraphy and relative magnitude of the paleosecular variation in the Phanerozoic: Earth Science Reviews, v. 12, p. 35–64.

Jones, D. L., et al, 1972, Southeastern Alaska—A displaced continental fragment?: U.S. Geological Survey Professional Paper 800-B, p. B211–B217.

———, 1977, Wrangellia—a displaced terrane in northwestern North America: Canadian Journal of Earth Sciences, v. 14, p. 2565–2577.

Lowrie, W., and W. Alvarez, 1981, One hundred million years of geomagnetic polarity history: Geology, v. 9, p. 392–397.

MacKevett, E. M., 1978, Geological map of the McCarthy Quadrangle, Alaska: U.S. Geological Survey Miscellaneous Investigations Map I-1032, scale 1:25,000.

Monger, J. W. H., and E. Irving, 1980, Northward displacement of north-central British Columbia: Nature, v. 285, n. 5763, p. 289–294.

Moore, J. C., et al, 1983, Paleogene evolution of the Kodiak Islands, Alaska: consequences of ridge-trench interaction in a more southerly latitude: Tectonics, v. 2, n. 3, p. 265–293.

Newton, C. R., 1983, Paleozoogeographic affinities of Norian bivalves from the Wrangellian, Peninsular and Alexander terranes, western North America, *in* C. H. Stevens, Pre-Jurassic suspect terranes of the western Cordillera: Society of Economic Mineralogists and Paleontologists Symposium.

Odin, G. S., 1982, The Phanerozoic time scale revised: Episodes, v. 1982, n. 3, p. 3–9.

Packer, D. R., 1972, Paleomagnetism of the Mesozoic in Alaska: PhD Dissertation, University of Alaska, 160 p.

Panuska, B. C., 1983, Paleomagnetic data from the Cretaceous MacColl Ridge Formation and tectonic implications for the Wrangellia Terrane (Abs.): EOS, American Geophysical Union Transactions, v. 64, n. 45, p. 668

———, 1984, Paleomagetism of the Wrangellia and Alexander terranes and the tectonic history of southern Alaska: Unpublished PhD Dissertation, University of Alaska, 197 p.

———, and D. B. Stone, 1981, Late Paleozoic palaeomagnetic data for Wrangellia: resolution of the polarity ambiguity: Nature, v. 293, p. 561–563.

———, et al, 1984, A paleomagnetic pilot study of the Gravina-Nutzotin belt: U.S. Geological Survey Circular 868, p. 117–120.

Plumley, P. W. et al, 1983, Paleomagnetism of the Paleocene Ghost Rocks Formation, Prince William terrane, Alaska: Tectonics, v. 2, p. 295–314.

Pullaiah, G., et al, 1975, Magnetization changes caused by burial and uplift: Earth and Planetary Science Letters, v. 28, p. 133–143.

Silberling, N. J., and D. L. Jones, 1983, Paleontologic evidence for northward displacement of Mesozoic rocks in accreted terranes of the western Cordillera (abs.): Geological Association of Canada May Meeting.

Silver, L. T., and T. H. Anderson, 1983, Further evidence and analysis of the role of the Mojave-Sonora Megashear(s) in Mesozoic Cordilleran tectonics: Geological Society of America Abstracts with Programs, v. 15, n. 5, p. 273.

Stone, D. B., 1982, Triassic paleomagnetic data and paleolatitudes for Wrangellia, Alaska, *in* Short Notes on Alaska Geology, 1981: Alaska Division of Geological and Geophysical Surveys Geologic Report 73, p. 55–62.

———, and D. R. Packer, 1979, Paleomagnetic data from the Alaska Peninusula: Geologic Society of America Bulletin, Part I, v. 90, p. 545–560.

———, et al, 1982, Paleolatitudes versus time for southern Alaska: Journal of Geophysical Research, v. 87, p.

3697–3707.
Templeman-Kluit, D. J., 1979, Transported cataclasite, ophiolite and granodiorite in Yukon: Evidence of arc-continent collision: Geological Survey of Canada, Paper 79-14, 27 p.
Van der Voo, R., et al, 1980, Paleozoic paleomagnetism and northward drift of the Alexander terrane, southeastern Alaska: Journal of Geophysical Research, v. 85, p. 5281–5296.

Cenozoic Migration of Alaskan Terranes Indicated by Paleontologic Study

R. von Huene
G. Keller
T. R. Bruns
K. McDougall
U.S. Geological Survey
Menlo Park, California

Comparisons of microfossils from deep-sea cores, from samples of an exploratory drill hole, and from dredged rocks of the Gulf of Alaska, with coeval microfossil assemblages on the North American continent, provide constraints on the northward migration of the Yakutat block and the Prince William terrane during Tertiary time. The estimated paleolatitudes of microfauna and flora indicate that: (1) the Prince William terrane was attached to North America in its present position by middle Eocene time (40 to 42 Ma), consistent with models derived from paleomagnetic data, and (2) the adjacent Yakutat block was 30 ± 5° south of its present position in early Eocene (50 Ma), 20 ± 5° south in middle Eocene (40 to 44 Ma), and 15 ± 5° south in late Eocene time (37 to 40 Ma), thus requiring a northward motion of about 30° since 50 Ma. Moreover, the Yakutat block was at least 10° south of the Prince William terrane during Eocene time. These data are consistent with migration of the Yakutat block with the Pacific and Kula plates for at least the last 50 Ma.

The collision of the Yakutat block with North America resulted in subduction of the block coincident with uplift of the Kenai Mountains. The extension of the Kenai Mountains into the Kodiak area suggests that a southwest extension of the Yakutat block collided with the Kodiak margin and was completely subducted. The subducted extension of the Yakutat block could have connected the now subducting head of Zodiak deep-sea fan to a North American source of sediment during deposition of the fan.

INTRODUCTION

A comparison of microfossils from the gulf of Alaska, with coeval assemblages from California, Oregon, and Washington, provide constraints on past positions of allochthonous terranes. We summarize the paleontologic data in a number of informal and unpublished reports of the U.S. Geological Survey (USGS), the State of Alaska, the Deep Sea Drilling Project (DSDP), and the Middleton Island exploratory well that was released to us by Tenneco Oil Company. From this biostratigraphy there emerges a paleo-oceanographic history of two terranes in the western Gulf of Alaska.

Recent geological and geophysical studies around the Gulf of Alaska indicate a Cenozoic northward drift of large crustal fragments that are now lodged in the Alaskan continent (Stone and Parker, 1979; Plumley et al, 1982; Stone et al, 1982; Bruns, 1983a; Plafker, 1983; Moore et al, 1983). The movement histories proposed are based on incomplete data, and although all schemes agree on northward drift, each is different. The major differences involve the amounts of northward motion and the timing of this motion. The biostratigraphic data presented here provide additional constraints on the northward drift history suggested by geologic and paleomagnetic data.

The Prince William terrane and Yakutat block have moved north during the Cenozoic (Jones et al, 1981; Rogers, 1977). In the Kodiak group of islands, the Prince William terrane of Jones et al (1981) includes not only some insular terrane but also the shelf adjacent to the Kodiak group of islands and Kenai Peninsula (Fig. 1). The Yakutat block, to the north and east, is bounded on the northeast by the Fairweather fault and the Chugach–Saint Elias thrust fault system, on the west by the thrust faults paralleling Kayak Island and crossing the adjacent shelf, and on the south by the base of the continental slope. The biostratigraphic and paleoclimatic records presented here set constraints independent of plate tectonic models on some positions during the Cenozoic migration of these terranes.

Our study, based on new materials from the Middleton Island well, cores and dredge samples off Kodiak Island, and on a review of original materials from studies of other investigators, is reported in a more extensive paper by Keller et al (1984). We focus here on the tectonic consequences of that study and add some speculations with regard to terranes that may have been completely subducted just as the western part of the Yakutat block is presently being subducted beneath Alaska.

BIOSTRATIGRAPHIC AND PALEO-OCEANOGRAPHIC ANALYSES

Samples from the onshore and offshore regions of the Gulf of Alaska have been examined for planktonic and benthonic foraminifers, coccoliths, and diatoms to determine their age and paleoclimatic conditions of deposition. To obtain relative paleolatitudes, faunal assemblages were correlated with onshore marine sequences of California, Oregon, and Washington in areas

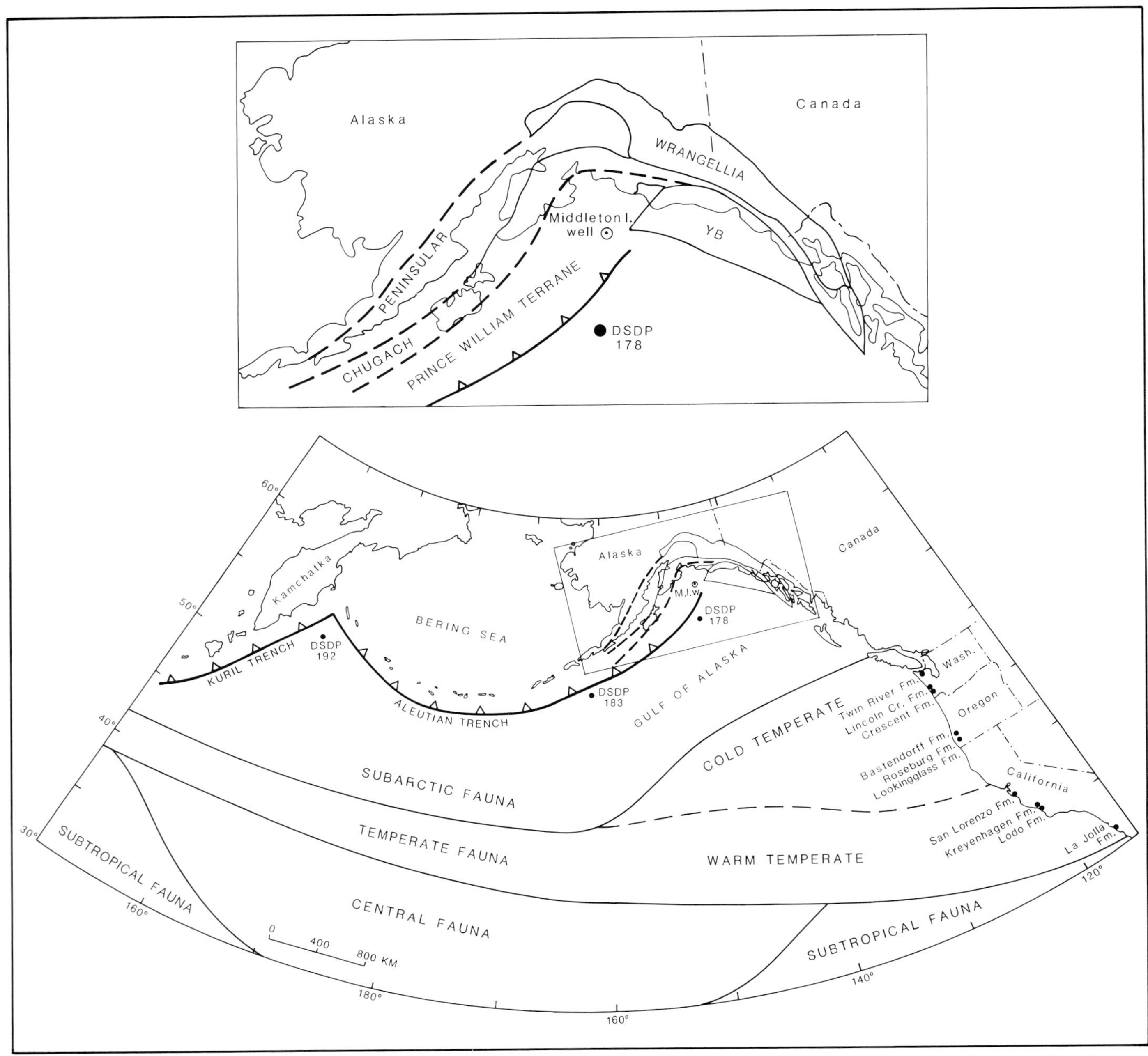

Figure 1—Location map of sites studied, terrane boundaries after Jones et al (1981), and geographic distribution of present faunal provinces. YB = Yakutat block.

that were stable since the sequences were deposited. In addition, general paleolatitudes are derived from comparison with the present distribution of planktonic faunal provinces (subtropical, temperate, subarctic; Fig. 1). The comparisons with present provinces are adjusted for the effects of major paleoclimatic changes. The main effect of globally warmer conditions, for instance, during early and middle Eocene time, was a northward shift of the faunal provinces; conversely, during globally cooler conditions of latest Eocene and Oligocene time, the subarctic fauna expanded southward. Therefore, our paleolatitude determinations based on planktonic faunal provinces may be 5 to 10° in error during times of climatic extremes.

Microfossil analyses were provided by several of our colleagues as well as from published sources. Coccoliths examined earlier and for this report were analyzed by Bukry (in Plafker et al, 1979, 1980; Poore and Bukry, 1978). Benthonic foraminifers were examined and reported by Rau (Rau et al, 1977; Rau, 1978; Plafker et al, 1979, 1980) and for this study by McDougall (Keller et al, in press). Planktonic foraminifers were reported for three dredge samples by Poore (Poore and Bukry, 1978) and were examined for this study by Keller. Diatoms originally studied by Koizumi (1973) for sites 192 and 183 and Schrader (1973) for site 178 were reexamined by Harper (1977) and by Barron (unpublished data). Paleontologic data are summarized in Figure 2 with respect to age, paleodepth and paleoenvironment; time scale is after Berggren et al (in press).

(Rau et al, 1977) and in the Middleton Island well, but these assemblages are commonly dominated by arenaceous forms characteristic of cooler shelf and slope faunas. Late Eocene assemblages from California (Mallory, 1959) and Japan (Ujie and Watanabe, 1960) also contain many of the same species as found in rocks of the Yakutat block; however, these more southern assemblages also contain warmer water species not present in Alaskan latitudes. These faunal comparisons suggest that the late Eocene to early Oligocene Yakutat block samples were deposited in an environment similar to samples from southwestern Washington and Oregon, or about 45 ± 5°N present latitude, in agreement with the latitude of deposition indicated by planktonic foraminifers and by plate tectonic reconstruction (Table 2).

PRINCE WILLIAM TERRANE: MIDDLETON ISLAND WELL

Microfossil assemblages from the Middleton Island well (present latitude 59°25′N) are generally sparse and poorly preserved in the upper part of the well; however, in the lower part definite age assignments could be made. Sixty-six samples were examined for coccolith and foraminifers. Twenty of these contained coccoliths, but only four samples have diagnostic or diverse assemblages adequate for good age correlations. Planktonic foraminifers are also poorly represented, whereas benthic foraminifers are more common. The Middleton Island well was studied earlier by Rau for benthic foraminifers (Rau et al, 1977) with the few samples available at that time. Correlations and faunal comparisons were made between the Middleton Island well and sediments of the Yakutat block described in Plafker et al (1979, 1980), Rau et al (1977), and Poore and Bukry (1978). All the samples were reexamined for planktonic foraminifers by Keller et al (1984).

Late Middle Eocene: 40 to 42 Ma

The basal sample of the Middleton Island well (3,658 to 3,642 m [12,001–11,949 ft]) contains the coccolith index species for the early late Eocene (38.5 to 40 Ma). Coccolith assemblages of sample 3,639 to 3,642 m (11,939–11,949 ft) contain forms that suggest the late middle Eocene. The overlying sample (3,633 to 3,637 m [11,919–11,932 ft]) contains forms that suggest a late Eocene age in agreement with the age determined from planktonic foraminifers. Coccolith stratigraphy therefore indicates that the middle to late Eocene boundary falls between sample 3,642 m (11,949 ft) and 3,636 m (11,929 ft) with a possible hiatus. Therefore, at its greatest depth the Middleton Island well penetrated late middle Eocene or earliest late Eocene sediment (40 to 42 Ma).

Benthic foraminifers are rare in the lower part of the Middleton Island well (3,658 to 3,258 m [12,001–10,689 ft]), but they also suggest a middle to late Eocene age. Rau et al (1977) reported a more diverse benthic fauna of questionable early to middle Eocene age, which could not be confirmed by Keller et al (1984).

The low-diversity microfossil assemblages in the Middleton Island well, as compared to coeval assemblages from dredge hauls from the continental slope of the Yakutat block, indicate that the Middleton Island assemblages were deposited in a significantly cooler water environment. This relation suggests that the Middleton Island late middle to early late Eocene fauna probably accumulated north of coeval sequences dredged from the Yakutat block.

Late Eocene: 36 to 38 Ma

Planktic foraminiferal assemblages in samples between 3,495 and 3,423 m (11,467 and 11,260 ft) are indicative of the late Eocene (36.5 to 38 Ma), although in the absence of the index species, it could be as young as early Oligocene. Coccolith assemblages are determinate late Eocene to early Oligocene. This sample is coeval with a dredge sample from the Yakutat block, as discussed earlier. The low faunal diversity and absence of warm-water species in both the Middleton Island well and the Yakutat sample suggest a similar cool environment indicative of high northern latitudes.

Latest Eocene and Early Oligocene: 34? to 38 Ma

Middleton Island well samples between 3,347 and 3,207 m (10,981 and 10,522 ft) contain planktonic foraminiferal assemblages indicative of late Eocene or early Oligocene age. Coccolith assemblages between 3,505 and 3,613 m (11,499 and 11,854 ft) can, at best, be assigned a late Eocene to early Oligocene age. This age is in good agreement with age determinations based on planktonic foraminifers.

Planktic microfossils are rare between 3,240 and 2,347 m (10,630 and 7,700 ft) but suggest that this interval is also of late Eocene or early Oligocene age. No planktonic microfossils are present between 2,347 and 890 m (7,700 and 2,920 ft); the latter sample contains latest Oligocene or early Miocene planktonic foraminifers.

Benthic foraminifers of late Eocene age are also observed between 3,246 and 1,329 m (10,650 and 4,360 ft). Mixed assemblages of Narizian and Refugian Stage species are present between 3,222 and 2,682 m (10,571 and 8,799 ft) with Refugian Stage species more common in the upper part. Similar mixed Narizian and Refugian Stage species assemblages were observed in coeval sediments of Washington and Oregon (McDougall, 1980). The late Eocene and early Oligocene age of the well section (Narizian to Refugian; Poore, 1980) agrees well with the age determinations based on planktonic microfossils. The benthic foraminiferal assemblages between 3,246 and 2,673 m (10,650 and 8,770 ft) contain many species characteristic of middle to lower bathyal depths with an influx of transported material from a nearby slope and shelf region.

The long-ranging, low-diversity planktic foraminifers from the late Eocene and early Oligocene section of the Middleton Island well (present latitude 59°25′N) are indicative of a cold water environment similar to coeval assemblages of the Lincoln Creek Formation of southwest Washington at 47°N present latitude. The cooler assemblages at the Middleton Island well, however, appear to have lived at a somewhat higher latitude than those of the Lincoln Creek Formation. We think the Middleton Island late Eocene sediment was deposited in an environment north of 50 ± 5°N present latitude (Table 1).

Coccolith assemblages also indicate a cold-water

environment as suggested by poor diversity and by the lack of tropical taxa. These cold-water assemblages are most similar to coeval assemblages from DSDP site 192 in the western North Pacific and also indicate that deposition occurred in colder waters than assemblages from southwestern Washington.

Oligocene (32? to 36 Ma)

Benthic species diagnostic of the Oligocene are present between 6,289 and 5,193 m (20,633 and 17,037 ft) (see also Rau et al, 1977). No planktonic microfossils are present. Benthic assemblages indicate deposition at middle to lower bathyal depths with continued transport from shelf and upper-slope regions.

Miocene to Pleistocene

Rare early Miocene species occur in sample 5,193–5,219 m (17,037–17,123 ft) and more commonly in sample 3,810–3,836 m (12,500–12,585 ft). Planktic foraminifers in sample 3,810–3,836 also suggest an early Miocene or very latest Oligocene age. Deposition continued in a lower to middle bathyal environment.

The upper part of the Middleton Island well, 3,001 to 939 m (9,846–3,081 ft), is of late Miocene to Pleistocene age (undifferentiated). Thus, a hiatus is present between the early and late Miocene deposits, a circumstance also observed in the nearby deep-ocean DSDP site 178 and noted by Lagoe (1983) in samples from Cape Yakataga. The presence of shallow-water benthic species, including numerous *Elphidiums*, suggest deposition at shelf depths or transport from inner shelf regions.

PACIFIC PLATE

DSDP cored sequences provide sufficient continuity to show the chronostratigraphic and paleoclimatic record and variations in the sedimentation patterns that result from large-scale changes in oceanographic conditions. Figure 3 illustrates the chronostratigraphic and lithologic records of deep-sea sequences of DSDP sites 192, 183, and 178, which are presently at high northern latitudes, and the Middleton Island well. The Yakutat block is not included because no similar cored sequence is available. The DSDP sites are on the Pacific plate and hence, during their northward migration, their relative positions on the plate remained constant through time. The related position of the Middleton Island well located on the Prince William terrane is less well known. Because of similar faunal and floral records (Fig. 3), the sediment at the Middleton Island well was probably deposited within the same general paleo-oceanographic and paleoclimatic regime as those of the DSDP sites.

Late Cretaceous

The oldest sediment overlying basement at site 192 (Hole A) is of Maestrichtian age (70 to 68 Ma). Coccolith assemblages indicate a tropical to subtropical environment; however, absence of siliceous microfossils suggest that deposition occurred north of the high productivity equatorial region (Worsley, 1973), or about 15 ± 5°N present latitude. This position is in good agreement with the predicted position of 19°N latitude based on plate tectonic reconstructions by Engebretson (1982, Table 2).

Sediment recovered from the Ghost Rocks Formation of the Kodiak Islands was considered to be of Late Cretaceous and Paleocene age by Byrne (1982). Based on paleomagnetic data, Plumley et al (1982) infer a Paleocene (62 Ma) paleolatitude of the Kodiak Islands about 17 ± 6° south of their present latitude. No sediment of Paleocene age was recovered in the north Pacific DSDP sites. At site 192 a hiatus marks the position of the uppermost Cretaceous, the Paleocene, and part of the early Eocene (68 to 51 Ma), and at site 183 early Eocene sediments rest on basement basalt (Fig. 3).

Eocene

Eocene sedimentation is disrupted by hiatuses at the early–middle Eocene boundary (49 to 46 Ma), middle–late Eocene boundary (43 to 41 Ma), and during late Eocene (39 to 37.5 Ma, Figs. 2,3). Early Eocene (51 to 49 Ma) coccolith assemblages at site 192 suggest that deposition occurred in the low-productivity central water mass (Fig. 1, Worsley, 1973), or at about 30 ± 5°N latitude and hence at a similar latitude to coeval assemblages of the Yakutat block. Plate tectonic reconstructions place site 192 and the Yakutat block at 31°N present latitude (Table 2), in agreement with the latitudinal limits determined from faunal evidence.

Middle Eocene sediment at site 192 indicates deposition within the central to temperate water masses, or about 40 ± 5°N latitude, as compared to 36°N latitude determined by plate reconstruction (Table 2). Late Eocene assemblages (37 to 40 Ma) are indicative of a cold subarctic environment, or about 45 ± 5°N latitude. Plate tectonic reconstruction suggests a somewhat lower position of 38°N present latitude (Table 2).

Cold subarctic assemblages are also present in the late Eocene sediment of site 183, although preservations of microfossils is poor. Wolfe (1977) reports pollen assemblages dominated by conifers, similar to coeval assemblages from the Alaskan mainland and cooler than coeval assemblages from Washington and Oregon. It was also observed earlier that foraminifers from the Middleton Island well indicated a position north of the Lincoln Creek Formation of Washington (i.e., north of 50°N latitude).

Thus, late Eocene to early Oligocene assemblages (34 to 37 Ma) from the Pacific (site 192), Prince William terrane (Middleton Island well), and the Yakutat block contain assemblages from the subarctic faunal province or about 50 ± 5°N present latitude. In contrast, plate tectonic reconstruction indicates that the late Eocene and early Oligocene position of sites 192 and 183 was about 10° further south, and the Yakutat block was about 12° further south (Bruns, 1983a; Table 2). This discrepancy may be explained by a southward expansion of the subarctic fauna in the latest Eocene and Oligocene resulting from initiation of polar glaciation (Keller, 1983a, 1983b; Keigwin and Keller, 1984). Therefore, the effect on faunal paleolatitude determinations would be a bias towards cooler, that is, higher latitudes at this time.

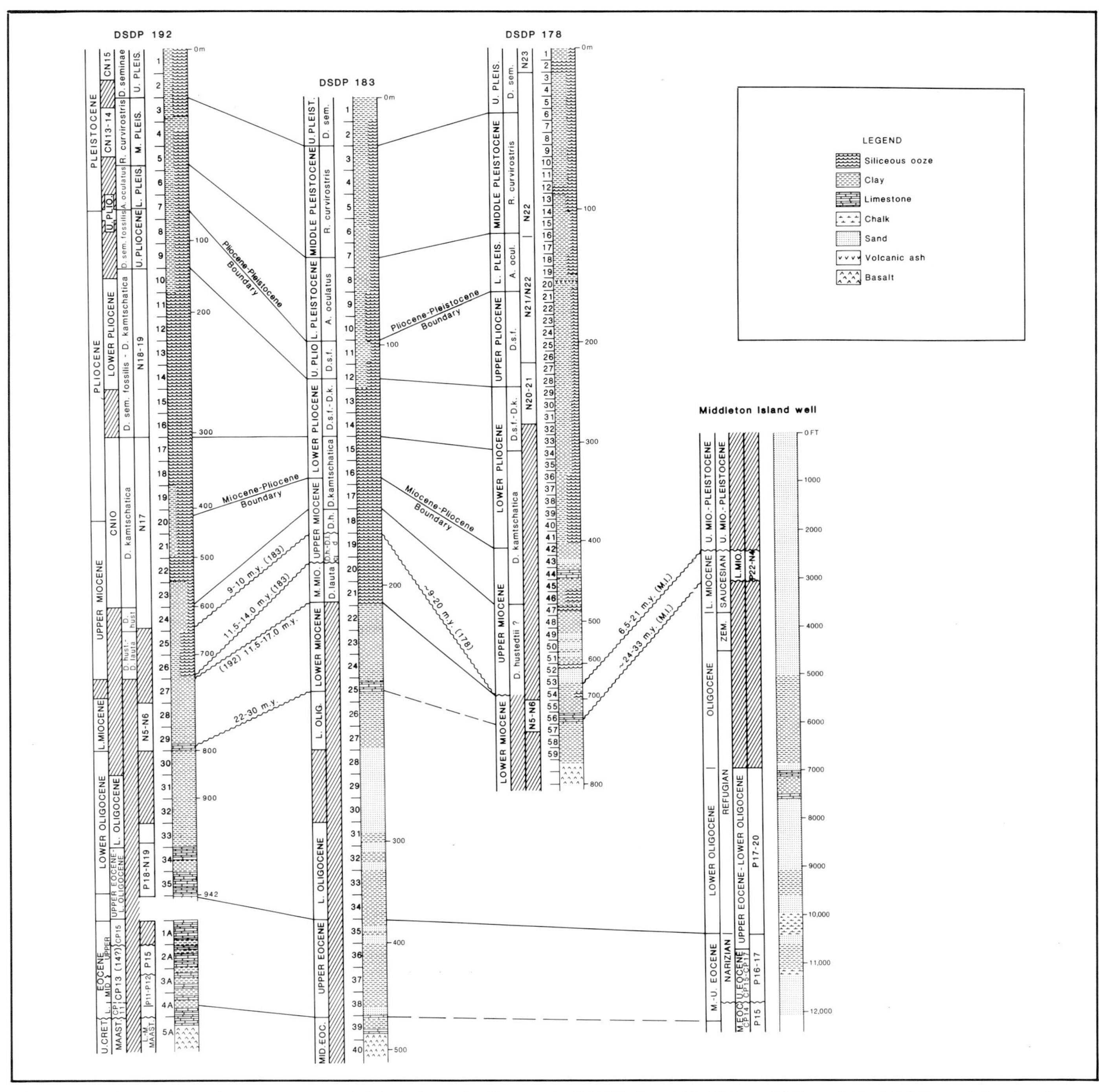

Figure 3—Biostratigraphic correlation and lithologies of sediments in DSDP sites 192, 183, and 178 and the Middleton Island well. Dashed zigzag lines mark hiatuses. Diagonal lines mark dissolution in sediments. The biostratigraphy at DSDP sites has been previously studied: nannofossils by Worsley (1983), diatoms for sites 192 and 183 by Koizumi (1973), Harper (1977), and Barron (unpublished data), foraminifers by Echols (1973) for sites 192 and 183 and by Keller (this report) and for site 178 by Ingle (1973). Absolute age scale after Berggren et al (in press).

Oligocene

Microfossil preservation in Oligocene sediment is poor, and a hiatus appears to represent the late Oligocene (30 to 20 Ma, Fig. 2). In addition, late Eocene to early Oligocene assemblages are commonly indistinguishable and indicate cold subarctic conditions. This circumstance suggests that by Oligocene time deposition at the North Pacific DSDP sites, the Prince William terrane, and the Yakutat block occurred in cold water north of Washington, or north of 50 ± 5°N present latitude.

Miocene to Pleistocene

Early Miocene sediment (23 to 20 Ma) is present at all North Pacific DSDP sites and the Middleton Island well (Fig. 2). Early Miocene assemblages are found in a limestone layer at sites 192, 183, and 178 and sediment enriched in carbonate at the Middleton Island well. Hiatuses are present throughout the Miocene sequences at sites 192 and 183, in the late Miocene (10 to 9 Ma), middle Miocene (14 to 11.5 Ma), and early Miocene (20 to 15 Ma). At site 178 and Middleton Island well, hiatuses occur between 9 to 20 Ma and 6.5 to 20 Ma, respectively.

Biostratigraphic control in the late Miocene to Pleistocene sequences is excellent, except for the Middleton Island well where correlation is more tenuous because of poor preservation of microfossils. Faunal and floral assemblages are typical of a subarctic to arctic environment.

Terrane Kinematics

The concept that the Prince William terrane and Yakutat block migrated northward *during the Cenozoic* has been addressed in paleomagnetic studies and plate tectonic reconstructions (Stone et al, 1982; Plumley et al, 1983; Moore et al, 1983; Bruns, 1983a) and are illustrated in Figures 4 and 5 with positions from our faunal data. Paleomagnetic data from the Prince William terrane (Plumley et al, 1983; Moore et al, 1983) show that early Paleocene (about 62 Ma) volcanic rocks of the Ghost Rocks Formation of Kodiak Island were emplaced at 25 ± 9° south of their present position (Figs. 4, 5). Moore et al (1983) present three models for movement of the Prince William terrane. Paleontologic data are consistent with these models but do not aid in differentiating between them. The Ghost Rocks Formation and the Middleton Island well are on the same terrane (Jones et al, 1981) and thus migrated northward together. Our paleontologic data indicate that the Middleton Island middle Eocene sediment was deposited about 8 ± 5° south of its current location.

Paleontological data indicate that the Yakutat block moved into the Gulf of Alaska after the Prince William terrane. The foraminiferal fauna of the Yakutat block are similar to coeval late early Eocene fauna of California at 36°N present latitude, to middle Eocene fauna of northern California at 40°N present latitude, and to late Eocene fauna of southwest Washington (Table 1) at about 45°N present latitude. Oligocene fauna (about 34 Ma) suggest a high-latitude depositional environment north of about 50 ± 5°N present latitude. Fauna of equivalent ages from the Prince William terrane and Yakutat block show that the Yakutat block was at least 10° south of the Prince William terrane during middle Eocene time and reached a similar latitude no earlier than middle Oligocene time.

Recent geological and geophysical studies of the Yakutat block indicate that it is currently colliding with, and accreting to, southern Alaska (Plafker et al, 1978; von Huene et al, 1979b; Bruns, 1979; Perez and Jacob, 1979; Lahr and Plafker, 1980; Bruns and Schwab, 1983) and that it has moved with the Pacific plate for at least Pliocene and Quaternary time (Schwab et al, 1980; Bruns, 1983a). One plate tectonic model for the movement history of the Yakutat block, which suggests that it moved with the Pacific and Kula plates since the Eocene (Fig. 5; Bruns, 1983a), agrees with the predicted position of the Yakutat block and the Eocene and Oligocene positions indicated by this study. Thus, within the limits of uncertainty in the latitude determinations, the paleontological data and terrane path reconstruction are consistent.

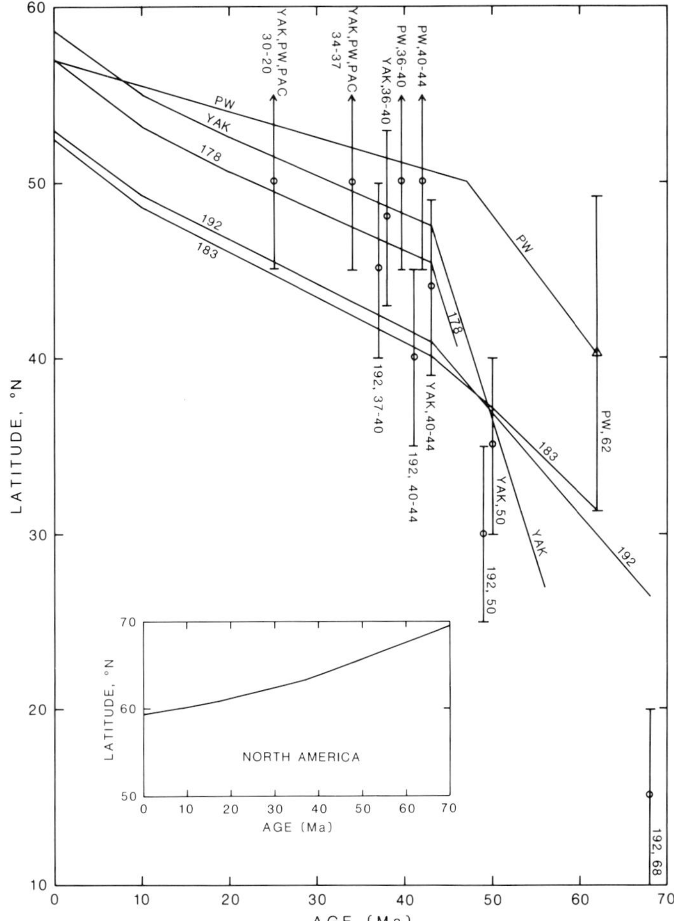

Figure 4—Plot of absolute paleolatitude versus age of sites studied in this report. All paleolatitudes have been adjusted for motion of Pacific and North America plates based on the movement of these plates relative to the fixed hotspot reference of Engebretson (1982). Travel paths relative to a fixed North America reference frame are shown in Figure 5. Circles with error bars indicate best estimate of paleolatitude positions determined from faunal data. Triangle with error bar marks paleomagnetic data of Plumley et al (1982, 1983). Labeled lines indicate latitude positions of indicated terranes and DSDP sites according to plate tectonic reconstructions of Engebretson (1982), Moore et al (1983), and Bruns (1983a). Inset shows southward drift of North America plate during last 70 Ma. Total closure between northward-moving terranes and North America plate can be determined by subtracting paleolatitudes of equivalent ages.

In summary, paleolatitudes determined from microfaunas and floras in the Gulf of Alaska coastal and offshore areas provide three main tectonic constraints: (1) The Prince William terrane was at a high northerly latitude near its present position by 40 to 42 Ma. This observation is consistent with paleomagnetic data from Kodiak Island (Plumley et al, 1983), provided the Prince William terrane traveled northward with the fast-moving Kula plate from 62 to 45 Ma and was incorporated into Alaska by about 45 Ma. (2) Fossil assemblages from the Yakutat block require northward motion of about 30° relative to a North American point of origin and also require that the Yakutat block was about 10° south of the Prince William terrane through Eocene and Oligocene time. These data are

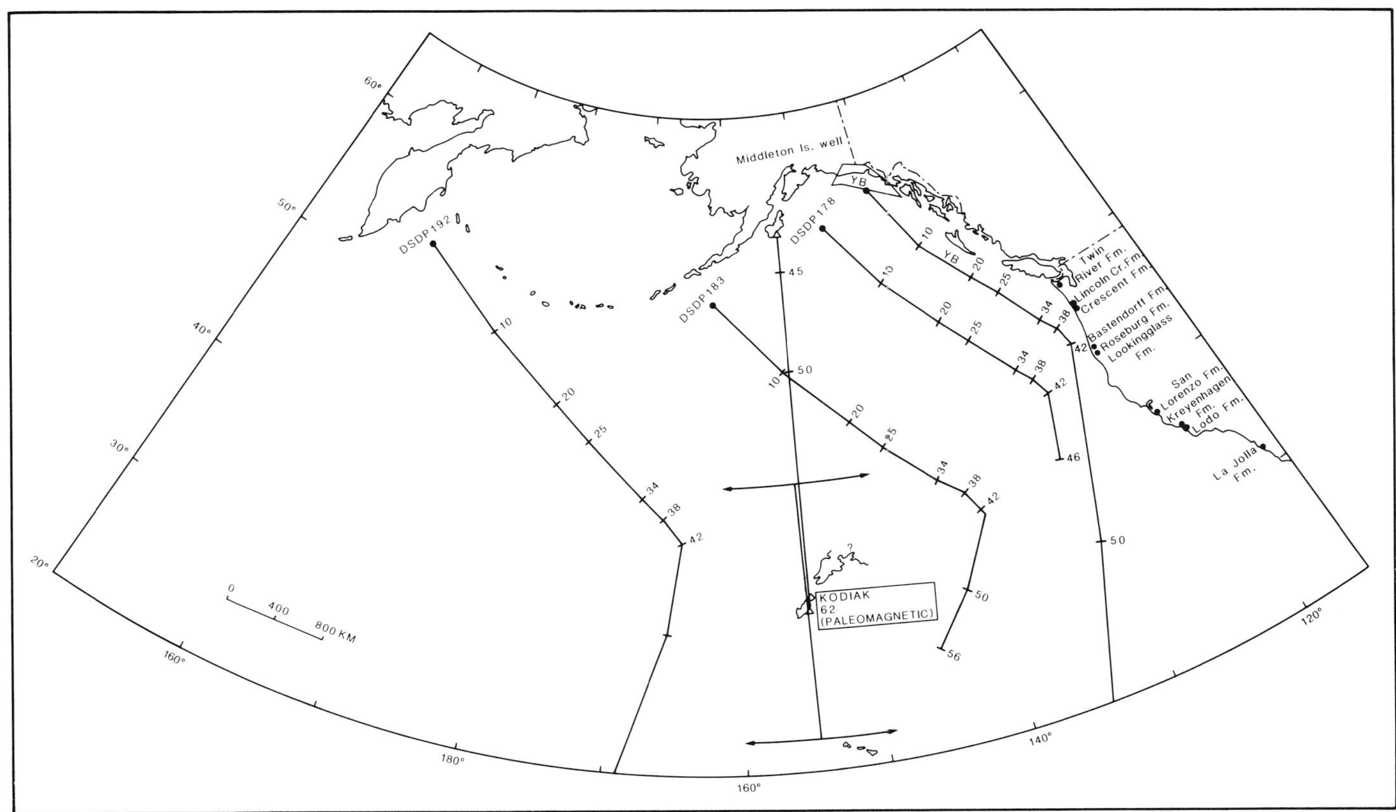

Figure 5—Backtrack positions of sample localities described in this paper (relative to a North America plate fixed in its current position). Heavy solid lines show movement paths according to Pacific–North America poles of Engebretson (1982), and models for Prince William terrane of Moore et al (1983) and for the Yakutat block (YB) of Bruns (1983a). Numbers along lines indicate positions in million years. Latitude position of Kodiak Island at 62 Ma based on paleomagnetic data of Plumley et al (1982, 1983).

consistent with motion of the Yakutat block with the Pacific and Kula plates during the past 50 Ma, as suggested by Bruns (1983a). (3) Data from the Pacific plate require about 40° of northward motion and are consistent with the plate tectonic reconstruction of Engebretson (1982) for Pacific plate motion based on hot-spot tracks.

IMPLICATIONS OF THE YAKUTAT BLOCK HISTORY

The 30° northward transport of the Yakutat block suggested by paleontologic data and the interaction of the block with the Alaskan continent are dynamic examples of processes inferred from ancient inactive terrane sutures. The Yakutat block has collided with Alaska on the north and has been subducting beneath Alaska on the west for the past 4 m.y. (Bruns, 1983a). Subduction of the block is shown by a distinctive magnetic anomaly that extends along the seaward margin of the Yakutat block and then continues about 280 km (174 mi) across the shelf west of the zone where the block is subducted. The anomaly reaches the Kenai Peninsula (Schwab et al, 1980), but its full extent has not been determined by magnetic surveying on land. Where the magnetic anomaly crosses the shelf, the amplitude decreases, indicating an increase in depth of the causative body. The increased depth indicates subduction of the Yakutat block. Thus, we infer a minimum length of the block now subducted beneath the late Mesozoic and early Tertiary rocks of the continent (Schwab et al, 1980; Bruns, 1983b). The minimum length of the subducted anomaly in the direction of plate convergence (258 km [160 mi]) divided by the rate of Pacific–North American relative plate convergence (6.6 cm/year [2.6 in./year]) gives a nearly 4 m.y. minimum age for initial subduction of the west end of the anomaly.

Another possible sign of subduction of a terrane is seen about 40 km (25 mi) west and 60 km (37 mi) south of the end of the magnetic anomaly in the passage that separates the Kenai from the Kodiak Mountains. Fisher et al (1983) have pointed out a 30 km (19 mi) long sequence of reflections that plunge from a depth of 11 km to 23 km (6.8–14 mi) in an arch that parallels the top of the Benioff zone (Fig. 6). Using their suggested velocities and the time interval from top to bottom of the reflective sequence indicates a layer from 2 to 5 km (1.2–3.1 mi) thick. One cause these authors suggest for the reflections is a sedimentary rock sequence that could have been detached from the subducting plate. We believe that sediment sequence of this thickness might represent part of a subducted terrane.

Subduction of the westernmost end of the Yakutat block at least 4 m.y. ago corresponds roughly with the initial uplift of the present Kenai Mountains. This uplift is marked in Cook Inlet by the introduction of detritus from the mountain range and the end of a sediment supply from the interior of the continent during the Pliocene, or about 5 m.y. ago (Kirschner and Lyon, 1973). Prior to the mountain-building phase, the Kenai area was an eroded

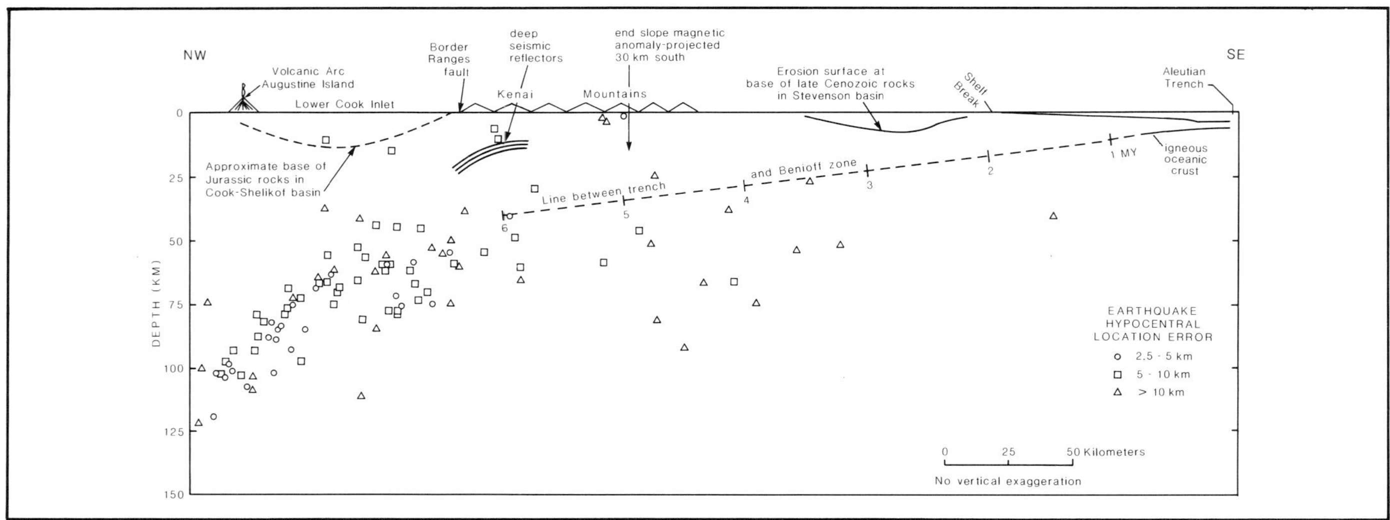

Figure 6—Diagrammatic section through the waterway between the Kenai Peninsula and Kodiak group of islands and normal to the trench showing some tectonic features of the Prince William terrane, the extent of the subducted continuation of the Yakutat block slope magnetic anomaly, and the deep layered seismic reflections. The component of convergence along the line of section is 4.7 cm/year (1.9 in/year), and the time in Ma since introduction of a point on the oceanic plate is shown by numbers along the line between the trench and the Benioff zone. Earthquake hypocenters located from records of a local network during the period 1971 to 1980 are projected from 20 km (12 mi) on either side of the line of section.

lowland. The concordant summits of the Kenai Mountains reflect a pre-Pliocene erosional surface that extended across the present mountainous areas. On the adjacent continental shelf a regional erosional surface forms the base of a reflective sequence in most seismic reflection records (Fisher and von Huene, 1980; Bruns, 1979). This extensive erosion surface was disrupted by mountain building and subsidence in the present land and continental shelf areas, respectively.

The Kenai–Kodiak Mountains are in a peculiar position and have no counterpart in the rest of the Aleutian arc-trench system. Forearc areas are generally not the sites of high mountains. Furthermore, the Benioff zone beneath the Kenai–Kodiak area has an anomalous shallow dip (4 to 5°) and the distance between the trench and arc (almost 400 km [249 mi]) is unusually great (Jacob et al, 1977). Thus, some process in addition to normal subduction of ocean crust might be considered.

The deep crustal mechanism by which subduction of a block causes the uplift of mountains is conjectural; however, the concurrent uplift of the Kenai Mountains, depression of the shelf, and the subduction of the Yakutat block suggest a genetic relation. If the Kenai Mountains were indeed formed by the collision and subduction of the Yakutat block, this relation could provide an explanation for some unresolved tectonic problems along the western Gulf of Alaska, and we present these ideas to stimulate further consideration.

If uplift of the Kenai Mountains is related to the collision and subduction of the Yakutat block, then it begs the question of the origin of the Kodiak Mountains, which are the southwest continuation of the Kenai Mountains. The Kodiak area has the unusually great arc-trench separation and the shallow-dipping Benioff zone noted beneath the Kenai Peninsula and Prince William Sound (von Huene et al, 1979a). Did the Yakutat block once extend southwest of its present boundaries on part of the ocean crust now subducted beneath the Kodiak group of islands? Could the subduction of a southern continuation of the Yakutat block be responsible for the Kodiak Mountains? Not only does the anomalous position of the Kodiak Mountains suggest such a subducted continuation of the block, but so does the Zodiak Fan.

The Zodiak Fan has been an unexplained feature of the western Gulf of Alaska since its discovery by Hamilton (1967). Stevenson et al (1983) have defined the extent of this huge fan complex and show that its proximal part is subducting just south of the Kodiak group of islands (Fig. 7). Thus, the feeder channel and associated continental margin are on the now subducted Pacific plate beneath the Kodiak area. Stevenson et al (1983) considered several models to connect Zodiak Fan to a North American source of sediment, but these remained equivocal.

We reconstruct a terrane convergence history using the plate motion model of Engebretson (1983). The subducted leading or northwest edge of the Yakutat block and its extension are positioned beneath the Kenai–Kodiak mountains at the time mountain building started. The present position is then derived by moving the leading edge with the Pacific plate during the past 5 m.y. (Fig. 7). The trailing edge position of the subducted block beneath the Kodiak area is poorly constrained by geophysical data but can be estimated from the position of the head of Zodiak fan, assuming its apex was at the base of the extension of the Yakutat block. This position also corresponds to the southwest end of the Kenai–Kodiak mountains. The shape of the Yakutat block extension was diagrammed based on these assumptions (Fig. 7).

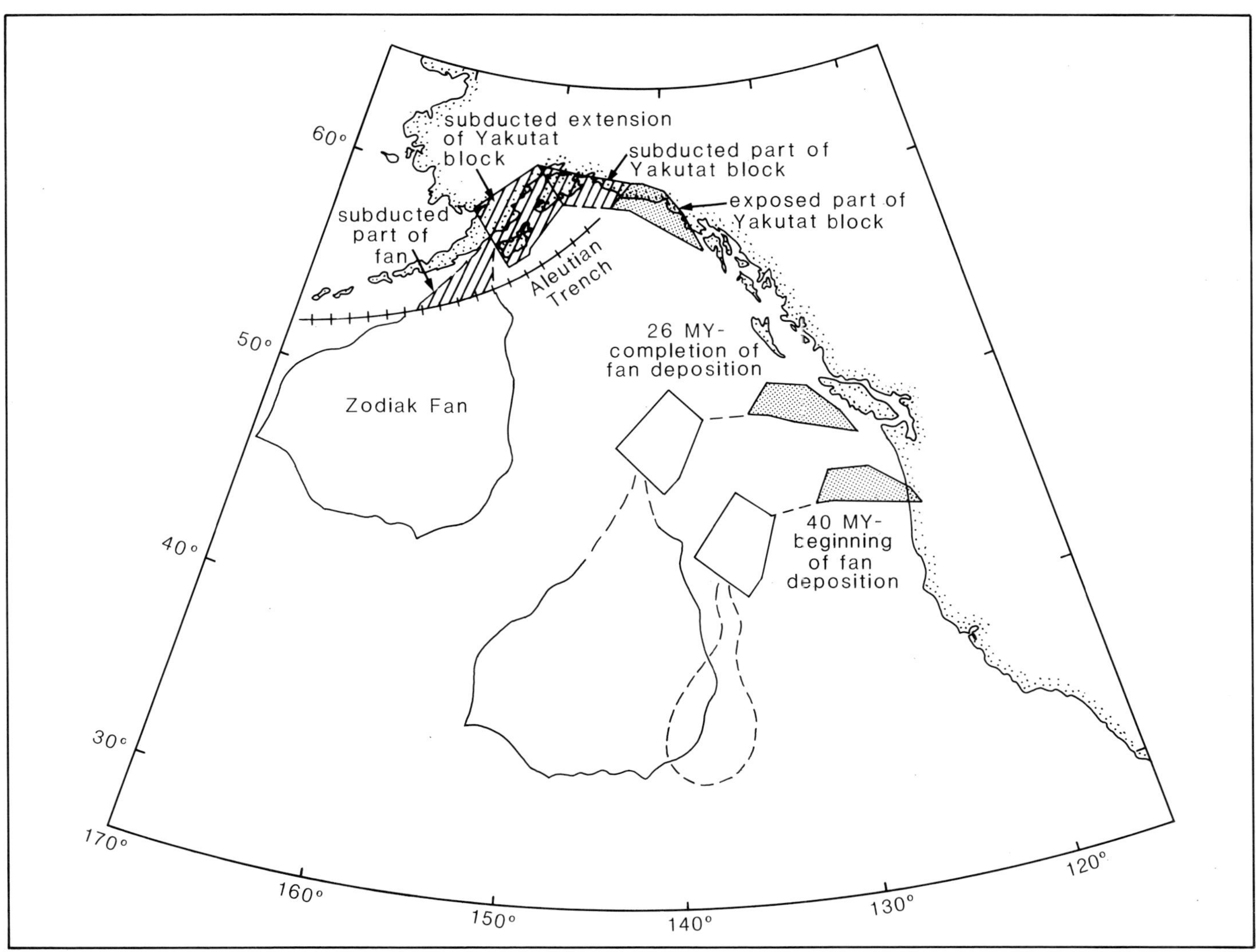

Figure 7—Reconstruction of a possible subducted extension of the Yakutat block beneath the Kodiak area. The area of the Pacific plate that subducted between 1 and 8 m.y. ago and may have extended between Zodiak Fan and the Yakutat block is indicated by hatchurs. Position of the Yakutat block, its possible southwest extention, and Zodiak Fan are shown when the fan first began to develop (40 Ma), when fan sedimentation ended (26? Ma), and in its present position (subducted parts hatchured). Migration of the Yakutat block and its subducted continuation along the magnetic anomaly marking the slope of the block is after Bruns (1983a).

Zodiak fan began to receive sediment about 40 m.y. ago, and construction of the fan ended about 26 m.y. ago (Stevenson et al, 1983). If the reconstructed Yakutat block and Zodiak fan are rotated back to the 40 m.y. and 26 m.y. positions, they are near the North American continent. Therefore, during deposition the fan might have been connected by the Yakutat block to drainages from Washington State and British Columbia (Fig. 7). A North American sediment source for the large amount of material in Zodiak fan is very appealing, but not without some possible problems. Stevenson has pointed out the possibility of a spreading ridge separating the Yakutat block from our postulated Yakutat block extension about 40 m.y. ago or the alternate possibility of a transform fault terminating the spreading ridge similar to the scheme of Bruns (1983a).

Our restoration is consistent with present plate motion models and we consider it an attractive working hypothesis because it involves an adequate source of sediment for Zodiak fan. It also provides an explanation for the episodic tectonic history of the Kodiak area (Fisher and von Huene, 1980) during a period of constant plate convergence. The Kodiak shelf was subaerially eroded in mid-Miocene time as shown by the widespread erosion surface in seismic records and the regional hiatus (Fig. 2). In contrast to the erosion surface are the basins and mountains of the present tectonic regime that began to develop in late Miocene time. Plate tectonic models show no change in relative motion since Oligocene time (Engebretson, 1983) and thus in this subduction-dominated tectonic system the collision and subduction of a terrane may have caused the change in tectonic regime.

REFERENCES

Armentrout, J. M., et al, 1980, Cenozoic stratigraphy and late Eocene paleoecology of southwestern Washington, *in* Geologic field trips in western Oregon and southwestern Washington: Oregon Department of Geology and Mineral Industries Bulletin 101, p. 79–120.

Bandy, O. L., 1956, Ecology of foraminifera in northeastern Gulf of Mexico: U.S. Geological Survey Professional Paper 274-G, p. 179–204.

Barron, J. A., 1980, Lower Miocene to Quaternary diatom biostratigraphy of Leg 57, off northeastern Japan, *in* Initial reports of the Deep Sea Drilling Project: Washington, DC, U.S. Gov't. Printing Office, v. 56–57, p. 641–685.

Beck, Jr., M. E., 1980, Paleomagnetic record of plate-margin tectonic processes along the western edge of North America: Journal of Geophysical Research, v. 85, p. 7115–7131.

Beck, R. S., 1943, Eocene foraminifera from Cowlitz River, Lewis County, Washington: Journal of Paleontology, v. 17, p. 584–614.

Berggren, W. A., et al, in press, Paleogene geochronology and chronostratigraphy, *in* Geochronology and the geological record: Geological Society of London Special Paper.

Blondeau, A., and E. E. Brabb, 1983, Large foraminifers of Eocene age from the coast ranges of California, *in* E. E. Brabb, ed., Studies in Tertiary stratigraphy of the California Coast Ranges: U.S. Geological Survey Professional Paper 1213, p. 41–48.

Blow, W. H., 1969, Late Middle Eocene to Recent planktonic foraminiferal biostratigraphy, *in* P. Bronnimann and H. H. Renz, eds., First international conference on planktonic foraminifera from the Oligocene-Miocene Cipero and Lengua Formations of Trinidad: U.S. National Museum Bulletin 215, p. 97–123.

Bruns, T. R., 1979, Late Cenozoic structure of the continental margin, northern Gulf of Alaska, *in* A. Sisson, ed., The relationship of plate tectonics to Alaska geology and resources: Anchorage, Proceedings of the Alaska Geological Society Symposium, 1977, p. 11–130.

———, 1983a, A model for the origin of the Yakutat block, an accreting terrane in the northern Gulf of Alaska: Geology, v. 11, p. 718–721.

———, 1983b, Structure and petroleum potential of the Yakutat segment of the northern Gulf of Alaska continental margin: U.S. Geological Survey Miscellaneous Field Studies Map MF-1430, 22 p. and 3 sheets, scale 1:5000,000.

———, and W. C. Schwab, 1983, Structure maps and seismic stratigraphy of the Yakataga segment of the continental margin, northern Gulf of Alaska: U.S. Geological Survey Miscellaneous Field Studies Map MF-1424, 3 sheets scale 1:250,000.

Byrne, T., 1982, Structural geology of coherent terranes in the Ghost Rocks Formation, Kodiak Islands, Alaska, *in* J. K. Leggett, and O. Blackwell, eds., Trench-forearc geology; sedimentation and tectonics on modern and ancient active plate margins: Geological Society of London, p. 229–242.

Chase, C. G., 1978, Plate kinematics: The Americas, East Africa, and the rest of the world: Earth and Planetary Science Letters, v. 37, p. 355–368.

Cushman, J. A., and G. D. Hanna, 1927, Foraminifera from the Eocene near Coalinga, California: California Academy of Science Proceedings 4th Series, v. 16, p. 205–228.

Drugg, W. S., 1959, Eocene stratigraphy of the Hoko River area, Olympic Peninsula, Washington: Master's Thesis, University of Washington, 192 p.

Echols, R. J., 1973, Foraminifera, Leg 19, Deep Sea Drilling Project, *in* Initial reports of the Deep Sea Drilling Project: Washington, DC, U.S. Gov't. Printing Office, v. 19, p. 721–736.

Engebretson, D. C., 1982, Relative motion between oceanic and continental plates in the Pacific basin: PhD Dissertation, Stanford University, p. 1–211.

Fisher, M. A., and R. von Huene, 1980, Structure of Upper Cenozoic strata beneath Kodiak shelf, Alaska: Bulletin of the American Association of Petroleum Geologists, v. 64, n. 7, p. 1014–1033.

———, et al, 1983, Possible seismic reflections from the downgoing Pacific plate, 275 km arcward from the eastern Aleutian Trench: Journal of Geophysical Research, v. 88, p. 5835–5849.

Hamilton, E. L., 1967, Marine geology of abyssal plains in the Gulf of Alaska: Journal of Geophysical Research, v. 72, p. 4189–4213.

Hanna, M. A., 1926, Geology of the La Jolla quadrangle, California: California University Publications, Department of Geological Sciences Bulletin, v. 16, p. 187–246.

Harper, H. E., 1977, Diatom biostratigraphy of the Miocene/Pliocene boundary in marine strata of the circum North Pacific: PhD Dissertation, Harvard University, p. 1–112.

Ingle, Jr., J. C., 1973, Neogene foraminifera from the northeastern Pacific Ocean, Leg 18, Deep Sea Drilling Project, *in* Initial reports of the Deep Sea Drilling Project: Washington, DC, U.S. Gov't. Printing Office, v. 18, p. 517–567.

Jacob, K. H., et al, 1977, Trench-volcanic gap along the Alaskan-Aleutian arc: facts and speculations on the role of terrigenous sediments, *in* M. Talwani and W. C. Pitman, III, eds., Island arcs, deep-sea trenches, and back-arc basins: American Geophysical Union, M. Ewing Series 1, p. 243–258.

Jones, D. L., et al, 1981, Map showing tectonostratigraphic terranes of Alaska, columnar sections, and summary description of terranes: U.S. Geological Survey Open-File Report 81-792, 2 p.

———, et al, 1983, Recognition, character and analysis of tectonostratigraphic terranes in western North America, *in* M. Hashimoto and S. Uyeda, eds., Accretion tectonics in the circum-Pacific regions: Tokyo, Advances in Earth and Planetary Sciences, Terra Scientific Publishing Company, p. 21–36.

Keigwin, L. D., and G. Keller, 1984, Middle Oligocene climatic change from equatorial Pacific DSDP Site 77B: Geology, v. 12, p. 1–16.

Keller, G., 1983a, Biochronology and paleoclimatic implications of middle Eocene to Oligocene planktonic foraminiferal faunas: Marine Micropaleontology, v. 7, p. 463–486.

———, 1983b, Paleoclimatic analyses of middle Eocene through Oligocene planktonic foraminiferal faunas: Paleogeography, Paleoclimatology, and Paleoecology, v. 42, p. 73–94.

———, et al, 1984, Paleoclimatic evidence for Cenozoic migration of Alaskan terranes: Tectonics, v. 3, n. 4, p. 473–495.

Kirschner, C. E., and C. A. Lyon, 1973, Stratigraphic and tectonic development of Cook Inlet petroleum province: American Association of Petroleum Geologists Memoir 19, p. 346–407.

Koizumi, I., 1973, The late Cenozoic diatoms of Sites 183–193, Leg 19, Deep Sea Drilling Project, *in* Initial reports of the Deep Sea Drilling Project: Washington, DC, U.S. Gov't. Printing Office, v. 19, p. 805–856.

Lagoe, M. B., 1983, Oligocene to Pliocene foraminifera from the Yakutat Reef section, Gulf of Alaska Tertiary Province, Alaska: Micropaleontology, v. 29, p. 202–222.

Lahr, J. C., and G. Plafker, 1980, Holocene Pacific-North America plate interaction in southern Alaska: implications for the Yakutaga seismic gap: Geology, v. 8, p. 483–486.

Lattanzi, R. D., 1979, Planktonic foraminiferal biostratigraphy of Exxon Company USA wells drilled in the Gulf of Alaska during 1977 and 1978: Journal of the Alaska Geological Society, p. 48–49.

Lyle, W., and J. Morehouse, 1978, Tertiary formations in the Kodiak Island area, Alaska, and their petroleum reservoir and source—rock potential: Alaska Department of Natural Resources, Division of Geological and Geophysical Surveys, Alaska Open-File Report 114, 47 p.

Mallory, V. S., 1959, Lower Tertiary biostratigraphy of the California Coast Ranges: Tulsa, OK, American Association of Petroleum Geologists, 416 p.

McDougall, K., 1980, Paleoecological evaluation of late Eocene biostratigraphic zonations of the Pacific coast of North America: Journal of Paleontology, v. 54, Society of Economic Paleontologists and Mineralogists Paleontology Monograph 2, 75 p.

Miles, G. A., 1977, Planktonic foraminifera of the Lower Tertiary Roseburg, Lookingglass and Flournoy Formations, southwest Oregon: PhD Dissertation, University of Oregon, Eugene, 359 p.

———, 1981, Planktonic foraminifers of the lower Tertiary Looking-glass and Flournoy Formations (Umpqua Group), southwest Oregon; Pacific Northwest Cenozoic Biostratigraphy, J. M. Armentrout, ed.: Geological Society of America Special Paper 184, p. 185–204.

Molnar, P., and T. Atwater, 1973, Relative motion of hot spots in the mantle: Nature, v. 246, p. 288–291.

Moore, J. C., et al, 1983, Paleogene evolution of the Kodiak Islands, Alaska: Consequences of Ridge-Trench Interaction in a more southerly latitude: Tectonics, v. 2, p. 265–294.

Murry, J. W., 1973, Distribution and ecology of living benthic foraminiferids: New York, Crane, Russak, and Company, Inc., 274 p.

Nuttall, W. L., 1930, Eocene foraminifera from Mexico: Journal of Paleontology, v. 4, p. 271–293.

Okada, H., and D. Bukry, 1980, Supplementary modification and introduction of code numbers to the low-latitude coccolith biostratigraphic zonation (Bukry, 1973, 1975): Marine Micropaleontology, v. 5, p. 321–325.

Perez, O. J., and K. H. Jacob, 1979, Tectonic model and seismic potential of the eastern Gulf of Alaska and Yakataga seismic gap: Journal of Geophysical Research, v. 85, no. B12, p. 7132–7150.

Phleger, F. B, 1960, Ecology and distribution of recent foraminifera: Baltimore, Johns Hopkins Press, 297 p.

Plafker, G., 1983, The Yakutat block: an actively accreting tectonostratigraphic terrane in southern Alaska: Geological Society of America Abstracts with Programs, v. 15, p. 406.

———, et al, 1979, Geologic implications of 1978 outcrop sample data from the continental slope: U.S. Geological Survey, Circular 804-B, p. 143–146.

———, et al, 1980, Preliminary geology of the continental slope adjacent to OCS Lease sale 55, eastern Gulf of Alaska, petroleum resource implications: U.S. Geological Survey Open-File Report 80-1089, 72 p.

Plumley, P. W., et al, 1982, Paleomagnetism of volcanic rocks of the Kodiak Islands indicates northward latitudinal displacement: Nature, v. 300, p. 50–52.

———, 1983, Paleomagnetism of the Paleocene Ghost Rocks Formation, Prince William Terrane, Alaska: Tectonics, v. 2, p. 295–314.

Poore, R. Z., 1980, Age and correlation of California Paleogene benthic foraminiferal stages: U.S. Geological Survey Professional Paper 1162-C, 8 p.

———, and E. E. Brabb, 1977, Eocene and Oligocene planktonic foraminifera from the upper Butano sandstone and type San Lorenzo Formation, Santa Cruz Mountains, California: Journal of Foraminiferal Research, v. 7, p. 249–272.

———, and D. Bukry, 1978, Preliminary report on Eocene calcareous plankton from the eastern Gulf of Alaska continental slope: U.S. Geological Survey Circular 804-B, p. 141–142.

———, et al, 1977, Lower Tertiary biostratigraphy of the northern Santa Lucia Range, California: U.S. Geological Survey Journal Research, v. 5, p. 735–745.

Rau, W. W., 1948, Foraminifera from the Porter Shale (Lincoln Formation), Grays Harbor County, Washington: Journal of Paleontology, v. 22, p. 152–174.

———, 1958, Stratigraphy and foraminiferal zonation in some of the Tertiary rocks of southwestern Washington: U.S. Geological Survey Oil and Gas Investigations Chart OC-57, 2 sheets.

———, 1964, Foraminifera from the northern Olympic Peninsula, Washington: U.S. Geological Survey Professional Paper 374-G, p. G1–G33.

———, 1966, Stratigraphy and foraminifera of the Satsop River area, southern Olympic Peninsula, Washington: Washington Division of Mines and Geology Bulletin 53, 66 p.

______, 1978, Unusually well preserved and diverse Eocene foraminifers in dredge samples from the eastern Gulf of Alaska continental slope: U.S. Geological Survey Circular 804-B, p. 139

______, 1981, Pacific Northwest Tertiary benthic foraminiferal biostratigraphic framework—An overview; *in* J. M. Armentrout, ed., Pacific Northwest Cenozoic biostratigraphy: Geological Society of America Special Paper 184, p. 67–84.

______, et al, 1977, Preliminary foraminiferal biostratigraphy and correlation of selected stratigraphic sections and wells in the Gulf of Alaska Tertiary Province: U.S. Geological Survey Open-File Report 77-747, 54 p.

Rogers, J. F., 1977, Implications of plate tectonics for offshore Gulf of Alaska petroleum exploration: Proceedings of the 9th Annual Offshore Technology Conference, p. 11–16.

Roth, G. H., 1974, Biostratigraphy and paleoecology of the Coaledo and Bastendorff Formations, southwestern Oregon: PhD Dissertation, Oregon State University, 270 p.

Schmidt, R. R., 1970, Planktonic foraminifera from the lower Tertiary of California: PhD Dissertation, University of Southern California, 339, p.

Schrader, H. J., 1973, Cenozoic diatoms from the northeast Pacific, Leg 18, *in* Initial reports of the Deep Sea Drilling Project: Washington, DC, U.S. Gov't. Printing Office, v. 18, p. 673–798.

Schwab, W. C., et al, 1980, Maps showing structural interpretation of magnetic lineaments in the northern Gulf of Alaska: U.S. Geological Survey Miscellaneous Field Studies Map MF-1245.

Serova, M. Y., 1976, The *Caucasina eocaenica kamchatica* Zone and the Eocene-Oligocene boundary in the northwestern Pacific *in* Progress in micropaleontology: New York, Special Publication Micropaleontology Press, p. 314–328.

Snavely, Jr., P. D., et al, 1978, Twin River Group (upper Eocene to lower Miocene)—defined to include the Hoko River, Makah, and Pysht Formations, Clallam County, Washington: U.S. Geological Survey Bulletin 1457-A, p. 111–120.

Stevenson, A. J., et al, 1983, Tectonic and geologic implications of the Zodiak fan, Aleutian Abyssal Plain, northeast Pacific: Geological Society of America Bulletin, v. 94, p. 259–273.

Stock, J. M., and P. Molnar, 1983, Some geometrical aspects of uncertainties in combined plate reconstructions: Geology, v. 11, p. 697–701.

Stone, D. B., and D. R. Parker, 1979, Paleomagnetic data from the Alaska Peninsula: Geological Society of America Bulletin, v. 90, p. 545–560.

______, et al, 1982, Paleolatitude versus time for southern Alaska: Journal of Geophysical Research, v. 87, p. 3697–3707.

Thoms, R. E., 1965, Biostratigraphy of the Umpqua Formation, southwestern Oregon: PhD Dissertation, University of California, Los Angeles, 219 p.

Tysdal, R. G., and G. Plafker, 1978, Age and continuity of the Valdez Group, southern Alaska, *in* N.F. Sohl and W. B. Wright, eds., Changes in stratigraphic nomenclature by the U.S. Geological Survey, 1977: U.S. Geological Survey Bulletin, 1457-A, p. A120–A124.

Ujie, B. H., and H. Watanabe, 1960, The Poronai foraminifera of the northern Ishikari Coal field, Hokkaido: Science Reports of the Tokyo University, v. 7, p. 117–136.

Vedder, J. G., et al, 1983, Stratigraphy, sedimentation and tectonic accretion of exotic terranes, southern Coast Ranges, California, *in* J. S. Watkins and S. L. Drake, eds., Studies in continental margin geology: American Association of Petroleum Geologists Memoir 34, p. 471–498.

von Huene, R. E., et al, 1979a, Cross section, Alaska Peninsula-Kodiak Island-Aleutian Trench: Geological Society of America Map and Chart Series MC-28A, scale 1:250,000.

______, 1979b, Continental margins of the eastern Gulf of Alaska and boundaries of tectonic plates, *in* J. S. Watkins and L. Montadert, eds., Geological and geophysical investigations of continental margins: American Association of Petroleum Geologists Memoir 29, p. 273–290.

Warren, A. D., and J. H. Newell, 1981, Calcareous plankton biostratigraphy of the type Bastendorff Formation, southwest Oregon, *in* J. M. Armentrout, ed., Pacific Northwest Cenozoic biostratigraphy: Geological Society of America Special Paper 184, p. 105–112.

Wolfe, J. A., 1977, Paleogene floras from the Gulf of Alaska region: U.S. Geological Survey Professional Paper 997, 100 p.

Worsley, T. R., 1973, Calcareous nannofossils, Leg 19 of the Deep Sea Drilling Project, *in* Initial reports of the Deep Sea Drilling Project: Washington, DC, U.S. Gov't. Printing Office, v. 19, p. 741–750.

Early Cretaceous Orogenic Belt in Northwestern Alaska: Internal Organization, Lateral Extent, and Tectonic Interpretation

Stephen E. Box
University of California
Santa Cruz, California

An Early Cretaceous orogenic belt can be traced along a sinuous 4,000 km (2,700 mi) long trend in southwestern, west-central, and northern Alaska and in northeastern U.S.S.R. A general transect in the tectonic transport direction across the grain of each segment of the belt reveals a structural stack consisting of (from top to bottom) an Early Cretaceous (Neocomian) andesitic volcanic terrane; a structurally complex terrane of older mafic igneous and cherty sedimentary rocks; a thrust belt consisting of (1) a variety of lithologies that experienced a late Early Cretaceous low-grade dynamothermal metamorphic event, and/or (2) unmetamorphosed, imbricated upper Paleozoic and Mesozoic slope- and shelf-type sedimentary rocks; and a deep basin of mid-Cretaceous (Albian and younger) clastic rocks derived from the latter three belts. Mid-Cretaceous rocks tie the orogenic belt to the North American and Eurasian cratons.

The general along-trend similarity of the Early Cretaceous histories of each of these terranes and of the timing and style of their juxtaposition suggest that this belt developed by a coherent tectonic process. Comparison of features of this orogenic belt with those developing from Neogene arc-continent collisions in the western Pacific suggests that the Alaskan–Siberian orogenic belt resulted from the Early Cretaceous collision of an active volcanic arc with North American continental lithosphere. The southern end of the belt (in southwestern Alaska) is apparently repeated by subsequent right-lateral faulting.

INTRODUCTION

Application of the terrane concept (Coney et al, 1980) to the geology of western North America has resulted in the discrimination of some 200 terranes whose early histories are considered to be unrelated. Some of these terranes are considered to be "exotic" to North America (i.e., Wrangellia; Jones et al, 1977), while others appear to be displaced fragments of North American provinces (i.e., Nixon Fork; Plumley and Coe, 1982; Blodgett, 1983). The mode of assembly of this collage continues to be a puzzle, although some combination of interplate convergence and strike-slip faulting is considered likely.

The internal organization of this collage varies from relatively ordered to extremely chaotic. The relative degree of internal organization is revealed by the size and lateral continuity of terranes within the collage, as well as by the consistency of interterrane relationships. Some areas, such as the central Alaska Range, are characterized by small, discontinuous terranes that lack consistent interterrane relationships (Jones et al, 1981). Other areas, such as the Pacific margin of Alaska, are underlain by large, linear terranes (i.e., Peninsular–Chugach–Prince William) that have consistent geometric relationships between them, although the time of arrival in their present position remains debatable (Moore et al, 1983).

This paper focuses on the internal organization of the orogenic belt in northern and west-central Alaska and the apparent continuations of this organization into northeastern U.S.S.R. and southwestern Alaska (see tectonic map of Fig. 1). The consistent internal organization of the collage in western Alaska and northeastern U.S.S.R. suggests that this long linear belt developed by a coherent tectonic process. These terranes (see terrane map of Howell et al, this volume, for the names of terranes encompassed by Fig. 1 of this report) are tied by sedimentary overlap and provenance-stitching in mid-Cretaceous time to both the North American and the Eurasian cratons (Coney et al, 1980; Churkin and Trexler, 1981) after experiencing an intense Early Cretaceous orogeny. The nature of the geologic record in western Alaska and in northeastern U.S.S.R. for the time period immediately prior to the mid-Cretaceous overlap is explored here in an effort to constrain the nature of the plate interactions that resulted in assembly of these juxtaposed terranes. Structural, stratigraphic, igneous, and metamorphic events of that age recorded in these terranes provide clues to the tectonic style of their accretion, and these features are emphasized in Figure 1. A comparison is made between features of the Alaskan–Siberian orogenic belt and those resulting from Neogene arc-continent collision in areas of the western Pacific, and a model for the evolution of the Cretaceous belt is proposed.

EARLY CRETACEOUS OROGENESIS IN NORTHERN AND WEST-CENTRAL ALASKA

The Early Cretaceous orogenic belt of northern and west-central Alaska (Fig. 1) is composed of a central,

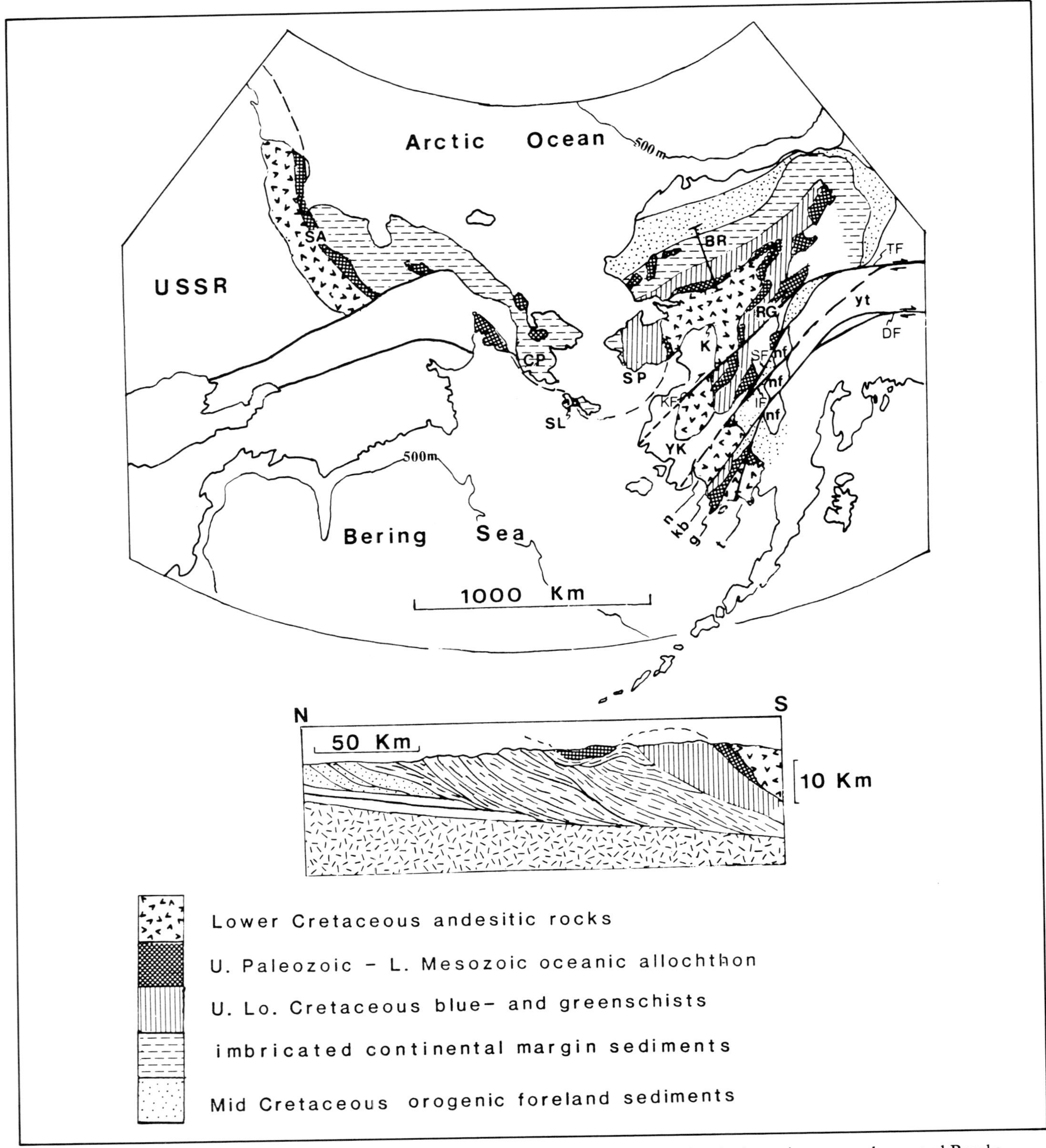

Figure 1—Generalized tectonic map of northwestern Alaska and northeastern U.S.S.R. (geologic section across the central Brooks Range modified from Mull, 1982). Early Cretaceous orogenic belt can be traced from the South Anyuy zone in northeastern U.S.S.R. eastward to the eastern Brooks Range of northern Alaska, then southwestward along the Ruby Geanticline to the southwest Alaskan coast. SA = South Anyuy zone, CP = Chukotsk Peninsula, SL = St. Lawrence Island, SP = Seward Peninsula, BR = Brooks Range, K = Koyukuk basin, RG = Ruby Geanticline, YK = Yukon-Kuskokwim delta, DF = Denali fault, IF = Iditarod-Nixon Fork fault, KF = Kaltag fault, SF = Susalatna fault, TF = Tintina fault. Terranes: n = Nyack, kb = Kilbuck, g = Goodnews, t = Togiak, nf = Nixon Fork, yt = Yukon-Tanana.

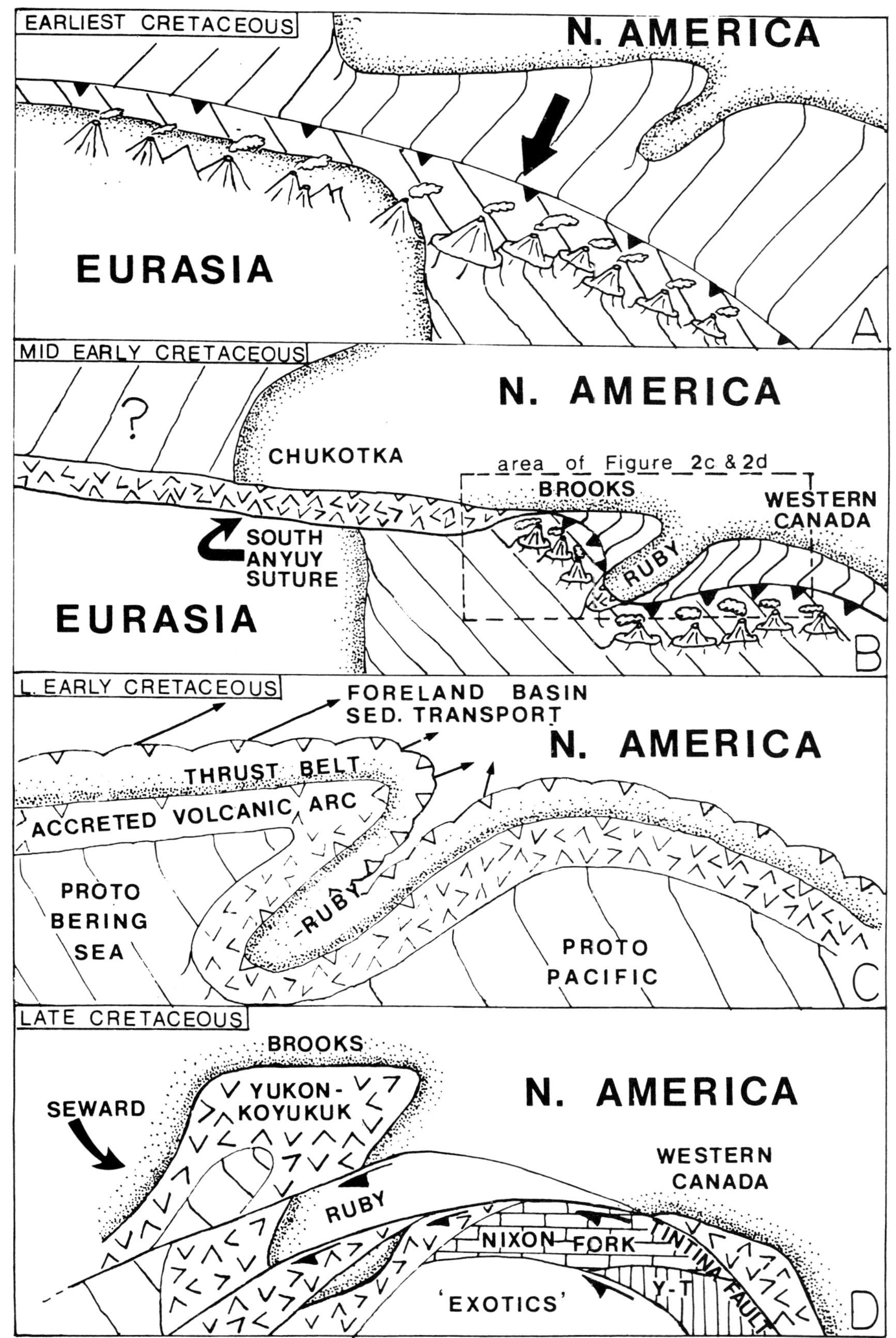

Figure 3—Proposed tectonic model for the development of the late Early Cretaceous orogenic belt in northwestern Alaska and northeastern U.S.S.R. and its subsequent disruption in southern Alaska by right-slip faulting.

REFERENCES

Blodgett, R. B., 1983, Paleobiogeographic affinities of Devonian fossils from the Nixon Fork terrane, southwestern Alaska, *in* C. H. Stevens, ed., Pre-Jurassic rocks in western North American suspect terranes: Society of Economic Paleontologists and Mineralogists Symposium Volume.

Box, S. E., 1982, Kanektok suture, SW Alaska: geometry, age, and relevance (Abs.): EOS, American Geophysical Union Transactions, v. 63, no. 45, p. 915.

______, 1983, Tectonic synthesis of Mesozoic histories of the Togiak and Goodnews terranes, SW Alaska: Geological Society of America Abstracts with Programs, v. 15, p. 406.

______, and Moore, J. C., 1981, Mesozoic tectonic framework, northern Bristol Bay, southwestern Alaska (Abs.): EOS, American Geophysical Union Transactions, v. 62, no. 45, p. 1038–1039.

Bundtzen, T. K., and W. G. Gilbert, 1983, Outline of the geology and mineral resources of Upper Kuskokwim Region, Alaska: Journal of the Alaska Geological Society, v. 3, p. 101–117.

______, and G. M. Laird, 1980, Preliminary geology of the McGrath-Upper Innoko River Area, western interior Alaska: Alaska Division of Geological and Geophysical Surveys, Open-File Report 134, 35 p.

Cardwell, R. K., and B. L. Isacks, 1978, Geometry of subducted lithosphere beneath the Banda Sea in Eastern Indonesia from seismicity and fault plane solutions: Journal of Geophysical Research, v. 83, p. 2825–2838.

Carter, D. J., et al, 1976, Stratigraphical analysis of island-arc continental margin collision in eastern Indonesia: Journal of the Geological Society of London, v. 132, p. 179–198.

Churkin, Jr., M., and J. H. Trexler, 1981, Continental plates and accreted oceanic terranes in the Arctic, *in* A. E. M. Nairn, et al, eds., The ocean basins and margins, v. 5, The Arctic Ocean: New York, Plenum Press, p. 1–20.

Coney, P. J., et al, 1980, Cordilleran suspect terranes: Nature, v. 288, p. 329–333.

Dempsey, W. J., et al, 1957, Profiles of Bethel Basin, Alaska: U.S. Geological Survey Open-File Map, scale 1:250,000.

Fujita, K., and J. T. Newberry, 1982, Tectonic evolution of northeastern Siberia and adjacent regions: Tectonophysics, v. 89, p. 337–357.

Gabrielse, H., in press, Major transcurrent displacements along the northern Rocky Mountain Trench and related lineaments in north-central British Columbia: Geological Society of America Bulletin.

Grantz, A., 1966, Strike-slip faults of Alaska: U.S. Geological Survey Open-File Report 66-267, 82 p.

Griscom, A., 1978, Aeromagnetic interpretation of the Goodnews and Hagemeister Islands quadrangles region, southwestern Alaska: U.S. Geological Survey Open-File Report 78-9-C, 20 p. and 2 sheets, scale 1:250,000.

Grybeck, D., et al, 1977, Geologic map of the Brooks Range, Alaska: U.S. Geological Survey Open-File Report 77-166-B, 2 sheets, scale 1:1,000,000.

Hamburger, M. W., et al, 1983, Seismotectonics of the northern Philippine Island arc, *in* D. E. Hayes, ed., The tectonic and geologic evolution of southeast Asian seas and islands, Part 2: American Geophysical Union Geophysical Monograph 27, p. 1–22.

Hamilton, W., 1979, Tectonics of the Indonesian region: U.S. Geological Survey Professional Paper 1078, 345 p.

Hoare, J. M., and W. L. Coonrad, 1959a, Geologic map of the Bethel quadrangle, Alaska: U.S. Geological Survey Map I-285, scale 1:250,000.

______, 1959b, Geologic map of the Russian Mission quadrangle, Alaska: U.S. Geological Survey Map I-292, scale 1:250,000.

______, 1978a, Geologic map of the Goodnews and Hagemeister Islands region, southwestern Alaska: U.S. Geological Survey Open-File Report OF-78-9-B, scale 1:250,000.

______, 1978b, Lawsonite in southwestern Alaska: U.S. Geological Survey Circular 772-B, p. 55–57.

______, 1983, Graywacke of Buchia Ridge and correlative Lower Cretaceous rocks in the Goodnews Bay and Bethel quadrangles, southwestern Alaska: U.S. Geological Survey Bulletin 1529-c, 17 p.

Holloway, N. H., 1982, North Palawan block, Philippines —its relation to Asian mainland and role in evolution of South China Sea: Bulletin of the American Association of Petroleum Geologists, v. 66, p. 1355–1383.

Jeletsky, J. A., 1975, Jurassic and Lower Cretaceous paleogeography and depositional tectonics of Porcupine Plateau, adjacent areas of northern Yukon and those of Mackenzie District: Geological Survey of Canada, Paper 74-16, 52 p.

Jones, D. L., and A. Grantz, 1964, Stratigraphic and structural significance of Cretaceous fossils from Tiglukpuk Formation, northern Alaska: Bulletin of the American Association of Petroleum Geologists, v. 48, p. 1462–1474.

______, and N. J. Silberling, 1979, Mesozoic stratigraphy—The key to tectonic analysis of southern and central Alaska: U.S. Geological Survey Open-File Report 79-1200, 37 p.

______, et al, 1977, Wrangellia—A displaced terrane in northwestern North America: Canadian Journal of Earth Sciences, v. 14, p. 2565–2577.

______, et al, 1981, Tectonostratigraphic terrane map of Alaska: U.S. Geological Survey Open-File Report 81-792, scale 1:2,500,000.

______, et al, 1983, Oceanic terranes of north-central and northern Alaska: Geological Society of America Abstracts with Programs, v. 15, p. 428.

Mayfield, C. F., et al, 1983, Stratigraphy, structure, and palinspastic synthesis of the western Brooks Range, northwestern Alaska: U.S. Geological Survey Open-File Report, 83-779, 58 p.

McCabe, R., et al, 1982, Geologic and paleomagnetic evidence for a possible Miocene collision in western Panay, central Philippines: Geology, v. 10, p. 325–329.

Miller, T. P., et al, 1972, Preliminary geologic map of the eastern Solomon and southeastern Bendeleben quadrangles, eastern Seward Peninsula, Alaska: U.S. Geological Survey Open-File Report, scale 1:250,000.

Moll, E. J., and W. W. Patton, Jr., 1981, Preliminary report on the Late Cretaceous and Early Tertiary volcanic and related plutonic rocks in western Alaska: U.S. Geological Survey Circular 844, p. 73–76.

Moore, J. C., et al, 1983, Paleogene evolution of the Kodiak Islands, Alaska: consequences of ridge-trench interaction in a more southerly latitude: Tectonics, v. 2, p. 265–293.

Mull, C. G., 1979, Nanushuk Group deposition and the Late Mesozoic structural evolution of the central and western Brooks Range and Arctic Slope, *in* T. S. Albrandt, ed., Preliminary geologic, petrologic, and paleontologic results of the study of Nanushuk Group rocks, North Slope, Alaska: U.S. Geological Survey Circular 794, p. 5–13.

———, et al, 1976, New structural and stratigraphic interpretations, central and western Brooks Range and Arctic Slope: U.S. Geological Survey Circular 733, p. 24–26.

Patton, Jr., W. W., 1973, Reconnaissance geology of the northern Yukon-Koyukuk province, Alaska: U.S. Geological Survey Professional Paper 774-A, p. A1–A17.

———, 1983, Yukon-Koyukuk basin—key to the tectonics of western Alaska: Geological Society of America Abstracts with Programs, v. 15, p. 408.

———, and B. Csejtey, Jr., 1979, Geologic map of St. Lawrence Island, Alaska: U.S. Geological Survey Open-File Report 79-945, scale 1:250,000.

———, and I. L. Tailleur, 1977, Evidence of the Bering Strait region for differential movement between North America and Eurasia: Geological Society of America Bulletin, v. 88, p. 1298–1304.

———, et al, 1977, Preliminary report on the ophiolites of northern and western Alaska, *in* R. G. Coleman, and W. P. Irwin, eds., North American ophiolites: Oregon Department of Geological and Mineral Industries Bulletin 95, p. 51–58.

———, et al, 1980, Preliminary geologic map of the Medfra quadrangle, Alaska: U.S. Geological Survey Open-File Report 80-811-A, scale 1:250,000.

Plumley, P. W., and R. S. Coe, 1982, Paleomagnetic evidence for reorganization of the north Cordilleran borderland in Late Paleozoic time: Geological Society of America Abstracts with Programs, v. 14, p. 224.

Pollock, S. M., 1982, Structure, petrology, and metamorphic history of the Nome Group blueschist terrane, Salmon Lake area, Seward Peninsula, Alaska: Unpublished Masters's Thesis, University of Washington, 222 p.

Roeder, D., and C. G. Mull, 1978, Tectonics of the Brooks Range ophiolites, Alaska: Bulletin of the American Association of Petroleum Geologists, v. 62, p. 1696–1713.

Sainsbury, C. L., 1972, Geologic map of the Teller quadrangle, western Seward Peninsula, Alaska: U.S. Geological Survey Map I-685, scale 1:250,000.

———, et al, 1970, Blueschist and related greenschist facies rocks of Seward Peninsula, Alaska: U.S. Geological Survey Professional Paper 700-B, p. B33–42.

Seslavinskiy, K. B., 1970, Structure and development of the South Anyui fault trough, west Chukotka: Geotectonics, v. 4, p. 311–317.

———, 1979, The South Anyuy geosuture, western Chukotka: Transactions of the Academy of Sciences of the USSR, Earth Science Section, v. 249, p. 78–81 (American Geological Institute English translation).

Shew, N., and F. H. Wilson, 1981, Map and table showing radiometric ages of rocks in southwestern Alaska: U.S. Geological Survey Open-File Report 81-886, 26 p.

Shilo, N. A., and S. M. Til'man, 1981, The tectonic zones of northeastern USSR and the formation of continental crust, *in* A. E. M. Nairn, et al, eds., The ocean basins and margins, v. 5, The Arctic Ocean: New York, Plenum Press, p. 413–438.

Silver, E. A., and R. B. Smith, 1983, Comparison of terrane accretion in modern Southeast Asia and the Mesozoic North American Cordillera: Geology, v. 11, p. 198–202.

Suppe, J., 1981, Mechanics of mountain building and metamorphism in Taiwan: Geological Society of China Memoir No. 4, p. 67–85.

Templeman-Kluit, D., 1979, Transported cataclastite, ophiolite, and granodiorite in Yukon: evidence of arc-continent collision: Geological Survey of Canada Paper 79-14, 27 p.

Till, A. B., 1982, Granulite, peridotite, and blueschist—early tectonic history of the Seward Peninsula, Alaska (Abs.), *in* Western Alaska geology and resource potential: Program and Abstracts, Alaska Geological Society Symposium, p. 33–34.

Turner, D. L., et al, 1979, K–Ar geochronology of the southwestern Brooks Range, Alaska: Canadian Journal of Earth Sciences, v. 16, p. 1789–1804.

———, in press, Geologic and geochronologic studies of the Early Proterozoic Kanektok metamorphic complex of southwestern Alaska: Canadian Journal of Earth Sciences.

Voyevodin, V. N., et al, 1978, The eugeosynclinal complex of the Mesozoides of Chukchi Peninsula: Geotectonics, v. 12, p. 472–477.

Wallace, W. K., 1983, Major lithologic belts of southwestern Alaska and their tectonic implications: Geological Society of America Abstracts with Programs, v. 15, p. 406.

Young, F. G., et al, 1976, Geology of the Beaufort-Mackenzie Basin: Geological Survey of Canada Paper 76-11, 65 p.

Tectonostratigraphic Terranes in Southwest Oregon

M. C. Blake, Jr.
D. C. Engebretson
A. S. Jayko
D. L. Jones
U.S. Geological Survey
Menlo Park, California

Recent geologic and paleomagnetic studies suggest that southwest Oregon consists of a collage of small, fault-bounded, late Mesozoic tectonostratigraphic terranes. These terranes and their probable paleotectonic environments are the Gold Beach (oceanic island arc), Yolla Bolly (transitional from continental arc-continental margin?), Snow Camp (island arc?), Sixes River (transform melange), Pickett Peak (subduction or collision complex), and Western Klamath (island arc and marginal basin).

Fossils and radiometric dates from these terranes overlap in age spanning Late Jurassic and Early Cretaceous time. The similarities in age, combined with pronounced differences in lithostratigraphy, sandstone petrology, and deformational, igneous, and metamorphic history, suggest that these terranes formed in different paleoenvironments that were subsequently accreted to North America.

Western Klamath terrane was probably in its present position with respect to North America by Early Cretaceous time. The other terranes except Gold Beach were accreted and dispersed prior to the mid-Eocene and appear to have been involved in a post-accretion Early Tertiary collision event. The Gold Beach terrane may have accreted in post mid-Eocene time.

INTRODUCTION

Geologic and geophysical data summarized herein coupled with numerous regional studies by other workers (including Dott, 1971; Coleman, 1972; Ramp, 1975; Ramp et al, 1977; Roure, 1981; Roure and Blanchet, 1983; Smith et al, 1982; Garcia, 1982) have been used to distinguish six tectonostratigraphic terranes in the upper Mesozoic rocks of southwestern Oregon based on the internal stratigraphic criteria defined by Coney et al (1980) for Cordilleran suspect terranes.

Important points that we emphasize are: (1) Each terrane is characterized by an internal stratigraphic record that differs markedly from that of nearby, coeval terranes; (2) each terrane is fault bounded, and intermediate lithologic facies do not connect adjoining terranes; and (3) previous efforts to fit all the rocks from these disparate terranes into one vertical stratigraphic sequence were based on misconceptions concerning the age and structural relations of the separate terranes.

The six terranes we describe are the Gold Beach, Yolla Bolly, Sixes River, Pickett Peak, Snow Camp, and Western Klamath terranes (Fig. 1). In the Yolla Bolly and Western Klamath terranes, we have subdivided the terranes into subterranes to differentiate some of the more subtle differences within terranes but emphasize that these units may have once been continuous.

DESCRIPTION OF TERRANES

Gold Beach Terrane

The Gold Beach terrane, named for outcrops near the town of Gold Beach and adjacent coastal areas, consists of the Upper Jurassic Otter Point Formation (Koch, 1966; Dott, 1971) unconformably overlain by the Upper Cretaceous Houstenaden Creek unit of Bourgeois (1980b), Cape Sebastian Sandstone of Dott (1971), and Hunters Cove Formation of Dott (1971). Fossils of Aptian to Albian age (Evitt, 1980, personal communication) were collected by Roure from a tectonic zone between the Otter Point Formation and the Hunters Cove Formation. Probably, these fossils are from the Houstenaden Creek unit and, if so, narrow the age unconformity to Lower Cretaceous.

The Otter Point Formation consists largely of interbedded volcanic sandstone, mudstone, and conglomerate as well as breccia and tuff of andesitic composition (Figs. 2, 3). These rocks appear to have formed in deep water, adjacent to a Late Jurassic (Tithonian) island arc (Aalto and Dott, 1970; Dott, 1971; Coleman, 1972; Walker, 1977). The structure of the Otter Point ranges from little-deformed sedimentary rocks to intensely sheared broken formation or melange that contains blocks of pillow basalt, radiolarian chert, and serpentinite. The stratigraphic position of these rocks is not known. Metamorphism of the Otter Point sedimentary

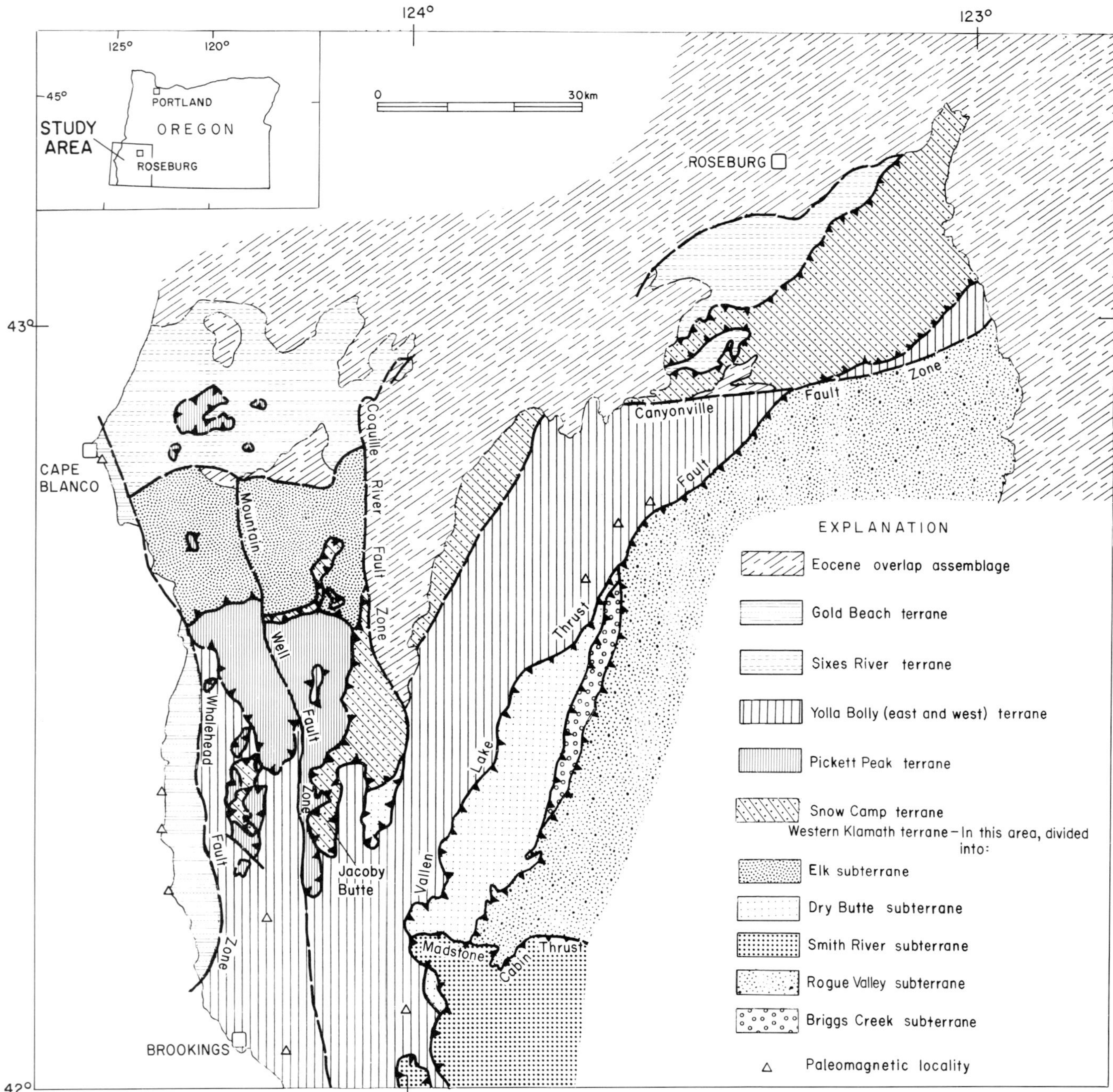

Figure 1—Tectonostratigraphic terrane map of southwestern Oregon including location of paleomagnetic sites.

rocks is limited to abundant veins of laumontite. Prehnite, pumpellyite, and epidote formed during ocean-floor metamorphism are present locally in the pillow lavas.

The overlying Cretaceous rocks grade from deep- to shallow-to-deep water facies up section and are predominantly quartzofeldspathic in composition (Dott, 1971; Bourgeois, 1980b). Bourgeois (1980b) suggested that the composition of the Cretaceous sandstone and conglomerate is incompatible with a Klamath Mountains source terrane. A well-exposed depositional contact between Cretaceous rocks and the Otter Point occurs locally, but in most places it is obscured by younger thrust faulting and folding (Dott, 1971).

Yolla Bolly Terrane

The Yolla Bolly terrane in Oregon (Silberling et al, 1984) consists of sedimentary and igneous rocks of the Upper Jurassic and Lower Cretaceous Dothan Formation. These rocks were subdivided by Widmier (1962) into a volcanic-rich coastal member and a sandstone-rich inland member. In addition to the pronounced difference in gross lithology described below, there are subtle differences in sandstone petrology, geochemistry, and textural grade that we used to subdivide these rocks into two subterranes, eastern and western. Megafossils and radiolarians found in both subterranes range in age from Tithonian to Hauterivian.

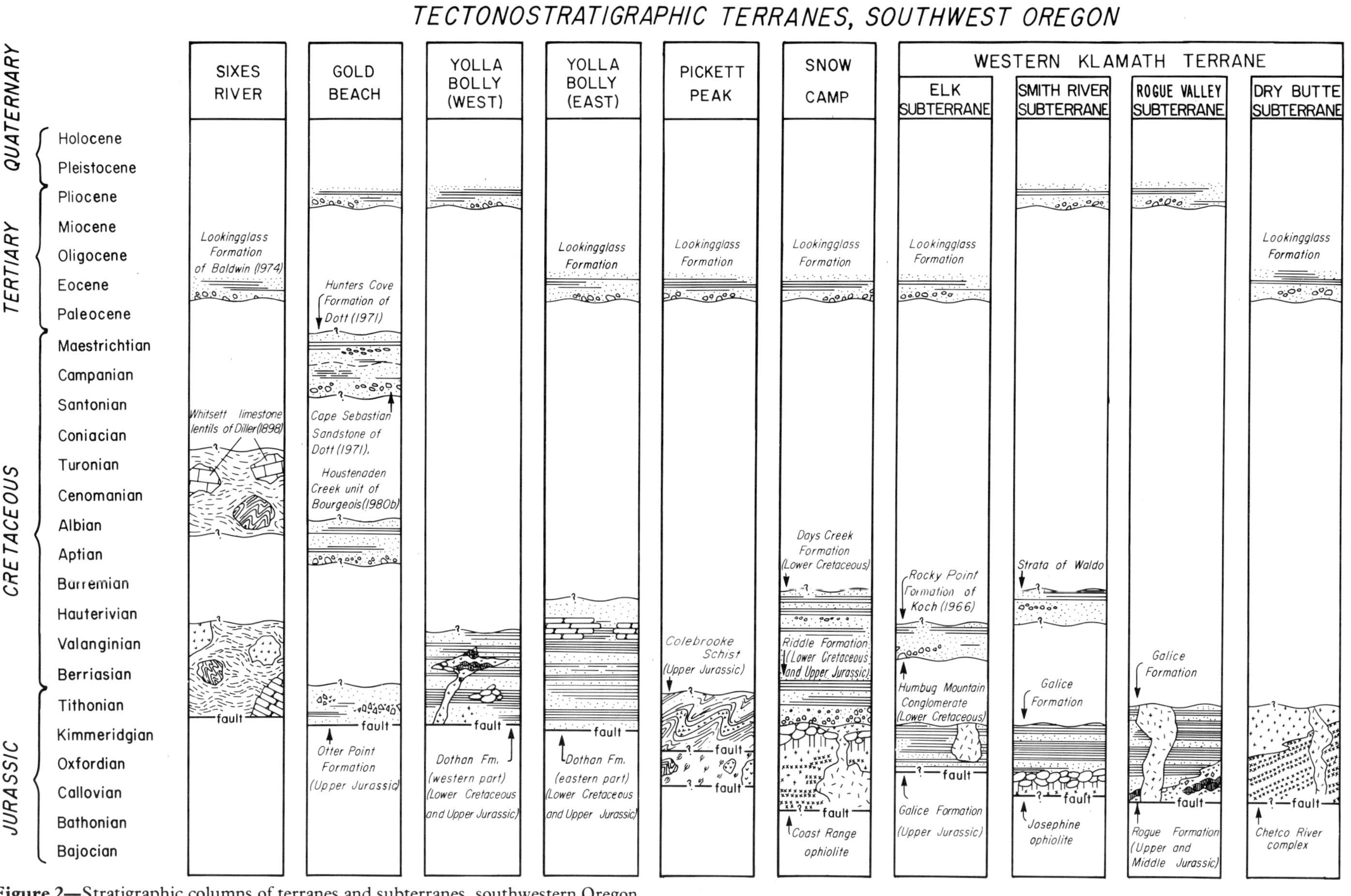

Figure 2—Stratigraphic columns of terranes and subterranes, southwestern Oregon.

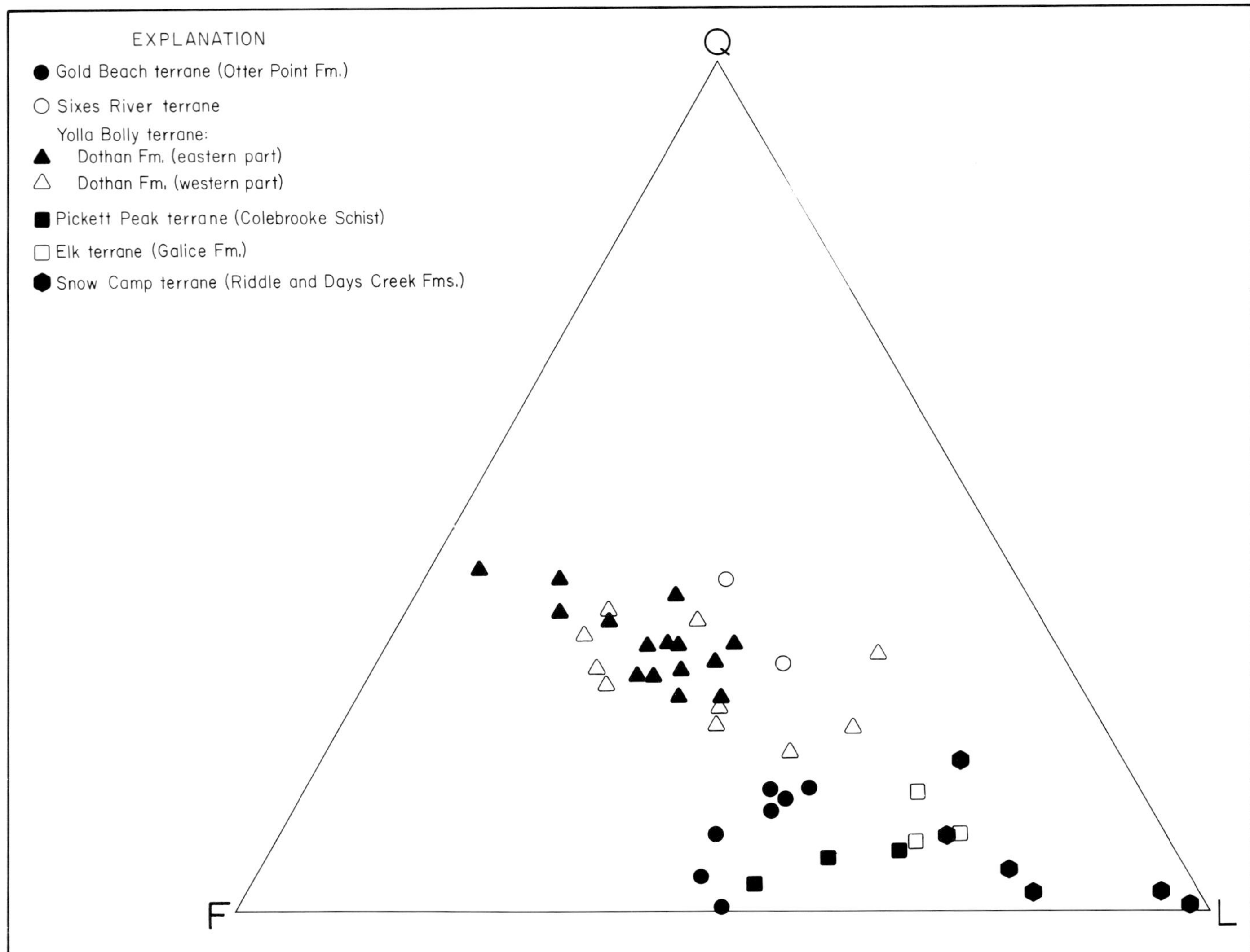

Figure 3—Modal analyses of sandstone (graywackes) from different terranes. Q = monocrystalline quartz; F = total feldspar; L = lithic fragments including chert. Approximately 500 grains were counted per sample.

Rocks in the western subterrane consist of sandstone, mudstone, and conglomerate turbidites with minor interbedded radiolarian chert and abundant tuffs, flows, and shallow intrusive rocks of basaltic to dacitic composition. Detailed petrographic and geochemical studies (Wiggins, 1980) indicate that these igneous rocks are calc-alkaline, volcanic-arc lavas, probably erupted along a continental margin. Structurally, the western subterrane is only moderately sheared ("broken formation") with serpentine-marked shear zones restricted to coastal exposures north of Brookings and near contacts with the adjacent Gold Beach terrane. Sandstone bodies are nonfoliated (textural zone 1, 1 Blake et al, 1967) and commonly contain pumpellyite. Wiggins (1980), however, identified both lawsonite and aragonite in intrusive and extrusive volcanic rocks, indicating that the rocks were metamorphosed to lower blueschist-facies conditions during accretion.

The eastern subterrane consists mainly of sandstone, mudstone, and conglomerate turbidites with moderately abundant radiolarian chert interbeds. Volcanic rocks are largely restricted to isolated bodies of pillow lava near the thrust fault contact with the overlying Western Klamath terrane. Sandstones are somewhat less volcanogenic and more quartzofeldspathic than in the western subterrane (Fig. 3; Table 4) and have abundant large detrital flakes of white mica. Graywackes are usually incipiently foliated (textural zone 2A) with pumpellyite, the characteristic metamorphic mineral in the southern part of the subterrane, and detrital K-feldspar, prehnite, pumpellyite, and laumontite veins to the northeast. The eastern subterrane is of lower metamorphic grade than the western subterrane.

It appears that the eastern and western subterranes may be lateral facies of a terrane that has arc affinities to the west and continental margin to the east.

Sixes River Terrane

North of the Sixes River and extending eastward to the vicinity of Roseburg (Fig. 1) is an area previously mapped as the Otter Point Formation (Lent, 1969; Dott, 1971; Beaulieu and Hughes, 1976) (see *Gold Beach*

Terrane). These rocks consist of highly sheared mudstone, sandstone, and conglomerate, together with relatively abundant tectonic blocks (knockers) of high-grade blueschist and eclogite (Coleman and Lanphere, 1971). In addition, blocks of Middle Cretaceous (Albian–Cenomanian) pelagic and coralline limestone associated with pillow lava are present near Roseburg (Whitsett limestone lentils of Diller, 1898). Sliter (1984) correlates these limestones with those found to the south near Laytonville, California (Alvarez et al, 1980) and also formed at low latitudes (Sliter, 1984). Common laumontite in the sandstone indicates zeolite facies metamorphism. The Sixes River terrane differs from the Otter Point Formation of the Gold Beach terrane by the presence of exotic blueschist and limestone blocks and typically quartzofeldspathic sandstone composition (Fig. 3; Lent, 1969). Concretions in the sheared matrix contain *Buchia fischeriana*, which is of slightly younger age than *Buchia piochii*, the characteristic fossil of the Otter Point Formation.

The Sixes River terrane is correlative with the Central melange terrane of northern California (Blake et al, this volume). It is inferred to have a complex history of accretion overprinted by deformations associated with zones of transform faulting.

Pickett Peak Terrane

The Pickett Peak terrane in Oregon (Blake et al, 1982b) consists of the Colebrooke Schist plus an underlying serpentinite melange containing rare blueschist and abundant greenstone knockers. The Colebrooke (Coleman, 1972) is made up predominantly of textural zone 2B-3A metagraywacke and mica schist (metapelite), plus minor metamorphosed tuffs and pillow lava and rare metachert. Metamorphic mineral assemblages are transitional between the lower blueschist and greenschist facies and are most commonly represented by quartz-albite-phengite-chlorite ± lawsonite ± calcite in metagraywacke, and albite-chlorite-actinolite-epidote ± crossite in metabasalt. The underlying serpentinite melange locally contains blocks of blueschist with glaucophane, lawsonite, and fine-grained garnet. These blocks are finer grained and lower grade than the blocks of blueschist and eclogite found in the Sixes River terrane and are interpreted as higher grade equivalents of the Colebrooke Schist.

No fossils have been found in the Colebrooke Schist or in the underlying melange. Radiometric dates of 125 to 138 m.y. (Dott, 1971; Coleman, 1972; Lanphere et al, 1978) indicate an Early Cretaceous metamorphic age.

Snow Camp Terrane

Extending northeast-southwest across the middle of the area (Fig. 1) is a dismembered ophiolite sequence with overlying sedimentary rocks here named the Snow Camp terrane. These rocks include serpentinized peridotite, gabbro, sheeted igneous complex, tholeiitic pillow basalt (Gullixson et al, 1980), and, along the Rogue River, a thick sequence of andesitic to dacitic volcanic rocks (Gray et al, 1982), plus a sheeted dike complex dated at 164 ± 2 m.y. by the Pb/U method (Saleeby, 1984). Original stratigraphic order of the ophiolite elements is rarely preserved because of imbricate thrusting. Overlying the volcanic rocks are flyschlike lithic (chert-rich) sedimentary rocks (Figs. 2, 3) of the Upper Jurassic and Lower Cretaceous Riddle Formation and the Lower Cretaceous Days Creek Formation. These sediments are unmetamorphosed but usually are involved in the imbricate thrusting seen in the ophiolite. Megafossils are common and radiolarian chert is unknown. This terrane has been correlated with the Coast Range ophiolite and overlying Great Valley sequence of California (Jones, 1973). Notable differences exist, however, in particular the presence of small intrusive bodies in the ophiolite, local development of schistosity, and some possible remnants of older (Paleozoic?) ophiolite; Dott (1971) obtained a K–Ar hornblende age of 285 ± 25 m.y. from gabbro that appears to be from this terrane.

Western Klamath Terrane

Middle and Upper Jurassic rocks along the western margin of the Klamath Mountains and in a large outlier to the west along the coast form five distinctive subterranes of the Western Klamath terrane. Each of these subterranes is characterized by a slightly different substrate, but most contain flyschlike sedimentary rocks (Galice Formation) of the same age but having different source areas, and somewhat different overlap assemblages (Fig. 2). It is likely that the subterranes were once part of a complex arc-marginal basin-continental margin sequence that is now highly disrupted.

The Elk subterrane consists of middle Upper Jurassic (Oxfordian–Kimmeridgian) metasedimentary and rare metavolcanic rocks [Galice and Rogue(?) Formations] intruded by Upper Jurassic plutons and unconformably overlain by Lower Cretaceous (Valanginian) sandstone, mudstone, and conglomerate of the Humbug Mountain Conglomerate and the Rocky Point Formations of Koch (1966). The metasediments and overlying Cretaceous sedimentary rocks appear to be similar to rocks of the Smith River subterrane and the Cretaceous overlap assemblage occurring at the southern end of the Klamath Mountains, near Red Bluff, California (Blake and Jones, 1977). The composition of metagraywacke from this subterrane is more lithic (chert-rich) than any of the other terranes with the exception of the Riddle–Days Creek sandstones of the Snow Camp terrane (Fig. 3; Table 1).

The Dry Butte terrane, here redesignated as the Dry Butte subterrane, was named by Loney and Himmelberg (1977) for the predominantly igneous complex that forms an arcuate belt along the western margin of the Klamath Mountains in Oregon. Included in it are metagabbro, metaperidotite, gabbro, quartz diorite, and amphibolite (Chetco River complex of Hotz, 1971). This subterrane is interpreted as a deeply eroded Middle and Upper Jurassic island arc complex based on the age of the quartz diorite (Hotz, 1971).

Thrust over the Dry Butte subterrane is the Briggs Creek subterrane, consisting of a highly deformed and recrystallized garnet-bearing amphibolite (= Briggs Creek Amphibolite of Garcia, 1979). Structurally on top of the amphibolite (Fig. 1) lies a fourth subterrane, the Rogue Valley subterrane. It includes the type Galice and Rogue

Table 1

	Gold Beach (6)	Yolla Bolly (west) (16)	Yolla Bolly (east) (21)	Pickett Peak (14)	Elk (8)	Snow Camp (2)
SiO_2	64.02	66.62	68.82	66.96	73.51	79.06
Al_2O_3	13.79	13.47	14.01	13.96	11.78	8.47
FeO^a	6.03	5.09	4.66	5.62	4.09	3.91
MgO	3.89	3.12	1.98	3.18	2.11	2.28
CaO	3.77	2.83	1.55	0.86	1.46	0.59
Na_2O	3.02	3.82	4.03	2.81	2.24	1.57
K_2O	1.03	1.28	1.15	1.30	1.52	0.92
H_2O	3.43	2.75	2.73	3.36	2.48	2.44
TiO_2	0.70	0.66	0.70	0.85	0.57	0.46
P_2O_5	0.18	0.16	0.25	0.29	0.20	0.11
MnO	0.18	0.10	0.12	0.07	0.05	0.18
	100.04	100.00	100.00	99.26	100.01	99.99
Sr^{87}/Sr^{86}	—	.7045	.7055	.7070	.7059	—

[a]Total iron as FeO.

Table 1—Average chemical analyses of graywackes and metagraywackes from several terranes and subterranes. Number in parentheses indicates the number of samples averaged. Includes data from Coleman (1972) and unpublished analyses.

Formations (Wells and Walker, 1953), which apparently formed above an ophiolitic basement. The Rogue Valley subterrane differs from the others by virtue of its greater amount of intercalated calc-alkaline volcanic rocks (Johnson, 1980; Garcia, 1982) and associated volcanogenic composition of the Galice sandstones. In addition, the recent discovery of Triassic radiolarians in chert presumably beneath the Rogue Formation (Roure and De Wever, 1983) indicates a much older age for the basal ophiolite than for the other basement rocks of this terrane.

To the south and structurally overlying the Dry Butte and Rogue River Valley subterranes along the Madstone Cabin thrust fault (Ramp, 1975) is the Smith River subterrane (Fig. 1) composed of the Josephine Peridotite (Loney and Himmelberg, 1976, 1977), together with overlying gabbro, volcanic rocks, and clastic flyschlike rocks (and very minor volcaniclastic rocks). Recent work (Harper, 1980) in northernmost California suggests that the Josephine Peridotite is the base of a relatively complete ophiolite sequence interpreted as having formed in a marginal basin. The overlying Galice Formation here, unlike its counterpart in the Elk subterrane, is more quartzofeldspathic in composition, apparently derived from a more continental source.

A regional penetrative metamorphism of the Galice sedimentary rocks associated with each of these subterranes is prehnite-pumpellyite to the lowest greenschist facies (Harper, 1983, personal communication) and has been dated at about 150 m.y. (Nevadan) (Lanphere et al, 1978). This metamorphic event is unique to the Western Klamath terrane.

STRATIGRAPHIC RELATIONSHIPS AND INFERRED TECTONIC SETTINGS

As shown in Figure 2, most of the subterranes of the Western Klamath terrane contain sedimentary rocks of the Galice Formation. The Dry Butte and Briggs Creek subterranes are exceptions, possibly representing deeper structural levels of the same terrane. We believe that these subterranes probably formed an island arc-marginal basin adjacent to the continental margin during Middle Jurassic time and were subsequently imbricated during the Nevadan orogeny. The timing of this event can be seen in the hiatus in sedimentation between the deformation and metamorphism of the Galice Formation and the deposition of the overlying Lower Cretaceous overlap assemblage [Humbug Mountain Conglomerate–Rocky Point Formation on the Elk subterrane and the somewhat younger sedimentary strata at Waldo (Fig. 2) in the Smith River terrane].

Within the Snow Camp terrane we see a basal ophiolite (locally overlain by arc volcanic rocks) identical in age to that of the Smith River subterrane but marked by continuous sedimentation throughout the entire interval of the Nevadan orogeny. The Colebrooke Schist of the Pickett Peak terrane is probably of about the same age as the other terranes but records a deformational and metamorphic history not seen elsewhere. The Yolla Bolly terrane is compositionally totally different from the Snow Camp and Gold Beach terranes and, furthermore, represents, in part, a probable continental arc. The Gold Beach terrane is clearly the upper part of an andesitic island arc with Lower and Upper Cretaceous continent-derived overlap deposits unlike anything in the other terranes.

The Sixes River terrane is very similar to the Central melange terrane of the northern California Coast Range and includes the same kind of pelagic limestone blocks as near Laytonville, which have been shown to have been derived from south of the equator (Alvarez et al, 1980). With the exception of the subterranes of the Western Klamath terrane, we see no evidence that the other terranes represent once-contiguous parts of the same continental margin, as proposed by Roure (1981) and Roure and Blanchet (1983).

STRUCTURAL RELATIONS

Major faults or fault zones separate all the terranes and subterranes of southwestern Oregon. Most boundary faults are low-angle thrusts, so that the assemblage of terranes constitutes stacks of subhorizontal nappes (Coleman, 1972; Blake et al, 1982; Roure and Blanchet, 1983).

Evidence for episodic accretion is seen in the timing of faulting and in the subsequent deformation of earlier structures. The earliest event involved imbrication of the Dry Butte–Briggs Creek, Rogue Valley, and Smith River subterranes and formation of the penetrative deformation that characterizes the Nevadan orogeny. This was followed by intrusion of Late Jurassic plutons and the deposition of the Lower Cretaceous overlap assemblage. It is possible that the Snow Camp terrane was the highest imbricate sheet during this thrusting; however, it is presently juxtaposed with the other terranes along younger Mesozoic or Cenozoic thrusts.

The timing of accretion of the Pickett Peak terrane remains a problem because of inadequate dating of these rocks. Its structural position suggests, however, that it was accreted subsequent to the previously described Nevadan(?) event, most likely in Early Cretaceous time.

The Yolla Bolly terrane, based on more complete data in northern California (Blake and Jones, 1983), probably was accreted during Middle Cretaceous time (~ 92 m.y.). The Sixes River terrane, because of its far-traveled Middle Cretaceous limestone blocks, must have arrived sometime during the Late Cretaceous or Early Tertiary.

This accretion was followed by renewed imbrication that involved rocks of the Lower Cenozoic Roseburg Formation of Baldwin (1974) (Perttu and Benson, 1980). The detailed structural studies of Roure (1981) show that the sense of over-thrusting was from east-southeast to west-northwest.

The last major tectonic event was the formation of the high-angle faults that offset the nappe structures. The westernmost of these, the Whalehead fault, may have had major right-lateral movement along it and may have been responsible for bringing in the Gold Beach terrane.

CENOZOIC OVERLAP ASSEMBLAGE

A thick sequence of middle(?) Eocene sandstone, mudstone, and conglomerate, which includes the Lookingglass and Flournoy Formations of Baldwin (1974), and the Tyee Formation overlaps the Mesozoic terranes as well as the lower(?) Eocene Roseburg Formation in the northern part of the map area (Baldwin, 1974; Carayon, 1984). These little-deformed overlying sedimentary units appear to be synorogenic flysch deposits that were derived in part from the uplifted Klamath Mountains to the south after accretion was completed. They clearly document that nearly all of the terranes, including the far-traveled limestone blocks in the Sixes River terrane, were at the latitude of southwest Oregon by about 50 m.y.

PALEOMAGNETIC DATA

Paleomagnetic directions have been obtained from reconnaissance sampling of the Gold Beach and Yolla Bolly terranes. Although the preliminary nature of the data precludes conclusions concerning precise paleolatitudes and rotations, the results do suggest that the studied terranes possess unique magnetic signatures that support the interpretation of different paleoenvironments and geologic histories (Table 2; Fig. 4).

Gold Beach Terrane

Mudstone, fine sandstone, and limey concretions were sampled from 23 sites at four localities of the Otter Point Formation within the Gold Beach terrane as shown in Figure 1. The sites display a wide range of structural attitudes, and significant improvement is seen in the between-site scatter after correcting the paleomagnetic directions to horizontal. Both normal and reverse polarities are observed.

Assuming that we are measuring a sedimentary

Table 2

Terrane	Number of Sites	Number of Samples	Before Tectonic Correction				After Tectonic Correction			
			D^a	I	A_{95}	K	D	I	A_{95}	K
Gold Beach	23	71	51	47	38.3	1.6	262	43	8.5	13.8
Yolla Bolly	16	66	140	59	12.4	9.8	124	37	11.2	11.8

[a] D, I Declination and inclination of remanent magnetization; A_{95}, radius of 95% confidence in degrees; K, precision parameter.

Table 2—Average of site mean directions from reconnaissance sampling of two terranes in southwest Oregon.

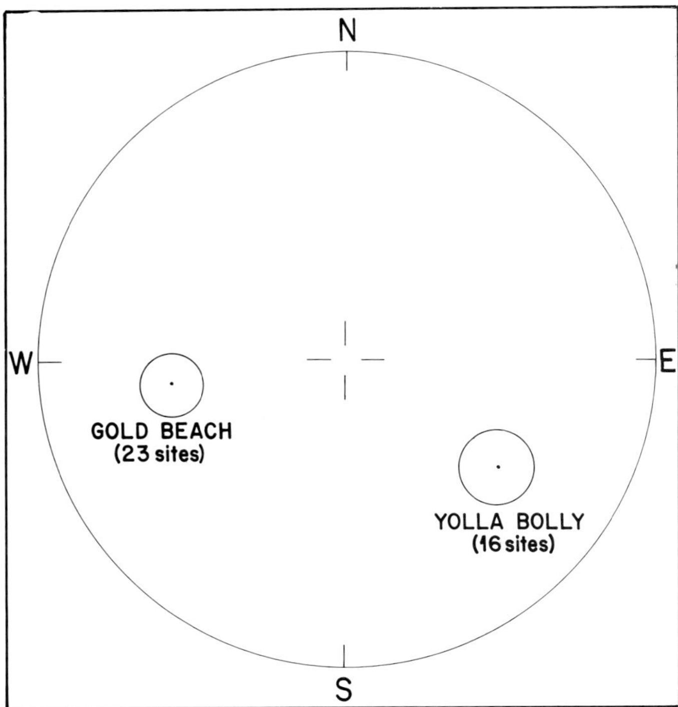

Figure 4—Paleomagnetic directions for Gold Beach and Yolla Bolly terranes after correction for tilt of beds.

remanent magnetization acquired approximately 150 m.y. ago, the data suggest that the rocks of this terrane have been displaced northward about 1,200 km (746 mi) if the magnetization was acquired in the northern hemisphere (Tables 2, 3). Although there is a large displacement indicated, it may not be significant considering the large uncertainty associated with the 150 m.y. pole (Irving and Irving, 1982). If future studies confirm the 150 m.y. wander pole with a greater degree of certainty, then this large apparent displacement will be significant. Alternatively, if the magnetization was acquired in the southern hemisphere, then the displacement would be on the order of 6,700 km (4,163 mi), which is significant even with the large 150 m.y. uncertainty. The depositional age of the rocks range from Tithonian to Valanginian, approximately 150 to 136 m.y. Comparison with the 150 m.y. pole was chosen because it gives the most conservative estimate of displacements. Both the northern and southern hemisphere options are equally valid. The Gold Beach terrane is also not geologically constrained to be at its present latitude until mid- to late Cenozoic, because it is not overlapped by the Eocene sediments.

Yolla Bolly Terrane

Preliminary results from 16 sites at 6 localities from fine-grained graywacke, chert, and pillow lava of the Dothan Formation are included in Table 1. Only a slight improvement in the dispersion is seen after the beds were restored to horizontal. It appears that there is a more significant improvement in the inclination dispersion, which suggests either undetected structures not accounted for or localized independent rotations about near vertical axes. Only one polarity has been recognized in the Yolla Bolly terrane thus far, and it is possible that we are seeing an effect of metamorphic overprinting rather than an original sedimentary remanent magnetization.

If this magnetization is primary and about 150 m.y. ago, then the apparent northward displacement is on the order of 1,600 km (994 mi) with a clockwise rotation of 150 ± 25°. Again, the large uncertainties associated with the 150 m.y. paleopole render this apparent displacement insignificant. If, on the other hand, the magnetization is an overprint acquired about 90 to 100 m.y. ago during regional metamorphism and the tilt factor is negligible, then the implied displacement with respect to the 90 m.y. pole is 3,600 km (2,237 mi) with a clockwise rotation of 154 ± 16°. The fact that all the sites represented one polarity direction during a time interval when the polarity flipped several times (Tithonian to Valanginian) suggests that the magnetization is an overprint. The regional structural fabric is one of subhorizontal thrust sheets (Coleman, 1972), which indicates that the tilt factor may be negligible. However, because the magnetization may be the result of an anisotrophy associated with deformation, the tectonic implications of these results are limited.

CONCLUSIONS

The data presented above suggest that southwest Oregon consists of a number of fault-bounded, late Mesozoic tectonstratigraphic terranes. The various subterranes of the Western Klamath terrane may have formed a complex island arc–marginal basin–continental margin sequence during Middle Jurassic time that was imbricated and accreted to North America during the Nevadan orogeny. From its stratigraphy and present outboard position, the Elk subterrane is believed to have been transported northward from a previous postaccretion position at the southern end of the Klamath Mountains in northern California.

The Snow Camp terrane is an accreted fragment of Jurassic island arc and overlap assemblage that is of similar age and lithology to parts of the Western Klamath terrane but lacks a penetrative deformation and metamorphism. It is likely that it was translated northward to its present position along strike-slip faults in a similar manner to the Elk subterrane during a pre mid-Eocene transform episode.

The Yolla Bolly rocks structurally underlie the rocks of the Western Klamath terrane in northern California and southwest Oregon and were probably accreted at about 90 m.y., corresponding to the timing of blueschist-facies metamorphism in northern California. Unlike northern California, the western subterrane records an episode of silicious igneous (arc?) activity.

The Pickett Peak terrane is completely allochthonous and is not rooted beneath the Western Klamath terrane, as to the south in northern California. Most likely it was dispersed northward after being metamorphosed and accreted to the south in California and bears no primary relationship to any of the other terranes.

The Sixes River terrane is a possible correlative of the Central melange terrane in northern California and contains scraps of similar exotic seamounts. It was probably formed by large-scale northward, strike-slip or oblique-slip motions related to passage of the Kula–

Figure 3g—Landslide-dominated topography typical of the sheared argillite matrix that forms the Central Franciscan belt. Looking west from Mendocino Pass; 100 m (328 ft)-size block of greenstone in center of photo.

Yolla Bolly terrane and dispersed slabs of the Pickett Peak terrane northward (Fig. 2). This faulting may have been associated with accretion of some of the exotic rocks in the Central belt, based on the fact that it appears to predate accretion of the Coastal terrane and because an Eocene overlap assemblage overlies similar faults in southwest Oregon (Blake et al, this volume).

A second set of high- to low-angle faults with both reverse and right-slip components trend about N 50° W across the earlier structures near Covelo (Fig. 2). Lower Cretaceous to Eocene strata in the Covelo area were apparently dispersed by this episode of faulting, and thus the faulting must be post-Eocene. Much of the deformation in the Central belt melange therefore appears to have formed by early thrust faulting followed by or concurrent with at least two periods of oblique right-slip faulting. South of the study area the second (N 50° W) set of oblique right-slip faults translated tens of kilometer-long slabs of the Coast Range ophiolite and Great Valley sequence northwestward away from the west side of the Great Valley province (McLaughlin, 1978; McLaughlin et al, 1983b) incorporating them into the Central belt.

Coastal Belt

The westernmost and youngest belt of the Franciscan assemblage, named the Coastal belt by Bailey and Irwin (1959), consists of several mappable terranes, the Coastal, Yager, and King Range, that are distinguished based on differences in lithology, age, sandstone composition, and structural state (Kramer, 1976; Bachman, 1981; McLaughlin et al, 1982, 1983b). The Coastal belt is separated from the Central belt melange by the Coastal belt thrust (Jones et al, 1978), an east-dipping structure locally folded into subvertical orientation and truncated by younger high-angle faults.

Figure 3h—Coherent arkosic sandstone and shale of the Coastal terrane near Leggett. Beds about 1 m (3.3 ft) thick.

Figure 3i—Disrupted graywacke and argillite (broken formation) of the Coastal terrane near Cape Mendocino; pocket knife for scale.

The Coastal terrane is characterized by largely arkosic sandstone, mudstone, and conglomerate that range in age from Late Cretaceous (c.f. 80 m.y.) to late Eocene (c.f. 40 m.y.). Laumontite veins are abundant, and fine-grained prehnite and pumpellyite can be seen in many thin sections. The Coastal terrane is structurally highly disrupted by brittle fracturing, shearing, and folding, with the fabric varying from partially disrupted sandstone-rich rocks in the eastern part (Fig. 3h) to melange in the western exposures (Fig. 3i). The chaotic section of the Coastal terrane contains far-traveled blocks of pillow basalt

Figure 3j—Coherent strata of the Yager terrane along Van Duzen River near Bridgeville; lens cap for scale.

Figure 3k—Folded and disrupted strata of the King Range terrane along Cooskie Creek, 10 km (6.2 mi) south of mouth of Mattole River.

and foraminiferal limestone, which probably are scraps of Late Cretaceous (Campanian to Maestrichtian) midocean sea mounts (Sliter, 1984; Harbert and others, 1983); as well as some rare blocks of medium-grade blueschist (Type III; Coleman and Lee, 1963).

The Yager Formation is the sole component of the Yager terrane (McLaughlin, 1983; McLaughlin et al, 1983b; Fig. 2). It differs from the Coastal terrane in that rocks of the Yager terrane consist largely of well-bedded (Fig. 3j), little-sheared, but locally highly folded mudstone-rich turbidites (Fig. 3k), with interbeds of quartzofeldspathic and arkosic, muscovite-bearing sandstone, locally with thick lenses of polymict conglomerate. In general, the Yager terrane is lithologically, structurally, and chronologically more homogeneous than the Coastal terrane. Shearing is only locally prominent near the western and eastern contacts with the Coastal terrane and Central belt; however, complex folding is widespread. Fossils (dinoflagellates and foraminifers) from the Yager terrane indicate that these rocks are no older than Paleocene and are as young as late Eocene. Thus, the age of the Yager terrane overlaps and is younger than the oldest rocks of the Coastal terrane. The Yager terrane has been interpreted as an uplifted and deformed slope-basin or trench-slope basin deposit that was deposited on older trench- and trench-slope deposits of the Coastal terrane (Underwood, 1983).

Penetrative shearing and deformation of the Coastal terrane and complex folding of the Yager terrane largely occurred between about 38 to 24 m.y. ago during or following the late Eocene and predated the middle and upper Miocene overlap assemblage of the Wildcat Group as used by Ogle (1953) and McLaughlin et al (1982). Emplacement of the Central belt over the Coastal belt along the Coastal belt thrust is also constrained to the interval between 38 and 24 m.y. also based on the overlap age of the Wildcat Group across the Coastal belt thrust of the Eel River basin. Rocks of the Wildcat Group are locally sheared into the Coastal terrane as the result of ongoing northeast-southwest-oriented convergence between the Pacific and North American plates at the Mendocino triple junction (McLaughlin et al, 1983b).

The youngest terrane of the Franciscan assemblage is the King Range terrane (McLaughlin et al, 1982). The King Range terrane is a composite of two subterranes (the Point Delgada and King Peak) of differing ages and lithologies, separated by steep faults that are stitched together by cross-cutting, adularia-bearing veins dated at 13.8 m.y. (McLaughlin et al, in press). The Point Delgada subterrane consists of basaltic pillow flows, pillow breccias, and tuffs, intruded by diabase sills. The igneous section is overlain by folded and sheared, pumpellyite-bearing, quartz-rich to arkosic sandstone and argillite that is devoid of detrital K-feldspar. Melange zones containing rare blueschist blocks cut the igneous and overlying sedimentary rocks. Radiolarians from red argillite interbedded with pillow breccias are of Late Cretaceous (Campanian or Coniacian) age.

The King Peak subterrane is characterized by complexly folded and broken calcareous argillite and interbedded quartzofeldspathic to volcaniclastic sandstone probably deposited in a lower slope or trench setting and associated locally with melange containing minor pods and lenses of pillow basalt, pelagic limestone, plus radiolarian and diatom-bearing chert (McLaughlin, 1983). Both the well-bedded pelagic and hemipelagic rocks and the melange blocks of pillow basalt-chert contain radiolarians and/or foraminifers of middle Miocene age or younger. A few poorly preserved foraminifers identified as Paleocene or Eocene in age are most likely reworked, although it is possible that part of the King Peak subterrane is Paleogene in age (McLaughin et al, 1982).

Aeromagnetic data (Griscom, 1980) indicates that the

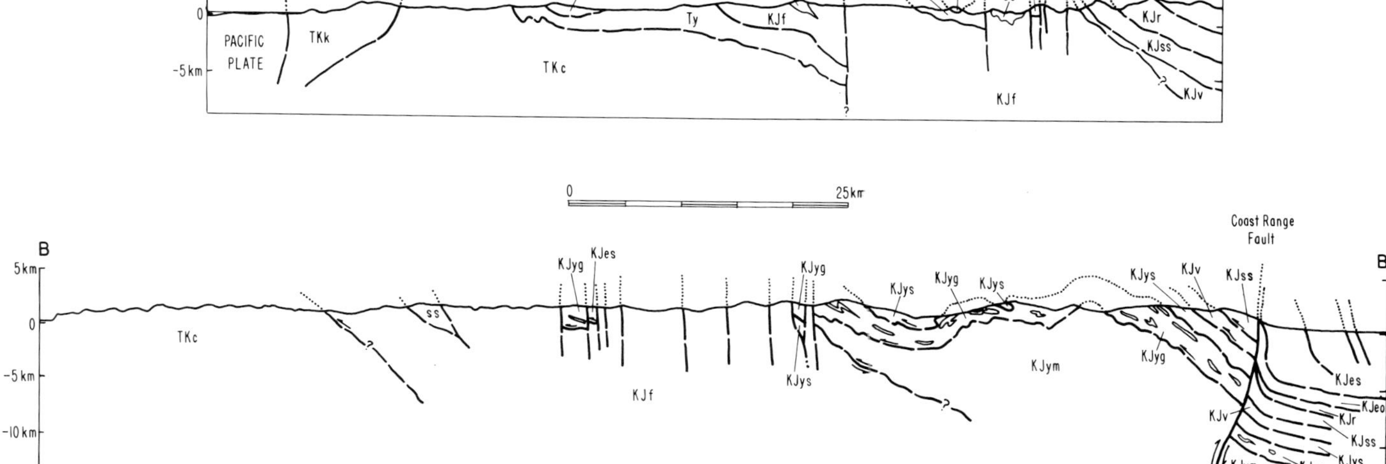

Figure 4—Structure sections across the northern California transect. See Figure 2 for locations. Map unit symbols (see Fig. 2) used may not necessarily conform to standards set forth by the U.S. Geological Survey.

King Range terrane dips southwestward, off the Coastal terrane, and thus was obducted (overthrust northeastward) onto the North American margin in post-middle Miocene time. This accretion is interpreted to result from northeast-southwest-oriented compression and uplift associated with late Neogene convergence between the Pacific and North American plates at the Mendocino triple junction (McLaughlin et al, 1982, 1983a) and postdates the stitching of the King Peak and Point Delgada subterranes at ~14 m.y.

STRUCTURAL RELATIONS

Near Red Bluff, the rocks of all three provinces come together in what has been called the Yolla Bolly junction (Blake et al, 1984). Both the Klamath Mountains and Great Valley rocks were formerly thought to be in thrust fault contact with the underlying Franciscan rocks, along the Coast Range thrust (Bailey et al, 1970), but more recent work (Worrall, 1981; Blake et al, 1984) suggests a much more complicated history. To the south of the Yolla Bolly junction, the bounding fault is nearly everywhere high-angle, and recent seismicity suggests that this structure is still active (unpublished report, California Department of Water Resources). The name Coast Range fault is retained for this segment (Worrall, 1981). To at least 15 km (9.3 mi) west of the Yolla Bolly junction, the bounding structure between the Franciscan assemblage and the Klamath Mountains province is a zone of high-angle faults with a large component of left-lateral strike-slip movement (Jones and Irwin, 1971; Worrall, 1981). Further west and north the contact is interpreted as a low-angle east-dipping thrust (Irwin et al, 1974).

Structural relations across the northern California Coast Ranges are summarized in the cross-sections (Fig. 4). The Late Cretaceous(?) low-angle thrust faults in the Eastern Franciscan belt are truncated by the Late Cenozoic(?) Coast Range fault on the east and Early to Middle Tertiary(?) north-trending high-angle faults on the west. The Central Franciscan belt is dominated by elongate slabs of graywacke and linear zones of "exotic" lithologies (pelagic limestone, eclogites, and high-grade blueschist). These linear features are inferred to be structurally controlled by at least two generations of high-angle faults with right-slip and reverse components. Earlier thrust relationships are preserved locally. The western margin of the Central Franciscan belt is underthrust by the rocks of the Coastal and Yager terranes. Farther west, the King Range terrane is obducted onto the Coastal terrane.

GEOLOGIC HISTORY

The following sequence of events is inferred to have occurred in the Coast Ranges of northern California. Figure 5 is an attempt to show schematically the relative timing of these events from the stratigraphic relations within each of the terranes.

1. Formation of the Coast Range ophiolite and the oceanic island arc represented by the Smith River subterrane at about 160 ± 5 m.y. (Hopson et al, 1981; Saleeby et al, 1982).

2. Timing of the Nevadan orogeny at about 150 ± 5 m.y. associated with isoclinal folding, lower greenschist-facies metamorphism, and development of a penetrative fabric in the Smith River subterrane, inferred to be related to collision of the arc with North America. This deformational and metamorphic event has not been recognized in the rocks of the Elder Creek terrane but could be expressed in the unconformity between the ophiolite and overlying mafic breccia. The pronounced Nevadan metamorphic cleavage seen in the Smith River subterrane, which is absent in the coeval Elder Creek terrane, requires either that the Elder Creek terrane was not in close proximity to the collision site or else was part

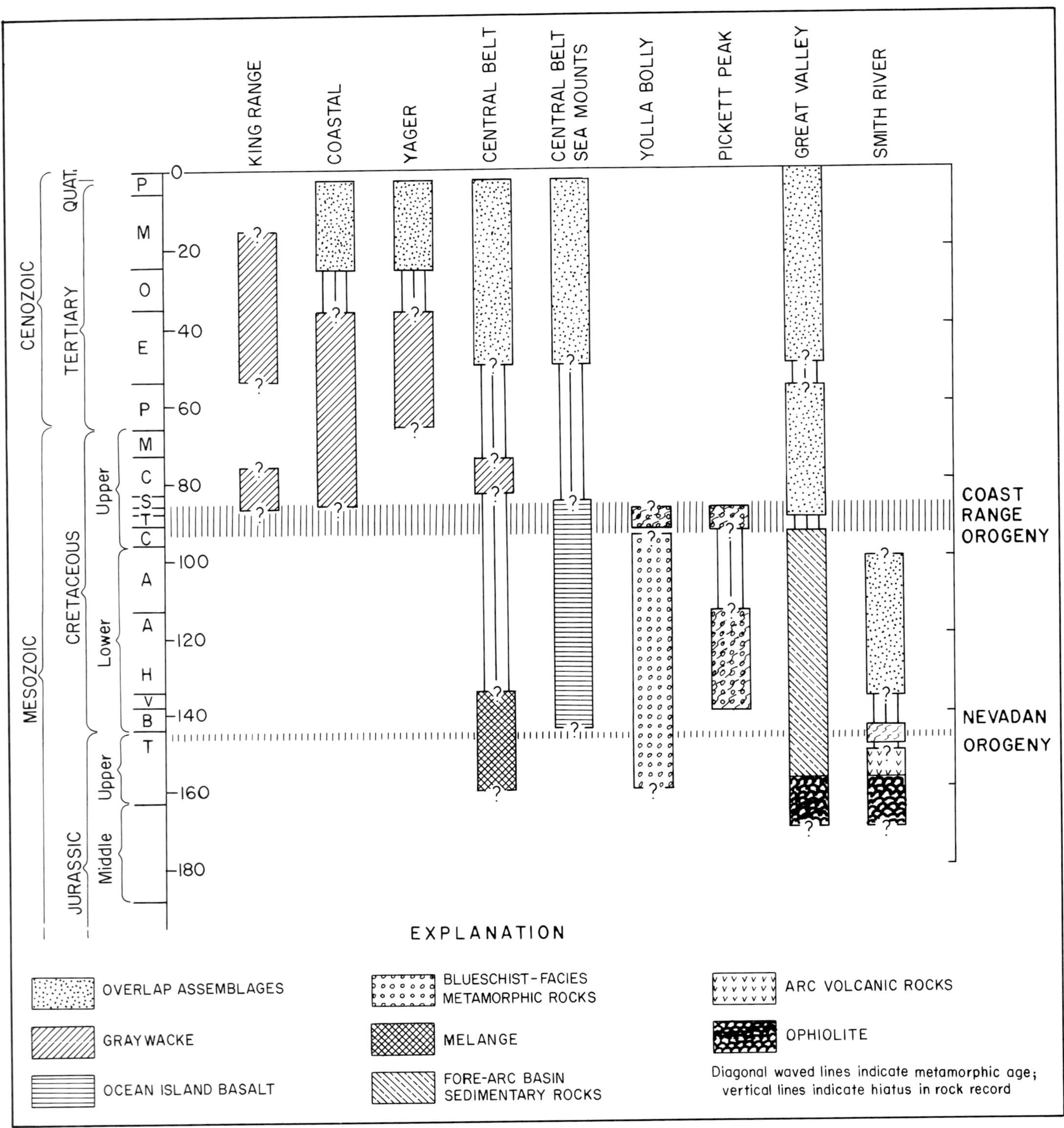

Figure 5—Terrane-time diagram.

of the upper plate during Nevadan deformation, thus avoiding metamorphism. We prefer the second model at the present time, because slabs of the Smith River subterrane that occur along the Coast Range fault are inferred to be derived from below the Elder Creek terrane south of the Yolla Bolly junction.

The origin of the high-grade blueschist, amphibolite, and eclogite blocks of the Central belt melange remains a mystery. Recent Pb/U dating of a blueschist block from a serpentinite diapir near Leech Lake Mountain (Fig. 2) at 162 ± 3 m.y. (Mattinson, 1981) suggests that these blocks predate the beginning of Franciscan assemblage subduction and may be older than the Nevadan metamorphic event. The age of the protolith and timing of metamorphism of the Pickett Peak terrane also remains an enigma, although the available radiometric data suggest an age of about 125 m.y. for the metamorphism (Lanphere et al, 1978; McDowell et al, in press).

3. Deposition of protolithic sediments of the Yolla Bolly terrane possibly began as early as Middle Jurassic time but definitely by the Late Jurassic and continuing up until the Middle Cretaceous. The age of the alkalic intrusive rocks is not known at this latitude but to the south of the study area in the Diablo Range, Pb/U ages on zircons from similar rocks were dated at 95 m.y. (Mattinson and Echeverria, 1980).

4. Timing of the Middle Cretaceous Coast Range orogeny. This event is associated with the accretion of the Yolla Bolly terrane to the continental margin and subsequent metamorphism. The 92 m.y. U/Pb metamorphic age on Yolla Bolly terrane rocks in the Diablo Range (Mattinson and Echeverria, 1980) correlates with the timing of a Turonian unconformity in the Great Valley sequence (McLaughlin et al, 1980; Blake and Jones, 1981).

5. Accretion and incorporation into the Franciscan assemblage Central Franciscan belt of far-traveled oceanic and sea-mount terranes (i.e., Laytonville Limestone of Alvarez et al, 1980) occurred sometime after the Middle Cretaceous (the depositional age of some of the younger deep-marine blocks) and was completed prior to about 50 m.y., based on the Eocene overlap assemblage in southwest Oregon. The accretion of the oceanic fragments may coincide with the truncation of the western margin of the Yolla Bolly terrane and northward dispersion of the Yolla Bolly, Pickett Peak, and Elder Creek terranes in the Central belt.

6. Deposition of the Yager and Coastal terranes during the Late Cretaceous to late Eocene followed by accretion and formation of the Coastal belt thrust prior to deposition of the middle Miocene and younger Wildcat Group of Ogle (1953).

7. Accretion of the King Range terrane (Late Cretaceous to Miocene) in post-middle Miocene time (probably less than 14 m.y.).

REFERENCES

Alvarez, W., et al, 1980, Franciscan complex limestone deposited at 17° south paleolatitude: Geological Society of America Bulletin, v. 91, p. 476–484.

Bachman, S. B., 1981, The Coastal Belt of the Franciscan: Youngest phase of northern California subduction, *in* J. K. Leggett, ed., Trench and forearc sedimentation and tectonics in modern and ancient subduction zones: Geological Society of London Special Publication, p. 401–418.

Bailey, E. H., and W. P. Irwin, 1959, K-feldspar content of Jurassic and Cretaceous graywacke of the northern Coast Ranges and Sacramento Valley, California: Bulletin of the American Association of Petroleum Geologists v. 43, p. 2797–2802.

———, et al, 1970, On-land Mesozoic ocean crust in California Coast Ranges, *in* Geological Survey Research 1970; U.S. Geological Survey Professional Paper 700-C, p. C70–C81.

Blake, Jr., M. C., and A. S. Jayko, 1983, Geologic map of the Yolla Bolly-Middle Eel Wilderness and adjacent roadless areas: U.S. Geological Survey Miscellaneous Field Studies Map MF-1595-A, scale 1:62,500.

———, and D. L. Jones, 1974, Origin of Franciscan melanges in northern California: Society of Economic Paleontologists and Mineralogists Special Paper 19, p. 255–263.

——— and ———, 1981, The Franciscan assemblage and related rocks in northern California: a reinterpretation, *in* W. G. Ernst, ed., Geotectonic development of California: Englewood Cliffs, NJ, Prentice-Hall, Inc., p. 307–328.

———, et al, 1967, Upside-down metamorphic zonation, blueschist facies, along a regional thrust in California and Oregon: U.S. Geological Survey Professional Paper 575C, p. 1–9.

———, et al, 1981, Geology of a subduction complex in the Franciscan assemblage of northern California: Oceanologica Acta, 1981 N° Sp, p. 167–172.

———, et al, 1982, Preliminary tectonostratigraphic terrane map of California: U.S. Geological Survey Open-File Report 82-593, 9 p., 3 maps, scale 1:750,000.

———, et al, 1984, Geologic map of the Red Bluff 1:100,000 quadrangle, California: U.S. Geological Survey Open-File Report 84-105, 33 p., 1 map, scale 1:100,000.

Brown, Jr., R. D., 1964, Thrust-fault relations in the northern Coast Ranges, California: U.S. Geological Survey Professional Paper 475D, p. 7–13.

Coleman, R. G., and D. E. Lee, 1963, Metamorphic aragonite in the glaucophane schists of Cazadero, California: American Journal of Science, v. 260, p. 577–595.

Griscom, A., 1980, Aeromagnetic and interpretation maps of the King Range and Chemise Mountain instant study areas, northern California: U.S. Geological Survey Miscellaneous Field Studies Map MF-1196-B.

Harbert, W. P., et al, 1983, Paleomagnetism of the Parkhurst Limestone, Coastal Belt Franciscan, Northern California: EOS, American Geophysical Union Transactions, v. 64, p. 868.

Harper, G. D., 1980, The Josephine ophiolite, remains of a Late Jurassic marginal basin in northwestern California: Geology, v. 8, p. 333–337.

Herd, D. G., 1978, Intracontinental plate boundary east of Cape Mendocino: Geology, v. 6, p. 721–725.

Hopson, C. A., et al, 1981, Coast Range ophiolite, western California, *in* W. G. Ernst, ed., The geotectonic development of California: Englewood Cliffs, NJ, Prentice-Hall, Inc., p. 418–510.

Irwin, W. P., 1960, Geologic reconnaissance of the northern Coast Ranges and Klamath Mountains, California: California Division of Mines Bulletin 179, 80 p.

———, 1981, Tectonic accretion of the Klamath Mountains *in* W. G. Ernst, ed., The geotectonic development of California: Englewood Cliffs, NJ, Prentice-Hall, p. 29–49.

———, et al, 1974, Geologic map of the Pickett Peak Quadrangle: U.S. Geological Survey Quadrangle Map GQ-1111, scale 1:62,500.

Jayko, A. S., 1983, Deformation in low-grade blueschists of the Eastern Franciscan belt, northern California (Abs.): Geological Society of America Abstracts with Programs, v. 15, p. 313.

Jones, D. L., 1975, Discovery of *Buchia Rugosa* of Kimmeridgian age from the base of the Great Valley sequence (Abs.): Geological Society of America Abstracts with Programs, v. 7, n. 3, p. 330.

______, and R. E. Imlay, 1969, Structural and stratigraphic significance of the Buchia zones in the Colyear Springs-Paskenta area, California: U.S. Geological Survey Professional Paper 647-A, 24 p.

______, and W. P. Irwin, 1971, Structural implications of an offset Early Cretaceous shoreline in northern California: Geological Society of America Bulletin, v. 82, p. 815–822.

______, et al, 1978, Distribution and character of Upper Mesozoic subduction complexes along the west coast of North America: Tectonophysics, v. 47, p. 207–222.

Kanter, L., et al, 1982, Paleomagnetism of continental and oceanic terranes of coastal California: EOS, American Geophysical Union Transactions, v. 63, n. 45, p. 915.

Kelsey, H. M., and D. K. Hagans, 1982, Major right-lateral faulting in the Franciscan assemblage of northern California in late Tertiary time: Geology, v. 10, n. 7, p. 387–391.

Kramer, J. C., 1976, Geology and tectonic implications of the Coastal Belt Franciscan, Ft. Bragg-Willits area, northern Coast Ranges, California: PhD Dissertation, University of California, Davis, 128 p.

Lanphere, M. A., 1971, Age of the Mesozoic oceanic crust in the California Coast Ranges: Geological Society of America Bulletin, v. 82, p. 3209–3212.

______, et al, 1978, Early Cretaceous metamorphic age of the South Fork Mountain Schist in the northern Coast Ranges of California: American Journal of Science, v. 278, p. 798–815.

Mattinson, J. M., 1981, U-Pb systematics and geochronology of blueschists: preliminary results: EOS, American Geophysical Union Transactions, v. 62, n. 45, p. 1059.

______, and L. M. Echeverria, 1980, Ortigalita Peak gabbro, Franciscan complex: U-Pb date of intrusion and high-pressure-low temperature metamorphism: Geology, v. 8, p. 589–593.

Maxwell, J. C., et al, 1981, Geologic cross-sections, northern California Coast Ranges to northern Sierra Nevada, and Lake Pillsbury area to southern Klamath Mountains: Geological Society of America Map and Chart Series MC-18N.

McDowell, F. W., et al, in press, Glaucophane schists and ophiolites of the northern California Coast Ranges: Ages, mineralogy, and their tectonic implications: Geological Society of America Bulletin, v. 95.

McLaughlin, R. J., 1978, Preliminary geologic map and structural sections of the central Mayacamas Mountains and the Geysers steam field, Sonoma, Lake, and Mendocino Counties, California: U.S. Geological Survey Open-File Report 78-389, scale 1:24,000.

______, 1983, Post middle Miocene accretion of Franciscan rocks, northwestern California: Reply to discussion by S. G. Miller and K. R. Aalto: Geological Society of America Bulletin, v. 94, p. 1028–1031.

______, et al, 1980, Structure of late Mesozoic rocks in the core of the Wilbur Springs antiform, northern Coast Ranges, California (Abs.): Geological Society of America Abstracts with Programs, v. 12, p. 119.

______, et al, 1982, Post-middle Miocene accretion of Franciscan rocks, northwestern California: Geological Society of America Bulletin, v. 93, p. 595–605.

______, et al, 1983a, Terrane boundary relations and tectonostratigraphic framework south Eel River Basin, northwestern California: American Association of Petroleum Geologists Meeting, Program and Abstracts, 58th Annual Meeting, Pacific Sections, American Association of Petroleum Geologists, Society of Economic Geologists, Society of Economic Paleontologists and Mineralogists, p. 112–113.

______, et al, 1983b, Tectonostratigraphic framework of the Franciscan assemblage and lower part of the Great Valley sequence in The Geysers-Clear Lake Region, California: EOS, American Geophysical Union Transactions, v. 64, p. 868.

______, et al, in press, Paragenesis and tectonic significance of base and precious metal occurrences along the San Andreas fault at Pt. Delgada, California: Economic Geology.

Ogle, B. A., 1953, Geology of Eel River Valley Area, Humboldt County, California: California Division of Mines Bulletin 164, 128 p.

Roure, F., and P. De Wever, 1983, Decouverte de radiolarites du Trias dans l'unite occidentale des Klamath, sud-ouest de l'Oregon, U.S.A.: Consequences sur l'age des peridotites de Josephine: Comptes Rendus de l'Academie des Sciences (Paris), v. 297, p. 161–164.

Saleeby, J. B., et al, 1982, Time relations and structural-stratigraphic patterns in ophiolite accretion, west central Klamath Mountains, California: Journal of Geophysical Research, v. 87, p. 3831–3848.

Silberling, N. J., et al, 1984, Lithotectonic terrane map of the North American Cordillera: U.S. Geological Survey Open-File Report 84-523, scale 1:2,500,000.

Sliter, W. V., 1984, Cretaceous low-latitude pelagic limestone from the Franciscan Assemblage, California, and the Sixes River terrane, Oregon (Abs.): Geological Society of America Abstracts with Programs v. 16, n. 5, p. 333.

Suppe, J. 1973, Geology of the Leech Lake Mountain-Ball Mountain region, California: University of California Publications in Geological Sciences, v. 107, p. 1–81.

Talley, K. L., 1976, Descriptive geology of the Redwood Mountain outlier of the South Fork Mountain Schist, Northern Coast Ranges, California: MS Thesis, Southern Methodist University, Dallas, 86 p.

Underwood, M. B., 1983, Depositional setting of the Paleogene Yager Formation, northern Coast Ranges of California; *in* D. K. Larue and R. J. Steel, eds., Cenozoic marine sedimentation, Pacific margin,

U.S.A.: Society of Economic Paleontologists and Mineralogists, Pacific Section, Symposium Volume, p. 81–101.

Worrall, D. M., 1981, Imbricate low-angle faulting in uppermost Franciscan rocks, South Yolla Bolly area, northern California: Geological Society of America Bulletin, v. 92, p. 703–729.

Metamorphic, Deformational, and Temporal Constraints on Terrane Assembly, Northern Klamath Mountains, California

L. Bruce Hill*
Stanford University
Stanford, California

An important interval of orogenic activity between 172 and 150 Ma ago is responsible for the assembly, metamorphism, and deformation of four terranes in the north-central Klamath Mountains of California. From structural, petrologic, and geochronological data, it is apparent that in an early pulse of tectonism, between 172 and 162 Ma ago, the Hayfork terrane and Marble Mountains terrane were juxtaposed along a low-angle thrust. The Condrey Mountain terrane then underthrust the Marble Mountains and Hayfork terranes sometime after 162 Ma. Later amalgamation of the Smith River terrane to the three-terrane assemblage (Marble Mountains, Hayfork, and Condrey Mountain terranes) occurred at about 150 Ma.

Observed deformational fabrics support three major episodes of regional folding. Early but apparently unrelated isoclinal folding events are associated with prograde metamorphism in the Marble Mountains and Condrey Mountain terranes. Tectonic mixing of the diverse mafic, sedimentary, and ultramafic lithologies of the Marble Mountains terrane occurred prior to this first dynamothermal event. A second stage of isoclinal folding affected the Hayfork terrane as well as the Hayfork and Condrey Mountain (?) terranes and was accompanied by low-grade recrystallization. A third folding event, regionally variable in style, is associated with the underthrusting of the Smith River terrane to the west and overprints all contacts and preexisting structural fabrics in the north-central Klamaths.

INTRODUCTION

The Klamath Mountains are situated within a long belt of possible accreted Mesozoic rocks extending along the cordillera from Alaska to Mexico (Coney et al, 1981). Within the Klamath Mountains numerous fault-bounded tectonostratigraphic terranes exhibit complex and unique geologic histories. Many of the original features of these terranes have been obscured by polyphase deformation and metamorphism. If the tectonic fabrics in terranes such as these can be understood in terms of the plate tectonic environments in which the rocks were deformed and metamorphosed, light will be shed upon processes involved in accretion of rock sequences to western North America in the Mesozoic.

Four thrust fault-bounded terranes of the northern Klamath Mountains are examined in this paper. Detailed studies of the metamorphic and deformational histories of these rocks, coupled with the geochronological constraints, are used to reconstruct temperature-depth-time paths for the terranes during Middle Jurassic times.

*Present Address: SOHIO Petroleum Company, 1801 California Street, Suite 3500, Denver, Colorado 80202

LITHOTECTONIC TERRANES

Four distinctive terranes have been identified in the area of Figures 1 and 2 in the north-central and western Klamath Mountains of California by Blake et al (1982). From west to east the terranes discussed in this paper are: (1) the Smith River terrane, (2) the Marble Mountains terrane, (3) the northern Hayfork terrane, and (4) the Condrey Mountain terrane.

The Marble Mountains Terrane

The Marble Mountains terrane, a northern subdivision of Irwin's (1972) western Paleozoic and Triassic belt, is a complex sequence of rocks that has experienced recurring cycles of volcanism, plutonism, deformation, and metamorphism. The Marble Mountains terrane can be subdivided into smaller lithologic units that include: (1) mafic dike complex(es), (2) dismembered ultramafics with associated (?) amphibolite, (3) metavolcanic-metasedimentary complexes, and (4) the Preston Peak Ophiolite. Common rock types of the Marble Mountains terrane include abundant mafic schist, metaperidotite, marble, minor chert, black schist, and quartzite. It has been suggested that this assemblage may be a tectonic melange (Coleman, 1983, personal communication). The

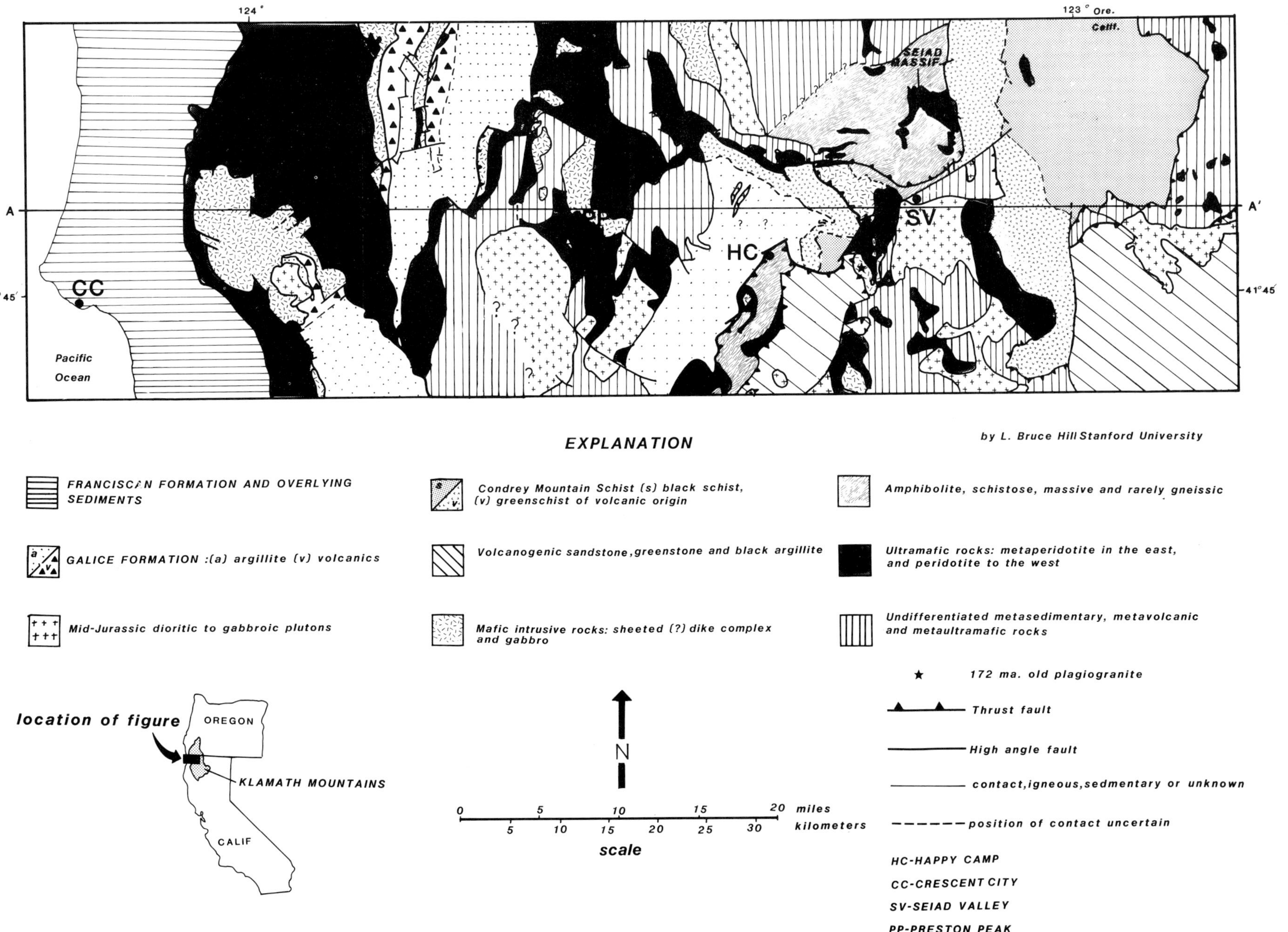

Figure 1—Interpretive geologic map of the north-central Klamath Mountains (geologic map based on CDMG Weed quadrangle [1:250,000] by D. Wagner and G. Saucedo, California Division of Mines and Geology Regional Geologic Map Series Map No. 3A, in preparation).

Table 1

Deformation Event	D1	D2	[a]D3	[a]D3
Folding Phase	F_1	F_2	F_3	F_4
Fold profile	transposed	isoclinal	chevron	chevron
Foliation	penetrative (S_1)	none	crenulation (local)	none
Size of folds	unknown	very large (kilometers)	small (m)– large (km)	small (m)
Abundance	—	rare	moderate	localized
Trend and plunge of fold axis	unknown	NE	NE	NW
Strike/dip of axial plane	unknown	unknown	NE–NW	NW–NE
Associated metamorphism	amphibolite facies	retrograde	none	none
Relationship to foliation development	synchronous	deforms S_1	deforms S_1 and retro-grade min	deforms S_1 and retro-grade min

[a]F_3 and F_4 apparently developed as a conjugate pair subsynchronously during D3 deformation.

Table 1—Folding summary: Marble Mountains terrane.

Pressures of metamorphism in the Hayfork terrane were probably low. Evidence of low Na(M4) in the Hayfork amphiboles suggests that pressures were 2 kbars or less. Additionally, the lack of pervasive metamorphic fabrics in the Hayfork metavolcanics indicates a static environment of recrystallization. Thus, the northern Hayfork terrane experienced metamorphism near the uppermost limit of the greenschist facies, at lower pressures and, to a lesser extent, temperatures than recorded by nearby metabasites of the Marble Mountains terrane. This relationship supports a break in metamorphic grade across the faulted contact between the two terranes (T_2).

Deformation

The first episode of deformation recognized in this terrane (Table 2) is evidenced by a penetrative cleavage, S_1, cutting original bedding in the argillites. Since minor fold hinges associated with this fabric have not been observed, folding, associated with S_1 in this terrane, is demonstrated by bedding-cleavage intersections that indicate a northeasterly trend for the F_1 axis. As mentioned above, recrystallization accompanying the development of the S_1 foliation took place at the greenschist-amphibolite transition.

The second folding event observed in the rocks of this terrane is F_2. This deformation is weak and is characterized by open folding without cleavage development. F_2 folding in this terrane accounts for broadly undulating antiform/synform pairs reported by Rawson (unpublished map, University of Oregon) and Peterson (1982). The northeasterly trending minor folds of this age are rare. S_1 foliations are largely homoclinal, and data indicate that the sequence is subhorizontal and not strongly affected by F_2.

F_3, the third episode of folding in the northern Hayfork terrane, is manifested in widely scattered, steeply dipping brittle kink bands and crenulations.

Condrey Mountain Terrane

Metamorphism

As indicated in Figure 4, the mafic schists of the Condrey Mountain terrane were crystallized under conditions transitional between the greenschist and blueschist facies. Actinolite-albite-epidote-chlorite-sphene-quartz ± pyrite is the common paragenesis in the metabasites and corresponds to the S_1 foliation. Absence of calcite suggests a low PCO_2 during metamorphism. Mapping by Donato et al (1980) as well as by Helper (1983) has revealed blue amphibole-bearing mafic schist within the Condrey Mountain window indicative of moderately high pressures of recrystallization (Ernst, 1979; Brown, 1977). Additionally, stilpnomelane and deerite, minerals found in other high-pressure terranes, have been recognized (Coleman, personal communication).

During a second episode of recrystallization, chlorite formed, probably at the expense of earlier formed actinolite and albite. Thin-section petrography F_2 folds indicates that the development of the S_2 foliation is the result of chlorite crystallization parallel to the axial plane of the F_2 folds.

Deformation

The layered greenschists and black phyllites east of Happy Camp (Fig. 1) were examined in detail resulting in

Table 2

Deformation Event	D1	D2	D3	[a]D4
Folding Phase		F_1	F_2	F_3
Fold profile		isoclinal (?)	open concentric	kink
Foliation		penetrative (S_1)	none	none
Size of folds		unknown	small (m)	small
Abundance		unknown	rare	rare
Trend and plunge fold axis		NE	NNE	—
Strike/dip of axial plane		subhorizontal	NE–NW	N–W
Associated metamorphism		upper greenschist	none	none
Relationship to foliation development		synchronous	deforms S_1	deforms S_1

[a]Kinking may be the result of Tertiary uplift in the area proposed by Coleman and Helper (1983).

Table 2—Folding summary: Hayfork terrane.

the formulation of a four-phase deformational history (Table 3). Structural studies of the Condrey Mountain terrane, further east within the Condrey Mountain window, have been completed by Donato et al (1980), Helper (1983), and Manning (unpublished data). The following discussion is based on data derived from the rocks correlated with the Condrey Mountain Schist to the west, near Happy Camp, California.

The four-folding phases exhibited by the rocks of the Condrey Moutain terrane (F_1 through F_4) are comparatively similar in style and orientation to the F_1 through F_3 sequence described for the Marble Mountains terrane above (compare Figs. 1, 3). The oldest recognized fabric is a transposed foliation (S_1) recrystallized in the greenschist to blueschist facies (Helper, 1983). Fold closures associated with this foliation have not been observed in the field and are presumed to have been obliterated during subsequent strain.

A second phase of isoclinal folding, F_2, has also been recognized in the greenschist and black phyllites of the Condrey Mountain terrane. Fold closures of this age are uncommon but well exposed, having similar-style fold profiles distinctly deforming the earlier S_1 schistosity. In general, axial planes strike west-northwest and dip to the north. Fold axes trend and plunge erratically northeast to northwest. A penetrative foliation is associated with F_2 folding, and the crystallization of chlorite along the axial surfaces of these folds is responsible for the development of the foliation, as mentioned above.

F_3 in the Condrey Mountain terrane is characterized by chevron to tight upright folds with northeasterly striking and northwesterly dipping crenulation to slip cleavages forming in the more tightly oppressed hinges. Fold axes vary in trend and plunge from north-northeast to north-northwest. This third episode of folding is ubiquitous throughout the Condrey Mountain terrane as indicated by data of Helper (1983). The structural trends of these F_3 folds are also continuous into the adjacent Marble Mountains terrane, and the thrust fault between the two terranes is clearly deformed by these folds as observed in the field. As a result, F_3 folds in both terranes are attributed to the same deformation event, D3. Moreover, it is clear that the two terranes were juxtaposed before the onset of D3 deformation.

A late fourth folding event is a steep cross-cutting kink banding that is probably related to the late kinking seen in the Marble Mountains and Hayfork terranes.

Smith River Terrane

Metamorphism

Snoke (1972) reports that the Galice Formation in the Gasquet quadrangle (California) has suffered low-grade metamorphism under P/T conditions no higher than greenschist facies. More detailed work by Harper (1980) suggests recrystallization in the pumpellyite-actinolite facies in volcanics overlying the Josephine Ophiolite as indicated by the assemblage chlorite-albite-quartz-pumpellyite-calcite-actinolite. Relict igneous fabrics have been recognized in these rocks, suggesting a somewhat static type of metamorphism at low temperatures.

This regional metamorphic event developed with

Table 3

Deformation Event	[a]D1(?)	[a]D2(?)	D3	[b]D4(?)
Folding Phase	F_1	F_2	F_3	F_4
Fold profile	transposed (S_1)	isoclinal (S_2)	chevron to tight	kink
Foliation	penetrative (dominant)	penetrative (subdominant)	local crenulation	none
Size of folds	—	large (km)	small (m)	very small
Abundance	—	rare	common	sporadic
Trend and plunge of fold axis	—	NNE	NE-NW	—
Strike/dip of axial plane	—	E-W/N	NE-NW	N-S/W
Associated metamorphism	greenschist-blueschist	retrograde	?	none
Relationship to foliation development	synchronous with S_1	synchronous with S_2	deforms S_2 and S_1	deforms S_2 and S_1

[a]Deformational events in the Condrey Mountain terrane may not be related in time or space to deformational events in the three other terranes (see text for explanation).
[b]F_4 kinking may be related to Tertiary doming of the Condrey Mountain terrane proposed by Coleman and Helper (1983).

Table 3—Folding summary: Condrey Mountain terrane.

"Nevadan" folding that forms pervasive tight to isoclinal F_1 folds and a steep S_1 penetrative cleavage in the argillites of the Galice Formation (Harper, 1980). Geochronological constraints developed by Saleeby et al (1982) place the age of the Josephine ophiolite and the concomitant deposition of the Galice Formation after 157 Ma. Therefore, this metamorphic and deformational event must have occurred no earlier than 157 Ma.

Deformation

The oldest structures recognized in the Galice Formation overlying the Josephine Ophiolite are isoclinal in profile with a strong slatey cleavage (Table 4). Original bedding is deformed by these F_1 folds. The age of the Galice flysch in this area has been constrained between 157 and 150 Ma (Saleeby et al, 1982). Therefore, this first deformation and any subsequent folding must be post-157 Ma old.

The second deformation described in these rocks is F_2 and is characterized by chevron, kink, and concentric style folds, apparently related to the Cretaceous Coast Range thrust (Harper, 1980).

INTERPRETATION OF FOLDING AND METAMORPHISM

In the ensuing discussion, a comparison of folding and metamorphism between the four terranes provides some baseline constraints concerning the possible interactions among the terranes.

Folding in the north-central Klamaths can be largely attributed to three important Jurassic deformational events, D1 to D3 (Tables 1 through 4). The Marble Mountains and Condrey Mountain terranes show evidence of all three (F_1 to F_4 folding), the northern Hayfork terrane two (F_1 and F_2), and the Smith River terrane one (F_1, see discussion below). These data might suggest that the Marble Mountains and Condrey Mountain terranes, having similar folding histories, were juxtaposed before the onset of the first folding cycle. Such a hypothesis is unlikely for two reasons: (1) the syndeformational metamorphism of the underlying Condrey Mountain terrane is of the high P/T type whereas the overlying Marble Mountains terrane exhibits syndeformational metamorphism of the low P/T variety, (2) the thrust contact between the two terranes is not apparently deformed by D1- or D2-related structures. Therefore, the first two deformations involving recrystallization in each terrane, although similar in style and orientation, may have formed in altogether different tectonic environments. However, since D3 structures overprint the contact between the terranes, D3 occurred after the two terranes were juxtaposed after 162 Ma ago. Hence, the Condrey Mountain terrane may have been tectonically removed from its high-pressure metamorphic environment and thrust against the base of the Marble Mountains terrane. A reasonable site for such a process might be beneath the hot hanging wall of an island arc, along a subduction-related zone of imbrication. In such a scenario, the Condrey Mountain terrane may represent an underplated fragment from the lower, cooler plate, while

Table 4[a]

Deformation Event	D1	D2	D3	[b]D4(?)
Folding Phase			F_1	F_2
Fold profile			tight to isoclinal	chevron to concentric
Foliation			S_1 slatey cleavage	none
Size of folds			large (km)	small (m)–large (km)
Abundance			uncommon	abundant
Trend and plunge of fold axis			SSE	SSE
Strike/dip of axial plane			NNW–ENE	NW–NE
Associated metamorphism			pumpellyite actinolite facies	none
Relationship to foliation development			synchronous	deforms S_1

[a]Data from Harper (1980).
[b]A possible Cretaceous or younger deformation not related to deformations in terranes to the east.

Table 4—Folding summary: Smith River terrane.

the Marble Mountains terrane may be a sliver from the amphibolite facies mafic and ultramafic rocks beneath the leading edge of an island arc.

The 162 Ma Wooley Creek batholith intrudes the Marble Mountains terrane–Hayfork terrane boundary (T_2) (Allen et al, 1982), indicating that D1 deformation and attendant metamorphism must have occurred prior to pluton intrusion. However, because of the 172 Ma igneous age of the dike complex (see next section), this first dynamothermal event is confined between 172 and 162 Ma.

A correlation of F_2 in the Marble Mountains terrane with F_1 in the Hayfork terrane is indicated for the following reasons: (1) F_2 fold axes in the Marble Mountains terrane have northeast trends and plunges as indicated by unrefolded F_2 fold hinges; F_1 fold axes in the Hayfork terrane may also have northeast trends and plunges as demonstrated by the orientation of bedding-cleavage intersections; (2) strikes and dips of foliations are concordant between adjacent parts of terranes; (3) progressive syn-F_1 metamorphism of the Hayfork terrane occurred at temperatures of about 450 to 500°C; similar temperatures are recorded by the metabasites of the Marble Mountains terrane during retrograde recrystallization; and (4) the thrust between the two terranes (T_2) is overprinted by the syn-F_1 prograde greenschist facies metamorphism. As a result, it is probable that F_1 in the Hayfork terrane and F_2 in the Marble Mountains terrane were generated during one regional tectonometamorphic event, D2. Thus, no evidence for D1 structures in the Hayfork terrane are apparent. This relationship suggests that the two terranes were juxtaposed after D1 but before D2.

In the Smith River terrane, F_1 folds are younger than 157 Ma as previously noted. Therefore, in correlating F_1 folds with structures in the three terranes to the east, D1 and D2 deformations, confined between 172 and 162 Ma, are eliminated. However, since D3 is post-162 Ma in age, it is possible that F_1 folding in the rocks of the Smith River terrane is the result of the regional D3 deformation. Accordingly, D3 may have formed during the "Nevadan orogeny" of Harper (1980), when the Smith River terrane was sutured to the three terranes to the east.

A temporal sequence of events may now be proposed for the early history of the four terranes: (1) post-D1 thrusting of the northern Hayfork terrane across the Marble Mountains terrane, (2) D2 deformation with greenschist facies metamorphism of the northern Hayfork terrane and retrograde metamorphism of the Marble Mountains terrane accompanied by simultaneous intrusion of calc-alkaline plutons at 162 Ma, (3) underplating of the two-terrane assemblage by the Condrey Mountain terrane prior to D3, and (4) thrusting of the Smith River terrane beneath the composite three-terrane assemblage to the east.

Table 5

Age	Sample	Interpreted Significance
[a]172 Ma Pb/U zircon	metamorphosed plagiogranite from mafic dike complex of the Marble Mountains terrane	age of oceanic crust from Marble Mountains terrane; maximum age of D1 deformation and amphibolite facies metamorphism
[b]170 Ma Pb/U zircon	leucocratic stock in the Condrey Mountain Schist (deformed)	minimum depositional age of the Condrey Mountain schist protolith; minimum age of F_1 in the Condrey Mountain terrane
[b]167 Ma K–Ar glaucophane	metavolcanic rock from the Condrey Mountain terrane	cooling age for first metamorphic event in the Condrey Mountain terrane
[c]165 Ma K–Ar hornblende	retrograded amphibolite block near Preson Peak in Marble Mountains terrane	minimum age for D1 related amphibolite facies metamorphism in the Marble Mountains terrane
[d,e]162 Ma Pb/U zircon	Slinkard Pluton (S) Wooley Creek Batholith (W.C.)	W.C. pluton cuts Marble Mountains/Hayfork terrane boundary (T_2), giving a minimum age of T_2 Slinkard is post-S_1/pre-F_3, thus 162 is a maximum age of F_3 and D3, and a minimum age of S_1 and D1 plutons are rootless above T_3; 162 is a maximum age of T_3 amphibolite xenoliths in (S); 162 is a minimum age of amphibolite facies metamorphism and D1
[c]150 Ma Pb/U zircon	undeformed plagiogranite dike in the Galice Fm. of Smith River terrane	minimum depositional age of the Galice Formation on the Josephine Ophiolite; a maximum age for D3 produced F_1 folds in the Galice
[f]150–156 Ma Kimmeridgian (Cox et al time scale)	*Buchia concentrica* fossils from the Galice Formation	maximum depositional age of the Galice Formation; maximum age for D3-related F_1 folding in the Smith River terrane

[a]Saleeby et al (1984).
[b]Coleman et al (1983).
[c]Reported by Saleeby et al (1982).
[d]Reported by Allen et al (1982).
[e]Reported by Barnes et al (1982).
[f]Reported by Harper (1980).

Table 5—Geochronological constraints pertinent to the timing of tectonomorphic events in the north-central Klamath Mountains.

GEOCHRONOLOGICAL CONSTRAINTS

Within the region of the four terranes described above, four stages of thrusting and folding and at least two stages of metamorphism were identified. Geochronological constraints summarized in Table 5 have been helpful in formulating a progressive sequence of structural and metamorphic events within an absolute time frame. Table 6 outlines the proposed sequence of events.

A new 172 (±1) Ma Pb/U age (Table 7, Saleeby et al, 1984) from a metamorphosed leucocratic differentiate of an amphibolite facies dike complex exposed south of Seiad Valley provides an important new geochronological constraint on the absolute timing of events in the north-central Klamaths. The date suggests a Bathonian igenous age for the formation of the ophiolite protolith and provides an upper limit on the age of metamorphism and deformation on the Marble Mountains terrane.

A 170 (± 2) Ma concordant Pb/U igneous crystallization age (Coleman et al, 1983) determined on zircon separates from a metamorphosed leucocratic stock intruding the Condrey Mountain Schist provides a minimum age for the protolith of the Condrey Mountain Schist. A K–Ar metamorphic age of 167 Ma (Coleman et al, 1983) gives a minimum age of the metamorphism of the Condrey Moutain terrane. Thus, the greenschist-blueschist event recorded in the Condrey Mountain terrane may have occurred between 167 Ma and 170 Ma ago.

A retrograded amphibolite block in the Preston Peak area of the Marble Mountains terrane gives a 165 Ma K–Ar age (Saleeby et al, 1982), indicating that the major amphibolite facies event was prior to that date but no earlier than the 172 Ma igneous age of the dike complex. Amphibolite xenoliths in the unmetamorphosed 162 Ma Slinkard Pluton give a minimum age for the high-grade metamorphism of the Marble Mountains terrane. Because of the 162 Ma Slinkard Pluton found in the Marble Mountains and Hayfork terranes was intruded while the

Table 6

Phase	Age	Event
0	pre–172 Ma	genesis of tectonostratigraphic terranes: *Marble Mountains terrane:* simatic crustal rocks and superimposed island arc intrusives *Hayfork terrane:* interarc basin assemblage with superimposed island arc intrusives *Condrey Mountain terrane:* subducted interarc basin (?) assemblage
1	172–165 Ma	tectonic mixing (T_1) of Marble Mountains terrane lithologies; amphibolite facies metamorphism of the Marble Mountains terrane and D1 deformation; blueschist-greenschist facies metamorphism of the Condrey Mountain terrane
2	165–162 Ma	T_2 thrusting of Hayfork across Marble Mountains; D2 deformation and metamorphism: *Hayfork terrane:* prograde upper greenschist *Marble Mountains terrane:* retrograde upper greenschist *Condrey Mountain terrane:* coeval (?) retrograde greenschist intrusion of calc-alkaline plutons
3	162–150 Ma	T_3 thrusting of the Marble Mountains/Hayfork composite terrane over the Condrey Mountain terrane
4	157–150 Ma	formation of the Josephine Ophiolite and concomitant deposition of the Galice Formation (both in the Smith River terrane)
5	150–145 Ma	assembly of the Smith River terrane with the three terranes to the east (Hayfork, Marble Mountains, and Condrey Mountain terranes) D3 deformation: isoclinal F_1 folding in the Smith River terrane transitional with upright F_3 folding in the Marble Mountains and Condrey Mountain terranes and upright F_2 folding in the Hayfork terrane

Table 6—An interpretive sequence of events in the assembly of terranes in the north-central Klamath Mountains.

Table 7

Concentration	Isotopic Composition	Ages (Ma)[b]		
^{238}U (ppm)	$^{206}Pb/^{204}Pb$	$^{206}Pb^{238}U$	$^{207}Pb/^{235}U$	$^{207}Pb/^{206}Pb$
239	1501	172 ± 2	172 ± 2	179 ± 10

[a]Sample collected by Hill, separations and age determination by Saleeby.
[b]Decay Constants:
$\lambda^{238}U = 1.5513 \times 10^{-10}$/year
$\lambda^{235}U = 9.8485 \times 10^{-10}$/year
$^{235}U/^{238}U = 137.8$

Table 7—Geochronological data for the concordant 172 ($\pm$1) Ma age determination by Saleeby et al (1984). Location of sample is shown in Figure 1.

surrounding amphibolite country rocks were still hot (country rocks are not thermally metamorphosed), pre-existing retrograde metamorphism of the Marble Mountains terrane and accompanying prograde metamorphism of the northern Hayfork terrane (?) probably occurred between 162 Ma and 165 Ma, during F_2 folding (?). The important earlier stage of prograde metamorphism of the Marble Mountains terrane as well as T_2 would be pre-165 Ma, perhaps coeval with metamorphism of the Condrey Mountain terrane between 167 Ma and 170 Ma.

In the interval between 157 and 150 Ma, the Josephine Ophiolite was generated according to Saleeby et al (1982) and the Galice Formation deposited during an interval of extensional tectonics (Harper, 1983a). The tectonic regime became compressional again between 150 and 145 Ma (Harper, 1983b) during the faulting of the Smith River terrane with the crystalline rocks to the east.

SUMMARY

The following sequence of tectonic and metamorphic events is proposed using the constraints developed in this paper (Table 5): (1) T_1 ductile imbrication of metabasites and metaperidotite and general premetamorphic mixing of lithologies; D1 deformation-F_1 folding and amphibolite facies metamorphism between 165 and 172 Ma in the Marble Mountains terrane; coeval (?) F_1 folding and greenschist-blueschist facies metamorphism between 167 and 170 Ma in the Condrey Mountain terrane; (2) T_2 thrusting of the northern Hayfork terrane over the Marble Mountains terrane; (3) F_2 folding of the Marble Mountains and Condrey Mountain (?) terranes, and F_1 folding of the Hayfork terrane; prograde greenschist-amphibolite transitional facies metamorphism in the northern Hayfork terrane; retrograde metamorphism in the Marble Mountains and Condrey Mountain (?) terranes between 162 and 165 Ma; subsynchronous intrusion of the 162 Ma old calc-alkaline plutons; (4) T_3 thrusting of the Marble Mountains–northern Hayfork terranes over the Condrey Mountain terrane; (5) deposition and formation of the Galice–Josephine assemblage of the Smith River terrane between 157 and 150 Ma; and finally, (6) juxtaposition of the Smith River terrane with the remaining three terranes to the east; D3 deformation and pumpellyite-actinolite facies metamorphism in the Smith River terrane, and coeval upright folding in the easterly terranes.

If the D3 correlation across the four terranes is correct, the differences in folding geometry continuous among the four terranes might be attributed to a regional scale strain gradient; D3-associated folding is isoclinal and most ductile in the west, becoming more brittle and upright to the east in the Marble Mountains, Condrey Mountain, and Hayfork terranes. Diachronous folding such as this has been described by Price (1972) who has demonstrated that profound changes in the style of folding from one place to another can primarily be a function of the material properties of the rock and may not necessarily indicate superimposed folding. These rock properties are in turn directly a function of pressure and temperature. Thus, a ductile flow regime generating isoclinal folds in one locality might grade into a more brittle regime producing an upright style of folding as is described for D3 in the east. This kind of diachronous deformation might be expected during the thrusting of the Smith River terrane, probably derived from a regime of high heat flow, beneath a cooler uplifted assemblage of terranes to the east. D3-related folding has therefore overprinted all four terranes between 145 Ma and 157 Ma. Accordingly, "Nevadan" deformation (Harper, 1983b) may in part be the result of this event.

In summary, F_3 deformation records a gradient in folding style interpreted to be the result of suturing a warm ductily deforming terrane (Smith River terrane) to a cooler, more brittly deforming assemblage of metamorphic terranes.

In conclusion, the four terranes described were consolidated into a complex geologic package during two important compressional orogenic episodes over an active orogenic period of about 20 Ma. The first stage may have begun at about 172 Ma with D1 ductile deformation and metamorphism that may have taken place at moderate levels of the crust, at the base of an island arc in the case of the Marble Mountains terrane (Hill, 1984), and in the adjacent (?) part of a trench in the case of the Condrey Mountain terrane. Later in the first episode, prior to 162 Ma, the Hayfork and Marble Mountains terranes were assembled, probably at higher levels of the arc, as indicated by the low-pressure metamorphism of the Hayfork terrane. Therefore, D1 and D2 deformation and T_2 thrusting are envisioned as one progressive tectonometamorphic event. The transitional blueschist-greenschist facies Condrey Mountain terrane may have then been transported from the footwall of the arc upward (?) and attached to the base of the Marble Moutains terrane by thrusting. In the second event, at about 150 Ma, the shallower levels of an interarc basin assemblage (Smith River terrane) were juxtaposed with the previously uplifted composite crystalline package represented by the Marble Mountains, Condrey Mountain, and northern Hayfork terranes. During this event diachronous folding developed across the accretionary orogen, effectively overprinting all previous tectono-metamorphic fabrics and terrane boundaries.

ACKNOWLEDGMENTS

The author gratefully acknowledges Dr. Robert G. Coleman who provided inexhaustible guidance as well as field support for this project through the U.S. Geological Survey. Dr. Juhn G. Liou is also thanked for his invaluable discussions on the metamorphism of the northern Klamath Mountains. Extensive comments on this manuscript were provided by Robert Coleman, Juhn Liou, Elizabeth Miller, Nick Mortimer, Craig Manning, and David Howell. Chevron U.S.A. is acknowledged for fellowship support during the 1983–1984 academic year at Stanford University.

REFERENCES

Allen, C. M., et al, 1982, Comagmatic nature of the Wooley Creek Batholith and the Slinkard Pluton and age constraints on tectonic and metamorphic events in the western Paleozoic and Triassic Belt, Klamath Mountains, Northern California (Abs.): Geological Society of America Abstracts with Programs, v. 4, n. 4, p. 145.

Apted, M. J., and J. G. Liou, 1983, Phase relations among greenschist, epidote amphibolite and amphibolite in a basaltic system: American Journal of Science, v. 283-A, p. 328–354.

Barnes, C. G., et al, 1982, Evidence for basal detachment of the western Paleozoic and Triassic Belt, northern Klamath Mountains, California (Abs.): Geological Society of America Abstracts with Programs, v. 14, n. 4, p. 147.

Blake, M. C., et al, 1982, Preliminary tectonostratigraphic terrane map of California: United States Geological Survey Open-File Report 82-593.

Brown, E. H., 1977, The crossite content of a Ca amphibole as a guide to pressure of metamorphism: Journal of Petrology, v. 18, pt. 1, p. 53–72.

Coleman, R. G., and M. D. Helper, 1983, The significance of the Condrey Mountain dome in the evolution of the Klamath Mountains, California and Oregon (Abs.):

Geological Society of America Abstracts with Programs, v. 15, n. 5, p. 294.

Coleman, R. G., et al, 1983, Geologic map of the Condrey Mountain Roadless Area, Siskyou County, California: United States Geological Survey Miscellaneous Field Studies Map MF-1549-A.

Coney, P. J., et al, 1981, Cordilleran suspect terranes: Nature, v. 288, p. 329–333.

Donato, M. M., et al, 1980, Geology of the Condrey Mountain Schist, northern Klamath Mountains, California and Oregon: Oregon Geology, v. 42, n. 7, p. 125–129.

______, 1981, Northward continuation of the Hayfork terrane, north-central Klamath Mountains, California (Abs.): Geological Society of America Abstracts with Programs, v. 13, n. 2, p. 52.

Ernst, W. G., 1979, Coexisting sodic and calcic amphiboles from high pressure metamorphic belts and the stability of barroisitic amphibole: Mineralogical Magazine, v. 43, p. 269–278.

Hanks, C. L., 1981, The emplacement history of the Tom Martin Ultramafic Complex and associated metamorphic rocks, north-central Klamath Mountains, California: Master's Thesis, University of Washington, 112 p.

Harper, G. D., 1980, Structure and petrology of the Josephine Ophiolite and overlying sedimentary rocks, northwestern California: PhD Dissertation, University of California, 260 p.

______, 1983a, A depositional contact between the Galice Formation and a late Jurassic ophiolite in northwestern California and southwestern Oregon: Oregon Geology, v. 45, n. 1, p. 3–7.

______, 1983b, The Nevadan Orogeny—collapse of a west-facing island arc/back arc/remnant arc complex (Abs.): Geological Society of America Abstracts with Programs, v. 15, n. 5, p. 294.

Helper, M. A., 1983, Deformation-metamorphism relationships in a regional blueschist-greenschist facies terrane, Condrey Mountain Schist, north-central Klamath Mountains, Northern California (Abs.): Geological Society of America Abstracts with Programs, v. 15, n. 5, p. 427.

Hill, L. B., 1984, A tectonic and metamorphic history of the north-central Klamath Mountains, California: PhD Dissertation, Stanford University, 246 p.

Howell, D. G., and D. L. Jones, 1983, The principles of terrane analysis and some key definitions: Circum-Pacific Terranes Conference Abstracts, August, 1983, Stanford University Press.

Irwin, W. P., 1972, Terranes of the western Paleozoic and Triassic Belt in the southwestern Klamath Mountains, California: United States Geological Survey Professional Paper 800-C, p. C103–C111.

Klein, C. W., 1975, Structure and petrology of the southeastern portion of the Happy Camp quadrangle, Siskiyou County, northwestern California: PhD Dissertation, Harvard University.

______, 1977, Thrust plates of the north-central Klamaths near Happy Camp: California Division of Mines and Geology Special Report 129, p. 23.

Lieberman, J. L., and J. M. Rice, 1983, Prograde metamorphism of marble and peridotite in the Seiad Ultramafic Complex, California (Abs.): Geological Society of America Abstracts with Programs, v. 15, n. 5, p. 436.

Liou, J. G., 1971, Synthesis and stability relations of prehnite, $Ca_2Al_3Si_3O_{10}(OH)_2$: American Mineralogist, v. 56, p. 507–531.

______, et al, 1974, Experimental studies of phase relations between greenschist and amphibolite in a basaltic system: American Journal of Science, v. 274, p. 613–632.

Loomis, T. P., and R. R. Gottschalk, 1981, Hydrothermal origin of mafic layers in alpine-type peridotites: evidence from the Seiad Ultramafic Complex, California, U.S.A.: Contributions to Mineralogy and Petrology, v. 76, p. 1–11.

Moody, J. B., et al, 1983, Experimental characterization of the greenschist/amphibolite boundary in mafic systems: American Journal of Science, v. 283, p. 48–92.

Peterson, S. W., 1982, Geology and petrology around Titus Ridge, north-central Klamath Mountains, California: Master's Thesis, University of Oregon, Eugene, 73 p.

Price, R. A., 1972, The distinction between displacement and distortion in flow, and the origin of diachronism in tectonic overprinting in orogenic belts: 24th International Geological Congress, section 3, p. 545–551.

Saleeby, J. B., et al, 1982, Time relations and structural-stratigraphic patterns in ophiolite accretion, west-central Klamath Mountains, California: Journal of Geophysical Research, v. 87, n. B5, p. 3831–3848.

______, 1984, Pb/U ages on thrust plates of the west central Klamath Mountains and Coast Ranges, northern California and southern Oregon: Geological Society of America Abstracts with Programs, Annual Meeting, Reno, Nevada.

Saxena, S. K., 1976, Two-pyroxene geothermometer: a model with an approximate solution: American Mineralogist, v. 61, p. 643–652.

Schiffman, P., and J. G. Liou, 1980, Synthesis and stability relations of Mg-Al pumpellyite, $Ca_4Al_5MgSi_6O_{21}(OH)_7$: Journal of Petrology, v. 21, n. 3, p. 441–474.

Snoke, A. W., 1972, Petrology and structure of the Preston Peak area, Del Norte and Siskiyou Counties, California; PhD Dissertation, Stanford University.

Spear, F. S., 1980, NaSi = CaAl exchange equilibrium between plagioclase and amphibole; an empirical model: Contributions to Mineralogy and Petrology, v. 72, p. 33–41.

______, 1981, An experimental study of hornblende stability and compositional variation in amphibolite: American Journal of Science, v. 281, p. 687–734.

Wright, J. E., 1982, Permo-Triassic accretionary subduction complex, south-western Klamath Mountains, northern California: Journal of Geophysical Research, v. 87, n. B5, p. 3805–3818.

Table 1 (continued)

Map no. (Fig. 1)	Pluton	Lithology	Method	Mineral	Age (m.y.)	Reference
29	Russian Peak	granodiorite	K–Ar	B	147	Romey, 1962; Evernden and Kistler, 1970
30	Russian Peak	granodiorite	K–Ar	B	144	Romey, 1962; Evernden and Kistler, 1970
31	Cobbles in conglomerate	trondhjemite	U/Pb	Z	455	Mattinson and Hopson, 1972
32	Unnamed	gabbro	U/Pb	Z	480	Mattinson and Hopson, 1972
33	"Craggy Peak"	gabbro	K–Ar	H	426	Lanphere et al, 1968
34	"Craggy Peak"	gabbro	K–Ar	H	447	Lanphere et al, 1968
35	"Craggy Peak"	trondhjemite	K–Ar	B	136	Lanphere et al, 1968
36	Castle Crags	granodiorite	K–Ar	B	167	Lanphere et al, 1968
	Castle Crags	granodiorite	K–Ar	H	175	Lanphere et al, 1968
	Castle Crags	granodiorite	K–Ar	B	135	Lanphere et al, 1968
	Castle Crags	granodiorite	K–Ar	B	136	Lanphere et al, 1968
37	Castle Crags	granodiorite	K–Ar	B	162	Lanphere et al, 1968
38	Sugar Pine	qtz-diorite	K–Ar	B	139	Lanphere et al, 1968
	Sugar Pine	qtz-diorite	K–Ar	H	137	Lanphere et al, 1968
39	Unnamed	gabbro	K–Ar	H	340	Lanphere et al, 1968
40	Deadman Peak	qtz-diorite	K–Ar	B	133	Holdaway, 1963; Evernden and Kistler, 1970
41	North Fork ophiolite	plagiogranite	U/Pb	Z	265–310	Ando et al, 1983
42	Caribou Mtn.	trondhjemite	K–Ar	B	133	Davis, 1961; Evernden and Kistler, 1970
43	Horseshoe Lake	diorite	K–Ar	B	133	Davis, 1961
44	Ironside Mtn.	syenodiorite	K–Ar	B	169	Lanphere et al, 1968
45	Denny intrusive unit	?	U/Pb	Z	165	Wright, 1981
46	Ironside Mtn.	syenodiorite	K–Ar	B	171	Lanphere et al, 1968
	Ironside Mtn.	syenodiorite	U/Pb	Z	163	Wright, 1981
47	Ironside Mtn.	diorite	U/Pb	Z	170	Wright, 1981, 1982
48	Ammon Ridge	diorite	K–Ar	?	148	Young, 1978
	Ammon Ridge	diorite	K–Ar	?	152	Young, 1978
49	Saddle Gulch	gabbro	K–Ar	H	189	Lanphere, 1977, personal communication
50	Glen Creek Complex	diorite	U/Pb	Z	144	Wright, 1981; Snoke et al, 1982
	Glen Creek Complex	gabbro	K–Ar	H	151	Irwin et al, 1974
51	Bear Wallow	diorite	U/Pb	Z	193	Wright, 1981
52	East Fork	diorite	K–Ar	B	164	Lanphere, 1983, personal communication
	East Fork	diorite	K–Ar	H	149	Lanphere, 1983, personal communication
53	"WR-1"	qtz-diorite	U/Pb	Z	193	Wright, 1981
54	"PCCK"	qtz-diorite	U/Pb	Z	198	Wright, 1981
55	"Beegum"	qtz-diorite	U/Pb	Z	207	Wright, 1981
56	Walker Point	gabbro	U/Pb	Z	169	Wright, 1981
57	Basin Gulch	diorite	U/Pb	Z	170	Wright, 1981, 1982
58	Shasta Bally	granodiorite	K–Ar	B	135	Lanphere et al, 1968
	Shasta Bally	granodiorite	K–Ar	H	131	Lanphere et al, 1968
59	Shasta Bally	qtz-diorite	K–Ar	B	134	Lanphere et al, 1968

(continued)

Table 1 (continued)

Map no. (Fig. 1)	Pluton	Lithology	Method	Mineral	Age (m.y.)	Reference
	Shasta Bally	qtz-diorite	K–Ar	H	133	Lanphere et al, 1968
	Shasta Bally	qtz-diorite	K–Ar	?	134	Lanphere and Jones, 1978
	Shasta Bally	qtz-diorite	U–Pb	Z	136	Lanphere and Jones, 1978
60	Shasta Bally	qtz-diorite	K–Ar	B	131	Curtis et al, 1958; Evernden and Kistler, 1970
61	Mule Mtn.	qtz-diorite	K–Ar	H	392	Albers et al, 1981
	Mule Mtn.	qtz-diorite	U/Pb	Z	400	Albers et al, 1981
62	Pit River	granodiorite	K–Ar	H	251	Lanphere et al, 1968
	Pit River	granodiorite	K–Ar	H	220	Zeller, 1965
	Pit River	granodiorite	U/Pb	Z	260	Albers, 1982, personal communication
63	Yellow Butte	qtz-monzonite	K–Ar	B	137	Hotz, 1971
	Yellow Butte	qtz-monzonite	K–Ar	H	138	Hotz, 1971

Table 1—List of isotopic ages of plutonic rocks of the Klamath Mountains. The K–Ar ages originally reported have been recalculated, where appropriate, by use of the conversion tables of Dalrymple (1979). Abbreviations used: K–Ar, potassium-argon; U/Pb, uranium/lead; B, biotite; H, hornblende; Z, zircon; m.y., million years.

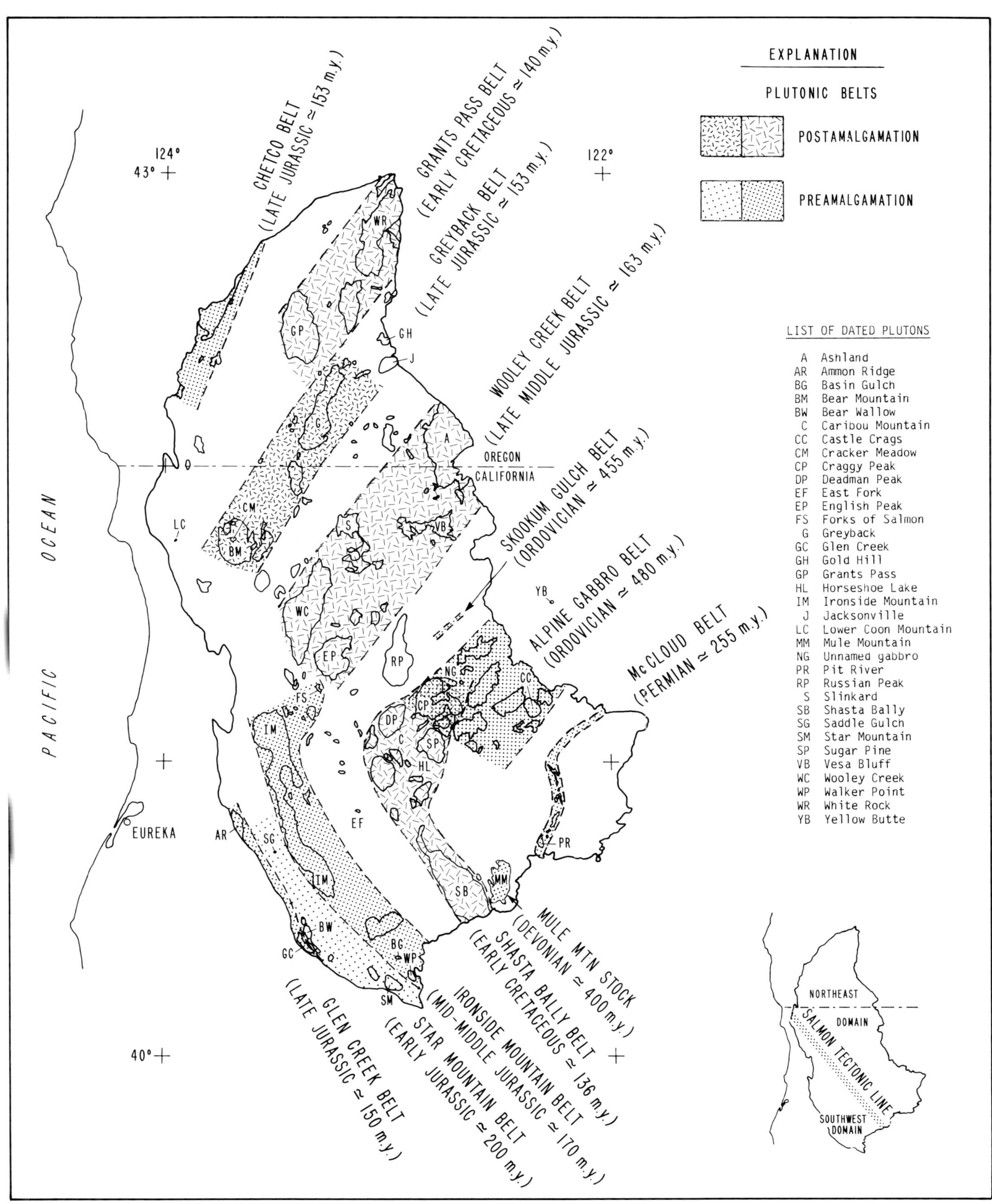

F gure 2—Map of Klamath Mountains province showing outlines of major plutons and the trends of the plutonic belts. Ultramafic ophiolitic rocks are not shown. Letter symbols on map correspond to names in list of dated plutons. The time scale of Harland et al (1982) was used for correlating the isotopic ages with geologic time.

Other plutons that are isotopically dated but not assigned to specific belts include the Russian Peak, East Fork, and Yellow Butte plutons. Neither the Russian Peak nor the East Fork plutons seem to belong to nearby belts, but both plutons are important because they place constraints on the time of suturing of their host terranes. The Russian Peak pluton (144 to 147 m.y.) cross-cuts the sutures between the Central Metamorphic and the Western Paleozoic and Triassic terranes. The East Fork pluton (149 to 164 m.y.) cuts the suture between the Central Metamorphic and North Fork terranes. The Yellow Butte pluton (138 m.y.), which is exposed through a window in Tertiary volcanic rocks of the Cascade Range about 19 km (12 mi) northeast of the Klamath Mountains province, may represent an extension of the Shasta Bally belt.

The Skookum Gulch belt is unusual and may not be a true plutonic belt. It consists of several bodies of silicic plutonic rocks in a matrix of Paleozoic schist and phyllite. The contact relations are not clear, but the plutonic rocks are thought possibly to be large tectonic blocks in a melange rather than local intrusions (Hotz, 1977). The plutonic rocks are tentatively considered early Paleozoic in age because of their similarity to cobbles of isotopically dated trondhjemite (455 m.y.; Mattinson and Hopson, 1972) that occur in a lower Paleozoic conglomerate at nearby Lovers Leap (Potter et al, 1977; Hotz, 1977; Lindsley-Griffin, 1977). The cobbles as well as the plutonic bodies in the Skookum Gulch melange may well have come from the Trinity ophiolite.

PREAMALGAMATION AND POSTAMALGAMATION PLUTONS

As a corollary to the general concepts of plate tectonics and the accretion process, the plutons are categorized as preamalgamation or postamalgamation on the basis of whether they intruded before or after their host terrane was joined to an adjacent terrane.[1] The preamalgamation plutons intrude only a single terrane. The postamalgamation plutons in some instances may intrude a single terrane but in other instances may intrude two or more contiguous terranes.

The preamalgamation plutons of the Klamath Mountains occur in two principal genetic settings: (1) The plutons are part of an ophiolite suite, as exemplified by the early Paleozoic gabbroic plutons that intrude the Trinity ultramafic sheet; and (2) the plutons are similar in age and composition to volcanic strata they intrude, forming comagmatic plutonic-volcanic pairs, and are thought to represent the intrusive phases of volcanic island arcs. The arc-related plutons include the Mule Mountain stock (related to the Devonian Balaklala Rhyolite), plutons of the McCloud belt (related to the Permian Dekkas Andesite), and the Early or Middle Jurassic plutons of the Ironside Mountain belt (related to the Hayfork Bally Meta-andesite). It should be noted that the plutons of the McCloud belt are preamalgamation in the sense that they are genetically related to intermediate-age (Permian) volcanic strata of a long-standing volcanic arc, but they are postamalgamation in relation to the Devonian suturing of the Central Metamorphic terrane to the early part of the Eastern Klamath terrane.

In contrast to the preamalgamation plutons, the postamalgamation plutons are significantly younger than their host rocks and seem to bear no genetic relation to them. They are assigned to belts mainly on the basis of their isotopic ages. These plutons are considered to form postamalgamation belts at places where some of the plutons cross-cut the suture between the host and an adjacent terrane, or where the suture is known to be older than the plutons on the basis of other regional considerations. Postamalgamation plutons seem to be absent southwest of the Shasta Bally belt in the southwest domain except for the small East Fork pluton, whereas belts of post-amalgamation plutons greatly predominate northwest of the Shasta Bally belt in the northeast domain.

The Wooley Creek and Ironside Mountain pluton belts are virtually coincident along trend, but the plutons of the Wooley Creek belt are appreciably less mafic and more quartzose than those of the Ironside Mountain belt. The plutons of both belts intrude presumably correlative volcanogenic strata of the Hayfork terrane (Donato et al, 1981) even though the belts are in separate domains (Fig. 2). The Ironside Mountain batholith is clearly truncated by terrane boundaries (Irwin, 1977). However, because the Wooley Creek, Slinkard, and Vesa Bluffs plutons cross terrane boundaries (Donato et al, 1982; Barnes, 1983; Mortimer, 1984), the plutons of the Wooley Creek belt are postamalgamation even though they appear on average to be only a few million years younger than the preamalgamation plutons of the Ironside Mountain belt (Fig. 3).

The plutons of the Klamath Mountains are grouped in Figure 3 according to the terranes in which they occur and, in most instances, according to the appropriate plutonic belt. The ages of small plagiogranite pods in the Jurassic Josephine ophiolite and the late Paleozoic North Fork ophiolite also are shown. They define the ages of the ophiolites and help to illustrate the time interval between the formation of the ophiolitic base of the terrane and the intrusion of the terrane by younger plutons. In the case of the Josephine ophiolite, the interval is only a few million years, and with the North Fork ophiolite the interval is more than 100 m.y. The time interval between the intrusion by the early Paleozoic plutons of the Alpine gabbro belt and the intrusion by the plutons of the Shasta Bally belt is approximately 300 m.y. This sequential increase in the time intervals between emplacement of the oldest and youngest plutonic rocks of the terranes follows the generally sequential west-east increase in the ages of the oldest rocks of the terranes.

DISPLACEMENT AND ROTATION OF TERRANES

The places of origin of most of the terranes are not known; they could have formed just west of the continental

[1]In an earlier report (Irwin, 1984), the plutons were categorized as preaccretion and postaccretion plutons, using the term accretion in a general sense, i.e., the joining of one terrane to another, rather than in the restricted sense of Jones et al (1983) that implies the joining of a terrane to a continental margin. For conformity with the terminology of this volume, the term accretion is used in the restricted sense, and the term amalgamation is used to imply the joining of terranes in an oceanic setting.

Mines and Geology Map Sheet, scale 1:250,000.
Wagner, D. L., and G. J. Saucedo, in press, Geologic map of the Weed quadrangle, 1:250,000, California: California Division of Mines and Geology Regional Geologic Map Series, Map 4A.
Wells, F. G., and D. L. Peck, 1961, Geologic map of Oregon west of the 121st meridian: U.S. Geological Survey Miscellaneous Investigations Map I-325, scale 1:500,000.
Wright, J. E., 1981, Geology and uranium-lead geochronology of the western Paleozoic and Triassic subprovince, southwestern Klamath Mountains, California: PhD Dissertation, University of California, Santa Barbara, 300 p.
———, 1982, Permo-Triassic accretionary subduction complex, southwestern Klamath Mountains, northern California: Journal of Geophysical Research, v. 87, n. B5, p. 3805–3818.
Young, J. C., 1978, Geology of the Willow Creek quadrangle, Humboldt and Trinity Counties, California: California Division of Mines and Geology Map Sheet 31, scale 1:62,500.
Zeller, E. J., 1965, Modern methods for measurement of geologic time: California Division of Mines and Geology Mineral Information Service, v. 18, n. 1, p. 9–16.

Structural and Metamorphic Aspects of Middle Jurassic Terrane Juxtaposition, Northeastern Klamath Mountains, California

N. Mortimer
Stanford University
*Stanford, California**

An approximately 18 km (11 mi) thick structural section of rocks assigned to the "western Paleozoic and Triassic belt" (TrPz belt) of Irwin (1966) is exposed in the northeastern Klamath Mountains of California. The division of these rocks into five terranes and identification of multiple episodes of deformation and metamorphism provide new information on the postdepositional history of the TrPz belt. Late Triassic deformation and blueschist facies metamorphism of the Fort Jones terrane is well established (Hotz et al, 1977; Borns, 1980). In this paper, evidence for a Middle Jurassic subgreenschist to amphibolite facies, low-pressure, progressive metamorphism affecting the North Fork, Salmon River, Hayfork, and Marble Mountains terranes is presented. A late Middle Jurassic igneous intrusion, the Vesa Bluffs pluton, cuts and contact metamorphoses all five TrPz belt terranes. Two major preintrusion deformations are recognized in the North Fork, Salmon River, Hayfork, and Marble Mountains terranes. The first was major faulting related to juxtaposition of the terranes along low-angle faults; the second (possibly continuous with the first) was a flattening deformation synchronous with the Middle Jurassic regional metamorphism. Postintrusion deformations were Late Jurassic open folding (the Nevadan orogeny) and regional doming and tilting of Neogene age. The Jurassic structural and metamorphic history of the terranes is consistent with assembly of the TrPz belt in a single, evolving Middle and Late Jurassic arc/subduction system. Gravity modeling indicates that low-angle Jurassic structures are essentially preserved intact in this part of the Klamaths.

INTRODUCTION

The Paleozoic and Mesozoic rocks of the Klamath Mountains province are exposed in the western United States between the Pacific Ocean and the Cascade volcanic arc (Fig. 1). Irwin (1966) divided the province into four major lithotectonic units called belts, bounded by east-dipping thrusts, and later subdivided the "western Paleozoic and Triassic belt" (TrPz belt) into smaller lithotectonic units he called terranes (Irwin, 1972). A recent summary of the geology of the Klamath Mountains is given by Irwin (1981).

This paper is concerned with the postdepositional history of the TrPz belt that occupies the central structural levels of the Klamath Mountains province (Fig. 2). In California, the TrPz belt has been divided into the Marble Mountains, Rattlesnake Creek, Hayfork, Salmon River, North Fork, and Fort Jones terranes by Blake et al, 1982. These terranes, like the larger lithotectonic units, are bounded by generally low-angle east-dipping faults and are interpreted by Davis (1968) and Irwin (1981) to constitute a series of stacked thrust plates (Fig. 2). To date, most of our knowledge regarding the geologic history of the TrPz belt has come from studies in the southern Klamath Mountains (e.g., Irwin, 1972; Davis et al, 1979; Wright, 1982; Ando et al, 1983). These and other studies have provided essential mapping and stratigraphic data but have tended not to emphasize aspects of structure and metamorphism. Ongoing and completed work in the northern California Klamaths (e.g., Hotz, 1967, 1977, 1979; Borns, 1980; Coleman et al, 1983; Kays and Ferns, 1980; Donato et al, 1982; Hill, 1984; Mortimer, 1984) is better able to address such aspects in part because of the higher grades of TrPz belt regional metamorphism to the north. This paper is the first published synthesis of structural and metamorphic data from a transect across the TrPz belt.

The issue of where the TrPz belt terranes originated (Davis et al, 1978; Wright, 1982) is addressed elsewhere (Mortimer, 1984); interpretations presented in this paper offer specific new constraints on the nature, timing, and location of TrPz belt deformation and metamorphism. Of more general interest, the data show how structural and metamorphic criteria can be used to constrain suturing and postsuturing events more precisely than intrusive and overlap relations. The timescale of Harland et al (1982) is used throughout this paper, and the old K–Ar ages of Lanphere et al (1968) have been recalculated using new decay constants to conform with this.

*Present address: Department of Geological Sciences, University of British Columbia, Vancouver, BC, Canada

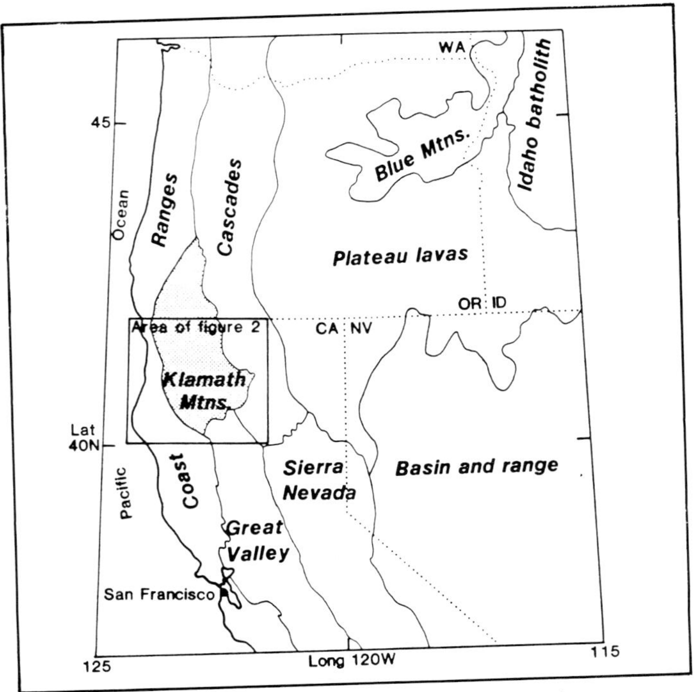

Figure 1—Location of the Klamath Mountains province.

GENERAL GEOLOGY

In the light of mapping in the northeastern Klamath Mountains from 1981–1983, a case can be made for division of the formerly undivided TrPz belt into five terranes that correlate with similarly juxtaposed terranes in the southern and central Klamath Mountains. The justification for this, along with relevant paleontologic, petrologic, and geochemical data, can be found in Mortimer (1984). Brief descriptions of the TrPz belt terranes as defined by Blake et al (1982) are given below.

The Fort Jones terrane was originally defined as the Stuart Fork Formation in the southern Klamath Mountains (Davis et al, 1965). In the northern Klamath Mountains, it consists of tectonically disrupted chert, argillite, marble, volcanic rocks, and gabbro (Hotz, 1967, 1977, 1979; Borns, 1980) that were metamorphosed to blueschist facies in Late Triassic time (Hotz et al, 1977; Borns, 1980). The North Fork terrane consists of tectonically imbricated Permian to Early Jurassic chert and argillite and Permian alkalic volcanic rocks and limestone (Irwin, 1972; Ando et al, 1983; Mortimer, 1984).

The Salmon River terrane is composed of tectonically disrupted ?Permo–Carboniferous tholeiitic volcanic rocks, diabase, gabbro, and harzburgite (Ando et al, 1983). The Salmon River terrane is considered by Irwin (1972) and Ando et al (1983) to be the basement portion of the North Fork terrane. The Hayfork terrane is a composite terrane (Wright, 1982) consisting of an eastern portion of Permian to Late Triassic or Early Jurassic chert, argillite, chert-argillite breccia, alkalic volcanic rocks, and late Paleozoic limestones (Irwin, 1981; Wright, 1982; Mortimer, 1984) and a western portion of Middle Jurassic volcanic and volcaniclastic rocks and their plutonic equivalents (Irwin, 1972; Fahan, 1982; Wright, 1982).

The Marble Mountains terrane consists of structurally disrupted peridotite, basic igneous rocks, siliceous sedimentary rocks, and marble that are metamorphosed to amphibolite facies (Rawson and Petersen, 1982; Donato et al, 1982; Hill, 1984). Rawson and Petersen (1982) suggest that the Marble Mountains terrane is correlative with the less metamorphosed Rattlesnake Creek terrane of the southern and central Klamath Mountains (Irwin, 1972; Wright, 1981) whose protolith ages are Late Triassic and Early Jurassic. The terranes structurally above (Central Metamorphic terrane) and below (Condrey Mountain terrane) the TrPz belt are not considered in this paper.

This structural and metamorphic synthesis draws directly on the data and interpretations of Hotz (1967, 1979) for the Marble Mountains terrane and Borns (1980) for the Fort Jones terrane. The author has collected samples from but not mapped these terranes. Interpretation of TrPz belt deformation is more straightforward when considered in a metamorphic framework. For this reason, metamorphism is dealt with separately and before deformation in the following sections.

METAMORPHISM

Prior to the subdivision of the TrPz belt in the northeastern Klamath Mountains, Hotz (1967, 1979) and Kays and Ferns (1980) proposed an east-to-west increase in metamorphic grade from greenschist to amphibolite facies in rocks that are now recognized as the Salmon River, Hayfork, and Marble Mountains terranes (Fig. 3). Hotz (1979) realized that the Triassic blueschist facies rocks of the Fort Jones terrane had undergone a different deformational and metamorphic history from the other (structurally underlying) Jurassic metamorphic TrPz belt rocks. This current study confirms and expands on Hotz's interpretation of two separate metamorphic events that are discussed separately below. The Triassic event has been investigated in detail by Borns (1980), thus in this paper more attention is devoted to the Jurassic metamorphism.

High-Pressure Late Triassic Event

The diagnostic blueschist facies minerals lawsonite and jadeite were first recognized in rocks of the Fort Jones terrane (Stuart Fork Formation) by Hotz (1973) although glaucophane schists were previously noted by Masson (1949). The age of metamorphism was established by Hotz et al (1977) who reported Late Triassic (214 to 222 m.y.) K–Ar and Ar–Ar (white mica) ages from three lawsonite-bearing schists near Yreka.

The blueschists are not isolated blocks in melange, but blueschist-facies metamorphism is coherent and regional in nature (Borns, 1980). Borns (1980) estimated P–T conditions during metamorphism to be 300 to 400°C and 9 to 11 kbar. The structurally underlying TrPz belt terranes do not contain blueschist-facies minerals and have not experienced this high-pressure Late Triassic metamorphism. The fault zone separating the Fort Jones terrane from the North Fork terrane is therefore a major structural and metamorphic break.

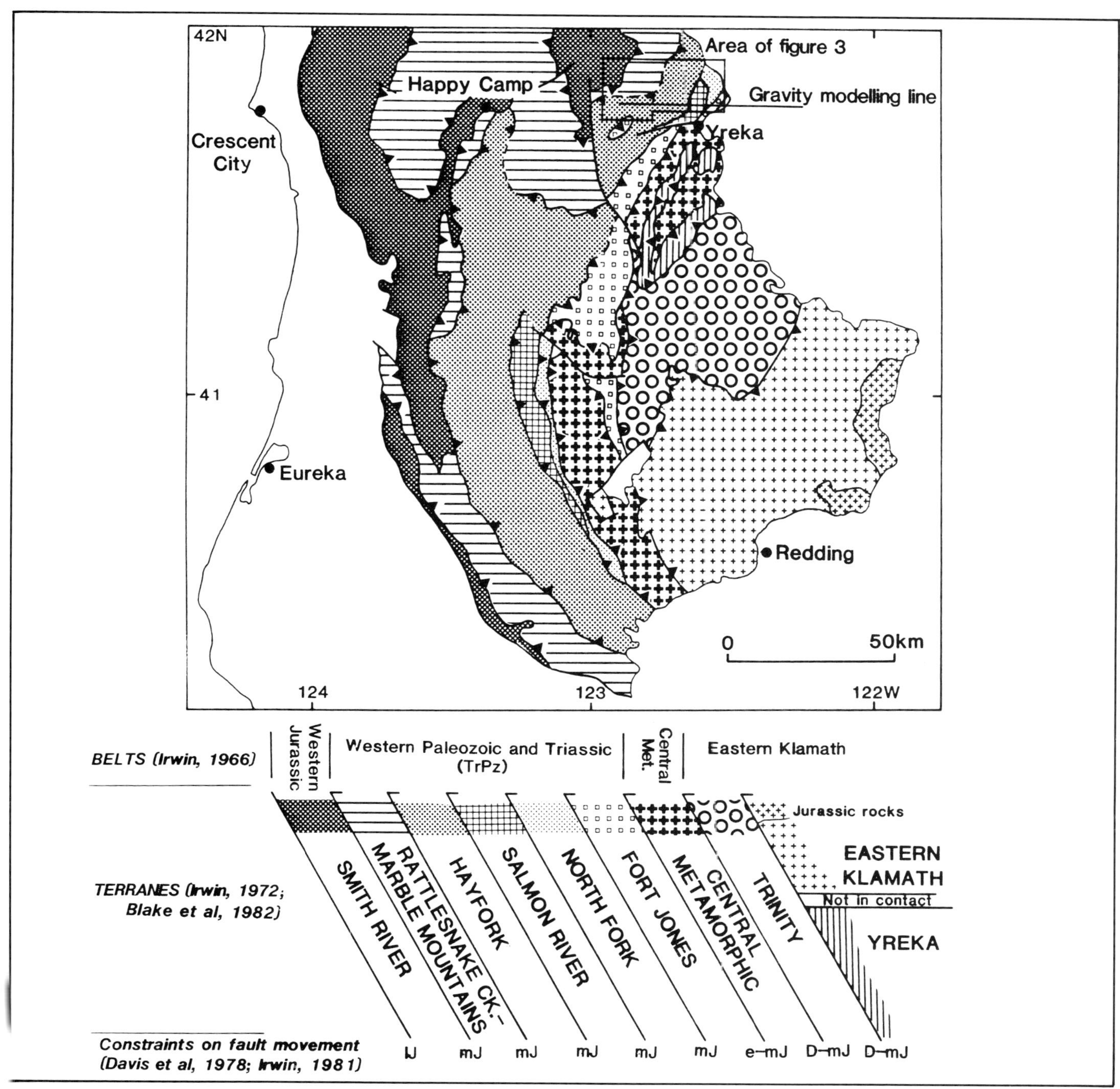

Figure 2—Major lithotectonic subdivisions of the California Klamath Mountains. Terrane boundaries in the northeast part of the province are slightly modified from Blake et al (1982). Plutons are omitted for clarity.

Low-Pressure Middle Jurassic Event

The North Fork, Salmon River, Hayfork, and Marble Mountains terranes consist of five lithologies: tholeiitic metabasites, alkalic metabasites, siliceous metasediments, calcareous metasediments, and metaperidotites. Metacarbonates in the area of Figure 3 are almost always pure calcite marbles and do not show any mineralogical variation with metamorphic grade. Metaperidotite assemblages were not used as indicators of metamorphic grade in this study because of the difficulty in distinguishing relict from metamorphic olivine and prograde from retrograde antigorite. Alkalic metabasites, which can be distinguished from the tholeiites on the basis of major element chemistry (Mortimer, 1984), are generally calcite bearing and contain the assemblage albite + chlorite + calcite + sphene ± epidote. This assemblage is stable over a wide range of P–T conditions and is indicative of high $P(CO_2)$ during metamorphism (Miyashiro, 1973). Locally high $P(CO_2)$ during metamorphism of these rocks is readily accounted for by abundant interbedded limestone and marble.

In contrast, tholeiitic metabasites (and a few alkalic

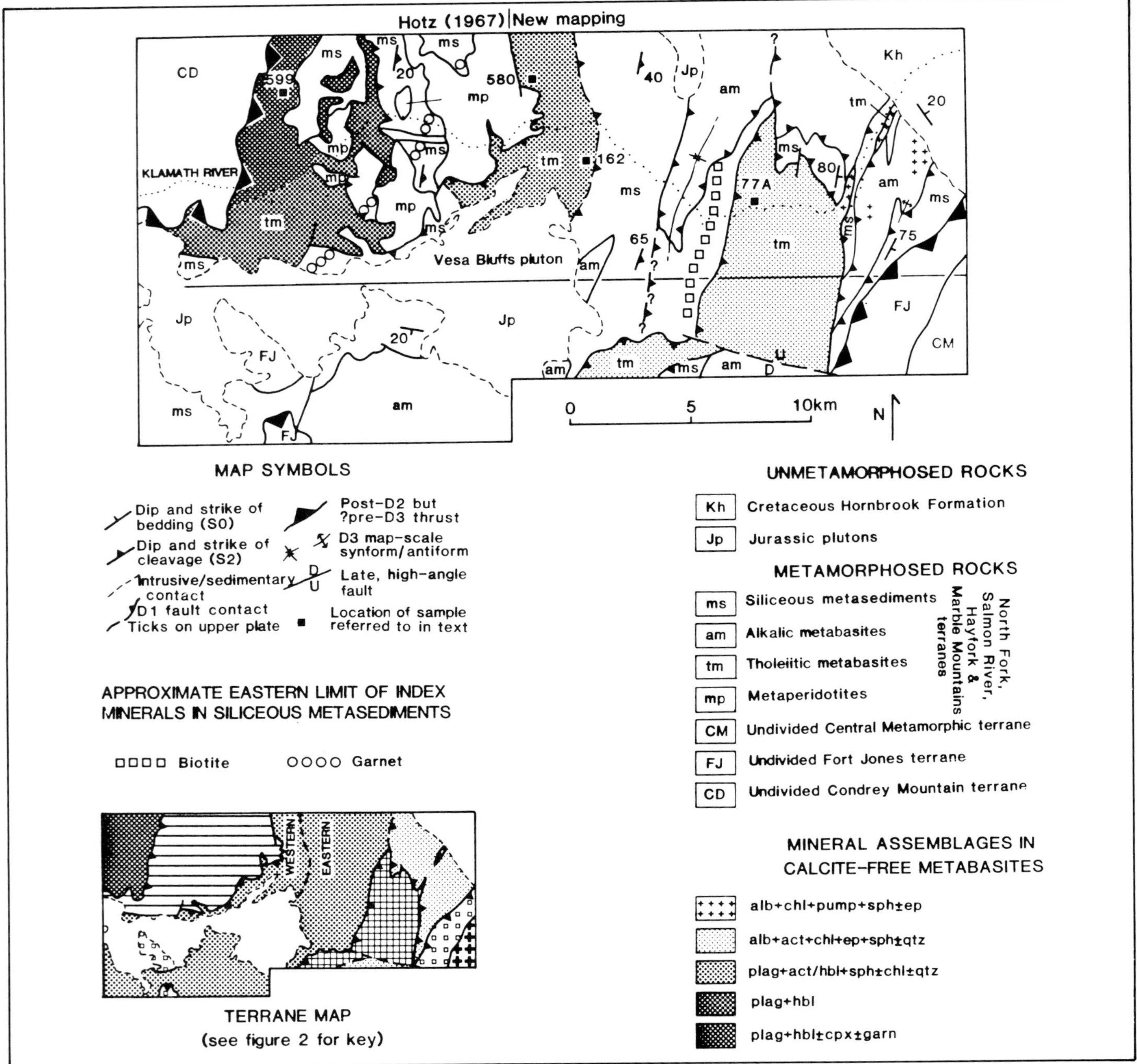

Figure 3—Simplified geologic map of the study area, emphasizing the distribution of metamorphic protoliths and progressive mineral change in calcite-free metabasites. Inset at lower left shows the major lithotectonic subdivisions of the area.

metabasites) are calcite-free and contain more diverse mineral assemblages than do the calcite-bearing metabasites (Fig. 4). Metasedimentary rocks also display progressive mineral changes across the study area. The following discussion of metamorphism is based on progressive mineral changes observed in calcite-free metabasites and siliceous metasediments (summarized in Figs. 3, 4).

The North Fork terrane is the least metamorphosed of the TrPz belt terranes and contains relict, premetamorphic textures and minerals. Examples include well-preserved radiolaria and fusulinids in cherts and limestones and relict palagonite, calcic plagioclase, augite, and hornblende in volcanic rocks. Cleavage is developed only in argillaceous lithologies. Metamorphic assemblages in calcite-free metabasites (Fig. 4) indicate subgreenschist facies metamorphism, a lower grade than recognized by Hotz (1979) and Kays and Ferns (1980).

The Salmon River terrane consists entirely of tholeiitic basalt and diabase. Original textures are again well preserved (e.g., chilled pillow rims and ophitic textures). Relict igneous augite is preserved only in the eastern portion of the terrane; westward augite is replaced by actinolite. Very rarely, albite and actinolite define a weak,

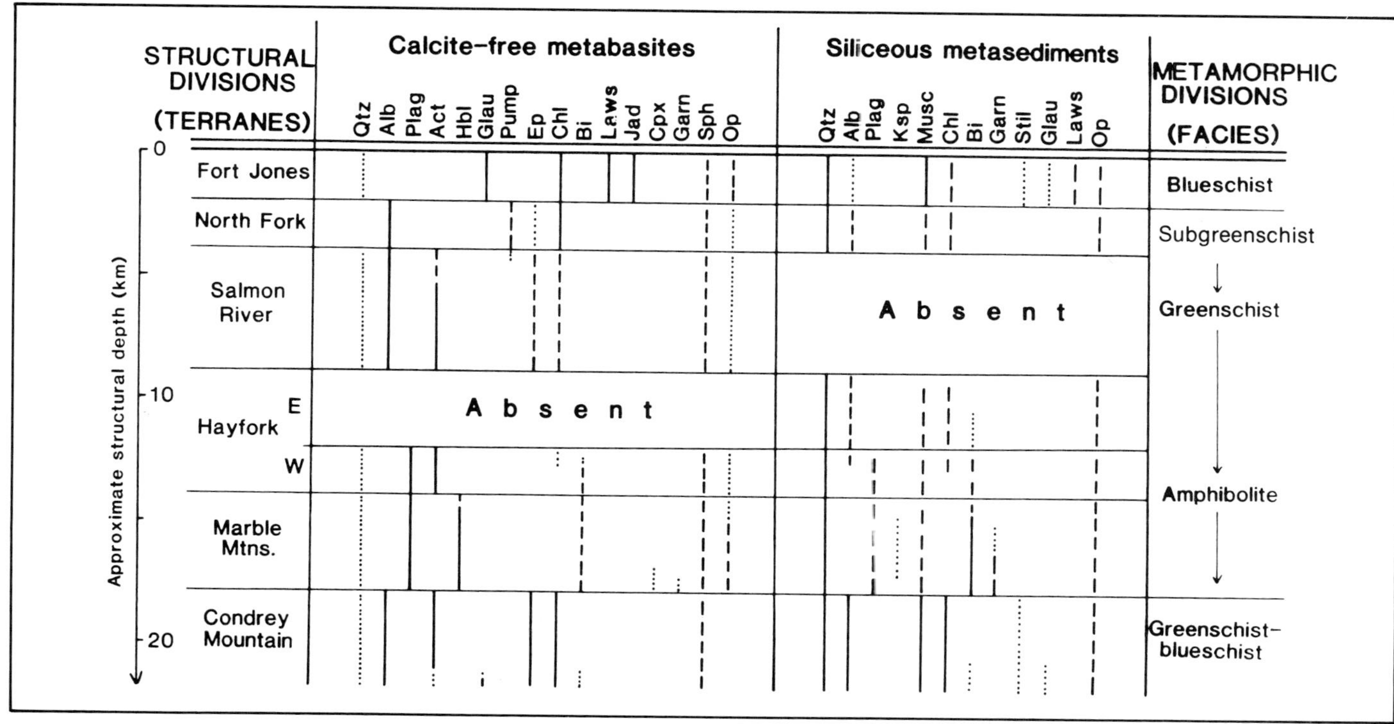

Figure 4—Variation of metamorphic mineral assemblages in the terranes of the area of Figure 3. Fort Jones terrane data from Borns (1980), Marble Mountains and Condrey Mountain terrane data from Hotz (1967). Abbreviations are as follows: Qtz = quartz, Alb = albite, Plag = plagioclase, Act = actinolite, Hbl = hornblende, Glau = glaucophane, Pump = pumpellyite, Ep = epidote, Chl = chlorite, Bi = biotite, Laws = lawsonite, Jad = jadeite, Cpx = non-jadeitic clinopyroxene, Garn = garnet, Sph = sphene, Op = opaques, Ksp = alkali feldspar, Musc = white mica, Stil = stilpnomelane.

planar fabric in zones a few centimeters wide, but for the most part the rocks are unfoliated. Mineral assemblages and the composition of metamorphic plagioclase and amphibole in the Salmon River terrane (Figs. 4, 5) indicate greenschist-facies metamorphism.

Traces of biotite first appear in metasediments of the eastern Hayfork terrane. In the western Hayfork terrane, static metamorphic textures overprint volcaniclastic bedding in calcite-free metabasites. In contrast to the Salmon River terrane metabasites, the metamorphic amphibole is actinolitic hornblende and the feldspar a calcic plagioclase (Fig. 5), indicative of transitional greenschist–amphibolite-facies conditions (Liou et al, 1974).

Hotz (1967) defined three metamorphic zones in rocks corresponding to the Marble Mountains terrane based on the pleochroic color of metamorphic hornblende in metabasites. Mineral assemblages and plagioclase and amphibole compositions in Marble Mountains terrane metabasites (Figs. 4, 5) indicate amphibolite facies metamorphism, a higher grade than Hayfork terrane metabasites. The occurrence of garnet and clinopyroxene in metabasites, and garnets in metasediments in the lowest structural levels of the Marble Mountains terrane mark the metamorphic culmination of TrPz belt rocks in the area of Figure 3. The TrPz belt is thrust (Hotz, 1979) over the structurally and metamorphically distinct Middle to Late Jurassic greenschist–blueschist-facies rocks of the Condrey Mountain terrane (Figs. 3, 4; Hotz, 1967, 1979; Helper, 1983).

Contact Metamorphism

The Vesa Bluffs pluton is one of a number of calc-alkaline plutons of latest Middle and earliest Late Jurassic age that intrude terranes of the Klamath Mountains (Hotz, 1971; Davis et al, 1978; Irwin, 1981; Wright and Sharp, 1982; Allen et al, 1982). The contact metamorphic aureole of the Vesa Bluffs pluton is generally no more than 200 m (650 ft) wide. Contact metamorphism is clearly later than regional metamorphism, a fact that is most apparent where the pluton has intruded greenschist- and blueschist-facies rocks. For example, within the contact aureole, eastern Hayfork terrane metasediments are upgraded to biotite hornfelses and greenschist-facies Salmon River terrane metabasites are upgraded to hornblende-plagioclase granofelses. In the Fort Jones terrane, glaucophane is replaced by intergrowths of albite and chlorite, and lawsonite is replaced by clinozoisite (Borns, 1980).

Discussion

With the exception of local retrograding, no evidence for more than one episode of regional metamorphism is seen in the rocks of the North Fork, Salmon River, Hayfork, and Marble Mountains terranes. Although outcrop of calcite-free metabasites and siliceous metasediments is discontinuous across the North Fork, Salmon River, Hayfork, and Marble Mountains terranes (Fig. 3), consideration of the metamorphic assemblages in each terrane (Fig. 4) suggests that metamorphic grade increases steadily westward (and down-structure) across these terranes. The fact that metamorphism overprints the

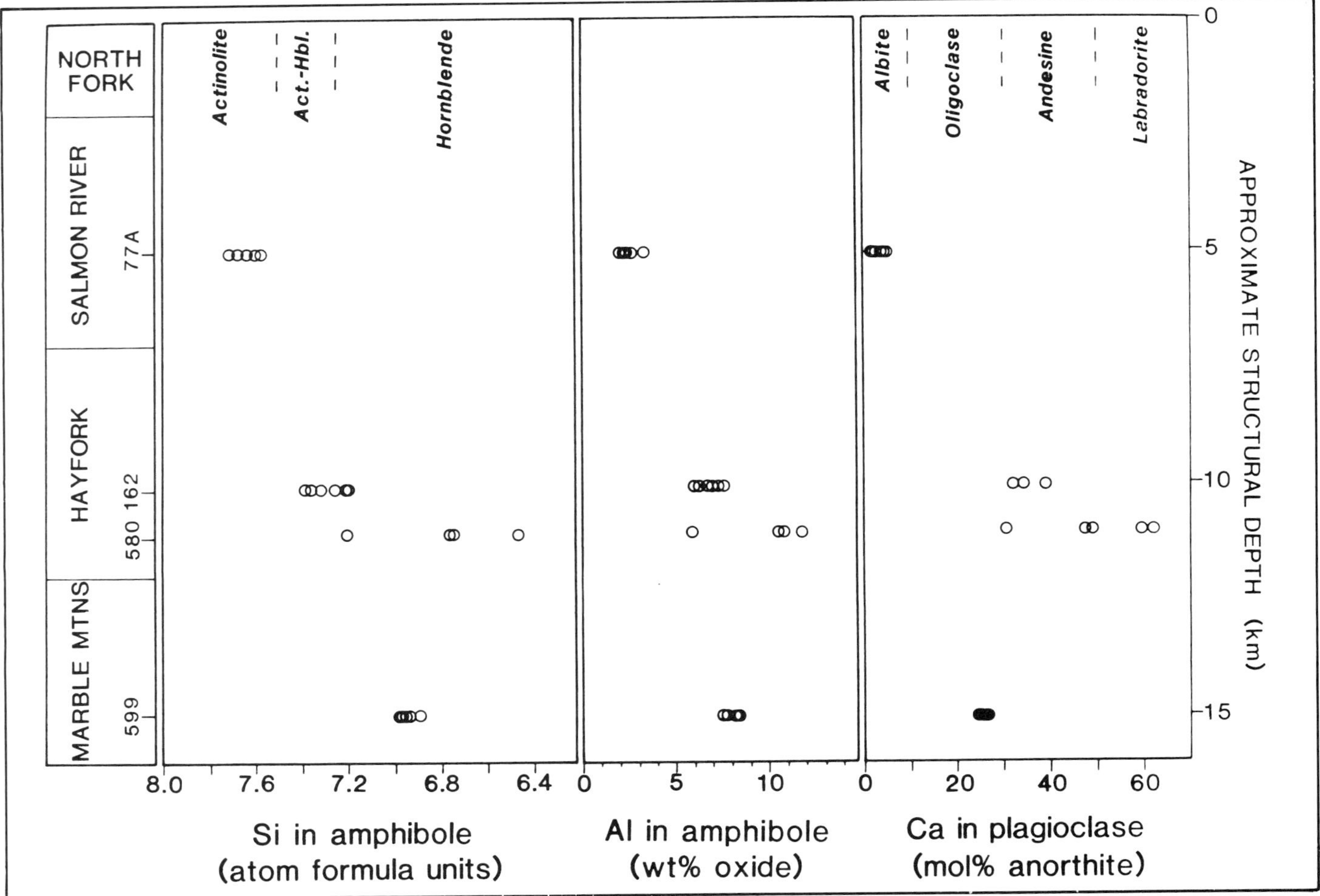

Figure 5—Variation in composition of metamorphic amphibole and feldspar in the Salmon River, Hayfork, and Marble Mountains terranes. See Mortimer (1984) for analyses.

terrane boundaries is significant because it requires that by the time of metamorphism, the four terranes had to have been juxtaposed. Additionally, the observed westward increase in grade rules out *major* tectonic disruption of the four terranes subsequent to regional metamorphism (cf. Hill, this volume).

A direct maximum age limit on this metamorphism is provided by fossils in the North Fork terrane; the youngest well-constrained radiolarian ages in the area of Figure 3 are Pliensbachian (194 to 200 m.y.; Jones, personal communication). If the 168 to 177 m.y. K–Ar ages on igneous hornblende from the western Hayfork terrane of the southern Klamath Mountains (Fahan, 1982) are applicable to correlative rocks in the study area, then metamorphism is confined to have taken place in a very short time; the minimum age limit of regional metamorphism is given by the 164 ± 5 m.y. K–Ar (hornblende) age from the Vesa Bluffs pluton in the area of Figure 3 (Lanphere et al, 1968), which intrudes the regionally metamorphosed rocks.

Evidence that this is a low-pressure metamorphic event include (1) the absence of mineral assemblages indicative of pumpellyite-actinolite facies (Coombs et al, 1976) or albite-epidote amphibolite facies (Liou et al, 1974; Apted and Liou, 1983); (2) presence (in the western Hayfork terrane) of the assemblage calcic plagioclase + actinolitic hornblende ± chlorite indicative of low-pressure greenschist-amphibolite transitions (i.e., below the intersection of the chlorite-out and epidote-out reactions in Fig. 6; Liou et al, 1974; Liou and Ernst, 1979); and (3) low Na(M4) contents of metamorphic amphibole in the buffering assemblage of Brown (1977) that are similar to Na contents of amphiboles from other low-pressure regimes (Fig. 7; Brown, 1977).

The P–T regime of this Middle Jurassic event can be crudely defined by a comparison with experimental phase equilibria studies on natural basaltic systems (Fig. 6). The low-temperature end of the P–T path defined in Figure 6 is constrained by the pumpellyite-bearing assemblages in the North Fork terrane and the upper end by the sporadic occurrence of clinopyroxene-bearing amphibolites in the Marble Mountains terrane.

In summary, two temporally and spatially separate regional metamorphic events can be recognized in the terranes of the TrPz belt in the northeastern Klamath Mountains. These are (1) Late Triassic, high-pressure, blueschist-facies metamorphism restricted to the Fort Jones terrane and (2) Middle Jurassic, low-pressure, subgreenschist to amphibolite-facies metamorphism superimposed on the North Fork, Salmon River, Hayfork, and Marble

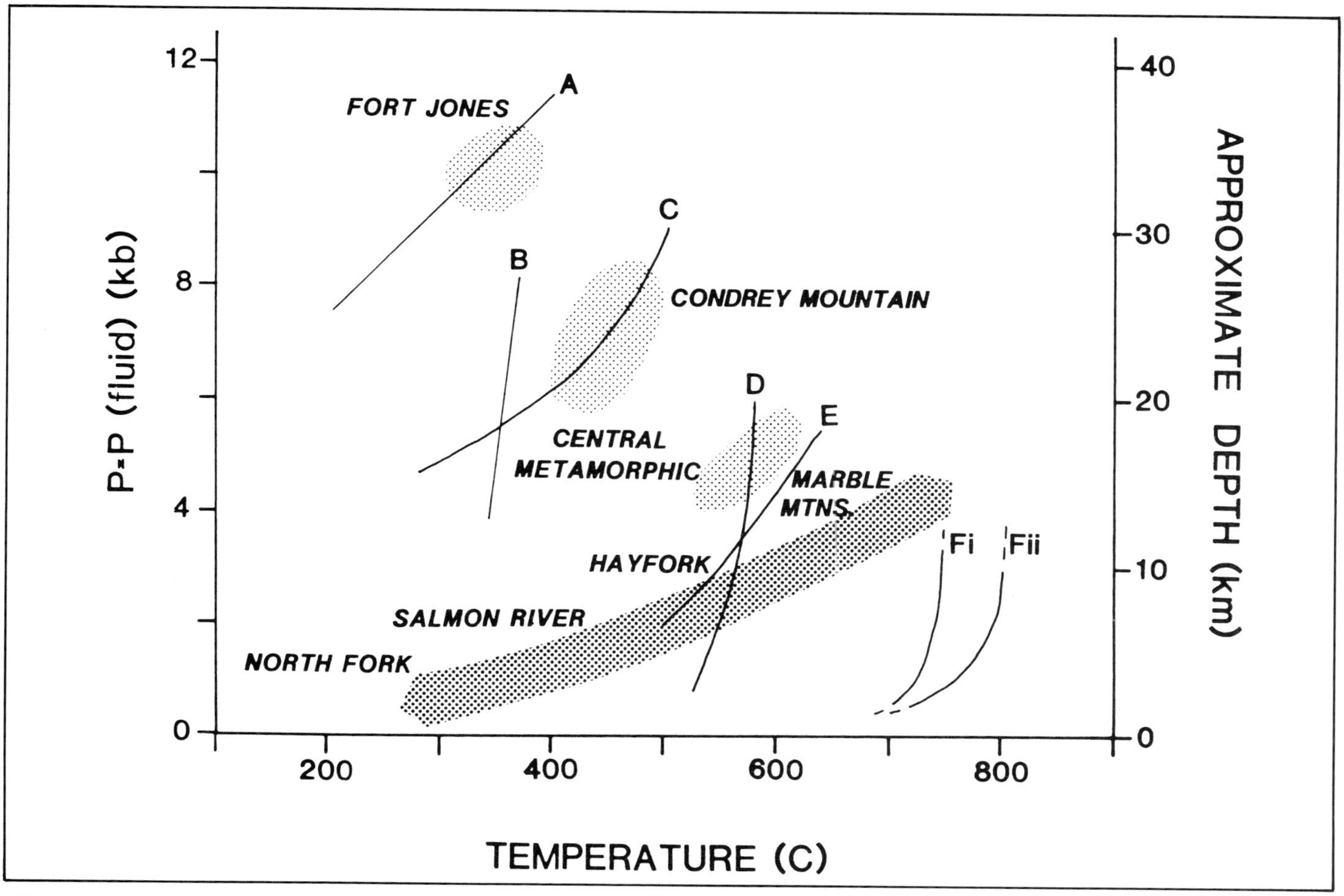

Figure 6—Approximate inferred P–T conditions for regional metamorphic events in the Klamath Mountains, constrained by experimentally determined metabasite phase equilibria. Data for Fort Jones terrane from Borns (1980). Lines A–F represent the following. A: impure jadeite in (Newton and Smith, 1967); B: pumpellyite out (Nitsch, 1971); C: glaucophane composition field (Ernst, 1979); D: chlorite out, QFM buffer (Liou et al, 1974; Apted and Liou, 1983); E: epidote out, QFM buffer (Liou et al, 1974; Apted and Liou, 1983); F: clinopyoxene in i) HM buffer, ii) QFM buffer (Spear, 1981). The Condrey Mountain terrane contains glaucophane but no lawsonite or jadeite (Hotz, 1967; Coleman et al, 1983); parts of the Central Metamorphic terrane contain albite-epidote amphibolite facies assemblages (Hotz, 1977; Cashman, 1980).

Mountains terranes. The Vesa Bluffs pluton intruded and contact metamorphosed all these terranes in latest Middle Jurassic time.

Metamorphism in the Southern TrPz Belt

Metamorphic relations in the TrPz belt 100 km (60 mi) to the south are somewhat different. There the Fort Jones terrane (Stuart Fork Formation) was, until recently, thought only to have been metamorphosed to the greenschist facies (Davis et al, 1965). Whole-rock K-Ar ages of 148, 158, and 245 m.y. were reported for the Fort Jones terrane by Lanphere et al (1968). Recently, rare relict lawsonite and glaucophane has been found in Fort Jones terrane rocks in the central Klamath Mountains (Jayko and Blake, 1984; Blake, personal communication, 1984) demonstrating an older high-pressure event in these rocks. The other TrPz belt terranes are metamorphosed to subgreenschist or greenschist facies, and local metamorphic gradients are spatially related to plutons (Fahan, 1982). Well-preserved radiolaria in the North Fork, Hayfork, and Rattlesnake Creek terranes (Irwin et al, 1978) demonstrate

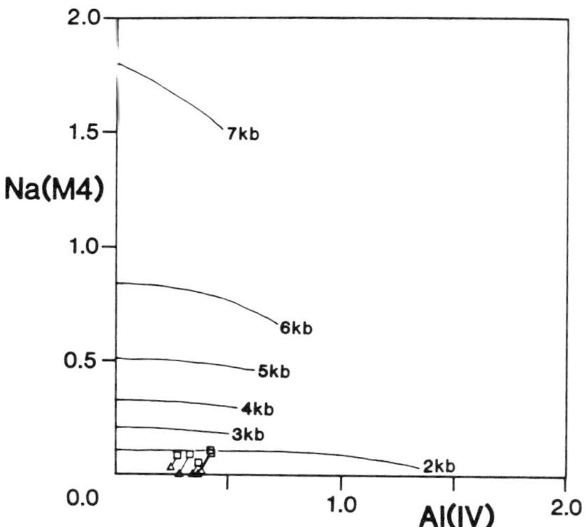

Figure 7—Amphibole compositional data from sample 77A plotted on the Al(IV) versus Na (M4) diagram of Brown (1977). Squares are analyses recalculated with full stoichiometric allocation of ferric iron; triangles are analyses recalculated with all ferrous iron (Papike et al, 1974).

that these metasediments were never metamorphosed to high grades.

The major differences in metamorphic grade between the northern and southern TrPz belt can be ascribed to two factors: (1) lateral thermal gradients during Middle Jurassic regional metamorphism such that higher grades were never regionally attained in the southern Hayfork and Rattlesnake Creek terranes, and (2) more extensive greenschist-facies recrystallization during Middle to Late Jurassic pluton intrusion in the southern TrPz belt than in the north. Jayko and Blake (1984) ascribe the obliteration of blueschist mineralogies in the Fort Jones terrane to a greenschist-facies overprint synchronous with pluton intrusion. Jurassic thermal events may also help explain the spread of whole-rock K–Ar ages in the southern part of the Fort Jones terrane.

STRUCTURE

An approximately 18 km (11 mi) section (structural thickness) of the TrPz belt is exposed in the area of Figure 3. Extreme lithologic heterogeneity, almost total absence of coherent stratigraphy, and general lack of facing directions and structural overprints in the rocks hinder identification and correlation of deformation events with any reasonable degree of certainty. Nonetheless, the case can be made that the North Fork, Salmon River, Hayfork, and Marble Mountains terranes have shared a common structural (as metamorphic) history since their mutual juxtaposition in Middle Jurassic time. Almost all deformation in the Fort Jones terrane is Triassic in age (Borns, 1980) and is not dealt with in this paper.

D1—Inter- and Intraterrane Faulting

At the surface, the faults bounding the terranes are steep to shallow east-dipping zones generally between 20 and 100 m (70–300 ft) wide, containing rock slices from contiguous terranes. Regionally they map out as low-angle, thrust-like faults (Irwin, 1966; Fig. 2 of this paper). The terranes are also internally faulted such that originally coherent stratigraphy is destroyed. This is especially obvious in the North Fork terrane where radiolarian cherts of different ages are mutually juxtaposed (Mortimer, 1984) and in the North Fork and Hayfork terranes where lenses of competent (volcanic) lithologies have become isolated in a matrix of argillaceous metasediments. Portions of the terranes are thus melanges (definition of Hsu, 1968). Argillaceous matrices are absent from the other terranes and most of the North Fork terrane but the style of faulting throughout the area is similar to melange deformation, with older rocks juxtaposed on younger and vice versa. The structural condition of most of the terranes is such that they are probably best described as "broken formations" (Hsu, 1968); the lithologic integrity of individual terranes being preserved despite extensive internal disruption.

Chaotic faulting is thus responsible for the inter- and intraterrane juxtaposition of North Fork, Salmon River, Hayfork, and Marble Mountains terrane lithologies prior to their shared regional metamorphism and is interpreted to be the earliest recognizable deformation in the area. Although specific cross-cutting relationships are observed (e.g., truncation of North Fork terrane faults at the Salmon River terrane boundary; Fig. 3), the general style of faulting between and within terranes is similar, and all this deformation is grouped under a single, nonpenetrative deformation event designated D1.

D2—Flattening and Isoclinal Folding

Throughout the area of Figure 3 a flattening foliation, axial planar to rare isoclinal folds, is variably developed. It is manifested as a penetrative slaty cleavage in argillites, a weak penetrative cleavage in marbles and siliceous argillites, pinch-and-swell structure in cherts, and micaceous cleavage in schists. Low-grade metabasites are not penetratively deformed but most amphibolite-facies metabasites are well foliated and lineated (Hotz, 1967, 1979). Correlation of the flattening fabrics is based on the fact that the flattening foliation is everywhere defined by prograde regional metamorphic minerals (Fig. 4) that probably formed during the same metamorphic event (see above). The flattening fabric in all four terranes can therefore be attributed to the same (synmetamorphic) deformation, designated D2 (with flattening fabric S2).

Isoclinal folding accompanied S2 formation, but the extent of folding is difficult to assess. Where seen together in outcrop, S0 (bedding) and S2 are almost always parallel (see also Fig. 8). S2 is parallel to the axial planes of rare 10 cm (4 in.) amplitude isoclinal folds in marble of the Hayfork and Marble Mountains terranes. In ribbon cherts of the North Fork and eastern Hayfork terranes, bedding and cleavage dip east with cleavage sometimes slightly steeper than bedding (Fig. 9a). Of the 15 facing directions found (using radiolarian age sequences, graded volcaniclastic beds and pillows), 4 were overturned (one each in the North Fork and Salmon River terranes and two in the Hayfork terrane; Mortimer, 1984) implying overturning though on an unknown scale.

Kays and Ferns (1980), working in the Marble Mountains terrane north of the area of Figure 3, interpret D1 faulting to be synchronous with metamorphism, D2 isoclinal folding, and axial planar cleavage development. Although D1 faults and D2 cleavage are indeed subparallel throughout the study area, cleavage was not observed to be better developed near faults. Thus the interpretation of a D2 cleavage superimposed on already disrupted rocks (albeit shortly after D1) is preferred.

D3—Open Folding

Postmetamorphic centimeter to 10 m (.4 in–30 ft) scale, open to tight, upright to overturned, north-south trending folds affect S0 and S2 in metasediments. No overprints were found that might suggest more than one postmetamorphic mesoscopic folding, and the folds are here loosely grouped in a single deformation event, D3. The extreme variability in style is in part a function of rock type; argillites and thin-bedded sediments are crenulated and crumpled on a cm scale (Fig. 9b). Thicker bedded sediments, including ribbon cherts, show meter scale buckling, and massive siliceous argillites fold on a larger scale still. Unfoliated metabasites and metaperidotites do not show recognizable mesoscopic folds. Where best developed, in the ribbon cherts of the North Fork terrane,

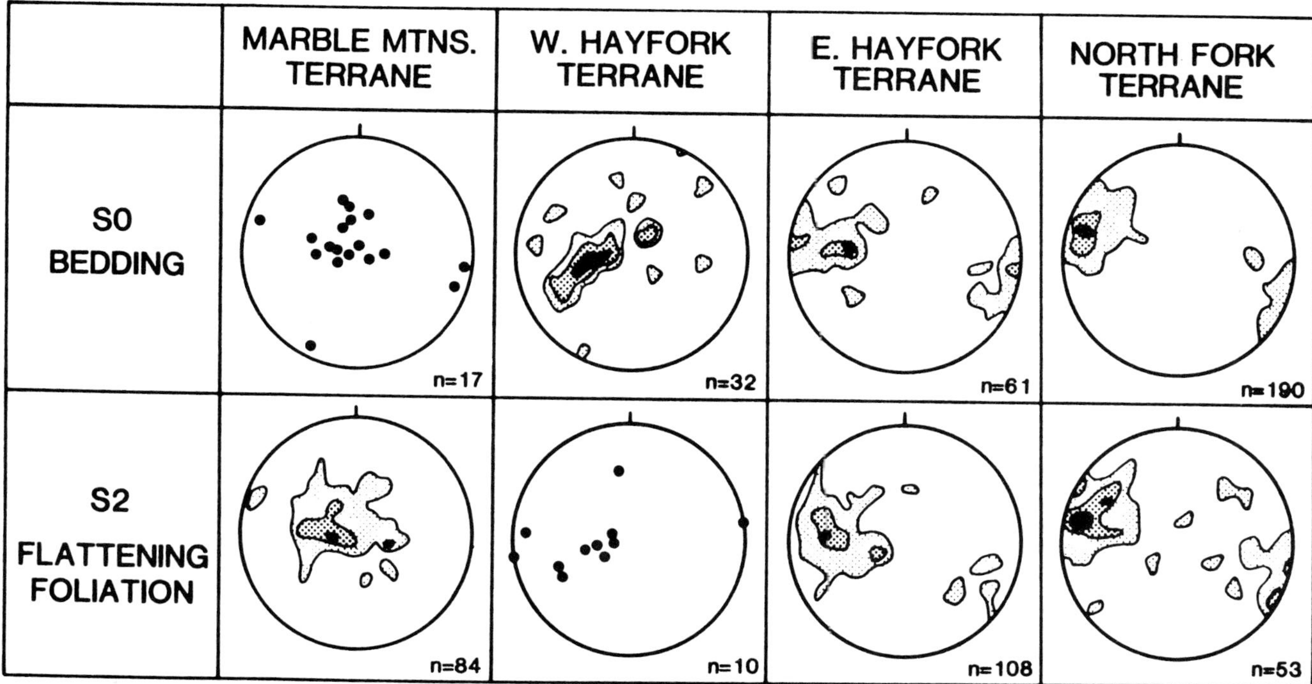

Figure 8—Equal-area stereonets of poles to planar fabric data from the area of Figure 12. Marble Mountains terrane data from Hotz (1967). Contours at 2, 6, and 10% per 1% area.

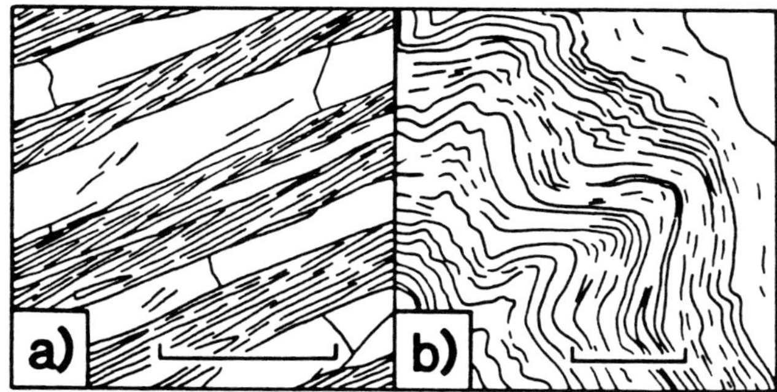

Figure 9—(a) Bedding/S2 cleavage intersection in ribbon chert of the eastern Hayfork terrane, (b) F3 folds in interbedded marble, volcaniclastic rocks, and argillite (S2 parallel to bedding) of the western Hayfork terrane. Scale bars are 10 cm (3.9 in.) long.

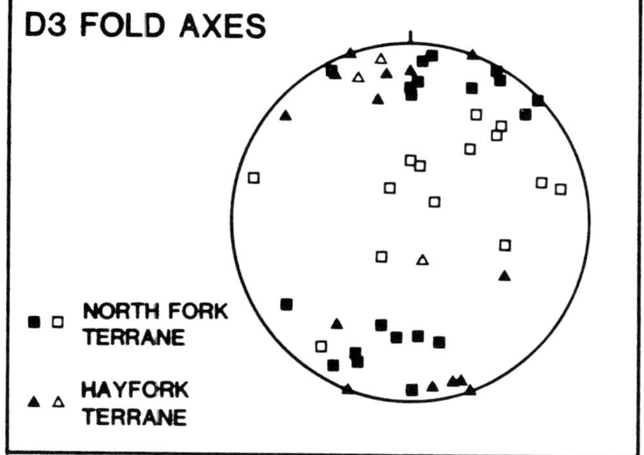

Figure 10—Equal-area stereonet of D3 fold axes from the North Fork and Hayfork terranes. Open symbols are axes measured within 50 m (160 ft) of major volcanic outcrops. Closed symbols are axes measured from the central portions of metasediment outcrops.

the folds are seen in outcrop to be noncylindrical. Some scatter in the north-south trends of fold axes can also apparently be accounted for by strain heterogeneity resulting from the proximity of volcanic rocks (Fig. 10).

A north-south trending upright synform (wavelength approximately 1,500 m [5,000 ft]) in the eastern Hayfork terrane is the best example of a map scale D3 structure (Fig. 3). Other possible examples are folding of the easternmost chert/volcanic contact in the North Fork terrane (Fig. 3) and still smaller synforms and antiforms in metasediments of the North Fork and Hayfork terranes (Mortimer, 1984). Given that certain D1 faults have been demonstrably folded, the apparent steepening and reactivation of other D1 faults can also probably be attributed to D3 deformation.

D4—Regional Tilting and Doming

Evidence for Neogene doming of the northern Klamath Mountains, accompanied by high-angle faulting, has been presented by Barnes and Rice (1983), Coleman and Helper (1983), and Mortimer and Coleman (1984). The gradual change in orientation of S0 and S2 from north-striking and gently dipping in the Marble Mountains terrane to more northeast-striking and steeply dipping in the North Fork terrane (Fig. 5) may be an expression of this doming. This would be consistent with Jurassic structures remaining shallow near the center of the dome and being steepened on its eastern flank.

The unmetamorphosed Upper Cretaceous Hornbrook Formation (Nilsen et al, 1983) unconformably overlies the North Fork and Fort Jones terranes in the area of Figure 3. It contains no mesoscopic folds but has a regional east-northeast dip of some 20 to 30° that was probably acquired during Neogene deformation. All Jurassic structures in the Klamaths basement are truncated by the unconformity, and the Hornbrook Formation thus definitively constrains the age of pre-Neogene deformation and metamorphism in the area of Figure 3 as pre-Late Cretaceous.

GRAVITY MODELING

Some indication of the deep structure of the area is obtained by modeling of Bouguer gravity anomalies (Fig. 11). The interactive computer program GRAVMODEL, developed by Eureka Resources Inc., was used to match the observed gravity profile along the line shown in Figure 2 with those computed for geologically reasonable models. Gravity data were compiled by Eureka from a number of sources including Kim and Blank (1973).

The bodies defined in the preferred gravity model of Figure 11 represent terranes exposed along or near the section line. The densities assigned to the Condrey Mountain and Marble Mountains terranes are from Barnes et al (1982) and the Vesa Bluffs pluton from Hotz (1971); the other bodies are proportional averages of the author's own density measurements and/or published rock densities (Clark, 1966; Hotz, 1977, 1979).

Absolute thicknesses and dips of terranes are not well constrained in this simple gravity model. However, two points that can be made are that (1) instead of maintaining their often steep surface dips, the terrane boundaries must flatten at depth and dip gently east, and (2) the western gravity maximum (over the outcrop of the Vesa Bluffs pluton) is probably best explained by buried high-density metaperidotite bodies in the Marble Mountains terrane (these are common where the terrane is exposed at the surface). No high-angle boundaries between terranes are required at depth. This is in keeping with the regional style of low-angle east-dipping faults seen throughout the Klamaths province (Irwin, 1981).

TIMING AND CORRELATION OF JURASSIC EVENTS

Timing of D1 and D2 is constrained by the same criteria used to date the regional metamorphism; they occurred prior to pluton intrusion at 164 ± 5 m.y. and after the deposition of Pliensbachian sediments in the North Fork terrane and, if correlations with the southern Klamath Mountains are correct, after the 168 to 177 m.y. eruption of rocks in the western Hayfork terrane (Fahan, 1982). D1, D2, and regional metamorphism thus took place in a relatively short time in the Middle Jurassic (Fahan and Wright, 1983). Timing of D3 is more difficult to constrain. Cross-cutting relations between D3 and the Vesa Bluffs pluton are not seen, but open folding definitely occurred after Middle Jurassic metamorphism and before deposition of the Hornbrook Formation in Turonian time (Nilsen et al, 1983). This age range may be narrowed by making assumptions based on regional relationships: Hill (1984, this volume), working near Happy Camp (Fig. 2), has correlated postmetamorphic open folds in the Hayfork, Marble Mountains, Condrey Mountain, and Smith River terranes with each other. This deformation is tightly constrained to be between 145 and 150 m.y. old in the Galice Formation of the Smith River terrane (Harper, 1983) and is generally recognized to be the Nevadan orogeny. The D3 folds in the area of Figure 3 are probably equivalent to these Nevadan folds.

Pluton intrusion at about 164 ± 5 m.y. constrains the TrPz belt metamorphism, plutonism, D1, and D2 to be strictly pre-Nevadan events (cf. Irwin et al, 1978). Many K–Ar mineral ages for TrPz belt rocks do cluster between 144 and 151 m.y. (Lanphere et al, 1968), thus spanning the age of the Nevadan orogeny, but, in the author's opinion, record nothing more than the interval during which TrPz belt hornblendes and biotites were slowly cooling down through their blocking temperatures from pre-Nevadan regional metamorphism. Uplift of the region was complete by Late Cretaceous time as paleosols developed on the metamorphic basement are overlain by the basal beds of the Hornbrook Formation (Nilsen et al, 1983).

SUMMARY, DISCUSSION, AND SPECULATION

Juxtaposition of the North Fork, Salmon River, Hayfork, and Marble Mountains terranes took place along low-angle, east-dipping faults with much inter- and intraterrane disruption (D1) in Middle Jurassic time. Following this, a ductile flattening deformation (D2) synchronous with low-pressure subgreenschist to amphibolite facies regional metamorphism was superimposed on the terranes. A slice of Late Triassic blueschist-facies crust, the Fort Jones terrane, was then thrust over this Jurassic metamorphic package. During latest Middle Jurassic time, the whole of the TrPz belt was intruded by calc-alkaline plutons (Davis et al, 1978; Wright, 1981; Allen et al, 1982). Timing of open folding (D3) is poorly constrained but can reasonably be assigned to the (latest Jurassic) Nevadan orogeny. Upper structural levels of the TrPz belt were exposed at the surface before Turonian time. These events are summarized in Figure 12. The absence of deep high-angle discontinuities (Fig. 10) suggests that major strike-slip faults, recognized elsewhere in the North American Cordillera, are absent in this part of the Klamaths. Thus, apart from regional Neogene doming and minor block faulting, low-angle Jurassic structures are still preserved intact.

The preferred interpretation of the above sequence of Middle and Late Jurassic events is that they took place in a single arc/subduction system. In this model, terrane juxtaposition by low-angle faulting occurred in the high, brittle levels of the subduction zone and was followed by transport to appropriate (though still relatively low-pressure) metamorphic depths. Subsequent events were

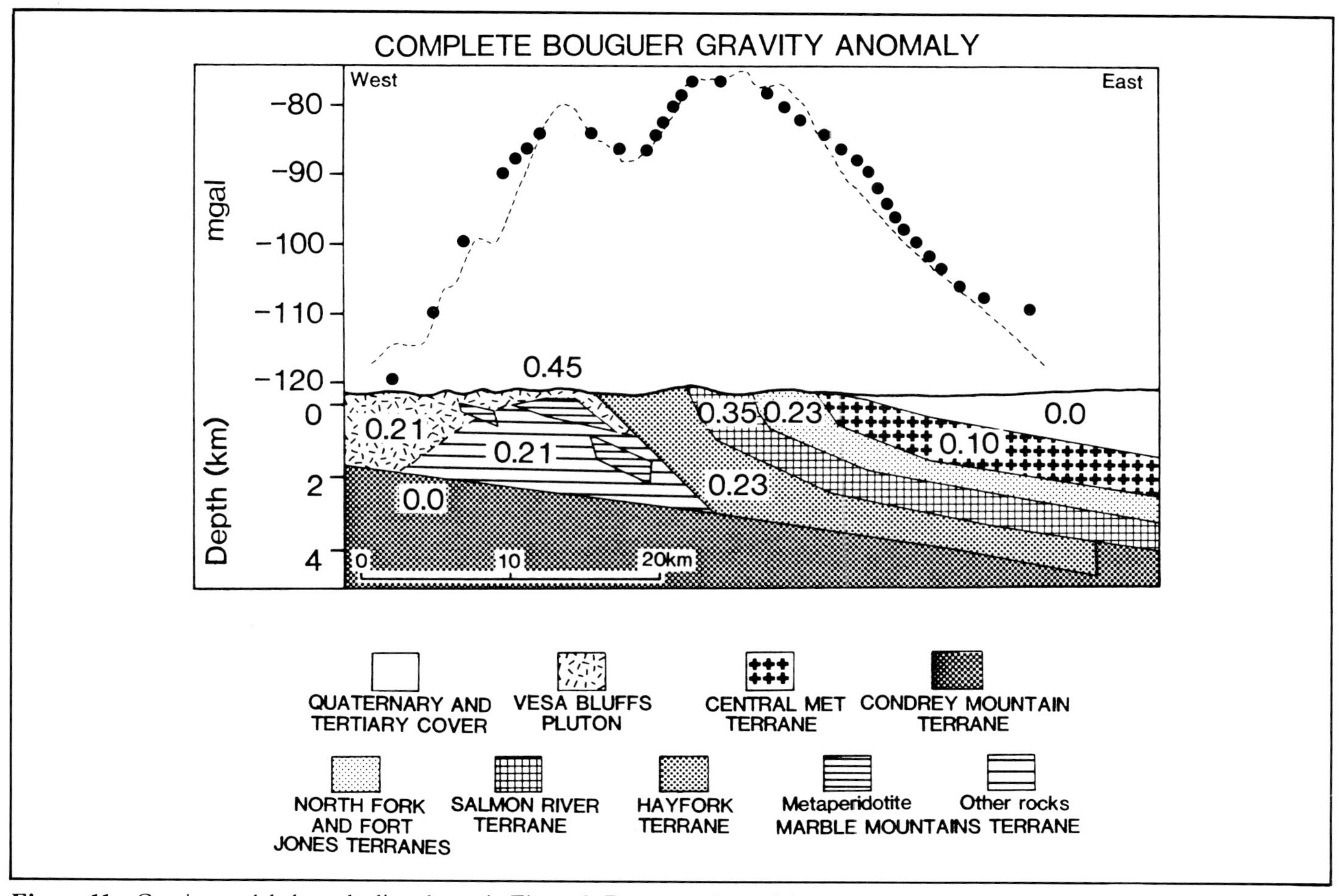

Figure 11—Gravity model along the line shown in Figure 2. Dots are values of the observed gravity anomaly, dashed line is computed anomaly. Numbers are positive density contrasts between the modeled bodies (in g/cc). The contact between the Vesa Bluffs pluton and Marble Mountains terrane cannot be determined using the gravity model; its position in the cross section is therefore questionable.

intra-arc thrusting followed by pluton intrusion, cooling, and uplift to the surface. This particular model is attractive because of its relative simplicity and consistency with current observations and ideas about style of deformation, metamorphic facies, and igneous activity in arc/subduction systems (e.g., Shiki and Misawa, 1982; Ernst, 1974; Ringwood, 1977).

In addition, most Klamath investigators would agree that a Middle Jurassic arc was constructed across rocks of an already assembled TrPz belt (e.g., Davis et al, 1978; Wright, 1982; Wright and Sharp, 1982; Saleeby, 1983; Fahan and Wright, 1983), their main evidence being the intrusion of plutons. Structural and metamorphic interpretations presented here support such an idea and can be used to infer the specific position of the TrPz belt in the arc/subduction system through time.

ACKNOWLEDGMENTS

I would like to thank Bob Coleman for encouragement and advice throughout this project; Juan de la Fuente, geologist with the Klamath National Forest, for logistical support in the field; and Mel Erskine of Eureka Resource Associates Inc., for assistance with gravity modeling. Earlier drafts of the manuscript were improved by comments from Bob Coleman, Juhn Liou, Elizabeth Miller, Meghan Miller, Mary Donato, and Bruce Hill. Financial support from the Daniel Pidgeon Fund of the Geological Society of London and NSF grant EAR 82-13347 (Coleman)is gratefully acknowledged.

REFERENCES

Allen, C. A., et al, 1982, Comagmatic nature of the Wooley Creek batholith and the Slinkard pluton and age constraints on tectonic and metamorphic events in the western Paleozoic and Triassic belt, Klamath Mountains, N. California (Abs.): Geological Society of America Abstracts with Programs, v. 14, n. 4, p. 145.

Ando, C. J., et al, 1983, The ophiolitic North Fork terrane in the Salmon River region, central Klamath Mountains, California: Geological Society of America Bulletin, v. 94, p. 236-252.

Apted, M. J., and J. G. Liou, 1983, Phase relations among greenschist, epidote-amphibolite, and amphibolite in a basaltic system: American Journal of Science, v. 283A, p. 328-354.

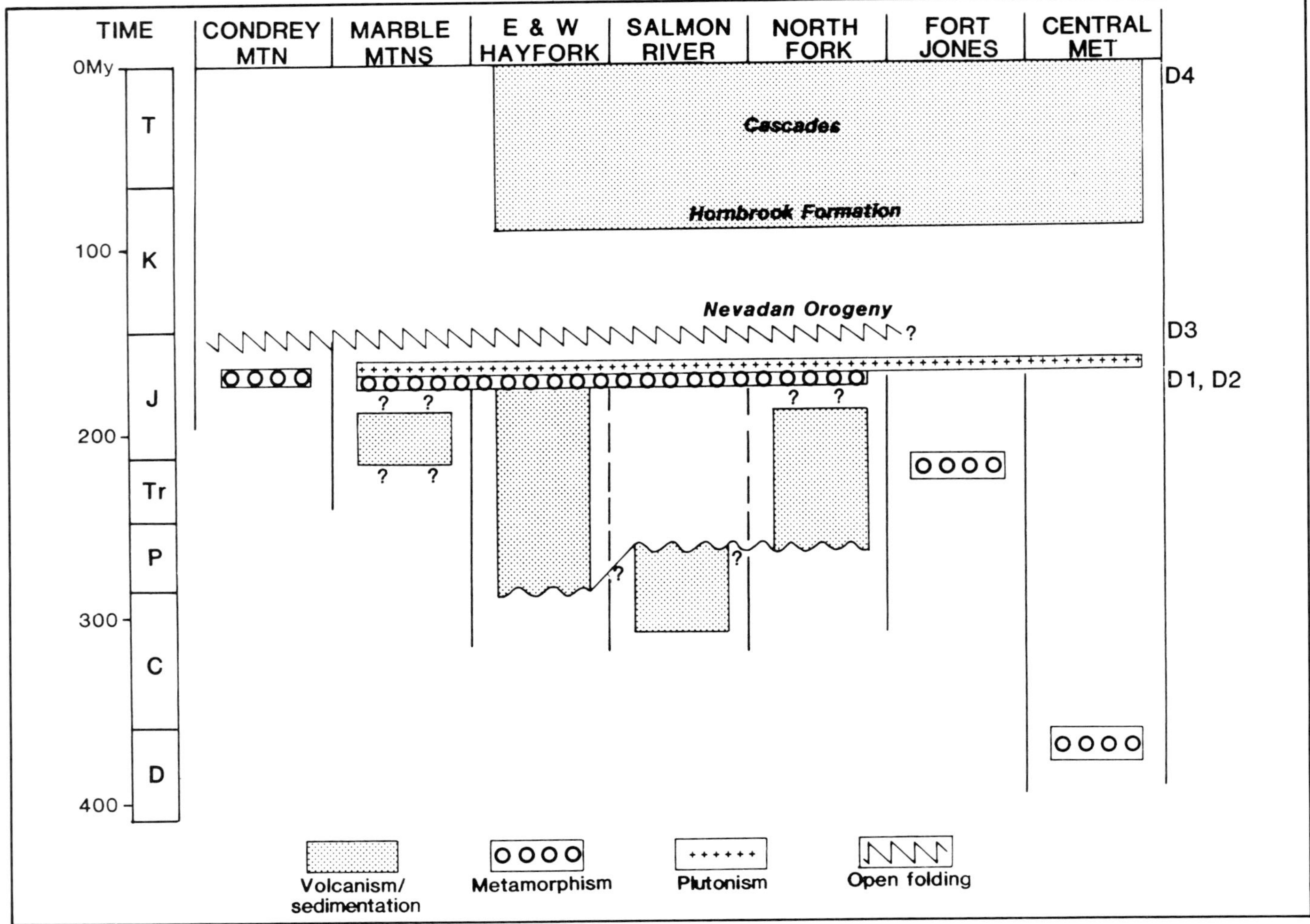

Figure 12—Comparison of the depositional, plutonic, and metamorphic records of seven Klamath terranes. Timescale of Harland et al (1982). Data from Hotz et al (1977), Cashman (1980), Wright (1981), Fahan (1982), Irwin et al (1978), Ando et al (1983), and Hill (1984). Symbols in the right margin indicate the time of the TrPz belt deformation events discussed in the text.

Barnes, C. G., and J. M. Rice, 1983, Titled plutons in the Klamath Mountains (Abs.): Geological Society of America Abstracts with Programs, v. 15, n. 5, p. 314.

______, et al, 1982, Evidence for basal detachment of the western Paleozoic and Triassic belt, northern Klamath Mountains, California (Abs.): Geological Society of America Abstracts with Programs, v. 14, n. 4, p. 147.

Blake, Jr., M. C., et al, 1982, Preliminary tectonostratigraphic terrane map of California: U.S. Geological Survey Open-File Report 82-593.

Borns, D. J., 1980, Blueschist metamorphism of the Yreka-Fort Jones Area, Klamath Mountains, Northern California: PhD Dissertation, University of Washington, 155 p.

Brown, E. H., 1977, The crossite content of Ca-amphibole as a guide to the pressure of metamorphism: Journal of Petrology, v. 19, p. 53–72.

Cashman, S. M., 1980, Devonian metamorphic event in the northeastern Klamath Mountains, California: Geological Society of America Bulletin, v. 91, pt. I, p. 453–459.

Clark, S. P., 1966, Handbook of physical constants: Geological Society of America Memoir 97.

Coleman, R. G., and M. A. Helper, 1983, The significance of the Condrey Mountain Dome in the evolution of the Klamath Mountains, California and Oregon (Abs.): Geological Society of America Abstracts with Programs, v. 15, n. 5, p. 294.

______, et al, 1983, Geologic map of the Condrey Mountain Roadless area, Siskiyou County, California. United States Geological Survey Miscellaneous Field Studies Map MF-1540-A.

Coombs, D. S., et al, 1976, Pumpellyite-actinolite facies schists of the Taveyanne Formation near Loeche, Valais, Switzerland: Journal of Petrology, v. 17, p. 440–471.

Davis, G. A., 1968, Westward thrust faulting in the south-central Klamath Mountains, California: Geological Society of America Bulletin, v. 79, p. 911–934.

______, et al, 1965, Structure, metamorphism and plutonism in the south-central Klamath Mountains, California: Geological Society of America Bulletin, v. 76, p. 933–966.

______, et al, 1978, Mesozoic construction of the

Cordilleran "collage," central British Columbia to central California; *in* Mesozoic paleogeography of the western United States: Society of Economic Paleontologists and Mineralogists, Pacific Coast Paleogeography Symposium 2, p. 1–32.

______, et al, 1979, Cross section of the central Klamath Mountains, California: Geological Society of America Map and Chart Series, MC-28I.

Donato, M. M., et al, 1982, Geologic map of the Marble Mountains Wilderness, Siskiyou County, California: U.S. Geological Survey Miscellaneous Field Map, MF-1452A.

Ernst, W. G., 1974, Metamorphism and ancient continental margins, *in* C. A. Burk and C. L. Drake, eds., The geology of continental margins: New York, Springer-Verlag, 1009 p.

______, 1979, Coexisting sodic and calcic amphiboles from high-pressure belts and the stability of barroisitic amphibole: Mineralogical Magazine, v. 43, p. 269–278.

Fahan, M. R., 1982, Geology and geochronology of a part of the Hayfork terrane, Klamath Mountains, northern California: MS Thesis, University of California, Berkeley, 127 p.

______, and J. E. Wright, 1983, Plutonism, volcanism, folding, regional metamorphism and thrust faulting: contemporaneous aspects of a major middle Jurassic orogenic event within the Klamath Mountains, northern California (Abs.): Geological Society of America Abstracts with Programs, v. 15, n. 5, p. 272–273.

Harland, W. B., et al, 1982, A geologic time scale: Cambridge, Cambridge University Press, 131 p.

Harper, G. D., 1983, The Nevadan Orogeny—Collapse of a west-facing island arc/back arc/remnant arc complex (Abs.): Geological Society of America Abstracts with Programs, v. 15, n. 5, p. 294.

Helper, M. A., 1983, Deformation-metamorphism relationships in a regional blueschist-greenschist facies terrane, Condrey Mt. Schist, north-central Klamath Mountains, N. Calif. (Abs.): Geological Society of America Abstracts with Programs, v. 15, n. 5, p. 427.

Hill, L. B., 1984, Tectonic and metamorphic history of the north-central Klamath Mountains, California: PhD Dissertation, Stanford University, 248 p.

Hotz, P. E., 1967, Geologic map of the Condrey Mountain quadrangle and parts of the Seiad Valley and Hornbrook quadrangles, California: U.S. Geological Survey Geological Quadrangle Map GQ-618.

______, 1971, Plutonic rocks of the Klamath Mountains, California and Oregon: U.S. Geological Survey Professional Paper 684B, 20 p.

______, 1973, Blueschist metamorphism in the Yreka-Fort Jones area, Klamath Mountains, California: U.S. Geological Survey Journal of Research, v. 1, n. 1, p. 53–61.

______, 1977, Geology of the Yreka Quadrangle, Siskiyou County, California: U.S. Geological Survey Bulletin 1436, 72 p.

______, 1979, Regional metamorphism in the Condrey Mountain Quadrangle, north-central Klamath mountains, California: U.S. Geological Survey Professional Paper 1086, 25 p.

______, et al, 1977, Triassic blueschist from northern California and north-central Oregon: Geology, v. 5, n. 11, p. 659–663.

Hsu, K. J., 1968, Principles of melanges and their bearing on the Franciscan-Knoxville paradox: Geological Survey of America Bulletin, v. 79, p. 1063–1074.

Irwin, W. P., 1966, Geology of the Klamath Mountains province: California Division of Mines and Geology Bulletin 190, p. 19–38.

______, 1972, Terranes of the Western Paleozoic and Triassic Belt in the Southern Klamath Mountains, California: U.S. Geological Survey Professional Paper 800C, p. C103–C111.

______, 1981, Tectonic accretion of the Klamath Mountains; *in* W. G. Ernst, ed., The geotectonic development of California, Rubey v. I: Englewood Cliffs, NJ, Prentice Hall, Inc., 706 p.

______, et al, 1978, Radiolarians from pre-Nevadan rocks of the Klamath Mountains, California and Oregon, *in* Mesozoic paleogeography of the western United States: Society of Economic Paleontologists and Mineralogists, Pacific Coast Paleogeography Symposium 2, p. 303–310.

Jayko, A. S., and M. C. Blake, 1984, Geologic map of part of the Orleans Mountain Roadless area, Siskiyou and Trinity Counties, California: U.S. Geological Survey Miscellaneous Field Studies Map MF-1600-A.

Kays, M. A., and M. L. Ferns, 1980, Geologic field trip guide through the north-central Klamath Mountains: Oregon Geology, v. 42, n. 2, p. 23–35.

Kim, C. K., and H. R. Blank, Jr., 1973, Bouguer gravity map of California, Weed Sheet: California Division of Mines and Geology, scale 1:250,000.

Lanphere, M. A., et al, 1968, Isotopic age of the Nevadan orogeny and older plutonic and metamorphic events in the Klamath Mountains, California: Geological Society of America Bulletin, v. 79, p. 1027–1052.

Liou, J. G., and W. G. Ernst, 1979, Oceanic ridge metamorphism of the East Taiwan Ophiolite: Contributions to Mineralogy and Petrology, v. 68, p. 335–348.

______, et al, 1974, Experimental studies of the phase relations between greenschist and amphibolite in a basaltic system: American Journal of Science, v. 274, p. 613–632.

Masson, P. H., 1949, Geology of the Gunsight Peak district, Siskiyou County, California: MA Thesis, University of Californis, Berkeley, 74 p.

Miyashiro, A., 1973, Metamorphism and metamorphic belts: New York, John Wiley & Sons, Inc., 492 p.

Mortimer, N., 1984, Petrology and structure of Permian to Jurassic rocks near Yreka, Klamath Mountains, California: PhD Dissertation, Stanford University, 84 p.

______, and R. G. Coleman, 1984, A Neogene structural dome in the Klamath mountains, California and Oregon, *in* T. H. Nilsen, ed., Geology of the Upper Cretaceous Hornbrook Formation, Oregon and California, Society of Economic Paleontologists and Mineralogists, Pacific Section, v. 42, p. 179–186.

Newton, R. C., and J. V. Smith, 1967, Investigations concerning the breakdown of albite at depth in the earth: Journal of Geology, v. 75, p. 268–286.

Nilsen, T. H., et al, 1983, Geologic map of the outcrop of the Hornbrook Formation, Oregon and California: U.S. Geological Survey Open-File Report 83-373.

Nitsch, K. H., 1971, Stabilitatsbeziehungen von Prehnit und Pumpellyit-haltigen Pargenesen: Contributions to Mineralogy and Petrology, v. 30, p. 240–260.

Papike, J. J., et al, 1974, Amphiboles and pyroxenes: characterization of other than quadrilateral components and estimation of ferric iron from microprobe data (Abs.): Geological Society of America Abstracts with Programs, v. 6, p. 1053–1054.

Rawson, S. A., and S. W. Petersen, 1982, Structural and lithologic equivalence of the Rattlesnake Creek terrane and high-grade rocks of the Western Paleozoic and Triassic belt, north-central Klamath Mountains, California (Abs.): Geological Society of America Abstracts with Programs v. 14, n. 4, p. 226.

Ringwood, A. E., 1977, Petrogenesis in island arc systems,*in* M. Talwani and W. C. Pitman, III, eds., Island arcs, deep sea trenches and back arc basins: American Geophysical Union, Maurice Ewing Series 1, p. 311–324.

Saleeby, J. B., 1983, Accretionary tectonics of the north American Cordillera: Annual Review of Earth and Planetary Sciences, v. 11, p. 45–73.

Shiki, T., and Y. Misawa, 1982, Forearc geological structure of the Japanese islands, *in* J. K. Leggett, ed., Trench-forearc geology: Geological Society of London Special Publication 10, p. 63–73.

Spear, F. S., 1981, An experimental study of hornblende stability and compositional variability in amphibolite: American Journal of Science, v. 281, p. 697–734.

Wright, J. E., 1981, Geology and U-Pb geochronology of the western Paleozoic and Triassic subprovince, Klamath Mountains, Northern California. PhD Dissertation, University of California, Santa Barbara, 300 p.

______, 1982, Permo-Triassic accretionary subduction complex, southwestern Klamath Mountains, Northern California: Journal of Geophysical Research, v. 87, n. B5, p. 3805–3818.

______, and W. D. Sharp, 1982, Mafic and ultramafic intrusive complexes of the Klamath-Sierran region, California: Remnants of a Middle Jurassic arc complex (Abs.): Geological Society of America Abstracts with Programs, v. 14, n. 4, p. 245–246.

Salinian Block U/Pb Age and Isotopic Variations: Implications for Origin and Emplacement of the Salinian Terrane*

James M. Mattinson
Eric W. James
University of California
Santa Barbara, California

The Salinian composite terrane is a large allochthonous strip of crust in western California that is underlain by a basement complex comprising Cretaceous granitic rocks and high-grade metamorphic rocks that are locally Precambrian but chiefly of unknown age. Details of Paleogene movements on the major crustal breaks that bound the block—the San Andreas and Sur-Nacimiento fault zones—are controversial. U/Pb isotopic ages and common Pb systematics of the metamorphic and plutonic rocks elucidate the magmatic history of the terrane, the petrogenesis of the magmas, and place constraints on the movement history of the block. Plutonic activity began ca. 105 to 120 m.y. ago in the northwestern part of Salinia and migrated southeastward over a period of ca. 40 m.y. The trend of ages, plus geographic variations in common Pb systematics, inheritance of zircons, and metamorphic basement isotopic characteristics indicate that, in mid-Late Cretaceous time, Salinia was part of a west-facing continental arc, underlain, at least in part, by Precambrian basement similar in age (ca. 1.7×10^9 years) to crystalline rocks on the adjacent craton. The youngest (ca. 75 to 80 m.y. old) plutonic rocks in Salinia must have been emplaced *prior* to the departure of the Salinian terrane from its source region and prior to its juxtaposition with terranes to the southwest that are underlain by Franciscan and Great Valley rocks.

INTRODUCTION

The Salinian composite terrane of California (Vedder et al, 1982), a large allochthonous terrane of granitic and metamorphic basement, is clearly a major and critical element in the Mesozoic and Cenozoic tectonic framework of western North America. Yet, as discussed below, its origin and much of its geologic history remain enigmatic. The Salinian composite terrane is underlain by a basement complex comprising Cretaceous granitic rocks and high-grade metamorphic rocks that are locally Precambrian but chiefly of unknown age.

The terrane is bounded by two major crustal breaks—the San Andreas fault zone on the northeast and the Sur-Nacimiento fault zone on the southwest. Along both these structures, the granitic and metamorphic rocks of the Salinian terrane are sharply juxtaposed against the predominantly "oceanic" rocks of the Franciscan assemblage.

For many years following the classic paper of Hill and Dibblee (1953), the Salinian terrane has been regarded as allochthonous "only" in terms of perhaps 500 to 600 km (310–375 mi) of right-lateral movement along the San Andreas fault and its precursors. Thus, it should have been possible to "match up" the Salinian terrane with its parental terrane to the South. Such a "simple" explanation for the block is appealing in terms of its relatively straightforward tectonic interpretation and, indeed, has been supported by a number of workers (e.g., Hamilton, 1978; Nilsen, 1978; Page, 1981; Dickinson, 1982, 1983). Other workers including Ross of the U.S. Geological Survey (USGS), who is certainly the major contributor to the petrology of Salinian granitic and metamorphic rocks, point to serious problems with this interpretation, particularly with regard to the metamorphic rocks of the Salinian terrane. (See Ross, 1978, for a concise summary.) Thus, it may be necessary to look farther afield for the source of the Salinian terrane; perhaps much farther south, as suggested by preliminary paleomagnetic data (Champion et al, 1980, 1984; McWilliams and Howell, 1982; Wilson et al in Page, 1982; Kanter and Engebretson, 1982; Kanter et al, 1982) or even to the north as suggested on the basis of isotopic patterns by Kistler (1978) and Kistler and Peterman (1978).

Thus, while the most recent phase of tectonic emplacement of the Salinian terrane clearly is related to the Neogene opening of the Gulf of California and accompanying few hundreds of kilometers of right-lateral offset along the San Andreas fault zone, there is considerable controversy regarding the timing, extent, and nature of earlier phases of emplacement of the terrane. Current (and conflicting) tectonic models for the

*Research supported by National Science Foundation grant EAR80-08215, a UCSB faculty research grant, and Penrose Grant, 2848-81

emplacement of the Salinian terrane in its present position invoke additional north-northwestward translations of a few hundreds or thousands of kilometers, and possible major westward-directed thrusting (e.g., compare Dickinson, 1983; Page, 1982; Silver, 1982, 1983; Vedder et al, 1982). Equally problematical is the nature of movements along the Sur–Nacimiento fault system, which bounds the "outboard" side of the terrane. These tectonic problems are discussed in more detail later.

The granitic and metamorphic nature of the Salinian basement permits the application of a variety of isotopic techniques to determine ages and petrogenesis. Isotopic ages and initial ratios provide some important constraints on the origin and time of emplacement of the Salinian terrane. First, the absence of plutons, or of any evidence of high temperature metamorphism in the adjacent Franciscan assemblage, clearly indicates that pluton emplacement, metamorphism, and even the high-temperature phase of postplutonic, postmetamorphic thermal history of the Salinian terrane all predate much of the tectonic transport of the Salinian terrane, and certainly its amalgamation with the Sur–Obispo composite terrane (McWilliams and Howell, 1982) and final emplacement against North America. Thus, accurate ages for pluton emplacement, metamorphism, and thermal history provide an upper limit on the age of amalgamation and emplacement of the Salinian terrane. Second, initial Pb isotopic ratios and the isotopic characteristics of zircons provide strong clues to the age and nature of the source of the magmas that fed the Salinian plutons. Such data also shed light on the provenance of the protoliths of the metamorphic rocks. Finally, the variations of ages and initial isotopic ratios on a large scale, throughout the Salinian terrane, place constraints on the tectonic setting, and even the polarity of the setting in which the Salinian terrane evolved prior to transport and "docking" at its present location.

BACKGROUND

Tectonic Setting

The general tectonic setting of the Salinian terrane and its relationship to the San Andreas and Sur–Nacimiento fault zones are shown in Figure 1. Early workers (e.g., Taliaferro, 1943) suggested that movements on these fault zones were predominantly vertical, with little or no significant horizontal component. A major advance in our understanding of the tectonic setting of the Salinian block was the demonstration of large-scale (hundreds of kilometers) right-lateral displacement along the San Andreas fault zone (Hill and Dibblee, 1953; Crowell, 1962) and the recognition of the role of that fault as a segment of the boundary between the North American and Pacific plates (Wilson, 1965). Despite considerable effort, however, the details of the timing and magnitude of movements along the San Andreas and Sur–Nacimiento fault zones remain controversial. At the present time, there are two major schools of thought regarding the pre-Neogene tectonics of the Salinian terrane. One school, drawing heavily on recent paleomagnetic evidence, holds that the Salinian terrane has been displaced northward a total of 2,500 to 5,600 km (1,550–3,500 mi) since Late Cretaceous time. The second school rejects the paleomagnetic data and holds that Salinia is allochthonous "only" to the extent of several hundred kilometers. The basis for these ideas is examined in more detail below.

Certainly the Neogene movement history of (and within?) the Salinian terrane is only the latest manifestation of the extensive translations associated with plate interactions between the Pacific realm and the North American continental block. This is made clear by the mounting evidence for northward translations of thousands of kilometers for the "Wrangellia" terranes (e.g., Jones et al, 1977) and some parts of the Franciscan complex (Blake and Jones, 1978; Alvarez et al, 1979). In fact, the concept of wholesale accretion to the western margin of the North American continent of far-traveled allochthonous terranes must be regarded as one of the most significant tectonic concepts of the past decade. The question here is the extent to which the Salinian terrane has participated.

First, the need for important pre-Neogene movement of the Salinian terrane relative to North America has long been recognized. In the simplest sense the problem here is in reconciling apparent offsets of at least 500 to 600 km (310–375 mi) for parts of the Salinian terrane, relative to the Sierran batholith belt, with well-demonstrated Neogene offsets of only about 300 km (186 mi) along the San Andreas fault zone. The "extra" offset has been attributed to either a proto- or ancestral San Andreas fault (e.g., Suppe, 1970; Silver et al, 1971; Gastil et al, 1972; Garfunkel, 1973; Armstrong and Suppe, 1973), to slivering of the Salinian block by internal faults (Johnson and Normark, 1974; Ross and Brabb, 1973; Dibblee, 1976), to major offsets (400 km [250 mi]) on faults other than the San Andreas, some of which transect the Salinian block (Howell, 1975), or to west-directed thrusting (Silver, 1982, 1983).

Some of the faults within the Salinian terrane are well documented (e.g., the Rinconada fault with 40 to 60 km [25–37 mi] of right slip, Dibblee, 1976; Graham, 1978; and the San Gregorio–Hosgri fault with 80 to 115 km [50–70 mi] of right slip, Silver, 1974; Hall, 1975; Graham and Dickinson, 1978). Lithologic differences across these faults have led Ross (1977, 1978) to subdivide Salinia into four sub-blocks, each bounded by faults (Fig. 2). These sub-blocks include: (1) a central block including the La Panza Range, most of the Santa Lucia Range, the Gabilan Range, and the Ben Lomond area; (2) a western block including the western Santa Lucia Range west of the Palo Colorado Fault; (3) a southeastern block (part of the "Tujunga terrane" of Blake et al, 1982), including the terrane between the Red Hills–San Juan–Chimeneas fault and the San Andreas fault; and (4) a northern block including all the Salinian terrane north of the Zayante–Vergeles fault.

Recent paleomagnetic work in the Salinian and adjacent terranes raises the possibility of pre-Neogene translations in the thousands of kilometers, however. Paleomagnetic studies of Late Cretaceous sedimentary rocks at Pigeon Point (Champion et al, 1980, 1984; Wilson et al in Page, 1982) suggest 2,500 to 5,600 km (1,550–3,500 mi) of post-Late Cretaceous northward translation of the Salinian terrane. Work on the Paleocene sedimentary rocks of San Pedro Point (Champion et al, 1984 and in

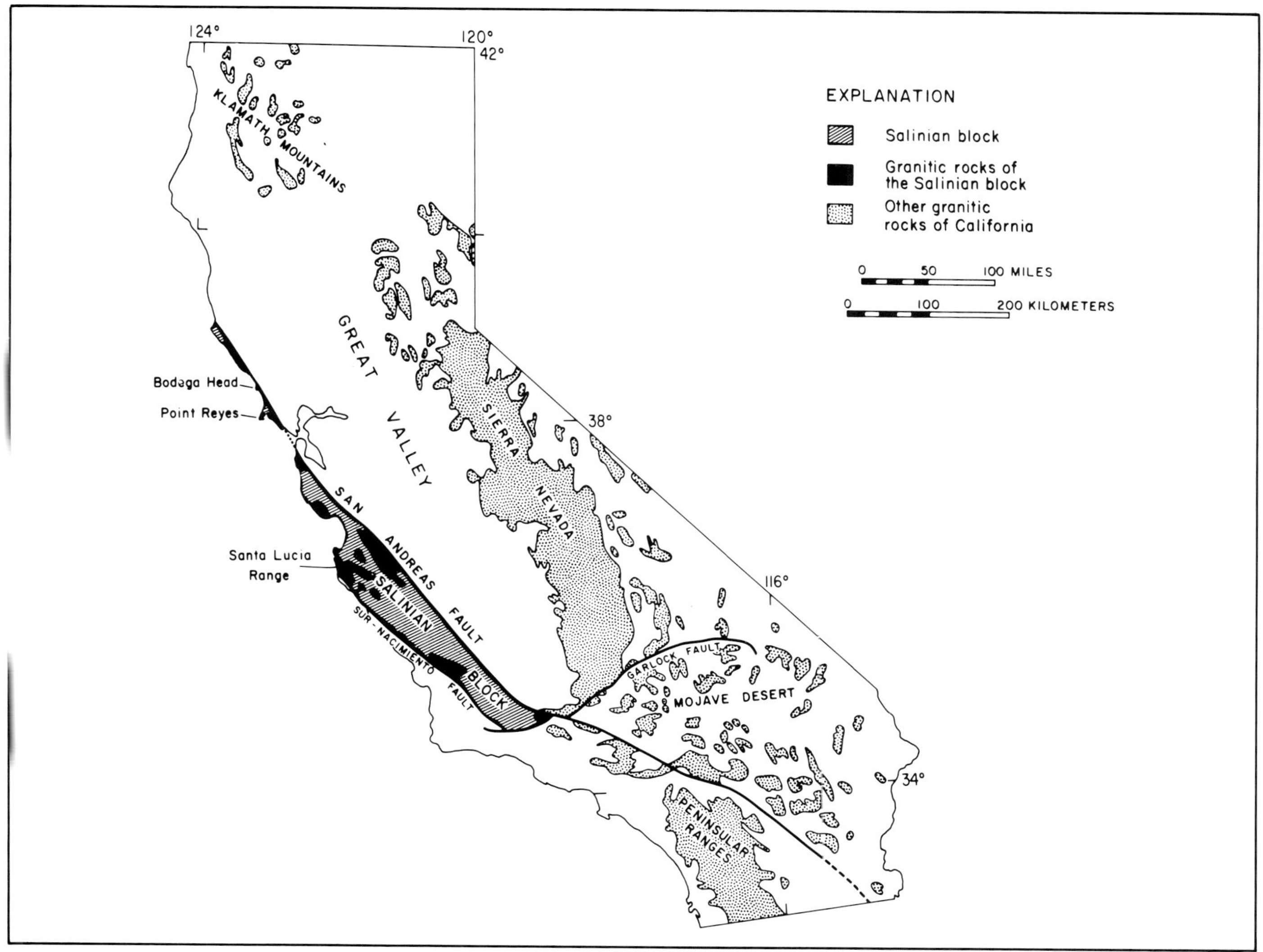

Figure 1—Simplified map of California showing the Salinian composite terrane, its major areas of granitic rock, and other granitic rocks of California.

McWilliams and Howell, 1982) suggests about 2,100 km (1,300 mi) of northward translation since the Paleocene. These interpretations involve major structural corrections, however, and many of the original data points have been discarded, primarily because of problems with secondary magnetization. Thus, the interpretations should be viewed with caution. Nevertheless, additional supporting evidence for large translations comes from preliminary paleomagnetic work on Salinian plutonic rocks (Kanter et al, 1982), paleomagnetic work on the Stanley Mountain terrane, which lies outboard of Salinia and may have been amalgamated with it as early as Late Cretaceous time (McWilliams and Howell, 1982), preliminary paleomagnetic results from spilites of the Gualala basin (Kanter and Engebretson, 1982; Kanter et al, 1982), and petrographic work on Salinian granitic rocks suggesting that Salinia does not "fit" with the southern Sierra Nevada (Ross, 1978; but also see below). A comprehensive paper on the tectonics of the Salinian terrane that wholeheartedly accepts the paleomagnetic data is that of Page (1982). Page (1970, 1981, 1982) also has made major contributions to the problems of the nature of the Sur–Nacimiento fault zone and its role in the disappearance of the now-missing western margin of the Salinian magmatic belt (discussed briefly later).

Other recent tectonic models reject the paleomagnetic evidence for thousands of kilometers of movement for Salinia relative to the craton. Dickinson (1982, 1983) suggests a more "traditional" translation of the Salinian terrane from a position just south of the southernmost Sierra Nevada but uses a roughly equal amount of left slip (500 to 600 km [310–375 mi]) on the Sur–Nacimiento fault system to juxtapose the outboard Franciscan terranes against the granitic and metamorphic rocks of Salinia. Some support for the "Salinian–Sierran Connection" is found in recent work by Ross. Ross of the USGS, who has done so much work in the Salinian terrane, has recently been working in the southern tip of the Sierra Nevada. Ross has previously emphasized the lack of matching rock types between the Salinian block and the Sierras or the Mojave, but now (Ross, in press) he notes some possible strong correlations between granitic rocks of the Gabilan Range and those of the Sierran tail. Ross (1976) earlier

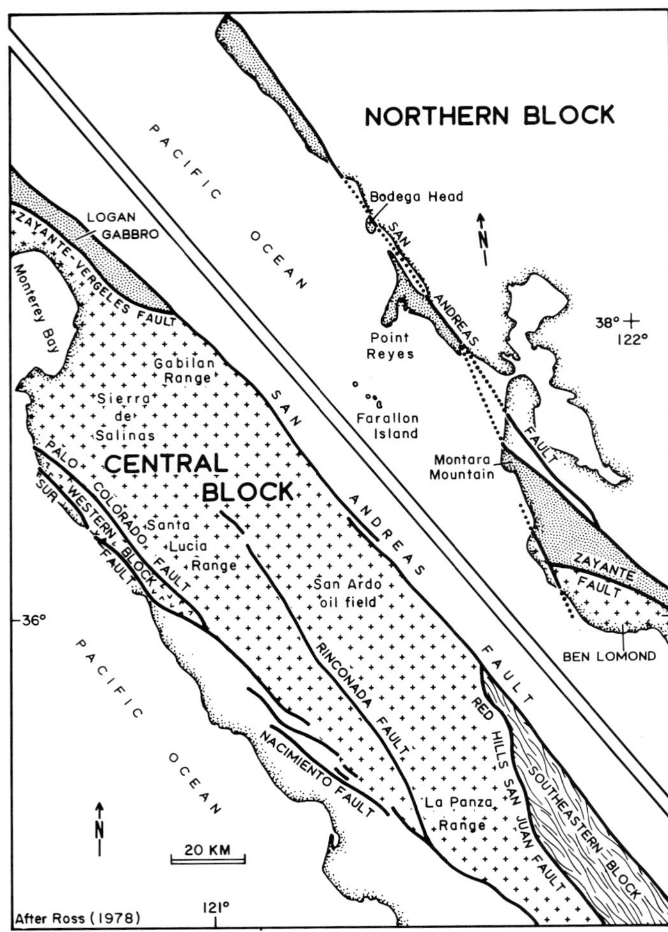

Figure 2—The four sub-blocks of Salinia. (After Ross, 1978. Copyright © 1978 The Society of Economic Paleontologists and Mineralogists. Used with permission.)

proposed a possible correlation between the schist of Sierra de Salinas in the Salinian terrane and the schist of Portal–Ritter Ridge, southeast of the Sierran tail, on the northeast side of the San Andreas fault. We are in the process of testing these possible correlations with a combination of field, petrologic, and isotopic investigations.

A totally different model has been proposed by Silver (1982, 1983). Silver's model involves a two-stage movement history for the Salinian terrane. First, west-directed Paleogene thrusting transported Salinia to the west some 125 to 150 km (78–93 mi) from the east side of the Peninsular Ranges–southern Sierran Cretaceous arc, overriding the intervening western part of the batholithic belt, the arc-trench gap (Great Valley sequence), and trench (accreted Franciscan terranes) in the process. Second, Neogene San Andreas right slip (~ 300 km [186 mi]) translated the Salinian terrane (in this model a klippe) to its present position. This intriguing model solves several Salinian dilemmas, including the apparent need for pre-Neogene San Andreas movements and the "missing" western part of the Salinian igneous arc and arc-trench gap (discussed below). If the southernmost tip of the Sierras is part of Silver's thrust package, then the correlations suggested by Ross (in press) between the Gabilan Range and the southernmost Sierras, discussed above, would also be accommodated. Needless to say, the models of both Dickinson and Silver have no place for the possible thousands of kilometers of poleward displacement suggested by the paleomagnetic data for the Salinian terrane and adjacent terranes.

The outboard boundary of Salinia, the enigmatic Sur–Nacimiento fault zone, deserves further attention before we leave the topic of Salinian tectonics. As with the relationship of Salinia to North America, the nature and significance of the Sur–Nacimiento fault zone are controversial. Furthermore, this critical boundary has received less attention than has the San Andreas fault zone. Page (1970, 1981, 1982) has called attention to the fundamental problem regarding the juxtaposition of Salinian granitic–high-grade metamorphic basement against roughly coeval low-grade accretionary Franciscan rocks; that is, the absence of the western part of the Salinian arc (i.e., western foothills belt equivalent) and arc-trench gap deposits (i.e., Great Valley equivalent), amounting to some 100 to 250 km (62–155 mi) in width (Page, 1982). In earlier work, Page (1970) suggested that the Sur–Nacimiento fault zone might be a fossil subduction zone and invoked tectonic erosion or rasping of the continental margin as a possible mechanism for the removal of the now missing western portion of Salinia. In more recent papers, Page (1981, 1982) favors "mega-transport" (i.e., large-scale strike-slip) for removal of the missing fragments of Salinia, probably to more northerly latitudes.

Dickinson (1983), as discussed earlier, postulates left slip (i.e., southerly displacement) along the Nacimiento fault during Cretaceous time, whereas Silver's model (1982, 1983) permits but does not require major movements along the western boundary of Salinia to explain the "missing" part of the arc and arc-trench gap. Instead, the "missing" part of Salinia would have been left behind during the tectonic extraction of a Salinian "mega-flake" from the east side of the Cretaceous batholith belt of southern and central California.

Summary of Basement Geology

Metamorphic Rocks

The metamorphic framework of the Salinian composite terrane comprises three lithologically and geographically distinct suites of metamorphic rocks (Ross, 1977). Each suite is confined to a particular sub-block or set of sub-blocks within Salinia (Fig. 2). The suites are: (1) "Sur-Series" type metamorphic rocks of the central, western, and northern sub-blocks; (2) high-grade gneisses, migmatites, and schists of the southeastern sub-block (Barrett Ridge Slice or "Tujunga terrane"); and (3) the schist of Sierra de Salinas in the central sub-block. These suites are discussed briefly below.

The first suite of metamorphic rocks, the "Sur Series" of the Santa Lucia Range (Trask, 1926) and their presumed equivalent elsewhere in the block, are chiefly quartzo-feldspathic granofels and gneiss, biotite-feldspar quartzite, aluminous and graphitic schists, marble, para (?) amphibolite, and calc-silicate rocks. Thus, the protoliths of these rocks appear to have been chiefly of sedimentary, continental shelf origin. At the present time, the only hint of an age for this metamorphic suite is the report by Bowen and Gray (1959) of possible coral or crinoid-like "fossils" in marble of the Gabilan Range, suggesting a Paleozoic age. That these are in fact fossils at all is disputed by Page (1981, p. 378) and questioned by Ross (1977; and 1978, personal communication). Metamorphism of the suite largely accompanies plutonism (Compton, 1966; Weibe, 1970; James, 1984) and thus is about mid-Cretaceous in age (Mattinson, 1978). Despite considerable study, no satisfactory correlations between these rocks and any rocks of the east side of the San Andreas fault have been made (Ross, 1977, 1978).

The second suite of metamorphic rocks, from the southeastern sub-block (Barrett Ridge Slice of Ross, 1977; or Tujunga terrane of Blake et al, 1982), comprises predominantly high-grade gneisses, migmatities, and schists, with local thin-bedded amphibolite, marble, calc-hornfels, and quartzite. The metamorphic rocks crop out in a large block in the Mt. Pinos–Mt. Abel area, the southeasternmost basement in the Salinian composite terrane, and in two very small areas at Barrett Ridge and Red Hills. Mattinson (1983) reports an age of ca. 1.7×10^9 years for the Barrett Ridge gneiss. Ross (1977) suggests the possibility that the high-grade rocks of the Barrett Ridge Slice may correlate with the high-grade rocks of the San Gabriel Mountains.

The third and possibly most unusual metamorphic terrane in the Salinian terrane is underlain by the schist of the Sierra de Salinas in the central sub-block. This is a rather homogeneous biotite quartzofeldspathic schist with minor quartzite, marble, and amphibolite. On the basis of chemistry, Ross (1976) suggests that the schist was derived from a thick, uniform pile of graywacke. The schist crops out on both sides of the Salinas Valley and has been penetrated by wells as far south as the San Ardo oil field (Ross, 1976). The Sierra de Salinas schist is quite unlike any of the other metamorphic rocks in the Salinian terrane. Instead, Ross (1976) correlates the schist with the Pelona and Orocopia schists in the Mojave Desert, especially the Pelona schist of Portal and Ritter Ridge and Quartz Hill. Haxel and Dillon (1978) concur with the general lithologic similarity between the Sierra de Salinas schist and the Pelona–Orocopia–Rand schist but note several differences: The distinctive layering that characterizes the Pelona–Orocopia schist is not well developed in the Sierra de Salinas schist, the Sierra de Salinas schist is metamorphosed to higher grade than is the Pelona–Orocopia, and the Sierra de Salinas schist is intruded by large Cretaceous plutons that are lacking in the Pelona–Orocopia schist. They conclude that the two schist terranes may have had similar, perhaps correlative protoliths, but distinctly different metamorphic histories. Specifically, the Sierra de Salinas schist is intruded by ca. 80 to 85-m.y.-old plutons (Mattinson, 1982a) that are evidently unmetamorphosed. Thus, the metamorphism of the schist is mid-Late Cretaceous or older. In contrast, the last metamorphism of the Pelona–Orocopia is dated at between 50 and 60 m.y. ago (Ehlig, 1968; Ehlig et al, 1975; Conrad and Davis, 1977). The Rand schist may have an older metamorphic age, however, inasmuch as it is intruded by a stock dated by Kistler and Peterman (1978) as 74-m.y.-old (K–Ar); and by Silver et al (1984) as 73 ± 3 m.y. old (zircon U-Pb). In fact, in a recent regional overview, Burchfiel and Davis (1981) correlate the Rand and Sierra de Salinas schists and consider them to have been metamorphosed some 20 to 30 m.y. earlier than the Pelona–Orocopia schist. Burchfiel and Davis (1981) raise the interesting possibility that all of these schists were derived from Franciscan rocks, an idea originally proposed by Yeats (1968). According to this hypothesis, the Sierra de Salinas–Pelona–Orocopia–Rand schists represent Franciscan rocks that were underthrust, partly on a low-angle subduction zone, beneath the western margin of North America, with the Sierra de Salinas–Rand schist segment being underthrust prior to 80 m.y. and the Pelona–Orocopia schist segment being underthrust between 50 and 60 m.y. ago. However, recent work by Mukasa et al (1984) indicates that the protolith of the Orocopia Schist in the Chocolate Mountains is older than ca. 163 m.y., based on U/Pb dating of zircon from a metadiorite dike that intrudes the schist. Preliminary evidence from the Sierra de Salinas schist (Mattinson, 1982a) clearly indicates that the protolith of that schist was derived predominantly from a Precambrian source.

Plutonic Rocks

The plutonic rocks of the Salinian composite terrane may be conveniently discussed in terms of the same tectonic sub-blocks (after Ross, 1978) used earlier in the discussion of tectonics and metamorphic rocks. The most exposures of plutonic rocks are in the central block that includes the Gabilan Range, the largest area of extensive granitic outcrops in the entire Salinian terrane, the Ben Lomond area, the La Panza Range, and the eastern part of the Santa Lucia Range. The granitic rocks exposed in this sub-block are rather typical of the Cordilleran batholithic belt in a general way. They are calc-alkaline to calcic, with variation diagram trends in the range of the Sierra Nevada and Peninsula Ranges trends (Ross, 1972), and appear to have been emplaced at mesozonal depths. The Salinian plutons contrast with the western parts of the Sierra Nevada and Peninsular Ranges batholiths in that diorite and true trondhjemite are rare. Another significant difference between the Salinian plutons and the Sierran and Peninsular Ranges plutons is the lack of any metallic mineralization in the former (Ross, 1978), although in very recent work, Ross (in press) notes that mineralization drops off drastically in the southernmost Sierra, which, as discussed earlier, is a candidate for correlation with the Gabilan Range.

An anomalous suite of rocks at Logan, between the northern and central sub-blocks, includes anorthositic and quartz gabbros of possible ophiolitic affinities. These rocks have been correlated with gabbro near Eagle Rest Peak in the San Emigdio Mountains by Ross (1970), who also proposes that gabbroic clasts in sedimentary rocks at Gualala in the northern sub-block were originally derived from near Eagle Rest Peak.

Plutonic rocks in the northern sub-block comprise only rather small isolated outcrop areas at Montara Mountain, the Farallon Islands, Point Reyes, and Bodega

Head, and submerged outcrops at Cordell Bank. These rocks are essentially identical petrographically to the plutonic rocks of the central sub-block discussed above.

The plutonic rocks of the western sub-block comprise a suite of deep-seated synmetamorphic granitoids including charnokitic tonalite and cumulate ultramafites that are associated with lower granulite-grade wall rocks (Compton, 1960; Weibe, 1970). These rocks are thus unique in terms of their metamorphic history and their thermal history (Mattinson, 1978) within the Salinian block. In terms of emplacement ages and origin of their magmas, however, the plutons of the western sub-block are essentially identical to the mesozonal plutons of the northern sub-block (see Mattinson, 1978; and discussions below).

Plutonic rocks of the southeastern sub-block at Barrett Ridge and Red Hills occur as small intrusions in high-grade quartzofeldspathic gneiss of Precambrian age (Mattinson, 1983). The intrusions include gneissic granodiorite, alaskite granite, pegmatite, and fine-grained dike rocks.

ISOTOPIC DATA

Methods

The data presented below result from several years work throughout the Salinian composite terrane by Mattinson and a detailed study of two areas (Ben Lomond and Montara Mountain) by James. We have determined, or are in the process of determining, U/Pb isotopic ages for zircon, sphene, and apatite from most of the plutonic bodies in the Salinian terrane. We have measured initial Pb isotopic compositions from the same set of samples. In addition, we have a small amount of age and isotopic data for the metamorphic rocks of the Salinian block. In many of the plutonic rocks and the metamorphic rocks, the presence of inherited zircons and/or discordance resulting from subsequent disturbances complicates the determination of emplacement, source area, and/or provenance ages. In many of these cases the analysis of several fractions of zircon, plus other mineral species, pinpoints clearly the age of emplacement and/or source area. In other cases, large uncertainties are necessarily involved. Detailed analytical results and discussion will be published elsewhere. For the purposes of this paper we will summarize only the results. For this discussion, we will continue to use the fourfold subdivision of Salinia of Ross (1978): *northern block* including the Farallon Islands, Cordell Bank, Point Reyes, Bodega Head, and Montara Mountain; *central block* including the Ben Lomond area, the Santa Lucia Range east of the Palo Colorado fault, the Sierra de Salinas, the Gabilan Range, and the La Panza Range; *western block* including the Santa Lucia Range west of the Palo Colorado fault; and the *southeastern block* including the Barrett Ridge and Red Hills areas, and the Mount Abel-Pinos area. However, much of our data is summarized on a base from which the slip along faults within and bounding the sub-blocks has been removed (Fig. 3).

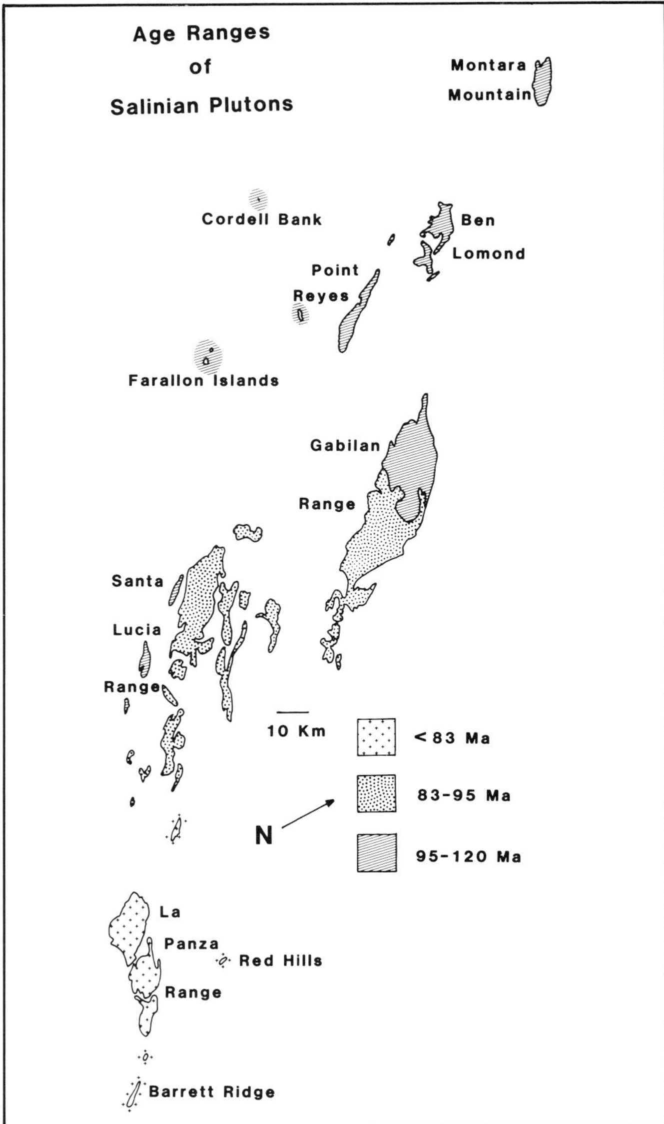

Figure 3—Age ranges of Salinian plutons, 110 km (68 mi) of right slip removed on the San Gregorio-Hosgri fault and 60 km (37 mi) removed on the Rinconada (Dibblee, 1976; Dickinson, 1983).

Emplacement Ages

In general, emplacement ages for Salinian plutons range from a maximum of 110 to 120 m.y. to a minimum of ca. 75 m.y., for a total span of some 40 m.y. The distribution of ages tends to be fairly systematic, with the highest ages in the northwest and the lowest in the southeast (Fig. 3). Plutons of the northern block of Ross (1978) have emplacement ages predominantly in the range of 95 to 110 m.y., with a few possibly as old as 120 m.y. These include the granitic rocks of Bodega Head and Point Reyes (Mattinson, 1978), Cordell Bank and the Farallon Islands (Mattinson, in preparation), and Montara Mountain (James, 1984, and in preparation).

Plutons of the large ($\sim$ 55 $\times$ 280 km [35 $\times$ 175 mi]) central block show a range of ages. Plutons from the northern part of the central block, in the Ben Lomond area (James, 1984, and in preparation), the Monterey area (Mattinson, in preparation), and the northern half of the Gabilan Range (Mattinson, 1982a, and in preparation) were emplaced during the same time interval as those in the northern block, i.e., 95 to 120 m.y. Plutons from the southern part of central block, including the southern half of the Gabilan Range and the La Panza Range, are substantially younger, mostly in the 80 to 90 m.y. range (Mattinson, 1982a, and in preparation; Ehlig and Joseph, 1977; Joseph and Ehlig, 1981).

Plutonic rocks from the narrow western block of Ross (1978) are charnokitic tonalites (Compton, 1960), quite distinctive from all other plutonic rocks in the Salinian block, as discussed earlier. Nevertheless, these rocks are identical in emplacement age to the rocks of the northern block and the northern part of the central block: about 105 m.y. (Mattinson, 1978).

The scattered plutons of the southeastern block (Barrett Ridge Slice) are the youngest yet found in the Salinian composite terrane—in the range of 75 to 80 m.y. (Mattinson, 1983, and in preparation).

Cooling Ages

Isotopic ages from a variety of minerals reflect "cooling ages" that approximate plutonic emplacement ages only for relatively shallow, rapidly cooled intrusive bodies. Examples include K–Ar ages of biotite and hornblende and U/Pb ages of sphene (titanite) and apatite. Ages of these minerals from Salinian plutons are younger than emplacement ages inferred from zircon data by as little as 1 to 2 m.y. to as much as 25 m.y. Moreover, as is well known, the different minerals have different "blocking" or "retention" temperatures and thus commonly record different ages from the same sample. Thus, apatite U/Pb and biotite K–Ar ages are typically lower than sphene U/Pb and hornblende K–Ar ages (e.g., see Mattinson, 1978 and 1982b). Such ages may reflect relatively straightforward postintrusive cooling, cooling after reheating during a strong subsequent thermal event (e.g., nearby emplacement of younger plutons), or partial opening of the isotopic systems in response to a mild subsequent event.

Within the Salinian composite terrane, cooling ages based on U/Pb determinations from apatite and sphene generally show consistent patterns. In general, of course, all the cooling ages are somewhat younger than the inferred emplacement ages for the same pluton; but the details are revealing. Cooling ages for the oldest plutons tend to show the greatest "offset" from emplacement ages, whereas cooling ages from the youngest plutons show the smallest offset from emplacement ages. For example, plutons with emplacement ages of 80 to 85 m.y. in the central sub-block and those with emplacement ages of 75 to 80 m.y. in the southeastern sub-block have cooling ages only about 1 to 4 m.y. younger than the emplacement ages (Mattinson, 1983, and in preparation). These younger plutons of the southern Gabilan Range and southeastern sub-block are among the youngest recognized in the Salinian block, and evidently: (1) were emplaced at relatively shallow levels and/or were rapidly unroofed, and (2) were not subjected to any subsequent significant thermal disturbances.

Plutons with emplacement ages of ca. 105 m.y. from the northern sub-block have cooling ages ca. 10 to 12 m.y. younger than the emplacement ages (Mattinson, 1978; James, 1984). This may reflect the present exposure of a deeper crustal level (and hence, slower cooling after emplacement) for these rocks compared to the younger plutons discussed above. Alternatively, the plutons from the northern sub-block may have been subjected to a subsequent thermal pulse strong enough to raise temperatures above ~ 500°C (932°F), the temperature necessary to reset the U/Pb system in sphene (Mattinson, 1978, 1982b) and the K–Ar system in hornblende (e.g., Harrison et al, 1979).

The charnokitic plutons of the western sub-block have emplacement ages of ca. 105 m.y. but cooling ages of only ca. 75 to 80 m.y. (Mattinson, 1978). Wiebe (1970) suggests a depth of 15 to 20 km (9–12 mi) for this terrane, based on mineral assemblages in metamorphic rocks associated with the plutons. Based on this, Mattinson (1978) suggested that the plutons may have remained at depth (and at high temperatures) for some 25 m.y. after their emplacement, followed by rapid uplift, erosion, and cooling. The subsequent determination of younger ages for plutons elsewhere in Salinia (but not recognized yet in the western sub-block) suggests an alternative that would alleviate the need for maintaining very high temperatures (>500°C [932°F]) for such a long time: Emplacement of plutons in the western sub-block at 80 m.y. ago, just below the present level of exposure, could have provided a thermal pulse sufficient to reset the sphene, apatite, and biotite ages. There is no direct evidence at the present time to choose between these possibilities, however.

Inherited Zircons

The presence of inherited zircons in granitic rocks is simultaneously a blessing and a curse. An inherited zircon component, especially if of mixed provenance and coupled with postintrusive Pb loss, can severely complicate the U/Pb systematics of a zircon suite. In some cases, only an approximation of the emplacement age can be made, or reliance must be placed on other methods (e.g., U/Pb on sphene or Rb/Sr whole-rock determinations). On the other hand, inherited zircons in granitic rocks can be profound indicators of petrogenesis and the age of the source area of the granitic magma, or at least of a substantial component of contamination of the magma. The very presence of an inherited component of zircon in an igneous rock indicates the involvement of preexisting continental crustal rocks at some stage in the petrogenesis of the magma, either via anatexis or significant contamination in the region of magma formation, transport, or emplacement. Moreover, in cases of relatively "well-behaved" U/Pb systematics, the age of the source material/contaminating component can be determined with some accuracy.

The majority of Salinian plutons contain inherited zircons. The use and evaluation of this characteristic as a tracer is necessarily somewhat qualitative, but regional trends within the block are clear. Inherited zircons are at a minimum in the northern sub-block, the western sub-block, and the northern part of the central sub-block.

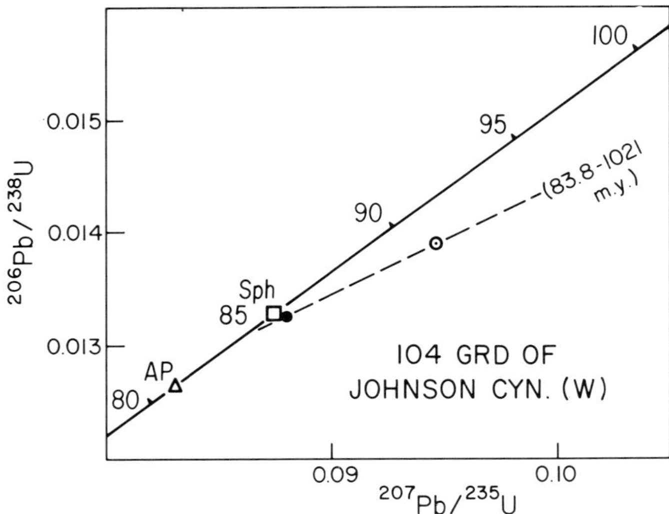

Figure 4—Small portion of a concordia plot showing inherited zircons in a Gabilan Range granodiorite. Closed circle = very fine-grained zircon; open circle = coarse-grained zircon: open square = sphene; open triangle = apatite. The fine-grained zircon contains only a very minor inherited component and closely approximates the emplacement age of the granite. The coarse fraction contains a significant inherited component, evidently of Precambrian age.

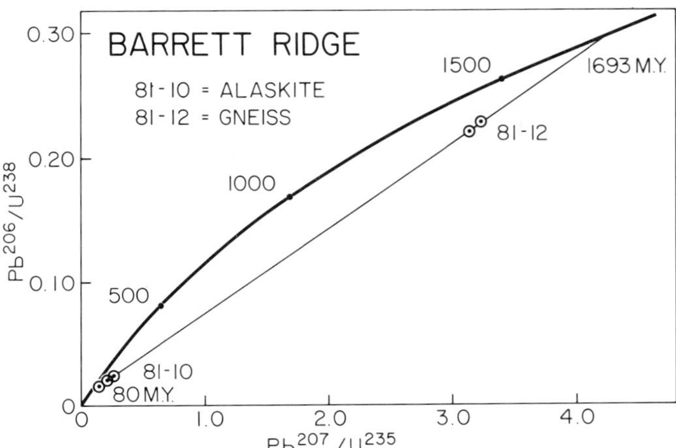

Figure 5—Condordia diagram for zircons from the Barrett Ridge gneiss and intruding alaskite pluton. The chord shown is fitted to all the data. The plutonic zircons alone still define a chord with similar upper and lower intercepts.

Plutons at Bodega Head, Cordell Bank, Montara Mountain, Ben Lomond, and the western Santa Lucia Range are largely devoid of measureable inherited zircons (Mattinson, 1978, and in preparation; James, 1984, and in preparation). Plutons from Point Reyes and the Farallon Islands contain small but conspicuous components of inherited zircon (Mattinson, 1978, and in preparation).

Plutons from most of the central block, including all of the Gabilan Range and a limited sampling of the La Panza Range, all include inherited zircons. The inherited component in these plutons ranges from minor to very prominent (Fig. 4). Where the inherited component is prominent, the upper intercept ages can be well determined. These ages range from early Paleozoic to Precambrian.

Plutons in the southeastern sub-block have zircon systematics that are dominated by inherited components. These populations are very well behaved, defining highly linear arrays on concordia diagrams. The upper intercepts are in the range of 1.7×10^9 years (Fig. 5), close to the age of the Precambrian Barrett Ridge gneiss exposed in the area (Mattinson, 1983, and in preparation).

Common (Initial) Pb Systematics

The isotopic composition of Pb incorporated in a pluton at the time of its formation (magma genesis) is a useful petrogenetic indicator. In general, magmas derived from depleted mantle sources (i.e., those that give rise to oceanic basalts) are relatively nonradiogenic, especially with regard to the $^{206}Pb/^{204}Pb$ ratios. Magmas derived from continental crustal rocks are generally highly radiogenic, whereas "typical" island arc sources yield intermediate isotopic values. Ancient high-grade metamorphic rocks yield still a fourth isotopic signature, with relatively high $^{207}Pb/^{204}Pb$ ratios for a given $^{206}Pb/^{204}Pb$ ratio. This "granulitic" signature is the product of early evolution in a high U/Pb ("Mu") environment, followed by a depletion of U during metamorphism.

Initial Pb isotopic compositions for Salinian plutons are all quite highly evolved and show systematic variations throughout the Salinian block. These are summarized in Figure 6. Plutons from the northwestern part of Salinia (northern block, western block, and northern central block) are the least radiogenic but still plot with the most radiogenic samples from the central Sierra Nevada (Chen and Tilton, personal communication), as shown on Figure 6. The most northwesterly Salinian samples (Cordell Bank, Farallon Islands, and Montara Mountain) tend to be the least radiogenic of the group.

Plutons from the central block, as represented by the Gabilan Range, are significantly more radiogenic than are the northwestern samples discussed above. This tends to parallel the greater degree of continental crustal involvement shown by inherited zircons in the Gabilan Range, as discussed earlier.

Finally, plutons from the southeastern sub-block appear to define a mixing relationship between "normal" Salinian Pb's and Pb characteristic of the Precambrian Barrett Ridge gneiss, which bears a "granulitic" signature. This is especially evident in the plot of $^{208}Pb/^{204}Pb$ versus $^{206}Pb/^{204}Pb$ (Fig. 6) and is consistent with the evidence for an important inherited Precambrian component in these plutons.

Metamorphic Rocks

Our isotopic work on Salinian metamorphic rocks is still rather limited. We have data for the Barrett Ridge gneiss of the southeastern sub-block (Mattinson, 1983, and in preparation), the schist of Sierra de Salinas of the central sub-block (Mattinson, 1982a, and in preparation), and the Santa Cruz schist of the central sub-block (James, 1984).

Zircon data from the Barrett Ridge gneiss (Fig. 5)

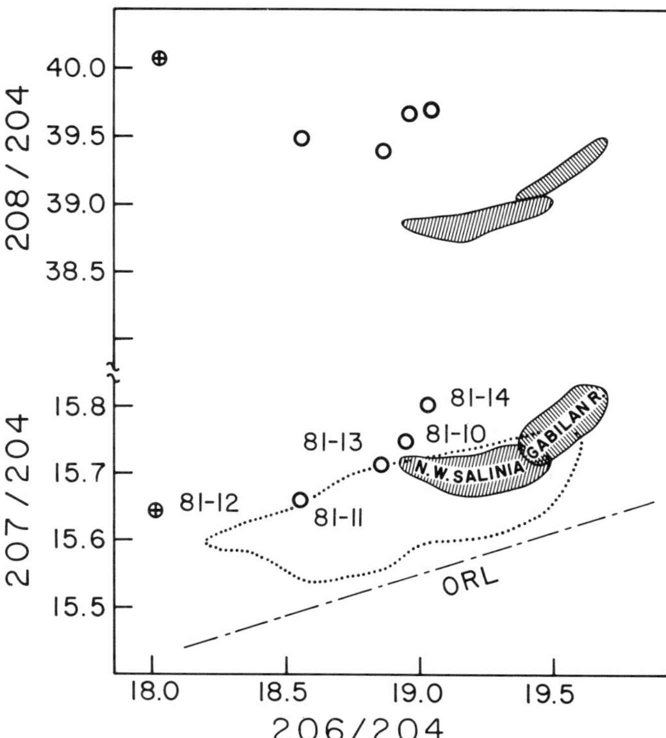

Figure 6—Initial (common) Pb data for Salinian plutons and metamorphic rocks. Sample 81-12 is the Barrett Ridge gneiss; 81-10 through 81-14 are intrusive rocks of southeastern Salinia, some of which intrude the gneiss. The fields for the northern and western sub-blocks and for the Gabilan Range are crosshatched. The large field indicated by the dotted line is for Sierran granodiorites (Chen and Tilton, personal communication) and the line marked "ORL" is a regression line for oceanic basalts (Tilton, personal communication).

clearly indicate a Precambrian age of ~ 1.7×10^9 years for the gneiss (Mattinson, 1983). This age probably approximates the age of the protolith of the gneiss. An early (Precambrian) high-grade metamorphism is suggested by the common Pb data (Fig. 6), which require an early period of Pb evolution in a high $^{238}U/^{204}Pb$ (Mu) environment, followed by U (but not Th) depletion and a long period of "arrested" Pb development in the resulting low Mu environment. Work in progress may help better quantify the timing of these events. The Late Cretaceous cross-cutting plutons produced a resetting of U/Pb apatite ages in the gneiss but had no obvious widespread textural or mineralogical effects on the gneiss.

Isotopic data from the schist of Sierra de Salinas in the southern Gabilan Range show both similarities and contrasts to the data for the Barrett Ridge gneiss discussed above. The zircon fractions from the schist are highly discordant and show some scatter on concordia with Precambrian Pb/Pb ages and middle to early Paleozoic U/Pb ages. Some of the zircons are rounded, suggesting a detrital origin. Common Pb from the schist shows no hint of the "granulitic" signature that characterizes the Barrett Ridge gneiss, instead closely resembling the Pb from the intruding Salinian plutons. U/Pb ages of apatite from the schist record the intrusion of the nearby Late Cretaceous plutons. These data only put broad limits on the age of the schist: The protolith was, at least in part, derived from a source that was rich in Precambrian zircons; and metamorphism was pre-Late Cretaceous based on the ages of the postmetamorphic intrusions. Additional isotopic experiments, planned and in progress, may place tighter constraints on the time of deposition of the protolith of the schist and the time of its metamorphism.

Preliminary isotopic data for zircons from the Santa Cruz schist of the Ben Lomond area are rather similar to the data for the schist of Sierra de Salinas. The zircons yield highly discordant ages, with Precambrian upper intercept ages and Mesozoic lower intercept ages that probably reflect, at least in part, resetting owing to Cretaceous intrusions in the area.

CONCLUSIONS

The patterns of age and isotopic data within the Salinian composite terrane suggest the following conclusions: (1) Mesozoic magmatic activity spanned some 40 m.y., from about 110 to 120 to 75 to 80 m.y. (2) The locus of igneous activity swept across the terrane from west to east over this interval, with the oldest plutons in the northern sub-block and the youngest in the Barrett Ridge slice. (3) The variations in common Pb systematics and inheritance of zircons parallel the age distribution, with the degree of inheritance and the "continental signature" of the common Pb ratios increasing towards the southeast. (4) These variations indicate that the Salinian magmatic arc evolved in a west-facing continental margin setting. (5) Direct measurements at Barrett Ridge and evidence from inherited zircons in many of the plutons point to basement ages in the range of ~ 1.7×10^9 years, compatible with known ages of autochthonous basement to the southeast, on the craton (e.g., Silver, 1971). (6) The age of the youngest Salinian plutons limits the age of juxtaposition of Salinia with rocks of the Franciscan and Great Valley sequences across the Sur–Nacimiento to no earlier than 75 to 80 m.y.

REFERENCES

Alvarez, W., et al, 1979, Franciscan limestone deposited at 17° south paleolatitude (Abs.): Geological Society of America Abstracts with Programs, v. 11, p. 66.

Armstrong, R. L., and J. Suppe, 1973, Potassium-argon geochronometry of Mesozoic igneous rocks in Nevada, Utah, and southern California: Geological Society of America Bulletin, v. 84, p. 1375–1392.

Blake, Jr., M. C., 1982, Map of geologic terranes of California: U.S. Geological Survey Open-File Report 82-593, scale 1:500,000.

———, and D. L. Jones, 1978, Allochthonous terranes in northern California?— a reinterpretation, *in* D. G. Howell and K. A. McDougall, eds., Mesozoic paleogeography of the western United States: Society of Economic Paleontologists and Mineralogists, Pacific Section, Pacific Coast Paleogeography Symposium 2, p. 397–400.

Bowen, O. E., and C. H. Gray, Jr., 1959, Geology and economic possibilities of the limestone and dolomite deposits of the northern Gabilan Range, California: California Division of Mines Special Report 56, 40 p.

Burchfiel, B. C., and G. A. Davis, 1981, Mojave Desert and environs, *in* W. G. Ernst, ed., The geotectonic development of California, Rubey, v. I: Englewood Cliffs, NJ, Prentice Hall, Inc., p. 217-252.

Champion, D., et al, 1980, Paleomagnetism of the Cretaceous Pigeon Point Formation and the inferred northward displacement of 2500 km for the Salinian block, California: EOS, American Geophysical Union Transactions, v. 61, p. 948.

______, 1984, Paleomagnetic and geologic data indicating 2500 km of northward displacement for the Salinian and related terranes, California: Journal of Geophysical Research, v. 89, p. 7736-7752.

Compton, R. R., 1960, Charnokitic rocks of Santa Lucia Range, California: American Journal of Science, v. 58, p. 609-636.

______, 1966, Granitic and metamorphic rocks of the Salinian block, California Coast Ranges: California Division of Mines and Geology Bulletin 190, p. 277-287.

Conrad, R. L., and T. E. Davis, 1977, Rb/Sr geochronology of cataclastic rocks of the Vincent trust, San Gabriel Mountains, southern California (Abs.): Geological Society of America Abstracts with Programs, v. 9, p. 403-404.

Crowell, J. C., 1962, Displacement along the San Andreas fault, California: Geological Society of America Special Paper 71, 61 p.

Dibblee, Jr., T. W., 1976, The Rinconada and related faults in the southern Coast Ranges, California, and their tectonic significance: U.S. Geological Survey Professional Paper 981, 55 p.

Dickinson, W. R., 1982, Offset of geologic terranes in coastal California by successive or alternating episodes of sinistral and dextral strike slip (Abs.): Geological Society of America Abstracts with Programs, v. 14, n. 4, p. 160.

______, 1983, Cretaceous sinistral strike slip along Nacimiento fault in coastal California: Bulletin of the American Association of Petroleum Geologists, v. 67, n. 4, p. 624-645.

Ehlig, P. L., 1968, Causes of distribution of Pelona, Rand, and Orocopia Schist along the San Andreas and Garlock faults, *in* W. R. Dickinson and A. Grantz, eds., Proceedings of conference on geologic problems of San Andreas fault system: Stanford University Publications, Geological Sciences, v. 11, p. 294-305.

______, and S. E. Joseph, 1977, Polka dot granite and correlation of La Panza quartz monsonite with Cretaceous batholithic rocks north of Salton Trough, *in* D. G. Howell, et al, eds., Cretaceous geology of the California Coast ranges west of the San Andreas Fault: Society of Economic Paleontologists and Mineralogists, Pacific Section, Pacific Coast Paleogeography Field Guide 2, p. 91-96.

______, et al, 1975, Tectonic implications of the cooling ages of the Pelona Schist (Abs.): Geological Society of America Abstracts with Programs, v. 7, p. 315.

Garfunkel, Z., 1973, History of the San Andreas fault as a plate boundary: Geological Society of America Bulletin, v. 84, p. 2409-2430.

Gastil, R. G., et al, 1972, The reconstruction of Mesozoic California: 24th International Geologic Congress, Montreal, 1972, Proceedings Section 3, p. 217-229.

Graham, S. A., 1978, Role of the Salinian block in the evolution of the San Andreas fault system: Bulletin of the American Association of Petroleum Geologists, v. 62, p. 2214-2231.

______, and W. R. Dickinson, 1978, Evidence for 115 kilometers of right slip on the San Gregorio-Hosgri fault trend: Science, v. 199, p. 179-181.

Hall, Jr., C. A., 1975, San Simeon-Hosgri fault system, coastal California: economic and environmental implications: Science, v. 190, p. 1291-1294.

Hamilton, W., 1978, Mesozoic tectonics of the western United States, *in* D. G. Howell and K. S. McDougall, eds., Mesozoic paleogeography of the western United States: Society of Economic Paleontologists and Mineralogists, Pacific Section, Pacific Coast Paleogeography Symposium 2, p. 33-70.

Harrison, T. M., et al, 1979, Geochronology and thermal history of the Coast Plutonic Complex, near Prince Rupert, B. C.: Canadian Journal of Earth Sciences, v. 16, p. 400-410.

Haxel, J., and J. Dillon, 1978, The Pelona Orocopia Schist and the Vincent Chocolate Mountain thrust system, southern California, *in* D. G. Howell, and K. A. McDougall, eds., Mesozoic paleogeography of the western United States: Society of Economic Paleontologists and Mineralogists, Pacific Section, Pacific Coast Paleogeography Symposium 2, p. 453-469.

Hill, M. L., and T. W. Dibblee, Jr., 1953, San Andreas, Garlock and Big Pine faults, California—a study of the character history and tectonic significance of their displacements: Geological Society of America Bulletin, v. 64, p. 443-458.

Howell, D. G., 1975, Hypothesis suggesting 700 km of right slip in California along northwest-oriented faults: Geology, v. 3, n. 2, p. 81-84.

James, E. W., 1984, U/Pb ages of plutonism and metamorphism in part of the Salinian block, Santa Cruz and San Mateo Counties, California (Abs.): Geological Society of America Abstracts with Programs, v. 16, p. 291.

Johnson, J. D., W. R. Normark, 1974, Neogene tectonic evolution of the Salinian block, west-central California: Geology, v. 2, p. 11-14.

Jones, D. L., et al, 1977, Wrangellia—a displaced terrane in northwest North America: Canadian Journal of Earth Sciences, v. 14, p. 2565-2577.

Joseph, S. E., and P. L. Ehlig, 1981, Isotopic correlation of La Panza Range granitic rocks, southern Salinian block, with similar rocks in the central and eastern Transverse Ranges: Geological Society of America Abstracts with Programs, v. 13, p. 63.

Kanter, L. R., and D. C. Engebretson, 1982, Preliminary paleomagnetic results from the Gualala block,

———, et al, 1982, Paleomagnetism of continental and oceanic terranes of coastal California: EOS, American Geophysical Union Transactions, v. 63, n. 45, p. 915.

Kistler, R. W., 1978, Mesozoic paleogeography of California: a viewpoint from isotope geology: *in* D. G. Howell and K. A. McDougall, eds., Mesozoic paleogeography of the western United States: Society of Economic Paleontologists and Mineralogists, Pacific Section, Pacific Coast Paleogeography Symposium 2, p. 75–84.

———, and Z. E. Peterman, 1978, Reconstruction of crustal blocks of California on the basis of initial strontium isotopic compositions of Mesozoic granitic rocks: U.S. Geological Survey Professional Paper 1071, 17 p.

Mattinson, J. M., 1978, Age, origin, and thermal histories of some plutonic rocks from the Salinian block of California: Contributions to Mineralogy and Petrology, v. 67, p. 233–245.

———, 1982a, Granitic rocks of the Gabilan Range: U-Pb isotopic systematics and implications for age and origin (Abs.): Geological Society of America Abstracts with Programs, v. 14, n. 4, p. 184.

———, 1982b, U-Pb "blocking temperatures" and Pb loss characteristics in young zircon, sphene, and apatite (Abs.): Geological Society of America Abstracts with Programs, v. 14, n. 7, p. 558.

———, 1983, Basement rocks of the southeastern Salinian block, California: U-Pb isotopic relationships (Abs.): Geological Society of America Abstracts with Programs, v. 15, n. 5, p. 414.

McWilliams, M. O., and D. G. Howell, 1982, Exotic terranes of western California: Nature, v. 297, p. 215–217.

Mukasa, S. B., et al, 1984, A Late Jurassic minimum age for the Pelona-Orocopia schist protolith, southern California (Abs.): Geological Society of America Abstracts with Programs, v. 16, p. 323.

Nilsen, T. H., 1978, Late Cretaceous geology of California and the problem of the proto-San Andreas fault, *in* D. G. Howell and K. A. McDougall, eds., Mesozoic paleogeography of the western United States: Society of Economic Paleontologists and Mineralogists, Pacific Section, Pacific Coast Paleogeography Symposium 2, p. 559–573.

Page, B. M., 1970, Sur-Nacimiento fault zone of California: Continental margin tectonics: Geological Society of America Bulletin, v. 81, p. 667–690.

———, 1981, The southern Coast Ranges, *in* W. G. Ernst, ed., The geotectonic development of California, Rubey v. I: Englewood Cliffs, NJ, Prentice Hall, Inc., p. 329–417.

———, 1982, Migration of Salinian composite block, California, and disappearance of fragments: American Journal of Science, v. 282, p. 1694–1734.

Ross, D. C., 1970, Quartz gabbro and anorthositic gabbro: Markers of offset along San Andreas fault in the California Coast Ranges: Geological Society of American Bulletin, v. 81, p. 3647–3662.

———, 1972, Petrographic and chemical reconnaissance study of some granitic and gneissic rocks near the San Andreas fault from Bodega Head to Cajon Pass, California: U.S. Geological Survey Professional Paper 698, 92 p.

———, 1976, Metagraywacke in the Salinian block, central Coast Ranges, California—and a possible correlative across the San Andreas fault: U.S. Geological Survey Journal of Research, v. 4, n. 6, p. 683–696.

———, 1977, Pre-intrusive metasedimentary rocks of the Salinian block, California—a tectonic dilemma: *in* J. H. Stewart, et al, eds., Paleozoic paleogeography of the western United States: Society of Economic Paleontologists and Mineralogists, Pacific Section, Pacific Coast Paleogeography Symposium 1, p. 371–380.

———, 1978, The Salinian block—a Mesozoic granitic orphan in the California Coast Ranges: *in* D. G. Howell and K. A. McDougall, eds., Mesozoic paleogeography of the western United States: Society of Economic Paleontologists and Mineralogists, Pacific Section, Pacific Coast Paleogeography Symposium 2, p. 509–522.

———, in press, Basement rocks of the Salinian block and southernmost Sierra Nevada and possible correlations across the San Andreas, San Gregorio-Hosgri, and Rinconada-Reliz-King City fault zones: U.S. Geological Survey Professional Paper 1317.

———, and E. E. Brabb, 1973, Petrography and structural relations of granitic basement rocks in the Monterey Bay area, California: U.S. Geological Survey Journal of Research, v. 1, p. 273–282.

Silver, E. A., 1974, Structural interpretation from free-air gravity on the California continental margin, 35°–40°N (Abs.): Geological Society of America Abstracts with Programs, v. 6, p. 253.

———, et al, 1971, Tectonic development of the continental margin off central California: Geological Society of Sacramento Annual Field Trip Guidebook, p. 1–10.

Silver, L. T., 1971, Problems of crystalline rocks of the Transverse Ranges (Abs.): Geological Society of America Abstracts with Programs, v. 3, n. 2, p. 193–194.

———, 1982, Evidence and a model for west-directed Early to mid-Cenozoic basement overthrusting in southern California (Abs.): Geological Society of America Abstracts with Programs, v. 14, n. 7, p. 617.

———, 1983, Paleogene overthrusting in the tectonic evolution of the Transverse Ranges, Mojave and Salinian regions, California (Abs.): Geological Society of America Abstracts with Programs, v. 15, n. 5, p. 438.

———, et al, 1984, Some observations on the tectonic history of the Rand Mountains, Mojave Desert, California (Abs.): Geological Society of America Abstract with Programs, v. 16, p. 333.

Suppe, J. E., 1970, Offset of late Mesozoic basement terrains by the San Andreas fault system: Geological Society of America Bulletin, v. 81, p. 3253–3258.

Taliaferro, N. L., 1943, Geologic history and structure of the central Coast Ranges, California: California Division of Mines and Geology Bulletin 118, p.

119–163.
Trask, P. D., 1926, Geology of the Point Sur quadrangle, California: California University, Department of Geological Science Bulletin, v. 16, n. 6, p. 119–186.
Vedder, J. G., et al, 1982, Stratigraphy, sedimentation, and tetonic accretion of exotic terranes, southern Coast Ranges, California, *in* J. S. Watkins and C. L. Drake, eds., Studies in continental margin geology: American Association of Petroleum Geologists Memoir 34, p. 471–496.
Wiebe, R. A., 1970, Pre-Cenozoic tectonic history of the Salinian block, western California: Geological Society of America Bulletin, v. 81, p. 1837–1842.
Wilson, J. T., 1965, A new class of faults and their bearing on continental drift: Nature, v. 207, p. 343–347.
Yeats, R. S., 1968, Rifting and rafting in the southern California borderland, *in* W. R. Dickinson and A. Grantz, eds., Proceedings of conference on geologic problems of San Andreas fault system: Stanford University Publications, Geological Sciences, v. 11, p. 307–322.

Modeling the Motion Histories of the Point Arena and Central Salinia Terranes

L. R. Kanter*
M. Debiche
Stanford University
Stanford, California

Paleomagnetic data from several formations of the Point Arena and central Salinia terranes of coastal California indicate that both terranes originated far to the south of their present positions. The thick sequences of arkosic clastic rocks on both terranes indicate that they probably remained adjacent to North America during most of their travels. Geologic and paleomagnetic constraints are used to model motion histories for each terrane that are compatible with the relative motions between North America and Pacific Basin plates calculated by Engebretson (1982). The terranes moved poleward (northward) with the Pacific plate during the Neogene, and they were fixed with respect to North America during the Middle Tertiary. Pre-Neogene poleward motion is modeled as a translation along the North American continental margin driven by oblique subduction of the Farallon plate beneath North America. Restoration of the Mexico–Central America continental margin to a more northerly trend is necessary to achieve the rates of northward translation required by the paleomagnetic data.

Although the two terranes were moved by the same plate interactions, their detailed motion histories are somewhat different. The paleomagnetic data indicate that the Point Arena terrane, presently located north of Salinia, originated south of central Salinia and passed outboard of it in the Miocene.

INTRODUCTION

The Salinian terrane is an elongate region with granitic and prebatholithic metamorphic rocks west of the San Andreas fault in central California. It is in fault contact with Franciscan rocks to both east and west and is conspicuously out of place with respect to the California batholithic belt (Fig. 1).

The plutonic compositions, initial strontium ratios, and metasedimentary rocks of Salinia indicate that it is a disrupted fragment of a continental arc, the Cordilleran arc being the closest and most likely source. The traditional view that Salinian granitic rocks originated in the California batholith belt between the Sierra Nevada and the Peninsular Ranges has recently been expanded upon by Dickinson (1983). Other workers, however, have described geologic relations that indicate that the rocks of Salinia and the Sierra Nevada are not correlative (Ross, 1978; Page, 1982; Vedder et al, 1983).

The Point Arena terrane is located just 22 km (14 mi) northwest of the northernmost granitic outcrop of Salinia (Blake et al, 1982). It is separated from Franciscan rocks to the east by the San Andreas fault, a structural position analogous to that of Salinia. Although the Point Arena and Salinia terranes have several similar formations, a few major geologic differences cause them to be defined as separate but probably related terranes.

Paleomagnetic data are now available from three rock units in each of the two terranes (Kanter, 1983; Champion et al, in press). Both terranes appear to have originated far to the south of their present positions and have paleomagnetic records of progressive poleward motion through time. The paleolatitude curves for the terranes, in combination with geologic data, provide constraints for modeling terrane motions. Neogene poleward motion is related to the San Andreas transform system. The paleomagnetic data from both terranes also require significant amounts of pre-Neogene poleward motion.

It is likely that both terranes remained adjacent to North America during pre-Neogene motion, judging from their thick, terrigenous-sourced Cretaceous–Paleogene sedimentary sequences. Pre-Neogene motion could have occurred in two ways: as strike-slip faulting slightly inboard of a compressional plate boundary (Nilsen and Clarke, 1975) or along a transform boundary between the North America and Kula plates (Atwater, 1970).

Poleward transport of the terranes by the Kula plate would require major transform offset of the eastern part of the Kula–Farallon ridge to bring the Kula–Farallon–North America triple junction far enough south to pick up the Point Arena and Salinia terranes. If this could be accomplished, the calculated rates would be ample to produce the transport needed (Engebretson, 1982; Page

*Present address: ARCO Exploration Co., P.O. Box 2819, Dallas, Texas

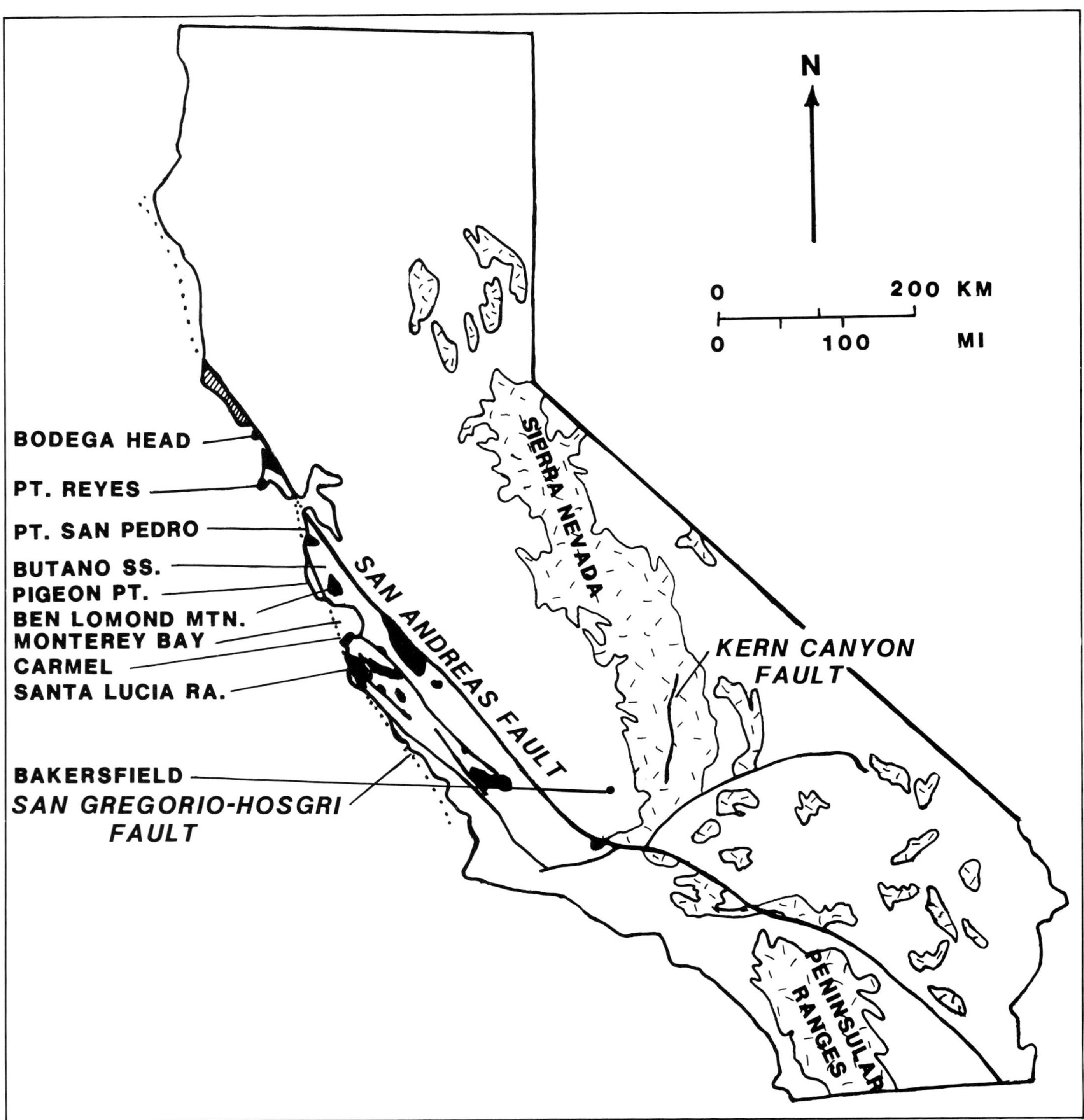

Figure 1—Map of central California showing granitic rocks of Salinia (black), Sierra Nevada batholith, and the Point Arena terrane.

and Engebretson, 1984). This is the mechanism that Page (1982) called on for transport of Salinia.

The Farallon plate lay west of North America throughout most of the Mesozoic and Early Tertiary and was probably being subducted obliquely beneath North America for most of this time (Engebretson, 1982). Oblique dextral subduction of the Farallon plate is a likely driving force for poleward transport of the Salinian and Point Arena terranes along the North American continental margin. Oblique subduction can be resolved into two components, one perpendicular and one parallel to the continental margin (Fig. 2). The latter component can provide a mechanism for poleward transport as is the case today in the Sumatra region (Fitch, 1972; Page et al, 1979).

Relative motion between the Farallon and North America plates peaked in Paleocene time at about 160 km/m.y. (100 mi/m.y.), the direction of relative motion being about N 55° E (Engebretson, 1982). The rate of dextral translation parallel to the continental margin is extremely sensitive to the azimuth of the Paleocene

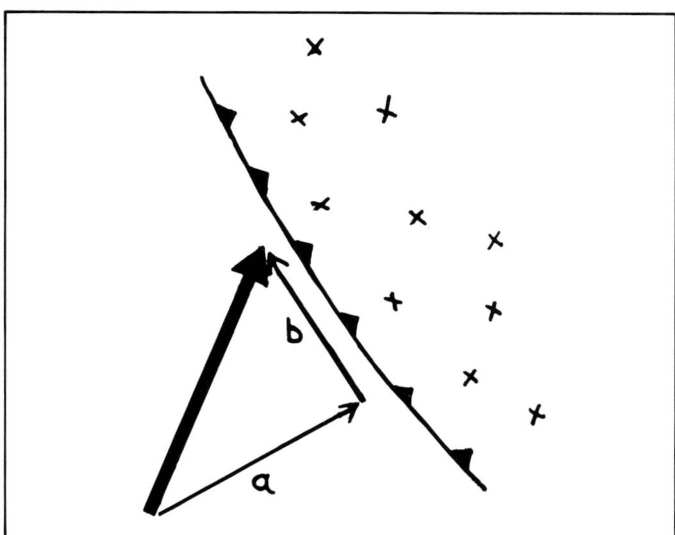

Figure 2—Vector diagram showing resolution of oblique subduction into components perpendicular (a) and parallel (b) to the continental margin. Trench shown by barbs.

continental margin. If it was the same as it is today (about N 25° W), the maximum dextral motion at a site with present-day coordinates 19.5° N, 252° E was only about 20 km/m.y. (12 mi/m.y.) from 48 to 56 Ma.

Some evidence exists for counterclockwise motion of parts of Mexico and Central America since the Early Tertiary (Dickinson and Coney, 1980; Urrutia-Fucugauchi, in press), suggesting that the orientation of the continental margin of southern North America may have been more northerly in the Late Cretaceous and Early Tertiary. The maximum dextral translational velocity produced by Farallon–North America relative motion along a due north-trending continental margin is 86 km/m.y. (53 mi/m.y.) during the same 48 to 56 Ma period, more than four times greater than that for a margin with the same trend as the present one.

Paleomagnetic results from terranes are compared to the North America apparent polar wander path of Harrison and Lindh (1982) except during the lower-middle Eocene, when the pole of Diehl et al (1983) is used. All confidence limits are standard errors unless otherwise noted. The timescale used is that of Harland and others (1983).

GEOLOGY OF THE TERRANES

The oldest rocks of Salinia are undated metasedimentary rocks of continental affinity (Compton, 1966; Ross, 1977). The plutonic basement consists of granodiorite, quartz monzonite, tonalite, and lesser amounts of rocks with more mafic and more felsic compositions. Pluton intrusion ages fall mostly in the range 100 to 110 Ma but are as young as 82 Ma in the Gabilan Range (Mattinson, 1982). Potassium-argon ages (75 to 90 Ma) reflect a major episode of uplift and cooling.

The plutonic rocks are overlain by arkosic uppermost Cretaceous and Paleogene shallow- to deep-water clastic strata whose facies patterns, geometries, and paleocurrents indicate deposition in a continental borderland setting like that offshore southern California today (Nilsen and Clarke, 1975; Howell and Vedder, 1978). These are overlain by Neogene sedimentary and volcanic rocks.

The stratigraphic sequence of the Point Arena terrane includes uppermost Cretaceous and Paleocene and Eocene turbidites broadly similar to those of Salinia, overlain by Neogene clastic rocks and basalt (Wentworth, 1966). No granitic basement is exposed in the terrane, however, and the structural base of the section is formed by undated spilitic pillow basalts unlike anything seen in Salinia. This is one of the main reasons for differentiating the Point Arena terrane from Salinia (Fig. 3).

Salinia is a composite terrane internally disrupted by Neogene faults (Page, 1982), at least one of which (San Gregorio–Hosgri) has more than 100 km (62 mi) of right-lateral offset (Graham and Dickinson, 1978). The motion histories of various parts of Salinia are therefore not necessarily the same. The Butano Sandstone and the rocks at Point San Pedro lie on a portion of Salinia that is close to the San Andreas fault, inboard of the known intraterrane Neogene faults, and will be called here central Salinia. The Pigeon Point Formation lies outboard of the San Gregorio–Hosgri fault and the intra-Salinian faults of the Santa Lucia Range region and thus has experienced a Neogene history slightly different from that of central Salinia. By removing the effects of Neogene intraterrane motion, data from the Pigeon Point Formation can be used with that from the Butano Sandstone and the Point San Pedro rocks to determine the movement history of central Salinia.

The Point Arena terrane, defined as a terrane distinct from Salinia on the basis of geologic differences, is a more outboard terrane. It lies west of the San Andreas/San Gregorio–Hosgri fault system, thereby suggesting a Neogene history like that of the Pigeon Point Formation. Its paleomagnetic data indicate a pre-Neogene history different from that of central Salinia.

TERRANE DISPLACEMENTS

Paleolatitudes and paleolatitude discrepancies for central Salinia, Point Arena, and the Sierra Nevada are listed in Table 1 and presented graphically in Figure 4. Each data point is represented by a rectangle whose height is the displacement range and whose width is the age range. Five points from the San Andreas fault time-displacement curve, shown as small, black rectangles, indicate displacement across the San Andreas fault only (Graham, 1976). The larger, open rectangles represent paleomagnetic data from the central Salinia and Point Arena terranes. Where age limits are uncertain, as for the German Rancho Formation, or unknown, as for the Point Arena spilites, they are dotted. The paleomagnetic displacements are total displacements relative to the North American craton in a southerly (antipoleward) direction going backwards in time, and the limits are ± one standard error. Several of the rock units have experienced Neogene motion on the San Gregorio–Hosgri fault and faults within the Salinian terrane, the total poleward displacement being about 180 km (112 mi) (Graham, 1976; Graham and Dickinson, 1978; Kanter, 1983). For these units the displacement limits after

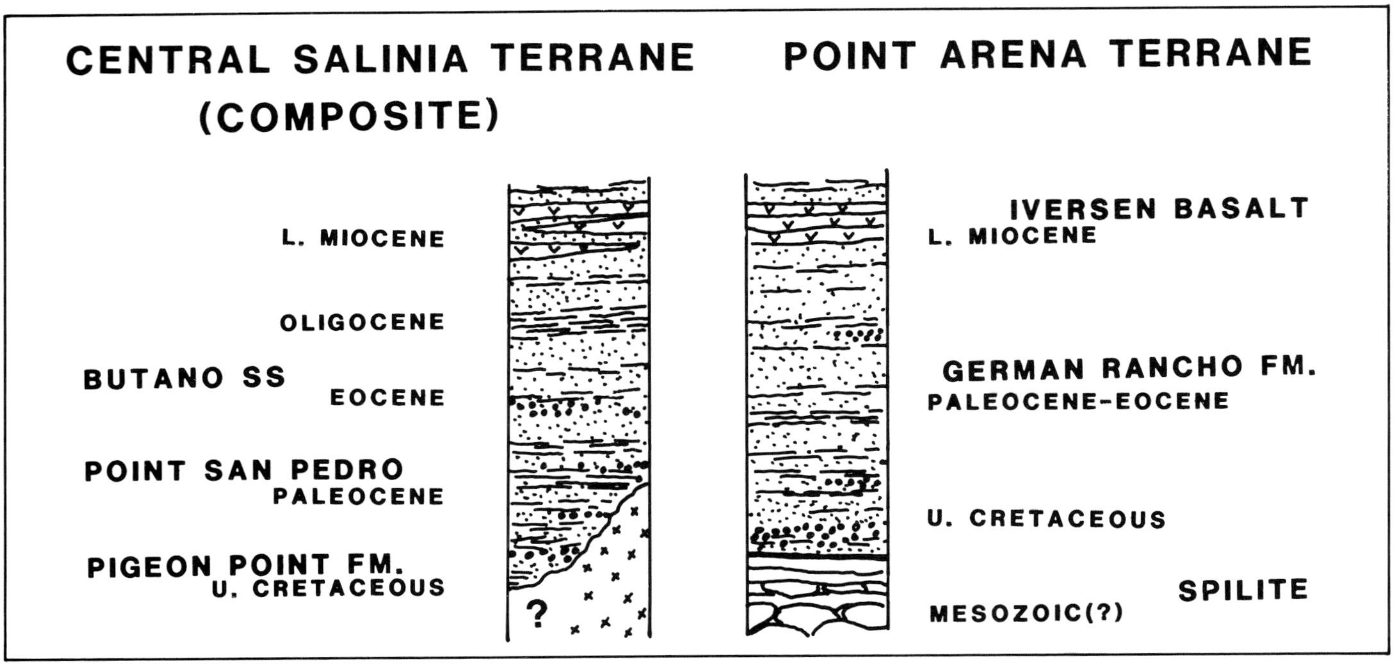

Figure 3—Schematic stratigraphic sections for central Salinia and Point Arena terranes. Rock units with measured paleolatitudes are labeled. Central Salinia section is a composite.

Table 1

Unit	Age	λ_0^a	λ_x^b	Paleo. Discrep.	Displacement Km	Displacement Mi
Butano Sandstone	42–55	34.6 ± 2.4	39.9 ± 1.2	5.3 ± 2.7	590 ± 300	367 ± 186
Point San Pedro[c]	65–66	25.0 ± 2.0	42.9 ± 1.9	17.9 ± 2.8	1,990 ± 310	1,237 ± 193
Pigeon Point Formation[c]	83–65	20.8 ± 2.7	45.8 ± 2.3	25.0 ± 3.5	2,780 ± 390	1,728 ± 242
Iversen Basalt	21–25	29.3 ± 7.4	38.8 ± 1.5	9.5 ± 7.6	1,050 ± 840	652 ± 522
German Rancho Formation	38–65	25.0 ± 1.4	41.4 ± 1.2	16.3 ± 1.8	1,810 ± 200	1,125 ± 124
Spilite[d]	80–160	5.1 ± 1.7	42.0 ± 2.3	36.9 ± 2.9	4,100 ± 320	2,548 ± 199
Central Sierra Nevada[e]	96–83	42.3 ± 3.8	49.8 ± 3.3	7.5 ± 4.9	830 ± 540	516 ± 336
Southern Sierra Nevada[f]	80–86	36.2 ± 3.2	46.8 ± 3.0	10.6 ± 4.4	1,180 ± 490	733 ± 304

[a]λ_0 is measured paleolatitude.
[b]λ_x is expected paleolatitude from cratonal pole.
[c]Data from Champion et al, in press.
[d]λ assuming an age of 130 Ma.
[e]Data from Frei et al, in press.
[f]Data from Kanter and McWilliams, 1982.
Other data from Kanter, 1983.

Table 1—Paleolatitude discrepancy and displacement with respect to the North American craton.

this motion is removed are shown by dashed lines.

Figure 4 shows that all of the units studied paleomagnetically have undergone more motion than the approximately 300 km (186 mi) of slip documented for the San Andreas fault. The Eocene Butano Sandstone of central Salinia has been offset only about 300 km (186 mi) from the Point of Rocks Sandstone east of the San Andreas fault (Clarke and Nilsen, 1973), but its paleolatitude implies a total poleward displacement of 590 ± 300 km (367 ± 186 mi) since Eocene time (Kanter, 1983). Paleomagnetic studies of the central Sierra Nevada (Frei et al, 1984) and southernmost Sierra Nevada (Kanter and McWilliams, 1982) also show poleward displacement (Table 1), implying that all of central California may have undergone northward motion by Neogene strike-slip along faults located east of the Sierra Nevada. According to some estimates, a total of about 100 km (62 mi) of Neogene northward displacement has occurred on a few faults in the

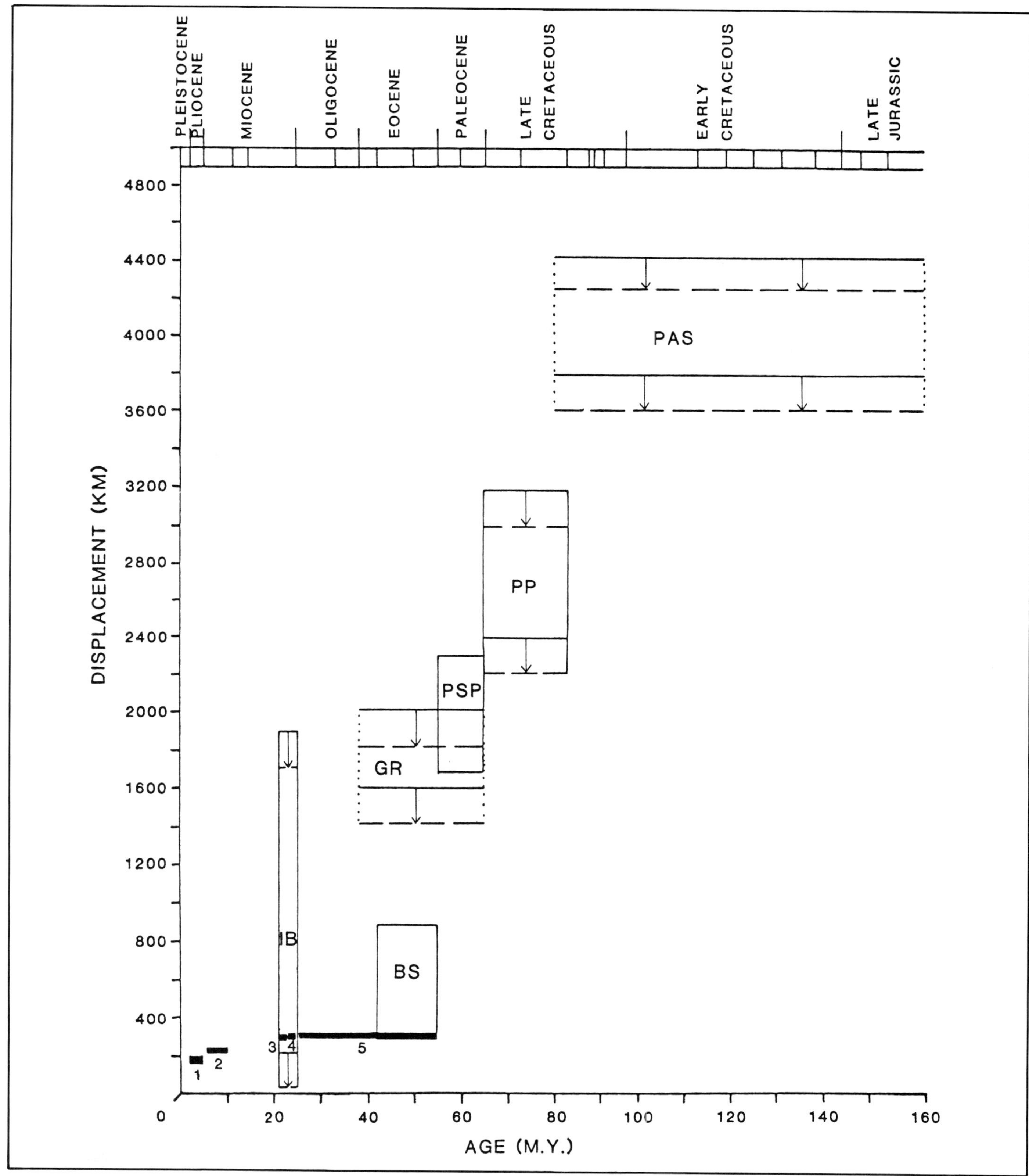

Figure 4—Time-displacement plot for central Salinia and the Point Arena terrane. Black rectangles are from the San Andreas time-displacement curve. Points 1 through 4 are points 1, 2, 5, and 6 from Graham (1976). Point 5 is from Nilsen and Link (1975) and Nilsen (personal communication). Open rectangles show age range and poleward displacement limits for rock units studied paleomagnetically. Central Salinia: BS = Butano Sandstone; PP = Pigeon Point Formation; PSP = rocks at Point San Pedro. Point Arena terrane: GR = German Rancho Formation; IB = Iversen Basalt; PAS = spilite. For open rectangles displacement is measured relative to the North American craton. For rock units that have been translated along the San Gregorio-Hosgri and other intra-Salinian faults, dashed lines with arrows show effect of removal of 180 km (112 mi) of Neogene poleward displacement.

southwestern Great Basin (Albers, 1967; Stewart et al, 1968; 1970; Stewart, 1983). In addition the Kern Canyon fault zone within the Sierra Nevada batholith runs nearly due north and has about 15 km (9 mi) of right-lateral separation (Ross, personal communication). The paleomagnetic result from the Butano Sandstone suggests that the amount of post-Eocene displacement along faults of this type is about 300 km (186 mi).

The total poleward displacement indicated paleomagnetically for the Paleocene rocks at Point San Pedro (Champion et al, in press) is 1,990 ± 310 km (1,237 ± 193 mi). Salinia moved 1,400 ± 430 km (870 ± 267 mi) toward the Early Tertiary pole between deposition of the Point San Pedro rocks and the Butano Sandstone.

The Upper Cretaceous Pigeon Formation was deposited nearly 2,800 km (1,740 mi) to the south. After accounting for its additional 180 km (112 mi) of Neogene poleward motion along intra-Salinian faults, we can say that central Salinia was about 2,600 ± 390 km (1,616 ± 242 mi) south of its present position in the uppermost Cretaceous. The terrane experienced 2,010 ± 400 km (1,249 ± 249 mi) of motion along a pre-Eocene "proto-San Andreas" zone.

The Point Arena terrane, outboard of the San Gregorio–Hosgri fault, has a Neogene history like that of the Pigeon Point Formation and slightly different from that of central Salinia. It was offshore of southern California, west of central Salinia, and probably approximately due west of Bakersfield by earliest Miocene, when the Iversen Basalt and similar rocks in the Mindego Formation of central Salinia were erupted (Table 1). This tie is based on chemical and stratigraphic similarities (Graham and Peabody, 1981), palinspastic reconstruction, and the paleolatitude determined for the Iversen Basalt (Kanter, 1983).

The locus of deposition for the lower to middle Eocene and possibly Paleocene rocks of the German Rancho Formation was 1,810 ± 200 km (1,125 ± 124 mi) south of the present position and 1,220 ± 360 km (758 ± 224 mi) south of the site of partly coeval Butano Sandstone deposition. Thus, significant pre-Neogene poleward displacement appears to be indicated for the German Rancho Formation, in contrast to the Butano Sandstone of central Salinia, which experienced little or no pre-Neogene displacement.

The undated spilites at the structural base of the Point Arena terrane formed in equatorial latitudes. If one assumes an age of 130 Ma, a paleolatitude discrepancy of 4,100 ± 320 km (2,548 ± 199 mi) is inferred. The calculated paleolatitude discrepancy is large regardless of the age assumed for these rocks.

REASONABLE MOTION HISTORIES FOR THE TERRANES

If the paleomagnetic data from the Butano Sandstone are ignored, it is possible to fit a single straight line through all the pre-Neogene rock units in Figure 4, suggesting that no displacement has occurred between the Point Arena and central Salinia terranes. However, careful examination of the available geologic, chronologic, and paleomagnetic data indicates that the two terranes have had different motion histories.

The basements of the terranes are geologically distinct and their Paleogene sedimentary sequences, while broadly similar, cannot be correlated. The paleomagnetic data from the German Rancho Formation of the Point Arena terrane suggest it was deposited in the Early Tertiary significantly farther south than was the Butano Sandstone of central Salinia. The geologic, palinspastic, and admittedly loose paleomagnetic correlations of the lower Miocene Iversen Basalt with rocks of the Mindego Formation of central Salinia suggests proximity of the two terranes in the early Neogene, however.

The age of the German Rancho Formation is crucial and somewhat uncertain. Wentworth (1966) reports Paleocene and Eocene fossils from the formation (Fig. 4); however, Ingle (1983, personal communication) reports that forams listed by Wentworth from the two localities where 11 of the 15 German Rancho paleomagnetic sites were located are early to middle Eocene and possibly latest Paleocene in age, as are most of the fossil sites on Wentworth's map. Ages are not available for the other German Rancho paleomagnetic sampling localities. Thus, while the rocks sampled could possibly have an age range as wide as that indicated on Figure 4 (55 to 38 Ma), it appears more likely that most, if not all, of the paleomagnetic samples have ages in the range latest Paleocene to middle Eocene (58 to 42 Ma), which is little different from the age range of the Butano Sandstone (55 to 42 Ma).

Even though the mean age of the German Rancho paleomagnetic samples may be somewhat older than that of the samples from the Butano Sandstone, the Early Tertiary paleomagnetic results from both terranes cannot be fit with a straight line that is consistent with our best estimates of the ages of the units that were sampled.

For all of these reasons it seems likely that some amount of motion of the Point Arena terrane relative to central Salinia occurred after the middle Eocene and before the formation of the Iversen Basalt, although the amount of motion need not be large because considerable leeway is provided by the error limits.

The relative motion probably occurred along a fault near the continental margin. In regions where strike-slip motion accompanies oblique subduction at present, the strike-slip appears to be localized along one main zone near the trench. It was along such a fault that the Late Cretaceous–Eocene motion of both terranes occurred (Fig. 5a). The relative motion between Point Arena and central Salinia probably occurred by activation of a new splay of this fault that passed outboard of central Salinia about 50 Ma, abandoning its more easterly branch in a manner similar to the abandonment of the Pilarcitos fault in the Neogene San Andreas fault system (Fig. 5b).

The branch of the ancient fault that passes between the Point Arena and central Salinian terranes should be present on the continental shelf between the southern end of the Point Arena terrane and the northern end of central Salinia. The granitic exposures at Bodega Head and Point Reyes lie between these two points. There are no paleomagnetic data available from the tiny exposure at

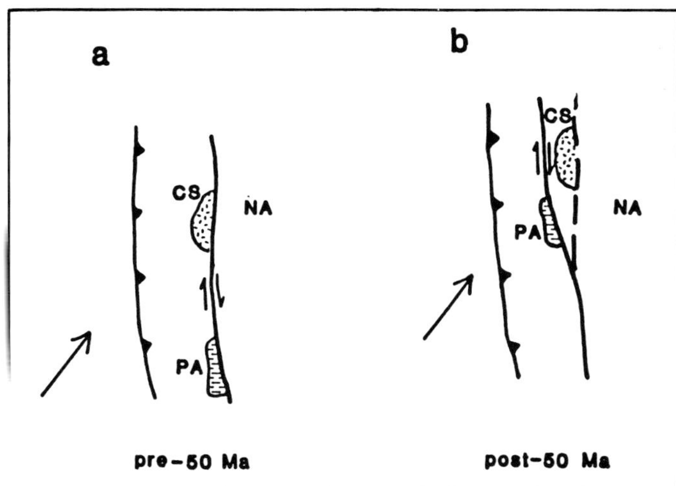

Figure 5—Sketch of hypothesized faults active during pre-Neogene transport of terranes. Barbs show trench. NA = North America; CS = central Salinia; PA = Point Arena terrane. (**a**) Pre-50 Ma. (**b**) Post-50 Ma. Dashed line shows inactive branch of fault.

Bodega Head, and the granitic rocks at Point Reyes appear to be mostly remagnetized, so it is not possible to determine whether these regions share the Paleogene motion history of the Point Arena terrane. Several workers, however, have noted similarities between the Point Reyes sedimentary section and sections near Carmel and along the south flank of Ben Lomond Mountain, which may tie Point Reyes to the Monterey Bay region in the Tertiary (Clark, 1966; Galloway, 1977). This would argue for location of the fault between Point Reyes and the Point Arena terrane. This hypothesis is also consistent with the different basement of the Point Arena terrane. The trace of the ancient fault may be in part reactivated; it is almost certainly disrupted by Neogene faults on the continental shelf.

Geologically reasonable estimates of cumulative poleward displacements were determined working backward from the present using geologic and paleomagnetic data as guides. The displacements and corresponding paleomagnetic data are presented in Table 2. Displacement relative to a fixed North American craton is considered for three regions: the Point Arena terrane, central Salinia, and the Sierra Nevada–Great Valley block. Point Arena and central Salinia are separated by the San Gregorio–Hosgri and intra-Salinian faults as well as by the hypothetical ancient fault discussed previously. Central Salinia and the Sierra Nevada–Great Valley block are separated by the San Andreas fault and the more easterly branch of the ancient fault, and Sierra Nevada–Great Valley is separated from the craton by faults located east of the Sierra Nevada. Figure 6 is a cartoon showing the terrane motions through time.

Paleomagnetic data for the Butano Sandstone of central Salinia indicate a residual post-Eocene poleward displacement of 310 ± 300 km (193 ± 186 mi) after restoration of 300 km (186 mi) of offset along the northwest-trending San Andreas fault (Table 1). The 400 km (249 mi) Neogene displacement assumed for Sierra Nevada–Great Valley relative to the craton (Table 2) is slightly greater than the mean residual poleward displacement but is well within the one sigma error limits. The results from older rocks of the Sierra Nevada batholith lend support to the concept of Neogene motion of Sierra Nevada–Great Valley and allow (but do not require) an earlier episode of motion for this block of several hundred kilometers. The Neogene displacement of central Salinia with respect to the craton is the sum of Neogene San Andreas fault motion (300 km [186 mi]) and the displacement of Sierra Nevada–Great Valley with respect to the craton. The Neogene displacement of Point Arena is 180 km (112 mi) greater than that of central Salinia, based on an assumed displacement of this amount along the San Gregorio–Hosgri fault and other intra-Salinian faults.

Going back in time, estimates of displacements are made for each region whenever there is a paleomagnetic determination available. Where no data are available for a particular terrane, the displacements listed are extrapolations. All of the postulated displacements lie within one standard error of the paleolatitude discrepancies with the exception of Point San Pedro (60 Ma), which is 1.6 standard errors from the paleomagnetic data determined by Champion et al (in press). A displacement lying outside the one sigma error limit is necessary to give a rate of motion within the upper limit of rates produced by the oblique subduction of the Farallon plate beneath North America at this time.

MODELING TERRANE MOTION HISTORIES

The motion histories of central Salinia and the Point Arena terrane can also be modeled by assuming that the terranes moved with specified plates and then calculating the positions of the terranes and their paleomagnetic inclinations at different times. These predicted paleomagnetic inclinations can then be compared to the paleomagnetic inclinations measured from the terranes (Engebretson, 1982).

This modeling uses the relative plate motions determined by Engebretson (1982) with North America held fixed. His calculations assume that the hotspots are fixed with respect to each other. The predicted paleomagnetic directions are calculated assuming a dipole field.

Terranes can have three types of relative motion: They can move with North America, which produces no relative motion between the two; they can ride completely on an oceanic plate, assuming the full motion of that plate with respect to North America; or they can move parallel to the continental margin during oblique subduction of an oceanic plate. In the latter case the terrane moves with some fraction of the component of the oceanic plate's relative motion that is parallel the continental margin. Models for both terranes are listed in Table 3. For each model a terrane trajectory showing the modeled path of the terrane is given in Figure 6.

The pre-Neogene position of each terrane was determined by considering offsets on the San Andreas, San Gregorio–Hosgri, and intra-Salinian faults. California from the Sierra Nevada westward was assumed to have undergone an additional 400 km (249 mi) of Neogene northward motion along faults located east of the Sierra Nevada, as discussed in the previous section.

Table 2

	PA Point Arena	CS Central Salinia	SNGV Sierra Nevada Great Valley	Craton
Present	0	0	0	0
Early Miocene 20 Ma	830 (515)[c]	650 (404)[b]	400 (249)[a]	0
Early to Middle Eocene 48 Ma	1,210 (752)	650 (404)[d]	400 (249)	0
Paleocene–Eocene 54 Ma	1,610 (1,001)[e]	1,050 (652)	400 (249)	0
Paleocene 60 Ma	1,960 (1,218)	1,500 (932)[f]	?500 (311)	0
Latest Cretaceous 75 Ma	2,860 (1,778)	2,400 (1,492)[g]	?600 (373)	0
Late Cretaceous 90 Ma	3,700 (2,300)	2,600 (1,616)	?600 (373)[h]	0
Cretaceous?	4,200 (2,610)[i]		?600 (373)	0

[a]Assumed displacement along faults east of Sierra Nevada.
[b]Displacement relative to a based on known offset along northwest-trending San Andreas fault.
[c]Displacement relative to b based on 180 km (112 mi) north-south separation between Iversen Basalt of PA and Mindego Formation of CS. Consistent with displacement (1,050 ± 840 km [652 ± 522 mi]) inferred from paleomagnetism of Iversen basalt.
[d]Consistent with displacement (590 ± 300 km [367 ± 186 mi]) inferred from paleomagnetism of Butano Sandstone.
[e]Consistent with displacement (1,810 ± 200 km [1,125 ± 124 mi]) inferred from paleomagnetism of German Rancho Formation.
[f]Consistent within 1.6 sigma with displacement (1,990 ± 310 km [1,237 ± 193 mi]) inferred from paleomagnetism of Point San Pedro rocks. Larger displacement would imply unreasonably rapid movement between 60 and 54 Ma.
[g]Consistent with displacement 2,600 ± 390 km (1,616 ± 242 mi), which is amount inferred from paleomagnetism of Pigeon Point Formation less 180 km (112 mi) of poleward displacement along the San Gregorio-Hosgri and intra-Salinian faults.
[h]Estimate based on paleomagnetic results from the central Sierra Nevada (Frei et al, in press).
[i]Consistent with the paleomagnetic results from the spilites of the Point Arena terrane assuming that their age is 130 Ma.

Table 2—Cumulative displacement in a southerly direction from present position displacements in km (mi).

Neogene motions of both terranes were modeled by assuming that the terranes moved with the Pacific plate, much as Baja California is moving today. The duration of Pacific plate transport was varied in successive runs to achieve the Neogene motion required for each terrane. An alternative possibility is that the terrane moved parallel the continental margin at a velocity less than or equal to the tangential component of Pacific–North America motion.

Before the inception of Pacific plate transport, we assume that the terranes were attached to North America for varying periods of time during the Middle Tertiary. Pre-Neogene poleward displacement was modeled by strike-slip motion subparallel to the North America continental margin as a result of oblique convergence of the Farallon and North America plates.

For the Point Arena terrane the timing of this motion is loosely constrained to be prior to 25 Ma (pre-Iversen Basalt), probably beginning in Campanian time (80 Ma). The motion is constrained by the measured paleolatitude for the German Rancho Formation at 54 Ma.

For central Salinia, the timing of pre-Neogene poleward displacement is very tightly constrained and quite difficult to model adequately. Granitic rocks were intruded into continental crust 110 to 100 Ma, possibly continuing up to 82 Ma. Beginning in Campanian time the granitic rocks were rapidly uplifted and eroded and arkosic sedimentation occurred in discrete, deep sedimentary basins in a borderland province, continuing through at least middle Eocene time. By about 50 Ma, central Salinia was adjacent to the western San Joaquin Valley. Measured paleolatitudes exist for rocks with mean ages of about 75 Ma (Pigeon Point Formation), 60 Ma (Point San Pedro rocks), and 48 Ma (Butano Sandstone).

Transport parallel to the continental margin is modeled as occurring between 80 and 50 Ma. Before this time central Salinia is modeled as being part of the Mexico–Central America Cordilleran arc (Fig. 7). Because the timing of central Salinian motion does not afford much opportunity for adjustment, the main factor that can be changed is the trend of the continental margin. Complete efficiency of transport was assumed to obtain the maximum amount of transport. In the modeling program the trend of the continental margin was approximated by a small circle on the globe (Table 3).

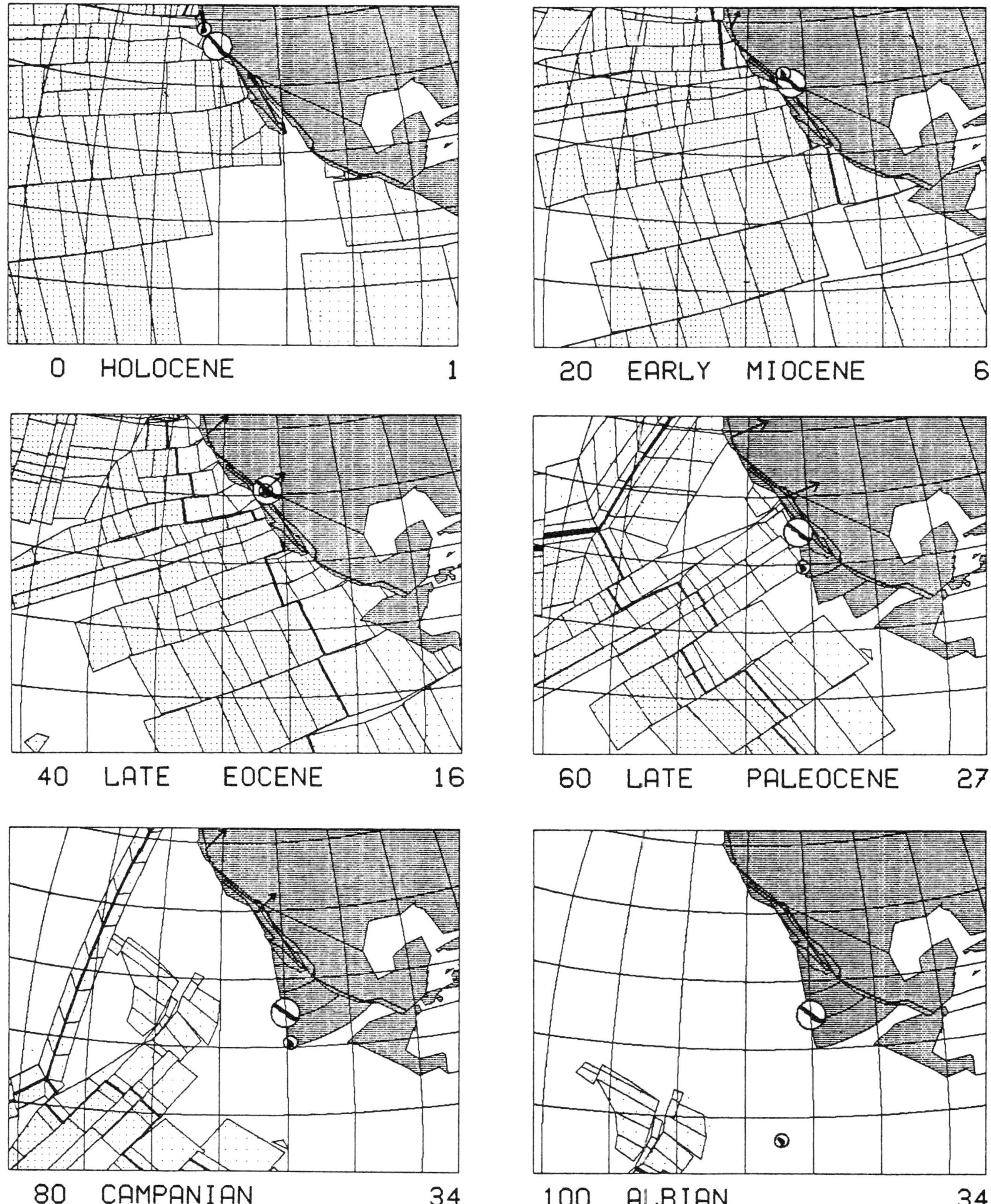

Figure 6—Position of central Salinia (in large circle) and Point Arena (in small circle) terranes through time. Age (m.y.) in lower left of each diagram; anomaly number in lower right. Size and shape of Point Arena terrane is distorted for clarity. North America fixed coordinates with a 10° latitude-grid. North America plate in gray. Heavy line is western coastline; lighter line is continental margin. Pacific plate with transforms and isochrons shown in small stipple. Farallon plate and remnants with inferred transforms and isochrons shown by large stipple. Arrows represent local velocities of oceanic plates relative to North America. Rotated continental margin visible from 100 Ma to 60 Ma.

Table 3

Central Salinia Model 6	
Plate	Interval (Ma)
Pacific	0–17
North America	17–50
Farallon[a]	50–80
North America	80–110
Point Arena Model	
Plate	Interval (Ma)
Pacific	0–22
North America	22–38
Farallon[b]	38–80
Farallon	80–160

[a]The continental margin was approximated by a small circle of radius 70.440° centered at (15° N 322° E). The terrane moved along this circle at a rate equal to the projection of the Farallon convergence vector on the circle.
[b]Same as in above except that the radius of the small circle was 69.56°.

Table 3—Models.

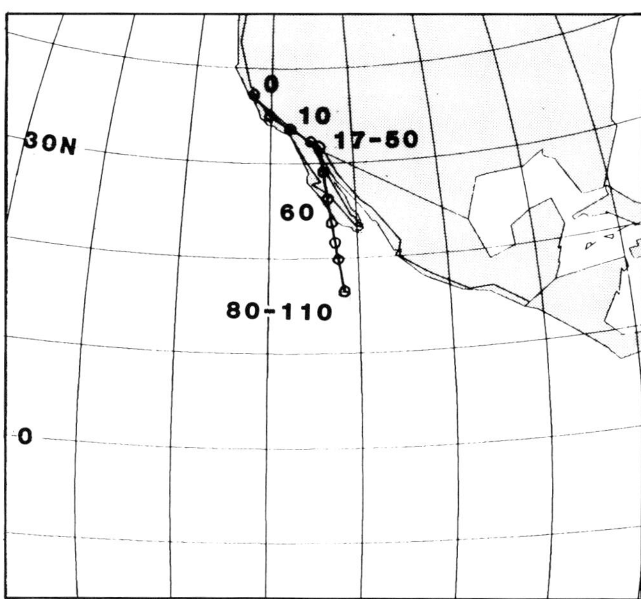

Figure 7—Trajectory of a point on central Salinia assuming the plate transport history of Salinia Model 6 (Table 3). Circles represent position at 5 m.y. intervals. Trajectory is in North America fixed coordinates, with 10° latitude-longitude grid.

The location and trend of the Mexico-Central America continental margin during the Late Cretaceous and Early Tertiary are not well known. For the models discussed here, the margin was assumed to trend approximately north-south, the southern end of the margin being located well to the west of its present position. This orientation is more favorable for poleward translation than is the present orientation of the margin. If from 80 to 50 Ma Salinia moved poleward along this margin at a velocity equal to the northward component of Farallon-North America convergence, the modeled paleolatitudes fit those determined paleomagnetically within the error limits of the latter. The observed paleolatitudes cannot be explained by Farallon-North America convergence along a margin trending northwest as it does today (Fig. 8).

For Point Arena Model 9 we use the same continental margin trend for the Point Arena terrane. Because the spilites of this terrane appear to be oceanic and are very different from the granitic rocks of Salinia, we consider them to be a true piece of Farallon oceanic crust. Before 80 Ma, the approximate age of the oldest Point Arena sedimentary rocks, the spilites ride as part of the Farallon plate (Fig. 9). Their age is not known, so a wide range of possible ages is shown in Figure 10. For calculating the amount of poleward displacement, an age of 130 Ma was assumed. Point Arena Model 9 fits the German Rancho Formation and Iversen Basalt paleomagnetic data quite well with translation parallel to the continental margin lasting between 80 and 38 Ma, well within the lenient age constraints for this terrane. If the hypothesis of the origin of the spilites as basement of the terrane is correct, this model suggests that the age of the spilites is 90 to 130 Ma.

CONCLUSIONS

We conclude that poleward displacements of the central Salinia and Point Arena terranes as inferred from paleomagnetic data can be modeled by assuming that during the Neogene these terranes were transported along the continental margin by the Pacific plate. For a time in the Middle Tertiary, the terranes were probably attached to North America. In the Paleogene and Late Cretaceous they appear to have been transported along a north-trending continental margin by northeasterly convergence of the Farallon plate relative to North America. Successful modeling of the terrane motions using the mechanism of pre-Neogene Farallon-North America plate interaction requires a continental margin trend for Mexico and Central America that is considerably more northerly than the present one. If Kula-North America interaction can be invoked to transport the terranes for a period of time, less of an adjustment in continental margin trend is necessary.

REFERENCES

Albers, J. P., 1967, Belt of sigmoidal bending and right-lateral faulting in the western Great Basin: Geological Society of America Bulletin, v. 78, p. 143–156.

Atwater, T., 1970, Implications of plate tectonics for the Cenozoic evolution of western North America: Geological Society of America Bulletin, v. 81, p. 3513–3536.

Blake, M. C., et al, 1982, Preliminary tectonostratigraphic terrane map of California: U.S. Geological Survey Open-File Report 82-593, 10 p.

Champion, D. E., et al, in press, Paleomagnetic and geologic data indicating 2500 km of northward displacement for the Salinian and related terranes, California: Journal of Geophysical Research, v. 89.

Clark, J. C., 1966, Tertiary stratigraphy of the Felton-Santa Cruz area, Santa Cruz Mountains, California: PhD Dissertation, Stanford University, 184 p.

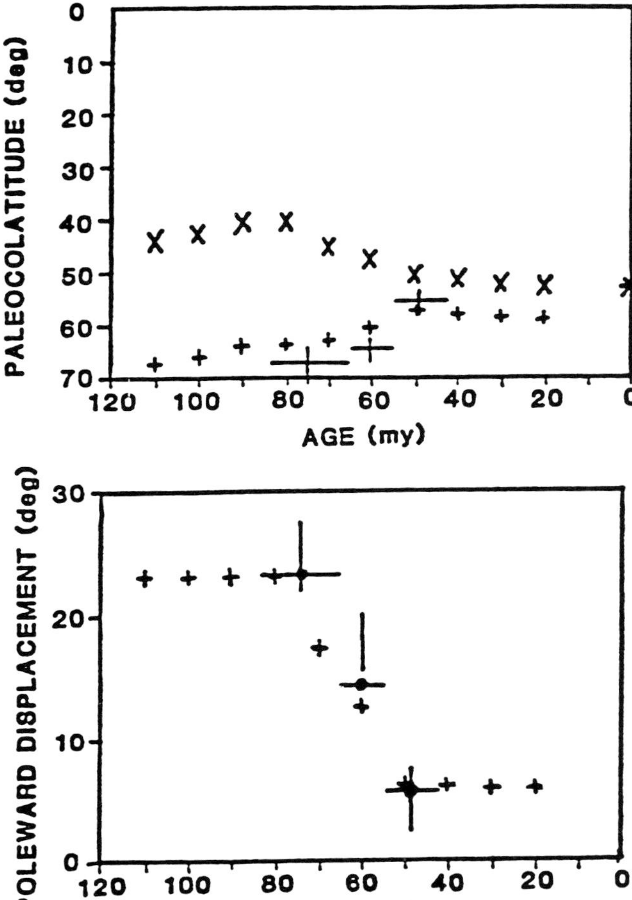

Figure 8—Paleocolatitude and poleward displacement for central Salinia paleomagnetic data and Salinia Model 6. In upper graph X's are paleocolatitude for cratonal North America and +'s are paleocolatitudes predicted by the model. Age and colatitude limits on paleomagnetic data shown by crosses. In lower graph poleward displacement relative to North America is shown for model and data by +'s and crosses, respectively.

Clarke, Jr., S. H., and T. H. Nilsen, 1973, Displacement of Eocene strata and implications for the history of offset along the San Andreas fault, central and northern California, *in* A. Nur and R. Kovach, eds., Proceedings of conference on tectonic problems of San Andreas fault system: Stanford University Publications, Geological Sciences, v. 13, p. 358-367.

Compton, R. R., 1966, Granitic and metamorphic rocks of the Salinian block *in* E. H. Bailey, ed., Geology of northern California: California Division of Mines and Geology Bulletin 190, p. 277-287.

Dickinson, W. R., 1983, Cretaceous sinistral strike slip along Nacimiento fault in coastal California: Bulletin of the American Association of Petroleum Geologists, v. 67, p. 624-645.

______, and P. J. Coney, 1980, Plate tectonic constraints on the origin of the Gulf of Mexico, *in* R. H. Pilger, Jr., ed., The origin of the Gulf of Mexico and the early opening of the central North Atlantic Ocean: School of Geoscience, Louisiana State University, Baton Rouge, p. 27-36.

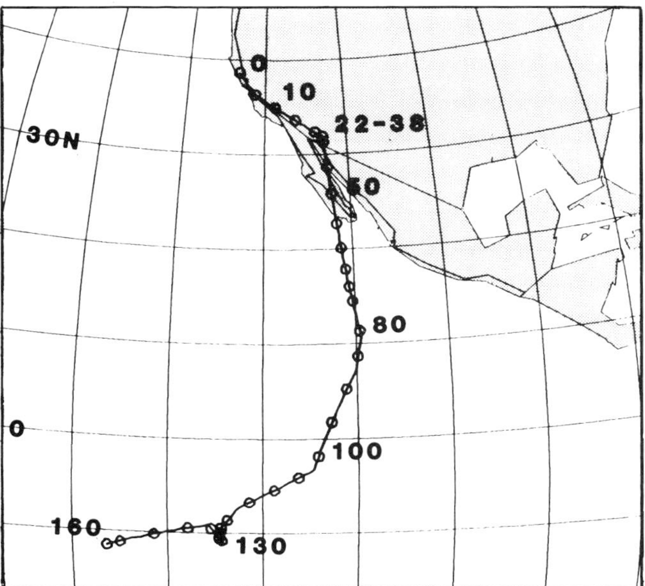

Figure 9—Trajectory of a point on the Point Arena terrane assuming the plate transport history of Point Arena Model 9. Conventions as in Figure 7.

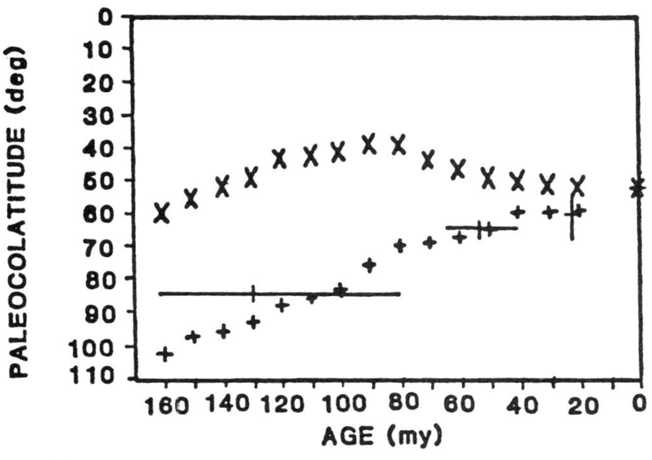

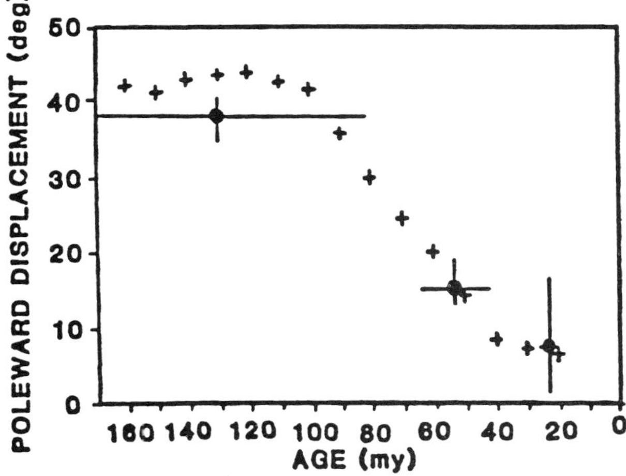

Figure 10—Paleocolatitude and poleward displacement for Point Arena paleomagnetic data and Point Arena Model 9. Conventions as in Figure 8.

Diehl, J. F., et al, 1983, Paleomagnetism of the late Cretaceous-early Tertiary north-central Montana alkalic province: Journal of Geophysical Research, v. 88, p. 10593-10610.

Engebretson, D. C., 1982, Relative motions between oceanic and continental plates in the Pacific basin: PhD Dissertation, Stanford University, 211 p.

Fitch, T. J., 1972, Plate convergence, transcurrent faults, and internal deformation adjacent to southeast Asia and the western Pacific: Journal of Geophysical Research, v. 77, p. 4432–4460.

Frei, L. S., et al, 1984, Paleomagnetic results from the central Sierra Nevada: Constraints on reconstructions of the western United States: Tectonics, v. 3, p. 157-177.

Galloway, A. J., 1977, Geology of the Point Reyes Peninsula, Marin County, California: California Division of Mines and Geology Bulletin 202, 72 p.

Graham, S. A., 1976, Tertiary sedimentary tectonics of the central Salinian block of California: PhD Dissertation, Stanford University, 510 p.

———, and W. R. Dickinson, 1978, Evidence for 115 kilometers of right slip on the San Gregorio- Hosgri fault trend: Science, v. 199, p. 179-181.

———, and C. E. Peabody, 1981, New evidence for major strike-slip along the San Gregorio-Hosgri fault of the San Andreas tranform system (Abs.): Geological Society of America Abstracts with Programs, v. 13, p. 463.

Harland, W. B., et al, 1983, A geologic time scale: Cambridge, Cambridge University Press, 128 p.

Harrison, C. G. A., and T. Lindh, 1982, A polar wandering curve for North America during the Mesozoic and Cenozoic: Journal of Geophysical Research, v. 87, p. 1903-1920.

Howell, D. G., and J. G. Vedder, 1978, Late Cretaceous paleogeography of the Salinian block, California, *in* D. G. Howell and K. A. McDougall, eds., Mesozoic paleogeography of the western United States: Society of Economic Paleontologists and Mineralogists, Pacific Section, p. 33-70.

Kanter, L. R., 1983, Paleomagnetic constraints on the motion history of Salinia: PhD Dissertation, Stanford University, 156 p.

———, and M. O. McWilliams, 1982, Rotation of the southernmost Sierra Nevada, California: Journal of Geophysical Research, v. 87, p. 3819-3830.

Mattinson, J. M., 1982, Granitic rocks of the Gabilan Range, California: U-Pb systematics and implications for age and origin (Abs.): Geological Society of America Abstracts with Programs, v. 14, p. 184.

Nilsen, T. H., and S. H. Clarke, Jr., 1975, Sedimentation and tectonics in the early Tertiary borderland of central California: U.S. Geological Survey Professional Paper 925, 64 p.

———, and M. H. Link, 1975, Stratigraphy, sedimentology, and offset along the San Andreas fault of Eocene to lower Miocene strata of the northern Santa Lucia Range and the San Emigdio Mountains, central California, *in* D. W. Weaver, et al, eds., Conference on future energy. Horizons of the Pacific Coast: Society of Economic Paleontologists and Mineralogists, Pacific Section, and Society of Exploration Geophysicists, Pacific Section, p. 367–400.

Page, B. G. N., et al, 1979, A review of the main structural and magnetic features of northern Sumatra: Journal of the Geological Society of London, v. 136, p. 569-579.

———, and D. C. Engebretson, 1984, Correlation between the geologic record and computed plate motions for central California: Tectonics, v. 3, p. 133-156.

Ross, D. C., 1977, Pre-intrusive metasedimentary rocks of the Salinian block, California—a tectonic dilemma, *in* J. H. Stewart, et al, eds., Paleozoic paleogeography of the western United States: Society of Economic Paleontologists and Mineralogists, Pacific Section, p. 371-380.

———, 1978, The Salinian block—a Mesozoic granitic orphan in the California Coast Ranges, *in* D. G. Howell and K. A. McDougall, eds., Mesozoic paleogeography of the western United States: Society of Economic Paleontologists and Mineralogists, Pacific Section, p. 509-522.

Stewart, J. H., 1983, Extensional tectonics in the Death Valley area, California: Transport of the Panamint Range structural block 80 km northwestward: Geology, v. 11, p. 153-157.

———, et al, 1968, Summary of regional evidence for right-lateral displacement in the western Great Basin: Geological Society of America Bulletin, v. 79, p. 1407-1414.

———, et al, 1970, Reply to discussion on summary of regional evidence for right-lateral displacement in the western Great Basin: Geological Society of America Bulletin, v. 81, p. 2175-2180.

Urrutia-Fucugauchi, J., in press, Paleomagnetism and tectonic evolution of the southern Cordillera: Geology, v. 12.

Vedder, J. G., et al, 1983, Stratigraphy, sedimentation, and tectonic accretion of exotic terranes, southern Coast Ranges, California, *in* J. S. Watkins and C. L. Drake, eds., Studies in continental margin geology: American Association of Petroleum Geologists Memoir 34, p. 471–498.

Wentworth, C. M., 1966, The upper Cretaceous and lower Tertiary rocks of the Gualala area, northern Coast Ranges, California: PhD Dissertation, Stanford University, 197 p.

Geometry and Tectonic Setting of Sea-Floor Spreading for the Josephine Ophiolite, and Implications for Jurassic Accretionary Events along the California Margin

Gregory D. Harper
State University of New York
Albany, New York

Jason B. Saleeby
California Institute of Technology
Pasadena, California

Elizabeth A. S. Norman*
University of Utah
Salt Lake City, Utah

Ophiolites with Jurassic petrogenetic ages occur in the western Sierra Nevada and Klamath Mountains, California Coast Ranges, and probably underlie much of the California Great Valley. The Josephine ophiolite of the western Klamath Mountains represents one of the most complete, intact, and best studied of the Jurassic ophiolites. A fracture zone history of the ophiolite is indicated by a thin crustal section, highly fractionated ferrobasalts, and a distinctive olistostrome locally overlying the ophiolite interpreted as the sediment fill of a fracture zone valley. Orientations of sheeted dikes, transform remnants, and upper mantle flow fabrics all indicate a sea-floor spreading direction and transform motion parallel to the continental margin. Likewise, some dike orientations and fracture zone features of the Sierra Nevada and Coast Range ophiolites suggest spreading and transform motion parallel to the continental margin. The Great Valley is a major morphotectonic feature that may have originated as a large transform valley that in later Cretaceous time became a forearc basin. Geophysical lineaments of the Valley along with fracture zone features of the northern Coast Range ophiolite and southern Sierra Nevada suggest a left-stepping family of fracture zones in and around the margin of the Great Valley. It is suggested that Jurassic ophiolites of California record major sinistral-sense transform motion and related oblique spreading oriented parallel to the continental margin. Such motion may be linked to the Mojave-Sonora megashear and reflects decoupling between North American and Pacific basin plates during rapid northwestward motion of the North American plate.

INTRODUCTION

Ophiolites with igneous ages clustering around 160 m.y. are one of the most striking features of California geology. These include the Josephine ophiolite in northwestern California, the Smartville ophiolite in the Sierra Nevada foothills, and the Coast Range ophiolite exposed primarily along the western edge of the California Great Valley. Many workers have suggested a marginal basin origin for these ophiolites based on the widespread occurrence of coeval arc volcanic and plutonic complexes in California and southwestern Oregon (Fig. 1) and the occurrence of arc-derived detritus in sediments overlying the ophiolites (Evarts, 1977; Xenophontos and Bond, 1978; Harper, 1980; Saleeby et al, 1982). The Josephine ophiolite represents one of the most intact and best studied ophiolites in the North American Cordillera (Dick, 1976, 1977; Harper, 1980; Saleeby et al, 1982). Geochemical data and stratigraphic relations support its origin in a marginal basin adjacent to an active island arc. Geometric relations in the orientations of sheeted dikes, fracture zone remnants, and mantle flow fabrics consistently indicate a spreading direction parallel to the trend of the continental margin (approximately north-south). The 160 m.y. sea-floor spreading ages that are so widespread in California

*Present Address: Chevron U.S.A., Denver, Colorado

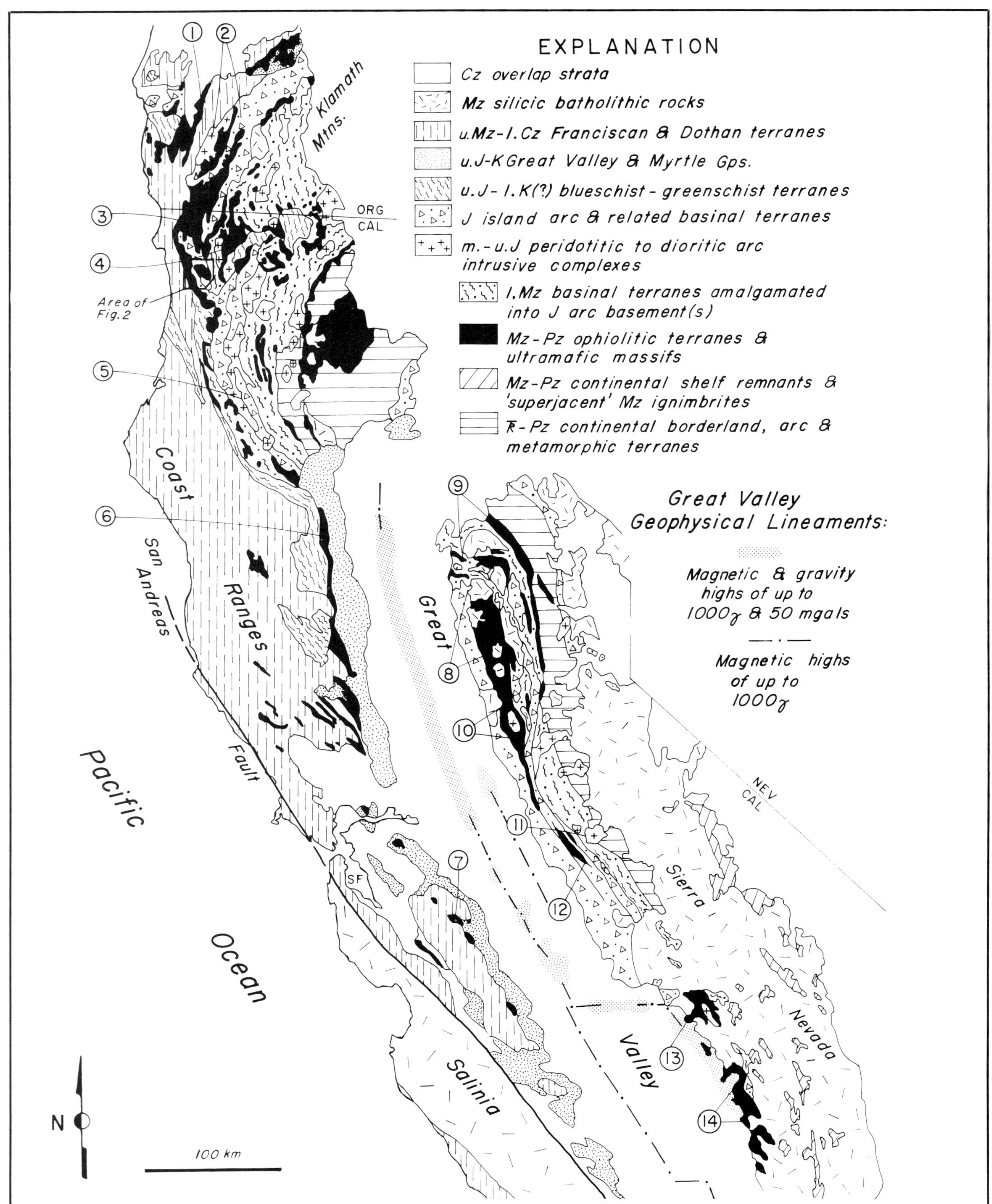

Figure 1—Generalized map showing distribution of ophiolitic, Mesozoic island arc, and related terranes of central California to southwestern Oregon region (modified from Irwin, 1979, and references given in Table 1). Great Valley gravity-magnetic lineaments are from synthesis presented in Cady (1975). Numbers on map refer to locations of successions depicted in Figure 3.

correspond in time with major transform motion of the Mojave–Sonora megashear (Silver and Anderson, 1974, 1983) and oblique components in relative motion along the California margin (Gordan et al, 1981; Engebretson, 1982). The possibility of transform faulting along the California margin in conjunction with the spreading generation of marginal basin crust should be considered very closely. Such processes will simultaneously disperse previously accreted terranes, accrete juvenile crust, and produce structural breaks that may nucleate plate convergence upon changes in relative plate motions.

This paper will focus on the following: (1) the spreading geometry and fracture zone history of the Josephine ophiolite; (2) time and space relations between 160 m.y. ophiolites and Middle-to-Late Jurassic volcanic and plutonic arc rocks of the region between central California and southern Oregon; (3) major Jurassic structural features of northern and central California that suggest or are compatible with transform motion parallel to the margin; and (4) a model for Jurassic transform and spreading geometry and consideration of such a model in the light of regional tectonics and plate kinematic patterns.

Figure 1 is a generalized map of the region from central California to southern Oregon showing major belts of accretionary terranes. Text and diagrams below will focus on the Jurassic ophiolitic terranes, island arc and related basinal terranes, and on Jurassic arc-related plutons. But first, for a regional frame of reference, several other groups of terranes should be noted. These include terranes of the Franciscan and Dothan that were accreted against the Jurassic ophiolitic and arc terranes in mid-Cretaceous to Early Tertiary time and a belt consisting of several Paleozoic terranes in the eastern Klamaths and along the northern and axial Sierra Nevada. In the central and southern axial Sierra, such Paleozoic terranes can be linked genetically to North American sial (Saleeby et al, in press); however, they are likely to have been in a strike-slip dispersal mode throughout much of Mesozoic time. Time relations in the amalgamation of such parautochthonous North America fragments to Paleozoic borderland terranes of the northern Sierra Nevada and eastern Klamaths are poorly understood, although stitching relations by mid-Jurassic peridotitic to dioritic arc plutons provide an important constraint relative to 160 m.y. ophiolites and related arc terranes. Furthermore, a complex sequence of early Mesozoic basinal terranes and related late Paleozoic to early Mesozoic ophiolite fragments formed part of the amalgam that was stitched by the mid-Jurassic plutons. The amalgam of early Mesozoic basinal terranes, Paleozoic borderland, and parautochthonous North American shelf terranes, and the Jurassic arc and ophiolitic terranes were accreted to North America during the Late Jurassic Nevadan orogeny. Cretaceous batholithic rocks form a major stitching belt across the Nevadan orogen. This stitched orogen formed the North American "autochthon" against which Franciscan and Dothan terranes were accreted. Late Jurassic–Early Cretaceous blueschist-greenschist terranes shown on Figure 1 may represent remnants of the Nevadan metamorphic core and perhaps some of the earliest Franciscan–Dothan accretion-related metamorphic events.

Figure 1 also shows the locations of key stratigraphic or structural successions diagnostic of the various Jurassic ophiolitic and arc terranes that are illustrated in Figure 3. Informal names for the successions and references are given in Table 1. The stratigraphic succession at Location 3 is the Josephine ophiolite, which underlies an area of over 800 sq km (309 sq mi) and includes one of the largest ultramafic massifs in the Cordillera. A complete ophiolite sequence is exposed at this location, and much of the ultramafic massif extends into southern Oregon. Attention will focus first on the Josephine ophiolite with emphasis on the geometry and tectonic setting of its sea-floor spreading genesis.

SPREADING GEOMETRY OF THE JOSEPHINE OPHIOLITE

The orientation of sheeted dikes in ophiolites has been used to estimate the trend of the spreading axis (e.g., Pallister, 1981). The strike of the dikes is assumed to have been parallel to the ridge crest; this assumption is supported by the fact that fissures observed at spreading centers are statistically parallel to the spreading axis (Ballard and Van Andel, 1977; Luyendyk and Macdonald, 1977; Van Andel and Ballard, 1979). Several hundred sheeted dikes were measured in the Josephine ophiolite (Harper, 1982) and consistently yield east-west strikes after correction for folding. The dikes dip to the south at 20 to 40° after structural correction; this dip is probably the result of rotations at the spreading center about a subhorizontal axis by listric faulting (Verosub and Moores, 1981; Harper, 1982). Thus, the sheeted dikes suggest that the Josephine ophiolite formed along spreading centers oriented normal to the western Jurassic belt of the Klamaths and presumably the continental margin; i.e., the spreading direction was parallel to the continental margin. The spreading geometry deduced from the sheeted dikes is also substantiated by high-temperature flow fabrics in the peridotite; in general, these fabrics indicate mantle flow in a direction normal to the trend of the sheeted dikes (Harding and Bird, 1983). Such a flow pattern is expected beneath spreading ridges as shallow hot asthosphere ascends and spreads and is supported by seismic anisotropy often observed in modern oceanic upper mantle (Christensen and Salisbury, 1975).

A possible complication in using structural relations within the ophiolite to infer spreading directions is that part of the Klamath Mountains have apparently undergone clockwise rotations of as much as 70° about a vertical axis in Tertiary time (Simpson and Cox, 1980; Fagin and Gose, 1983; Schultz and Levi, 1983). Plutons from the northern part of the range appear to have rotated approximately 75° clockwise, but paleomagnetic data for plutons from the central Klamaths are ambiguous (Schultz and Levi, 1983). Perhaps the best indication of rotations is structural trends in the Galice Formation that lies depositionally on the ophiolite. Galice cleavage and bedding surfaces generally dip steeply to the east and bend from northwest in the southern Klamaths to northeast in the northern Klamaths. If we assume that the change in structural trend is due to the aforementioned rotations, then the northwestern

Klamaths have undergone a large clockwise rotation (as indicated by paleomagnetic data), whereas the central and southwestern Klamaths probably have not rotated appreciably. Such an analysis can be extended into the western Sierra Nevada where paleomagnetic data have shown that units correlative with and having similar structural trends to the Galice Formation of the southwestern Klamaths have undergone little or no rotation (Bogen, 1983; Frei and Cox, 1983). Because the structural trends are north-northwest in the present study area (Loc. 3), a clockwise rotation of up to 20 to 30° is permitted by this line of reasoning. Restoration of such a rotation brings the deduced spreading direction of the Josephine ophiolite into even closer parallelism with the California margin and the major Juassic ophiolite belts located to the south. Such a rotation is used palinspastically in Figure 5, discussed below.

The spreading geometry deduced above is further substantiated by the recognition of a fossil transform valley within the Josephine ophiolite. Transform faults and related fracture zones represent some of the most complex structural features of oceanic domains. The Josephine transform valley is filled with the Lems Ridge olistostrome; this chaotic deposit has not been previously described. Its importance along with the importance of fracture zone features elsewhere in California is one emphasis of this paper, and thus the Lems Ridge olistostrome will be discussed in some depth.

EVIDENCE FOR A FRACTURE ZONE IN THE JOSEPHINE OPHIOLITE

The strongest evidence for a fracture zone in the Josephine ophiolite is the local occurrence of the Lems Ridge olistostrome containing clasts derived from the ophiolite (Harper et al, 1983). The olistostrome has a maximum thickness of approximately 700 m (2,300 ft) and occurs on an east-dipping limb of a large fold where it has been brought to the surface along a reverse fault (Fig. 2). The olistostrome is a complex unit consisting of pebbly mudstone, massive volcaniclastic greenstone, and intercalated graywacke and slaty argillite beds. It is well exposed in road cuts on Lems Ridge and on the South Fork of the Smith River.

The Lems Ridge olistostrome conformably overlies pillow lavas of the Josephine ophiolite, and in one outcrop pebbly mudstone occurs between pillows. The olistostrome is overlain by slates and metagraywackes of the Late Jurassic (Late Oxfordian–Early Kimmeridgian) Galice Formation; the contact is gradational with graywacke and slaty argillite interbedded with volcaniclastic greenstone and pebbly mudstone (Norman, 1984).

Elsewhere, the Josephine ophiolite is conformably overlain by the Galice Formation consisting of a thin pelagic sequence of chert and slaty argillite that is overlain and locally interbedded with graywacke, slate, and minor pebble conglomerate. Graywackes in the basal few hundred meters are characteristically rich in volcanic rock fragments and feldspar, whereas graywackes higher in the section (the bulk of the Galice Formation) are rich in chert and argillite clasts (Harper, 1983). Graywackes within and directly overlying the Lems Ridge olistostrome are petrographically similar to the volcanic-rich graywackes of the lower Galice Formation. This indicates that the Lems Ridge olistostrome is equivalent to the lower Galice Formation and, because of its much greater thickness, suggests that it was deposited in a deep trough.

The most common clast types in the pebbly mudstone are sandstones and shales. The sandstone clasts are diverse and include feldspathic lithic wackes, quartzofeldspathic wackes, and rare orthoquartzite containing a few percent microcline. Other clast types include quartz-mica schist, phyllite, porphyritic andesite, silicic volcanics, pumice, marble, greenstone, metadiabase, metagabbro, chert, and rare ultramafics. Some of the clasts are as much as 15 m (50 ft) in diameter; the larger clasts are mostly metagabbro, greenstone, gray-green chert, and, less commonly, marble.

The volcaniclastic greenstone portions of the Lems Ridge olistostrome include tuffs, volcaniclastic sandstones, and breccias. Where sedimentary textures are not obscured by recrystallization, the clasts can be observed to consist of predominantly volcanic rock fragments (> 85%) and crystals of plagioclase, clinopyroxene, and hornblende. Larger clasts, up to a meter in length, occur sporadically and include argillite, metagabbro, and serpentinite. In addition, large isolated outcrops 15 to 20 m (50–65 ft) long of metagabbro and greenstone occur associated with the volcaniclastic greenstone and are either very large clasts or talus deposits formed on the sea floor. The contact between the volcaniclastic greenstone and pebbly mudstone varies from planar to highly irregular, probably the result of soft sediment deformation.

The volcaniclastic greenstone appears to have been derived from an active volcanic arc. The major and rare-earth chemistry determined for one sample indicates that it is andesitic and light-rare earth enriched, typical of calc-alkaline volcanic rocks (Norman, 1984).

The clasts of ultramafic rocks, metagabbro, metadiabase, and greenstone within the Lems Ridge olistostrome were apparently derived from the Josephine ophiolite. Many of the metagabbro and ultramafic clasts are protomylonites and mylonites. Except for the tectonitic overprint, these rocks are petrographically similar to the rocks from the Josephine ophiolite. The protolith of two ultramafic clasts, determined from the composition of porphyroclasts, is a clinopyroxene cumulate, a common rock type in the Josephine ophiolite. The occurrence of ultramafic rocks indicates that lower crustal portions of the ophiolite were exposed on the sea floor.

One of the strongest lines of evidence that the Josephine ophiolite was the source of many clasts in the Lems Ridge olistostrome is the abundance of pebble- to boulder-size clasts of gray-green chert. Except for their massive nature, the cherts resemble those that overlie the ophiolite elsewhere. Radiolarians separated from several of the chert clasts are Jurassic in age and similar to those from cherts directly overlying pillow lavas of Josephine ophiolite to the north (Jones, 1983, personal communication; Harper et al, 1983).

The Lems Ridge olistostrome is similar to inner fan channel deposits of submarine fans (Walker and Mutti, 1973). However, there are several features of the olisto-

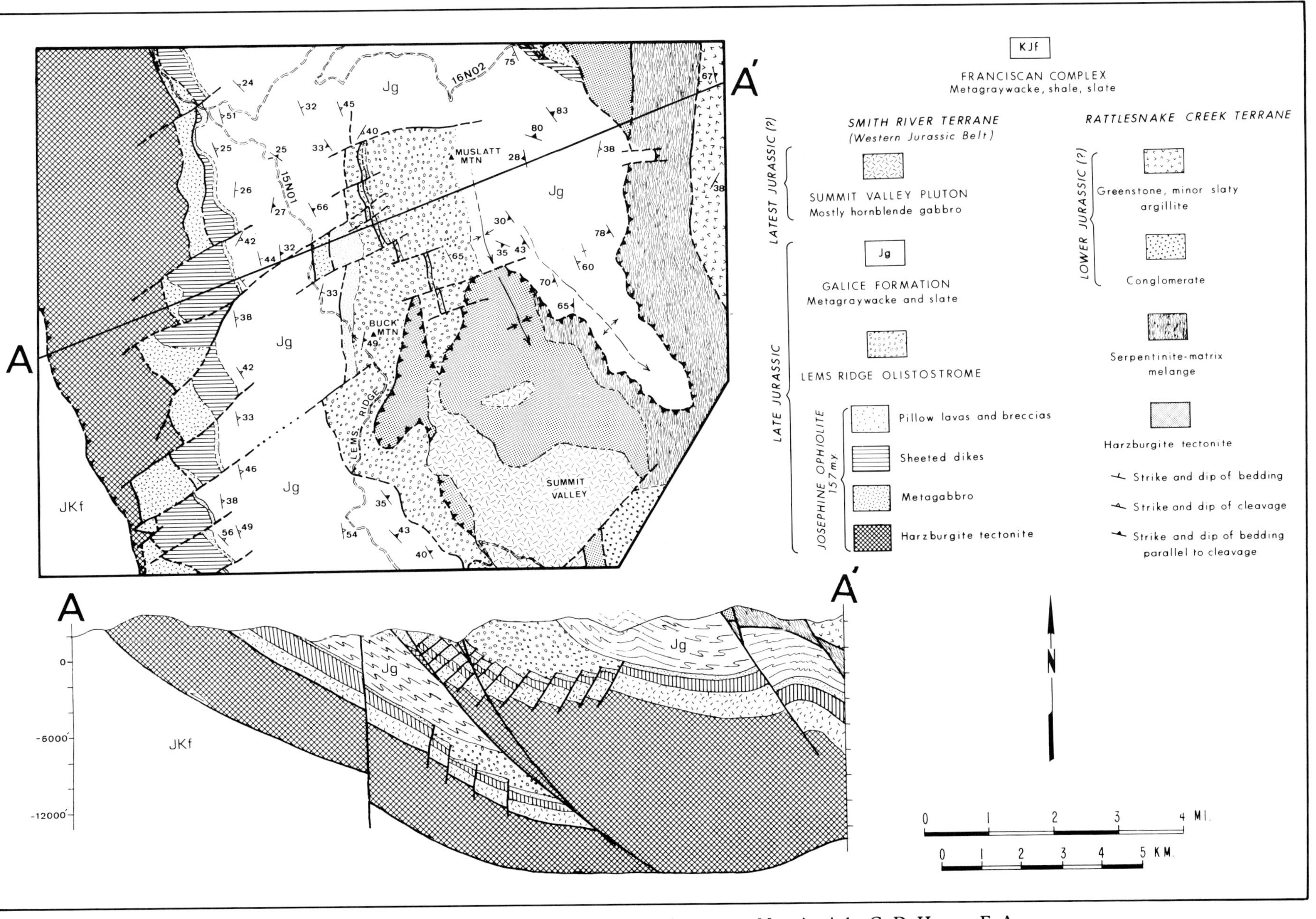

Figure 2—Geologic map of part of the western Klamath belt showing the Lems Ridge olistostrome. Mapping is by G. D. Harper, E. A. Norman, and C. M. Gorman.

strome that suggest it was deposited in a fracture zone valley (i.e., a tectonic depression rather than an erosional valley):

1. Inner fan channel deposits are deposited in channels eroded into slope deposits consisting primarily of mudstone (Walker and Mutti, 1973). In contrast, the Lems Ridge olistostrome directly overlies the Josephine ophiolite.

2. Ophiolite-derived clasts are common, and some are very large. Lower crustal rocks such as metagabbros and serpentinites are commonly dredged from modern fracture zones that offset slow spreading centers (e.g., Bonatti and Honnorez, 1976; Prinz et al, 1976).

3. Many of the ophiolite clasts have cataclastic fabrics, consistent with exposure to the sea floor by faulting.

Modern fracture zones are characterized by deep depressions, the depth of which is dependent on the amount of offset of the spreading centers (Choukroune et al, 1978; Fox et al, 1980). Based on its thickness, the Lems Ridge olistostrome must have filled a depression that was at least 700 m (2,300 ft) deep.

The crustal thickness of the Josephine ophiolite is also suggestive of generation near a fracture zone. Several studies indicate that oceanic crust may be very thin near fracture zones (Fox et al, 1980; Stroup and Fox, 1981; Sinha and Louden, 1983; Corimer et al, 1983). The crustal thinning can be attributed to a decreased magma supply caused by the juxtaposition of relatively cold lithosphere against the spreading center (Nelson, 1981). The crustal thickness near fracture zones may be as little as 1 or 2 km (.6 or 1.2 mi) (Fox et al, 1980; Stroup and Fox, 1981).

The maximum crustal thickness of the Josephine ophiolite, measured from the base of the ultramafic cumulates to the top of the pillow lavas, is only approximately 3.6 km (2.2 mi). In many areas, however, the crustal thickness is only about 1 km (.6 mi) (e.g., Section A–A′, Fig. 2); in these areas, pillow lavas are less than 100 m (330 ft) thick and sheeted dikes may be less than 200 m (650 ft) thick, compared to their maximum thicknesses of 400 and 1,500 m (1,300 and 4,900 ft), respectively.

The crustal thickness of modern backarc basins is variable but is generally between 5 and 8 km (3 and 5 mi) (Ambos and Hussong, 1982), similar to oceanic crust. Thus, the crustal thickness of the Josephine ophiolite is much less than backarc basin or oceanic crust, unless widespread serpentinization of the lower crust is invoked (e.g., Nichols et al, 1980). The very thin crustal section of the Josephine ophiolite, especially west of the Lems Ridge olistostrome (Fig. 2), is possibly the result of generation at a spreading center adjacent to a large fracture zone.

The geochemistry of dikes and lavas of the Josephine ophiolite is also supportive of a fracture zone origin. Lavas dredged near oceanic fracture zones are systematically more differentiated than those erupted along normal ridge segments (Hekinian and Thompson, 1976), especially near the tips of propagating rifts (Sinton et al, 1983). The highly fractionated lavas are ferrobasalts enriched in incompatible elements such as Ti, Zr, and rare earths. This effect is possibly the result of low magma supply rates coupled with moderate cooling rates allowing closed-system fractionation (Sinton et al, 1983). Highly fractionated lavas occur in the upper parts of the Josephine ophiolite. These lavas are ferrobasalts and are strongly enriched in Ti, Zr, and rare earths (e.g., Fig. 4, note Ti-rich samples). The ferrobasalts have up to 13.0 wt % FeO* (total iron as FeO) and 2.4 wt % TiO_2. The geochemistry of lavas directly underlying the Lems Ridge olistostrome is also supportive of a fracture zone origin; dikes and lavas from elsewhere in the Josephine ophiolite are similar to island arc tholeiites but are in some respects transitional to MORB (Fig. 4; Harper, 1984). Pillow lavas from beneath the olistostrome have lower Ti/Cr and higher Ti/V, both of which make them more similar to MORB (Fig. 4); the different chemistry of these lavas might be due to unusual melting conditions in the vicinity of a fracture zone. A second possibility is that the pillow lavas beneath the olistostrome represent more MORB-like lavas on the opposite side of the fracture zone.

The outcrop pattern of the Lems Ridge olistostrome may be used to deduce the orientation of the fracture zone valley and thus the spreading direction. The olistostrome crops out in a north-south belt extending for 13 km (8 mi) (Fig. 2) and must also extend further north in the subsurface as it is downfaulted on its northern end. Note that the olistostrome is absent above the ophiolite both to the west and to the east (Fig. 2); this indicates that the olistostrome filled a north-south depression, suggesting a fracture zone oriented north-south. The fracture zone orientation deduced from the outcrop pattern of the Lems Ridge olistostrome is consistent with the spreading geometry deduced from both sheeted dikes and mantle flow fabrics. This is treated as a strong reference point below in the consideration of regional spreading geometries within other Jurassic ophiolites of California. Attention will now focus on the local tectonic setting of Josephine spreading.

TECTONIC SETTING OF JOSEPHINE SPREADING

Numerous workers have suggested a backarc (marginal) basin origin for the Josephine ophiolite (Dick, 1977; Vail, 1977; Harper, 1980; Saleeby et al, 1982). Several lines of evidence indicate spreading within an island arc setting.

1. The Josephine ophiolite is thrust over a coeval calc-alkaline arc complex comprised of the Rogue Formation and related plutonic rocks (Loc. 2). The Rogue is overlain by the Galice Formation, which is virtually identical in age and petrography to "Galice" that depositionally overlies the Josephine ophiolite (Harper, 1983).

2. Metagraywackes of the Galice Formation overlying the Josephine ophiolite contain abundant andesitic detritus.

3. The Josephine ophiolite and overlying Galice Formation are cut by 150 m.y. calc-alkaline dikes and sills, indicating that arc magmatism occurred within the basin approximately 7 m.y. after formation of the ophiolite.

4. The lavas and dikes have a strong geochemical arc component (Harper, in press). This is indicated by plots using "immobile" trace elements, particularly the Ti/Cr plot (Fig. 4). In some respects, however, the lavas and dikes are transitional between island arc tholeiites (IAT) and mid-ocean ridge basalts (MORB) that are typical of modern backarc basins: (1) the lavas and dikes have Ti/V ratios that mostly fall between 20 and 28, transitional

between IAT and MORB (Fig. 4), and (2) the dikes and lavas have higher Ni contents at a given Ti/Cr ratio than IAT. In addition, Sr and Nd isotopes for a metagabbro indicate derivation from a depleted mantle source characteristic of modern oceanic crust or oceanic island arcs (Shaw and Wasserburg, in press), and the Pb isotopic composition is intermediate between oceanic arcs and MORB (Chen and Shaw, 1982). As noted above, pillow lavas underlying the Lems Ridge olistostrome are much more similar to MORB.

5. The crystallization sequence inferred from the cumulate sequence and from phenocryst assemblages is different from MORB, and the abundance of magnesian orthopyroxene in the cumulate sequence indicates an affinity to island arc magmas (Hawkins and Evans, 1983).

Thus, the regional stratigraphic relations, geochemistry, and petrography all indicate an origin of the Josephine ophiolite within an island arc setting. The unusual transitional chemistry of the dikes and lavas is somewhat problematic in that modern backarc basin basalts are typically similar to MORB (Hawkins, 1977), although transitional basalts have been dredged from the East Scotia Sea (Tarney et al, 1981). The transitional chemistry of Josephine dikes and lavas is considered to be the result of either formation of the ophiolite during the earliest phases of spreading or formation along short spreading segments normal to the trend of the arc as postulated in the model below.

Another manifestation of the 160 m.y. spreading is the Preston Peak dike complex that comprises one of the main components of a polygenetic ophiolitic terrane that structurally overlies the Josephine ophiolite and Galice Formation (Snoke, 1977; Saleeby et al, 1982). The dike complex consists of mafic dikes, sills, and diabase breccia that were built over and intruded into an older ultramafic tectonite and amphibolite basement (Loc. 4). The ultramafic rocks vary from coarsely recrystallized massive peridotite to well-foliated mylonitic rocks. Scattered inclusions of amphibolite are concordantly interlayered with the ultramafic tectonites. These basement rocks are cross-cut by dikes petrographically similar to the overlying dike complex. The basement underwent a severe deformation-metamorphic history and serpentinization prior to injection of mafic dikes and construction of the dike complex.

The dike complex consists primarily of diabase that grades upward into diabase breccia, both of which show a lower greenschist-facies static metamorphism. The intrusive section is approximately 1 to 2 km (.6 to 1.2 mi) thick and consists primarily of dikelike intrusions having chilled margins. The diabase breccia is approximately 500 m (1,640 ft) thick and has been interpreted either as a near-vent pyroclastic deposit (Snoke, 1977) or as a talus deposit formed along a submarine fault scarp (Saleeby et al, 1982). Above the breccia lies a thin sequence of siliceous agrillite.

The Preston Peak complex has been called an ophiolite, but it clearly does not represent oceanic crust generated at a spreading ridge. Based on major, trace, and rare-earth element data (Snoke et al, 1977), as well as clinopyroxene phase chemistry (Snoke and Whitney, 1979), it has been suggested that the ophiolite represents a primitive island arc tholeiite assemblage built across an older unroofed basement. Alternatively, it may be a rift-edge facies formed at the edge of the marginal basin in which the Josephine ophiolite was formed (Saleeby et al, 1982).

The Preston Peak dike complex has been assigned an age of approximately 160 m.y. based on a 159 ± 2 m.y. Pb/U zircon age determined on a late-stage quartz diorite dike and a 165 m.y. reset K–Ar amphibole age on an amphibolite having clear static textural overprinting with retrograde mineral assemblages. The Preston Peak dike complex and Josephine ophiolite are coeval within the analytical error of the dating techniques. This, along with their close spatial relationship, suggests a close genetic link. In addition, new analytical data for the Josephine ophiolite (Harper, 1984) indicate a similar chemistry between the Preston Peak dike complex and dikes and lavas of the Josephine ophiolite. Both the Preston Peak and Josephine samples are peculiar in that they have Cr concentrations lower than MORB, yet have Ni concentrations similar to MORB but higher than arc tholeiites (Fig. 4). Thus, both rock suites are considered to be transitional between MORB and island-arc tholeiites; this distinctive chemistry is considered further evidence for a genetic link between the Josephine and Preston Peak ophiolites.

The Preston Peak ultramafic assemblage has been correlated by Norman et al (1983) with the Late Triassic to Early Jurassic Rattlesnake Creek terrane of Irwin (1972). Such rocks occur along the western margin of the belt of early Mesozoic basinal and ophiolitic terranes that were amalgamated to later become the Jurassic arc basement (Fig. 1). The essential features of this amalgam are its basinal depositional sequences consisting of chert, argillite, flysch, limestone blocks, and pillow lava, and the occurrence of late Paleozoic and early Mesozoic ophiolite fragments (Irwin, 1972; Snoke, 1977; Wright, 1982; Ando et al, 1983).

Structurally overlying the Rattlesnake Creek terrane is the western Hayfork terrane (Loc. 5). The western Hayfork appears to represent the remnants of an island arc volcanic center that is intruded by cosanguinous peridotitic to dioritic plutons (Wright, 1981; Harper and Wright, in press). Interesting time and space relations exist between the western Hayfork terrane, the peridotitic-dioritic arc plutonic belt, and the Josephine ophiolite. Arc volcanism and magmatism ceased in the western Hayfork terrane and in the arc plutonic belt at the time of intrusion of the Preston Peak dike swarm and spreading generation of the Josephine ophiolite (Saleeby et al, 1982; Harper and Wright, in press). Furthermore, the locus of arc volcanism shifted to the Rogue sequence at this time. These relations suggest that Josephine spreading disrupted the locus of Jurassic arc activity in a fashion that is comparable to modern fringing arc systems of the western Pacific where extinct "remnant" arcs are created by rifting and sea-floor spreading resulting in new localization of an active arc segment (Karig, 1972). Such an arc–interarc basin–remnant arc triad is apparent by sequentially viewing sections 2 through 5 in Figure 3. However, the Josephine spreading geometry suggests that the western edge of the Hayfork–Rattlesnake Creek remnant arc was a boundary transform.

Table 1

1. Wild Rogue ophiolite	Ramp and Gray, 1980; Saleeby, in press b
2. Rogue arc complex	Garcia, 1982; Saleeby, in press b
3. Josephine ophiolite	Harper, 1980, 1984; Saleeby et al, 1982
4. Preston Peak dike complex	Snoke, 1977; Saleeby et al, 1982
5. Western Hayfork arc sequence	Wright, 1982; Harper and Wright, in press
6. Northern Coast Range ophiolite	Bailey et al, 1970; Hopson et al, 1981; Blake and Saleeby, unpublished data
7. Southern Coast Range ophiolite	Evarts, 1977; Sharp and Evarts, 1982; Bailey et al, 1970; Hopson et al, 1981
8. Smartville ophiolite	Xenophontos and Bond, 1978; Saleeby and Moores, in press
9. Feather River dike swarm and older ophiolitic slabs	Hietanen, 1981; Saleeby and Moores, in press
10. Folsom dike swarm and Bear Mtns. ophiolitic complex	Clark, 1964; Behrman, 1978; Springer, 1980; Saleeby, 1982
11. Sonora dike swarm and Calaveras complex	Schweickert et al, 1977; Sharp and Saleeby, 1979; Sharp, 1980
12. Penon Blanco arc sequence	Morgan, 1976; Saleeby, 1982
13. Kings River ophiolite	Saleeby, 1978, 1982
14. Kaweah ophiolitic melange	Saleeby, 1979, 1982

Table 1—Informal names and references for stratigraphic and structural successions in ophiolitic and island arc terranes shown in Figure 3.

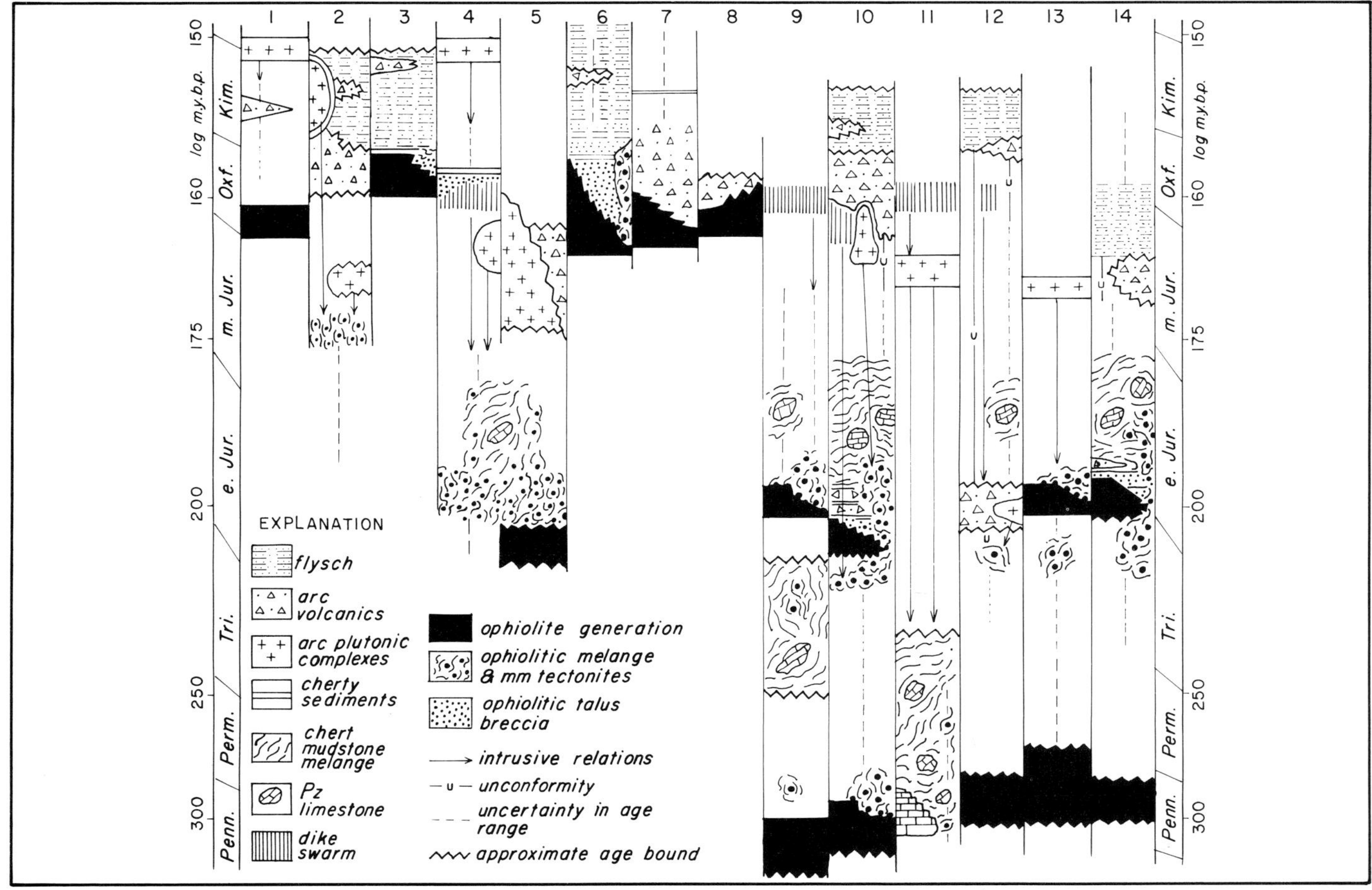

Figure 3—Stratigraphic or structural seccessions diagnostic of ophiolitic and island arc terranes of the western Sierra Nevada, Klamath Mountains, and Coast Ranges. Locations are shown on Figure 1 and references along with informal names are given in Table 1. Log time scale roughly reflects limitations in time resolution because of polymetamorphism and structural disruption, primarily in older assemblages. Rock bodies in a given succession that are not shown in direct contact or connected by unconformity or intrusive symbols are presently in tectonic contact.

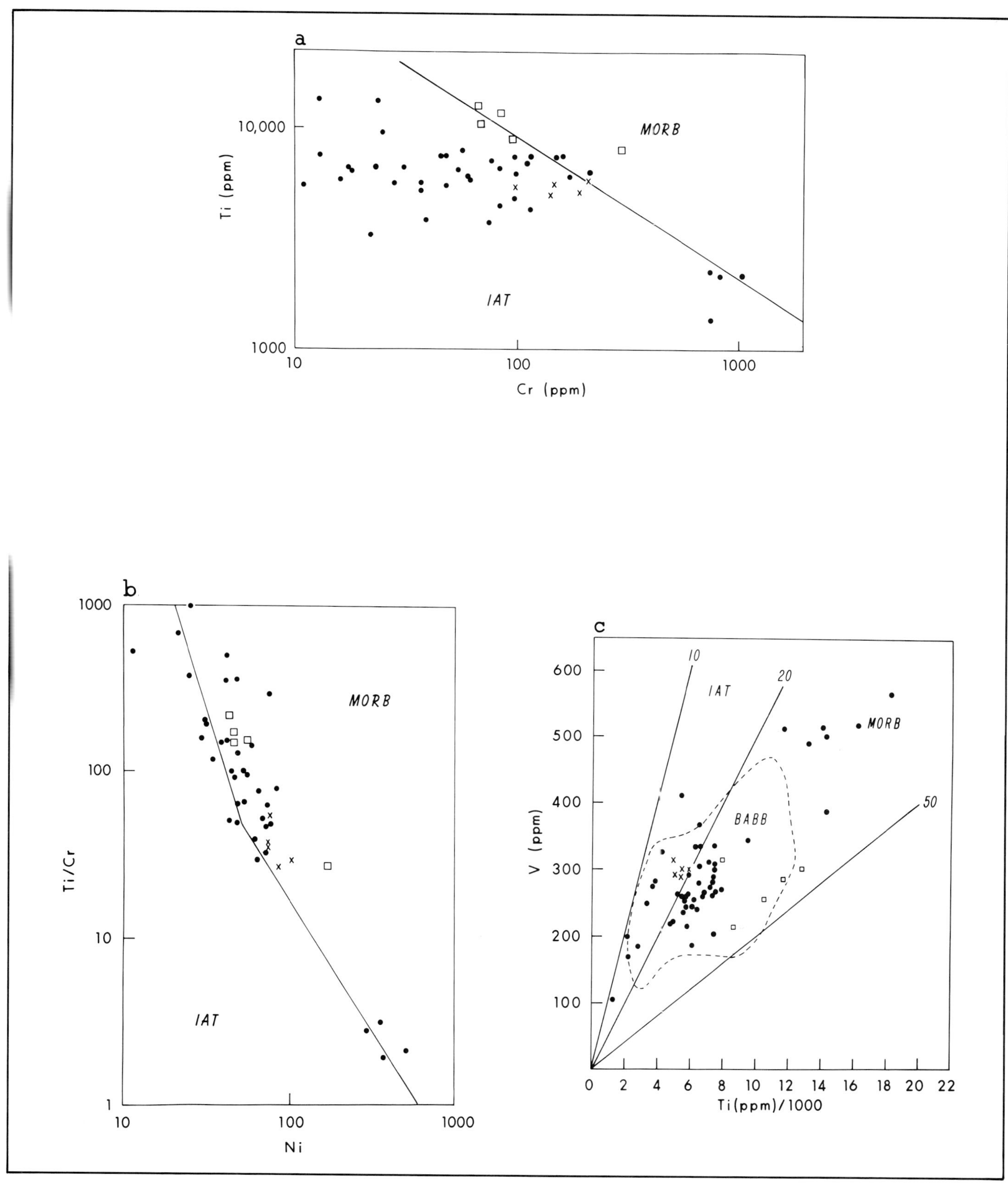

Figure 4—Plots of trace elements for dikes and lavas of the Josephine ophiolite (solid circles). Josephine pillow lavas beneath the Lems Ridge olistostrome (squares), and the mafic complex of the Preston Peak complex (X). Data are from Harper (in press and unpublished data) for the Josephine ophiolite and from Snoke et al (1977) for the Preston Peak complex. (**a**) Ti versus Cr plot after Pearce (1975). (**b**) Ti/Cr versus Ni plot after Beccaluva et al (1979). (**c**) Ti versus V plot after Shervais (1982). MORB = mid-ocean ridge basalt, IAT = island-arc tholeiite, BABB = backarc basin basalt.

The diabase (talus?) breccia of the Preston Peak complex and possibly the structurally complex serpentinite melange of the Rattlesnake Creek terrane (Wright, 1981) may be vestiges of such a boundary structure. Superimposed thrusting has unfortunately obscured possible primary relations.

In summary, the spreading geometry of the Josephine ophiolite, its petrochemistry, and its stratigraphic, temporal, and spatial relations with adjacent arc terranes suggest that its sea-floor spreading generation split a fringing island arc into remnant and active arc segments. Furthermore, spreading and transform trends were parallel to the continental margin. A similar geometry is suggested by structural and spatial relations in the Coast Range ophiolite belt.

FRACTURE ZONE AND ARC SEGMENTS OF THE COAST RANGE OPHIOLITE

The Coast Range ophiolite extends primarily along the eastern margin of the Coast Ranges (Fig. 1) where it forms the basement for the Great Valley Sequence. These rocks lie structurally above terranes of the Franciscan complex along the Cretaceous to Early Tertiary Coast Range thrust (Bailey et al, 1970). Pb/U isotopic ages on the Coast Range ophiolite cluster around 160 m.y. (Hopson et al, 1981). The Coast Range belt consists of two distinctly different segments that are separated by the main drainage system of the Great Valley where it funnels into the San Francisco Bay (Blake and Jones, 1981).

The northern ophiolite segment consists of normal ophiolitic components that were disrupted on the sea floor with the development of major fault scarps that reached deep plutonic levels resulting in talus breccias that became interbedded with ophiolitic pillow lavas (Loc. 6; Hopson et al, 1981; Blake and Saleeby, unpublished data). Blocks of mafic metamorphic tectonite formed contemporaneously with ophiolite melange zones that remained diapirically active through much of Cretaceous time. The northern Coast Range ophiolite has characteristics that are typical of large oceanic fracture zones and is here interpreted as such. The present orientation of the northern Coast Range fracture zone is similar to the transform trend deduced in the Josephine ophiolite. The older (Upper Kimmeridgian) facies of the Great Valley Group interfingers with the ophiolitic talus breccias and is compositionally similar to Galice flysch that overlies the Josephine ophiolite (Harper and Saleeby, unpublished data). Isotopic studies of similar heavy mineral populations are in progress in order to test a possible genetic link between the two flysch sequences. Paleocurrent data suggest that submarine fan channels of the northern Great Valley Sequence followed a north to south orientation along the trend of the Coast Range ophiolite (Ojakangas, 1968). Perhaps the lower intervals of the Great Valley Sequence were funneled along a transform valley system controlled by the northern Coast Range fracture zone.

The southern Coast Range ophiolite segment differs markedly from the northern segment. Here much, and perhaps the entire, igneous sequence represents arc volcanism and plutonism. The arc affinity of the igneous sequence is clearly shown by geochemical data (Bailey and Blake, 1974; Evarts, 1977; Shervais and Kimbrough, 1983; Natland, in press) and is consistent with field observations of abundant silicic and intermediate intrusives and volcanics. The structuring of the arc igneous sequence into an ophiolite stratigraphy suggests an intra-arc rifting environment. The paucity of well-ordered, extensive sheeted dikes indicates a poorly organized rift system. One of the few and best developed sheeted sequences occurs at the northern end of the southern segment where an original east-west orientation is suggested by paleomagnetic data (Williams, 1983); unlike the Josephine ophiolite, however, these dike orientations cannot be interpreted in light of a complete intact ophiolite sequence. The southern Coast Range segment is best displayed in the Diablo Range, and thus this segment will be referred to below as the Diablo rifted arc.

The along-strike spatial relations between the northern Coast Range fracture zone and the Diablo rifted arc mimic the relations between the Josephine ophiolite and the Rogue arc. Such relations suggest longitudinal splitting and transform dispersal of the arc segments by sea-floor spreading parallel to the continental margin. Direct relations between the western Klamath belt and the Coast Range belt cannot be made, although their relative orientations are consistent with one another. The Wild Rogue ophiolite situated west of the western Klamath belt (Loc. 1) also represents a rift basin formed adjacent to an active arc. Perhaps this along with other ophiolite remnants of southwesternmost Oregon are a continuation of the Coast Range belt. The along-strike relations of arc and transform-related ophiolite segments also occur within the Sierra Nevada foothills. However, here a more complex and longer lived interaction of arc and transform tectonics is recorded.

ARC AND FRACTURE ZONE TERRANES OF THE WESTERN SIERRA NEVADA

The western metamorphic belt of the Sierra Nevada consists of interleaved ophiolitic, marine basinal, and island arc terranes that were progressively amalgamated and then finally accreted in Triassic through Late Jurassic time. Most of the pre-Middle Jurassic terranes, together with Paleozoic rocks thought to be parautochthonous to North America, were stitched by Middle Jurassic periodotitic to dioritic arc plutons (Wright and Sharp, 1982; Saleeby et al, in press). Major Late Jurassic faults intervene between some plutons, and thus the details of the stitching pattern are unknown. Ophiolitic terranes of the western Sierra occur in two distinctly different structural and petrogenetic settings.

1. The Smartville ophiolite (Loc. 8) represents an intact 160 m.y. rifted arc sequence (Xenophontos and Bond, 1978) that has been thrust eastward over early Mesozoic basinal and ophiolitic melange terranes (Moores and Day, 1983). A complete ophiolite section is not present, but the outstanding feature is great swarms of 160 m.y. sheeted dikes that both cut and feed into basaltic pillow lavas and basaltic to locally dacitic volcaniclastic rocks. Cumulate rocks are rare, and depleted mantle rocks are not observed.

2. An older, more complex ophiolite belt extends semicontinuously along the entire length of the foothills metamorphic belt. It consists of a polygenetic mixture of ophiolitic melange and ultramafic ± mafic tectonite massifs (Saleeby, 1982) and will be referred to below as the foothills ophiolite belt. Apparent igneous ages within the belt cluster near 300 and 200 m.y., and metamorphic dates on amphibolite-facies tectonites reveal a complex pattern of events with perhaps four distinct metamorphic peaks between 300 m.y. and 155 m.y. dispersed, and in several places superimposed, along the belt (Behrman, 1978a; Saleeby, 1982; Saleeby and Sharp, unpublished data). The 300 m.y. and 200 m.y. protolith assemblages are intermixed and in places cannot even be resolved adequately by Pb/U zircon data owing to polymetamorphism. Initial Pb, Sr, and Nd isotopic compositions are complex and record derivation from multiple mantle reservoirs including ultradepleted oceanic mantle and a distinctly different, possibly subcontinental mantle (Saleeby and Chen, 1978; Saleeby, 1982, and unpublished data; Shaw and Wasserburg, in press). The foothills ophiolite belt constitutes basement for Jurassic arc rocks and is intimately associated with early Mesozoic basinal terranes (Morgan, 1976; Behrman, 1978b; Saleeby, 1979, 1982). Ophiolite exposures in the central foothills (Locs. 10 and 12) are in the cores of large Nevadan anticlines.

Numerous papers have been written on the foothills ophiolite belt from a perspective that it represents a major fracture zone assemblage upon which the Jurassic arc nucleated (Behrman, 1978a; Saleeby, 1978, 1979, 1981, 1982, in press a). Our intention is not to reiterate such discussions here but to draw attention to some of the cogent points.

1. In a number of locations the prevailing Late Jurassic (Nevadan) northwest structural trends of the foothills can be shown to have followed preexisting structures within the ophiolitic basement. These older fracture zone orientations are also parallel to 160 m.y. fracture zone trends of the Coast Ranges and western Klamaths and to major geophysical lineaments of the Great Valley.

2. Mesozoic basinal terranes of the western Sierra occur in exceedingly narrow belts, are in large part olistostromal, and often contain locally derived ophiolitic debris. Such terranes may represent a family of transform valley deposits.

3. The intermixing of 300 and 200 m.y. protolith assemblages appears in part to be primary. The younger assemblage was evidently injected into or along the edge of the older, previously disrupted assemblage, and in places injection occurred in a hot, high-stress environment (Saleeby, 1982). Such tectonic injection probably represents sea-floor spreading adjacent to, or leaking within, a preexisting fracture zone.

4. The foothills ophiolite belt coincides with the locus of a major truncation (or apparent truncation) in the Paleozoic to early Mesozoic North American shelf (Hamilton and Myers, 1967; Saleeby, 1981). Together the ophiolite belt, truncation zone, and fragments of the parautochthonous North America are considered remnants of a major early Mesozoic, and perhaps older, boundary transform between plates of the Pacific basin and the western margin of the North American plate (Saleeby, 1981, in press a). Mesozoic island arc sequences, ignimbrite fields, and batholithic complexes were superimposed across and probably migrated along the boundary zone. This fundamental structure is referred to below as the Sierran boundary transform.

The history of island arc activity in the western Sierra Nevada overlaps with the arc history of the western Klamaths and the southern Coast Ranges. Most notable is the belt of Middle Jurassic peridotitic to dioritic plutons in the Klamaths and Sierras (Fig. 1). The Smartville rifted arc is identical in age to the Diablo rifted arc. Callovian to Lower Kimmeridgian arc volcanics of the western Sierra (Locs. 8, 10, 12, 14) coincide in age with the western Hayfork and Rogue arc sequences of the western Klamaths (Locs. 2, 5). However, basement continuity apparently exists between the Middle and Late Jurassic arc sequences in the western Sierra, whereas the Josephine rift basin intervenes between the Rogue and western Hayfork arcs in the Klamaths. Older, Early Jurassic arc activity is recorded in the central Sierra foothills by the Penon Blanco Formation (Loc. 12, Morgan, 1976; Saleeby, 1982) and in scattered fault-bounded slices in the northern Sierra Nevada (Saleeby and Moores, in press). The older arc-rocks of the western Sierra may have analogs in the Rattlesnake Creek terrane of the western Klamath Mountains (Wright, 1982).

Lower to Middle Jurassic arc sequences rest depositionally on Paleozoic and Triassic basement amalgams of the eastern Klamaths and northern Sierra (Irwin, 1966; McMath, 1966; Clark, 1976). Paleogeographic proximity between these eastern arc sequences and the western Jurassic arc sequences of the Sierra and Klamaths is suggested by two points: (1) Middle Jurassic peridotitic to dioritic plutons appear to stitch basement amalgams of the eastern and western arc sequences (Harper and Wright, in press); and (2) clasts of the distinctive Permian McCloud Limestone occur within Middle Jurassic arc rocks of the western Sierra (Loc. 10; Berman and Parkison, 1978; Saleeby, 1983) and within the Hayfork terrane of the western Klamaths (Harper and Wright, in press). The McCloud Limestone is one of the distinctive formations of the eastern Klamath amalgam. It is not suggested here that the eastern arc sequences are simply continuous lateral equivalents of the western arc sequences. Rather, a complex pattern of transform and oblique rifts is suggested within and along the locus of Jurassic arc activity, and the original spatial relations have been further complicated by the Nevadan orogeny and post-Nevadan dispersion. Attention will now focus on our interpretation of Jurassic sea-floor spreading along the California margin, and how it relates to transform dispersion of arc segments and to plate kinematic patterns.

JURASSIC SEA-FLOOR SPREADING AND TRANSFORM MOTION ALONG THE CALIFORNIA MARGIN

An important finding displayed graphically in Figure 3 is the regional synchronism in ophiolite genesis and basaltic dike intrusion into older accreted terranes at 160 m.y. throughout the region shown in Figure 1. Furthermore, the Sierran-Klamath belt of Middle Jurassic peridotitic to dioritic plutons ceased its main activity at 160 m.y., apparently as a response to the rifting and sea-floor spreading. This reponse is distinctly pre-Nevadan, even though the Nevadan orogeny followed closely in time (ca. 150 m.y.). Parallel relations exist in the southeast California–Sonora, Mexico, region where the continental magmatic arc was disrupted at about 160 m.y. as rapid transform motion of the Sonora–Mojave megashear commenced (Silver and Anderson, 1974, 1983, and personal communication, 1983; Anderson and Schmidt, 1983). Do these parallel events merely reflect a random sampling of events packaged into a mosaic of unrelated terranes? We suggest not and present the following model primarily in hopes of stimulating discussion and future research on processes of transform faulting and ocean crust generation within the Cordilleran orogen.

Figure 5 is a model of the 160 m.y. spreading and transform geometry for the central California to southern Oregon region. The transform geometry of the Josephine ophiolite and the northern Coast Range ophiolite is strongly influential in the model. One of the main interpretations presented regards the structure and origin of the Great Valley basement. Gravity, magnetic, seismic, and basement core data synthesized by Cady (1975) indicate that the Valley is underlain by oceanic-type basement. In Figure 1 the traces of major magnetic and gravity anomalies are shown. Source rock models of the anomalies generated by Suppe (1978) and Griscom (in Saleeby et al, in press) suggest that the linear anomalies arise from steep edges or inflections in gently west-dipping ophiolitic crust. Such crust is disrupted beneath the Coast Ranges and is apparently uplifted and exposed as the Coast Range ophiolite along the Coast Range thrust and younger high-angle faults. The major Great Valley anomalies extend for over 300 km (186 mi) down the axis of the northern and central Valley until a transverse anomaly appears to offset the main anomaly over to the western Sierra foothills (Zietz et al, 1969). South of this transverse anomaly, the main anomalies coincide with the Kings–Kaweah fracture zone assemblage; the last phase of spreading along or within the Kings–Kaweah fracture zone was Early Jurassic (about 200 m.y., Loc. 13, 14).

It is suggested that the origin of the Great Valley geophysical lineaments is revealed by the Kings–Kaweah ophiolite belt, and that these lineaments represent major fracture zones in Jurassic oceanic crust. Two major phases of Jurassic spreading occurred along the Great Valley trend—200 and 160 m.y. The extent of Early Jurassic spreading and transform motion is unknown because of superimposed arc magmatism and dispersal by the 160 m.y. spreading and transform episode. In Figure 5 it is suggested that the transverse lineament between the main mid-Valley and Kings–Kaweah lineaments is the remnant of a short spreading-ridge segment. Repeated dispersal arising from Early Jurassic and/or 160 m.y. spreading along this ridge segment may account for the marked change between the central and southern Sierra metamorphic belt. In the south, very little of the metamorphic belt remains (Fig. 1). To the north, a major Middle to Late Jurassic volcanic arc remains in a deformed but relatively intact state. The paucity of Callovian to Kimmeridgian arc-related strata in the southern foothills suggests major dispersal arising from 160 m.y. spreading. If so, a local boundary between 160 m.y. crust and the older Kings–Kaweah transform lies buried beneath the Valley fill west of the ophiolite belt. Such a boundary could step westward at the transverse lineament and continue northward as the main axial Great Valley lineament. The lower amplitude southern continuation of the axial lineament could be a fracture zone within 160 m.y. crust, akin to the northern Coast Range fracture zone. South of the San Francisco Bay area, the possible continuation of the northern Coast Range transform is shown as a boundary zone against the Diablo rifted arc. This suggests that a normal rift edge between juvenile oceanic crust and the rifted arc lies concealed where the drainage of the Great Valley funnels into the San Francisco Bay area. Dike orientations reflecting this normal rift edge are shown provisionally in Figure 5 after Williams (1983). As noted earlier, an analogous pattern in longitudinal dispersion of arc segments is suggested in the structural and stratigraphic relations between the Josephine ophiolite and the Rogue arc sequence. The Wild Rogue ophiolite (Loc. 1) probably represents yet another remnant of a rift basin formed adjacent to a Late Jurassic arc segment. Dike orientations for this ophiolite are shown tentatively based on the predominance in orientations observed in the Rogue River canyon where the ophiolite belt was first described (Ramp and Gray, 1980) and dated (Saleeby, in press b).

The Great Valley basement is suggested to be primarily the remnants of a composite leaky transform system much like the Cayman Trough (Perfit and Heerzen, 1978) or the Andaman Sea (Curray et al, 1979). The Cayman Trough is of particular interest because a small spreading ridge segment has inserted itself and is spreading along the axis of a large preexisting transform deep that has been instrumental in dispersing active and extinct arc segments. An analogous situation is suggested in Figure 5 with the Great Valley rift. Igneous accretion along the Great Valley appears to have nucleated along the preexisting Sierran boundary transform; the eastern oceanic wall of the boundary transform contains fragments of late Paleozoic fracture zone crust against which, and perhaps into which, Early Jurassic ophiolitic crust was generated (Locs. 11, 13, 14). Direct contacts between 160 m.y. ophiolite and this older Sierran fracture zone complex are not observed. This older fracture zone complex constituted much of the Sierran foothills arc basement at 160 m.y. Where 160 m.y. intra-arc rifting occurred within the older fracture zone framework, systems of sheeted and isolated dikes intruded preferentially along the preexisting fracture zone fabric; examples include the Feather River and Folsom dike swarms, which were emplaced into older ophiolitic melange (Locs. 9, 10), and perhaps the Smartville ophiolite (Loc. 8). In contrast, the Sonora dike swarm (Loc. 11) was emplaced into the mid-Jurassic arc plutonic belt and its annealed metamorphic wall rocks (Sharp, 1980); here the dikes trend generally east-northeast, indicating extension parallel to the axis of the Sierra Nevada. These relations suggest that a very limited

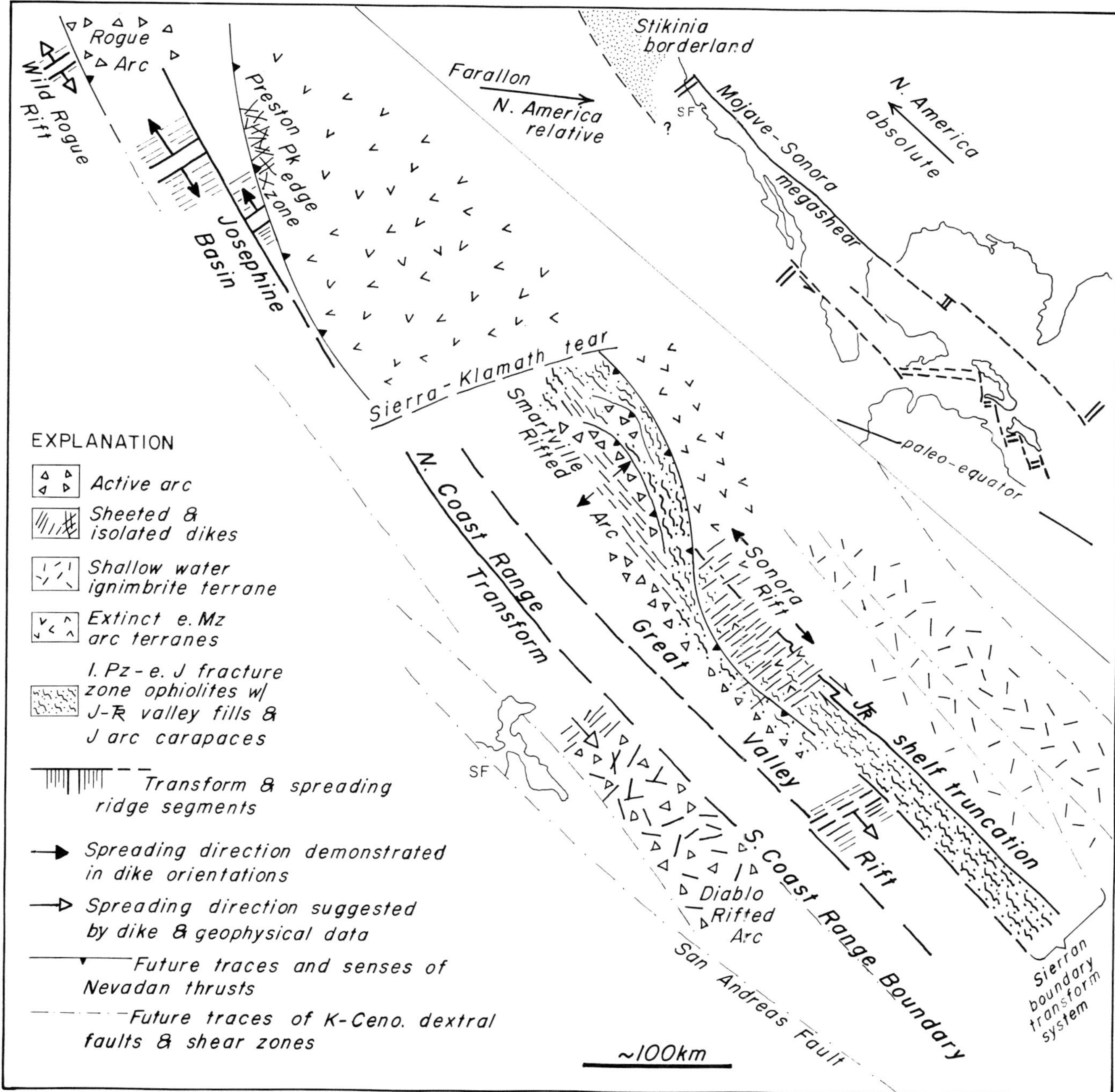

Figure 5—Sketch map showing proposed spreading and transform geometry for 160 m.y. ophiolites of California and southwest Oregon and their spatial relations with coeval and older Mesozoic arc terranes. Also shown are the traces of "future" Late Jurassic thrust faults. Ancillary map in upper right is Mercator projection of Mojave–Sonora megashear (after Anderson and Schmidt, 1983) showing its relation to 160 m.y. rift and transform features of main figure. Stikinia borderland region is taken after Davis et al (1978), Monger and Irving (1980), and Saleeby (1983). Absolute motion of North America is after Gordan et al (1982); relative motion between North America and Farallon plate is after Engebretson (1982).

component of 160 m.y. transform motion occurred along the Sierran foothills, perhaps an amount equivalent to the amount of extension represented in the Sonora dike swarm. As shown in Figure 5, the main transform motion is suggested to have occurred along Great Valley lineaments and the northern Coast Range transform.

Figure 5 also shows the general distribution of the Early to Middle Jurassic arc sequences that were built over the Paleozoic borderland amalgam. As noted above, arc volcanism and plutonism ceased in these regions during or prior to the commencement of the 160 m.y. rift-transform system. In the southeast, an extensive ignimbrite terrane was active over the North American sialic edge during the 160 m.y. rifting event. It should be noted that blocks and clasts of quartzite and silicic tuff occur within the Jurassic chert-argillite-sandstone olistostrome fill of the Kaweah transform valley (Loc. 4). The ignimbrite terrane developed in an environment dominated by shallow-marine deposition. A number of large caldera complexes and major slump and olistostrome complexes have been identified (Fiske and Tobisch, 1978; Saleeby et al, 1978; Busby-Spera, 1983). The ignimbrite terrane apparently formed over the actively fragmenting North American shelf edge.

Figure 5 gives an impression of a left-stepping family of transforms between the Josephine ophiolite, northern Coast Ranges, axial Great Valley and Kings–Kaweah belts (lineaments). The ancillary map in the upper right shows 160 m.y. spreading and transform geometry suggested for the Mojave–Sonora megashear (Silver and Anderson, 1974, 1983, personal communication, 1983; Anderson and Schmidt, 1983). The map is modified from Anderson and Schmidt (1983), who constructed it by a Mercator projection from the pole of rotation defined by the megashear. Temporal and spatial relations and the sinistral sense of movement all suggest a link between the two systems. The Josephine–Great Valley rift system and megashear together define a zone of major decoupling between the North American plate and plates of the Pacific basin. This may be viewed in the context of the late Middle to Late Jurassic rapid northwestward absolute motion of North America suggested by Gordan et al (1981) with a linkage to spreading in the equatorial Atlantic. The model is also consistent with sinistral-sense oblique convergence between North American and the Farallon plates suggested by Engebretson (1982). The decoupling zone disrupted both fringing island arc terranes of central California and southern Oregon and the continental arc terrane of southeastern California and Sonora. The petrotectonic significance of the southern Sierra ignimbrite field is uncertain; it could be a direct response to rifting along the sialic edge, or it could represent a major extensional cycle within the continental arc.

THE NEVADAN OROGENY AND POST-NEVADAN DISPERSION

The Nevadan orogeny was originally defined as a Jurassic deformation event in the Sierra Nevada region by Blackwelder (1914). Numerous workers in subsequent years included batholith emplacement and related metamorphism as part of the Nevadan orogeny, which in effect expanded its time span to include much of the Mesozoic. The Nevadan orogeny was redefined by Bateman and Clark (1974) as a short-lived, intense deformation event in Late Jurassic time. An approximate timing of 150 m.y. fits Bateman and Clark's (1974) definition and time relations in the western Klamaths discussed in Saleeby et al (1982) and is adopted here. The main manifestations of the Nevadan orogeny in the Sierra–Klamath belts are the development of through-going fault systems along the western margins of the ranges and regional folding with the development of slaty cleavages, primarily in argillaceous strata. Nevadan metamorphism is generally lower- to subgreenschist facies. Two notable exceptions are at Location 10 where amphibolite facies conditions were reached because of primary heat within the volcanic arc (Saleeby, 1982) and in a large window (Condrey Mountain) exposing blueschist–greenschist rocks east of Locations 3 and 4 in the Klamath Mountains (Klein, 1977). The Condrey Mountain window is of interest because it exposes possible Josephine basin-affinity rocks that were thrust eastward beneath the extinct mid-Jurassic arc and its basement amalgam. Schists similar to those in the Condrey Mountain window occur in the northern Coast Ranges and, in particular, along the western perimeter of the Klamath province. A fundamental question is whether these schists represent additional thrust sheet(s) of Josephine basin rocks that were extracted from the Nevadan metamorphic core and reshuffled during early Franciscan–Dothan accretion.

Eastward underthrusting of the Josephine basin beneath older terranes of the Klamath Mountains is supported by structural studies (Irwin, 1977; Snoke, 1977; Klein, 1977) and seismic refraction and magnetic data (Fuis et al, in press). This sense of thrusting is depicted on Figure 5 with underthrusting nucleating along the boundary structure between the Josephine basin and the extinct Middle Jurassic arc terrane. The Preston Peak dike complex and perhaps ophiolitic melange of the Rattlesnake Creek terrane represent possible remnants of this boundary zone. The opposite sense of thrusting is depicted in the northwestern Sierra where the Late Jurassic arc terrane and its fracture zone basement are thrust eastward over the extinct mid-Jurassic arc and its basement amalgam (Moores and Day, 1983). To the south the sense of thrusting is irresolvable.

The change in Nevadan vergence from the Klamaths to the northern Sierra suggests a major Nevadan-age tear between the two ranges. Such a tear could have nucleated along a spreading ridge segment similar to that hypothesized for the southern Great Valley transverse lineament. Alternatively, the change in vergence could have been directly influenced by longitudinally propagating rifts in the Smartville rifted arc (including Locs. 8, 9, 10). In this case, buoyant crust consisting of warm attenuated active arc material and bounding and subjacent fracture zone material preferentially moved tectonically over the extinct arc and its Paleozoic borderland amalgam.

The Nevadan orogeny is widely viewed as a collisional orogenic event (Schweickert and Cowan, 1975; Moores and Day, 1983). The questions that must be considered closely are what collided with what, where were they when they collided, and to what extent has post-collision dispersion modified them. These questions must be viewed in a larger context than the area shown in Figure 1. Basement amalgams for the Jurassic arc terranes of the western Sierra–Klamath belt can be compared quite closely to the much larger Stikinian amalgam of British Columbia (Davis et al, 1978; Saleeby, 1983). This large amalgam began to accrete to North America in mid-Jurassic time, probably in California paleolatitudes, and subsequently was dispersed northward in Cretaceous and Tertiary time (Monger and Irving, 1980). It seems likely that the rift geometry depicted in Figure 5 occurred within or near the southern termination of the actively accreting Stikinian amalgam. This amalgam is shown as a Stikinian borderland in the ancillary sketch of Figure 5. Late Jurassic flysch sequences notably enriched in early Mesozoic chert detritus (Loc. 2, 3, 6, 10, 12, 14) are likely to have been derived from Stikinian orogenic highlands. Perhaps there is a general provenance link between the flysch deposits and the initial Late Jurassic molasse of the Bowser Basin that blankets a large portion of Stikinia and is rich in early Mesozoic chert detritus (Eisbacher, 1981). The collapsing Stikinia borderland may have driven short-lived consumption of 160 m.y. rift basin crust and convergence across transform links resulting in what California workers call the Nevadan orogeny.

Modern cases of active spreading and transform faulting within a collapsing arc framework are nicely displayed in the Bismark, Soloman, and Coral Seas of the Melanesian reentrant. Here the collapse is being driven by the impact of the Ontong–Java Plateau on the outer fringes of the system. This brings up another important question concerning the possible impact of the Wrangellian amalgam with the outer fringes of the Stikinian borderland causing its collapse. The major Oxfordian unconformity over northern Wrangellia (Jones et al, in press) is perhaps a result of such an impact, which propagated by basin closure(s) inward to precipitate the Nevadan orogeny. This view is posed as a question for regional workers to consider.

Finally, it must be emphasized that the configuration of the continental margin depicted in Figure 1 reflects post-Nevadan dispersion to a much greater extent than Nevadan or pre-Nevadan tectonics. The great northward displacements of the Stikinia and Wrangellia amalgams are evidence for such dispersion, which may have in effect removed most of the "Nevadan" orogen from the California region. Fragments of 160 m.y. rifting and spreading products may be lodged in northwestern Washington as scattered 160 m.y. arc-affinity ophiolitic nappes which arrived in mid- to Late Cretaceous time (Whetten et al, 1978, 1980). These may also attest to the importance of post-Nevadan dispersion.

Notable uncertainties in the relations depicted in Figure 5 relate to post-Nevadan dispersion. The first concerns the locus of the megashear through the northern Mojave–Sierra Nevada region. An observable link with the Josephine–Great Valley rift system cannot be made confidently owing to Cretaceous and Cenozoic dextral faults within and east of the Sierran metamorphic framework and batholith. Furthermore, geochemical and paleomagnetic data suggest at least 300 km (190 mi) and allow up to 1,000 km (620 mi) of post-Nevadan northward displacement of the Sierran ± Klamath metamorphic belt (Frei and Cox, 1983; Saleeby et al, in press). Dextral-sense structures within the Sierra Nevada thought to have formed by right-oblique convergence from Late Triassic through Cretaceous time (Saleeby, 1981) are resolved into differing Jura–Triassic and mid- to Late Cretaceous episodes by additional field and geochronological studies. Finally, paleomagnetic data from the Diablo segment of the Coast Range ophiolite are ambiguous and allow its genesis between paleoequatorial and California latitudes (Beebe and Luyendyk, 1983, and personal communication; Williams, 1983). In the case of an equatorial origin, the Coast Range ophiolite could be more directly linked to the southern megashear system (ancillary map, Fig. 5) suggested by Anderson and Schmidt (1983) to have extended through the present Mexican volcanic belt. This would imply that the southern Coast Range boundary structure (Fig. 5), or perhaps the main Great Valley lineament, functioned as a tremendously rapid dextral transform in Early Cretaceous time in the northward transport of the Coast Range ophiolite to its mid-Cretaceous forearc basin site. An equatorial origin for the Coast Range ophiolite might also explain the different sedimentary and structural histories of the Josephine and northern Coast Range ophiolites; notably, the latter did not experience the Nevadan orogeny and did not receive epiclastic sediments until the Upper Kimmeridgian (Hopson et al, 1981).

The three uncertainties discussed above draw attention to some of the first-order problems in Cordilleran accretionary tectonics. These include a rigorous understanding of the paleomagnetic data and in particular the interplay of physical and geological processes in rock magnetization and a better structural and stratigraphic understanding of post-Jurassic dispersion and its impact on earlier accretionary events. It is hoped that this paper has helped focus attention on these and other important problems.

ACKNOWLEDGMENTS

Acknowledgment is made to the donors of the Petroleum Reseach Fund, administered by the American Chemical Society, for the support of field and laboratory study of the Galice Formation by Harper. Funding for analytical work was provided by the Research Committee of the University of Utah and is gratefully acknowledged. Support for field and geochronological studies in west-coast ophiolite and island arc terranes from N. S. F. grants EAR-7925998 and EAR-8206382 and the Alfred P. Slowan Foundation is gratefully acknowledged by Saleeby.

REFERENCES

Allen, C. M., et al, 1982, Comagmatic nature of the Wooley Creek batholith and the Slinkard pluton and age constraints on tectonic and metamorphic events in the western Paleozoic and Triassic belt, Klamath Mountains, N. California (Abs.): Geological Society of America Abstracts with Programs, v. 14, p. 145.

Ambos, E. L., and D. M. Hussong, 1982, Crustal structure of the Mariana Trough: Journal of Geophysical Research, v. 87, p. 4003–4018.

Anderson, T. H., and V. A. Schmidt, 1983, The evolution of Middle America and the Gulf of Mexico-Caribbean Sea region during Mesozoic time: Geological Society of America Bulletin, v. 94, p. 941-966.

Ando, C. J., et al, 1983, The ophiolitic North Fork terrane in the Salmon River region, central Klamath Mountains, California: Geological Society of America Bulletin, v. 94, p. 236-252.

Bailey, E. H., and M. C. Blake, 1974, Major chemical characteristics of Mesozoic Coast Range ophiolite in California: U.S. Geological Survey Journal of Research, v. 2, p. 637-658.

______, et al, 1970, On-land Mesozoic oceanic crust in California Coast Ranges: U.S. Geological Survey Professional Paper 700-C, p. C70–C81.

Ballard, R. D., and T. H. Van Andel, 1977, Morphology

and tectonics of the inner rift valley at lat 36° 50′ N on the Mid-Atlantic Ridge: Geological Society of America Bulletin, v. 88, p. 507–530.

Bateman, P. C., and L. D. Clark, 1974, Stratigraphic and structural setting of the Sierra Nevada batholith: Pacific Geology, v. 8, p. 78–89.

Beccaluva, L., et al, 1979, Geochemical discrimination between ocean floor and island arc tholeiites; application to some ophiolites: Canadian Journal of Earth Sciences, v. 16, p. 1874–1882.

Beebe, W., and B. P. Luyendyk, 1983, Preliminary paleomagnetic results from the Llanada remnant of the Coast Range ophiolite, San Benito County, California: EOS, American Geophysical Union Transactions, v. 64, p. 686.

Behrman, P. G., 1978a, Pre-Callovian rocks west of the Melones fault zone, central Sierra Nevada foothills, *in* D. G. Howell and K. A. McDougall, eds., Mesozoic paleogeography of the western United States: Society of Economic Paleontologists and Mineralogists, Pacific Section, Pacific Coast Paleogeography Symposium 2, p. 337–348.

______, 1978b, Paleogeography and structural evolution of a middle Mesozoic volcanic arc-continental margin; Sierra Nevada foothills, California: Unpublished PhD Dissertation, University of California, Berkeley, 301 p.

______, and G. A. Parkison, 1978, Paleogeographic significance of the Callovian to Kimmeridgian strata, central Sierra Nevada foothills, California, *in* D. G. Howell and K. A. McDougall, eds., Mesozoic paleogeography of the western United States: Society of Economic Paleontologists and Mineralogists, Pacific Section, Pacific Coast Paleogeography Symposium 2, p. 349–360.

Blackwelder, E., 1914, A summary of the orogenic epochs in the geologic history of North America: Journal of Geology, v. 22, p. 633–654.

Blake, Jr., M. C., and D. L. Jones, 1981, The Franciscan assemblage and related rocks in northern California: a reinterpretation, *in* W. G. Ernst, ed., The geotectonic development of California, Rubey v. 1: Englewood Cliffs, NJ, Prentice-Hall, Inc., p. 305–328.

Bogen, N. L., 1983, Magnitude of crustal extension across the northern Great Basin (40° N): paleomagnetic constraints: EOS, American Geophysical Union Transactions, v. 64, p. 685.

Bonatti, E., and J. Honnorez, 1976, Sections of the earth's crust in the equatorial Atlantic: Journal of Geophysical Research, v. 81, p. 4104–4116.

Busby-Spera, C. J., 1983, Large-volume rhyolite ash-flow eruption and submarine caldera collapse in the lower Mesozoic southern Sierra Nevada, California: EOS, American Geophysical Union Transactions, v. 64, p. 873.

Cady, J. W., 1975, Magnetic and gravity anomalies in the Great Valley and western Sierra Nevada metamorphic belt, California: Geological Society of America Special Paper 168, 56 p.

Cater, F. W., and F. G. Wells, 1953, Geology and mineral resources of the Gasquet quadrangle, California-Oregon: U.S. Geological Survey Bulletin 995-C, p. 79–133.

Chen, J. H., and H. F. Shaw, 1982, Pb-Nd-Sr isotopic studies of California ophiolites (Abs.): Geological Society of America Abstracts with Programs, v. 14, p. 462.

Choukroune, P., et al, 1978, In situ structural observations along transform fault A in the FAMOUS area, Mid-Atlantic Ridge: Geological Society of America Bulletin, v. 89, p. 1013–1029.

Christensen, N. I., and M. H. Salisbury, 1975, Structure and constitution of the lower oceanic crust: Reviews of Geophysics and Space Physics, v. 13, p. 57–86.

Clark, L. D., 1964, Stratigraphy and structure of part of the western Sierra Nevada metamorphic belt, California: U.S. Geological Survey Professional Paper 410, 70 p.

______, 1976, Stratigraphy of the north half of the western Sierra Nevada metamorphic belt, California: U.S. Geological Survey Professional Paper 923, 26 p.

Corimer, M. H., et al, 1983, New evidence for anomalously thin oceanic crust along the Kane fracture zone: EOS, American Geophysical Union Transactions, v. 64, p. 314.

Curray, J. R., et al, 1979, Tectonics of the Andaman Sea and Burma, *in* J. S. Watkins et al, eds., Geological and geophysical investigations of continental margins: American Association of Petroleum Geologists Memoir 29, p. 189–198.

Davis, G. A., et al, 1978, Mesozoic construction of the Cordilleran "collage," central British Columbia to central California, *in* D. G. Howell and K. A. McDougall, eds., Mesozoic paleogeography of the western United States: Society of Economic Paleontologists and Mineralogists, Pacific Section, Pacific Coast Paleogeography Symposium 2, p. 1–32.

Dick, H. J. B., 1976, The origin and emplacement of the Josephine peridotite of southwestern Oregon: Unpublished PhD Dissertation, Yale University, 409 p.

______, 1977, Partial melting in the Josephine peridotite; I. The effect on mineral composition and its consequences for geobarometry and geothermometry: American Journal of Science, v. 277, p. 801–832.

Dott, Jr., R. H., 1971, Geology of the southwestern Oregon coast west of the 124th meridian: Oregon Department of Geological and Mineral Industries Bulletin, v. 69, 63 p.

Eisbacher, G. H., 1981, Late Mesozoic-Paleogene Bowser Basin molasse and Cordilleran tectonics, Canada: Geological Association of Canada Special Paper 23, p. 125–151.

Engebretson, D. C., 1982, Relative motions between oceanic and continental plates in the Pacific basin: Unpublished PhD Dissertation, Stanford University.

Evarts, R. C., 1977, The geology and petrology of the Del Puerto ophiolite, Diablo Range, central California Coast Ranges, *in* R. G. Coleman and W. P. Irwin, eds., North American ophiolites: Oregon Department of Geology and Mineral Industries Bulletin 95, p. 121–139.

Fagin, S. W., and W. A. Gose, 1983, Paleomagnetic data from the Redding section of the eastern Klamath belt: Geology, v. 11, p. 505–508.

Fahan, M. R., and J. E. Wright, 1983, Plutonism, volcanism, folding, regional metamorphism and thrust faulting: contemporaneous aspects of a major Middle Jurassic orogenic event within the Klamath Mountains, northern California (Abs.): Geological Society of America Abstracts with Programs, v. 15, p. 272.

Fiske, R. S., and O. T. Tobisch, 1978, Paleogeographic significance of volcanic rocks of the Ritter Range pendant, central Sierra Nevada, California: *in* D. G. Howell and K. A. McDougall, eds., Mesozoic paleogeography of the western United States: Society of Economic Paleontologists and Mineralogists, Pacific Section, Pacific Coast Paleogeography Symposium 2, p. 209–222.

Fox, P. J., et al, 1980, Evidence for crustal thinning near fracture zones: implications for ophiolites, *in* A. Panayiotou, ed., Ophiolites: Proceedings of the International Ophiolite Symposium: Geological Survey of Cyprus, p. 161–168.

Frei, L. S., and A. Cox, 1983, The Sierra Nevada batholith as an eastern boundary of accreted terranes since the Late Cretaceous: Circum-Pacific Terrane Conference Abstracts, v. 1, Stanford University Press.

Fuis, G. S., et al, in press, A geologic interpretation of seismic-refraction results in northeastern California: Journal of Geophysical Research.

Garcia, M. O., 1979, Petrology of the Rogue and Galice Formations, Klamath Mountains, Oregon: identification of a Jurassic island arc sequence: Journal of Geology, v. 86, p. 29–41.

———, 1982, Petrology of the Rogue River island-arc complex, southwest Oregon: American Journal of Science, v. 282, p. 783–807.

Gordan, R. G., et al, 1981, Paleomagnetic Euler Poles for the absolute motion of North America during the Mesozoic and late Paleozoic: EOS, American Geophysical Union Transactions, v. 62, p. 853.

Gray, G. G., 1982, Northward continuation of the Rattlesnake Creek terrane, north-central Klamath Mountains, California (Abs.): Geological Society of America Abstracts with Programs, v. 14, p. 167.

Hamilton, W., and W. B. Myers, 1967, The nature of batholiths: U.S. Geological Survey Professional Paper 554-C.

Harding, D. J., and J. M. Bird, 1983, Accreting plate margin structures in the Josephine ophiolite tectonite and cumulate sequences: EOS, American Geophysical Union Transactions, v. 64, p. 846.

Harper, G. D., 1980, The Josephine Ophiolite—remains of a Late Jurassic marginal basin in northwestern California: Geology, v. 8, p. 333–337.

———, 1982, Evidence for large-scale rotations at spreading centers from the Josephine ophiolite: Tectonophysics, v. 82, p. 25–44.

———, 1983, A depositional contact between the Galice Formation and a Late Jurassic ophiolite in northwestern California and southwestern Oregon: Oregon Geology, v. 45, p. 3–9.

———, 1984, The Josephine ophiolite, northwestern California: Geological Society of America Bulletin, v. 95, p. 1009–1026.

———, and J. E. Wright, in press, Middle to Late Jurassic tectonic evolution of the Klamath Mountains, California: Tectonics.

———, et al, 1983, The Lems Ridge olistostrome—sediment fill of an ancient fracture zone: Geological Society of America Abstracts with Programs, v. 15, p. 427.

Hawkins, J. W., 1977, Petrologic and geochemical characteristics of marginal basin basalts, *in* M. Talwani and W. C. Pitman, III, Island arcs, deep sea trenches, and back-arc basins: American Geophysical Union, Washington, DC, p. 355–365.

———, and C. E. Evans, 1983, Geology of the Zambales Range, Luzon, Philippine Islands: Ophiolite derived from an island arc-backarc basin pair, *in* D. E. Hayes, ed., The tectonic and geologic evolution of southeast Asia seas and islands, Part 2: Washington, DC, American Geophysical Union, p. 124–138.

Hekinian, R., and G. Thompson, 1976, Comparative geochemistry of volcanics from rift valleys, transform faults and aseismic ridges: Contributions to Mineralogy and Petrology, v. 57, p. 145–162.

Hietanen, A., 1981, Petrologic and structural studies in the northwestern Sierra Nevada, California: U.S. Geological Survey Professional Paper 1226-A-C, 59 p.

Hopson, C. A., et al, 1981, Coast Range ophiolite, western California, *in* W. G. Ernst, ed., The geotectonic evolution of California, Rubey v. 1, Englewood Cliffs, NJ, Prentice-Hall, Inc., p. 418–510.

Irwin, W. P., 1966, Geology of the Klamath Mountains province, *in* E. H. Bailey, ed., Geology of northern California: California Division of Mines and Geology Bulletin, v. 190, p. 19–38.

———, 1972, Terranes of the western Paleozoic and Triassic belt in the southern Klamath Mountains, California: U.S. Geological Survey Professional Paper 800-C, p. 103–111.

———, 1977, Ophiolitic terranes of California, Oregon, and Nevada: Oregon Department Geology Mineral Industries Bulletin, v. 95, p. 75–92.

Jones, D. L., et al, in press, Collision tectonics in the Cordillera of western North America; examples from Alaska: Transactions of the Royal Society of London, Series A.

Karig, D. E., 1972, Remnant arcs: Geological Society of America Bulletin, v. 83, p. 1057–1068.

Kimbrough, D. L., and J. W. Shervais, 1983, Geochemistry of the Coast Range ophiolite: A composite paleooceanic basement terrane in western California: EOS, American Geophysical Union Transactions. v. 64, p. 327.

Klein, C. W., 1977, Thrust plates of the north-central Klamath Mountains near Happy Camp, California: California Division of Mines and Geology Special Report 129, p. 23–26.

Luyendyk, B. P., and K. C. Macdonald, 1977, Physiography and structure of the inner floor of the FAMOUS rift valley: observations with a deep-towed instrument package: Geological Society of America Bulletin, v. 88, p. 648–663.

McMath, V. E., 1966, Geology of the Taylorsville area, northern Sierra Nevada, California, *in* E. H. Bailey, ed., Geology of northern California: California

Division of Mines and Geology Bulletin, v. 190, p. 173-183.

Monger, J. H. W., and E. Irving, 1980, Northward displacement of north-central British Columbia: Nature, v. 285, p. 289-294.

Moores, E. M., 1970, Ultramafics and orogeny, with models of the U.S. Cordillera and the Tethys: Nature, v. 228, p. 837-842.

______, and H. W. Day, 1983, An overthrust model for Mesozoic Sierran tectonics: Geological Society of America Abstracts with Programs, v. 15, p. 293.

Morgan, B. A., 1976, Geology of Chinese Camp and Moccasin quadrangles, Tuolumne County, California: U.S. Geological Survey Miscellaneous Field Studies Map MF-840, scale 1:24,000.

Natland, J. H., in press, California Coast Range ophiolite remnants: fragments of an island arc crustal segment: Journal of Geology.

Nelson, K. D., 1981, A simple thermal-mechanical model for mid-ocean ridge topographic variation: Geophysical Journal of the Royal Astronomical Society, v. 65, p. 19-30.

Nichols, J., et al, 1980, Seismic velocity structure of the ophiolite at Point Sal, southern California, determined from laboratory measurements: Geophysical Journal of the Royal Astronomical Society, v. 63, p. 165-185.

Norman, E. A. S., 1984, Petrology and structure of the Summit Valley area, Klamath Mountains, California: Unpublished MS Thesis, University of Utah, Salt Lake City.

______, et al, 1983, Northern extension of the Rattlesnake Creek terrane (Abs.): Geological Society of America Abstracts with Programs, v. 15, p. 314-315.

Ojakangas, R. W., 1968, Cretaceous sedimentation, Sacramento Valley, California: Geological Society of America Bulletin, v. 79, p. 445-461.

Pallister, J. S., 1981, Structure of the sheeted dike complex of the Samail ophiolite near Ibra, Oman: Journal of Geophysical Research, v. 86, p. 2661-2672.

Pearce, J. A., 1975, Basalt geochemistry used to investigate past tectonic environments on Cyprus: Tectonophysics, v. 25, p. 41-67.

Perfit, M. R., and B. C. Heerzen, 1978, The geology and evolution of the Cayman Trench: Geological Society of America Bulletin, v. 89, p. 1155-1174.

Prinz, M., et al, 1976, Ultramafic and mafic dredge samples from the equatorial Mid-Atlantic Ridge and fracture zones: Journal of Geophysical Research, v. 81, p. 4087-4103.

Ramp, L., and F. Gray, 1980, Sheeted dikes of the Wild Rogue Wilderness, Oregon: Oregon Geology, v. 42, p. 119-124.

Saleeby, J. B., 1978, Kings River ophiolite, southwest Sierra Nevada foothills, California: Geological Society of America Bulletin, v. 89, p. 617-636.

______, 1979, Kaweah serpentinite melange, southwest Sierra Nevada foothills, California: Geological Society of America Bulletin, v. 90, p. 29-46.

______, 1981, Ocean floor accretion and volcano-plutonic arc evolution of the Mesozoic Sierra Nevada, California, *in* W. G. Ernst, ed., The geotectonic development of California, Rubey v. 1: Englewood Cliffs, NJ, Prentice-Hall, Inc., p. 132-181.

______, 1982, Polygenetic ophiolite belt of the California Sierra Nevada—Geochronological and tectonostratigraphic development: Journal of Geophysical Research, v. 87, p. 1803-1824.

______, 1983, Accretionary tectonics of the North American Cordillera: Annual Review of Earth and Planetary Sciences, v. 15, p. 45-73.

______, in press a, The California margin, a hybrid boundary transform since when?, *in* Ocean basins and margins, v. 7: New York, Plenum Publishing Co.

______, in press b, Pb/U zircon ages from the Rogue River area, western Jurassic belt, Klamath Mountains, Oregon: Geological Society of America Abstracts with Programs, v. 16.

______, and J. H. Chen, 1978, Preliminary report on initial lead and strontium isotopes from ophiolitic and batholithic rocks, southwestern foothills, Sierra Nevada, California: U.S. Geological Survey Open-File Report 78-701, p. 375-376.

______, and E. M. Moores, in press, Continent-Ocean transect: Geochronological and structural profiles across the northern (c1) and southern (c2) Sierra Nevada metamorphic belt (Abs.): Geological Society of America Abstracts with Programs, v. 16, n. 6.

______, et al, 1978, Early Mesozoic paleotectonic-paleogeographic reconstruction of the southern Sierra Nevada region, *in* D. G. Howell and K. A. McDougall, eds., Mesozoic paleogeography of the western United States: Society of Economic Paleontologists and Mineralogists, Pacific Section, Pacific Coast Paleogeography Symposium 2, p. 311-336.

______, et al, 1982, Time relations and structural-stratigraphic patterns in ophiolite accretion, west central Klamath Mountains, California: Journal of Geophysical Research, v. 87, p. 3831-3848.

______, et al, in press, Explanatory pamphlet for the Continent-Ocean Transect: Corridor 2, central California offshore to the Colorado plateau.

Schultz, K. L., and S. Levi, 1983, Paleomagnetism of Middle Jurassic plutons of north-central Klamath Mountains (Abs.): Geological Society of America Abstracts with Programs, v. 5, p. 427.

Schweickert, R. A., and D. S. Cowan, 1975, Early Mesozoic tectonic evolution of the western Sierra Nevada, California: Geological Society of America Bulletin, v. 86, p. 1329-1336.

Sharp, W. D., 1980, Ophiolite accretion in the northern Sierra: EOS, American Geophysical Union Transactions, v. 61, p. 1122.

Shaw, H. F., and G. J. Wasserburg, in press, Isotopic constraints on the origin of Appalachian mafic complexes: American Journal of Science.

Shervais, J. W., 1982, Ti-V plots and the petrogenesis of modern and ophiolitic lavas: Earth and Planetary Science Letters, v. 59, p. 101-118.

______, and D. L. Kimbrough, 1983, The Coast Range ophiolite of California; a complete island arc-oceanic crust terrane (Abs.): Geological Society of America Abstracts with Programs, v. 15, p. 685.

Silver, L. T., and T. H. Anderson, 1974, Possible left-lateral early to middle Mesozoic disruption of the southwestern North American craton margin (Abs.): Geological Society of America Abstracts with Programs, v. 6, p. 955.

———, 1983, Further evidence and analysis of the role of the Mojave-Sonora megashear(s) in Mesozoic Cordilleran tectonics (Abs.): Geological Society of America Abstracts with Programs, v. 15, p. 273.

Simpson, R. W., and A. Cox, 1980, Paleomagnetic evidence for tectonic rotation of the Oregon Coast Range: Geology, v. 5, p. 585–589.

Sinha, M. C., and K. E. Louden, 1983, Lateral changes in crustal structure at the Oceanographer fracture zone: EOS, American Geophysical Union Transactions, v. 64, p. 314.

Sinton, J. M., et al, 1983, Petrologic consequences of rift propagation on oceanic spreading ridges: Earth and Planetary Science Letters, v. 62, p. 193–207.

Snoke, A. W., 1977, A thrust plate of ophiolitic rocks in the Preston Peak area, Klamath Mountains, California: Geological Society of America Bulletin, v. 88, p. 1641–1659.

———, and S. E. Whitney, 1979, Relict pyroxenes from the Preston Peak ophiolite, Klamath Mountains, California: American Mineralogist, v. 64, p. 865–873.

———, et al, 1977, The Preston Peak ophiolite, Klamath Mountains, California, an immature island arc: Petrochemical evidence: California Division of Mines and Geology Special Report 129, p. 67–79.

———, et al, 1982, Significance of mid-Mesozoic peridotitic to dioritic intrusive complexes, Klamath Mountains-western Sierra Nevada, California: Geology, v. 10, p. 160–166.

Stroup, J. B., and P. J. Fox, 1981, Geologic investigations in the Cayman Trough: evidence for thin oceanic crust along the Mid-Cayman Rise: Journal of Geology, v. 89, p. 395–420.

Suppe, J., 1978, Cross section of southern part of northern Coast Ranges and Sacramento Valley, California: Geological Society of America Map and Chart Series MC-28B.

Tarney, J., et al, 1981, Geochemical aspects of back-arc spreading in the Scotia Sea and western Pacific: Philosophical Transactions of the Royal Society of London, Series A, v. 300, p. 263–285.

Vail, S. G., 1977, Geology and geochemistry of the Oregon Mountain area, southwestern Oregon and northern California: Unpublished PhD Dissertation, Oregon State University, 159 p.

Van Andel, T. H., and R. D. Ballard, 1979, The Galapagos Rift at 86° W: 2. Volcanism, structure and evolution of the rift valley: Journal of Geophysical Research, v. 84, p. 5390–5406.

Verosub, K. L., and E. M. Moores, 1981, Tectonic rotations in extensional regimes and their paleomagnetic consequences for oceanic basalts: Journal of Geophysical Research, v. 86, p. 6335–6349.

Walker, R. G., and E. Mutti, 1973, Turbidite facies and facies associations, *in* G. V. Middleton and A. H. Bouma, eds., Turbidites and deep water sedimentation: Society of Economic Paleontologists and Mineralogists, Pacific Section, p. 119–157.

Wells, F. G., and G. W. Walker, 1953, Geology of the Galice quadrangle, Oregon: U.S. Geological Survey Geologic Quadrangle Map GQ-25, scale 1:62,500.

Whetten, J. T., et al, 1978, Ages of Mesozoic terranes in the San Juan Islands, Washington, *in* D. G. Howell and K. A. McDougall, eds., Mesozoic paleogeography of the western United States: Society of Economic Paleontologists and Mineralogists, Pacific Section, Pacific Coast Paleogeography Symposium 2, p. 117–132.

———, et al, 1980, Allochthonous Jurassic ophiolite in northwest Washington: Geological Society of America Bulletin, v. 91, p. 359–368.

Williams, K. M., 1983, The Mount Diablo ophiolite, Contra Costa County, California: Unpublished MS Thesis, San Jose State University.

Wright, J. E., 1981, Geology and uranium-lead geochronology of the western Paleozoic and Triassic subprovince, southwestern Klamath Mountains, California: Unpublished PhD Dissertation, University of California, Santa Barbara, 300 p.

———, 1982, Permo-Triassic accretionary subduction complex, southwestern Klamath Mountains, northern California: Journal of Geophysical Research, v. 87, p. 3805–3818.

———, and W. D. Sharp, 1982, Mafic-ultramafic intrusive complexes of the Klamath-Sierran region, California: Remnants of a Middle Jurassic arc complex (Abs.): Geological Society of America Abstracts with Programs, v. 14, p. 245–246.

Xenophontos, C., and G. C. Bond, 1978, Petrology, sedimentation, and paleogeography of the Smartville terrane (Jurassic)—bearing on the genesis of the Smartville ophiolite, *in* D. G. Howell and K. A. McDougall, eds., Mesozoic paleogeography of the western United States: Society of Economic Paleontologists and Mineralogists, Pacific Section, p. 291–302.

Zietz, I., et al, 1969, Aeromagnetic investigation of crustal structure for a strip map across the western United States: Geological Society of America Bulletin, v. 80, p. 1703–1714.

Subduction, Magmatic Arcs, and Foreland Deformation

Warren Hamilton
U.S. Geological Survey
Denver, Colorado

Subduction is usually assumed to consist of the sliding down a slot, fixed in the mantle, of oceanic lithosphere that rolls over a platen also fixed in position; and the overriding plate is commonly assumed to be shortened compressively across the entire width of the magmatic arc and the belt of foreland deformation. Both of these assumptions are disproved by the actual characteristics of modern convergent-plate systems in which the subducting plate is of normal oceanic lithosphere and the Benioff seismic zone has a moderate to steep inclination. Actual hinges retreat into incoming oceanic plates as overriding oceanic (island arc) or continental plates advance. Subducting slabs sink more steeply than the inclinations of Benioff seismic zones, which mark positions, not trajectories, of slabs. The apparent dip is a product of the velocity of advance of the overriding plate and of the velocity of sinking (and hence of age and density) of the subducting slab. Extreme compressive shear imbricates the surficial accretionary wedge pushed in front of the overriding plate, but that plate itself commonly is not deformed compressively. Slight to severe extension, not shortening, occurs across most modern magmatic arcs, perhaps because the steeply sinking subducting slabs displace underlying mantle downward, resulting in extension of the mantle and lithosphere above the slabs.

Accretionary wedges have dynamic-equilibrium surfaces whose slopes decrease upward exponentially. Poorly consolidated materials on the subducting plate are scraped off at the toe of a wedge, the shallowest materials being accreted against the toe, the deeper being accreted beneath the toe a little farther back. Soft-sediment deformation characterizes this regime. Although many students of fossil wedges infer an olistostromal origin for much such deformation, the rarity of olistostromes visible in seismic-reflection profiles across modern trenches makes such interpretations in general unlikely. Consolidated sediments and shreds of the subducting lithosphere plate itself are underplated against the bottom of the wedge. The base of the wedge is dragged back and the wedge thus thickened by the subducting plate beneath, but the wedge simultaneously flows forward and thins by gravitational spreading. The result is internal imbrication of the wedge, and the maintenance of a dynamic profile as the wedge grows both laterally and vertically. Clastic sedimentation in trenches is primarily in the form of longitudinal turbidites, which may flow as far as 3,000 km (1,900 mi) from their source rivers. The source terranes need bear little similarity to nearby parts of the overriding plate against which the turbidites are plated in an accretionary wedge. (A lack of understanding of such relationships has led to unsound speculations regarding development of the Franciscan accretionary wedge of California.) The ratio within accretionary wedges of broken formation and melange to strata coherent at outcrop scale varies greatly, and presumably reflects the ratio of plate-convergence velocity to sediment supply at the toe of the wedge. High-pressure metamorphism to blueschist and even eclogite facies occurs in that part of the wedge which is beneath the overriding plate, but the dynamics of this metamorphism, and particularly of the manner in which its products are returned to shallow crustal levels, are poorly understood.

A narrow strip of oceanic upper lithosphere commonly is attached to the front of an overriding continental plate. Where this is the case, the understuffing of accretionary melange raises the thin, oceanic leading edge of the plate to form an outer-arc ridge, and this produces behind that leading edge, partly by elastic flexure, an outer-arc basin. Such a ridge and basin may be displayed in bathymetry as well as structure (for example, off modern Sumatra and Java and in the paleobathymetry and longitudinal deposition of the Lower Cretaceous part of the "Valley Facies" of California) or may appear as a bathymetric shelf underlain by structural ridge and basin (such as modern Chile and southern Alaska and much of the Upper Cretaceous and Paleogene parts of the "Valley Facies"). Where the leading edge of a continental plate is instead of continental crust, as was the case in the Klamath Mountains region of California and Oregon during Cretaceous time, ridge and basin are not formed. Only the leading edges of modern overriding plates are deformed by shear and compression; outer-arc basins are not shortened except at their outer edges.

Subduction of oceanic lithosphere beneath a continental plate often begins as a consequence of a plate collision. Such a collision adds an oceanic island arc or a continental fragment to a continental plate. Convergence between megaplates continues, but the continental plate is too low in density to be subducted, so a new subduction system breaks through outboard of the continental plate as enlarged by the subduction. In the case of California, the Late Jurassic collision with North America of a composite island arc was followed by inauguration of subduction of Pacific lithosphere eastward beneath the continent as enlarged by the addition of that arc. A narrow strip of the oceanic lithosphere that formed in Middle and Late Jurassic time behind the eastward-migrating arc before it collided with North America was also left attached to the continent. This strip is preserved as the ophiolitic basement of the Valley Facies, exposed in the Coast Ranges and buried beneath the Great Valley. Deposition of pelagic sediments and early distal turbidites on that basement preceded the collision; later deposition and development and filling of the outer-arc basin in the manner noted previously followed the collision. Much of the ophiolitic material present as shreds within the Franciscan accretionary wedge, west of and beneath the Great Valley ophiolite, also represents fragments of oceanic lithosphere produced behind migrating Jurassic island arcs.

A magmatic arc is primarily a product of the subducting slab, hence migrates as the position and configuration of that slab change. Extension tends to be localized within the magmatic-arc part of an overriding plate because the vertical zone represented by the arc is heated and weakened by rising magma. Oceanic island arcs are characterized by normal faulting. In the extreme form of arc extension, hinge retreat is rapid, the two sides of an arc split apart along the current magmatic belt, the back side lags behind as the leading, continuously magmatic part migrates forward to track the subducting slab, and a marginal ocean basin is produced by "backarc spreading." The extensive ophiolites exposed on land are mostly of this marginal-basin, migrating-arc type, rather than of lithosphere formed at midocean ridges. (Voluminous cumulate dunite, voluminous cumulate rich in orthopyroxene, thick crust, much plagiogranite, and various chemical signatures distinguish migrating-arc ophiolite from oceanic lithosphere formed at spreading ridges.) Tectonic accretion of such arc ophiolite to continents occurs where arcs collide with continents; the ophiolite can represent either the onramping leading edge of the arc, or, as in the Great Valley case, the lithosphere trailing behind the arc. The hypothetical process of "obduction," whereby oceanic crust is conjectured to be thrust onto continental crust in the sense opposite to that of subduction, has yet to be documented anywhere.

Modern oceanic island arcs migrate, collide, reverse polarity, bend, fold back on themselves in plan, and generally behave in complex and fast-changing ways. Collisions are commonly oblique and progressive, and are not simultaneous along their lengths. Island arcs with long steady-state histories are exceptional, and even those migrate and change their lengths and shapes. Island-arc terranes accreted to continents are often products of aggregation of diverse arcs at sea before final continental collision. The modern Philippine Islands and the Mesozoic island-arc terranes of the Klamath Mountains of California are examples of such aggregates of many arcs each.

The Middle Tertiary component of Basin–Range extension is a continental analog of the spreading of an oceanic plate above a subducting slab and was approximately orthogonal to the continental margin and bounding trench of the time. This early part of Basin– Range spreading occurred during and shortly after widespread arc magmatism through the same region, and thermal softening of the crust by the introduction of mantle heat in the magmas likely facilitated extension. Late Tertiary and Quaternary extension of the province has been, by contrast, obliquely northwestward and represents distributed slipping, in the San Andreas direction, between North American and Pacific lithosphere plates. In both middle and late Cenozoic stretching regimes, the upper crust has responded by the rotational collapse of brittle fault blocks; the middle crust has been widened by the slipping apart of great crustal lenses along ductile shear zones; and the lower crust has probably been extended by more continuously ductile shear. (The middle-crust structures are widely exposed in the region, whereas deep-crustal ones are not known to be exposed.) The combined effect of Cenozoic stretching has been a doubling of the width of the Basin–Range province.

An active example of spreading of continental crust above a subducting slab of oceanic lithosphere is given by North Island, New Zealand. There, a magmatic belt of hybrid arc-and-extension type, adjacent to the magmatic arc proper, is undergoing rapid extension, and normal faulting is widespread. The North Island magmas erupt through a thick accretionary wedge of terrigenous clastic sediments, by reaction with which the magmas are greatly modified. The general surface of the belt is below sea level in the north and little above it in the south. Surface altitude, and hence crustal thickness, apparently is decreasing with time despite the addition of mantle magma to the crust, so extension is outpacing such magmatism. The volcanic rocks are bimodal—largely silicic, but partly basaltic. The entire package of thick crust formed of ensimatic terrigenous clastic sediments plus mostly silicic magmas of transitional arc/extensional type is in many ways analogous to that of middle Paleozoic New England, for which a similar setting is inferred.

At the other extreme of ratios of extension to the mantle component of magmatism, the magmatic arc of the central Andes stands high above sea level, and apparently magmatism outbulks extension; but even here, premagmatic rocks display only minor compressional deformation, and normal faulting is present. Volcanic rocks of the central Andes resemble the granitic rocks of the Cretaceous Sierra Nevada batholith of California in composition and frequency distribution. The oceanic plate being subducted beneath the central Andes has a gentle dip and may be dragging on the base of the overriding continental plate. Crustal shortening is commonly assumed to be occurring in the overriding plate, but this has yet to be proved.

The walls of batholithic plutons—the upper-crustal parts of continental magmatic arcs—commonly display isoclinal folding, with or without severe disruption. Such deformation is generally local and an accompaniment of contact metamorphism, hence is a product of the rise and spread of the plutons and not of shortening of the lithosphere plate. Isoclinal folding is produced by discontinuous laminar flow, not by accordion compression. (Superposed coaxial folds in metamorphic rocks are commonly taken as evidence for multiple episodes of deformation, but more generally are instead products of continuing deformation within single episodes.) The multiple episodes of severe, prebatholithic crustal shortening and metamorphism that some structural geologists have inferred from structures in the metamorphic aureoles of central and eastern Sierra Nevada plutons are disproved by the lack of appropriate deformation in the pre-Cretaceous rocks of ranges adjacent to the east. Modern stratovolcanoes atop tracts of sedimentary strata tend to be ringed by concentric compressional structures produced by the spreading of the volcanoes and their subjacent plutons. Both deep and shallow deformation are driven by gravity.

A foreland thrust belt records cratonward imbrication of a preexisting stratal wedge. The back of the wedge is somehow thickened, and the wedge is partly pushed and partly spreads gravitationally, imbricating upslope toward the craton because the top of the wedge slopes downward toward that craton. The active thrust belt of the central Andes has a surface that rises more or less continuously westward to the high Cordillera Oriental, which I infer to owe its altitude to crustal thickening by voluminous intrusion along the present magmatic arc. (For a long period ending about 5 m.y. ago, the central Andean magmatic arc lay primarily farther west, in the Cordillera Occidental, but then that locus was abandoned. I infer the eastern Cordillera to have been inflated since that time and to mark the present magmatic arc, which will break through more extensively at the surface when the crust to be traversed is heated adequately.) The analogous Sevier Belt, the dominantly Cretaceous foreland thrust belt of North America, formed synchronously and in structural continuity with great Cretaceous batholiths farther west. Great shortening of continental crust is inferred by many geologists to have accompanied both Andean and Sevier foreland thrusting, but I do not regard this as proved. I infer instead that the crustal thickening that drove at least much of the thrusting was due to the gravitational rise and spread of batholithic plutons at the west sides of the imbricated wedges.

A correlation between the amount of foreland thrusting and the surface altitude of the magmatic arc is likely (and, hence, a direct correlation between thrusting and the volume of mantle-derived magma, and an inverse correlation between thrusting and plate extension across the magmatic arc). Severe extension, and no thrusting, is underway in North Island, New Zealand. There must be a great Paleogene batholith beneath the vast ignimbrite terrane of the Sierra Madre Occidental of Mexico, but only moderate surface altitude was attained, and little if any thrusting resulted. The high-standing magmatic arcs of the modern central Andes, the Paleogene Himalaya (see subsequent discussion), and, presumably, the Cretaceous North American Cordillera, produced major thrust belts.

In southeastern California and southwestern Arizona, a short distance northeast of the major Cretaceous batholithic belt, no stratal wedge was present in Cretaceous time, and no foreland thrust belt was produced; but the terrane there of middle-crustal, top-to-the-northeast synmetamorphic shearing may be analogous mechanically. Farther southeast, in Mexico, a thrust belt was produced in appropriate strata.

The preceding discussion of thrust belts is concerned with subduction systems involving normal oceanic lithosphere. Where collision of light crustal masses is instead involved in zones of plate convergence, severe crustal shortening can be the result. Crustal thickening due to collision can also produce a foreland thrust belt. Such was the case in Neogene New Guinea, where the collision of an island arc from the north ramped continental-margin strata southward onto the Australia–New Guinea craton.

The latest Cretaceous and early Paleogene Laramide deformation of the previously cratonic crust of the western United States was an event of crustal shortening. Right-slip, north-trending structures in central New Mexico give way northward and northwestward through the Central Rocky Mountains to an increasingly broad belt of basins and basement uplifts separated by great basement thrust faults and other compressive structures. The Laramide belt is a plate boundary of distributed convergence, whereby the Colorado Plateau region rotated 4° or so clockwise, relative to the continental interior, about an Euler pole in or near West Texas. COCORP reflection profiling in Wyoming can be interpreted to indicate that the great basement thrust faults of the upper crust flatten and splay downward into ductile shear zones that split the middle crust into lenses. The zone of distributed crustal shortening must continue, presumably northwestward, from the northwest limit of the cratonic Rocky Mountains to the edge of the North American plate as it existed in early Paleogene time. The Idaho batholith region was eroded to a general depth of about 15 km (10 mi) during latest Cretaceous through early Eocene time, and the uplift recorded by this erosion perhaps was due to Laramide crustal shortening on yet-unrecognized structures.

Laramide compression of the craton was synchronous with a marked change in absolute motion of North America. From about 80 to 40 m.y. ago, the continent moved rapidly westward, away from Europe. A corresponding flattening of the subducting slab, as its hinge migrated rapidly westward, is shown by the eastward sweep of arc magmatism, far inland from its earlier Cretaceous position in the Sierra Nevada batholith. The bottom of the continental lithosphere was eroded tectonically at this time, and the great thrust fault developed at its base is exposed in numerous places in southern California (where it is given various names—Vincent, Rand, Orocopia, etc.—in its various exposures). The fault truncates crystalline rocks of the continental crust, including the base of the Sierra Nevada batholith and other Cretaceous magmatic-arc terranes, at what were then depths of about 20 to 40 km (12–25 mi), as indicated by the uppermost amphibolite to middle granulite facies of the plutonic rocks. Beneath the thrust are metamorphosed terrigenous clastic (trench?) sediments, and subordinate oceanic crustal, upper mantle, and abyssal-pelagic sediments, which are exposed as far as about 200 km (120 mi) inland from the palinspastic position of the leading edge of the continental plate of the time. (This demonstrates that sediments, as dense metamorphic rocks, can be subducted far beneath a continent.) In the eastern Peninsular Ranges of south-central California, large lenses of middle-crust plutonic rocks, separated by thick ductile shear zones, were thrust relatively westward over one another near the end of Cretaceous time, apparently in response to the relatively eastward motion of the oceanic plate a short distance beneath. Presumably, drag of the base of the continent against the gently dipping subducting oceanic slab retarded the westward drift of the western part of the continental plate and resulted in the rotation shown by the Laramide compression of the craton.

The regional uplift during late Cenozoic time of much of western North America, from the eastern Great Plains to the western Cordillera, is perhaps due to the thermal and mineral-phase expansion of the abandoned, subducted slab that extended far east beneath the continent during latest Cretaceous and Paleogene time.

The southwest part of the North American craton was deformed previously, during Pennsylvanian time, by the Wichita–Uncompahgre–Oquirrh family of basins and basement uplifts, trending west-northwestward from Oklahoma to Nevada. Only at the Wichita end are the bounding structures defined clearly; there, major crustal overthrusting, and possibly also oblique left slip, is shown. The various Pennsylvanian basins and uplifts are similar dimensionally to those of Laramide deformation, and similarly may represent a zone of distributed crustal shortening between a southwestern plate and the interior of the continent. This compression presumably was related to the final closing of an ocean and the consequent collision of North and South America.

Middle and late Cenozoic clockwise rotation of the Colorado Plateau by a few degrees relative to the continental interior, about an Euler pole in north-central Colorado, is indicated by the southward-increasing crustal expansion of the Rio Grande rift system. This extension was primarily of late Oligocene and early Miocene age, but it continued during later Neogene time. An analogous amount of crustal shortening between rotated plateau and continental interior must be taken up in Wyoming, northwest Colorado, and northern Utah, although appropriate compressive structures are presently proved only in the eastern Uinta region.

Uncommonly thick continental crust, or the presence of several tens of kilometers of continental crust beneath deep-crustal rocks exposed by 30 km (20 mi) or so of erosion, is often ascribed to the subduction of one continent beneath another. The modern example commonly cited for the operation of this process is that of the hypothetical Tertiary sliding beneath Tibet of continental crust that previously was the northward extension of the Indian continent. Although the present Tibetan crustal thickness of about 70 km (45 mi) might be explained geometrically in these terms, the conjecture receives no support from the geology along the Indus-Tsangpo suture between India and Tibet. That suture is marked by ophiolite and melange and includes marine strata as young as Eocene. The very presence of these oceanic materials is all but inexplicable in terms of the undersliding hypothesis—and, even worse for the conjecture, the ophiolite dips south, not north, wherever it has actually been observed. My own inferences are that the final subduction of oceanic lithosphere that lay between Tibet and India was southward beneath India, and that the Tertiary migmatites and midcrustal granites of the high Himalaya are the deeply eroded roots of the resulting magmatic arc on the Indian plate. (There are no arc-magmatic rocks of proved post-Cretaceous age on the Tibetan side of the suture.) The hypothesis of crustal doubling by subduction of one continent beneath another thus is probably invalid in the one place where a modern example is commonly inferred. Extensive crowding aside of continental crust north of India is clearly shown by geologic structural patterns, and there may also have been much shortening of Tibetan crust by buckling and thrusting. Chinese paleontologists claim (published documentation is inadequate) that much of the uplift of the Tibetan Plateau has occurred within Neogene time. If this is correct, then part of the explanation for that uplift may be that the continental Mohorovicic discontinuity partly represents plagioclase-out density-phase changes, within rocks of mostly intermediate and mafic compositions, downward to rocks of upper-granulite and eclogite facies mineralogy. Such a phase change would migrate up or down as a result of changes in temperature or pressure caused by a varying thermal regime, or by tectonic thickening, or by tectonic or erosional thinning.

The concepts summarized briefly here rely on reports published by many geoscientists and mostly have been developed, documented, and applied in the reports listed in the Reference section.

The concepts of formation and tectonic accretion of "terranes" employed by some of the authors in this volume are derived in substantial part from a 1969 paper of mine (Mesozoic California and the underflow of Pacific mantle: Geological Society of America Bulletin, v. 80, p. 2409–2429). Some of those 1969 concepts are now obsolete in important ways. I urge those authors who are still recycling those concepts to update via the newer references listed in the following section.

REFERENCES

Hamilton, W., 1978, Mesozoic tectonics of the western United States: Society of Economic Paleontologists and Mineralogists, Pacific Section, Pacific Coast Paleogeography Symposium 2, p. 33–70.

______, 1979, Tectonics of the Indonesian region: U.S. Geological Survey Professional Paper 1078, 345 p. (Reprinted with minor corrections, 1981.)

______, 1980, Complexities of modern and ancient subduction systems: Studies in Geophysics—Continental Tectonics, Washington, DC, National Academy of Sciences, p. 33–41.

______, 1981a, Plate-tectonic mechanism of Laramide deformation: University of Wyoming Contributions to Geology, v. 19, n. 2, p. 87–92.

______, 1981b, Crustal evolution by arc magmatism: Philosophical Transactions of the Royal Society of London, Series A, v. 301, n. 1461, p. 279–291.

______, 1981c, Tectonic map of the Indonesian region, 1:5,000,000: U.S. Geological Survey Map I-875-D (2nd ed., slightly revised).

______, 1982, Structural evolution of the Big Maria Mountains, northeastern Riverside County, south-eastern California, *in* E. G. Frost and D. L. Martin, eds., Mesozoic-Cenozoic tectonic evolution of the Colorado River region, California, Arizona, and Nevada: San Diego, Cordilleran Publishers, p. 1–27.

______, 1983, Geological and ecological studies of Qinghai-Xizang Plateau (Review): Science, v. 219, p. 1315–1316.

Genetic Relationship between Lower Mesozoic Continental Strata of the Colorado Plateau and Marine Strata of the Western Great Basin: Significance for Accretionary History of Cordilleran Lithotectonic Terranes

Robert Lupe*
N. J. Silberling
U.S. Geological Survey
Denver, Colorado

The Auld Lang Syne Group incorporates the most westerly lower Mesozoic rocks in northern Nevada whose deposition can be related to the continent. Chinle and Auld Lang Syne strata are interpreted as remnants of related alluvial and marine deposition that represent a Late Triassic overlap of the Golconda allochthon in central Nevada and thus place an upper age limit on accretion of this allochthon to North America.

Three successive major influxes of fine-grained terrigenous clastic sediment from an eastern source are recorded in the marine Upper Triassic to Lower Jurassic rocks of the Auld Lang Syne Group in northwest Nevada. These events may correspond to major cycles of fluvial deposition observed in the nearest significant outcrops of Upper Triassic rocks to the east, those of the Chinle Formation 500 km (310 mi) away in east-central Utah. A genetic relationship between these widely separated sedimentary sequences, suggested by the possibility of equating cycles of fluvial delivery of sediment to cycles of deposition in the marine environment, is also supported by similarities in their compositions and by the nature of a single, isolated exposure of the Chinle in the intervening region of northeastern Nevada. By extrapolation from the relatively well-dated marine rocks in northwestern Nevada, deposition of the Monitor Butte Member and higher parts of the Chinle Formation in Utah may not have begun until Norian (mid Late Triassic) time and may have continued into earliest Jurassic time with important consequences for dating nonmarine fossils and paleomagnetic poles from the red beds of the Colorado Plateau.

INTRODUCTION

The purpose of this paper is to point out a basis for establishing a sediment-source relationship between Lower Mesozoic strata of the Auld Lang Syne Group in northwestern Nevada and the Chinle Formation of the Colorado Plateau (Fig. 1) and to discuss the implications of this relationship for interpreting the age of nonmarine strata of the western interior and the accretionary history of Cordilleran lithotectonic terranes.

Strata of the Auld Lang Syne and Chinle are, respectively, shallow- to deep-marine deposits formed along the Late Triassic continental margin and nonmarine red beds representing the upper part of an extensive alluvial plain. A genetic relationship between these widely separated sets of rocks was postulated previously. Silberling and Wallace (1969, p. 40) suggested that the derivation of the large volume of terrigenous clastic sediments within the Auld Lang Syne Group in northwest Nevada was by river "drainage from some large portion of the eastern Great Basin, . . . Colorado Plateau, and Rocky Mountain provinces." Supporting this conjecture, Poole (1961, p. 140), Stewart et al (1972, p. 76), O'Sullivan (1977, p. 145), and Lupe (1977, p. 369), among others, demonstrated strong westward to northwestward fluvial transport during Chinle deposition that would have carried sediment from the region into Nevada. Although a genetic relationship between the Auld Lang Syne Group and Chinle Formation is thus reasonable, demonstration of such a relationship has remained elusive because the principal exposures of these two rock units not only represent exclusively different depositional settings but also are about 500 km (310 mi) apart (Fig. 1). Moreover, combined with the almost total absence of outcropping Upper Triassic strata within the intervening region of the eastern Great Basin, nothing else about the geologic record in this region provides evidence for inference about its Late Triassic paleography.

Recently, however, well-defined depositional cycles in the Chinle Formation of east-central Utah have been

*Present Address: Amerada Hess Corporation, Denver, Colorado.

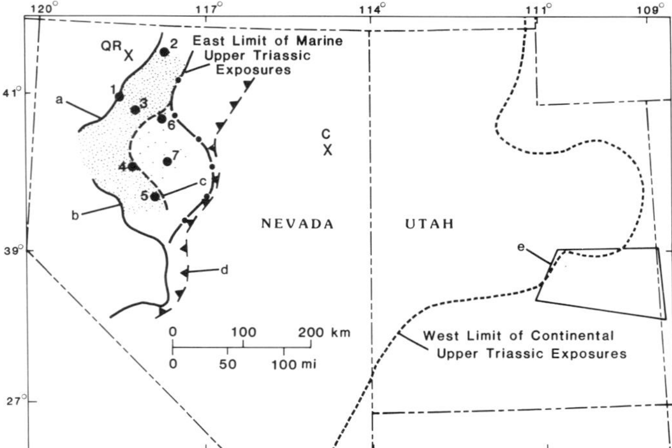

Figure 1—Index map showing the location of the Auld Lang Syne Group in northwest Nevada and the Chinle Formation in east-central Utah. The Auld Lang Syne is lightly stippled where it overlies the Star Peak Group as part of the so-called shelf terrane and heavily stippled where it forms the only pre-Tertiary strata of the so-called basinal terrane in northwest Nevada. The boundary between the shelf and basinal terranes is the Fencemaker Thrust system. Numbered localities within the outcrop area of the Auld Lang Syne correspond to numbered columns in Figure 2. a = northwest limit of exposures of Auld Lang Syne Group; b = trace of hypothetical structures juxtaposing Auld Lang Syne Group and other lower Mesozoic facies (Speed, 1978); c = west limit of Triassic Star Peak Group platform carbonate rocks; d = east limit of Golconda allochthon; e = principal study area of Chinle exposures. QR = exposures of the Quinn River Formation; C = exposures of the Chinle Formation near Currie, Nevada.

recognized by Lupe (1977, p. 365; 1979). Comparison of these depositional cycles, now recognized over a wide area of Chinle outcrop, with the episodic and cyclic nature of terrigenous sediment supply recorded in the Auld Lang Syne Group, provides a new basis for genetically relating the continental Chinle deposition to that of the marine Auld Lang Syne along the Late Triassic Pacific margin some hundreds of kilometers farther west (Fig. 2).

This postulated genetic relationship is supported by compositional compatibility between the finer grained clastic rocks of the Chinle and the Auld Lang Syne, as well as by the characteristics of a single, isolated, small outcrop of Chinle near Currie in northeast Nevada (Fig. 1), about midway in the gap between Upper Triassic exposures in northwestern Nevada and those in east-central Utah.

CHINLE DEPOSITIONAL CYCLICITY

The Chinle Formation includes a wide variety of nonmarine sedimentary rock types including conglomerates, sandstones, mudstones, and carbonate rocks formed in environments ranging from high-energy fluvial to low-energy lacustrine and soil-forming situations. Outside of the study area in east-central Utah (Fig. 1), where the cyclic nature of Chinle deposition has been studied most intensively (Lupe, 1977, 1979), Chinle exposures are widely distributed in eastern Utah, western Colorado, northwestern New Mexico, northern Arizona, and southernmost Nevada. Throughout most of its outcrop area the Chinle overlies nonmarine red beds of the Moenkopi Formation, deposited during late Early Triassic and possibly also earliest Middle Triassic time. The Chinle is overlain by poorly dated continental deposits of the Glen Canyon Group that are in turn overlain by partly marine Middle Jurassic (Bajocian and younger) strata of the Carmel Formation and its lateral equivalents.

The stratigraphically lowest rocks conventionally included in the Chinle Formation are interpreted as paleosols developed mainly in the upper beds of the Moenkopi Formation, or where the Moenkopi was eroded in pre-Chinle time, in the Lower Permian Cutler Formation (Finch, 1959, p. 151; Schultz, 1963, p. 43; Johnson, 1964, p. 775; Stewart et al, 1972, p. 89). In the east-central Utah study area, these paleosols constitute the Temple Mountain Member of the Chinle, a kaolinitic silcrete commonly 10 or more meters (33 or more feet) thick. Farther south, equivalent paleosols are the "mottled strata" (Stewart et al, 1972, p. 15) that form the base of the Chinle Formation beneath the Shinarump Member (Fig. 2). The conglomeratic and sandy Shinarump is envisaged as part of the same passive, stagnant depositional regime represented by the Temple Mountain and "mottled strata." Considering its broad areal distribution, the Shinarump is remarkably thin, averaging about 10 m (33 ft) in thickness (Stewart et al, 1972, p. 19), and it has a complexly discontinuous, patchy distribution, commonly occurring as channel-fill deposits. The anomalously coarse-grained rocks of the Shinarump include conglomerate of grain-supported, well-rounded, resistant clasts and quartzose sandstone, and the dominant interstitial clay is kaolinite (Schultz, 1963, p. 29), all of which point to a history of reworking and surface exposure. The upper contact of the Shinarump can be interpreted as an erosional unconformity (Stokes, 1950; Young, 1964; Stewart et al, 1972, Fig. 7) and may represent a significant break in deposition. The pulses of energetic local deposition and sedimentary winnowing interspersed with periods of intense weathering and leaching recorded in the Shinarump suggested to Stokes (1950) its origin as pediment alluvium at a time when little sediment was moving into or out of the region.

In contrast to the relatively passive deposits and paleosols forming the basal parts of the Chinle, the stratigraphically higher and major part of the formation can be interpreted as the result of successive phases of active, regionally integrated, fluvial deposition. Within the southeastern Utah study area, three distinct fining-upward fluvial-lacustrine cycles constitute the Monitor Butte and higher parts of the formation (Fig. 2). High-energy fluvial processes abruptly began each cycle, at the base of which occur relatively coarse-grained sediments of proximal and distal braided streams. In the middle and upper parts of each cycle, deposition was in lower energy environments such as rivers in lower flow and their associated floodplains and abandoned channels and lakes. In many places, soils cap the cycles and indicate temporary cessation of sedimentation (Lupe, 1977, 1979).

This cyclic deposition resulted at least in part from changes in the amount of water flowing through the alluvial system. A large pulse of renewed flow introduced each cycle, resulting in deposition on the floodplain of thick,

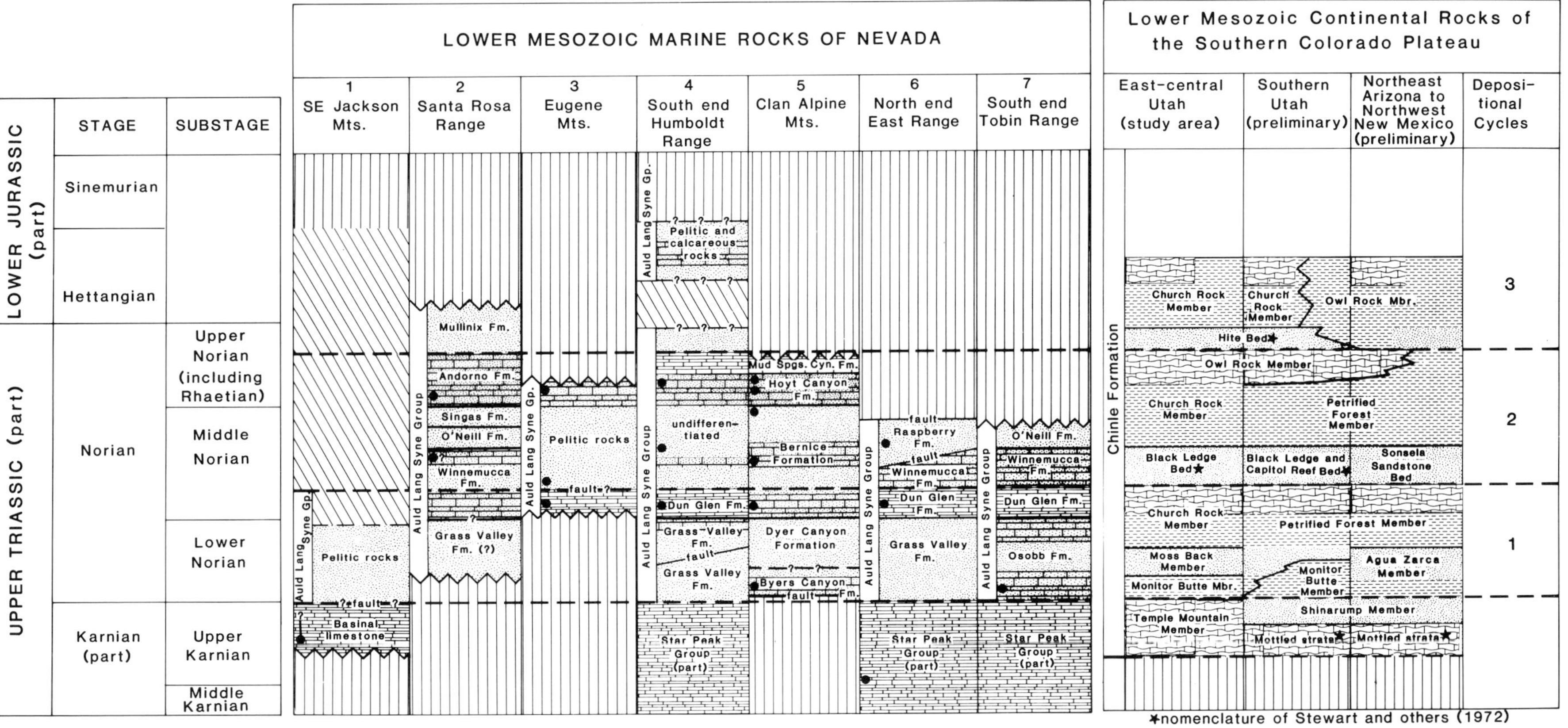

Figure 2—Correlation of representative rocks assigned to the Auld Lang Syne Group and its inferred genetic equivalents in northwest Nevada with depositional cycles and members of the Chinle Formation in east-central Utah. Units in the Clan Alpine Mountains (col. 5) are those of Speed (1978), who did not include them in the Auld Lang Syne Group. Heavy, dashed lines separate cycles; numbered columns for lower Mesozoic rocks in western Nevada correspond to numbered localities on Figure 1. Occurrences of age-diagnostic ammonites and halobiid or monotid bivalves in the Auld Lang Syne are marked by solid circles. In the columns for Nevada, terrigenous clastic rocks are stippled; calcareous clastic rocks, or carbonate rocks interbedded with terrigenous clastic rocks are shown by stippled rectangular pattern; and predominantly carbonate units are shown by solid rectangular pattern. Vertical ruling indicates absence of strata; diagonal ruling indicates strata possibly present but not positively identified. In the columns for the southern Colorado Plateau, rocks that are predominantly sandstone are stippled and those predominantly mudstone are dashed. Paleosols are shown by wavy pattern with short vertical lines.

high-energy deposits of proximal and distal braided streams as well as more distal deltaic and lacustrine deposits that reflect large volumes of relatively low-energy or standing water. As deposition during a pulse in stream flow continued, the proximal braided streams prograded over the formerly more distal sediments of lower energy environments. Then, as energy in the system decreased, sediments were deposited in overbank-lacustrine and low-energy fluvial environments. Decrease in transport energy was not necessarily accompanied by decrease in water supply in the middle parts of cycles. Finally, after a period toward the end of a cycle when essentially only overbank-lacustrine sediments were deposited, conditions became so stable that widespread soils developed. This implies little sediment introduction or removal towards the ends of cycles and, hence, a drastic reduction in water supply. This sequence of events was repeated three times during Chinle time, and each time the cycle-producing, depositional mechanism possessed less energy than previously.

Cyclic deposition of the Chinle extends over a much larger area than that of the study area outlined on Figure 1. In addition to the section in the east-central Utah study area, sections of the Chinle farther south, in southern Utah and in northern Arizona and New Mexico, are also shown in Figure 2. Although interpretation of these southern sections is preliminary, they also appear to record the same cyclic deposition and predominantly west-northwest fluvial sediment transport (Stewart et al, 1972, p. 88, 96). North of the east-central Utah study area, in northern Utah and in southern Wyoming and Idaho, cyclic deposition is not clearly shown in continental Upper Triassic rocks. Nevertheless, the large area in the middle and southern parts of the Colorado Plateau that was characterized by cyclic fluvial deposition during Chinle time should have contributed volumetrically significant pulses of terrigenous-clastic sediment to the region farther northwest and west.

The composition of the Chinle reflects two different sediment sources: (1) the crystalline Precambrian rocks of the Uncompahgre and Front Range Highlands of the Ancestral Rocky Mountains generally to the east, and (2) contemporaneous siliceous volcanic rocks and tuffs from the region of the Mogollon Highland to the south (Stewart et al, 1972). Sandstones in the fluvial, cyclic parts of the Chinle are dominated by quartz grains but are variably feldspathic (generally containing more abundant potash feldspar than plagioclase) and micaceous (Cadigan in Stewart et al, 1972, p. 56). In the more southern parts of the Colorado Plateau montmorillonite and mixed-layer montmorillonitic clays are abundant in the Chinle (Schultz, 1963), and, along with actual lapilli and rhyolite fragments, they reflect a siliceous volcanic source area to the south. Farther north, in areas such as the east-central Utah study area that were more influenced by drainage from the Ancestral Rockies, this montmorillonitic component is less conspicuous, and the clay fraction of some analyzed samples is wholly made up of chlorite and illite.

On the basis of plant and vertebrate fossils contained within it, the Chinle has long been correlated in a general way with the Newark Supergroup of the North American eastern seaboard and with the Keuper of the European Germanic Basin (Stewart et al, 1972, p. 86). Dating of the Chinle, as well as these other continental deposits, in terms of the marine Triassic standard has remained imprecise. Recently, three Upper Triassic floral zones have been established by Ash (1980) for North America, and the lower two of these are recognized in the Chinle. The lowest zone, the *Eoginkgoites* Zone, characterizes the Shinarump and Temple Mountain Members of the Chinle, the part of the formation regarded herein as predating cyclic fluviatile deposition. In terms of marine stages of the Upper Triassic, this zone is regarded by Ash as middle Carnian in age.

The *Dinophyton* Zone, the middle of the three Upper Triassic floral zones established by Ash, is reported by him from the Monitor Butte Member in western New Mexico, as well as from the better known localities in the lower part of the Petrified Forest Member in Arizona. The "late Carnian, ?Norian" age assigned by Ash to the *Dinophyton* Zone is derived from its occurrence in parts of the Newark Supergroup in the eastern United States that can be dated by a chain of palynologic correlations as younger than the middle part of the marine Carnian in Alpine Europe and older than the Rhaetian (as typified in the Alps by the Kössen beds that overlap in age with the upper Norian as used on Figure 2 of the present paper). Opportunities for palynological sampling of sequences in Alpine Europe that can be directly dated by age-diagnostic marine invertebrates of latest Carnian and early and middle Norian age are few, if any, and hence dating spore and pollen assemblages representing this part of the Upper Triassic rests on interpolation beween younger and older floras (Schuurman, 1979). As a consequence, paleobotanical dating of the initial Chinle fluvial cycle that began with the Monitor Butte Member more closely than latest Carnian to middle Norian is not currently possible.

The age of still higher parts of the Chinle, such as those making up the second and third major depositional cycles in the east-central Utah study area, remains poorly constrained in terms of direct paleontologic evidence obtained from the Chinle itself.

CHINLE FORMATION IN NORTHEAST NEVADA

In the broad expanse between Upper Triassic outcrops in east-central Utah and those in northwest Nevada, one limited area of exposure (locality C, Fig. 1), about 10 km (6.2 mi) north of Currie in northeast Nevada, lends support to a genetic relationship between the Chinle Formation and the Auld Lang Syne Group. Scattered outcrops of Upper Triassic rocks at this locality originally were mapped as Chinle by Scott (1954). The Chinle here rests on impure limestone and calcareous siltstone assigned to the Thaynes Formation, which is capped by a conspicuous unit of algal-laminated dolomicrite about 10 m (33 ft) thick. Ammonites that occur near the top of the Thaynes in this area are of the *Haugi* Zone—the highest zone of the Lower Triassic. The apparently highest part of the Chinle section dips under, and presumably underlies, a thick sequence of mature quartzose sandstone equated with the Aztec Sandstone of Triassic (?) and Jurassic age (a correlative of the Navajo Sandstone) by Stewart and Carlson (1978).

Although so poorly exposed in low, rolling, alluviated hills that the Chinle section cannot be pieced together,

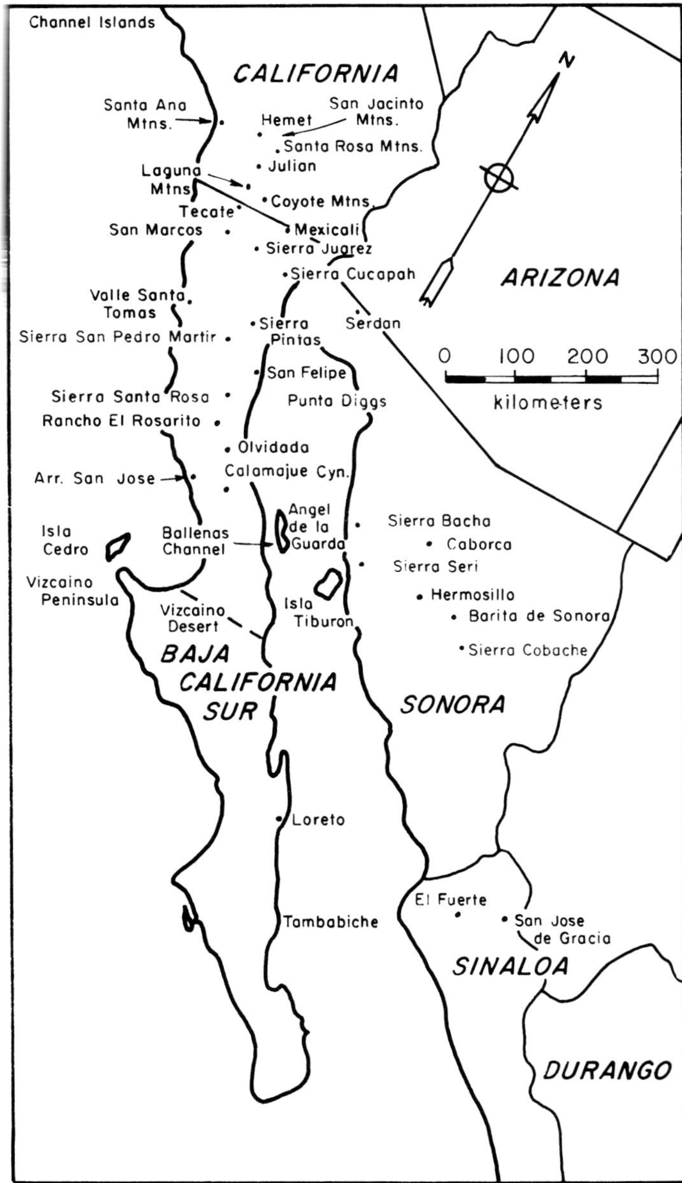

Figure 2—Locality map.

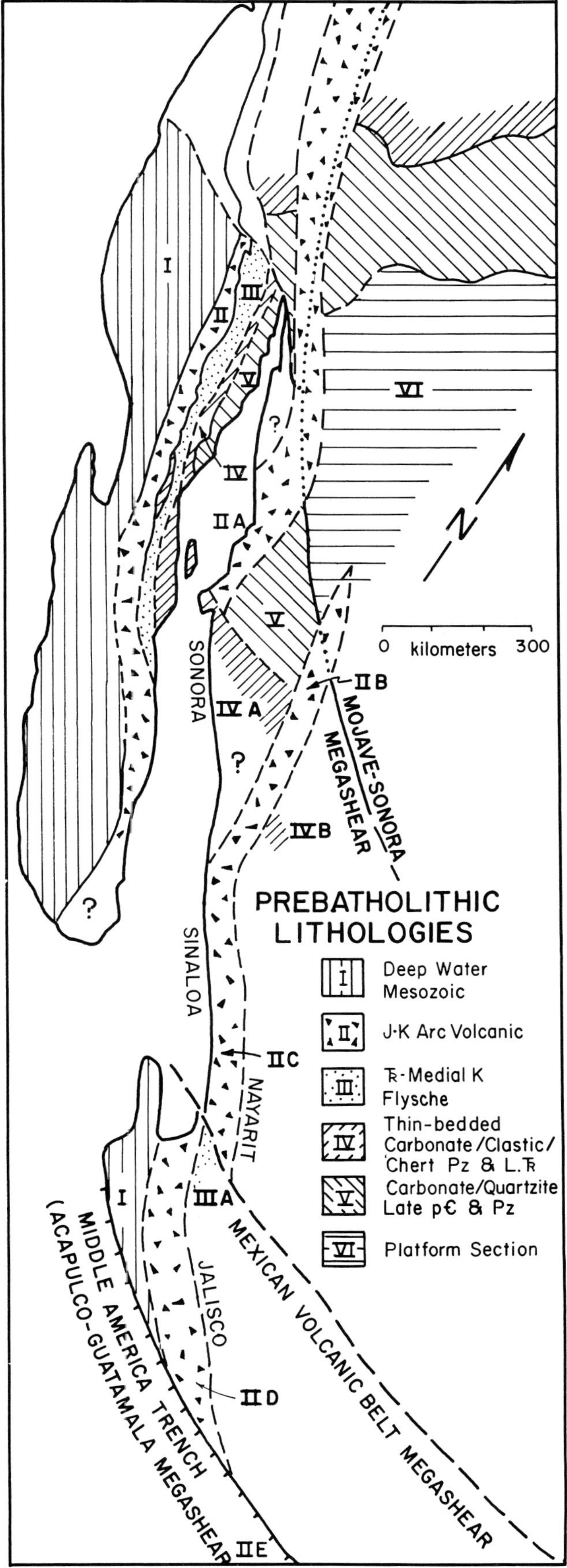

Figure 3—Regional framework for terranes of Peninsular California.

volcanic detritus. Reef limestones are common in the Cretaceous strata of both Baja California and Sinaloa. The volcanic strata are intruded by calcic to calc-alkaline granitic rocks ranging in age from 120 Ma to Paleogene.

In western Sonora (IIA, Fig. 3), medial Cretaceous and Upper Jurassic volcanic strata are interbedded with marine carbonate and clastic strata (Hardy, 1981; Beauvais and Stump, 1976). In eastern Sonora is a belt of interbedded carbonate and andesitic volcanic rocks, first described by King (1939) and recently confirmed by Rangin (1982) after a very critical field review (IIB, Fig. 3).

Mullen (1978) describes these rocks as they appear in northernmost Sinaloa. The continuation of these rocks southward in Sinaloa (IIC, Fig. 3) is described by Bonneau (1971). He finds that in Sinaloa these strata are Albian–Cenomanian rather than Albian–Aptian.

A belt of Cretaceous volcanic and sedimentary rocks in western Nayarit (IID, Fig. 3) is reported by Gastil (1979) and Gastil et al (1979). At Barra Navidad (latitude 19° 10′), these are interbedded with carbonate rock. Work by Jensky (1975) suggests that these arc volcanic strata are in part younger than Cenomanian.

PENINSULAR FLYSCH TERRANE

A nearly continuous belt of metamorphosed flysch can be followed from the northernmost end of Peninsular California southward to the Agua Blanca Fault (latitude 31° 30′ N) (III, Fig. 1). These northern flysch sequences, the Bedford Canyon Formation of the Santa Ana Mountains, the French Valley Formation of the Hemet area, and the type Julian Schist of San Diego County, have been dated respectively as Triassic and Jurassic (Criscione et al, 1978), Upper Triassic (?) (Gastil et al, 1975), and Triassic (?) (Hudson, 1922). Strata in northwestern Baja California Norte are locally well-preserved but have yielded no fossils.

South of the Agua Blanca Fault, thick flysch sections have been mapped near Rancho El Rosarito (latitude 30° 30′ N), in the Mina Olvidada area (latitude 30° 00′ N), and in the Arroya Calamujue area (latitude 30° 25′ N). In these sections the only fossils are from the 30th parallel (Phillips, 1984) and indicate an Early Cretaceous, probably Albian–Aptian age (Durham, 1984; personal communication).

The strata attributed to this terrane range from proximal turbidites with boulder conglomerate to distal turbidites with intercalated chert beds. Quartzite conglomerates (Phillips, 1984) and Precambrian zircons (Bushee et al, 1963) suggest a cratonal provenance for the flysch sequences, whereas minor amounts of andesite-rhyolite volcanic material in the Cretaceous sections suggest proximity to the contemporaneous arc. Some 6.5+ km (4 mi) of stratigraphic thickness has been measured on the 30th parallel.

It is tempting to relate all of these deposits to a single Triassic to Cretaceous apron of continentally derived debris, but it is entirely possible that the Triassic–Jurassic strata of southern California and the Lower Cretaceous strata of the 30th parallel belong to unrelated terranes. Flysch deposits, reportedly including a Jurassic ammonite, are found in western Nayarit (IIIA, Fig. 3; Gastil et al, 1979).

PALEOZOIC AND LOWER TRIASSIC ROCKS OF THE 30th PARALLEL AND CALAMUJUE CANYON

A several kilometer thick section along the 30th parallel includes highly variable lithologies of quartzite-carbonate, fine-grained clastics, thin-bedded carbonate rocks, bedded chert, and olistostrome blocks. The Paleozoic portion of this section has yielded few identified fossils, the most significant being early Permian *Parafusulina kummeli* (Stevens, in Gastil and Miller, 1983). Strata below the recognized Permian contain unidentified crinoidal debris, corals, and byrozoan fragments (Delattre, 1984).

Above the Permian rocks are fine-grained clastics, interbedded chert-carbonate-sandstone-pebble conglomerate, and thin intervals of organic-rich, micritic limestone. These dark limestones contain Early Triassic (Smithian) conodonts and ammonites (Gastil et al, 1981; Gastil and Miller, 1983; Buch, 1984). The entire section has undergone upper greenschist to lower amphibolite grade regional metamorphism.

Recent conodont discoveries in Calamujue Canyon (latitude 29° 30′) indicate the presence of clastic and carbonate strata of Lower Mississippian age.

THE BALLENAS TERRANE

Rocks grouped in this terrane occur along the Ballenas Channel of the Gulf of California from latitude 29° 30′ to 29° 40′ N, on the southwestern coast of Isla Angel de la Guarda, on the southern tip of Isla Tiburon, and at one point along the adjacent coast of Sonora (Gastil and Krummenacher, 1976, 1977) (for place names refer to Fig. 2; for generalized stratigraphy refer to Fig. 4). The only identified fossils are from horizons of bioclastic (crinoidal) limestone that yield conodont fragments of Lower Devonian age. This section consists largely of thin-bedded graywacke, chert, and carbonate rocks, with some olistostromal breccias. Parts of this section can be termed carbonate-rich flysch. There are also thicker carbonate rock units, grading to quartz arenite. One 100 m (328 ft) thick unit consists primarily of volcanic/volcaniclastic (basaltic?) strata. Other (possibly overlying) sections contain considerable metadolostone and quartzite, bedded chert, and pillow basalt.

Rocks of the Ballenas Channel appear to be greenschist facies, except where in contact with small bodies of tonalite and gabbro. Locally the rocks have been tightly folded but do not display the extreme flattening or strong downdip lineation found in some of the terranes described above. Fold axes are subhorizontal, and recumbent and thrust structures occur locally. It is possible that this terrane correlates with rocks in the Sierra Pintas (latitude 31° 35′ to 31° 50′) and with rocks east of Hermosillo, Sonora (Noll, 1981; Poole et al, 1983).

ROCKS OF THE SIERRA PINTAS

McEldowney (1970) mapped the northern Sierra Pintas and made the first discovery of Paleozoic fossils in Peninsular California. They are crinoid columnals, rugose

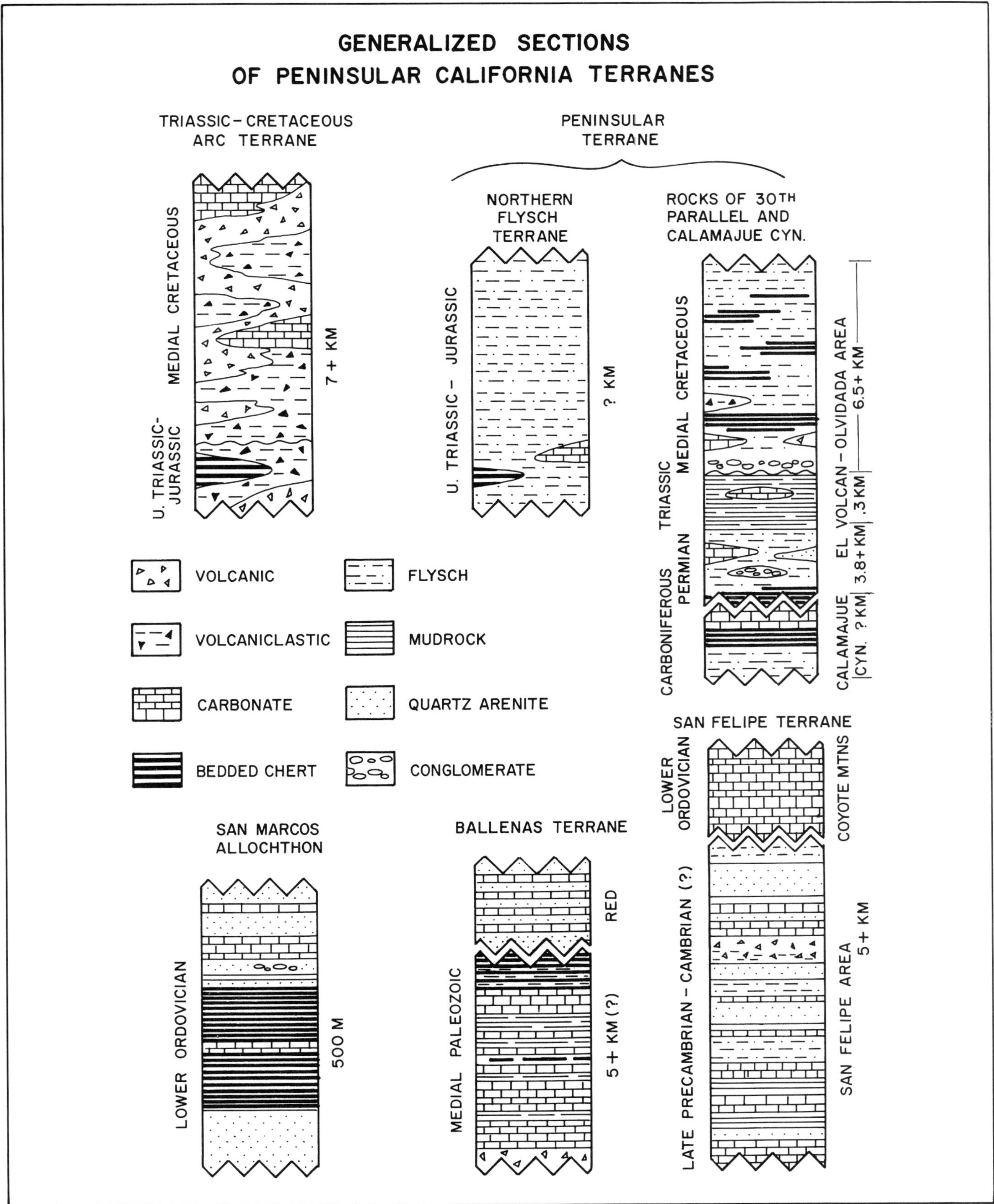

Figure 4—Generalized stratal sections for the terranes of Peninsular California.

corals, and brachiopods. No precise age has been determined, but Stevens (personal communication) has identified a coral as an upper Paleozoic lophophyllid. In the area where these fossils were found, the strata consist of fine clastics, chert, a crinoidal carbonate bed about 3 m (10 ft) thick, and sandstones and conglomerates consisting largely of basaltic debris with admixtures of carbonate rock and chert, both as clasts and as interbeds. In the northwestern part of the range, the section is largely flysch with a thick section of bedded chert. In the southern part of the range, the section includes a variety of carbonate rocks, bedded chert, and pebbly mudstones. No granitic intrusions are exposed, and much of the range appears to be greenschist facies. The northern basement rocks are recumbently folded on near horizontal N 80° west-trending fold axes, in sharp contrast with adjacent parts of the peninsula.

Most of the range is covered by Tertiary volcanic strata, and considering the contrasts of both lithology and structure between the northern and southern Sierra Pintas, it is possible that more than one terrane is represented. There are lithologic similarities both to the rocks of the Ballenas Terrane and to the more highly metamorphosed rocks of the Sierra Juarez, to the northwest.

Rocks of the El Fuerte Formation and San Jose de Garcia, northernmost Sinaloa, are reportedly calcareous flysch and chert deposits, similar to the Ballenas terrane of Baja California and Sonora (Mullen, 1978; Roldan, 1983, personal communication).

SAN FELIPE TERRANE

Near the town of San Felipe, the rocks range from lower greenschist grade along the coast to lower amphibolite grade a few kilometers inland. Anderson (1982) mapped these rocks in detail and constructed a composite stratigraphic section. He concludes that there is a strong resemblance to the Upper Precambrian/Lower Cambrian section of the Caborca area in western Sonora. Prominent units are the thick ultrapure quartz arenite (La Provedora–Zabriskie equivalent?) and a metabasalt unit (Puerto Blanco equivalent?). Unfortunately, no fossils have been discovered.

Although the rocks of the Sierra Cucapach are more metamorphosed than those of either San Felipe or the Coyote Mountains, they include at least two distinct metasedimentary sequences. The section exposed along the Mexicali–San Felipe Highway is thinly interlaminated metamorphosed dolostone/argillite, with some thicker argillite units. The assemblage exposed on the west side of the range includes interbedded metacarbonate and metaclastics rocks, a single unit of coarsely recrystallized ultrapure metaquartzite, and a section of "finely bedded" amphibolite. This association of laminated amphibolite with carbonate and metaquartzite strata is also found in the Coyote Mountains (latitude 32° 50′), west of Punta Diggs (latitude 30° 55′), and west of the Sierra Santa Rosa (latitude 30° 55′). These amphibolites, like those at San Felipe, may correlate with the basalt and basaltic conglomerate at the base of the Cambrian in the Caborca area (Eells, 1972; Stewart et al, in press).

Conodonts of Early Ordovician age were first reported from carbonate-quartzite rocks of the eastern peninsula by Miller and Dockum (1983). This locality is in a section of the Coyote Mountains estimated to be 1,200 m (3,937 ft) thick and consisting of 90% dolomitic marble. This section may overlie (stratigraphic-top indicators have yet to be recognized) a section of mixed clastics and subordinate carbonate rock, including at least one prominent ultrapure quartz arenite and one pebbly arkose member. It is tempting to speculate that the Lower Ordovician conodont-bearing carbonate rock is equivalent to part of the southern Great Basin Pogonip Group and that "underlying" strata include correlatives of the Zabriskie and Wood Canyon Formations. Engle (1982, personal communication) reports that conodonts believed to be of similar age have subsequently been recovered from carbonate-quartzite strata of the Borrego and Santa Rosa Mountain area, northeastern San Diego and Riverside Counties. Similar carbonate quartzite-rich sections are known from the San Jacinto Mountains south to the southeastern end of the Sierra San Pedro Martir.

Throughout the San Felipe Terrane (Figs. 1, 4) units are thick-bedded, fine clastic rocks are subordinate, and carbonate rock predominates over quartzite. Most rocks are amphibolite facies metamorphic grade and in many areas are intimately intruded by Cretaceous age tonalite and granodiorite.

SAN MARCOS ALLOCHTHON

Lothringer (1984) made a detailed study of an allochthonous block (blocks) measuring 5.5 by 1 km (3.4 by .62 mi), 44 km (27 mi) south of Tecate. These blocks are part of a melange, and we consider them to be giant olistoliths. The two major lithologies are ultrapure quartzite (>99% SiO_2) and bedded chert (grading to siliceous shale) with thin argillaceous partings. Calcium carbonate marble is a third but relatively minor section component. Upper Lower Ordovician conodonts have been recovered from a crinoidal limestone (Gastil and Miller, 1981, 1983). Algal (?) and oolitic structures were observed in other carbonate rock layers. Cross-bedding occurs but is rare in both the quartzite and carbonate rocks. Granule-size quartz grains are common in the quartzite, and a well-rounded pebble to cobble conglomerate of chert and quartzite clasts occurs locally. A few meters of carbonaceous slate is found at one locality.

We interpret the giant olistoliths of Rancho San Marcos to have been deposited into a flysch basin during the Mesozoic (?). They were intruded, folded, and metamorphosed to zeolite and greenschist grades during the Upper Cretaceous. The resistant quartzite layers are bent into open folds, with less resistant flysch showing a strong axial plane foliation.

The combination of carbonate rock and massive clean quartzite with thick-bedded chert sections also suggests a correlation to the Valmy, Vinini, and other "transitional" rocks of north-central Nevada and east-central California (Ketner, 1984, personal communication). This allochchonous block occurs within the Triassic–Cretaceous flysch of the Peninsular Terrane.

A block of carbonate rocks, covering only a few acres, crops out a few kilometers northwest of the San Marcos blocks. A considerable area of allochthonous rock exists some 6 km (3.7 mi) to the southwest (Friet, 1984, San Diego State University Senior Report). It is possible that many of the carbonate rock exposures within the Peninsular flysch terrane of western Baja California are also allochthonous blocks.

There are several contrasts between the Lower Ordovician rocks of San Marcos and the Lower Ordovician rocks of Coyote Mountain (San Felipe Terrane). One is the small ratio of carbonate rock to metaquartzite and the relative abundance of bedded chert at San Marcos. A second is the fact that the conodont fauna of Coyote Mountain has "Midcontinent" faunal affinities, common to the miogeoclinal rocks of the Great Basin, whereas the San Marcos fauna is similar to Early Ordovician "North Atlantic" faunas, found in the deeper water facies rocks of the Roberts Mountain allochthon. These contrasts suggest that the Coyote Mountain and Rancho San Marcos localities, respectively, correlate with the miogeoclinal facies (as found in southern Nevada) and the "inner arc basin" facies (as found in the Roberts Mountain allochthon of west-central Nevada).

ROCKS OF WESTERN SONORA

The Caborca area (Eells, 1972; Stewart et al, in press), consists of Precambrian gneiss and granite overlain by Late Precambrian to Devonian miogeoclinal rocks, platform deposits of Carboniferous to Early Jurassic age, and arc volcanic strata of Late Jurassic age. The Paleozoic rock of the Caborca area is bounded on the northeast by the Mojave–Sonora megashear and on the southwest by the Late Cretaceous batholith.

Southeast of Hermosillo, in the Sierra Cobachi and near the Barita de Sonora Mine, Rebeico quadrangle (IV, Fig. 3), are rocks of Ordovician to Devonian age that more resemble the "inner arc basin" facies of the Roberts Mountains allochthon (Ross, 1977; Poole et al, 1977). It is the recognition of these strata that led Dickinson (1981) to bring the tectonic elements of northern California and Nevada all the way around to Sonora and inspired Silver and Anderson (1983) to add Megashear II. These rocks are included in the Ballenas Terrane (IVA, Fig. 3).

Near the Gulf of California, prebatholithic rocks of Sonora have been metamorphosed and intruded much as they are in the Peninsula. Some metamorphosed strata are clearly a continuation of strata in the Caborca area, but others are thinly bedded carbonate/flysch and/or interbedded with andesite/rhyolite volcanic rocks. Between latitudes 29° 30′ N and 29° 50′ Anderson and Silver (1969) reported volcanic rocks of latest Jurassic and medial Cretaceous ages (IIA, Fig. 3). Throughout the Sierra Bacha, Sierra Seri, and eastern Isla Tiburón, andesitic to rhyolitic volcanic rocks are associated with metacarbonate and other metasedimentary strata of uncertain age (Gastil and Krummenacher, 1977).

AS-YET UNIDENTIFIED TERRANES

Although much of what has already been described may appear to the reader as "unidentified," Peninsular California includes additional large areas of metamorphic rock for which correlation is even less certain.

Extensive areas of highly foliated and recrystallized granitic rock occur in the Santa Rosa Mountains of Alta California, north of the town of Julian, throughout the Laguna Mountains of San Diego County, the Coyotes, the Cucupahs, and the northern Sierra San Pedro Martir. In many cases these rocks are so foliated and recrystallized as to be of uncertain protolith (Hirsch, 1984, MS Thesis, San Diego State University). The fact that these rocks include xenoliths of metasedimentary rock tempts the observer to identify them as earlier (probably Mesozoic) intrusives into Paleozoic and/or Mesozoic country rock, and for many this is probably the correct explanation. However, there are reasons to be suspicious. Many of these rocks, unlike most plutons of the Cretaceous batholith, are S type (Todd and Shaw, 1979) and have heavy mineral concentrates that are predominantly monazite or xenotime rather than zircon. Xenoliths in the granitic gneiss near Julian are largely calc-silicate rock, although calc-silicate rocks are rare in the host Julian Schist. We cannot yet dismiss the possibility that some of these gneisses are Precambrian basement.

Until recently, the largely metaclastic rocks found throughout the Sierra Juarez and the northern half of the Sierra San Pedro Martir (Santa Eulalia Formation of Woodford and Harriss, 1938) were carelessly referred to as "Julian Schist"—the amorphous "country rock" of the batholith. The discovery of Ordovician conodonts in the Coyote Mountains now forces us to make new assessments. The Coyote Mountain miogeoclinal section appears to give way progressively to carbonate-poor strata southwestward across the eastern foothills of the main Peninsular Range. However, patches of metacarbonate rock occur throughout the Sierra Juarez, and there are also intervals of recrystallized bedded chert. The type Julian Schist and the Mesozoic(?) flysch of northwestern Baja California do not contain carbonate rock or bedded chert. We are forced to speculate whether we are looking at two very different assemblages of dominantly metaclastic rock, one that is early to middle Paleozoic, possibly even older, and one that is Mesozoic. We have included these possibly older dominantly flysch rocks in the Ballenas Terrane.

Because these unidentified rocks are almost entirely of sillimanite metamorphic grade, it is unlikely that the puzzle will be unraveled by fossil discoveries. A solution will require the systematic description of protoliths in both the weakly and strongly metamorphic terranes and techniques of geochemical fingerprinting.

TERRANE BOUNDARIES

We will consider terrane boundaries in sequence from west to east across the peninsula. As stated earlier, we will not attempt here to delineate boundaries within the composite continental borderland terrane.

At Arroya San Jose, arc volcanics of the medial

Cretaceous Alisitos Formation unconformably overlie Upper Triassic and Lower Jurassic volcanic arc strata (Minch, 1969). In the Vizcaino peninsula, clasts up to a meter in diameter, believed to be derived from the Alisitos arc, form conglomerates of Turonian age (Robinson, 1979; Patterson, 1979). Thus, by Turonian time, it appears that the pre-Cretaceous arc and the arc-derived molasse of the borderland were part of the same terrane as the Jurassic–Cretaceous arc.

Despite considerable search, no depositional contacts have been found between the arc volcanic rocks of the western peninsula and the adjacent flysch belt to the east. In the two places where direct contact is observed, the contact is a fault. In the Santa Ana Mountains, harzburgite and melange are exposed in the fault zone (Davis, 1981, personal communication). Along the southwestern edge of the Sierra San Pedro Martir there is not only a contrast in lithology but in metamorphic grade and deformational style, with the zone of mylonitization being a kilometer wide. The deformation style in the flysch suggests reverse faulting of the interior of the peninsula over the arc, rather than strike-slip faulting.

On the 30th parallel, however, Lower Cretaceous flysch rests unconformably on Triassic and Permian strata. There is a temptation to conclude from this that the entire belt of Upper Triassic–Cretaceous flysch rocks is in depositional contact with Lower Triassic and Paleozoic rocks to the east. For this reason, these two sequences have been grouped together as the Peninsular Terrane.

The proposed correlation of carbonate-quartzite rocks of the San Felipe area with the Caborca rocks (cratonal North America) (Fig. 3) argues against major strike-slip separation between the two areas (beyond the 300 km [186 mi] of post-Miocene translation). Silver and Anderson's (1983) Megashear II contradicts this proposed correlation. Any reconstruction that brings Peninsular California from far to the south ignores the fact that there are no reported strata similar to the Caborca rocks in southern Mexico or Central America.

Although many of the Jurassic–Cretaceous volcanic rocks of western Sonora are in low-angle fault contact with the underlying Caborca rocks, the evidence adequately demonstrates that these volcanic rocks were deposited on the same kinds of rocks that they now rest on. Large metamorphosed quartzite and carbonate bodies in the southern Sierra Bacha and Sierra Seri suggest that the Caborca rocks are exposed nearly all the way to the Gulf coast (Figs. 2, 3).

DISCUSSION

Except for paleomagnetic evidence from gabbroic and granitic rocks of the western peninsula (Beck and Plumbley, 1979; Beck, 1980; Erskine and Marshall, 1980), there would be no reason to consider terranes of Peninsular California exotic to North America. In the Vizcaino Peninsula borderland rocks appear to have been in depositional contact with the peninsular arc since at least Turonian time. The arc is in fault contact with the flysch, but most probably by reverse faults resulting from the postemplacement uplift of the batholith's axis. The flysch rests unconformably on Triassic and upper Paleozoic strata near Mina Olvidada. And the Lower Paleozoic carbonate-quartzite rocks of the eastern edge of the peninsula appear continuous with those on the opposite side of the Gulf (Fig. 3). Thus, one can argue that the best evidence at hand, skimpy as it is, favors in situ origins for Turonian and younger rocks (save for 300 km [186 mi] on the San Andreas Plate Boundary) and for all rocks east of the Jurassic–Cretaceous arc (except for 800 km [497 mi] of mid-Jurassic translation on the Mojave–Sonora megashear).

Faunal and lithologic similarities exist between the "inner arc basin" Ordovician and Devonian rocks of northwestern Nevada and east-central California, and similar age rocks found in Baja California, and between the Permian and Triassic rocks of the 30th parallel and those of the Inyo Mountains, California. Eight hundred kilometers (497 mi) of left-lateral slip along the Mojave–Sonora Megashear I places Baja California adjacent to these proposed correlatives. It is not our purpose to propose these correlations as additional evidence for the megashear but rather to indicate that the distribution is not inconsistent with the megashear. It is of course still possible to make the stratigraphic correlations without strike-slip juxtaposition.

Several questions are left unanswered. From what location did the giant olistoliths of Ordovician quartzite and bedded chert slide into a Mesozoic(?) flysch basin at Rancho San Marcos? If the Ordovician rocks in the eastern part of the peninsula were part of the miogeocline, was there also an exotic cratonal terrane to the west? On Isla Cedros, Boles and Landis (1984) found that upper Paleozoic olistostromal blocks of shallow-water carbonate rock and quartzite were transported from the north and deposited into a Jurassic clastic basin. Then there is the Sierra Pintas with medial (?) Paleozoic basaltic strata and a structural style foreign to the peninsula, located almost astride the belt of miogeoclinal rocks. Are all of these suspect rocks part of an exotic terrane sutured onto western North America at some time near the end of the Paleozoic and moved northeastward during the Jurassic?

If we assume that paleomagnetic indications of transport from far to the south for much of the Peninsula is correct, are there any scenarios left by which this evidence can be reconciled? To answer this we should first define the total extent of Cretaceous and older rocks that have shared this hypothetical translation. If all of the peninsular granitic rocks moved together, then there is no choice but to place the candidate "suture" in or east of the Gulf of California. In which case proposed correlations across the Gulf of California are either erroneous or result from the fortuitous existence of a subparallel, left-lateral Megashear II that effectively cancels out right-lateral movement for pre-Jurassic rocks. If the granitic rocks in the eastern part of the peninsula do not show this translation, we are dealing with a Sumatran-type intra-arc fault that was active during the emplacement of the batholith for which the evidence has been largely destroyed by the continued emplacement and final uplift of the granitic rocks. The paleomagnetically hypothesized provenance of the Mesozoic arc terrane (locality IIE) is along the Middle American Trench.

In summary, the prebatholithic rocks of Peninsula California can be considered as an integral part of western North America, but unexplained relationships still exist, and exotic provenance for some terranes cannot be ruled out.

ACKNOWLEDGMENTS

Research has been assisted by grants from the National Science Foundation, the National Geographic Society, and the San Diego State University Foundation. I have benefited by discussions with Jim Boles, Tom Moore, Calvin Stevens, Keith Ketner, John Stewart, Barney Poole, Raul Madrid, and many others, and from the published ideas of the above, as well as L. T. Silver, Tom Anderson, and William Dickinson. Assistance has been given by Bruce Wardlow, Celestina Gonzales de Lecuanda, Gary Webster, and of course by many graduate and undergraduate students, only some of whom have been acknowledged in the text. My colleague, Richard Miller, has made a major contribution in providing the conodont identifications.

REFERENCES

Anderson, P. V., 1982, Prebatholithic stratigraphy of the San Felipe area, Baja California, Mexico: MS Thesis, San Diego State University, 100 p.

______, 1984, Prebatholithic strata of San Felipe: the recognition of Cordilleran miogeosynclinal deposits in northeastern Baja California, Mexico (Abs.): Society of Economic Paleontologists and Mineralogists, Pacific Section Meeting at San Diego, p. 29–30.

Anderson, T. H., and L. T. Silver, 1969, Mesozoic magmatic events of the northern Sonora coastal region, Mexico (Abs.): Geological Society of America Abstracts with Programs, v. 1, p. 3.

______ and ______, 1981, The role of the Mojave-Sonora Megashear in the tectonic evolution of Sonora, Mexico (Abs.): Geological Society of America Abstracts with Programs, v. 13, p. 47.

Beauvais, L., and T. E. Stump, 1976, Corals, molluscs, and paleogeography of Late Jurassic strata of the Pozo Serna, Sonora, Mexico: Paleogeographics, v. 19, p. 275–301.

Beck, Jr., M. E., 1980, Paleomagnetic record of plate-margin tectonic processes along the western edge of North America: Journal of Geophysical Research, v. 85, p. 7115–7131.

______, and P. W. Plumbley, 1979, Late Cenozoic subduction and continental margin truncation along the middle America Trench: Discussion: Geological Society of America Bulletin, pt. I, v. 90, p. 792–794.

Boles, J. R., and C. D. Landis, 1984, Jurassic sedimentary melange and associated facies, Baja California, Mexico: Geological Society of America Bulletin, v. 94, p. 513–521.

Bonneau, M., 1971 (1969), Una nueva area Cretacica fossilifera en el Estado de Sinaloa: Sociedad Geologica Mexicana Boletin, v. 32, p. 159–167.

Buch, P., 1984, Upper Permian (?) to Lower Triassic stratigraphy, east of El Marmol, northeastern Baja California, Mexico, *in* V. Frizzel, ed., Symposium volume on Baja California: Society of Economic Paleontologists and Mineralogists, Pacific Section, p. 31–36.

Bushee, J., et al, 1963, Lead-alpha dates for some basement rocks of southwestern California: Geological Society of America Bulletin, v. 74, p. 803–806.

Criscioni, J. J., et al, 1978, The age and sedimentation/diagenesis for the Bedford Canyon Formation and the Santa Monica Formation in southern California: a Rb/Sr evaluation, *in* D. G. Howell and K. A. MacDougall, eds., Mesozoic paleogeography of the western United States: Society of Economic Paleontologists and Mineralogists, Pacific Section, Pacific Coast Paleogeography Symposium 2, p. 385–396.

Crocker, J., and M. Campbell, 1984, Prebatholithic stratigraphy of the Ballenas Channel area, Baja California Sur (Abs.): Society of Economic Paleontologists and Mineralogists, Pacific Section Meeting at San Diego.

Delattre, M. P., 1984, Lower Permian metasedimentary rocks of Zamora, northeastern Baja California, Mexico, *in* V. Frizzel, ed., Symposium volume on Baja California: Society of Economic Paleontologists and Mineralogists, Pacific Section Meeting at San Diego, p. 23–29.

Dickinson, W. R., 1981, Plate tectonic evolution of the southern Cordillera, *in* W. R. Dickinson and W. D. Payne, eds., Relations of tectonics to ore deposits in the southern Cordillera: Arizona Geological Society Digest, v. 14, p. 113–135.

Eells, J. L., 1972, Geology of the Sierra de la Berruga, northwestern Sonora, Mexico: MS Thesis, California State University, San Diego, 77 p.

Erskine, B. G., and M. Marshall, 1980, A paleomagnetic and rock magnetic investigation of the northern Peninsular Ranges Batholith, Southern California: EOS, American Geophysical Union Transactions, v. 61, p. 948.

Gastil, R. G., 1979, Reconnaissance geology of west-central Nayarit, Mexico: summary: Geological Society of America Bulletin, v. 90, p. 15–18.

______, 1981, The tectonic history of Peninsular California and adjacent Mexico, *in* W. G. Ernst, ed., The geotectonic development of California, Rubey Symposium v. I: San Francisco, Freeman and Company, p. 284–305.

______, and D. Krummenacher, 1976, Reconnaissance map of coastal Sonora between Puerto Lobos and Bahia Kino: Geological Society of America Map and Chart Series MC-16.

______ and ______, 1977, Reconnaissance geology of coastal Sonora between Puerto Lobos and Bahia Kino: Geological Society of America Bulletin, v. 88, p. 189–198.

______, and R. H. Miller, 1981, Lower Paleozoic strata on the Pacific Plate of North America: Nature, v. 292, n.

5826, p. 828-830.

______ and ______, 1983, Prebatholithic terranes of southern and Peninsular California, U.S.A. and Mexico; status report, *in* C. Stevens, ed., Suspect terranes symposium volume: Society of Economic Paleontologists and Mineralogists, p. 49-61.

______ and ______, 1984, Prebatholithic paleogeography of Peninsular California, *in* V. Frizzel, ed., Symposium volume on Baja California: Society of Economic Paleontologists and Mineralogists, Pacific Section, p. 9-16.

______, et al, 1975, Reconnaissance geology of the state of Baja California: Geological Society of America Memoir 140, 170 p.

______, et al, 1978, Mesozoic history of peninsular California and related areas east of the Gulf of California, *in* D. G. Howell and K. A. MacDougall, eds., Mesozoic paleogeography of the western United States: Society of Economic Paleontologists and Mineralogists, Pacific Section, p. 107-116.

______, et al, 1979, Reconnaissance geology of west-central Nayarit, Mexico: Geological Society of America Map and Chart Series MC-24, scale 1:200,000.

______, et al, 1981, Lower Triassic strata near El Volcan, Baja California, Mexico (Abs.): Geological Society of America Abstracts with Programs, v. 13, n. 2, p. 57.

Hardy, L. R., 1981, Geology of the Central Sierra de Santa Rosa, Sonora, Mexico, *in* L. Ortlieb and Q. J. Roldon, eds., Geology of northwestern Mexico and southern Arizona: Field Guides and Papers for Geological Society of America Cordilleran Section Meeting in Hermosillo, p. 73-98.

Hudson, F. S., 1922, Geology of the Cuyamaca region of California with special reference to the origin of the nickeliferous pyrrhotite: California University, Department of Geology Science Bulletin, v. 13, p. 175-252.

Jensky, W. A., 1975, Reconnaissance geology and geochronology of the Bahia de Banderas area, Nayarit and Jalisco, Mexico: Master's Thesis, University of California, Santa Barbara, 80 p.

King, R. E., 1939, Geological reconnaissance in northern Sierra Madre Occidental of Mexico: Geological Society of America Bulletin, v. 50, p. 1625-1727.

Lothringer, C. J., 1984, Geology of a Lower Ordovician allochthon, Rancho San Marcos, Baja California, Mexico, *in* V. A. Frizzell, Jr., ed., Geology of the Baja California Peninsula, Pacific Section: Society of Economic Paleontologists and Mineralogists. v. 39, p. 17-22.

McEldowney, R. C., 1970, Geology of the northern Sierra Pinta, Baja California, Mexico: MS Thesis, San Diego State University, 78 p.

Miller, R. H., and M. S. Dockum, 1983, Ordovician conodonts from metamorphosed carbonates of the Salton Trough, California: Geology, v. 11, p. 410-412.

Minch, J. A., 1969, A depositional contact between the pre-batholithic Jurassic and Cretaceous rocks in Baja California, Mexico (Abs.): Geological Society of America Abstracts with Programs for 1969, pt. 3, p. 42-43.

Mullan, H. S., 1978, Evolution of part of the Nevadan Orogeny in northwestern Mexico: Geological Society of America Bulletin, v. 89, p. 1175-1188.

Noll, J. H., 1981, Geology of the Picacho, Colorado area, northern Sierra de Cobache, Central Sonora, Mexico (Abs.): Geological Society of America Abstracts with Programs, v. 3, n. 2, p. 99.

Patterson, D. L., 1979, The Valle Formation—physical stratigraphy and depositional model, southern Vizcaino Peninsula, Baja California Sur, *in* P. L. Abbott and R. G. Gastil, eds., Baja California geology: Field Guide and Papers, Geological Society of America Meeting, San Diego, p. 73-82.

Peiffer-Rangin, F., 1979, Les zones isopiques du Paleozoique inferieur du nord-ouest Mexicain, temoins du relais entre les Appalaches et la cordillere ouest-americaine: Compte Rendu Acadamie Sciences Paris, Series D, v. 288, p. 1517-1519.

Phillips, J. R., 1984, "Middle" Cretaceous metasedimentary rocks of La Olvidada, northeastern Baja California, Mexico, *in* V. Frizzel, ed., Symposium volume on Baja California: Society of Economic Paleontologists and Mineralogists, Pacific Section, p. 37-41.

Poole, F. G., et al, 1977, Silurian and Devonian paleogeography of western United States, *in* J. H. Stewart, et al, eds., Paleozoic paleogeography of the western United States: Pacific Coast Paleogeography Symposium 1, Pacific Section, Society of Economic Paleontologists and Mineralogists, p. 39-66.

______, 1983, Bedded barite deposits of middle and late Paleozoic age in central Sonora, Mexico (Abs.): Geological Society of America Abstracts with Programs, v. 15, n. 5, p. 299.

Rangin, C., 1982, Contribution a l'etude geologique du system cordillerain du nord-ouest de Mexique: Memoires des Sciences de la Terre et Marie Curie, 588 p.

Ross, R. J., 1977, Ordovician paleogeography of western United States, *in* J. H. Stewart, et al, eds., Paleozoic paleogeography of the western United States: Pacific Coast Paleogeography Symposium 1, Pacific Section, Society of Economic Paleontologists and Mineralogists, p. 19-38.

Robinson, J. W., 1979, Structure and stratigraphy of the northern Vizcaino Peninsula, *in* P. L. Abbott and R. G. Gastil, eds., Baja California geology: Field Guide and Papers, Geological Society of America Meeting, San Diego, p. 77-82.

Silver, L. T., and T. H. Anderson, 1983, Further evidence and analysis of the role of the Mojave-Sonora megashear(s) in Mesozoic Cordilleran tectonics (Abs.): Geological Society of America Abstracts with Programs, v. 15, n. 5, p. 273.

Stewart, J. H., 1982, Regional relations of Proterozoic Z and Lower Cambrian rocks in the western United States and northern Mexico, *in* J. D. Cooper, et al, eds., Geology of selected areas in the San Bernardino Mountains, western Mojave Desert, and southern Great Basin, California: Geological Society of America, Cordilleran Section, Annual Meeting Guidebook and Volume, p. 171-186.

———, et al, in press, Upper Proterozoic and Cambrian rocks in the Caborca Region Sonora, Mexico—physical stratigraphy, biostratigraphy, paleocurrent studies, and regional relations: U.S. Geological Survey Professional Paper 1309.

Todd, V. R., and S. E. Shaw, 1979, Structural metamorphic, and intrusive framework of the Peninsular Ranges batholith in southern San Diego County, California, *in* P. L. Abbott and V. R. Todd, eds., Mesozoic crystalline rocks: Manuscripts and road logs for Geologic Society of America Annual Meeting in San Diego, p. 177–232.

Woodford, A. O., and T. P. Harriss, 1938, Geological reconnaissance across Sierra San Pedro Martir, Baja California: Geological Society of America Bulletin, v. 49, p. 1297–1336.

Tectonostratigraphic Terranes of the Vizcaino Peninsula and Cedros and San Benito Islands, Baja California, Mexico

David L. Kimbrough
University of California
Santa Barbara, California

Mesozoic rocks on the Vizcaino Peninsula and Cedros and San Benito Islands consist of at least seven distinctive tectonostratigraphic terranes. The largest is the Vizcaino terrane, which represents a fragment of a low-latitude Upper Triassic-Lower Jurassic oceanic arc consisting of ophiolite basement and overlying arc-derived strata that formed the site of later Jura-Cretaceous magmatism associated with a tectonically active continental margin insular arc. A serpentinite matrix melange terrane containing high P/T tectonite clasts occurs in low-angle fault contact beneath Upper Triassic ophiolite of the Vizcaino terrane in a small area. On Cedros Island, the Choyal terrane is a Middle to Late Jurassic volcanic arc assemblage that consists of volcanic-plutonic basement sequences, including ophiolite, that are overlain by Middle Jurassic volcaniclastic strata and Middle to Late Jurassic shelf-slope sediments that record a continental collision. Blueschist facies and lower grade subduction complex assemblages occur as a series of thrust sheets tectonically beneath the Choyal terrane and in a fault block at the southwest corner of the island; these rocks are collectively referred to as the Cedros terrane but consist of several sequences that may represent distinct subterranes. The San Benito Islands are divided into three fault-bounded terranes: an intact sequence of graywacke and argillite of unknown age, a Late Jurassic-Early Cretaceous assemblage of sheared graywacke + chert + basalt, and a serpentinite matrix melange with high P/T tectonite blocks. The Vizcaino-Cedros-San Benito terranes resemble Mesozoic terranes of the California Coast Ranges, the western Sierra Nevada foothills, and the western Klamath Mountains of California and probably evolved along strike from one another within the same plate margin settings. Paleomagnetic inclination data and stratigraphic relations indicate that if the Vizcaino terrane represents a far-traveled allochthonous unit, then much of this displacement may have occurred in a narrow time interval in the Late Cretaceous. The timing of possible large scale displacement of the Choyal terrane is restricted to Late Jurassic to Late Cretaceous.

INTRODUCTION

Oblique convergence between the North American and Pacific realms during the Mesozoic and Cenozoic resulted in a complex series of accretionary and truncation events involving fragments of continental and oceanic volcanic arcs, displaced continental fragments, and intervening marginal ocean basins all of various ages (Coney, 1981; Saleeby, 1983). These fault-bounded fragments have been termed tectonostratigraphic terranes (Coney et al, 1980) and comprise the bulk of the western Cordillera from Alaska to Mexico. Most of these accreted terranes are considered allochthonous with respect to the North American craton.

Mesozoic ophiolite, volcanic arc, and subduction complex terranes are extensively exposed on the Vizcaino Peninsula and on Cedros and San Benito Islands (Cohen et al, 1963; Jones et al, 1976; Kilmer, 1977; Rangin, 1978; Rangin et al, 1982; Klienast and Rangin, 1982; Barnes, 1984; Hickey, 1984; Kimbrough, 1984; Boles and Landis, 1984; Moore, 1984, and this volume). These rocks crop out within a 250 × 40 km (155 × 25 mi) northwest-trending belt that underlies the Pacific continental margin of Baja California (Figs. 1, 2). The Vizcaino-Cedros-San Benito terranes represent portions of convergent plate margin regimes that were accreted to the continental margin by late Mesozoic time. Magmatism associated with ophiolite and volcanic arc terranes proceeded semicontinuously from Upper Triassic to Lower Cretaceous time. Unmetamorphosed and structurally intact stratigraphic sequences that depositionally overlie ophiolite and volcanic arc basement substrates provide an unusually clear record of arc magmatism, tectonics, and sedimentation. Rocks that correlate with parts of the Franciscan Complex of California occur on Cedros and San Benito Islands and possibly the Vizcaino Peninsula and record continued late Mesozoic subduction of oceanic crust beneath a continental margin.

Subsequent to their continental accretion in the Middle or Late Jurassic (Kimbrough, 1982; Boles and Landis, 1984), the Vizcaino and Choyal terranes formed the forearc basement substrate of mid- to Late Cretaceous west-facing continental margin arc, represented to the east by the Alisitos Group and Peninsular Range Batholith of Baja California (Silver et al, 1979; Gastil et al, 1981; Beggs, 1984; Patterson, 1984a). The Valle Formation (Cenomanian-Santonian) represents forearc basin submarine fan deposits

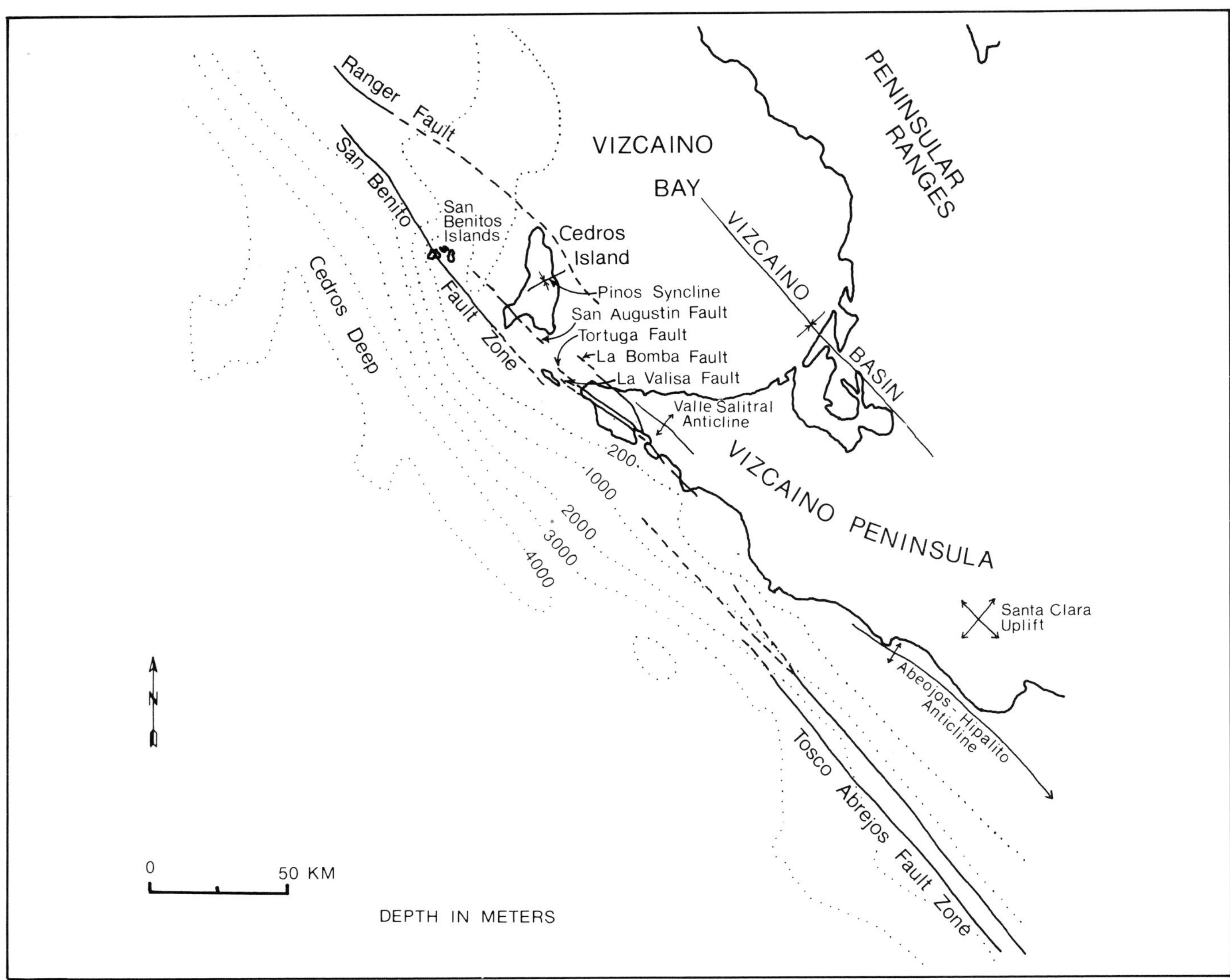

Figure 1—Bathymetry and major structural features of the Baja California borderland in the Vizcaino Peninsula region.

derived from the Alisitos arc (Patterson, 1984b) and conformably overlies ophiolite and arc terranes on the Vizcaino Peninsula and Cedros Island. The ophiolite and arc terranes were never deeply buried by these forearc basin sediments, nor were they thermally overprinted by later volcanic arc magmatism. This shallow forearc basin tectonic setting accounts for the good preservation of these terranes.

Mesozoic paleo-oceanic terranes similar to the Baja California terranes described here comprise a volumetrically important and widespread component of the Cordilleran collage (Saleeby, 1983). Southward from the Transverse Ranges of Southern California, however, these terranes have very limited exposure in the largely submerged borderland region of Southern and Baja California (Howell and Vedder, 1981). The Vizcaino–Cedros–San Benito terranes comprise by far the largest and best preserved of these exposures and occur in a localized region of uplifted borderland from 800 to 1,000 km (497–621 mi) southward of the Transverse Ranges. Lithologically similar rocks occur 300 km (186 mi) south of the Vizcaino Peninsula in the Magdelena Bay region but are now considered as distinct tectonostratigraphic terranes (Blake et al, 1984).

REGIONAL GEOLOGICAL SETTING

Although the Pacific margin of Baja California now occupies a relatively quiescent intraplate setting, it locally preserves the bathymetric, structural, and geophysical characteristics of an active convergent margin (Riveroll, 1978). The Cedros Deep is a flat-floored elongate basin that parallels the foot of the continental slope, only 40 km (25 mi) to the west of the San Benito Islands, and was an active trench until at least 15 m.y. ago (Atwater, 1970; Riveroll, 1978; Fig. 1). Seismic reflection data show a strong reflector that dips to the east, beneath the toe of the continental slope, and is interpreted as oceanic crust that was partially subducted along the now inactive Cedros Deep trench. Riveroll (1978) reports gravity anomalies associated with the Cedros Deep and outer shelf edge that are similar to anomalies associated with active convergent margins. The belt of rock that contains the Mesozoic age

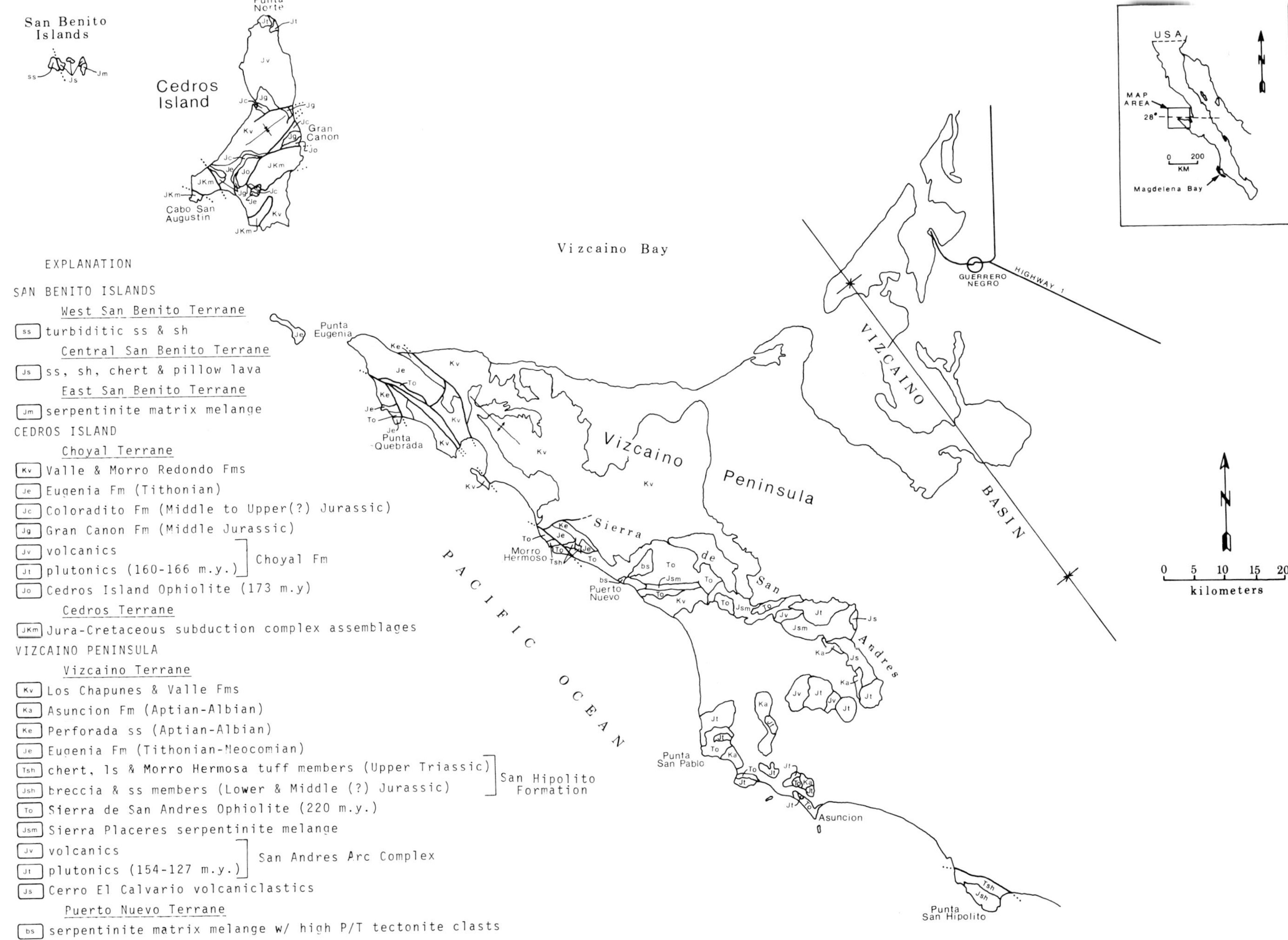

Figure 2—Simplified geologic map showing tectonostratigraphic terranes of the Vizcaino Peninsula and Cedros and San Benito Islands.

Vizcaino–Cedros–San Benito terranes occupies a structural position along the outer arc ridge of this fossilized Miocene age subduction zone.

The Vizcaino–Cedros–San Benito belt is bounded on its northeast side by the northwest-trending Vizcaino Basin. This basin contains from 3,000 to 5,000 m (9,843–16,404 ft) of Valanginian to middle Eocene sedimentary rocks of the Upper Cretaceous Valle Formation and Early Tertiary Bateque Formation with Quaternary cover sediments (Madrid-Solis, 1979; Helenes, 1984). The Vizcaino basin covers any major crustal sutures or post-accretion fault zones that intervene between the Vizcaino–Cedros–San Benito terranes and the Peninsular Range Batholith which intrudes lower Paleozoic North American platform sequences of the Baja California Peninsula to the east (Gastil and Miller, 1984). The Vizcaino basin is similar in many respects to the Great Valley of Central California that separates the Franciscan Complex and Coast Range Ophiolite of the Coast Ranges on its west side, from the Sierra Nevada batholith to the east that is intruded into lower Paleozoic shelf and slope-rise facies strata that were deposited across Precambrian sialic crust of the North American craton (see references in Saleeby, 1983).

The Vizcaino–Cedros–San Benito belt is bounded on its southwest side by extensions of the San Benito shear zone to the north (Krause, 1965) and the Tosco–Abreojos fault zone to the south (Spencer and Normark, 1979; Fig. 1). These fault systems are marked by basement ridges and scarps that trend approximately N 35° W along the slope and outer shelf for approximately 800 km (497 mi) and may have taken up a major part of the Pacific–North American transform plate boundary motion between 4.5 and 12 to 14 m.y. (Spencer and Normark, 1979). Extensions of these fault systems adjacent to the Vizcaino Peninsula are difficult to recognize because of the steep irregular slope and flat narrow continental shelf (Fig. 1). However, major northwest-trending Late Tertiary folds and faults that occur on the Vizcaino Peninsula and Cedros Island are probably related subparallel elements of the Late Tertiary San Benito–Tosco–Abreojos fault system. Around the tip of the Vizcaino Peninsula, the northwest-trending La Bamba, Bahia Tortugas, and La Valisa faults have an accumulated 50 km (31 mi) of post-Miocene right-lateral displacement (Robinson, 1975).

THE VIZCAINO TERRANE

Introduction

The oldest recognized rocks of the Vizcaino terrane are remnants of an Upper Triassic ophiolite. These rocks are exposed discontinuously in a northwest-trending belt along the southwestern edge of the Vizcaino Peninsula, from Punta San Hipolito to Punta Quebrada, and include a well-preserved ophiolite massif in the northwest part of the Sierra de San Andres (Moore, 1983; Barnes, 1984; Fig. 2). The sedimentary cover over this ophiolite basement is preserved at Punta San Hipolito and in the Morro Hermoso area (Fig. 3). These strata are assigned to the San Hipolito Formation and contain various lithofacies deposited within the same Upper Triassic oceanic volcanic arc setting (Barnes, 1984). Reef-associated slump blocks of limestone in the San Hipolito Formation suggest a tropical setting for this oceanic arc (Stanley, 1979).

A Late Jurassic–Early Cretaceous volcanic arc complex intruded and was built above the Triassic ophiolite basement on the Vizcaino Peninsula (Kimbrough, 1982; Barnes, 1982; Moore, 1983). Volcanic arc rocks are dominated by andesite flows and breccias with intercalated and overlying volcanogenic sedimentary rock (Rangin, 1978; Moore, 1983, 1984). Plutons mainly of tonalitic and granodioritic compositions intrude both the Jura–Cretaceous arc rocks and disrupted Triassic ophiolite basement and have zircon U-Pb ages that range from 154 m.y. to 127 m.y. (Kimbrough, 1982; Barnes, 1982). Late Jurassic and Early Cretaceous marine volcanogenic strata of the Vizcaino Group (Barnes, 1982, 1984) are associated with this superimposed arc sequence. Facies relations, the presence of relatively siliceous volcaniclastics including andesitic and dacitic pyroclastic flows, plus a small but significant component of continentally derived detritus indicate that the source of Vizcaino Group sediments was an active insular arc adjacent to a continental craton (Barnes, 1982; Hickey, 1984). Tectonic disruption of this arc setting is recorded by major unconformities that locally expose subarc intrusives (Barnes, 1984) and a serpentinite matrix melange that contains fragments of both the Triassic ophiolite and Jurassic arc volcanic/plutonic rocks (Moore, 1983).

Sierra de San Andres Ophiolite

The Upper Triassic Sierra de San Andres Ophiolite and its depositionally overlying sedimentary strata is best exposed in the western part of the Sierra de San Andres where it forms a gently northwest-plunging antiform that is cut by numerous high-angle faults (Moore, 1983; Fig. 2). A serpentinite matrix melange with blueschist- and greenschist-facies metasedimentary and metaigneous tectonite clasts occurs structurally beneath the ophiolite, as seen in a small window at the core of the antiform (Moore, 1983). This melange constitutes the Puerto Nuevo terrane, described below.

Although disrupted by faulting, the ophiolite consists of a well-ordered stratigraphic sequence that consists of (from bottom to top): harzburgite-dunite tectonite, cumulate olivine + plagioclase + clinopyroxene gabbro, noncumulus hornblende gabbro and minor plagiogranite, dike and/or sill complex, and pillow lava (Moore, 1983). The Morro Hermosa tuff member of the San Hipolito Formation, discussed below, is in depositional contact on pillow lava of the ophiolite sequence.

Plagiogranite from the ophiolite has been dated at two widely separated localities on opposing flanks of the antiform in the western Sierra de San Andres. In Arroyo San Cristobal, hornblende quartz diorite intrudes noncumulus hornblende gabbro and has yielded a zircon U-Pb isotopic age of approximately 220 m.y. (Kimbrough, 1982). South of Puerto Escondito, a small lens-shaped mass of albitite intrudes noncumulus hornblende gabbro and dike complex. A U-Pb mineral isochron on feldspar and sphene separates from the albitite also indicates an age around 220 m.y. (Barnes and Mattinson, 1981). These U-Pb ages are interpreted as the crystallization age of the entire igneous portion of the ophiolite suite in the western Sierra de San Andres.

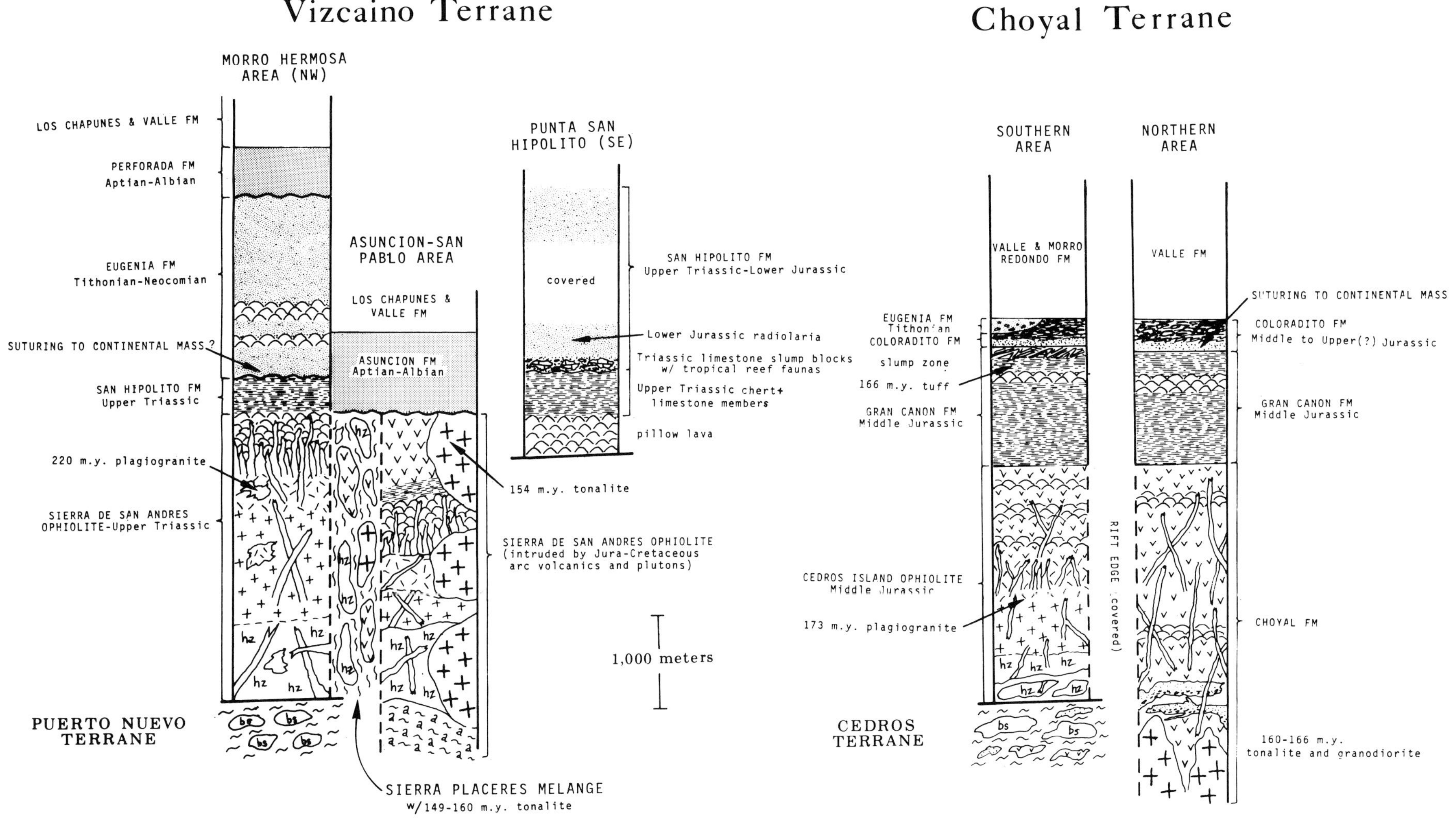

Figure 3—Generalized stratigraphy of the Vizcaino and Choyal terranes.

Upper Triassic ophiolitic rocks also occur in a small area in the northern part of the Vizcaino Peninsula near Punta Quebrada (Fig. 2). Here, an intact sequence of well-preserved pillow lava contains interpillow limestone with Carnian to upper-middle Norian radiolarian assemblages (Barnes, 1982). Conformably beneath these pillow lavas are tectonized metagabbroic rocks and minor serpentinite that are intruded by diabasic dikes.

A great mass of pillow lava cut by dikes and irregular bodies of micrograbbro crops out at Punta Asuncion. Westward and inland along the coast from Punta Asuncion, these rocks are underlain by dike complex and noncumulus gabbro (Moore, 1983). These rocks strike toward the Punta San Roque headland where there is a strongly deformed assemblage of sheared ultramafic rocks and gabbro and foliated fine-grained amphibolite. The volcanic cover of this dismembered ophiolite is overlain by tuffaceous chert and limestone (Moore, 1983) that is undated but similar in appearance to Upper Triassic strata of the San Hipolito Formation at Punta San Hipolito, described below. Jura–Cretaceous plutons associated with the San Andres arc complex, described below, intrude this dismembered ophiolitic basement (Kimbrough, 1982; Moore, 1983).

The Punta Quebrada and Asuncion–San Pablo ophiolite assemblages, as well as pillow lava at the base of the San Hipolito Formation at Punta San Hipolito, are considered correlative to the well-preserved ophiolite in the western Sierra de San Andres and are collectively referred to as the Sierra de San Andres ophiolite. Moore (1983), however, recognizes petrologic differences between the Asuncion–San Pablo and Sierra de San Andres ophiolite remnants and considers them as parts of separate terranes (Moore, 1983, and this volume).

San Hipolito Formation

A 2,400 m (7,874 ft) thick homoclinally dipping section of Upper Triassic and Jurassic marine sedimentary rocks is exposed in the area of Punta San Hipolito in the southwestern part of the Vizcaino Peninsula (Fig. 2). These rocks comprise the type locality for the San Hipolito Formation and have been divided into four informal members in this area (Finch and Abbott, 1977). The basal portion of the section is a 450 m (1,476 ft) thick sequence of tuffaceous chert, limestone, and interbedded tuff deposited on pillow lava (Fig. 3). The basal chert member of the formation gives way upsection to the limestone member, which may reflect shoaling of the basin floor.

Exceptionally rich, well-preserved radiolarian assemblages in the chert member and the hemipelagic pelecypod *Monotis* cf. *M. subcircularis* in the upper part of the limestone member indicate an Upper Triassic age (late Carnian to late Norian) for these units (Pessagno et al, 1979). The upper limit of the *Capnodoce* radiolarian zone (Carnian–Norian) occurs within the chert member. This zone is widespread in the western Cordillera and is already described from the Rattlesnake Creek terrane of the Klamath Mountains, the Suplee–Izee area of eastern Oregon, and from the Cache Creek Group of British Columbia (Pessagno et al, 1979).

The hemipelagic pelecypod *Halobia* sp. also occurs in the basal chert member of the formation and in interpillow areas in the underlying basalts suggesting a short time interval between extrusion of the pillow lava and blanketing by sediment. The pillow lava at the base of the section is interpreted as the upper portion of an ophiolite correlative to the Sierra de San Andres Ophiolite, 80 km (50 mi) to the northwest. The basal chert member (up to 240 m [787 ft] thick) contains thinly laminated tuffaceous chert with flame structures and low-angle cross laminae that indicate current reworking. Thin green volcaniclastic sandstone beds are intercalated within the chert. The overlying limestone member (up to 210 m [689 ft] thick) consists of thinly bedded, light-gray tuffaceous limestone with abundant poorly preserved radiolaria. Crystal lithic tuff beds and volcanic breccia lenses up to 6 m (20 ft) thick are abundant within the limestone member. Chaotic soft sediment slump folds and auto breccias occur in the upper part of the chert member and in the limestone member.

A submarine erosion surface with up to 100 m (328 ft) of relief was cut into the limestone member and filled by fossiliferous limestone megabreccia blocks with a sandy volcaniclastic matrix (breccia member of Finch and Abbott, 1977). An ammonite collected from a limestone block is probably an arcestid such as *Arcestes* and indicates a Middle or Late Triassic age (Silberling, 1979, written communication to Gastil). Stanley (1979) has identified other Triassic fossils in the limestone blocks and notes the similarity of the fossil assemblages to other Triassic coral reefs in western North America that indicate a warm, shallow, tropical marine environment. These limestone blocks probably represent slump blocks from reef accumulations that fringed volcanic islands of a low-latitude Triassic oceanic arc.

The limestone breccia member is overlain conformably by approximately 1,850 m (6,070 ft) of immature volcaniclastic sandstones (sandstone member of Finch and Abbott, 1977). Thin limestone and chert beds are intercalated within the basal portion of the sandstone member. A limestone interbed approximately 250 m (820 ft) above the limestone breccia member yielded a well-preserved Lower Jurassic (upper Pliensbachian) radiolarian assemblage (Whalen and Pessagno, 1984). The upper part of the sandstone member at Punta San Hipolito therefore probably extends at least into the Middle Jurassic.

In the Morro Hermosa area, 80 km (50 mi) northwest of Punta San Hipolito, pillow lava along the northwest edge of the Sierra de San Andres Ophiolite is depositionally overlain in several places by 200 to 300 m (656–984 ft) thick sections of fine-grained zeolitized tuffs, tuffaceous sandstone, siltstone, limestone, and tuffaceous chert. Interpillow limestone at the base of the sequence contains Carnian (?) lower-to upper-middle Norian radiolaria (Barnes, 1982). Chert and tuffaceous chert near the base of the section contain similar age radiolarian assemblages and conodonts (Barnes, 1982). *Monotis* sp., a Late Triassic index fossil, was also found on bedding surfaces in one horizon near the base of the section (Barnes, 1982). These rocks make up the Morro Hermoso tuff unit of Rangin (1978).

The Morro Hermoso tuffs are now recognized as part of the San Hipolito Formation (Barnes and Mattinson, 1981; Barnes, 1984), as originally suggested by Finch and

Abbott (1977) on the basis of lithologic similarity to the type section of the formation at Punta San Hipolito 80 km (50 mi) to the southeast. The Morro Hermoso tuffs are dominated by fine-grained zeolite tuffs and lesser volcaniclastic sandstone and siltstone (Barnes, 1982). This dominantly volcanogenic unit, although fine grained, apparently represents a more proximal volcanogenic facies compared to the tuffaceous chert and limestone facies exposed at the base of the formation at Punta San Hipolito, described above.

A recent timescale compilation by Harland et al (1982) indicates close agreement of the 220 m.y. radiometric ages from the ophiolite with the biostratigraphic ages from inter-pillow limestone of the ophiolite and the Morro Hermosa tuffs. This suggests that formation of the ophiolite occurred in close proximity to active arc volcanism, perhaps in a marginal basin (Kimbrough, 1982). Petrologic data from volcanic rocks in the western Sierra de San Andres and the Asuncion area support an arc-related origin for the ophiolite (Moore, 1983).

San Andres Arc Complex

The southeastern half of the Sierra de San Andres consists of mafic pillow lava, andesite flows and breccias, and volcanogenic sedimentary rocks that are intruded by tonalitic and granodioritic plutons (Rangin, 1978; Moore, 1983, 1984; Fig. 2). The plutons and metavolcanic rocks immediately north of Mesa de las Auras and plutons that intrude the probable Upper Triassic ophiolitic basement in headland areas from Asuncion to Punta San Pablo are part of this complex. These Jurassic volcanic and plutonic rocks, referred to here as the San Andres arc complex, apparently intruded and built above the Upper Triassic ophiolite basement of the Vizcaino terrane. The San Andres arc complex has been correlated to the magmatic arc rocks on Cedros Island (Rangin, 1978; Barnes, 1984; Moore, 1984). However, differences in stratigraphy and the age of plutonism, described below, suggest these arc assemblages may represent separate terranes.

Zircon U–Pb age data on plutons associated with the San Andres arc complex indicate minimum emplacement ages between 127 m.y. and 154 m.y. (Kimbrough, 1982; Barnes, 1982). These data provide a minimum age for the arc volcanic rock and sediments they intrude and indicate that the plutonic rocks, which are mainly biotite-hornblende tonalite and hornblende-biotite granodiorite, were emplaced over a considerable span of time, perhaps as much as 25 m.y.

Albian age marine strata of the Asuncion Formation depositionally overlie plutonic and volcanic rock of the San Andres arc complex 15 km (9.3 mi) northwest of Asuncion (Barnes, 1982; Fig. 2, 3). This is the only locality on the Vizcaino Peninsula where fossiliferous strata are seen in depositional contact on Jurassic arc basement and is the only available stratigraphic age constraint on flow rocks and intrusives of the San Andres arc complex. Tonalite beneath the erosional unconformity yielded a concordant zircon U–Pb age of 154 m.y., implying that uplift and erosion of the arc occurred between 154 m.y. and Albian time (Barnes, 1982).

A thick sequence of pre-Upper Jurassic volcaniclastic turbidites consisting of intercalated conglomerate, sandstone, and argillite occurs in the southwestern portion of the Sierra de San Andres and has been informally referred to as the Cerro El Calvario volcaniclastics (Moore, 1983; 1984; Fig. 2). These volcaniclastic rocks are intruded by tonalite dated at 154 m.y. by K–Ar and U–Pb methods (Troughton, 1974; Barnes, 1982). Moore (1984) tentatively correlates these strata with the sandstone member of the San Hipolito Formation on the basis of lithologic and sedimentologic similarity.

In the central part of the Sierra de San Andres, a wide serpentinite matrix basement melange, the Sierra Placeres melange of Moore (1983), separates the Upper Triassic Sierra de San Andres Ophiolite in the northwest part of the range from the Jurassic age San Andres arc complex in the southeast part of the range. The melange matrix consists of schistose serpentinite derived from tectonite peridotite of the Triassic ophiolite basement. Large blocks and slabs of San Andres arc volcanic and plutonic rock and ophiolite peridotite (up to several kilometers in length) float in a serpentinite matrix and are elongate parallel with the matrix foliation (Moore, 1983). The melange shear zone is best developed from Puerto San Jose to Cerro el Calvario and has a steeply dipping northeast–southwest-trending structural grain. Large dikelike bodies of foliated meta-gabbro occur within the serpentinite matrix, subparallel to the matrix foliation, and may represent synkinematic intrusions contemporaneous with basement deformation and melange development.

A sample for U–Pb age dating was collected from a large plutonic block within the Sierra Placeres melange in the central Sierra San Andres just east of Asuncion road. A single slightly discordant zircon fraction from this sample indicates a crystallization age in the range 149 to 160 m.y., indicating that the melange development, at least in part, postdates this age (Kimbrough, 1982). Aptian–Albian age strata conformably overlie the melange and postdate its development (Moore, 1983). The wide area of melange development, its steeply dipping structure, and its composition and age suggest that it represents a zone of wrench faulting within the Late Jurassic–Early Cretaceous San Andres arc complex, perhaps in response to a component of oblique subduction (Kimbrough, 1982).

A thick sequence of Upper Jurassic–Lower Cretaceous volcanogenic sedimentary rock is exposed around the tip of the Vizcaino Peninsula at Punta Eugenia, near Punta Quebrada, and in the Morro Hermoso area (Fig. 2). These strata are considered a part of the San Andres arc complex (Barnes, 1982; Hickey, 1984). Late Jurassic strata in the Punta Eugenia area are intruded by several small andesite plugs, the largest of which has a K–Ar date of 125 m.y. (Robinson, 1975). Hickey (1984) has divided the strata in the Punta Eugenia area into two distinct stratigraphic units, the Tithonian to Neocomian Eugenia Formation (2,000 m [6,562 ft]) and the Aptian–Albian Perforada sandstone (500 m [1,640 ft]). These two units are interpreted as a proximal clastic apron that was deposited on steep submarine slopes of a fan-delta system adjacent to an active insular volcanic arc. Paleocurrent transport indicators in the Eugenia Formation show a consistent trend of sediment transport toward the west (Barnes, 1984; Hickey, 1984). The unconformity between the Eugenia Formation and the Perforada sandstone is related to basement uplifts in Lower Cretaceous time (Barnes, 1984; Hickey, 1984). The Jura-Cretaceous strata in the Morro Hermosa area are also part of the Eugenia Formation but are more strongly volcanogenic and contain abundant intercalated pillow lava horizons (Barnes, 1984) in contrast to the Punta Eugenia sections described by Hickey (1984). The Eugenia Formation and Perforada sandstone record major volcanic activity (Tithonian–Neocomian), degradation, unroofing, and eventual drowning (Albian) by forearc submarine fan deposits of the Valle Formation (Barnes, 1984; Hickey, 1984).

Late Jurassic Eugenia Formation strata unconformably overlie the Morro Hermoso tuff member of the San Hipolito Formation near Morro Hermoso and the small exposure of Upper Triassic pillow lava near Punta Quebrada (Barnes, 1984). These relations indicate that the basement substrate beneath the Eugenia Formation is Upper Triassic ophiolite. The base of the Eugenia Formation is nowhere else exposed on the Vizcaino Peninsula.

The sediment provenance of Eugenia Formation strata is reflected by sandstone petrology and clast suites within conglomerate beds. Barnes (1982) has documented a stratigraphic succession in the detrital composition of the Eugenia Formation sandstones characterized from base to top by volcanogenic, transitional, and quartzofeldspathic petrofacies. This sequence is interpreted as a record of uplift and unroofing of the San Andres–Cedros arc complex (Barnes, 1982). Hickey has additionally recognized a metamorphiclastic petrofacies in a restricted area that is characterized by a high content of continentally derived detritus including metasedimentary fragments, polycrystalline quartz, and chert.

Conglomerate horizons in the Eugenia Formation contain volcanic clasts as the dominant component; plutonic clasts are a minor (1–10%) but ubiquitous component (Hickey, 1984). A distinctive peraluminous biotite granite clast collected from Aptian–Albian strata north of Punta Quebrada contains a strongly discordant zircon population. Concordia modeling indicates a crystallization age around 150 m.y. and an inherited component of Precambrian zircon (approximately 1.3 b.y.). Similar biotite granite clasts occur throughout most of the Eugenia Formation and Perforada sandstone and reach up to 2 m (6.6 ft) in diameter. This material appears to have been derived from a proximal source area. Granitic rocks of this type or age have not been described from anywhere within the Baja California Peninsula (Silver et al, 1979; Gastil et al, 1981). The absence of a suitable source terrane for this apparently exotic rock type may indicate translation of the Vizcaino terrane relative to the North American craton since Aptian–Albian time and prior to the opening of the Gulf of California 5 m.y. ago.

Moore (1983, and this volume) divides the Vizcaino terrane, as described here, into two Upper Triassic ophiolite terranes separated by the Sierra Placeres melange. Differences in the petrology of the ophiolite remnants in the Asuncion–San Pablo area versus the western Sierra de San Andres, along with differences in the overlying Jura-Cretaceous stratigraphy, are the main basis for this subdivision. However, the lithologic similarity of the ophiolite sequences in these areas, and the presence of biostratigraphically similar Upper Triassic volcanogenic and pelagic strata on both sides of the Sierra Placeres melange, grouped together as members of the San Hipolito Formation (Barnes, 1984), suggest that rocks on either side of the Sierra Placeres melange zone are closely related. The Sierra Placeres melange appears to be a Late Jurassic–Early Cretaceous structure related to San Andres arc complex magmatism.

CHOYAL TERRANE

On Cedros Island a Middle Jurassic oceanic arc/ophiolite complex and overlying Jura-Cretaceous strata comprise the Choyal terrane (Fig. 2). The Middle Jurassic Gran Canon Formation and the Middle to Late(?) Jurassic Coloradito Formation that depositionally overlie volcanic arc and ophiolite basement are stratigraphic elements not recognized on the Vizcaino Peninsula (Fig. 3). Volcanic-arc derived strata of the Gran Canon Formation are older than dated volcanogenic strata associated with the San Andres Arc Complex of the Vizcaino terrane. Olistostromes in the Coloradito Formation contain exotic megaclasts of chert, limestone, and orthoquartzite found only on Cedros Island. However, Tithonian conglomerate and sandstone on Cedros Island has been correlated to the Eugenia Formation on the Vizcaino Peninsula (Kilmer, 1977), and the Valle Formation is also widely exposed. These two formations constitute a possible overlap assemblage that may indicate proximity of the Vizcaino and Choyal terranes at least since the Late Cretaceous. Nevertheless, the absence of Middle Jurassic rocks on the Vizcaino Peninsula that are clearly correlative to the extensive Middle Jurassic history recorded by the Choyal terrane requires that these two areas, at least tentatively, must be considered separate tectonostratigraphic terranes.

The northern part of Cedros Island, from Arroyo El Choyal northward, is composed of a thick pile of mafic to silicic submarine volcanic rocks that are intruded by andesitic dike swarms and several small granodiorite and tonalite plutons and stocks. The volcanic rocks comprise a compositionally diverse suite of subalkaline lavas that have both tholeiitic and calc-alkaline affinities (Kimbrough, 1982). Zircon ages from the plutons intruding the volcanics range from 160 m.y. to 166 m.y. (Kimbrough, 1982). This basement complex has been interpreted as a well-preserved portion of a fossil volcanic arc eruptive center (Jones et al, 1976; Rangin, 1978; Kimbrough, 1982) and is referred to as the Choyal Formation.

In the southern half of Cedros Island, rocks representing most levels of an idealized ophiolite section (Cedros Island Ophiolite) occur in fault-bounded blocks that are in fault contact with an adjacent blueschist facies and lower grade metamorphic complex (Kilmer, 1977, 1979; Rangin, 1978; Kimbrough, 1982). The plutonic rocks of the ophiolite appear to be floored by mantle derived harzburgite-dunite tectonite and roofed by submarine volcanic rocks that are in turn depositionally overlain by tuffaceous radiolarian chert (Kimbrough, 1982). Sheeted dikes are not preserved, but abundant subparallel mafic dikes, which intrude hornblende gabbros along the east coast of the island between Arroyo San Carlos and Gran Canon, may have been sheeted at a higher stratigraphic level that has been removed by faulting. Plagiogranite intruded into noncumulus hornblende gabbro has yielded a zircon U-Pb age of 173 m.y. that is interpreted as the emplacement age for the entire igneous portion of the ophiolite suite (Kimbrough, 1982). Ophiolite emplacement is related to extensional rifting within arc basement (Kimbrough, 1982).

The Cedros Island Ophiolite and the Choyal Formation are separated by a down-faulted northeast-southwest-trending syncline (Pinos syncline of Kilmer, 1979). Locally, volcanogenic strata of the Gran Canon Formation depositionally and concordantly overlie volcanic flows and breccias of both the ophiolite and the Choyal Formation, on the southeast and northwest limbs of the Pinos syncline, respectively (Kimbrough, 1984). The Cedros Island Ophiolite and the Choyal Formation together formed a topographically rugged, submarine volcanic substrate upon which overlying Jurassic strata of the Gran Canon, Coloradito, and Eugenia Formations were deposited.

The Gran Canon Formation (300-1,200 m [984–3,937 ft]) consists of volcaniclastic and epiclastic sediment with minor pillow lava and conglomerate intercalated in its upper portion (Kimbrough, 1984). These volcanogenic sediments are inferred to have been derived from an undissected magmatic arc. Sedimentary structures, composition, unfossiliferous nature, stratigraphy, and tectonic setting combine to suggest that Gran Canon Formation strata were deposited in a relatively deep-water arc-flanking or intra-arc basin (Kimbrough, 1984). There is no evidence for a component of terrigenous sediment until the onset of Coloradito Formation sedimentation, discussed below.

Although megafossils are scarce within the Gran Canon Formation, bivalves collected from a fine-grained tuff bed in the middle part of the formation were identified as Bositra buchi (Roemer) (Imlay, 1968, written communication to Kilmer). Bositra is common in Toarcian to Callovian age rocks and occurs in abundance in Callovian strata of the Bedford Canyon Formation in southern California (Jones et al, 1976). Zircon from a hornblende crystal-lithic tuff in the upper part of the Gran Canon Formation has yielded a concordant U-Pb isotopic age of 166 m.y. (Kimbrough, 1982). This radiometric age corresponds to late Middle Jurassic using the time scale of Harland et al (1982) and is consistent with the presence of Bositra in the section.

The Coloradito Formation conformably overlies the Gran Canon Formation and is well exposed on both flanks of the Pinos syncline and in the south part of the island. The formation (300–400 m [984–1,312 ft]) consists of basal thin-bedded sandstone flysch and overlying units of black silty argillite, tuff, pebbly sandstone, and conglomerate (Kilmer, 1979; Boles and Landis, 1984). The argillite units locally have a sheared scaly matrix type deformation and contain a spectacular assemblage of exotic rock types including Paleozoic limestones, orthoquartzites, Triassic ribbon cherts, and volcaniclastics that range from meter size clasts to large slab-like blocks 50-100 m (164–328 ft) long by 8 to 10 m (26–33 ft) thick (Kilmer, 1977; Boles and Landis, 1984).

The Coloradito Formation is assigned a Middle Jurassic age based on a 159 m.y. hornblende K-Ar date on andesitic tuff in the upper one-third of the formation (Gastil et al, 1978) and on upper Callovian ammonites from graywacke within the formation (Rangin, 1978). It is possible, however, that these ages were determined from reworked blocks.

Tuff beds in the Coloradito Formation record continued arc volcanism while the sheared argillite and exotic blocks record an abrupt influx of mature continental detritus. The source of the exotic blocks is unclear, but they have great significance for tectonic and paleogeographic reconstructions of the Choyal terrane. Boles and Landis (1984) suggest that the exotic blocks could have been derived from North American Paleozoic platform sequences in eastern Baja California that are intruded by the Peninsular Range Batholith (Gastil and Miller, 1984). The source of these clasts is still unclear, however.

The Eugenia Formation on Cedros Island was first mapped and described by Kilmer (1977), who correlated it to the Eugenia Formation on the Vizcaino Peninsula. Boles and Landis (1984) have interpreted the Eugenia Formation on Cedros Island (200 m [656 ft]) as a submarine channel sequence, probably related to the upper part of a submarine channel fill. The basal 10 to 20 m (33–66 ft) of the formation in the southern part of the island contains megablocks of Triassic green to black banded radiolarian chert, siliceous sandstone, siltstone, and limestone, set in a poorly exposed matrix of poorly sorted belemnite-bearing conglomeratic sandstone. This basal unit is overlain by a sequence of generally massive amalgamated conglomerate units. A Late Jurassic radiolarian fauna (early Tithonian, Zone 3) is reported from siltstone within the lower 50 m (164 ft) of the formation in this area (Boles and Landis, 1984). The Valle Formation depositionally overlies both the Eugenia and Coloradito Formations on Cedros Island with mild angular discordance.

Boles and Landis (1984) interpret the Coloradito and Eugenia Formations as a slope-channel facies deposited adjacent to a continental crustal block of high relief. The sudden appearance of a coarse exotic sediment in debris flows and extensive soft sediment slump folding in the directly underlying Gran Canon Formation strata apparently records tectonism and sedimentation associated with the continental accretion of the Choyal terrane at around the Middle/Upper Jurassic boundary (Boles and Landis, 1984; Kimbrough, 1984).

SUBDUCTION COMPLEX TERRANES

Introduction

Rocks that are widely exposed on Cedros and San Benito Islands show structural and lithologic features that typify parts of the Franciscan Complex in California (Cohen et al, 1963; Suppe and Armstrong, 1972; Jones et al, 1976; Kilmer, 1977, 1979; Rangin, 1978; Rangin et al, 1981; Klienast and Rangin, 1982). All of the above authors recognized the similarity of rocks on Cedros and San Benito Islands to parts of the much better known

Franciscan Complex rocks exposed in the Coastal Ranges of Northern and Central California.

Melanges on Cedros Island are similar to those of the California Coast Ranges that separate coherent units of well-bedded sedimentary rock. There is a characteristic mixture of oceanic basement rocks with terrigenous and pelagic sedimentary rocks, all of which have been variably recrystallized under high P/T metamorphic conditions. Both serpentinite matrix and sedimentary matrix melanges are present on Cedros and San Benito Islands, some with tectonic inclusions of coarse-grained high P/T tectonites.

Coherent units of sedimentary strata on both Cedros and San Benito Islands are dominated by turbiditic beds of texturally immature quartzofeldspathic sandstone and argillite, with or without intercalated volcanics, chert, and limestone (Cohen et al, 1963; Rangin, 1978; Kimbrough, 1982). Biostratigraphic ages from coherent sedimentary strata and melange blocks, based on radiolarian faunas from chert, range from upper Pliensbachian–Toarcian to Early Cretaceous (Rangin, 1978, 1981; Kimbrough, 1982) similar to the ages reported from the Franciscan Complex of California (Pessagno, 1977). Locally on Cedros Island, a large melange block contains a thick section of Lower Jurassic to possible Early Cretaceous pure red ribbon radiolarian chert in depositional contact on pillow lava (Kimbrough, 1982).

San Benito Terranes

The San Benito Islands are a group of three islands situated on the edge of the continental shelf only 40 km (25 mi) east of the Cedros Deep (Fig. 1). West Island, the largest of the three, is cut by the northwest-trending San Benito shear zone, a major right-lateral fault zone that can be traced for several hundred kilometers along the San Benito ridge and possibly connects with the Tosco–Abreojos fault to the south (Krause, 1965; Spencer and Normark, 1979).

Mapping by Cohen et al (1963) separates three distinct fault-bounded map units on the islands (Fig. 2) that are considered separate terranes: (1) The West San Benito terrane is an intact sequence of graywacke and argillite that occurs west of the San Benito shear zone; no ages have been reported; (2) the Central San Benito terrane is a sheared graywacke + chert + basalt association to the east of the San Benito shear zone that contains Late Jurassic and Early Cretaceous radiolarian assemblages (Rangin et al, 1981); and (3) the East San Benito terrane is a serpentinite matrix melange with coarse-grained high P/T blueschist and amphibolite tectonic blocks that also occurs to the east of the San Benito shear zone. Two K–Ar analyses on hornblende and white mica from samples of schist collected out of the melange yield ages of 148 m.y. (Suppe and Armstrong, 1972).

Cedros Terrane

Rocks correlative to the Franciscan Complex are exposed in three separate areas in the southern part of Cedros Island and cover approximately 80 sq km (31 sq mi). These rocks are collectively referred to here as the Cedros terrane even though they are probably subdivisible into several separate subterranes (Rangin, 1978; Klienast and Rangin, 1982; Kimbrough, 1982).

The Cedros massif (Kimbrough, 1982) and the adjacent Punta Prieta Ridge are diapirs of variably recrystallized high P/T metasedimentary and metaigneous rocks and serpentinite that have domed upwards through the Middle Jurassic arc/ophiolite basement of the Choyal terrane (Kimbrough, 1982). The Cedros massif is comprised of several thrust sheets composed of strongly deformed and recrystallized metasedimentary sequences, coherent metavolcanic units, and serpentinite and sedimentary matrix melange containing blocks of metagraywacke, metabasalt, glaucophane schist, recrystallized limestone, and radiolarian chert (Rangin, 1978). These roughly tabular narrow fault-bounded units have variable metamorphic grades, lithologies, and structural features that suggest a complex postmetamorphic structural history. The thrust sheets of the Cedros massif, and also perhaps Punta Prieta ridge, resemble schuppen complexes described from the Franciscan Complex in the Goat Mountain area (Suppe and Foland, 1978), around Pachecho Pass (Cowan, 1974), and on Catalina Island (Platt, 1975).

A prominent map unit of glaucophane schist metasedimentary rock within the Cedros massif is excellently exposed along the east coast of the island. Float blocks of metachert probably derived from this unit have yielded two white mica K–Ar ages of 110 ± 2 m.y. and 109 ± 2 m.y. (Suppe and Armstrong, 1972). These schists appear similar to the Catalina schist exposed on Catalina Island and the Palos Verde Peninsula, which yields similar K–Ar ages.

Franciscan-type rocks are also exposed in the southwest corner of Cedros Island on the nearly peninsular cape of Cabo San Augustin. This area is separated from the rest of the island by the San Augustin Fault, a possible extension of the San Benito shear zone (Kilmer, 1979) or a related subparallel fault. Coherent units of quartzose sandstone, shale, chert, volcanics, and limestone are separated by broad tracts of spectacularly exposed sedimentary matrix melange that contains blocks of rounded blueschist, graywacke, volcanic rocks, gabbro, quartz-rich plutonic rocks, radiolarian chert, limestone, orthoquartzite, and ultramafic rocks (Kimbrough, 1982). Graywacke from coherent units and melange matrix contains lawsonite and jadeitic pyroxene (Rangin, 1978; Klienast and Rangin, 1982). Smaller blocks in the melange often have phaccoidal shapes and are engulfed in a pervasively sheared argillaceous matrix. Red radiolarian chert from a melange block has yielded a Late Jurassic, probably zone 2A, radiolarian assemblage (Jones et al, 1976).

The San Augustin melange also contains large blocks of recrystallized fossiliferous limestone, one of which is reported to be a crinoidal limestone with Permian fusilinids (Rangin, 1978). Permian fossils from the Cabo San Augustin limestone blocks are under investigation by Rangin at the University of Paris, and samples appear to have faunas related to those of Permian reef limestones that occur in Sonora, Mexico (Blake, 1981, personal communication). Paleozoic limestone blocks are not recognized as a component of the Franciscan Complex of California.

Puerto Nuevo Terrane

On the Vizcaino Peninsula, a serpentinite matrix melange (the Puerto Nuevo melange of Moore, 1983) occurs structurally beneath the harzburgite-dunite member of the

Sierra de San Andres ophiolite at the core of an anticlinal structure in the ophiolite basement, as outlined earlier (Jones et al, 1976; Moore, 1983). This melange is well exposed along the Puerto Nuevo road and is referred to as the Puerto Nuevo terrane.

The melange occurs structurally beneath nonschistose harzburgitic peridotite that is cut by impressive swarms of rodingitized gabbroic and diabasic dikes. Rounded blocks of high P/T tectonites, ranging mainly from 1 to 20 m (3.3–66 ft) in diameter, are dispersed as tectonic clasts within schistose antigorite serpentinite matrix in a zone subparallel to the base of the ophiolite (Moore, 1983). The subhorizontal geometry of the Puerto Nuevo melange suggests that the Sierra de San Andres ophiolite is a thin, unrooted, crystalline sheet that has been underthrust by the melange.

A variety of coarse-grained blueschist- and greenschist-facies blocks occur in the Puerto Nuevo melange, including glaucophane-epidote-aragonite metabasalts and crossite-stilpnomelane metacherts (Jones et al, 1976) and glaucophane-quartz, garnet-glaucophane-quartz and glaucophane-white mica quartz schists. Most of the high P/T tectonites have strong schistose fabrics and some show evidence of superposed folds. Some blocks are strongly mylonitized and others exhibit retrograde reaction rinds on their outer margins.

Blueschist facies rocks of the Puerto Nuevo melange may be related to similar high-grade Franciscan Complex metamorphic rocks on Cedros and San Benito Islands, but alternatively may be derived from a pre-Jurassic blueschist facies terrane related to an earlier pre-Franciscan regime of plate convergence or collision along a continental margin. In California, probable Middle and Late Triassic blueschist facies metamorphic rocks occur structurally inboard of the Franciscan Complex, associated with Triassic and older ophiolite belts of the Sierra Nevada foothills and the western Klamath Mountains. The association of the Puerto Nuevo melange with the Late Triassic age Morro Hermosa ophiolite raises suspicion about the metamorphic ages of blueschist blocks in the melange. Radiometric age dating is needed here.

MOVEMENT HISTORY OF TERRANES

Critical to understanding the paleogeographic significance of the Baja California terranes described here is an understanding of their movement history with respect to each other and to the stable North American craton. Karig et al (1978) and Coney et al (1980) suggest that the Vizcaino region has been tectonically transported relative to the main part of Baja California. Boles and Landis (1984) and Barnes (1984) regard the Vizcaino and Choyal terranes as essentially in place. Patterson (1984b) presents paleomagnetic evidence that indicates the Valle Formation (Cenomanian–Santonian) on the Vizcaino Peninsula and Cedros Island, and Alisitos arc rocks that occur to the east along the axis of the Baja California Peninsula, were at the same paleolatitude in Late Cretaceous. The evidence further suggests that the Alisitos arc and the Valle depositional basin have moved very little, perhaps 3° northward, with respect to the North American craton. Recent discoveries show that the Alisitos arc/Peninsular Range Batholith was intruded into North American Paleozoic platform sequences exposed along the eastern margin of this Late Cretaceous arc complex and is thus tied to the stable craton (Gastil and Miller, 1984). This later discovery is consistent with; and lends support to, the paleomagnetic evidence presented by Patterson (1984b). If the Vizcaino and Choyal terranes represent exotic fragments of far-traveled tectonic settings, then the evidence reviewed above suggests that most of their northward motion occurred prior to the Late Cretaceous. Further, exotic continental granite detritus found in strata of the Vizcaino terrane extend upward into Aptian–Albian strata. If this granitic detritus was derived from a far-removed continental margin, it implies that the major relative motion between the Vizcaino terrane and the North American craton must postdate the deposition of the Aptian–Albian strata. The timing of possible northward displacement of the Vizcaino terrane may therefore be restricted to a narrow interval in the mid-Cretaceous. McWilliams and Howell (1982) have presented paleomagnetic evidence for the Salinia and Stanley Mountain terranes of California that suggest a large component of northward relative motion along the western margin of North America during the Late Cretaceous and Early Tertiary. The possible Late Cretaceous displacement of the Vizcaino and Choyal terranes could therefore be related to the displacement suggested for these California terranes to the north.

Another important question is the possibility of large-scale relative displacement between the Vizcaino terrane and the Choyal terrane on Cedros Island. The rationale for distinguishing these two terranes was discussed earlier. It is not clear that the continental suturing events inferred for the Choyal terrane and the Vizcaino terrane represent the same tectonic event, for example an arc-continent collision, along the same segment of continental margin. Resolution of this uncertainty calls for paleomagnetic investigation.

CORRELATIVE CORDILLERAN TERRANES TO THE NORTH

The Vizcaino–Cedros–San Benito terranes represent the best exposure of Mesozoic paleo-oceanic assemblages along a 1,500 km (932 mi) segment at the southern end of the western Cordilleran accreted belt. The evolution of these terranes and their relationships to the much more widely exposed Mesozoic terranes to the north is important from a regional standpoint in reconstructing the Mesozoic paleogeography of the western Cordillera. The subduction complex assemblages of Cedros and San Benito Islands have obvious correlatives with older elements of the Franciscan Complex, many of which were discussed earlier. The following discussion is restricted to possible correlatives of the Vizcaino and Choyal terranes.

The Upper Triassic ophiolite and overlying San Hipolito Formation of the Vizcaino terrane has possible correlatives in Jura-Triassic ophiolites and arc assemblages of the western Klamath Mountains and western Sierra Nevada foothills (see Saleeby, 1983, and references therein). The presence of the Capnodoce radiolarian zone in the San Hipolito Formation, the Rattlesnake Creek ophiolite assemblage of the western Klamath Mountains, the Suplee-Izee area of eastern Oregon (Cache Creek-affinity), and the Cache Creek Group of British Columbia strengthens this correlation. The association of the Sierran and Klamath terranes with distinctive upper Paleozoic Cache Creek or Cache Creek-affinity basement (Saleeby, 1983) is an important difference, however, between the Sierran-Klamath Jura-Triassic arc complexes and the Triassic rocks of the Vizcaino terrane.

The Middle and Upper Jurassic history of arc volcanism, basement rifting, and continental accretion recorded by the Choyal terrane on Cedros Island bears remarkable resemblance to Jurassic ensimatic arc assemblages and ophiolites in the western Klamath Mountains, western Sierra Nevada foothills, and the Coast Ranges of California. The Smartville and Josephine ophiolites (Xenophontos and Bond, 1978; Harper, 1980) were apparently generated during rifting events associated with tectonic extension across arc terranes. The Cedros ophiolite is similarly ascribed to rifting within an active arc terrane and appears to preserve an edge zone assemblage generated during the initial stages of tectonic extension and marginal or intra-arc basin development. Further, stratigraphic relationships indicate that the Jurassic volcanic arc terranes and rift basin ophiolites of the Klamaths and Sierran foothills were generated in proximity to the paleocontinental margin (Behrman and Parkison, 1978) and were subsequently collapsed against the continental margin shortly after their formation (Saleeby, 1982). Continental accretion of the Choyal terrane is apparently recorded by the Coloradito Formation, and this accretionary event probably took place around the Middle/Upper Jurassic boundary, or 10 to 15 m.y. following the rifting event inferred from the Cedros Island ophiolite. The timing of rifting and subsequent collapse against the continental margin is thus similar to the timing of these events inferred for the Klamath and Sierran foothills arc/ophiolite terranes. The Choyal terrane probably formed along strike from the Jurassic age ophiolite/arc terranes of California within the same tectonic setting.

ACKNOWLEDGMENTS

Field excursions and conversations with numerous people and particularly T. E. Moore, J. J. Hickey, D. A. Barnes, J. R. Boles, C. A. Landis, and J. M. Mattinson helped to develop the ideas presented in this review. I thank D. G. Howell and J. R. Boles for critical reviews that improved earlier versions of this manuscript. Laboratory work was supported by NSF grant EAR 80-08215 to Mattinson. Fieldwork on Cedros Island was supported in part by a Geological Society of America Penrose grant and a Harold T. Stearn Fellowship.

REFERENCES

Atwater, T., 1970, Implications of plate tectonics for the Cenozoic tectonic evolution of western North America: Geological Society of America Bulletin, v. 81, p. 3513–3536.

Barnes, D., 1982, Basin analysis of volcanic arc derived, Jura-Cretaceous sedimentary rocks, Vizcaino Peninsula, Baja California Sur, Mexico: PhD Dissertation, University of California, Santa Barbara, 240 p.

——, 1984, Volcanic arc derived, Mesozoic sedimentary rocks, Vizcaino Peninsula, Baja California Sur, Mexico, *in* V. A. Frizzell, ed., Geology of the Baja California Peninsula: Society of Economic Paleontologists and Mineralogists, Pacific Section, v. 39, p. 119–130.

——, and J. Mattinson, 1981, Late Triassic–Early Cretaceous age of eugeoclinal terranes, western Vizcaino Peninsula, Baja California Sur, Mexico (Abs.): Geological Society of America Abstracts with Programs, v. 13, p. 43.

Beggs, J. M., 1984, Volcaniclastic rocks of the Alisitos Group, Baja California, *in* V. A. Frizzell, ed., Geology of the Baja California Peninsula: Society of Economic Paleontologists and Mineralogists, Pacific Section, v. 39, p. 43–52.

Behrman, P. G., and G. A. Parkison, 1978, Paleogeographic significance of the Callovian to Kimmeridgian strata, central Sierra Nevada Foothills, California, *in* D. G. Howell, and K. A. McDougall, eds., Mesozoic paleogeography of the western United States: Society of Economic Paleontologists and Mineralogists, Pacific Section, Pacific Coast Paleogeography Symposium 2, p. 349–360.

Blake, Jr., M. C., et al, 1984, Tectonostratigraphic terranes of Magdelena Island, Baja California Sur, *in* V. A. Frizzell, ed., Geology of the Baja California Peninsula: Society of Economic Paleontologists and Mineralogists, Pacific Section, v. 39 p. 183–192.

Boles, J. R., and C. A. Landis, 1984, Jurassic sedimentary melange and associated facies, Baja California, Mexico: Geological Society of America Bulletin, v. 95, p. 513–521.

Cohen, L. H., et al, 1963, Geology of the San Benito Islands, Baja California, Mexico: Geological Society of America Bulletin, v. 74, p. 1355–1370.

Coney, P. J., 1981, Accretionary tectonics in western North America, *in* W. R. Dickinson and W. D. Payne, eds., Relations of tectonics to ore deposits in the southern Cordillera: Arizona Geological Society Digest, p. 23–38.

——, et al, 1980, Cordilleran suspect terranes: Nature, v. 288, p. 329–333.

Cowan, D. S., 1974, Deformation and metamorphism of the Franciscan subduction zone complex northwest of Pacheco Pass, California: Geological Society of America Bulletin, v. 85, p. 1623–1634.

Finch, J. W., and P. L. Abbott, 1977, Petrology of a Triassic marine section, Vizcaino Peninsula, Baja California Sur, Mexico: Sedimentary Geology, v. 19, p. 253–273.

Gastil, G., and R. H. Miller, 1984, Prebatholithic paleogeography of the Peninsula California and adjacent Mexico, *in* V. A. Frizzell, ed., The geology of the Baja California Peninsula: Society of Economic Paleontologists and Mineralogists, Pacific Section, v. 39, p. 9–15.

——, et al, 1978, Mesozoic history of peninsular California and related areas east of the Gulf of California, *in* D. G. Howell and K. A. McDougall, eds., Mesozoic paleogeography of the western United States: Society of Economic Paleontologists and Mineralogists, Pacific Section, Pacific Coast Paleogeography Symposium 2, p. 107–116.

______, et al, 1981, The tectonic history of peninsular California and adjacent Mexico, *in* W. G. Ernst, ed., The geotectonic development of California, Rubey v. 1: Englewood Cliffs, NJ, Prentice-Hall, Inc., p. 284–305.

Harland, W. B., et al, 1982, A geologic time scale: Cambridge University Press, 131 p.

Harper, G. D., 1980, The Josephine ophiolite-remnant of a Late Jurassic marginal basin in northwestern California: Geology, v. 8, p. 333–337.

Helenes, J., 1984, Dinoflagellates from Cretaceous to Early Tertiary rocks of the Sebastian Vizcaino Basin, Baja California, *in* V. A. Frizzell, ed., The geology of the Baja California Peninsula: The Society of Economic Paleontologists and Mineralogists, Pacific Section, v. 39, p. 89–106.

Hickey, J., 1984, Stratigraphy and composition of a Jura-Cretaceous volcanic arc apron, Punta Eugenia, Baja California Sur, Mexico, *in* V. A. Frizzell, ed., The geology of the Baja California Peninsula, Society of Economic Paleontologists and Mineralogists, Pacific Section, v. 39, p. 149–160.

Hopson, C. A., et al, 1981, Coast Range Ophiolite, western California, *in* W. G. Ernst, ed., The geotectonic development of California, Rubey v. 1: Englewood Cliffs, NJ, Prentice-Hall, Inc., p. 418–510.

Howell, D. G., and J. G. Vedder, 1981, Structural implications of stratigraphic discontinuities across the southern California borderland, *in* W. G. Ernst, ed., The geotectonic development of California, Rubey v. 1: Englewood Cliffs, NJ, Prentice-Hall, Inc.

Jones, D. L., et al, 1976, The four Jurassic belts of northern California and their significance to the geology of the southern California borderland, *in* D. G. Howell, ed., Aspects of the geologic history of the California continental borderland: American Association of Petroleum Geologists, Pacific Section, Miscellaneous Publication 24, p. 343–362.

Karig, D., et al, 1978, Late Cenozoic subduction and continental margin truncation along the northern middle America trench: Geological Society of America Bulletin, v. 89, p. 265–276.

Kilmer, F. H., 1977, Reconnaissance geology of Cedros Island, Baja California, Mexico: Southern California Academy of Science Bulletin, v. 76, p. 91–98.

______, 1979, A geological sketch of Cedros Island, Baja California, *in* P. L. Abbott and R. G. Gastil, ed., Baja California geology: Field Guide and Papers, Geological Society of America Meeting, San Diego, p. 11–28.

Kimbrough, D. L., 1982, Structure, petrology and geochronology of Mesozoic paleooceanic basement terranes on Cedros Island and the Vizcaino Peninsula: Unpublished PhD Dissertation, University of California, Santa Barbara, 395 p.

______, 1984, Paleogeographic significance of the Middle Jurassic Gran Canon Formation, Cedros Island, Baja California Sur, *in* V. A. Frizzell, ed., The geology of the Baja California Peninsula: Society of Economic Paleontologists and Mineralogists, Pacific Section, v. 39, p. 107–117.

Klienast, J. R., and C. Rangin, 1982, Mesozoic blueschists and melanges of Cedros Island (Baja California, Mexico): a consequence of nappe emplacement or subduction?: Earth and Planetary Science Letters, v. 59, p. 119–138.

Krause, D. C., 1965, Tectonics, bathymetry and geomagnetism of the southern continental borderland west of Baja California, Mexico: Geological Society of America Bulletin, v. 76, n. 6, p. 617–650.

Madrid-Solis, A., 1979, Petroleum exploration in Baja California: a new productive province, *in* P. L. Abbott and R. G. Gastil, eds., Baja California: Field Guide and Papers, Geological Society of America Meeting, San Diego, p. 121–126.

McWilliams, M. O., and D. G. Howell, 1982, Exotic terranes of western California: Nature, v. 297, n. 5863, p. 215–217.

Minch, J. C., et al, 1976, Geology of the Vizcaino Peninsula, *in* D. G. Howell, ed., Aspects of the geologic history of the California continental borderland: American Association of Petroleum Geologists, Pacific Section, Miscellaneous Publication 24, p. 135–195.

Moore, T. E., 1983, Geology, petrology, and tectonic significance of the Mesozoic paleooceanic terranes of the Vizcaino Peninsula, Baja California Sur, Mexico: PhD Dissertation, Stanford University, 376 p.

______, 1984, Sedimentary facies and composition of Jurassic volcaniclastic turbidites at Cerro El Calvaria, Vizcaino Peninsula, Baja California Sur, Mexico, *in* V. A. Frizzell, ed., The geology of the Baja California Peninsula: Society of Economic Paleontologists and Mineralogists, Pacific Section, v. 39, p. 131–148.

Patterson, D. L., 1984a, Los Chapunes and Valle sandstones: Cretaceous petrofacies of the Vizcaino Basin, Baja California Sur, Mexico: Society of Economic Paleontologists and Mineralogists, Pacific Section, v. 39, p. 161–172.

______, 1984b, Paleomagnetism of the Valle Formation and the Late Cretaceous paleogeography of the Vizcaino Basin, Baja California, Mexico: Society of Economic Paleontologists and Mineralogists, Pacific Section, v. 39, p. 173–182.

Pessagno, Jr., E. A., 1977, Upper Jurassic radiolaria and radiolarian biostratigraphy of the California Coast Ranges: Micropaleontology, v. 23, p. 56–113.

______, et al, 1979, Upper Triassic radiolaria from the San Hipolito Formation, Baja California: Micropaleontology, v. 25, n. 2, p. 160–197.

Platt, J. P., 1975, Metamorphic and deformational processes in the Franciscan Complex, California: Some insights from the Catalina schist terrane: Geological Society of America Bulletin, v. 86, p. 1337–1347.

Rangin, C., 1978, Speculative model of Mesozoic geodynamics, central Baja California to northwestern Sonora (Mexico), *in* D. G. Howell and K. A. McDougall, eds., Mesozoic paleogeography of the western United States: Society of Economic Paleontologists and Mineralogists, Pacific Section, Pacific Coast Paleogeography Symposium 2, p. 85–106.

———, et al, 1981, Geochemistry of the Mesozoic bedded cherts of Central Baja California (Vizcaino-Cedros-San Benito): implications for paleogeographic reconstruction of an old oceanic basin: Earth and Planetary Science Letters, v. 54, p. 313-322.

Riveroll, G. C., 1978, A marine geophysical study of Vizcaino Bay and the continental margins of Western Mexico: PhD Dissertation, Oregon State University.

Robinson, J. W., 1975, Reconnaissance geology of the northern Vizcaino Peninsula, Baja California Sur, Mexico: MS Thesis, San Diego State University, 114 p.

Saleeby, J., 1982, Polygenetic ophiolite belt of the California Sierra Nevada: Geochronological and tectonostratigraphic development: Journal of Geophysical Research, v. 87, n. 3, p. 1803-1824.

———, 1983, Accretionary tectonics of the North American Cordillera: Annual Review of Earth and Planetary Science, v. 15, p. 45-73.

Silver, L. T., et al, 1979, Some petrological, geochemical and geochronological observations of the Peninsular Ranges batholith near the international border of the U.S.A. and Mexico, *in* P. L. Abbott and V. R. Todd, eds., Mesozoic crystalline rocks: Peninsular Ranges batholith and pegmatites and Point Sal ophiolite: San Diego State University Publication, p. 83-110.

Spencer, J. E., and W. R. Normark, 1979, Tosco-Abreojos fault zone: A Neogene transform plate boundary within the Pacific margin of southern Baja California, Mexico: Geology, v. 7, n. 11, p. 554-557.

Stanley, G. D., 1979, Paleoecology, structure, and distribution of Triassic coral buildups in western North America: Article 65, The University of Kansas Paleontological Contributions.

Suppe, J., and R. L. Armstrong, 1972, Potassium-argon dating of Franciscan metamorphic rocks: American Journal of Science, v. 272, p. 217-233.

Troughton, G. H., 1974, Stratigraphy of the Vizcaino Peninsula near Asuncion Bay, Territorio de Baja California, Mexico: MS Thesis, San Diego State University, 83 p.

Whan, P. A., and E. A. Pessagno, Jr., 1984, Lower Jurassic radiolaria, San Hipolito Formation, Vizcaino Peninsula, Baja California Sur, *in* V. A. Frizzell, ed., Geology of the Baja California Peninsula: Society of Economic Paleontologists and Mineralogists, Pacific Section, p. 53-66.

Xenophontos, C., and G. C. Bond, 1978, Petrology, sedimentation and paleogeography of the genesis of the Smartville terrane (Jurassic)—bearing on the genesis of the Smartville ophiolite, *in* D. G. Howell and K. A. McDougall, eds., Mesozoic paleogeography of the western United States: Society of Economic Paleontologists and Mineralogists, Pacific Section, Pacific Coast Paleogeography Symposium 2, p. 291-302.

The Mexican Thrust Belt

Maria Fernanda Campa U.
PEMEX
Mexico, D. F.

The Mexican Thrust Belt divides into three major transverse geological provinces. From north to south they are: the Basin and Range province, the Northern province, and the Southern province. Each province is composed of several structural domains and/or tectonostratigraphic terranes. The Basin and Range province consists of two structural domains that border the southwestern end of the North American craton. The Northern province consists of four structural domains informally named the Thrust Front, the Altiplano Central, the Thrust Back, and the West Margin. The Southern province is divided into a western composite terrane (Guerrero) and an eastern branch, which is overprinted by the Thrust Front, Thrust Back, and West Margin structural domains.

The boundary between the Northern and Southern thrust belt provinces separates Paleozoic tectonostratigraphic terranes of eastern Mexico from Mesozoic terranes of the Cordillera of western and southern Mexico. The spatial distribution of terranes is controlled by the position of allochthonous continental blocks.

The Mexican Thrust Belt was formed mainly by compressional deformation related to the Laramide orogeny. However, several subsequent tectonic events, the most conspicuous being extension of the Basin and Range province, have also added to deformation. The Motagua-Polochic sinistral fault system is the southern terminus of the Mexican Thrust Belt and represents the boundary between the North American and Caribbean plates.

The eastern part of the Mexican Thrust Belt represents the western limit for hydrocarbon production in Mexico. Mesozoic limestones that are potential hydrocarbon reservoirs lie east of the thrust belt.

INTRODUCTION

The eastern edge of the North American Cordillera is defined by a major thrust belt, which has a general north-south trend and bifurcates southward at the latitude of Yellowstone Park (Fig. 1) (Stewart, 1976; Coney, 1978). The extent of thrusting southward into Arizona and New Mexico remains poorly known mainly because of the structural complexity and the paucity of field studies. The continuation of the thrust belt into Mexico is likewise difficult to define accurately (Silver and Anderson, 1974; Stewart, 1977; Drewes, 1978; Peiffer, 1979).

Nevertheless, the structural trend of the Sierra Madre Oriental in northeastern Mexico (Fig. 2) is thought to be linked to the Cordilleran thrust belt (de Cserna, 1956; Tardy, 1980; Suter, 1982). This study attempts to clarify the general structure of the Mexican Thrust Belt and also to determine its genetic relationship with the entire thrust belt of the North American Cordillera (Fig. 1).

TECTONOSTRATIGRAPHIC TERRANES OF MEXICO

The generalized tectonostratigraphic terrane map shows the distribution of the major basement terranes of Mexico (Fig. 3). Most basement rocks of Mexico are covered by younger sedimentary or volcanic sequences. The sequences include Mesozoic sedimentary rocks related to the opening of the Gulf of Mexico that cover most of eastern Mexico, and Cenozoic volcanic rocks that include the ignimbrite province of the Sierra Madre Occidental and the younger volcanoes of the Trans Mexican volcanic arc.

The basement rocks beneath the overlying sequences consist of a collage of terranes and autochthonous structural domains that form three geological zones:

1. A northern zone that is the southward extension into Mexico of North American Precambrian basement and associated Paleozoic and Mesozoic overlap sequences, some of which are autochthonous, and others that are interpreted to be composite displaced terranes.

2. An eastern zone that borders the Gulf of Mexico, composed mainly of upper Paleozoic rocks that are heterogeneous in character but have a common origin as material accreted to North America during the Late Paleozoic Appalachian-Ouachita Marathon orogeny.

3. A southwestern zone that makes up Mexico's broad Pacific margin, which is characterized by a heterogeneous assemblage composed mainly of submarine volcanic and Mesozoic sedimentary rocks and small terranes that contain pre-Mesozoic rocks. The Mexican Thrust Belt separates the eastern zone of Paleozoic autochthonous strata from the southwestern collage of terranes that represents Mesozoic and younger accretion.

THE MEXICAN THRUST BELT

The Mexican Thrust Belt is a relatively continuous morphological and structural zone and is interpreted as an orogenic belt that originated from continental compressive deformation during the Laramide orogeny (Figs. 2, 3).

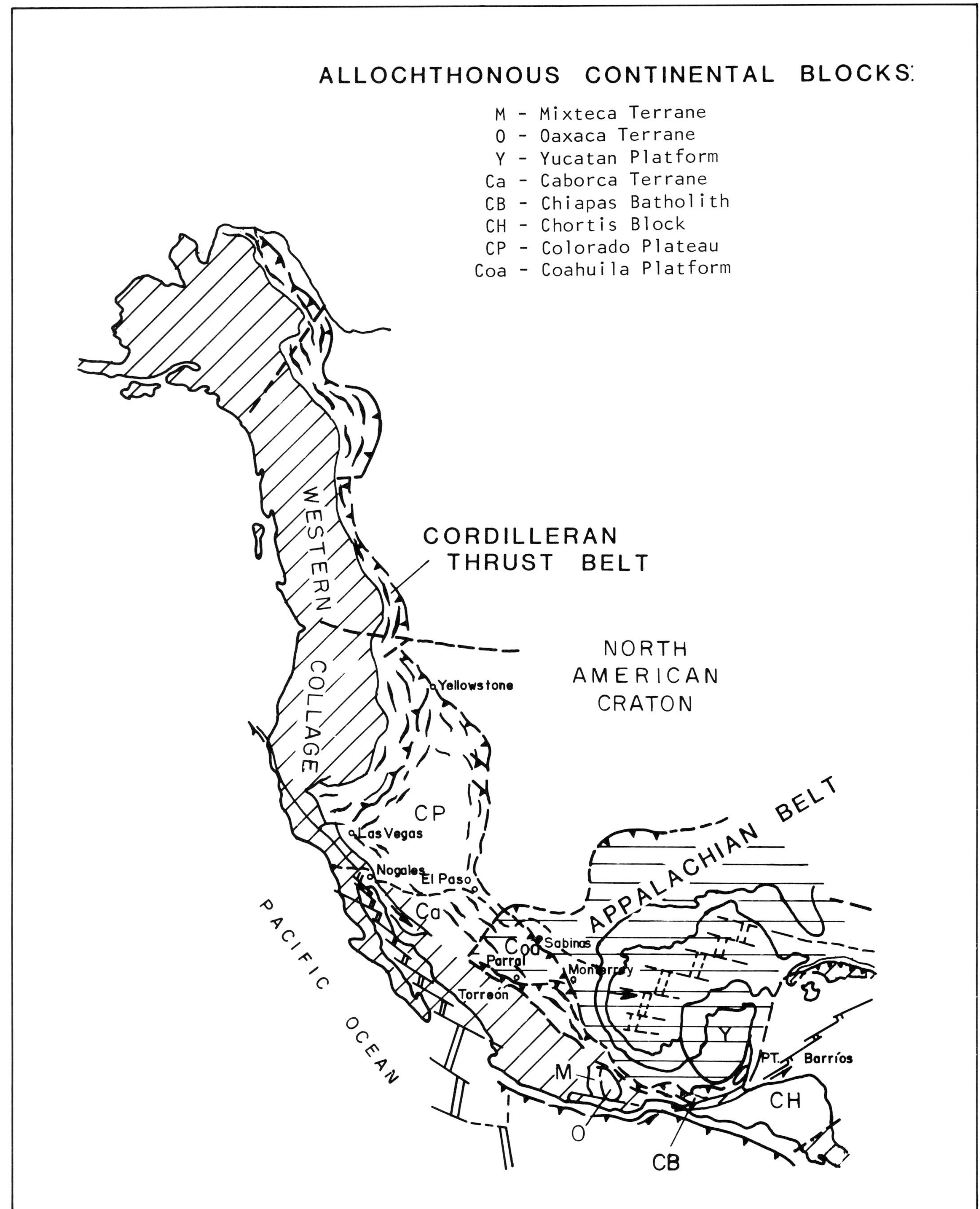

Figure 1—Major structural elements of the Cordilleran thrust belt and terranes of the Mexican Thrust Belt.

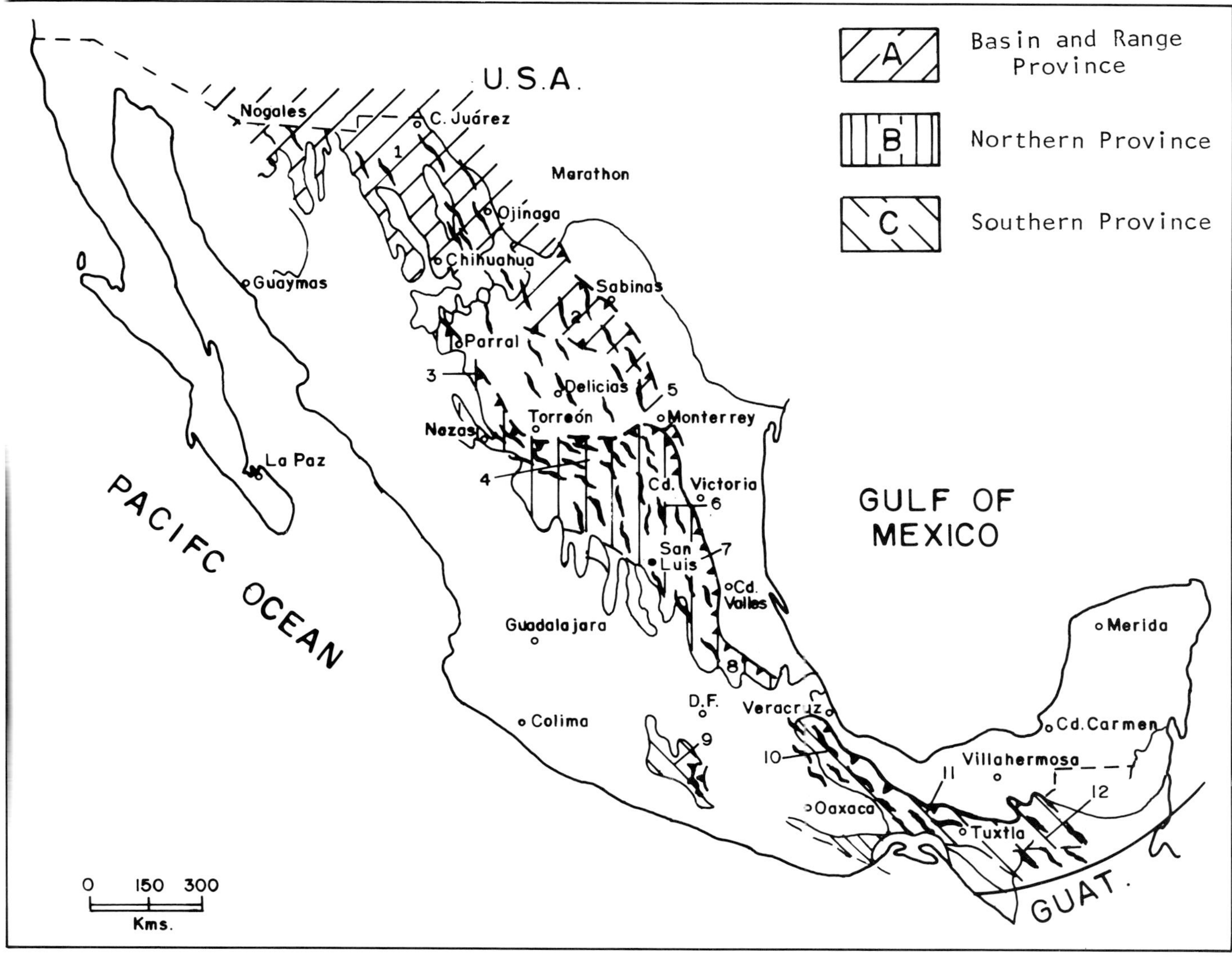

Figure 2—Location of provinces and sectors of the Mexican Thrust Belt

A Basin and Range Province
1. Chihuahua–Sonora Basin
2. Sabinas Basin

B Northern Province
3. San Pedro El Gallo Sector
4. Transversal Sector
5. Curvature of Monterrey Sector
6. Victoria Sector
7. Valles–San Luis Platform Sector
8. Huayacocotla Sector

C Southern Province
9. Guerrero Sector
10. Sierra Mazateca Sector
11. Tehuantepec Isthmus Sector
12. Sierra de Chiapas Sector

However, part of the Mexican Thrust Belt located between the southwestern United States (Stewart, 1977) and northwestern Mexico was affected by late Cenozoic extensional deformation, which produced an alternating series of elongated mountain ranges (horsts) and basins (grabens) filled with alluvial sediments.

The Mexican Thrust Belt is described systematically in this paper from the northwest to the southeast using structural and stratigraphic "sectors" and transects shown in Figures 4, 5, and 6. Geologic units shown in one sector may continue into adjacent sectors, while others show abrupt changes. The boundaries between terranes shown in the transects are considered to be faults (Figs. 7, 8). Figure 2 illustrates how the Mexican Thrust Belt is geologically divided into three provinces: (1) the Basin and Range province, (2) the Northern province, and (3) the Southern (bifurcated) province.

Basin and Range Province

The Basin and Range province in north-central Mexico is a region partially covered by continental volcanic rocks of the Sierra Madre Occidental and by basin-filling strata composed of thick alluvial sediments of Plio-Pleistocene age (Fig. 2). The province is divided into two separate basins, the Chihuahua–Sonora basin to the west (De Ford, 1969) and the Sabinas basin to the east (Flores, 1980). Each

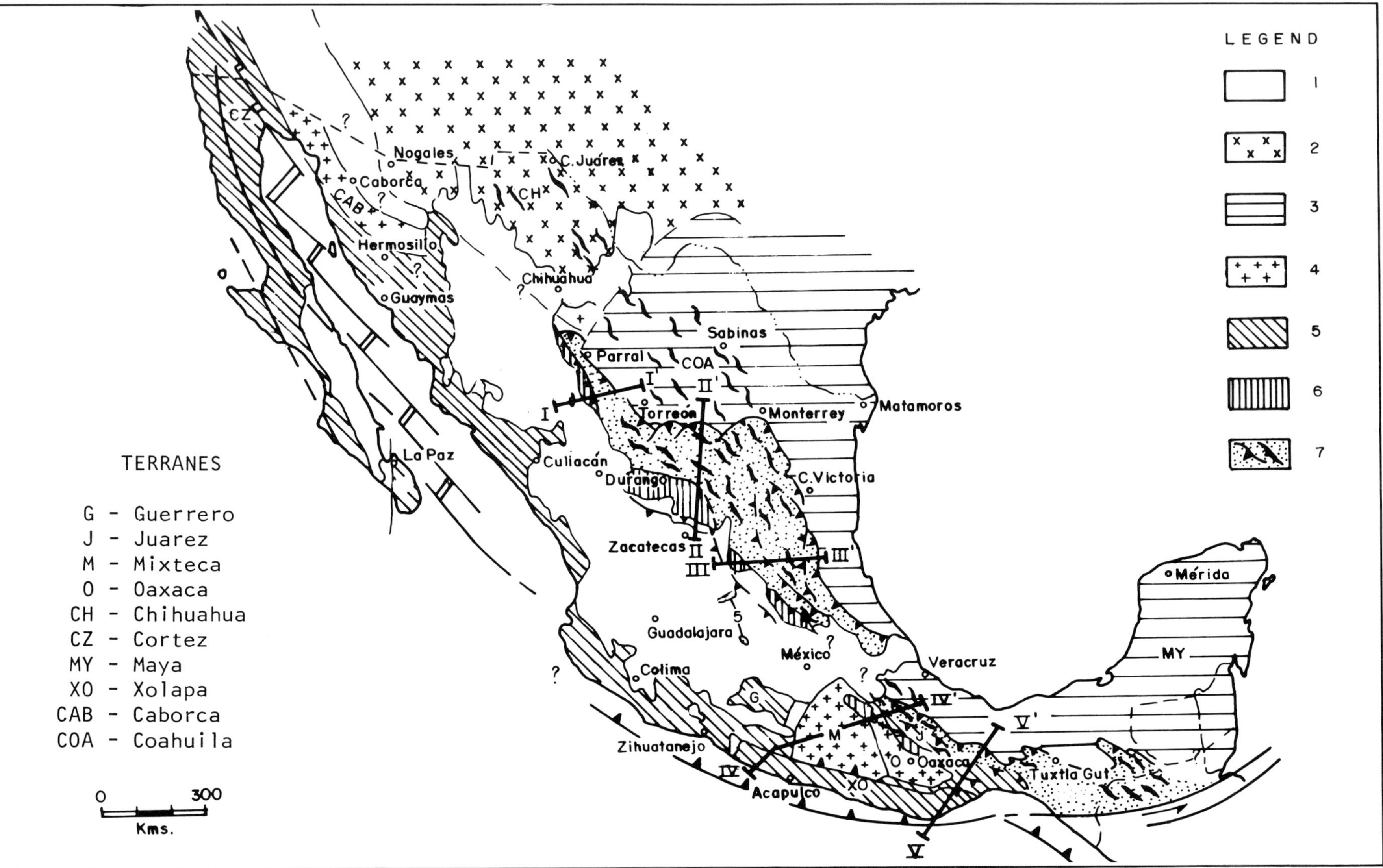

Figure 3—Tectonostratigraphic terranes, location of the Mexican Thrust Belt, distribution of prominent overlap sequences, and location of transects shown in Figures 7 and 8.

Units:

1. Cenozoic volcanic rocks that overlie basement
2. Southwestern end of the North American Craton
3. Eastern Appalachian basements and Gulf of Mexico Mesozoic transgressive overlap sequence
4. Continental allochthonous blocks immersed in western collage
5. Western Cordilleran collage of island arc and other fragments of igneous oceanic crust and melange
6. Late Jurassic to Early Cretaceous turbidite belt
7. Mexican Thrust Belt

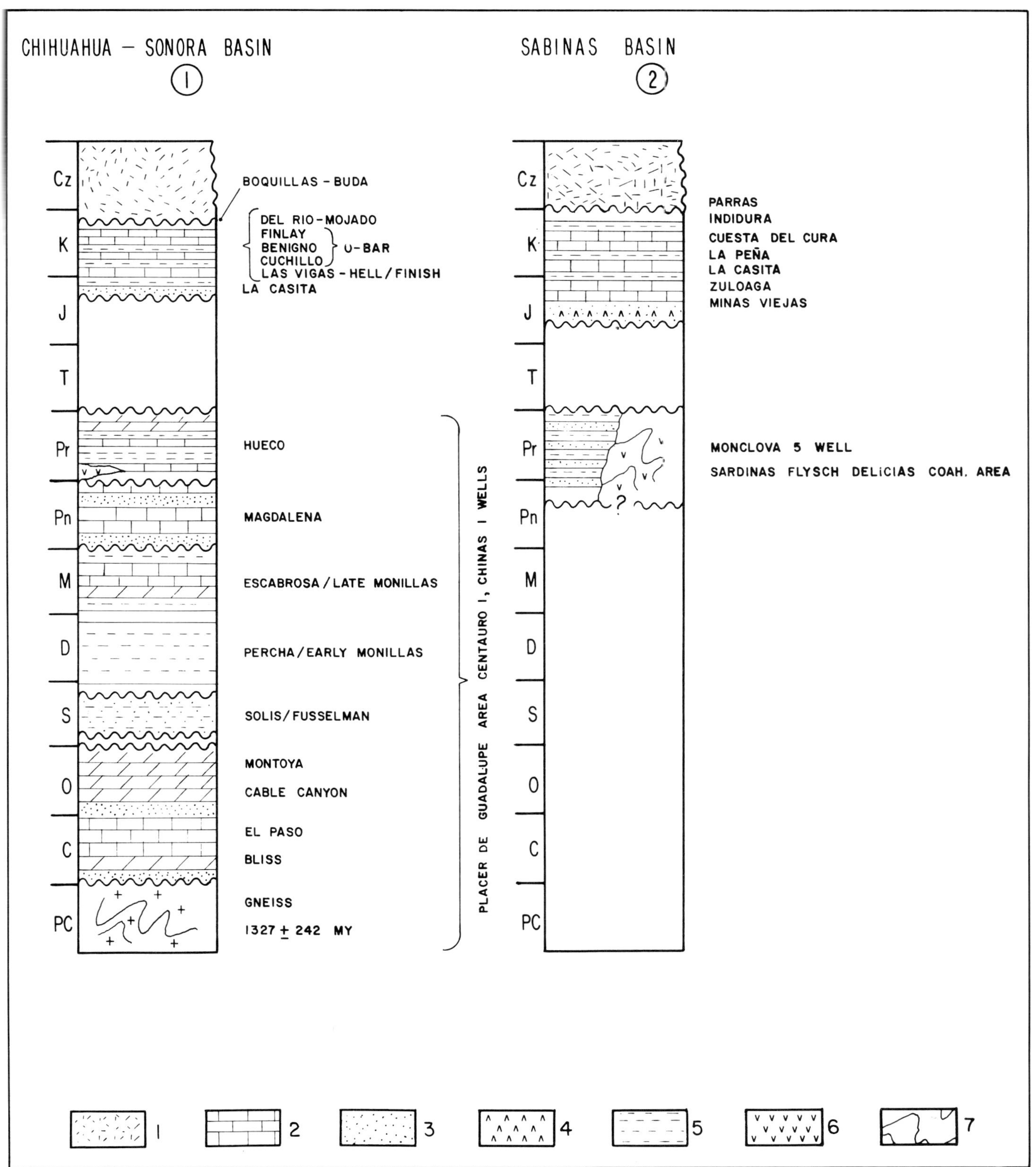

Figure 4—Tectonic units of the Basin and Range Province

1. Interbedded ignimbrite, andesite, and rhyolite
2. Limestone
3. Sandstone and conglomerate
4. Evaporite deposits
5. Shale
6. Volcanic rocks
7. Metamorphic rocks

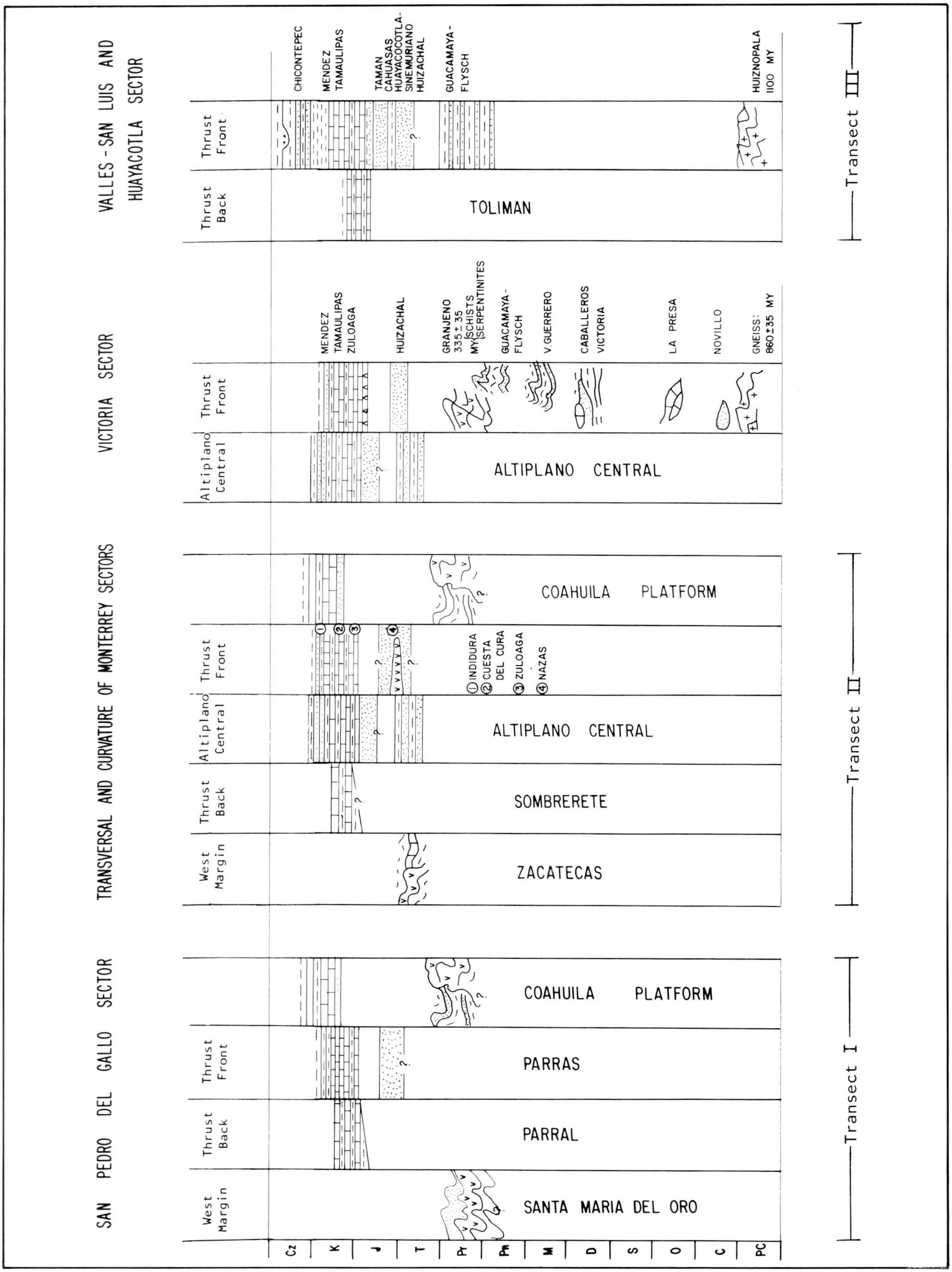

SAN PEDRO DEL GALLO SECTOR

West Margin — SANTA MARIA DEL ORO

Thrust Back — PARRAL

Thrust Front — PARRAS

COAHUILA PLATFORM

Transect I

TRANSVERSAL AND CURVATURE OF MONTERREY SECTORS

West Margin — ZACATECAS

Thrust Back — SOMBRERETE

Altiplano Central — ALTIPLANO CENTRAL

Thrust Front — ① INDIDURA ② CUESTA DEL CURA ③ ZULOAGA ④ NAZAS

COAHUILA PLATFORM

Transect II

VICTORIA SECTOR

Altiplano Central — ALTIPLANO CENTRAL

Thrust Front — MENDEZ TAMAULIPAS ZULOAGA; HUIZACHAL; GRANJENO 335 ± 35 MY SCHISTS SERPENTINITES; GUACAMAYA-FLYSCH; V. GUERRERO; CABALLEROS VICTORIA; LA PRESA; NOVILLO; GNEISS: 860±35 MY

VALLES-SAN LUIS AND HUAYACOTLA SECTOR

Thrust Back — TOLIMAN

Thrust Front — CHICONTEPEC; MENDEZ TAMAULIPAS; TAMAN CAHUASAS HUAYACOCOTLA-SINEMURIANO HUIZACHAL; GUACAMAYA-FLYSCH; HUIZNOPALA 1100 MY

Transect III

Cz K J T Pr Pn M D S O C PC

Figure 5—Columns of the tectonostratigraphic terranes of the Northern province; transect locations shown on Figures 3, 7, 8, and 9.

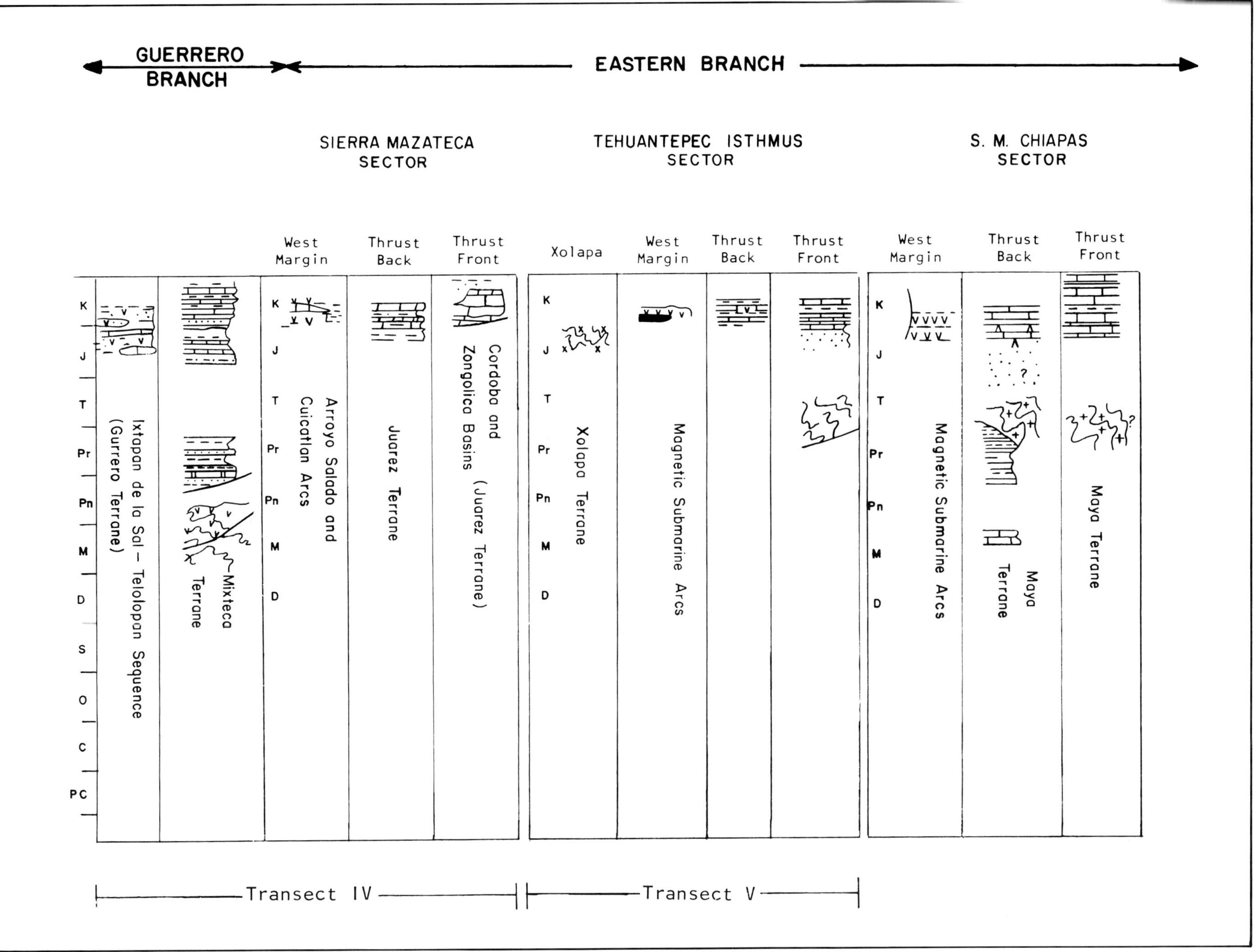

Figure 6—Columns of the tectonostratigraphic terranes of the Southern province; transect locations shown on Figures 3, 10, and 12.

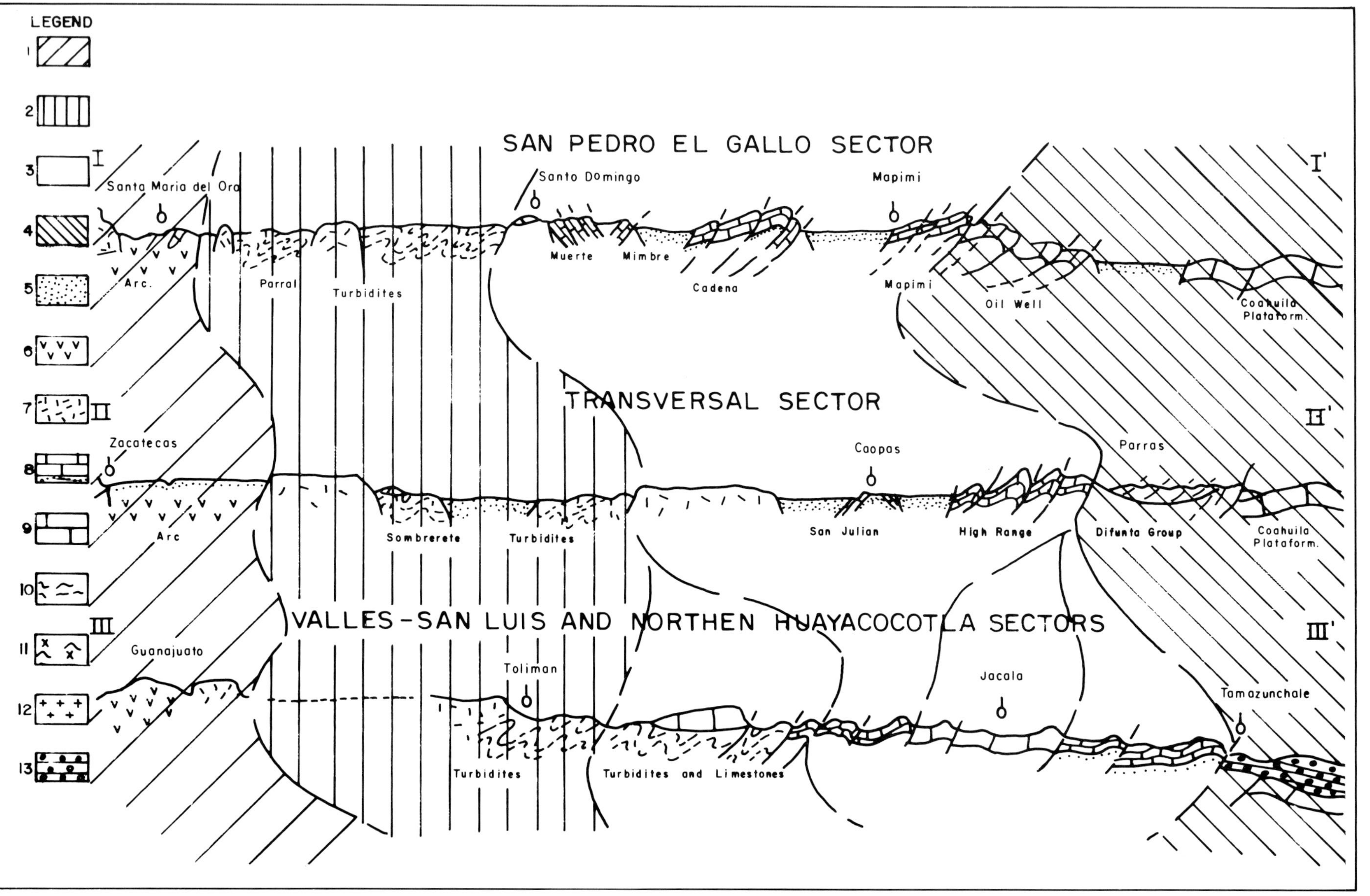

Figure 7—Structure sections across the thrust belt and the tectonostratigraphic terranes in the Northern province.

Units:

1. Western Cordilleran collage, consists of island arc and other fragments of igneous oceanic crust and melange
2. Late Jurassic to Early Cretaceous turbidite belt
3. Eastern Gulf of Mexico, Mesozoic transgressive sequence
4. Coahuila Platform
5. Quaternary alluvium
6. Submarine volcanic deposits
7. Cenozoic continental volcanic rocks
8. Limestone and red beds
9. Platform limestones
10. Turbidites
11. Metamorphic rocks
12. Granitic rocks
13. Terrigenous marine strata

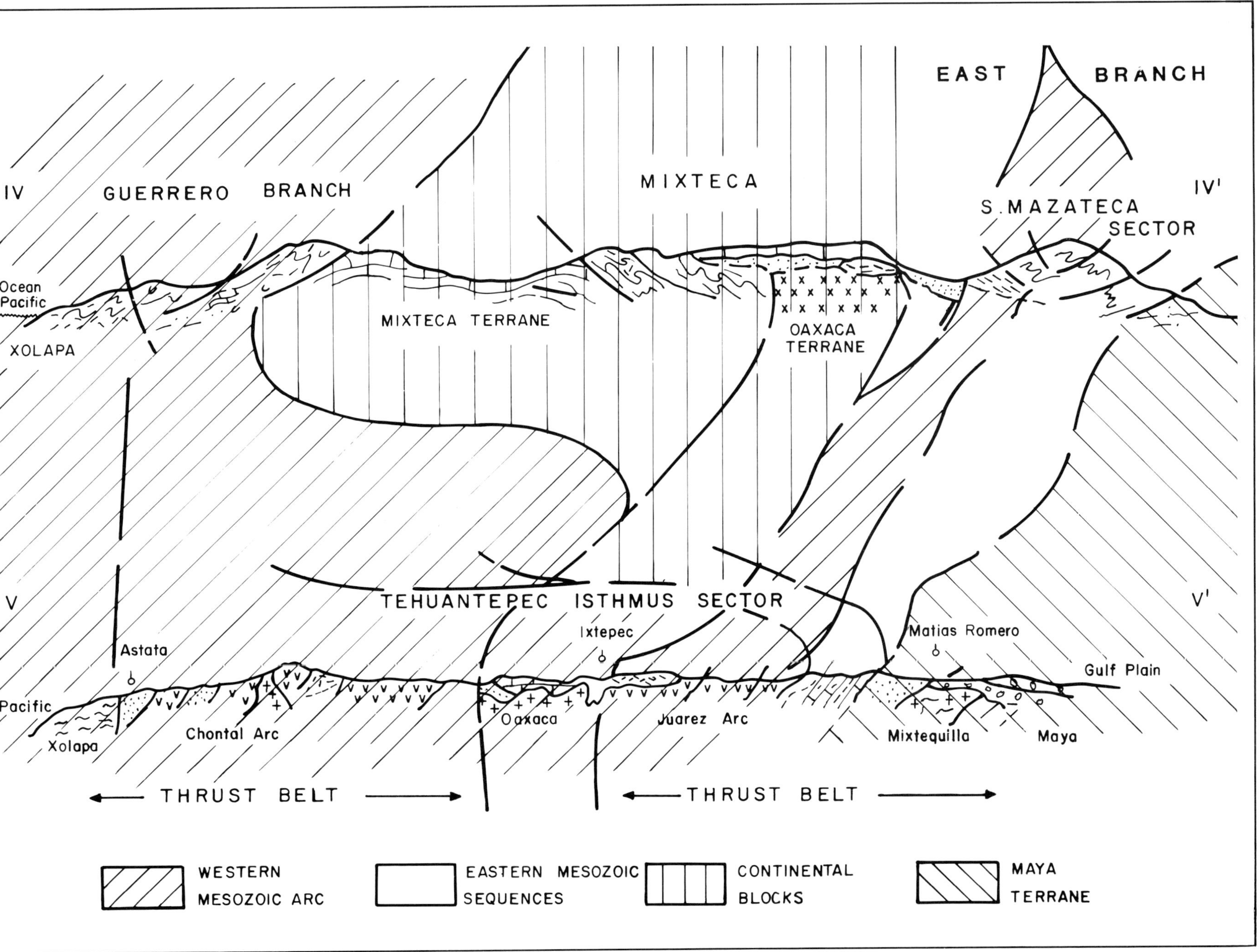

Figure 8—Structure sections across the thrust belt and the tectonostratigraphic terranes in the Southern province. See Figure 7 for explanation of symbols.

basin is distinguished by a unique type of basement and structural deformation (Malpica and De la Torre, 1980). The rocks of the Chihuahua–Sonora basin consist of Paleozoic limestone, sandstone, and shale (Fig. 4). The fauna and lithologies correlate with Paleozoic sequences of Arizona and New Mexico (De Ford, 1969; Pierce, 1976). Precambrian rocks crop out in scattered localities in northeastern Sonora and have been penetrated in exploratory wells drilled in Chihuahua (López, 1979). The rocks are similar to the Precambrian basement of the North American craton. Granitic and metamorphic rocks that are similar to Paleozoic outcrops near Delicias, Coahuila (Mexico), and near Marathon, Texas, have been discovered by drilling in the Monclova oil field area (Fig. 1).

Rocks of the Sabinas basin that crop out near the town of Delicias, Coahuila (Bose, 1921), consist of Permian flysch and late Paleozoic granitic rocks that are conformably overlain by Early Cretaceous limestone. Because the Sabinas basin basement rocks lie on the same structural trend as the Marathon uplift of Texas (Denison et al, 1971), the Sabinas basement is considered to be a continuation of the Appalachian–Ouachita–Marathon belt (Bridges, 1965; 1970; Flawn et al, 1961).

In the Chihuahua–Sonora basin, recumbent folds and thrust faults with northwest-trending axes verge towards the northeast (Flawn and Diaz, 1959). The Sabinas basin contains recumbent folds that also have northwest-trending axes. Reverse faulting is also associated with the Sabinas basin folding. The faults display northeastward displacements in the eastern part of the basin against the Tamaulipas Platform and display southwest displacements in the western part of the basin against the Coahuila Platform. The structural pattern of the Sabinas basin is interpreted as having originated by intracontinental deformation, whereas the Chihuahua–Sonora structural pattern could be interpreted as the continuation of fault trends in Arizona with folding and thrusting against the southwestern edge of the Colorado Plateau (Drewes, 1978).

Northern Province

The Northern province is dominated by a high mountain range and is divided into four structural domains. From east to west they are: the Thrust Front, the Altiplano Central, the Thrust Back, and the West Margin (Tardy, 1976; Figs. 5, 7).

Thrust Front

The Mesozoic stratigraphy of the Thrust Front domain is shown in Figures 5 and 7. Transect II in Figures 5 and 7 illustrates red beds (Nazas Formation) that underlie Late Jurassic limestones and quartz sandstones (Burkhardt, 1930), which in turn grade into Early Cretaceous basin limestone and then into Late Cretaceous turbidites (Pantoja, 1972). The red beds are Jurassic continental strata interlayered with volcanic deposits and can be correlated with the same type of red beds that extend around the Gulf of Mexico and that are also present in Arizona (Damon et al, 1981). Crystalline basement that underlies the Mesozoic sequence has recently been reported near Ciudad Victoria, (Fig. 2) where late Paleozoic schist and serpentinite (330 ± 35 m.y.), Grenvillian gneiss (1,140 ± 80 m.y.), and late Paleozoic flysch reportedly crop out (Carrillo, 1961; de Cserna et al, 1977). The composite basement is highly deformed into eastward-verging thrust sheets; the schist and serpentinite form the upper basement sheet, which is thrust over the gneiss and over metasedimentary rocks that contain a Devonian–Mississippian fauna as well as exotic blocks of quartzite and marble; the latter sequence is in turn thrust over a Permian flysch sequence and exhibits a similar eastward vergence (Figs. 5, 7).

The extent of Paleozoic and Grenvillian basement rocks under the remainder of the thrust belt is unknown. Scattered outcrops of basement suggest that tectonic blocks are also involved in the imbrication within the Thrust Front domain.

Mesozoic strata are cut by thrust faults in the ranges of the Thrust Front domain and also overlie most of the Western Margin domain of the Victoria sector (Fig. 5). In the Victoria sector, the Mesozoic cover is deformed into anticlines and synclines that overlie Late Jurassic evaporites that have been deformed into domes (Carrillo, 1965). The southeast end of the Northern province is similar to that of the Victoria sector (Fig. 5) except for a peculiar area of thick and highly deformed Early Jurassic shales interlayered between the red beds. The two red bed units, separated by shales of Sinemurian age, are considered to be Late Triassic (Huizachal Formation) and Middle Jurassic in age (Cahuasas Formation). The Lower Jurassic (Liassic) strata may be coeval with the so-called Pacific realm of H. P. Laubscher, University of Basel (1981, personal communication), which contains a bivalve fauna that is similar to a fauna in sequences in the Andean Cordillera of Chile and Argentina (A. V. Hillebrandt, Geology and Paleontology Institute of Berlin, 1982, personal communication). The Liassic black shales interlayered between the red beds in the Huayacocotla anticlinorium as well as those involved in the imbricated deformation are interpreted as a leading thrust sheet between the tectonically broken red beds.

Altiplano Central

The Mesozoic stratigraphy of the Altiplano Central domain is similar to that of the Thrust Front terrane (Fig. 5). A sequence of pelitic rocks, partly metamorphosed to phyllite, is similar to pelitic rocks that crop out in the Sierra de Charcas, near the city of San Luis Potosi (Fig. 2). The pelitic strata contain Late Triassic faunal assemblages (Martinez, 1972). An exploratory oil well (Tapona 1) penetrated more than 4,000 m (13,123 ft) of phyllite. The folds and thrusts in the Altiplano Central terrane are affected by the basin and range structure and form asymmetrical anticlines and synclines.

Thrust Back

Long and narrow northwest-oriented fold axes of the Thrust Back domain follow the western limit of the Altiplano Central (Figs. 5, 7). The belt consists of a homogeneous sequence of marine turbidites of Late Jurassic to Early Cretaceous age (Flores, 1982; Eguiluz and Campa, 1982). The turbidite belt is interpreted as abyssal sediments deposited on a passive continental margin. Close to the town of Nazas the turbidite belt is thrust eastward over the Thrust Front domain (Fig. 2) (Eguiluz and Campa, 1982). Near the southeastern limit, similar deposits are thrust over the Valles–San Luis Platform (Fig. 7, Transect III) (Carrillo and Suter, 1982).

Western Margin

Most of the Western Margin domain is covered by Cenozoic continental volcanic deposits. However, there are scattered outcrops of complexly deformed and metamorphosed shale, sandstone, and limestone that are interlayered with submarine volcanic rocks (Figs. 5, 7).

The very complex composite terrane that forms the Western Margin is interpreted to be composed of island arc submarine volcanics and melange that formed along a convergent oceanic margin. The age of the composite terrane ranges from late Paleozoic in the Santa Maria del Oro area (Fig. 7, Transect I) (Burkhardt, 1930; McGehee, 1976) to Late Triassic in the Zacatecas area, is probably Mesozoic in Guanajuato, and is Early Cretaceous in the Tlalpujahua, Michoacan area (Fig. 3). The oceanic terranes are thought to be a southward continuation of the Cordilleran belt of suspect terranes discussed by Coney et al (1980) and Campa and Coney (1983).

Southern Province

Southward from the Transverse Volcanic Arc, the Mexican Thrust Belt bifurcates into two belts around the Mixteca and Oaxaca terranes (Fig. 3) (Campa and Coney, 1983). Both branches are characterized by northwest-oriented fold axes and the northeastward displacement of the thrust belt (Figs. 6, 8). The eastern branch consists of autochthonous Mesozoic sedimentary rocks and the western or informally named Guerrero composite terrane, consists of accreted oceanic rock (Campa et al, 1974; Carfantan, 1981).

Guerrero Terrane

The Guerrero branch follows a north-south trend (Figs. 3, 8) around the western margin of the Mixteca terrane (Campa and Coney, 1983). In the Teloloapan–Ixtapan de la Sal region (Campa et al, 1974), the branch is made up of highly deformed metasedimentary and metavolcanic deposits that have a Late Jurassic to Early Cretaceous marine fauna. Crystalline basement is not exposed in the Guerrero branch (Figs. 6, 8). Toward the east, the metamorphosed volcanic-sedimentary sequence is thrust eastward over Cretaceous limestone and turbidites and covers part of the Mixteca terrane, whose basement is composed of late Paleozoic metamorphic rocks, called the Acatlan Complex (Fig. 8, Transect IV) (Ortega, 1978).

The deformation in imbricated sheets is dominated by eastward vergence. The southeast area of the Guerrero branch is broken abruptly by an east-west oriented reverse fault with southward displacement; the branch is also terminated by middle Cenozoic granitic intrusives (Salinas, 1982). The reverse fault and the granitic rocks are considered to be the boundary of the exotic Xolapa terrane, which trends parallel to the Pacific Coast (de Cserna, 1956; Campa and Coney, 1983). Thus, the western branch of the Mexican Thrust Belt involves the western part of the Guerrero terrane and the eastern cover of the Mixteca terrane (Fries, 1960). The Mixteca and the Oaxaca terranes of southern Mexico are probably allochthonous continental blocks (Ortega et al, 1977).

Eastern Branch

The Eastern branch of the bifurcated Southern province parallels the Gulf Coast southward to the Tehuantepec Isthmus (Fig. 8, Sierra Mazateca) where the trend flexes toward the Pacific Coast and curves around the west edge of the Yucatan Platform (Sierra de Chiapas) and finally passes into northwestern Guatemala (Clemons et al, 1974).

The rugged mountainous topography is similar to that of the Northern province Thrust Front domain. The Eastern branch forms a long and narrow belt that trends southeastward until it suddenly widens into Sierra Madre de Chiapas. The branch then bends eastward and extends to the Maya mountains in Belize, where the Mexican Thrust Belt terminates against the active Motagua–Polochic sinistral fault system.

The Eastern branch of the Southern province is divided into three longitudinal structural domains: (1) Thrust Front, (2) Thrust Back, and (3) Western Margin (Fig. 6, Transects IV and V).

Thrust Front

The mountain ranges of the Thrust Front are made up of Early Jurassic and probably Triassic red beds, Late Jurassic to Early Cretaceous platform limestone, and Late Cretaceous turbidites (Fig. 9). The Mesozoic strata overlie two different basement types in Chiapas (Fig. 6): late Paleozoic flysch and limestone basement that crops out close to the Guatemala border, and the granitic and metamorphic complex of the Chiapas and Mixtequita batholiths that extends across Chiapas to the Tehuantepec Isthmus (Fig. 9, Unit 6). The batholiths are generally considered to be Permo-Triassic in age, although in some areas the Chiapas batholith has been dated as old as Precambrian. The Thrust Front is characterized by anticlines and synclines, which are thrust in sheets that verge northeastward over the Maya terrane (Gonzalez, 1976; Mossman and Viniegra, 1976; Campa and Coney, 1983).

Thrust Back

The Thrust Back and Thrust Front trend parallel to the Sierra Mazateca but are terminated in the Tehuantepec Isthmus by batholithic rocks (Fig. 9). The stratigraphic sequence in the Thrust Back domain (Fig. 6) consists chiefly of thin interstratified limestone and shale containing a Late Jurassic to Early Cretaceous fauna. Along the Santo Domingo River, the Mesozoic strata overlie red beds and locally overlie blocks of late Paleozoic schist (Charleston, 1980). The structures and the mountainous topography are very similar to the ranges of the Thrust Front in the Southern province but differ significantly from the structure of the Thrust Back of the Northern province. The main structures of the Thrust Back (Southern province) are anticlines and synclines that have been thrust northeastward over the Thrust Front (Fig. 8).

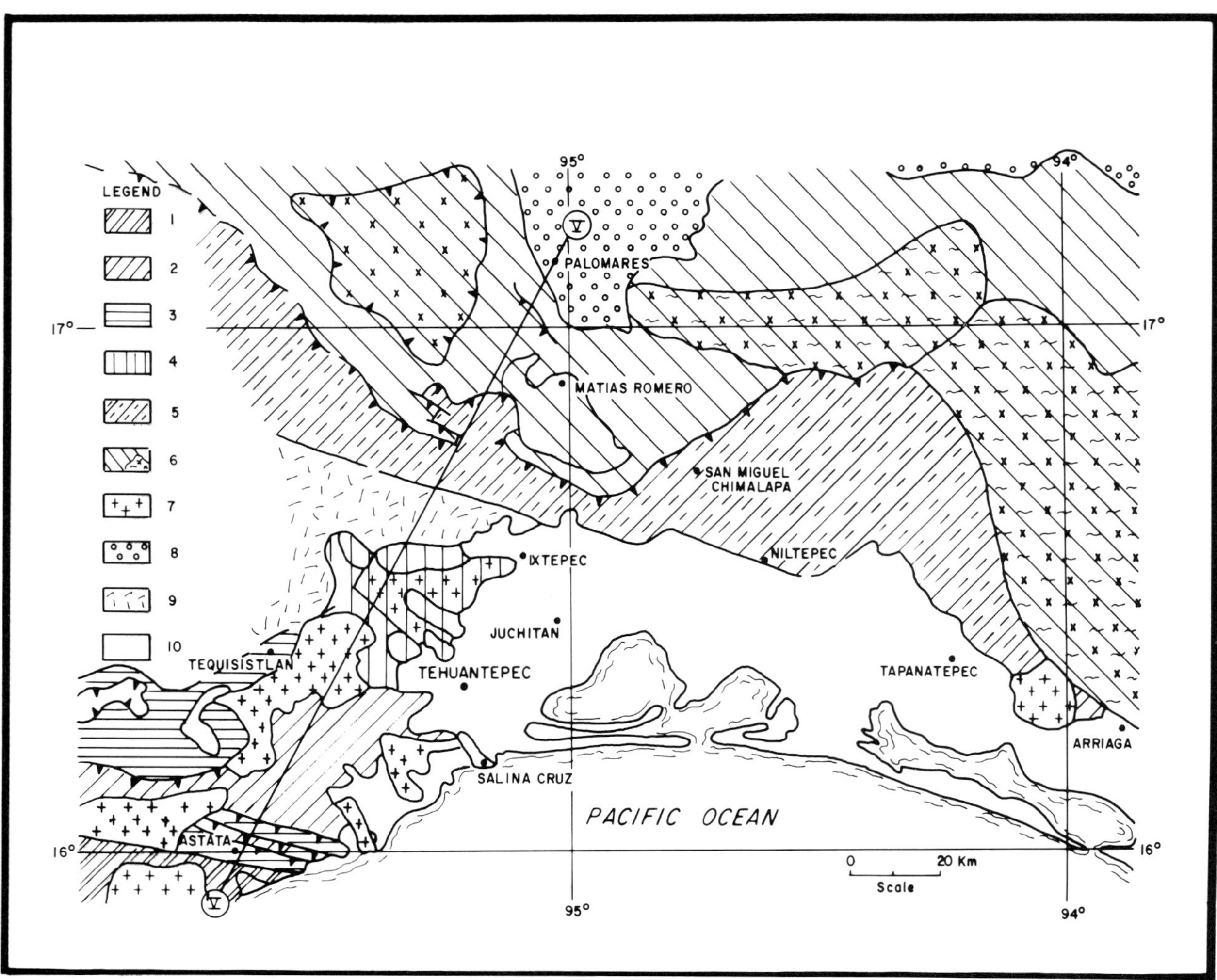

Figure 9—Tectonostratigraphic Map of the Tehuantepec Isthmus, Southern province.

1. Xolapa terrane, migmatitic-gabbroic rocks probably of Jurassic age
2. Chontal terrane, island arc Mesozoic sequences
3. Late Cretaceous turbidite belt
4. Early Cretaceous limestone and red beds that overlie the Oaxaca terrane
5. Isolated sequences of arc and ultrabasic rocks probably of Mesozoic age
6. Permo-Triassic granitic-metamorphic batholith of Chiapas and Mixtequita
7. Cenozoic granitic pluton of the Isthmus region
8. Gulf of Mexico, Cenozoic marine regressive sequence
9. Volcanic deposits, probably Cenozoic in age
10. Quaternary alluvium and sand bar deposits

Western Margin

The Western Margin of the Southern province consists of a complex sequence of limestone and shale interlayered with submarine lavas whose age is poorly known, although local faunas suggest an Early Cretaceous age. The uppermost thrust sheets are composed of phyllite, greenschist, serpentinite, and peridotite that are probably part of an ophiolite sequence of unknown age. Major faults and batholithic rocks define the southeastern limits of the Western Margin. The structural grain is dominated by anticlines and synclines that are thrust northeastward and overlie both the Thrust Back and Thrust Front domains.

CONCLUSIONS

The Mexican Thrust Belt extends southward from the Cordilleran thrust belt and was mainly created by compressional and transverse shear stresses that formed recumbent folds and thrust sheets. The principal stresses were oriented northeast-southwest and mainly affected preexisting continental blocks, namely the Coahuila Platform and the Altiplano Central, Mixteca, Oaxaca, and Yucatan terranes (Figs. 1, 3). Detailed investigations throughout the Mexican Thrust Belt are needed to clarify the geometry of regional patterns. The Mexican Thrust

Belt separates a region of eastern thrust and fold belt from the western Cordillera, which is interpreted to be a collage of terranes. The northern part of the Mexican Thrust Belt (Basin and Range Province of this paper) is interpreted to be the immediate continuation of the North American craton and of the structures present in the southwestern United States. The southeastern end of the Mexican Thrust Belt continues into the Altas Cuchumatanes Mountains of Guatemala and terminates against the active Motagua-Polochic fault, which is a sinistral fault system, and represents the main tectonic boundary between the North American and Caribbean Plates (Figs. 1, 3).

Timing of Deformation

The youngest sedimentary rocks involved in the major episodes of Mexican Thrust Belt deformation are Late Cretaceous. Thus, the folding and thrust faulting of the Cordilleran thrust belt took place in early Cenozoic time (Laramide orogeny), except for the segment between Yellowstone, Wyoming, and Las Vegas, Nevada, which was deformed earlier during the Sevier orogeny (Fig. 1). The Basin and Range extensional deformation is post-Laramide (middle to late Cenozoic time) and is probably related to an extensional response of competent cratonic blocks after the compressive Laramide event to the west (Fig. 1). Deformation during Plio-Pleistocene time as well as Recent time has also left an imprint on the structure of the Mexican Thrust Belt.

Direction and Displacement

The Mexican Thrust Belt is characterized by fold axes that are generally oriented northwest-southeast and fault planes that dip southwest. Overall displacement direction is toward the east-northeast. Near the northern end of the Mexican Thrust Belt shown in Figure 5, Transect III, the displacement towards the east has been estimated to be approximately 40 km (25 mi) (Tardy, 1980).

Near the city of Monterrey, Mexico, the tectonic displacement is estimated to be 100 km (62 mi) towards the northeast, and in the same area, the minimum amount of shortening is calculated to be 30 to 50% (Padilla, 1982).

At the Tehuantepec Isthmus a metamorphic block (La Mixtequilla) was estimated to have been displaced 200 km (124 mi) towards the northeast (Fig. 9, Unit 6) (PEMEX, unpublished geophysical data).

Continental Implications

The northern and southern ends of the Cordilleran thrust belt were deformed and disrupted during Plio-Pleistocene time. Figure 1 shows a break in the thrust belt between the southwestern United States and northwestern Mexico, a position close to the southern end of the North American craton. In this region the extensional Basin and Range structures are Middle Tertiary in age.

The Cordilleran thrust belt branches into two segments between Yellowstone Park, Wyoming, and Monterrey, Mexico. The eastern branch follows the Laramide front, which is broken by the Rio Grande Graben, following southward through the Sabinas Basin. Deformation of the western branch took place early during the Sevier orogeny in northwestern Mexico.

Between Las Vegas, Nevada, and El Paso, Texas, and southward near the Mexican cities of Parral, Torreon, and Monterrey, the thrust belt appears genetically related to the Laramide orogeny and rejoins with the continental Laramide eastern front. The two branches of the thrust belt surround the Colorado Plateau and the Coahuila Platform continental blocks (Fig. 1).

The west margin of the North American continent has been deformed by several tectonic events that are not clearly differentiated at present. One of the important events that formed the Mexican Thrust Belt is the compression associated with the Laramide orogeny; another event relates to the extensional forces that formed the Basin and Range province of the southwestern United States and northwestern Mexico; the youngest tectonic event is neotectonic and remains presently active.

Hydrocarbon Potential

The Mesozoic sedimentary rocks that constitute the Mexican Thrust Belt show a general transgressive depositional sequence that might be genetically related to a passive margin that formed during the opening of the Gulf of Mexico (Fig. 3, Unit 3). Along the eastern side of the thrust belt there are shallow marine carbonates that grade upward and westward into a sequence of deeper water turbidites (Fig. 5). This belt of limestone flysch is then thrust over the platform deposits and defines the border separating the eastern and western regions of allochthonous terranes (Fig. 3). The turbidite belt is interpreted to be the western limit of stratigraphic and/or structural traps for oil in Mesozoic rocks. The subsurface Mesozoic rocks are potentially oil productive to the east, in spite of complex structural deformation.

ACKNOWLEDGMENTS

I wish to express my gratitude for the work of many people who assisted me over a period of several years especially J. Aubouin, M. Tardy, and J. C. Carfantan from France; M. Suter and R. Laubscher from Switzerland; R. Prince from Canada; P. Damon and H. Drewes from the United States; S. Eguiluz, R. Flores, V. H. Garduno, R. Gonzalez, R. Padilla, and J. T. Castro-Mora from Mexico, who contributed with their work in thrust belt problems to make this paper possible. I am also very grateful to the Group of the Map of Accretionary Terranes of the North American Cordillera Project formed by Norm Silberling and David Jones of the U.S. Geological Survey; Jim Monger of the Geological Survey of Canada; and Peter J. Coney, from the University of Arizona at Tucson.

REFERENCES

Bose, E., 1921, On the Permian of Coahuila, northern Mexico: American Journal of Science, v. 1, p. 187-194.

Bridges, L. W., 1965, Stratigraphy of Mina Plumosas-Placer de Guadalupe area: West Texas Geological Society Guidebook, Publication 64-50, p. 50-59.

______, 1970, Paleozoic history of the southern Chihuahua tectonic belt: West Texas Geological Society and the University of Texas at Austin Symposium in Honor of Professor R. K. De Ford, p. 69.

Burkhardt, C. 1930, Etude synthetique sur le Mesozoique mexicain; Memoirs Socièté Paleontologists, Suisse, n. 49-50, p. 280.

Campa, F. M., and P. J. Coney, 1983, Tectono-stratigraphic terranes and mineral resource distributions in Mexico: Canadian Journal of Earth Science, v. 20, p. 1040–1051.

______, et al, 1974, La secuencia Mesozoica volcánica sedimentaria metamorgizada de Ixtapan de la Sal, Mexico-Telolopan, Gro.: Boletín de la Sociedad Geológica Mexicana, v. 35, p. 7–28.

Carfantan, J. C., 1981, Paleogeography and tectonics of the Sierra de Juárez isthmus of Tehuantepec area and its relations with other terranes of southern Mexico and Central America (Abs): Geological Society of America Abstracts with Programs, v. 13, n. 2, p. 48.

Carrillo, B., 1961, Geologia del Anticlinorio Huizachal Peregrina al NW de Cd. Victoria, Tamps: Boletín de la Asociatión Mexicana de Geólogos Petroleros, v. 13, n. 1, y. 2, p. 98.

______, 1965, Estudio geológico de una parte del Anticlinorio de Huayacocotla: Boletín de la Asociación Mexicana de Geólogos Petroleros, v. 17, n. 5-6, p. 23.

Carrillo, M., and M. Suter, 1982, Tectónica de los alrededores de Zimapán, Hidalgo y Querétaro, Libro guia de la Excursión Geológica Mexicana, p. 1–20.

Charleston, S., 1980, Stratigraphy and tectonics of the Rio Santo Domingo area, state of Oaxaca, Mexico (Abs.): Twenty-Sixth Congress Geólogos Internacional Program and Abstracts, p. 99.

Clemons, R. E., et al, 1974, Stratigraphic nomenclature of recognized Paleozoic and Mesozoic rocks of western Guatemala: Bulletin of the American Association of Petroleum Geologists, v. 58, p. 313–320.

Coney, P. J., 1978, Mesozoic–Cenozoic Cordilleran plate tectonics: Geological Society of America Memoir 152.

______, et al, 1980, Cordilleran suspect terranes: Nature, v. 288, p. 239–333.

Damon, P. E., et al, 1981, Age trends of igneous activity in relation to metallogenesis in the southern Cordillera, *in* W. R. Dickinson and W. D. Payne, eds., Relation of tectonics to ore deposits in the southern Cordillera: Arizona Geological Society Digest, v. 14, p. 137–154.

de Cserna, Z., 1956, Tectónica de la Sierra Madre Oriental de Mexico entre Torreón y Monterrey: Publicaciones del XX Congreso Geológico Internacional.

______, et al, 1977, Aloctono del Paleozoico Inferior en la región de Cd. Victoria, Tamps: UNAM Instituto de Geología Revista 1, p. 33–43.

De Ford, R. K., 1969, Some keys to the geology of northern Chihuahua, *in* The border region: New Mexico Geological Society 20th Field Conference, 61 p.

Denison, R. E., et al, 1971, Basement rock framework of part of Texas, southern New Mexico and northern Mexico, *in* K. Seewld and D. Sundeen, eds., The geologic framework of the Chihuahua tectonic belt: West Texas Geological Society, Middland, p. 3–14.

Drewes, H., 1978, The Cordilleran orogenic belt between Nevada and Chihuahua: Geological Society of America Bulletin, v. 89, p. 641–657.

Eguiluz, S., and M. F. Campa, 1982, El Geosinclinal Mexicano en el Sector de San Pedro El Gallo, Durango: VI Convención Nacional de la Sociedad Geológica Mexicana, Resumenes, p. 3.

Flawn, P. T., and T. Diaz, 1959, Problems of Paleozoic tectonics in north-central and northeastern Mexico: Bulletin of the American Association of Petroleum Geologists, v. 43, p. 224–230.

______, et al, 1961, The Ouachita system: Texas University Bureau of Economic Geology Publication 6120, 401 p.

Flores, L. R., 1980, Análisis Tectónico estructural del Golfo de Sabinas, a partir de datos de subsuelo, superficie y satélite: IMP Subdirección de Exploración, Proyecto C-1097, unpublished.

______, 1982, Análisis estratotectónico del norte y noreste de Mexico, un nuevo método: VI Convención Nacional, Sociedad Geológica Mexicana, Resumenes, p. 44.

Fries, C., 1960, Geologia del estado de Morelos y de partes adyacentes de Mexico y Guerrero, región central meridional de Mexico: Boletín del UNAM Instituto de Geología, n. 60, p. 236.

Garrison, J. R., et al, 1980, Rb-Sr isotopic study of the ages and provenance of Precambrian granulite and Paleozoic greenschists near Cd. Victoria, Mexico, *in* The origin of the Gulf of Mexico and early opening of the central North Atlantic Ocean: Proceedings of a Symposium, Department of Geology, Louisiana State University, Baton Rouge, p. 37–49.

González, A. J., 1976, Resultados obtenidos en la exploración de la plataforma de Córdoba y principales campos productores: Boletín de la Sociedad Geológica Mexicana, v. 37, n. 2.

López, R. E., 1979, Geología de Mexico: Ediciones Mexico, 3 v.

Malpica, C. R., and L. G. De la Torre, 1980, La integracíon estratigráfica del Paleozoico de Mexico, Partes I, II y III y anexos: IMP Subdireccíon de Tecnología de la Exploración, Proyecto C-1079, unpublished.

Martinez, P. J., 1972, Exploración Geológica del área. El Estribo San Francisco, S. L. P. (hojas K-8 y K-9): Boletín de la Asociación Mexicana de Geólogos Petroleros, v. 24, n. 7-9, p. 327–402.

McGehee, R. V., 1976, Las rocas metamorficas del Arroyo de la Pimita, Zacatecas, Zac: Boletín de la Sociedad Geológica de Mexico, v. 37, p. 1–10.

Mossman, R. W., and F. Viniegra, 1976, Complex fault structures in Veracruz, Province of Mexico: Bulletin of the American Association of Petroleum Geologists, v. 60, n. 3, p. 379–388.

Ortega, G. F., et al, 1977, Lithologies and geochronology of the Precambrian craton of southern Mexico (Abs.): Geological Society of America Abstracts with Programs, v. 9, p. 1121–1122.

Padilla, S. R., 1982, Geologic evolution of the Sierra Madre Oriental between Linares, Concepcion del Oro, Saltillo and Monterrey, Mexico: PhD Dissertation, University of Texas at Austin.

Pantoja, A. J., 1972, La formación Nazas del Levantamiento de Villa Juárez, Estado de Durango: II Convención Nacional de la Sociedad Geológica Mexicana, Mazatlán, Memoria, p. 194–196.

Peiffer, R. F., 1979, Les zones isopiques du Paleozoique inferieur dur nordouest Mexicain, temoins du relais entre les Appalaches et al Cordillere West Americaine: Comptes Rendus Academie Sciences du Paris Series D, v. 288, p. 1517–1519.

Pierce, H. W., 1976, Elements of Paleozoic tectonics in Arizona: Arizona Geological Society Digest, v. 10, p. 37–58.

Salinas, P. J. C., 1982, Las relaciones de las rocas metamórficas con la cobertura sedimentaria Mesozoica en la Sierra Madre del Sur, regiones de la Montana Guerrero y la Mixteca Oxaquena: VI Convención Nacional de la Sociedad Geológica Mexicana, Resumenes, p. 89.

Silver, L. T., and T. H. Anderson, 1974, Possible left-lateral early to middle Mesozoic disruption of the southwestern North American craton margin (Abs.): Geological Society of America Abstracts with Programs, v. 6, p. 955–956.

Stewart, J. H., 1976, Late Precambrian evolution of North America: plate tectonics implication: Geology, v. 4, p. 11–15.

———, 1977, Basin-range structure in western North America, a review: Geological Society of America Memoir 152, 31 p.

Suter, M., 1982, Deformaciones laramidicas y potencial petrolero del borde oriental de la plataforma de Valles–San Luis Potosi: V Convencion Nacional de la Sociedad Geológica Mexicana, Resumenes, p. 82.

Tardy, M., 1976, La happe de parras: un trait essentiel de la structure laramienne du secteur transverse de la Sierra Madre Oriental, Mexique: Bulletin Societé Geologique du France, (7), XVII, p. 77–87.

———, 1980, Contribution a l'étude geologique de la Sierra Madre Orientale du Mexique: Doctoral Dissertation, Université Pierre et Marie Curie de Paris, 445 p.

blocks consisting of serpentinized harzburgite and dunite, cumulate gabbro (containing clinopyroxene, plagioclase, olivine, and orthopyroxene as cumulate phases), noncumulate uralite gabbro that is locally intruded by dikes and stocks of plagiogranite, a sheeted dike complex, and pillow lava and breccia. The ophiolite has been tectonically thinned internally by a combination of low-angle and high-angle faults but has an estimated minimum thickness of about 3,000 m (9,843 ft).

The age of the ophiolite has been isotopically determined from sphene from an albitite dike and zircon from a small stock of plagiogranite, both of which yield concordant U-Pb ages of 220 m.y. (Barnes and Mattinson, 1981; Kimbrough, 1982). These isotopic ages are Norian (Late Triassic) (Palmer, 1983) and are in approximate agreement with fossil ages from the overlying sedimentary unit (see below). The upper part of the ophiolite displays prehnite-pumpellyite to greenschist facies assemblages that are interpreted to result from ocean-floor hydrothermal metamorphism. The presence of tuffaceous rocks depositionally above the ophiolite suggests that the complex was formed adjacent to an active volcanic arc, possibly in a marginal basin. Trace element geochemistry of the volcanic and hypabyssal rocks of the ophiolite are consistent with, but do not prove, a marginal basin origin (Fig. 4).

At Morro Hermoso, the Sierra de San Andres ophiolite is overlain by 200 to 300 m (656–984 ft) of thin-bedded, varicolored altered vitric tuff with minor interstratified tuffaceous chert, fine-grained volcaniclastic sandstone, and pink limestone. This sequence has been informally termed the Morro Hermoso Formation by Rangin (1976, 1978) and the Puerto Escondido tuff member of the San Hipolito Formation by Barnes (1982), who correlates it to all four members of that formation of the Vizcaino Sur terrane. The tuffs contain radiolarian assemblages and the megafossil *Monotis* that indicate a Carnian to Norian (Late Triassic) age.

The basal contact of this tuffaceous sequence on the ophiolite is nearly everywhere faulted but at Puerto Escondido may be depositional. The interstratified sandstones are entirely volcanogenic and consist of grains composed of plagioclase, clinopyroxene, amphibole, glass shards, volcanic quartz, and, locally, pumice fragments (Barnes, 1982). The tuffs contain abundant authigenic minerals including laumontite and pumpellyite (Barnes, 1982). The metamorphic assemblages indicate that the Upper Triassic part of the Vizcaino Norte terrane has undergone zeolite facies burial metamorphism.

The uppermost unit of the Vizcaino Notre terrane is the Eugenia Formation, which consists largely of volcanogenic conglomerate, graywacke, and shale (Robinson, 1975; Minch et al, 1976; Boles, 1978; Barnes, 1982). The Eugenia Formation crops out over a wide area in the northern part of the Vizcaino Peninsula and has been divided into lower and upper parts by Boles and Hickey (1979).

The lower part is about 1,500 km (4,921 ft) in thickness and contains megafossils and radiolarian assemblages that are as old as Tithonian (Late Jurassic) and as young as Valanginian (Early Cretaceous). This part consists largely of sandy siltstone and thin-bedded sandstone with thick conglomerate packages where examined by the author at

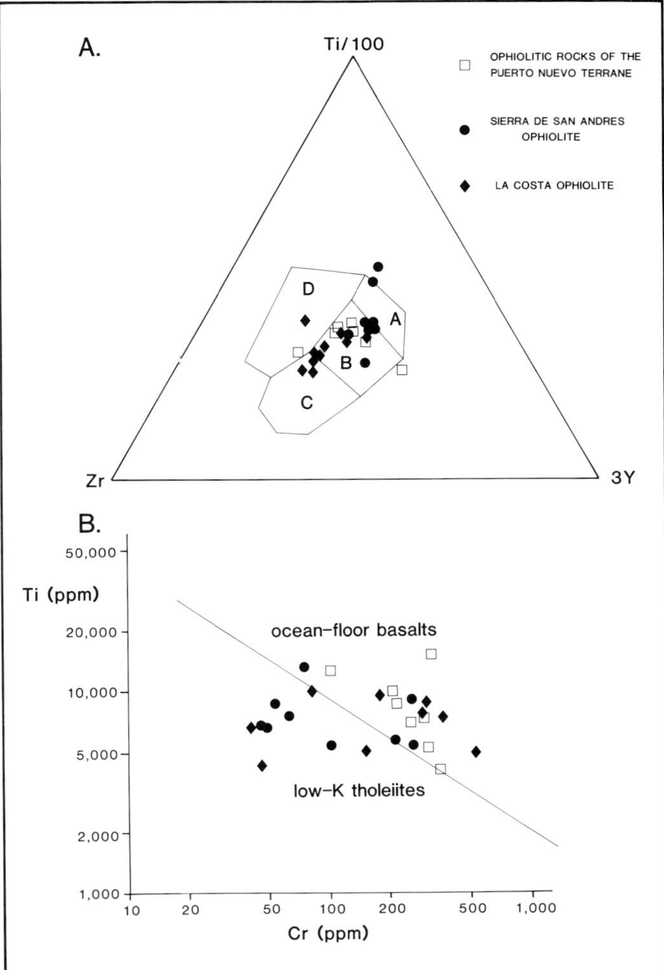

Figure 4—Hypabyssal and volcanic samples from the ophiolitic rocks of the Vizcaino Norte terrane (Sierra San Andres ophiolite), Vizcaino Sur terrane (La Costa ophiolite), and the Puerto Nuevo terrane plotted on A, the Ti-Zr-Y plot of Pearce and Cann (1973), and B, the Ti-Cr diagram of Pearce (1975). Ocean-floor basalts plot in field B, island-arc basalts in fields A and B, calc-alkalic basalts in fields B and C, and within-plate basalts in field D of Figure 5a. The plots suggest that the ophiolitic lithologies of the Puerto Nuevo terrane may have affinities with ocean-floor basalts whereas the ophiolites of the other two terranes may have a mixed ocean-floor and island-arc origin.

Morro Hermoso. The basal contact of the Eugenia with the Upper Triassic tuffs is everywhere a fault or at best enigmatic, although Barnes (1982) reports that it is unconformable. Because the lowermost part of the Eugenia is shale-rich and poorly consolidated and may have easily been disrupted, it is possible that the contact is a faulted unconformity.

Boles and Hickey (1979) and Barnes (1982) report that the lower part of the Eugenia Formation consists mostly of sediment-gravity-flow deposits with some traction flow deposits that accumulated as a debris apron in deep water along a shelfless, mostly volcanic, structural high that was located to the east. The conglomerates of the Eugenia are angular and poorly sorted and contain clasts that are largely of volcanic arc (basalt to dacite) and intraformational origin. Other clast types include xenolith-rich tonalite, peraluminous two-mica granite, rare quartz-pebble conglomerate, and black and green chert (Hickey, 1983, personal communication). One clast of two-mica, peraluminous granite has yielded a discordant U-Pb isochron on zircon that indicates a crystallization age of 147 m.y. and contamination prior to crystallization by country rocks that had an age of 1.3 b.y. (Kimbrough, 1982). Granite of this age is unknown in situ in Baja California, and the clasts can only have been derived from a cratonal source area such as that exposed in Sonora or southern Mexico (Ortega-Gutierrez, 1981).

The sandstones of the Eugenia Formation consist largely of felsic, intermediate, and mafic volcanic rock fragments, but locally contain a small but persistent percentage of quartz, some metamorphic lithic fragments, and generally about 1% potassium feldspar (Boles, 1978; Hickey, 1983, personal communication). Barnes (1982) reports that the stratigraphically lowest sandstones lack quartz and potassium feldspar grains and that these components increase in abundance up-section. Secondary mineral assemblages within the sandstones have been interpreted by Boles (1978) and Barnes (1982) to indicate mild thermal diagenesis at temperatures less than 100° C.

At Punta Quebrada and Morro Hermoso, the Eugenia Formation contains an intercalated sequence of clinopyroxene-plagioclase microporphyric pillow lava of basaltic-andesitic composition that is as much as 650 m (2,133 ft) in thickness. These volcanic rocks are locally still glassy and have major and trace element chemistries indicative of an island-arc-tholeiitic composition (Moore, unpublished data).

The upper part of the Eugenia Formation is about 1,000 m (3,281 ft) in thickness and consists predominantly of well-sorted medium- to fine-grained sandstone characterized by abundant biotite and quartz and as much as 12% potassium feldspar grains. These strata lie unconformably on the lower part of the Eugenia (Hickey, 1983, personal communication) and contain Aptian-Albian (Middle Cretaceous) foraminifera (Boles and Hickey, 1979). This part of the Eugenia contains graded beds and Bouma sequences that are interpreted by Boles and Hickey (1979) to have been deposited on a submarine fan that was shed from the east. These strata have many similarities to the Middle Cretaceous Valle Formation (see Overlap sequence) that overlies the Eugenia Formation and should probably have their own nomenclature (Hickey, 1983, personal communication).

Vizcaino Sur Terrane

The Vizcaino Sur terrane crops out in the coastal hills and interior mountains of the southern Vizcaino Peninsula (Fig. 1). It consists, in ascending order, of the Upper Triassic La Costa ophiolite (Moore, 1983), the Upper Triassic to Lower Jurassic San Hipolito Formation, pre-Upper Jurassic volcanic and volcaniclastic rocks in the Cerro El Calvario area, and Middle Jurassic to Lower Cretaceous intrusive rocks.

The base of the sequence consists of an ophiolitic sequence that crops out along the coast between Punta San Pablo and Punta San Hipolito (Fig. 5). The ophiolite, although tectonized, has a consistent steep southerly dip and is composed, in ascending order, of serpentinized harzburgite, cumulate gabbro (with plagioclase, clinopyroxene, olivine, and orthopyroxene as cumulate phases), noncumulate uralite gabbro, sheeted dikes, and pillow lava and breccia.

The ophiolite structurally overlies foliated amphibolite of unknown age at Punta San Pablo. The amphibolite has a minimum thickness of 750 m (2,461 ft) and consists of a homogenous assemblage of hornblende, plagioclase, and sphene that retain relict ophitic microgabbro textures within less-deformed phacoids. The amphibolite may have resulted from contact dynamothermal metamorphism during obduction of the ophiolite but is thicker and lacks the reverse mineralogical zonations observed in amphibolites at the base of many ophiolite sequences (Coleman, 1977). Alternatively, the amphibolite body may have resulted from contact metamorphism by an underlying, unseen intrusive body. However, it contains a penetrative fabric that is cross-cut by a pluton dated at 140 m.y. by Kimbrough (1982). These features may argue that the amphibolite represents another terrane of pre-Cretaceous age.

Although complicated by extensive low- and high-angle faults, the thickness of the ophiolite is estimated to be about 6,000 m (19,685 ft). No isotopic age dates have been obtained directly from this ophilitic sequence, but radiolaria obtained from interpillow chert at its top are Carnian-Norian (Late Triassic) in age (Barnes, 1982), and plutonic rocks that are intrusive into the gabbroic parts of the ophiolite are as old as 151 m.y. The ophiolite is therefore interpreted to be no younger than Late Triassic.

Trace element geochemistry of hypabyssal and volcanic rocks from the ophiolite are somewhat ambiguous but can be interpreted to indicate a marginal basin origin for the complex (Fig. 4). The ophiolite is depositionally overlain by tuffaceous chert that suggests proximity of the ophiolite to a volcanic arc at the time of its formation. The ophiolite has undergone regional propylitic alteration and is locally amphibolitized by Jurassic intrusive rocks.

The ophiolite is directly overlain by the type section of the San Hipolito Formation (Mina, 1957; Finch and Abbott, 1977). The San Hipolito consists of four informal members which are, in ascending order, the chert member, the limestone member, the breccia member, and the sandstone member. The two lowest members contain the megafossils *Monotis* and *Halobia* and a radiolarian assemblage that indicate a Carnian-Norian (Late Triassic) age (Pessagno et al, 1979). Radiolarian of Pliensbachian and/or Toarcian age have been reported by Whalen (1983) from near the base of the sandstone member, indicating that part of the formation is at least as young as Early Jurassic.

The basal chert member consists mostly of laminated green tuffaceous chert that contains flame structures and low-angle cross-stratification, with lesser amounts of red radiolarian chert and minor volcaniclastic sandstone. The chert member grades upward into the limestone member that consists of about 60% fossiliferous tuffaceous limestone and 35% interstratified volcaniclastic sandstone. Sandstones of the two lowest members have similar compositions that average Q_4 F_{37} L_{59} and consist essentially of plagioclase and

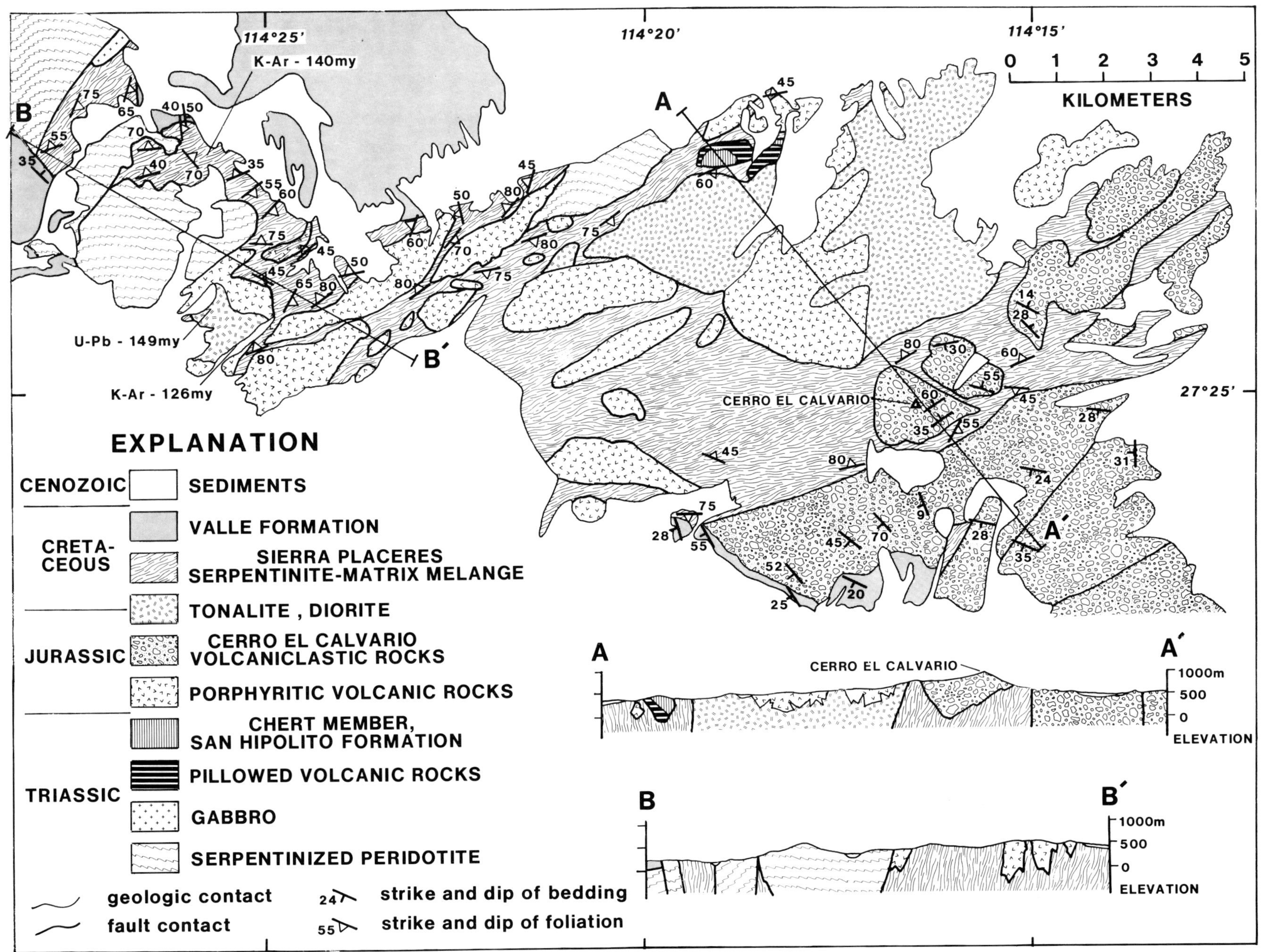

Figure 5—Simplified geologic map of the Sierra Placeres melange.

intermediate volcanic lithic fragments with abundant pyroxene, and hornblende and glass shards. All quartz grains are of volcanic origin (Finch and Abbott, 1977).

The overlying breccia member consists of sediment-gravity-flow deposits of granule to boulder conglomerate and breccia that contain clasts composed almost entirely of intermediate and mafic volcanic rock, mostly plagioclase and clinopyroxene porphyries (ankaramite), and clasts of shallow-water limestone. Barnes (1982) reports that the breccia member lies on an erosional surface with relief of as much as 50 m (164 ft). It is overlain by a thick sequence of fine-grained, thin-bedded volcaniclastic sandstone, locally with intrastratified conglomerate, that composes the sandstone member. The top of the San Hipolito Formation is not exposed. Sandstones of the two upper volcaniclastic members have the composition Q_0 F_{29} L_{71} and consist essentially of plagioclase and intermediate volcanic detritus including rare volcanic quartz grains and abundant grains of pyroxene and amphibole (Finch and Abbott, 1977).

Volcaniclastic sedimentary rocks that are similar to the upper members of the San Hipolito Formation crop out in the southern Sierra de San Andres at Cerro El Calvario. These strata overlie pillow lava of probable volcanic arc affinity (Moore, unpublished data) and are intruded by granitic rocks that yield U-Pb isotopic ages on zircon of 154 m.y. (Barnes, 1982), indicating a pre-Late Jurassic age. The rocks consist of interstratified coarse-grained conglomerate, breccia, sandstone, and argillite that were deposited as sediment-gravity flows. The conglomerates contain clasts that consist mostly of plagioclase, clinopyroxene, and hornblende porphyries, including ankaramite, and intraformational lithologies. They also contain rare carbonate and hypabyssal intrusive rocks. Sandstones from the sequence have an average composition of Q_0 F_{41} L_{59} and are characterized by abundant volcanic lithic grains, oscillatory-zoned plagioclase, pyroxene and hornblende, and rare quartz grains. Based on their similar lithology, composition, sedimentology, and age, the volcaniclastic strata of the Cerro El Calvario area are interpreted to be correlative with the Lower Jurassic volcaniclastic strata of the two upper members of the San Hipolito Formation (Moore, 1984).

The mafic to intermediate volcanic composition of the sandstones and conglomerates and the absence of any continentally derived detritus in the San Hipolito Formation and related strata have been interpreted to result from deposition as epiclastic and pyroclastic debris shed from a nearby active island arc during the Late Triassic and Early Jurassic (Finch and Abbott, 1977). These strata have been subsequently altered and contain low-grade metamorphic minerals including albite, laumontite, quartz, epidote, and abundant calcite.

The ophiolite and overlying Triassic and Lower Jurassic sedimentary rocks of the Vizcaino Sur terrane are intruded by a number of tonalite plutons. U-Pb isotopic ages of the plutons range from 154 to 130 m.y. (Barnes, 1982; Kimbrough, 1982). Mafic phases commonly include subequal amounts of green-brown hornblende and red-brown biotite, but biotite dominates in several of the plutons. Locally, many of the plutons display hornfels and amphibolite contact metamorphic zones that are typically about 100 m (328 ft) in width. Although one tonalite pluton displays a well-developed synkinematic structural fabric, most of the plutons have massive textures except for the effects of deformation under brittle conditions after crystallization.

Sierra Placeres Melange

A serpentinite-matrix melange, herein called the Sierra Placeres melange, separates the Vizcaino Norte and Vizcaino Sur terranes (Figs. 1, 5). This melange crops out in the Sierra de San Andres between San Jose de Castro and Cerro El Calvario in the interior of the Vizcaino Peninsula. The melange trends east-northeast and is as wide as 15 km (9.3 mi). It is characterized by enveloping abundant blocks of ophiolitic and island arc affinity (Fig. 6) and can be distinguished from the Puerto Nuevo melange by its lack of blueschist, greenschist, and amphibolite blocks displaying a well-developed penetrative deformation or polydeformational history.

Blocks contained within the Sierra Placeres melange are as much as 8 km (5 mi) in length. Enclosed rocks include hornblende gabbro and diorite, hornblende tonalite, weakly foliated amphibolite, massive volcanic and hypabyssal rocks with abundant large plagioclase and pyroxene phenocrysts (ankaramite), and quartz-poor volcaniclastic conglomerate, sandstone, and shale that are interpreted to be of island-arc affinity. The melange also contains blocks of serpentinized harzburgite, layered peridotite, pillow basalt, and radiolarian chert and tuffaceous chert that are interpreted to be of ophiolitic affinity. One block, over 1 km (.6 mi) in length, contains a succession consisting, in ascending order, of about 100 m (328 ft) of pillow lava with diabase dikes, 30 m (98 ft) of tuffaceous radiolarian chert, and about 30 m (98 ft) of siliceous tuff that is interpreted to represent the uppermost part of an ophiolite.

A U-Pb isotopic age of 149 m.y. ($^{207}Pb/^{206}Pb$ age of 161 m.y.) on zircon has been obtained from a block in the melange that contains hornblende tonalite which is intrusive into massive porphyritic volcanic rocks (Kimbrough, 1982). K-Ar ages on hornblende from blocks consisting of hornblende porphyry and amphibolite are 126 $\pm$ 4 m.y. and 140 $\pm$ 21 m.y., respectively (Krummenacher, 1980, personal communication). Radiolarian assemblages from chert blocks within the melange have been identified as Early Jurassic (Jones and Murchey, 1983, personal communication).

The serpentinite matrix of the melange is composed of coarse-grained, moderately indurated chrysotile that generally is strongly foliated. This foliation wraps around enclosed blocks, but between inclusions it has an average east-northeast and near-vertical orientation. Map-scale elongate blocks are oriented parallel to the strike of the foliation and generally have steeply dipping contacts with the melange.

Rangin (1978), on the basis of the enclosed chert blocks, suggested that the Sierra Placeres melange is a disrupted ophiolite. However, the lack of coherent stratigraphy, presence of inclusions of island arc affinity, and a consistent penetrative fabric in the serpentinite matrix suggest that the melange resulted from tectonic mixing along a major fault zone. The lithologies,

Figure 6—View along strike of the Sierra Placeres melange showing block of massive serpentinized harzburgite enveloped in a steeply dipping foliated serpentinite matrix.

compositions, and ages of the blocks included within the Sierra Placeres melange are similar to those of the Vizcaino Norte and Vizcaino Sur terranes, which lie adjacent to the melange, but not the Puerto Nuevo terrane. The blocks are therefore interpreted as tectonic fragments derived from the adjacent Vizcaino Norte and Vizcaino Sur terranes.

The nature and amount of the displacement across the melange is unknown. The near-vertical orientation of elongate blocks and matrix foliation suggests that the melange was a zone of strike-slip or dip-slip faulting. The large differences in the stratigraphy across the melange suggest that the zone represents a large amount of displacement, possibly indicative of large-scale strike-slip movement. The width of the Sierra Placeres melange (as much as 15 km [9.3 mi]) is greater than that commonly associated with strike-slip faults (Blake and Howell, 1982, personal communication). The large width may, however, be interpreted as an apparent width across a low-angle fault such as a thrust fault. The near-vertical orientation of structures within the melange may be produced, in this scenario, by later upward diapiric movement of the matrix serpentinite. This interpretation suffers, however, because it is uncertain which of the two adjacent terranes is the upper plate. The best guess is that the Vizcaino Norte terrane overlies the Vizcaino Sur terrane because of its very low-grade metamorphic overprint.

The age of the Sierra Placeres melange can be constrained fairly closely. Isotopic ages from blocks enclosed in the melange are as young as about 130 m.y. and provide a maximum age of its formation. The melange is depositionally overlain by sedimentary strata of Albian age (see below). These sedimentary rocks are unaffected by the structures contained within the melange and provide a minimum age of deformation of about 110 m.y. (Palmer, 1983). The Sierra Placeres melange is therefore interpreted to have formed between 110 and 130 m.y. (Early Cretaceous).

Overlap Sequence

The Vizcaino Norte and Vizcaino Sur terranes, but not the Puerto Nuevo terrane, are depositionally overlain by the Middle Cretaceous Valle Formation. The Valle Formation, over 2,000 m (6,562 ft) in thickness (Robinson, 1975), covers the older rocks across most of the Vizcaino Peninsula and has been divided by Rangin (1978) into two parts. The lowermost part consists of thin-bedded mudstone, siltstone, and fine-grained sandstone turbidites of Albian to Cenomanian age. The coarser grained upper part consists of sandstone and conglomerate turbidites of mostly Turonian age. Detritus in the Valle Formation is largely resedimented debris derived from a mixed intermediate volcanic, metamorphic, and batholithic source located to

Figure 7—Photograph showing sheared serpentinite of the Sierra Placeres melange (spm) unconformably overlain by shallow-marine sandstone of the Valle Formation (Kv), also called the Asuncion formation by Barnes (1982).

the east (Robinson, 1975; Patterson, 1979). The sequence is generally interpreted as a deep-sea fan complex constructed in a forearc basin associated with the Cretaceous Allisitos arc (Patterson, 1979).

In the northern part of the Peninsula, the basal contact of the Valle is commonly a low-angle fault, especially near the Sierra de San Andres where it may have undergone gravity sliding during the Cenozoic uplift of the range. Robinson (1975) describes locations north of Bahia Tortugas where the Valle rests on an angular unconformity on the Eugenia Formation of the Vizcaino Norte terrane. Barnes (1982) reports, however, that the ages of the two units are nearly equivalent and that the relationship is conformable at one of these locations.

To the south, the lowest part of the Valle Formation rests unconformably on tonalite, volcaniclastic rock, and cumulate gabbro of the Vizcaino Sur terrane. It also rests unconformably on the Sierra Placeres melange (Fig. 7). The basal strata of the Valle Formation (*sensu lato*) in this area are Aptian to Cenomanian marginal-marine and deep-water deposits and consist of bioclastic limestone, olistostromes, stratified breccia and conglomerate, fossiliferous ungraded beds of fine-grained sandstone, and rare tuff beds (Barnes, 1982). These clastic rocks contain debris of obvious local derivative such as ankaramite, cumulate gabbro, serpentinite, and amphibolite. The strata indicate that the basement rocks of the Vizcaino Peninsula were exposed to erosion during the Middle Cretaceous.

Barnes (1982) has informally named this basal sequence the Asuncion formation and shows it to have a thickness of at least 800 m (2,625 ft). These basal strata are conformably and gradationally overlain by at least 500 m (1,640 ft) of thin-bedded sandstone, siltstone, and mudstone turbidites of Cenomanian age followed by at least 750 m (2,461 ft) of coarser grained turbidites of Cenomanian to Turonian age (Patterson, 1979; Barnes, 1982). Thus, the Valle Formation blankets both the Vizcaino Norte and Vizcaino Sur terranes and the Sierra Placeres melange complex that separates these two terranes.

GEOLOGIC IMPLICATIONS

In the previous sections, I have described the geology of the Vizcaino Peninsula as consisting of three principal terranes. Their similarities and differences are summarized in Table 1. The composite Puerto Nuevo terrane can be distinguished from the other two by its penetratively and polydeformed elements and its moderately low-temperature and high-pressure pumpellyite-actinolite, blueschist-greenschist, epidote amphibolite, and eclogite facies

Table 1

Characteristics	Puerto Nuevo Terrane	Vizcaino Norte Terrane	Vizcaino Sur Terrane
Distribution	exposed in structural window under the Vizcaino Norte terrane	Northern Vizcaino Peninsula	Southern Vizcaino Peninsula
Type	composite	noncomposite	noncomposite
Age	pre-Cretaceous	Late Triassic to Early Cretaceous	Late Triassic to Early Cretaceous
Lithologic characteristics	blocks in melange	coherent	coherent
Metamorphism	subduction zone (low T-high P)	ocean-floor (high T-low P) and subsequent burial (zeolite facies)	ocean floor (high T-low P) and subsequent island arc (high T-low P) thermal metamorphism
Deformation	Penetrative	Brittle	Brittle
Sandstone compositions	quartz-poor graywacke and quartzofeldspathic sandstone (metamorphosed)	quartz-poor graywacke to quartzofeldspathic sandstone	quartz-poor volcaniclastic sandstone
Provenance of sandstone	volcanic-plutonic arc	volcanic-plutonic arc and craton	volcanic arc
Paleogeographic setting at time of origin	midocean ridge and volcanic arc	island arc–marginal basin system	island arc–marginal basin system

Table 1—Summary of characteristics of the three principal terranes of the Vizcaino Peninsula.

metamorphism. The Vizcaino Norte and Vizcaino Sur terranes are similar in that both contain a basal ophiolite and overlying tuffaceous pelagic rocks of Late Triassic age. However, contrasts in the ratio of volcaniclastic to pelagic and biogenic detritus in their Triassic sequences suggest that the Vizcaino Norte and Sur terranes may not have been contiguous but may have been separated by significant distances. Moreover, the Vizcaino Sur terrane contains a thick sequence of volcanic and volcaniclastic strata that was intruded by Middle Jurassic to Lower Cretaceous granitic rocks, whereas the Vizcaino Norte terrane lacks any rocks of Early and Middle Jurassic age but contains a thick Upper Jurassic to Lower Cretaceous sequence of volcanogenic sandstone with contributions from plutonic and cratonal source areas. Notably, both terranes seem to lack strata of Hauterivian and Barremian (middle Early Cretaceous) age, although a single ammonite of Hauterivian age was reported from a structurally complicated area near Morro Hermoso (Miller, U.S. Geological Survey, 1982, personal communication).

Barnes (1982) suggests that the temporal and stratigraphic dissimilarities between the Vizcaino Norte and Vizcaino Sur terranes can be explained by facies changes and erosional events that are time transgressive. However, the presence of a major zone of dislocation, represented by the Sierra Placeres melange, between the northern and southern terranes of the Vizcaino Peninsula, shows that any reconstruction of events must include substantial amounts of tectonic shortening between the terranes prior to the Late Cretaceous. For this reason, the original relative paleogeographic positions of the terranes are unknown. As there is now considerable evidence for large-scale tectonic transport of paleo-oceanic terranes of Mesozoic age now emplaced along the western margin of North America (Jones et al, 1977; Alvarez et al, 1980), these two geologic provinces with different middle Mesozoic geologic histories can be considered as separate tectonostratigraphic terranes.

Despite the differences between the Vizcaino Norte and Vizcaino Sur terranes, the similar age and composition of their basal ophiolites suggest that they may have originated in the same general part of the paleo-Pacific ocean (Panthalassa). Although not conclusive, the weight of the geochemical, petrologic, and stratigraphic data suggests that both ophiolites formed adjacent to contemporaneous volcanic arcs and in a position to receive sediment from those arcs, possibly as parts of one or more marginal basins. Because the Triassic sedimentary rocks that overlie the ophiolites completely lack continental detritus and are composed entirely of volcanogenic and biogenic debris, the marginal basin(s) must have formed in a position removed from a continental source. At present, the nearest known outcrop of Triassic volcanic and plutonic rocks that may have provided detritus to the Vizcaino is in the Caborca-Magdalena area of northwestern Sonora (Rangin, 1978; Gastil et al, 1978; Anderson and Silver, 1969). Alternatively, the arc complex(es) that provided the tuffaceous sediment to the marginal basin(s) may have since been disrupted and parts removed by the later tectonic events that have affected the Vizcaino Peninsula terranes.

By the Early Jurassic, island-arc volcanism was initiated on the Vizcaino Sur terrane, implying that a reorganization of convergent plate boundaries had occurred. Coarse-grained volcanic detritus accumulated on the oceanic basement of the Vizcaino Sur terrane and resulted in the construction of a volcanic arc on the La Costa ophiolite. The volcanic debris is poor in quartz and consists essentially of pyroxene porphyry (ankaramite) and hornblende porphyry of basaltic-andesite and andesite composition. The paucity of quartz and absence of any continentally derived debris indicate that this Lower Jurassic island arc was intraoceanic and removed from a source of continental detritus. Island arc volcanism, represented by the intrusion of hornblende- and biotite-bearing diorites and tonalite plutons into the older volcanic and ophiolitic basement of the Vizcaino Sur terrane, continued during the Late Jurassic to Early Cretaceous. The more siliceous late phase of this arc may have provided the volcanic detritus that composes the Eugenia Formation of the Vizcaino Norte terrane as suggested by Barnes (1982). However, no sedimentary or volcanic rocks of Late Jurassic to Early Cretaceous age occur in the Vizcaino Sur terrane nor have any plutonic rocks of the same age been recognized in the Vizcaino Norte terrane. Thus, there are no transitional facies between the two terranes, implying that if they were part of the same arc complex, there have been significant amounts of shortening since its inception.

Importantly, the Early Cretaceous part of the Eugenia Formation contains clasts of quartz-pebble conglomerate and granitic rocks that indicate isotopic contamination by Precambrian sources. These clasts represent the earliest appearance of detritus derived from a cratonal source in the terranes of the Vizcaino Peninsula and may record the approach of the Vizcaino Norte terrane toward a craton during the Late Jurassic and Early Cretaceous. However, there is no evidence of a compressional event of Nevadan age at about 150 to 156 m.y. in the Vizcaino Peninsula as suggested by Rangin (1983). This may suggest that the approach of the Vizcaino Norte terrane was by transform rather than by convergent motion and did not result in a collision with the North American continental margin as proposed for the Nevadan orogeny by Schweickert and Cowan (1975).

By the end of the Early Cretaceous, the Vizcaino Norte and Vizcaino Sur terranes were tectonized and emplaced against one another along the Sierra Placeres melange. This melange is therefore a suture between the terranes. Deformation within the terranes is brittle in style and metamorphism low in grade, implying that emplacement occurred at shallow structural levels. Simultaneously, arc volcanism ended in the Vizcaino Peninsula and was initiated along the Alisitos arc in the interior of the Baja California peninsula during the Albian. This arc is composed in large part of silicic volcanic and volcaniclastic rocks, implying that it developed on a continental margin similar to the modern Andean arc. Most workers have generally suggested that this more eastern arc provided the volcanic and plutonic detritus that composes the Valle Formation which overlaps the suture between the Vizcaino Norte and Vizcaino Sur terranes (Rangin, 1978; Gastil et al, 1978; Barnes, 1982) (Fig. 8). Thus, by the Albian (late Early Cretaceous) a second reorganization of convergent plate motions had taken place and resulted in the emplacement (obduction?) of the Vizcaino Norte and Vizcaino Sur terranes and the initiation of the Andean-type Alisitos arc.

In contrast to the history of the above two terranes, the geologic history and age of emplacement of the Puerto Nuevo terrane is unclear. The metaigneous and metasedimentary components of the Puerto Nuevo terrane have a pre-Cretaceous age and generally are of oceanic affinity. This terrane records Late Triassic to Early Cretaceous subduction metamorphism that may have been paired with one of the island-arc complexes described above. However, there is no evidence of a genetic link between the Puerto Nuevo terrane and the Vizcaino Norte and Vizcaino Sur terranes. Likewise, there is no control on the time of emplacement of the Puerto Nuevo terrane beneath the Vizcaino Norte terrane. It seems likely that the Vizcaino Norte terrane was obducted onto the Puerto Nuevo terrane during the same reorganization event that juxtaposed the Vizcaino Norte and Vizcaino Sur terranes. However, blueschist and associated metamorphic detritus is not known in the Sierra Placeres melange, the basal, proximally derived part of the Valle Formation (*sensu lato*), or any other pre-Quaternary rocks of the Vizcaino Peninsula. Additionally, the Puerto Nuevo melange is not overlapped by any pre-Quaternary sedimentary units. Thus, the Puerto Nuevo terrane may have been emplaced as late as the Tertiary, possibly as boundins that have migrated along strike-slip faults as proposed by Karig (1980).

The nature and age of accretion of the Vizcaino composite terrane onto the margin of what is presently Baja California is unknown because this contact is covered by the Quaternary fill of the Vizcaino basin. Mina (1957) reported the presence of Upper Cretaceous turbidites east of the basin and correlated them with the Valle Formation west of the basin. Helenes (1981), however, has contrasted the Tertiary sections exposed on the eastern and western sides of the basin and pointed out stratigraphic differences that could be attributed to strike-slip faulting. Although the significance of these differences is not clear, they may indicate that emplacement of the Vizcaino composite terrane into its present position in Baja California may not have been completed until after the Miocene.

CONCLUSION

The Vizcaino Peninsula consists of three principal tectonostratigraphic terranes that are composed largely of rocks of ophiolitic and island-arc affinity. The ophiolites of the Vizcaino Norte and Vizcaino Sur terranes are interpreted to have originated as parts of one or more marginal basins of Late Triassic age. The age and site of

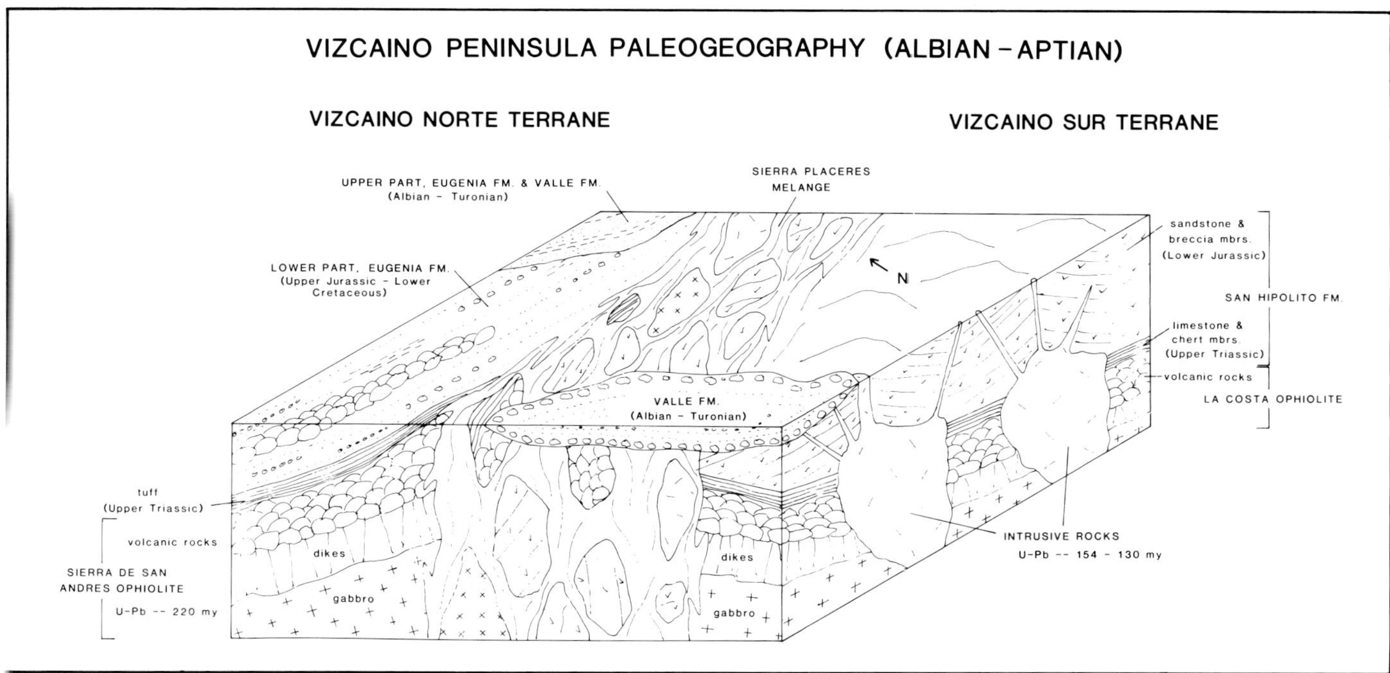

Figure 8—Block diagram showing paleogeographic reconstruction of the Vizcaino Norte and Vizcaino Sur terranes during the Albian (late Early Cretaceous). Position of the Puerto Nuevo terrane at that time is unknown and is therefore not shown in the diagram.

origin of the composite Puerto Nuevo terrane is less clear, but it consists of an ophiolitic subterrane that may have been formed at a midocean ridge and a more silicic volcanic-plutonic subterrane that may represent fragments of a volcanic-arc complex.

Gastil et al (1978) and Rangin (1978) previously suggested that the ophiolites of the Vizcaino Norte and Vizcaino Sur terranes were formed as a marginal basin behind a west-facing arc that fringed the southern margin of the North American craton. However, the Upper Triassic to at least Lower Jurassic sedimentary rocks that cover these ophiolites do not contain detritus that could have been shed from a continental area, such as abundant quartz, tourmaline, garnet, staurolite, muscovite, vein quartz, metamorphic rock fragments, nor any other evidence of close proximity to a craton. The covering sedimentary rocks of the ophiolites were instead derived entirely from active island-arc sources or are biogenic in origin. Thus, there exists no evidence that these terranes were tied to the continental margin of the North American craton as a fringing island arc–backarc system until well after their basal ophiolites were formed. Although poorly dated, the metamorphosed and disrupted Puerto Nuevo terrane similarly consists of rocks entirely of oceanic affinity that contain no evidence of linkage to a continental margin at the time of their origin or metamorphism.

Although the fringing arc model of Gastil et al (1978) and Rangin (1978) cannot be entirely eliminated, it seems more likely that the ophiolites and covering sediments in these terranes originated as parts of marginal basins in Panthalassa (the paleo-Pacific Ocean) at locations that may have been very distant from the continental margin of North America and underwent a history of tectonic transport toward the North American craton. Continental detritus contained within the Upper Jurassic to Lower Cretaceous Eugenia Formation of the Vizcaino Norte terrane represents the oldest contribution of cratonal debris to the Vizcaino terranes and may signal the approach of that terrane toward the American craton. By the Early Cretaceous, the Vizcaino Norte and Vizcaino Sur terranes were deformed and sutured together along the Sierra Placeres melange and probably were accreted to the margin of North America, although not necessarily at the present latitude of the Vizcaino Peninsula. This indicates that the terranes may have been transported with one or more parts of the ocean floor at Panthalassa for at least as much as 80 m.y. prior to their accretion to the margin of North America.

Plate motions calculated by Engebretson (1982) and Stone et al (1982) suggest that thousands of kilometers of Panthalassa have been subducted along the southern margin of the North American craton since the Triassic. Because most of the Farallon and all of the Kula plates have been destroyed by this convergence, the paleogeography of the boundary between these two terranes and the North America craton are not fully understood. Jones et al (1978) suggested that this part of Panthalassa may have had a complicated paleogeography that consisted of a number of island arcs, marginal or interarc basins, transform boundaries, and small oceanic plates much like that of the present configuration of the modern western Pacific Ocean basin. The oceanic character and long period of time between origin of their basal ophiolites and the first appearance of cratonal debris suggested that the terranes of the Vizcaino Peninsula represent preserved allochthonous fragments of one or more such microplates of Panthalassa. Without paleomagnetic evidence or knowledge of the paleogeography of Panthalassa, however, the details of the history of transport and emplacement of these fragments by such processes as orthogonal or oblique convergence, rifting, transform motion, or a combination of these processes, may not be resolvable.

ACKNOWLEDGMENTS

The ideas presented here benefited greatly from the work of, and discussions with, my fellow students of the geology of the Vizcaino Peninsula, including Gordon Gastil, David Kimbrough, Javiar Helenes, James Hickey, Debra Patterson, and David Barnes. Financial support from the National Science Foundation (Grant EAR 80-08527 to J. G. Liou), the Geological Society of America (Research Grants 2503-79 and 2697-80), and the Shell Fund of Stanford University is greatly appreciated. The manuscript was reviewed and improved by the help of J. G. Liou and B. M. Page and typed by Irene Rogers, Henriette Phillips, Mary Milan, Rita Parsons, Terry Higgins, and Margo Maguire at the Branch of Western Regional Geology, U.S. Geological Survey.

REFERENCES

Alvarez, W., et al, 1980, Franciscan Complex limestone deposited at 17° south paleolatitude: Geological Society of America Bulletin, v. 91, p. 476–484.

Anderson, T. H., and L. T. Silver, 1969, Mesozoic magmatic events of the northern Sonora coastal region (Abs.): Geological Society of America Abstracts with Programs, v. 1, p. 3–4.

Barnes, D. A., 1982, Basin analysis of volcanic arc-derived Jura-Cretaceous sedimentary rocks, Vizcaino Peninsula, Baja California Sur, Mexico: PhD Dissertation, University of California, Santa Barbara, 249 p.

______, and J. M. Mattinson, 1981, Late Triassic-Early Cretaceous age of eugeoclinal terranes, western Vizcaino Peninsula, Baja California Sur, Mexico (Abs.): Geological Society of America Abstracts with Programs, v. 13, p. 43.

Boles, J. R., 1978, Basin analysis of the Eugenia Formation (Late Jurassic), Punta Eugenia area, Baja California, *in* D. G. Howell and K. A. McDougall, eds., Mesozoic paleogeography of the western United States: Society of Economic Paleontologists and Mineralogists, Pacific Section, Pacific Coast Paleogeography Symposium 2, p. 493–498.

______, and J. J. Hickey, 1979, Eugenia Formation (Jura-Cretaceous) Punta Eugenia area, *in* P. L. Abbott and R. G. Gastil, eds., Baja California geology: Department of Geological Sciences, San Diego State University, Fieldtrip Guidebook for the 1979 Geological Society of America Annual Meeting, San Diego, p. 65–71.

Coleman, R. G., 1977, Ophiolites: ancient oceanic lithosphere?: New York, Springer-Verlag, 229 p.

Coney, P. J., et al, 1980, Cordilleran suspect terranes: Nature, v. 288, p. 329–333.

Engebretson, D. C., 1982, Relative motions between oceanic and continental plates in the Pacific basin: PhD Dissertation, Stanford University, 211 p.

Ernst, W. G., and Y. Seki, 1967, Petrologic comparison of the Franciscan and Sanbagawa metamorphic terranes: Tectonophysics, v. 4, p. 463–478.

Finch, J. W., and P. L. Abbott, 1977, Petrology of a Triassic marine section, Vizcaino Peninsula, Baja California Sur, Mexico: Sedimentary Geology, v. 19, p. 253–273.

Gastil, R. G., and R. H. Miller, 1981, Lower Paleozoic strata on the Pacific plate of North America: Nature, v. 292, p. 828–829.

______, et al, 1978, Mesozoic history of peninsular California and related areas east of the Gulf of California, *in* D. G. Howell and K. A. McDougall, eds., Mesozoic paleogeography of the western United States: Society of Economic Paleontologists and Mineralogists, Pacific Section, Pacific Coast Paleogeography Symposium 2, p. 107–116.

Helenes, E. J., 1981, Tertiary evolution of the Sebastian Vizcaino Peninsula, Baja California, Mexico (Abs.): Geological Society of America Abstracts with Programs, v. 13, p. 60.

Hickey, J. J., 1984, Stratigraphy and composition of a Jura-Cretaceous volcanic arc apron, Punta Eugenia, Baja California Sur, Mexico, *in* V. A. Frizzell, ed., The geology of the Baja California peninsula: Los Angeles, Society of Economic Paleontologists and Mineralogists, p. 149–160.

Jones, D. L., et al, 1976, The four Jurassic belts of northern California and their significance to the geology of the southern California borderland, *in* D. G. Howell, ed., Aspects of the geologic history of the California borderland: American Association of Petroleum Geologists, Pacific Section, Miscellaneous Publication 24, p. 343–362.

______, et al, 1977, Wrangellia—a displaced terrane in northwestern North America: Canadian Journal of Earth Sciences, v. 14, p. 2565–2577.

______, et al, 1978, Microplate tectonics of Alaska—significance for the Mesozoic history of the Pacific Coast of North America, *in* D. G. Howell and K. A. McDougall, eds., Mesozoic paleogeography of the western United States: Society of Economic Paleontologists and Mineralogists, Pacific Section, Pacific Coast Paleogeography Symposium 2, p. 71–74.

______, et al, 1982, Character, distribution, and tectonic significance of accretionary terranes in the central Alaska Range: Journal of Geophysical Research, v. 87, p. 3709–3717.

Karig, D. E., 1980, Material transport within accretionary prisms and the "Knocker" problem: Journal of Geology, v. 88, p. 27–39.

Kimbrough, D. L., 1982, Structure, petrology, and geochronology of Mesozoic paleooceanic terranes on Cedros Island the Vizcaino Peninsula, Baja California Sur, Mexico: PhD Dissertation, University of California, Santa Barbara, 395 p.

Mina, F., 1957, Bosquejo geologico del Territorio Sur de la

Baja California: Boletín de la Asociatión Mexicana de Geólogos Petroleros, v. 9, p. 139-269.

Minch, J. C., et al, 1976, Geology of the Vizcaino Peninsula, *in* D. G. Howell, ed., Aspects of the geologic history of the California continental borderland: American Association of Petroleum Geologists, Pacific Section, Miscellaneous Publication 24, p. 136-195.

Moore, T. E., 1979, Geologic summary of the Sierra de San Andres ophiolite, *in* R. G. Gastil and P. L. Abbott, eds., Baja California geology: Department of Geological Sciences, San Diego State University, Fieldtrip Guidebook for the 1979 Geological Society of America Annual Meeting, San Diego, p. 95-106.

______, 1983, Geology, petrology, and tectonic significance of paleooceanic terranes of the Vizcaino Peninsula, Baja California Sur, Mexico: PhD Dissertation, Stanford University, 376 p.

______, 1984, Sedimentary facies and composition of Jurassic volcaniclastic turbidites at Cerro El Calvario, Vizcaino Peninsula, Baja California Sur, Mexico, *in* V. A. Frizzell, ed., The geology of the Baja California peninsula: Los Angeles, Society of Economic Paleontologists and Mineralogists, p. 131-148.

Ortega-Gutierrez, F., 1981, Metamorphic belts of southern Mexico and their tectonic significance: Geofisica Internacional, v. 20, p. 177-202.

Palmer, A. R., 1983, The decade of North American geology 1983 geologic time scale: Geology, v. 11, p. 503-504.

Patterson, D. L., 1979, The Valle Formation—physical stratigraphy and depositional model, southern Vizcaino Peninsula, Baja California Sur, *in* R. G. Gastil and P. L. Abbott, eds., Baja California geology: Department of Geological Sciences, San Diego State University, Fieldtrip Guidebook for the 1979 Geological Society of America Annual Meeting, San Diego, p. 73-76.

______, 1984a, Los Chapunes and Valle sandstones: Cretaceous petrofacies of the Vizcaino basin, Baja California, Mexico, *in* V. A. Frizzell, ed., The geology of the Baja California peninsula: Los Angeles, Society of Economic Paleontologists and Mineralogists.

______, 1984b, Paleomagnetism of the Valle Formation and the Late Cretaceous paleogeography of the Vizcaino basin Baja California, Mexico, *in* V. A. Frizzell, ed., The geology of the Baja California peninsula: Los Angeles, Society of Economic Paleontologists and Mineralogists.

Pearce, J. A., 1975, Basalt geochemistry used to investigate past tectonic environments on Cyprus: Tectonophysics, v. 25, p. 41-67.

______, and J. R. Cann, 1973, Tectonic setting of basic volcanic rocks determined using trace element analysis: Earth and Planetary Science Letters, v. 19, p. 290-300.

Pessagno, Jr., E. A., et al, 1979, Upper Triassic radiolaria from the San Hipolito Formation, Baja California: Micropaleontology, v. 25, p. 160-197.

Rangin, C., 1976, Le complexe ophiolitique de Basse Californie, une paleocroute oceanique ecaille (Peninsule de Vizcaino, Baja California, Mexique): Bulletin Sociètè Géologique France, Series 7, v. XVIII, n. 6, p. 1677-1685.

______, 1978, Speculative model of Mesozoic geodynamics, central Baja California to northeast Sonora (Mexico), *in* D. G. Howell and K. A. McDougall, eds., Mesozoic paleogeography of the western United States: Society of Economic Paleontologists and Mineralogists, Pacific Section, Pacific Coast Paleogeography Symposium 2, p. 85-106.

______, 1983, Geodynamic significance of Late Triassic to Early Cretaceous volcanic sequences of the Vizcaino Peninsula and Cedros Island, Baja California, Mexico: Geology, v. 11, p. 552-556.

Robinson, J. W., 1975, Reconnaissance geology of the northern Vizcaino Peninsula, Baja California Sur, Mexico: MS Thesis, San Diego State University, 114 p.

Schweikert, R. A., and D. S. Cowan, 1975, Early Mesozoic tectonic evolution of the western Sierra Nevada, California: Geological Society of America Bulletin, v. 86, p. 1329-1336.

Stone, D. B., et al, 1982, Paleolatitudes versus time for southern Alaska: Journal of Geophysical Research, v. 87, p. 3697-3709.

Troughton, G. H., 1974, Stratigraphy of the Vizcaino Peninsula near Asuncion Bay, Territorio de Baja California, Mexico: MS Thesis, San Diego State University, 83 p.

Whalen, P. A., 1983, Lower Jurassic radiolaria, San Hipolito Formation, Vizcaino, Baja California Sur (Abs.): Bulletin of the American Association of Petroleum Geologists, v. 67, p. 568-569.

Tectonostratigraphic Terranes, Pacific Northwest Quadrant

Tectonic Evolution of Kamchatka and the Sea of Okhotsk and Implications for the Pacific Basin

Bruce F. Watson*
Kazuya Fujita
Michigan State University
East Lansing, Michigan

Kamchatka Peninsula can be divided into approximately ten tectonostratigraphic terranes that are primarily of oceanic origin. Most of these terranes are remnants of island arcs and oceanic crust that amalgamated in Mesozoic and Cenozoic time. Little reliable evidence exists for pre-Mesozoic crust anywhere in Kamchatka although small blocks of Paleozoic material may have been introduced through transcurrent motion. Accretion in western Kamchatka started about 100 Ma ago, while the current interrelations in eastern Kamchatka developed about 60 Ma ago. Many of the major events that affected eastern Kamchatka since 75 Ma ago can be related to changes in plate motion between the Pacific and Eurasian plates and the development of the Aleutian arc. The Sea of Okhotsk is bounded on the north by Mesozoic subduction zones and is suggested to be formed primarily of oceanic crust with an oceanic plateau in its northern half and island arcs in its southern half.

INTRODUCTION

Kamchatka Peninsula is located in the northeastern U.S.S.R. along the northwestern margin of the Pacific Ocean basin. It is bounded on the east by the Kurile-Kamchatka trench and on the west by the relatively shallow part of the Sea of Okhotsk. The terranes of eastern Kamchatka are the basement for the volcanic arc resulting from present-day subduction.

The majority of recent Soviet investigators of the region have agreed on an island arc or volcanic arc origin for most of Kamchatka (Avdeiko, 1971; Gnibidenko et al, 1974; Erlich, 1979; Savostin et al, 1983; and others). In addition, a difference has long been noted between the stratigraphy and tectonics of eastern, central, and western Kamchatka (Vlasov et al, 1965; Marakhanov and Potap'ev, 1981; and others). It is only recently, however, that sufficient data have become available to begin a detailed examination of the region.

The geology and tectonics of Kamchatka are very important to both the study of northeastern Siberia and the northern Pacific because of its structural position between them. The study of Kamchatka may allow us to interrelate the motions of Eurasia and the plates in the Pacific basin and to constrain relative motion models now being developed for the Pacific (for example, Engebretson, 1982).

In this paper, we examine the geology and development of Kamchatka in the context of plate tectonics and tectonostratigraphic terranes (Jones et al, 1983). We summarize some of the controversies on the age of Kamchatka, present a preliminary tectonostratigraphic terrane map, and suggest a possible model for the evolution of Kamchatka and the Sea of Okhotsk that is consistent with available data. In the absence of accessibility, both to the outcrops themselves and to most of the primary literature sources, we have used literature on the geology of Kamchatka in translation (see Fujita et al, 1982), supplemented by a large number of Soviet geologic maps and several Russian sources. We realize that there are severe limitations to this method. In particular, there is the lack of an even and detailed data base and an inability, as noted by Hamilton (1970), to verify questionable or ambiguous data. In Soviet literature, it is also difficult to distinguish between factual data and extrapolations.

In Figure 1, we present a terrane map of Kamchatka. More detailed geologic information, both on lithologies and sources, is presented in Watson (1985) and is omitted here. A relatively detailed geologic map of Kamchatka has been published by Krasniy (1964).

METAMORPHIC TERRANES OF KAMCHATKA

The ages of the oldest formations now exposed in Kamchatka have been the subject of much debate among Soviet geologists. Early mapping identified several metamorphic outcrops that were presumed to be Precambrian on the basis of their metamorphic grade (Mokrousov and Marchenko, 1964). The oldest reliable ages from macropaleontological data are Jurassic to Early Cretaceous for parts of the Kvakhon (Kv; these abbreviations refer to

*Present Address: Superior Oil, 12401 Westheimer, Houston, Tx 77077

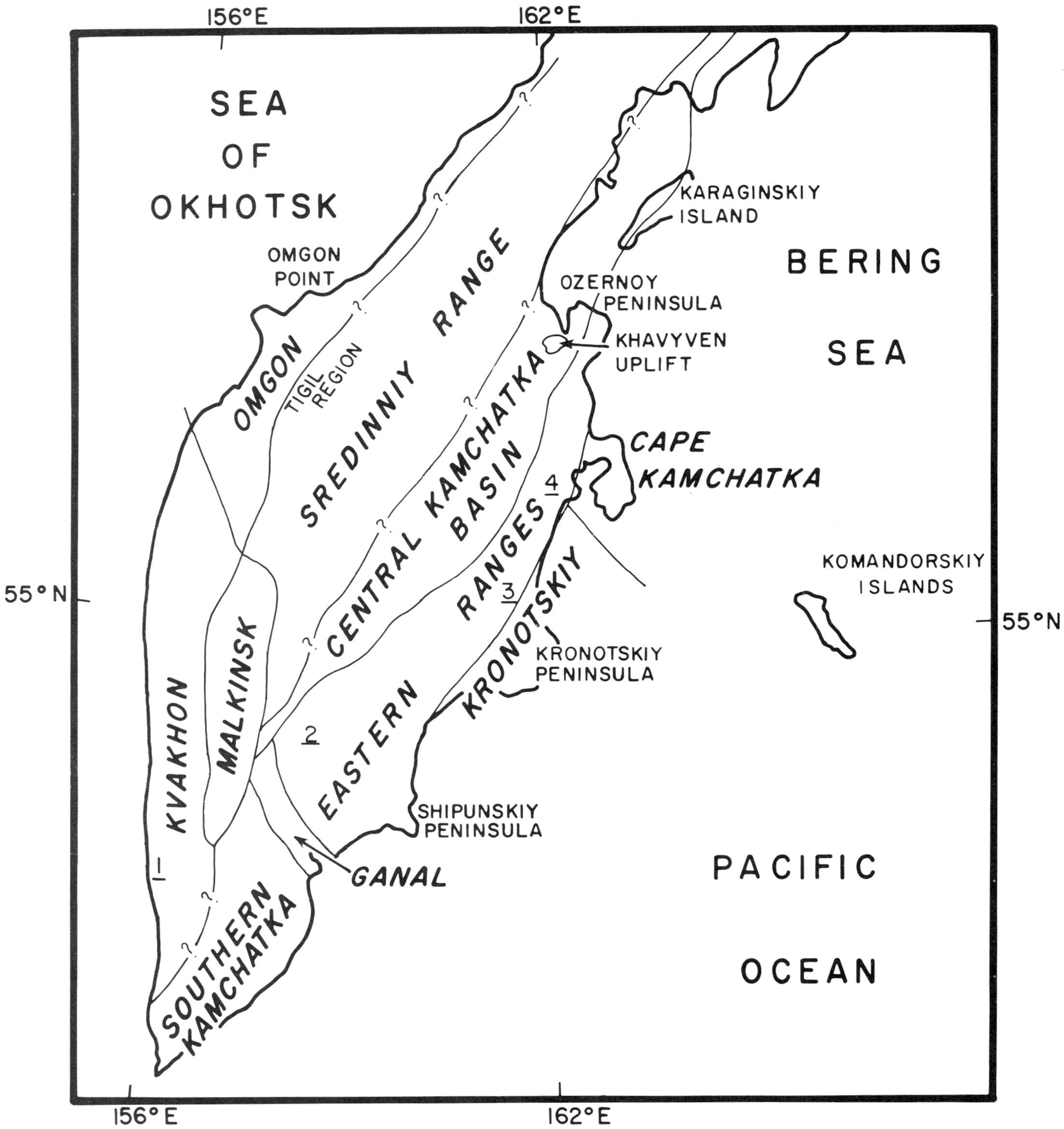

Figure 1—Index and preliminary tectonostratigraphic terrane map of Kamchatka Peninsula. Underscored numbers denote locations of (1) Ust' Bolsheretsk, (2) Valaginskiy Range, (3) Tumrok Range, and (4) Kumroch Range.

Figs. 2, 3, and 5) suite (Sidorchuk and Khanchuk, 1981) of the Kvakhon terrane (Fig. 2), and Late Cretaceous for the bulk of the peninsula (Aksenovich et al, 1964; Krasniy, 1966; Voronkov and Smirnov, 1971; German et al, 1978). Radiometric age determinations and the discovery of Paleozoic spores, however, have caused considerable uncertainty (see Parfenov, 1970).

Malkinsk Terrane

The Malkinsk terrane (Fig. 1), which consists entirely of the so-called Central Massif, is the most controversial in Kamchatka. The situation is complicated by differing geologic names assigned to various formations by different authors and the extrapolation of some formations from neighboring terranes. The lowest unit (Kolpakova series,

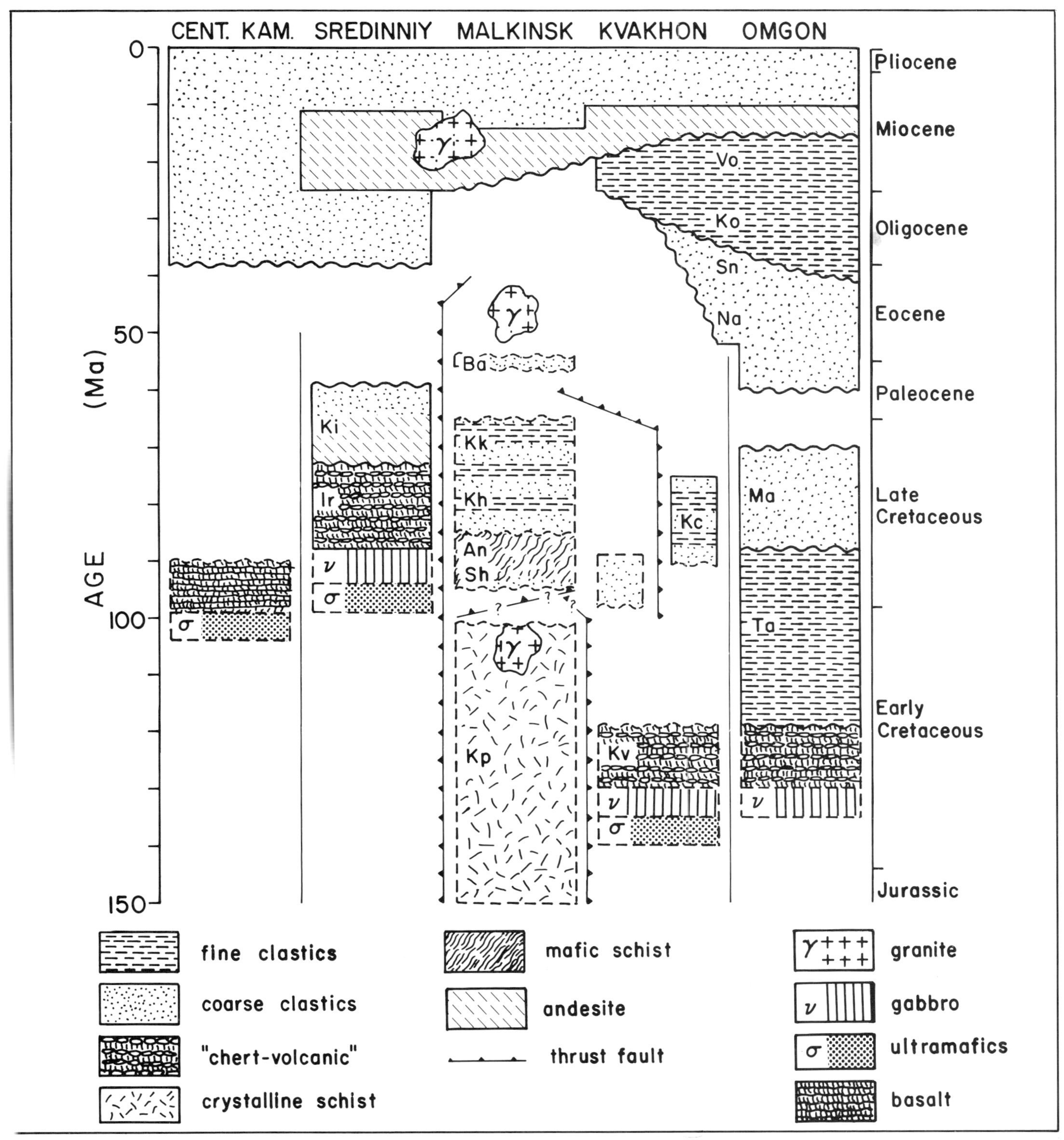

Figure 2—Stratigraphic columns for the terranes of western Kamchatka. Two letter codes within the columns identify specific formation names. Abbreviations are given in the text.

Kp; Fig. 2) is composed of crystalline schist, gneiss, and amphibolite, which have been intruded by now-metamorphosed granites. This is overlain by the Malkinsk series, which is also intruded by granites. For the most part, the Kolpakova series has undergone two phases of metamorphism, while the Malkinsk series has only undergone one (Shul'diner et al, 1980). The lowest unit of the Malkinsk series is the Shikhtinsk (Sh) suite, which is composed of crystalline schist, gneiss, metamorphosed pelite, and basal coarse clastics. This is overlain (concordantly ?) by the Andrianovka (An) suite of metamorphosed mafic rocks. The two suites are sometimes referred to as the Kamchatka series (Mokrousov and Marchenko, 1964); the label is also sometimes applied to

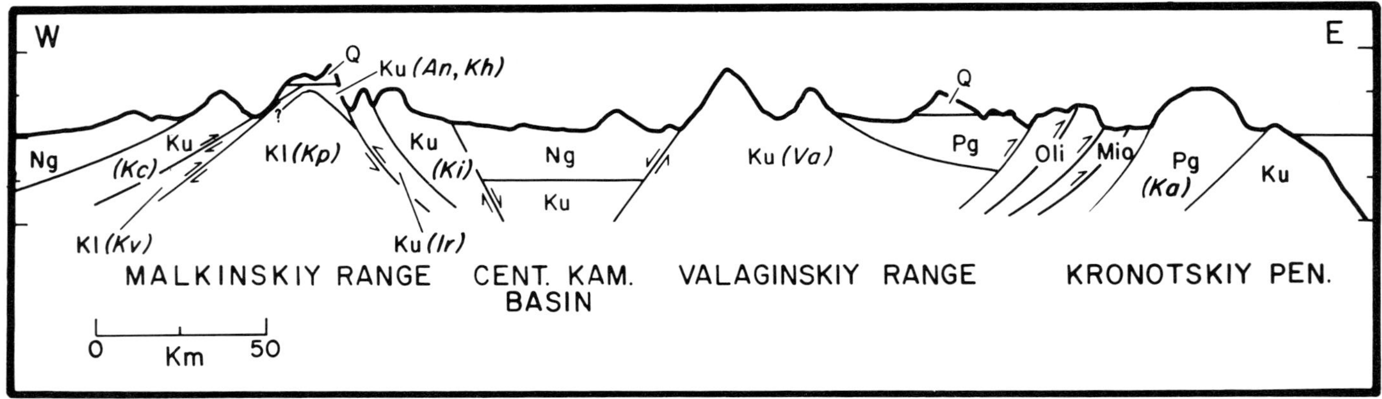

Figure 3—Schematic cross section across Kamchatka at the latitude of Kronotskiy Peninsula. Formation ages are given with formations identified in parentheses. Abbreviations are given in the text.

the Shikhtinsk suite alone (Gnibidenko et al, 1974; Marakhanov and Potap'ev, 1981). Overlying these series are the Kheivan (Kh) suite of phyllites and metamorphosed sandstones and the Khimkinsk (Kk) suite of metavolcanics. Unconformably overlying the Malkinsk series in a few localities is the coarse clastic Baraba (Ba) suite, which has a Paleocene flora (Aprelkov et al, 1964).

Although earlier Soviet literature indicates that the Kvakhon suite and Kikhchik (Kc) series overlies the Malkinsk series, which in turn are overlain by the Irunei (Ir) and Kirganik (Ki) series, recent data suggest that these two groups of formations are in tectonic contact with the underlying deposits (Shul'diner et al, 1980; Sidorchuk and Khanchuk, 1981; Zhegalova, 1981). The Kvakhon suite is composed of tuffaceous rocks with a melange and sheets of ultramafics at the base (Shul'diner et al, 1980; Sidorchuk and Khanchuk, 1981) and is paleontologically dated as being of Late Jurassic to Early Cretaceous age. The Kvakhon suite has undergone blueschist metamorphism, and pebbles of the metamorphosed rocks are found in overlying terrigenous sediments that contain Early to Late Cretaceous flora (Sidorchuk and Khanchuk, 1981). The presumed younger Kikhchik series is composed of turbidites and mafic schists and overlies (tectonically ?) both the Kvakhon suite and the Malkinsk series. We consider the Kikhchik series and the Kvakhon suite to be part of the Kvakhon terrane, which appears to be an arc assemblage overlying oceanic crust that has been obducted over the Malkinsk terrane, possibly in several thrust sheets, in mid-Cretaceous time (Fig. 3).

The Irunei series (Fig. 2) is found on the eastern side of the Central Massif and consists of mafic extrusives, tuffs, and siliceous rocks with tuffs dominant in the upper part of the section (Shul'diner et al, 1980). Krasniy (1964) maps ultramafics and gabbros along the thrust fault contact between the Irunei and Malkinsk rocks. Faunal remains are rare in central Kamchatka; however, a Cenomanian *Inoceramus* has been identified from the tuffs of the Irunei, just east of the Central Massif (Aksenovich et al, 1964), and *Inoceramus schmidti* has been found in Irunei-like rocks south of the Ganal Range (German et al, 1978). Overlying the Irunei is the Kirganik series of calc-alkaline volcanics, cherty rocks, and terrigenous sediments in which early Paleogene spores and pollen have been identified (Shapiro, 1981). These two series form the Sredinniy Range terrane, which appears to represent a Late Cretaceous to Early Paleogene arc sequence formed on oceanic crust.

Radiometric dates from the Malkinsk terrane are primarily K-Ar determinations that range from isolated figures of 314 (compare with Ganal Range age below) and 250 Ma, to a continuous spread from 192 Ma to 30 Ma with a maximum number around 50 Ma (over 50 determinations; Erlich, 1979; Shul'diner et al, 1979). In addition, a 470 Ma date on a pegmatite intruding metamorphosed deposits is reported by Parfenov (1970); the dating method is unknown. The radiometric dates are primarily on granitic intrusions in the Malkinsk terrane, but a number are from the "basement" with a maximum number at 55 Ma and 105 Ma (Fig. 4). The 105 Ma maximum lies in the time between the deposition of the Kvakhon suite and the age of the overlying clastics with pebbles, suggesting that this metamorphic event may be due to the obduction of the Kvakhon terrane. The younger 50 Ma maximum is synchronous with an absence of sedimentation in the Malkinsk, Sredinniy, and Omgon terranes and is here related to the obduction of the Sredinniy Range onto the Malkinsk (Fig. 3) or to the suturing of the Irunei, Malkinsk, and Eastern Ranges terranes in the early Paleogene.

The evidence for a Precambrian age for parts of the Malkinsk terrane is twofold. First, Kuz'min and Chukhonin (1980) report 1.3 Ga ages from the gneiss of the Kolpakova series. The dating was performed using $^{207}Pb/^{206}Pb$ ratios determined from a few milligrams of zircon using an unvalidated method of determining isotopic composition by thermionic emission (Chukhonin, 1978). Second, Mokrousov and Marchenko (1964) report Late Precambrian sporomorphs from the metamorphosed Kheivan suite. These sporomorphs were all identified by Timofeev, whose correlation ability using these supposed remains has been questioned by Schopf (1969). Thus, the claims for a Precambrian age for the Malkinsk terrane remain equivocal.

Of greater interest, however, are Devonian and Carboniferous spores that have been reported from the Kheivan suite (Sivertseva and Smirnova, 1974). A

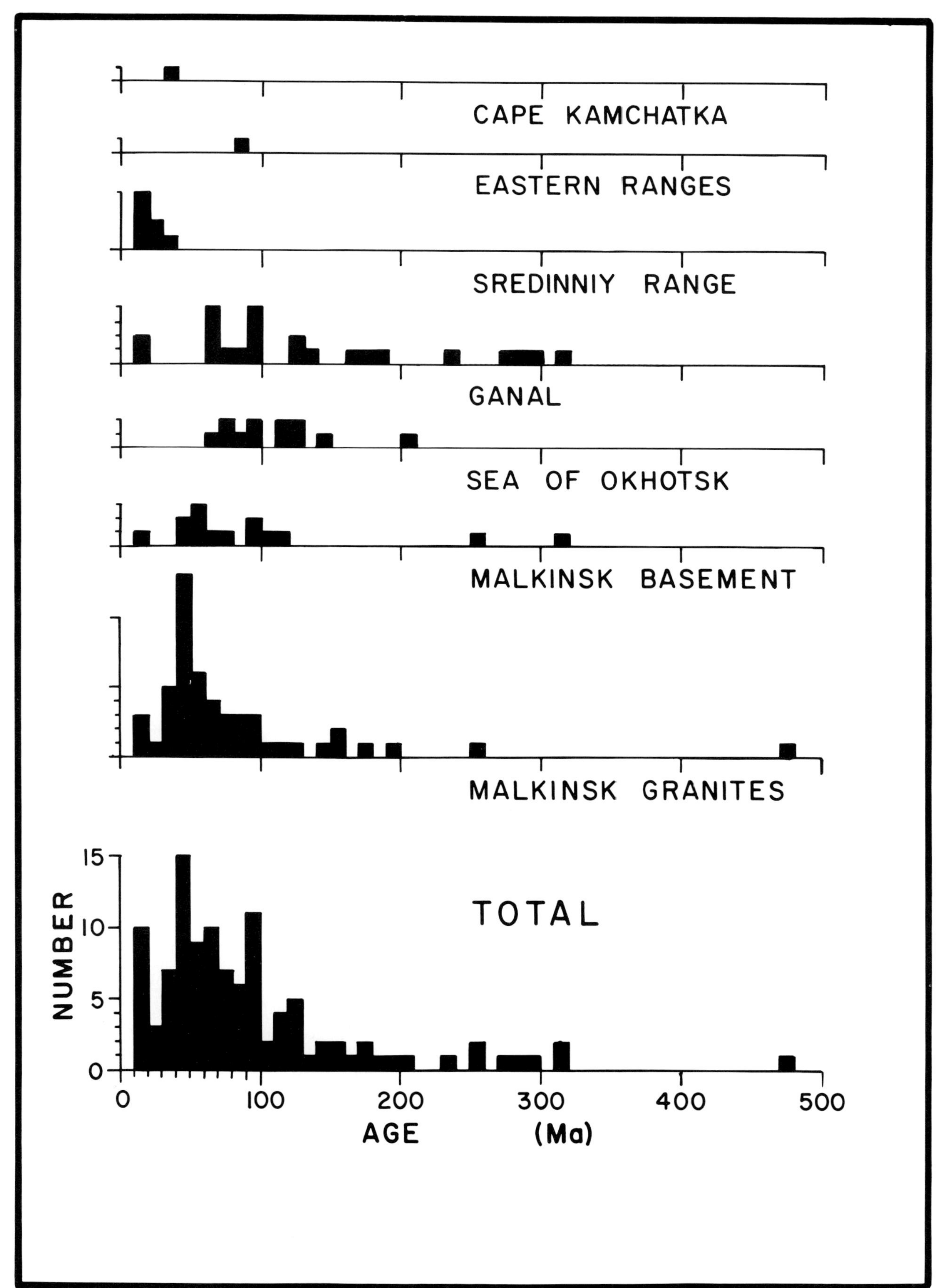

Figure 4—Histogram of radiometric (K–Ar) age determinations in and near Kamchatka.

relatively good assemblage has been reported from black shales. In the same paper, however, the authors report obtaining a similar assemblage from a biotite schist. This casts some doubt on the reliability of the report since such a high degree of metamorphism would tend to destroy any evidence of spores (Tschudy, 1969). The reliability of this report is also questioned because of the contradictory and confusing reports of Devonian, Carboniferous, Triassic, and Jurassic spores and Triassic flora from the Kikhchik series, all of which are reported to be poorly preserved (Aksenovich et al, 1964). If these Devonian spores are really present, either in situ or weathered from now-missing blocks, it is possible that there are some fragments of Devonian material in the Malkinsk terrane or its margins. Such fragments of older material are found in the Mesozoic subduction complexes of northwestern Kamchatka, the Koryaks, and southwestern Japan (Fujita and Newberry, 1983; Taira et al, 1983). Taira et al (1983) have suggested that the blocks of Paleozoic material in the Kurosegawa zone of southwestern Japan, which lies outboard of a Jurassic accretionary complex, may be due to large-scale transcurrent motion or oblique subduction that has juxtaposed older rocks seaward of the younger accretionary prism. It is therefore possible that the older blocks in the northeast Siberian accretionary zones may have had a similar origin since the relative motions of possible plates in the Pacific basin (Pacific, Izanagi, and Farallon) all were north-northeast with respect to Eurasia in the Mesozoic (Engebretson, 1982).

Thus, we suggest that the Malkinsk terrane is of Mesozoic age and developed along a convergent margin. The blueschist and tuffs in the Kvakhon suite suggest that the eastern margin of the Kvakhon terrane represented an island arc in Early Cretaceous time. The metamorphism in the Malkinsk terrane is of two types and ages (Shul'diner et al, 1980). The first, which affected the Kolpakova series, is a high-temperature, high-pressure type and is 100 Ma or older in age. This can be related to the obduction of the Kvakhon suite and/or metamorphism in a deeper part of the subduction zone. If the Paleozoic blocks are present, the basement of the Malkinsk terrane may record earlier episodes of terrane amalgamation that were brought to their present location in the Mesozoic. The younger metamorphism of approximately 50 Ma age is primarily related to the intrusion of granites. The metamorphic grade is concentric about the intrusions and one can trace unmetamorphosed sediments into metamorphosed regions (Lebedev et al, 1967).

One additional observation is that except for the grade of metamorphism, the original composition of the Andrianovka and Kvakhon suites, and the Kikhchik series and the Kheivan suite, are relatively similar. The Shikhtinsk suite is described as being composed of a basal arkose and gritstone beds with pebbles of the underlying rocks metamorphosed to a higher grade than the overlying suites. It is possible that this represents a tectonic contact zone, as drawn by Zhegalova (1981), and that the Andrianovka and Kheivan suites represent oceanic crust that has been obducted onto the Kolpakova series. In this event, the Kolpakova series would be the base onto which subsequent thrusts were obducted.

Ganal Terrane

Two formations are found in the Ganal Range, the Ganal (Ga) and Stenovaya (Sv) suites, which are separated by a southwestward-dipping thrust fault (German et al, 1978). The lower Ganal suite, which is thrust over the Stenovaya, is composed of amphibolite and mafic schists metamorphosed to the epidote-amphibolite facies level (Fig. 5). The suite contains no faunal remains; however, K–Ar age dates ranging from 314 to 12 Ma have been reported with maxima at about 90 and 65 Ma (Fig. 4). The older dates (about 200 Ma) are on gabbroic rocks and are said by the authors to be potentially unreliable. Petrologic analysis of the Ganal suite indicates that the parent rocks were alkaline to tholeiitic basalts of oceanic origin (Rozen and Markov, 1973). Subordinate blueschist assemblages are found along the thrust contact with the Stenovaya suite (Dobretsov and Kuroda, 1970; German et al, 1978), suggesting that the contact is a major terrane boundary. To the south, the Ganal suite is in normal fault contact with rocks equivalent to the Irunei series in which *Inoceramus schmidti* (Late Cretaceous) remains have been identified (German et al, 1978). The metamorphism of the Ganal is not continued into the Irunei rocks (Shul'diner et al, 1979) and implies that the contact postdates metamorphism.

The supposedly overlying Stenovaya suite consists of felsic to mafic schists and metasediments (German et al, 1978). The parent materials are reported to be quartz keratophyre and spilite and suggest an island arc origin. The metamorphic grade in this series decreases northward from the epidote-amphibolite to greenschist facies as one moves away from the contact with the Ganal suite. The faunally dated basal sediments of the Valaginskiy Range, north of the Ganal Range, are composed of spilite, diabase, tuff, and clastics (Serova et al, 1970; Gnibidenko et al, 1974), which are petrologically similar to the Stenovaya. Thus, it is suggested that the Stenovaya suite is equivalent to the Valaginskiy Range sequence and that the thrust between the Ganal and Stenovaya suites represents the suture between the Ganal and Eastern Ranges terranes.

One Rb–Sr age date of 487 Ma is known from the "least altered, plagiogranite porphyry" stock intruding the Stenovaya suite (German et al, 1978). If our correlation in the previous paragraph is correct, this stock would be intruding Late Cretaceous formations and the date is highly anomalous. The stock has also been K–Ar dated at 60 Ma and lies within a band of Miocene granites that also intrude the Stenovaya. In light of these facts and correlations, we question the 487 Ma date and suggest that perhaps some excess radiogenic strontium has migrated from surrounding or deeper rocks as noted in Hamilton (1965). We thus interpret the Ganal terrane as a remnant of ocean floor that was obducted onto the Eastern Ranges, perhaps about 60 Ma ago, and then accreted onto the Sredinniy Range terrane (Irunei series) about 50 Ma ago.

Central Kamchatka Basin Terrane

The Khavyven uplift is located in the northernmost part of the Central Kamchatka basin and is composed of a core of greenschists and microquartzites (metamorphosed chert?) and contains mafic and ultramafic rocks (Fig. 2). The formations have been evenly metamorphosed and have

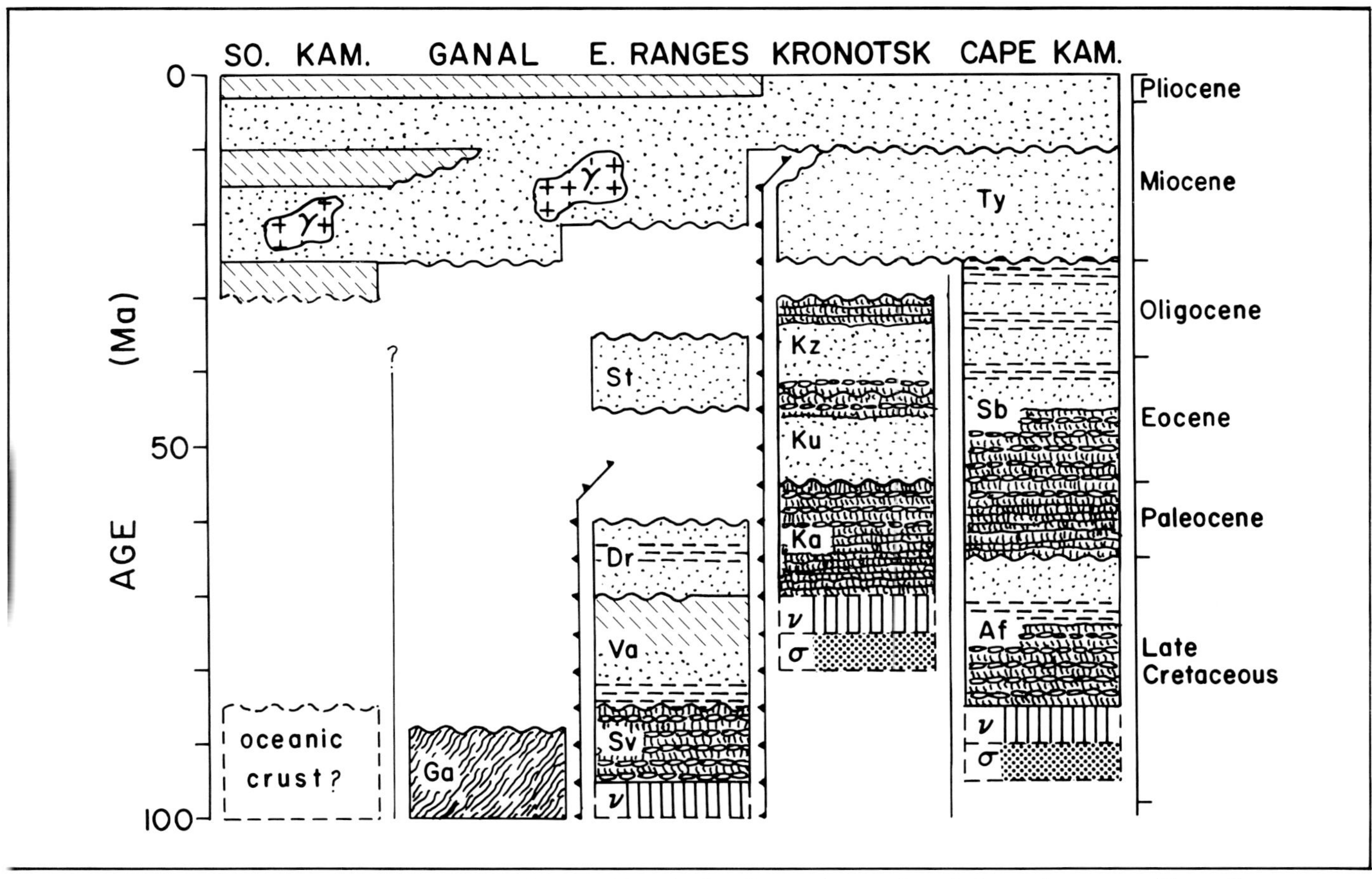

Figure 5—Stratigraphic columns for the terranes of eastern Kamchatka. Conventions and symbols are the same as Figure 2.

petrologic similarities to the lower parts of the Stenovaya suite. The uplift appears, therefore, to be formed of oceanic rocks. K-Ar dates of 122 to 92 Ma are known (Gnibidenko and Marakhanov, 1973) and the basement rocks are probably of Cretaceous or Jurassic age.

Cape Kamchatka Terrane

In the southeastern part of Cape Kamchatka, there is the so-called Olenegorskiy pluton composed of gabbroic and ultramafic rocks (Vysotskiy and Gracheva, 1981). The rocks have compositions of ocean floor basalts and have been K-Ar dated as being 900 Ma. However, this outcrop is overlain by radiolaria dated mafic extrusives, tuffs, and siliceous rocks, with clastics at the top, of the Late Cretaceous Afrika (Af) series (Markov et al, 1972; Khotin, 1972). Overlying this series is the faunally dated Paleogene Stolbovsk (Sb) series (Fig. 5).

The Olenegorsk rocks are clearly the remnant of obducted oceanic crust, and the overlying rocks suggest a Cretaceous age. Mafic rocks are known to have excess argon, and other Soviet age dates have proven to be anomalously old relative to known geology (Dalrymple and Lanphere, 1969; Salop, 1977). Thus, we believe no rocks older than Cretaceous outcrop in the Cape Kamchatka terrane.

TERRANES OF EASTERN KAMCHATKA

Although the details of the process remain unclear, there is no question that the development of the terranes in eastern Kamchatka occurred in Late Cretaceous to Tertiary time (Fig. 5). The Eastern Ranges terrane consists of the Valaginskiy, Tumrok, and Kumroch Ranges. A similar stratigraphy exists along the length of this belt and indicates it is one terrane. The basement consists of spilites, keratophyres, tuffs, and cherts, with associated gabbros and ultramafics (Markov et al, 1969; Gnibidenko et al, 1974). The upper section of the Valaginsk (Va) series is composed of clastics that gradually change up-section to calc-alkaline volcanics (Markov et al, 1969; Florenskiy and Florenskiy, 1972; Rotman et al, 1973; Shapiro, 1981). Fragments of *Inoceramus* date the older parts of the series as Late Cretaceous (Serova et al, 1970). Dunite, harzburgite, pyroxenite, and serpentinized gabbro are found throughout the ranges (Rotman et al, 1973) and are similar in composition to ophiolites elsewhere. A K-Ar age of 80 Ma has also been obtained from the ultramafics (Rotman et al, 1973) and supports the interpretation that the Eastern Ranges represent a Late Cretaceous island arc built on oceanic crust. Savostin et al (1983) interpret the Eastern Ranges as part of the Irunei arc of our Sredinniy Range terrane. Because the entire Mesozoic is covered by extrusive volcanics of the Miocene Sredinniy Ranges arc, it is possible that the Eastern Ranges are a displaced section of that arc that was separated by backarc spreading. The relatively old ages from the Khavyven uplift and the lack of younger island arc volcanics from the Eastern Ranges, however, favor a two-arc interpretation; the issue is not resolved.

In the Paleocene, the Eastern Ranges arc accreted onto the Ganal terrane and western Kamchatka. Volcanism ceased and sandstones and shales alone were deposited (Drozdovsk suite, Dr) in eastern Kamchatka (Serova et al, 1970). This event can be explained as a shift in the relative motion vectors between Kamchatka and the Pacific basin from perpendicular to the peninsula to parallel (see Fig. 7). Fujita and Newberry (1983) have suggested that the Olyutorsk terrane accreted onto the Koryak highlands about this time, emplacing the Vyvenka and Vatyna ophiolites. These complexes are contiguous with ophiolites identified on Karaginskiy Island and Ozernoy Peninsula and thus the latter were probably emplaced at about the same time. This event also correlates well with the development of the Komandorskie Islands and uplift on Cape Kamchatka, suggesting that the Aleutian arc developed at this time.

Tertiary deposits on Cape Kamchatka consist of basalt, andesite tuff, and chert (Stolbovsk series) overlain by a thick clastic sequence containing tuffaceous horizons (Tyushevka series, Ty). A similar stratigraphy is found on the Komandorskie Islands as well (Shmidt, 1978) and suggests that the two regions are part of the same terrane. Pre-Miocene structures and deposits on Cape Kamchatka and the Eastern Ranges, however, are distinctly different and are separated by a major thrust (Markov et al, 1969). Thus, we infer that Cape Kamchatka did not develop in situ relative to the rest of Kamchatka but was thrust into position in Miocene time. This is also supported by the presence of post-Paleocene andesitic volcanism in northernmost Kamchatka, at the latitude of Karaginskiy Island (Korzhenevskiy, 1962), and a gravity low offshore of Karaginskiy Island that trends into the neck of Cape Kamchatka (Shapiro, 1976), which may be the remnant of an Oligocene to Miocene subduction zone.

The development of the Aleutian subduction zone would help to explain the change in the convergence direction between Kamchatka and the Pacific basin. The new subduction zone, which strikes east-west, would pull the subducting plate in a more northerly direction, resulting in transcurrent motion along the eastern margin of Kamchatka.

Reinitiation of convergence between Kamchatka and the Pacific in late Oligocene time led to the formation of the Sredinniy Range volcanic arc that overlaps the Kvakhon, Malkinsk, Eastern Ranges, Ganal, Southern Kamchatka, and Sredinniy Range terranes. Both granitic intrusions and extrusive volcanics are found throughout this belt, implying that all of these terranes had accreted together by this time (Figs. 2, 5). Subduction under this arc closed the strait between Cape Kamchatka and the Eastern Ranges.

The nature of the Southern Kamchatka terrane is not clear; however, the basement is inferred to be oceanic crust on the basis of xenoliths found in volcanic ejecta from this region and from high gravity and magnetic anomalies that indicate high-density basement (Shul'diner et al, 1979; Marakhanov and Potap'ev, 1981).

Paleogene clastics of the Stanislavsk (St) suite overlay the volcanics of the Eastern Ranges. East of these exposures is an accretionary wedge of late Paleogene to Miocene age that is now cut by many thrust faults (Fig. 3; Shapiro and Selivertsov, 1975). Further east is the Grechishkin thrust that separates the wedge from the Kronotskiy terrane. The Kronotskiy terrane (Fig. 5) is composed of submarine volcanics (Cape Kamenistiy suite, Ka), sediments, and cherts (Kubovsk suite, Ku), capped by alkali basalt flows (Kozlovsk suite, Kz) of Late Cretaceous to Oligocene age (Sadreev and Dolmatov, 1965; Suprunenko, 1976). Serpentinized peridotite, pyroxenite, and gabbro are present at the base of the exposure. Mafic flows are also found on Shipunskiy Peninsula, although their relationship to the Kronotskiy and Eastern Ranges terranes is not clear. Petrologic analysis of the basalts exposed on Kronotskiy Peninsula indicates that they are similar to Hawaiian tholeiites (Rotman and Markovskiy, 1968), and thus we suggest that the terrane is an accreted seamount province. The duration of volcanic activity in the Kronotskiy terrane is relatively long and therefore may not be related to the Emperor Seamount chain which lies nearby. Alternatively, the Kronotskiy terrane may have been formed on a hot spot located on the Kula-Pacific ridge axis that ceased spreading in Tertiary time (Engebretson, 1982). Accretion of the Kronotskiy terrane to the Eastern Ranges terrane occurred in middle Miocene time and led to the development of the subduction zone in its present location.

Thus, eastern Kamchatka records convergent plate motions in the Late Cretaceous, transcurrent relative motion in the Paleogene, and convergence again in the Neogene. Comparison of these motions with the relative motions between Eurasia and the Kula and Pacific plates in these intervals indicate that the geology observed in Kamchatka can be explained by relative motion between the Eurasian and Pacific plates, alone. The Pacific-Eurasia and Kula-Eurasia relative motions both yield convergence prior to and after the Paleogene (Engebretson, 1982). However, relative motions between the Kula and Eurasian plates are convergent in the Paleogene but are conservative between Eurasia and the Pacific. The latter is what we infer from the geology; therefore, we suggest that the Kula plate did not extend this far west in the Paleogene and that the Pacific plate did (see Figs. 7, 8).

TERRANES OF WESTERN KAMCHATKA

The terranes that contributed to the development of western Kamchatka are the Central Kamchatka Basin, Sredinniy Range, Malkinsk, Omgon, and Kvakhon (Figs. 1, 2). The Central Kamchatka Basin is probably composed of trapped or backarc formed oceanic crust. High gravity and magnetic anomalies suggest that the basement is at not too great a depth (Rivosh, 1964); this is also suggested by the exposures of the Khavyven uplift discussed previously. Oligocene to Quaternary volcanics and clastics fill the basin (Rivosh, 1964; Krasniy, 1964), but the exact nature of this basin is unknown because of the extensive Neogene and younger cover.

The Kvakhon terrane has already been described in its easternmost exposures. It extends from the edge of the Malkinsk terrane, under the Bolsheretsk depression, and into the Sea of Okhotsk. It is bounded on the north by a series of northwest-trending faults known as the "diagonal geosuture" (Rotman, 1964). The basement mafic rocks apparently underlie most of the region as evidenced by amphibolized gabbros found in a drillhole at a depth of 534 m (1,752 ft) near Ust' Bolsheretsk (Fig. 1; Gnibidenko et al, 1974). This gabbro is correlated to a refractor (5.2 km/sec [17,000 ft/sec]) that dips gently to the west and somewhat more steeply to the north (Marakhanov and Potap'ev, 1981). Near the diagonal suture, the depth to this refractor reaches 4 km (2.5 mi) and is believed to be overlain by Cenozoic sediments (Smirnov, 1971). The Paleogene is absent throughout most of the southern part of the terrane and is only locally present near the diagonal suture (Fig. 5). The Paleogene sediments are primarily coarse clastics (Napana, Na, and Snatol, Sn, suites), when present, which generally fine upwards and become more marine. The Miocene is represented by marine fine clastic deposits (Voyampolka, Vo, and Kovacha, Ko, series).

North of the diagonal suture, we have identified two terranes, the Omgon and the Sredinniy Range. The Omgon terrane is believed to be floored by oceanic crust as evidenced by outcrops of gabbros (Krasniy, 1964) and "chert-volcanic" deposits underlying the mid-Cretaceous turbidites of the Tal'niki (Ta) and Maynachsk (Ma) suites (Voronkov and Smirnov, 1971; Avdeiko, 1971). These turbidites have been faunally dated and are overlain by coarser clastics. The Cretaceous of the Omgon terrane is unconformably overlain by sediments identical to those described for the Kvakhon terrane except that the Paleogene deposits are extensive. The deposition began in the Eocene with coarse clastics, and the transition to finer clastics occurred earlier than south of the diagonal suture. The deposits are continuous across the suture by late Eocene time, indicating that the two blocks were contiguous and in their present relative positions.

The Omgon and Sredinniy Range terranes are separated by a zone of steeply folded rocks and thrust faults (Aprelkov and Zhegalov, 1972; Dimitriyeva, 1981). Blueschists have also been described from this boundary zone (Dobretsov and Kuroda, 1970; Sidorchuk and Khanchuk, 1981) which extends northward into the Kamchatka isthmus. The Sredinniy Range terrane is floored in the Tigil region by rocks that have been associated with the Irunei series and may be equivalent to them. These deposits underlie the entire Sredinniy Range terrane and are overlain by the extrusive volcanics of the Late Cretaceous to Paleogene Irunei volcanic arc (Shapiro, 1981).

The geology described above can be interpreted as follows. First, the basalts of the Kvakhon and Omgon terranes are both ocean floor and island arc types (Rotman and Markovskiy, 1968; Sidorchuk and Khanchuk, 1981). They are both approximately the same age, have blueschist assemblages on their eastern margins, and are overlain by similar sediments in the Tertiary. We suggest that they are of the same genetic origin, but since the Kvakhon terrane obducted onto the Malkinsk terrane, it was raised above base level and generated more coarse clastics than the Omgon terrane and had a longer hiatus in the Late Cretaceous and Paleogene. The diagonal suture may be a result of the southern block obducting while the northern one did not, or it may be due to the obduction of the Irunei series of the Sredinniy Range. Since Eocene deposits overlap both the Omgon and Sredinniy Range terranes, the suturing of the terranes must have occurred before this time; this is consistent with data presented earlier.

The Irunei island arc was active in the Late Cretaceous, probably for a period of 20 to 25 Ma. Calc-alkaline volcanics are found both east of the Malkinsk terrane and in the Kamchatka isthmus; the deposits are probably continuous but have been overlain by the Miocene volcanic deposits of the Sredinniy Range.

SEA OF OKHOTSK

The evolution of the shallow oceanic part of the Sea of Okhotsk has long been an enigma. For the most part, Soviet workers have assumed the presence of a large Precambrian block within the northern part (for example, Khain and Seslavinsky, 1973) and have postulated that clastic debris was shed from the Precambrian block to form the deposits of western Kamchatka. The available geologic and geophysical data, however, suggest that the Sea of Okhotsk is not formed of continental crust and that the region was separate from Eurasia until earliest Paleogene time.

In the northeastern corner of the Sea of Okhotsk, where the structures of Kamchatka and the Okhotsk–Chukotsk volcanic belt meet, lie Taygonos Peninsula and the Penzhina Range (Fig. 6). Taygonos Peninsula has been divided into several northeast–southwest-striking zones, each of which are fault bounded (inset to Fig. 6; Leonenko, 1976). The stratigraphy of the zones is complex, but they suggest convergent plate motions through the Mesozoic. Subsequent to the accretion of the Omolon terrane onto Eurasia in the Jurassic (Fujita and Newberry, 1982), subduction began under the Viskichun (Vi) terrane (Fig. 6), which is an island arc sequence 4 to 5 km (13,000–16,000 ft) thick (Nekrasov, 1971). Starting around 150 Ma, granitic intrusions developed and spread throughout all of Taygonos Peninsula by about 115 to 91 Ma (Albian) ago (K–Ar determinations; Firsov, 1965). The Viskichun terrane accreted to Eurasia in Late Jurassic (?) time (Fujita, 1978) and subduction continued under the Neyneg (Nn) terrane (accretionary wedge?). Along the northwest shore of Penzhina Gulf is the Pribrezhnaya (Pr) terrane consisting of ultramafics overlain by spilites and cherts. A K–Ar date of 183 Ma has been determined on the ultramafics (Nekrasov, 1971), and the overlying deep-sea sediments have been dated by radiolaria as Late Jurassic to Cretaceous (Alekseyev, 1981). Thus, we suggest that the oceanic crust is of Middle Jurassic age. The entire peninsula is overlain by a thin veneer of Aptian to Coniacian continental coal-bearing facies (Pergament, 1958).

On the eastern side of Penzhina Bay (Fig. 6), in northwesternmost Kamchatka, ultramafics and deep-marine rocks are found in the Penzhina Range (Kuyul Massif; see inset, Fig. 6) and as thrust sheets at the top of Mt. Dlinnaya (Kolyasnikov and Krasniy, 1981; Alekseyev,

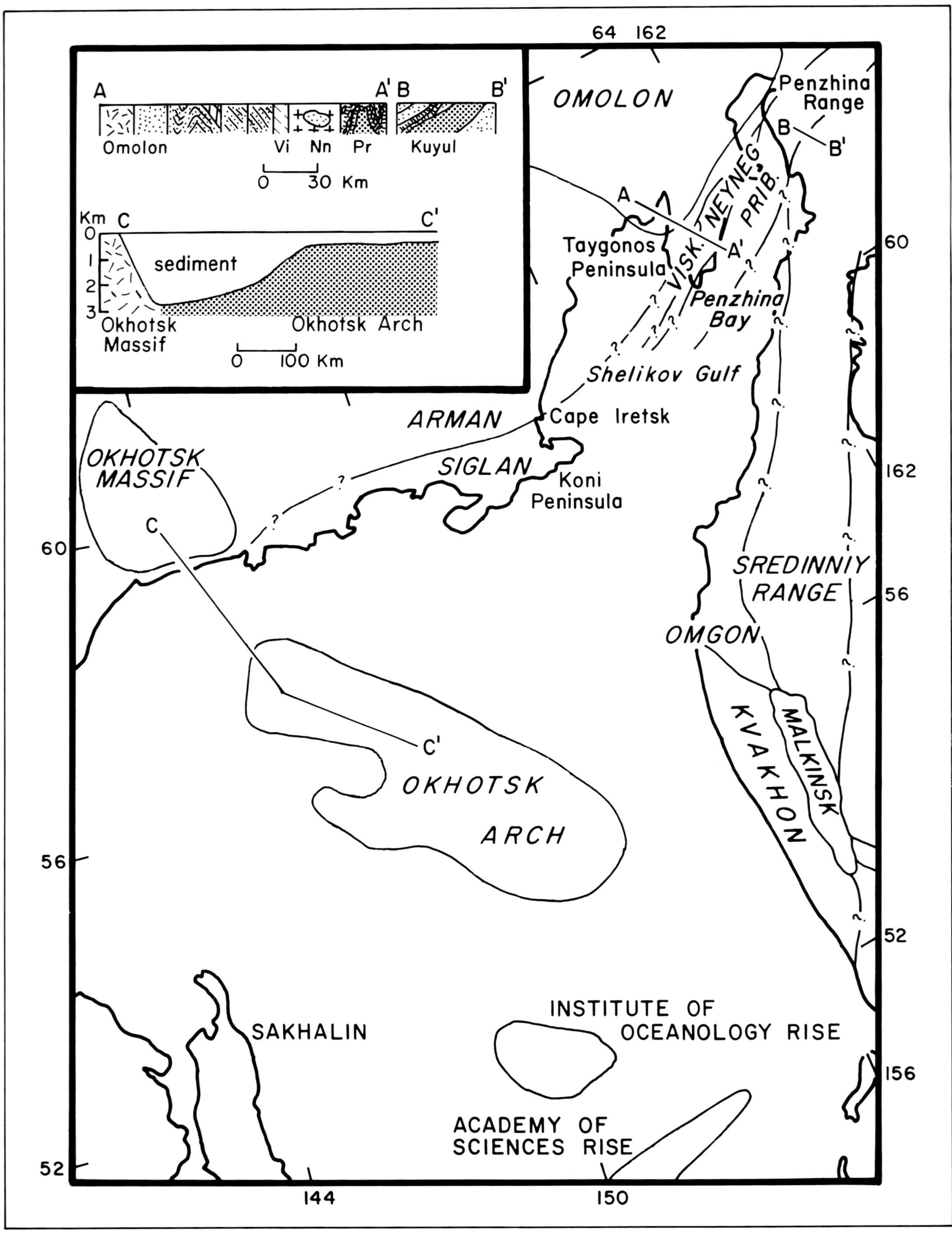
A
A'
B
B'
Omolon
Vi
Nn
Pr
Kuyul
0 30 Km
Km
C
C'
0
1
2
3
sediment
Okhotsk Massif
Okhotsk Arch
0 100 Km
64 162
OMOLON
Penzhina Range
NEYNEG
PRIB.
VISK.
Taygonos Peninsula
Penzhina Bay
Shelikov Gulf
ARMAN
Cape Iretsk
SIGLAN
Koni Peninsula
OKHOTSK MASSIF
C
60
162
56
SREDINNIY RANGE
OMGON
KVAKHON
MALKINSK
C'
OKHOTSK
ARCH
56
52
INSTITUTE OF OCEANOLOGY RISE
SAKHALIN
156
ACADEMY OF SCIENCES RISE
52
144
150

981). Fragments of the Penzhina Range ophiolites are found in the Tylakryl group of Hauterivian to Barremian age (125 Ma) and K–Ar radiometric age dates of 132 to 112 Ma are known from the Kuyul massif (Alekseyev, 1981). Thus, we suggest that both the Penzhina and Pribrezhnaya ophiolite complexes were emplaced about 125 Ma ago. Isolated blocks of faunally dated Paleozoic rocks and discordant K–Ar ages of 400 to 150 Ma are known from the Penzhina Range (Firsov and Dobretsov, 1969; Beliy and Gel'man, 1980; Alekseyev, 1981). This, again, is similar to discoveries from the Kurosegawa zone of Japan (Nozawa, 1977) and may represent the results of transcurrent motion or oblique subduction (Taira et al, 1983).

Along the north-central coast of the Sea of Okhotsk, two different terranes are found near Magadan (Fig. 6). The northern Arman terrane is composed of miogeosynclinal sandstones and shales with volcanics present only in the uppermost sections. The southern Siglan terrane is composed primarily of Late Triassic to Jurassic island arc deposits (Yudin and Izmailov, 1966; Zaborovskaya, 1978). Since the lithologies and ages are similar, the Siglan and Viskichun terranes probably represent parts of the same arc (the Uda–Murgal terrane of Howell et al, 1983). In addition, geophysical surveys between the Koni and Taygonos Peninsulas suggest the continuation of the Pribrezhnaya terrane offshore as well (Belyayev et al, 1966).

The suture between the Siglan and Arman terranes is the Chelomdzha–Yamsk fault along which some possible remnants of oceanic crust are preserved. At Cape Iretsk, a tabular body of peridotite and gabbro is known. In addition, isolated outcrops of ultramafics are known along the strike of the fault between the Okhotsk massif and Taygonos Peninsula (Umitbayev, 1977). The fault is sutured by granites radiometrically (K–Ar) dated between 130 and 100 Ma (Yudin and Izmailov, 1966).

Overlying the Cretaceous subduction zone complexes are the continental arc deposits of the Okhotsk–Chukotsk volcanic belt which was active from Aptian through Danian time (for example, Osipov, 1976). This represented an Andean type margin that subducted crust of the Sea of Okhotsk.

Recent seismic reflection surveys have provided considerable information on the aquatic part of the Sea of Okhotsk (Gnibidenko and Khvedchuk, 1982). Island arc type rocks have been dredged off of the Academy of Sciences and the Institute of Oceanology rises and have been dated as being Late Cretaceous and older (Burk and Gnibidenko, 1977). Although some of these dredge samples may be ice-rafted debris, the Academy of Sciences rise, at least, can be structurally connected to the Malkinsk or Kvakhon terranes from seismic refraction data (Marakhanov and Potap'ev, 1981) and is probably a genuine subduction complex.

Figure 6—Index and terrane map of the Sea of Okhotsk and its margins. Terranes are identified by slanted letters. VISK denotes Viskichun, PRIB denotes Pribrezhnaya. Inset shows cross sections across Taygonos Peninsula (A–A′), Penzhina Range (B–B′), and the Okhotsk massif–Okhotsk Arch (C–C′). Lithologies in sections A–A′ and B–B′ are as in Figure 2. Terrane abbreviations for section A–A′ are given in the text.

The north-central part of the Sea of Okhotsk is underlain, in part, by the Okhotsk Arch (Gnibidenko and Khvedchuck, 1982), a large, flat-topped section of elevated acoustic basement with an upper-layer refraction velocity of 5.2 km/sec (17,000 ft/sec) (Kosminskaya et al, 1963). Although Gnibidenko and Khvedchuk (1982) suggest it is connected to the Precambrian Okhotsk massif (K–Ar dates of 1.23 to 1.88 Ga and Rb–Sr date of 3.7 ± 0.5 Ga; Salop, 1977), it is clear from the seismic reflection data that the Okhotsk Arch terminates prior to the coastline and that there is a deep sedimentary basin between them (section C-C′ in Fig. 6; Figure 5 in Gnibidenko and Khvedchuk, 1982). In addition, the Okhotsk–Chukotsk volcanic belt crosses the Okhotsk massif with no change in trend, which would be unlikely if the Okhotsk massif continued some 400 km (250 mi) out into the Sea of Okhotsk.

From its size (500 × 200 km [300 × 125 mi]) and seismic velocity, we suggest that it could be an oceanic plateau similar to Hess or Shatskiy rise (compare with Hussong et al, 1979; Gettrust et al, 1980). The arrival of the Okhotsk Arch into the Okhotsk–Chukotsk subduction zone resulted in the termination of subduction and the accretion of the Sea of Okhotsk to Eurasia. Subsequently, sediment from the regions surrounding the Sea filled in the basinal parts.

DISCUSSION

Kamchatka appears to be divisible into a number of tectonostratigraphic terranes that display evidence of an oceanic origin, primarily island arcs. A model for the tectonic evolution of the region is presented below that has similarities to the present Philippine Sea and the Marianas in its earlier phases.

In the Late Jurassic, the Omolon massif accreted onto Eurasia (Fujita and Newberry, 1982). Subduction was initiated along the southern margin of the newly expanded Eurasian continent along the Viskichun–Siglan arc system. About the same time, the Kvakhon arc developed far offshore, effectively trapping the Sea of Okhotsk and the Okhotsk Arch oceanic plateau. A similar origin has been suggested for the Bering Sea (Cooper et al, 1976). The origin of the basal parts of the Malkinsk terrane is not known, but it too may have developed as an island arc; its relation to the Kvakhon terrane is not known at this time (Fig. 7a). The situation existing in the Early Cretaceous, therefore, would be similar to the present Philippine Sea with Eurasia representing China, the Viskichun arc analogous to the Ryukyus, the Kvakhon arc analogous to the Marianas, and the Okhotsk Arch being a larger version of the Daito rise. Parts of the margins of the trapped ocean basin were probably transform, like the southern end of the Marianas. Subduction was directed under the Kvakhon arc and blueschist metamorphism occurred.

About 120 to 110 Ma ago, the Siglan–Viskichun arc accreted to the Eurasian mainland. Since transcurrent motions may have dominated between the Pacific basin and parts of the Kvakhon arc, the Malkinsk terrane was juxtaposed against the Kvakhon terrane (Fig. 7b), and at about 110 to 100 Ma, the Kvakhon terrane was obducted, possibly in several thrust sheets, onto the Malkinsk terrane

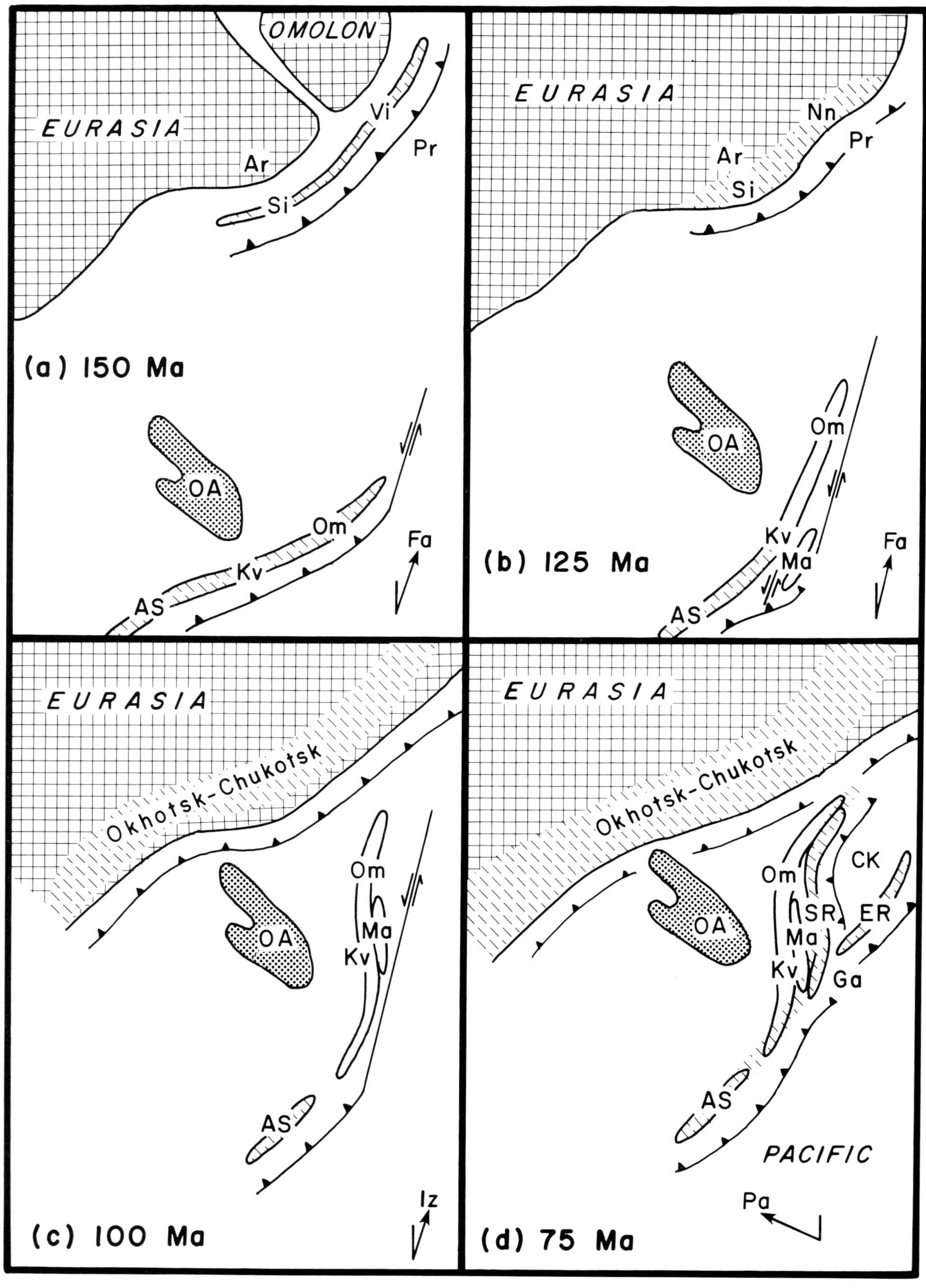
OMOLON
EURASIA
Vi
Pr
Ar
Si
(a) 150 Ma
OA
Om
Kv
AS
Fa
EURASIA
Nn
Pr
Ar
Si
OA
Om
Kv
Ma
(b) 125 Ma
AS
Fa
EURASIA
Okhotsk-Chukotsk
Om
Ma
OA
Kv
AS
Iz
(c) 100 Ma
EURASIA
Okhotsk-Chukotsk
CK
Om
SR
ER
OA
Ma
Kv
Ga
AS
PACIFIC
Pa
(d) 75 Ma

(Fig. 7c). Subduction then started in the Okhotsk–Chukotsk belt, consuming the proto-Sea of Okhotsk. About this time the northern end of the Omgon terrane may have collided with Eurasia, emplacing the Penzhina ophiolite.

Subduction started about Santonian time outboard of the Malkinsk terrane and formed the Irunei volcanic arc on oceanic crust. Simultaneously, subduction also started under the Eastern Ranges arc; the Central Kamchatka Basin terrane was formed either as a result of backarc spreading or by being trapped. This situation prevailed until about 60 Ma when the Okhotsk Arch collided with Eurasia, terminating the subduction of the Sea of Okhotsk (Fig. 7d). The Eastern Ranges and Ganal terranes collided with each other and, shortly thereafter, with the now greatly expanded Eurasian plate (including western Kamchatka). Trapped in between, the Sredinniy Range terrane was thrust onto the Malkinsk and Omgon terranes.

Volcanism and convergence ceased in the Kamchatka region at about 55 Ma and transcurrent motion began off the eastern coast (Fig. 8a). The Olyutorsk terrane accreted about this time and the Aleutian arc developed. About 30 Ma ago, subduction resumed in the Kamchatka region, forming the Sredinniy Range volcanic arc and closing the strait between Cape Kamchatka and the Eastern Ranges. About 10 Ma ago, the Kronotskiy terrane was accreted and subduction shifted to its present locus (Fig. 8b).

CONCLUSIONS

The distribution of terranes and tectonic evolution of Kamchatka can be explained in terms of the formation and accretion of three island arcs: the Kvakhon, Irunei, and Eastern Ranges. The basement of Kamchatka is probably no older than the Jurassic although exotic blocks of Paleozoic material may have been introduced through transcurrent or oblique motions. These older blocks may have been basement terranes dispersed from older continental margins.

The Sea of Okhotsk appears to be primarily oceanic except for the northern region which is underlain by an oceanic plateau. Parts of the southern shallow region may be built on an island arc. There are apparently similarities between the geology of the Malkinsk terrane and the Hidaka terrane of Hokkaido (Arsent'ev et al, 1978; and others) indicating they may have been contiguous. Further study is needed.

The geologic interpretation, presented here, of Kamchatka representing a convergent plate margin in the Mesozoic and Neogene and a conservative margin in the Paleogene fits well with the relative motion vectors computed by Engebretson (1982) between the Eurasian and Pacific plates and poorly with motions between the Kula and Eurasian plates. This provides some encouragement that the evolution of the Pacific Ocean basin and its margins can be interrelated.

More data are needed, especially in western, northern, and southern Kamchatka. We hope, however, that the broad outlines presented here are a basis from which more detailed or alternate models may be developed.

Figure 7—Sketch map showing the tectonic evolution of the Sea of Okhotsk and Kamchatka during the late Mesozoic. Ar = Arman; AS = Academy of Sciences Rise; CK = Central Kamchatka Basin; ER = Eastern Ranges; Ga = Ganal; Nn = Nenyneg; OA = Okhotsk Arch; Pr = Pribrezhnaya; Si = Siglan; and Vi = Viskichun terrane. At lower left of each section the relative motion between the Farallon (Fa), Izanagi (Iz), or Pacific (Pa) plates and Eurasia is shown. (Engebretson, 1982. Used with the permission of the author.) Grid pattern shows areas of continental crust, diagonal dashed lines denote active volcanic arcs, and stippling shows ocean plateaus.

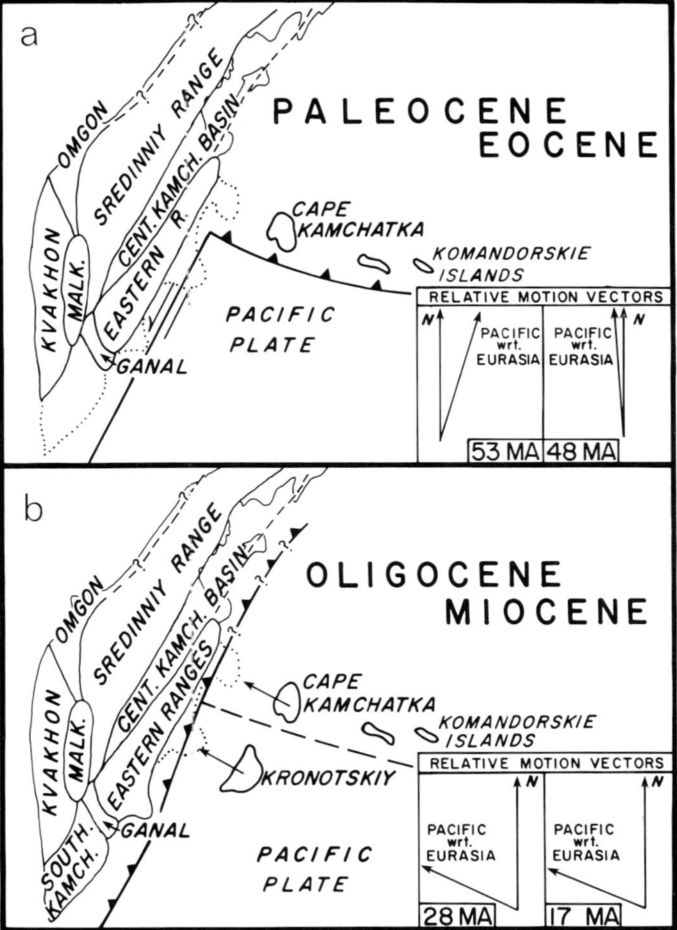

Figure 8—Sketch maps showing the tectonic evolution of Kamchatka in the (a) Early Tertiary and (b) mid-Tertiary. At lower right, relative motion vectors between Eurasia and the Pacific plates are shown. (Engebretson, 1982. Used with the permission of the author.)

ACKNOWLEDGMENTS

We thank Drs. D. G. Howell and M. Churkin, Jr., for encouragement, critical comments, and helpful suggestions on this paper. We are indebted to Dr. D. C. Engebretson for relative motion data, Dr. A. T. Cross for extensive discussions on palynology, and Dr. D. W. Scholl for comments on Soviet radiometric dates and unpublished information. Dr. D. Jones, M. J. Mellon, and Dr. R. G.

Gordon also provided helpful comments. This work was supported, in part, by NSF Grant EAR 80-25267. B. F. Watson was supported by the Department of Geological Sciences, Michigan State University, which is gratefully acknowledged.

REFERENCES

Aksenovich, A. V., et al, 1964, Precambrian and Paleozoic stratigraphy, *in* G. M. Vlasov, ed., Geology of the USSR, v. 31, Kamchatka and the Kurile and Komandorskie Islands: Moscow, Nedra Publishing House, p. 79–103 (in Russian).

Alekseyev, E. S., 1981, The Kuyul' serpentinite melange and the structure of the Talovsk-Mayna zone (Koryak highlands): Geotectonics, v. 15, p. 68–78.

Aprelkov, S. E., and Y. V. Zhegalov, 1972, Volcanic belts of Kamchatka: Geotectonics, v. 6, p. 123–126.

______, et al, 1964, Paleogene and Neogene stratigraphy of central Kamchatka, *in* G. M. Vlasov, ed., Geology of the USSR, v. 31, Kamchatka and the Kurile and Komandorskie Islands: Moscow, Nedra Publishing House, p. 154–177 (in Russian).

Arsent'ev, V. P., et al, 1978, Folded regions and recent platforms of east Europe and Asia: Novosibirsk, Nauka Publishing House, Siberian Branch, 319 pp. (in Russian).

Avdeiko, G. P., 1971, Evolution of geosynclines on Kamchatka: Pacific Geology, v. 3, p. 1–13.

Belyayev, I. V., et al, 1966, Tectonic zoning of Shelikov Gulf and adjacent areas (from geophysical data): Academy of Sciences of the USSR, Doklady, Earth Science Sections, v. 171, p. 109–112.

Beliy, V. F., and M. L. Gel'man, 1980, Meymechit in the Penzhina Range: Academy of Sciences of the USSR, Doklady, Earth Science Sections, v. 250, p. 128–131.

Burk, C. A., and G. S. Gnibidenko, 1977, The structure and age of acoustic basement in the Okhotsk Sea, *in* M. Talwani and W. C. Pitman, III, eds., Island arcs, deep sea trenches and back-arc basins: Washington, DC, American Geophysical Union, M. Ewing Series 1, p. 451–461.

Chukhonin, A. P., 1978, A mass-spectrometric study of the forms taken by lead in zircon: Geochemistry International, v. 15, p. 186–189.

Cooper, A. K., et al, 1976, Plate tectonic model for the evolution of the eastern Bering Sea Basin: Geological Society of America Bulletin, v. 87, p. 1119–1126.

Dalrymple, G. B., and M. A. Lanphere, 1969, Potassium-argon dating: San Francisco, W. H. Freeman, 258 pp.

Dimitriyeva, V. K., 1981, The tectonic nature of the so-called marginal West Kamchatka trough: International Geology Review, v. 23, p. 1266–1274.

Dobretsov, N. L., and I. Kuroda, 1970, Geologic laws characterizing glaucophane metamorphism in northwestern part of the folded frame of Pacific Ocean: International Geology Review, v. 12, p. 1389–1407.

Engebretson, D. C., 1982, Relative motions between oceanic and continental plates in the Pacific basin: PhD Dissertation, Stanford University, 211 p.

Erlich, E. N., 1979, Recent structures of Kamchatka and position of Quaternary volcanoes: Bulletin Volcanologique, v. 42, p. 13–43.

Firsov, L. V., 1965, Absolute age of granitoids of the Taygonos Peninsula: Academy of Sciences of the USSR, Doklady, Earth Science Sections, v. 162, p. 46–48.

______, and N. L. Dobretsov, 1969, Age of glaucophane metamorphism at the northwestern fringe of the Pacific Ocean: Academy of Sciences of the USSR, Doklady, Earth Science Sections, v. 185, p. 46–48.

Florenskiy, I. V., and P. V. Florenskiy, 1972, Role of paleovolcanic rocks in the structure of eastern Kamchatka: Academy of Sciences of the USSR, Doklady, Earth Science Sections, v. 205, p. 84–87.

Fujita, K., 1978, Pre-Cenozoic tectonic evolution of northeast Siberia: Journal of Geology, v. 86, p. 159–172.

______, and J. T. Newberry, 1982, Tectonic evolution of northeastern Siberia and adjacent regions: Tectonophysics, v. 89, p. 337–357.

______ and ______, 1983, Accretionary terranes and tectonic evolution of northeast Siberia, *in* M. Hashimoto and S. Uyeda, eds., Accretion tectonics in the circum-Pacific regions: Tokyo, Terra Scientific Publishing Company, p. 43–57.

______, et al, 1982, Bibliography of northeast Siberian geology and geophysics (2nd ed.): United States Geological Survey Open-file Report 82-616, 80 p.

German, L. L., et al, 1978, Metamorphic complexes of the Ganalsky Range, Kamchatka: Pacific Geology, v. 13, p. 49–64.

Gettrust, J. F., et al, 1980, Crustal structure of the Shatsky rise from seismic refraction measurements: Journal of Geophysical Research, v. 85, p. 5411–5415.

Gnibidenko, G. S., and I. I. Khvedchuk, 1982, The tectonics of the Okhotsk Sea: Marine Geology, v. 50, p. 155–198.

______, and V. I. Marakhanov, 1973, Structure of the east Kamchatka anticlinorium: Academy of Sciences of the USSR, Doklady, Earth Science Sections, v. 210, p. 74–77.

______, et al, 1974, Geology and deep structure of Kamchatka Peninsula: Pacific Geology, v. 7, p. 1–32.

Hamilton, E. I., 1965, Applied geochronology: London, Academic Press, 267 p.

Hamilton, W., 1970, The Uralides and the motion of the Russian and Siberian platforms: Geological Society of American Bulletin, v. 81, p. 2553–2576.

Howell, D. G., et al, 1983, Preliminary tectonostratigraphic terrane map of the circum-Pacific region, *in* D. G. Howell, et al, eds., Proceedings of the Circum-Pacific Terrane Conference: Stanford University Publications, Geological Sciences, v. 18, p. 227–242.

Hussong, D. M., et al, 1979, The crustal structure of the Ontong Java and Manihiki oceanic plateaus: Journal of Geophysical Research, v. 84, p. 6003–6010.

Jones, D. L., et al, 1983, Recognition, character, and analysis of tectonostratigraphic terranes in western North America, *in* M. Hashimoto and S. Uyeda, eds.,

Accretion tectonics in the circum-Pacific regions: Tokyo, Terra Scientific Publishing Company, p. 21–35.

Khain, V. E., and K. B. Seslavinsky, 1973, Some basic problems of structure and tectonic history of the north-western segment of the Pacific mobile belt, *in* P. J. Coleman, ed., The western Pacific: Island arcs, marginal seas, geochemistry: New York, Crane, Russak and Company, p. 389–406.

Khotin, M. Y., 1972, Siliceous rocks in the Late Cretaceous volcanic-tuffaceous-siliceous sequence at Kamchatskii Cape: Lithology and Mineral Resources, v. 7, p. 338–349.

Kolyasnikov, Y. A., and L. L. Krasniy, 1981, The tectonic position of the ultramafic massif on Mount Dlinnaya (northwest Kamchatka): Geotectonics v. 15, p. 79–82.

Korzhenevskiy, B. A., 1962, New data on geology of the northern Central Range, Kamchatka: Academy of Sciences of the USSR, Doklady, Earth Science Sections, v. 142, p. 74–75.

Kosminskaya, I. P., et al, 1963, Basic features of the crustal structure of the Sea of Okhotsk and the Kuril-Kamchatka zone of the Pacific Ocean from deep seismic sounding data: Academy of Sciences of the USSR, Izvestiya, Geophysics Series, p. 11–27.

Krasniy, L. I., 1964, Geologic map of the northwestern part of the Pacific mobile belt: Ministry of Geology of the USSR, scale 1:1,500,000.

______, 1966, Geological structure of the northwestern part of the Pacific-Ocean mobile belt: Moscow.

Kuz'min, V. K., and A. P. Chukhonin, 1980, Precambrian age of gneisses of the Kamchatka pluton: Academy of Sciences of the USSR, Doklady, Earth Science Sections, v. 251, p. 61–63.

Lebedev, M. M., et al, 1967, Metamorphic zones of Kamchatka as an example of the metamorphic assemblages of the inner part of the Pacific belt: Tectonophysics, v. 4, p. 445–461.

Leonenko, N. A., 1976, Structural zonation of the Taygonos Peninsula: International Geology Review, v. 18, p. 640–646.

Marakhanov, V. I., and S. V. Potap'ev, 1981, Structural regionalization of the Kamchatka tectonic zone: Moscow, Nauka Publishing House, 87 p. (in Russian).

Markov, M. S., et al, 1969, Junction of east Kamchatka structures and the Aleutian island arc: Geotectonics, v. 3, p. 314–319.

______, et al, 1972, Basement of the Cretaceous geosyncline on the Kamchatka Cape Peninsula (eastern Kamchatka): Geotectonics, v. 6, p. 246–249.

Mokrousov, V. P., and A. F. Marchenko, 1964, Precambrian and Paleozoic stratigraphy, *in* L. I. Krasniy, ed., Geology of the USSR, v. 31, Kamchatka and the Kurile and Komandorskie Islands: Moscow, Nedra Publishing House, p. 55–79 (in Russian).

Nekrasov, G. Y., 1971, The position of ultrabasics, basic effusives and radiolarian rocks in the development of the Taygonos Peninsula and the Penzha Range: Geotectonics, v. 5, p. 288–292.

Nozawa, T., 1977, Pre-Silurian basement, *in* K. Tanaka and T. Nozawa, eds., Geology and mineral resources of Japan (3rd ed.), v. 1, Geology: Kawasaki, Geological Survey of Japan, p. 18–19.

Osipov, A. P., 1976, Magmatic formations in the Okhotsk-Chukotka volcanogenic belt: International Geology Review, v. 18, p. 579–584.

Parfenov, L. M., 1970, Structure of Precambrian of East Asia, *in* Y. A. Kosigin, ed., Tectonic problems of Precambrian of continents: Moscow, Nauka Publishing House, p. 131–154 (in Russian)

Pergament, M. A., 1958, Upper Cretaceous rocks of northwestern Kamchatka: Academy of Sciences of the USSR, Doklady, Earth Science Sections, v. 120, p. 463–468.

Rivosh, L. A., 1964, Some geophysical data bearing on deep structure of the Central Kamchatka trough: International Geology Review, v. 6, p. 2009–2014.

Rotman, V. K., 1964, Diagonal suture of western Kamchatka: Academy of Sciences of the USSR, Doklady, Earth Science Sections, v. 159, p. 61–63.

______, and B. A. Markovskiy, 1968, Types of geosynclinal basalt magmas in Kamchatka: Academy of Sciences of the USSR, Doklady, Earth Science Sections, v. 182, p. 70–72.

______, et al, 1973, The Kamchatka ultrabasic province: International Geology Review, v. 15, p. 1015–1024.

Rozen, O. M., and M. S. Markov, 1973, Origin of amphibolites and the metamorphic melanocratic basement of island arcs (Ganal Range, Kamchatka): Geotectonics, v. 7, p. 136–142.

Sadreev, A. M., and B. K. Dolmatov, 1965, New data on the thickness and age of effusive-pyroclastic and tuffogenic-sedimentary formations of Kronotski Peninsula: Academy of Sciences of the USSR, Izvestiya, Geology Series, n. 7, p. 122–126 (in Russian).

Salop, L. J., 1977, Precambrian of the northern hemisphere: Amsterdam, Elsevier, 378 p.

Savostin, L., et al, 1983, Geology and plate tectonics of the Sea of Okhotsk, *in* T. W. C. Hilde and S. Uyeda, eds., Geodynamics of the western Pacific-Indonesian region: Washington, DC, American Geophysical Union, Geodynamics Series, v. 11, p. 189–221.

Schopf, J. M., 1969, Precambrian microfossils, *in* R. H. Tschudy and R. A. Scott, eds., Aspects of palynology: New York, Wiley-Interscience, p. 145–161.

Serova, M. Y., et al, 1970, *Rzehakina epigona* zone in volcanic terrigenous formations of the Eastern Range, Kamchatka: Academy of Science of the USSR, Doklady, Earth Science Sections, v. 190, p. 72–73.

Shapiro, M. N., 1976, Northeastern continuation of the East Kamchatka synclinorium: Geotectonics, v. 10, p. 76–78.

______, 1981, Relationship between geosynclinal and island-arc regimes (as exemplified by Kamchatka): Geotectonics, v. 15, p. 363–374.

______, and V. A. Seliverstov, 1975, Morphology and age of folded structures in eastern Kamchatka, at the latitude of the Kronotskiy Peninsula: Geotectonics, v. 9, p. 240–244.

Shmidt, O. A., 1978, Tectonics of the Komandorskie Islands and structure of the Aleutian ridge: Moscow,

Nauka Publishing House, 100 p. (in Russian).
Shul'diner, V. I., et al, 1979, The crystalline basement of Kamchatka: structure and evolution: Geotectonics, v. 13, p. 136–145.
———, et al, 1980, Two types of pre-Mesozoic metamorphism in the Central Range, Kamchatka: Academy of Sciences of the USSR, Doklady, v. 251, p. 96–99.
Sidorchuk, I. A., and A. I. Khanchuk, 1981, Mesozoic glaucophane schist complex from western slope of Central Range of Kamchatka: Geology and Geophysics, v. 255, n. 3, p. 150–156 (in Russian).
Sivertseva, I. A., and A. N. Smirnova, 1974, On the discovery of Paleozoic spores in metamorphosed deposits on Kamchatka: Soviet Geology and Geophysics, v. 15, n. 6, p. 106–107.
Smirnov, L. M., 1971, The tectonics of west Kamchatka: Geotectonics, v. 5, p. 186–192.
Suprunenko, O. I., 1976, Time of formation and development of the Kurile-Kamchatka deep-sea trench: Academy of Sciences of the USSR, Doklady, Earth Science sections, v. 227, p. 112–113.
Taira, A., et al, 1983, The role of oblique subduction and strike-slip tectonics in the evolution of Japan, *in* T. W. C. Hilde and S. Uyeda, eds., Geodynamics of the western Pacific-Indonesian region: Washington, DC, American Geophysical Union, Geodynamics Series, v. 11, p. 303-316.
Tschudy, R. H., 1969, Relationship of palynomorphs to sedimentation, *in* R. H. Tschudy and R. A. Scott, eds., Aspects of palynology: New York, Wiley-Interscience, p. 79–96.
Umitbayev, R. B., 1977, Structural position and some features of ultramafic rocks in the northern Okhotsk region: Geotectonics, v. 11, p. 210–214.
Vlasov, G. M., et al, 1965, Aspects of tectonics of Kamchatka: International Geology Review, v. 7, p. 788–802.
Voronkov, Y. S., and V. N. Smirnov, 1971, New data on the age of Cretaceous sediments of the Moroshechnyy Range, southwestern Kamchatka: Academy of Sciences of the USSR, Earth Science Sections, v. 200, p. 52–53.
Vysotskiy, S. V., and A. A. Gracheva, 1981, Precambrian age of the Olenegorskiy basement swell in the eugeosynclinal zone of eastern peninsulas of Kamchatka: Academy of Sciences of the USSR, Doklady, Earth Science Sections, v. 257, p. 101–103.
Watson, B. F., 1985, Tectonic evolution of Kamchatka Peninsula: MS Thesis, Michigan State University.
Yudin, S. S., and L. I. Izmailov, 1966, The Chelomdzha-Yamsk deep fault: Academy of Sciences of the USSR, Doklady, Earth Science Sections, v. 166, p. 93–95.
Zaborovskaya, N. B., 1978, Interior zone of the Okhotsk-Chukotsk belt in Taygonos: Moscow, Nauka Publishing House, 199 p. (in Russian).
Zhegalova, G. V., 1981, Melange in the massifs of the gabbro-norite-cortlandite complex of the Sredinniy (Middle) Range of central Kamchatka: Geotectonics, v. 15, p. 273–278.

Accreted Terranes of China

Ji Xiong
Zhongshan University
Guangzhou, People's Republic of China

Peter J. Coney
University of Arizona
Tucson, Arizona

China is composed of tectonostratigraphic terranes that exhibit a complicated history of accretion. These terranes are grouped into two domains: a Pacific domain and a Tethys domain. Most of the terranes in the Pacific domain were assembled between late Paleozoic and early Mesozoic time. The accretionary ages of terranes in the Tethys domain generally become younger from middle to late Paleozoic of the Altai terrane in the north to the accretion in Tertiary time of the Himalaya terrane to the south. The terranes are characterized by internal homogeneity and continuity of stratigraphy, tectonic style, and history. The terranes are bounded by faults. Most of the terranes are allochthonous. Paleomagnetic studies show that some of the terranes have traveled a long distance from near the equator to their present positions.

The main period of accretionary activity ended during the Tertiary in China. However, the post-accretionary deformation is still active. Complex left strike-slip faulting, folding, and thrust faulting result in continued intraplate deformation of terranes. These tectonic characteristics are caused by the interaction of Pacific, Indian, and Eurasian plates. The different rates of intraplate movement also cause local extension and development of fault-depression basins.

INTRODUCTION

Terrane analysis suggests that China has grown by repeated collisions of terranes of variable ages and types. These terranes represent a wide spectrum of geologic environment, including continental fragments, scraps of oceanic basins, volcanic island arcs, deep-sea sediments, and ophiolites of various ages; some are of unknown origin (Fig. 2). The accretionary history is complex. It began in early Paleozoic time with the amalgamation of the Qilianshan terranes and continued up to the collision of India in Tertiary time.

Thirty-three tectonostratigraphic terranes have been recognized using the principles of terrane analysis (Coney et al, 1980; Jones et al, 1982b, 1983). These terranes are varied in size and shape. Each terrane is fault bounded and is characterized by an individual sequence that records a unique geological history different from that of adjoining terranes.

This paper describes the character, distribution, and accretion of terranes and casts light on the tectonic evolution of China.

DISTRIBUTION AND CHARACTER OF TERRANES

A generalized distribution of terranes in China is shown in Figure 1. Representative stratigraphic columns for each terrane are shown in Figure 2. These terranes are grouped into two domains: a Pacific domain and a Tethys domain. Terranes generally located to the east of the cities of Kunming, Chengdu, and Yinchuan are the terranes of the Pacific domain. They were mostly carried by "Pacific" plates to their present positions. Those located to the west are the terranes of the Tethys domain.

Terranes of the Pacific Domain

The terranes of the Pacific domain are for the most part northeast elongated and bounded by northeast-trending faults (neocathaysian and cathaysian faults; Fig. 1).

The diamond-shaped Yangzi terrane of the Pacific domain covers most of the territory of the Yangtze River drainage. It has a crystalline basement formed during late Proterozoic capped by Sinian to Middle Triassic marine deposits and Late Triassic terrestrial deposits. The Permian section is a thick pile of basalt. Devonian and Carboniferous deposits are mostly lacking. Paleomagnetic records suggest that it was located at about 2° N latitude during Permian time (McElhinny et al, 1981).

The Huanan terrane is composed of Precambrian to Silurian metamorphic rocks intercalated with volcanic rocks. This sequence was tightly folded during the Silurian Caledonian orogeny. It amalgamated with the Yangzi terrane during the Silurian to form a composite terrane. During the late Paleozoic both terranes were in the same tectonic setting. A series of shallow marine sediments were deposited. The Triassic on both terranes consists of continental sediments. Triassic granitic plutons (γ_5^2) stitch the Huanan and Yangzi terranes. From the Jurassic to Cretaceous, the Huanan terrane underwent intensive igneous intrusions.

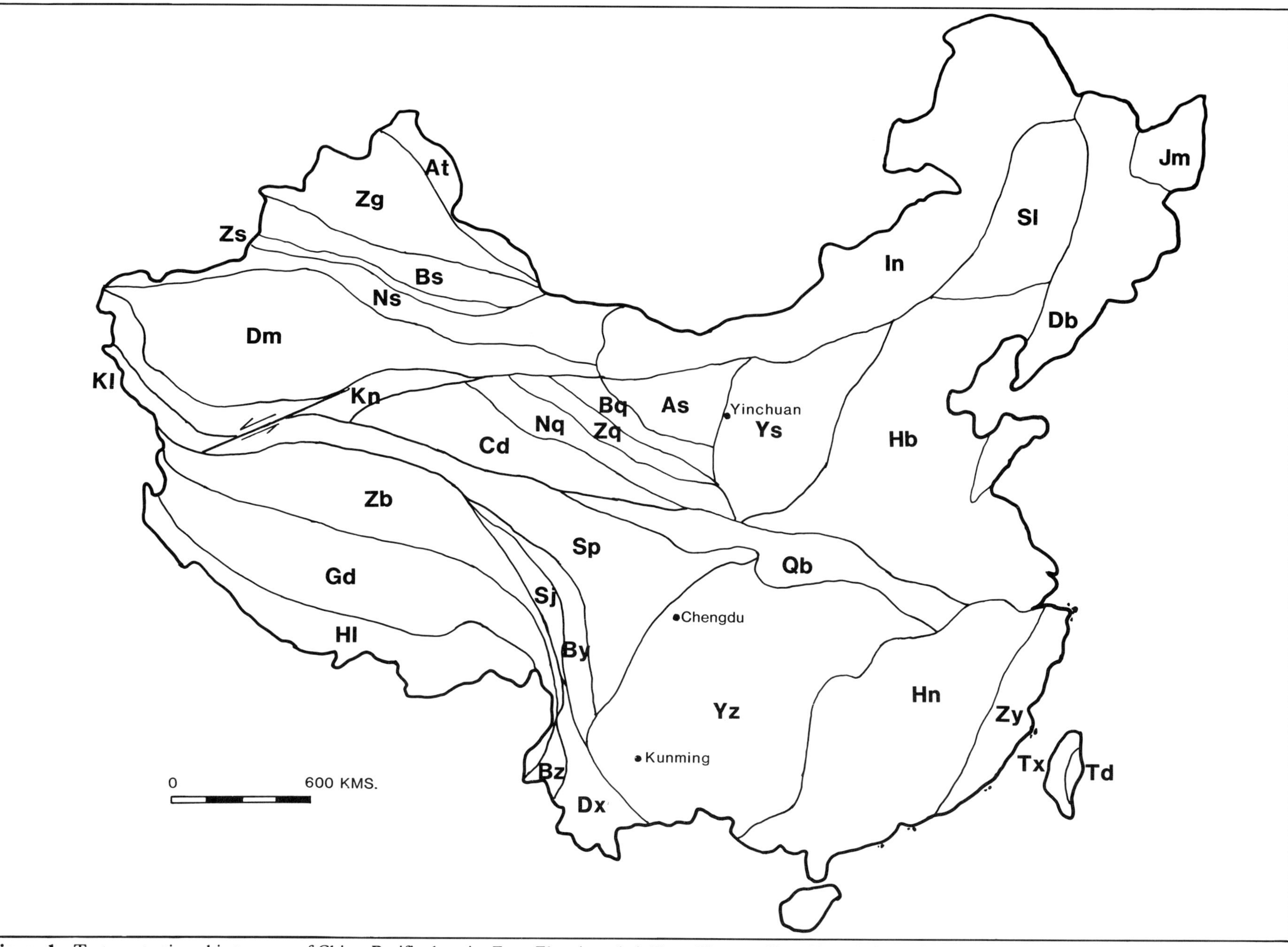

Figure 1—Tectonostratigraphic terranes of China. Pacific domain: Zy = Zheminyanhai; Hn = Huanan; Yz = Yangzi; Qb = Qinba; Hb = Huabei; Db = Dongbei; Jm = Jiamusi; Sl = Songliao; In = Inner-Mongolia; Ys = Yishan; Tx = Taixi; Td = Taidong. Tethys domain: As = Alashan; Bq = Beiqilianshan; Zq = Zhongqilianshan; Nq = Nanqilianshan; Cd = Chaidamu; Sp = Songpan-Ganzi; By = Baiyu-Yidun; Si = Sanjian; Zb = Zangbei; Gd = Gangdisi-Nianqing; Hl = Himalaya; Bz = Baoshan-Zhenkang; Dx = Dianxinan; Kl = Kalakunlun; Kn = Kunlun; Dm = Dalimu; Ns = Nantianshan; Zs = Zhongtianshan; Bs = Beitianshan; Zg = Zhungaer; At = Aertai.

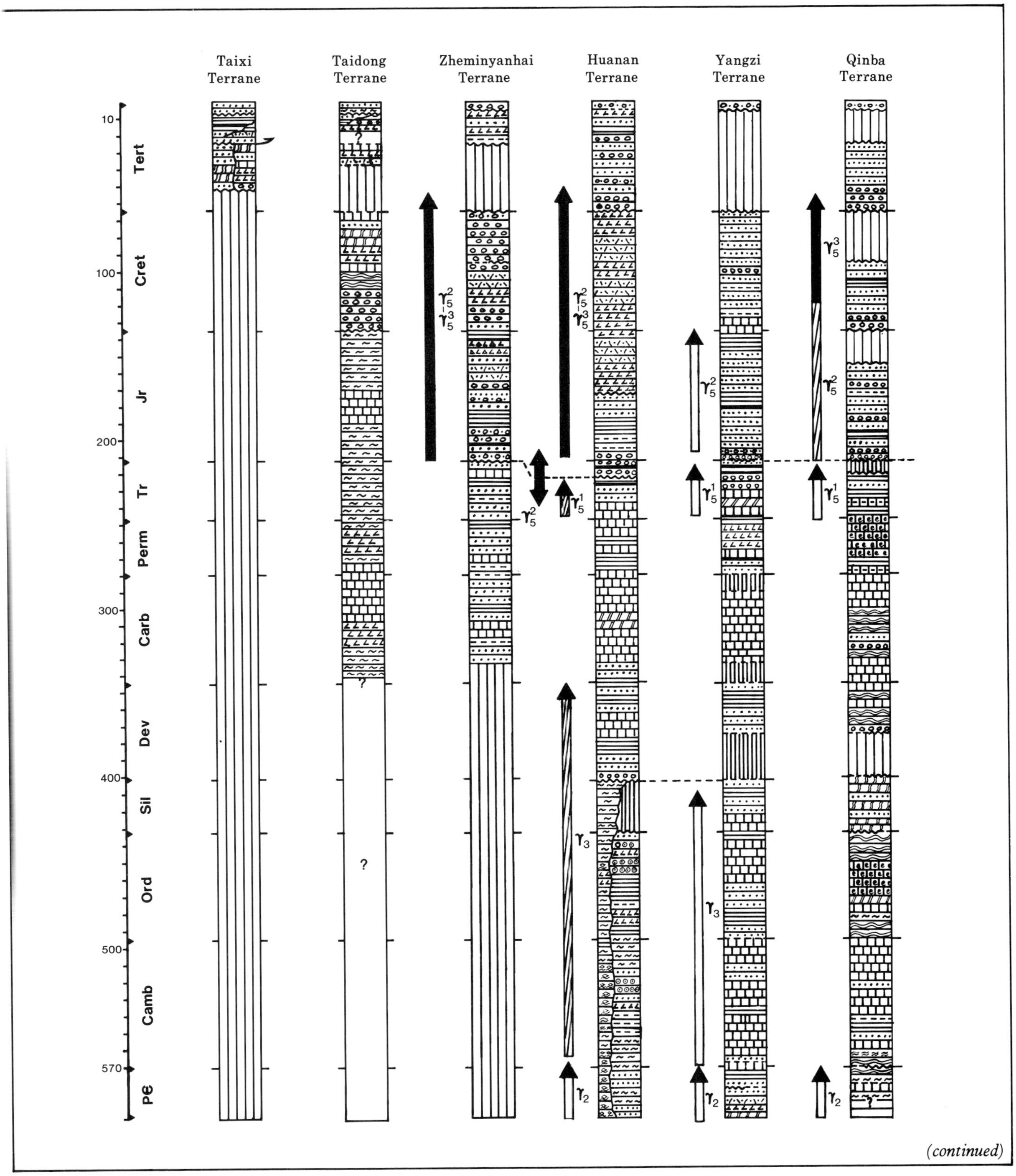

Figure 2—Tectonostratigraphic columns for each of the terranes of China. PC = Precambrian; Camb = Cambrian; Ord = Ordovician; Sil = Silurian; Dev = Devonian; Carb = Carboniferous; Perm = Permian; Tr = Triassic; Jr = Jurassic; Cret = Cretaceous; Tert = Tertiary. Key to terrane symbols is found on page 356.

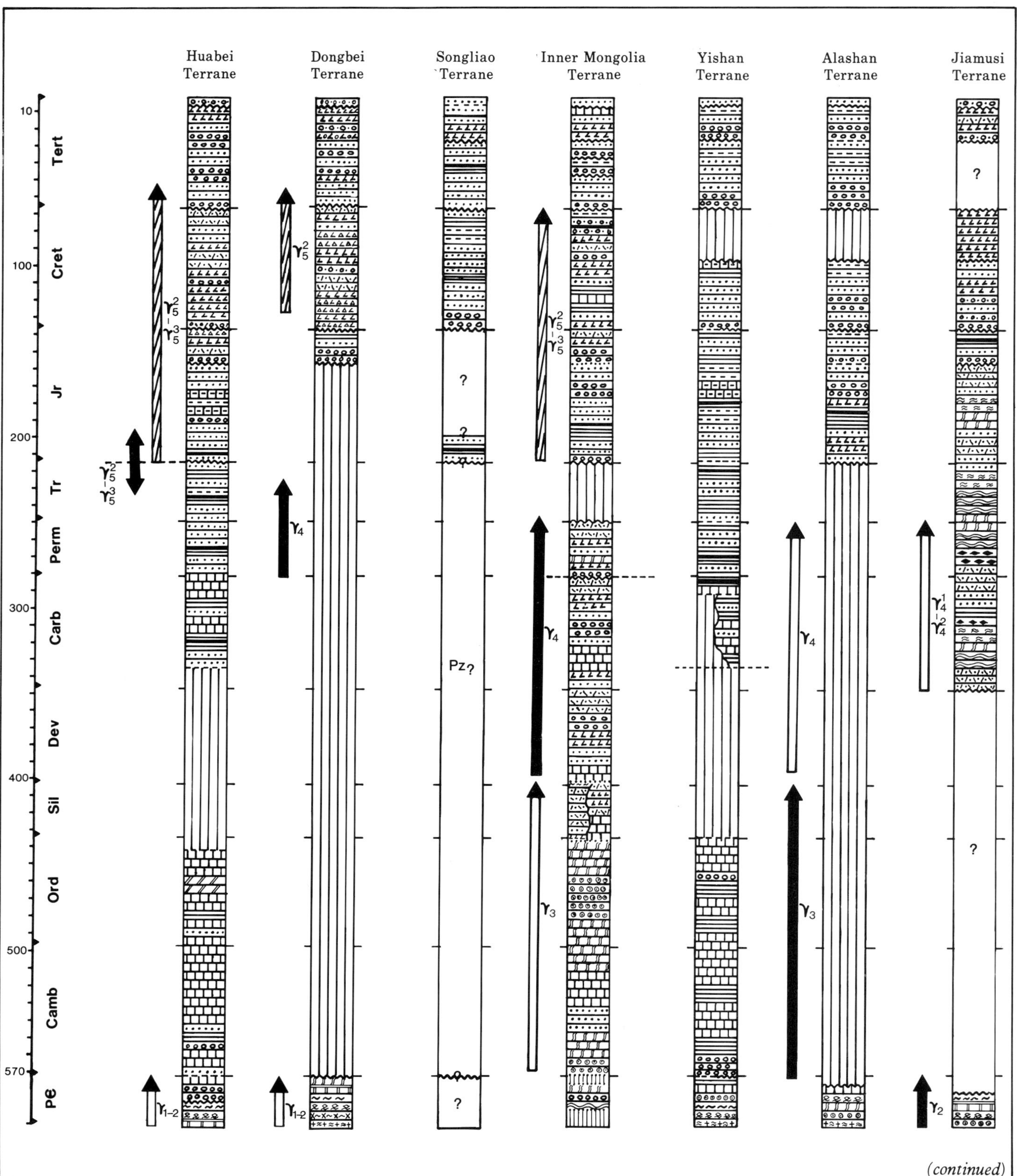

Figure 2—Continued.

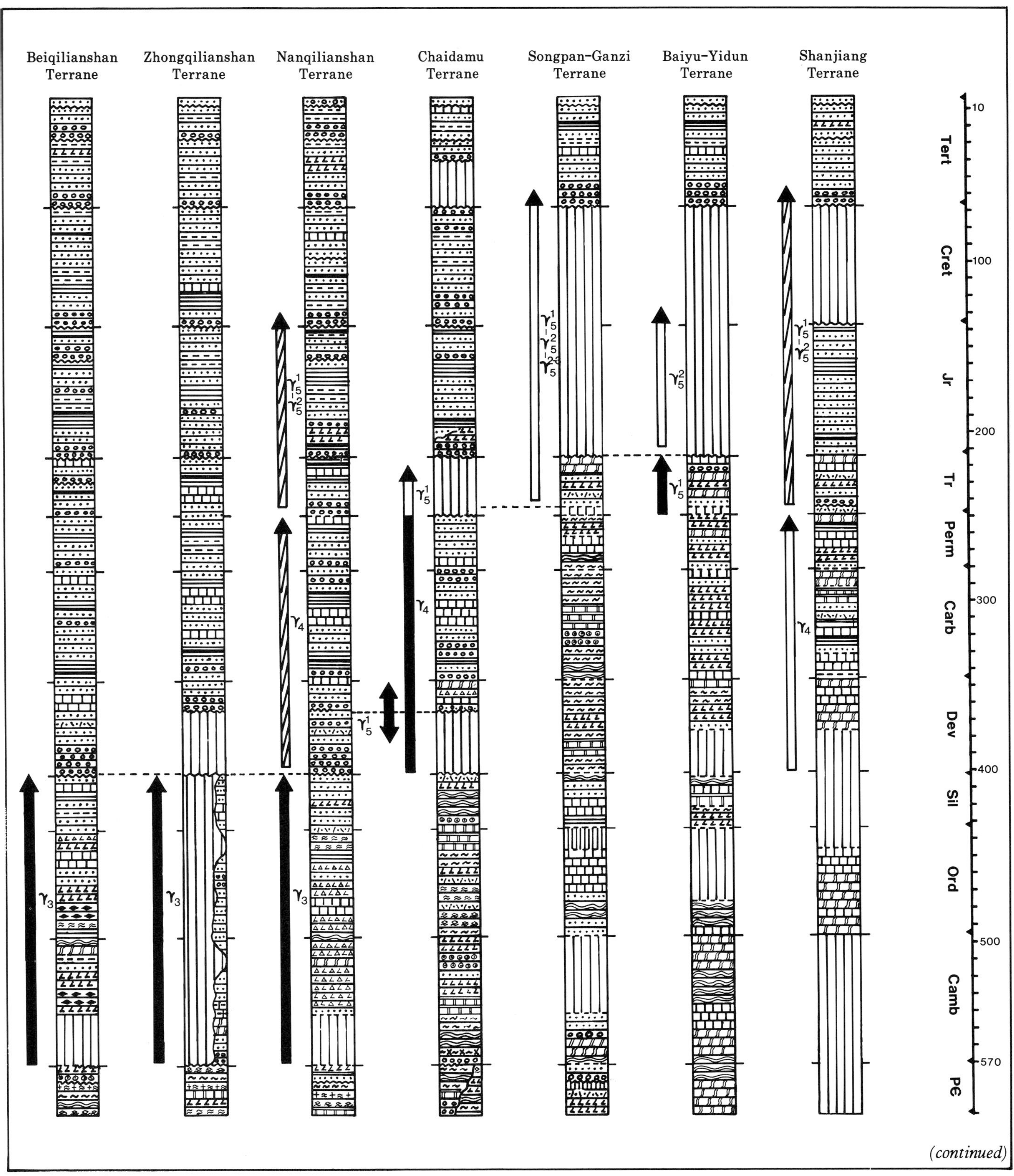

Figure 2—Continued.

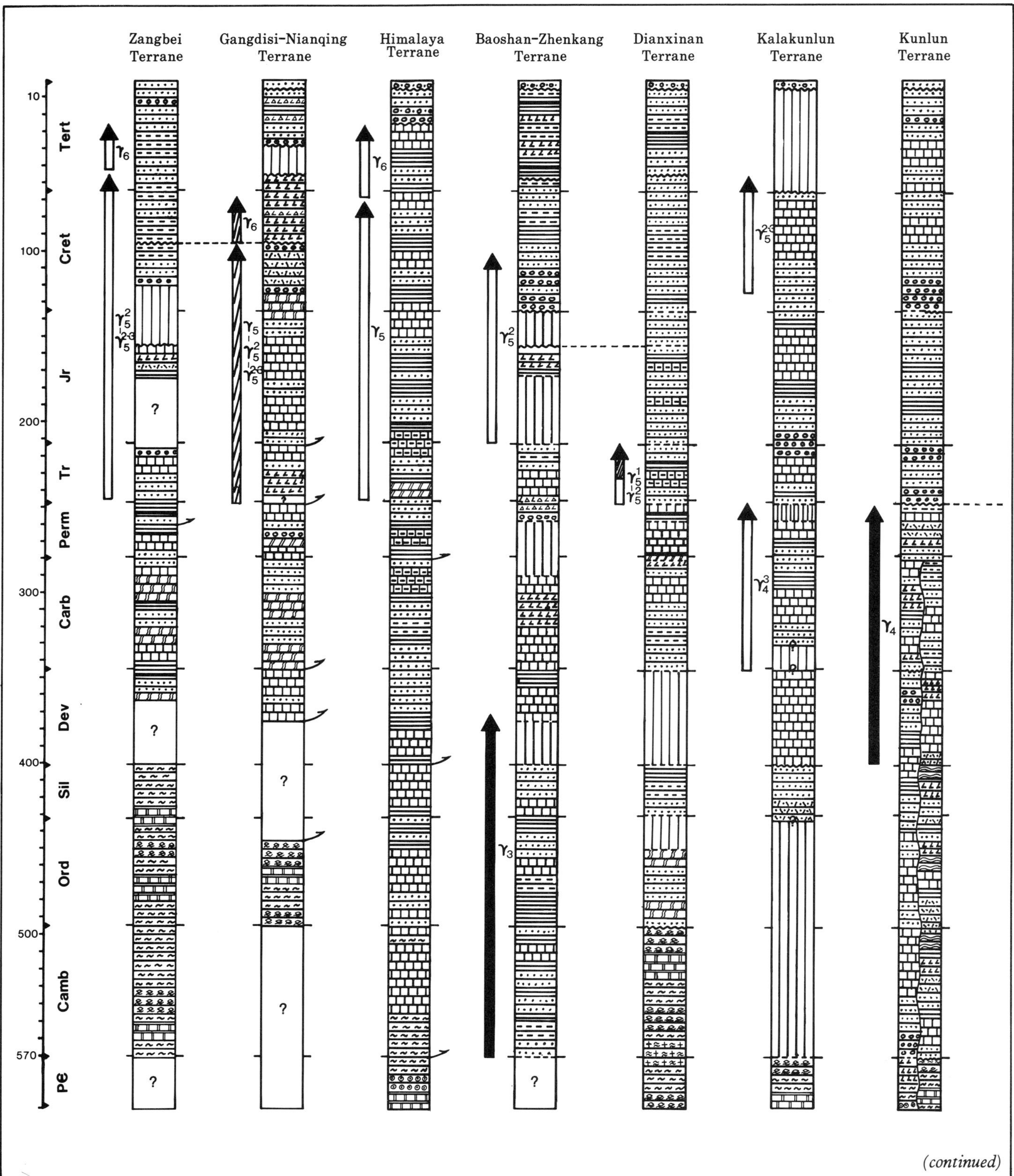

Figure 2—Continued.

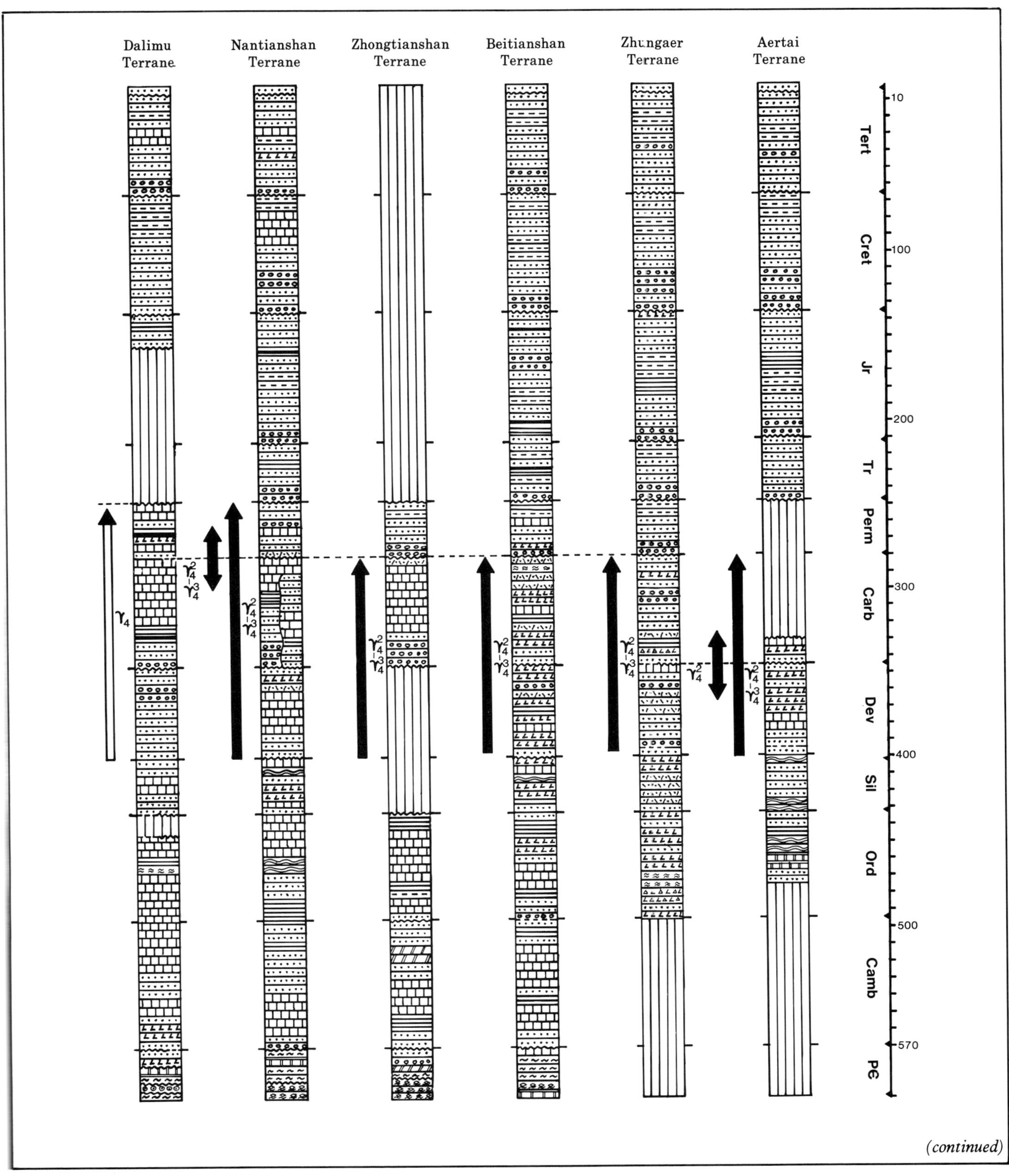

Figure 2—Continued.

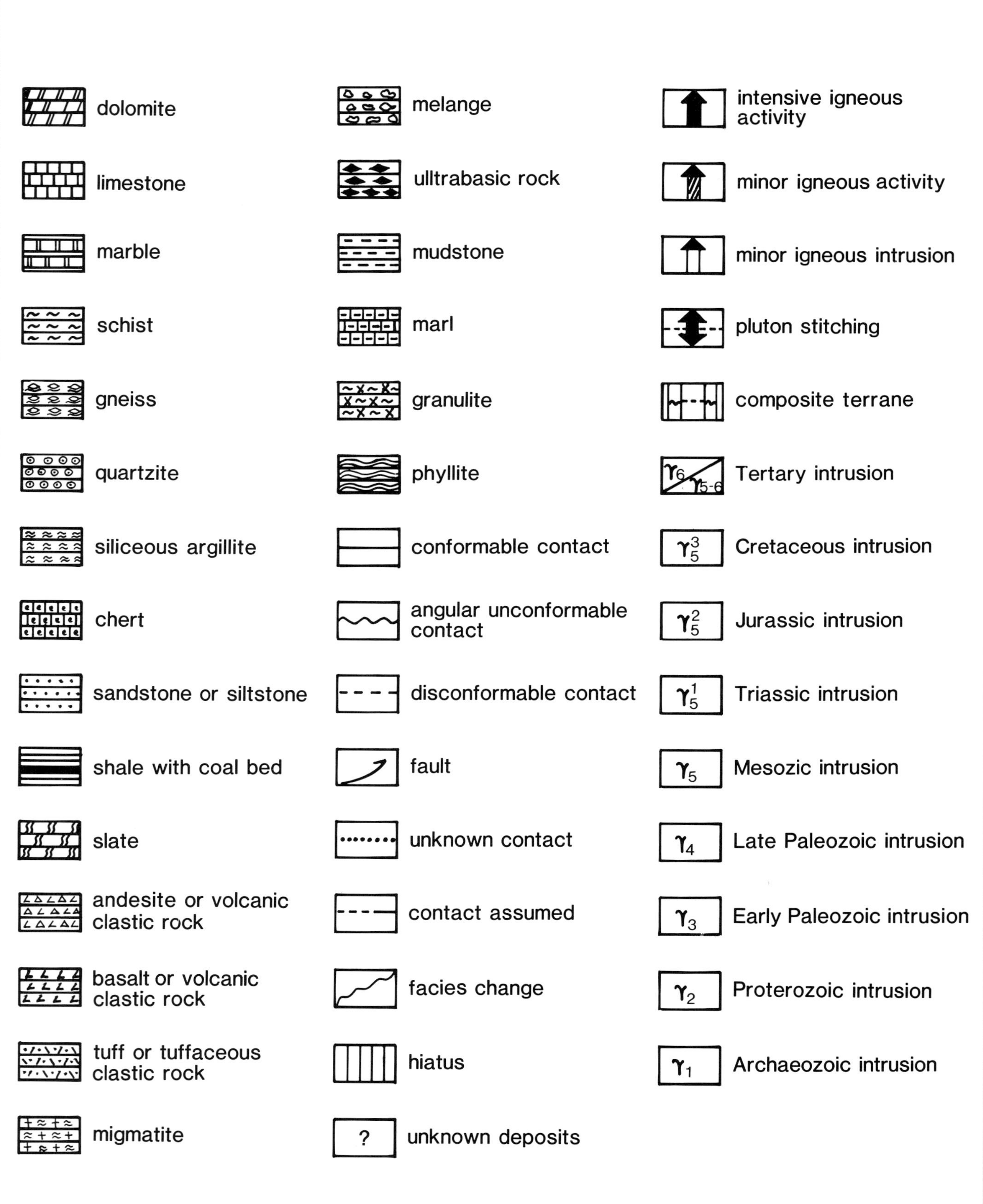

Figure 2—Continued.

The Huabei terrane was situated at about 10° N latitude during the Permian period (McElhinny et al, 1981) and was separated from Yangzi–Huanan superterrane. It is a microcontinent bearing the oldest basement in China, which consolidated during the Zhongtiao orogeny 1.7 b.y. (Tectonic Group of Geology and Mineral Resources Institute, 1978). Late Ordovician, Silurian, Devonian, and Lower Carboniferous rocks are generally absent. From Middle Carboniferous to Lower Permian the section is composed of strand and continental coal-bearing deposits.

The Songliao terrane is possibly a continental block that may have Precambrian crystalline basement. It also carries a thick sequence of Cretaceous lacustrine deposits. The Songliao terrane borders on the Inner-Mongolia terrane to the west. The Inner-Mongolia terrane has a late Paleozoic island arc assemblage built on Precambrian to Lower Paleozoic metamorphic rocks. The Songliao terrane borders on the Dongbei terrane to the east whose Precambrian basement is overlain by Late Jurassic continental deposits. The Yishan terrane contains Precambrian basement covered by a thin sequence of Cambrian to Ordovician marine rocks. No Silurian, Devonian, and Lower Carboniferous rocks are known.

The only west-northwest–east-southeast elongated terrane of the Pacific domain is the Qinba terrane, which contains a Precambrian basement covered by Paleozoic and early Mesozoic marine limestone and cherty limestone. The smallest terranes are Taidong terrane and Taixi terrane, which make up the island of Taiwan.

Terranes of the Tethys Domain

Terranes of the Tethys domain are mostly elongated west-northwest–east-southeast changing to north-northwest–south-southeast trends to the southeast (Fig. 1).

The Dalimu terrane is the largest terrane in this domain. It is a microcontinent bearing a crystalline basement formed during the late Proterozoic Dalimu orogeny. The northern part of it is amalgamated to the Tianshan composite terrane, which is composed of Beitianshan, Zhongtianshan, and Nantianshan subterranes. These three terranes are quite different in geologic history. The Beitianshan subterrane has an extremely thick sequence of Ordovician to Carboniferous island arc volcanic rocks, whereas the Nantianshan subterrane is mainly composed of late Paleozoic shallow-marine deposits containing volcanic detritus. The Zhongtianshan terrane may be a continental block separating these two terranes. The Zungaer terrane contains Ordovician to Devonian island arc volcanic rocks without Cambrian deposits or a Precambrian basement. The northwest tip of this domain is the Altai terrane whose Devonian island arc volcanic rocks are underlain by Middle Ordovician to Silurian metamorphic rocks.

The southern part of the Dalimu terrane is in juxtaposition with terranes of volcanic island arc affinity and continental fragments. The Himalaya terrane is situated at the southernmost region of the Tethys domain and is bounded by the Yaluzangbujiang River suture to the north. It is a continental fragment of Gondwanaland. Paleomagnetic data from the Jiangzi slate suggest a paleolatitude of 23° S during Late Jurassic time (Zhu, 1981). The Gangdisi–Nianqing terrane is bounded by the Yaluzangbujiang River ophiolite to the south and by the Bangonghu–Nujiang fault to the north. A few patches of Ordovician to Silurian migmatite gneiss and crystalline schist exposed in this terrane may represent the oldest rocks in this terrane (Ren et al, 1980). They are capped by upper Paleozoic shallow marine rocks. The Triassic and Cretaceous sequence contains volcanic rocks. The strata older than Early Cretaceous were intensively compressed with folds overturned southwards. The paleomagnetic data from the Late Jurassic Lasha limestone indicate a paleolatitude of 1° N (Zhu, 1981).

ACCRETIONARY HISTORY

The accretionary history of China is extremely complicated and incompletely known. However, we can still shed some light based on what is known or inferred.

Accreted Terranes of the Pacific Domain

The accreted terranes that make up the Pacific domain were carried in by the "Pacific" plates to their present positions. Most of the terranes were assembled between late Paleozoic and early Mesozoic time (Fig. 3). Yangzi terrane amalgamated with Huanan terrane during Silurian time to form a composite terrane that continued to travel northwest. During Permian time this superterrane was located at 2° N (McElhinny et al, 1981). In Late Triassic time the superterrane finally hit the Qinba terrane, which also collided with the Huabei terrane during the same period. Jurassic to Cretaceous plutons (γ_5^2–γ_5^3) stitch the two terranes. Ophiolites, flysch, subduction zone melange, and a paired metamorphic belt in the Qinba terrane indicate a north-dipping subduction zone (Li, 1975; Wu, 1980). A linear belt of Late Triassic and Early Cretaceous granites and granodiorites, which trends east from the Qinling through the Dabieshan to Huaiyang massif (Klimetz, 1983) may be related to this subduction. The Yangzi–Huanan superterrane coalesced with the Songpan–Ganzi terrane of the Tethys domain during Late Triassic to Early Jurassic time (Klimetz, 1983). The Zhemingyanhai terrane accreted against the Huanan terrane during Late Triassic time. The small occurrence of metamorphosed intermediate-basic volcanic rocks, layered mafic, and ultramafic rocks with jasper deposits along the terrane boundary fault may be the marks of the suture (Luo, 1979; Li et al, 1981).

The Huabei terrane was juxtaposed with the Yishan terrane by Middle to Late Carboniferous as indicated by the similarity of Late Carboniferous continental deposits. Since Permian time, a relative rotation of 120° has occurred between the Yangzi–Huanan superterrane and the Huabei terrane during their voyage (McElhinny et al, 1981).

The Songliao terrane may be a continental fragment that may have been caught by the accretion between the Inner-Mongolia terrane to its west and the Dongbei terrane to its east. The Dongbei terrane may have docked against the Huabei terrane and triggered the Tancheng–Lujiang grand fault during the Late Triassic. The isotopic age of 190 to 230 m.y. obtained from igneous rocks of northern

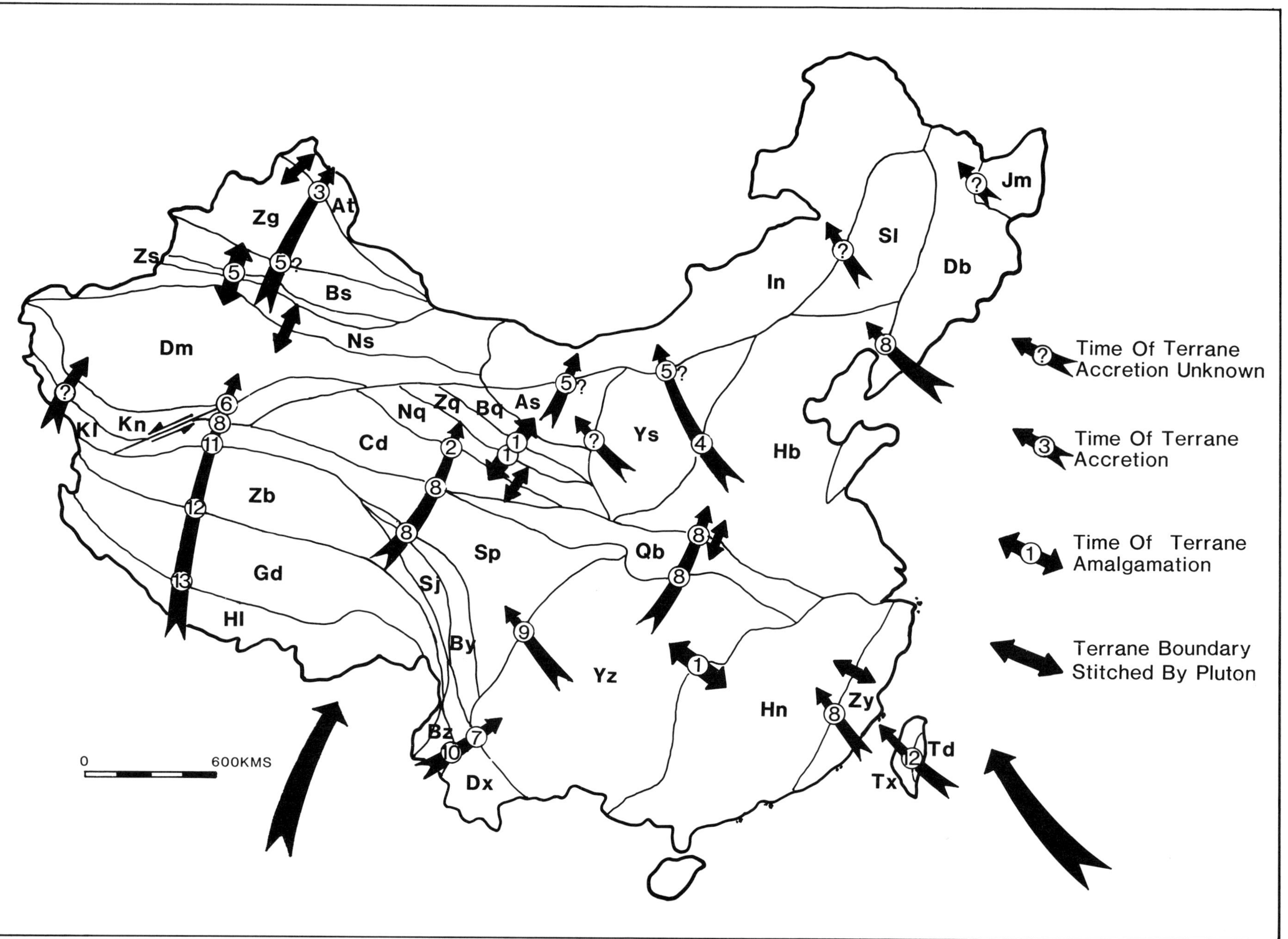

Figure 3—Time of terrane accretion in China. 1 = Silurian; 2 = Devonian; 3 = Early Carboniferous; 4 = Middle–Late Carboniferous; 5 = Permian; 6 = Late Permian; 7 = Early–Middle Triassic; 8 = Late Triassic; 9 = Late Triassic–Early Jurassic; 10 = Middle Jurassic; 11 = Cretaceous; 12 = Late Cretaceous; 13 = Tertiary.

Jiangsu Province and the metamorphic rock developed in the Zhongbaling range of the Anhui Province may represent a heating event of the macroshear and the early initial volcanic history (Weng and Wang, 1981) that were caused by the collision. Yet, the evolutionary history of Jiamusi terrane still remains a mystery.

Accreted Terranes of the Tethys Domain

The accretionary ages of the terranes of the Tethys domain generally become younger from the middle to late Paleozoic of the Altai terrane in the north to the accretion in Tertiary time of the Himalaya terrane to the south (Fig. 3). The Zhungaer terrane coalesced with the Aertai terrane during Early Carboniferous. The Tianshan composite terrane was grafted onto the northern margin of Dalimu terrane during Middle Permian.

The accretionary history of the Beiqilianshan, Zhongqilianshan, and Nanqilianshan terranes of Tethys domain is extremely complex. During the Late Silurian, double subduction occurred between Zhongqilianshan and Beiqilianshan terranes. The southern subduction was along the northern margin of Zhonqilianshan terrane. A high P/T metamorphic belt was produced by the subduction beneath the Zhongqilianshan terrane. Along the northern side, the Paleo-Ocean plate subducted underneath a north China paleocontinent. The intensive folding, faulting, and small early Paleozoic intrusions along this margin were the products of this subduction (Li et al, 1978, 1981). The Nanqilianshan terrane was amalgamated into the Zhongqilianshan terrane, and the Zhongqilianshan terrane combined with Beiqilianshan terrane during Silurian time to form the Qilianshan composite terrane. The Chaidamu terrane docked against the Qilianshan composite terrane during Devonian time. Subduction then shifted to the southern margin of the Chaidamu terrane. The late Paleozoic folds and Variscan granite and granodiorite intrusion along the southern margin of Chaidamu were produced by this subduction (Li et al, 1978).

The Qinba terrane accreted against the eastern edges of the Chaidamu terrane and the Qilianshan composite terrane as well as the southern Huabei terrane during Late Triassic time. The Songpan–Ganzi terrane hit the Chaidamu and Qinba terranes during Late Triassic time. After these continued accretions, the Qilianshan composite terrane joined together with the north China paleocontinent and the Dalimu terrane to further enlarge the paleo-continent.

The Tibetan plateau region south of Dalimu terrane appears to have evolved via a progressive Late Triassic through Early Tertiary southward stacking of rifted northern Gondwana crustal elements against the southern margin of the western China (Powell and Conaghan, 1975; Sengör, 1979; Bally et al, 1980; Andrews-Speed and Brookfield, 1982; Klimetz, 1983). The Dianxinan terrane accreted against the Yangzi terrane during Early to Middle Triassic (Duan and Zhao, 1981). The Baoshan–Zhenkang terrane docked against Dianxinan terrane during Middle Jurassic time. The Himalayan microcontinent that was rifted by the opening of Neo-Tethys (Sinha-Roy, 1982) finally hit the paleocontinent during Tertiary time. All of China was assembled by this time from continuous accretion.

POSTACCRETIONARY DEFORMATION

The main period of accretionary activity ceased during the Tertiary in China. However, postaccretionary complex strike-slip faulting, folding, and thrust faulting are still very active, resulting in continued intraplate deformation of terranes. These tectonic characteristics are caused by the interaction of the Pacific, Indian, and Eurasia plates.

The Tancheng–Lujiang grand fault that is the boundary fault between the Huabei and the Dongbei terranes underwent three stages of movement. Not only is it a geotectonic boundary but also it is an important metallogenic controlling line (Yang and Zhu, 1979). In the late Middle Jurassic to Early Cretaceous, it was compressional with sinistral shear. The northward movement of the Kula plate during this time period (Hilde et al, 1977) caused the eastern coast of Tancheng–Lujiang grand fault to move northwards. In the Late Cretaceous to Paleogene it displayed extensional characteristics. The rift was filled by a thick pile of Late Cretaceous lava, volcanic rocks, conglomerates, and sands. The continental sediments were shed from the terranes adjacent to the basin. The extension is probably related to the opening of the sea of Japan during Late Cretaceous to Early Tertiary time as a result of the Kula–Pacific ridge subduction (Hilde et al, 1977). In the Neogene to Quaternary, the Tancheng–Lujiang grand fault is compressional with dextral shear characteristics. The earthquake fault plane solutions and data of postquake deformation measurements indicate that the majority of the earthquakes are strike-slip. The change in the direction of motion of the Pacific plate from north-northwest to west-northwest 43 to 45 m.y.B.P. (Hilde et al, 1977; Butler and Coney, 1981) may cause the dextral shear.

In the Himalayas and in much of China north of them, all the relative motion between the India–Australia and Asian plate is absorbed as intraplate telescoping and translation where the crust appears to have been shortened by 800 km (497 mi) or more (Molnar and Tapponnier, 1975; Powell and Conaghan, 1975; Jones et al, 1982b). Seismicity is widespread. Not only are there numerous small earthquakes north of the Himalayas, but four of the seven great earthquakes occurred far north of the Himalayas (Molnar and Tapponnier, 1975). There is no typical plate boundary here. The continued northward movement of the India–Australia plate causes a series of large-scale left-lateral strike-slip grand faults that are displacing much of China toward the Pacific Ocean (Yan et al, 1981) to accommodate shortening of the crust (Chang et al, 1978). This also causes the rejuvenation of the paleo-orogenic belts 1,000 km (621 mi) north of the Yaluzangbujiang suture and deformation of a large region 3,000 km (1,864 mi) north of the suture.

The Qinghai–Tibet plateau has been pushed northward since Pleistocene time by the Indian plate as shown by paleomagnetic data. In eastern China, earthquake focal mechanism analysis and paleomagnetic studies also provide evidence that the continent is moving from south to north relative to the Pacific Ocean (Ren et al, 1980).

The Weihe basin of northern margin of the middle Qinba terrane is a postaccretionary fault-depression basin caused by local extension related to different rates of intraplate movement.

CONCLUSION

It is now clear that China is made up of terranes of different sizes and ages. The timing of accretion ranges from Paleozoic to Cenozoic. Paleomagnetic studies (McElhinny et al, 1981; Zhu, 1981) demonstrate that some of the terranes have traveled distances of up to several thousand kilometers before they accreted to China.

The terranes of China are composed of continental fragments, oceanic basins, volcanic arcs, and deep-sea sediments. Some are of unknown origin. Most of the terranes are known to be fault bounded. A few of them are separated by ophiolite belts and sutures. Some igneous activity in a few terranes seems to be intimately related to subduction. The movement of "Pacific" plates may have played an important role for the collision of terranes of the Pacific domain. Continued northward movement of the India-Australia plate causes the displacement of terranes in much of the Tethys domain and rejuvenation of a large region north of the Himalayan suture zone. The interaction of the Indian plate, Eurasian plate, and Pacific plate remain very active since accretion, causing strike-slip faulting, folding, and thrust faulting that result in continued intraplate terrane deformation.

REFERENCES

Andrews-Speed, C. P., and M. E. Brookfield, 1982, Middle Paleozoic to Cenozoic geology and tectonic evolution of the northern Himalaya: Tectonophysics, v. 82, p. 253–275.

Bally, A. W., et al, 1980, Notes on the geology of Tibet and adjacent areas—Report of the American plate tectonics delegation to the People's Republic of China: U.S. Geological Survey Open-File Report 80-501.

Ben-Avraham, Z., 1978, The evolution of marginal basins and adjacent shelves in east and southeast Asia: Tectonophysics, v. 45, p. 269–288.

______, and S. Uyeda, 1973, The evolution of the China basin and the Mesozoic paleogeography of Borneo: Earth and Planetary Science Letters, v. 18, p. 365–376.

______, et al, 1981, Continental accretion: From oceanic plateaus to allochthonous terranes: Science, v. 213, n. 3, p. 47–54.

Burrett, C. F., 1973, Plate tectonics and the fusion of Asia: Earth and Planetary Science Letters, v. 21, p. 181–189.

Butler, R. F., and P. J. Coney, 1981, A revised magnetic polarity time scale for the Paleocene and early Eocene and implication for Pacific motion: Geophysical Research Letters, v. 8, n. 4, p. 301–304.

Chang, C. F., et al, 1978, Tectonic evolution, classification of tectonic belts, and upward reason research of Himalaya, *in* Scientific papers of geology for international exchange: Beijing, China, Publishing House of Geology, p. 198–211 (in Chinese).

Chen, G. D., et al, 1978, On some features of tectonics of China *in* Scientific papers of geology for international exchange: Beijing, China, Publishing House of Geology, p. 119–132 (in Chinese).

Chen, S. Q., et al, 1981, Features of gravity and magnetic anomalies in central and northern parts of the south China Sea and their geological interpretation: Scientia Sinica, v. XXIV, n. 9, p. 1271–1284.

Coney, P. J., 1981, Accretionary tectonics in western North America: Arizona Geological Society Digest, v. XIV, p. 23–37.

______, et al, 1980, Cordilleran suspect terranes: Nature, v. 288, n. 5789, p. 329–333.

Duan, X. H., and H. Zhao, 1981, The Ailaoshan-Tengtiaohe fracture—The subduction zone of an ancient plate: Acta Geologica Sinica, n. 4, p. 258–266.

Fan, D. F., 1978, Outline of the tectonic evolution of southwestern China: Tectonophysics, v. 45, p. 261–267.

Feng, Y. M., and B. Q. Zhu, 1980, Melange and tectonic development of the west Qinling Mountains: Acta Geologica Sinica, n. 1, p. 34–43.

Gansser, A., 1980, The significance of the Himalayan suture zone: Tectonophysics, v. 62, p. 37–52.

Geologic Outline of China, 1978, Compilation group of geologic maps of China, Academica Sinica, *in* Scientific papers on geology for international exchange: Beijing, China, Publishing House of Geology, p. 33–56 (in Chinese).

Geology and Tectonic Characteristics of Tianshan, China, 1978, Compilation group of Geological Bureau of Xinjiang, *in* Scientific papers on geology for international exchange: Beijing, China, Publishing House of Geology, p. 188–197 (in Chinese).

Guo, L. Z., et al, 1983, On the formation and evolution of the Mesozoic-Cenozoic active continental margin and island arc tectonics of the western Pacific Ocean: Acta Geologica Sinica, n. 1, p. 11–21.

Hattori, I., 1982, The Mesozoic evolution of the Mino terrane, central Japan: A geologic and paleomagnetic synthesis: Tectonophysics, v. 85, p. 313–340.

Hilde, T. W. C., et al, 1977, Evolution of the western Pacific and its margin: Tectonophysics, v. 38, p. 145–165.

Ho, C. S., 1979, Geologic and tectonic framework of Taiwan: Geological Society of China Memoir 3, p. 57–72.

Howell, D. G., et al, 1978, Preliminary tectonostratigraphic terrane map of the circum-Pacific region: The American Association of Petroleum Geologists.

Ji, X., 1983, A preliminary study on characteristics and classification of tectonostratigraphic terranes: Journal of Geology and Geochemistry, v. 8, p. 16–19 (in Chinese).

______, in press, Suspect terranes in China, *in* Circum-Pacific Terranes Conference: Stanford University, Publications, Geological Sciences.

Jones, D. L., et al, 1981, Map showing tectonostratigraphic terranes of Alaska, columnar sections and summary description of terranes: U.S. Geological Survey Open-File Report, 81-792.

______, et al, 1982a, Character, distribution, and tectonic significance of accretionary terrane in central Alaska range. Journal of Geophysical Research, v. 87, n. 85, p. 3709-3717.

———, et al, 1982b, The growth of western North America: Scientific American, v. 247, n. 5, p. 70–84.
———, et al, 1983, Recognition, character, and analysis of tectonostratigraphic terranes in western North America, *in* M. Hashimoto and S. Uyeda, eds., Accretion tectonics in the circum-Pacific regions: Tokyo, Terra Scientific Publishing Company, p. 21–35.
Juan, V. C., 1975, Tectonic evolution of Taiwan: Tectonophysics, v. 26, p. 197–212.
Klimetz, M. P., 1983, Speculation on the Mesozoic plate tectonic evolution of eastern China: Tectonics, v. 2, n. 2, p. 139–166.
Larson, R. L., and C. G. Chase, 1972, Late Mesozoic evolution of the west Pacific Ocean: Geological Society of America Bulletin, v. 83, p. 3627–3644.
Leith, W., 1982, Rock assemblages in central Asia and the evolution of the southern Asian margin: Tectonics, v. 1, n. 3, p. 303–318.
Li, C. Y., 1975, Tectonic evolution of some mountain ranges in China, as tentatively interpreted on the concept of plate tectonics: Acta Geophysica Sinica, v. 18, n. 1, p. 52–74.
———, and Y. Q. Tang, 1983, Some problems on subdivision of paleo-plate in Asia: Acta Geologica Sinica, n. 1, p. 1–10.
———, et al, 1978, Tectonic evolution of Qinling and Qilianshan, *in* Scientific papers on geology for international exchange: Beijing, China, Publishing House of Geology, p. 33–42 (in Chinese).
———, et al, 1981, The metallogeny and plate-tectonics of China: Acta Geologica Sinica, n. 3, p. 195–204.
Luo, Z. L., 1979, On the occurrence of Yangzi old plate and its influence on the evolution of lithosphere in the southern part of China: Scientia Geologica Sinica, n. 2, p. 127–138.
McElhinny, M. W., et al, 1981, Fragmentation of Asia in the Permian: Nature, v. 293, p. 212–216.
Miyashiro, A., 1981, Tectonic and petrologic aspects of Asia: Geological Society of China Memoir 4, p. 1–31.
Molnar, P., and P. Tapponnier, 1975, Cenozoic tectonics of Asia: Effects of a continental collision: Science, v. 189, n. 4201, p. 419–426.
——— and ———, 1977, The collision between India and Eurasia: Scientific American, v. 236, n. 4, p. 30–42.
Powell, C. M., and P. J. Conaghan, 1975, Tectonic models of the Tibetan plateau: Geology, v. 3, n. 12, p. 727–731.
Ren, J. S., et al, 1980, Geotectonic evolution of China: Publishing House of China (in Chinese).
Ridd, M. F., 1980, Possible Paleozoic drift of SE Asia and Triassic collision with China: Journal of the Geological Society of London, v. 137, p. 635–640.
Sengör, A. M. C., 1979, Mid-Mesozoic closure of Permo-Triassic Tethys and its implications: Nature, v. 279, p. 590–593.
Silver, E. A., and R. B. Smith, 1983, Comparison of terrane accretion in modern southeast Asia and the Mesozoic North American Cordillera: Geology, v. 11, p. 198–202.
Sinha-Roy, S., 1982, Interaction of Tethyan blocks and evolution of Asian fold belts: Tectonophysics, v. 82, p. 277–297.
Tapponnier, P., et al, 1982, Propagating extrusion tectonics in Asia: New insights from simple experiments with plasticine: Geology, v. 10, n. 12, p. 611–616.
Tectonic Framework of China, 1978, Tectonic group of geology and mineral resources, Academia Sinica, *in* Scientific papers on geology for international exchange: Beijing, China, Publishing House of Geology, p. 69–86 (in Chinese).
Tectonic Outline of China, 1977, Guangzhou Seismologic Brigade, State Seismologic Bureau: Seismologic Publishing House.
Tsai, Y. B., et al, 1977, Tectonic implications of the seismicity in the Taiwan region: Geological Society of China Memoir 2, p. 13–41.
Wang, Q., and X. Y. Liu, 1976, Paleo-oceanic crust of the Chilienshan region, western China and its tectonic significance: Scientia Geologica Sinica, n. 1, p. 42–55.
Watts, A. B., et al, 1977, Sea-floor spreading in marginal basins of the western Pacific: Tectonophysics, v. 37, p. 167–181.
Weng, S. J., and W. Q. Wang, 1981, Tectonism and magmatism of lower Yangtze valley: Bulletin of the Nanjing Institute of Geology and Mineral Resources, v. 2, n. 3, p. 10–18.
Wu, H. Q., 1980, The glaucophane-schists of eastern Qinling and northern Qilian Mountain in China: Acta Geologica Sinica, n. 3, p. 195–207.
Yan, J. Q., et al, 1981, Recent tectonics in the Qinghai-Xizang plateau: Acta Geologica Sinica, v. 24, n. 4, p. 386–393.
Yang, G. Q., and Y. Zhu, 1979, Aeromagnetic survey for geological reconnaissance work in China: Acta Geophysica Sinica, v. 22, n. 4, p. 379.
Zhang, Q. W., and H. Z. Huang, 1982, The evolution of magmato-tectonic activation of the Meso-Cenozoic era in eastern China: Acta Geologica Sinica, n. 2, p. 111–122.
Zhu, Z. W., 1981, Paleomagnetism and continental drift of Tibet plateau: Acta Geophysica Sinica, n. 1, p. 40–49.

Mesozoic Accretion and Collision Tectonics of Northeastern Asia

L. M. Parfenov
Geological Institute, Siberian Branch
Academy of Sciences, Yakutsk, Lenin 39, USSR

B. A. Natal'in
Institute of Tectonics and Geophysics
Academy of Sciences, Khabarovsk, Kim-Yu-Chen 65, USSR

Mesozoic foldbelts of two ages are recognized in northeast Asia. Older Mesozoic (Late Jurassic–Neocomian) foldbelts include the Verkhoyano–Chukotsk region that lies northeast of the Siberian platform and the Mongolo–Okhotsk foldbelt that is located southeast of the Siberian platform (Fig. 1). Foldbelts of younger Mesozoic (Coniacian to Maestrichtian) age include the Koryaksk and most of the Sikhote–Alinsk foldbelts, which are roughly parallel to the modern Pacific boundary. To the east the earlier foldbelts are truncated by Cenozoic foldbelts and magmatic arcs (Fig. 1).

INTRODUCTION

The Mesozoic foldbelts consist of tectonostratigraphic terranes composing volcanic island arcs and strata of the continental margin (convergent and passive) affinity. These terranes provide information about fossil convergent boundaries of Late Precambrian, Paleozoic, and Mesozoic age (Parfenov et al, 1978, 1979, 1981; Parfenov and Natal'in, 1981) and outline the tectonic evolution of all of northeast Asia. The pre-Mesozoic tectonic development of the region, however, is poorly understood because most of the Precambrian and Paleozoic rocks are from isolated fragments within Mesozoic foldbelts. For this reason, the pre-Mesozoic tectonic evolution of this area is not discussed in this report. The paleotectonic reconstructions presented below are based on palinspastic reconstructions of the fault systems and foldbelts that affect the area.

ACCRETIONARY STRUCTURE

The structure of the region is characterized by a mosaic of mainly Early Precambrian sialic terranes ("megablocks" or "median massifs" in the terminology of many Russian geologists) that collided with each other and with the larger continental mass of the Siberian platform during the Mesozoic. These sialic terranes include the Okhotsk, Omolonsk, Bureinsk, and Khankaisk median massifs, which are as much as several thousand kilometers across (Fig. 1). The sialic terranes are characterized by an Early Precambrian basement overlain by a variety of younger formations of variable thickness. The terranes are separated from one another by foldbelts that are interpreted to have resulted from the interaction of island arcs and continental margins with the Kula plate as it moved northward with respect to Eurasia and North America in the Jurassic and Cretaceous.

The Koryaksk and Sikhote–Alinsk foldbelts consist of accreted sequences of island arc and convergent continental margin-type strata juxtaposed against passive margin sequences (Figs. 2, 3). Following initiation of subduction, forearc basin–accretionary wedge systems formed on and adjacent to the accreted terranes and migrated oceanward through time. Rocks of the associated volcanic-plutonic belts can be shown to be contemporaneous with deformation in the accretionary prisms. The imbricated thrusts within the accretionary wedges indicate oceanward vergence (Fig. 4). The relative positions of the various volcanic-plutonic belts, forearc basins, and accretionary wedges have remained essentially unchanged.

The accretionary structure of the Sikhote–Alinsk foldbelt is overprinted and complicated by large left-lateral faults that are oriented subparallel to the trend of the belt and the former continental margin. This fault system was longstanding and gradually migrated oceanward (Fig. 5). Formation of the faults was related to oceanic plate motion roughly parallel to the continental margin. Their tectonic position and amount of horizontal displacement are comparable to that of the San Andreas fault system in California.

The Mongolo–Okhotsk and South-Anyuisk foldbelts separate continental terranes and may have resulted from collision of adjacent continental terranes (Figs. 6, 7). The foldbelts include the same tectonic elements as the accretionary systems but are extremely disrupted and locally metamorphosed to greenschist and lawsonite-glaucophane-schist facies. They are characterized by regionally developed schistosity and transposition structures that are locally expressed as secondary tectonic banding. Volcanic and plutonic belts are superimposed on the rigid sialic basement of the bounding megablocks or continental terranes adjacent to the foldbelts and represent magmatic arcs that were developed in association with

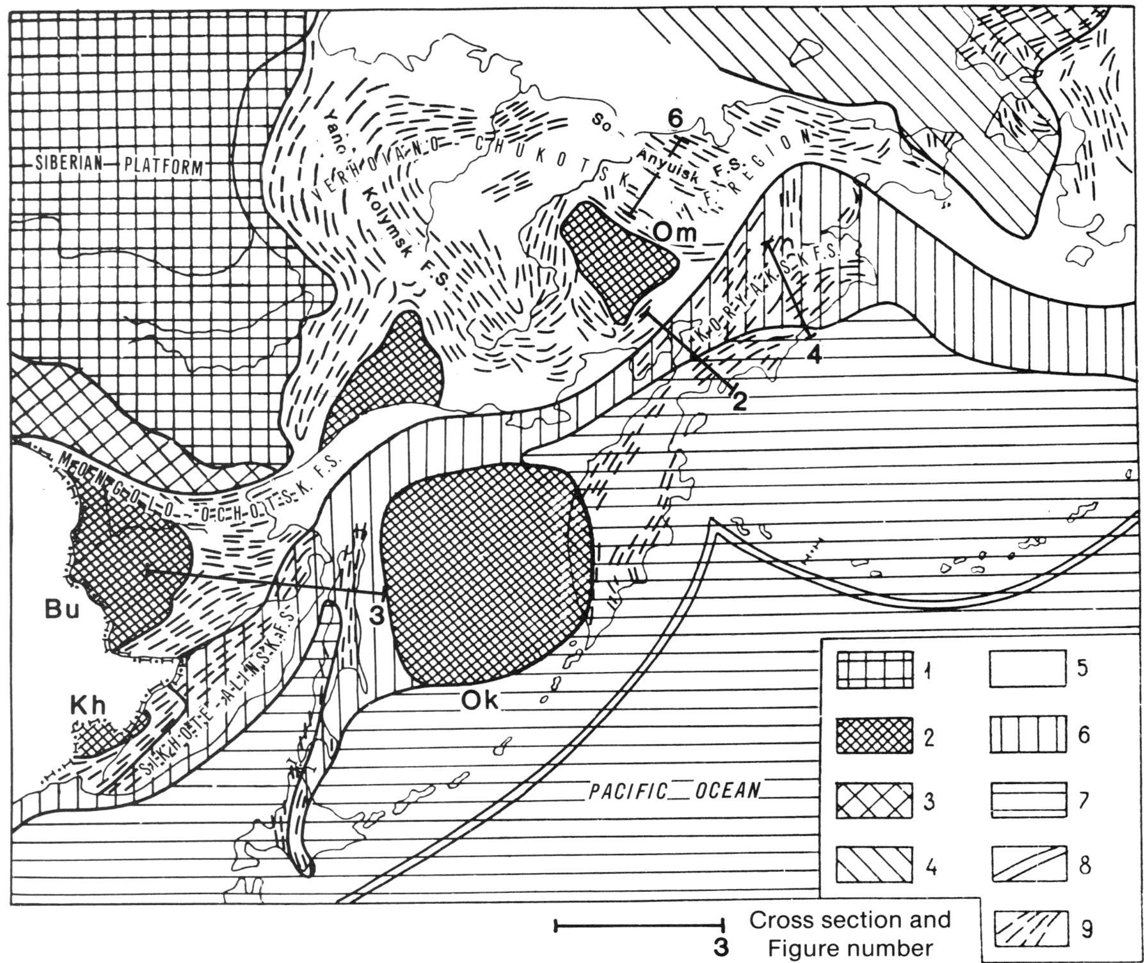

Figure 1—Map showing the foldbelts of northeastern U.S.S.R. and Alaska. (1) Siberian platform; (2) microcontinental sialic terranes (median massifs): Ok = Okhotsk, Om = Omolonsk, Bu = Bureinsk, Kh = Khankaisk; (3) late Archean–early Proterozoic Stanovoy foldbelt; (4) middle Paleozoic Brooks–Wrangel foldbelt; (5) Late Jurassic–Neocomian foldbelts; (6) latest Cretaceous foldbelts; (7) Cenozoic foldbelts, including island arcs and adjacent ocean floor; (8) trenches; (9) structural trends. All cross sections are shown with figure number.

subduction at the time of collision of these terranes. The accretionary-wedge and forearc-basin complexes related to this subduction event later became zones of extreme dislocation that obscured their earlier structural relationships. The edges of the rigid sialic terranes exhibit considerable structural disruption that can be traced in some places for as much as 100 km (62 mi) toward the interior of the terranes, for example in the Aldano–Stanovoy region (Figs. 7, 10).

The Mongolo–Okhotsk and South-Anyuisk collisional foldbelts are intruded and partly overlain by volcanic-plutonic rocks including the Umlecano–Ogodzhinsk magmatic belt. This volcanic-plutonic belt stretches for 100 km (62 mi) and cuts the deformed edges of the adjacent continental terranes and associated molasse deposits.

The Yano–Kolymsk foldbelt (Fig. 1) has considerable width, variable trend, and consists of a thick and widespread nonmagmatic sedimentary sequence. The foldbelt resulted from the collision of the Siberian platform with an island arc (including the Alazeya volcanic arc) that was developed on a passive continental margin (Fig. 8). The Momsko–Polousnensk uplift, west of the Alazeya arc, may have represented an accretionary wedge that was associated with this collision. The age of the folding and thrusting in the Yano–Kolymsk system decreases toward the Siberian platform, which may reflect progression in age of the collisional event (Fig. 9).

PALEOTECTONIC RECONSTRUCTION

The following is a scenario for the Mesozoic tectonic evolution of northeast Asia. In the early Mesozoic, several large continental masses of pre-early Mesozoic age can be recognized. These are: the Eastern Siberian continent (Siberian Platform), the Chukotsk microcontinental terrane, and the Bureinsk-Khankaisk microcontinental terrane (Fig. 10). The existence of the Okhotomorsk microcontinental terrane is inferred, based on the presence

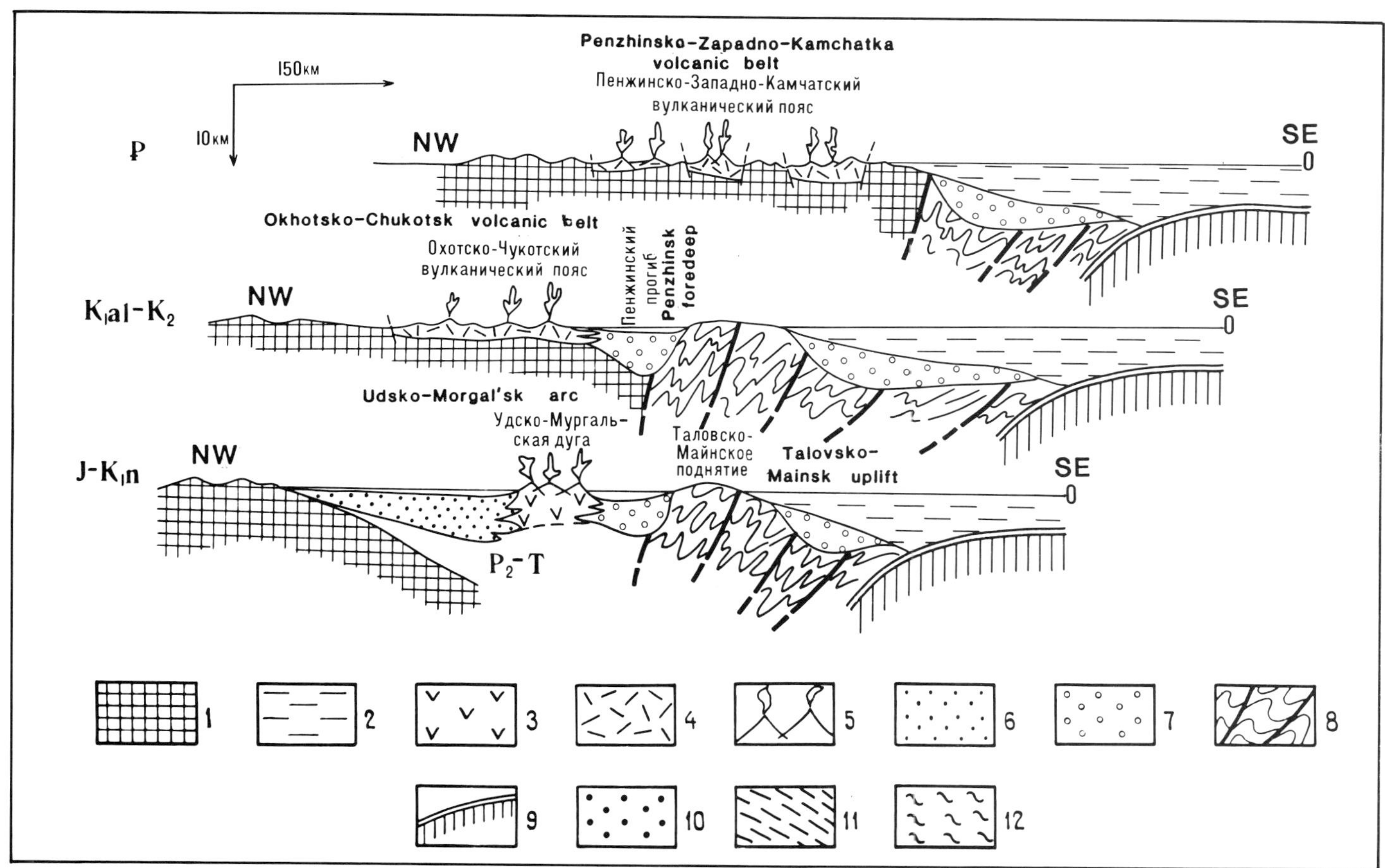

Figure 2—Paleotectonic sections illustrating the tectonic evolution of the Penzhinsk segment of the Okhotsk–Chukotsk belt and the adjacent part of the Koryaksk foldbelt. The Jurassic to Early Cretaceous (J-K_1n) tectonostratigraphic terranes are related to the Udsko–Murgal'sk island arc. The continental margin was formed in Aptian–Albian (K_1al–K_2) after the Kolymsk folding event that affected the entire Verkhoyano–Chukotsk region (Fig. 1). Relative to the earlier volcanic arc, the Okhotsk–Chukotsk volcanic-plutonic belt developed in a more continentward position. Petrochemical data suggest that this resulted from a flattening of the associated Benioff zone. The forearc basin can be discerned in the modern structure of the region and is superimposed on the forearc basin of the Udsko–Murgal'sk arc. The accretionary wedge was considerably expanded by the attachment of the early Mesozoic volcanic arc complexes of the Kankaren Range. The extensive forearc basin represented by the Upper Cretaceous olistostrome-bearing terrigenous complexes were formed on the arc side of the trench slope. During the Paleogene (P), the active continental margin was displaced considerably oceanward from its earlier position. (1) continental blocks; (2) ocean water; (3) volcanic arcs; (4) volcanic-plutonic belts; (5) active volcanos; (6) backarc basin; (7) forearc basin; (8) accretionary wedges; (9) oceanic crust; (10) deposits of the upper shelf; (11) deposits of the lower shelf and continental slope; (12) oceanic deposits.

of ancient metamorphic rocks in western Kamchatka and some geophysical data from the Sea of Okhotsk. The margin of the Eastern Siberian continent during the early Mesozoic is delineated on the north by the Oloysk and Alazeysk arcs, on the southeast by the Udsko–Murgalsk island arc, and on the south by the granodioritic batholithic belts of the Stanovoy Range. The microcontinental terrane margins were mainly passive in character. The presence of ophiolites within the intervening accretionary zones indicates that these microcontinental terranes were separated from the Siberian craton by oceanic crust.

The convergence and collision of the sialic microcontinental terranes with the Siberian craton resulted in mid-Jurassic–Early Cretaceous folding (Fig. 11). Collision of the Eastern Siberian continent and the Alazeya arc occurred at the end of Middle and the beginning of Late Jurassic (Fig. 9). The Bureinsk–Khankaisk microcontinental terrane collided with the Siberian craton during the Late Jurassic producing the Mongolo–Okhotsk foldbelt (Fig. 7). The closing of the South-Anyui ocean basin and the formation of the South-Anyuisk foldbelt occurred in mid Early Cretaceous time (Fig. 6).

By the late Early Cretaceous, the continental margins of northeastern Asia had become similar in shape to its modern margins (Fig. 12). During the late Early Cretaceous to Late Cretaceous, the Okhotsk–Chukotsk volcanic-plutonic belt (Fig. 2) marked the site of an active continental margin along the southeastern margin of the Siberian craton. The Sikhote–Alinsk volcanic arc (Fig. 3) is present on the east of Sikhote–Alinsk and may be the southern continuation of this magmatic belt. The northeast-trending strike-slip faults of the Sikhote–Alinsk region represent

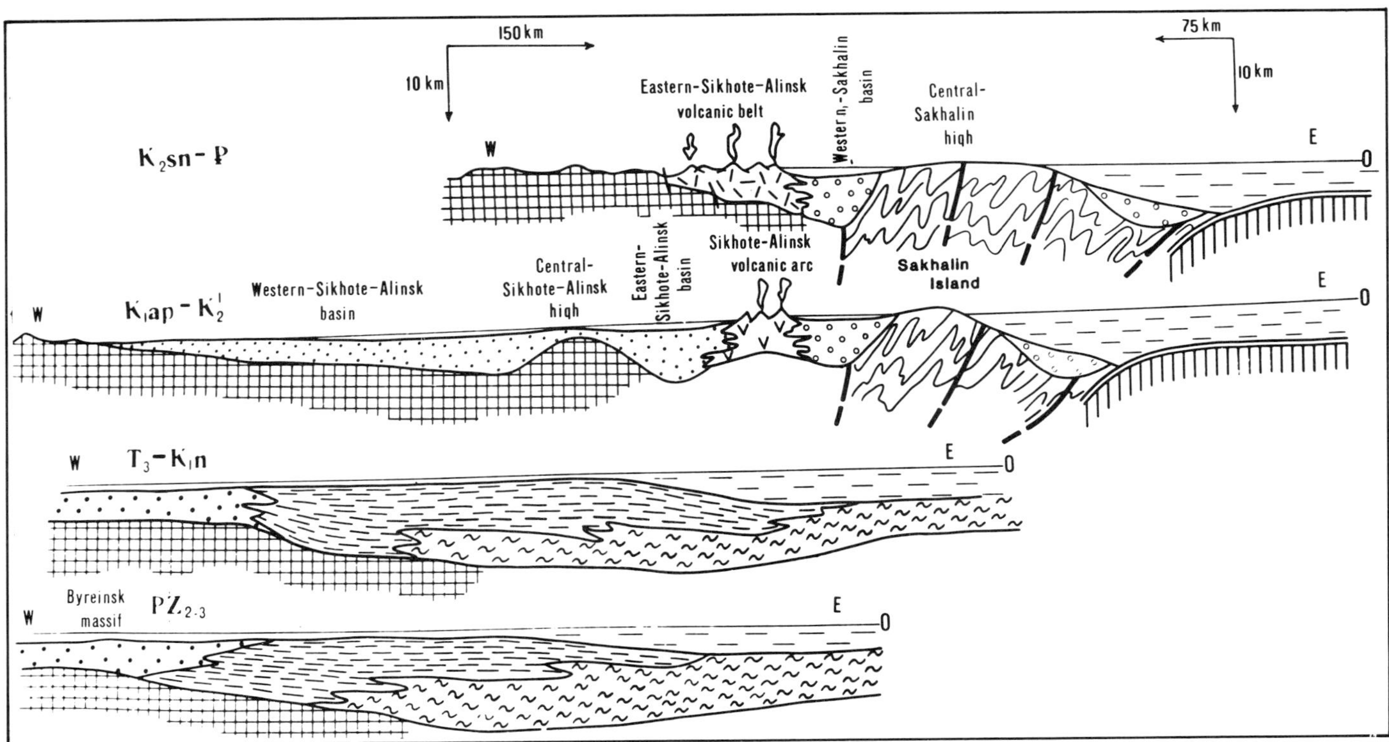

Figure 3—Paleotectonic sections illustrating the tectonic evolution of the Sikhote-Alinsk-Sakhalinsk region. A passive continental margin developed along the eastern margin of the Bureinsk massif from the end of middle Paleozoic (PZ_{2-3}) through the Neocomian (T_3–K_1m). The Sikhote-Alinsk volcanic arc formed in the Aptian–Turonian (K_1ap–K_2l). Fragments of this volcanic arc occur as debris within the younger formations. Contemporaneous backarc basins are represented by thick terrigenous sequences that compose most of the Sikhote-Alinsk Range. Although somewhat disrupted, a forearc basin sequence that overlies an ophiolite is exposed west of Sakhalin. A corresponding accretionary zone consisting of upper Paleozoic–early Mesozoic tholeiitic and alkaline basalts, cherts, graywackes, glaucophane-bearing schists, and eclogites is exposed on Eastern Sakhalin. During Senonian–Paleogene (K_2sn-P), the volcanic arc was replaced by an Andean-type margin that includes the Eastern Sikhote-Alinsk volcanic-plutonic belt and its forearc basin, including the earlier deposits of the Aptian–Turonian island arc. Oceanic and flyschoid sediments of the late Mesozoic accretionary forearc basin were added to the Aptian–Turonian accretionary belt during the Senonian to Paleogene. Symbols are the same as for Figure 2.

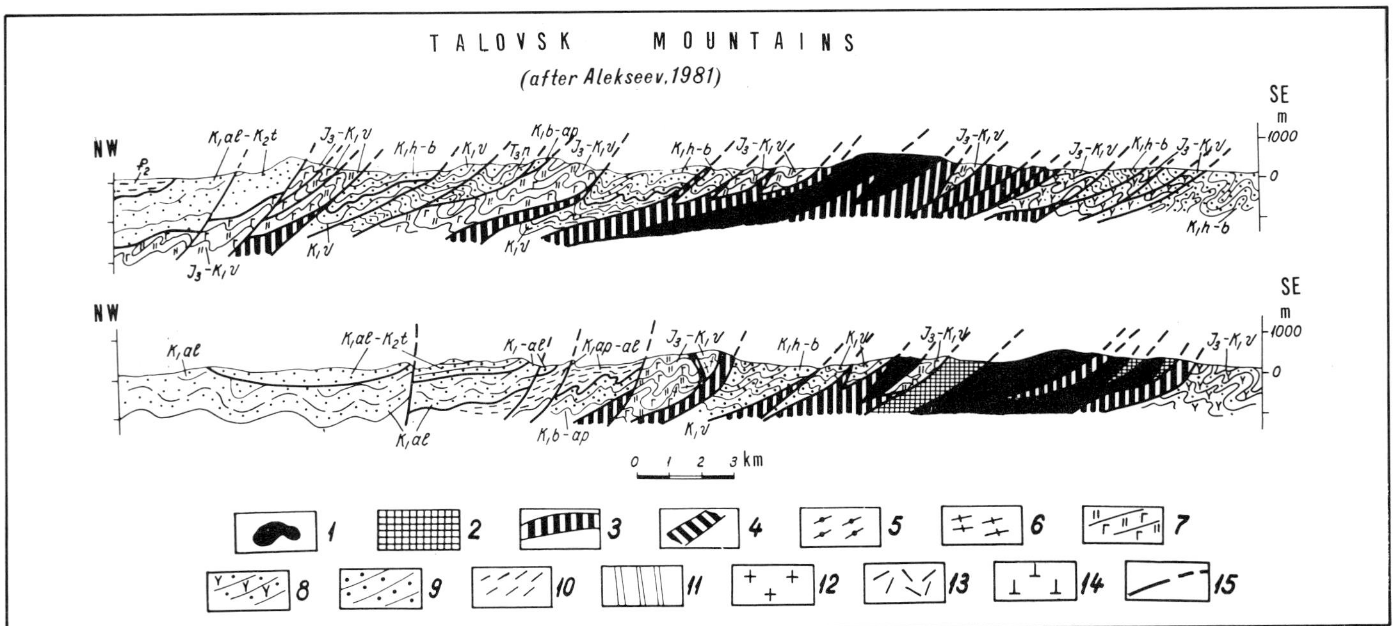

Figure 4—Generalized geological sections across the Talovsko-Mainsk accretionary wedge within the Koryaksk upland. The accretionary wedge associated with the Udsko-Murgal'sk arc was not significantly modified by later deformational events. (1) ultramafic rock; (2) gabbroic rock; (3) serpentinite melange; (4) gabbro and anorthosites; (5) amphibole-two-pyroxene schists; (6) schist; (7) chert and basalt; (8) chert, tuff, shale, and sandstone; (9) sandstone; (10) siltstone and argillite; (11) sheeted dikes; (12) granite; (13) silicic volcanic rocks; (14) intermediate and mafic volcanic rocks; (15) faults.

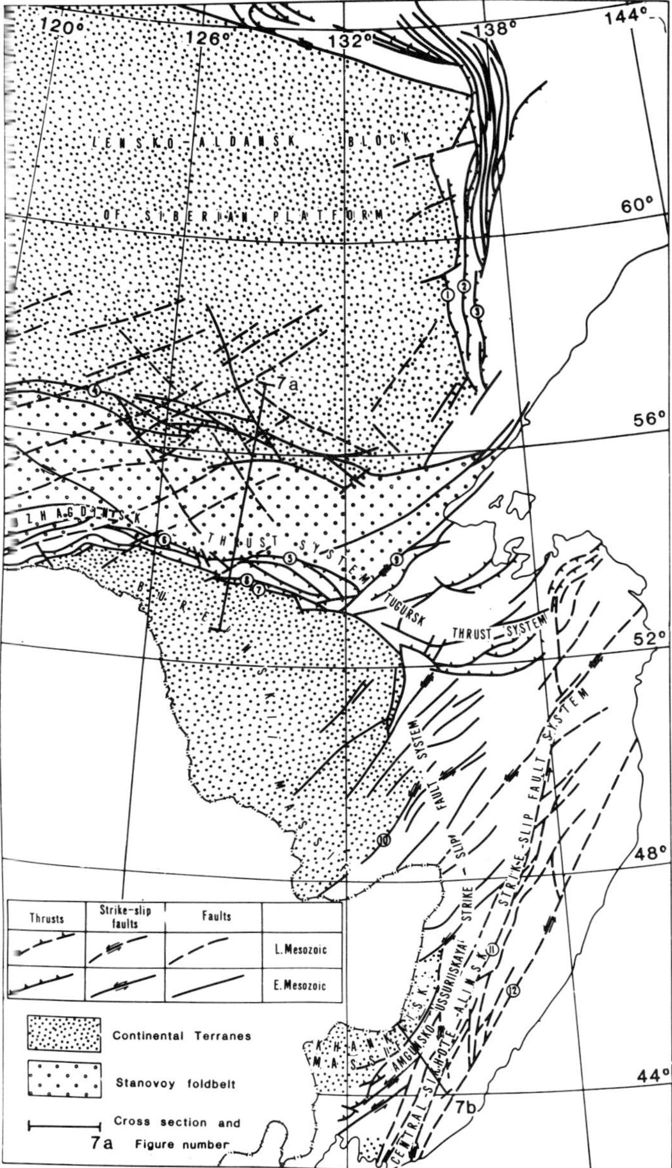

Figure 5—Mesozoic fault systems in the southern part of the Soviet Far East (after Parfenov et al, 1981). The oldest faults (Early Cretaceous) are mainly left-lateral strike-slip faults of the Angunsko-Ussuriisk system that displaced in a stepwise fashion the eastern edges of the Bureinsk and Khankaisk massifs. The en echelon fault pattern is characteristic of this system. The Central-Sikhote-Alinsk system of extensive left-lateral strike-slip faults was formed in the Middle Cretaceous. Horizontal displacements of both systems were taken up by the Turgusk thrust system. The faults are labeled as follows: (1) Nel'kano-Kyllaksk; (2) Guvindinsk; (3) Chelatsk; (4) South Chul'mansk; (5) Lansk; (6) North Tukuringrsk; (7) Ninni-Sagayansk; (8) South-Tukuringrsk; (9) Uligdansk; (10) Kukansk; (11) Central- Sikhote-Alinsk; (12) Vostochnyi.

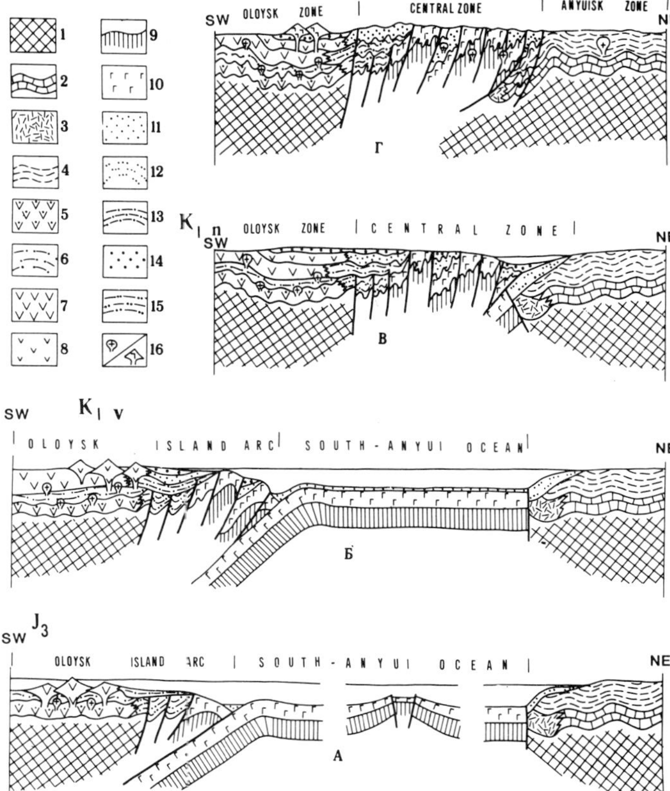

Figure 6—Paleotectonic sections illustrating the tectonic evolution of the South-Anyuisk foldbelt. The Eastern-Siberian craton (left part of the diagram) and Chukotsk (right part of the diagram) continental terrane were separated in early Mesozoic by the South-Anyui Ocean in which there was active spreading until the end of Jurassic. The Chukotsk block had a passive margin consisting of the thick terrigenous series of the Chukotsk miogeosynclinal system superimposed on the relics of a Carboniferous convergent boundary. The Eastern-Siberian continental block was bordered by the Late Jurassic Oloysk island arc. This volcanic arc was superimposed on the similar, but more complexly deformed Upper Triassic–Middle Jurassic terranes. Collision and suturing of the two continental blocks and formation of a collisional system occurred at the end of Neocomian (K_1n). (1) Pre-Riphean folded basement of the Chukotsk cratonic block and Paleozoic complexes of the Alazeisko-Oloysk system (Eastern Siberian continental block); (2) middle Paleozoic clastic and carbonate sequences; (3) middle Paleozoic volcaniclastic sequences; (4) Triassic clastic sequences; (5) Upper Triassic calc-alkalic volcanic rocks and graywacke; (6) early Middle Jurassic flysch; (7) Upper Jurassic and (8) Neocomian calc-alkalic and alkalic volcanic rocks, graywacke, and conglomerate; (9) ultramafic and gabbroic rocks and plagiogranite; (10) Upper Jurassic basalt, chert, schist, and graywacke; (11) Berriasian–Valanginian flysch; (12) shale, sandstone; (13) Upper Jurassic flysch; (14) Hauterivian graywacke; (15) Upper Triassic flysch; (16a) granitic rocks; (16b) volcano.

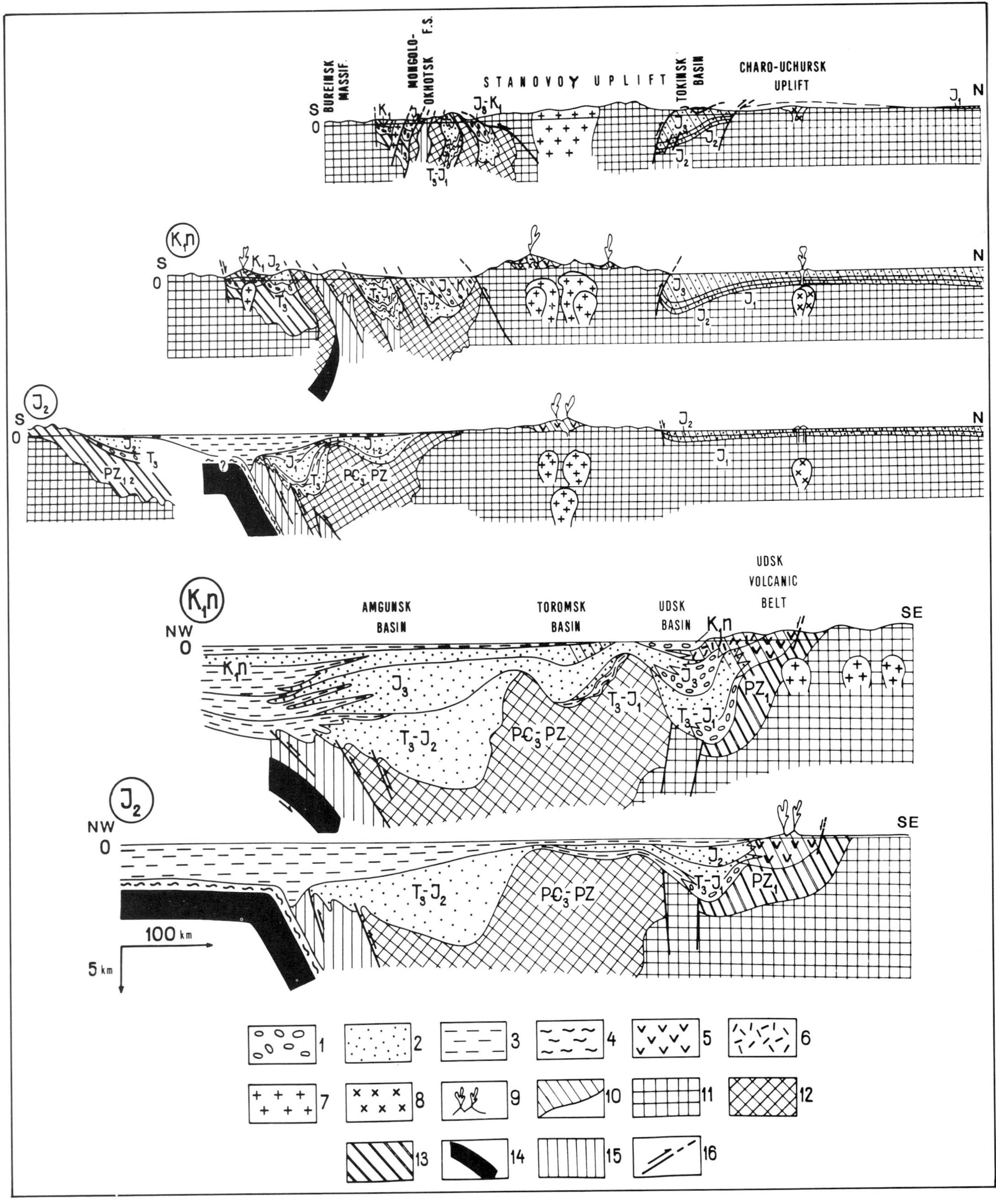
BUREINSK MASSIF
MONGOLO-OKHOTSK F.S.
STANOVOY UPLIFT
TOKINSK BASIN
CHARO-UCHURSK UPLIFT
S
N
NW
SE
AMGUNSK BASIN
TOROMSK BASIN
UDSK BASIN
UDSK VOLCANIC BELT
100 km
5 km

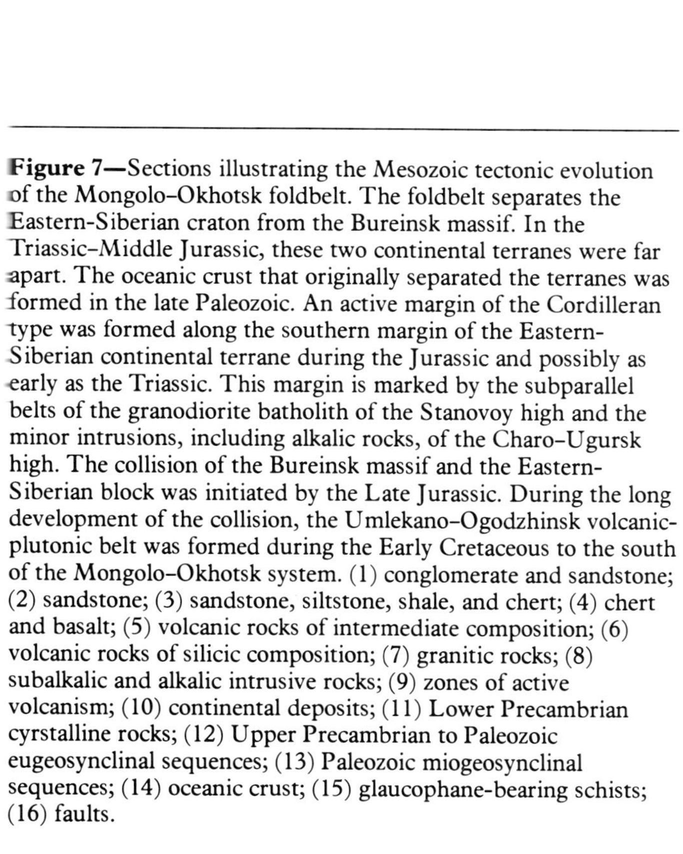

Figure 7—Sections illustrating the Mesozoic tectonic evolution of the Mongolo-Okhotsk foldbelt. The foldbelt separates the Eastern-Siberian craton from the Bureinsk massif. In the Triassic-Middle Jurassic, these two continental terranes were far apart. The oceanic crust that originally separated the terranes was formed in the late Paleozoic. An active margin of the Cordilleran type was formed along the southern margin of the Eastern-Siberian continental terrane during the Jurassic and possibly as early as the Triassic. This margin is marked by the subparallel belts of the granodiorite batholith of the Stanovoy high and the minor intrusions, including alkalic rocks, of the Charo-Ugursk high. The collision of the Bureinsk massif and the Eastern-Siberian block was initiated by the Late Jurassic. During the long development of the collision, the Umlekano-Ogodzhinsk volcanic-plutonic belt was formed during the Early Cretaceous to the south of the Mongolo-Okhotsk system. (1) conglomerate and sandstone; (2) sandstone; (3) sandstone, siltstone, shale, and chert; (4) chert and basalt; (5) volcanic rocks of intermediate composition; (6) volcanic rocks of silicic composition; (7) granitic rocks; (8) subalkalic and alkalic intrusive rocks; (9) zones of active volcanism; (10) continental deposits; (11) Lower Precambrian cyrstalline rocks; (12) Upper Precambrian to Paleozoic eugeosynclinal sequences; (13) Paleozoic miogeosynclinal sequences; (14) oceanic crust; (15) glaucophane-bearing schists; (16) faults.

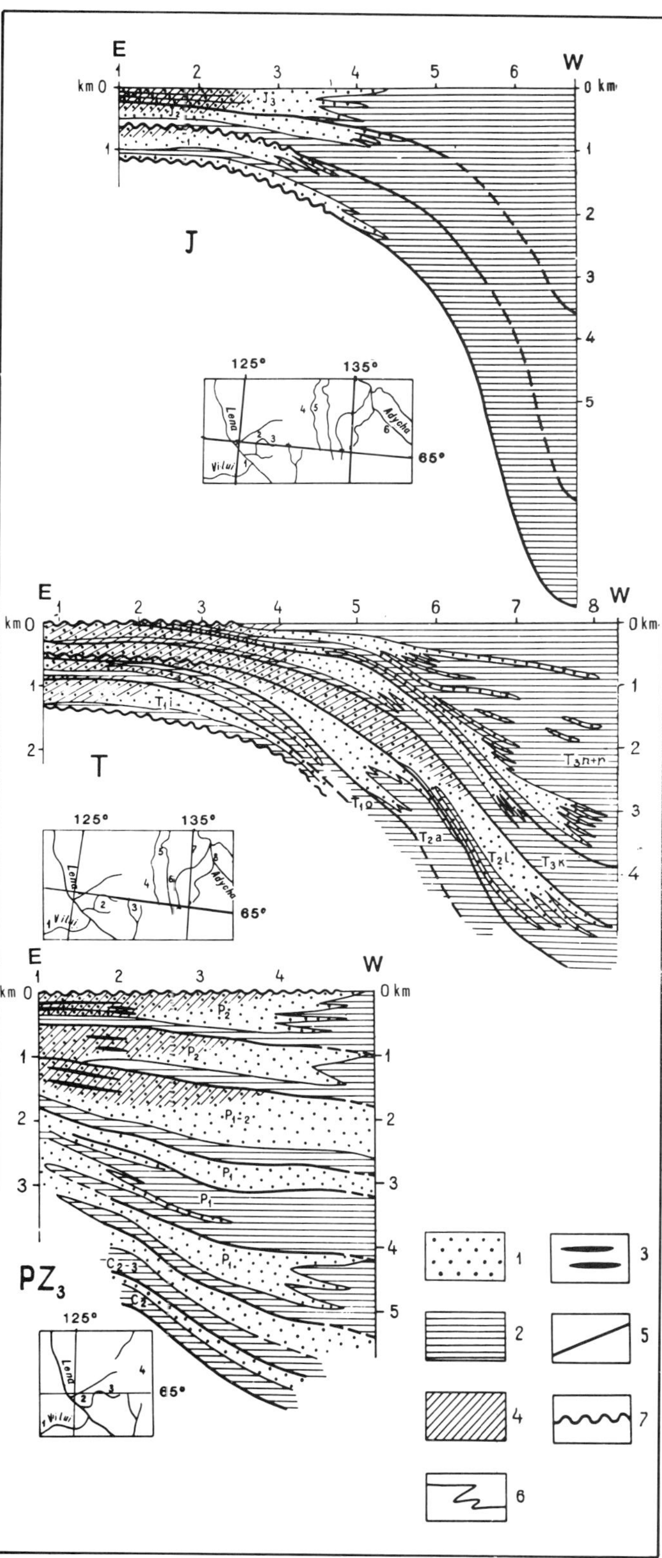

Figure 8—Lithologic features of the Verkoyansk complex of the Late Paleozoic–Triassic passive continental margin (Arkhipov and Parfenov, 1980; with additions by the authors). (1) sandstone; (2) siltstone, shale, argillite; (3) coal; (4) continental deposits; (5) time lines; (6) facies changes; (7) unconformities.

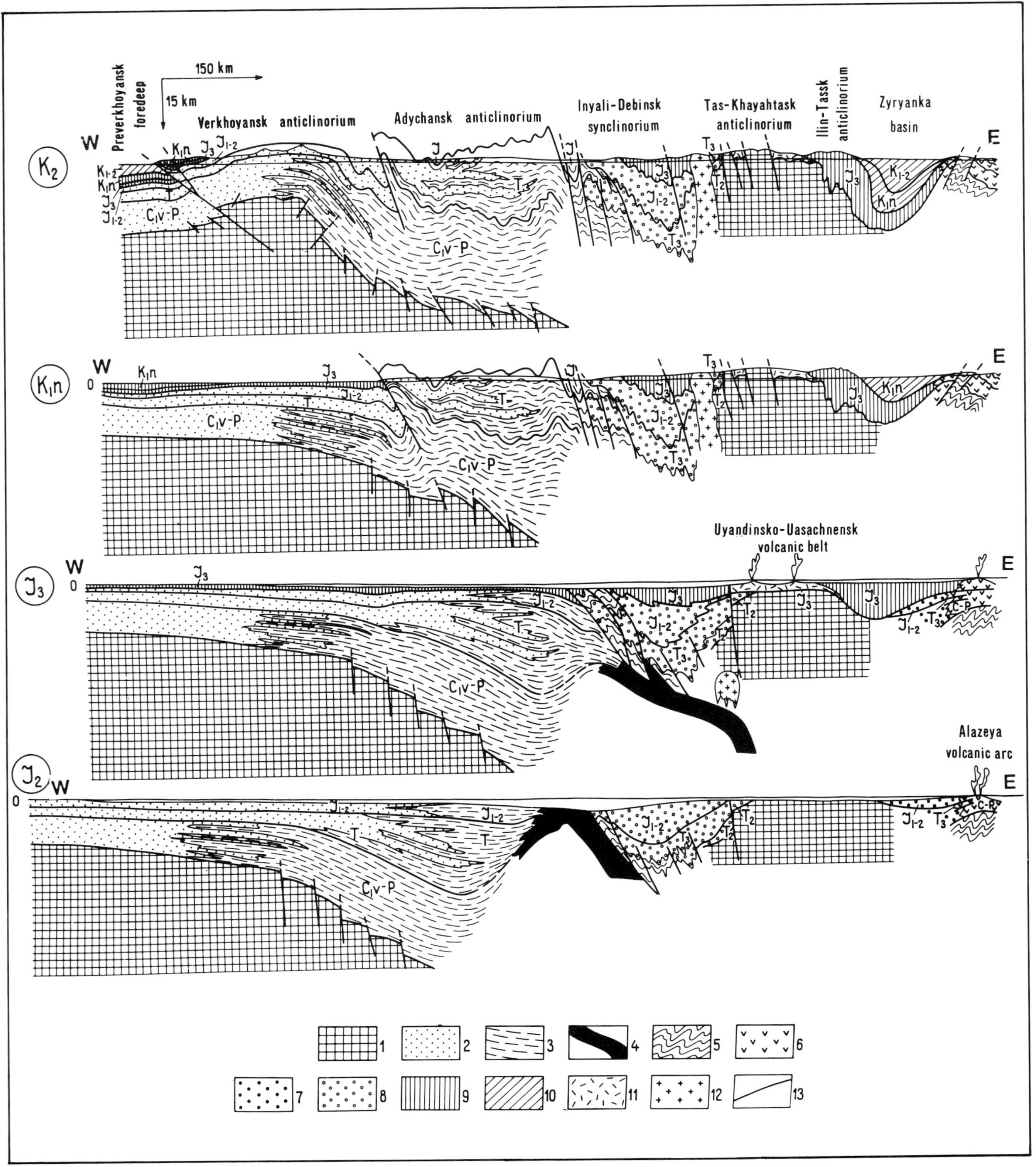

Figure 9—Paleotectonic sections illustrating the tectonic evolution of the central parts of the foldbelts bordering the Siberian platform on the northeast. (1) continental terranes with different stratigraphic sequences and ages; (2) sandstone; (3) siltstone and shale; (4) oceanic crust; (5) schist; (6) volcanic arc complexes; (7) forearc basins; (8) accretionary forearc basins; (9) molasse; (10) upper molasse; (11) volcanic rocks of various composition; (12) granites; (13) faults.

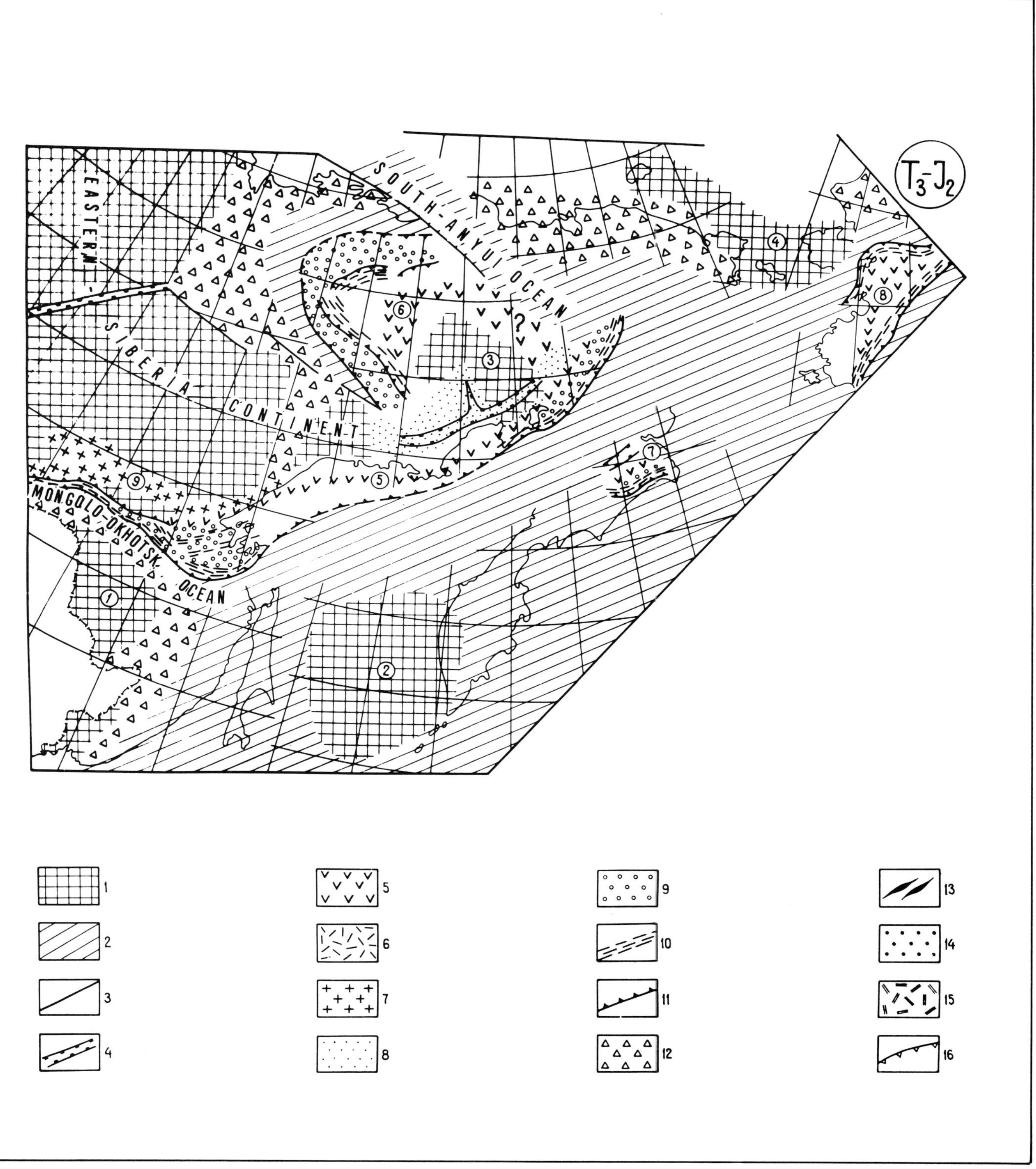

Figure 10—Late Triassic–Middle Jurassic paleotectonic reconstruction. (1) continents and microcontinental terranes; (2) oceanic crust; (3) transform faults; (4) rift zones; (5) volcanic arcs; (6) volcanic-plutonic belts; (7) granodiorite batholithic belts; (8) backarc basins; (9) forearc basins; (10) accretionary wedges; (11) position of subduction zones; (12) passive continental margins; (13) Cretaceous foldbelts; (14) intermontane troughs and rift basins behind active continental margins of Cordilleran type; (15) epicollisional volcanic-plutonic belts; (16) thrust. ①–④ microcontinents: ① Bureinsk–Khankaisk, ② Okhotomorsk, ③ Omolonsk, ④ Chukotsk; ⑤–⑧ volcanic arcs: ⑤ Udsko–Murgal'sk, ⑥ Alazeisk, ⑦ Kankaren, ⑧ Yukon–Koyukuk; ⑨ Stanov belt of granodioritic batholiths.

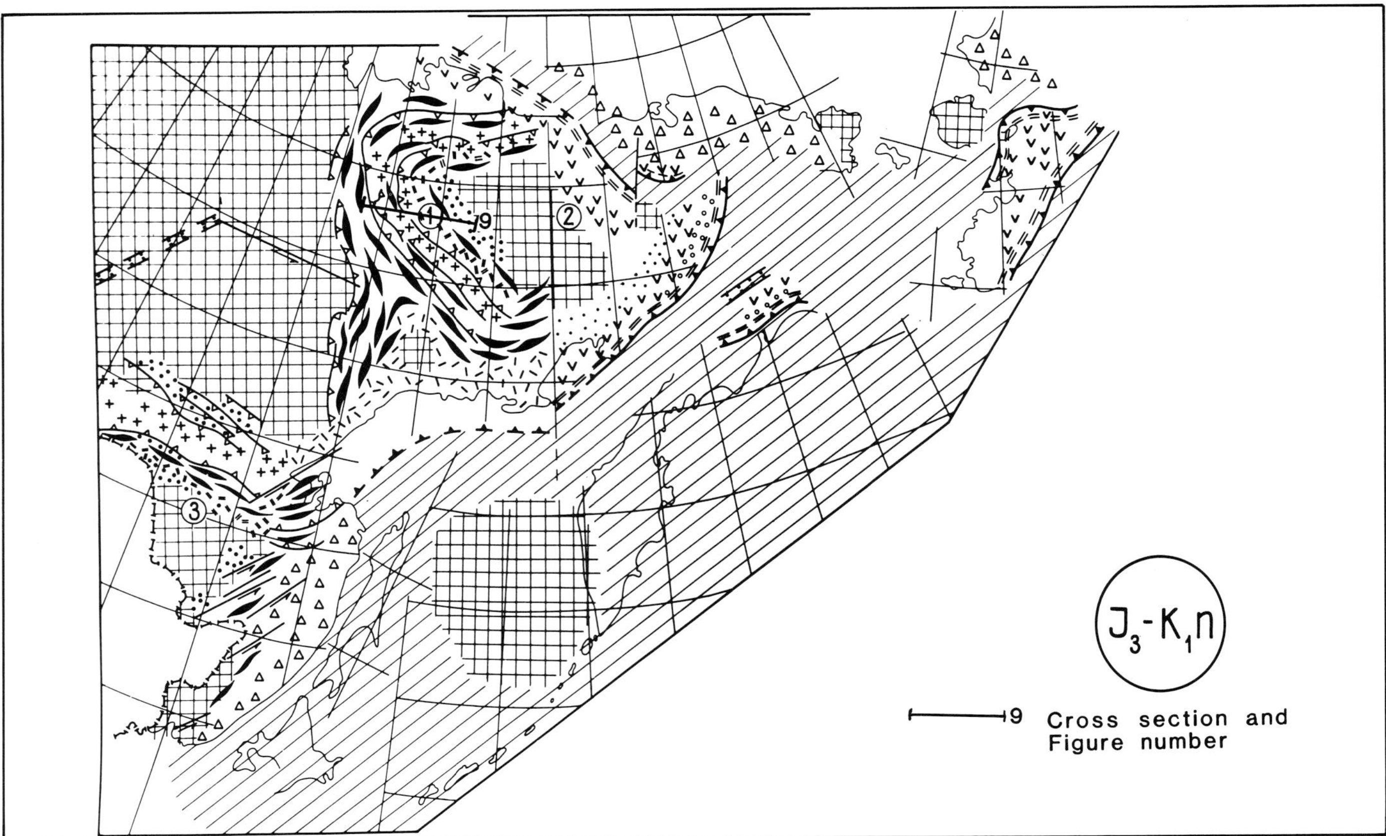

Figure 11—Late Jurassic–early Lower Cretaceous (Neocomian) paleotectonic reconstruction. Symbols are the same as for Figure 10. The diagram is labeled as follows: (1) Uyandinsko-Yasachnensk foldbelt; (2) Kolymo-Omolonsk massif; (3) Umlekano-Ogodzhinsk foldbelt.

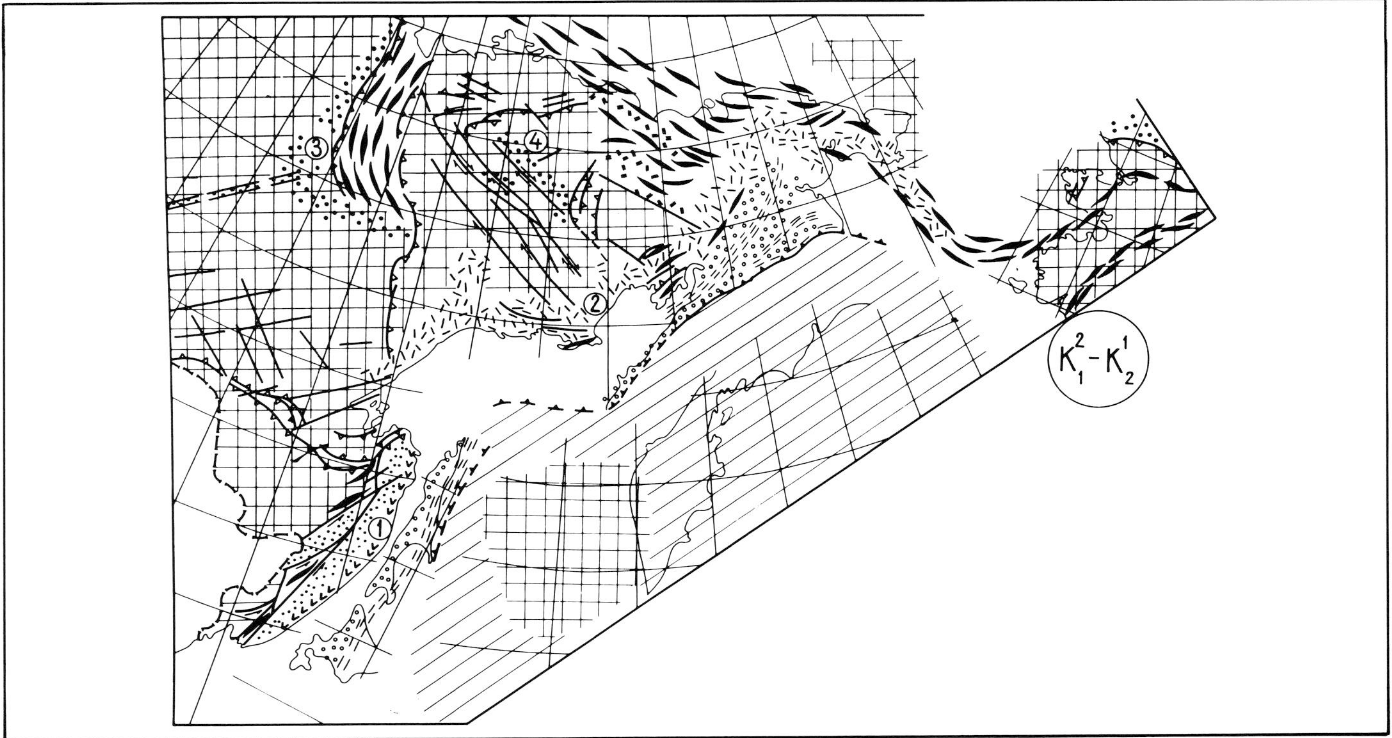

Figure 12—Paleotectonic reconstruction of late Lower Cretaceous–early Upper Cretaceous (Aptian–Senonian). Symbols are the same as for Figure 10. The diagram is labeled as follows: (1) Sikhote-Alinsk volcanic arc; (2) Okhotsko-Chukotsk volcanic-plutonic belt; (3) Preverkhoyansk foredeep; (4) Zyryansk basin.

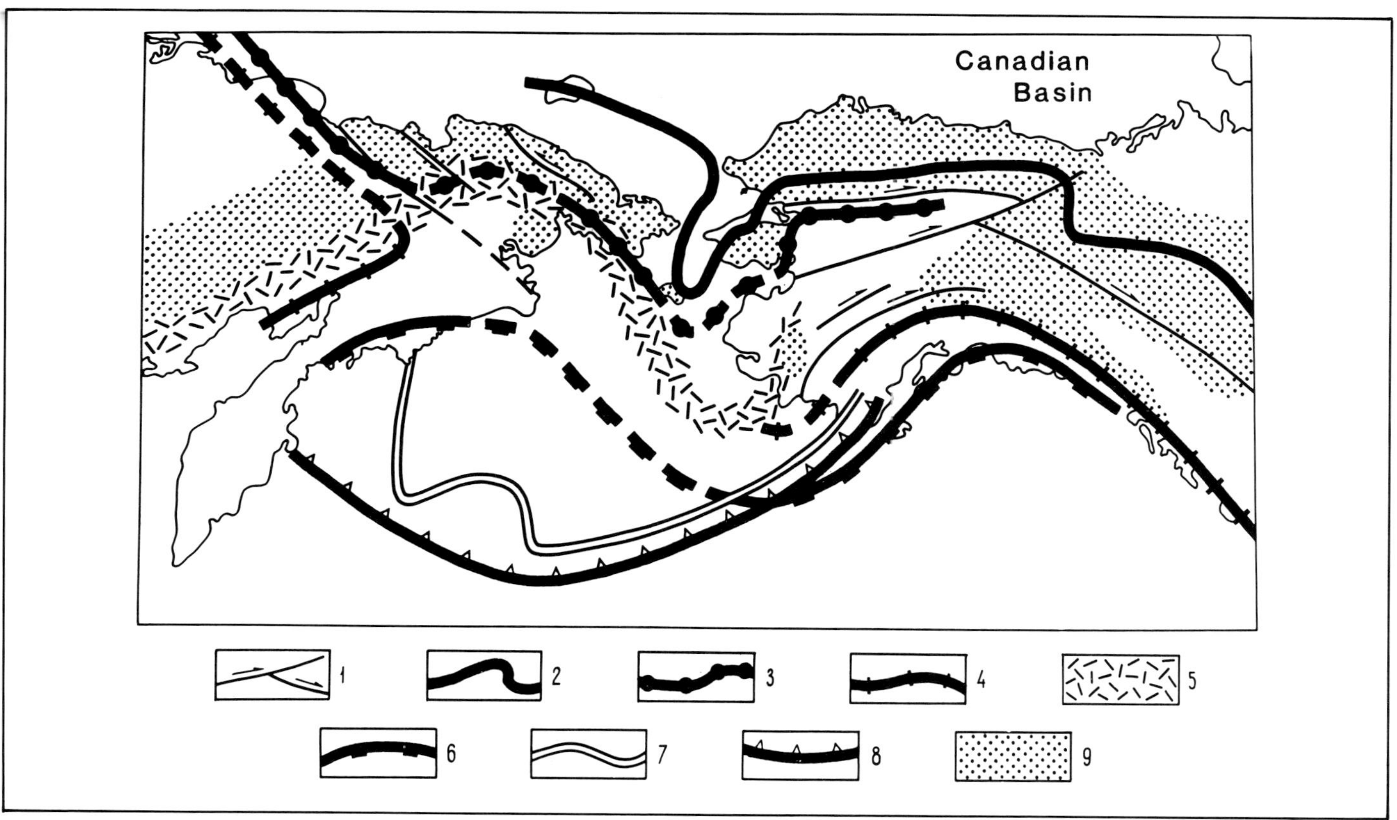

Figure 13—Features of the Beringovomorsk (Bering Sea) orocline. (1) strike-slip faults; (2) zone of Paleozoic folding; (3) Late Jurassic-Neocomian eugeosynclinal foldbelts; (4) Mesozoic volcanic arcs; (5) Okhotsko-Chukotsk volanic-plutonic belt and its possible extension into Alaska; (6) latest Cretaceous eugeosynclinal foldbelt; (7) Paleogene volcanic arc; (8) Neogene volcanic arc; (9) Mesozoic continents and microcontintents.

transform faults that were active after collision of the Bureinsk-Khankaisk microcontinental terrane with the southeastern margin of the Siberian craton. The faults resulted from the northwestward movement of the Kula plate and the Okhotomorsk microcontinental terrane toward the Okhotsk-Chukotsk belt.

After accretion of the above terranes to the Eurasian continent, they were extensively folded and faulted during the late Mesozoic. These later tectonic events were mainly caused by the opening of the North Atlantic and Canadian Ocean basins and reflect the presence of a zone of compression between Eurasia and North America (Fig. 13).

ACKNOWLEDGMENTS

T. E. Moore, W. K. Wallace, K. D. Small, D. G. Wetherell, and M. Churkin, Jr., of ARCO Alaska, Inc., revised the original manuscript to clarify the terminology. Some geographic names and coordinates were added to the figures to help readers unfamiliar with the region.

REFERENCES

Alexandrov, A. A., 1978, Nappes and imbricated-thrust structures of the Koryakskoye upland: Moscow, Nauka, 122 p. (in Russian).

Alekseev, E. S., 1981, Kuyulskiy serpentine melange and the structure of the Talovsko-Mainskaya zone (Koryakskoye upland): Geotectonics, n. 1, p. 105–120 (in Russian).

Arkhipov, Y. V., and L. M. Parfenov, 1980, The Adychanskaya zone of gentle deformation: Reports of the USSR Academy of Sciences, v. 250, n. 1, p. 155–158 (in Russian).

Nekrasov, G. E., 1978, New data on the tectonic structure of Pekul'ney Range (left bank of the Anadyr' River): Reports of the USSR Academy of Sciences, v. 238, n. 6, p. 1433–1436 (in Russian).

Parfenov, L. M., and B. A. Natal'in, 1981, Rules governing the structure and tectonic evolution of Mesozoic and Cenozoic foldbelts of the northwest part of the Pacific margin: Geology and Geophysics, n. 7, p. 3–15 (in Russian).

———, et al, 1978, Geodynamics of north-eastern Asia in the Mesozoic and Cenozoic and the nature of its volcanic belts: Journal of the Physics of the Earth, v. 26, Suppl., p. S503–S525.

———, et al, 1979, Tectonic zonation and tectonostratigraphic evolution of northeastern Asia: Moscow, Nauka, 240 p.

———, et al, 1981, Tectonic evolution of the active continental margins of the northwestern Pacific Rim: Geotectonics, n. 1, p. 85–104 (in Russian).

Rifting, Drifting, and Crustal Accretion in the Taiwan Sector of the Asiatic Continental Margin*

W. G. Ernst
University of California
Los Angeles, California

C. S. Ho
Energy and Mining Research Service Organization
Taipei, Taiwan

J. G. Liou
Stanford University
Stanford, California

Permian and younger rocks record the effects of rifting and passive margin deposition, sea-floor spreading, convergence and oblique closure, volcanic and plutonic arc construction, arrival of ophiolitic material, and the suturing of exotic trench-argillite and andesitic arc assemblages in central and eastern Taiwan. Large portions of the sialic crust evidently formed approximately in situ, then were deformed and thrust landward during subsequent tectonic events, but far-traveled terranes and allochthonous fragments of terranes played a substantial role in accretion. Recognized and suspected exotics include: (1) lower or mid-Mesozoic amphibolites + serpentinites (now high-rank metaophiolites), situated anomalously in the inboard upper Mesozoic Tailuko metamorphosed miogeoclinal and/or continental slope belt (itself possibly but not definitely far traveled); (2) the outboard upper Mesozoic Yuli high-pressure, low-temperature metamorphic trench-argillite melange complex; (3) Mio-Pliocene tectonic blocks of blueschist metaophiolite, which have been emplaced in the east-central part of the Yuli terrane; (4) olistostromal debris supplied to the Pliocene Lichi Melange of the yet further eastward Coastal Range, from preexisting Miocene oceanic crust of the South China Sea; and (5) the Neogene calc-alkaline Luzon arc that began to collide with the Asiatic sialic crust in Plio-Pleistocene time. The Cenozoic slate series represents an additional, largely fault-bounded parautochthonous terrane, which formed as the Tertiary miogeoclinal cover along the Asiatic continental margin to the west of the approaching Luzon arc and was itself thrust westward during the Plio-Pleistocene arc collision.

INTRODUCTION

Taiwan is a relatively well-exposed, thoroughly studied portion of a Permian and younger orogenic belt. Because it is relatively small, one might expect that complications in this mountain chain would be only moderate. However, intricacies revealed by geologic, structural, petrologic, and geophysical investigations suggest caution in accepting unreservedly any plate tectonic scenario, at present, as overly simplified.

Our attempt to unravel the history of rifting, drifting, and crustal accretion of Taiwan will be based on its current plate tectonic setting, the geologic relationships of its several lithotectonic belts, and recognizable or interpretative processes that have shaped the nature of the island; the various episodes of metamorphism are emphasized in placing constraints on the development of this portion of the crust. We apply the tectonostratigraphic terrane concept (Coney et al, 1980; Jones et al, 1983) as an aid in understanding the geologic growth of Taiwan. We hope that this review will stimulate new investigations that in turn will produce more realistic refinements in the future.

Current Plate Tectonic Setting

As illustrated in Figure 1, Taiwan occupies a prominent salient along the leading edge of the Asiatic lithospheric plate (Biq, 1971, 1981; Chai, 1972; Bowin et al, 1978; Hamilton, 1979; Ernst, 1983a). Present-day differential motions of the impinging plates are well documented by earthquake foci (Tsai, 1978; Hamburger et al, 1983). The more easterly Philippine Sea plate is moving northwestward (Seno, 1977) relative to Asia. Convergence is accommodated along its northern boundary by subduction of the Philippine Sea plate beneath the Ryukyu volcanic arc, which has been constructed on oceanic crust directly seaward from the Asiatic continental margin. The Tatun andesitic volcano field at the northern tip of Taiwan probably represents the western termination of the Ryukyu arc. Because of curva-

* Institute of Geophysics and Planetary Physics Publication No. 2493.

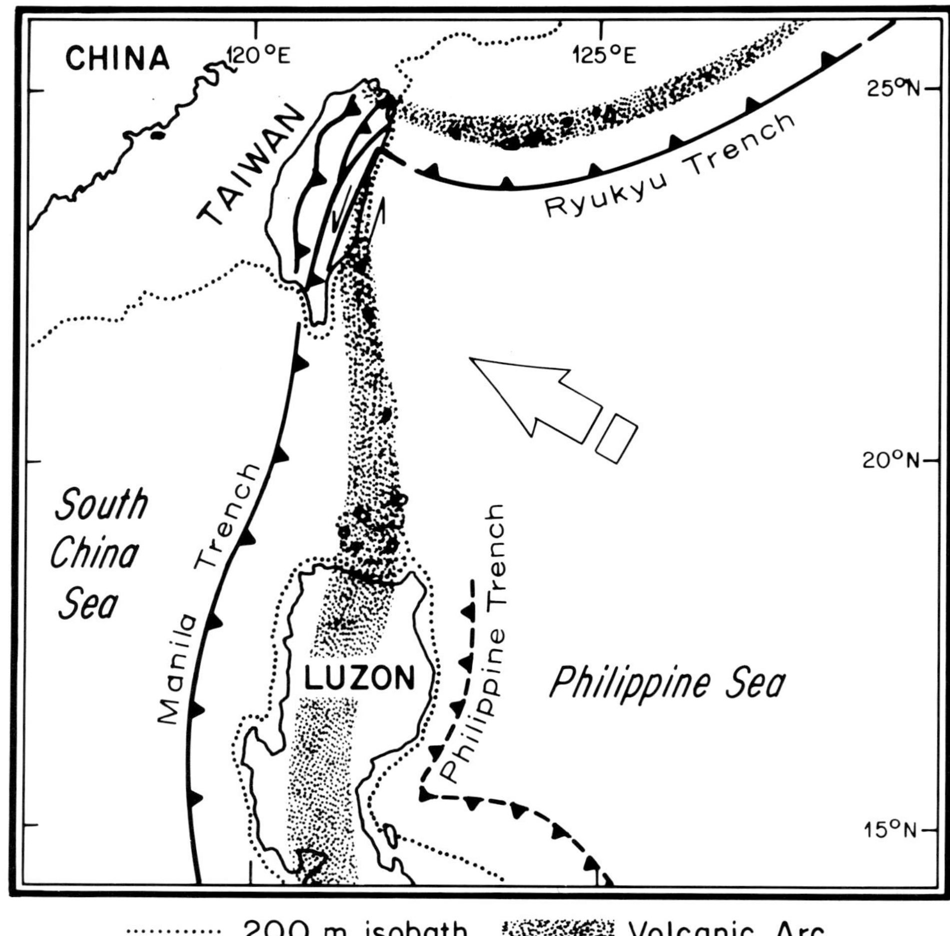

Figure 1—Regional plate tectonic setting of Taiwan at the intersection of Ryukyu and Manila trenches with the continental margin of Asia, largely after Hamilton (1979). The arrow indicates the approximate motion of the Philippine Sea plate relative to Asia (7 cm/year [3 in./year] according to Seno, 1977). The Philippine Sea plate is descending obliquely beneath the Ryukyu arc but is being thrust over the South China Sea and Chinese continental margin. Thus, relative to the Asiatic plate, the Manila trench is moving concomitantly westward with time. The Philippine trench appears to be propagating northward (Lewis and Hayes, 1983).

ture of the surface expression of this convergent plate boundary, the western extension of the Ryukyu trench currently appears to be experiencing chiefly right-lateral transform motion.

South of Taiwan, convergence is accommodated along the western boundary of the Philippine Sea plate by eastward underflow of the South China Sea and incipient subduction of the Chinese continental margin. The northern part of the Philippine Islands, a northward-extending bathymetric high, and the Coastal Range of eastern Taiwan constitute the Luzon calc-alkaline volcanic arc, which has been produced by this eastward subduction of the Asiatic lithospheric plate; modern igneous activity is confined to the southern portion of the arc, but volcanism occurred in the more northerly Coastal Range during Miocene and Pliocene time. The island of Taiwan marks a sector where the Asiatic continent adjacent to the South China Sea mantle lithosphere in large part has not followed the latter as it descended beneath the Luzon arc. Instead, the edge of the continent has decoupled from the subducting slab and may be described as having collided with the arc; in the process, this eastern edge of Asia is being structurally compressed, thickened, and uplifted.

Taiwan therefore consists of portions of two major tectonic entities: (1) the Asiatic sialic crust including pre-Tertiary basement, and (2) the accreted Philippine Sea Luzon arc. Overlying the basement of the continental margin lies an imbricated, westward-verging sedimentary prism of Cenozoic age. In part deposited on the Asiatic continental margin and in part on the now-consumed South China Sea oceanic crust, this passive margin section is largely allochthonous: Portions of this unit have been thrust westward up to about 160 km (100 mi) towards the mainland during continued convergence (Suppe, 1976; Suppe and Wittke, 1977). Its contact with the Luzon volcanic arc occurs along the Longitudinal Valley of eastern Taiwan, a modern intraorogenic left-lateral strike-slip zone (Allen, 1962; Biq, 1965; Hsu, 1976a). General relations are sketched in Figure 2.

Prior to the collision of the Coastal Range and the Asiatic continental margin that commenced about 4 m.y. ago (Chi et al, 1981), oceanic crust of the South China Sea

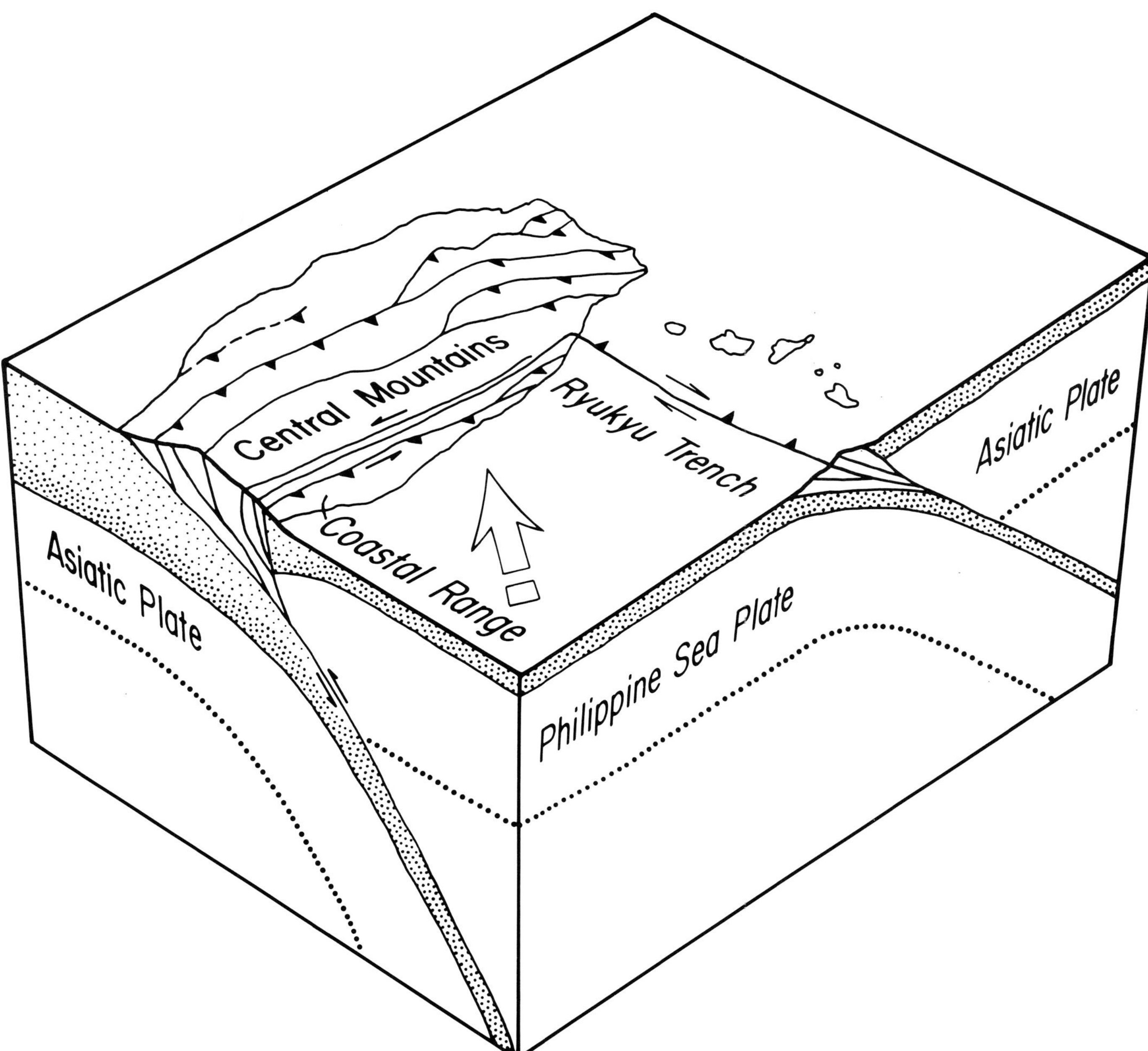

Figure 2—Schematic diagram, only approximately to scale, showing present plate tectonic configuration of Taiwan (Ernst, 1983a). Whereas the convergent plate junction located seaward from the Ryukyu arc is a well-defined linear feature, the eastward underflow of the Asiatic plate beneath the Philippine Sea and surmounting Coastal Range (the Luzon calc-alkaline arc) is being choked off because of accumulation of the imbricated passive margin wedge (+ forearc basin) and distributory thrust fault system present throughout the main part of Taiwan. The Longitudinal Valley marks the juxtaposition of exotic Coastal Range (including its forearc basin) with the miogeoclinal wedge and its Asiatic basement; a minimum of 150 km (90 mi) left-lateral slip has taken place along this strike-slip fault system since the Pliocene onset of the collisional event (Tsai, 1978).

lay along the eastern edge of the Asiatic lithospheric plate (as it still does today southwest of Taiwan). Consumption of this intervening oceanic crust allowed (1) impaction of the Chinese sialic crust, initially in northeasternmost Taiwan, with the northern end of the calc-alkaline Luzon arc, (2) incipient subduction of the Asiatic continental crust, and (3) imbrication and westward thrusting of the miogeoclinal wedge. Distributory shear apparently has moved westward with time; in a sense, this has resulted in the temporary offloading of the sutured, sheared units onto the nonsubducted Philippine Sea plate. Impaction has partly been accommodated by sinistral strike-slip as well as by landward thrusting. Thus, a broad zone of deformation is envisioned, rather than a discrete, well-defined plate boundary; the Longitudinal Valley fault zone simply constitutes a major strand of a rather broad distributory shear system. As convergence continued, the zone of collision has extended southwestward as more sialic material apparently became involved in the eastward underflow, and the South China Sea has been progressively consumed. Portions of the Asiatic basement were also transported continentalward (see Fig. 4). Ultimately, buoyancy forces should choke off the Manila trench subduction zone, and a new west-dipping convergent plate junction will be initiated, probably, as conjectured by Bowin et al (1978), as a northward extension of the reactivated Philippine trench (see also Cardwell et al, 1980; Lewis and Hayes, 1983). This process ultimately should result in transference of the entire Luzon arc to the then stabilized Asiatic plate, including surviving remnants of the South China Sea, and the westward underflow of the Philippine Sea lithosphere beneath Taiwan (Ernst, 1983a).

GENERAL GEOLOGY AND PETROLOGY

The various lithotectonic belts of Taiwan are illustrated in Figure 3 (Ho, 1982). The interpretation of individual suspect terranes will be presented in the sections dealing with crustal evolution. From west to east the lithotectonic realms are: (1) the Coastal Plain; (b) the Western Foothills; (c) the Hsuehshan Range; (d) the Backbone Range; (e) the Basement Complex; and (f) the Coastal Range. Units (b), (c), and (d) comprise the imbricated, allochthonous Cenozoic cover strata and minor intercalated mafic volcanics of the sequence of slates and argillites structurally overlying (e) the pre-Tertiary Tananao Schist Complex, representing the Asiatic continental crust; (f) is the Neogene Luzon calc-alkaline arc plus associated sediments as well as the Luzon trough forearc basin fill of the Coastal Range. With the exceptions of (a) and (f), all sections have been at least weakly recrystallized; moreover, the Coastal Range contains the East Taiwan Ophiolite, which was subjected to oceanic metamorphism prior to its incorporation in the Luzon arc.

Various diastrophic events have been previously documented in Taiwan. Ho (1982) referred to two principal mountain-building phases: the late Mesozoic multistage Nanao orogeny involving the pre-Tertiary basement, and the Plio-Pleistocene Penglai (= Taiwan) orogeny, resulting in the deformation, progressive metamorphism, and uplift of most of the island. Yen (1976; 1981, personal communication), in addition, divided the pre-Tertiary deformation into earlier and later Mesozoic stages and distinguished between the Neogene orogenies in the Coastal Range on the one hand, and the rest of Taiwan on the other.

(a) Coastal Plain

Of Pleistocene and Recent age, the Coastal Plain consists of terrace gravels, alluvium, and well-bedded but poorly consolidated clastic sediments. This uplifted but otherwise practically undeformed section passes westward into epicontinental sediments of the Taiwan Strait; eastern, thicker portions have been deposited upon or are tectonically overlain locally by allochthonous strata of the Western Foothills.

(b, c, d) Western Foothills and Slate Series

Shallow marine, detrital units of the Western Foothills (a few coal-bearing units occur in the north) are largely Miocene to Pleistocene in age; the Hsuehshan Range contains Eocene and Oligocene quartzose, carbonaceous metasandstones, argillites, and slates; and the Backbone Range consists chiefly of Paleocene, Eocene, and Miocene slaty and phyllitic metaclastic rocks (e.g., see Yen, 1973). Informally known in aggregate as the slate series, this nonturbiditic sedimentary prism consist of continental shelf deposits on the west, which gradually give way eastward to units that were deposited on the Asiatic continental slope and rise. The slate series contains intercalated, somewhat alkalic mafic lavas and tuffaceous units as well as diabasic dikes. Toward the east, the sedimentary sequence thickens to approximately 10 km (6 mi). The dominantly clastic debris is more sandy in the northern and west-central parts. This cover series contains detritus lithologically similar to the basement, hence probably was locally derived.

Virtually the entire section consists of a stack of folded, imbricate thrust sheets that appear to root in the vicinity of the Longitudinal Valley. Simplified, balanced geologic cross sections of portions of northern and west-central Taiwan (Suppe, 1980a, 1980b) are presented in Figures 4 and 5, respectively. The transported nature of this foreland fold-and-thrust belt is evident; the cumulative shortening ranges up to approximately 160 km (100 mi). However, on the east, basal and intraformational conglomerates contain pebbles of the underlying pre-Cenozoic basement (Yen et al, 1956; Suppe et al, 1976), so at least some of the basal units of the cover series are autochthonous. In general, tops face east, but overall the section youngs in the opposite direction, suggesting the progressive westward upward-climbing transport of older, right-side-up units over younger, structurally lower strata. Metamorphism has affected all these units, with grade increasing eastward (Chen, 1981; Liou, 1981a). Deformation and recrystallization of rocks as young as Plio-Pleistocene fix the time of the Penglai orogeny.

The Kenting Melange of southernmost Taiwan (see Fig. 3) appears to represent a Mio-Pliocene deep-water olistostrome deposited within a basin spatially separate from that of the slate series immediately prior to the arrival of the Luzon arc (Page and Lan, 1983).

(e) Tananao Schist

The pre-Cenozoic basement of Taiwan (Yen, 1954a, 1954b, 1960; Ho, 1975, 1979) crops out as a narrow belt immediately west of the Longitudinal Valley. It bears evidence of multiple deformation (Yen, 1967, 1976; Wang, 1979). As indicated in Figure 3, two principal lithotectonic zones have been recognized, separated by the poorly exposed Shoufeng fault (Yen, 1963). In the vicinity of the Southern Cross-Island Highway, this break appears to be a west-dipping imbricate fault zone (Stanley et al, 1981), but elsewhere its presence and attitude are equivocal.

On the west lies the Tailuko belt, consisting of Permian and Lower Mesozoic pelitic schists, quartzofeldspathic gneisses, platform carbonate units, intercalated mafic and tuffaceous metasediments, greenschists and amphiboles + minor lenses of serpentinite (Fuh, 1962; Liou et al, 1981). In the north, this belt has been locally intruded by later Mesozoic granitic rocks (Wang-Lee, 1981). The occurrence of exclusively plutonic s-type

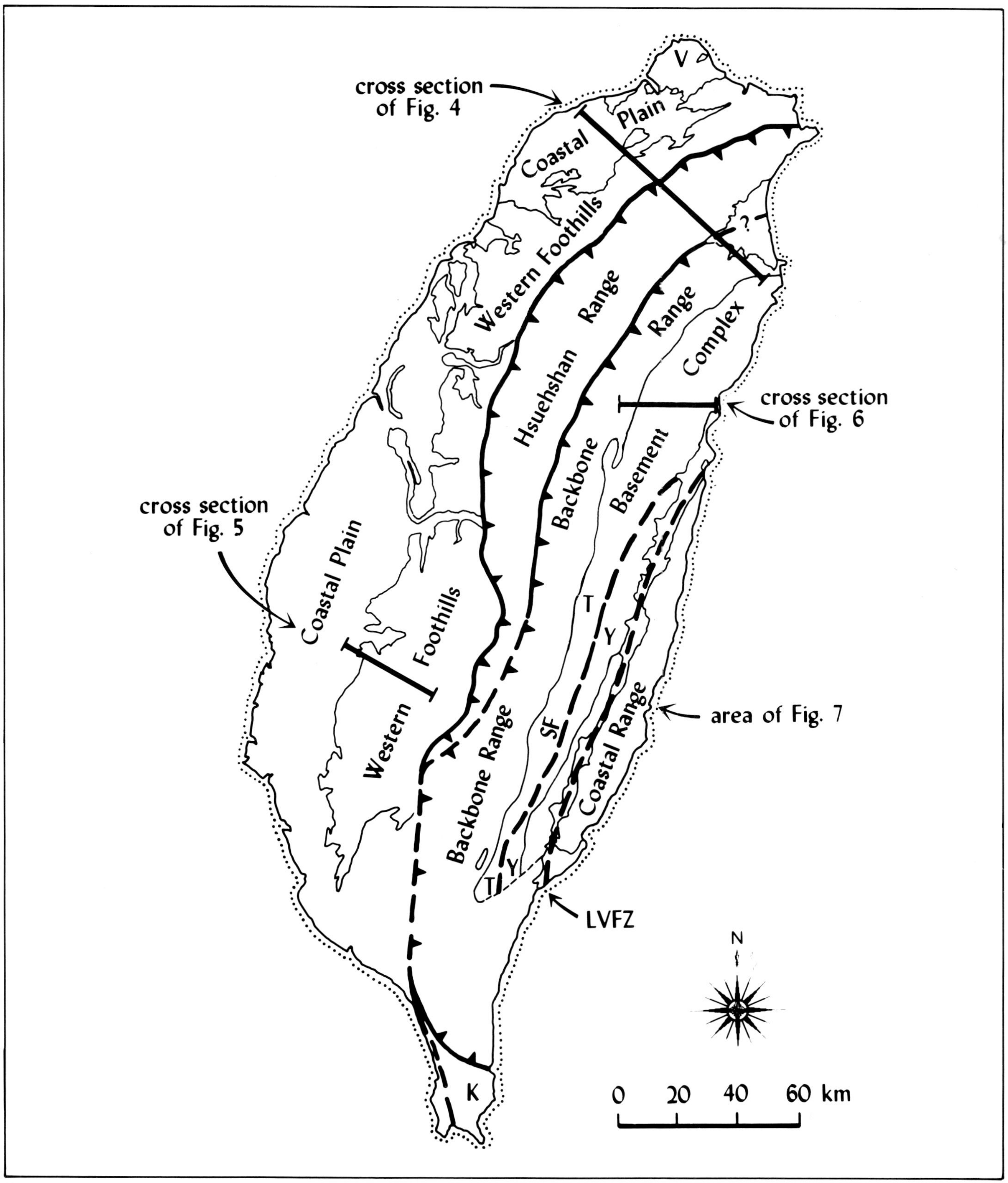

F gure 3—Lithotectonic belts of Taiwan, after Ho (1975). (Copyright © 1975 The Ministry of Economic Affairs, Republic of China. Used with permission.) The Tatun volcanic group is indicated by V, the Kenting Melange by K. The Coastal Range of eastern Taiwan is separated from the pre-Cenozoic basement terrane by the Longitudinal Valley, the site of a major left-lateral strike-slip fault zone (LVFZ). The Tananao Schist terrane is divided into two belts by the hypothetical Shoufeng fault (SF); these are the Tailuko belt (T), and the Yuli terrane (Y). Successively more westerly lithotectonic realms of the Cenozoic cover series have been juxtaposed by an east-dipping series of thrusts (barbs on upper plates); only a few of these are shown (none for the Western Foothills belt).

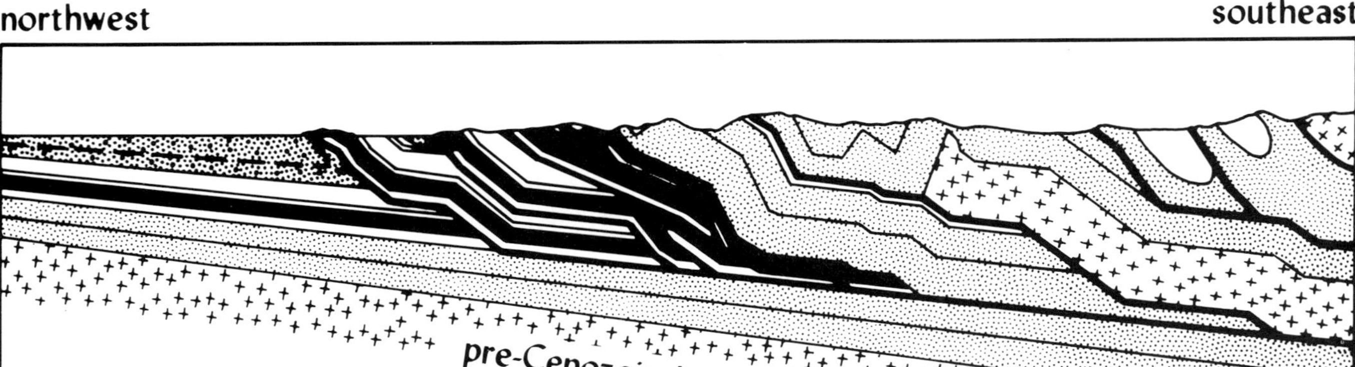

Figure 4—Simplified structural cross section of northern Taiwan. (After Suppe, 1980a. Copyright © 1980 The Geological Society of China. Used with permission.) No vertical exaggeration. Location of cross section is shown in Figure 3. Superjacent series rocks are metaclastic lithologies (heavy dots on west indicate unmetamorphosed shales).

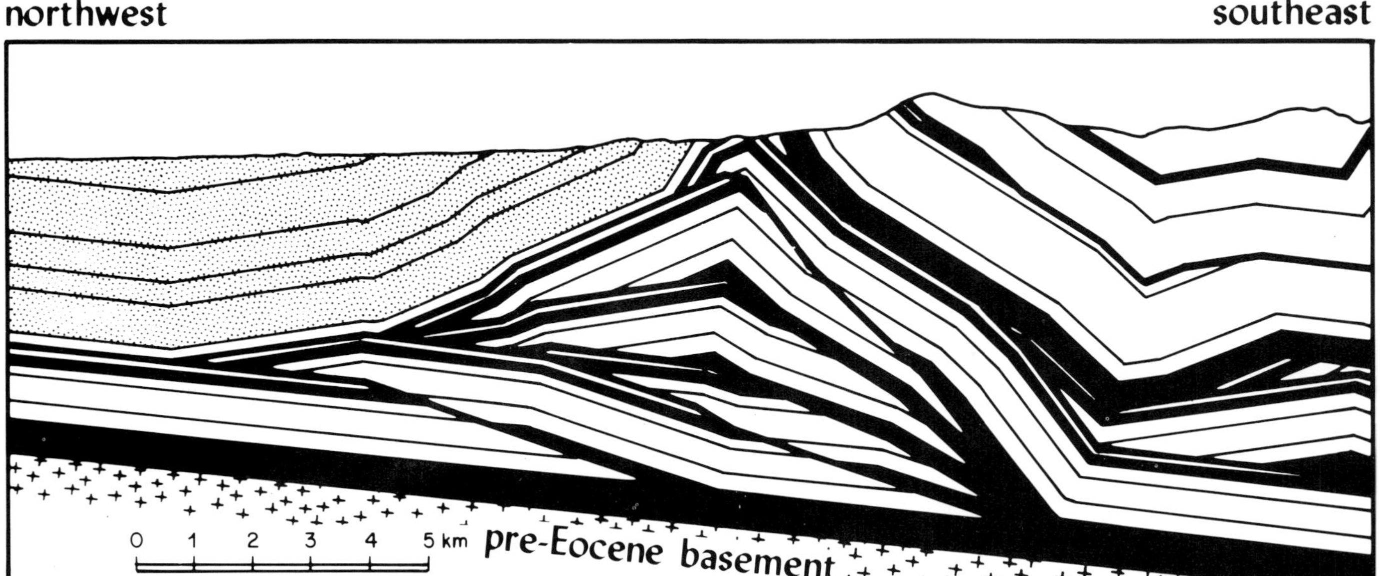

Figure 5—Geologic cross section of Western Foothills belt of southern Taiwan. (After Suppe, 1980b. Copyright © 1980 by the Chinese Petroleum Corporation. Used with permission.) No vertical exaggeration. Important compressive deformation appears to die out westward. Location of cross section is shown in Figure 3. Superjacent series rocks are metaclastic lithologies (heavy dots on west indicate young, unmetamorphosed units).

granitic rocks indicates that relatively deep levels of this sialic arc were remobilized in the later Mesozoic. A cross section through the Tailuko belt transecting east-central Taiwan (Ernst, 1983b) is illustrated in Figure 6. The Tailuko belt contains traces of several pre-Cenozoic metamorphic events in addition to the Plio-Pleistocene greenschistic overprinting (Liou et al, 1981; Lo and Wang-Lee, 1981; Ernst et al, 1981; Ernst, 1983a). These include: (1) production of amphibolite facies assemblages in earlier Mesozoic time; (2) emplacement of calc-alkaline granitoids and localized thermal progradation of country rocks to upper amphibolite facies during the mid(?) Mesozoic; and (3) Late Cretaceous deformation of the granitic rocks and retrograde formation of lower grade mineral assemblages (Yen and Rosenblum, 1964). The highest grade metamorphics occur adjacent to the felsic plutons, and both are confined to the northernmost portion of the belt (Chu, 1981; Wang-Lee, 1981). Although the timings of the various recrystallization events are poorly constrained, the relative age sequence and association with stages of intrusion and deformation appear to be clear. The Cenozoic metamorphic overprinting must be Pliocene or younger because of the occurrence of greenschist-facies metadiabase dikes cutting Miocene slates.

East of the Tailuko belt lies a monotonous melange of pre-Cenozoic pelitic and mafic plus ultramafic schists of oceanic affinities (Liou, 1981b), collectively termed the Yuli belt; carbonate strata and granitoids are conspicuous by their absence from this terrane. The provenance of the clastic sediments is not well constrained. Rocks of the Yuli belt exhibit dominantly high-rank greenschist-facies mineral compatibilities, chiefly ascribable to the Plio-

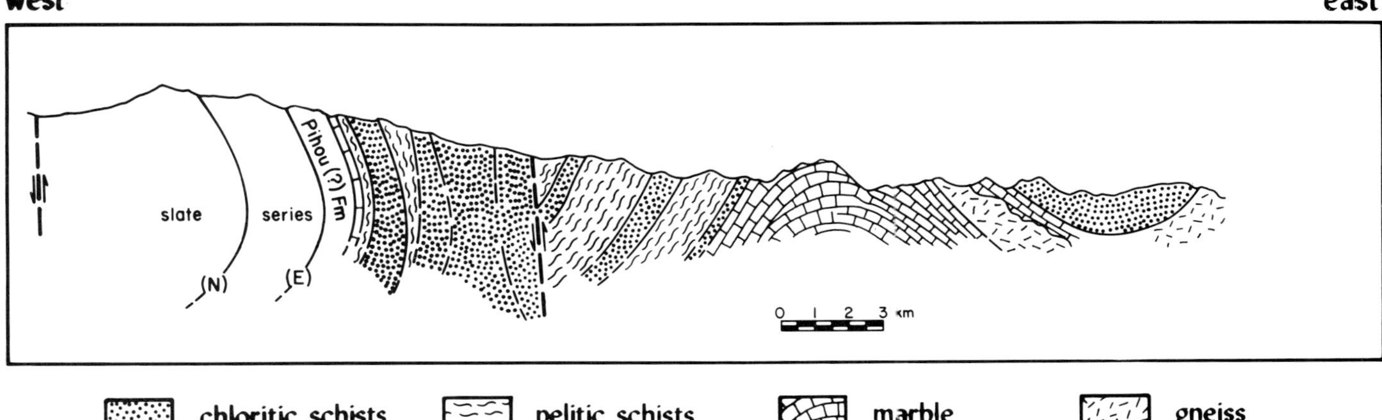

Figure 6—Schematic, interpretative cross section of the Tailuko Gorge area. (Ernst, 1983b. Journal of Metamorphic Geology, v. 1, p. 305–325. Copyright © 1983 Blackwell Scientific Publications Ltd. Used with permission.) No vertical exaggeration. Location of the cross section is shown in Figure 3. The line of section does not encounter the Shoufeng fault zone and the Yuli terrane. The slate series is Tertiary in age; the Pihou(?) Formation is either Cretaceous or Tertiary. Basement units to the east of these units are pre-Cenozoic in age of deposition.

Pleistocene overprinting, as noted in the eastern Mu-kua Chi area (Ernst and Harnish, 1983). However, melange matrix and the enclosed ophiolitic tectonic blocks (associated with serpentinite lenses) in the east-central portion of the belt attest to the Late Cretaceous crystallization of relatively high-pressure barroisitic amphibolites and the late Miocene production of blueschists (Yen, 1966; Liou et al, 1975; Lan and Liou, 1981; Liou, 1981b; Jahn et al, 1981).

The problematic Shoufeng fault has been compared with the Median Tectonic Line of southwestern Japan (Miyashiro, 1961). Similarly, it juxtaposes a continental-ward calc-alkaline magma-intruded, thermally recrystallized, largely miogeoclinal and/or continental slope section (Tailuko belt) against a more seaward trench-argillite lithotectonic zone lacking felsic igneous rocks but containing "cold" alpine-type serpentinized peridotites and mafic igneous bodies (Yuli terrane). In general, open folding appears to characterize the Tailuko belt, whereas oceanward vergence is typical of the Yuli terrane (Yen, 1967; Biq, 1971; Stanley et al, 1981). Thus, these two Mesozoic belts possess some of the contrasts that characterize landward circum-Pacific volcanic/plutonic arcs and outboard subduction complexes (Ernst, 1977a).

(f) Luzon Arc

The Neogene Coastal Range of eastern Taiwan consists of calc-alkaline volcanic units and associated strata that are lithologically distinct from—and are everywhere in fault contact with—belts lying to the west of the Longitudinal Valley. The structure of the Coastal Range is complicated by low-angle (flat to east-dipping) imbricate thrust faults; in general, the oldest rocks are exposed towards the center of the range as the cores of north–northeast-trending anticlines, with younger units disposed peripherally. Overall geologic relationships are presented in Figure 7. The stratigraphy (Hsu, 1956, 1976b; Chi et al, 1981; Ho, 1982) is summarized below.

The oldest rocks that crop out in the Coastal Range are lower Miocene andesitic volcanics, agglomerates, tuffs and associated volcanogenic sediments of the Tuluanshan Formation (base not exposed), and hypabyssal andesitic/dioritic plugs of the Chimei Complex (not separately indicated in Fig. 7). The latter possesses whole-rock K–Ar dates as old as 22 m.y. (Ho, 1969), and lenses of reef limestone near the top of the Tuluanshan Formation yield early-mid Miocene foraminiferal dates (Chang, 1967). More recent biostratigraphic work (Chi et al, 1981) indicates that, locally, the upper part of the Tuluanshan is middle or even late Miocene in age. This calc-alkaline unit laterally interfingers with and passes stratigraphically upwards into turbiditic, volcanogenic conglomerates, sandstones, and lesser amounts of siltstones and mudstones of the Takangkou Formation. Dominantly a flysch unit, the Takangkou also contains intercalated calc-alkaline volcanics. Low portions of the section contain detritus exclusively of volcanic arc provenance, but towards the top, clastic debris from the Asiatic passive margin (Cenozoic cover series) begins to appear (Teng, 1979, 1980). This upper Miocene–Pliocene unit (Chi et al, 1981) in turn passes laterally and vertically into a Plio-Pleistocene muddy olistostrome, the Lichi Melange (Wang, 1976; Page and Suppe, 1981). The olistostrome contains detritus of several distinct source regimes: the immediately adjacent, more easterly calc-alkaline (Luzon) arc; slaty material of the Cenozoic cover series overlying the western sialic basement (Asiatic continental margin); and East Taiwan Ophiolite (crust + uppermost mantle of the South China Sea) from the suture zone. The andesitic and passive margin strata are present in the Lichi Melange exclusively as pebbles and cobbles, but the mafic + ultramafic detritus occurs both as conglomeratic lenses and as slide blocks up to 1 km (.6 mi) in length and several hundred meters in thickness (Liou et al, 1977; Ernst, 1977b). The larger blocks exhibit stratigraphic relationships which, in aggregate, allow reconstruction of the nature and history of the East Taiwan Ophiolite prior to its incorporation in the olistostrome (Liou et al, 1977). The mafic and ultramafic members of the suite formed during early Miocene time, probably at a spreading center; after a hiatus marked by thermal relaxation and

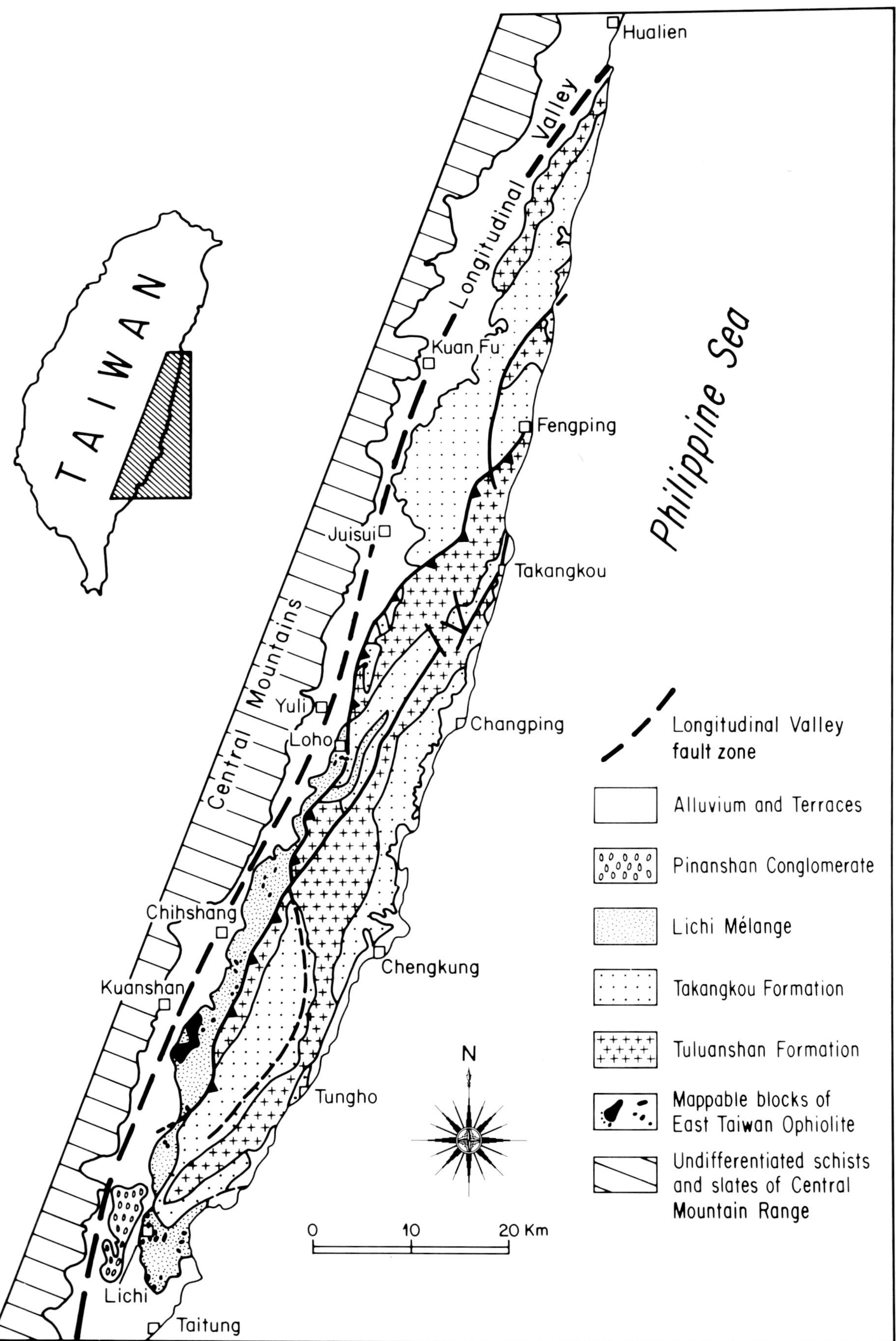

Figure 7—Regional geology of the Coastal Range. (Modified from Liou et al, 1977. Copyright © 1977 Energy Mining Research and Service Organization. Used with permission.)

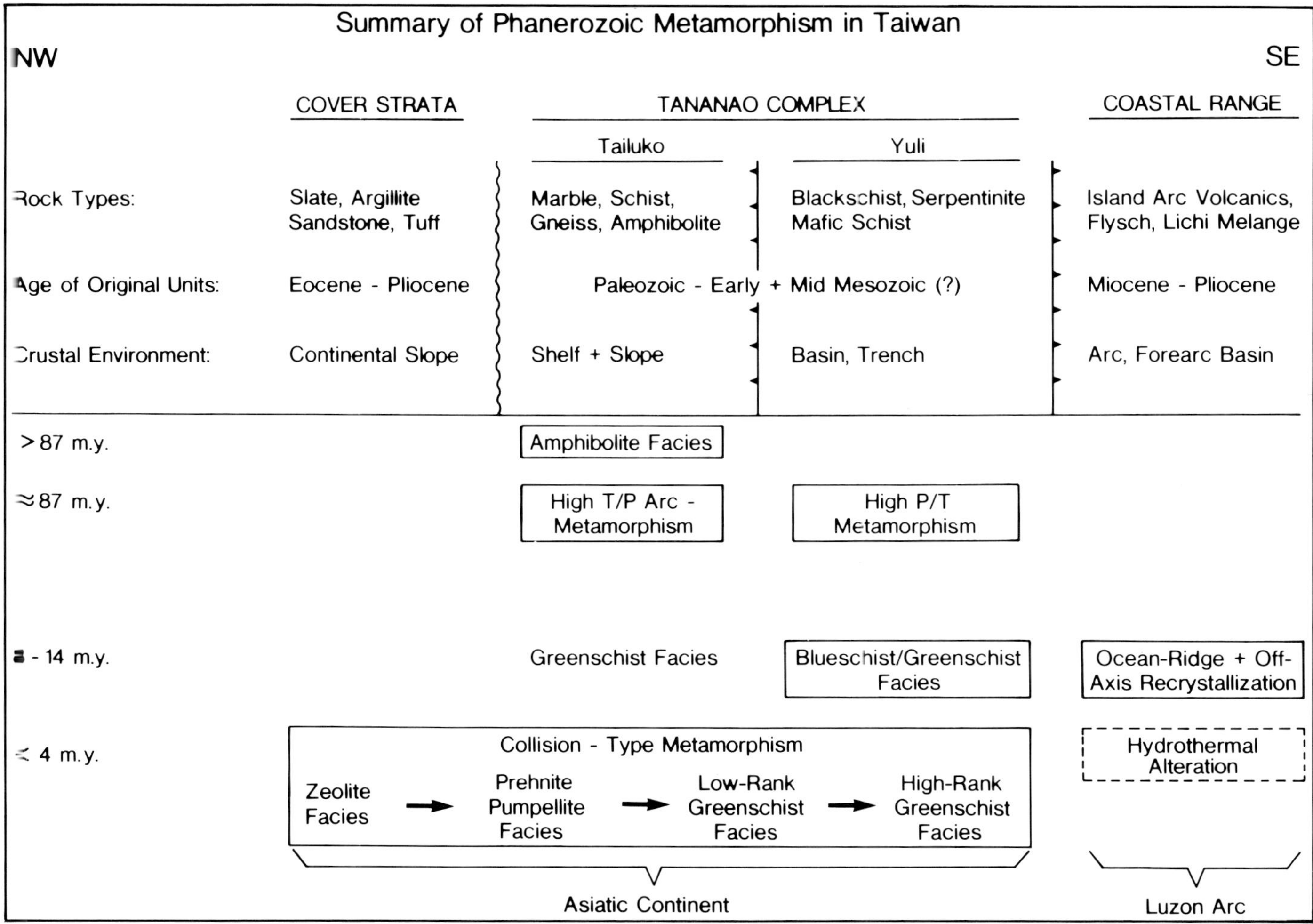

Figure 8—Summary of metamorphic episodes in Taiwan, largely after Yen (1976), Liou (1981a, 1981b), Ho (1982), and Ernst (1983a). Boxes enclose well-documented mineral assemblages. Tectonic and unconformable contacts are shown by barbed and wavy lines respectively.

deposition of red, deep-sea clays (Suppe et al, 1977), off-axis basaltic lavas covered the plutonic series during the middle Miocene (Suppe et al, 1981). Overlying the Lichi Melange, locally marked by an angular unconformity, are fluvial gravels of the upper Pleistocene Pinanshan Conglomerate; this deposit contains abundant detritus from the metamorphic pre-Cenozoic basement of the Asiatic continental margin.

Except for hydrothermal alteration associated with calc-alkaline volcanism represented by the Chimei Complex and the Tuluanshan Formation, Miocene and younger strata of the Coastal Range are completely unmetamorphosed. The Lichi Melange retains a virtually unrecrystallized clay matrix (Page and Suppe, 1981). Thus, metamorphic lithologies present in the East Taiwan Ophiolite were produced during oceanic-ridge and sea-floor recrystallization in the South China Sea (Liou, 1979; Liou and Ernst, 1979), prior to incorporation in the Lichi Melange as olistostromal debris.

METAMORPHIC EPISODES

The various recrystallization events recorded in the rocks of Taiwan are closely related to differential plate motions and to large-scale tectonic processes. The recognizable and distinctly different metamorphic stages are illustrated in Figure 8, where the generalized natures of the protoliths are also indicated. This figure summarizes the main recrystallization events described in previous sections and, by inference, provides constraints regarding lithospheric plate dynamics and reorganization in the Taiwan sector of the Asiatic continental margin.

CRUSTAL EVOLUTION

We will now attempt to unravel the plate tectonic history of Taiwan in light of the tectonostratigraphic terrane concept, starting with the more recent events and working backwards through time. Obviously, the interpretation of earlier events becomes progressively more speculative. What should be clear from the scenario now to be presented is that this Mesozoic and younger mountain belt, although small in regional extent, exhibits first-order complications which, although partially understood, are beyond the limits of our present state of knowledge to decipher completely.

Plio-Pleistocene Closure of the South China Sea

Collision of the Coastal Range with the Asiatic continental margin was a consequence of eastward subduction of the South China Sea oceanic crust-capped plate beneath the Luzon arc (Taylor and Hayes, 1983). Impaction was initiated in northeasternmost Taiwan and has since propagated southwards (Chai, 1972; Karig, 1973). The gradual approach of the calc-alkaline arc towards the Asiatic continent is recorded by the appearance of slate series clasts in upper portions of the Takangkou (upper Miocene) and Lichi (Pliocene) units; widespread exposure of the pre-Tertiary basement occurred later, as revealed by abundant detritus in the (Pleistocene) Pinanshan Conglomerate. The Luzon arc, marking the western edge of the Philippine Sea plate, is certainly far traveled relative to the Asiatic margin, as demonstrated by the lithologic dissimilarities tectonically juxtaposed across the Longitudinal Valley. In all ways, the Coastal Range fits the definition of a tectonostratigraphic terrane (Jones et al, 1983).

The East Taiwan Ophiolite, which occurs as blocks in the Lichi Melange near the suture zone, represents exotic fragments of the now-consumed South China Sea (Ernst, 1977b; Suppe et al, 1981). These slide blocks and olistostromal debris represent portions of a largely vanished oceanic terrane. The latter material was allochthonous to the Coastal Range.

Another consequence of the attempted subduction of the Asiatic continental crust was the imbrication and westward thrusting of the passive margin cover series and its Mesozoic basement (e.g., see Fig. 4; also Suppe, 1981). The eastward intensification of the Plio-Pleistocene recrystallization and deformation attests to progressively more profound depths of underflow, now recovered (Ernst, 1983a). Technically, inasmuch as its contacts with older lithologic units are chiefly faults, the slate series may be regarded as a separate terrane. It is clearly parautochthonous, having been transported westward because of about 160 km (100 mi) of shortening (Suppe, 1981). However, its nearly in situ origin along the eastern margin of the Asiatic continent is well-constrained, and some basal portions lie with depositional contact over the Tananao Schist belt, hence it cannot be described as devoid of genetic correlation to the underlying Tananao basement.

Cenozoic Rifting of the Asiatic Margin

Magnetic anomalies within the South China Sea document sea-floor spreading as old as mid-Oligocene, according to Taylor and Hayes (1983); these authors have speculated that block faulting of the Asiatic margin, which preceded production of the oceanic crust, may have been initiated in latest Cretaceous or Paleocene time. In any case, a Paleogene Atlantic-type continent-ocean transition provided the catchment basin for the Tertiary slate series and associated minor intercalated, slightly alkalic basalts and diabases; the latter probably are related to the rifting of the continental margin. Sea-floor spreading ceased in the South China Sea near the end of middle Miocene time (Taylor and Hayes, 1983), shortly after formation (and ridge + sea-floor metamorphism) of the East Taiwan Ophiolite. The Neogene closure of part of this basin: (1) had already begun to generate the calc-alkaline arc on the nonsubducted Philippine Sea lithospheric plate; (2) produced the Mio-Pliocene blueschist tectonic blocks of the Yuli terrane; and (3) culminated in a Plio-Pleistocene-Recent continent-island arc collision. The latter resulted in the aborted subduction and consequent westward relative transport of the foreland fold-and-thrust belt, including slivered portions of its basement of Tananao Schist and the superjacent slate series.

The blueschists attest to westward(?) subduction of segments of the South China Sea and reflect metamorphic portions of this poorly understood terrane in fragmentary form. As previously described, the passive margin slate section is parautochthonous, being neither strictly in place nor truly exotic. To refer to it as a far-traveled terrane would seem to be unjustified because the detritus (including basal and intraformational conglomerates) indicates that the cover series was derived from, and deposited on, the eastern edge of the Mesozoic basement (Yen et al, 1956; Suppe et al, 1976).

Inferred Mesozoic Rifting and Subduction of the Asiatic Margin

Prior to the Cenozoic opening of the South China Sea during Paleogene rifting, the Asiatic margin evidently was the site of a Mesozoic convergent plate junction of the circum-Pacific type (Miyashiro, 1961, 1967). This situation is indicated by the existence of the landward, calc-alkaline, igneous-metamorphic Tailuko belt, and the seaward, serpentinized peridotite-invaded trench-argillite sequence of the eugeoclinal Yuli belt (Ernst, 1983a). The Shoufeng fault marks the tectonic contact between these belts along what must represent a major suture. Lack of apparent correspondence of structures, lithologies, and provenance across this break allows the speculation that the more easterly, high-pressure, low-temperature melange belt might be allochthonous relative to the Asiatic continental margin. We therefore regard the Yuli belt as a suspect terrane (e.g., see Coney et al, 1980).

Prior to the construction of this paired belt assemblage in late Mesozoic time, mafic rocks + minor serpentinites evidently were tectonically inserted into a miogeoclinal section consisting of Permian and lower Mesozoic, well-ordered platform carbonates, and interlayered pelitic and psammitic strata. Whether the Tailuko belt was brought as a microcontinental fragment to its present location at the edge of the Asiatic craton through drifting and accretion or whether it was assembled essentially in place is not known. The former model could account for the southeastward projection of sialic basement which constitutes Taiwan, whereas the latter model better accounts for the miogeoclinal nature and age relationships of the pre-Cenozoic strata along the Asiatic margin. The associated metamorphosed mafic + ultramafic lenses appear to be ophiolitic and may represent lower or mid-Mesozoic oceanic crust + hydrated mantle obducted into a preexisting passive margin section (Liou et al, 1981). This suggests that latest Paleozoic rifting and spreading was succeeded in mid-Mesozoic time by the onset of convergence; underflow culminated in the late Mesozoic construction and assembly of the arc and trench belts.

We thus recognize: (1) an authochthonous(?) inboard sequence consisting of the upper Paleozoic–lower to mid-Mesozoic Atlantic margin-type section, converted in situ to a metamorphosed and granitoid-intruded Tailuko basement complex in later Mesozoic time; (2) tectonically emplaced exotic ophiolitic lenses in the Tailuko belt of what may represent portions of a lower or mid-Mesozoic paleo-Pacific oceanic plate; and (3) an outboard upper Mesozoic high-pressure, low-temperature Yuli melange suspect terrane.

SUMMARY

The Phanerozoic crustal evolution of Taiwan has involved a complex interplay of rifting, drifting, and arc collision. Late Mesozoic and younger stages in this development are illustrated diagrammatically in Figure 9. Much of the present island evidently has been assembled more or less in place by accretion of miogeoclinal and/or continental slope strata following attenuation and truncation of the sialic crust during sea-floor spreading. Additionally, calc-alkaline plutonic rocks were emplaced during convergent plate motion. Exotic and suspect terranes and far-traveled ophiolitic scraps have been sequestered in this sector of the Asiatic margin by convergent or oblique underflow. As illustrated in Figure 10, these foreign arrivals appear to include: (1) lower to mid-Mesozoic amphibolite + serpentinite lenses in the Tailuko belt; (2) the upper Mesozoic Yuli trench-argillite + melange complex; (3) Mio-Pliocene blueschist tectonic blocks situated in the east-central part of the Yuli terrane; (4) the Neogene Luzon arc; and (5) the Miocene East Taiwan Ophiolite, supplied as olistostromal debris to the Lichi Melange of the Coastal Range. It is possible that the Tailuko belt is far traveled, but relationships remain ambiguous. The slate series is best described as parautochthonous. In aggregate, these allochthonous, parautochthonous and suspect terranes and exotic oceanic fragments constitute an important aspect of the growth of this portion of the Asiatic continental crust.

ACKNOWLEDGMENTS

Aspects of this paper were discussed at the Second International Meeting on Circum-Pacific Tectonostratigraphic Terranes, which took place at Stanford, California, during late August 1983. It is based on research by numerous Chinese geologists as well as on our own work and that of our colleague, John Suppe. Support during the formulation of this synthesis was provided partly by the University of California, Los Angeles, and partly by the National Science Foundation through grant EAR80-17295/Ernst. Helpful reviews were obtained from C. Biq, B. M. Page, and T. P. Yen, for which the authors are very grateful.

REFERENCES

Allen, C. R., 1962, Circum-Pacific faulting in the Philippines-Taiwan Region: Journal of Geophysical Research, v. 67, p. 4795–4812.

Biq, C., 1965, The East Taiwan Rift: Petroleum Geology of Taiwan, n. 4, p. 93–106.

______, 1971, A fossil subduction zone in Taiwan: Proceedings of the Geological Society of China, v. 14, p. 146–154.

______, 1981, Collision, Taiwan-style: Geological Society of China Memoir 4, p. 91–102.

Bowin, C., et al, 1978, Plate convergence and accretion in Taiwan–Luzon region: Bulletin of the American Association of Petroleum Geologists, v. 62, p. 1645–1672.

Cardwell, R. K., et al, 1980, The spatial distribution of earthquakes, focal mechanism solutions, and subducted lithosphere in the Philippine and northeastern Indonesian islands: American Geophysical Union Monograph 23, p. 1–35.

Chai, B. H. T., 1972, Structural and tectonic evolution of Taiwan: American Journal of Science, v. 272, p. 389–422.

Chang, L.-S., 1967, A biostratigraphic study of the Tertiary in the Coastal Range, eastern Taiwan, based on smaller Foraminifera (I. southern part): Proceedings of the Geological Society of China, n. 10, p. 64–76.

Chen, C.-H., 1981, Change of x-ray diffraction pattern of K-micas in the metapelites of Taiwan and its implication for metamorphic grade: Geological Society of China Memoir 4, p. 475–490.

Chi, W. R., et al, 1981, Stratigraphic record of plate interactions in Coastal Range, eastern Taiwan: Geological Society of China Memoir 4, p. 155–194.

Chu, H. T., 1981, The discovery of sillimanite in northern Taiwan: Geological Society of China Memoir 4, p. 491–496.

Coney, P. J., et al, 1980, Cordilleran suspect terranes: Nature, v. 288, p. 329–333.

Ernst, W. G., 1977a, Mineral parageneses and plate tectonic settings of relatively high-pressure metamorphic belts: Fortschritte der Mineralogie, v. 54, p. 192–222.

______, 1977b, Olistostromes and included ophiolitic debris from the Coastal Ranges of eastern Taiwan: Geological Society of China Memoir 2, p. 97–114.

______, 1983a, Mountain building and metamorphism: a case history from Taiwan, *in* K. J. Hsü, (ed.), Mountain building: London, Academic Press, p. 247–256.

______, 1983b, Mineral parageneses in metamorphic rocks exposed along Tailuko Gorge, Central Mountain Range, Taiwan: Journal of Metamorphic Geology, v.

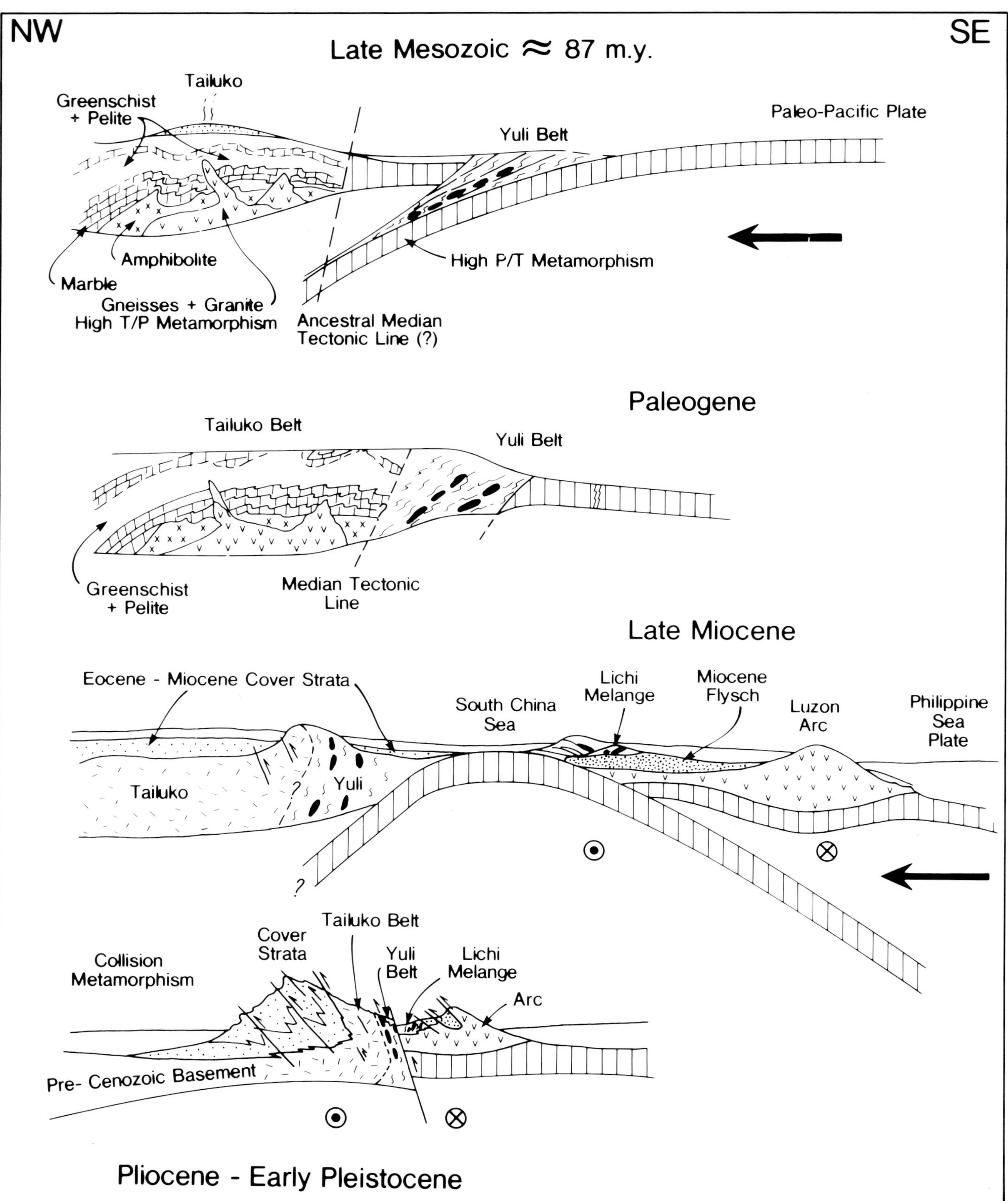

Figure 9—Schematic illustration of four stages in the accretionary growth of Taiwan over the past approximately 87 m.y. In lower two cross sections, ⊗ and ⊙ signify relative plate motions possessing large northward and southward components, respectively. Note that a late Miocene Molucca Sea type of double subduction (Hamilton, 1979) must have given way in Pliocene time to exclusively eastward underflow of the Asiatic plate, including progressively eastward zones slate series–Tailuko belt–Yuli terrane, for the latter to have developed the biotitic greenschists during attempted underflow and collision with the Luzon arc in Plio-Pleistocene time.

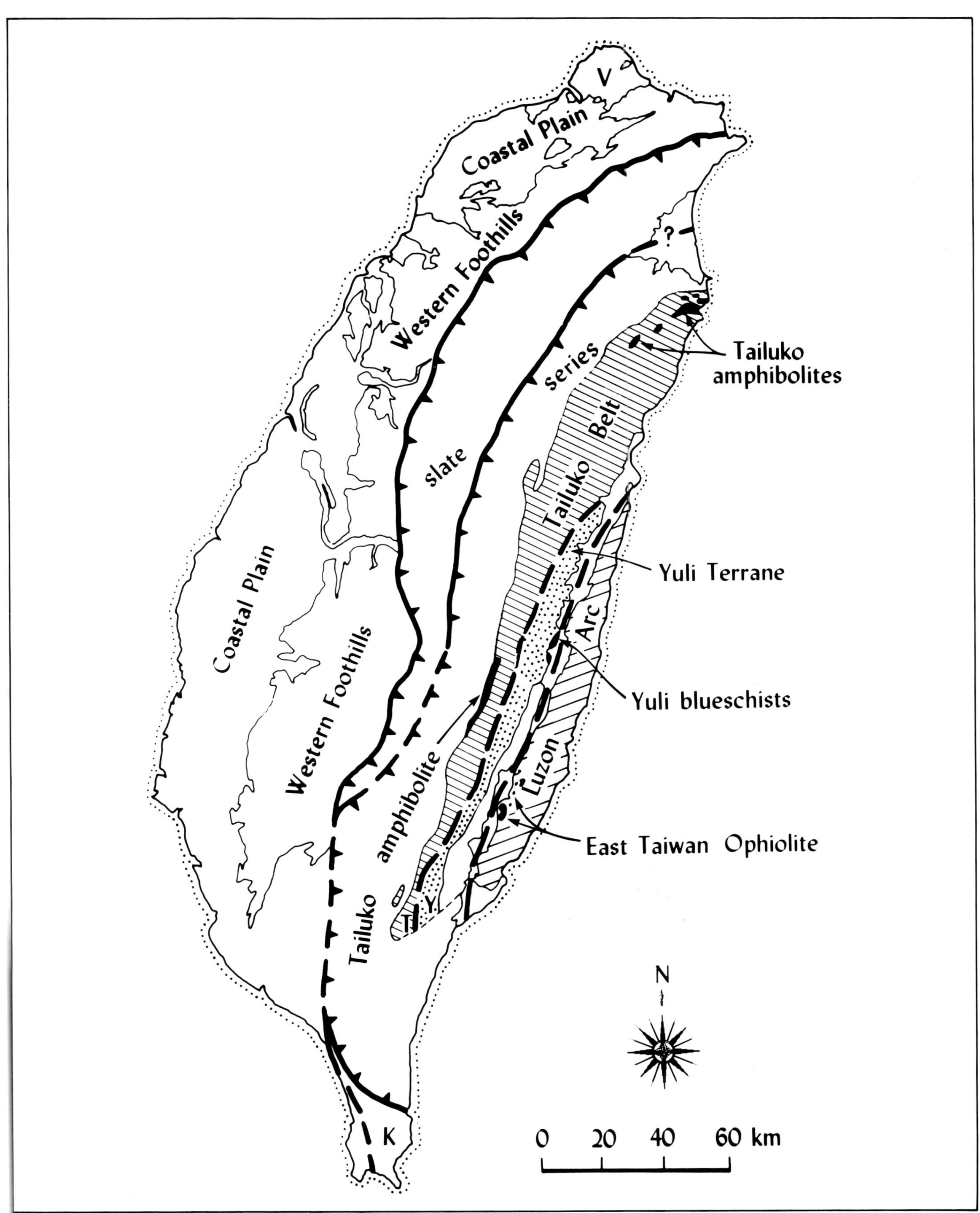

Figure 10—Preliminary tectonostratigraphic terrane map for Taiwan. The southernmost mafic + ultramafic body labeled Tailuko amphibolite is principally greenschist. See text for discussion.

1, p. 305-325.

———, and D. Harnish, 1983, Mineralogy of some Tananao greenschists facies rocks, Mu-kua Chi area, eastern Taiwan: Proceedings of the Geological Society of China, v. 26, p. 99-112.

———, et al, 1981, Multiple metamorphic events recorded in Tailuko amphibolites and associated rocks of the Suao-Nanao area: Geological Society of China Memoir 4, p. 391-441.

Fuh, T. M., 1962, Metamorphic rocks of the Nanao district, Taiwan, with special reference to the origin of orthogenesis: Geological Society of China Memoir 1, p. 113-132.

Hamburger, M. W., et al, 1983, Seismotectonics of the northern Philippine island arc: American Geophysical Union Monograph 27, p. 1-22.

Hamilton, W., 1979, Tectonics of the Indonesian Region: U.S. Geological Survey Professional Paper 1078, 345 p.

Ho, C. S., 1969, Geological significance of potassium-argon ages of the Chimei igneous complex of eastern Taiwan: Bulletin of the Geological Survey of Taiwan, n. 23, p. 1-52.

———, 1975, An introduction to the geology of Taiwan: explanatory text of the geologic map of Taiwan: Taipei, Ministry of Economic Affairs, Republic of China, 153 p.

———, 1979, Geologic and tectonic framework of Taiwan: Geological Society of China Memoir 3, p. 57-72.

———, 1982, Tectonic evolution of Taiwan: explanatory text of the tectonic map of Taiwan: Taipei, Ministry of Economic Affairs, Republic of China, 126 p.

Hsu, T. L., 1956, Geology of the Coastal Range, eastern Taiwan: Bulletin of the Geological Survey of Taiwan, n. 8, p. 39-63.

———, 1976a, Neotectonics of the Longitudinal Valley, eastern Taiwan: Bulletin of the Geological Survey of Taiwan, n. 25, p. 53-62.

———, 1976b, The Lichi Melange in the Coastal Range framework: Bulletin of the Geological Survey of Taiwan, n. 25, p. 87-96.

Jahn, B. M., et al, 1981, High-pressure metamorphic rocks of Taiwan: REE geochemistry, Rb-Sr ages and tectonic implications: Geological Society of China Memoir 4, p. 497-520.

Jones, D. L., et al, 1983, Recognition, character, and analysis of tectonostratigraphic terranes in western North America, *in* M. Hashimoto and S. Uyeda (eds.), Accretion tectonics in the circum-Pacific region: Tokyo, Terra Scientific Publishing Company, p. 21-35.

Karig, D. E., 1973, Plate convergence between the Philippine and Ryukyu Islands: Marine Geology, v. 14, p. 153-168.

Lan, C. Y., and J. G. Liou, 1981, Occurrence, petrology and tectonics of serpentinites and their associated rodingites in the Central Range, Taiwan: Geological Society of China Memoir 4, p. 343-389.

Lewis, S. D., and D. E. Hayes, 1983, The tectonics of northward propagating subduction along eastern Luzon, Philippine Islands: American Geophysical Union Monograph 27, p. 57-78.

Liou, J. G., 1979, Zeolite facies metamorphism of basaltic rocks from the East Taiwan Ophiolite: American Mineralogist, v. 64, p. 1-14.

———, 1981a, Recent high CO_2 activity and Cenozoic progressive metamorphism in Taiwan: Geological Society of China Memoir 4, p. 551-582.

———, 1981b, Petrology of metamorphosed oceanic rocks in the Central Range of Taiwan: Geological Society of China Memoir 4, p. 291-342.

———, and W. G. Ernst, 1979, Oceanic ridge metamorphism of the East Taiwan Ophiolite: Contributions to Mineralogy and Petrology, v. 68, p. 335-348.

———, et al, 1975, Petrology of some glaucophane schists and related rocks from Taiwan: Journal of Petrology, v. 16, p. 80-109.

———, et al, 1977, The East Taiwan Ophiolite, its occurrence, petrology, metamorphism and tectonic setting: Mining Research and Service Organization Special Report 1, 212 p.

———, et al, 1981, Geology and petrology and some poly-metamorphosed amphibolites and associated rocks in northeastern Taiwan: Geological Society of America Bulletin, v. 92, pt. I, p. 219-224, pt. II, p. 609-748.

Lo, C.-H., and C. M. Wang-Lee, 1981, Mineral chemistry in some gneissic bodies, the Hoping-Chipan area, Hualien, eastern Taiwan: Proceedings of the Geological Society of China, v. 24, p. 40-55.

Miyashiro, A., 1961, Evolution of metamorphic belts: Journal of Petrology, v. 2, p. 277-311.

———, 1967, Orogeny, regional metamorphism, and magmatism in the Japanese islands: Meddelser fra Dansk Geologie Forening Kobenhavn, v. 17, p. 390-446.

Page, B. M., and C. Y. Lan, 1983, The Kenting Melange and its record of tectonic events: Geological Society of China Memoir 5, p. 227-248.

———, and J. Suppe, 1981, The Pliocene Lichi Melange of Taiwan: its plate tectonic and olistostromal origin: American Journal of Science, v. 281, p. 193-227.

Seno, T., 1977, The instantaneous rotation vector of the Philippine Sea Plate relative to the Eurasian Plate: Tectonophysics, v. 42, p. 209-226.

Stanley, R. S., et al, 1981, A cross section through the southern Central Mountains of Taiwan: Geological Society of China Memoir 4, p. 443-474.

Suppe, J., 1976, Decollement folding in southwestern Taiwan: Petroleum Geology of Taiwan, n. 13, p. 25-35.

———, 1980a, A retrodeformable cross section of northern Taiwan: Proceedings of the Geological Society of China, v. 23, p. 46-55.

———, 1980b, Imbricate structure of western foothills belt, south-central Taiwan: Petroleum Geology of Taiwan, n. 17, p. 1-16.

———, 1981, Mechanics of mountain-building and metamorphism in Taiwan: Geological Society of China Memoir 4, p. 67-90.

———, and J. H. Wittke, 1977, Abnormal pore-fluid pressures in relation to stratigraphy and structure in the active fold-and-thrust belt of northwestern Taiwan: Petroleum Geology of Taiwan Meng Volume, n. 14, p. 11-24.

———, et al, 1976, Observations of some contacts between basement and Cenozoic cover in the Central Mountains, Taiwan: Proceedings of the Geological Society of China, v. 19, p. 59-70.

———, et al, 1977, Paleogeographic interpretation of red shales within the East Taiwan Ophiolite: Petroleum Geology of Taiwan, n. 14, p. 109-120.

———, et al, 1981, Paleogeographic origins of the Miocene East Taiwan Ophiolite. American Journal of Science, v. 281, p. 228-246.

Taylor, B., and D. E. Hayes, 1983, Origin and history of the South China Sea Basin: American Geophysical Union Monograph 27, p. 23-56.

Teng, L. S., 1979, Petrographic study of the Neogene sandstones of the Coastal Range, eastern Taiwan (I. Northern Part): Acta Geologica Taiwanica, n. 20, p. 129-155.

———, 1980, Lithology and provenance of the Fanshuiliao Formation, northern Coastal Range, eastern Taiwan: Proceedings of the Geological Society of China, v. 23, p. 118-129.

Tsai, Y. B., 1978, Plate subduction and the Plio-Pleistocene orogeny in Taiwan: Petroleum Geology of Taiwan, n. 15, p. 1-10.

Wang, C. S., 1976, The Lichi formation of the Coastal Range and arc-continental collision in eastern Taiwan: Bulletin of the Geological Survey of Taiwan, n. 25, p. 73-86.

Wang, Y., 1979, Some structural characteristics of Tertiary basement on Taiwan: Geological Society of China Memoir 3, p. 139-145.

Wang-Lee, C., 1981, Field trip guide to eastern section of the cross-island highway, *in* Seminar Plate Tectonics Metamorphic Geology, p. 9-22, Taipei, National Science Council Field Guidebook, 71 p.

Yen, T. P., 1954a, The gneisses of Taiwan: Bulletin of the Geological Survey of Taiwan, n. 5, p. 1-100.

———, 1954b, The green rocks of Taiwan: Bulletin of the Geological Survey of Taiwan, n. 7, p. 1-46.

———, 1960, A stratigraphic study on the Tananao Schist in northern Taiwan: Bulletin of the Geological Survey of Taiwan, n. 12, p. 53-66.

———, 1963, The metamorphic belts within the Tananao Schist terrain of Taiwan: Proceedings of the Geological Society of China, v. 6, p. 72-74.

———, 1966, Glaucophane schist of Taiwan: Proceedings of the Geological Society of China, v. 9, p. 70-73.

———, 1967, Structural analysis of the Tananao schist of Taiwan: Bulletin of the Geological Survey of Taiwan, n. 18, 110 p.

———, 1973, The Eocene sandstones in the Hsuehshan Range terrain, northern Taiwan: Proceedings of the Geological Society of China, v. 16, p. 97-110.

———, 1976, Geohistory of Taiwan: Proceedings of the Geological Society of China, v. 19, p. 52-58.

———, and S. Rosenblum, 1964, K-Ar ages of micas from the Tananao Schist terrains of Taiwan—A preliminary report: Proceedings of the Geological Society of China, v. 7, p. 80-81.

———, et al, 1956, Some problems on the Mesozoic formation of Taiwan: Bulletin of the Geological Survey of Taiwan, n. 8, p. 1-14.

Paleomagnetic Evidence for Accretion and Bending Tectonics of the Hida and the Circum-Hida Terranes, Central Japan

Kimio Hirooka
Satoshi Uchiyama
Tetsuhiro Date
Hiroyuki Kanai
Toyama University
Toyama, Japan

Tadashi Nakajima
Fukui University
Fukui, Japan

The paleomagnetic results from the Triassic sedimentary rocks in the circum-Hida terrane show their low latitudinal origin. On the other hand, the paleolatitudes derived from the Jurassic sedimentary rocks covering both the Hida and the circum-Hida terranes are consistent with those from the Triassic to Jurassic granitic rocks reported previously and in this paper. This result reconfirms the earlier conclusion that the two terranes were amalgamated in the middle latitudinal region by the Jurassic. The Mesozoic paleomagnetic declinations derived from the southern part of the Hida and the circum-Hida terranes face to the east and deviate from north-directing declinations observed generally in the other areas. This clockwise rotation of the paleomagnetic declinations may be ascribed to the postaccretionary bending of the Hida and the circum-Hida terranes.

INTRODUCTION

The understanding of the geotectonic evolution of the circum-Pacific region is one of the most difficult and important subjects in analyzing and synthesizing Mesozoic and Cenozoic accretion tectonics. The understanding of Mesozoic tectonism in the central part of Honshu Island, Japan, is also necessary in documenting the evolution of Japan, where several tectonostratigraphic terranes of different ages form a complex Paleozoic and Mesozoic collage (Hattori, 1982; Hirooka et al, 1983; Mizutani and Hattori, 1983), and much attention is currently focused there in order to weave the tectonic evolution of central Japan (Chihara and Komatsu, 1982; Sasajima, 1981).

Seven pre-Tertiary major tectonostratigraphic terranes (belts) are differentiated in the central part of Honshu Island (Fig. 1). The Hida terrane is generally located in the northernmost part of central Japan. This terrane is characterized by metamorphic rocks collectively called the Hida gneiss. This gneiss is considered to be Precambrian in origin. The Hida terrane is intruded by many plutonic rocks from Triassic to Jurassic age, called Funatsu granite. The Hida terrane is surrounded successively by northeast–southwest-trending narrow terranes (belts) such as the circum-Hida, the Mino, the Ryoke, the Sambagawa, the Chichibu, and the Shimanto terranes (belts) (Geological Survey of Japan, 1977). A brief summary of these terranes is offered by Hattori (1982). For example, the Ryoke is a high-temperature metamorphic derivative of the Mino terrane, the Sambagawa is a high-pressure metamorphic belt of unknown origin, and the Chichibu is a Mesozoic dismembered complex composed of early Paleozoic to middle Mesozoic rocks. No systematic paleomagnetic result has been reported from these three terranes. The Shimanto is considered to be a Cretaceous to Paleogene subduction complex composed of continental-derived sedimentary rocks and ocean-floor sedimentary rocks (Taira, 1981).

We have been analyzing the paleomagnetism of granitic and sedimentary rocks of the Hida, the circum-Hida, the Mino, and the Shimanto terranes. In our previous paper (Hirooka et al, 1983), we reported paleomagnetic results obtained from the Jurassic to Cretaceous granitic rocks of the Hida terrane and from the Permian to Cretaceous siliceous shales and cherts of the Mino and the Shimanto terranes. The paleolatitudes calculated from the paleomagnetic inclinations of the

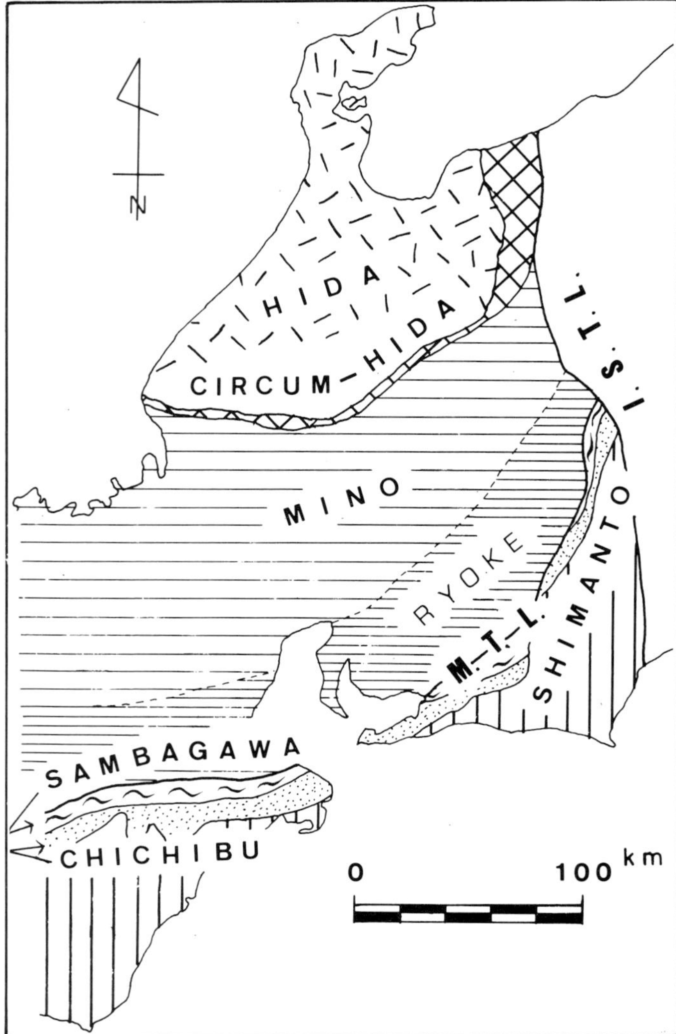

Figure 1—Terrane map of central Japan. The Hida, the circum-Hida, the Mino, the Sambagawa, the Chichibu, and the Shimanto terranes are mutually independent in origin. The Ryoke is a metamorphic derivative of the Mino. I.S.T.L. is the Itoigawa–Shizuoka tectonic line separating southwest Japan from northeast Japan. M.T.L. is the median tectonic line separating the Inner (northern) Zone from the Outer (southern) Zone in southwest Japan.

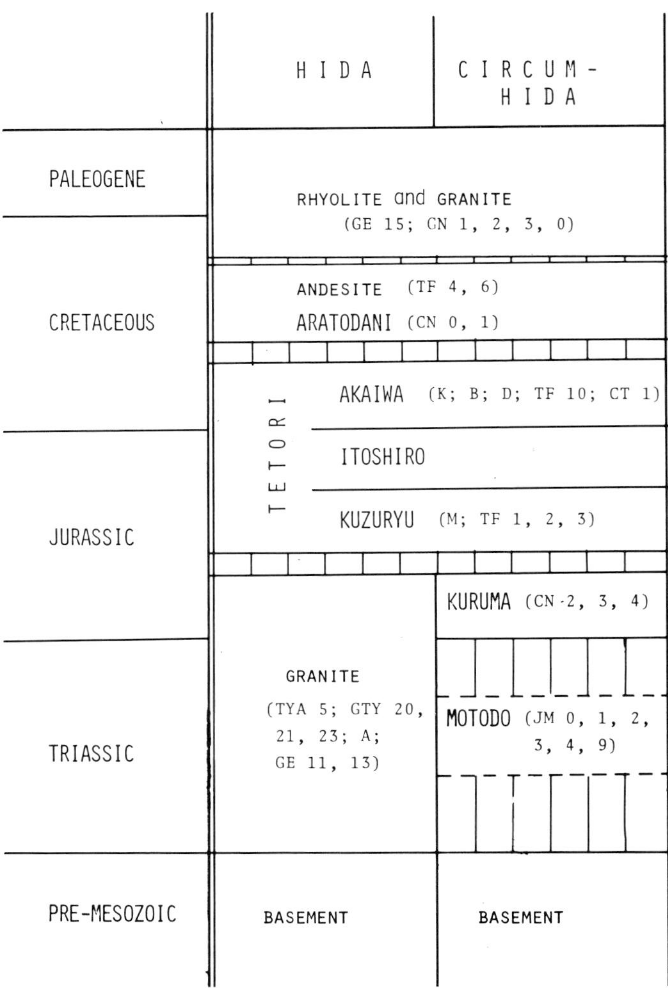

Figure 2—Mesozoic stratigraphic relation between the Hida and the circum-Hida terranes, central Japan. The letters and numbers in parentheses show paleomagnetic sample sets derived from individual formations.

Mesozoic pelagic sedimentary rocks in the Mino and the Shimanto terranes are very low, causing us to conclude that these rocks were deposited and acquired their remanent magnetization in equatorial regions. The paleomagnetic results from the Triassic to Jurassic and the Late Cretaceous granitic bodies in the Hida terrane, on the other hand, suggest that this terrane was situated during the Mesozoic around 31° N, which is almost identical to the present latitude.

As for the Hida terrane, we reported only the paleomagnetic data derived from granitic rocks in our previous paper (Hirooka et al, 1983). Granitic bodies, whose original horizontal planes can not be defined, provide no clear information about the tectonic modification such as tilting or folding after they were magnetized, so that we intended to determine accurately the paleolatitude of the Hida terrane again based on the paleomagnetic measurements of the middle Mesozoic Kuruma Group and the Tetori Group (Fig. 2). Both groups are sedimentary sequences covering the Hida and the circum-Hida terranes and enable us to apply tilting correction of the direction of magnetization by restoring the strata to the horizontal. Moreover, we attempted to detect the origin of the curved shape of the circum-Hida terrane as shown in Figure 3. In this paper, we report many paleomagnetic results from the sedimentary and volcanic rocks in the Hida and the circum-Hida terranes and substantiate the accretion tectonics and the post-accretionary bending that occurred in both terranes.

STRATIGRAPHY AND PALEOMAGNETIC SAMPLING

In the Hida terrane, paleomagnetic samples were collected from the Late Jurassic to Early Cretaceous sedimentary sequence called the Tetori (Maeda, 1961) (Fig. 3). The Tetori Group is divided into three subgroups, that is, the Kuzuryu, the Itoshiro, and the Akaiwa (Fig. 2). The

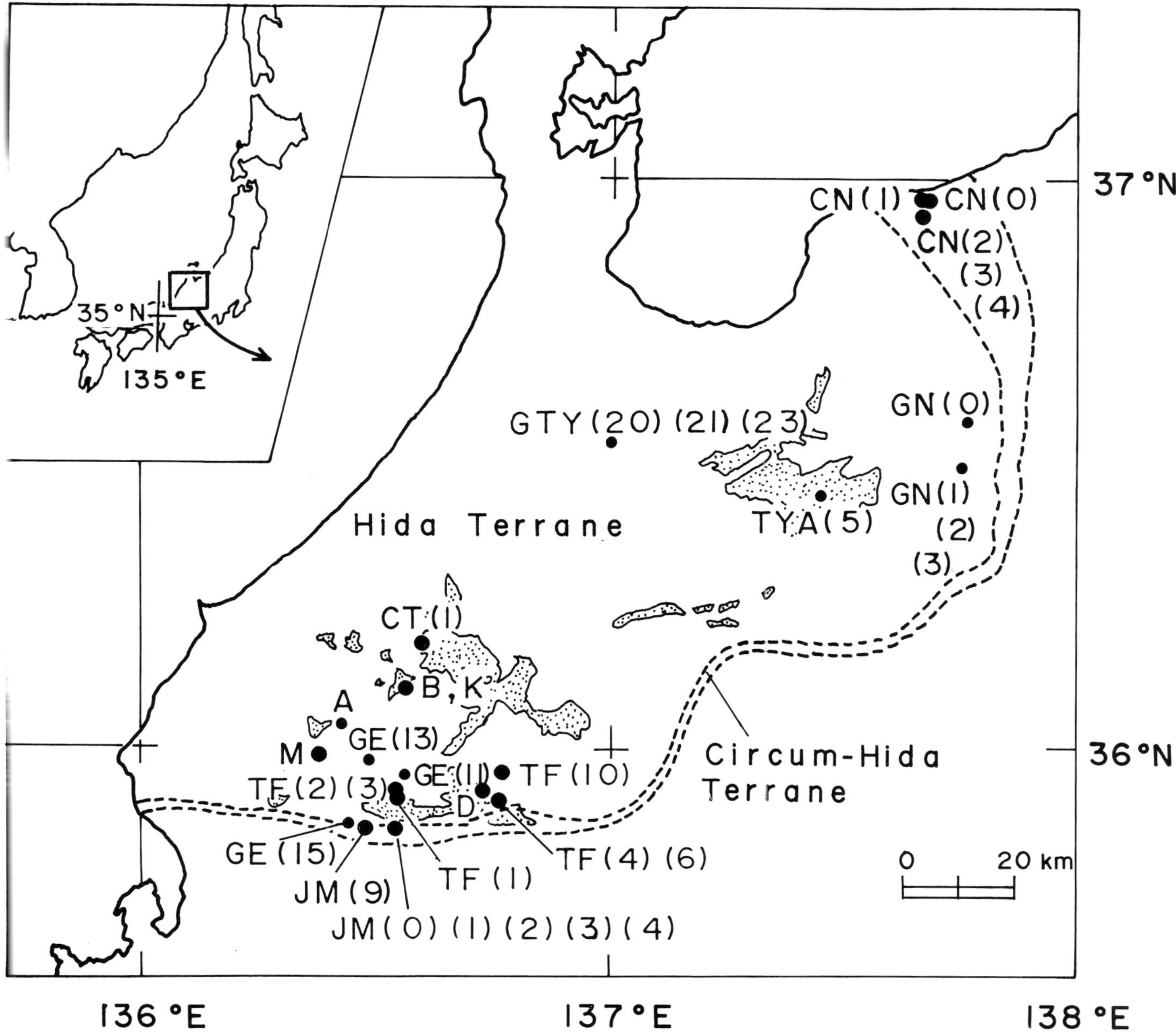

Figure 3—Paleomagnetic sampling sites in the Hida and the circum-Hida terranes. The Jurassic to Cretaceous sedimentary and volcanic rocks are onlapping sequence over the Hida and the circum-Hida terranes. For example, the Tetori Group is distributed in the stippled areas. Small solid circles, granite; large solid circles, sedimentary rock.

Kuzuryu Subgroup was deposited in the Middle to Late Jurassic in a shallow-sea to deltaic and partly subaerial environment, and its distribution is limited in a small segment near the southwestern part of the Hida terrane. The Itoshiro and the Akaiwa Subgroups are of the Late Jurassic to Early Cretaceous in age and are composed of brackish-water to lacustrine sedimentary rocks that are widely distributed in the southern part of the terrane. Of the nine sets of paleomagnetic samples from the Tetori Group, four sets (M; TF 1, 2, and 3) are from the Kuzuryu Subgroup and five sets (K; B; D; TF 10; CT 1) from the Akaiwa Subgroup. Rocks at sites TF 4 and 6 are layered andesite and tuff that unconformably overlie the Tetori Group. The age of these rocks is considered to be Middle to Late Cretaceous, because they are overlain by the Late Cretaceous rhyolite (Yamada et al, 1982).

Together with many small blocks of the Paleozoic ages, Mesozoic sedimentary formations are distributed in the narrow circum-Hida terrane, which is a fault contact with the Hida terrane. The paleomagnetic samples in the circum-Hida terrane were collected from the Triassic Motodo Formation (Omura, 1968), the Early Jurassic Kuruma Group (Kobayashi et al, 1957), and the Middle Cretaceous Aratodani Formation (Takizawa, 1980, 1984). The Motodo Formation is distributed in the southwestern part of the circum-Hida terrane and the samples were derived from six sites (JM 0, 1, 2, 3, 4, 9). The Kuruma Group and the Aratodani Formation are located in the northern segment

Table 1

Site	Locality Lat(°N)	Lon(°E)	N	D(°E)	I(°)	α_{95}(°)	k	ODF (Oe)	ODT (°C)	Bedding Strike(°)	Dip(°)
Hida terrane											
M	35°59′	136°23′	8	31.7	51.9	11.8	22.9	100	—	N37W	35N
TF(1)	35°54′	136°33′	11	9.8	66.9	17.3	8.0	100	—	N37W	23N
TF(2)	35°55′	136°33′	10	76.2	78.8	8.8	31.1	100	—	N19W	27N
TF(3)	35°55′	136°33′	22	89.9	57.3	10.2	10.2	—	450	N34W	32N
K	36°06′	136°34′	9	4.7	57.8	5.6	84.2	100	—	N61E	8N
B	36°06′	136°34′	6	347.9	59.9	8.1	69.1	150	—	N75W	27N
D	35°55′	136°44′	8	217.2	52.4	12.8	19.8	100	—	N57W	85N
TF(10)	35°57′	136°46′	10	44.2	40.9	5.0	94.4	NRM	—	N63W	28S
CT(1)	36°11′	136°36′	7	51.3	58.9	4.7	169.8	200	—	N77E	22S
TF(4)	35°54′	136°46′	10	219.9	74.3	2.1	558.6	150	—	N39W	39N
TF(6)	35°54′	136°46′	10	106.8	86.2	3.5	187.7	100	—	N26W	31N
Circum-Hida terrane											
JM(0)	35°51′	136°32′	9	82.1	49.2	8.5	37.9	150	—	N81E	56S
JM(1)	35°51′	136°32′	10	84.6	46.2	6.1	64.2	200	—	N81E	56S
JM(2)	35°51′	136°32′	10	73.4	44.7	4.4	121.3	100	—	N81E	56S
JM(3)	35°51′	136°32′	10	66.8	47.2	3.5	189.3	100	—	N81E	56S
JM(4)	35°51′	136°32′	10	67.8	57.7	10.4	22.5	150	—	N81E	56S
JM(9)	35°51′	136°29′	10	82.1	60.6	13.6	13.6	150	—	N59W	57S
CN(2)	36°56′	137°40′	10	36.8	67.5	16.3	9.8	250	—	N75E	42N
CN(3)	36°56′	137°40′	10	29.1	74.4	9.9	24.8	600	—	N25E	12N
CN(4)	36°56′	137°40′	10	36.5	76.2	9.3	28.0	300	—	N25E	12N
CN(0)	36°58′	137°41′	6	150.5	68.5	10.1	45.3	400	—	N51E	58N
CN(1)	36°58′	137°40′	10	183.4	85.7	2.4	396.5	300	—	N51E	58N

Table 1—Localities of samples, paleomagnetic data, and bedding attitudes of sedimentary rocks treated in this paper. Lat, latitude; Lon, longitude; N, number of samples; D and I, mean paleomagnetic declination and inclination; α_{95} and k, Fisher's circles of confidence of 95% and precision parameter; ODF and ODT, optimum demagnetizing field (1 Oe = 0.1 mT) and temperature.

of the terrane. The paleomagnetic samples were collected at five sites designated as CN 2, 3, and 4 in the Kuruma Group and as CN 0 and 1 in the Aratodani Formation.

In the Hida terrane, paleomagnetic measurements were made for not only sedimentary rocks but also Jurassic and Paleogene granitic rocks whose ages were determined geologically or isotopically (Kawano and Ueda, 1966). The results are listed in Table 3. In Tables 1 and 2, all site-sets are classified into "Hida terrane" and "circum-Hida terrane" according to their site-positions. This operation does not always mean that they are proper to the respective terranes. In fact, the Tetori Group overlies not only the Hida but also the circum-Hida terrane.

PALEOMAGNETIC MEASUREMENTS AND RESULTS

Measurements of the magnetization of the samples were carried out by employing a spinner magnetometer. After measuring the natural remanent magnetization (NRM), three or four pilot samples from every site were selected and submitted to the alternating field (A. F.) and the thermal demagnetizations. First, eight demagnetization steps such as 50, 100, 150, 200, 250, 300, 400, and 600 Oe peak fields or, in some cases, 100, 150, 200, 250, 300, 400, 500, and 600 Oe peak fields were applied to the pilot samples, and the optimum demagnetization steps were defined as the field in which Fisher's precision parameter, k (Fisher, 1953), reached a maximum. Second, the thermal demagnetization experiments of several steps such as 120, 200, 280, 410, 450, 500, 550, and 600°C or 200, 300, 350, 400, 450, and 500°C were carried out for a few pilot samples to ascertain the thermal stability of the magnetic remanences.

In all cases except site TF 3, the magnetization directions shifted along the similar migration paths that were observed in the A. F. demagnetization or hardly moved, so that we used the results of A. F. demagnetization as the paleomagnetic data. As an example, the results of the A. F. and the thermal demagnetizations for site JM 2 are shown in Figure 4. Based on these observations, the remaining samples were demagnetized at the optimum and neighboring two steps. As the paleomagnetic data, we accepted the demagnetization results that gave the largest precision parameter. In the case of TF 3 with very scattered NRM directions (Fig. 5), the A. F. cleaning did not work well; the thermal treatment was very effective. Accordingly, the remanent directions erased through the 450°C cleaning were accepted as the paleomagnetic data of this site. The paleomagnetic results revealed through the procedures above are summarized in Table 1. Consequently, we obtained stable paleomagnetic directions from six sites

Table 2

Site	Dc(°E)	Ic(°)	PL(°N)	Age
Hida terrane				
M	39.3	18.4	9.5	M.Jr
TF(1)	29.9	47.3	28.5	M-L.Jr
TF(2)	77.4	51.8	32.4	M-L.Jr
TF(3)	78.9	27.3	14.5	M-L.Jr
K	359.0	50.9	31.6	E.Cr
B	358.9	34.7	19.0	E.Cr
D	30.1	42.5	24.6	E.Cr
TF(10)	72.9	64.5	46.3	E.Cr
CT(1)	92.1	64.0	45.7	E.Cr
TF(4)	59.0	66.2	48.6	M.Cr
TF(6)	69.0	56.1	36.6	M.Cr
Circum-Hida terrane				
JM(0)	125.4	24.5	12.9	Tr
JM(1)	123.2	21.7	11.3	Tr
JM(2)	118.2	28.3	15.1	Tr
JM(3)	119.3	33.5	18.3	Tr
JM(4)	131.8	35.2	19.4	Tr
JM(9)	177.0	47.2	28.4	Tr
CN(2)	5.9	31.8	17.2	E.Jr
CN(3)	34.4	63.3	44.8	E.Jr
CN(4)	43.1	64.4	46.2	E.Jr
CN(0)	315.5	53.1	33.7	M.Cr
CN(1)	317.6	35.2	19.4	M.Cr

Table 2—Summary of paleomagnetic direction after bedding correction. Dc, Ic, and PL, mean declination, inclination, and paleolatitude, respectively. Tr, Jr, and Cr, Triassic, Jurassic, and Cretaceous, respectively. E, M, L, Early, Middle, and Late, respectively.

Table 3

Site	Locality Lat(°N)	Locality Lon(°E)	N	D(°E)	I(°)	α_{95}(°)	k	ODF (Oe)	PL(°N)	Age
TYA(5)	36°27′	137°27′	10	34.7	54.1	12.9	14.9	150	34.6	Jr
GTY(20)	36°32′	137°00′	8	320.6	59.1	17.5	11.0	100	39.9	Jr
GTY(21)	36°32′	137°00′	6	3.2	52.7	15.9	18.6	50	33.3	Jr
GTY(23)	36°32′	137°00′	8	3.9	58.3	10.9	27.0	200	39.4	Jr
A	36°01′	136°25′	9	354.3	59.3	4.4	138.3	200	40.2	Jr
GE(11)	35°56′	136°34′	8	46.5	61.4	10.0	31.9	100	42.5	Jr
GE(13)	35°58′	136°29′	9	357.9	67.0	19.9	7.6	100	46.7	Jr
GE(15)	35°51′	136°27′	10	3.2	56.8	12.0	17.1	200	37.4	Cr
GN(1)	36°30′	137°46′	9	335.9	39.1	20.2	7.5	150	22.1	51 Ma
GN(2)	36°30′	137°46′	10	1.0	52.5	5.6	75.0	100	33.1	51 Ma
GN(3)	36°30′	137°46′	4	331.3	41.1	15.0	38.4	100	23.5	51 Ma
GN(0)	36°35′	137°47′	6	8.6	46.2	14.1	23.4	100	27.5	41 Ma

Table 3—Summary of paleomagnetic directions obtained from Jurassic, Cretaceous, and Paleogene granitic bodies intruded into the Hida terrane. Lat and Lon, latitude and longitude, respectively; N, number of samples; D and I, declination and inclination, respectively; α_{95} and k, Fisher's oval of 95% confidence and precision parameter, respectively; ODF, optimum demagnetization field (Oe); PL, paleolatitude; Jr, Cr, Jurassic and Cretaceous, respectively.

N
N
N
N
AF
NRM
100 Oe
(ODF)
250 Oe
300 Oe
JM (2)
Thermal
NRM
270 °C
460 °C
590 °C

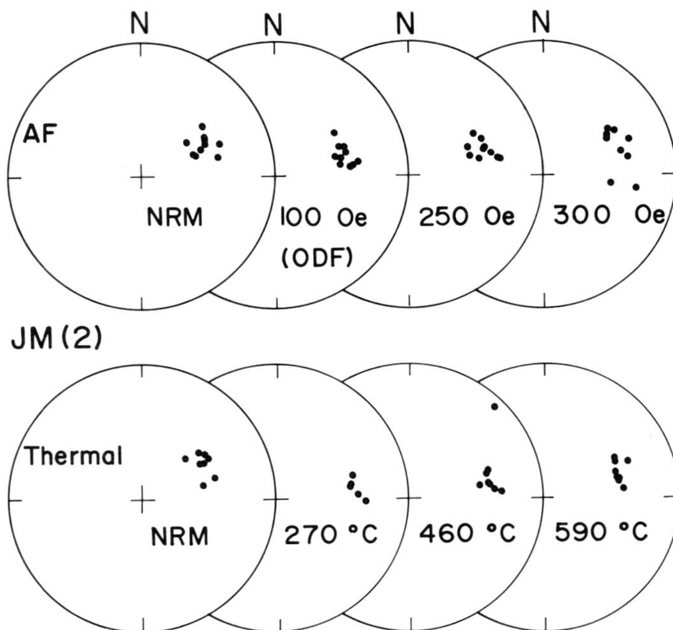

Figure 4—One of the typical behaviors of the magnetic remanence through the successive alternating and thermal demagnetization. No significant migration of magnetic directions is observed. Since many demagnetization experiments for other sample sets show similar behavior, the stable magnetizations were separated through the alternating field demagnetization. Projection on the lower hemisphere of equal-area net.

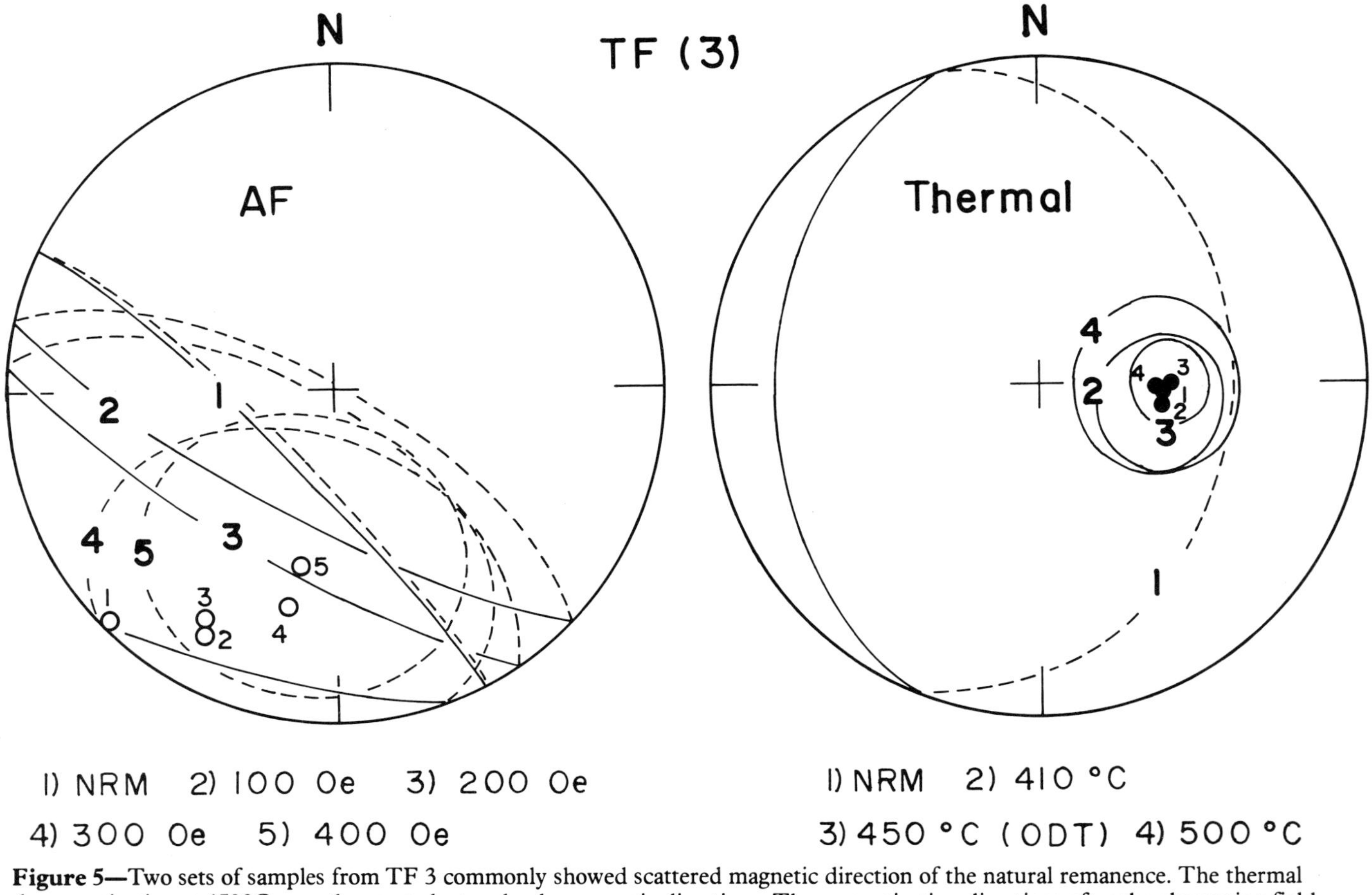

TF (3)
N
AF
N
Thermal
1) NRM 2) 100 Oe 3) 200 Oe
4) 300 Oe 5) 400 Oe
1) NRM 2) 410 °C
3) 450 °C (ODT) 4) 500 °C

Figure 5—Two sets of samples from TF 3 commonly showed scattered magnetic direction of the natural remanence. The thermal demagnetization at 450°C gave the most clustered paleomagnetic directions. The magnetization directions after the alternating field demagnetization generally gave smaller precision parameters, *k*, of Fisher (1953) than the thermal demagnetization at 450°C, and the result of 450°C cleaning was accepted as the stable one. Open circles, upper hemisphere; solid circles, lower hemisphere; equal-area projection; 95% confidence around the means is shown by ovals.

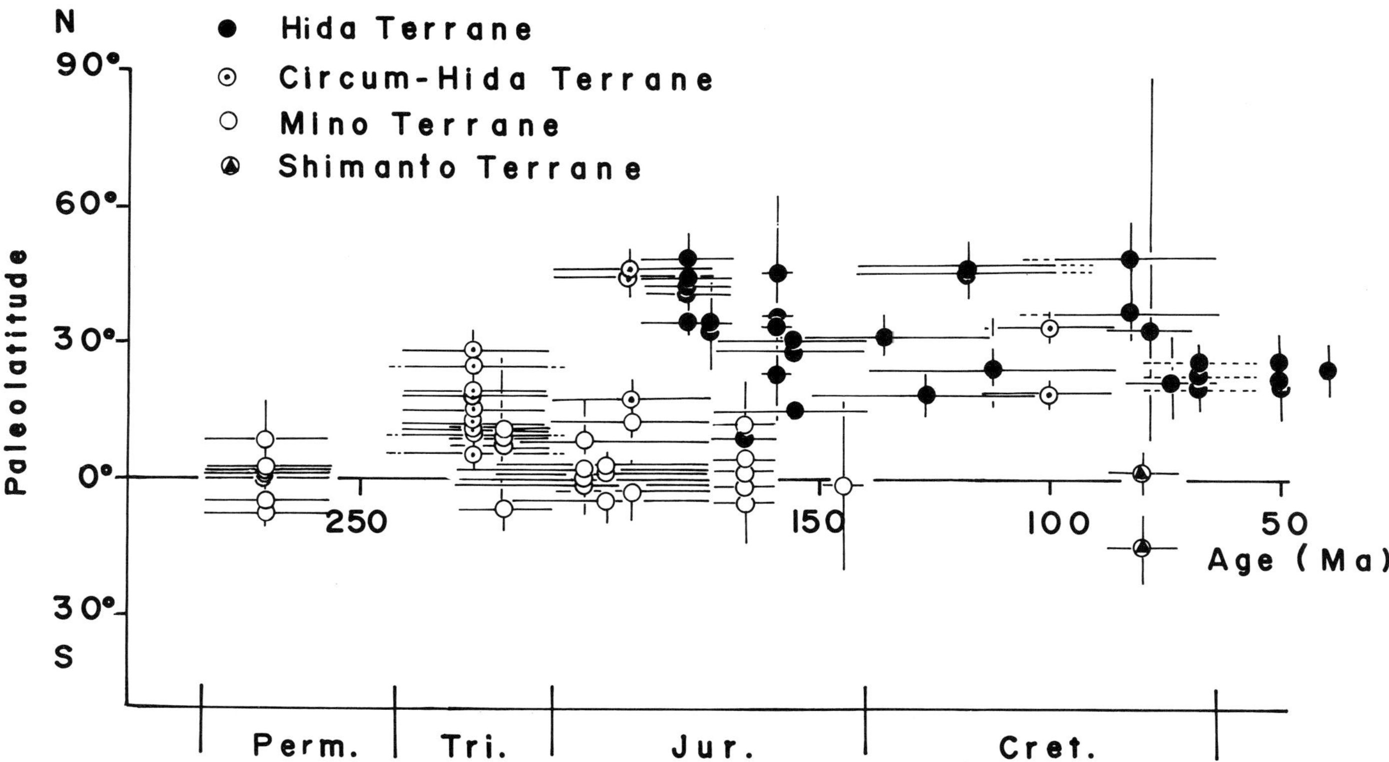

Figure 6—Paleolatitude versus age of rocks in the Hida, the circum-Hida, the Mino, and the Shimanto terranes, compiled from Hattori and Hirooka (1977, 1979), Hirooka et al (1983), and the data described in this paper. The successive accretion to the Hida terrane of the circum-Hida, the Mino, and some rocks in the Shimanto after their northward drift can be concluded based on this illustration.

of the Triassic sedimentary rocks in the circum-Hida terrane, from 16 sites of the Jurassic to Early Cretaceous sedimentary rocks covering the Hida and the circum-Hida terranes, and from 12 sites of granitic rocks in the Hida terrane.

We consider that the paleomagnetic remanences that were confirmed to be stable by both the A. F. and thermal demagnetizing experiments were acquired prior to the tilting of beds, and the paleomagnetic directions can be corrected by the dips around the strikes. The results are listed in Table 2 along with the paleolatitudes calculated from the inclinations.

DISCUSSION AND CONCLUSION

The paleolatitudes derived from the Motodo Formation generally are clearly lower than the present latitude (~ 36° N) of the sampling sites. These results are in good agreement with our previous results (Hirooka et al, 1983). On the other hand, other results from the Jurassic to Cretaceous sedimentary and volcanic rocks and the Paleogene granitic rocks suggest their middle latitudinal origin, although a few results such as those from M, B, TF 3, CN 1, and CN 2 show rather low latitudes. This observation is also consistent with our previous paleomagnetic data derived from Mesozoic granitic rocks in this region, and lead us to conclude that the Hida and the circum-Hida terranes were unified by the beginning of the Jurassic and that both terranes have been locating thereafter in the middle latitude as a composite terrane. This conclusion is supported by the geologic synthesis of Hattori (1982) and Mizutani and Hattori (1983), who clarified the Late Triassic to Early Jurassic accretion of the circum-Hida terrane to the Hida terrane; the former migrated from the low latitudinal regions. Unfortunately, there are no data with respect to the Paleozoic to Triassic paleolatitudes of the Hida terrane.

To the south of the circum-Hida terrane, there exist the Mino and the Shimanto terranes. The paleomagnetic data that were published by Hattori and Hirooka (1977, 1979) and Hirooka et al (1983) revealed the southern origin of the pelagic sedimentary and volcanic rocks in these terranes. The paleolatitudes calculated from these available data and those reported in this paper are summarized in Figure 6. The northward drift of the circum-Hida and the Mino terranes and several pelagic sedimentary rocks in the Shimanto terrane is clearly recognized. They must have been accreted to the Hida terrane, which has been situated around the present position successively since the Middle Mesozoic.

Apart from the discussion on the paleomagnetic inclinations, paleomagnetic declinations offer other important information on the geotectonism. From the paleomagnetic results from the Triassic to Jurassic and some Early Cretaceous sedimentary rocks of the southern part of the Hida and the circum-Hida terranes, we recognize an important tendency that their declinations mostly swing systematically to the east (Fig. 7). The paleomagnetic declinations from the Triassic to Jurassic granitic rocks (Hirooka et al, 1983) and the Jurassic sedimentary rocks, for example, CN 2, 3, and 4, generally direct to the north or within 45° from the north. This

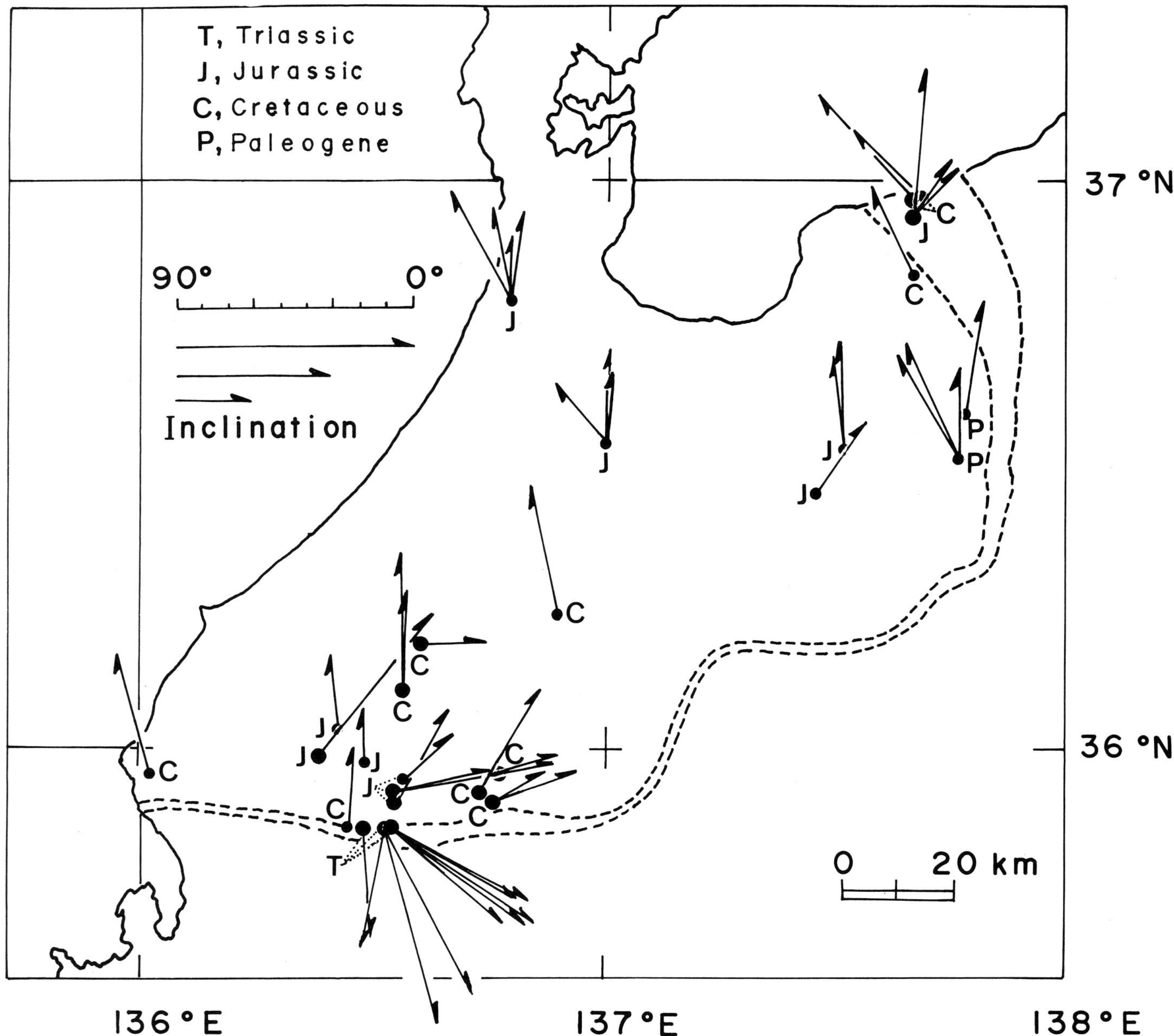

Figure 7—Paleomagnetic directions from the Triassic to Cretaceous rocks in the Hida and the circum-Hida terranes, central Japan, reported in this paper and by Hirooka et al (1983). The arrows with large and small solid circles show the paleomagnetic directions derived from the Mesozoic sedimentary rocks and the Early Mesozoic and the Late Mesozoic and Paleogene granitic rocks of this region, respectively. Many of the Jurassic to Early Cretaceous paleomagnetic declinations of the rocks in and along the southern circum-Hida terrane are deviated systematically to the east, suggesting the Middle to Late Mesozoic clockwise rotation of this part. J, C, and P in the figure show Jurassic, Cretaceous, and Paleogene, respectively.

observation suggests a clockwise rotation of the southern part of the Hida and the circum-Hida terranes. It is concluded that the present bowlike shape of the Hida and the circum-Hida terranes resulted from the demonstrated bending tectonics.

ACKNOWLEDGMENTS

The authors express their sincere thanks to Dr. I. Hattori of Fukui University and Drs. N. Yamada, F. Takizawa, and K. Wakita of the Geological Survey of Japan who offered their unpublished field data. Dr. H. Sakai of Toyama University assisted the paleomagnetic measurements. Financial support for this study was met partly by the Grant-in-Aid for Scientific Researches from the Ministry of Education, Culture, and Science (Prob. No. 58460050).

REFERENCES

Chihara, K., and M. Komatsu, 1982, The recent study in the Hida marginal zone, especially the Ōmi-Renge and the Jō-etsu Belts—a review: Memoirs of the Geological Society of Japan 21, p. 101–116 (in

Japanese with English abstract).
Fisher, R. A., 1953, Dispersion on a sphere: Proceedings of the Royal Astronomical Society of London, v. 217A, p. 295–305.
Geological Survey of Japan, 1977, Geology and mineral resources of Japan, v. 1, Geology: 430 p.
Hattori, I., 1982, The Mesozoic evolution of the Mino terrane, central Japan; a geologic and paleomagnetic synthesis: Tectonophysics, v. 85, p. 313–340.
———, and K. Hirooka, 1977, Paleomagnetic study of the greenstones in the Mugi-Kamiaso area, Gifu Prefecture, central Japan: Journal of the Japanese Association of Mineralogists, Petrologists and Economic Geologists, v. 72, p. 340–353.
——— and ———, 1979, Paleomagnetic results from the Permian greenstones in central Japan and their geologic significance: Tectonophysics, v. 57, p. 211–235.
Hirooka, K., et al, 1983, Accretion tectonics inferred from paleomagnetic measurements of Paleozoic and Mesozoic rocks in central Japan, *in* M. Hashimoto and S. Uyeda, eds., Accretion tectonics in the circum-Pacific regions: Tokyo, Terra Scientific Publishing Company, p. 179–194.
Kawano, Y., and Y. Ueda, 1966, K–Ar dating on the igneous rocks in Japan (V)—granitic rocks in southwestern Japan: Journal of the Japanese Association of Mineralogists, Petrologists and Economic Geologists, v. 56, p. 191–211 (in Japanese with English abstract).
Kobayashi, T., et al, 1957, On the Lower Jurassic Kuruma Group: Journal of the Geological Society of Japan, v. 63, p. 182–194 (in Japanese with English abstract).
Maeda, S., 1961, On the geological history of the Mesozoic Tetori Group in Japan: Journal of the College of Arts and Sciences, Chiba University, v. 3, p. 369–426 (in Japanese with English abstract).
Mizutani, S., and I. Hattori, 1983, Hida and Mino: tectonostratigraphic terranes in central Japan, *in* M. Hashimoto and S. Uyeda, eds., Accretion tectonics in the circum-Pacific regions: Tokyo, Terra Scientific Publishing Company, p. 169–178.
Omura, A., 1968, Sedimentological study of the Motodo Formation in the vicinity of Nishitani-Mura, Ōno-Gun, Fukui Prefecture, central Japan: Journal of the Geological Society of Japan, v. 74, p. 217–231 (in Japanese with English abstract).
Sasajima, S., 1981, Pre-Neogene paleomagnetism of Japanese Islands (and vicinities), *in* M. W. McElhinny and D. S. Valencio, eds., Paleoreconstruction of the continents: American Geophysical Union Geodynamic Series 2, p. 115–128.
Taira, A., 1981, Formative process of the Shimanto Belt, Kagaku, v. 51, p. 516–523 (in Japanese).
Takizawa, F., 1980, Mesozoic formations of the north-eastern part of the Hida Marginal Belt: Research Report of the Hida Marginal Belt, n. 1, p. 58–68 (in Japanese).
———, 1984, Upper limit of the Kuruma Group and its covered sediments in the Hida Marginal Belt: Abstracts of Annual Meeting of the Geological Society of Japan (Waseda), p. 202 (in Japanese).
Yamada, N., et al, 1982, Pre-Nohi volcanism in and around the circum-Hida Tectonic Belt: Abstracts of Annual Meeting of the Geological Society of Japan (Niigata), p. 203.

Jurassic Geologic Framework in the Japanese Islands

Yasuji Saito
National Science Museum
Tokyo, Japan

There are three lithologic units in the Japanese Jurassic sedimentary rocks: (1) trench and abyssal pelagic sediments in subduction complex and their metamorphic equivalents, (2) shallow- to nonmarine mollasselike deposits developed along the inner belts of pre-Jurassic terranes, and (3) shallow-water coherent deposits in and along the outer belts of pre-Jurassic terranes. Contained within the Jurassic subduction complex are oceanic blocks of late Early Carboniferous to Triassic age. The subduction complex represents two pulses of Early to Middle and Late to latest Jurassic, corresponding to subductions bringing about coeval magmatism of the peri-continental type. The thick mollasselike deposits resulted from the collision of certain buoyant features against a continental margin. The juxtaposition of the above three units, the occurrence of chaotically mingled sediments, and the complicated structures of the complex are accounted for by a tectonic accretion process related to oblique subduction.

INTRODUCTION

More than 20 geotectonic units have been recognized in the Japanese Islands (Tanaka and Nozawa, 1977; Hirokawa, 1978). They are tectonically juxtaposed, and most boundaries are recognized as faults or fault zones (Fig. 1). On the basis of recent biostratigraphic works on radiolaria-bearing sedimentary rocks, Taira et al (1983) have grouped them into three collective units that have several longitudinal zones of disruption ascribed to tectonism within strike-slip fault zones. The three units include pre-Jurassic ("older") terranes, Jurassic, and Cretaceous to Tertiary subduction complexes. In the "older" terranes, Jurassic rocks are sporadically developed and are shallow- to nonmarine sediments entirely different in origin from deeper lithofacies of the Jurassic subduction complex. Such lithologic contrast among the Jurassic deposits is of fundamental importance in understanding the tectonic development of the Japanese Islands as well as the formation of the Jurassic subduction complex. After reference to the recent works published in Japan, a brief comparative sketch of the pre-Cretaceous strata is presented.

OUTLINE OF PRE-CRETACEOUS GEOLOGIC UNITS

Distributions of lithofacies that characterized belts and terranes are schematically summarized in Figure 2. Geology for each belt or terrane, particularly the Jurassic component, is briefly summarized and interpreted below.

Hida

The Hida terrane consists mainly of quartz-feldspathic gneiss, marble, amphibolite, and granulite, metamorphic rocks derived from Precambrian to Paleozoic protoliths. Crystalline schists (210 to 250 Ma) of intermediate-pressure type are developed in the eastern margin known as the Unazuki zone. An association of these rocks having continental characteristics is considered to have occupied a marginal part of the Sino-Korea massif (Hiroi, 1981). Although radiometric ages show several metamorphic events, the final metamorphism is Middle Jurassic (170 to 180 Ma), and this is concordant with intrusives of the associated granitic rocks (Nozawa, 1977). This composite terrane is unconformably overlain by a thick pile of shallow- to nonmarine conglomerate and sandstone of Early Jurassic to Cretaceous age (Kobayashi, et al, 1957).

Hida Marginal

The Hida marginal is a zone of tectonic melange (Komatsu and Chihara, 1982) including blocks of glaucophane schist (310 to 360 Ma), metagabbro (about 400 Ma), Ordovician to Devonian shallow-water sedimentary rocks, and late Early Carboniferous to Middle Permian limestone formed on mafic volcanics in a matrix of serpentinite. Jurassic nonmarine clastic strata unconformably overlie the tectonic melange, suggesting amalgamation of the Hida and Hida marginal by Jurassic time (Mizutani and Hattori, 1983).

Sangun and Joetsu

The Sangun is characterized by high-pressure metamorphic rocks, consisting of pelitic, psammitic, quartz, and green schists (Hashimoto, 1964; Nishimura et al, 1977). Radiometric ages of micas from pelitic rocks range from 170 to 240 Ma. Late Triassic nonmarine and Early to Middle Jurassic shallow-water clastic rocks unconformably overlie the metamorphic terrane. Judging from the distribution and ages of the metamorphism, the terrane

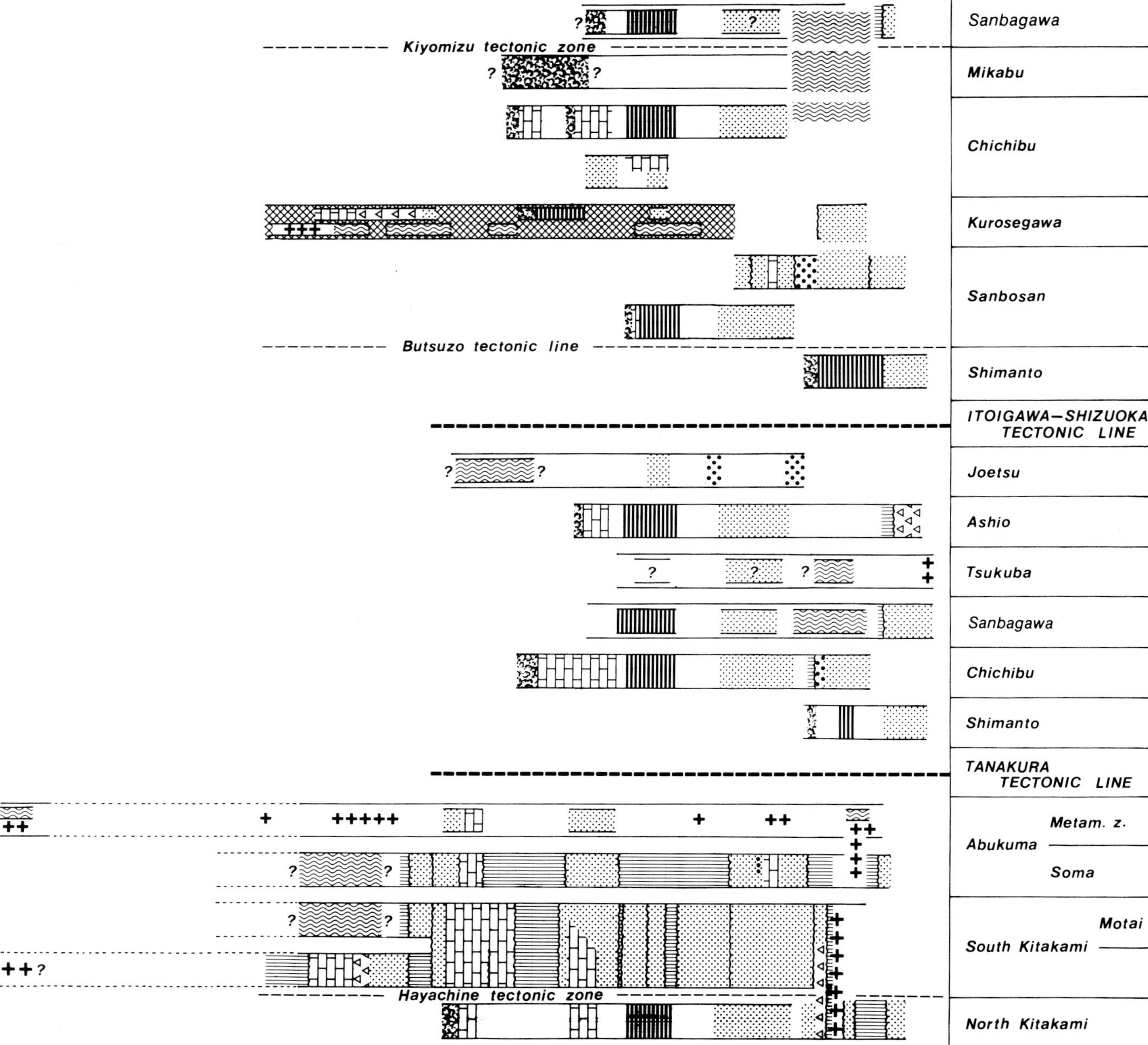

Figure 1—Geotectonic units of the Japanese Islands.

may be further divided. In the Joetsu region, pelitic and mafic metamorphic rocks of probably high-pressure type are sporadically developed, in association with ultramafic rocks (Hayama et al, 1969). Here, too, are thick sequences of Late Triassic and Early Jurassic shallow- to nonmarine clastic rocks.

Chugoku

The Chugoku has two modes of rock association. The northern exposure of the terrane is characterized by rocks of widely different ages and compositions, containing Early Carboniferous to Permian reef complexes built on submarine volcanos, coeval chert, and mudstone with subordinate sandstone (Ota, 1968; Hase et al, 1974). Unconformably overlying these strata are Triassic nonmarine conglomerate and sandstone. Triassic chert and Early Jurassic mudstone are also developed (Hayasaka et al, 1983). The southern part is composed mainly of Middle Permian to Triassic chert (Toyohara, 1976) and Early to Late Jurassic mudstone (Hayasaka et al, 1983). Rocks of different origins are chaotically mixed in both the northern and southern parts. The reef complexes and chert are interpreted as accreted oceanic materials (Kanmera and Nishi, 1983).

Maizuru

This zone consists of metamorphosed mafic and ultramafic rocks of late Paleozoic age, interpreted as disrupted ophiolite (Ishiwatari, 1978), associated with Middle Permian to Middle Triassic shallow-water clastic

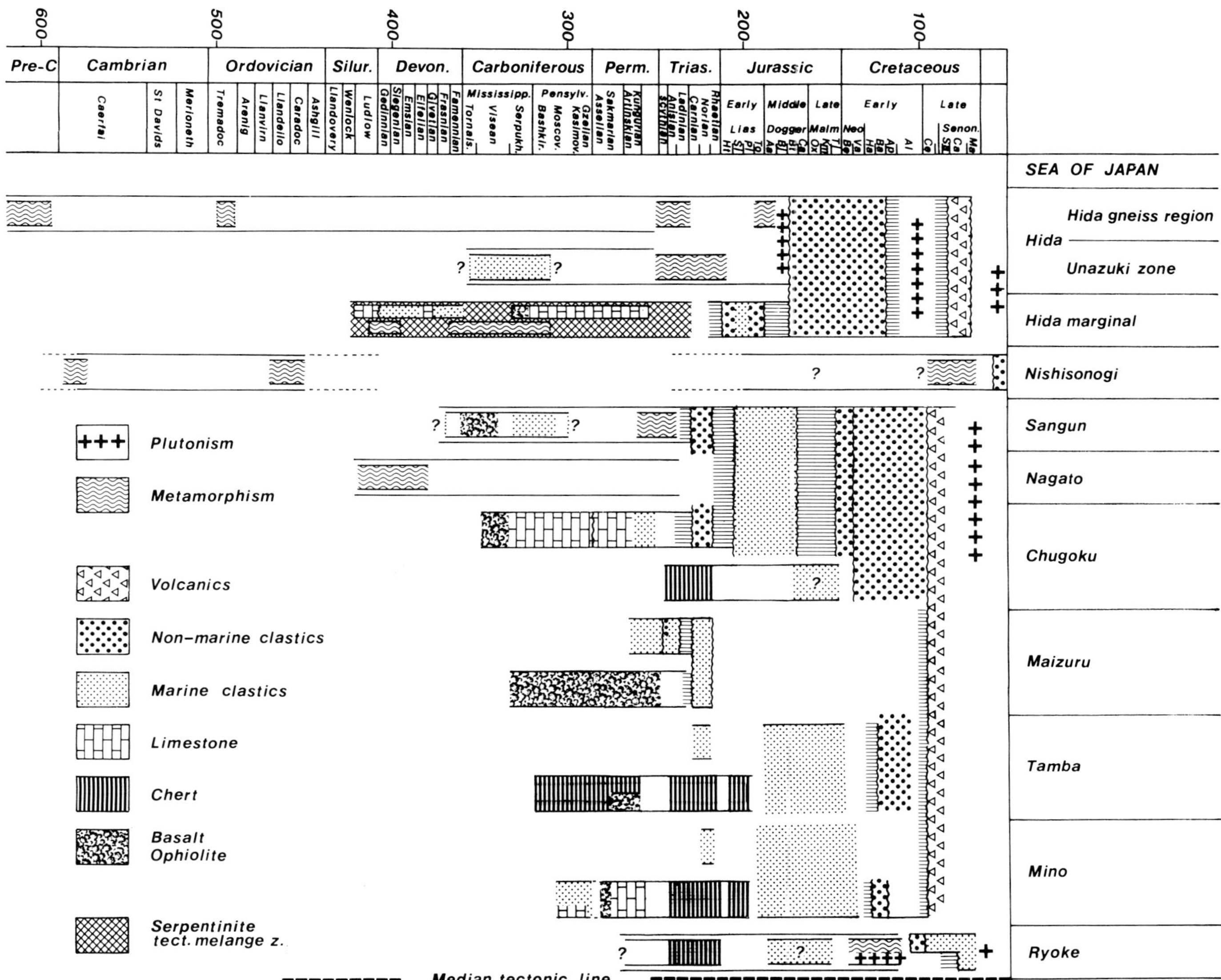

Figure 2—Distributions of lithofacies characterizing belts and terranes in the Japanese Islands.

rocks. Gabbroic rocks have radiometric ages from 240 to 280 Ma and metamorphic rocks from 200 to 330 Ma.

Tamba

The Tamba terrane consists of a structurally complex assemblage of different rock types including Late Carboniferous to Permian chert and minor limestone together with various amounts of mafic volcanics, Middle Triassic to Early Jurassic chert, coeval sandstone and mudstone, and Middle to Late Jurassic mudstone and olistostrome (Isozaki and Matsuda, 1980). These rock assemblages can be divided into two gross units based on rock composition and biostratigraphy (Ishiga, 1983). An older unit comprises Early to Middle Jurassic mudstone facies including large blocks or slabs of Late Carboniferous to Triassic chert, minor limestone and basalt, and are followed by Late Jurassic sandstone lithofacies. A younger unit consists of Late Jurassic mudstone lithofacies containing large blocks of Middle Triassic to Early Jurassic chert. The older and younger units are interpreted as Middle and Late Jurassic subduction complexes, respectively. They are unconformably overlain by Cretaceous nonmarine sedimentary and volcanic rocks.

Mino and Ashio

These are lithologically and biostratigraphically similar to the Tamba terrane. In the northern and western parts of the Mino, large blocks of Late Carboniferous to Permian limestone associated with mafic volcanics are embedded in Jurassic mudstone (Mizutani, 1981; Mizutani et al, 1981; Yamamoto, 1983). In the southeastern part, Middle Triassic to Early Jurassic chert is widely developed (Mizutani and Koike, 1982; Matsuda and Isozaki, 1982) along with Middle to Late Jurassic mudstone (Yao et al, 1980; Adachi, 1982; Matsuda and Isozaki, 1982) with minor conglomerate that contains Precambrian gneissic clasts (Shibata and Adachi, 1974). These strata compose a complicated structure consisting of many thrust sheets (Igo, 1979). Paleomagnetic data for the Mino terrane indicate that the rocks with oceanic affinity were formed in low latitude (Hattori, 1982; Mizutani and Hattori, 1983). In the Ashio region, Middle to Late(?) Jurassic radiolarians are reported from mudstone (Yao, 1979; Sashida et al, 1982) in which are embedded blocks and slabs of Middle Permian limestone built on mafic volcanics and Triassic chert. These are unconformably covered with Cretaceous(?) to Paleogene acidic volcanics.

Ryoke and Tsukuba

The Ryoke and Tsukuba terranes comprise low-pressure metamorphic belts consisting mainly of mica-rich schists and gneisses (Ono, 1977). Because Middle to Late Triassic conodonts are found from chert embedded in mudstone matrix of the Ryoke belt (Toyohara, 1976), the original rocks may have been Jurassic subduction complex. They are unconformably covered with a thick sequence of clastic sedimentary rocks or acidic volcanics of Late Cretaceous (Suyari, 1973). This means that the metamorphism is Early Cretaceous, although radiometric ages are from 60 to 80 Ma.

Sanbagawa, Mikabu, and Nishisonogi

These are regional metamorphic belts of high-pressure type, consisting of pelitic, psammitic, quartz, and green schists (Banno, 1964). Late Triassic conodonts found from low-grade metamorphic rocks (Suyari et al, 1980) suggest that the original rocks were formed probably as a part of the same Jurassic subduction complex as the Chichibu belt mentioned below. Metamorphism is seemingly related to Jurassic to mid-Cretaceous subduction although radiometric ages reflect only young ages spanning 70 to 102 Ma. The Mikabu belt, occurring at the southern part of the Sanbagawa, is a disrupted ophiolite composed of weakly metamorphosed cumulus peridotite, gabbro, pillow lava, and chert (Nakamura, 1971). It tectonically overlies weakly metamorphosed sedimentary rocks of the Chichibu belt. The Nishisonogi metamorphics are composed of pelitic, quartz, and green schists of high-pressure type. The radiometric ages of muscovite from the pelitic schist, ranging from 70 to 80 Ma, are close to the mean of the radiometric ages of the Sanbagawa metamorphism.

Chichibu and Sambosan

These belts consist of rocks varying in ages and compositions. One unit of the Chichibu is characterized by limestone, mafic volcanics, chert, and clastic rocks of the late Early Carboniferous to Permian (Maruyama and Yamasaki, 1978) and Triassic sandstone with a minor amount of limestone. These strata crop out along the Kurosegawa Tectonic Zone that runs through the central part of the Chichibu belt with an east-west trend. Another unit of these belts is composed of Triassic chert and Jurassic mudstone. From the northern zone of the Chichibu, Early to Middle Jurassic radiolarians are found (Sunouchi et al, 1982), and in its southern zone and the Sambosan belt Early to latest Jurassic radiolarians are reported (Taira et al, 1979; Matsuoka, 1982; Murata et al, 1982). These reflect a Jurassic subduction complex, and a continuous sequence from Triassic to Late Jurassic found in the southern part of the Chichibu (Matsuoka, 1983) suggests that the Jurassic oceanic plate sequence was covered by trench-filling terrigenous turbidites of Late Jurassic age.

Kurosegawa

This is a tectonic melange zone containing blocks of high-pressure schists (200–240, 320–330, and 350–390 Ma), garnet-amphibolite, granulite, garnet-biotite gneiss, sheared granite (400–450 Ma), Silurian limestone, Devonian shallow-water shale and acidic volcanics including welded tuff, and Carboniferous mafic volcanics and chert in a matrix of serpentinite (Maruyama, 1981). Paleomagnetic data for the acidic tuff and limestone indicate formation in low latitude requiring subsequent northward movement (Shibuya et al, 1983). This melange zone reflects disruption and dispersion within a strike-slip mobile zone (Taira et al, 1983).

Abukuma

This is a metamorphic terrane having probably continental characteristics, consisting of greenschist, amphibolite, and mica-rich gneiss of low-pressure type, associated with granitic rocks. The protolith is probably Paleozoic or older rocks and includes intermediate-pressure metamorphic rocks because of the existence of relict kyanite (Kano et al, 1973). Radiometric ages are mostly from 100 to 110 Ma, corresponding to the intrusion of granite in northeast Japan (Maruyama, 1979).

South Kitakami

The South Kitakami represents a peculiar region in Japan in having a well-stratified shallow-water sequence ranging from Silurian to Jurassic (Onuki, 1981), although Late Carboniferous rocks are absent. In the western part of this region, glaucophane-bearing rocks and ultramafic rocks, interpreted as disrupted ophiolite, are developed (Maekawa, 1981). Lithologic characteristics of the sedimentary and volcanic rocks show a continental affinity, in contrast to oceanic affinity of the other belts. Silurian to Devonian rocks, similar to those of the Kurosegawa, may have been a part of the Yangtze Massif in China. Therefore, it is considered to be allochthonous to the other pre-Cretaceous terranes of the Japanese Islands (Saito and Hashimoto, 1982). This region is bordered on its northeastern side by the serpentinite belt of a tectonic melange (Horikoshi, 1979).

North Kitakami

The North Kitakami consists of a thick sequence of deformed clastic rocks with chert, numerous mafic volcanics, and minor limestone. Possibly correlative strata continue to the north and underlie the main part of the western part of Hokkaido. Although limestone blocks occurring in the western part of this region are Carboniferous to Permian, most chert sequences are Triassic, and mudstone of the eastern part is mostly Middle to Late Jurassic. Hence, as a whole, the rock association is interpreted to be derived from Jurassic subduction complex (Minoura, 1983). The complex is intruded by mid-Cretaceous granitic rocks and overlain by mid-Cretaceous neritic conglomerate and sandstone.

LITHOLOGIC CONTRAST AMONG THE JURASSIC SEDIMENTS IN JAPAN

Based on many recent works on radiolarian biostratigraphy, the following geological aspects can be emphasized regarding the Jurassic system of the Japanese Islands.

1. The pre-Cretaceous geology of the Japanese Islands traditionally has been interpreted in the context of a late Paleozoic to early Mesozoic geosynclinal system. Recent biostratigraphic work has indicated, however, that all rock units or terranes in the islands are mingled with the Jurassic terrigenous sediments that comprise embedded "exotic" limestone, mafic volcanic rock, and chert ranging from late Early Carboniferous to Triassic. The chaotically mixed or structurally complex sedimentary facies are the products of an accretion process resulting from subduction (Taira et al, 1983). The tectonic units formed by the accretion process are Chugoku, Tamba, Mino, Ashio, Chichibu, Sambosan, and South and North Kitakami. Their metamorphic equivalents are the Sanbagawa and Ryoke belts.

2. The Jurassic rocks of the Japanese Islands occur in three modes: (1) subduction complex and their metamorphic equivalents mentioned above, (2) nonmarine to shallow-water mollasselike conglomerate and sandstone, sporadically developed as thick piles on the "older" terranes such as Sangun, Hida, Hida marginal, and Joetsu belts, and (3) shallow-water coherent units of dominantly sandstone and mudstone, developed in the South Kitakami and along the Kurosegawa Tectonic Zone.

3. Although the Jurassic subduction complex consists of, as a whole, Early Carboniferous to the latest Jurassic rocks, the complex may be subdivided into two units: an older and a younger subunit. The older unit composes Early to Middle Jurassic mudstone and olistostromal lithofacies containing blocks or slabs of Late Carboniferous to Triassic chert, limestone, and mafic volcanic rocks. The younger unit consists mainly of Late Jurassic mudstone and olistostromal lithofacies with blocks of Triassic to Early Jurassic chert. Locally tectonic blocks represent a continuous sequence ranging from Triassic to Late Jurassic.

4. The older complex may correspond to subduction forming magmatism along a continental margin developed in southeast China, the Korean Peninsula, and southwest Japan (Hida) of 190 to 160 Ma; whereas the younger subduction complex may correspond to magmatism occurring in southeast China and the Korean Peninsula from 160 to 130 Ma age (Takahashi, 1983).

5. The oceanic blocks or slabs in the Jurassic subduction complex are considered to represent an oceanic plate sequence developed on subducting slab. The oldest rocks known are late Early Carboniferous limestone formed on sea mounts, but the main constituents are chert ranging from Late Carboniferous to Early Jurassic. Pelagic shale intercalation in the cherts tends to increase upwards and to grade into siliceous shale and finally into coarse-grained clastic sediments.

6. The thick mollasselike deposits of Early Jurassic in the "older" terranes represent formation of deep inland basins and uplift of land mass, which are possibly due to deformation and would be caused by a collision of buoyant sea mounts or microcontinents against a continental margin. Large reef complexes in the inner side of the subduction complex and disrupted Silurian to Devonian rocks in the Hida marginal zone are possible candidates of the colliding masses.

7. In contrast with the inland basin deposits paleontologically characterized by not only the "cosmopolitan" and South Pacific elements but also by partially boreal faunal groups, the shallow-water deposits of the outer side include limestones with reef-forming organisms that indicate deposition in warmer environments. The Late Jurassic paleofloral provinces (Kimura, 1979) support the environmental contrast between the inner and outer sides. The inner-side flora belong to a Siberian paleofloral area, and the outer-side one to Kimura's Indo-European paleofloral area. The former has been considered to be temperate and the latter to be subtropical or tropical.

8. In southwest Japan, from the continental to Pacific sides, geologic units involving Jurassic strata are arranged in subparallel belts. Shallow- to nonmarine lithofacies crop out in inland depression. Strata of a subduction complex and its metamorphic equivalents lie to the south, and a younger Jurassic subduction complex that includes fault slices of shallow-water clastic lithofacies lies closest to the Pacific. Lying outboard of the youngest Jurassic sequence is the Shimanto belt, a Cretaceous subduction complex. The complex juxtaposition of the Jurassic sedimentary rocks of different origins can be explained by the accretion process related to the oblique subduction (Taira et al, 1983).

REFERENCES

Adachi, M., 1982, Some consideration on the *Mirifusus baileyi* assemblage in the Mino terrain, central Japan: News of Osaka of Micropaleontology, n. 5, p. 211–225 (in Japanese).

Banno, S., 1964, Petrologic studies on Sanbagawa crystalline schists in the Bessi–Ino district, central Shikoku, Japan: Journal of the Faculty of Science of the University of Tokyo, Sec. 2, v. 15, p. 203–319.

Hase, A., et al, 1974, The upper Paleozoic formation in and around Taishaki-dai, Chugoku massif, southwest Japan, with special reference to the sedimentary facies of limestone: Geologic Reports of Hiroshima University, v. 19, p. 1–39 (in Japanese).

Hashimoto, M., 1964, A review of the petrology of the Sangun metamorphic rocks, Japan: Bulletin of the National Science Museum, v. 7, p. 323–337 (in Japanese).

Hattori, I., 1982, The Mesozoic evolution of the Mino terrane, central Japan: A geologic and paleomagnetic synthesis: Tectonophysics, v. 85, p. 313–340.

Hayama, Y., et al, 1969, The Joetsu metamorphic belt and its bearing on the geologic structure of the Japanese Islands: Memoir of the Geological Society of Japan, n. 4, p. 61–82.

Hayasaka, Y., et al, 1983, Discovery of Jurassic radiolarians from the Kuga and Kanoashi groups in the western Chugoku district, southwest Japan: Journal of the Geological Society of Japan, v. 89, p. 527–530 (in

Japanese).

Hiroi, Y., 1981, Subdivision of the Hida metamorphic complex, central Japan, and its bearing on the geology of the Far East in pre-Sea of Japan time: Tectonophysics, v. 76, p. 317-333.

Hirokawa, O., ed., 1978, Geological map of Japan, 1:1,000,000: Geological Survey of Japan, Kawasaki, 2nd ed.

Horikoshi, E., 1979, Downward extension of the Kurosegawa tectonic zone based on the distribution of Quaternary volcanoes: Journal of Geological Society of Japan, v. 85, p. 427-434 (in Japanese).

Igo, H., 1979, Conodont biostratigraphy and restudy of geological structures at the eastern part of the Mino belt: Professor M. Kanuma Commemoration Volume, p. 103-113 (in Japanese).

Ishiga, H., 1983, Two suites of stratigraphic succession within the Tamba Group in the western part of the Tamba belt, southwest Japan: Journal of the Geological Society of Japan, v. 89, p. 443-454 (in Japanese).

Ishiwatari, A., 1978, A preliminary report on the Yakuno ophiolite in the Maizuru zone, inner southwest Japan: Journal of the Association for the Geological Collaboration in Japan (Earth Science), v. 32, p. 301-310 (in Japanese).

Isozaki, Y., and T. Matsuda, 1980, Age of the Tamba Group along the Hozugawa "anticline," western hills of Kyoto, southwest Japan: Journal of Geoscience, Osaka University, v. 23, p. 115-134.

Kanmera, K., and H. Nishi, 1983, Accreted oceanic reef complex in southwest Japan, *in* M. Hashimoto and S. Uyeda, eds., Accretion tectonics in the circum-Pacific regions: Tokyo, Terra Scientific Publishing Company, p. 195-206.

Kano, H., et al, 1973, The plutonic and metamorphic history of the Abukuma belt from the viewpoint of polymetamorphism, *in* K. Hide, ed., The Sanbagawa Belt: Hiroshima, Hiroshima University, p. 289-296.

Kimura, T., 1979, Late Mesozoic palaeofloristic provinces in East Asia: Proceedings of the Japan Academy, Series B, v. 55, p. 425-431.

Kobayashi, T., et al, 1957, On the Lower Jurassic Kuruma Group: Journal of the Geological Society of Japan, v. 63, p. 182-194 (in Japanese).

Komatsu, M., and K. Chihara, 1982, Tectonic melange and olistostrome in the Omi-Renge zone of the Hida marginal belt, *in* Symposium on tectonic melange zone: Geological Society of Japan Abstracts, p. 32-37 (in Japanese).

Maekawa, H., 1981, Geology of the Motai Group in the southwestern part of the Kitakami Mountains: Journal of the Geological Society of Japan, v. 87, p. 543-554 (in Japanese).

Maruyama, S., 1981, The Kurosegawa zone in the Ino district to the north of Kochi City, central Shikoku: Journal of the Geological Society of Japan, v. 87, p. 569-583.

———, and M. Yamasaki, 1978, Paleozoic submarine volcanoes in the high P/T metamorphosed Chichibu system of eastern Shikoku, Japan: Journal of Volcanic and Geothermal Research, v. 4, p. 199-216.

Maruyama, T., 1979, Rb-Sr geochronological studies in the granitic rocks of the southern Abukuma plateau, *in* H. Kano, ed., The basement of the Japanese Islands: Akita, Professor H. Kano Commemoration Volume, p. 523-558 (in Japanese).

Matsuda, T., and Y. Isozaki, 1982, Radiolarians around the Triassic-Jurassic boundary from the bedded chert in the Kamiaso area, southwest Japan: News of Osaka Micropaleontology, n. 5, p. 93-101 (in Japanese).

Matsuoka, A., 1982, Middle and Late Jurassic radiolarian biostratigraphy in the Sakawa and the Niyodo areas, Kochi Prefecture, southwest Japan: News of Osaka Micropaleontology, n. 5, p. 237-253 (in Japanese).

———, 1983, The conformable relationship between chert beds and clastic beds in the Triassic-Jurassic sequence of the southern subbelt of the Chichibu belt, Kochi Prefecture: Journal of the Geological Society of Japan, v. 89, p. 407-410 (in Japanese).

Minoura, K., 1983, Geology of the North Kitakami Belt: Chikyu (Earth Monthly), v. 5, 480-487 (in Japanese).

Mizutani, S., 1981, A Jurassic formation in the Hida-Kanayama area, central Japan: Bulletin of the Mizunami Fossil Museum, n. 8, p. 147-190.

———, and I. Hattori, 1983, Hida and Mino: Tectonostratigraphic terranes in central Japan, *in* M. Hashimoto and S. Uyeda, eds., Accretion tectonics in the circum-Pacific regions: Tokyo, Terra Scientific Publishing Company, p. 169-178.

———, and T. Koike, 1982, Radiolarians in the Jurassic siliceous shale and in the Triassic bedded chert of Unuma, Kagamihara City, Gifu Prefecture, central Japan: News of Osaka Micropaleontology, n. 5, p. 117-134 (in Japanese).

———, et al, 1981, Jurassic formations in the Mino area, central Japan: Proceedings of the Japan Academy, Series B, v. 57, p. 194-199.

Murata, M., et al, 1982, Late Mesozoic radiolarians fauna from the Sakaguchi Formation: News of Osaka Micropaleontology, n. 5, p. 327-337 (in Japanese).

Nakamura, Y., 1971, Petrology of the Toba Ultrabasic complex, Mie Prefecture, central Japan: Journal of the Faculty of Science of the University of Tokyo, Sec. 2, v. 18, p. 1-51.

Nishimura, Y., et al, 1977, Sangun belt, with special reference to the stratigraphy and metamorphism, *in* K. Hide, ed., The Sanbagawa Belt: Hiroshima, Hiroshima University, p. 252-282 (in Japanese).

Nozawa, T., 1977, Radiometric age map of Japan: Kawasaki, Geological Study of Japan.

Ono, A., 1977, Petrologic study of the Ryoke metamorphic rocks in the Takato Shiojiri area, central Japan: Journal of the Japanese Association of Mineralogists, Petrologists and Economic Geologists, v. 72, p. 453-468 (in Japanese).

Onuki, Y., 1981, Geology around the Kitakami River: Sendai, Hase Geological Survey, 224 p.

Ota, M., 1968, The Akiyoshi Limestone Group; a geosynclinal organic reef complex: Bulletin of the Akiyoshi-dai Science Museum, v. 5, p. 1-44 (in Japanese).

Saito, Y., and M. Hashimoto, 1982, South Kitakami Region: An allochthonous terrane in Japan: Journal of Geophysical Research, v. 87, p. 3691-3696.

Sashida, K., et al, 1982, On the Jurassic radiolarian assemblages in the Kanto district: News of Osaka Micropaleontology, n. 5, p. 51-66 (in Japanese).

Shibata, K., and M. Adachi, 1974, Rb-Sr whole-rock ages of Precambrian metamorphic rocks in the Kamiaso conglomerate from central Japan: Earth and Planetary Science Letters, v. 21, p. 277-287.

Shibuya, H., et al, 1983, A paleomagnetic study on the Silurian acidic tuff in the Yokokurayama lenticular body of the Kurosegawa tectonic zone, Kochi Prefecture, Japan: Journal of the Geological Society of Japan, v. 89, p. 307-309 (in Japanese).

Sunouchi, H., et al, 1982, Occurrence of Jurassic radiolarians from siliceous claystone in the northern belt of the Chichibu terrane, north of Ino Town, Kochi Prefecture, and its significance: Journal of the Geological Society of Japan, v. 88, p. 975-978 (in Japanese).

Suyari, K., 1973, On the lithofacies and the correlation of the Izumi Group of the Asan mountain range, Shikoku: Science Report of the Tohoku University, Series 2, v. 6, p. 489-495.

———, et al, 1980, Discovery of the late Triassic conodonts from the Sanbagawa metamorphic belt proper in western Shikoku: Journal of the Geological Society of Japan, v. 86, p. 827-828 (in Japanese).

Taira, A., et al, 1979, New observations on the Sambosan Group in the western Kochi Prefecture: Geological News, n. 302, p. 22-35 (in Japanese).

———, et al, 1983, the role of oblique subduction and strike-slip tectonics in the evolution of Japan, *in* T. W. C. Hilde and S. Uyeda, eds., Geodynamics of the western Pacific-Indonesian region: American Geophysical Union Geodynamic Series 11, p. 303-316.

Takahashi, M., 1983, Space-time distribution of late Mesozoic to early Cenozoic magmatism in East Asia and its tectonic implications, *in* M. Hashimoto and S. Uyeda, eds., Accretion tectonics in the circum-Pacific regions: Tokyo, Terra Scientific Publishing Company, p. 69-88.

Tanaka, K., and T. Nozawa, 1977, Geology and mineral resources of Japan: Kawasaki, Geological Survey of Japan, 430 p.

Toyohara, F., 1976, Geologic structures from Sangun-Yamaguchi zone to "Ryoke zone" in eastern Yamaguchi Prefecture: Journal of the Geological Society of Japan, v. 82, p. 99-111 (in Japanese).

Yamamoto, H., 1983, Occurrence of Late Jurassic radiolarians of the *Mirifusus baileyi* assemblage from New Village, Gifu Prefecture, central Japan: Journal of the Geological Society of Japan, v. 89, p. 595-596 (in Japanese).

Yao, A., 1979, Triassic and Jurassic radiolarians from Honshu geosynclinal deposits: Abstracts of the Annual Meeting of the Geological Society of Japan, p. 148 (in Japanese).

———, et al, 1980, Triassic and Jurassic radiolarians from the Inuyama area, central Japan: Journal of Geoscience, Osaka City University, v. 23, p. 135-154.

———, et al, 1982, Triassic and Jurassic radiolarian assemblages in southwest Japan: News of Osaka Micropaleontology, n. 5, p. 27-43 (in Japanese).

Tectonostratigraphic Terranes of Japan that Bear on the Tectonics of Mainland Asia

Z. M. Zhang
Stanford University
Stanford, California

The Japanese Islands are divided into six terranes: the Abean, Akiyoshi, Sakawa, Shimanto, Hidaka, and Chishima terranes. Tectonic lines traditionally regarded as important are not boundary faults in this terrane division scheme. Precambrian rocks exist in the Hida belt, which was, however, not a portion of the Sino-Korean craton. Ophiolites are considered to be basement for individual accretionary terranes in Japan. Mafic-ultramafic intrusives are not regarded as ophiolites.

The Japanese Islands reveal a pattern of oceanward growth of the continental crust. The present tectonic configuration of Japan is an end product of a long-term interaction between the Eurasian continent and a primary ocean, Panthalassa. The accretionary process ceases at one end of a terrane and shifts to the other end. This phenomenon is common in accretionary terranes of different ages and results whenever the divergent plate boundary is not parallel to the convergent plate boundary. The paleogeography of Japan during most of the Paleozoic indicated an affinity to the Yangtze craton, but Japan became close enough to have flora communication with the Sino-Korean craton during Late Permian. The late Paleozoic terrane of Japan was a part of the accretionary system along the eastern Asian continent. Direct extensions of Japanese Mesozoic terranes, however, could not be found in mainland Asia. The two phases of extension in eastern Asia during the Cretaceous–Paleogene led to the formations of a series of continental basins in the inner belt and of a string of marginal seas in the outer belt. Felsic magmatism accompanying the two phases of extension in eastern Asia could be widespread over an area 1,000 km (620 mi) wide from the coast inland. This extension must be controlled by certain deep-seated mechanisms or thin-skin tectonism that is similar to the present western North America, rather than by ridge descent.

INTRODUCTION

Japan lies within a festoon of island arcs characteristic of the western Pacific. The ocean side of these island arcs marks the present continental margin of Eurasia. This region provides a key in the study of the interaction between oceanic crust and continental crust on the globe. In order to approach the plate tectonic relationship of the Japanese Islands and mainland Asia, it is necessary to begin with a terrane analysis of Japan. From the principles and definitions used in terrane analysis (Coney et al, 1980; Jones et al, 1983; Howell and Jones, 1984), it is obvious that the more allochthonous and, especially, exotic terranes an area embraces the more difficult is the task of tracing its plate tectonic evolution. Extreme telescoping of terranes and deformation in the foreland (postaccretionary consolidation) and complicated strike-slip faulting, thrusting, and folding result in the breakup of some terranes (postaccretionary dispersion) in Japan (Taira, 1984), making plate tectonic analysis even harder. In regions where paleomagnetic, paleontologic, lithologic, and stratigraphic data are not sufficient, it is difficult to evaluate how many of the terranes are allochthonous or exotic. However, easier to answer are the questions of when a terrane or a group of terranes were accreted onto a preexisting continental block in Japan, how far a contemporaneous accretionary event could be traced within the Japanese Islands and in mainland Asia, and what is the effect of a particular accretionary event upon the foregoing terranes. The purpose of this paper is to put forward a preliminary division scheme of Japanese terranes, which is somewhat different from the latest ones by Japanese geologists (Geological Survey of Japan, 1977; Saito, 1984; Taira, 1984; Ozawa and Kanmera, 1984), and to offer some controversial ideas concerning the tectonics between the Japanese Islands and mainland Asia.

TECTONOSTRATIGRAPHIC TERRANES OF JAPAN

The Japanese Islands formed at the junction of four developing island arcs, namely, the Chishima (Kuril) arc connecting Kamchatka Peninsula to the north, the Honshu arc in the proper part of Japan, the Ryukyu arc connecting Taiwan to the south, and the Izu–Mariana arc, which extends southward from the apex of the Honshu arc. Before the formation of these island arcs, various terranes consisting of volcano-sedimentary formations had amalgamated to form the basement of Japan. These terranes experienced multiple-stage folding, faulting, metamorphism, and magmatism. The present-day tectonic framework of Japan is an end product of a long-term

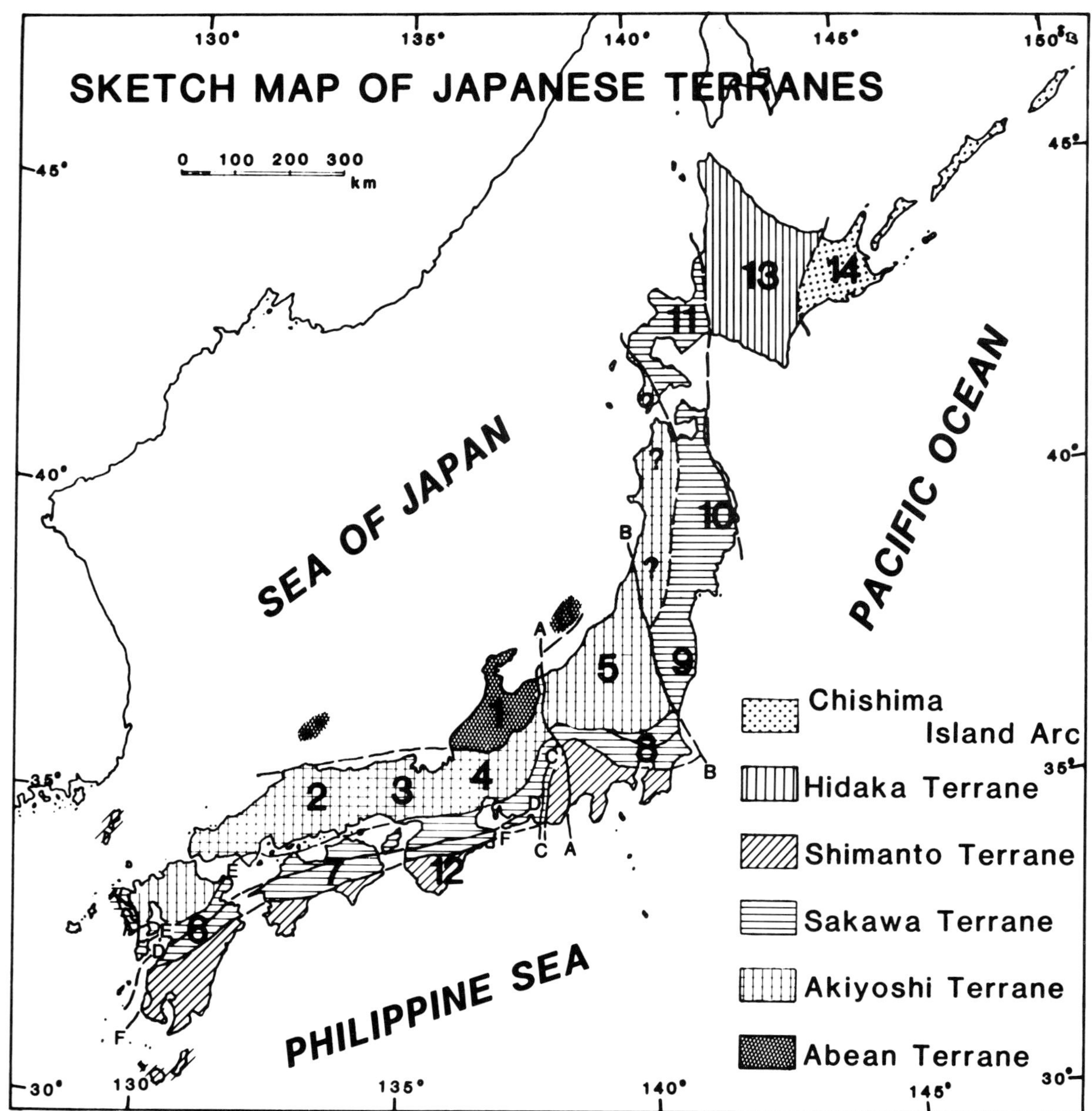

Figure 1—Sketch map of Japanese terranes. Terranes are shown by different patterns. Numbers indicate the locations of tectonostratigraphic columns in Figures 3 and 4. A–A = Itoigawa–Shizuoka tectonic line; B–B = Tanakura tectonic line; C–C = Akaishi tectonic line: D–D = median tectonic line; E–E = Oita-Kumamoto tectonic line; F–F = Butsuzu tectonic line.

interaction among the Eurasian plate, Pacific plate, and the now vanished Panthalassic ocean plate, e.g., the Kula plate (Uyeda and Miyashiro, 1974).

More than 20 tectonic units are recognized in the pre-Neogene geology of Japan (e.g., Geological Survey of Japan, 1977; Saito, 1984), and these tectonic units constitute six terranes (Fig. 1). They are the Abean, Akiyoshi, Sakawa, Shimanto, Hidaka, and Chishima terranes. The periods of terrane accretion appear to become younger oceanward (Geological Survey of Japan, 1977; Ozawa and Kanmera, 1984). The Sanriku–Jagan terrane (Sugimura and Uyeda, 1973) is the youngest, just developing offshore northeastern Japan. The major tectonism, plutonism, and metamorphism related to each of the accretionary events are shown in Figure 2. A few tectonic units that accreted onto a preexisting continental block in the Japanese Islands are treated as a single terrane because the postaccretionary dispersion was not severe enough to eliminate the tectonic relationship among those tectonic units.

The Abean terrane had undergone several stages of tectonism, which are named as Shizu, Setamai, and Abean orogenies, respectively, at the end of the early Visean, at the end of Late Carboniferous, and during the Permian. Each stage was characterized by faulting, folding and/or metamorphism, and subsequent uplifting. High P/T and low P/T metamorphism related to the accretionary process could be found in the Hida marginal belt and the Hida belt. The Akiyoshi terrane was accreted during the Late Triassic. Accretion was completed in the middle Carnic in

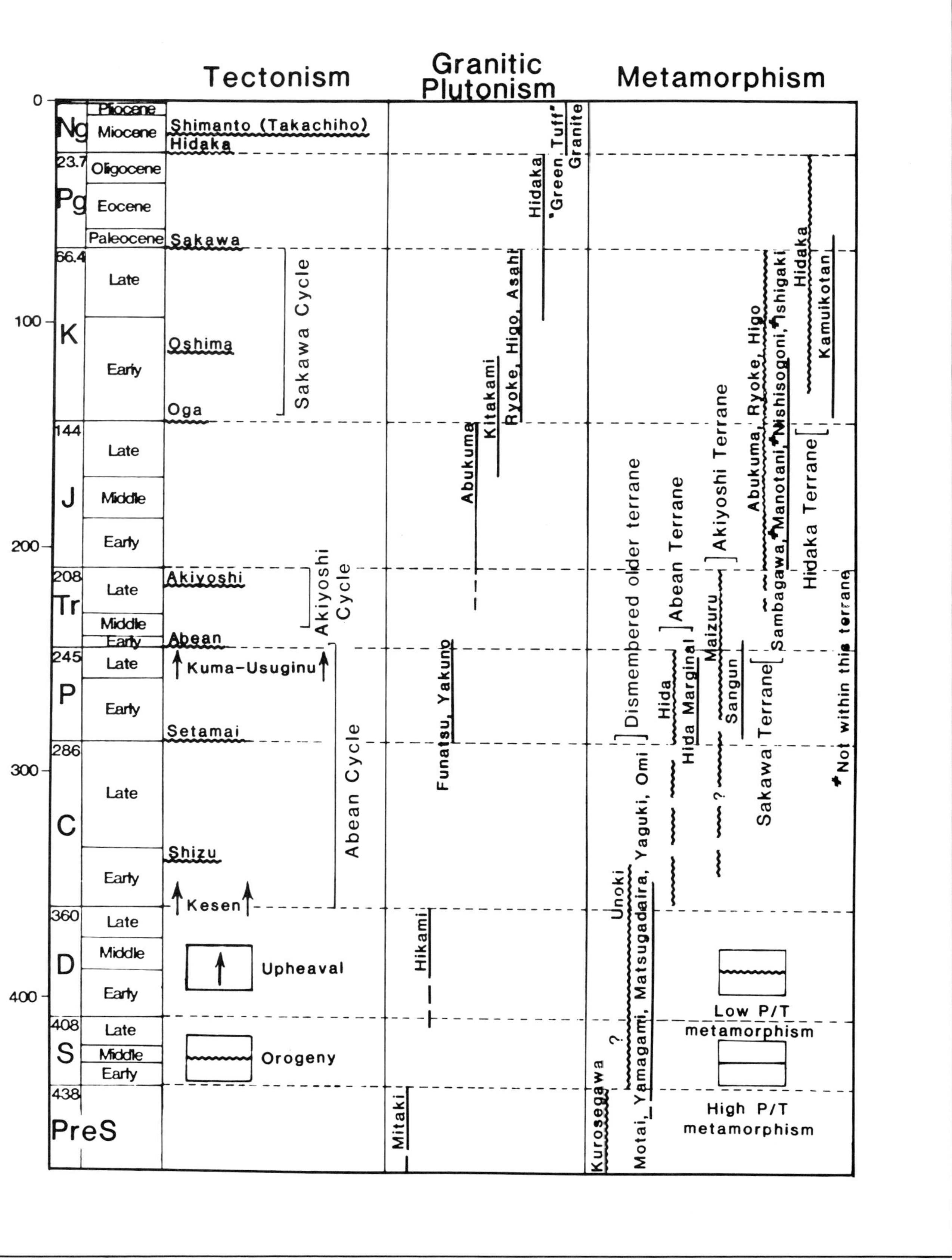

Figure 2—The main tectonic, plutonic, and metamorphic events in Japanese terranes (data mainly from Geological Survey of Japan, 1978). Geologic time scale by million years is on the left.

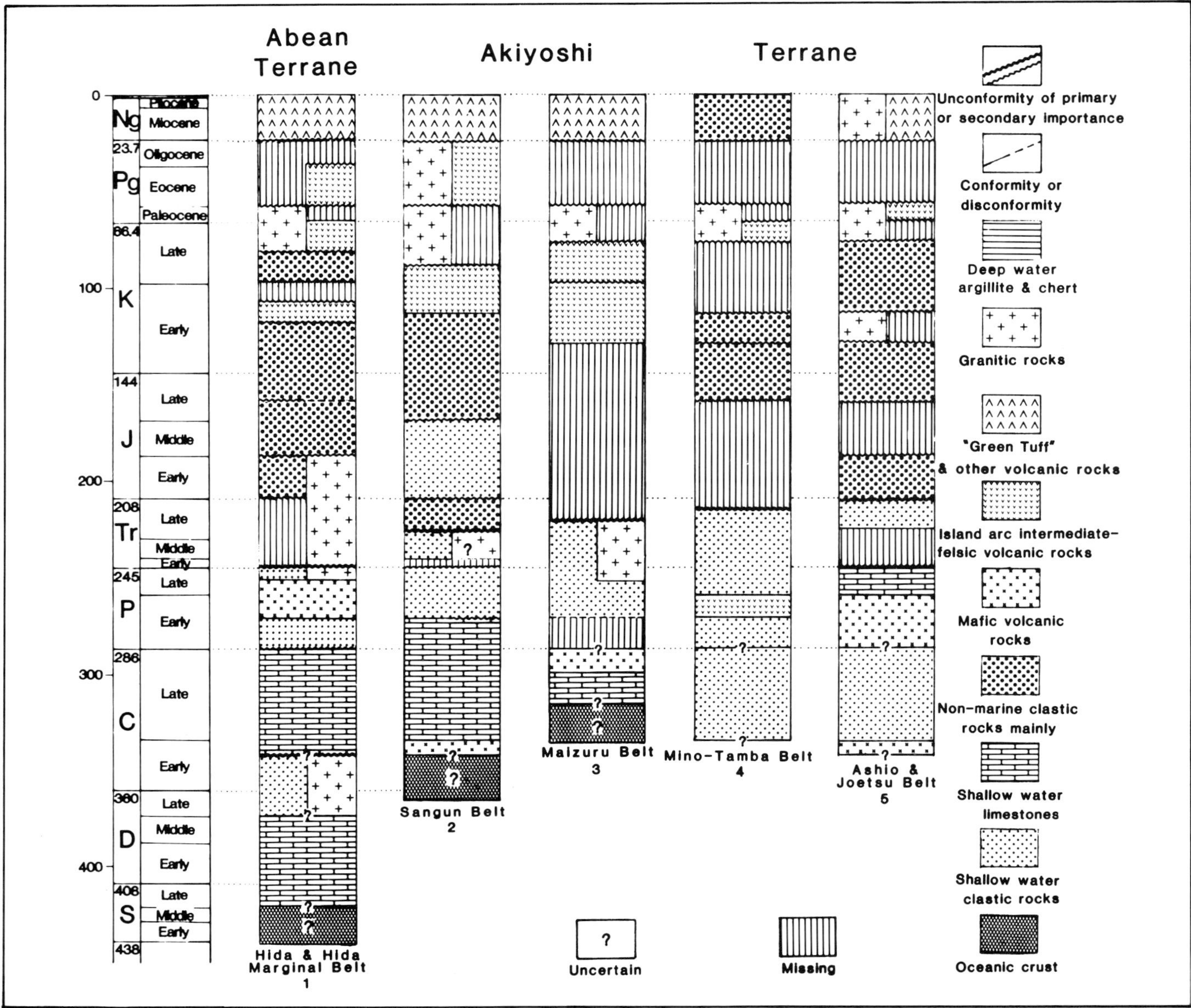

Figure 3—Tectonostratigraphic columns of Abean terrane and Akiyoshi terrane, showing sedimentary formations and tectonic, extrusive, and acidic intrusive events. Numbers show locations on Figure 1. Geologic time scale by million years is on the left. Legend is on the right. Data mainly from Geological Survey of Japan (1977).

the northern Kyushu and Chugoku Belts, at the end of Carnic in the Maizuru Belt, and at the end of Noric in Mino-Tamba and North Kanto Belts as shown in the tectonostratigraphic columns in Figure 3. Saito (1984) considered that the Mino-Tamba and Ashio Belts are Jurassic accretionary complexes. As a result of the Akiyoshi accretion, the Sangun and Maizuru metamorphics constitute a paired metamorphism belt. The Sakawa terrane was accreted on the older terranes during Jurassic and Cretaceous through the Oga, Oshima, and Sakawa tectonisms (Fig. 2). The northeastern section including Abukuma, Kitakami, and Outer Kitakami was attached before the Aptian age, while the southwestern section including the central Kyushu, the northern Shikoku, and the Kanto areas—that is, the Ryoke, Sanbagawa, Mikabu, Chichibu, and Sambosan Belts (Saito, 1984; Geological Survey of Japan, 1977)—were attached at the end of Cretaceous (Fig. 4). Subduction-related magmatism was widespread. The Sanbagawa and Ryoke metamorphics constitute a paired metamorphic belt. The northern subbelt of the Shimanto terrane is a Cretaceous accretionary complex, and the subsequent Eocene–Oligocene and early Miocene accretionary process led to the formation of the southern subbelt. The Hidaka terrane is a composite terrane. The Kamuikotan belt in the west was created by a westward subducting marginal sea during Cretaceous, and the Hidaka belt in the east was created by the subsequent eastward subduction of the same marginal sea during Early Tertiary (Miyashiro, 1977). The Chishima terrane is a portion of the Japanese island arcs that developed since the early Miocene, but its basement is similar to the Shimanto terrne. Several aspects about the present division scheme of Japanese terranes are emphasized below.

1. Major and minor division: The Japanese Islands

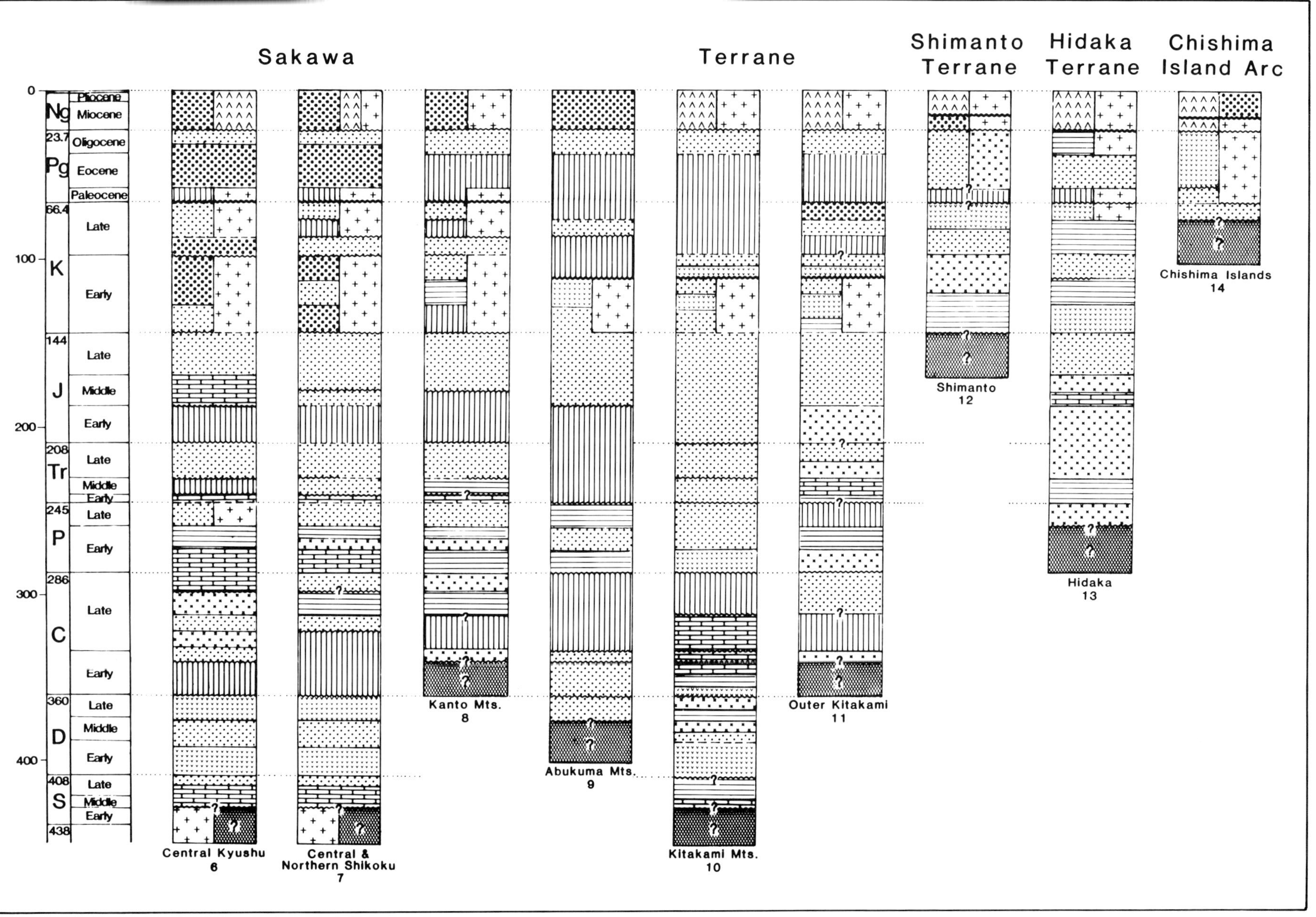

Figure 4—Tectonostratigraphic columns of Sakawa terrane, Shimanto terrane, Hidaka terrane, and Chishima terrane, showing sedimentary formations and tectonic, extrusive, and acidic intrusive events. Numbers show locations on Figure 1. Geologic time scale by million years is on the left. Legend is the same as in Figure 3. Data mainly from Geological Survey of Japan (1977).

have been traditionally divided into southwest Japan and northeast Japan by the Itoigawa–Shizuoka tectonic line as to the pre-Neogene geologic evolution. In view of the fact that similar pre-Neogene structures of southwestern Japan can be traced east of that line, some geologists consider that the Tanakura tectonic line should be the dividing line between southwest and northeast Japan (Geological Survey of Japan, 1977; Taira, 1984). Actually, extensions of the pre-Neogene structural units of southwest Japan could be found in the Kitakami Mountains and southwestern Hokkaido, even if there are some geologic dissimilarities on both sides of the tectonic line. Considering the principle of terrane analysis, both the Itoigawa–Shizuoka and the Tanakura tectonic lines are not believed to be of first-order importance. They are, nevertheless, the major postaccretionary structural lines.

The median tectonic line in southwest Japan is regarded as a dividing line between the inner zone and the outer zone in the area by most Japanese geologists (Geological Survey of Japan, 1977). It separates a high P/T Sanbagawa metamorphic belt from a low P/T Ryoke metamorphic belt, both of which are roughly contemporaneous products of a single period of accretion. This tectonic line is, therefore, not a boundary between two terranes but a secondary postaccretionary structural line within a single terrane.

2. Existence of Precambrian basement: Whether Precambrian rocks exist or not in the Japanese Islands has long been disputed. During the last two decades, more and more data have confirmed that Precambrian basement rocks actually occur in Japan as earlier mentioned by Minato et al (1965).

Radiometric determinations suggest that Precambrian rocks exist in the "Kamioka" granite and Miyakawa granite of the Hida area (Geological Survey of Japan). The occurrence of Precambrian continents within or at adjacent areas of Japan in certain geologic periods have been inferred from the discovery of Precambrian detritus, such as in the Hida Mountains (Geological Survey of Japan), the Mino Belt (Geological Survey of Japan), and in Mie Prefecture (Geological Survey of Japan). Metamorphic rocks are locally considered to be rejuvenated from Precambrian basement by later tectonothermal events; e.g., in the Hida Mountains, in the Kurosegawa belt, in the Nagato area of westernmost Honshu, and in Nagasaki Prefecture of Kyushu (see Minato et al, 1979). Precambrian rocks may also exist in the Oki Islands, the Noto Peninsula (Minato et al, 1979), the Abukuma Mountains, and the Kitakami Mountains (Geological Survey of Japan, 1977), but no direct evidence is available.

3. Tectonostratigraphic position of ophiolite: Coleman (1984) emphasizes that ophiolites could form in a variety of tectonic settings, but a twofold tectonic classification is most workable, namely, passive margin ophiolite (Tethyan type) and accretionary margin ophiolite (Cordilleran type). Ophiolites of Japan, either subduction-zone ophiolites or island arc ophiolites as mentioned by Miyashiro (1977), could be similar to Coleman's accretionary margin ophiolites. These ophiolite bodies are often the oldest rocks in the tectonostratigraphic sequence and are commonly considered to be basement for individual accretionary belts. The tectonostratigraphic columns in Figures 3 and 4 show the oceanic crust as basement for each terrane. However, some mafic-ultramafic complexes are reported to have intrusive contacts within the individual terranes (Geological Survey of Japan, 1977), and the radiometric dates show them to be clearly younger. These are the so-called ophiolites of Miyashiro (1977) and are found in many island arc terranes as mafic-ultramafic intrusions. However, they cannot be considered to have developed in ocean-spreading centers (ophiolites) since they do not exhibit the cogenetic mafic-ultramafic assemblages or stratigraphy normally associated with ocean crust formed at a spreading center. Snoke et al (1982) have clearly shown that these mafic-ultramafic intrusions develop in the Mesozoic arc terranes of the Klamath and Sierra Nevada Mountains of the western United States Cordillera and are intrusive into preexisting and older ophiolite basement.

4. Nature of Abean terrane: The Abean terrane consists of the Hida belt and the Hida marginal belt that rims the Hida belt from the south. Although Precambrian rocks do exist in the Hida belt, the author could not agree with Ozawa and Kanmera (1984) and Saito (1984) that the Hida belt was a portion of the Sino-Korean craton. Except for some Precambrian rocks of granulite facies, the Hida belt consists mainly of Hida metamorphic rocks and the associated Funatsu granite. The age of the metamorphic rocks is not well understood. The Rb–Sr whole-rock isochron age indicates mostly Paleozoic. In general, the statistical chart of the K–Ar and Rb–Sr age datings of the Hida metamorphic rocks show peaks around 500, 350, 240, and, especially, 180 m.y. (Geological Survey of Japan, 1977). The Hida belt seems not to be a Precambrian craton. Most protoliths of the Hida metamorphic rocks are considered to be greywacke intercalated with minor amounts of limestone, mudstone, and volcanic rocks (Geological Survey of Japan, 1977). The frequent variation of lithofacies suggests an unstable sedimentary environment or active continental margin. The Hida marginal belt consists of serpentine melange with exotic blocks of blueschists and unmetamorphic rocks. This diversity of rock suggests a tectonic collage similar to the Kurosegawa and Nagato belts or to those in the Tethyan tectonic belt of the Mediterranean (Ganser, 1974).

SOME TECTONIC PROBLEMS IN EASTERN ASIA AND RELATIONSHIP BETWEEN JAPAN AND MAINLAND ASIA

Manner of Continental Growth

The present structural pattern of Japanese Islands comes into being through five stages of evolution, that is, the Abean, Akiyoshi, Sakawa, and Shimanto–Hidaka accretion since late Paleozoic and island arc development since early Miocene. Generally speaking, the sedimentary formation of active continental margin nature, the volcanism, plutonism, deformation, and metamorphism related to accretionary processes have been shifted oceanward. Saito (1984) had the similar opinion that from the Sea of Japan to the Pacific Ocean, accretion had proceeded through Triassic(?), Jurassic, Cretaceous, and Paleogene progressively; whereas Hida, Abukuma, and

South Kitakami are considered to be continental affinity. Postaccretionary consolidation and dispersion do occur during the tectonic evolution of Japan; their effects, however, have not disturbed the general tendency of the continent's oceanward growth (Matsuda and Uyeda, 1971; Saito, 1984; Taira, 1984). Apparently, accretion and post-accretionary dispersion might have taken place alternatively in an area to yield a more complicated tectonic configuration. Underestimation of the effects of postaccretionary dispersion is misleading (Howell et al, 1984). But to what degree terrane analysis has proved the conception of the continent's oceanward growth to be unreasonable is still controversial among geologists. The author holds that the general tendency of a continent's growth is still oceanward although it is not necessarily a continuous process. Continental fragments, remnant volcanic arcs, oceanic islands and sea mounts, oceanic plateau and continentlike structures of mixed origin might be accreted on a continent to give a more complicated pattern of continental growth (Howell et al, 1984). However, the above-mentioned features occupy only a minor amount of the accretionary assemblages or their ages follow the same tendency of younging oceanward in most of the cases.

Panthalassa

The general oceanward migration of accretionary belts during Phanerozoic can be recognized not only in Japan but also elsewhere in the world, including southern China from the Yangtze craton to the Coast Range of eastern Taiwan (1 in Fig. 5), eastern Australia from the West Australian craton through Tasmania to northern Queensland , Antarctica from the East Antarctic craton through Ross-Filcher to the Antarctic Peninsula, and North America from the North American craton through the Klamath Mountains to the Coast Range (Blake et al, 1984; Hill, 1984). Obviously, there have been continuous or intermittent interactions between an oceanic plate with its adjacent continental plates since the beginning of Phanerozoic to the present time. In other words, Panthalassa and younger oceanic plates existed for the whole Phanerozoic period. As the successor of Panthalassa, the Pacific Ocean is not a secondary ocean like the Atlantic and Indian Oceans, which began to open since early Mesozoic. Instead, it is a primary ocean that has never closed completely.

Tectonism Along Strike of a Terrane

Despite the postaccretionary disturbance, it is common that the accretionary period revealed by tectonism, magmatism, and metamorphism becomes younger along the strike of an accretionary terrane. It is called tectonic shift and is an effect of diagonal convergence of plates. In other words, whenever the directions of divergent and convergent plate boundaries are not parallel to each other, subduction activity will cease along the convergent boundary from one end to another. Such features have been documented from the Cenozoic North American continental margin along the Juan de Fuca plate boundary and Cocos plate boundary (Dickinson and Snyder, 1979), the Mesozoic Akiyoshi and Sakawa terranes of Japan (see above), the late Paleozoic North Tienshan accretionary foldbelt of China (see Fig. 1 of Zhang et al, 1984), and the late Paleozoic Junggar–Hegen suture between the Cathaysian plate and the Siberian plate (2 in Fig. 5).

Affinity of Japanese Islands to Asian Cratons

As shown in Figures 3 and 4, Upper Ordovician/Silurian up to lower Visean strata occur in the Abean terrane and Sakawa terrane. Those strata, however, are totally lacking in the Sino-Korean craton. The peculiar solitary coral *Sugiyamaella* sp. originally described from the Japanese Tournaisian has been reported from the Qilian Mountains (3 in Fig. 5) and Inner Mongolia (4 in Fig. 5) of China (Guo, 1976). The most characteristic coral genus for Visean, *Kueichowphyllum* sp., is found from Japan through southern China, Viet Nam, central Asia, to Turkey, forming a faunal province of tropical sea, the Kuichowphyllum Sea as proposed by Hill (1948). According to Asama (1956), the Japanese flora of Permian age belong to the Cathaysian phytogeographic province because *Gigantopteris whitei* Halle, which is a characteristic element of the Lower Permian Shanxi Formation of North China, is found in the Kitakami Mountains. From these data, it is concluded that the Japanese terranes had closer affinity with the Yangtze craton rather than with the Sino-Korean craton during Silurian to Early Carboniferous. However, near the end of Permian when the Sino-Korean craton was combined with the Yangtze craton to make up the main part of the Cathaysian continent (Zhang et al, 1984), the Japanese terranes started to have flora communication with the Cathaysian phytogeographic province.

Late Paleozoic Active Continental Margin Along Mainland Asia

Existence of a late Paleozoic accretionary foldbelt along the maritime province (5 in Fig. 5) in southeastern China is advocated by some Chinese geologists (Li et al, 1982; Zhang et al, 1983; Zhang et al, 1984) but is seriously doubted by Huang (1984, personal communication). In some isolated areas of the maritime province, the Devonian to Lower Carboniferous active continental margin deposits are covered unconformably by Middle Carboniferous clastics. In the northern and central parts of Hainan Island (6 in Fig. 5), the same unconformity occurs roughly between Tournaisian and Visean. These areas constitute a late Paleozoic accretionary terrane that attached to the Yangtze craton. In the Changshan Mountains (7 in Fig. 5), the Devonian to Lower Carboniferous was active continental margin deposits, whereas unfolded and unmetamorphosed reef limestones appeared in late Visean. The Changshan area is another contemporaneous accretionary terrane that was attached to the Cambodian Massif from the north. The Qimantage and Burhanbuda Mountains in western China (8 in Fig. 5) were folded by an orogeny at approximately the same time. This terrane coeval with the above two was accreted onto the southward grown Sino-Korean craton. It is not clear whether these three terranes were connected to each other or not. Relatively strong tectonism of the same period also influenced the basement of the Mesozoic–Cenozoic mobile belt in the Far East of the Soviet Union

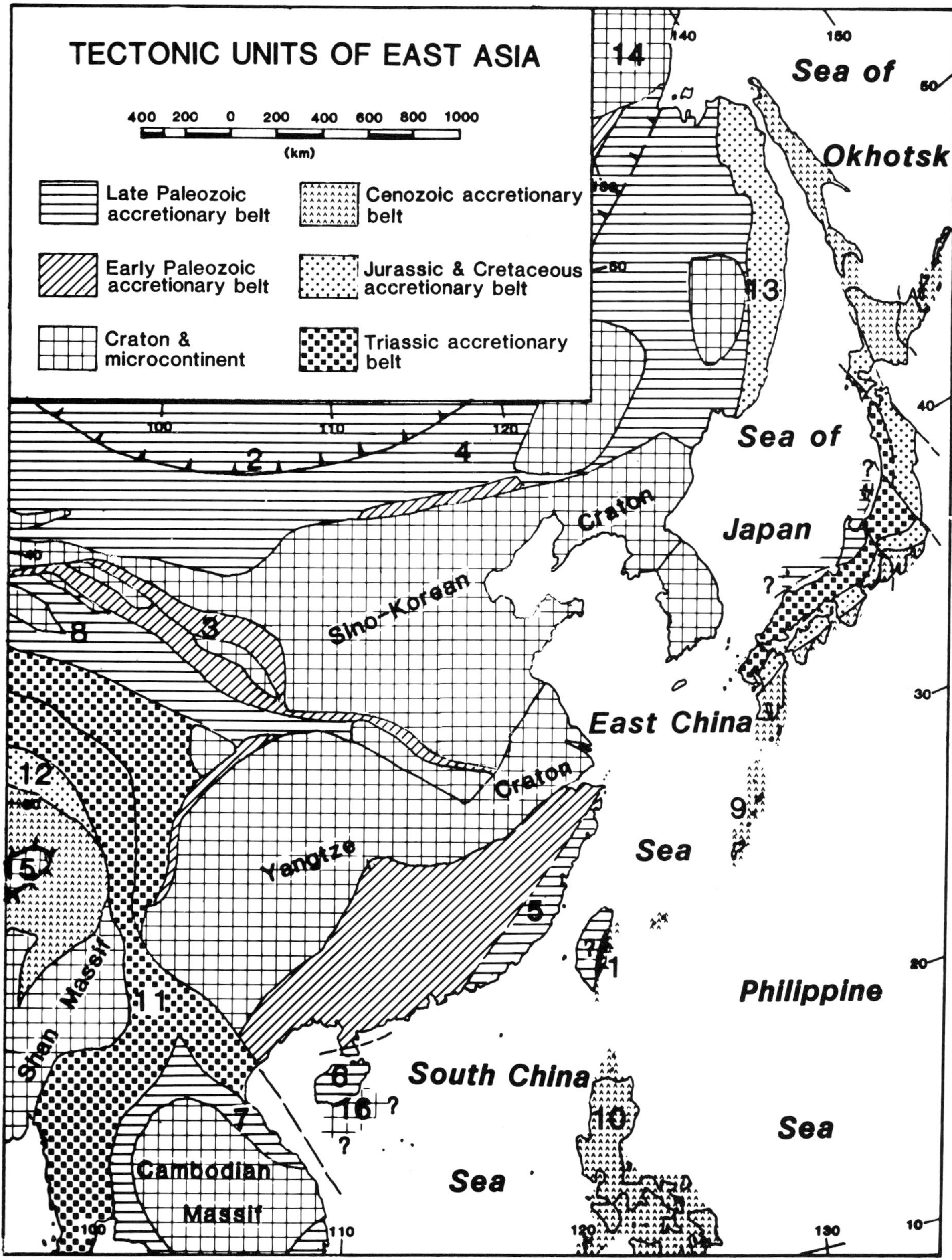

Figure 5—Tectonic units of east Asia. Cratons, microcontinents, and accretionary belts are shown by different patterns. 1 = Coast Range of eastern Taiwan; 2 = Junggar-Hegen suture; 3 = Qilian Mountains; 4 = Inner Mongolia; 5 = maritime province in southeastern China; 6 = northern and central parts of Hainan Island; 7 = Changshan Mountains; 8 = Qimantage and Burhanbuda Mountains; 9= Ryukyu Islands; 10 = the Cordillera Central of Luzon Island; 11 = Kekexili-Jinsha-Ailao region; 12 = North Tibet-West Yunnan; 13 = Nadanhada Mountains; 14 = Siberian craton; 15 = Indian craton; 16 = South China Sea craton.

from the Verkhoyan Mountains through the Kolyma River drainage to the Chukotsk Peninsula, but the whole area had not been converted into a stable environment. The Onimaru Transgression of late Visean was widespread in Japan, and about the same time, the Sino-Korean Craton was submerged for the first time after a long period of uplifting and erosion since the Middle Ordovician. The simultaneity of the above-mentioned terranes and the simultaneity of tectonism on stable craton and active continental margin reveals that the accretionary process had influenced not only the accretionary terrane itself but its foreland. It is also demonstrated that all the presently separated terranes might have had some spatial connection to each other before.

A younger accretionary terrane by the end of Permian might be overlapping on or extend side by side with the foregoing Early Carboniferous terranes before the opening of marginal seas along mainland Asia. The late Paleozoic terrane now extends from the Abean terrane through Ryukyu (Amamio, Okinawa, and Ishigake Island) (9 in Fig. 5) and the Central Range of Taiwan into the Cordillera Central of Luzon Island (10 in Fig. 5).

The Indosinian-Akiyoshi Orogeny

The Late Triassic Indosinian orogeny of China coeval with the Akiyoshi orogeny of Japan was very strong in the Kekexili-Jinsha-Ailao region (11 in Fig. 5) where an accretionary terrane was formed. The same terrane can be traced to Southeast Asia and to the west of the Pamir Mountain Knot. That was the Triassic Paleo-Tethys. The effect of this orogeny could be found also in the Mesozoic–Cenozoic mobile belt in the Far East of the Soviet Union. It appeared to be stratigraphic break or weak angular unconformity there. Relatively gentle folds, fault-bounded depressions, and granitic intrusions could be found in the Sino-Korean Craton and Yangtze Craton, but no contemporaneous accretionary terranes exist along the eastern margins of these cratons to bridge the Akiyoshi terrane in Japan and the Indosinian terrane in southwestern China and Southeast Asia (see Fig. 5). Whether it is an effect of postaccretionary dispersion, the available paleomagnetic data are insufficient. The Mesozoic sediments of Hida marginal belt, Mino belt, and Shimanto belt yield low to very low paleolatitudes. Hida belt seems stable in paleomagnetic direction during the same time (e.g., Hirooka et al, 1984; Kodama and Taira, 1984). The Mesozoic sediments of the Hida marginal belt and other terranes, for which paleomagnetic study has been done, are not in situ but appear as exotic blocks. In addition, the matrix including these exotic blocks shows much higher paleolatitude as mentioned by Maruyama (1984, personal communication). Then the Japanese terranes may have been formed at the paleolatitudes similar to their present latitudes. Long-distance drifting or postaccretionary dispersion of these terranes would be impossible, and the discontinuity of the Indosinian terrane with the Akiyoshi terrane seems to be a primary phenomenon.

The Yanshanian-Sakawa Orogeny

The Oga orogeny at the end of the Jurassic, the Oshima orogeny after Neocomian, and the Sakawa orogeny at the end of Cretaceous (Fig. 2) may be equivalents of the first, second, and third phases of the Yanshanian orogeny in China. The corresponding accretionary terranes formed through the Yanshanian orogenies could be found in North Tibet-West Yunnan (12 in Fig. 5) and Nadanhada Mountains of northeast China (13 in Fig. 5) but not along the eastern margins of the Sino-Korean Craton and the Yangtze Craton (see Fig. 5). Within the cratons, however, the Yanshanian orogeny is reflected by widespread magmatism and moderate folding, faulting, and thrusting of cover strata. The Neocomian coal-bearing series and continental volcanic deposits were commonly involved into folds and faults together with the Middle and Late Jurassic. Therefore, the remarkable unconformity is not between the Jurassic and Cretaceous, as some Chinese stratigraphers thought, but between the Neocomian and Aptian.

After the tectonism at the end of Neocomian, eastern Asia entered a stage of crustal extension. As a result, a number of continental sedimentary basins formed that are commonly bounded by reverse faults and cut randomly by normal faults (Fig. 6). Most of the basins continued to develop in the Paleogene and Neogene.

The Formation of Marginal Seas

The Cretaceous–Paleogene magmatism that brought widespread granitic rocks and the associated felsic volcanic rocks in Japan also played an important role in eastern mainland Asia (Fig. 7). The magmatism has been going on along with the first-phase and the second-phase extension along the continental margin. The large-scale rejuvenation of the sialic crust in a 1,000 km (620 mi) wide zone indicates a mechanism deep in the mantle. The ridge descent model (Uyeda and Miyashiro, 1974) for marginal sea formation seems unable to explain the simultaneity of formation of all the marginal seas in the western Pacific. The first phase and the second phase of extension that led to the occurrence of sedimentary basins on land and marginal seas near the ocean respectively in Early Cretaceous and Late Cretaceous–Paleogene were going on in northwest-southeast to west-northwest–east-southeast directions, and the resultant depressions arrange northeast-southwest to north-northeast–south-southwest directions in eastern Asia. That is the so-called Cathaysian tectonic trending by Lee (1939). Both the continental sedimentary basins in the inner side and the marginal seas in the outer side are oil-and gas-bearing.

CONCLUDING REMARKS

The Japanese Islands are divided into six terranes with a pattern of oceanward growth of continental crust. The present tectonic configuration of Japan is an end product of a long-term interaction between the Eurasian continent and a primary ocean, Panthalassa. Similar situations might have existed in other continents surrounding Panthalassa.

Accretionary process ceases at one end of a terrane and shifts to the other end. This phenomenon is common in accretionary terranes of different ages and results whenever the divergent plate boundary is not parallel to the convergent plate boundary.

Extension of the late Paleozoic accretionary terrane of

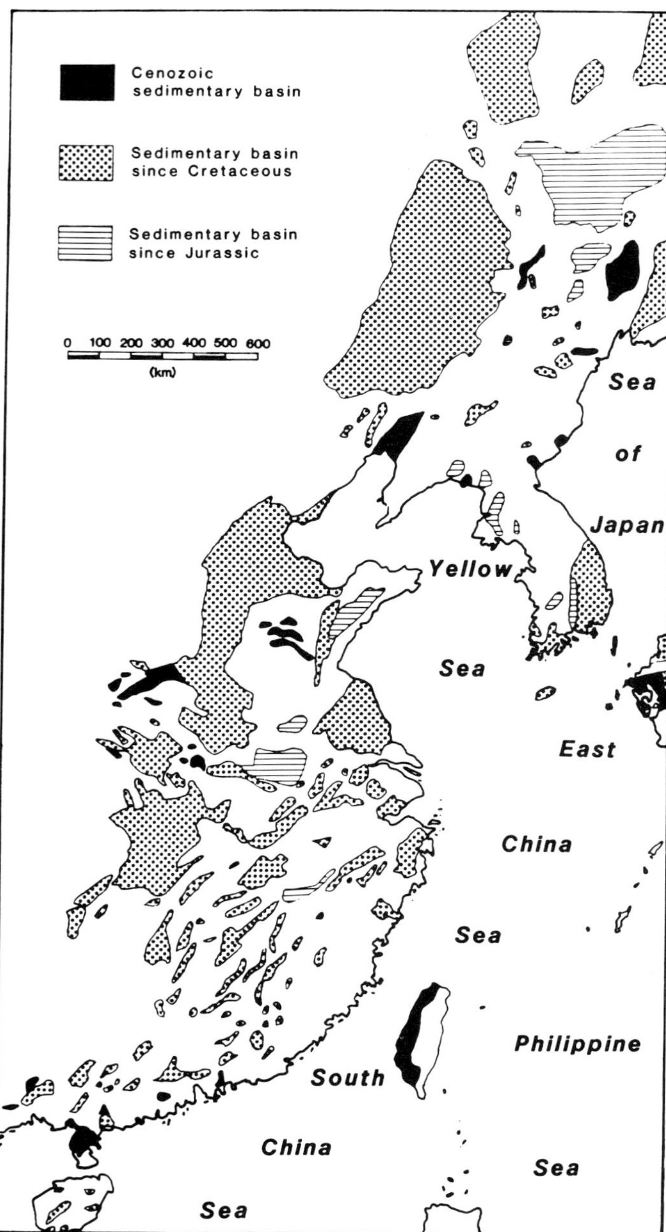

Figure 6—Mesozoic-Cenozoic sedimentary basins in eastern Asia, shown by different patterns. The Mesozoic sediments are mainly continental facies in mainland Asia.

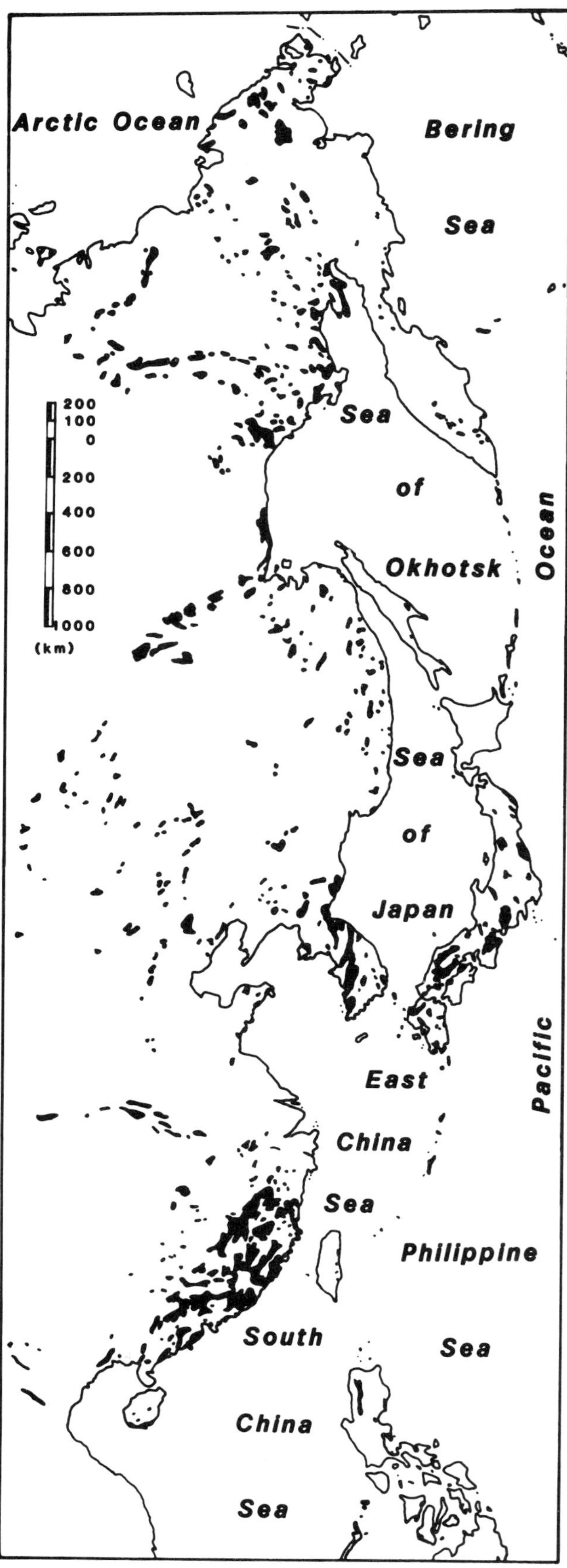

Figure 7—Distribution of late Mesozoic and Cenozoic granitic intrusions in eastern Asia.

Japan can be found in mainland Asia. The Mesozoic accretionary terranes of Japan, however, had no direct connection with those in the Indochina Peninsula and Southeast Asia. It reveals that two independent tectonic domains, the Tethyan and circum-Pacific, started to develop since the beginning of Mesozoic.

Two phases of extension in eastern Asia during Cretaceous-Paleogene led to the formations of a series of continental basins in the inner belt and of a string of marginal seas in the outer belt. Felsic magmatism accompanied the extension, which was controlled by thin-skin tectonism similar to the present western North America.

ACKNOWLEDGMENTS

Sincere thanks are due to a number of Japanese geologists who encouraged the author to write this paper by giving all-embracing and exciting discussions about China and mainland Asia during the Terrane Conference at Stanford in September of 1983. The author is very grateful to Professors Robert G. Coleman, Juhn G. Liou, Allan V. Cox, and Elizabeth L. Miller, and Drs. David G. Howell and David L. Jones for instructive suggestions and advice in preparation of the paper. The author's special acknowledgment is also given to Professor S. Maruyama who thoroughly reviewed the manuscript and helped to update the data.

REFERENCES

Asama, A., 1956, Permian plants from Maiya in northern Honshu, Japan: Proceedings of the Japanese Academy, v. 37, n. 7, p. 469–471.

Blake, Jr., M. C., et al, 1984, Preliminary tectonometamorphic terrane map of California, *in* D. G. Howell, et al, eds., Proceedings of the Circum-Pacific Terrane Conference: Stanford University Publications, Geological Sciences, p. 32.

Coleman, R. G., 1984, Ophiolites and their significance in terrane accretion and dispersion, *in* D. G. Howell, et al, eds., Proceedings of the Circum-Pacific Terrane Conference: Stanford University Publications, Geological Sciences, p. 66–67.

Coney, P. J., et al, 1980, Cordilleran suspect terranes: Nature, v. 288, n. 27, p. 329–333.

Dickinson, W. R., and W. S. Snyder, 1979, Geometry of triple junctions related to San Andres transform: Journal of Geophysical Research, v. 84, p. 561–572.

Ganser, A., 1974, The ophiolitic melange, a world-wide problem on Tethyan examples: Eclogae Geological Helvetiae, v. 67/3, p. 479–507.

Geological Survey of Japan, 1977, Geology and mineral resources of Japan: Tokyo, Sumitomo Printing and Publishing Co., Ltd., 430 p.

______, 1978, Geological map of Japan (1/1,000,000), Kawasaki, 2nd ed.

Guo, S. Z., 1976, Rugosa in atlas of palaeontology of North China (1) Inner Mongolia: Beijing, Publishing House of Geology, p. 74–75.

Hamilton, W., 1984, Subduction, magmatic arcs, and foreland deformation, *in* D. G. Howell, et al, eds., Proceedings of the Circum-Pacific Terrane Conference: Stanford University Publications, Geological Sciences, p. 105–107.

Hill, D., 1948, The distribution and sequence of Carboniferous coral faunas: Geological Magazine, v. 85, n. 3, p. 121–148.

Hill, L. B., 1984, Metamorphic and deformational constraints on terrane assembly, northern Klamath Mountains, California, *in* D. G. Howell, et al, eds., Proceedings of the Circum-Pacific Terrane Conference: Stanford University Publications, Geological Sciences, p. 108–110.

Hirooka, K., et al, 1984, Paleomagnetic evidence of accretion and tectonism of the Hida and the circum-Hida Terranes, central Japan, *in* D. G. Howell, et al, eds., Proceedings of the Circum-Pacific Terrane Conference: Stanford University Publications, Geological Sciences, p. 115–117.

Howell, D. G., and D. L. Jones, 1984, Tectonostratigraphic terrane analysis and some vernacular, *in* D. G. Howell, et al, eds., Proceedings of the Circum-Pacific Terrane Conference: Stanford University Publications, Geological Sciences, p. 6–9.

______, et al, 1984, Terrane accretion and the growth of continents, *in* D. G. Howell, et al, eds., Proceedings of the Circum-Pacific Terrane Conference: Stanford University Publications, Geological Sciences, p. 118.

Huang, T. K., et al, 1980, The geotectonic evolution of China: Beijing, Publishing House of Academy, 124 p.

Jones, D. L., et al, 1982, The growth of western North America: Scientific American, v. 247, n. 5, p. 70–85.

______, et al, 1983, Recognition, character, and analysis of tectonostratigraphic terranes in western North America, *in* M. Hashimoto and S. Uyeda, eds., Accretion tectonism in the circum-Pacific regions: Tokyo, Terra Scientific Publishing Co., p. 21–35.

Kimura, T., 1973, The old "Inner" Arc and its deformation in Japan, *in* P. J. Coleman, ed., The western Pacific island arcs, marginal seas, geochemistry, University of Western Australia Press, p. 255–274.

Kodama, K., and A. Taira, 1984, Paleomagnetism of the Shimanto Belt, southwest Japan, *in* D. G. Howell, et al, eds., Proceedings of the Circum-Pacific Terrane Conference: Stanford University Publications, Geological Sciences, p. 137.

Lee, J. S., 1939, The geology of China: London, Thomas Murby and Co., 528 p.

Li, C.Y., et al, 1982, Tectonic map of Asia (1/8,000,000): Beijing, Cartographic Publishing House.

Matsuda, T., and S. Uyeda, 1971, On the Pacific-type orogeny and its model—extension of the paired belts concept and possible origin of marginal seas: Tectonophysics, v. 11, p. 5–27.

Minato, M., et al, 1965, The geologic development of the Japanese Islands: Tokyo, Tsukiji Shokan Co., Ltd.

______, et al, eds., 1979, Variscan geohistory of northern Japan: the Abean Orogeny: Tokyo, Tokai University Press, 427 p.

Miyashiro, A., 1977, Subduction-zone ophiolites and island-arc ophiolites, *in* K. Saxena and S. Bhattacharji, eds., Energetics of geological processes: New York, Heidelberg, Berlin, Springer-Verlag, p. 188–214.

Ozawa, T., and K. Kanmera, 1984, Tectonic terranes of late Paleozoic rocks and their accretionary history in the circum-Pacific region viewed from fusulinacean paleobiogeography, *in* D. G. Howell, et al, eds., Proceedings of the Circum-Pacific Terrane Conference: Stanford University Publications, Geological Sciences, p. 158–159.

Saito, Y., 1984, Pre-Tertiary geotectonic units in Japan, *in* D. G. Howell, et al, eds., Proceedings of the Circum-

Pacific Terrane Conference: Stanford University Publications, Geological Sciences, p. 164–166.
Seno, T., and S. Maruyama, 1984, Paleogeographic reconstruction and origin of the Philippine Sea: Tectonophysics, v. 102, n. 1–4, p. 53–84.
Snoke, A. W., et al, 1982, Significance of mid-Mesozoic peridotitic to dioritic intrusive complexes, Klamath Mountains—western Sierra Nevada, California: Geology, v. 10, n. 3, p. 160–166.
Sugimura, A., and S. Uyeda, 1973, Island arcs: Japan and its environs. Development in geotectonics: Amsterdam, London, New York, Elsevier Scientific Publishing Co.
Sugisaki, R., and T. Tanaka, 1971, Magma types of volcanic rocks and crustal history in the Japanese pre-Cenozoic geosynclines: Tectonophysics, v. 12, p. 393–413.
———, et al, 1972, Late Paleozoic geosynclinal basalt and tectonism in the Japanese Islands: Tectonophysics, v. 14–15, p. 35–56.
Taira, A., 1984, Plate tectonic evolution of Japan, *in* D. G. Howell, et al, eds., Proceedings of the Circum-Pacific Terrane Conference: Stanford University Publications, Geological Sciences, p. 190–191.
Takahashi, M., 1983, Space-time distribution of Late Mesozoic to Early Cenozoic magmatism in East Asia and its tectonic implications, *in* S. Uyeda, ed., Accretion tectonics in the circum-Pacific region: Tokyo, Terra Scientific Publishing Co., p. 69–88.
Uyeda, S., and A. Miyashiro, 1974, Plate tectonics and the Japanese islands; a synthesis: Geological Society of America Bulletin, v. 185, p. 1159–1170.
Zhang, W. Y., et al, 1983, The marine and continental tectonic map of China (1/5,000,000). Beijing, Science Press.
Zhang, Z. M., et al, 1984, An outline of the plate tectonics of China: Geological Society of America Bulletin, v. 95, p. 295–312.

Terranes of the Central Philippines*

Robert McCabe
Texas A & M University
College Station, Texas

Jose N. Almasco**
Graciano Yumul†
Philippine Bureau of Mines and Geo-Science
Manila, Philippines

The Philippine Archipelago is a collection of volcanic arcs, ophiolite fragments, and rifted continental blocks that were welded into their current configuration prior to the late Miocene. Based on reported paleontologic dating and stratigraphic data, we recognize at least five tectonic elements in the central portion of the Philippines that can be classified as terranes.

These elements are: (1) the Central Philippine arc terrane, a stratigraphic terrane composed of a Lower Cretaceous to Recent island arc derived debris; (2) the Mindoro-Panay disrupted terrane, a heterogeneous terrane whose origin is related to the Miocene collision between the Philippine arc and the North Palawan continental terrane; (3) North Palawan continental terrane, a stratigraphic terrane composed of late Paleozoic and Mesozoic continental-derived sequences overlain by Cenozoic marine sediments; (4) the South Palawan disrupted terrane, an Early Tertiary to Miocene heterogeneous terrane overlain by Late Neogene shallow-water sediments; and (5) the Sulu-Zamboanga disrupted terrane, a poorly known terrane containing believed Mesozoic arc rocks and ophiolites. In addition to these five terranes, two additional terranes are identified and classified as suspect. These are: (1) East Luzon-Samar-Mindanao disrupted terrane, a heterogeneous terrane composed of Late Cretaceous marine sediments and Late Cretaceous(?) to Early Tertiary(?) ophiolitic rocks that sit over a metamorphic basement; and (2) the Cagayan arc terrane, a volcanic ridge located in the central part of the Sulu Sea that is composed of Miocene volcanics and limestones.

The assembly of these terranes is poorly understood. The North Palawan continental terrane is believed to have collided with the Central Philippines arc in the Miocene (Hamilton, 1979; McCabe et al, 1982a). The effects of this collision are the Mindoro-Panay disrupted terrane, a heterogeneous terrane that is composed of metamorphosed island arc rocks and ophiolites. Later deformational events have resulted in a series of left-lateral strike-slip faults. The Verde fracture zone between Mindoro and southwest Luzon is suggested to be one of these faults.

Late Miocene and Plio-Pleistocene paleomagnetic results from northeast Mindoro, Luzon, Marinduque, Negros, and northeastern Mindanao suggest that the entire Philippine archipelago has acted as a single tectonic unit during the past 5 m.y. Inclination values from Cretaceous to Oligocene rocks from the Philippines suggest that during the pre-Neogene the entire Philippine arc was located near equatorial latitudes. Neogene data are all close to the present inclination value, suggesting that the entire Philippine arc translated northward between the Early Neogene and Late Tertiary, as originally suggested by Hsu (1972). Declination values show the effects of the Miocene collision between Palawan and the Central Philippine arc. These results imply that caution must be used in defining terrane boundaries based on paleomagnetic declination values.

INTRODUCTION

Many of the world's orogenic belts are now recognized as being composed of numerous exotic blocks called terranes. Although many of these terranes share a common Neogene (or earlier) sedimentary and volcanic cover with the autochthonous rocks from a region, stratigraphic, paleontologic, and paleomagnetic studies suggest many of these blocks originated elsewhere and were translated to their current geographic position. For example, studies in the North American Cordillera (Jones et al, 1977, 1982; Irving, 1979; Coney et al, 1980; Beck, 1976, 1980) indicate that much of the western margin of North America is made up of a mosaic of numerous and apparently unrelated tectonostratigraphic terranes that were translated northward into their present location.

The processes responsible for the preaccretion assembly of these terranes is becoming a major topic of research. To date, most of this work is concentrated along

*Texas A & M Geodynamics Program Contribution Number 47.
**Present Address: Department of Geological Sciences, University of California at Santa Barbara, Santa Barbara, California
†Present Address: Department of Geology, University of Tokyo, Tokyo, Japan

the North American margin. Although these North American studies benefit from relatively well-constrained geologic data, numerous difficulties are encountered. These problems include later postassembly deformational events, poor preservation of initially intact stratigraphic sequences, and the difficulty of trying with partially exposed fragments of pre-Cenozoic stratigraphies to reconstruct the size, shape, and the kinematic history of these blocks.

Perhaps a way to overcome the difficulties of the older North American margin is to work with possible modern-day analogs found in Southeast Asia. In fact, recent workers (Silver and Smith, 1983; Hamilton, 1979; Saleeby, in press; Karig, 1983) have pointed out that many of the processes that occurred in the Mesozoic assembly of North America may have modern-day analogs in this region. As such, Southeast Asia appears to be an excellent laboratory for examining the assembly of terranes, and it also offers insights into the processes that lead to the initial isolation and the translation of these blocks prior to their eventual accretion.

One area of particular interest is the Philippine Archipelago, located along the western edge of the Philippine Sea Plate. This island chain is believed to result from the complex Cenozoic convergence between the Eurasian Plate, the Australian Plate, the Philippine Sea Plate, and the Pacific Plate. Figure 1 shows a simplified tectonic map of the Philippines and the surrounding regions. The figure shows that most of the margins of the Philippines are characterized by zones of active subduction, the two most prominent being the east-dipping Manila Trench and the west-dipping Philippine Trench.

In this region, we find examples of: (1) present-day amalgamation in the southern portion of Molucca Sea (Hatherton and Dickinson, 1968; Silver and Moore, 1978; Hamilton, 1977). This collision zone continues northward as a progressively older zone of Late Neogene amalgamation in Talaud (Hamilton, 1979; Moore, et al, 1981b; Moore and Silver, 1983) and between eastern and western Mindanao (Silver and Moore, 1978; Moore and Silver, 1983; Hamilton, 1979; Cardwell et al, 1980; Hawkins et al, this volume); (2) middle to late Miocene amalgamation of the North Palawan terrane to the western portion of the Philippine arc (Hamilton, 1979; Holloway, 1981; Taylor and Hayes, 1980; McCabe et al, 1982a; Karig, 1983); (3) Pliocene accretion of the North Luzon Ridge with Taiwan (Murphy, 1973; Karig, 1973; Suppe, 1980; Bowin et al, 1978). As such, this region appears ideal for studying the tectonic and stratigraphic processes involved in the consolidation of terranes.

Hamilton (1973, 1977, 1979) pointed out that the southern portion of this arc system is made up of at least six different tectonic units (which by the present definition would be classified as terranes). Hawkins et al (1982) distinguished three terranes in Mindanao. In addition, Karig (1983) recently identified at least six different tectonostratigraphic terranes in the northern portion of this arc system. In this paper we will examine a portion of the Philippine arc system, limiting our investigation to the terranes that make up the central portion of the archipelago (region enclosed in block in Fig. 1). Some of the terranes in this region are continuous with the tectonic stratigraphic units described by Hamilton, the Mindanao terranes classified by Hawkins et al, and with the northern Philippine terranes described by Karig.

TERRANE DEFINITIONS

Because the terrane concept is relatively new, some ambiguities still exist as to what a researcher defines as a terrane. The terrane definitions that we will use in this paper are from Jones et al (1983), Coney et al (1980), and Howell, Jones, and Schermer (this volume). In addition to these definitions, we define an additional type of terrane as a break-away terrane.

We suggest the use of the term break-away terrane to be used for a distinct stratigraphic package that shares an earlier history with another terrane, but later tectonic processes, such as opening of a marginal basin or transform faulting, separated this terrane from its place of origin. The important criteria for the identification of a break-away terrane is its common older history with another terrane and a younger stratigraphic succession distinct from the related terrane. Examples of break-away terranes are: (1) the North Palawan continental terrane that shares a pre-Middle Tertiary history with the southern China platform (Hamilton, 1979; Taylor and Hayes, 1980; Holloway, 1981); (2) the Sula Spur that is believed rifted from New Guinea (Hamilton, 1979); (3) the Palau Kyushu–West Mariana Ridge–Mariana Arc, all of which share a common pre-Oligocene history (Karig, 1971).

The recognition of break-away terranes is of prime importance to later tectonic reconstructions. If the place of origin of these terranes is recognized, it serves as a reference frame that can be used to constrain the trajectories of both the break-away terrane and other terranes that later become amalgamated to it. For example, although paleomagnetic evidence suggests that many of the terranes that make up the Philippines came from southerly latitudes (Hsu, 1972; Uyeda and McCabe, 1983; Fuller et al, 1983), the original locations of these terranes, with respect to each other, are still unknown. The only terrane for which we feel confident of its place of origin is the North Palawan continental terrane (Taylor and Hayes, 1980, 1983; Holloway, 1981). Therefore, in the future, when all of the terranes of the Philippine arc are identified, the Cenozoic location of the North Palawan continental terrane will serve as a reference point for understanding the relative trajectory of the other terranes from their place of origin to their present location.

One other term that we will use is suspect terrane (Coney et al, 1980). Although our meaning may be slightly different than that proposed by these authors, we will assign the term suspect terrane to a stratigraphic package that appears to be allochthonous to an established terrane where it is juxtaposed but with no identified distinct bounding faults. Based on this description, we identify two suspect terranes. These two terranes are: (1) East Luzon–Samar–Mindanao disrupted terrane, which later work may show to be a marginal portion of the central Philippine arc terrane; (2) the Cagayan arc terrane, which may have been built on top of both the North Palawan continental terrane and the South Palawan disrupted terrane.

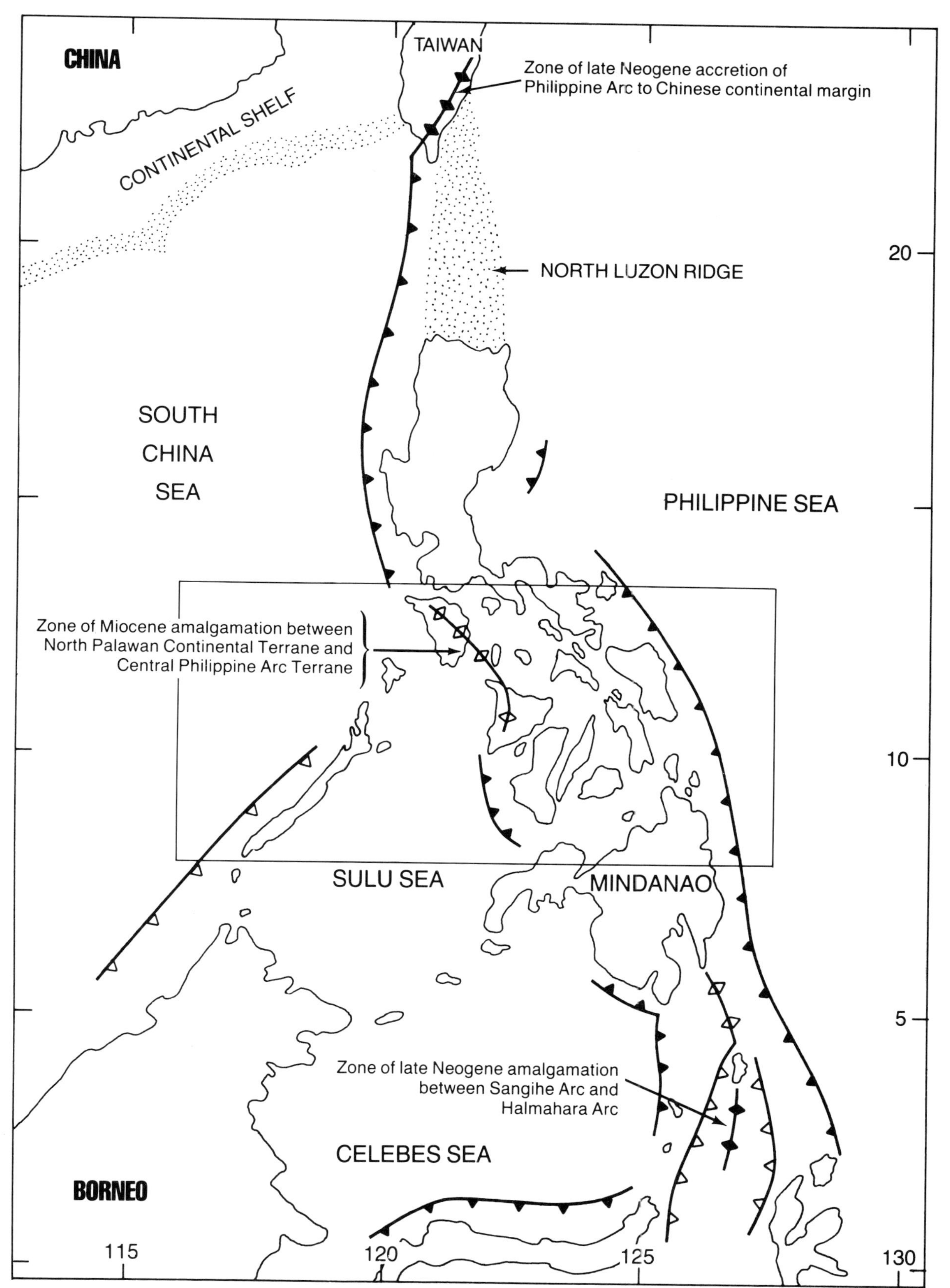

Figure 1—Simplified tectonic map of the western margin of the Philippine Sea Plate. Shown are zones of Neogene amalgamation in central Philippines and Molucca Sea. Also shown is zone of accretion of Philippines to Asian continental margin. Insert shows location of Figure 2. Single-sided hash marks represent zones of subduction, hash marks on upper plate. Diamonds show zones of collisions; solid ones are Pliocene to Recent collisions and open ones are Miocene in age.

DESCRIPTION OF TERRANES

The entire central Philippine region is composed of between five and eight (and perhaps more) distinct terranes that were stitched together between Eocene and late Miocene times. Geologic data from portions of this region are limited to only a few investigations; therefore, terrane boundaries are poorly constrained. Figure 2 shows the approximate boundaries of some of these terranes. These boundaries are based on Hashimoto (1981), Balce et al (1981), Hamilton (1979), the 1:1,000,000 scale map of the Philippines (Philippines Bureau of Mines, 1964), and our own field work in this region. It is hoped that this report will serve as a guide for other workers in the region, and we anticipate that later detailed studies will improve these boundaries and in all likelihood recognize other terranes. In fact, another paper from this volume (Hawkins et al, this volume) recognizes that the terrane that we label as the East Luzon–Samar–Mindanao is a composite terrane in the northeastern portion of Mindanao. In addition, these workers identify other terranes in southeastern Mindanao (a region that is not discussed in this paper).

Central Philippine Arc Terrane

This terrane is one of the largest in Southeast Asia, extending from the Molucca Sea to the eastern portion of Taiwan where it is accreting to the Chinese continental shelf. The Central Philippine arc terrane (CPAT) consists of island arc-related sequences that range in age from Lower Cretaceous to Recent. This terrane shows both periods of volcanic-plutonic activity and periods of volcanic quiescence. At present, it is impossible to comment on the polarity of the subduction systems that produced much of this terrane. The best stratigraphic section of this terrane is observed on the island of Cebu. Here an undated chlorite amphibolite schist (the Tunlob schist) forms the basement. Kinkel et al (1956) reports that this schist is unconformably overlain by Lower to Upper Cretaceous sequences. Kitamura et al (1968), however, argue that the schist is part of Cretaceous section and that the metamorphism of the schist reflects shearing at deeper levels. We originally suggested that this older portion of Cebu represents a distinctive terrane (reported in Howell et al, 1983); but, based on both the above arguments (Kinkel et al, 1965; Kitamura et al, 1968), we will classify these older rocks as part of the CPAT. The rest of the Cretaceous section that sits upon the schist is composed of folded volcanics (Cansi volcanics) that are metamorphosed up to the lowest greenschist facies, limestones (Tuburan limestone), and volcanic-derived sediments (Pandan Formation) (Hashimoto et al, 1978; Hashimoto and Balce, 1977; Balce et al, 1981). The lower portion of the Cretaceous section has been intruded by a 107 m.y. old diorite (Rb–Sr method; Momongan, personal communication). Deposited conformably upon the Upper Cretaceous section is a nearly complete Cenozoic section of volcanics, clastic sediments, coal seams, and limestones (Hashimoto, 1981; Balce et al, 1981).

Although a Lower Cretaceous section on Marinduque Island has yet to be found, this island, like Cebu, contains a similar section of Late Cretaceous to Recent diorites, volcanics, coralline limestones, and related sediments, suggesting that they share a common history. We suggest that the rest of the islands from the CPAT (with the possible exception of Bohol) are considerably younger in age than either Marinduque or Cebu. We believe that these younger islands are the products of Middle to Late Tertiary volcanic and plutonic activity that occurred along the margin of the older Marinduque–Cebu Cretaceous core. This volcanic and plutonic activity may have resulted from Late Tertiary to Early Neogene subduction proposed to have occurred along the western margin of the central Philippines (Hamilton, 1979; McCabe et al, 1982a; DeBoer et al, 1980; Uyeda and McCabe, 1983). Presently, the islands of Masbate, Guimaras, Negros, Panay, and Mindanao are believed to be of Cretaceous/Paleogene age (Philippines Bureau of Mines, 1964, 1981; Hashimoto, 1981). The assignment of this age is based on one questionable K–Ar age date from Guimaris of 59 m.y. (reported in Wolfe, 1981) and on a similarity between these islands' basement stratigraphy of island arc debris to the Cretaceous section of Cebu (also composed of island arc debris). However, unlike Cebu and Marinduque, these other islands do not have fossil dates older than Eocene. In fact, with the exception of one Eocene fossil date from Negros (Hashimoto, 1981) all of the fossil dates reported from this region are post-Eocene in age. On these islands, the mapped Cretaceous–Paleogene sections are mostly intermediate composition, arc-related volcanic breccias and tuffs that lie in contact with shallow-level diorites and quartz diorites that are usually heavily mineralized with sulfides. Recent Rb–Sr age dating of the Negros pluton gives an earliest Oligocene age (38 m.y.; Momongan, 1982, personal communication). In addition to this age, ages from volcanic-related sedimentary strata that sit stratigraphically above this volcanic-plutonic basement are late Oligocene to early Miocene in age (Gonzales, 1963; Balce et al, 1981; and our own field notes). On the bases of: (1) the 38 m.y. age of the Negros diorite; (2) the fact that the oldest fossil age in the region is Eocene and most fossils reported from the region are post-Eocene; (3) the intermediate chemistry of both the volcanics (most are andesitic) and the intrusives (dioritic); and (4) our field observations in Panay and Negros (which show that the volcanics lie directly above the pluton), we suggest that these Cretaceous/Paleogene metamorphosed sections on Panay, Negros, Guimaras, Masbate, and Mindanao are not of Mesozoic or Early Tertiary age but are Middle Tertiary volcanic ejecta and related sediments that are genetically related to the plutonic body which is also of Middle Tertiary age. Hamilton and Myers (1967) have hypothesized similar genetic relationships for many of the plutonic bodies of North America.

The timing of the plutonic activity in the CPAT is similar in age (Wolfe, 1981) and in composition (Balce et al, 1981; McCabe, 1979–1981 unpublished field notes) to the Central Cordillera and the Sierra Madre of northern Luzon. Based on the similarities in composition and stratigraphy, we suggest that the northern Luzon terranes of Karig (1983) are part of the CPAT. It should be noted that during the Neogene the northern Luzon terranes of Karig (1983) were likely offset westward with respect to the CPAT by strike-slip faulting. Age data (Wolfe, 1981) and geologic descriptions of the Central Cordillera of Mindanao

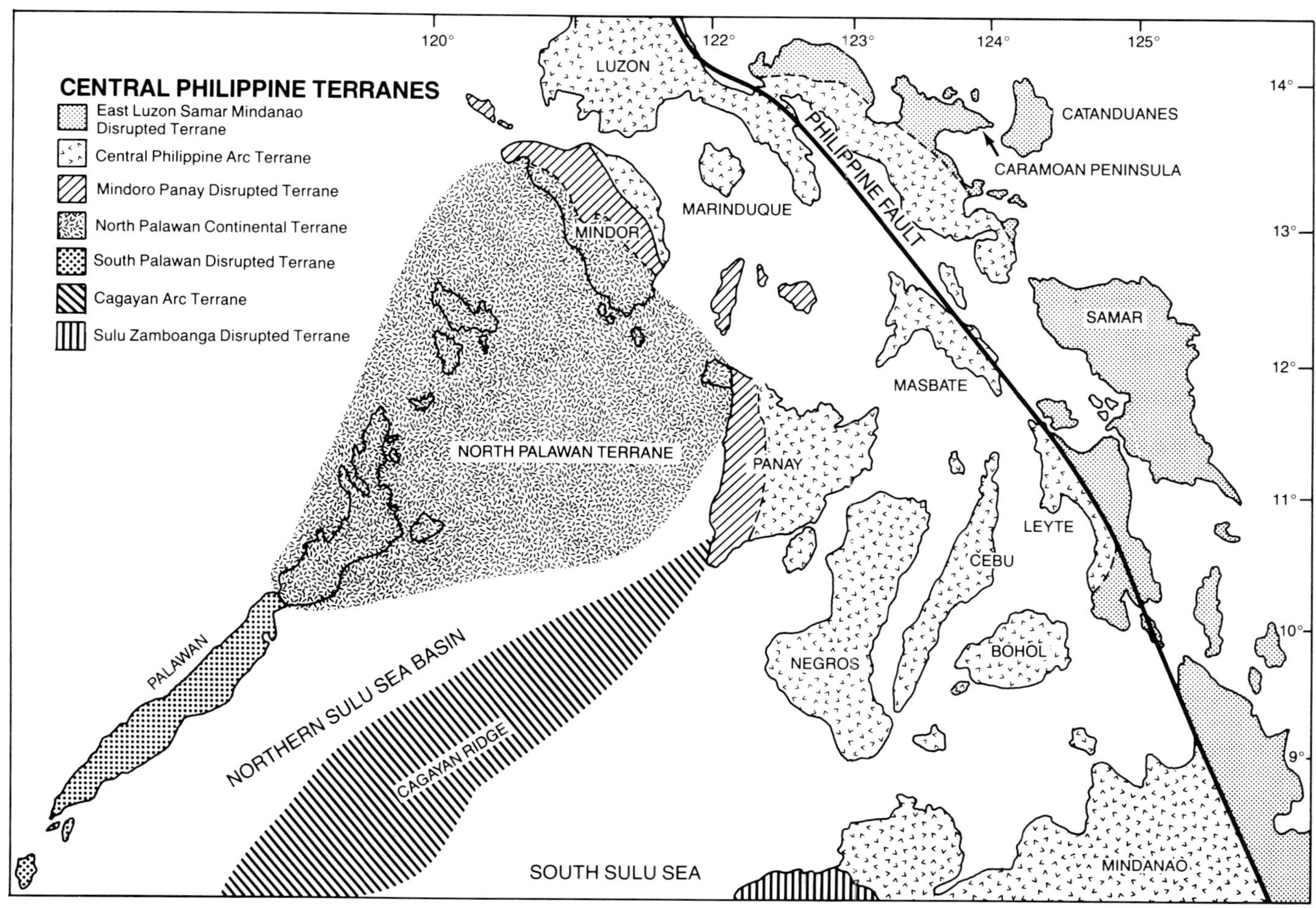

Figure 2—Terrane map of central Philippines as described in text. Terrane boundaries are only to be regarded as approximate. Boundary of CPAT and ELSMDT is shown with respect to westward occurrence of ophiolites. No displacement is shown on Philippine Fault. Although it is likely that this fault offsets CPAT with respect to ELSMDT, lack of data in region makes it impossible to predict the amount of this offset.

(Ranneft et al, 1960; Balce et al, 1981; Hamilton, 1979) also suggest that this region is the southern extension of the CPAT.

Mindoro-Panay Disrupted Terrane

The Mindoro-Panay disrupted terrane (MPDT) forms a thin linear belt that is located along the central portion of Mindoro and the western margin of Panay Island. The boundaries of this terrane are noted by faults. In Panay, a thrust fault in the eastern portion of the Antique Range marks the eastern limit of this terrane. The western limit on Panay is marked by a fault that trends north-south near the western coast of the island. On Mindoro, the eastern boundary is marked by the East Mindoro fault zone identified by Karig (1983). Field work in Puerto Galera (McCabe, Malicse, and Yumul, 1984, unpublished field notes) suggests that the northern boundary of this terrane is marked by a N 80° W trending thrust fault that passes directly through the town of Puerto Galera and appears to trend offshore into the Verde Island passage. If the nearly N 80° E trend of the structure observed in Puerto Galera is continued to the west (Fig. 3), it appears to trend toward Lubang and Ambil Island. This possible westward continuation of this structure suggests that it may be a portion of the Verde Passage suture identified by Karig (1983). The western margin of this terrane complex will be discussed in the following section.

On Panay Island, this terrane contains an assortment of Late Jurassic(?)/Early Cretaceous (McCabe et al, 1982a) ophiolitic fragments that are tectonically mixed with lower Miocene forearc deposits into a polymict melange (Hamilton, 1979). This entire section is overlain by flat-lying Late Neogene coralline limestones (Santos-Ynigo, 1949; Diegor, 1980; McCabe, unpublished field notes).

On the island of Mindoro, this disrupted terrane is more complex than it is on Panay. The apex of the middle to late Miocene collision event (Hamilton, 1979; McCabe et al, 1982a) is probably located in this region. To date, very little field work is reported from Mindoro. The work that has been done reports that this region includes quartzo-feldspathic schist, gneisses, marbles, mafic and ultramafic igneous rocks, pelitic schist, and siliceous intrusives (Philippines Bureau of Mines, 1964; Andal and Caagusan, 1968). Our preliminary field work in the northern portion of the region and more detailed studies in the southern and central portions of the island by Sarewitz question the occurrences of some of the reported high-grade rocks in the region (Sarewitz, 1984, personal communication). Although

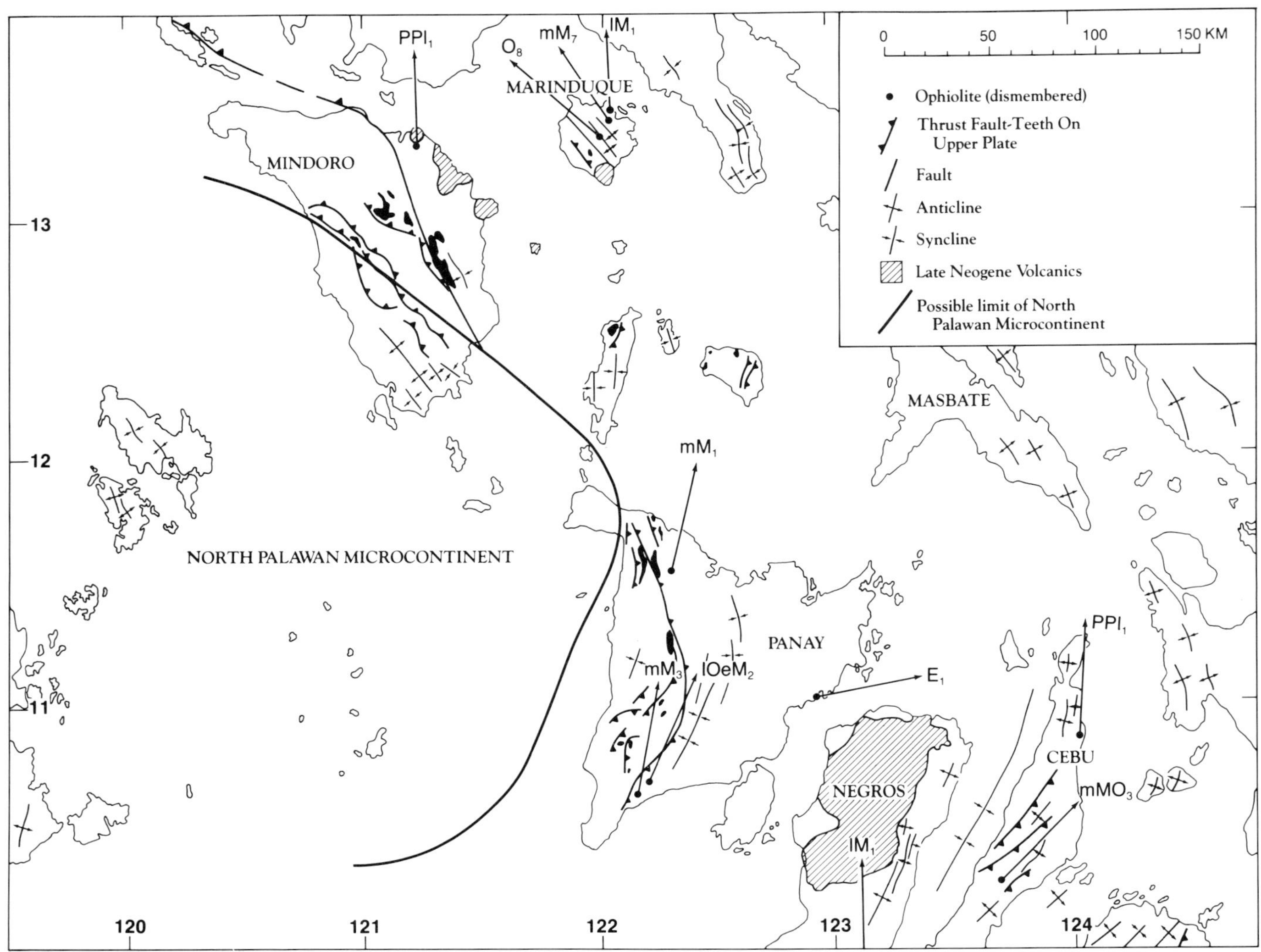

Figure 3—Figure showing the Oligocene and Neogene paleomagnetic directions from Central Philippine arc. Also shown are structures around central Philippines. Note that the pronounced bending of fold axis around collision zone is consistent with paleomagnetic directions. Location of volcanic gap between central Mindoro and Negros is also shown. Location of Verde fracture zone is approximated between Mindoro and Luzon (island just above map). The extension of this fault to the east of Mindoro is poorly controlled and may join with the East Mindoro fault of Karig (1983).

Andal and Caagusan (1968) describe the northern region as composed of gneisses, schist, and marbles, based on our field studies in this region it appears that much of the north coast is composed of phyllites and marbles. Our preliminary petrographic work on rocks from this region shows rocks that contain albite, epidote, chlorite, actinolite, white micas, and stilpnomelane mineral assemblages, indicating that the region is characterized by low-grade greenschist facies rocks. Field observations in the central portion of the island by Sarewitz (1984, personal communication) and Japan Metal Mining Agency (1983) suggest that much of this region is characterized by phyllitic rocks and slates with local occurrences of mica schist. Also found associated with these low-grade metamorphics in the central portion of the island are numerous occurrences of dismembered ophiolites. Some of these ophiolite fragments are in close association with amphibolites (some of which are garnet bearing). Owing to the limited amount of work that has been carried out in this region, the structural relationships between these ophiolite fragments and the metamorphics are unknown.

Karig (1983) labeled this region as the Mindoro metamorphic terrane. Karig noted the basic composition of these rocks and suggested that they are of island arc origin. Hashimoto and Sato (1968) suggest that these metamorphic rocks lay stratigraphically underneath the Jurassic Mansalay Formation, therefore suggesting a pre-Jurassic age. However, recent field studies in this region (Sarewitz, personal communication; Japan Metal Mining Agency, 1983 and personal communication) have yet to identify any evidence that would support Hashimoto and Sato's claim. Until field evidence substantiates the field relationships between the Mindoro metamorphics and the Mansalay Formation, the pre-Jurassic age must be regarded as suspect.

Because the age of the metamorphics is unknown, we speculate below on three possible origins for the Mindoro metamorphics. In the future, detailed field studies will be useful in distinguishing among these three possibilities.

1. The Mindoro metamorphics may be slices of the CPAT terrane that were tectonically mixed with ophiolitic fragments during the amalgamation of the North Palawan continental terrane with the CPAT. This origin (which we prefer) would be consistent with the fact that these rocks have an island arc origin (Karig, 1983) and contain diorites and quartz-diorites with Oligocene and Eocene K–Ar age dates (Japan Metal Mining Agency, 1983). Similar aged diorites are also noted on Negros and most likely Panay (discussed above). Other evidence that supports such an origin are ophiolites with Lower Cretaceous ages on Panay (McCabe et al, 1982a; Hashimoto et al, in preparation). If this scenario is correct, it would suggest that these ophiolites may represent the basement of the Cretaceous CPAT and not pieces of a Mesozoic ocean basin that has been totally destroyed by subduction. This model might suggest that the ophiolites are arc related and are not pieces of obducted crust. Recently, Bloomer and Hawkins (1983), Hawkins and Evans (1983), Hawkins et al (1984), and Silver et al (1983b) have shown that many of the Southeast Asian ophiolites may have such an origin.

2. A second origin for these rocks was suggested by Holloway (1981). Instead of requiring the forearc of the CPAT as the source, this model suggests that the rocks are a portion of the North Palawan continental terrane that was mixed during metamorphism.

3. The third model for the origin of the Mindoro metamorphics was suggested by Karig (1983). This model suggests that the Mindoro metamorphics are a distinct terrane that was sutured to the CPAT prior to the amalgamation with North Palawan.

North Palawan Continental Terrane

The northern portion of Palawan, the Calamian Island group (northeast of Palawan), the southwestern portion of Mindoro, a small peninsula in northwestern Panay, and possibly portions of the relatively unstudied Tablas Islands (north of western Panay) all are part of the North Palawan continental terrane (NPCT), a break-away terrane whose age and geologic history are totally unique with respect to the rest of the Philippine arc. In contrast to the other portions of the Philippine arc, late Paleozoic and Mesozoic rocks with probable continental affinities (Hashimoto and Sato, 1973) are found in this region. The anomalous position of these rocks and the intense structural deformation observed in the western portion of Panay and Mindoro lead Hamilton (1979) to suggest that a Miocene collision occurred between this block (which he suggested was rifted off China) and the Philippine arc. This conclusion has been supported by later marine geophysical studies (Taylor and Hayes, 1980, 1983), stratigraphic studies (Holloway, 1981), and paleomagnetic and geologic studies (McCabe et al, 1982).

The pre-Cenozoic stratigraphy of the NPCT is characterized by lightly metamorphosed (Hashimoto and Sato, 1968) to unmetamorphosed (Fontaine et al, 1983) upper Paleozoic through Mesozoic quartz-rich clastic sediments and limestones (Teves, 1953; Fontaine, 1979; Koike et al, 1967; Easton and Melandres, 1963; Andal, 1966). These pre-Cenozoic sediments are intruded by granitoids of proposed Paleogene age (Philippines Bureau of Mines, 1964). These older rocks are overlain by Eocene to early Oligocene nonmarine to deep-marine shales, silts, and carbonates and upper Oligocene to middle Miocene deep-water limestones (Holloway, 1981; Taylor and Hayes, 1980; Hashimoto, 1981). Both Taylor and Hayes (1980) and Holloway (1981) suggest that the Cenozoic sediments on the NPCT record a history of rifting and drifting of a continental margin similar to the continental break-up model presented by Falvey (1974).

The upper Miocene sediments found in southwestern Mindoro suggest that a marked change in tectonic regime occurred around this period. The late Miocene strata of this region are reported to be characterized by coarse clastic sands and boulder conglomerates that contain clasts of both the MPDT and the NPCT (Andal, 1966; Hamilton, 1979; unpublished data from Phillips Petroleum; Karig, 1983). The inclusion of clasts from both these terranes suggests that these terranes were province linked by late Miocene. This interpretation is also supported by paleomagnetic studies suggesting that the collision-related rotations of this region were over by latest Miocene (McCabe et al, 1982; 1984b).

On Figure 2, we define the western limit of the NPCT as the Ulugan Bay fault in the central portion of Palawan Island (Hamilton, 1979; Taylor and Hayes, 1980; Holloway, 1981; Karig, 1983). The eastern limit of the NPCT is more difficult to define. Here we will define this boundary as the eastern limit of rocks that only have a NPCT affinity. On Panay, we define this eastern limit to lie in a small north-south oriented valley that occurs just east of the small northwestern Peninsula of Panay (Buruanga Peninsula). The Buruanga Peninsula is characterized by metamorphic rocks of believed Paleozoic age. This age assignment is based on Permian limestones identified from Carabao Island located just north of the Buruanga Peninsula (Andal, 1966). East of this peninsula, the pre-Pliocene rocks are characterized by Mesozoic ophiolitic rocks and Miocene aged tuffaceous sediments.

As stated above, the amalgamation-related deformation on Mindoro is much more severe than on Panay. Because of this, the geology of central Mindoro is more difficult to understand. Therefore, the boundary between the NPCT and the MPDT is subject to debate. Hamilton (1979) infers the contact as the easternmost exposure of rocks with a North Palawan affinity. Taylor and Hayes (1980) and Holloway (1981) extended the Palawan block slightly eastward of Hamilton's limit to include the entire belt of highly deformed metamorphic rocks that make up the backbone of central Mindoro. Karig (1983) based the eastern edge of the NPCT on geophysical and bathymetric data from the southern end of the Manila Trench. From these data, Karig suggested that a plate boundary extends southward from the Manila Trench to southern Mindoro where it comes onland and continues as a belt of young and active faults and gravity lows. He further suggested that this plate boundary continues further south where it eventually becomes continuous with the northern extension of the Negros Trench. Based on the possible occurrence of a plate boundary between Mindoro and Negros, Karig divided the zone between the continuation of the Manila

Trench and eastern Mindoro into two separate terranes, the Mindoro metamorphic terrane and Southwest Mindoro collision zone.

Other workers in the region (Hamilton, 1979; Cardwell et al, 1980; DeBoer et al, 1980; McCabe et al, 1982) noted the lack of defined seismic zone in this area and argued for the nonexistence of plate boundary in this region. This is further supported by: (1) the very diffuse shallow earthquake pattern observed in the area and the total lack of thrusting focal mechanism solutions (Cardwell et al, 1980); (2) numerous east–southeast-trending strike-slip faults in the offshore area between Panay and Mindoro (unpublished Phillips Petroleum data); (3) the location of a volcanic gap from central Mindoro to northern Negros (McCabe et al, 1982); and (4) the subdued gravity signature of the region (DeBoer et al, 1980). Instead of the need for a rigid plate boundary in this region to accommodate the convergence associated with the Manila Trench, we suggest that this convergence is taken up by a diffuse zone of deformation that is spread over the entire collision region. Some of this convergence may be taken up by numerous small left-lateral strike-slip faults. As noted above, such faults are found in the region. In addition, we suggest that the Verde fracture zone (DeBoer et al, 1980), located between Mindoro and Luzon (Fig. 3), is an active left-lateral strike-slip fault that may have developed by reactivation of the pre-Pliocene Verde Passage suture of Karig (1983). This fault is characterized by an east–west-trending shallow zone of seismicity with rather poorly defined left-slip focal mechanism solutions (Cardwell et al, 1980). Seismic data from the region suggest that in eastern Mindoro, this fault curves to a more southeasterly strike. The development of late collision strike-slip faults are common to both continental collisions (Molnar and Tapponnier, 1975) and island arc collisions (Silver et al, 1983; McCabe, 1984a). Earlier, Newmark (1980) suggested using a slip-line model (Tapponnier and Molnar, 1976) for the development of the central Philippines.

Here, we follow the arguments of Hamilton (1979) and define the easternmost extension of the North Palawan terrane as being the contact between the continental affinity Paleozoic and Mesozoic rocks and the rocks that show arc affinities. By this definition, the contact between these two terranes would lie along the contact between the highly deformed Mindoro metamorphics, which have island arc affinities (Karig, 1983), and the quartz-rich Jurassic Mansalay Formation, a believed continental-derived sandstone in southern Mindoro. It should be pointed out that the continental origin of the Mansalay is questioned by Sarewitz (1984, personal communication) who points out that this sandstone is composed of plagioclase and is low in potassium feldspars. We contend that although the area is complexly deformed (by both collision-related tectonics and collision-related strike-slip faulting), the Mansalay Formation and the rocks to the west of this are still part of the NPCT that have been amalgamated to the Philippine arc. Evidence that is critical to this discussion but still lacking includes detailed structural data from Mindoro and the nature of the contacts between the various stratigraphic packages that make up the island. In addition, the age of the protoliths of the Mindoro metamorphics and the timing of the metamorphism of these rocks also needs to be better constrained.

South Palawan Disrupted Terrane

In contrast to the Paleozoic and Mesozoic sequences of North Palawan, the geology of the island south of the Ulugan Bay fault is quite different (Hamilton, 1979; Hashimoto, 1981; Balce et al, 1981; and many others). We name this region the South Palawan disrupted terrane (SPDT). This terrane is composed of ophiolites, amphibolites, and greenschist rocks that crop out along southeast-dipping thrust sheets (Hutchinson, 1975). These rocks of oceanic affinity are tectonically mixed with presumed continental-derived sequences of quartz and potassium feldspar bearing arkosic sandstones, shales, and mudstones. Fossil age data from the limestones in these continental sequences yield lower Eocene fauna (Samaniego, 1964) and may contain fossils as young as early Miocene (Casasola, 1956; Hamilton, 1979). All of these pre-middle Miocene rocks are strongly deformed, sheared, and boudinaged, which led Hamilton (1979) to classify this region as a melange. Unconformably overlying these strongly deformed rocks are essentially undeformed Late Neogene sedimentary and limestone sequences (Balce et al, 1981).

The SPDT is bounded on the northwest by a Palawan Trough, a bathymetric depression that has been described as a Tertiary to Early Neogene paleosubduction zone (Hamilton, 1973, 1979; Ludwig et al, 1979; Taylor and Hayes, 1982). Lying to the north of the Palawan Trough are the Reed Banks and other shoaling regions collectively referred to as the Dangerous Ground and called the Reed terrane by Nur (1983). Nur's Reed terrane also includes the NPCT. While these terranes (Reed terrane and NPCT) share a common pre-Cenozoic history, the Cenozoic stratigraphic column of North Palawan (Saldivar-Sali et al, 1981) is slightly different from that of the Reed Banks (Taylor and Hayes, 1980), which suggests that these two bodies may not share a common Cenozoic rifting and collision history. In addition, the structural relationship between these two bodies is unknown. Therefore, we separate these as two distinct structural bodies. Based on both seismic and drill core data, recent tectonic models of this area infer that a Miocene collision event occurred between the Reed terrane and SPDT (Hamilton, 1979; Ludwig et al, 1979; Taylor and Hayes, 1980, 1982). Hamilton (1979) and Ludwig et al (1979) suggested that the thrusting of ophiolites in the SPDT resulted from this collision. In contrast to a plate tectonic model for the evolution of this area, recently Hinz and Schluter (1983) presented a model for the evolution of the SPDT as an overthrust belt that resulted from compression between the northward-migrating Cayagan Ridge and the Reed Banks.

Sulu-Zamboanga Disrupted Terrane

This terrane, which begins just at the lower edge of Figure 2, will only be briefly discussed. The northeastern boundary shown in Figure 2 is defined by a zone of southward-directed thrust sheets that bring Late Tertiary sediments over ophiolites (Philippines Bureau of Mines, 1964). Ranneft et al (1960) describe the basement of this terrane as composed of sheared serpentinite, altered diabase, chlorite schist, phyllite, slate, and clastic sediments of unknown age that are overlain by Neogene sediments. Hamilton (1979) and Hawkins et al (1982) characterized this area as a melange, suggesting that this terrane may be a disrupted terrane. If so, this terrane is probably a continuous feature from the area shown in Figure 2 to the island of Tawi-Tawi, located at the southwestern edge of the Sulu arc, and perhaps comes onshore in northeast Borneo.

East Luzon–Samar–Mindanao Disrupted Terrane (Suspect)

This terrane, which is continuous with the Eastern metamorphic belt of Karig (1983), is located along the entire eastern margin of the archipelago. Hashimoto (1981) defined this belt as the Eastern ultramafic belt, but because both ophiolitic and arc-derived sequences are found in this region, we will not follow his classification scheme. The East Luzon–Samar–Mindanao disrupted terrane (ELSMDT) is characterized by a heterogeneous mixture of bedded Upper Cretaceous limestones, clastic sediments, and volcanics (Kitamura et al, 1968; Hashimoto et al, 1975; Wright et al, 1981; Moore et al, 1981a; Hawkins et al, this volume) that overlie an undated and possibly pre-Late Cretaceous metamorphosed basement (Irving, 1950; Hashimoto, 1981; Karig, 1983). The pre-Cenozoic sequences are reported to be unconformably overlain by Early to Middle Tertiary arc-derived volcanics, limestones, and clastic sediments (Balce et al, 1981; Hashimoto, 1981; Karig, 1983; Corby, 1951). Along the entire belt occurrences of ophiolitic rocks are reported that in many areas are in thrust contact with the pre-Oligocene rocks. Based on stratigraphic relationships, these ophiolitic rocks are reported to be Middle Tertiary in age (Melandres and Comsti, 1951; Santos-Ynigo and Esguerra, 1961), but this age should be regarded with extreme caution until it is confirmed by either radiometric age dating or paleontologic studies. Associated with these ophiolitic rocks, garnet amphibolites are reported from northeast Mindanao (Santos-Ynigo and Esguerra, 1961; Hawkins et al, 1981; Wright et al, 1981; Hawkins et al, this volume) and southeastern Luzon (Geary, 1984, personal communication). Hashimoto (1981) argued that the pre-Eocene rocks of this belt were metamorphosed by a Middle Tertiary event. Karig (1983), however, pointed out that the timing of the metamorphism of this region is subject to debate and further suggested that probably more than one metamorphic event has occurred. The pre-late Cenozoic strata of the region are overlain by Oligocene and Neogene sequences of arc volcanics, related clastic sediments, and limestones (Dumapit, 1973; Santos et al, 1979; Hashimoto, 1981; Balce et al, 1981). These late Cenozoic sequences may represent an overlap basin in that they are continuous with late Cenozoic sediments from the juxtaposed CPAT. If this is an overlap basin, it places an upper limit on the time of amalgamation of these terranes.

We classify this terrane as a suspect terrane. We feel that two criteria differentiate this terrane from the CPAT:

1. The ELSMDT is characterized by numerous ophiolitic exposures. Ultramafic fragments are only found in two rather restricted localities in the CPAT (along the base of two Miocene thrust faults in Marinduque and Cebu) while ophiolitic rocks in the CPAT are only found in the Central Cordillera of Mindanao (Philippines Bureau of Mines, 1964; Hawkins et al, 1982).

2. The Early Tertiary rocks from the ELSMDT are reported to have a regional metamorphic fabric (Hashimoto, 1981).

The exact western limit of this terrane is unknown. If a bounding fault exists between the ELSMDT and the CPAT, it is covered by a thick Neogene overlap basin assemblage of Late Neogene volcanics and related sediments. Furthermore, the structure of this bounding fault may have been deformed by movement along the large left-lateral Philippine fault, making any recognition of the paleosuture difficult. Moore and Silver (1983) suggest that the Philippine fault was reactivated along this proposed suture zone. In Figure 2, we define the contact between the ELSMDT and the CPAT as the western limit of ultramafic rocks near the Philippine fault.

Cagayan Arc Terrane (Suspect)

The Cagayan Ridge–Northern Sulu Basin is an inactive volcanic arc-forearc basin that is located in the northwestern portion of the Sulu Sea. This suspect terrane runs parallel with both the island of Palawan to the north and the Sulu–Zamboanga Ridge to the south.

The Northern Sulu Basin contains a thick sedimentary section that in places contains more than 7,500 m (24,606 ft) of sediment (Murauchi et al, 1973; Mascle and Biscarret, 1979). The sediments in this basin have been suggested as related to the Tertiary Crocker Formation, which is a well-exposed Oligocene sedimentary sequence exposed in Sabah (Beddoes, 1976). The basin is suggested to represent the volcanic forearc of the volcanic Cagayan Ridge (Hamilton, 1979). The Cagayan Ridge, which divides the shallower Northern Sulu Basin (water depths less than 2,000 m [6,562 ft]) from the Southern Sulu Basin (maximum depths 5,000 m [16,404 ft]), is a volcanic arc that Hamilton (1979) suggested was related to pre-Late Neogene subduction under the southern portion of Palawan. The ridge itself reaches the surface on the small island of Cagayan (near the Sabah coast), the Cuyo Islands (in the central portion of the basin), and perhaps the southwest portion of the island of Panay. All three of these areas are characterized by volcanic rocks that are overlain by reef limestones.

The arc is composed of Miocene volcanic rocks (two 15 m.y. K–Ar age dates [Rangin, 1984, personal communication]) that are distinct from any of the juxtaposed terranes. As such, we are forced to classify the CAT as a terrane. The southeastern boundary of this terrane is identified by bathymetry where the volcanic ridge drops off rapidly to the floor of the South Sulu Sea Basin. The northwestern boundary of this terrane is unknown (shown as ? in Fig. 2). If a structural boundary exists between this ridge and the terranes of Palawan, it is presently buried under the thick Neogene sedimentary blanket that covers most of the North Sulu Sea Basin (Murauchi et al, 1973; Hamilton, 1979). Because of the lack of a well-defined northwestern boundary fault and because this volcanic ridge lies along the margin of three distinct terranes (SPDT, NPCT, and MPDT) and may have been

Table 1

Location	Sites	Age	Rock Type	Dec.*	Inc.	α_{95}	Ref.
Luzon (western portion)							
Zambales	10	Eocene	VS	115.3	7.3	10.8	1
	1	Oligocene	S	81.5	9.0	—	1
	11	Late Neogene	V&S	356.1	17.1	4.9	1,2
Luzon (southeast portion)							
Bicol	19	Late Neogene	V&S	0.7	22.6	7.6	2,3
Marinduque	8	Oligocene	V	314.2	8.4	33.0	3
	5	Miocene	VI	317.7	16.2	25.8	3
	2	middle Miocene	VI	313.7	22.4	—	1
	2	late Miocene	VS	3.4	18.3	—	1,2
Mindoro	1	Pliocene	V	357.2	16.8	—	4
Cebu	4	Cretaceous	V	119.6	−5.3	—	5
Panay	6	late Oligocene–middle Miocene	VS	20.9	22.1	8.0	6
Negros	1	late Miocene	S	358.8	22.6	—	4

*Abbreviations: V = volcanic rocks and dikes; S = sedimentary rocks; I= intrusives; Dec. = declination; Inc. = inclination; α_{95} = radius of 95% confidence circle (only given if more than four sites reported); Ref. = Reference:
1. Fuller et al, 1983 2. McCabe et al, 1982b
3. Hsu, 1972 4. McCabe (unpublished data)
5. Noritomi and Almasco, 1981 6. McCabe et al, 1982a

Table 1—Paleomagnetic data from central Philippines (between 16°N and 10°N).

built on top of these three terranes after suturing, we classify this terrane as suspect.

Marginal Basins

The Philippine Islands are surrounded by four marginal seas: the South China Sea, the Celebes Sea, the Philippine Sea, and the Sulu Sea. Of these four basins, the Sulu Sea is quite distinct in that it is one of the smallest marginal seas in Southeast Asia and by the fact that it is totally enclosed by Late Tertiary to Neogene volcanic arcs (Fig. 1). Seismic-reflection profiles across this basin suggest that during the Neogene the Sulu Sea has been subducting along the Negros Trench and the Sulu Trench (Hamilton, 1979). Very little is known about this basin at present. Krause (1966) and Murauchi et al (1973) report that this basin is underlain with oceanic crust. Based on the relatively high heat flow (2.12 ± .25 HFU [Watanabe et al, 1977]), the depth of greater than 4,000 m (13,123 ft), and ages of the South China Sea and Celebes Sea, Weissel (1980) suggests an Oligocene age of this basin. However, since neither magnetic anomaly identification nor drill core data presently exist from the South Sulu Sea, this age must be used with a caution.

Because the Sulu Sea is a marginal basin, we will not classify the Sulu Sea as a terrane. However, two reasons suggest that such a classification may indeed be merited. First, it does satisfy the definition of a terrane as a fault-bounded, distinct stratigraphic package. Second, its position is unique in that it is locked on four sides with landmasses. Such a geometric arrangement makes it difficult to imagine any process that can totally subduct this basin; therefore, in the future some portion of this basin will probably become incorporated to Asia when the entire Philippine arc is accreted to the Asian landmass.

PALEOMAGNETISM

Paleomagnetic studies have become a useful tool of terrane analysis (Jones et al, 1977; Irving, 1979; Beck, 1976, 1980). Not only is paleomagnetism valuable for classifying the possible trajectories of terranes and their latitudinal origin, but careful use of paleomagnetic data, when combined with geologic data, can be very effective in testing the affinities of possibly related terranes.

For the past 6 years we have been doing paleomagnetic studies in the Philippine arc. Table 1 lists all of the reported paleomagnetic data collected by us and others from the central Philippines, south of 16° N latitude. Although this data set is incomplete, some basic trends are starting to emerge. The pre-Miocene data from Cebu, Marinduque, and Zambales (in western Luzon) are characterized by relatively low inclinations (less than 10°) with respect to present predicted dipole inclination value (approximately 20°) for the region (Table 1). In contrast, Neogene data show inclinations similar to the expected present-day value. Although some of the data set (especially Hsu, 1972) show relatively large α_{95} values, the fact that pre-Neogene data, which is from three separate and distant islands, collected by three separate research groups, show shallower than present field inclination values suggests that before the Miocene both the CPAT and Zambales terrane lay south of their present-day position and began a northward migration that was completed by middle(?) Miocene. Although the data set suggests that this northward translation occurred fairly rapidly after the Oligocene, the age control of the Late Tertiary and Early Neogene sites is not of good enough quality to determine when this translation occurred. From this data set, we infer that the CPAT and the Zambales ophiolite terrane of Karig

(1983) were all located within 5° of the equator until Oligocene–early Miocene. Unfortunately, paleomagnetic data are unable to constrain longitudinal variations. Therefore, with this data set alone we are unable to predict any genetic relationship between the Zambales ophiolite terrane and CPAT.

Looking at the declination values of rocks from the central Philippines, we suggest that workers in older accreted terranes must use declination values with extreme caution when doing terrane analysis. Figure 3 shows the Oligocene through Neogene values reported from the Central Philippine arc terrane. The figure shows that Oligocene to middle Miocene magnetic directions from Panay (McCabe et al, 1982a) are distinguishable from the present field by the fact that they are rotated 20° clockwise. In contrast, sites of Miocene/Oligocene age from Marinduque (Hsu, 1972; Fuller et al, 1983; McCabe, 1984b) are deflected counterclockwise from the present field direction. Late Miocene to Recent directions (McCabe, 1984b) from Negros and Mindoro (Table 1) and from Marinduque (McCabe et al, 1982b; Fuller et al, 1983) are indistinguishable from the present field. This paleomagnetic data set (clockwise southeast of the NPCT and counterclockwise northeast of the NPCT) have been interpreted by McCabe et al (1982a) to be the result of the middle to late Miocene docking of the NPCT to the CPAT. This collision-related bending is further supported by structures within the CPAT, which are also shown on Figure 3. On Negros, Panay, and Cebu (islands located southeast of the collision zone), fold axes of mapped Miocene structures trend to the northeast and Miocene thrusting in the region is toward the northwest. Islands located to the northeast of the Palawan Block, Mindoro and Marinduque, show similar age folds trending northwest while Miocene thrusting is toward the southwest.

Collision-related bending of a terrane appears to be common in Southeast Asia. Paleomagnetic data from Ryukyu arc, southern Marianas, Bonins, Sulawesi, and the northeastern margin of the Indian Plate show many examples of this type of behavior (McCabe, 1984a). Therefore, although inclination data are useful in constraining paleolatitudes of terrane, similarities or differences in declination values have little meaning in terrane analysis, unless they are carefully compared to geological and structural data.

DISCUSSION

The Philippine Archipelago is composed of a number of distinct terranes. Although these terranes are presently joined into one composite terrane, each behaved as an independent entity during some part of its history. All of these terranes became assembled by late Miocene time.

Of the various terranes composing the Philippine composite terrane, five of them, the Central Philippine arc terrane, the North Palawan continental terrane, the South Palawan disrupted terrane, the Mindoro–Panay disrupted terrane, and the Sulu–Zamboanga disrupted terrane, are all discrete tectonic elements, each of which records a unique geologic history. With the exception of the Mindoro–Panay disrupted terrane (which will be discussed in more detail below), the geologic history of the other four terranes is so different from its neighbor that it is not explained by simple rigid position stratigraphic models. Therefore, the present-day position of these terranes requires some type of independent tectonic movement for each of these bodies prior to their assembly.

We classify the Cagayan arc terrane and the East Luzon–Samar–Mindanao disrupted terrane as suspect. Later work may show that both are discrete terranes; however, because one (Cagayan arc terrane) lies at water depths greater than 1,000 m (3,281 ft), and the contact between the East Luzon–Samar–Mindanao disrupted terrane and the Central Philippine arc terrane is covered by a Neogene overlap sequence, we are unable to classify either of these packages as distinct terranes. Although future geophysical investigations may locate a fault in the offshore region between Palawan and the Cagayan Ridge, location of such a boundary between the East Luzon–Samar–Mindanao disrupted terrane and the CPAT may have been obscured by Neogene movement on the left-lateral Philippine fault.

The fifth terrane, the Mindoro–Panay disrupted terrane, is also classified as a unique stratigraphic package. We suggest that this terrane is the result of the collision between the CPAT and NPCT. Karig (1983) proposed that this terrane originated elsewhere and was translated by strike-slip faulting to its present position along what he defines as the West Luzon shear zone. We agree with Karig that strike-slip faulting has played a dominant role in shaping this area. However, we suggest that these faults are either syncollision or postcollision features that developed in response to inhibition or slowing of subduction because of the entry of the NPCT into the trench. We suggest a subduction-related collision origin for this region.

As mentioned earlier, the MPDT of western Panay is characterized by Lower Cretaceous ophiolites that are thrust from west to east over pre-late Miocene forearc deposits (Santos-Ynigo, 1949; Hamilton, 1979; McCabe, unpublished field work, 1981; United Nations Development Program and Philippines Bureau of Mines, 1983). Glaucophane schist is also reported locally (Mitchell, personal communication). On Mindoro, which we assume to be near the apex of the collision, the MPDT is characterized by ophiolites, polyphase deformed metamorphic rocks (Hashimoto and Sato, 1968) with possible island arc protoliths (Karig, 1983), amphibolites (some garnet bearing) of unknown affinity, and perhaps slices of NPCT that are tectonically mixed and thrust westward.

Based on the above relationships, we argue for convergence associated with this region between the early and middle Miocene, prior to the docking of the NPCT to the CPAT. Convergence is supported by paleomagnetic results that require a northward drift of the CPAT relative to the determined paleolatitude of the South China Sea, which Taylor and Hayes (1983) showed has remained at about the same latitude since middle Oligocene. The mechanical processes that result from amalgamation of two terranes are poorly understood. However, the above geologic evidence strongly suggests that metamorphism and thrust faulting with possible shortening of the forearc region has played an important role in the genesis of the

MPDT. We further suggest that the rocks found in this terrane are composed of thrust and highly sheared packets from deeper levels of the CPAT. These rocks were intensely deformed and thrust westward over the leading edge of the low-density NPCT as a result of the initial subduction of this block.

If this model for the evolution of this boundary is correct, we can infer that the intensity of deformation associated with the docking is inversely related to the distance from the collision. On Cebu, Negros, and Marinduque, together with eastern Panay, islands that are located a considerable distance from the collision, the collision-related deformation is characterized by the development of folds and minor thrust faults (Fig. 3). In western Panay, which is located just below the collision zone, folding is more complex and thrust faulting is pronounced. On Mindoro, which is located at the supposed apex of the collision zone, the collision-related deformation involves intense folding and thrusting and is characterized by rocks that suffered metamorphic recrystallization to the greenschist facies.

It is interesting to speculate that the Lower Cretaceous ophiolite sequence found on Panay, which is the same age as the presumed basement arc rocks in Cebu (Kitamura et al, 1968), may represent the initial oceanic basement that the arc was built upon and not an obducted piece of the proto-South China Sea located at the leading edge of the North Palawan continental terrane. Such an origin is consistent with recent observations by Hawkins and others (Hawkins and Evans, 1983; Bloomer and Hawkins, 1983; Hawkins et al, 1984) who point out that many of the ophiolitic rocks found in arc regions around the Marianas and the Philippines do not appear to be related to obducted portions of the downgoing plate. Silver et al (1983b) have documented that ophiolitic rocks found in the Sulawesi collision zone also appear to be derived from the arc region. These speculations, if correct, would suggest that many of the ophiolites that are common to collision zones in Southeast Asia (e.g., New Guinea, Taiwan, Hokkaido, and around the Izu area of southern Honshu Island) are arc derived and were thrust oceanward and tectonically mixed with forearc-derived sedimentary and volcanic sequences as a result of the collision.

CONCLUSIONS

1. The Central Philippine arc is a composite terrane that is made up of no fewer than five, and most likely more, distinct tectonostratigraphic terranes. These terranes, which were assembled prior to the late Miocene, are:

a. Central Philippine arc terrane—a stratigraphic terrane, which may extend from Taiwan to the Molucca Sea. This terrane is made up of Early Cretaceous to Holocene products of two or more volcanic arcs. The older arc served as a site of nucleation for the growth of the younger arcs.

b. Mindoro–Panay disrupted terrane—a heterogeneous mixed terrane that resulted from the Miocene collision of the Central Philippine arc terrane and the North Palawan continental terrane.

c. North Palawan continental terrane—a stratigraphic terrane composed of Paleozoic and Mesozoic continental-derived sequences overlain by Tertiary deep-water and shallow-water sediments. We also describe this terrane as a break-away terrane in that the pre-Cenozoic stratigraphy is similar to South China and the Cenozoic stratigraphy records a distinct history from South China.

d. South Palawan disrupted terrane—this poorly studied terrane appears to be a heterogeneous mixture containing Paleocene sediments and ophiolites.

e. Sulu–Zamboanga disrupted terrane—a relatively unstudied terrane composed of sheared serpentinite, altered diabase, chlorite schist, phyllites, slate, and clastic sediments of unknown age.

Besides these five terranes, two other terranes are identified as being suspect. These are:

f. Cagayan arc terrane—a Neogene submarine ridge that is believed related to Miocene volcanic arc activity. This terrane is identified as suspect in that it may represent a volcanic ridge that was deposited on top of the South Palawan disrupted terrane, North Palawan continental terrane, and the Mindoro–Panay disrupted terrane.

g. East Luzon–Samar–Mindanao disrupted terrane—a heterogeneous terrane composed of Late Cretaceous arc deposits, ophiolites (perhaps arc related), and metamorphic rocks. Although this terrane is differentiated from the Central Philippine arc terrane, later studies may show that it is a marginal facies to the Central Philippine arc.

2. The amount of paleomagnetic data from the area is limited for the most part to the Central Philippine arc terrane. These data show, as originally suggested by Hsu (1972), that all pre-Neogene sites are characterized by shallower than present field directions and suggest that much of this area lay near equatorial latitudes until the Neogene. This is similar to the findings of Fuller et al (1983) that Zambales has a shallow pre-Neogene paleolatitude. Neogene to Recent sites from the Philippines all show paleolatitudes that are close to present-day inclination values. The data suggest the entire Philippine arc migrated northward during the Late Tertiary and Early Neogene period, suggesting the presence of a convergent plate boundary between this area and the South China Sea during the Late Tertiary to Early Neogene.

3. Paleomagnetic data also show large amounts of differential rotations within the Central Philippine arc terrane. These data warn that declination data should be used with caution in relating the paleoaffinity of two stratigraphies that are believed related.

4. Although a strike-slip origin for the metamorphics of the Mindoro–Panay disrupted terrane has been suggested, the presence of ophiolites, blueschist, garnet amphibolites, and thrust faults in the Mindoro–Panay disrupted terrane implies a Miocene subduction-related collisional origin for this terrane.

5. The ophiolites thrust in western Panay are similar in age to the CPAT basement. This age relationship and the fact that these rocks are thrust oceanward suggest that this is an example of an arc related ophiolite and not an obducted piece of some totally subducted Mesozoic ocean basin.

6. The left-lateral Verde fracture zone (DeBoer et al, 1980), the East Mindoro fault zone (Karig, 1983), and the

occurrence of numerous strike-slip faults in the offshore regions between Mindoro, Panay, and Palawan are suggested to have developed in either the late Miocene or Pliocene in response to the stresses developed from the collision. As such, these faults may be analogous to similar faults noted in other collision areas such as eastern China (Molnar and Tapponnier, 1975), Sulawesi (Hamilton, 1979; Silver et al, 1983a), and Izu (Nakamura, 1984, personal communication).

ACKNOWLEDGMENTS

We would like to thank the Philippines Bureau of Mines and Geosciences and G. Balce, who provided much of the transportation used in this study. Thanks are given to D. Karig, K. Burton, G. Balce, F. Zanoria, W. Diegor, and M. Fuller for numerous discussions of the geology of the region. Special thanks are given to D. Sarewitz and E. Geary, who taught me (R.M.) much about the geology of Mindoro and who provided numerous comments on a draft of this paper. Last, useful criticisms were given by Eli Silver, S. Cisowski, S. Uyeda, J. Hawkins, G. Moore, W. Hamilton, and D. Howell. These reviews greatly improved this manuscript.

REFERENCES

Andal, D. R., 1966, A report on the discovery of fusulinids in the Philippines: Philippine Geologist, v. 20, p. 14-22.

——, and N. Caagusan, 1968, Geology of the iron deposits of northern Mindoro: Paper presented at the 2nd Geological Convention, Geological Society of the Philippines, p. 109-120.

Balce, G. R., et al, 1981, Metallogenesis in the Philippines: Geological Survey of Japan Report 261, p. 125-148.

Beck, M. E., 1976, Discordant paleomagnetic pole positions as evidence of regional shear in the western cordillera of North America: American Journal of Science, v. 276, p. 694-712.

——, 1980, Paleomagnetic record of plate-margin tectonic processes along the western edge of North America: Journal of Geophysical Research, v. 85, p. 7115-7131.

Beddoes, L. T., 1976, The Balabac Sub-basin, southwestern Sula Sea, Philippines: Proceedings of the Offshore South East Asia Conference, SEAPEX.

Bloomer, S. H., and J. W. Hawkins, 1983, Gabbroic and ultramafic rocks from the Mariana Trench—an island arc ophiolite, *in* D. E. Hayes, ed., Geophysical Monograph Series 27: American Geophysical Union p. 294-317.

Bowin, C., et al, 1978, Plate convergence and accretion in the Taiwan-Luzon region: Bulletin of the American Association of Petroleum Geologists, v. 62, p. 1645-1672.

Cardwell, R. K., et al, 1980, The spatial distribution of earthquakes, focal mechanism solutions, and subducted lithosphere in the Philippines and northeastern Indonesian islands, *in* D. E. Hayes, ed., The tectonic and geologic evolution of Southeast Asian seas and islands: American Geophysical Union Geophysical Monograph 23, p. 1-35.

Casasola, A. G., 1956, Geological reconnaissance of southern Palawan: Philippine Geologist, v. 10, p. 76-88.

Coney, P., et al, 1980, Cordilleran suspect terranes: Nature, v. 288, p. 329-333.

Corby, G. W., 1951, Geology and oil possibilities of the Philippines: Manila, Department of Agriculture and Natural Resources Technical Bulletin 21, 363 p.

DeBoer, J., et al, 1980, The Bataan orogene: eastward subduction, tectonic rotations, and volcanism in the western Pacific (Philippines): Tectonophysics, v. 67, p. 251-282.

Diegor, W., 1980, Some aspects in the geology, mineralization, and geotectonics of southwest Panay: Philippine Bureau of Mines and Geosciences Annual Geological Survey Division Seminar, 19 p.

Dumapit, P. T., 1973, Ground water geology of Bicol Peninsula: Journal of the Geological Society of the Philippines, v. 27, p. 24-44.

Easton, W. H., and M. M. Melandres, 1963, First Paleozoic fossil from Philippine Archipelago: Bulletin of the American Association of Petroleum Geologists, v. 47, p. 1871-1886.

Falvey, D. A., 1974, The development of continental margins in plate tectonic theory: Australian Petroleum Exploration Association Journal, v. 14, p. 95-106.

Fontaine, H., 1979, Note on the geology of Calamian Islands, North Palawan, Philippines: ESCAP-CCOP Newsletter, v. 6, p. 40.

——, et al, 1983, Marine Jurassic in Southeast Asia: United Nations ESCAP, CCOP Technical Bulletin, v. 16, p. 1-30.

Fuller, M. D., et al, 1983, Paleomagnetism of Luzon, *in* D. E. Hayes, ed., Geophysical Monograph Series 27: American Geophysical Union, p. 79-94.

Gonzales, B. A., 1963, Foraminiferal analysis on measured sections along the Tarao and Tanion Rivers, southwestern Iloilo: Philippines Bureau of Mines and Geosciences Report of Investigation 46, 35 p.

Hamilton, W., 1973, Tectonics of the Indonesian region: Geological Society of Malaysia Bulletin, v. 6, p. 3-10.

——, 1977, Subduction in the Indonesian region, *in* M. Talwani and W. C. Pitman III, eds., Island arcs, deep sea trenches and back arc basins, Maurice Ewing Series, v. 1: American Geophysical Union, p. 15-31.

——, 1979, Tectonics of the Indonesian region: U.S. Geological Survey Professional Paper 1078, 345 p.

——, and W. B. Myers, 1967, The nature of batholiths: U.S. Geological Survey Professional Paper 554-C, 29 p.

Hashimoto, W., 1981, Geologic development of the Philippines, *in* T. Kobayashi, et al, eds., Geology and palaeontology of Southeast Asia, v. 22: University of Tokyo Press, p. 83-170.

——, and G. Balce, 1977, A new correlation scheme for the Philippine Cenozoic formations: Proceedings of the 1st International Congress of Pacific Neogene Stratigraphy, Tokyo, 1976, p. 119-132.

______, and T. Sato, 1968, Contribution to the geology of Mindoro and neighboring islands, the Philippines, *in* T. Kobayashi, et al, eds., Geology and palaeontology of Southeast Asia, v. 5: University of Tokyo Press, p. 192–210.

______ and ______, 1973, Geologic structure of North Palawan and its bearing on the geological history of the Philippines, *in* T. Kobayashi, et al, eds., Geology and paleontology of Southeast Asia, v. 13: University of Tokyo Press, p. 145–161.

______, et al, 1975, Cretaceous systems of Southeast Asia, *in* T. Kobayashi, et al, eds., Geology and paleontology of Southeast Asia, v. 15: University of Tokyo Press, p. 219–287.

______, et al, 1978, Larger foraminifera from the Philippines, VI. Larger foraminifera found from the Lutak Hill Limestone, Pandan Valley, cental Cebu, *in* T. Kobayashi, et al, eds., Geology and paleontology of Southeast Asia, v. 19: University of Tokyo Press, p. 73–80.

Hatherton, T., and W. R. Dickinson, 1969, The relationship between andesitic volcanism and seismicity in Indonesia, the lesser Antilles, and other island arcs: Journal of Geophysical Research, v. 74, p. 5301–5310.

Hawkins, J. W., and C. A. Evans, 1983, Geology of the Zambales Range, Luzon, Philippine Islands: ophiolite derived from an island arc-backarc basin pair, *in* D. E. Hayes, ed., Geophysical Monograph Series 27: American Geophysical Union, p. 95–123.

______, et al, 1981, East Mindanao ophiolite belt: petrology of metamorphosed basal cumulate rocks (Abs.): EOS, American Geophysical Union Transactions, v. 62, p. 1086.

______, et al, 1982, Petrologic-tectonic evolution of volcanic arc, ophiolite and metamorphic rocks, central Mindanao, Philippines (Abs.): EOS, American Geophysical Union Transactions, v. 63, p. 113.

______, et al, 1984, Evolution of intra-oceanic arc-trench systems: Tectonophysics, v. 102, p. 175–205.

Hinz, K., and H. U. Schluter, 1983, Geology of the dangerous ground South China Sea and the continental margin of SW Palawan: Results of SONNE cruises SO-23 and SO-27, Sachbericht zum Forderungsvorhaben 03 R 3379: Hannover, Bundesanstalt fur Geowissenschaften und Rohstoffe, 17 p.

Holloway, N. H., 1981, The North Palawan block, Philippines: Its relation to the Asian mainland and its role in the evolution of the South China Sea: Geological Society of Malaysia Bulletin, v. 14, p. 19–58.

Howell, D. G., et al, 1983, Tectonostratigraphic terrane map of circum-Pacific region: U.S. Geological Survey Open-File Report 83-716, 18 p.

Hsu, I., 1972, Magnetic properties of igneous rocks in the northern Philippines: PhD Dissertation, Washington University, St. Louis, MO, 165 p.

Hutchinson, C. S., 1975, Ophiolites in Southeast Asia: Geological Society of America Bulletin, v. 86, p. 797–806.

Irving, E., 1979, Paleopoles and paleolatitudes of North America and speculations about displaced terrains: Canadian Journal of Earth Sciences, v. 16, p. 669–694.

Irving, E. M., 1950, Review of Philippine basement geology and its problems: Philippine Journal of Science, v. 79, p. 267–307.

Japan Metal Mining Agency, 1983, Report on geological survey of Mindoro Island: Phase II, unpublished report, 65 p.

Jones, D. L., et al, 1977, Wrangellia—A displaced terrane in north-western North America: Canadian Journal of Earth Sciences, v. 14, p. 2565–2577.

______, et al, 1982, Character, distribution, and tectonic significance of accretionary terranes in central Alaska Range: Journal of Geophysical Research, v. 87, n. 3709–3717.

______, et al, 1983, Recognition, character, and analysis of tectonostratigraphic terranes in western North America, *in* M. Hashimoto and S. Uyeda, eds., Accretion tectonics in the circum-Pacific regions: Tokyo, Terra Scientific Publishing Company, p. 21–35.

Karig, D. E., 1971, Structural history of the Mariana Island Arc system: Geological Society of America Bulletin, v. 82, p. 323–344.

______, 1973, Plate convergence between the Philippines and Ryukyu Islands: Marine Geology, v. 14, p. 153–168.

______, 1983, Accreted terranes in the northern part of the Philippine Archipelago: Tectonics, v. 2, p. 211–236.

Kinkel, A. R., et al, 1956, Copper deposits of the Philippines: Philippines Bureau of Mines Special Project Series Publication 16, 305 p.

Kitamura, N., et al, 1968, Preliminary notes on the geotectonics of the Eastern Philippine Arc and the Visaya Basin, *in* T. Kobayashi, et al, eds., Geology and palaeontology of Southeast Asia, v. 5: University of Tokyo Press, p. 186–191.

Koike T., et al, 1967, Fusulinid-bearing limestones pebbles found in the Agbahag Conglomerate, Mansalay, *in* T. Kobayashi, et al, eds., Geology and palaeontology of Southeast Asia, v. 4: University of Tokyo Press, p. 198–210.

Krause, D. C., 1966, Tectonics, marine geology, and bathymetry of the Celebes Sea–Sula Sea region: Geological Society of America Bulletin, v. 77, p. 813–832.

Ludwig, W. J., et al, 1979, Profiler-Sonobuoy measurements in the South China Sea Basin: Journal of Geophysical Research, v. 84, p. 3505–3518.

Mascle, A., and P. A. Biscarret, 1979, The Sula Sea: a marginal basin in South East Asia, *in* J. Watkins, et al, eds., Geological and geophysical investigations of continental margins: American Association of Petroleum Geologists Memoir 29, p. 373–381.

McCabe, R., 1984a, Implications of paleomagnetic data on the collision related bending of island arcs: Tectonics, v. 3, p. 409–428.

______, 1984b, Paleomagnetism of the Central Philippines and of Late Neogene rocks from Luzon: PhD Dissertation, University of Tokyo, 125 p.

______, et al, 1982a, Evidence against Late Neogene rotations of the island of Luzon: Paper presented at Southeast Asia Paleomagnetic Workshop held in Kuala

Lumpur, Malaysia, March, 1982.

———, et al, 1982b, Geologic and paleomagnetic evidence for a possible Miocene collision in western Panay, central Philippines: Geology, v. 10, p. 325-329.

Melandres, M. M., and F. A. Comsti, 1951, Reconnaissance geology of southeastern Davao: Philippine Geology, v. 5, p. 38–46.

Miranda, F. E., and B. S. Vargas, 1967, The geology and mineral resources of Catanduanes Province: Philippines Bureau of Mines Information Circular 62, 69 p.

Molnar, P., and P. Tapponnier, 1975, Cenozoic tectonics of Asia: effects of continental collision: Science, v. 189, p. 419–426.

Moore, G. F., and E. A. Silver, 1983, Collision processes in the Northern Molucca Sea, *in* D. E. Hayes, ed., Geophysical Monograph Series 27: American Geophysical Union, p. 360-372.

———, et al, 1981a, Geology of southern Mindanao, Philippines: EOS, American Geophysical Union Transactions, v. 62, p. 1086.

———, et al, 1981b, Geology of the Talaud Islands, Molucca Sea collision zone, northeast Indonesia: Journal of Structural Geology, v. 3, p. 467–475.

Murauchi, S., et al, 1973, Structure of the Sula Sea and the Celebes Sea: Journal of Geophysical Research, v. 78, p. 3437-3447.

Murphy, R. W., 1973, The Manila Trench–West Taiwan foldbelt: A flipped subduction zone: Geological Society of Malaysia Bulletin, v. 6, p. 27–42.

Newmark, R., 1980, Tectonic interaction in the Philippines: Calculated apparent slip along the Philippine Fault: Master's Thesis, University of California at Santa Cruz, 79 p.

Noritomi, K., and J. N. Almasco, 1981, Magnetic properties of rocks in and around the Atlas Porporhy Copper Mine, Cebu, Philippines: Journal of Mineralogy Collection, Akita University, Series A, v. 6, p. 45–77.

Nur, A., 1983, Accreted terranes: Reviews of Geophysics and Space Physics, v. 21, p. 1179-1185.

Philippines Bureau of Mines, 1964, Geological map of the Philippines, scale 1:1,000,000: 9 sheets.

Philippines Bureau of Mines and Geosciences, 1981, Geology and mineral resources of the Philippines, v. 1, Geology: Bureau of Mines Publication, 406 p.

Ranneft, T. S. M., et al, 1960, Reconnaissance geology and oil possibilities of Mindanao: Bulletin of the American Association of Petroleum Geologists, v. 44, p. 529–568.

Reyes, M. V., and E. P. Ordonez, 1970, Philippine Cretaceous smaller foraminifera: Journal of the Geological Society of the Philippines, v. 24, p. 1-67.

Saleeby, J. B., in press, Accretionary tectonics of the North American Cordillera: Annual Review of Earth and Planetary Sciences, v. 11, p. 45–73, 1983.

Samaniego, R. M., 1964, The occurence of *Globoratalia velascoensis* in the Philippines: Philippine Geologist, v. 1, p. 65–74.

Santos, R. R., et al, 1979, Preliminary report on the geologic reconnaissance of Surigao Peninsula and vicinity: Philippines Bureau of Mines Unpublished Geologic Report, 83 p.

Santos-Ynigo, L. M., 1949, Geology and pyrite deposits of southern Antique, Panay: Philippine Geologist, v. 4, p. 1-13.

———, and F. Esguerra, 1961, Geology and geochemistry of the nickeliferous laterites of Nonoc and adjacent islands, Surigao Province, Philippines: Philippines Bureau of Mines Special Project Series 18, 89 p.

Silver, E. A., and J. C. Moore, 1978, The Molucca Sea collision zone, Indonesia: Journal of Geophysical Research, v. 83, p. 1681-1691.

———, and R. B. Smith, 1983, Comparison of terrane accretion in modern Southeast Asia and the Mesozoic of the North American Cordillera: Geology, v. 11, p. 198-202.

———, et al, 1983a, Collision, rotation, and the initiation of subduction in the evolution of Sulawesi, Indonesia: Journal of Geophysical Research, v. 88, p. 9407-9418.

———, et al, 1983b, Ophiolite emplacement by collision between the Sula platform and the Sulawesi island arc: Journal of Geophysical Research, v. 88, p. 9419-9436.

Suppe, J., 1980, Mechanics of mountain building and metamorphism in Taiwan: Geological Society of China Memoir 4, p. 67–89.

Tapponnier, P., and P. Molnar, 1976, Slip line field theory and large scale continental tectonics: Nature, v. 264, p. 319–324.

Taylor B., and D. E. Hayes, 1980, The tectonic evolution of the South China Sea Basin, *in* D. E. Hayes, ed., Geophysical Monograph Series 23: American Geophysical Union, p. 89-104.

——— and ———, 1983, Origin and history of the South China Sea Basin, *in* D. E. Hayes, ed., Geophysical Monograph Series 27: American Geophysical Union, p. 23–56.

Teves, J. S., 1953, The pre-Tertiary geology of southern Oriental Mindoro: Philippine Geologist, v. 8, p. 1-36.

United Nations Development Program and Philippines Bureau of Mines and Geosciences, 1983, Cenozoic geological evolution of southwestern Panay and adjacent areas: Philippine Geologist, v. 37, p. 16–36.

Uyeda, S., and R. McCabe, 1983, A possible mechanism of episodic spreading of the Philippine Sea, *in* M. Hashimoto and S. Uyeda, eds., Accretion tectonics in circum-Pacific regions: Tokyo, Terra Scientific Publishing Company, p. 291-306.

Watanabe, T., et al, 1977, Heat flow in back-arc basins of the western Pacific, *in* M. Talwani and W. C. Pitman III, eds., Islands arcs, deep sea trenches and back arc basins, Maurice Ewing Series, v. 1: American Geophysical Union, p. 37.

Weissel, J., 1980, Evidence for the Eocene oceanic crust in the Celebes Basin, *in* D. E. Hayes, ed., Geophysical Monograph Series 23: American Geophysical Union, p. 37–47.

Wolfe, J. A., 1981, Philippine geochronology: Journal of the Geological Society of the Philippines, v. 35, p. 1-30.

Wright, E., et al, 1981, East Mindanao ophiolite belt: Petrology of volcanic series rocks: EOS, American Geophysical Union Transactions, v. 62, p. 1086.

Geology of the Composite Terranes of East and Central Mindanao

J. W. Hawkins
Scripps Institution of Oceanography
La Jolla, California

G. F. Moore
Tulsa University
Tulsa, Oklahoma

R. Villamor
Philippine Bureau of Mines and Geosciences
Surigao City, Republic of Philippines

C. Evans
Colgate University
Hamilton, New York

E. Wright
Scripps Institution of Oceanography
La Jolla, California

The Philippine Archipelago represents an evolving microcontinent formed by the amalgamation of many recognizable geologic terranes. Many of these terranes are, in turn, composites of several distinctive geologic systems that also constitute terranes. Remnants of island arc systems (volcanic, plutonic, and clastic rocks) are predominant but there are large fragments of depleted mantle peridotite, ophiolite from backarc basins or deep-sea floor, continental blocks, and marine sediments. Metamorphosed terrigenous, oceanic, and mantle material is also widespread. This tectonic collage has been thickened and overprinted by younger arc plutonic-volcanic series and by successor sedimentary basins.

We recognize two composite terranes in eastern Mindanao that were sutured together by mid-Tertiary (late Oligocene ?) time. Each includes remnants of arc and backarc basin rocks, mantle fragments, metamorphic rocks that appear to be related to arc deformation, and, locally, melange units. The western terrane in Mindanao is an extension of the Sangihe arc; the eastern terrane extends from Samar southward through Mindanao to the Halmahera arc. A Miocene and younger sedimentary successor basin (Agusan-Davao Trough) covers the presumed suture zone on Mindanao. Postamalgamation arc-volcanism has added new crust to the western terrane; the active Philippine Fault has dismembered parts of the eastern terrane.

The complex geologic history of Mindanao appears to be typical of the Philippine "super terrane" that has yet to be accreted to a continental mass. The Philippines give insight to an intermediate stage in the evolution from immature intraoceanic island arcs to the tectonic collages accreted to cratons as in western North America.

INTRODUCTION

The Philippine Archipelago represents an evolving microcontinent formed by the tectonic amalgamation of a wide variety of rock types, by emplacement of volcanic and plutonic rocks that have thickened and welded together the tectonically coalesced fragments, and by accumulation of thick sedimentary successor basins that have formed in situ. Within the complex tectonic collage of allochthonous blocks and fragments, there are stratigraphic, disrupted, and composite terranes (Jones et al, 1983). These include recognizable fragments of island arcs, sea mounts, backarc or deep sea-floor crust, mantle-derived peridotites showing varied degrees of depletion, metamorphosed ultramafic (mantle), mafic (oceanic), and silicic (continental) rocks, and a wide range of sedimentary rocks representing clastic material derived from volcanic-plutonic arcs, pelagic and near-shore sediments, and reef deposits. In a very simplified view, the Philippine Islands comprise a composite terrane formed of displaced crustal fragments, some of which may be derived from the Asian mainland; allochthonous terranes comprising oceanic and island arc crust and upper mantle; autochthonous volcanic-plutonic arcs and their derivative clastic rocks; and autochthonous sedimentary basins. The allochthonous units have been juxtaposed on lateral slip faults or sutured together by thrust faults or along zones of melange (Hamilton, 1979); many show evidence for extensive internal disruption. The recognition of belts of ophiolite, melange zones, and "fossil" volcanic-plutonic arcs has been important in demonstrating the composite nature of the archipelago (e.g., Hamilton, 1979; Karig, 1982, 1983; Hawkins and Evans, 1983; McCabe, this volume) and has given insights to the temporal variation in convergence zone trends and subduction polarity. Radiometric dating (e.g., Wolfe, 1981) and paleomagnetic studies (e.g., Fuller et al, 1983) have shown that the islands were formed by the coalescence (amalgamation) of a number of discrete geologic units formed at different times and at different geographic locations. The present location and relation of the islands to oceanic trenches and great faults has no direct bearing on the geologic setting of many of the constituent units at the time they were formed. The present configuration is but one scene in a dynamic tectonic evolutionary process involving accretion of tectonostratigraphic terranes and their disruption by lateral-slip faulting.

Recognition of discrete geologic units or "terranes" has proved useful in interpreting the history of complex areas such as northwestern North America (Monger et al, 1972; Coney et al, 1980; Jones et al, 1982, 1983; Schermer et al, 1984). Application of the terrane concept to the Philippines has helped to draw attention to the diversity of rock series recognizable and has demonstrated the allochthonous nature of these units (e.g., Moore et al, 1981b; Hawkins et al, 1981, 1982, 1984; Wright et al, 1981; Karig, 1982, 1983; Hawkins and Evans, 1983; McCabe, this volume).

Our main objective in this paper is to describe the geology of a complex belt of rocks in central and eastern Mindanao that has been assembled by collision of several different crust-mantle units along a zone extending south from Samar towards the Sangihe–Halmahera arcs (Fig. 1). This belt was referred to by Hamilton (1979) as the "older volcanic rocks and melange of Philippine and Halmahera arcs." We prefer to describe the belt as a collision complex inasmuch as melange constitutes only a relatively minor part of the belt, ophiolite and arc-derived clastic rocks are equally as important as volcanic material, and because the scale of individual rock units is on the order of hundreds of meters to tens of kilometers rather than on the scale of meters (or less) as implied by the term melange.

We present here a discussion of the geology of five areas in central and eastern Mindanao (Figs. 1–3) that characterize some of the complexity of the region and show both the problems and advantages in applying the terrane concept to the evolution of collision belts. The composite terranes we recognize each comprise an assemblage of disrupted terrane fragments and stratigraphic terranes. The eastern Mindanao composite terrane includes: (1) Pujada Peninsula (a disrupted terrane comprising a heterogeneous assemblage of backarc ophiolite, arc-derived clastic rocks, metamorphic rocks, clastic rocks, and carbonates); (2) Dinagat–Nonoc–Hanigad Island area (a dismembered "stratigraphic terrane" largely formed of island arc ophiolite); (3) Surigao area (a disrupted terrane comprising depleted mantle peridotite, metamorphic rocks, oceanic crust, arc volcanic and plutonic rocks, and arc-derived clastic rocks). The central Mindanao composite terrane (4), in Bukidnon Province, is a composite terrane formed of backarc or deep sea-floor ophiolite, metamorphic rocks, melange, arc-derived clastic rocks, limestone, and a modern arc-volcanic belt. A successor basin (5), the Agusan-Davao trough, formed after the Neogene collision between 1 and 4.

REGIONAL GEOLOGIC SETTING

Eastern Mindanao, Philippines, is a composite terrane formed as the result of a collision between two island arc systems during the mid-Tertiary (Hamilton, 1979). The suture zone between the collided arcs in southeastern Mindanao is believed to be defined by an ophiolite belt exposed on the Pujada Peninsula (Fig. 1; Moore and Silver, 1983). The Pujada ophiolite is part of a discontinuous belt of ophiolite, interspersed with arc-derived clastic rocks and volcanic-plutonic rocks of island arc origin, which extends from Samar southward through East Mindanao to the Talaud Islands (Moore et al, 1981a; Evans et al, 1983).

The occurrence of an ophiolite belt in southeastern Mindanao has been known for some time (Melendres and Comsti, 1951; Santos-Ynigo et al, 1961; Santos-Ynigo, 1965), and its significance was recognized by Hamilton (1979), who suggested that the Pujada Peninsula is the northern extension of the Molucca Sea collision zone. The Talaud Islands mark the southernmost exposures of rock on this postulated collision zone. The collision complex, serpentinized peridotite, and ophiolite rocks are exposed near Surigao and on Dinagat, Nonoc, and Hanigad Islands to the north giving a total length of at least 750 km (466 mi). Samar includes rocks of similar type and may add another 250 km (155 mi) to the belt. Hamilton (1979) has proposed that this collision belt is part of a great collision belt (Pacific Cordilleran system) that may act as a double orocline transforming the motion between east-directed movement in the southern Philippines and west-directed movement in Luzon (cf. his Figure 109). The Mollucca Sea collision zone separates the Sangihe Arc (west) from Halmahera (east) and corresponds to the Agusan–Davao Trough (mid- to later Neogene sedimentary basin) that separates the Central Cordillera of Mindanao (west) from the collision complex and Pujada Peninsula (east).

The Sangihe Island arc system can be traced from northern Indonesia, where the arc is still active, northward through south-central Mindanao to the Central Cordillera of Mindanao (Figs. 1, 2), where the arc recently has become inactive (Hamilton, 1979). Andesitic tuff, agglomerate, and volcaniclastic sedimentary rocks of probable Cretaceous to Paleogene age are reported to underlie the young volcanic rocks in central Mindanao (Metal Mining Agency, 1972).

Much of eastern Mindanao (Fig. 2) is covered with Cretaceous to Oligocene arc volcanic rocks (Ranneft et al,

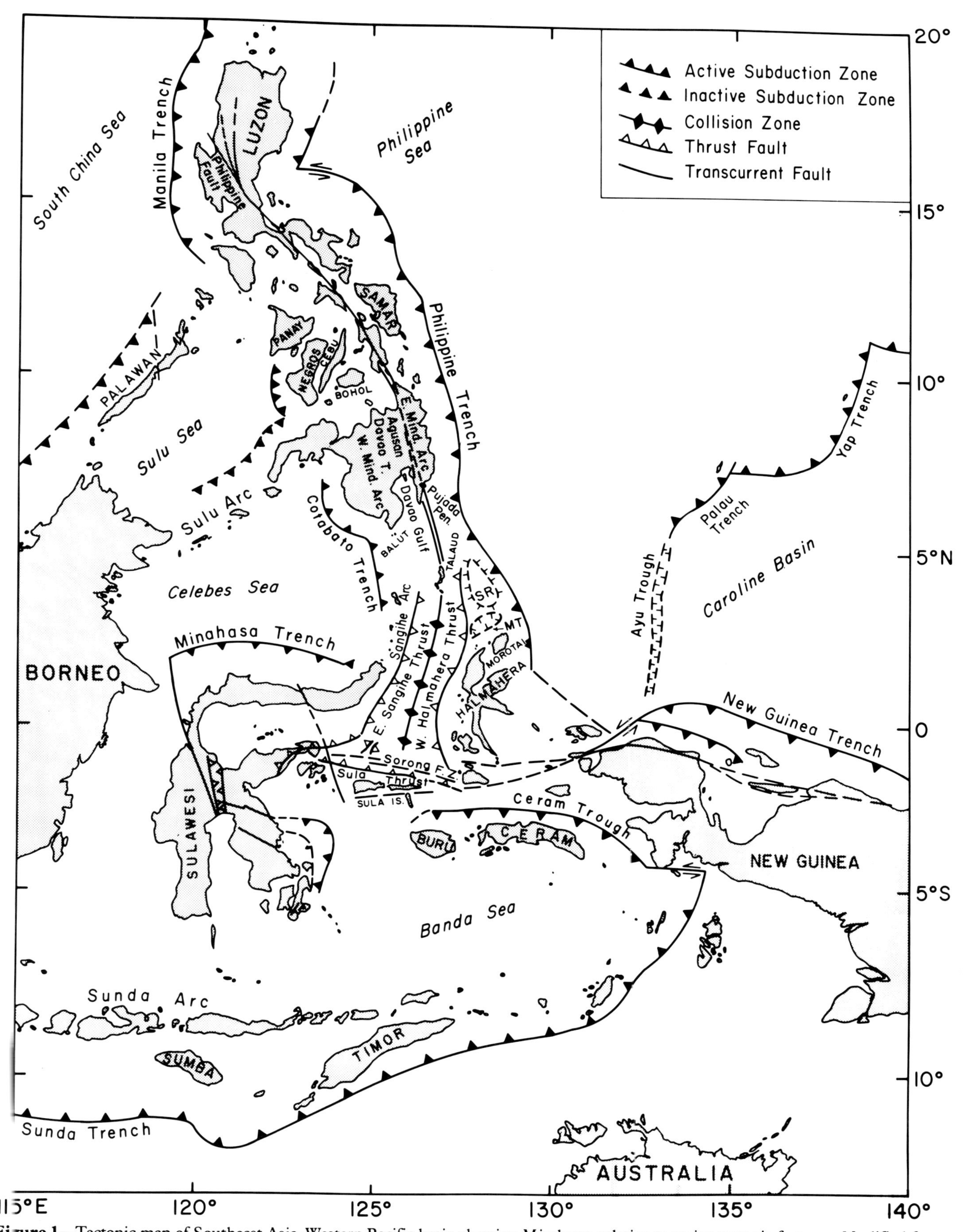

Figure 1—Tectonic map of Southeast Asia–Western Pacific basin showing Mindanao relative to major tectonic features. Modified from Hamilton (1979), Geologic Map of Philippines (1967), and Moore and Silver (1982).

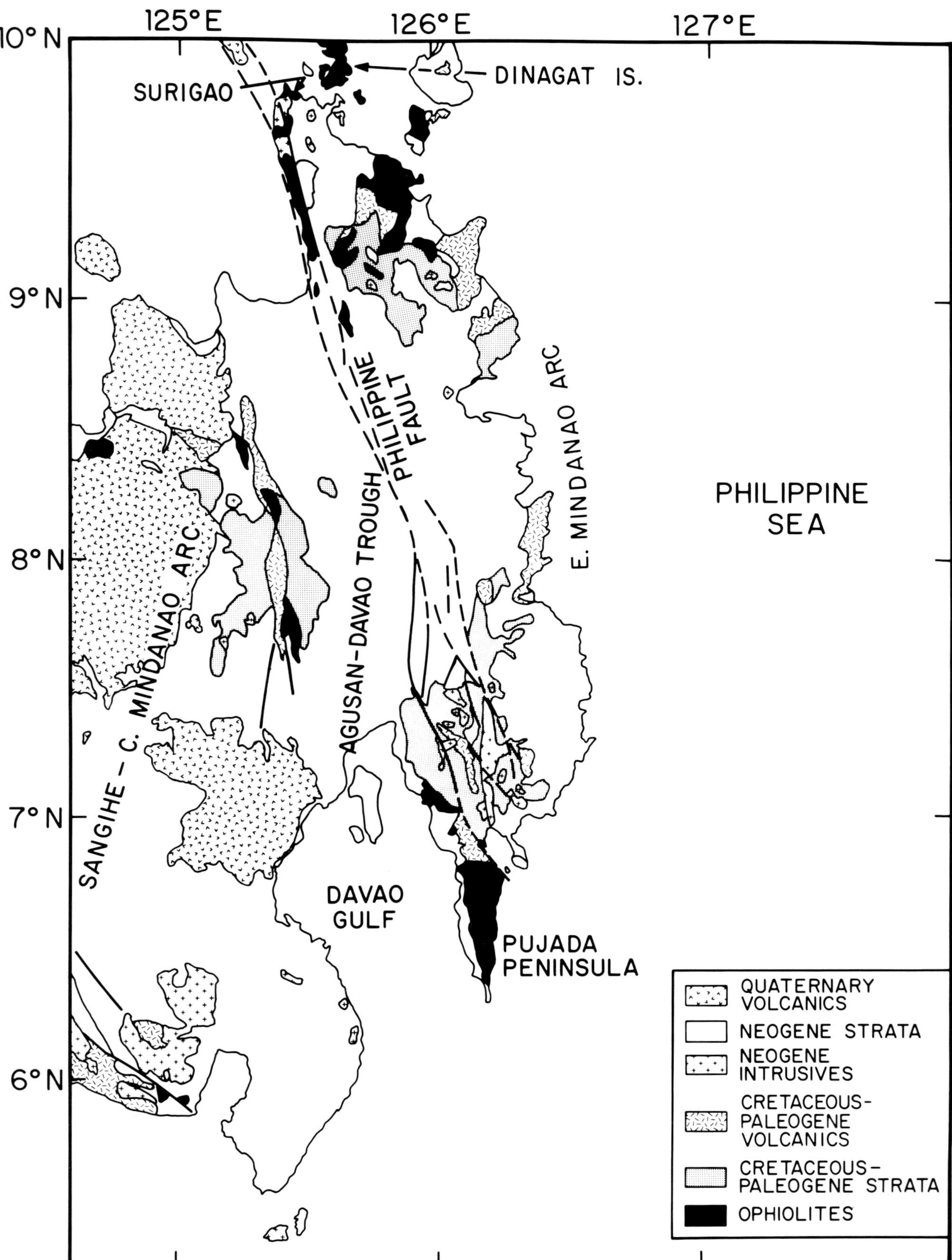

Figure 2—Central and eastern Mindanao showing major geographic and tectonic features. Agusan–Davao Trough successor basin separates central Mindanao composite terrane from eastern Mindanao composite terrane and covers the presumed suture zone. Philippine Trench, not shown, lies 100 km (62 mi) east of Mindanao.

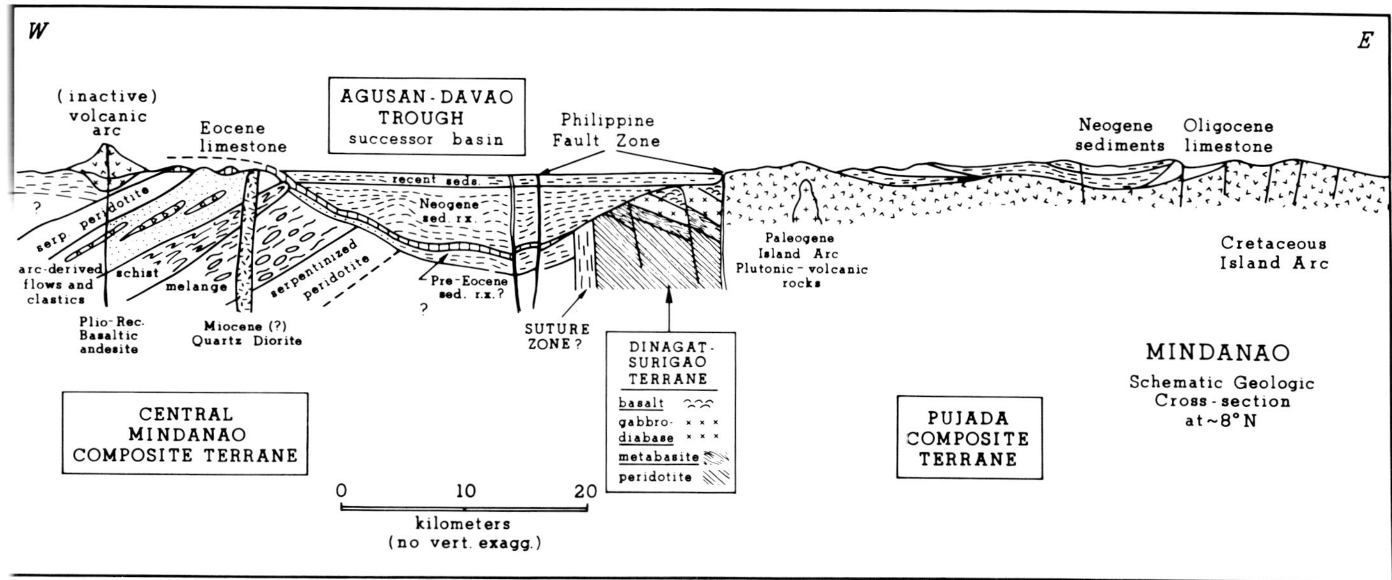

Figure 3—Schematic geologic cross section, at about 8° N, showing Agusan–Davao successor basin, probable location of suture zone between the terranes, and locus of Pliocene–Recent volcanoes. Details of structure and lithology in subsurface rocks based on inferences from exposures at surface and projected to depth. Subsurface details in Agusan–Davao Trough are based on seismic data (Murphy, 1981, written communication). Cross section is modified from a drawing by J. Murphy, AMOCO International.

1960; Metal Mining Agency, 1973; Matsumaru, 1974; Hashimoto, 1981; Wright et al, 1981). Similar Cretaceous and younger volcanic rocks have been found on Samar and Luzon Islands north of Mindanao (Reyes and Ordonez, 1970; Hashimoto et al, 1975; Hashimoto, 1981). These rocks have been interpreted as remnants of a Cretaceous to mid-Oligocene volcanic arc that extended from north of Samar through eastern Mindanao (Hamilton, 1979; Cardwell et al, 1980). The volcanic rocks are overlain by upper Oligocene limestones and Miocene coals and shales indicating cessation of volcanism by late Oligocene (Vergara and Spencer, 1957; Matsumaru, 1974).

The two arcs appear to have collided during the Neogene. The collision is still active south of Mindanao in the area of the Molucca Sea (Silver and Moore, 1978; Hamilton, 1979; McCaffrey et al, 1980). On Mindanao, the collision zone between the two arcs is occupied by the Agusan–Davao Trough (Figs. 1–3). The eastern margin of the Agusan–Davao Trough is composed of a complex of highly faulted and folded basement rocks that is exposed from Surigao southward to the Pujada Peninsula (Ranneft et al, 1960). The Philippine Fault Zone, which disrupts the eastern margin of the trough, presents a major tectonic complication in this region. Little is presently known about magnitude or age of offset along this fault, but it can be traced as a continuous feature in Mindanao from the Surigao region southward to the Pujada Peninsula (Ranneft et al, 1960; Allen, 1962; Fig. 1). The Philippine Fault Zone in Mindanao, and the area to the south, is still active as evidenced by the high concentration of seismicity in this region (Cardwell et al, 1980). Hamilton (1979) believes that the fault, although a major regional fault, does not exhibit strike-slip motion. During our field studies, we found evidence of en echelon folds and faults northeast of Pujada Peninsula that are consistent with left-lateral strike-slip along the fault. We were unable to define any strike-slip faults cutting Pujada Peninsula, but we suspect that some of the structural complexity in this region may be caused by strands of the Philippine Fault system that cut the Peninsula.

The tectonic grain of the Pujada Peninsula is north-northwest. This trend extends south to the Talaud Islands (Fig. 1), a Tertiary forearc terrane that also has a north-northwest structure (Moore et al, 1981a). The Philippine Fault Zone can be traced from Pujada Bay south-southeast along a bathymetric ridge to Talaud (Moore and Silver, 1983). The ridge is a horst block with fault strands on each side. These faults can be seen in two seismic profiles south of Mindanao and one profile just north of Talaud (Moore and Silver, 1983). The Philippine Fault Zone undoubtedly continues into the Talaud Islands area, but because of the complex geology on Talaud, the fault has not been recognized in the field. The volcanic rocks exposed on the Nanusa Islands east of Talaud may have been moved northward from the vicinity of Halmahera by motion along one of these strike-slip fault traces.

The structure of the Molucca Sea region changes drastically at Talaud. Bathymetry changes from a north-northwest trend to a north-northeast trend south of Talaud. The seismicity also changes: The active east-dipping subduction zone under Halmahera does not continue northward to the latitude of Talaud (Cardwell et al, 1980). Deformation in the collision zone south of Talaud is still active (Silver and Moore, 1978), whereas north of Talaud, deformation has slowed dramatically as indicated by the thick sediments of Davao Gulf region that are only moderately deformed (Cardwell et al, 1980). South of Talaud, the deformed collision complex is being thrust outward onto the flanks of the Sangihe and Halmahera arcs (Silver and Moore, 1978). These thrust faults can be traced northward only as far as the Talaud Islands. At the present time, the nature of the change in tectonic style at the Talaud Islands is unclear.

GEOLOGY OF THE PUJADA COMPOSITE TERRANE

The igneous and metamorphic rocks of the Pujada Peninsula and southeastern Mindanao constitute a distinctive terrane that may itself be a composite of two or more distinctive but dismembered microterranes (Fig. 4). For this discussion we will call it the Pujada composite terrane. The overall petrologic characteristics indicate an assemblage of rocks representing ophiolite, volcanic arc material, and their metamorphic equivalents (Tables 1, 2, 3).

Greenschist-facies metamorphosed mafic rocks are the lowest structural unit exposed on the Pujada Peninsula. Chlorite-actinolite schists forming the basement of the northern Pujada Peninsula are best exposed at Batobato Point (Fig. 4) and on the southern part of the peninsula. These schists (Magpapangi greenschist) underlie the amphibolite which in turn is overlain by peridotite (Villamor et al, 1984). Fault zones, presumed to be thrust faults, separate these rock units (Villamor et al, 1984). The degree of metamorphism of the schists exposed in the northern Pujada Peninsula appears to decrease to the east, and our impression is that these rocks grade into essentially unmetamorphosed rocks to the east in the area of Dawan where they are interbedded with basaltic volcanic rocks and thin-bedded red siliceous argillites (Melendres and Comsti, 1951). In the area northwest of Dawan it is clear that a fault juxtaposes the volcanic rocks and red sediments. In this area, the volcanic and sedimentary rocks are intersheared on a very small scale. Some of this intercalation may be a primary feature, the original material having been interlayered volcanic and sedimentary rocks. South of Dawan, along the coast of Pujada Bay, these sedimentary and volcanic rocks are strongly sheared and resemble melange. Pillow structures are preserved in the volcanic rocks at several localities.

The core of the Pujada Peninsula has thin discontinuous lenses of calcite and calc-silicate marble and metamorphosed mafic and ultramafic rocks (amphibolite and metaperidotite that comprise the lithologic types considered to be diagnostic of an ophiolite. Unmetamorphosed peridotite, gabbro, diabase, and basalt are also present and structurally overlie the metabasites.

The peridotite (Table 4) includes olivine and orthopyroxene cumulates and olivine-orthopyroxene-plagioclase cumulates. Serpentinization of the peridotite ranges from mild to nearly complete. Some peridotite samples show effects of extensive high-temperature shearing and recrystallization with thin streamers formed of a mosaic of tiny olivine crystals. Chromite forms individual grains and grain aggregates. The mafic rocks have the chemistry typical of midocean ridge or backarc basin basalts (Table 1). There are no distinctive chemical signatures to differentiate between these two origins in altered or weakly metamorphosed basalts. We can, however, distinguish between these rock types and basalts from sea mounts, island arcs, or continental settings.

Basalts from southeastern Mindanao (Fig. 2) and from the northwest side of Pujada Bay, near the Philippine Fault Zone (Fig. 4), have phenocrysts of plagioclase, hornblende, and clinopyroxene. The groundmass consists of plagioclase laths, pyroxene, and opaque minerals. At least some of these basalts have arc-tholeiitic chemistry (Wright et al, 1981, and Table 1). Thus, we have evidence in the Pujada terrane for crustal rocks from both an island arc terrane and a backarc or midocean ridge. We favor a backarc origin because of the nature of the associated rocks that indicate a large component of arc material.

Amphibolites from the northern Pujada Peninsula appear to have undergone retrograde metamorphism from amphibolite to greenschist facies. The amphibole is actinolite to calcic-Al-hornblende (see Table 4). The amphibole ranges from anhedral to euhedral and generally shows a preferred orientation of its long axis, which gives the rocks a lineated fabric. Epidote is common, forming clusters of small crystals. Garnets (almandine-grossularite-pyrope mixtures, Table 4) are also present in several of the amphibolites. Some of the garnets are euhedral, although some are poikiloblastic, and filled or rimmed with chlorite. Chlorite is also common as a replacement mineral on amphibole. Amphibolites from the central Pujada Peninsula are composed mostly of green hornblende in parallel blades, and plagioclase. Microprobe analyses indicate that they are all calcic amphiboles. No sodic amphiboles have been found. This is critical in establishing the P–T relations of metamorphism.

Petrology and Chemistry

The crystalline rocks of the Pujada composite terrane consist of a variety of types, which points to an affinity with oceanic lithosphere, volcanic arcs, and their metamorphic equivalents (Melendres and Comsti, 1951; Ranneft et al, 1960). Hamilton (1979) called attention to the significance of the ophiolite assemblage as an indication of former oceanic lithosphere. Ophiolite assemblages may form in several different tectonic settings (e.g., Hawkins, 1980; Hawkins and Evans, 1983) so we must look for distinctive characteristics to identify the type of ocean crust/mantle they represent. The Pujada terrane comprises rock types derived both from an island arc environment and from a backarc basin. These may have been part of a single convergent plate system, but now they have been juxtaposed by faults and thus they constitute different petrologic domains (stratigraphic terranes) within the larger terrane. For simplicity, we will refer to these as the backarc ophiolite and the island arc series. The backarc ophiolite forms much of the Pujada Peninsula; the arc series rocks are exposed, with arc-derived clastic rocks, along the eastern cordillera between 7° N and 10° N.

The basalts and diabases from the backarc ophiolite (Table 1) have TiO_2 concentrations in the range 1.16 to 1.63 wt%, Ni contents between 55 and 100 ppm, and FeO^*/MgO ratios between 1.00 and 1.60, where FeO^* indicates all iron represented as ferrous iron (see Fig. 5). These data are typical of fractionated midocean ridge basalts (MORB) or backarc basin basalts (BABB). In view of their close association with island arc volcanic rocks, we

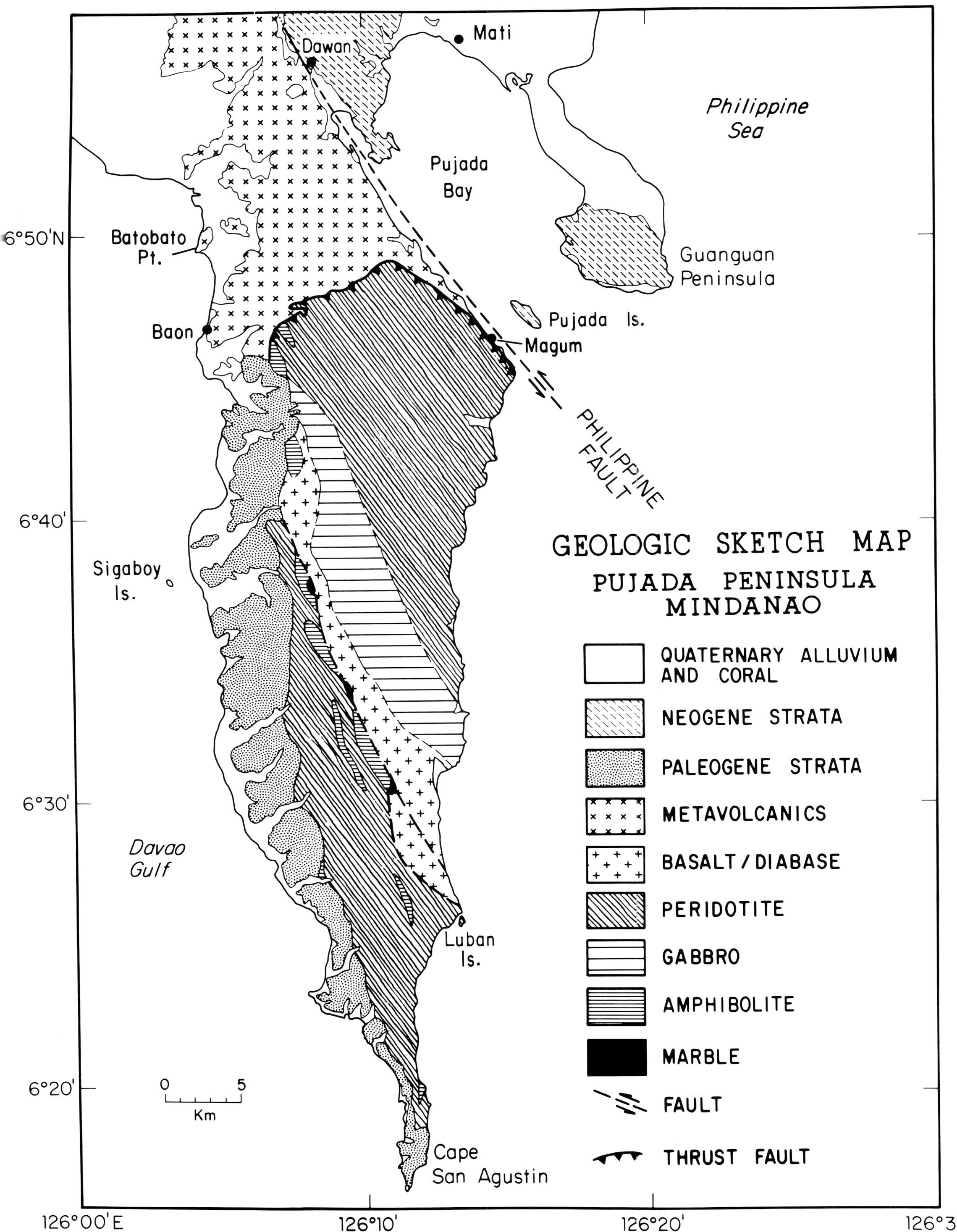

Figure 4—Geologic map of Pujada Peninsula. Based on Philippine Bureau of Mines (1967), Santos-Ynigo et al (1961), Melendres and Comsti (1951), and our field mapping.

Table 1

Sample	MIN 316	MIN 161	MIN 162	MIN 203	MIN 191	MIN 321	MIN 164	MIN 180	MIN 130
Type	B	D	D	D	D	D	G	G	G
Mg#	.646	.626	.607	.563	.575	.523	.623	.641	.802
					Weight Percent				
SiO_2	49.14	53.24	50.63	49.47	51.02	50.09	49.02	49.48	49.45
TiO_2	0.70	1.46	1.28	1.63	1.52	1.62	1.15	0.56	0.52
Al_2O_3	16.01	15.41	15.29	13.38	14.88	13.50	15.30	20.79	18.76
FeO*	8.90	9.00	9.71	11.32	10.18	11.62	9.49	4.98	3.93
MnO	0.18	0.17	0.17	0.24	0.21	0.19	0.18	0.10	0.07
MgO	9.12	8.46	8.40	8.18	7.72	7.16	8.78	6.41	8.94
CaO	12.67	9.27	11.60	11.80	11.76	11.35	14.11	14.02	17.32
Na_2O	2.72	3.42	2.73	2.78	2.13	3.18	1.80	2.65	1.05
K_2O	0.11	0.03	0.20	0.10	0.08	0.08	0.04	0.18	0.02
P_2O_5	0.08	0.01	0.11	0.13	0.07	0.21	0.05	0.19	0.02
Sum	99.74	100.47	100.19	99.07	99.61	99.04	99.92	99.36	100.15
					Trace Elements (ppm)				
Cr	384	262	—	234	—	117	195	—	—
Ni	141	100	85	71	56	63	90	78	161
V	251	253	275	287	284	349	284	189	236
Rb	1	1	1	1	2	1	1	1	1
Sr	145	99	121	131	106	152	124	185	213
Ba	31	2	7	11	11	8	12	27	25
Zr	56	97	88	111	110	146	72	49	21
Y	24	35	32	36	33	44	28	18	17
Nb	1	1	1	2	4	2	2	1	1

Abbreviations: B = basalt; D = diabase; G = gabbro; Mg# = Mg/(Mg + Fe).

Table 1—Backarc basin crustal rocks, Pujada Peninsula, eastern Mindanao; chemical composition.

interpret these as BABB. The element abundances and ratios clearly distinguish the backarc ophiolite basalts from the arc series.

Basalts and basaltic andesites of the island arc volcanic series (Table 2) contain 0.60 to 0.90 wt% TiO_2, Ni contents less than 27 ppm, usually in the 6 to 17 ppm range, and FeO*/MgO ratios between 1.75 and 3.30. These samples plot in the field for island arc basalts on element discriminant diagrams (Figs. 5–7). On some of these plots there is an overlap between arc and ocean-floor samples, but when all of the discriminants are considered, an arc origin is seen. The low Cr, Ni, Ti, Zr content and high Ba, Sr (Table 2) are distinctive characteristics of island arc basalts and basaltic andesites (e.g., Jakes and Gill, 1970).

As the FeO*/MgO ratios indicate, the island arc volcanic samples as a group are more differentiated than the BABB samples, and they show generally higher K_2O and SiO_2 contents. Sr values are high (> 300 ppm) in the arc samples as well. The BABB samples have 8% normative hypersthene and 8% normative olivine in contrast to the arc samples, which lack normative olivine and have about 16% normative hypersthene. The Y concentrations for the two groups are comparable (most fell in the 25–45 ppm range), which implies, once the effect of differentiation is subtracted, that the arc source was depleted in Y relative to the BABB source. The low concentration of Y, as well as Zr and Ti, is a characteristic of island arc tholeiites and helps to distinguish them from MORB or BABB.

In a plot of TiO_2 vs. Zr (Fig. 6), which utilizes the fields defined by Pearce and Cann (1973) for ocean-floor basalts (OFB), calc-alkaline basalts, and low-K tholeiites, the BABB samples plot as OFB, while the arc samples are in the calc-alkaline field. Similar distinctions are found in ternary plots of Ti-Zr-Y (Fig. 7) again using the fields of Pearce and Cann (1973). In general, arc tholeiites lie in the fields for calc-alkaline rocks in plots of this type, whereas BABB will plot in the ocean-floor basalt field.

Petrologic Discussion

The intensity of metamorphism in the rocks from the core of the peninsula is highly varied; it ranges from tectonized peridotite with granulite-facies mineral assemblages to serpentinites formed at low P and T. Some mafic rocks have amphibolite to greenschist-facies assemblages with well-developed lineation and schistosity while others have mineral assemblages and textures

Table 2

Sample	MIN 280	MIN 260	MIN 283	MIN 272	MIN 271	MIN 270	MIN 331-A
Type	B	B	BA	BA	BA	A	D
Mg#	.509	.406	.482	.478	.473	.421	.131
				Weight Percent			
SiO_2	50.85	48.04	55.16	53.87	54.45	59.71	70.34
TiO_2	0.82	1.05	1.11	0.80	0.80	0.65	0.16
Al_2O_3	14.97	20.88	15.14	18.04	20.41	18.99	15.59
FeO*	11.84	10.89	10.19	8.29	7.12	5.20	3.67
MnO	0.3	0.43	0.21	0.34	0.15	0.17	0.06
MgO	6.89	4.18	5.32	4.25	3.59	2.12	0.31
CaO	9.82	10.62	7.64	8.98	8.92	6.49	0.33
Na_2O	2.13	2.59	4.06	3.38	3.26	3.93	1.63
K_2O	1.03	0.26	0.06	1.14	0.20	2.45	5.86
P_2O_5	0.22	0.28	0.22	0.22	0.13	0.29	0.03
Sum	98.87	99.20	99.11	99.31	99.03	100.00	98.01
				Trace Elements (ppm)			
Cr	19	19	21	15	13	21	—
Ni	11	16	17	6	6	8	11
V	272	319	387	232	226	101	5
Rb	10	5	1	18	2	57	232
Sr	524	300	435	300	249	838	29
Ba	181	89	164	194	54	401	79
Zr	105	60	98	150	172	123	1177
Y	32	24	35	33	43	20	122
Nb	1	2	1	6	2	9	—

Abbreviations: B = basalt; BA = basaltic andesite; A = andesite; D = dacite; Mg# = Mg/(Mg + Fe).

Table 2—Island arc crustal rocks, Pujada Peninsula, and eastern Mindanao; chemical composition.

indicative of only low T hydrous static recrystallization. The diversity of textures and metamorphic facies strongly suggests that the core of the Peninsula is formed of an accumulation of slabs or blocks that have been derived from different levels of an oceanic lithosphere section and that have experienced different styles and intensities of metamorphism. At this time we have no basis to differentiate between different metamorphic cycles or a single event followed by tectonic disruption of a zoned metamorphic belt.

There is always a problem in determining the protolith of metamorphic rocks, but the chemical and mineralogic data for the Pujada samples appear to be distinctive enough to permit us to make a fairly reliable estimate. The serpentinites and metamorphosed ultramafic rocks were derived from depleted peridotite. The relict mineralogy of some samples clearly indicates that they were derived from harzburgite (OL + OPX) with chromite. This mineralogy is typical of depleted mantle peridotite; i.e., mantle that has been through a melting episode and represents the unmelted refractory residue beneath a spreading center, arc edifice, sea mount or some other magma generation site. Neither the textures nor the mineral composition [OL = Fo_{91}, OPX = $En_{90}Fs_9Wo_1$, CHR = Cr/(Cr + Al) = 0.43] are diagnostic of crystal accumulations found in stratified, fractionated mafic magma chambers, but this cannot be totally ruled out because the lowest (first to form) cumulate levels of some of these stratiform complexes resemble residual mantle material in composition.

The metamorphosed basaltic and gabbroic rocks exhibit a range in mineral and textural types (Table 4). Amphibolite-facies rocks include garnet-amphibole, garnet-amphibole-plagioclase, amphibole-zoisite (plagioclase), amphibole, calcite-amphibole, and amphibole-plagioclase assemblages. Sphene, chlorite, epidote, rutile, Fe-Ti oxides, and sodic plagioclase are present as accessory minerals or retrograde assemblages. Calcite marble forms discontinuous layers or lenses in the amphibolite terrane and clearly was derived from sedimentary material. The amphibole-bearing assemblages must have been derived from Ca, Al, Mg-rich rocks. The trace element (e.g., Ni, Cr, Sr) and minor element (e.g., Ti, Mn, Na, K) abundances of these samples argue for a basalt-gabbro source. The most likely parental rock would have been a (cumulate?) gabbro from the base of an oceanic or arc crustal section. Some of the mafic schists have chlorite-rich

Table 3

Sample	MIN 151	MIN 51	MIN 210	MDO 10b	MIN 50
Type	A	GA	A	A	GA
			Weight Percent		
SiO_2	44.52	43.75	52.05	49.70	47.95
TiO_2	1.44	1.58	1.01	0.99	2.58
Al_2O_3	17.32	18.30	14.33	14.44	14.47
FeO*	11.00	9.57	10.44	9.02	12.63
MnO	0.26	0.15	0.18	0.28	0.18
MgO	9.95	9.71	8.86	7.79	7.33
CaO	12.52	14.23	10.44	13.42	11.78
Na_2O	2.78	1.76	3.02	3.08	2.31
K_2O	0.24	0.08	0.15	0.34	0.10
P_2O_5	0.05	0.07	0.09	0.17	0.07
Sum	100.16	99.20	99.85	99.23	99.40
			Trace Elements (ppm)		
Ni	178	161	123	91	70
V	211	288	257	241	466
Rb	1	1	<1	6	3
Sr	348	597	145	118	155
Ba	17	48	21	—	0
Zr	273	99	38	66	146
Y	50	41	25	30	54
Nb	7	1	1	1	1

Abbreviations: A = amphibolite; hornblende, clinozoisite, sphene; GA = garnet amphibolite; hornblende, almandine, clinozoisite, oligoclase, sphene.

Table 3—Metamorphic rocks, Pujada Peninsula, eastern Mindanao; chemical composition.

assemblages and represent greenschist-facies rocks. These samples typically have strongly schistose textures. They probably represent the same rock types as the amphibolites, but they experienced lower T metamorphism. It is likely that they are the result of retrograde metamorphism of the amphibolite rather than a regional gradient in metamorphic intensity. This inference is based mainly on their distribution and the abundant evidence for tectonic transport and dislocation of all of the units. Low temperature-low pressure static metamorphism resulting from hydrothermal circulation formed vein-filling assemblages in some samples.

At this stage, it is not clear when or where the metamorphism took place. We have an assemblage of rocks representing depleted upper mantle and a crustal series of gabbro-basalt that could have come from the deep sea-floor (MORB-type crust) or backarc basin, and a young island arc (arc tholeiite-type crust). The metamorphic assemblages indicate that the rocks were sheared and recrystallized at high T and moderate P. The ultramafic rocks must have been close to their solidus temperature. If we assume that the ultramafic rocks are genetically related to the mafic schists, the simplest explanation is that the rocks were metamorphosed before parts of the crustal series had cooled below about 500°C and while the subjacent mantle was still at about 900 to 1000°C. This implies either that the metamorphism occurred close to the site of crustal generation and within 5 to 10 m.y. of the time of formation or that they have been brought up from great depths (e.g., 30 to 50 km [19–31 mi]) in the mantle. There is no indication of any high P mineral assemblages. Blueschist and eclogite minerals *have not* been identified in any of the microprobe analyses. If we postulate that the metamorphism formed while the crust/mantle series was still hot (and young), and if we consider the abundance of arc-derived clastic and igneous rocks, a possible explanation for the metamorphism may be constructed. Collision of arcs and backarcs, or arcs and "obducted" ocean crust, has been proposed to explain occurrences of similar rock series in other orogenic belts. The dynamo-thermal metamorphism of the Pujada series may have resulted from the thrusting of a slab of oceanic lithosphere into, or onto, an arc complex. The high temperature of this metamorphism suggests that young oceanic lithosphere was involved and an origin in a backarc basin would be consistent with these observations.

Sedimentary Rocks

North of the Pujada Peninsula, along the east coast of the Davao Gulf, there is an extensive terrane of weakly metamorphosed graywacke (Figs. 2–4). This graywacke belt trends to the southeast toward the Pujada Peninsula, and it extends nearly to the east coast of Mindanao east of Mati. The predominant exposures are thin-bedded volcaniclastic graywacke sandstones and shales. In the Hijo River area, calcarenite beds containing large foraminifera are interbedded with the graywackes. Ages of the fossils are Late Cretaceous, Paleocene, and Eocene. Planktonic forams from deep-water shales near the east coast also yield Eocene ages.

The Cretaceous to Miocene graywacke sandstones are composed of angular basaltic and andesitic volcanic rock fragments, ortho- and clino-pyroxene, plagioclase, amphibole, opaque minerals, minor quartz, and carbonate clasts. The volcanic rock fragments appear to have come from a hypersthene-clinopyroxene-bearing calc-alkaline (island arc) volcanic terrane. Microprobe data show that the clinopyroxene and hypersthene are typical of arc volcanic rocks. These graywackes are extremely hard and well indurated, and many are slightly metamorphosed. Epidote and chlorite are common, and several samples contain zeolites (laumontite identified by x-ray diffraction). The presence of laumontite indicates temperatures in excess of 200°C.

The Miocene–Pliocene sandstones exposed along the west coast of Pujada Peninsula (Fig. 4) are composed of angular fragments of andesitic volcanic rocks, hornblende, detrital chlorite and biotite, plagioclase, epidote, pyroxene, and ultramafic rocks. The source terranes for these rocks were an andesitic volcanic chain (Apo volcanics exposed across Davao Gulf?) and the Pujada ophiolite terrane. X-ray diffraction analyses of several samples yielded no low-grade metamorphic minerals.

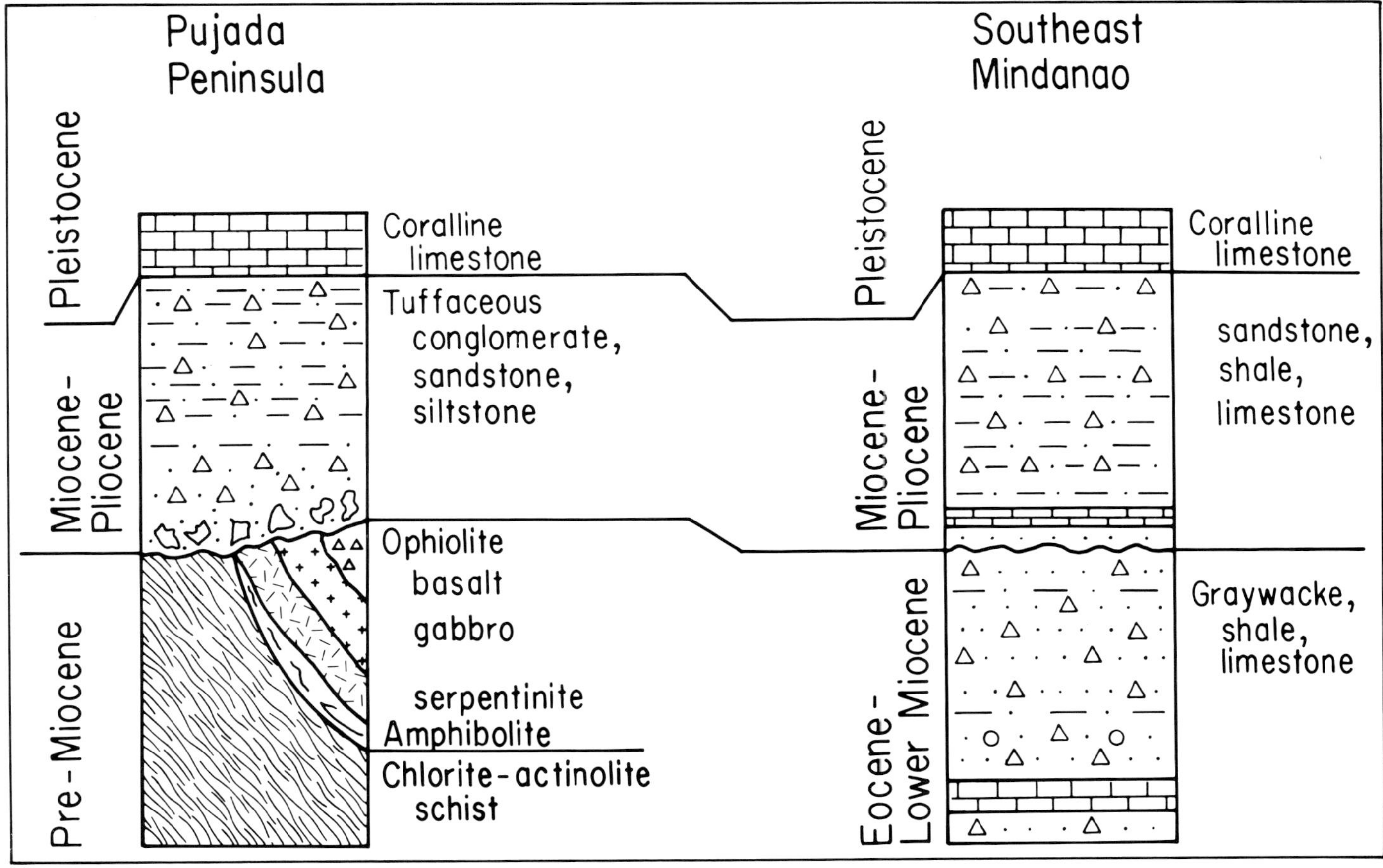

Figure 8—Schematic stratigraphic column for Pujada Peninsula.

GEOLOGY OF DINAGAT ISLAND TERRANE

Introduction

The Eastern Mindanao collision complex may be traced north toward Samar through Dinagat, Nonoc, Doot, and Hanigad Islands (Figs. 1, 2, 9). The northern end of Dinagat Island has exposures of upper mantle and lower crustal rocks (gabbro-diabase) that constitute part of a classic ophiolite assemblage. The southern part of Dinagat Island has garnet amphibolite and other metabasites which, by analogy with Pujada Peninsula rocks, probably were derived largely from cumulate mafic and ultramafic rocks from the basal crustal section of oceanic crust. Dunite, harzburgite, and clinopyroxene bearing harzburgite are present on Nonoc and Hanigad; the clinopyroxene-rich rocks have cumulate textures. Amphibolite (amphibole-oligoclase-epidote-sphene) is found on Hanigad Island. The Doot Island pillow basalts have abundant phenocrysts of olivine and clinopyroxene; they have heavy alteration by calcite but some aspects of their chemistry suggest an alkaline affinity, and they could be fragments of a sea mount. The sedimentary rocks on these islands are mainly fine-grained calcareous rocks that represent pelagic sediments deposited on the basaltic crust. A complete "stratigraphic section" from peridotite to basalt is not preserved, and we cannot be certain that all of the rocks are genetically related. However, the chemical and mineralogic characteristics of both the peridotite and the gabbro-diabase series indicate an arc tholeiitic assemblage, and we conclude that much of this ophiolite represents disrupted fragments of an island arc terrane. We give the informal name "Dinagat Island terrane" to the island group north of Surigao.

Northern Dinagat Island

The northern part of the island is formed largely of ultramafic rock and includes dunite, harzburgite, chromite-rich peridotite, and minor amounts of wehrlite, websterite, and clinopyroxenite. The peridotites show varied extent of serpentinization, and some samples are completely serpentinized. The peridotite mass has well-developed planar structures owing to preferred orientation of orthopyroxene and layers of chromite. The regional trend of the layering is northwesterly to nearly east-west; dip directions are variable with inclinations from 30° to vertical. Tightly folded chromite layers indicate intense ductile deformation. Some of the ultramafic rocks, especially those on the ridge crest of the island, have cumulate textures and planar structures that are the result of mineral stratification. We lack data to describe the orientation of this layering on a broad scale, but it appears to trend north-south and dips east at 50°. This suggests regional tilting down to the east, but we recognize that the igneous layering may not have been horizontal when formed.

The peridotite is cut by tabular intrusive bodies of

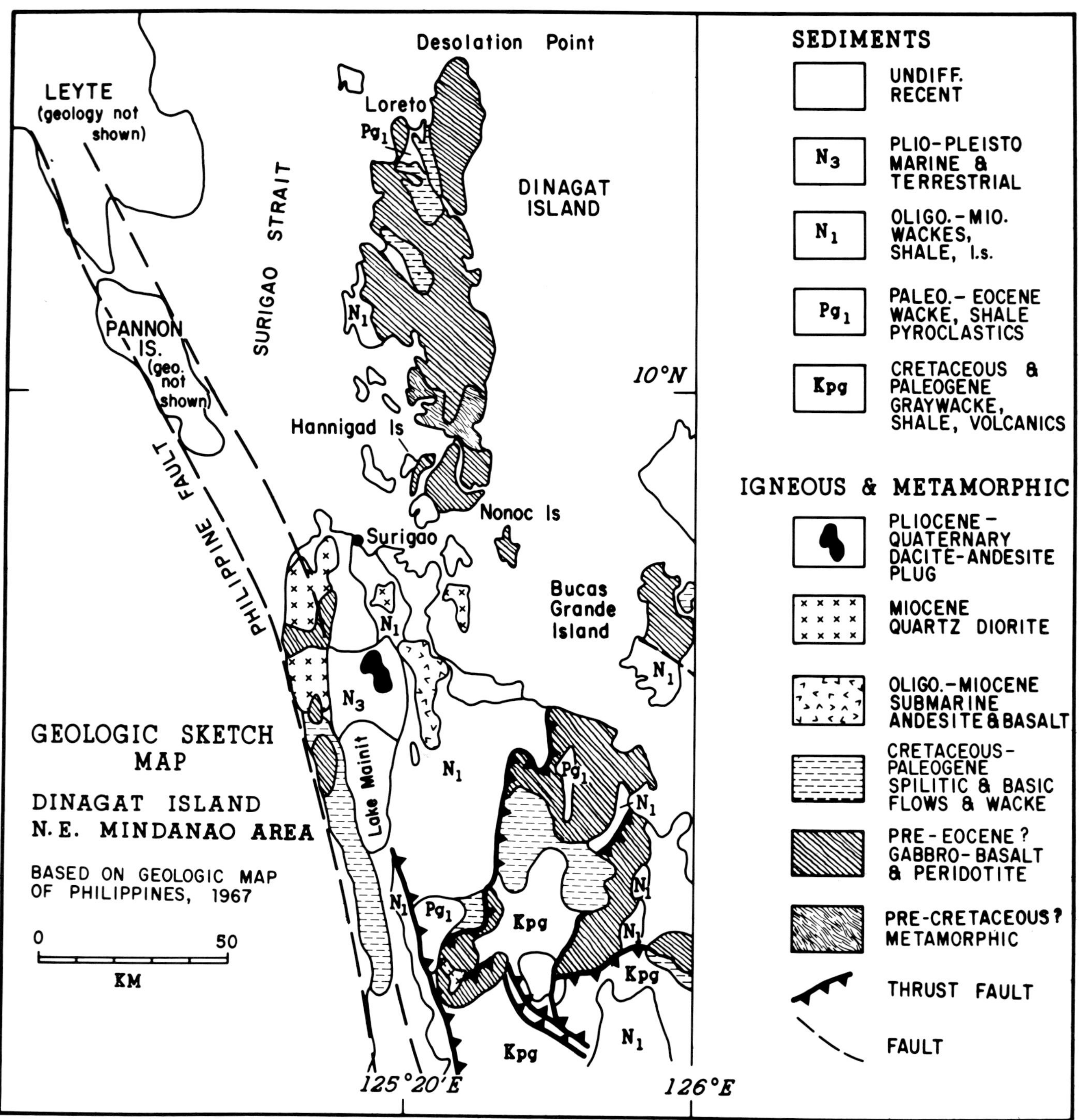

Figure 9—Geologic map of Dinagat Island–Surigao region showing major rock types and geologic structures. Based on Geologic Map of Philippines. (Copyright © 1967 The Philippine Bureau of Mines and Geosciences. Used with permission.)

diabase and microgabbo that are presumed to be feeder dikes to surface flows, although there are no known exposures of layered gabbro or basalt flows on the northern part of the island. Chromite is abundant in the peridotite, especially in the highly deformed dunite and harzburgite, and occurs as disseminated grains, in layers, pods, nodules, and orbicular structures.

Petrologic Summary

The peridotites are typical of ultramafic rocks found in other ophiolites (e.g., Hawkins and Evans, 1983). Harzburgite (olivine plus orthopyroxene) is the dominant rock type. Zones of dunite form layers nearly concordant with mineralogic layering in the harzburgite or form irregular discordant masses. The harzburgite has Mg-rich olivine ($Fo_{90-91.5}$) and enstatite (En_{89-90}); the dunite is even more Mg-rich (Fo_{93}). Chromite compositions in the harzburgite (Cr/[Cr + Al] = .50) are less enriched in Cr and Mg (less refractory) than in the dunite (Cr/[Cr + Al] = .75 — .80) and suggest that the dunite masses are the residue of more extensive melting than the harzburgite. The chromite compositions (Fig. 10) lie in the field

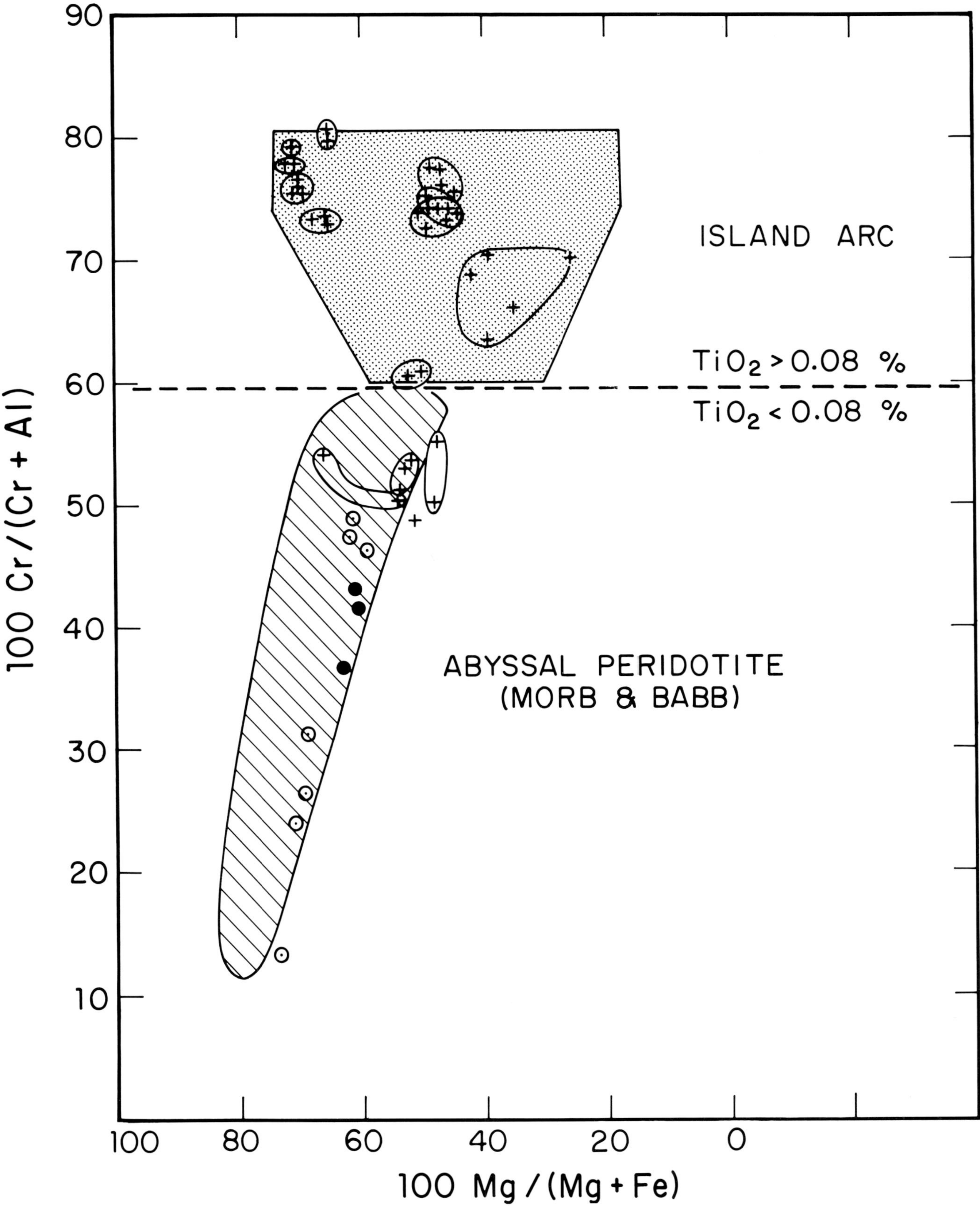

Figure 10—Chromite data for Mindanao peridotites (ratios are in atomic proportions). Fields are from Dick and Bullen (1984) and Hawkins and Evans (1983). Some Dinagat Island chromite plots close to the MORB-BABB field; these may be in pods of mantle less extensively melted than the main peridotite mass that has chromite typical of island arcs. Circled points are from Bukidnon Region, solid circles are from Pujada backarc ophiolite, and crosses are from Dinagat Island ophiolite. Data enclosed in circles are from the same sample.

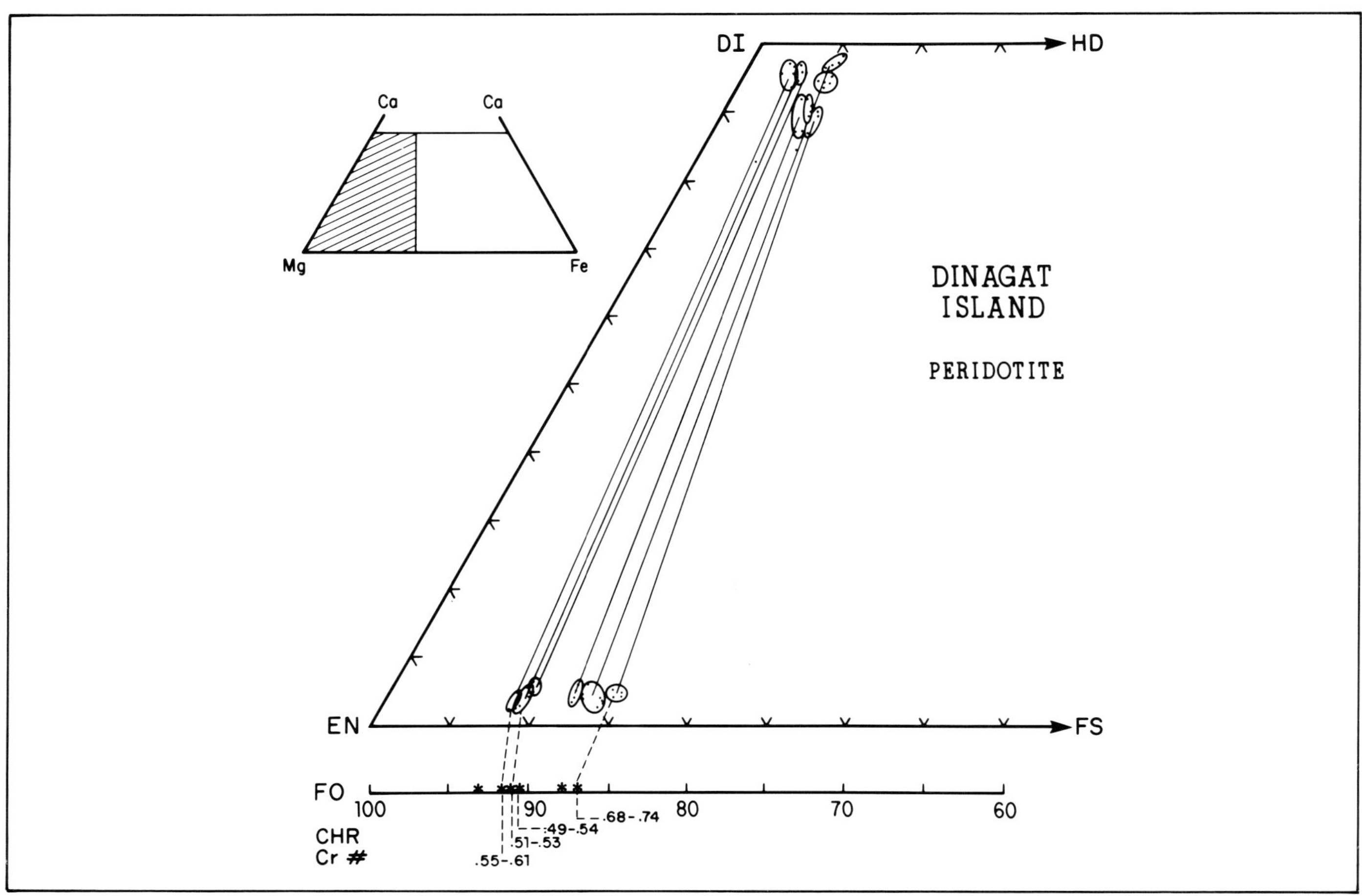

Figure 11—Microprobe data for olivine, chromite, and pyroxene from Dinagat Island peridotite. The lines link coexisting minerals. The high Mg# of silicates and high Cr# of spinels point to a depleted mantle composition and is consistent with an island arc setting. The samples with Fs 13–15 are from the cumulate textured section and represent lower crustal rocks. Inset shows location of this sketch relative to the pyroxene quadrilateral.

proposed for island arc settings (Hawkins and Evans, 1983; Evans, 1983; Dick and Bullen, 1984; Evans and Hawkins, in preparation). Some of the chromite data overlap the field proposed for ocean-floor rocks; these may be from pods of peridotite that were less extensively melted than the rocks with arc-like chromite data.

The internal structure of the peridotite is complex and resembles that of tectonized peridotite from other areas (e.g., Nicolas et al, 1971, 1980; Hawkins and Evans, 1983) The tight folding of mineralogic layering suggests high-temperature ductile deformation or "aesthenospheric flow" (Nicolas et al, 1971, 1980). The small-scale structures are typical of the transition we have studied in the Zambales Range, Luzon, at the boundary between depleted upper mantle peridotite and cumulate ultramafic rocks of the lower crustal section.

The cumulate-textured rocks of Dinagat include interlayered dunite, wehrlite (olivine plus clinopyroxene), and websterite (clinopyroxene plus orthopyroxene). In general, the cumulate-textured rocks have less Mg-rich minerals than the tectonized peridotite, e.g. Fo_{87} and En_{85} (Fig. 11). The sequence of crystallization appears to have been olivine and chromite-clinopyroxene-orthopyroxene. Plagioclase is not present but presumably formed last, and its absence may be due to crystal flotation. This crystallization sequence is typical of island arc magma series; it resembles the mineral sequence of the arc-series crust of the Zambales Range (Hawkins and Evans, 1983) and further supports the inference about tectonic setting based on the chromite data.

Mafic intrusive "dikes" cutting the peridotite give the best evidence for the nature of the tectonic setting in which the ophiolite formed. The diabase and microgabbro have island arc chemical characteristics (Table 5). For example, they have high silica, low Ti, Zr, Cr, and Ni for a given Mg content, normative hypersthene and quartz and distinctive element ratios. The presence of quartz and biotite in the micro-gabbro is also an important indicator of the similarity to arc material. We compare these to samples from island arc and other settings in Figures 5 through 7, using discriminant diagrams. They are similar to arc-tholeiitic series rocks from regions such as the Mariana arc and to samples from the Pujada Peninsula described in a separate section. The chemical data for these dikes are shown in Figure 12 where they are compared to "normal mid-ocean ridge basalt" and to island arc volcanic rocks. The Dinagat samples closely resemble basalts and basaltic andesites from the Mariana arc and are definitely different from N-MORB. This gives further support to the inferences made from mineralogy, bulk chemistry, and the

Table 5

Sample	DIN 5	DIN 3C	DIN 3B	DIN 10A	MIN 110	MIN 120
Type	D	D	G	D	B	BA
Mg#	.531	.488	.483	.405	.616	.490
			Weight Percent			
SiO_2	56.62	55.61	56.34	55.00	47.61	53.68
TiO_2	1.13	1.13	1.15	0.91	0.83	1.07
Al_2O_3	15.37	15.05	15.01	15.27	12.05	15.60
FeO*	8.93	10.39	10.32	12.15	10.44	10.20
MnO	0.21	0.21	0.21	0.21	0.22	0.24
MgO	5.67	5.55	5.40	4.63	9.39	5.50
CaO	7.60	5.84	7.14	8.68	16.73	7.77
Na_2O	3.17	4.92	3.52	2.60	1.34	4.84
K_2O	1.01	0.98	0.59	0.32	1.69	.20
P_2O_5	0.29	0.32	0.32	0.16	0.39	.17
Sum	100.00	100.00	100.00	99.93	100.69	99.27
			Trace Elements (ppm)			
Ni	58	67	81	288	129	221
V	383	370	354	411	337	279
Rb	16	13	8	2	1	26
Sr	381	281	236	290	96	673
Ba	421	309	197	79	11	450
Zr	60	79	83	24	67	49
Y	26	28	28	28	27	16
Nb	4	6	5	7	1	1

Dinagat Island samples DIN-5, 3C, 3B, 10A
Nonoc Island MIN-110
Doot Island MIN-120, pillow basalt
Abbreviations: D = diabasic textured andesite; G = micro-gabbro; B = basalt; BA = basaltic andesite; Mg# = Mg/(Mg + Fe).

Table 5—Dinagat, Nonoc, and Doot Islands, dikes and flows, chemical composition.

data for chromite in the peridotite intruded by the dikes.

Our data for crustal rocks on the southern end of the island are limited to the pillow basalts of Doot and Nonoc Islands, which are not distinctive enough to permit making a distinction between backarc basin crust or a deep seafloor origin. We favor a backarc basin origin because of the close proximity to the Dinagat Island arc-series rocks.

GEOLOGY OF THE SURIGAO AREA

The northeastern tip of Mindanao, near Surigao, is cut by the northwesterly trending Philippine fault that has dismembered and laterally translated large blocks and slivers of the eastern Mindanao collision complex. Rock types in this area (Fig. 9) include metamorphosed mafic rocks, peridotite, and gabbro similar to those exposed on the Pujada Peninsula and on the small islands off the north shore of Mindanao. This apparent continuity of distinctive rock types helps to link parts of the eastern Mindanao collision complex and gives support to the contention that it is a geologic terrane. The ultramafic rocks include dunite, pyroxene-rich peridotite (harzburgite), and serpentinite. Extensive Fe- and Ni-rich laterite covers much of the area (Esguerra, 1960, 1967). Gabbro dikes (probable arc-related magmas) cut the peridotite, and the peridotite is overlain by Eocene basalt and limestone (Esguerra, 1967). These in turn are overlain by Miocene clastic rocks that include conglomerate, coarse sandstone, and carbonaceous shale.

Metamorphic rocks are various metamorphosed mafic and ultramafic rocks including garnet amphibolite. The descriptions of these rocks, plus amphibolite samples we have studied, lead us to infer that they are equivalent to the metamorphic mafic rocks of Pujada Peninsula and Hanigad Island. That is, they probably represent metamorphosed cumulate rocks, formed at the base of ocean crust, and metamorphosed while the crust was still near solidus temperatures (young crust). We see no reason to consider these to be part of a distinct metamorphic terrane.

Our samples do not include any examples of melange, but Hamilton (1979) speculates that the "chaotic conglomerates" of the area may be post-Eocene melange comprising ultramafic rocks, fossiliferous Eocene limestone, and fragments of metamorphic rock. This interpretation seems consistent with our observations, and we speculate that the melange development may be related to the (mid-Tertiary?) collision between the eastern and central belts. We also recognize the possibility that the chaotic disruption of rock units could be a consequence of lateral slip along major faults and not necessarily related to plate convergence.

GEOLOGY OF THE BUKIDNON COMPOSITE TERRANE

The northward extension of the Sangihe volcanic arc marks the western boundary of the Agusan–Davao Trough; it forms part of Mindanao's Central Cordillera (Ranneft et al, 1960) comprising volcanic rocks, deformed clastic rocks, and crystalline "basement" (Figs. 2, 3). Mapping in Bukidnon Province by the Philippine Bureau of Mines (Villamor and Marcos, 1981) has delineated areas of metamorphosed igneous and sedimentary rocks; serpentinized peridotite; and mappable units of highly diverse rock types such as graywacke, siltstone, shale, volcanic rocks, pyroclastic rocks, and marble (probable melange) and deformed nearly monolithologic belts of sedimentary and volcanic rock (Fig. 13). The key to the original tectonic setting of these rocks is in their mineralogy and chemistry (Table 6) and is summarized in Table 7. We conclude that they were derived from an island arc setting and represent imbricated slabs and fragments of depleted mantle peridotite, island arc volcanic-plutonic rocks and clastic rocks derived from them, and rocks metamorphosed under low-temperature–moderately high-pressure conditions. The latter include derivatives of terrigenous clastic rocks and arc or sea-floor rocks; metamorphic conditions were transitional between lower greenschist and blueschist facies. These rocks represent several subterranes that have an earlier history of tectonic assembly. Plio-Pleistocene and Holocene basaltic-andesite from the modern volcanic arc caps the collision complex.

The internal structure of the rocks of this belt of deformed rocks (here informally called the Bukidnon collision complex) is not well known. The map pattern and attitudes of rock types suggest that the belt is formed of several imbricated layers or thrust slices that have been folded along northwesterly trending fold axes. Erosion has breached the folded thrust planes, exposing deeper layers in erosional windows. An alternative explanation may be that several large (tens of kilometers) elongated blocks of strongly folded rocks have been juxtaposed on northwesterly trending lateral-slip faults. We favor the first explanation in view of the collision history of the region of the Sangihe arc, but more mapping is needed before this can be proved.

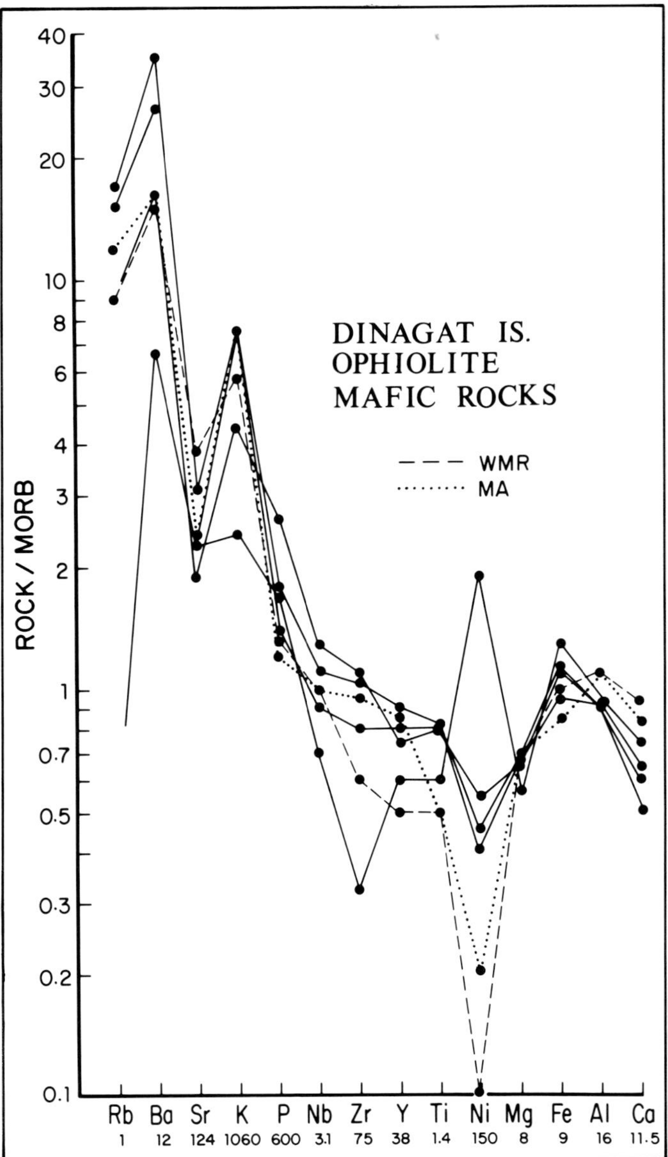

Figure 12—Element abundances in Dinagat Island basaltic andesite and micro-gabbro dikes normalized to element abundances in N-MORB (shown on abscissa). MORB-like samples should plot as straight line at 1 on ordinate. Data for Mariana Arc and West Mariana Ridge basaltic andesites closely parallel the Dinagat Island pattern and argue for an island arc origin of the Dinagat ophiolite.

Metamorphic Rocks

The rocks considered to be the oldest unit in the Bukidnon collision complex are schists, phyllites, and slates (Villamor and Marcos, 1981), but neither the age of metamorphism nor the age of the protolith are known. The metamorphic rocks include slate, phyllite, sericite schist, chlorite schist, and muscovite schist (Villamor and Marcos, 1981). In this report we give mineral data for a quartz-two mica schist collected by Villamor that has characteristics of metamorphic conditions transitional from lower greenschist to blueschist facies. The schist has a well-developed foliation formed by planar alignment of the mica and flattened lenticles of polycrystalline, mosaic-textured quartz grains. A second schistosity has developed along axial planes of flattened folds. The main minerals are quartz, colorless and green mica, and chlorite. Other minerals include porphyroblasts of albite, spessartine garnet, clinozoisite, Ti-magnetite, and graphite (?) (<3 modal percent of each). Apatite, tourmaline, and zircon are trace constituents (<1 modal percent of each). There are two micas (colorless and light green) that are phengite or solid solution series mixtures of muscovite and celadonite.

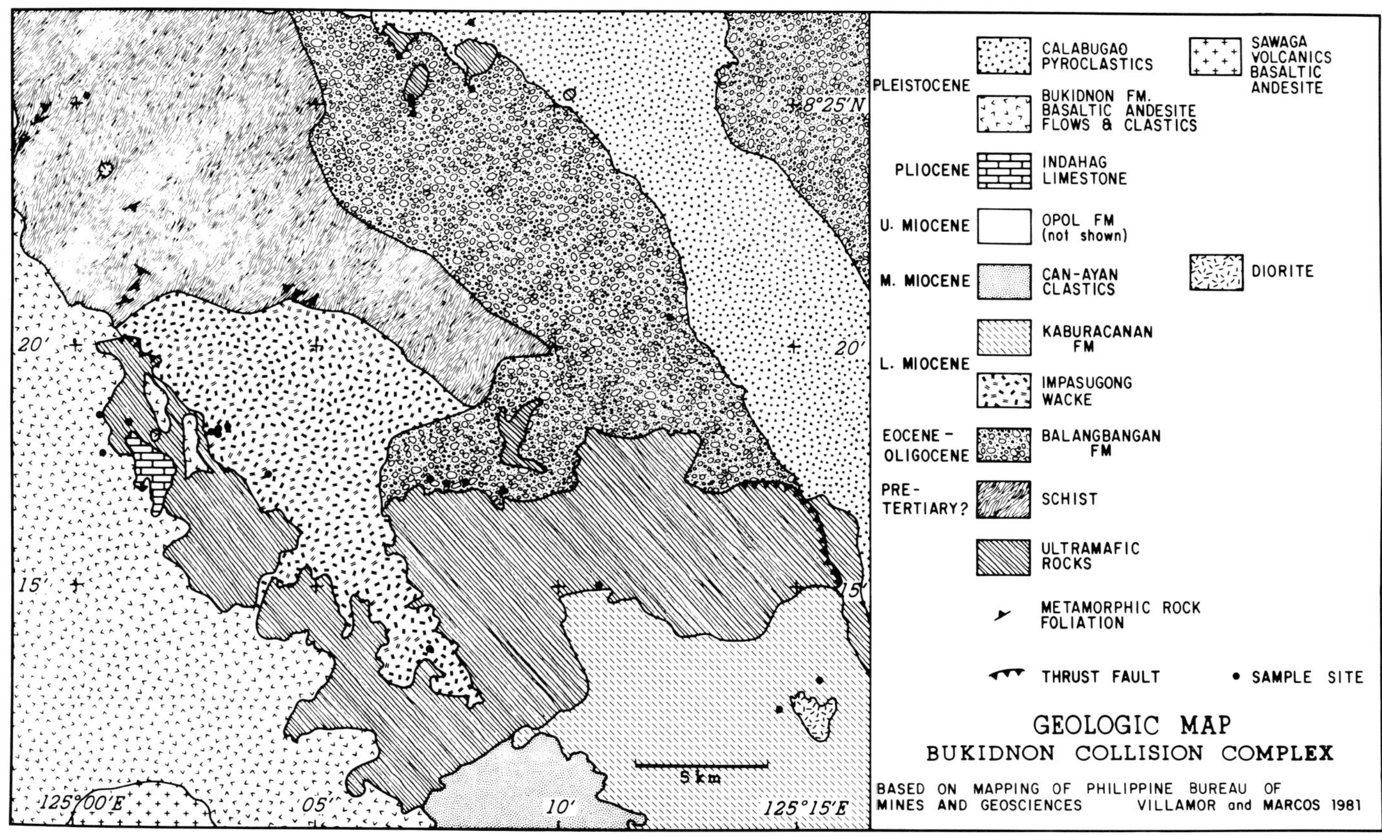

Figure 13—Geologic map of part of Bukidnon Province showing rocks of the central Mindanao collision complex. Ages shown are tentative and are based on comparisons with rocks from nearby areas. (After Villamor and Marcos, 1981. Copyright © 1981 The Philippine Bureau of Mines and Geosciences. Used with permission.)

Table 6

	1	2	3	4	5	6	7
SiO_2	48.59	24.99	36.14	36.51	45.20	55.87	—
TiO_2	0	0.08	0.24	0.05	0.44	0.07	0.05
Al_2O_3	27.34	21.49	20.57	20.63	10.57	2.13	46.38
FeO*	4.47	22.68	9.52	3.95	14.49	6.12	13.69
MnO	0.04	1.09	27.92	12.15	0.47	0.17	0.17
MgO	2.74	16.86	0.53	0.32	12.70	34.26	17.65
CaO	0.01	0	5.55	21.38	12.13	0.51	0.02
Na_2O	0.66	0	0	0	1.30	0	0
K_2O	9.91	0.02	0	0	0.36	0	0
Cr_2O_3	0	0	0.02	0	0.04	0.28	21.66
Sum	93.98	87.20	100.48	95.00	97.68	99.39	99.62

1. Phengite; quartz mica schist, basal spacing $d_{(001)}$ 9.9611 Å, optic angle, $2v\alpha$ 0–5°
2. Chlorite; quartz mica schist
3. Garnet; quartz mica schist; spessartine 63.3, pyrope 2.1, almandine 18.7, andradite 4.0, grossularite 11.9
4. Piedmontite; quartzite clast in melange unit
5. Actinolitic-hornblende; greenschist clast in melange unit (amphibole, oligoclase, epidote, sphene, magnetite)
6. Orthopyroxene, En_{90}, with Fo_{90-91} in serpentinized harzburgite, "backarc" mantle
7. Chromian spinel, serpentinized harzburgite, "backarc" mantle with opx #6 and Fo_{90-91}

Table 6—Microprobe analyses, key minerals in Bukidnon terrane collision complex.

Table 7

Rock type / minerals		Origin
A. Metamorphic Rocks		
1. Quartz-mica schist		Terrigenous quartzofeldspathic sediment
quartz	epidote	moderately high P fluid, low T
phengite	Ti-magnetite	(greenschist to blueschist)
Mg-chlorite	spessartine garnet	
Na-plagioclase	tourmaline	
2. Amphibole schist		
actinolitic-hornblende	chlorite	Intermediate composition
Na-plagioclase	sphene	volcanic or volcaniclastic rock
epidote		moderate P, low T
phengite		(greenschist)
1 and 2 in schist unit		
3. Piedmontite-rich laminated siliceous marble		Manganiferous pelagic sediment
quartz calcite piedmontite		moderate P, low T
4. Calc-silicate marble		
calcite	epidote	Marly limestone
grossularite garnet	quartz	moderate P, moderate T
amphibole		(lower amphibolite)
5. Amphibolite		
actinolitic hornblende	Fe-Ti oxide	Basalt
plagioclase	$An_{25-27}Ab_{71-74}Or_1$	moderate P, moderate T
epidote		(higher greenschist)
sphene		
3, 4, and 5 in "melange" unit (Balangbangan Fm.)		
B. Ultramafic Rocks		
1. Serpentinized harzburgite		
olivine	Fo_{90-91}	Depleted upper mantle
orthopyroxene	$En_{90}Fs_9Wo_1 - En_{89}Fs_{10}Wo_1$	
clinopyroxene	$En_{50}Fs_3Wo_{47} - En_{47}Fs_5Wo_{48}$	
chromite	Cr/(Cr + Al) 0.13 − 0.31	
	Fe″/Mg 0.36 − 0.45	
2. Serpentinite		
C. Volcanic Rocks		
1. Basaltic andesite (Impasugang Fm.)		Immature island arc
plagioclase	($An_{49}Ab_{50}Or_1$)	
clinopyroxene	$En_{42-46}Fs_{7-15}Wo_{41-47}$	
2. Andesite (Balangbangan Fm.)		
plagioclase	$An_{1-2}Ab_{98-99}Or_{0.5}$	
clinopyroxene	$En_{43-45}Fs_{12-14}Wo_{42-44}$	
amphibole, magnesio-hastingsite		
epidote		
Fe-Ti oxide		
3. Basaltic andesite (Calabagao Pyroclastics)		
plagioclase	$An_{61-73}Ab_{37-27}Or_1$	
clinopyroxene	$En_{43-45}Fs_{13-14}Wo_{42}$	
Fe-Ti oxide		
D. Graywacke (Balangbangan and Impasugang Fms.)		
plagioclase	amphibole	Clastic rocks from immature island arc
clinopyroxene	orthopyroxene	
chlorite	Fe-Ti oxides	
quartz	andesite clasts	

Table 7—Mineral assemblages and origin of rock types, Bukidnon collision complex, Mindanao.

collision zone, Indonesia: Journal of Geophysical Research, v. 83, p. 1681-1691.

Teves, J. S., et al, 1951, Reconnaissance geology of Agusan: Philippine Geologist, v. 5, p. 24-40.

Velde, B., 1965, Phengite micas: Synthesis, stability and natural occurrence: American Journal of Science, v. 263, p. 886-913.

Vergara, J. V., and F. D. Spencer, 1957, Geology and coal resources of BisligLingig region, Surigao: Philippine Bureau of Mines Special Project Series 14, p. 1-63.

Villamor, R., and D. M. Marcos, 1981, Preliminary report on the geology of Tankulan, Dumalaguing, Sumilao and Kalasungay quadrangles: Phillipine Bureau of Mines and Geosciences Regional Office X, Surigao City, 36 p.

———, et al, 1984, Report on the geology of Sigaboy, Tugabili and Cape San Augustin quadrangles, southern Pujada Peninsula, Davao Oriental: Philippine Bureau of Mines and Geosciences, Regional Office X, Surigao City, 39 p.

Wolfe, J. A., 1981, Philippine geochronology: Journal of the Geological Society of the Philippines, v. 35, n.1, p. 1-30.

Wright, E., et al, 1981, East Mindanao ophiolite belt: Petrology of volcanic series rocks: EOS, Transactions of the American Geophysical Union, v. 62, p. 1086.

Tectonostratigraphic Terranes, Pacific Southwest Quadrant

Suspect Terranes and Cambrian Tectonics in Northern Victoria Land, Antarctica

J. D. Bradshaw
S. D. Weaver
University of Canterbury
Christchurch, New Zealand

M. G. Laird
New Zealand Geological Survey
Christchurch, New Zealand

The early Paleozoic rocks of northern Victoria Land comprise at least three and possibly five suspect terranes. The key unit is the Bowers terrane, a narrow (35 km [22 mi]) fault-bounded strip of folded Cambrian sediments and volcanics that can be traced for 350 km (217 mi) from the Southern Ocean to the Ross Sea. To the west, amphibolitic paragneiss in the Lanterman Range and lower grade quartzose flysch in the Daniels and Morozumi Ranges are host to Cambro-Ordovician migmatites and granitoid plutons. The paragneiss and flysch-type successions do not occur together and are tentatively regarded as two suspect terranes. East of the Bowers terrane lie two further suspect terranes, an extensive nonschistose quartz-rich flysch sequence (Robertson Bay Group) and a narrow fault-bounded strip of schistose rocks that separates the Robertson Bay rocks from the Bowers terrane. The narrow strip has structural and metamorphic features not seen in either of the adjacent terranes.

The Bowers terrane contains three units, a lower volcanosedimentary unit 3.5 km (11,500 ft) thick overlain gradationally by 2.5 km (8,200 ft) of Middle and early Upper Cambrian mixed marine sediments and followed with a slight discordance by more than 4 km (13,000 ft) of fluviatile rocks, mainly quartzites. The volcanics are pillow lavas, hyaloclastites, and volcanic debris flows, predominantly basalt and andesite with subordinate dacite and rhyolite. The basalts have the chemistry of island arc tholeiites (low Ti, Zr, Y) and are primitive melts with high Mg, Cr, and Ni. Andesites show affinities with high-Mg andesites (or boninites) and were probably fractionated from basalt in central complexes below island volcanoes.

The volcanic rocks strongly suggest arc volcanism in an intraoceanic setting and therefore their current position seems highly anomalous. In particular, they lie between areas of broadly contemporaneous Cambrian rocks of continental or continent-derived character, they are faulted against paragneiss that is cut within 6 km (3.7 mi) by mature "S" type granitoids typical of arcs in areas of continental crust, and they are overlain by a generally regressive sequence that culminates in thick fluviatile quartzites.

It is difficult to account for all the features of the Cambrian geology of north Victoria Land with a simple convergent margin model, and we propose a model involving strike-slip modification of an arc that varied from intraoceanic to intracontinental along its length. There are marked similarities between the model and aspects of the current geology of New Zealand and particularly the Andaman-Burma sector of the Sunda arc. The adoption of an allochthonous terrane hypothesis is also significant with respect to the relationship of the Cambrian rocks of Australia and Antarctica in a united Gondwana continent. A speculative reconstruction of the Cambrian continental margin that incorporates our model is presented.

INTRODUCTION

The concept of terrane analysis, with its emphasis on the identification of units with distinctive geological histories, may prove valuable in Antarctica and a useful alternative to the older tendency to synthesize across great distances, a process which led to such unwieldy concepts as the Ross Geosyncline. However, in Antarctica many unit boundaries are concealed by major glaciers and may or may not be tectonic, so that there is a real danger of overenthusiastic terrane designation. Most units are, and must always remain, suspect terranes, and Antarctic terrane maps will be controversial. In north Victoria Land, however, where adjacent belts have rocks of distinctly different character, we feel that the terrane concept is valuable in focusing attention on outstanding problems, particularly because more conventional interpretations of the Cambrian of the region have severe problems.

MAJOR UNITS

Traditionally in northern Victoria Land, the 400 km (249 mi) wide belt of exposed Late Precambrian and Paleozoic sedimentary rocks between the Polar Plateau and the Ross Sea has been divided into three major units, plus some minor ones of uncertain affinity (see Gair et al, 1969). These comprise from west to east: (1) Wilson Group, metasediments and paragneiss of possible Late Precambrian or Cambrian age with migmatites and granitoids of the Cambro-Ordovician Granite Harbour

Intrusives; (2) Bowers Supergroup, folded Cambrian and ?Ordovician sediments and volcanics; and (3) Robertson Bay Group, extensive quartzose flyschlike sediments of ?early Paleozoic age cut by Devonian granitoids. All these units were considered to be components of the Ross Geosyncline deformed and metamorphosed during the Cambro-Ordovician Ross orogeny and locally by the Devonian Borchgrevink orogeny. A subdivision into three fault-bounded units termed the Rennick, Bowers Trough, and Robertson Bay tectonic zones was proposed by Grindley and Oliver (1983). Work in the early 1970s showed marked differences between the three belts (Laird et al, 1976; Laird, 1982; Bradshaw et al, 1982, Wodzicki et al, 1982); however, wide-ranging reconnaissance mapping reported by Tessensohn et al (1981) suggested that the differences had been overemphasized and metamorphic gradations and facies transitions had been overlooked. These differences of interpretation (see Bradshaw and Laird, 1983) arise in part from an examination of different sections and in part because the recognition of only three major belts is too simplistic.

Bowers Terrane

The distinctive Bowers terrane is the key unit and forms a belt 15 to 35 km (9–22 mi) wide extending for 350 km (217 mi) across northern Victoria Land from the Southern Ocean to the Ross Sea (Fig. 1). Lower Paleozoic stratified rocks in the terrane are united in the Bowers Supergroup, which exceeds 10 km (32,800 ft) in thickness and comprises Sledgers Group (oldest), Mariner Group, and Leap Year Group (Fig. 2).

The Sledgers Group, which is at least 3.5 km (11,500 ft) thick and notably volcanogenic, consists of two interfingering units, the Glasgow and Molar Formations. The base of the Sledgers Group is unknown except perhaps west of Reilly Ridge (Fig. 1) where there is polymict conglomerate ("Husky Conglomerate") with clasts of Lanterman terrane rocks (with retrograde greenschist mineralogy) and basaltic blocks very similar to those of the Glasgow Formation. An unconformity with Wilson Group is seen only in a fault sliver in the Lanterman Fault Zone, but the conglomerate is not known in continuity with the Glasgow Formation proper. Gibson (1984, personal communication) suggests that the conglomerate postdates the Glasgow Formation and is partly derived from it.

The Glasgow Formation includes flows of basalt, andesite, and (less commonly) rhyolite, and closely associated volcanic breccias. It is at least 2.5 km (8,200 ft) thick at Mt. Glasgow. Pillow lavas are common in the Edlin Neve–Sheehan Glacier area and at Mt. McCarthy; elsewhere, the formation is chiefly represented by volcanic debris flow deposits. Most of the formation appears to have accumulated under marine conditions.

The Molar Formation, which occurs in thick sequences (up to 1.3 km [4,300 ft]), interfingers with the Glasgow Formation at many localities. It consists dominantly of dark mudstone and thin beds of fine sandstone but with intercalated conglomerates. Clasts in the latter consist mainly of volcanic and intrusive rocks (about 70%) with basaltic fragments comprising about half of the igneous clasts; the remainder includes andesites, dacites, rhyolites, and granitoids. Clasts of metamorphic quartz and orthoquartzite have also been recognized. A 10 m (33 ft) thick laterally persistent limestone bed occurs within the Molar succession in the lower Carryer Glacier and on the western slopes of Mt. Soza. The lower Carryer exposures also show several thick mass flow units, one of which, near the top of the succession, consists of mudstone with large rafts of basalt and limestone up to 10 m (33 ft) in length. The limestone contains probable birdseye structures, which are a feature of marginal marine environments. The age of the Sledgers Group from sparse macrofossils is Middle Cambrian (Cooper et al, 1983).

The concordantly overlying Mariner Group (late Middle Cambrian to Late Cambrian) comprises at least 2.5 km (8,200 ft) of fossiliferous sandstone, calcareous mudstone, and limestone forming a largely nonvolcanogenic regressive sequence with shallow and marginal marine sediments at the top (Andrews and Laird, 1976).

The Leap Year Group rests on a marked erosion surface that truncates successively older horizons of the Mariner Group towards the northwest. In the lower Carryer Glacier and west of Mt. Soza, the Mariner Group has been completely removed by erosion and the Leap Year Group rests directly on the Sledgers Group. The Leap Year Group comprises the local and probably slightly older polymict Carryer Conglomerate, the quartzose Reilly Conglomerate, and the Camp Ridge Quartzite. The Group is predominantly fluviatile and reaches a thickness of at least 4 km (13,000 ft) in the Leitch Massif (Fig. 1) and could be as much as 7 km (23,000 ft) thick. No shelly fossils are known from the Leap Year Group, but its stratigraphic position, trace-fossils, and radiometric data are consistent with a Late Cambrian to Ordovician age (Adams et al, 1982).

Depositional Environment

Paleocurrent data for the Molar Formation indicate a dominant transport direction (within the finer grained sequences) towards the southeast, although there is a reversal of the trend in the Mt. McCarthy area. By contrast, coarser sandstones and conglomerates associated with channelized fining-upward sequences show a strong transport component from the southwest and northwest. The finer grained sediments are interpreted as axial deposits and the coarser grained mass-flow sediments as laterally derived deposits infilling channels on a basin slope with a volcanic plus high-grade metamorphic/granitoid source terrane lying to the southwest.

Paleocurrent data from the Mariner Group are sparser; however, trends both from the southwest and, in the head of the Mariner Glacier, from the southeast appear to dominate. Slumps and limestone clast-dominated debris flows in the Reilly Ridge area show downslope movement towards the northeast, again suggesting a basin margin towards the southwest. Paleocurrent trends in the Carryer Conglomerate also indicate a direction of transport towards the east or northeast. Data from the Camp Ridge Quartzite indicate that the transport direction was towards the northwest or north-northwest south of the Molar Massif, swinging round towards the northeast, north of this.

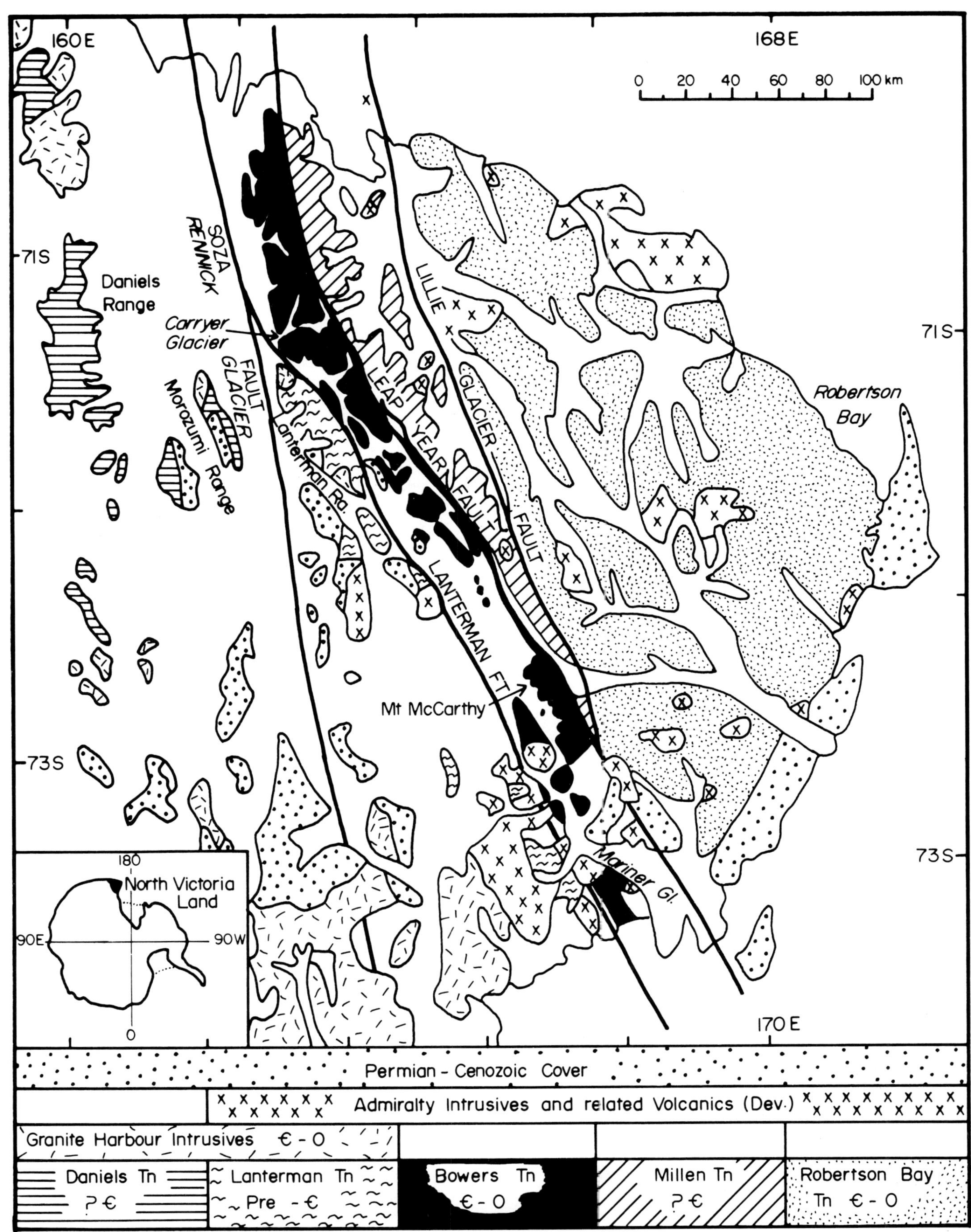

Figure 1—Terrane map of northern Victoria Land.

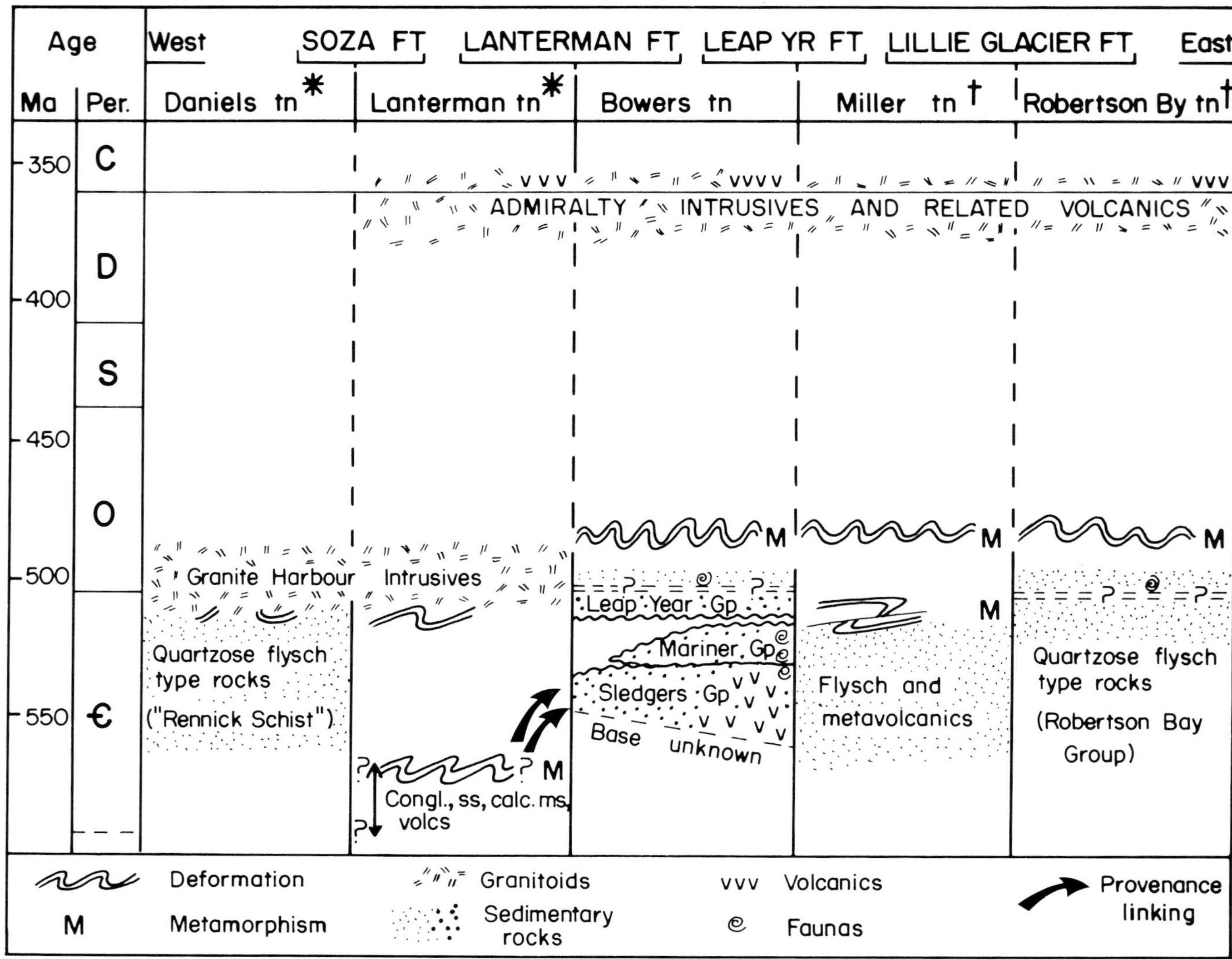

Figure 2—Terrane diagram for northern Victoria Land. Note that the case for distinction between the pairs of terranes marked * and †, respectively, is less powerful than that for the separation of the Bowers terrane from all other units.

Active subsidence of the Bowers zone allowed deposition of over 3.5 km (11,500 ft) of Sledgers Group comprising volcanic debris flows, subsidiary pillow lavas erupted from sea mounts, and interfingering sedimentary sequences. Polymict debris flows and channel fills indicate a continental source to the southwest. The mainly fine-grained and largely volcanic-free Mariner Group records the cessation of volcanism, slowing of subsidence, and the progradation of a sediment wedge indicated by a regressive sequence culminating in a tidal-flat environment (Andrews and Laird, 1976). An active basin margin and slope lay to the southwest, indicated by limestone-rich debris flows and channelized sequences. Convincing evidence of a northeast margin is lacking. Leap Year Group sediments suggest major paleogeographic changes with the uplift of granitoid-metamorphic source areas to the southwest and the southeast.

Structure

The Bowers terrane is characterized by folds and faults subparallel to the Leap Year Fault (Fig. 1). Although the Leap Year Group is unconformable on older rocks, the discordance is slight and post-Leap Year Group structures are dominant. The persistent Camp Ridge syncline (200 km + [124 mi] long) is an upright, gently plunging structure. To the west, axial surfaces dip northeast, and in the northeast they dip steeply to the southwest, thus forming a broad fan. Older, sometimes steeply plunging, folds in Mariner and Sledgers slates are in some cases synsedimentary and in other cases due to non-plane strain.

Western Terranes

West of the Bowers terrane the old Wilson Group appears to comprise two discrete units, the Lanterman terrane in the east and the Daniels terrane in the west.

The Lanterman terrane consists of paragneiss of at least amphibolite facies (Wodzicki et al, 1982) showing polyphase deformation with high strain and intruded by Cambro-Ordovician Granite Harbour intrusives. K-Ar isotopic ages mainly reflect this event (Adams et al, 1982; Kreuzer et al, 1981), and the age of metamorphism may be significantly older (Adams, 1983, personal communication).

Sedimentary protoliths include sandstone, conglomerate, mudstone, calcareous mudstone, and basic volcanics. Gibson (personal communication) believes that the conglomerates are younger and unconformable on the Lanterman metamorphic complex. The boundary with the Bowers terrane is the Lanterman Fault Zone (Bradshaw et al, 1982). The eastern part of the Lanterman terrane shows retrogressive metamorphism towards the fault zone, and major shears with abundant talc, magnesite, and actinolite occur parallel to it. Faults also occur within Bowers rocks. Provenance linking is indicated by blocks of conglomerate of Lanterman type in younger conglomerate in the Bowers terrane, which suggests proximity of parts of the Bowers and Lanterman terranes by Middle Cambrian at the latest.

The Rennick Glacier is the site of important Cretaceous and younger faulting that may follow the line of an older tectonic boundary. To the west in the Morozumi and Daniels Ranges, a quartzose flyschlike succession is host to a series of granites and migmatites of Granite Harbour type (Kleinschmidt, 1981; Plummer et al, 1983). In detail both the original sedimentary rocks and the structural/metamorphic history (Kleinschmidt and Skinner, 1981) differ from that of the Lanterman Range, and because it has been argued (Tessensohn et al, 1981) that the sediments are Robertson Bay Group, it seems reasonable to separate a Daniels suspect terrane.

The Lanterman terrane lies east of the Rennick Glacier in the Lanterman and Salamander Ranges and may extend to the south towards Terra Nova Bay where Skinner (1983) maps areas of Snowy Point paragneiss of probably Late Precambrian age. The Lanterman terrane appears to terminate against the Lanterman Fault Zone and is not known north of the Sledgers Glacier. The Daniels terrane west of the Rennick Glacier forms the Usarp Mountains, the Wilson Hills, and may extend to the northwest and include the "Berg Group" of Ravich et al, 1965.

Both Daniels and Lanterman terranes are intruded by mainly "S" type granitoids of the Granite Harbour Intrusives with maximum ages (Rb-Sr whole rock) of slightly over 500 Ma (Vetter et al, 1983; Adams, personal communication). K-Ar ages indicate a contemporaneous metamorphic event in Daniels metasediments (Adams, personal communication).

Eastern Terranes

The area previously mapped as Robertson Bay Group, which abuts the Bowers terrane along the Leap Year Fault (or fault zone) contains two belts of rock that Findlay and Field (in press) suggest are discrete terranes. Immediately to the east, between the Leap Year Fault and the Lillie Glacier Fault, the Millen terrane comprises schistose metasedimentary and metavolcanic rocks with early isoclinal folding and thrusting not seen in either adjacent terrane. The Millen terrane is only 40 km (25 mi) wide in the north and narrows southward. The age of the rocks is uncertain, but available K-Ar dates suggest that it is older than 500 Ma (Adams et al, 1982; Adams, personal communication). Further research is necessary to establish the affinities of the parent rocks of the Millen terrane. At present it is recognized mainly on the basis of a distinctive deformational history and the presence of more metavolcanic units than in typical Robertson Bay rocks.

East of the Lillie Glacier Fault the Robertson Bay terrane extends over 140 km (87 mi) eastwards to Robertson Bay. The terrane is underlain by a nonschistose flyschlike succession of quartzose sediments derived from a continental source and deposited on a submarine fan facing north or northwest (Wright, 1981; Field and Findlay, 1983). The Robertson Bay rocks are at lower metamorphic grade and show a simpler structural history than the Millen terrane rocks, with a single phase of northwest-trending upright folds.

The Robertson Bay Group has been considered to be Late Precambrian-Middle Cambrian (Wright, 1981; Cooper et al, 1983), an age consistent with K-Ar geochronology (Adams et al, 1982; Kreuzer et al, 1981). Recent isolation of earliest Ordovician (Tremadocian) faunas from limestone blocks in a debris flow within the Robertson Bay terrane is surprising (Burrett and Findlay, in press; Wright et al, in press). Although these rocks appear to be intercalated in the Robertson Bay Group, the possibility remains that they are a separate unit.

IGNEOUS GEOLOGY

In attempting to work out the geological development of northern Victoria Land, the igneous rocks have proved particularly helpful and are therefore described separately below.

Bowers Terrane Volcanics

The mainly Middle Cambrian Glasgow Formation is found throughout the length of the Bowers terrane. The formation comprises volcanic breccia, probably emplaced by submarine debris flows, pillow breccia, pillow lava, massive lava, and rare welded ignimbrite. The lower part of the succession is mainly andesite and dacite with subordinate basalt and rhyolite. In the west the upper part is similar, but in the east basalts predominate and andesites are subordinate.

Petrographically, many lavas have well-preserved primary mineralogy and texture. Metamorphic alteration, variable in intensity, has produced the assemblages albite + sericite + epidote + chlorite + calcite ± prehnite ± pumpellyite ± actinolite, indicating prehnite pumpellyite facies conditions. There is no indication that the metamorphism has an unrecognized influence on geochemistry.

Basaltic rocks are tholeiitic and are either aphyric or have phenocrysts of clinopyroxene, ± olivine pseudomorphs ± calcic plagioclase in an altered glassy mesostasis that sometimes shows quench textures. The clinopyroxene phenocrysts are endiopsideaugite ranging from $Ca_{39}Mg_{55}Fe_6$ to $Ca_{41}Mg_{43}Fe_{16}$. There is no olivine in the groundmass, but red-brown magnesiochromite is characteristic. The latter have a $Cr/Cr + Al = 0.79–0.70$, similar to several ophiolites but unlike those from MORB, which are typically 0.6–0.2 (Cameron et al, 1979; Coish and Church, 1979). Andesites have phenocrysts of intermediate plagioclase ± hornblende ± orthopyroxene.

The geochemistry of 48 samples of Glasgow Formation samples are presented and discussed in Weaver et al (1984). Briefly, the majority of the basalts are high in Mg, Cr, and Ni, have low FeO/MgO, and notably low Sr,

K, Rb, Nb, P, Ti, and Y. A single young flow in the southwest stands out with high K, Rb, P, Zr, Ti, and Y relative to the others, and several flows transitional between this composition and that of the majority have been identified among the youngest basalts in the north. Andesite and dacites are comparable with the majority of basalts with low Sr–Rb and Nb–Y. MgO is high in relation to SiO_2 in some andesites.

Comparison of the Glasgow Formation with many other complexes indicates that the basic rocks are island arc tholeiites of a primitive type probably developed in an intraoceanic setting. The andesites and dacites are of similar origin and derived by fractionation of low-Ti basalt in central complexes below a largely submarine arc. The late high-Ti basalt and transitional types may indicate incipient intra-arc rifting or marginal basin development just prior to cessation of arc development.

Granitic Rocks

Two suites of Paleozoic granitoids occur in northern Victoria Land. The Devonian Admiralty Intrusives are discordant plutons (see Grindley and Oliver, 1983) that cut all units at least as far west as the Lanterman terrane. This suite is related to a tectonic pattern younger than the one under discussion.

The older suite, the Granite Harbour Intrusives, is the local representative of Late Cambrian–Early Ordovician granitoids of varied type found throughout the Transantarctic Mountains. In northern Victoria Land these are restricted to the Daniels and Lanterman terranes and are not known within or east of the Bowers terrane.

In the Daniels Range a complex succession of pretectonic, syntectonic, and posttectonic granitoids (Plummer et al, 1983) are seen within an original continental crustal section at least 8 to 10 km (26,000–33,000 ft) thick (Babcock et al, 1983). The granites are predominantly mature "S" type granites (Plummer et al, 1983; Vetter et al, 1983; Wyborn, 1983) typical of continental arcs in crust of normal thickness, and it is therefore surprising that similar Granite Harbour rocks occur in the Lanterman terrane to within 6 km (4 mi) of the margin of the Bowers terrane with its primitive lavas.

SYNTHESIS AND CONCLUSIONS

Northern Victoria Land

Six salient points bear on the geotectonic synthesis of north Victoria Land and are summarized below.

1. The volcanic rocks of the Bowers terrane are primitive arc types developed in an intraoceanic or very thin continental crustal setting. Their juxtaposition with mature "S" type granitoids of the Lanterman and Daniels terranes is strongly anomalous and suggests regional telescoping or amalgamation of terranes.

2. The Bowers terrane volcanics show little variation over 300 km (186 mi) along the terrane, and it is therefore likely that the terrane is subparallel to the original arc (northwest-southeast). Perpendicular to the strike of the arc probable Cambrian continent-derived sediments extend 200 km (124 mi) northeastward to the margin of the Antarctic continent. To the southwest lies the Antarctic Shield. The Bowers terrane volcanics are clearly out of place.

3. The unfolded width of the Bowers terrane, not more than 60 km (37 mi), is small for a typical arc. The interfingering relationship of volcanics with quartzose sediments and conglomerate in the Bowers Group suggests access to a continental source and that the preserved portion is the continental or backarc side.

4. After the termination of volcanism there was further subsidence of the Bowers terrane to accommodate the 2.5 km (8,200 ft) Mariner Group. There is evidence of an active (?faulted) margin to the southwest but no basin margin to the northeast. The sequence is regressive. After tilting and erosion, at least 4 km (13,000 ft) of mainly fluviatile quartzites were deposited, again showing evidence of continental sources to the southwest and southeast.

5. The Millen terrane, between the Leap Year and Lillie Glacier Faults, shows a period of strong deformation that predates the northwest-trending upright folding common to the Bowers, Millen and Robertson Bay terranes (Findlay and Field, 1983).

6. The available K–Ar geochronology suggests a very similar history of metamorphism and uplift in all terranes from about 500 Ma or a little earlier, suggesting that all elements were close to the present relative position by that time. Integration of the scattered fossil ages and geochronology is made difficult by uncertainties about the age of the base of the Ordovician. Estimates range from 490 ± 9 Ma (Gale et al, 1979) to 505 ± 15 Ma (Harland et al, 1982). Odin (1982) believes that the base of the Cambrian could be as young as 530 ± 10 Ma. These uncertainties affect the interpretation of the significance of the Tremadocian localities and the relative age of the Bowers terrane volcanics to the older phases of the Granite Harbour Intrusives.

Given the existing constraints it is difficult, if not impossible, to explain the Cambrian development of north Victoria Land in terms of a simple convergent margin model. Development of orogenic belts involving large-scale strike-slip has been widely recognized both around the Pacific and within older continents (e.g., Badham, 1982), and seems likely to be the case here. If the sediments of the Daniels terrane are indeed Robertson Bay Group, as proposed by Tessensohn et al (1981), a hypothesis involving large-scale strike-slip movement seems inescapable.

The tectonic scheme outlined in Figure 3 is not a unique solution and will require modification or replacement as exploration proceeds. The essence of the model is a volcanic arc related to a convergent margin that changes in character from intraoceanic to continental along its length (Fig. 3), much as the Tonga–Kermadec arc changes as it intersects the New Zealand continent. In the late Middle Cambrian, oblique subduction is thought to have led to dextral strike-slip faulting within and between elements of the arc, and in particular to the displacement of the outer margin of the continent and its replacement by a primitive arc (Fig. 2). The suggestion (Gibson, 1984, personal communication) that the Husky Conglomerate is derived from the Bowers volcanics rather than basal to them is important in this context. Large-scale strike-slip movements subparallel to the arc are currently taking place in New Zealand and the Sunda Arc (Fig. 3). It is also interesting to note that backarc spreading in the Andaman Sea is taking place directly adjacent to the submarine volcanic arc and that such an arrangement might explain the chemically anomalous compositions of the youngest Bowers terrane lavas.

The nonvolcanic Mariner Group has a geology consistent with development on a subsiding arc adjacent to continental crust (Fig. 2). The unconformity and lithological contrast between the Mariner and Leap Year Groups suggest a changing tectonic setting and the development of major continental source areas to the south and west. The sedimentation of the Robertson Bay Group is almost certainly coeval with much of that of the Bowers Supergroup, but differences in environment suggest it cannot have been adjacent as it is now (Fig. 3). It is likely that the outer parts of the Bowers terrane were removed and replaced by submarine fan deposits over a period that overlapped with the deposition of the Leap Year Group (Fig. 3d). The chronology of events in the continental magmatic arc (granite Harbour Intrusives) has not been resolved, but it is possible that the early migmatites of the Daniels terrane are coeval with the Bowers terrane volcanics, whereas the younger plutons are similar in age to widespread deformation and metamorphism in all terranes.

The model proposed has similarities to the present northern part of the Sunda Arc (Fig. 3). Clearly the latter is much more complex and in particular is strongly influenced by the impingement of the Bengal Fan on the trench with the consequent growth of a massive forearc ridge. However, other aspects such as the relative position of delta, shelf, submarine fan, and volcanic arc, the dispersal patterns of continental and volcanic sediment, the dextral strike-slip tectonics, and the longitudinal change in volcanic arc setting show that our Cambrian model is reasonably actualistic.

Antarctica-Australasia

It is not possible to trace the north Victoria Land terranes outside the region discussed. Nevertheless, comparisons with the remainder of the Transantarctic Mountains to the south and southern Australia to the north suggest that broadly similar tectonic patterns occur. Summaries of the geology of these areas can be found in Cooper and Grindley (1982), Laird and Bradshaw (1982), Laird (1981, 1982), Grindley and Davey (1982), and in Oliver et al (1983), particularly section 2.

Late Precambrian and early Paleozoic rocks form the basement throughout the Transantarctic Mountains and are overlain by Devonian to Jurassic flat-lying sediments except in northern Victoria Land where volcanics and granitoids of Devono-Carboniferous age also occur. The present mountain range appears to cross-cut the major tectonic elements (Fig. 4), but the persistence of Late Cambrian–Early Ordovician granitoids for over 2,500 km (1,553 mi) from the Rennick Glacier to the Thiel Mountains suggests an extensive magmatic arc and convergent plate boundary. Cambro-Ordovician deformation and granitoids are the hallmark of the "Ross orogeny." This convergent margin appears to have extended into Australia where deformation and granitoids of the same age and type typify the "Delamerian orogeny" of eastern South Australia.

In Antarctica, the older Precambrian crust is known only at the Adelieland coast and in the western margin of the Central Transantarctic Mountains (Fig. 4). In the Late Proterozoic (Riphean–Vendian) a major sedimentary complex, including both immature turbidites and mature quartzite/carbonates, developed near this margin. Strong deformation, particularly in the east, calc-alkaline volcanism (Queen Maud Mountains), and possible Vendian granitoids (Queen Maud Mountains, southern Victoria Land, Terra Nova Bay, etc.; see Skinner, 1983a, 1983b) strongly suggest that the Beardmore orogeny represents a Vendian convergent margin.

Subsequently in the Early Cambrian a new rapidly subsiding trough appears to have developed within the Beardmore complex of the Central Transantarctic Mountains in which 8 km (26,000 ft) of shallow marine sediments including up to 5 km (17,000 ft) of carbonates were laid down (Byrd Group). Further to the east, clastics and silicic volcanics interpreted as arc rocks (Liv Group) (Stump, 1982; Borg, 1983) overlie an eastern rim of Precambrian rocks. Late Cambrian–Ordovician deformation and intrusion of granitoids is widespread in the (eastern) outer belt but is much less common to the west.

Prior to the Ross–Delamerian orogeny a similar pattern has been recognized in Australia. In South Australia varied Cambrian rocks represent a shelf and rapidly subsiding trough (Kanmantoo Trough 6 km+ [20,000 ft]) developed on the southern and eastern margin of the Gawler Craton. The trough was bounded on the east by the Precambrian Wilyama Block which in turn, on its eastern margin, was the site of mixed Cambrian sedimentary and volcanic rocks (Mt. Wright volcanic belt). The latter is thought to represent an arc near the margin of the Cambrian ocean (Scheibner, 1972; Powell, 1983). The continental margin probably passed southward to the east of the Glenelg Belt and Cambro-Ordovician Wando Granodiorite of Victoria.

As in Antarctica, the Cambrian appears to be marked by thick sequences developed in extensional basins behind a magmatic arc developed in Precambrian crust. However, it must be stressed that there is no true equivalent of the Beardmore orogeny in South Australia and that the Cambrian rocks of north Victoria Land form part of a distinctly different sedimentary-volcanic complex.

In summary, in the Cambrian the eastern margin of the Australian–Antarctica continent was a convergent plate boundary. Extensional troughs with thick sedimentary sequences developed along much of its length (Fig. 4). These troughs which were probably discontinuous and quasi-independent, include the Cambrian of the Pensacola Mountains, Queen Maud Mountains, Central Transantarctic Mountains, perhaps northwestern Victoria Land (Daniels terrane), and South Australia (Fig. 4). To the east, remnants of the discontinuous outer ridge include the eastern Queen Maud Mountains, southern Victoria Land, Terra Nova Bay, Lanterman terrane, and the Wilyama Block. Still further east lay the Cambrian ocean

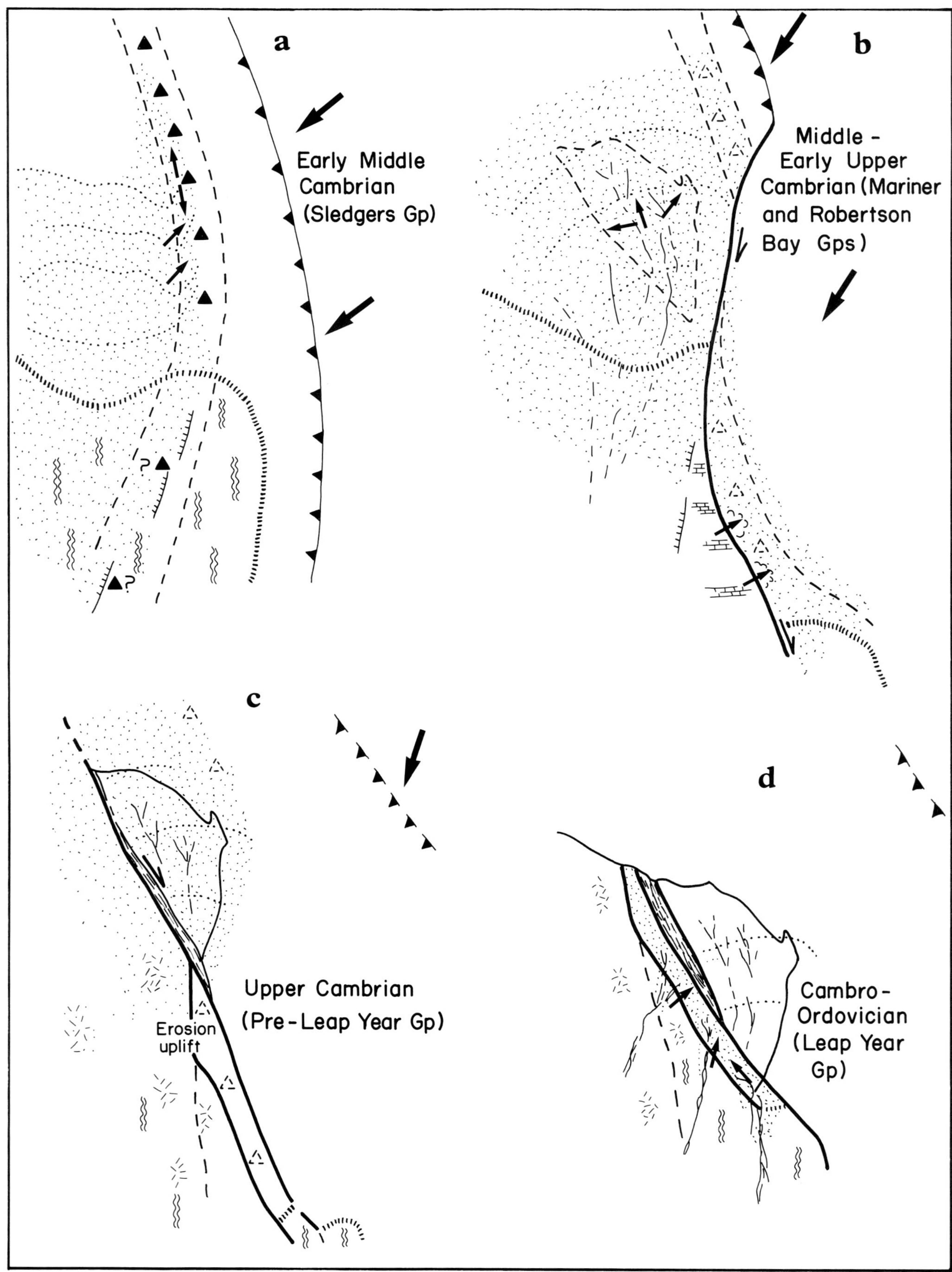

Figure 3—(a–d) Schematic model for the Cambrian evolution of northern Victoria Land.

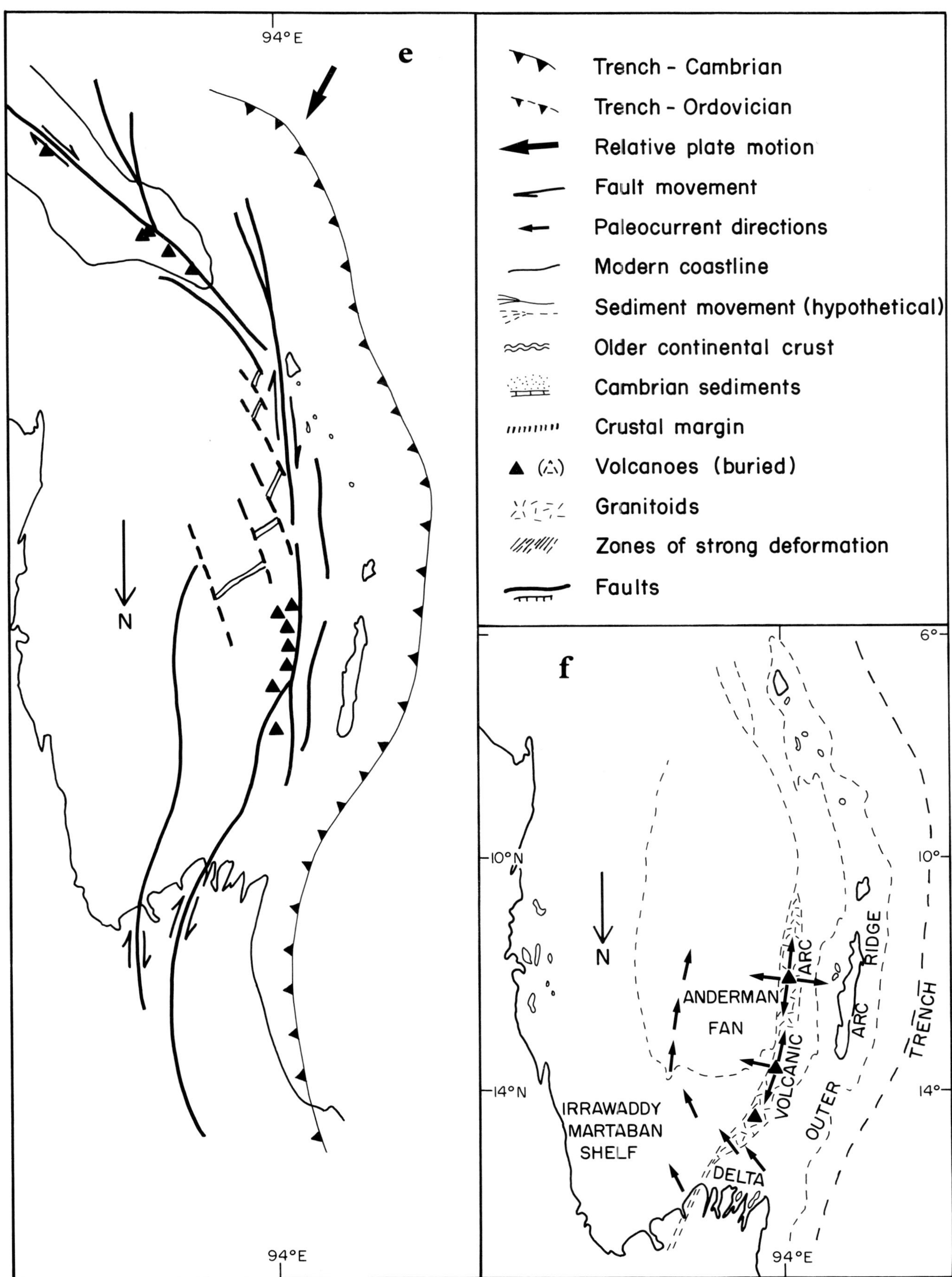

Figure 3—(e–f) Comparative figures of the neotectonics and major sedimentary environments and sediment dispersal patterns in the northern part of the Sunda arc (based on Curray et al, 1979, Karig et al, 1980, and Cas et al, 1980). These latter figures have south at the top. The scale of (f) is similar to that of the north Victoria Land diagrams.

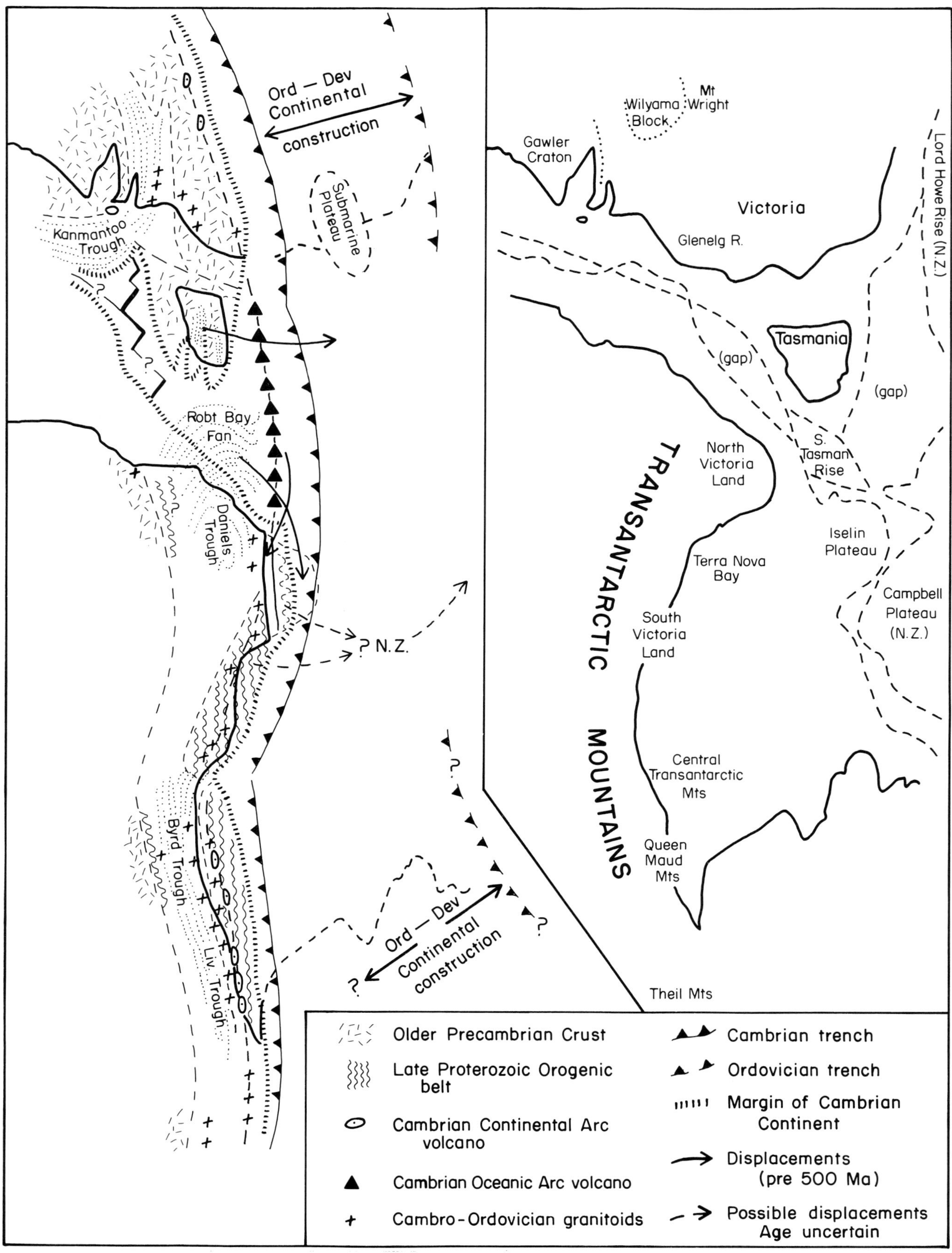

Figure 4—A speculative reconstruction of the Australia–Antarctic continental margin in the Cambrian. The Cambrian rocks of Tasmania and particularly New Zealand (off the eastern edge of 4b) now lie within continental crust developed during the Ordovician–Devonian.

basin now seen only as scattered remnants such as the Bowers arc, the basic igneous rocks (both arc and MORB types; Crawford, 1983) and pre-Ordovician clastics of Victoria (Cooper and Grindley, 1982), and the oceanic plateau or Proterozoic microcontinent believed to underlie central Victoria (Powell, 1983; Scheibner, this volume).

As Cooper and Grindley remark (1982, p. 72), reconstruction of early Paleozoic relations between Australia and Antarctica is difficult because of the paucity of convincing examples of geological links such as distinctive rock units or structural belts, and conversely the apparent similarity between regions not juxtaposed in any morphologically satisfactory reconstruction. The problem is even more severe if New Zealand is considered. Part of the difficulty stems from the use of a morphological fit as a basis for reconstruction because this overlooks the probability of large-scale displacement of terranes prior to breakup and the possibility of the line of separation following old terrane boundaries. Eastern Australia and New Zealand have been at or near convergent boundaries for much of the Phanerozoic and major pre-Carboniferous rearrangements were proposed nearly 10 years ago by Cooper (1975).

A loose assembly of Australia and Antarctica is used in Figure 4. Much of the Transantarctic sector seems to have been quiescent since the Ordovician, but Devonian arc-related magmatism is widespread in northeastern Victoria Land, Tasmania, Victoria, and east of the Ross Sea in Marie Byrd Land. Ordovician to Devonian terrane displacements might be expected in these areas. Late Precambrian–Cambrian rocks of Tasmania and New Zealand do show affinities with Australian and Antarctic terranes, and it is probable that the difficulties with Gondwana reassembly are due to displacements during the Cambrian–Devonian. Tasmanian rock types and a pattern of Cambrian crustal extension and rifting suggest that it is a more extreme development of the southern Australia pattern in which crustal extension led to complete isolation. Volcanism and deformation is widespread in the Late Cambrian, but the relationship of the crustal blocks to Cambrian subduction zones is unresolved. Cambro-Ordovician granitoids are absent. In New Zealand small scraps of Precambrian show late Proterozoic metamorphism and therefore resemble Antarctica more than Australia. The Cambrian comprises continent-derived clastics and arc-type volcanics that suggest proximity to a continent and a convergent margin. Like Tasmania Cambro-Ordovician granitoids are absent but unlike Tasmania there is no record of Cambro-Ordovician deformation. Both Tasmania and New Zealand Precambrian–Cambrian terranes appear to be crustal slivers that became detached and isolated by early Paleozoic tectonics.

ACKNOWLEDGMENTS

Field work in Antarctica was supported by Antarctic Division, Department of Industrial and Scientific Research, as part of the New Zealand Antarctic Research Programme. We thank B. D. Field, G. W. Grindley, and D. G. Howell for helpful comments on an earlier draft.

REFERENCES

Adams, C. J., et al, 1982, Potassium-argon geochronology of the Precambrian–Cambrian Wilson and Robertson Bay Groups and Bowers Supergroup, north Victoria Land, Antarctica, *in* C. Craddock, ed., Antarctic geoscience: Madison, University of Wisconsin Press, p. 543–548.

Andrews, P. B., and M. G. Laird, 1976, Sedimentology of a Late Cambrian regressive sequence (Bowers Group), northern Victoria Land, Antarctica: Sedimentary Geology, v. 16, p. 12–44.

Babcock, R. S., et al, 1983, Geology of the Daniels Range Intrusive Complex, northern Victoria Land, Antarctica (Abs.), *in* R. L. Oliver, et al, eds., Antarctic earth science: Canberra, Australian Academy of Science, p. 118.

Badham, J. P. N., 1982, Strike-slip orogens—an explanation for the Hercynides: Journal of the Geological Society of London, v. 139, p. 495–504.

Borg, S. G., 1983. Petrology and geochemistry of the Queen Maud Batholith, Central Transantarctic Mountains, with implications for the Ross Orogeny, *in* R. L. Oliver, et al, eds., Antarctic earth science: Canberra, Australian Academy of Science, p. 165–169.

Bradshaw, J. D., and M. G. Laird, 1983, Pre-Beacon geology of northern Victoria Land: a review, *in* R. L. Oliver, et al, eds., Antarctic earth science: Canberra, Australian Academy of Science, p. 98–101.

———, et al, 1982. Structural style and tectonic history in northern Victoria Land, *in* C. Craddock, ed., Antarctic geoscience: Madison, University of Wisconsin Press, p. 809–816.

Burrett, C., and R. H. Findlay, 1984, Cambrian and Ordovician conodonts from the Robertson Bay Group, Antarctica and their tectonic significance: Nature, v. 307, p. 723–725.

Cameron, W. E., et al, 1979, Boninites, komatiites and ophiolitic basalts: Nature, v. 280, p. 550–553.

Cas, R. A. F., et al, 1980, Ordovician palaeogeography of the Lachlan Fold Belt: a modern analogue and tectonic constraints: Journal of the Geological Society of Australia, v. 27, p. 19–31.

Coish, R. A., and W. R. Church, 1979, Igneous geochemistry of mafic rocks in the Betts Cove ophiolite, Newfoundland: Contributions to Mineralogy and Petrology, v. 70, p. 29–39.

Cooper, R. A., 1975, New Zealand and southeast Australia in the Early Paleozoic: New Zealand Journal of Geology and Geophysics, v. 18, p. 1–20.

———, and G. W. Grindley, 1982, Late Proterozoic to Devonian sequences of southeastern Australia, Antarctica and New Zealand and their correlation: Geological Society of Australia Special Publication 9, 103 p.

———, et al, 1983, Age and correlation of the Cambrian–Ordovician Bowers Supergroup, northern Victoria Land, *in* R. L. Oliver, et al, eds., Antarctic earth science: Canberra, Australian Academy of Science, p. 128–131.

Crawford, A. J., 1983, Tectonic development of the Lachlan Foldbelt and construction of the continental crust of southeast Australia: Geological Society of Australia Abstracts, v. 9, p. 30–32.

Curray, R. J., et al, 1979, Tectonics of the Andaman Sea and Burma, *in* J. S. Watkins, et al, eds., Geological and geophysical investigations of continental margins: American Association of Petroleum Geologists Memoir 29, p. 189–198.

Field, B. D., and R. H. Findlay, 1983, Sedimentology of the Robertson Bay Group, northern Victoria Land, *in* R. L. Oliver, et al, eds., Antarctic earth science: Canberra, Australian Academy of Science, p. 102–106.

Findlay, R. H., and B. D. Field, 1983, Tectonic significance of deformation affecting the Robertson Bay Group, northern Victoria Land, *in* R. L. Oliver, et al, eds., Antarctic earth science: Canberra, Australian Academy of Science, p. 107–112.

______ and ______, in press, Structural comparison of the Robertson Bay and Millen terrains, north Victoria Land, Antarctica, *in* Contributions to the geology of north Victoria Land: American Geophysical Union.

Gair, H. S., et al, 1969, Geologic map of Antarctica, Sheet 13, *in* V. C. Bushnell and C. Craddock, eds., Geologic map of Antarctica: Antarctic Map Folio Series 12, pl. XII.

Gale, N. H., et al, 1979, A Rb–Sr whole rock isochron for the Stockdale Rhyolite of the English Lake District and a revised mid Palaeozoic time scale: Journal of the Geological Society of London, v. 136, p. 235–242.

Grindley, G. W., and F. J. Davey, 1982, The reconstruction of New Zealand, Australia and Antarctica, *in* C. Craddock, eds., Antarctic geoscience: Madison, University of Wisconsin Press, p. 15–29.

______, and P. J. Oliver, 1983, Post-Ross Orogeny cratonisation of northern Victoria Land, *in* R. L. Oliver, et al, eds., Antarctic earth science: Canberra, Australian Academy of Science, p. 133–139.

Harland, W. B., et al, 1982, A geologic time scale: Cambridge, Cambridge University Press, 131 p.

Karig, D. E., et al, 1980, Structural framework of the fore-arc basin, N. W. Sumatra: Journal of the Geological Society of London, v. 137, p. 77–90.

Kleinschmidt, G., 1981, Regional metamorphism in the Robertson Bay Group area and in the southern Daniels Range, north Victoria Land, Antarctica, a preliminary comparison: Geologishes Jahrbuch, v. B41, p. 201–228.

______, and D. N. B. Skinner, 1981, Deformation styles in basement rocks of north Victoria Land: Geologishes Jahrbuch, v. B41, p. 155–199.

Kreuzer, A., et al, 1981, K–Ar and Rb–Sr dating of igneous rocks from north Victoria Land, Antarctica: Geologishes Jahrbuch, v. B41, p. 267–273.

Laird, M. G., 1981, Lower Paleozoic rocks of Antarctica, *in* C. H. Holland, ed., Lower Paleozoic rocks of the Middle East, eastern and southern Africa, and Antarctica; New York, Wiley Interscience, p. 257–314.

______, 1982, Lower Paleozoic rocks of the Ross Sea area and their significance in the Gondwana context: Journal of the Royal Society of New Zealand, v. 11, p. 425–438.

______, and J. D. Bradshaw, 1983, New data on the Lower Palaeozoic Bowers Supergroup, northern Victoria Land, *in* R. L. Oliver, et al, eds., Antarctic earth science: Canberra, Australian Academy of Science, p. 123–126.

______, et al, 1976, Reexamination of the Bowers Group (Cambrian), northern Victoria Land, Antarctica (preliminary note): New Zealand Journal of Geology and Geophysics, v. 19, p. 275–282.

______, et al, 1982, Stratigraphy of the Late Precambrian and Early Paleozoic Bowers Supergroup, northern Victoria Land, Antarctica, *in* C. Craddock, ed., Antarctic geoscience: Madison, University of Wisconsin Press, p. 535–542.

Odin, C. S., 1982, the Phanerozoic time scale revisited: Episodes 1982, n. 3, p. 3–9.

Oliver, R. L., et al, 1983, Antarctic earth science: Canberra, Australian Academy of Science, p. 697.

Powell, C. M., 1983, Tectonic relationships between the Late Ordovician and Late Silurian paleogeographies of southestern Australia: Journal of the Geological Society of Australia, v. 30, p. 353–373.

Plummer, C. C., et al, 1983, Geology of the Daniels Range, northern Victoria Land, Antarctica: a preliminary report, *in* Oliver, R. L., et al, eds., Antarctic earth science: Canberra, Australian Academy of Science, p. 113–117.

Ravich, M. G., et al, 1965, Oates Coast and George V Coast, *in* Pre-Cambrian of East Antarctica: Jerusalem, Israel Program for Scientific Translation, p. 354–405.

Scheibner, E., 1972, The Kanmantoo pre-cratonic province in New South Wales: Quarterly Notes of the Geological Survey of New South Wales, v. 7, p. 1–10.

Skinner, D. N. B., 1983a, The geology of Terra Nova Bay, *in* R. L. Oliver, et al, eds., Antarctic earth science: Canberra, Australian Academy of Science, p. 150–155.

______, 1983b, The granites and two orogenies of southern Victoria Land, *in* R. L. Oliver, et al, eds., Antarctic earth science: Canberra, Australian Academy of Science, p. 160–163.

Stump, E., 1982, Ross Supergroup in the Queen Maud Mountains, *in* C. Craddock, ed., Antarctic geoscience: Madison, University of Wisconsin Press, p. 565–569.

Tessensohn, F., et al, 1981, Geological comparison of basement units in north Victoria Land, Antarctica: Geologishes Jahrbuch, v. B41, p. 31–88.

Vetter, U., et al, 1983, Geochemistry, petrography and geochronology of the Cambro-Ordovician and Devono-Carboniferous granitoids of northern Victoria Land, *in* R. L. Oliver, et al, eds., Antarctic earth sciences: Canberra, Australian Academy of Science, p. 140–143.

Weaver, S. D., et al, 1984, Geochemistry of Cambrian volcanics of the Bowers Supergroup and implications for the Early Palaeozoic tectonic evolution of northern Victoria Land, Antarctica: Earth and Planetary Science Letters, v. 68, p. 128–140.

Wodzicki, A., et al, 1982, Petrology of the Wilson and Robertson Bay Groups and Bowers Supergroup, northern Victoria Land, Antarctica, *in* C. Craddock, ed., Antarctic geoscience: Madison, University of Wisconsin Press, p. 549–554.

Wright, T. O., 1981, Sedimentology of the Robertson Bay Group, north Victoria Land, Antarctica: Geologishes Jahrbuch, v. B41, p. 127–138.

———, et al, 1984, Newly discovered youngest Cambrian or oldest Ordovician fossils from the Robertson Bay terrane (formerly Precambrian), northern Victoria Land, Antarctica: Geology, v. 12, p. 301–305.

Wyborn, D., 1983, Chemistry of Palaeozoic granites of northern Victoria Land (Abs.), *in* R. L. Oliver, et al, eds., Antarctic earth science: Canberra, Australian Academy of Science, p. 144.

Accretion and Dispersal Tectonics of the Southern New England Fold Belt, Eastern Australia

Peter A. Cawood*
Evan C. Leitch
University of Sydney
New South Wales, Australia

Major structural divisions of the southern New England Fold Belt (eastern Australia) are fault-bounded blocks the character and distribution of which developed during climactic Permian orogenesis. Using stratigraphic and structural data, the blocks can be restored to their preorogenic positions, reversing movements on transcurrent faults and unwinding major folds. Major rocks sequences plotted on this reconstruction define a relatively simple pattern of five tectonostratigraphic units that record the accretionary history of the fold belt. The Tamworth terrane comprises two superimposed arc-fringe-forearc basin sequences, one of early Paleozoic and the other of middle-late Paleozoic age. To its east lies the Wisemans Arm terrane, a thin sliver of Devonian forearc basin strata that is in turn bounded to the east by early Paleozoic subduction complex rocks of the Woolomin terrane. In belts further east lie two further subduction complex terranes, the Cockburn terrane and the Texas terrane of middle and late Paleozoic age, respectively. Although direct pre-Permian linkages are lacking, provenance studies indicate that all terranes developed in close association; no exotic elements have been identified. Permian fragmentation and dispersal of the terranes resulted from a complex sequence of rift, compressional, and strike-slip movements that were accompanied by clastic sedimentation, metamorphism, protrusion of ultramafic rocks, and voluminous silicic magmatism. It is only in the mid-Carboniferous that clear linkages can be recognized between the southern New England Fold Belt and the earlier cratonized Lachlan Fold Belt to its west.

INTRODUCTION

The New England Fold Belt of New South Wales and Queensland, a meridionally oriented structure some 1,600 km (994 mi) long, developed during Paleozoic times close to the margin of the Gondwana continent. For much of this time the margin was a convergent plate boundary, and the fold belt is dominated by tholeiitic and calc-alkaline igneous rocks and associated sediments and subduction-accreted oceanic rocks. Our discussion treats only the southern section of the fold belt, that part south of a lobe of Mesozoic sediments of the Surat Basin (Fig. 1).

Traditionally, this southern part of the fold belt has been divided into a number of "structural blocks" (Leitch, 1974) that owe their distinctive character to a combination of differences in rock sequence, level of crustal exposure, metamorphic grade, and intensity of deformation (Fig. 1). These blocks are bounded by major faults and granite plutons, but they do not comprise discrete terranes in the sense of Jones et al (1983), because juxtaposed blocks contain common lithostratigraphic associations with similar and related geologic histories. The present block structure is the result of major transcurrent faulting and large-scale folding during Permian orogenesis that disrupted and fragmented a less complex arrangement of tectonostratigraphic elements. We consider that these earlier elements comprise the terranes of the southern New England Fold Belt and that the presently observed blocks are the products of their dispersal. Changes in tectonic style associated with the commencement of terrane dispersal resulted in the accumulation of the distinctive linkage assemblage of sedimentary and igneous rocks that provides important constraints in reconstructing the predispersal pattern of terranes.

This paper identifies and defines the major terranes in New England, providing a stratigraphic and structural analysis of each, and outlines evidence provided by provenance linking, overlap assemblages, and pluton stitching in determining interterrane relationships. In addition, the timing and causes of terrane dispersion and the overall relationship of these terranes to other crustal masses in eastern Australia is assessed.

TERRANES

Five major tectonostratigraphic units, fault-bounded and each with a different geological history, are recognized

*Present Address: Department of Earth Sciences, Memorial University of Newfoundland, St. John's, Newfoundland, Canada.

(Fig. 2b). These are: (1) Tamworth terrane; (2) Wisemans Arm terrane; (3) Woolomin terrane; (4) Cockburn terrane; and (5) Texas terrane. As a result of Permian dispersal, these units have been fragmented and scattered among the various blocks of the New England Fold Belt (compare Figs. 1 and 2). By reversing the major fault movements between blocks and unbending oroclinal folds, the terranes reassemble into a relatively simple series of parallel linear belts (Fig. 2b). This reconstruction is not a complete depiction of the fold belt immediately prior to Permian dispersal because no attempt has been made to remove the affects of either penetrative Permian deformation within the blocks or second-order folding and faulting. The reconstruction is based on the following adjustments to the presently observed pattern of blocks:

1. The Hastings block has been restored to a position along strike from the Tamworth block (Leitch, 1980; Cawood, 1982a).
2. The Yarrowitch block has been moved southwest along the Kilburnie fault, the curved trace of which leads to counterclockwise rotation of the block during restoration to a position north of the Hastings block (cf. Flood and Shaw, 1977).
3. The oroclinal bend in the southern part of the Tamworth block has been straightened, with concomitant rotation of the Manning, Yarrowitch, and Hastings blocks to maintain physical continuity of these divisions (Cawood, 1982a).
4. The major fold affecting the Bonshaw and Coffs Harbour blocks recognized by Flood and Fergusson (1982) has been unwound so that these blocks are aligned approximately parallel to the trend of the Tamworth and Macdonald blocks. The Warwick block has been rotated in concert with these movements, restoring it to a position north of the Tamworth block.
5. The Tabulam block has been restored to a position along strike from the Tamworth block as indicated by Flood and Fergusson (1982).
6. The early Paleozoic slice of the Port Macquarie block has been moved south to a position along the trend of the Macdonald, Manning, and Yarrowitch blocks.

The resulting structure (Fig. 2a) is twice as long as at present (Fig. 1), but this is probably exaggerated by several features unallowed for in our restoration. Thus, the Armidale block is the locus of emplacement of a major mass of Permian-Triassic granitic rocks (Fig. 1) and was probably extended at this time, and the Coffs Harbour and Bonshaw blocks were probably lengthened by interlayer slip during oroclinal bending. It is also possible that the contact between the Armidale and Macdonald blocks was originally a major sinistral transcurrent fault that merged with the northern extension of the Peel fault. Before displacement on this hypothetical fracture, the Armidale block may have lain further south, adjacent to the Yarrowitch and Manning blocks.

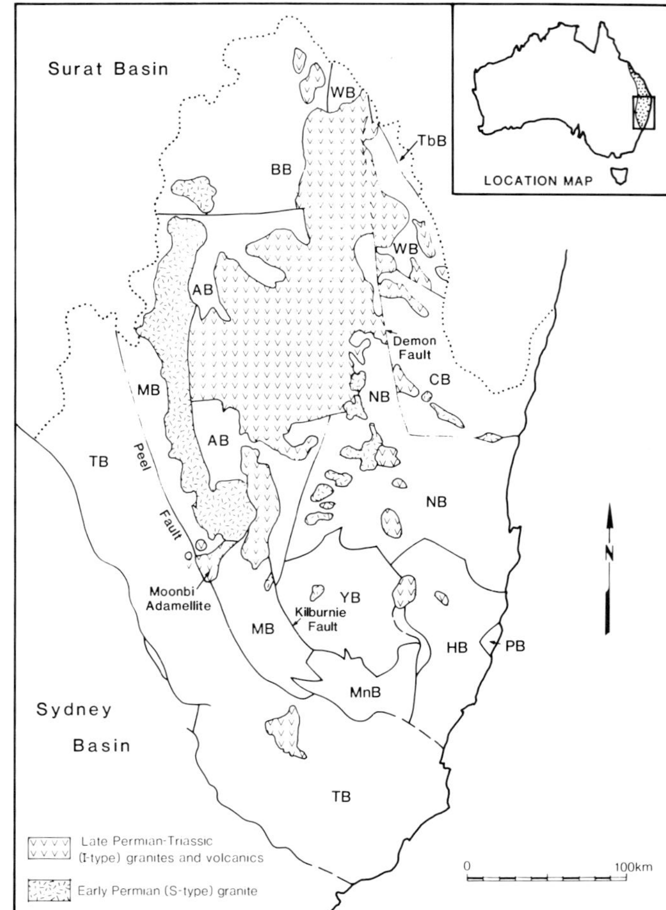

Figure 1—Major structural divisions of the southern New England Fold Belt. AB = Armidale block; BB = Bonshaw block; CB = Coffs Harbour block; HB = Hastings block; MB = Macdonald block; MnB = Manning block; NB = Nambucca Slate belt; PB = Port Macquarie block; TB = Tamworth block; TbB = Tabulam block; WB = Warwick block; YB = Yarrowitch block.

Tamworth Terrane

The Tamworth terrane holds the key to understanding the interrelationship of terranes within the New England Fold Belt. The reassembling of the Tamworth, Hastings, Tabulam, and Warwick blocks, the dispersed constituents of this terrane (Figs. 1, 2), into a single linear structure constrains and determines the reconstruction of all other terranes within the fold belt. Also, the terrane contains an uncomplicated and relatively complete sequence of clastic rocks that is well dated and spans the known history of the fold belt. This allows the establishment of a detailed record of temporal changes in provenance, which through provenance linking allows the time of accretion of other terranes to be determined. The character of the terrane and the basis for considering the Tamworth, Hastings, Tabulam, and Warwick blocks part of a single terrane are outlined in detail below. Summary lithological columns are presented in Figure 3.

The stratigraphic record of the terrane is best preserved in the Tamworth block (or belt; Korsch, 1977), which consists largely of stratified rocks of Middle Cambrian to Early Permian age (Fig. 3). Early Paleozoic rocks are restricted to a small area along the eastern margin of the block where some 1,350 m (4,429 ft) of Middle Cambrian–Early (?)Ordovician siltstone, sandstone, conglomerate, and tuff derived from a western source and

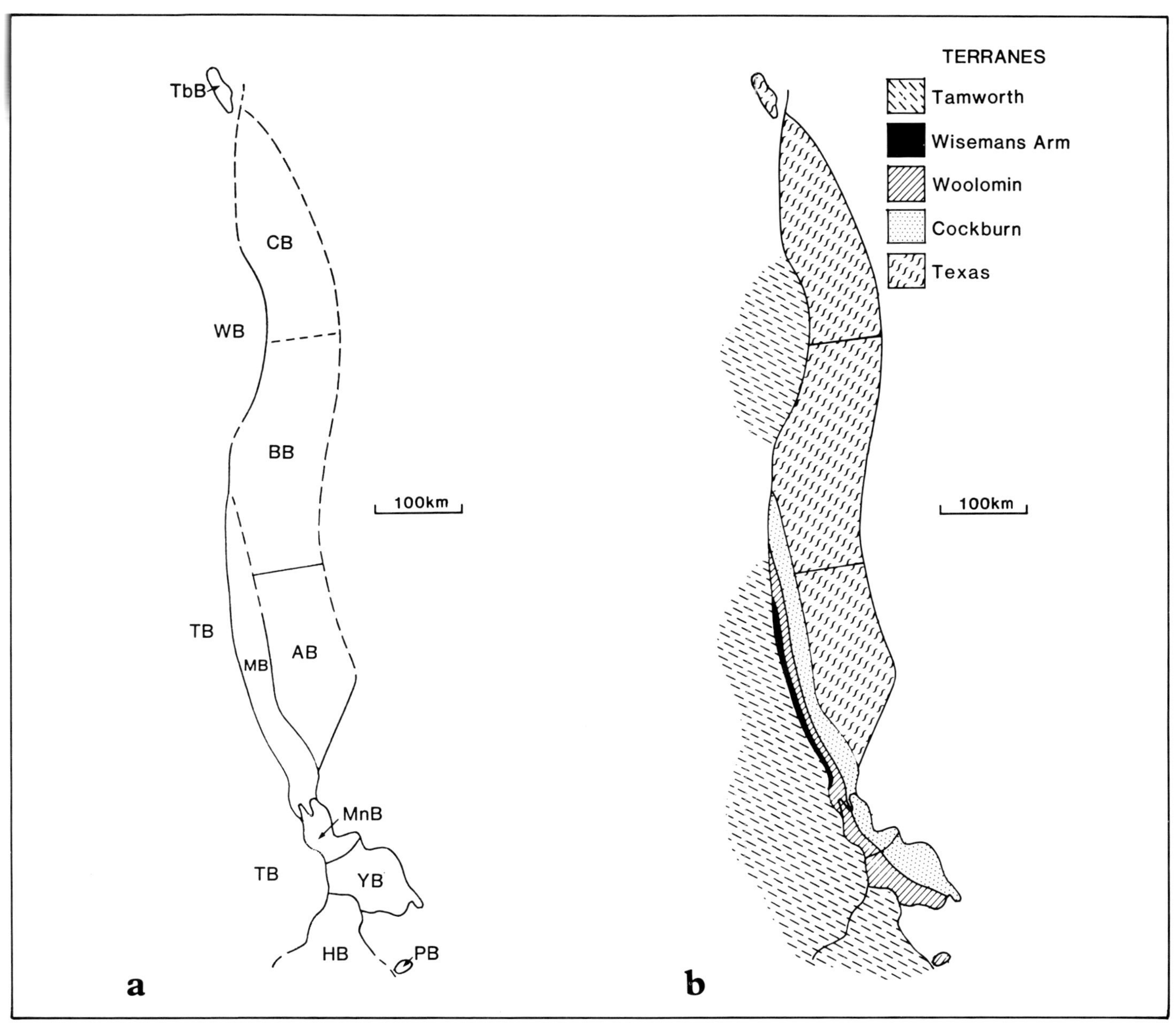

Figure 2—(**a**) Position of crustal divisions of the southern New England Fold Belt prior to Permian dispersal. See Figure 1 for key to blocks. (**b**) Pre-Permian distribution of terranes. The Nambucca block contains the Early Permian overlap assemblage and hence it is not shown on this diagram.

deposited in a submarine-fan complex are unconformably overlain by Middle Ordovician debris-flow breccias and limestone that range up to 100 m (328 ft) in total thickness (Cawood, 1976). Silurian strata are absent. Earliest Devonian rocks are shallow-water limestone, conglomerate, sandstone, siltstone, and tuff that rest unconformably on the early Paleozoic material. Clastic material was derived from a volcanic chain to the west, and extrusive and high-level intrusive andesitic rocks (now keratophyres) are associated with the sedimentary rocks.

Late Early–Late Devonian sections range up to about 3,000 m (9,843 ft) thick and are made up of a diverse sequence of conglomerate, siltstone, sandstone, tuff, limestone, basalt lavas, and associated dolerite dykes and sills. There is an overall tendency for facies associations to become of deeper water type up-section in eastern parts of the belt, with shallow-marine rocks being succeeded by thick redeposited sequences, whereas in the west, shallow-water rocks accumulated throughout most of the Devonian. All the clastic sediments were derived from the west.

Earliest Carboniferous rocks are of similar facies to those of Late Devonian age, with thick (up to 2,500 m [8,202 ft]) redeposited clastics in the east and shallow-marine sandstone, conglomerate, and limestone farther west (Mory, 1982). A western volcanic chain remained the major sediment source. By Early Visean times, rocks of shallow-water type were accumulating throughout much of the Tamworth block, and andesitic, dacitic, and rhyolitic ash flows extended across it (Roberts and Engel, 1980). Rapid differential subsidence gave rise to small basins that were filled by redeposited volcaniclastic materials (Skilbeck, 1982).

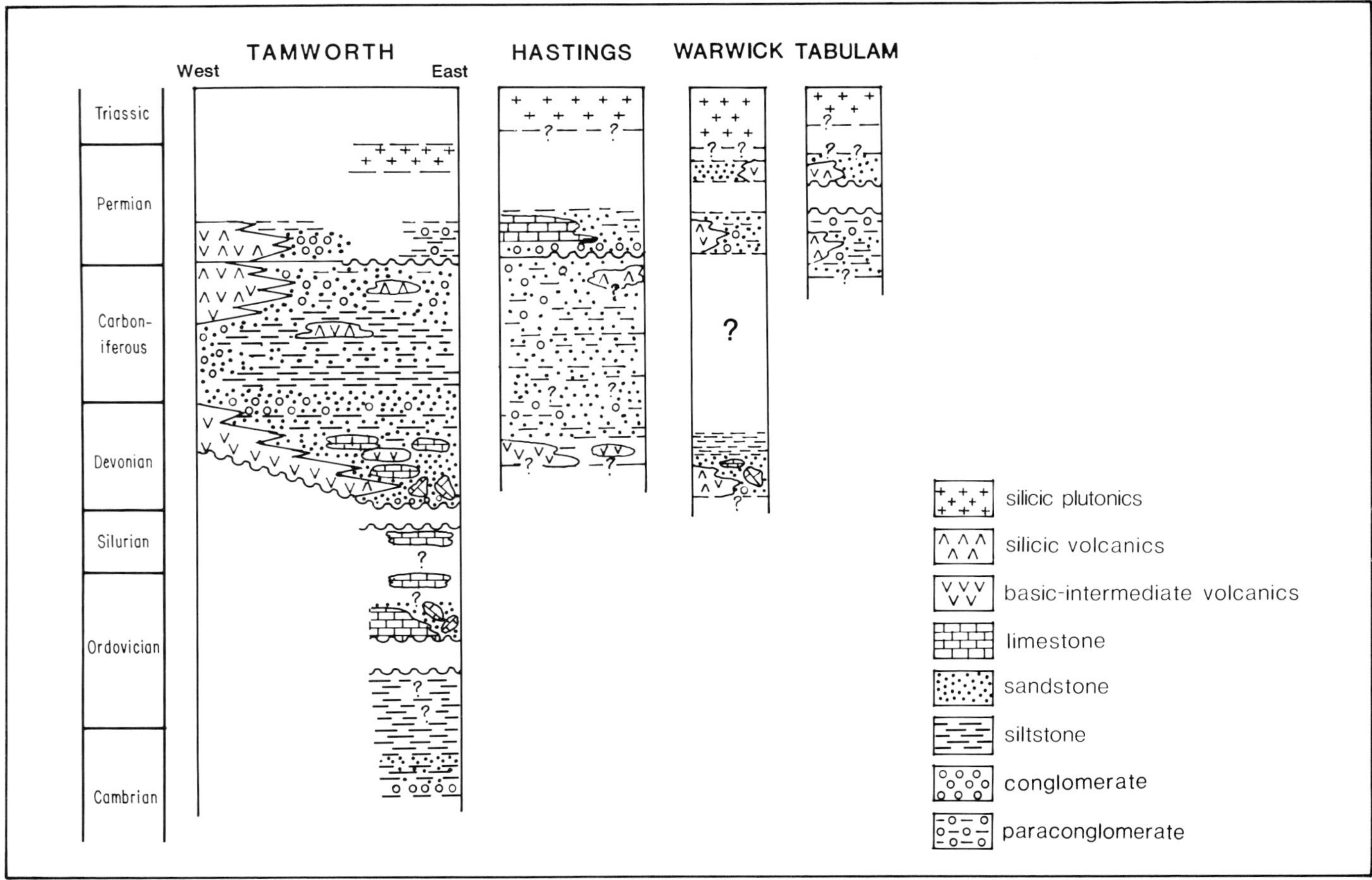

Figure 3—Time-space plot for structural divisions that collectively comprise the Tamworth terrane.

Late Permian rocks in the west comprise mafic and silicic volcanics, sandstone, conglomerate, tuff, and coal ranging up to about 1,400 m (4,593 ft) in thickness, whereas in the east, sandstone, siltstone, paraconglomerate, and limestone of this age form sequences up to about 1,000 m (3,281 ft) thick.

The stratigraphy of the Tamworth terrane in the Hastings block is incompletely established, but the oldest rocks are basalt and andesite pillow lavas and intercalated siltstone and tuff of probable Devonian age that are overlain by an upward-coarsening sequence of redeposited siltstone, sandstone, and paraconglomerate about 4,000 m (13,123 ft) thick. The sediments are quartz-poor volcaniclastics of basaltic and andesitic provenance. Younger (Carboniferous) strata occur in the northern part of the block where they comprise sediments of andesitic, dacitic, and rhyolitic derivation, of turbidite type in their lower part, but of shallow-water facies higher in the sequence (Lindsay, 1969). These are overlain by Early Permian paraconglomerate, limestone, and siltstone. Further south, dacite breccias and flows are possibly of Carboniferous age.

The oldest rocks of the Tabulam block are Late Carboniferous shallow-marine strata, predominantly siltstone and sandstone but including paraconglomerate and minor silicic volcanic rocks and tuff (Thomson, 1976), all of which accumulated in a shelf environment below wave base. Clasts in the conglomerate include large amounts of granitic material, silicic volcanic debris, and fine-grained sedimentary rocks including chert.

The Tamworth terrane in the Warwick block is characterized by an intimate association of basalt and andesite flows, breccias and shallow intrusives, quartz-poor volcanic-derived clastics, chert, siltstone, coralline limestone, and limestone breccias that form a structurally complex sequence several thousand meters thick. Most limestone masses are of Early Devonian age, but at least one block contains an Ordovician fauna (Wass and Dennis, 1977).

Restructuring of the Tamworth Terrane

Leitch (1980) pointed out that the geologic histories of the Hastings and adjoining blocks (Fig. 1) were inconsistent with their present spatial relations. He suggested that the Hastings block was either an element exotic to the rest of the New England Fold Belt or that it was a displaced section of the Tamworth block, moved at least 100 km (62 mi) north on a transcurrent fault along its western margin. Cawood (1982a) favored the latter hypothesis and proposed that the block had been displaced from a position along strike from the Tamworth block by a combination of oroclinal bending and sinistral faulting in the Late Permain. Evidence that supports a close association between the Tamworth block and the Hastings block during Devonian and Carboniferous times includes:

1. Although the age of the early part of the Hastings block sequence remains to be established, it is strikingly similar to that in the southeastern part of the Tamworth block where Devonian basalt pillow lavas are overlain by thin-bedded redeposited sandstone and siltstone and then coarse sandstone-dominated redeposited rocks. Like the Hastings block strata, these rocks are of mafic igneous provenance, are quartz-poor, and contain detrital clinopyroxene.

2. Detrital mineralogy and facies associations in Carboniferous sedimentary rocks of the two blocks are similar, and the occurrence of Carboniferous(?) dacites in the Hastings block finds parallels in the Tamworth block.

Both blocks are only mildly deformed and have suffered only burial metamorphism.

Flood and Fergusson (1982) argued that a great asymmetrical drag fold affected the northeastern area of Paleozoic strata in Figure 1, resulting in the Tabulam block being displaced from an original position north of, and along strike from, the Tamworth block. Unfolding of this structure placed the Warwick block west of the Bonshaw and Coffs Harbour blocks (Fig. 2). The sedimentary facies and provenance of the strata in the Tabulam block are consistent with this reconstruction, for although McCarthy et al (1974) highlighted the presence of granitic debris, several conglomerate horizons of similar age in the Tamworth block contain a high proportion of silicic plutonic debris. Several interpretations of the Warwick block in this restored position are possible. Flood and Fergusson (1982) suggested the Warwick block comprises an exotic terrane, accreted to the New England Fold Belt prior to the development of the subduction complex of the more eastern blocks, whereas Leitch (1975) argued that the rocks most closely resembled the Devonian of the Tamworth block and treated them as a displaced fragment of this element. Although no detailed biogeographic study of Devonian coral and conodont faunas in the Warwick block has been carried out, they appear to be broadly similar to those elsewhere in Australia (Strusz, 1976; Telford, 1972), arguing against a highly exotic origin. This and lithologic similarities lead us to group the Warwick block with the other blocks forming the Tamworth terrane.

The tectonic setting of the Tamworth terrane is clearly arc related. Tholeiitic and calc-alkaline igneous rocks and their degradation products have built up a Paleozoic sequence that consists of a complex interdigitation of terrestrial and shallow- and deep-marine epiclastic rocks, ash-fall and ash-flow tuffs, lavas, breccias, and small intrusive masses. Clastic rocks exceed primary igneous material in volume, and paleocurrent data show that the locus of volcanic activity mostly lay to the west.

There is little direct evidence from within the terrane as to whether the rocks accumulated adjacent to a forearc or a backarc margin. No Late Proterozoic calc-alkaline volcanics are known from eastern Australia or adjacent sectors of Gondwanaland, and the Cambrian strata are probably little younger than arc inception. Because backarc formation usually postdates the commencement of arc activity by several tens of millions of years, these rocks are more likely to have accumulated in a forearc than a backarc basin. A forearc position throughout the Paleozoic is supported by inferred interterrane relationships.

Wisemans Arm Terrane

A narrow fault-bounded sliver of interbedded quartz-poor volcaniclastic sandstone, siltstone, conglomerate, and tuff, associated with which are large olistoliths of limestone, siltstone, and chert, make up the Wisemans Arm terrane. Rocks of this terrane, of Early Devonian age, are known only from the western side of the Macdonald block. They were termed the Wisemans Arm Formation by Leitch and Cawood (1980) and comprise a sequence of redeposited epiclastic sediments, ash-fall tuffs, and slide blocks. Although the rocks are extensively faulted, there is no evidence of imbricate repetition, and no intercalated volcanic rocks have been identified. Seemingly unfaulted sequences range up to at least 1,800 m (5,906 ft) in thickness.

The rocks are clearly relatively deep-water marine deposits. Most detritus was derived from an arc source, but chert olistoliths and some volcanic debris come from rocks of ocean-floor type (Leitch and Cawood, 1980; Cawood, 1982c). This suggests deposition in the outer part of a forearc basin that was bound oceanwards by an uplifted accretionary subduction complex.

Woolomin Terrane

The Woolomin terrane is made up of an early Paleozoic basalt-chert-siltstone association found in the Manning, Macdonald, Yarrowitch, and Port Macquarie blocks. Although these rocks have been accorded different formational names in each block, and those in the Yarrowitch Block have undergone multiple deformation and regional metamorphism up to amphibolite grade (Morand, 1982), they are all of very similar geological character and form a near-continuous linear belt when Permian movements are reversed (Fig. 2b).

Detailed investigations in the Macdonald block show that the terrane is made up of a series of repeated slices, mostly less than 1,000 m (3,281 ft) thick, in which basalt is overlain by chert and then siltstone and tuff (Cawood, 1982b). The basal basaltic horizon, which consists of pillowed and massive flows with midocean ridge magmatic affinities (Cawood, 1984) always has a sheared lower contact with structurally underlying slices. Conformably overlying chert is radiolarian-bearing and often contains concordant manganese-rich pods. Siltstone, which forms partings and laminae within the chert, progressively increases in amount up-section where it is interstratified with green and grey tuffaceous rocks. Basalts intercalated in this part of the sequence are within-plate alkaline varieties characterized by the presence of pink titanium-rich calcic clinopyroxene phenocrysts. Lithofeldspathic sandstone of mafic volcanic provenance occurs close to the top of a few slices.

Both the internal structure of the Woolomin terrane and the nature of its constituent rocks favor its interpretation as a subduction complex. Ocean-floor rocks dominate the terrane, suggesting that detritus from the associated arc was largely trapped in the forearc basin and rarely gained access to the trench.

Cockburn Terrane

A distinctive assemblage of chert, siltstone, sandstone, and mafic igneous rocks that form a continuous belt along the eastern parts of the Macdonald, Manning, and Yarrowitch blocks in their predispersal positions (Fig. 2) comprises the Cockburn terrane. The overall structure of these rocks is believed to be similar to that of the Woolomin terrane: a stack of thrust slices younging consistently to the west with the basal unit of the slices commonly consisting of basalt of ocean-floor type (Cawood, 1984). However, in contrast to the Woolomin terrane, sandstone and fine conglomerate are present in the upper part of many slices, the basal mafic rocks are thicker and in some slices basalt passes downwards into massive dolerite, and intercalated alkaline volcanic rocks are more prominent.

A Devonian age is favored for the Cockburn terrane. Fossil evidence of age is sparse, but the petrography of sandstones within the unit is similar to the petrography of Devonian clastic rocks in the Tamworth terrane, and distinct from those of Carboniferous age.

Geological relationships and the character of both the sedimentary and igneous rocks indicate that the Cockburn terrane comprises part of an accretionary subduction complex. The abundance of coarse volcaniclastic rocks in the upper part of thrust slices suggests that, at the time this terrane was developing, large amounts of arc-derived detritus were being fed into the associated trench.

Texas Terrane

The Texas terrane consists of apparently thick sequences of slaty siltstone and redeposited feldspathic and lithofeldspathic sandstone derived from intermediate and silicic volcanic rocks, chert masses, altered mafic volcanic rocks, and recrystallized silicic ash-fall tuff that make up the greater part of the Armidale, Bonshaw, and Coffs Harbour blocks (Figs. 1, 2).

Investigations of the structure of rocks in the northwestern part of the Coffs Harbour block indicate the presence of transected folds and of melange horizons as well as less-disrupted areas of stratified rocks (Fergusson, 1982a, 1982b, in press). Fergusson interpreted the structure in terms of an accretionary subduction complex, and although the structure further south appears to be more coherent (Korsch, 1981a), it is likely that all of the terrane is of this origin. Flood and Fergusson (1982) traced major lithological boundaries from the Coffs Harbour block into the Bonshaw block and showed that these blocks form part of a major dextral fold, one limb of which is partly replaced by the Demon fault (Fig. 1). Although the structure of the Armidale block is complex and little understood, it seems to be a southern continuation of the rocks on the western side of the Bonshaw block and hence a continuation of the Carboniferous subduction complex.

TERRANE LINKAGES

Three types of linkages can be established between the New England terranes: linkages inferred from provenance studies, linkages derived from sedimentary sequences that extend across terrane boundaries, and linkages established by stitching by granitic plutons (Fig. 4).

Provenance Linkages

Provenance studies are considered by some to provide more equivocal ties than either overlap relationships or pluton stitching, but we have found them to be of much greater value in establishing the history of terrane development. Rocks as old as Cambrian make up the terranes whereas the oldest overlapping sequences recognized are earliest Permian and the oldest granites are probably slightly younger. Without the linkages indicated by shared clastic sources, which can be demonstrated at least back to the Early Devonian, much uncertainty would remain concerning terrane history.

Within the Tamworth terrane, clastic sedimentary rocks are almost all of igneous provenance, and a number of studies (Crook, 1960a, 1960b; Chappell, 1968; Leitch and Willis, 1982; Cawood, 1983) have revealed the detailed nature of their detrital components and temporal provenance changes useful in correlation both within and beyond the terrane.

Two cycles of basic-silicic volcanic activity are recorded in the strata of the Tamworth terrane. The first extends from the Middle Cambrian to Early Devonian with provenance changing from a mixed basaltic and andesitic source in the Cambrian to more strongly andesitic in the Middle Ordovician and andesitic and dacitic in the Early Devonian. Calcic clinopyroxene is the only detrital ferromagnesium phase present within rocks of this age range Its low Al and Ti content and restricted range of Fe:Mg ratios indicate that the source volcanics crystallized from subalkaline magmas with mildly calc-alkaline affinities (Cawood, 1983).

The commencement of the second volcanic cycle is marked by a reversion to more mafic volcanism in the distributive province and detritus changes from basaltic and andesitic in the Devonian to andesitic, dacitic, and rhyolitic in the Carboniferous. Detrital calcic clinopyroxene compositions and whole-rock sandstone geochemistry of the Devonian sediments indicate derivation from volcanics of island arc tholeiite affinity (Chappell, 1968; Leitch, 1974; Cawood, 1983). The incoming of detrital hornblende, biotite, and potassium feldspar in Carboniferous strata (Crook, 1960b) marks a change in the composition of the source volcanics to calc-alkaline, in accord with the calc-alkaline affinities deduced by Wilkinson (1971) for Carboniferous volcanics from the western and southern margin of the Tamworth block. Plutonic debris occurs at isolated levels throughout this second volcanic cycle and represents erosion through the volcanic carapace into the roots of the magmatic arc (Leitch and Willis, 1982).

No close connection can be demonstrated between the provenance of the early Paleozoic sandstones of the Tamworth terrane and the provenance of the rare sandstones of the Woolomin terrane. Both were derived from a volcanic arc source, but the Woolomin rocks contain a higher percentage of detrital quartz, a higher abundance of vitric volcanic grains, and more Ti-rich detrital clinopyroxene crystals (Cawood, 1983). However, by the Early Devonian both terranes were providing debris for the Wisemans Arm terrane and hence must have been closely associated by this time.

Cawood (1982c) showed that the modal composition of

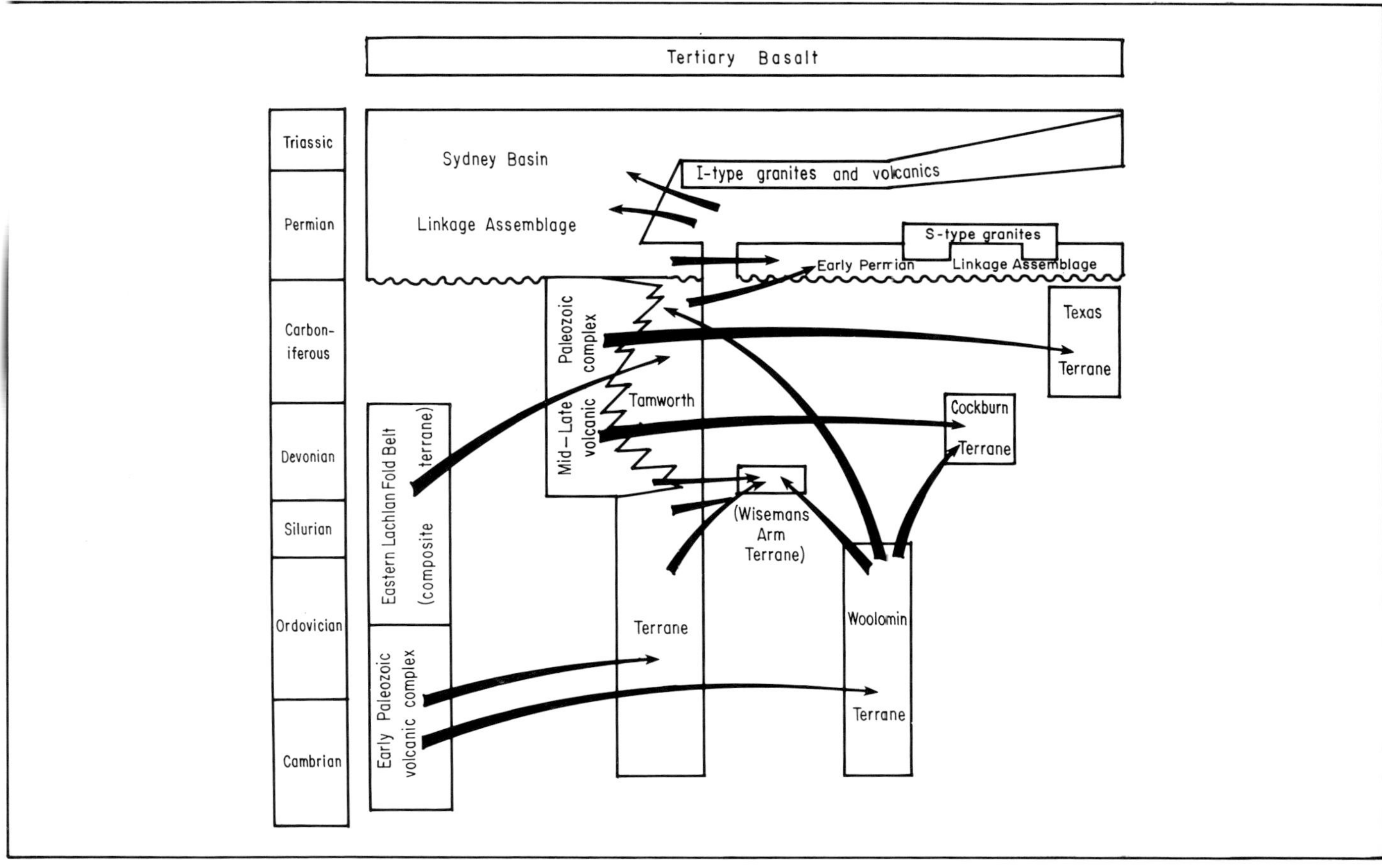

Figure 4—Terrane linkages in the southern New England Fold Belt. Arrows signify linkages indicated by sedimentary petrography.

sandstones from the Wisemans Arm terrane overlaps with the composition of Early Devonian Tamworth terrane rocks, and that a group of Mn-rich detrital calcic clinopyroxenes in the sandstones are identical with detrital calcic clinopyroxenes in Early Devonian Tamworth terrane rocks. The Wisemans Arm rocks also contain a group of Ti-rich calcic clinopyroxenes (Cawood, 1982c), distinct from any in the Tamworth terrane but of similar composition to those found in the basaltic rocks of the Woolomin terrane. The latter terrane was also the source for the chert olistoliths of the Wisemans Arm terrane (Leitch and Cawood, 1980).

Study of the provenance of Cockburn terrane rocks is yet to be completed, but results so far available indicate strong affinities with the Devonian rocks of the Tamworth terrane. The sandstones are quartz-poor and contain large amounts of detrital plagioclase and vitric and felsitic volcanic rock fragments. Detrital calcic clinopyroxene is widespread, but neither detrital hornblende nor biotite, which are present in terranes containing Carboniferous strata, have been identified in any of the sedimentary rocks of this terrane. Metabasalt and chert clasts similar to the rocks of the Woolomin terrane suggest an additional linkage.

Detailed petrographic studies of sandstones from the Texas terrane by Korsch (1978, 1981b) have shown that they have a distinctive character. Compared with older (pre-Carboniferous) sandstones, these rocks contain higher amounts of quartz, pyroxene is much less common, and lithic grains of silicic volcanic rocks are more common. Hornblende and biotite are widespread minor detrital components, and muscovite, tourmaline, zircon, and epidote are encountered, as are metamorphic fragments and chert. As pointed out by Korsch (1981b), these rocks have a magmatic arc provenance, and they match closely the provenance of Carboniferous sediments in the Tamworth terrane.

Overlap Sequences

Early Permian strata occur stratigraphically overlying and/or faulted within each of the five terranes. They constitute a distinctive lithologic association and generally occur in sequences at least 1,000 m (3,281 ft) thick that include paraconglomerate of mass-flow origin and dark micaceous siltstone (e.g., Mayer, 1972; Leitch, 1974; Korsch, 1977). Lenses of bioclastic limestone occur in the upper part of many sequences. Silicic ash-fall tuffs and intercalated basaltic and silicic volcanic rocks are widespread. Detritus within the clastic rocks comprises recycled sedimentary material similar to that of the Woolomin and Cockburn terranes as well as an igneous component comparable to the Late Carboniferous and Early Permian volcanics of the Tamworth terrane.

We treat these rocks collectively as an overlap assemblage, for although later dispersal movements along terrane boundaries and other faults have destroyed their physical continuity, they are all members of the same lithological association.

The association is widespread in the Nambucca block where several writers have suggested they comprise part of an accretionary subduction complex. Supporting evidence for such an interpretation is lacking, and the nature of the strata accords better with deposition in a rapidly downsinking graben flanked by rocks of older terranes (Cawood, 1982a). In fact, the time of accumulation of the Early Permian strata is transitional between the termination of the convergent-plate-margin setting that was responsible for terrane generation and accretion and the establishment of an oblique or strike-slip margin in which the terranes were fragmented and disrupted into the present series of crustal blocks. The importance of this overlap assemblage lies, thus, in providing a minimum age for accretion of all five terranes.

Pluton Stitching

A major suite of I-type plutons and volcanics were emplaced in central and eastern New England during the Late Permian and Triassic (Fig. 1; age range 255–180(?) Ma; Shaw and Flood, 1981). They cut across the boundaries between a number of the major blocks, providing a minimum age for the termination of terrane dispersion. Rocks of the Tamworth terrane in the Tamworth block are stitched to those of the Wisemans Arm, Woolomin, and Cockburn terranes by the 255 Ma Moonbi Adamellite (Fig. 1), whereas Tamworth terrane strata of the Hastings block are stitched to Woolomin terrane rocks of the Yarrowitch block by a Triassic silicic volcanic complex (Fig. 1) and those of the Warwick block are stitched to Texas terrane material by Late Permian granite plutons.

A belt of S-type plutons lies along the boundary between the Macdonald and Armidale blocks, stitching together rocks of the Woolomin, Cockburn, and Texas terranes. They are older (average age 275 Ma; Shaw and Flood, 1981) than the I-type bodies and are restricted to the three accretionary subduction-complex terranes.

TECTONIC INTEGRITY OF TERRANE RECONSTRUCTION

The reconstruction in Figure 2 derives from the removal of major translational and rotational movements indicated by documented structures and rock sequences. One check on the consistency of the reconstruction, and on the closely linked development of the terranes indicated by provenance studies, is a comparison of the arrangement of the terranes with that anticipated from their tectonic character.

The existence of a long-lived, largely calc-alkaline volcanic chain lying at various times within or immediately west of the Tamworth terrane, provides good evidence that the New England Fold Belt evolved in the vicinity of a convergent plate margin, and this terrane has been widely interpreted as an arc-margin–forearc basin realm (e.g., Leitch, 1975; Crook, 1980). Two major episodes of convergence are indicated by the sequence within the Tamworth terrane: an Early Paleozoic one and another of Devonian to Carboniferous (Early Permian?) age, the break coinciding with a major unconformity that has cut out any Silurian strata. The Woolomin terrane is interpreted as an accretionary subduction complex that formed during the earlier underthrusting, and the Cockburn and Texas terranes as similar complexes relating to the latter episode. The progressively younger age of the subduction complex terranes from west to east, all lying west of the Tamworth terrane on our reconstruction, is the only arrangement consistent with this tectonic scheme. The position of the Wisemans Arm terrane on the reconstruction accords with its interpretation as an outer-forearc-basin deposit and suggests that it lies on basement of Woolomin terrane rocks. (If this is eventually demonstrated, then the Wisemans Arm terrane will need no longer be considered a separate tectonostratigraphic unit but should be incorporated into the Woolomin terrane.) Uplift indicated by the presence of Woolomin terrane olistoliths in the Wisemans Arm rocks may have accompanied renewed subduction accretion at the start of the second convergent episode in the Devonian (Leitch, 1978a).

PERMIAN DISPERSAL AND ACCOMPANYING EVENTS

The net result of Permian dispersal is the presently observed arrangement of blocks, but insufficient data are available to allow full specification of movement paths, movement history, and causative stress fields. Tensional or transtensional stresses are indicated by the character of Early Permian sedimentary sequences and accompanying volcanic rocks. Thick redeposited Early Permian sections accumulated in a basinal complex formed by rifting of the earlier accreted rocks. Midocean-ridgelike basalt was extruded in the Nambucca Slate Belt (Asthana, personal communication), indicating incipient sea-floor spreading (cf. Scheibner, 1976), and sinistral movement on faults at a high angle to the axis of rifting led to displacement of the Yarrowitch and Hastings blocks (Cawood, 1982a, Fig. 2). Emplacement of granite plutons derived from the fusion of sedimentary rocks (Flood and Shaw, 1977), and the rise of ultramafic material, now largely represented by sheared serpentinite masses (Leitch, 1980) commenced at this time.

Late in the Early Permian the development of a strongly compressional regime throughout the southern New England Fold Belt is indicated by widespread folding accompanied by cleavage formation and regional metamorphism in rocks east of the Peel fault. Detailed structural studies in the Nambucca Slate Belt indicate that a wrench regime evolved in the Late Permian (Leitch, 1978b). Dextral strike-slip movements displaced the Tambulam block to the southeast, accompanied by major bending of the rocks of the Bonshaw and Coffs Harbour blocks (Flood and Fergusson, 1982). Emplacement of voluminous granite bodies, the product of melting of an igneous protolith (Shaw and Flood, 1981), was possibly controlled by this regime (Evans and Roberts, 1980).

The relationship between Permian terrane dispersal and plate movements is controversial, partly owing to the absence of definitive data on the magmatic affinity of Early Permian volcanics in the western part of the fold belt.

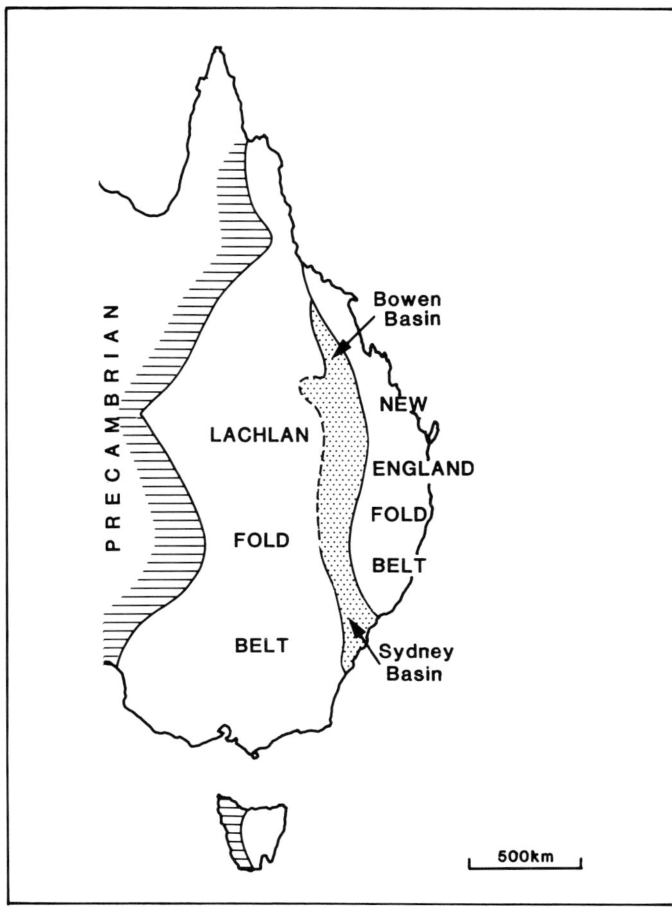

Figure 5—Major pre-Mesozoic tectonic units of eastern Australia.

These rocks, mainly basalts and rhyolites, are either a continuation of the Carboniferous magmatic arc activity and hence evidence of continued plate convergence (Leitch, 1975; Scheibner, 1976) or a bimodal rift suite signifying a major change in plate interaction. Harrington (1982) and Korsch (1982) argue the latter case and suggest that by the Early Permian the convergent boundary had been replaced by a zone of strike-slip movement with associated pull-apart rifts. By the Late Permian the plate margin had stepped eastward and the New England Fold Belt was a broad zone of transcurrent faulting and granite emplacement (Cawood, in press).

NEW ENGLAND FOLD BELT-LACHLAN FOLD BELT ACCRETION

The New England Fold Belt is separated from Precambrian shield areas of Australia by the Lachlan Fold Belt (Fig. 5), a broad zone of Cambrian to Late Devonian stratified rocks and granite last affected by strong deformation during the Carboniferous. The contact between the fold belts is masked by younger rocks, notably those of the Sydney Basin, which provide an earliest Permian linkage. Earlier relationships are less clear with two competing hypotheses; one views the New England Fold Belt as an arc assemblage formed outboard of a Lachlan continental margin, perhaps in an exotic location (e.g., Crook, 1980), whereas the other (e.g., Scheibner, 1976; Cawood, 1980) considers the two belts to have evolved in close association, with temporary isolation following opening of an interarc basin in the Devonian.

Early Paleozoic connections are based mainly on broad paleogeographic considerations. Thus, Cawood (1980) argued that Ordovician andesites, the oldest exposed rocks in the eastern part of the Lachlan Fold Belt, are a young part of the arc that was the source of the Cambrian and Early Ordovician clastics of the Tamworth terrane and was associated with the subduction accretion of the Woolomin terrane.

Devonian isolation is indicated by contrasts in the provenance of sedimentary rocks of this age in the two fold belts. These are most striking in the Late Devonian when quartz-rich craton-derived sandstones accumulated in the Lachlan, and quartz-poor volcanolithic sandstones were deposited in the New England Fold Belt. Complete separation of the depositional areas is strongly indicated at this time, suggesting either earlier dispersal (if Cawood's arguments are accepted) or continued independent development.

Clasts of a variety of intermediate and silicic plutonic rocks, and of silicified siltstone, hornfels, slate, and schist, occur in glaciogene conglomerates of Late Visean–Namurian age in the western part of the Tamworth terrane (McKelvey and White, 1964; White, 1968). These were derived from the west and are the first influx of clearly cratonic detritus. There appears to us a close temporal relationship between this event, a change in the overall composition of rocks extruded from the volcanic chain that bordered the Tamworth belt from andesitic to dacitic and rhyolitic, and folding and metamorphism in the eastern Lachlan Fold Belt (Kanimblan orogeny). Perhaps all are the product of the final accretion of the New England Fold Belt to the Lachlan (Leitch, 1974), although we cannot preclude the possibility of Late Carboniferous strike-slip movement along the suture (cf. Korsch and Harrington, 1981).

Detailed discussion of the relationship of the Lachlan Fold Belt to the Precambrian masses further west is beyond the scope of this paper. Widespread Late Devonian strata of broadly similar facies and provenance to those in the eastern Lachlan Fold Belt overlap from the Lachlan onto the shield, providing a linkage predating the Carboniferous New England accretion.

CONCLUSIONS

The accretionary history of the southern New England Fold Belt lasted from at least the Middle Cambrian until the latest Carboniferous. Five terranes developed during this period, three of which are subduction complexes and two that are arc-fringe and forearc basin elements. Although direct linkages are no older than Permian, comparisons of the detrital mineralogy of the terranes indicate that for at least parts of their histories they evolved in close mutual association, with rocks of similar age in different terranes sharing a common provenance. Subduction zone accretion was discontinuous, for although it is unclear whether the growth of individual subduction terranes was progressive or episodic, the Woolomin terrane is separated by a clear time-break from the Cockburn terrane, and a hiatus probably separates the latter from the Texas terrane.

Despite a long history of convergent plate interaction, we have identified no major exotic elements, neither oceanic plateau, arc fragment, nor microcontinental block. The absence of the last possibly reflects the rarity of microcontinents prior to the breakup of Gondwanaland. (However, detailed paleomagnetic and paleobiogeographic studies that have been most valuable in recognizing exotic terranes elsewhere have yet to be carried out in the New England Fold Belt.) The maintenance of the convergent system in a relatively constant position throughout most of the Paleozoic is probably a function of the absence of large accreted exotic elements. Conversely, it can be speculated that the arrival of a major terrane in the Early Permian, perhaps an oceanic plateau, led to the outstepping of the convergent margin (Packham and Leitch, 1974). Any such mass would now form part of the Lord Howe Rise, displaced by the opening of the Tasman Sea.

Thrust slices within the accretionary subduction complexes show a sequence of basaltic oceanic crust variously overlain by pelagic sedimentary rocks, ocean island basalt, and arc-derived clastics. This sequence records the movement of ocean floor an unknown but possibly great distance from ocean ridge to convergent margin. Whereas the lower section of each slice probably formed in regions remote from the developing fold belt and hence can be appropriately termed exotic, they are linked to the proximal arc clastics by intermediate lithofacies, and hence cannot be considered independent terranes.

Accretion of the New England Fold Belt to the Australian craton in the middle of the Carboniferous was accompanied by a change in the composition of volcanism along the western margin of the fold belt and by folding and metamorphism further west. There is little evidence of deformation within the fold belt related to this event. It is yet to be determined whether the fold belt evolved independently of the tectonic elements further west prior to the Carboniferous or whether accretion merely reversed the affects of an earlier (Silurian or Early Devonian) episode of dispersal.

Breakup and the onset of terrane dispersal commenced in the Early Permian and involved rifting, compressive, and strike-slip movements. Most deformation and regional metamorphism, as well as the emplacement of granite plutons and ultramafic masses, occurred during dispersal rather than accretion.

ACKNOWLEDGMENTS

We thank Dave Howell for his review of the manuscript, and Sheila Binns and Len Hay for their assistance in its preparation.

REFERENCES

Binns, R. A., 1966, Granitic intrusions and regional metamorphic rocks of Permian age from the Wongwibinda District, north-eastern New South Wales: Journal of the Proceedings of the Royal Society of New South Wales, v. 99, p. 5-36.

Cawood, P. A., 1976, Cambro-Ordovician strata, northern New South Wales: Search, v. 7, p. 378-379.

———, 1980, The geological development of the New England Fold Belt: Unpublished PhD Dissertation, University of Sydney.

———, 1982a, Tectonic reconstruction of the New England Fold Belt in the Early Permian: an example of development at an oblique-slip margin, *in* P. G. Flood and B. Runnegar, eds., New England geology, Voisey Symposium Volume: Armidale, N. S. W., University of New England, p. 25-34.

———, 1982b, Structural relations in the subduction complex of the Paleozoic New England Fold Belt, eastern Australia: Journal of Geology, v. 90, p. 381-392.

———, 1982c, Correlations of stratigraphic units across the Peel Fault System, *in* P. G. Flood and B. Runnegar, eds., New England geology, Voisey Symposium Volume: Armidale, N. S. W., University of New England, p. 55-61.

———, 1983, Modal composition and detrital clinopyroxene geochemistry of lithic sandstones from the New England Fold Belt (east Australia): a Paleozoic fore-arc terrain: Geological Society of America Bulletin, v. 94, p. 1199-1214.

———, 1984, A geochemical study of metabasalts from a subduction complex in eastern Australia: Chemical Geology, v. 43, p. 29-47.

———, in press, The development of the S. W. Pacific margin of Gondwana: Correlations between the Rangitata and New England Orogens: Tectonics.

Chappell, B. W., 1968, Volcanic greywackes from the Upper Devonian Baldwin Formation, Tamworth-Barraba district, New South Wales: Journal of the Geological Society of Australia, v. 15, p. 87-102.

Crook, K. A. W., 1960a, Petrology of Tamworth Group, Lower and Middle Devonian, Tamworth-Nundle district, New South Wales: Journal of Sedimentary Petrology, v. 30, p. 353-369.

———, 1960b, Petrology of Parry Group, Upper Devonian-Lower Carboniferous, Tamworth-Nundle district, New South Wales: Journal of Sedimentary Petrology, v. 30, p. 538-552.

———, 1980, Fore-arc evolution in the Tasman Geosyncline: the origin of the southeast Australian continental crust: Journal of the Geological Society of Australia, v. 27, p. 215-232.

Evans, P. R., and J. Roberts, 1980, Evolution of central eastern Australia during the late Paleozoic and early Mesozoic: Journal of the Geological Society of Australia, v. 26, p. 325-340.

Fergusson, C. L., 1982a, Structure of the Late Palaeozoic Coffs Harbour Beds, northeastern New South Wales: Journal of the Geological Society of Australia, v. 29, p. 25-40.

———, 1982b, An ancient accretionary terrain in eastern New England—evidence from the Coffs Harbour block, *in* P. G. Flood and B. Runnegar, eds., New

England geology, Voisey Symposium Volume: Armidale, N. S. W., University of New England, p. 63–70.

———, in press, The Gundahl Complex of the New England Fold Belt, eastern Australia: a tectonic melange formed in a Palaeozoic subduction complex: Journal of Structural Geology.

Flood, P. G., and C. L. Fergusson, 1982, Tectonostratigraphic units and structure of the Texas-Coffs Harbour region, *in* P. G. Flood and B. Runnegar, eds., New England geology, Voisey Symposium Volume: Armidale, N. S. W., University of New England, p. 71–78.

Flood, R. H., and S. E. Shaw, 1977, Two 'S-Type' granite suites with low initial $^{87}Sr/^{86}Sr$ ratios from the New England Batholith, Australia: Contributions to Mineralogy and Petrology, v. 61, p. 163–173.

Harrington, H. J., 1982, Tectonics and the Sydney Basin: Advances in the study of the Sydney Basin: Programme and Abstracts, 16th Symposium, Department of Geology, University of Newcastle, p. 15–19.

Jones, D. L., et al, 1983, Recognition, character, and analysis of tectonostratigraphic terranes in western North America, *in* M. Hashimoto and S. Uyeda, eds., Accretion tectonics in the circum-Pacific regions: Tokyo, Terra Scientific Publishing Co., p. 21–35.

Korsch, R. J., 1977, A framework for the Palaeozoic geology of the southern part of the New England Geosyncline: Journal of the Geological Society of Australia, v. 25, p. 339–355.

———, 1978, Petrographic variations within thick turbidite sequences: an example from the late Palaeozoic of eastern Australia: Sedimentology, v. 25, p. 247–265.

———, 1981a, Deformational history of the Coffs Harbour block: Journal of the Proceedings of the Royal Society of New South Wales, v. 114, p. 17–22.

———, 1981b, Some tectonic implications of sandstone petrofacies in the Coffs Harbour association, New England Orogen, New South Wales: Journal of the Geological Society of Australia, v. 28, p. 261–269.

———, 1982, Early Permian tectonic events in the New England Orogen, *in* P. G. Flood and B. Runnegar, eds., New England geology, Voisey Symposium Volume: Armidale, N.S.W., University of New England, p. 35–42.

———, and H. J. Harrington, 1981, Stratigraphic and structural synthesis of the New England Orogen: Journal of the Geological Society of Australia, v. 28 p. 205–226.

Leitch, E. C., 1974, The geological development of the southern part of the New England Fold Belt: Journal of the Geological Society of Australia, v. 21, p. 133–156.

———, 1975, Plate tectonic interpretation of the Paleozoic history of the New England Fold Belt: Geological Society of America Bulletin, v. 86, p. 141–144.

———, 1978a, Olistostromes and the onset of subduction: Bulletin of the Australian Society of Exploration Geophysicists, v. 9, p. 157–158.

———, 1978b, Structural succession in a Late Palaeozoic slate belt and its tectonic significance: Tectonophysics, v. 47, p. 311–323.

———, 1980, The Great Serpentine Belt of New South Wales: diverse mafic-ultramafic complexes set in a Palaeozoic arc, *in* A. Panayiotou, ed., Ophiolites: Proceedings of the International Ophiolite Symposium, Cyprus, 1979, p. 637–648.

———, 1981, Rock units, structure and metamorphism of the Port Macquarie block, eastern New England Fold Belt: Proceedings of the Linnean Society of New South Wales, v. 104, p. 273–292.

———, and P. A. Cawood, 1980, Olistoliths and debris flow deposits at ancient consuming plate margins: an eastern Australian example: Sedimentary Geology, v. 25, p. 5–22.

———, and S. G. A. Willis, 1982, Nature and significance of plutonic clasts in Devonian conglomerates of the New England Fold Belt: Journal of the Geological Society of Australia, v. 29, p. 83–89.

Lindsay, J. F., 1969, Stratigraphy and structure of the Palaeozoic sediments of the Lower Macleay Region, north-eastern New South Wales: Journal of the Proceedings of the Royal Society of New South Wales, v. 102, p. 41–55.

Mayer, W., 1972, Diamictite sedimentation in the Nowendoc-Mt. George area, northeastern New South Wales: Geological Society of Australia Joint Special Group Meeting, Canberra Abstracts, p. G14–G17.

McCarthy, B., et al, 1974, Late Carboniferous-Early Permian sequence north of Drake, N.S.W., dates, deposition and deformation of Beenleigh block: Queensland Government Mining Journal, v. 75, p. 324–329.

McKelvey, B. C., and A. H. White, 1964, Geological map of New England, 1:100,000, Horton Sheet (No. 290), with marginal text: Armidale, N.S.W., University of New England.

Morand, V. J., 1982, Structure and metamorphism in the central part of the Tia Complex, *in* P. G. Flood and B. Runnegar, eds., New England geology, Voisey Symposium Volume: Armidale, N. S. W., University of New England. p. 95–104.

Mory, A. J., 1982, The Early Carboniferous palaeogeography of the northern Tamworth Belt, New South Wales: Journal of the Geological Society of Australia, v. 29, p. 357–366.

Packham, G. H., and E. C. Leitch, 1974, The role of plate tectonic theory in the interpretation of the Tasman Orogenic Zone, *in* The Tasman Geosyncline—a symposium: Geological Society of Australia, Queensland Division, p. 129–155.

Roberts, J., and B. A. Engel, 1980, Carboniferous palaeogeography of the Yarrol and New England Orogens, eastern Australia: Journal of the Geological Society of Australia, v. 27, p. 167–186.

Scheibner, E., 1976, Explanatory notes on the tectonic map of New South Wales, scale 1:1,000,000: Geological Survey N.S.W., 283 p.

Shaw, S. E., and R. H. Flood, 1981, The New England Batholith, eastern Australia: geochemical variations in time and space: Journal of Geophysical Research, v. 86, p. 10530–10544.

Skilbeck, C. G., 1982, Carboniferous depositional systems of the Myall Lakes district, northern New South Wales, *in* P. G. Flood and B. Runnegar, eds., New England geology, Voisey Symposium Volume: Armidale, N.S.W., University of New England. p. 121–132.

Strusz, D. L., 1967, *Chlamydophllum, Iowaphyllum* and *Sinospongophyllum* (Rugosa) from the Devonian of New South Wales: Palaeontology, v. 10, p. 426–435.

Telford, P. G., 1972, Lower Devonian conodonts from the Silverwood Group, southeast Queensland: Proceedings of the Royal Society of Queensland, v. 83, p. 61–68.

Thomson, J., 1976, Geology of the Drake, 1:100,000 sheet: Sydney, Geological Survey N.S.W., 185 p.

Wass, R., and D. M. Dennis, 1977, Early Palaeozoic faunas from the Warwick-Stanthorpe region, Queensland: Search, v. 8, p. 207–208.

White, A. H., 1968, The glacial origin of Carboniferous conglomerates west of Barraba, New South Wales: Geological Society of America Bulletin, v. 79, p. 675–686.

Wilkinson, J. F. G., 1971, The petrology of some vitrophyric calcalkaline volcanics from the Carboniferous of New South Wales: Journal of Petrology, v. 12, p. 587–619.

Suspect Terranes in the Tasman Fold Belt System, Eastern Australia

Erwin Scheibner
Geological Survey of New South Wales
Sydney, Australia

The orogenic, composite, substantially Paleozoic Tasman Fold Belt System (Tasmanides) builds most of eastern Australia. This system represents the eastern margin of Gondwanaland, and it has developed as part of the tectonically complex southwest Pacific region.

This is the first systematic attempt to define suspect and stratotectonic terranes in eastern Australia. Only recently the existence of suspect terranes in this region was suggested by Scheibner (1982), Harrington (1983), and Powell (1983a, 1983b). Hard evidence in the form of reliable paleomagnetic and bioprovincial data is meager.

By earliest Cambrian time, a well-developed west-Pacific type active plate margin existed in eastern Australia. Hence continental breakup and sea-floor spreading, coupled with separation and dispersal of microcontinents and possibly some plate convergence, must have occurred during pre-Cambrian times. Precambrian complexes now constitute the basement of, and internal massifs within, the Tasmanides. The dispersed Precambrian microcontinental blocks became cores of separate tectonostratigraphic terranes. Collisional movement of these Precambrian basement blocks (microcontinental) and changes in the style of B-subduction (Mariana vs. Chilean) can explain the episodes of orogeny that punctuated development of the Tasman Fold Belt System, and these orogenic episodes also represented episodes of terrane accretion. The episodes of terrane accretion were followed, or sometimes overlapped, by episodes of terrane dispersion. Episodes of terrane dispersion were an integral part of episodic rearrangement of the active plate margin. Each such episode was characterized by creation of extensional features (volcanic rifts, marginal and interarc basins, and other basins) and subsequent closure of these and their inversion. During these processes individual terranes were further displaced relative to each other.

INTRODUCTION

The concept of tectonostratigraphic terranes (Jones et al, 1983) is gaining wide recognition as terrane analysis is being done in more orogenic regions. This concept is a natural outgrowth of plate tectonics. It is applicable to those regions that developed at active plate margins, especially where subduction of major oceanic plates has occurred. An important factor in the development of terranes seems to be oblique plate convergence (transpression) or divergence (transtension).

In orogenic belts for which active plate margin settings can be documented, all the terranes beyond the autochthonous miogeoclinal or paratectonic zone fringing a cratonal foreland should be considered "suspect" (cf. Coney et al, 1980; Williams and Hatcher, 1982). Experience has shown that diverse crustal fragments outboard of autochthonous paratectonic zones are often structurally uncoupled and therefore "suspect." They are "suspect" in that they originated away from the areas where they now occur. This means that these diverse crustal fragments could be allochthonous tectonostratigraphic terranes.

The existence of faulted boundaries to terranes is inherent in the definition and implies displacement between terranes, but the amount of displacement is not quantified. The boundaries between terranes are mostly sutures of various kinds. Some are suture zones with ophiolites marking plate convergence zones; some sutures only contain slivers of Alpine-type ultramafics, indicating deep penetration of these dislocations; and some sutures are cryptic, with mismatches in stratigraphic relations between adjoining terranes. Some terranes are continental or oceanic crustal blocks in which the basement/cover relationship is preserved; in others, thrusting at various crustal levels has caused the formation of nappes or lithospheric flakes.

Failure to recognize major tectonic junctions between adjacent terranes has led to many gross oversimplifications and misconceptions in the interpretation of orogenic belts (cf. Barber, this volume).

The Tasman Fold Belt System or Tasmanides exposed in eastern Australia is a composite, mainly Paleozoic, orogen. It formed part of Eastern Gondwanaland and has developed as part of the tectonically complex Southwest Pacific region. This region has been the site of continuous orogenic activity throughout Phanerozoic time (Packham, 1973). An active plate margin setting was a characteristic of eastern Australia during Paleozoic time (Oversby, 1971; Packham, 1973; Scheibner, 1972a, 1972b, 1974; Solomon and Griffiths, 1972). While the existence of an active plate margin setting for the Tasmanides has been recognized for over a decade, the presence of suspect terranes in this region has only recently been suggested (Scheibner, 1982; Harrington, 1983; Powell, 1983a, 1983b). A systematic attempt to define suspect and tectonostratigraphic terranes in the whole Southwest Quadrant of the circum-Pacific region has just been made (Scheibner in Howell et al, 1983). Hard evidence for these terranes in the form of reliable paleomagnetic and bioprovincial data is meager.

Here we will concentrate only on the Tasmanides, and mainly their southern part.

SUSPECT TERRANES IN THE TASMAN FOLD BELT SYSTEM

In tectonic analyses and syntheses of eastern Australia, usually only the Paleozoic development is considered. However, orogenic facies, ophiolites, and orogenic extrusives and intrusives indicate that by the earliest Cambrian a well-developed active converging plate margin existed here (Oversby, 1971; Scheibner 1972a, 1972b; Solomon and Griffiths, 1972; Crawford and Keays, 1978). Hence breakup, sea-floor spreading associated with separation, and dispersal of microcontinents must have occurred during pre-Cambrian time.

The Proterozoic rocks of the Australian craton should reflect the history of this pre-Cambrian passive and active plate formation.

Late Proterozoic Development

During the late Proterozoic, the Australian craton was covered by extensive platformal strata (the Central Australian Platform Cover, Geological Society of Australia, 1971; Plumb, 1979). Late accumulations of shallow water to continental sediments and lesser intraplate volcanics occurred in basins (Adelaide, Officer, Amadeus, cf. Fig. 1) that were associated either with intraplate rifting or compression (cf. Lambeck, 1983). Very thick accumulations were localized over the most recently cratonized mobile belts (Plumb, 1979). At a later stage, especially during the latest Proterozoic glaciations, large areas of platform cover extended from western Tasmania to western Australia (Plumb, 1979, Figure 9).

The major late Proterozoic rift zones were discordant to the eastern margin of the craton and the subsequent Paleozoic Tasman active plate margin. The remains of these rifts are located opposite the salients in the Tasmanides, indicating some deeper crustal relationships (cf. Fig. 1; more detail in Scheibner, in preparation). These late Proterozoic rift zones are similar in many respects to aulacogenes (Olenin, 1967; Preiss et al, 1981; Rutland, 1976; Scheibner, 1972b, 1974; von der Borch, 1980).

Controversy exists regarding the origin and the tectonic development of the Adelaide Fold Belt. An earlier interpretation of Sprigg (1952) suggested formation of the Adelaidean sediments as a miogeosynclinal (miogeoclinal) continental terrace, implying existence of a continental margin immediately to the southeast. The subsequent aulacogene model (see above) has been disputed by Preiss et al (1981), who argued that it does not represent a failed arm of a triple junction. They suggested a model in which the Adelaide Rift developed as a protracted multiple rifted arch system in a passive continental margin setting (cf. Veevers and Cotterill, 1978). von der Borch (1980) modified the earlier general aulacogene model by suggesting a Cambrian triple junction between the intracratonic Central Flinders Zone (aulacogene) and the ancient continental margin to the south.

While accepting the modified aulacogene and continental margin rift model of von der Borch (1980), we could speculate that the prolonged late Proterozoic rifting and extra-arch basin formation (Preiss et al, 1981; von der Borch, 1980) were related to continental breakup further east than the subsequent Cambrian breakup that localized the ensimatic Kanmantoo Trough at the continental margin.

The Proterozoic rocks that occurred further east could have formed in passive or active plate margin settings. They now constitute an unknown proportion of the basement and inliers (internal massifs) in the Tasmanides. The character and extent of the Precambrian basement in the Tasmanides is subject to controversy. Views range from a completely ensialic setting (Rutland, 1976), to an interspersed ensialic and ensimatic setting (Scheibner, 1974, 1976), to a mostly ensimatic setting (Crook, 1980). The late Proterozoic Central Australian Platform Cover, and especially the Adelaide and Amadeus rift zones that developed adjacent to the active plate margin, could be considered equivalent to miogeoclinal or paratectonic belts in other orogens. The eastern limit of these belts is the western boundary of the Tasman Fold Belt System (Fig. 1). All the terranes to the outboard of these miogeoclinal belts should be considered suspect.

Early Paleozoic Development

Whether or not the Kanmantoo Trough and similar features were separate terranes might be disputed. The Kanmantoo Trough was probably ensimatic but it was filled with sediments that accumulated in continental slope and rise tectonic settings (von der Borch, 1980). These Atlantic-type continental margin complexes were involved in early Paleozoic plate convergence and were strongly telescoped and deformed during the Late Cambrian closure. In some areas faults separate these orthotectonic complexes from the paratectonic belts (Adelaide Fold Belt), yet the facies transition from the neighboring paratectonic belts was only slightly disrupted (von der Borch, 1980). So the interpretation of the continental margin complexes as separate tectonostratigraphic terranes probably will remain a subject of dispute; however, consideration should be given to the fact that complexes further outboard are progressively more allochthonous (e.g., the eastern part of the Kanmantoo–Glenelg terrane).

Suspect and Tectonostratigraphic Terranes in the South Australian–Victorian–New South Wales Segment of the Tasman Fold Belt System

The presently recognized terranes are shown in Figure 2. A flow chart illustrating the time/space relationship of these terranes is shown in Figure 4. Cartoons of lithospheric sections in natural scale illustrate the hypothetical tectonic development (Fig. 5).

The Kanmantoo–Glenelg Terrane

The Kanmantoo–Glenelg terrane is a composite regionally metamorphosed terrane composed of ?late Proterozoic to Cambrian sediments, slate and graywacke (some turbidites), volcanics, and ?ophiolites. Basement consisted of thinned continental crust (Gawler–Willyama orogenic domain; cf. Geological Society of Australia, 1971)

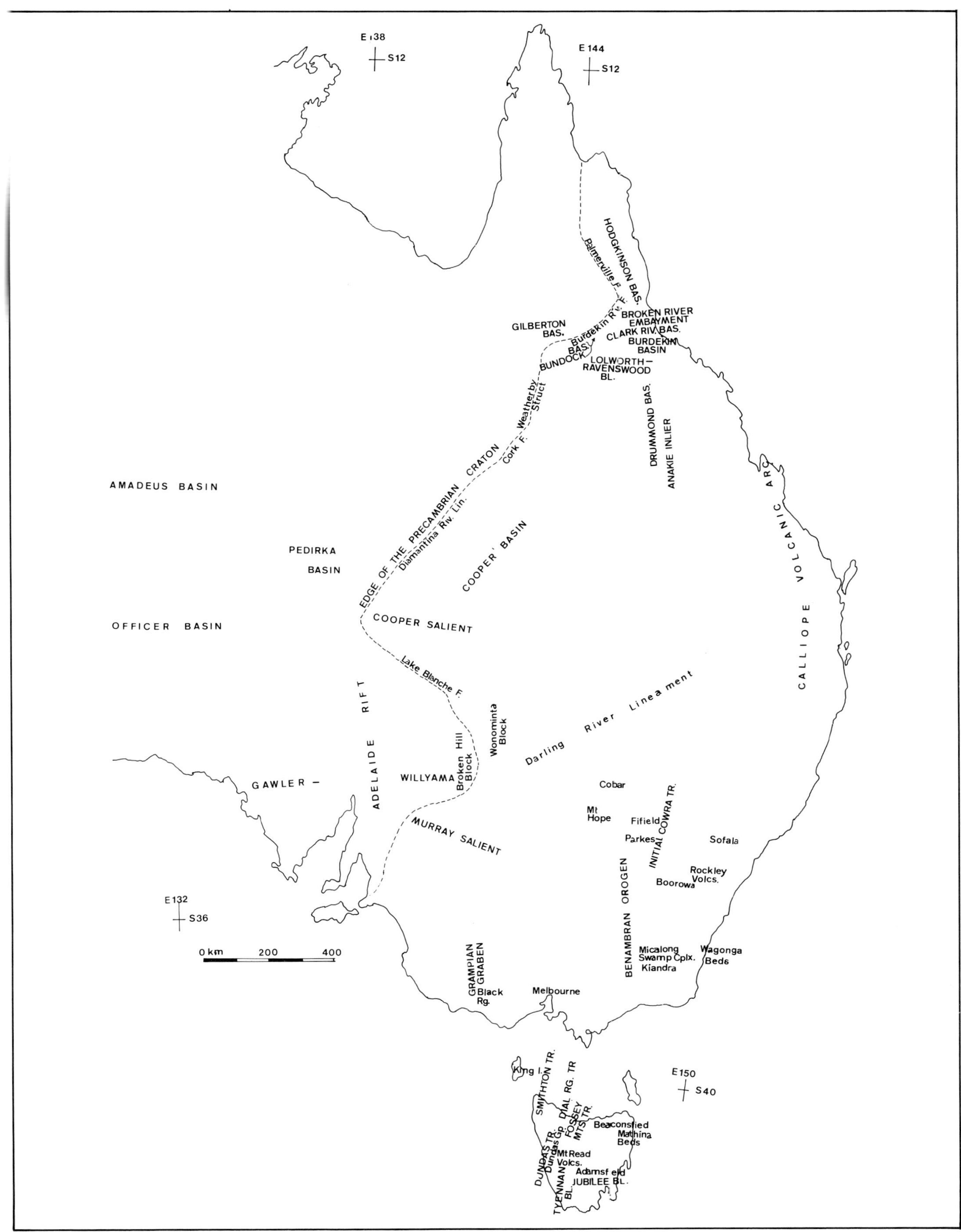

Figure 1—Locality diagram.

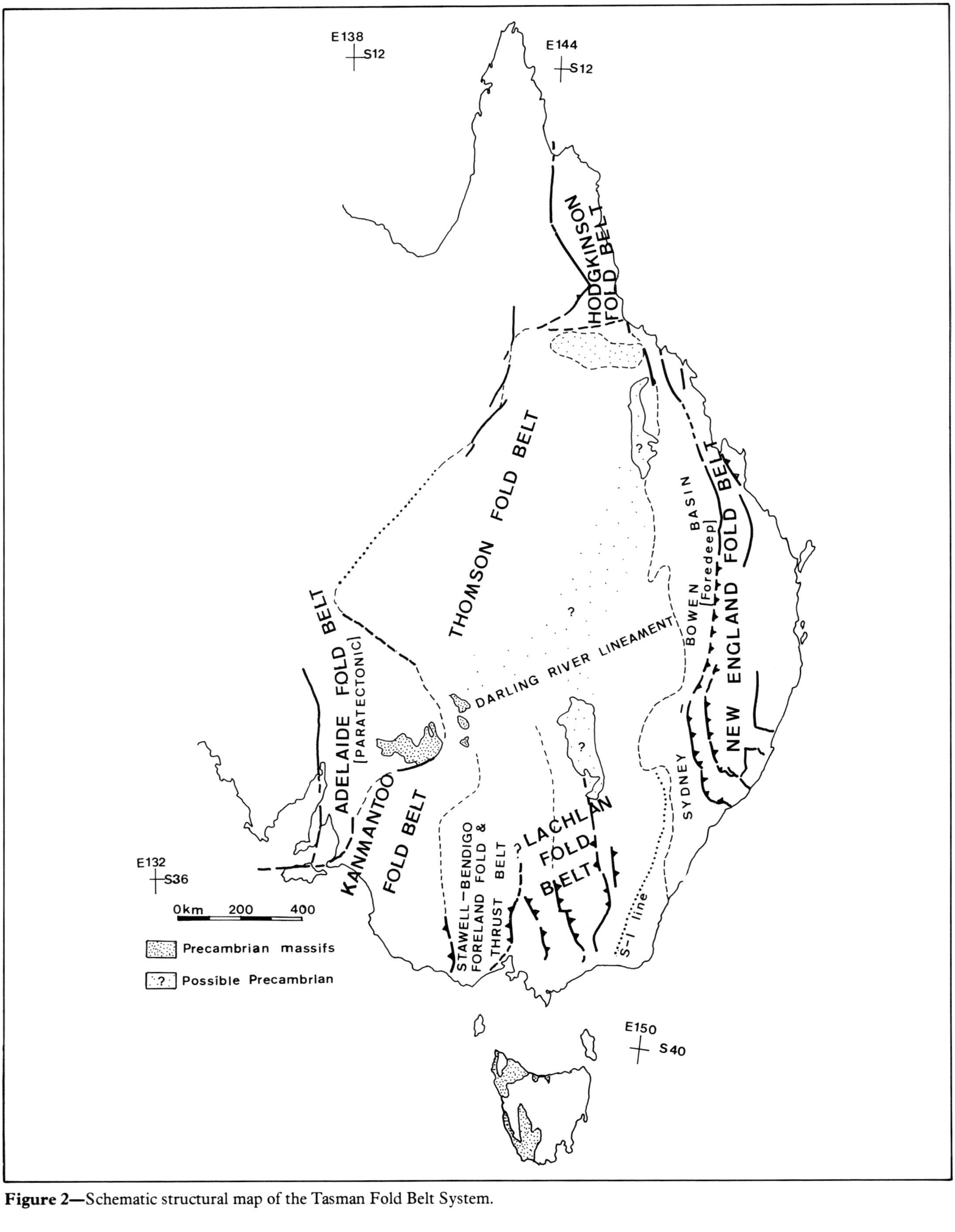

Figure 2—Schematic structural map of the Tasman Fold Belt System.

in the west and north and was probably replaced by oceanic crust further east. According to von der Borch (1980), sedimentological evidence indicates that separation in the Kanmantoo Trough area (innermost edge of this terrane) occurred by earliest Cambrian time. The earlier suggestion (Scheibner, 1972a, 1974) of the ensimatic character of this trough was based on the substantial thickness of over 6 km (3.7 mi) (Daily and Milnes, 1973) of rapidly deposited deep-water sediments. Direct evidence for oceanic crust here is missing. Further east, greenstones (mainly mafic volcanics) of uncertain origin occur in the Black Ranges in Victoria (cf. VandenBerg, 1978, and in Cooper and Grindley, 1982). These greenstones could have been part of the late Proterozoic oceanic crust (including basement of a volcanic arc) outboard of the continental margin up to the western limit of the Victorian microcontinent that now forms the basement of the Melbourne terrane (3) and adjacent terranes 2 and 4 (Fig. 3).

The eastern boundary of the Kanmantoo-Glenelg terrane is the Woorndoo Fault (cf. VandenBerg in Cooper and Grindley, 1982), and to the west of the fault, in the Stavely Belt, there are Early Cambrian or older andesitic breccia and serpentinite, with lesser rhyolite, andesite, basalt, chert, and lapilli tuff together with intercalated volcaniclastic sediments. These rocks comprise the Mt. Stavely complex. According to Crawford (1983), there are continental margin andesites at Mt. Stavely, while high-Mg low-Ti andesites occur at Mt. Dryden.

Orogenic deformation and metamorphism started in the Middle Cambrian and culminated with the emplacement of Ordovician granites (mainly S-type) (cf. Daily in Cooper and Grindley, 1982). This is the time of terrane accretion. During the Late Silurian to Early Devonian, some terrane dispersion in Victoria is indicated by the formation of the Grampian graben (rift) filled with acid volcanics and rapidly deposited (about 6 km [3.7 mi]) continental quartzose sandstone, red siltstone, and mudstone (cf. VandenBerg in Cooper and Grindley, 1982). In New South Wales, Late Cambrian to Ordovician continental to shallow-water sediments represent molassic complexes (Scheibner, 1974; Webby in Cooper and Grindley, 1982). Some Early Devonian sediments and volcanics (Cootawundy Beds) are similar to the fill of the Grampian graben. Permian and younger cover sediments together with the molassic complexes represent overlap sequences.

The Stawell-Bendigo Terrane

The Stawell-Bendigo terrane is an allochthonous terrane composed of Cambrian to Ordovician, partly metamorphosed sediments (turbidites), which were thrust eastward either during the Silurian or the Middle Devonian (terrane accretion). Structurally, this terrane is the Stawell-Bendigo Foreland Fold and Thrust Belt, the allochthonous thin-skin tectonic character of which was only recently recognized by Cox et al (1983).

The eastern front of this terrane is the Heathcote Greenstone Belt. Besides low-Ti lavas (boninites) and some ultramafics, apparently superposed midocean ridge basalt lavas (Crawford, 1983) are also present. Fossils in associated sediments (shales and cherts) indicate an Early Cambrian to Ordovician age, and at Lancefield the Riddell Grits (Early Gisbornian upwards) drape across the Cambrian sequence (VandenBerg, personal communication). The interpretation suggested here is that this greenstone belt is at the front and sole of the terrane. The fact that the Late Ordovician sequence passes into the Silurian (VandenBerg, personal communication), would imply submarine thrusting along listric thrust faults. The greenstones and associated sediments represent immediate basement at least for part of this terrane. In thrust contact is the deeper basement, formed by the Victorian microcontinent (Proterozoic rocks).

The above rocks were intruded and stitched by Early and Late Devonian S- and I-type granites with which felsic volcanics are associated. Some of this igneous activity was associated with limited terrane dispersion (rifting). Cover rocks are Permian and younger in age.

Melbourne Terrane

The Melbourne terrane is formed by Early to Middle Cambrian andesites, volcaniclastics, shale, some cherts, and rare limestone lenses. Middle Ordovician (Early Ordovician on Mornington Peninsula) to Early Devonian sediments (mostly turbidites) (cf. VandenBerg, 1978) were deposited in a basin (Melbourne Trough) that was at the back of the Kanmantoo orogen and was a foreland basin for the Benambran orogen. These rocks form an autochthonous cover on the Victorian microcontinent (Proterozoic rocks). The neighboring terranes (2 and 4 on Fig. 3) were accreted onto this terrane during Late Ordovician-Early Silurian Benambran orogeny. The fill of the Melbourne Trough was not deformed during this accretionary event. Orogenic deformation occurred during Middle Devonian Tabberabberan orogeny.

Late Devonian granites and associated felsic volcanics plus sediments are partly associated with limited terrane dispersion of the order of a kilometer to tens of kilometers and represent transitional tectonic complexes. Cover rocks are Permian and younger in age.

Howqua-Tabberabbera Terrane

The Howqua-Tabberabbera terrane is here considered to be an allochthonous (thrust westward) terrane composed of Ordovician sediments (turbidites) in the east and in the west the Cambrian Mt. Wellington Greenstone Belt that is in faulted contact with neighboring areas (cf. VandenBerg, in Cooper and Grindley, 1982). This belt comprises dismembered ophiolites and low-Ti mafic lavas (Crawford, 1983), together with sediments of Cambrian, Ordovician, and Silurian-Early Devonian age. The interpretation here is that this greenstone belt is at the front and sole of the Howqua-Tabberabbera terrane. Accretion occurred during the Late Ordovician-Early Silurian Benambran orogeny. This terrane structurally could be interpreted as the foreland fold and thrust belt of the Benambran orogen (the older part of the Lachlan Fold Belt) that comprises terranes 5, 6, 7, and 8 (Fig. 3).

The greenstones and associated sediments are the immediate basement. Deeper basement is formed by the Victorian microcontinent (Proterozoic rocks) probably in thrust relationship.

Silurian and Early Devonian shallow-water sediments and some Devonian granites could be related to limited,

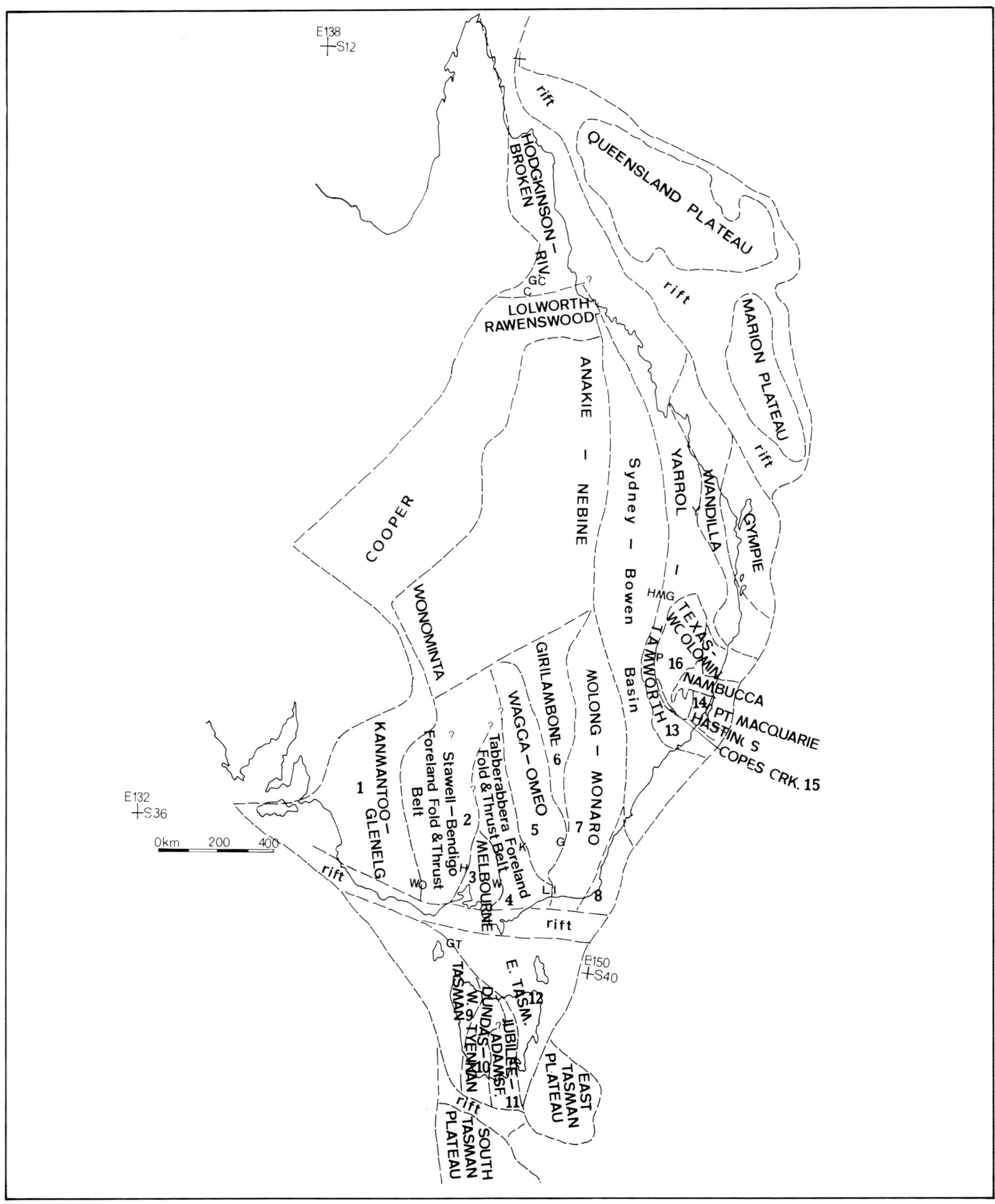

Figure 3—Suspect terranes in the Tasman Fold Belt System. Numbers identify terranes in Figures 5, 7, and 8. C = Clark River Fault; G = Gilmore Suture; GC = Gray Creek Fault Zone; GT = Gambier-Tamar Fracture Zone; H = Heathcote Greenstone Belt and M. William Fault; HMG = Hunter-Mooki-Goondiwindi Thrust System; K = Kiewa Thrust (Suture); LI = Long Plain-Indi Fault Zone; P = Peel Fault Zone (Thrust); Y = Yarrol Fault Zone; W = M. Wellington Greenstone Belt; WO = Woorndoo Fault.

quantitatively unknown, terrane dispersion, and similarly the Late Devonian Molassic sediments and associated felsic and rare mafic (bimodal) volcanics (the last part of a transitional tectonic province, cf. Geological Society of Australia, 1971). The Tabberabberan (Middle Devonian) and Kanimblan (Carboniferous) orogenies affected this terrane.

Permian and younger rocks represent cratonic cover.

Wagga–Omeo Terrane

The Wagga–Omeo terrane is a high-T regional metamorphic terrane composed of Ordovician quartzwackes and slates (turbidites), cherts, and some basic volcanics (?dismembered ophiolites). Metasediments locally reached hornblende amphibolite facies, and there are migmatites associated with anatectic granites. This metamorphism and deformation was caused by the Benambran orogeny, which probably resulted from the collision of the next easterly terrane (Molong–Monaro, 6 on Fig. 1) with the Victorian microcontinent (Melbourne terrane). Terrane accretion occurred during this collision, and the Wagga–Omeo terrane was thrust westward along the Kiewa thrust or suture. Anatectic, S-type Silurian granites occur in this terrane but do not occur further west in the structurally lower terranes.

Some terrane dispersal of the order of a kilometer to tens of kilometers and rifting occurred during the Silurian and Early Devonian (VandenBerg, in Cooper and Grindley, 1982). The Silurian rocks (felsic and andesitic volcanics and sediments) are strongly cleaved and foliated and are extensively metamorphosed to greenschist facies. However, locally they are flat lying and unaltered associated with A-type granitoids (VandenBerg, personal communication). While the Early Devonian rocks (felsic volcanics and sediments) in the south form broad open synclinal structures, in the north around Cobar they are strongly deformed. In several belts close to the Lachlan River, volcanic rifts were filled with bimodal, but prevailingly felsic, volcanics intruded by A- and S-type comagmatic granites.

Late Devonian to Early Carboniferous continental sediments, red beds, and volcanics with rare basic volcanics (bimodal) represent transitional tectonic complexes. Terminal deformation occurred during the Carboniferous Kanimblan orogeny. Permian and younger rocks represent the cover.

Girilambone Terrane

The Girilambone terrane (?correlative of the Anakie–Nebine terrane in the Thomson Fold Belt) is a metamorphic terrane composed of Girilambone Group, which consists of pre-Silurian quartz- and graywackes alternating with slates (turbidites), cherts, basic volcanics, and ultrabasics (?dismembered ophiolites). There is no fossil evidence for the age of the Girilambone Group. These rocks are intruded by Early to Middle Silurian granites and could be as old as Late Precambrian to Cambrian. To the east they seem to be basement to the neighboring Ordovician volcanic arc (Molong–Monaro terrane). Some authors interpret this and the previous terranes as one geotectonic unit (Pogson, 1982) formed in a backarc basin setting during the Ordovician. Data indicate that the Wagga–Omeo and Girilambone terranes were deformed and accreted during the Benambran orogeny. The boundary between them is the continuation of the Gilmore Suture. Shallow-water Devonian sediments and felsic volcanics are associated with rifting and reflect some later terrane dispersal. A peculiar feature of this terrane is the circular gabbro-peridotite, ?Alaskan-type intrusives of Early Devonian age that occur mainly around Fifield. Late Devonian to Carboniferous continental sediments represent molassic complexes. The Carboniferous Kanimblan orogeny was the terminal paroxysm. Post-Carboniferous rocks represent cover rocks.

Molong–Monaro Terrane

The Molong–Monaro terrane is a composite terrane comprised of ?Late Cambrian to Ordovician arc volcanics (partly shoshonitic) in the west, and forearc basin quartz-rich turbidites, slates, and cherts in the east. In the east in New South Wales and in the southwest in Victoria, flyschlike sedimentation continued into the Early Silurian. Basement in the west was Girilambone Group rocks, plus probably Proterozoic complexes, building a hypothetical microcontinent termed the Molong microcontinent here (cf. Fig. 5). The eastern limit of the Molong microcontinent is the S-I line of White and Chappell (1977). This line is the boundary between an area with mixed S- and I-type granites* in the west as opposed to only I- and A-type granites to the east. The basement immediately east of the S-I line under the forearc basin sediments was probably oceanic, and it is suggested that it was subsequently tectonically underplated (underthrust during subduction-collision) by a hypothetical Proterozoic oceanic plateau, probably a volcanic arc. Partial melting of this underplated terrane would explain the geochemistry of igneous rocks east of the S-I line (Beams, 1980; Williams et al, 1983.)

Accretion of the Molong–Monaro terrane occurred during the Early Silurian Benambran orogeny, when the Molong and Victorian microcontinents collided, closing the intervening Wagga Marginal Sea. The Gilmore Suture forms the boundary between these two terranes (Scheibner, 1982). The boundary between terranes 6 and 7 is placed along the Parkes Thrust and its northerly and southerly projections.

Middle to Late Silurian (Tumut Trough) and Middle Silurian to Middle Devonian (Hill End Trough) sediments (including turbidites) and volcanics represent basin formation and rifting associated with terrane dispersal. Initial volcanic rifts or volcanotectonic depressions are characterized by bimodal but dominantly felsic subaerial and submarine volcanism. At least locally (Tumut Trough), rifting progressed through breakup, crustal separation, and associated sea-floor spreading as indicated by the Coolac ophiolite suite (Ashley et al, 1979). S- and I-type granites

*Chappell and White (1974) used the terms S- and I-type to refer to organic granitoids derived from sedimentary and igneous source rocks, respectively. A-type felsic granitoids are thought to have been derived from crust that had previously produced I-type magmas, so that source rocks were residual from that prior melting (Collins et al, 1982).

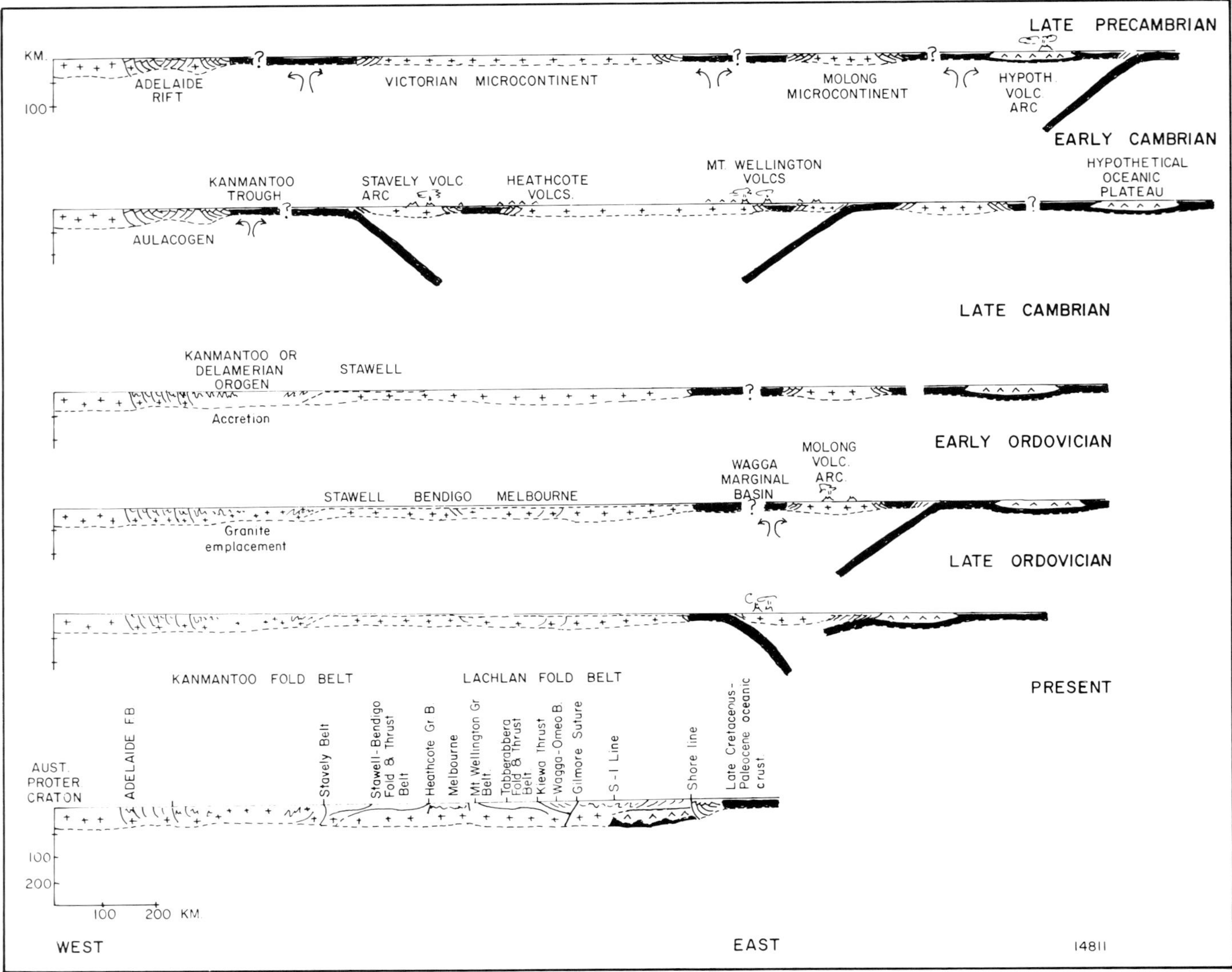

Figure 4—Cartoons showing development of the Kanmantoo Fold Belt. Question marks indicate decoupling and an unknown amount of oceanic crust.

and associated volcanics were emplaced in tectonic highs between the troughs and basins. These tensional features were progressively closed during the Devonian Bowning and Tabberabberan orogenic events. Molassic complexes followed, starting in the late Middle and Late Devonian. Locally, volcanic rifting occurred, as indicated by bimodal volcanics, followed by molassic, mostly continental sediments, including red beds.

Final deformation occurred during the Carboniferous Kanimblan orogeny, followed by emplacement of post-kinematic I-type and sometimes A-type granites (Collins et al, 1982). Permian and younger rocks form the cover.

Narooma Terrane

The Narooma terrane is an allochthonous terrane composed of Ordovician, mainly quartz-rich turbidites, cherts, and some volcanics, all representing outer arc slope and trench (?abyssal plain) strata (Powell, 1983a, 1983b). Terrane accretion occurred during the Silurian Benambran orogeny. Late Devonian to Carboniferous molassic complexes comprising bimodal volcanics and shallow-water and continental sediments followed. This rifting appears to have followed the western boundary of this terrane. Post-kinematic Carboniferous granites have an A-type character (Collins et al, 1982). Permian and younger rocks represent cover.

TECTONIC INTERPRETATION

The relationship and tectonic development of the described terranes is shown schematically in Figure 4.

Late Proterozoic to Ordovician Tectonic Development

The rock record indicates that by the earliest Cambrian time a well-developed west-Pacific type active plate margin existed in the area of present eastern Australia. Hence continental breakup, sea-floor spreading coupled with separation and dispersal of microcontinents, and possibly some plate convergence, must have occurred

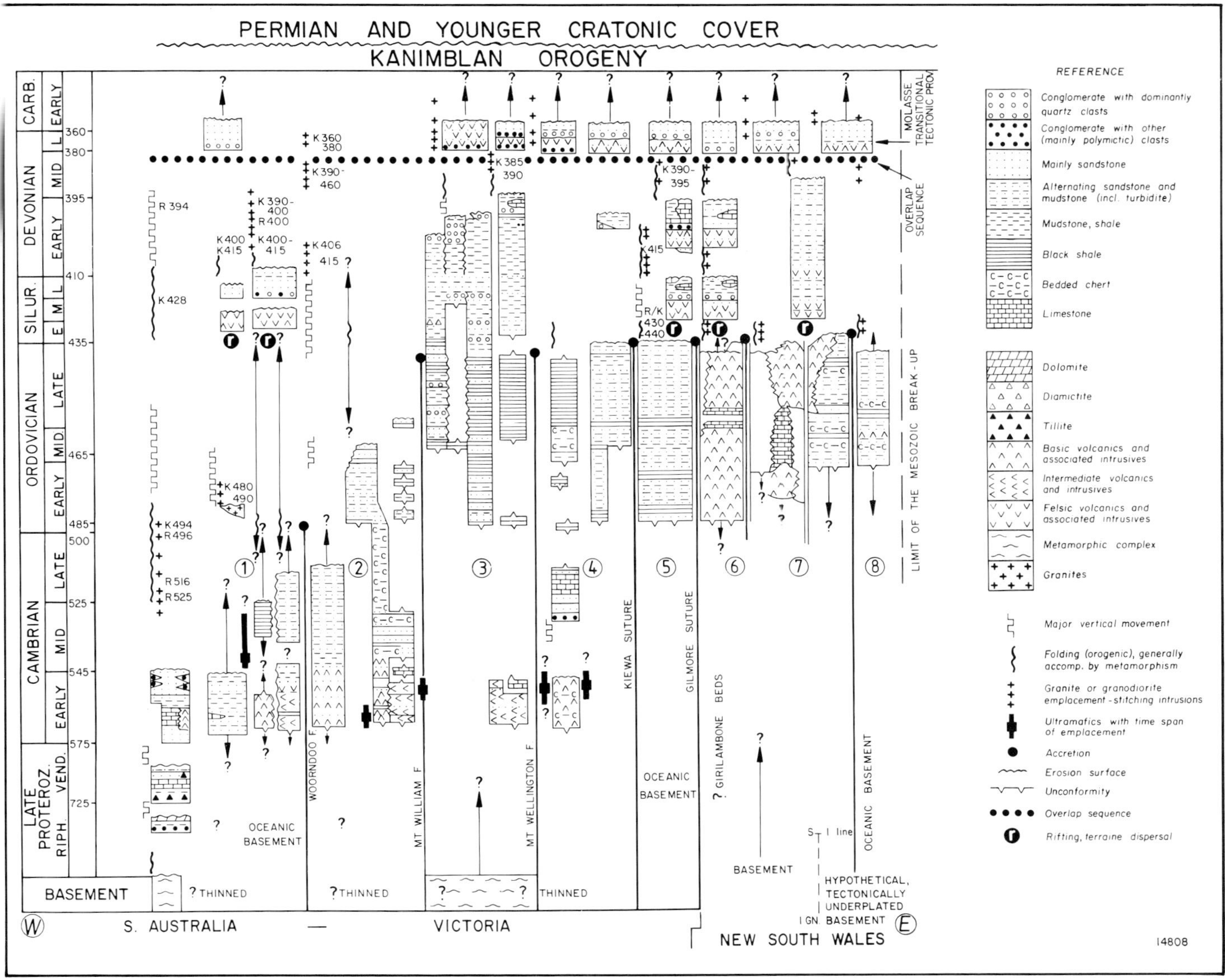

Figure 5—Flow chart of terrane accretion for the southern part of the Tasman Fold Belt System. Numbers in circles indicate terranes as shown on Figure 3. Data adopted from Cooper and Grindley (1982), slightly modified.

during pre-Cambrian times. These Precambrian complexes now constitute part of the basement and internal massifs (inliers) in the Tasmanides. The dispersed Precambrian microcontinental blocks (Victorian and Molong) became cores of separate tectonostratigraphic terranes.

The following paleogeographic reconstruction is interpreted from available data for the latest Precambrian. The rifted and extended Australian craton with an extra-arch basin (Preiss et al, 1981) or rift (aulacogene) occurred in the area of the present Adelaide Fold Belt. Outboard to the east was a hypothetical marginal sea of unknown size separating the craton from the Victorian microcontinent. Further east was another marginal sea separating the Victorian and Molong microcontinents. Outboard of the plate margin, embedded in the Ur-Pacific oceanic plate, was a hypothetical oceanic plateau formed by an intraoceanic volcanic arc (Fig. 4).

During the earliest Cambrian, the ensimatic Kanmantoo Trough formed adjacent to the Adelaide Rift, which became an aulacogene (von der Borch, 1980). The Kanmantoo Trough was rapidly filled with continental slope and rise sediments (von der Borch, 1980), and it was part of a wider (and older) marginal sea.

Several belts of Early Cambrian greenstones, including volcanic arc rocks, low-Ti lavas (boninites), and apparently superimposed midoceanic ridge basaltic (MORB) lavas (Crawford, 1983) occur in central Victoria, in an area underlain by the Victorian microcontinent. According to Crawford (1983), subduction of a backarc basin spreading center under an intraoceanic volcanic arc would cause formation of boninites and MORB lavas. The still active spreading center would provide the heat source to generate low-Ti lavas from hydrous, shallow sub-backarc basin and subarc mantle. Continued activity of the subducted spreading center could provide the superimposed MORB lavas. Another possibility is that the boninites were formed during the extension that caused formation of the ensimatic Kanmantoo Trough.

While some of the Cambrian volcanic arc complexes are autochthonous under the Ordovician–Devonian

Melbourne Trough sequence (VanderBerg in Cooper and Grindley, 1982), it is here suggested that the others (the Mt. Wellington and Heathcote Belt) are allochthonous.

One possible explanation of the two western greenstone belts (Stavely and Heathcote) is that they originated by assumed eastward subduction under the Victorian microcontinent and formation of boninites during rifting of the volcanic arc (cf. Fig. 4). Another possibility is that the Stavely Volcanic Arc and the Heathcote belt are unrelated and were accreted onto the Victorian microcontinent after the Early Cambrian. The large size (about 400 km [249 mi]) of the Victorian microcontinent assumed on the basis of distribution of crustal-derived granites would require westerly subduction and subsequent rifting to explain the Mt. Wellington greenstone belt on the east.

Diachronous orogenic deformation at the Australian active plate margin commenced in Middle Cambrian and continued into the Ordovician (Delamerian orogeny of Thomson, 1969). The inner marginal sea, including the Kanmantoo Trough, was inverted by collisional movement of the Victorian microcontinent. Retrograde metamorphism affected the Precambrian craton (for example the 520 ± 40 Ma thermal pulse in the Broken Hill Block; Harrison and McDougall, 1981) and the paratectonic cratonic cover was deformed. This is then deformation of the Adelaide Fold Belt in South Australia (Thomson, 1969; Preiss et al, 1981). Terrane stitching orogenic granites (Late Cambrian to Early Ordovician, cf. Milnes et al, 1977) were emplaced not only in the inverted Kanmantoo–Glenelg terrane, but also in the old craton at the margin (cf. Thomson, 1969). The orogenic belt that resulted from the deformation of the Early Cambrian active plate margin is referred to as the Kanmantoo Fold Belt (Scheibner, 1972a, 1972b, 1976).

The late orogenic (molassic) Middle Cambrian to Ordovician sediments of the Kanmantoo orogen were mostly shallow marine to continental (cf. Webby, in Cooper and Grindley, 1982), but a wide foreland basin developed on the Victorian microcontinent above the earlier orogenic complexes (Cox et al, 1983). This basin included the Stawell, but mainly the Ballarat (Bendigo) and Melbourne Troughs (VandenBerg, in Cooper and Grindley, 1982).

Starting possibly in the latest Cambrian, but definitely in earliest Ordovician (Sherwin, 1979), the Molong Volcanic Arc developed east of, and completely detached from, the Victorian microcontinent. It represents a separate terrane. Most authors of published models (Oversby, 1971; Packham, 1973; Scheibner, 1972b, 1974; Cas et al, 1980) have agreed on a model in which the Ordovician volcanics in New South Wales formed a volcanic island arc bordered on the west by a marginal sea (Wagga Marginal Sea) and an open ocean (Monaro Slope and Basin) on the east. Such a model was mainly based on the interpretation that the Ordovician volcanics are similar to those that characterize volcanic island arcs.

Recently, results of geochemical investigations of the Ordovician volcanics by L. Wyborn (1977) and Owen and Wyborn (1979) have become available. The results are from the southern part of the volcanic arc around Kiandra. These authors detected three types of igneous suites: minor tholeiitic basalts (Jagungal Volcanics); island arc tholeiites (Gooandra Suite); and shoshonites (Nine Mile Suite). Owen and Wyborn (1979) considered the geochemistry of other volcanics from the Molong Volcanic Arc and suggested that the Walli Andesite at Clifenden, the Kenyu Formation north of Boorowa, and the Sofala Volcanics at Sofala are all shoshonitic. These authors concluded that the controversy surrounding the generation and diverse setting of Cenozoic shoshonites precludes any well-founded speculation on the plate tectonic setting of the Ordovician Molong Volcanic Arc. Owen and Wyborn (1979) speculated that if Johnson et al's (1978) arguments are valid, then arc volcanism can occur by uplift of modified mantle without contemporaneous subduction. If this can also be applied to magmas less enriched in large-ion lithophile elements than the shoshonites, then the Gooandra Volcanics (island arc tholeiites) could also have originated without contemporaneous subduction. According to Owen and Wyborn (1979), all that can be safely concluded is that mantle under the Molong Volcanic Arc was modified. This modification could have originated from the asthenosphere alone, or been due to subduction, and this subduction could have preceded the volcanism by a considerable time interval.

L. Wyborn (1977) rejected the subduction model completely and suggested that the discussed rocks formed during Ordovician rifting and fracturing of a Late Precambrian continental crust in a marginal plateau environment.

New paleontological data (Kilpatrick and Fleming, 1980; Sherwin, 1979) indicate that sedimentation in the Wagga Marginal Basin and volcanism on the Molong Volcanic Arc started in the earliest Ordovician or even latest Cambrian. The previously published models have to be amended in this respect. The volcanism lasted for about 65 Ma or more. The above-discussed geochemical data and conclusions are pertinent to the later half of the development of the Molong Volcanic Arc, and it can be speculated that perhaps geochemical data are only available from the matured volcanic arc. I think that the balance is tipped towards an orogenic converging plate margin setting, rather than a passive diverging margin as suggested by L. Wyborn (1977).

The real problem is the uncertainty about the origin of modern shoshonitic suites (cf. Johnson et al, 1978). In the Southwest Pacific, modern shoshonites occur mostly in areas where there is some older continental-type crust. Coulon and Thorpe (1981) came to the conclusion that shoshonites form where the continental crustal thickness exceeds 20 km (12 mi).

It is here concluded that there must have been some earlier continental-type crust in the areas of Ordovician shoshonitic volcanism on the Molong Volcanic Arc. This crust could have originated during the previous orogenic episode, but it may well have been the separated microcontinent (Molong microcontinent) discussed earlier. It was also mentioned that the Girilambone Group rocks and correlatives (Jindalee Group) could represent part of the Molong Volcanic Arc basement on the west. On the east the Late Ordovician Rockley Volcanics interfinger with the flyschlike Triangle Group, and no older basement is known. However, as mentioned in the terrane description, based on the presence of S-type granites, the Molong microcontinent could have extended east to the S-I line of White et al (1976).

Most authors envisaged westerly dipping subduction under the Molong Volcanic Arc. However, the evidence is not completely conclusive. The shape of the arc has been distorted subsequently during tensional and compressional events. The Parkes and Molong segments of the volcanic arc are separated by a belt of intensively deformed undated flyschlike rocks that have been interpreted as fill of a possible early interarc basin (Initial Cowra Trough of Scheibner, 1974). If this proves correct, this tensional feature would indicate operation of Mariana-type B-subduction. The backarc basin (Wagga Marginal Sea) would be expected also to be extensional, perhaps partly floored with oceanic crust, but certainly including extended earlier continental crust. Part of this marginal basin formed during pre-Ordovician time and part during the Ordovician, possibly as a backarc basin.

Using the calculations of median volcanic arc values of Dickinson (1973) and accepting 65 Ma as the minimal duration of the Molong Volcanic Arc, the arc-trench gap would have been about 130 to 150 km (81–93 mi) wide, and the width of the arc over 80 km (50 mi). This means that in the model of westward subduction, the Monaro Slope and Basin sediments (Scheibner, 1974) would mainly represent forarc basin fill (this interpretation supports Crook, 1980), and only in the South Coast area of New South Wales could we expect to find accretionary wedge rocks. Perhaps the Wagonga Formation, which contains some basic volcanics, represents this accretionary wedge (Scheibner, 1974, 1982). Recently, Powell (1983a, 1983b) recognized that the Ordovician belt at Narooma has stripy cleavage, which he interpreted as characterizing an accretionary prism setting (outer arc slope). These rocks are here referred to as the Narooma terrane. Some uncertainty in the westward subduction model is presented by the eastward younging of Ordovician arc rocks. The oldest rocks occur west of Parkes and the youngest on the east, around Sofala and Rockley. If the above distribution is real and the conclusions of Dickinson (1973) about progressive migration of magmatic arcs away from the trench is correct, then subduction could have been towards the east. In the southern part of the volcanic arc, L. Wyborn (1977) noted eastward progression of volcanics from tholeiitic through island arc tholeiitic to shoshonitic. This would also support eastward subduction.

I am inclined to interpret the data as indicating westward subduction followed by a flip of subduction eastward, allowing subduction of any oceanic lithosphere (Wagga Marginal Basin) west of the Molong Volcanic Arc. The fill of this marginal sea was deformed, metamorphosed, and gave rise to the Wagga–Omeo metamorphic belt. The marginal sea and volcanic arc complexes were sutured along the Gilmore Suture (Scheibner, 1982). This is best documented in the south where the Gilmore and Long Plain–Indi Fault Zones form the boundary. These faults are reverse faults modified from westerly dipping thrusts by later deformation. The volcanic arc complex has underthrust the marginal sea complex. The Gilmore Suture is complex and slices of the volcanic arc occur west of the main dislocation. The Gilmore Fault Zone or Suture has many Late Silurian to Devonian intermediate-basic-ultrabasic (circular gabbro-peridotite or Alaskan-type) intrusions associated with it, and it is an important gold-bearing metallogenic feature (Suppel and Degeling, 1982). With less certainty it can be extended northward towards the area east of Cobar and Louth. North of the Lachlan River Lineament the Gilmore Suture may have two strands, the eastern one defined by the Alaskan-type intrusions within the Girilambone terrane.

The Wagga–Omeo metamorphic belt on the west is bounded by another collisional suture zone: the Kiewa Thrust. Along this east-dipping thrust the high-T/low-P Ordovician metamorphics (Omeo) intruded by syn-kinematic anatectic S-type granites are in contact with mildly metamorphosed rocks of the Tabberabbera subzone (VandenBerg, 1978). This subzone in turn was thrust westward and could be classified as the west-verging Tabberabbera Foreland Fold and Thrust Belt. The Mt. Wellington Greenstone Belt is at the western front and sole of this belt. All this deformation occurred during the Late Ordovician–Early Silurian Benambran orogeny and, surprisingly, sedimentation in the Melbourne Trough (foreland basin) was continuous during this time.

It is suggested that during the same time there was eastward thrusting of the Stawell–Bendigo Foreland Fold and Thrust Belt (Cox et al, 1983) and a weak metamorphic event in the Kanmantoo Fold Belt (Milnes et al, 1977).

In brief, the following model could explain the discussed data. During the earliest Ordovician, and probably earlier, a west-facing Mariana-type B-subduction developed east of the Molong microcontinent. The Molong Volcanic Arc formed on this microcontinent. The backarc basin (Wagga Marginal Basin) was partly floored by oceanic crust that formed during the Late Cambrian to Early Ordovician subduction and/or was a remnant from an earlier marginal sea between the Victorian and Molong microcontinents. The Monaro Forearc Basin was to the east of the volcanic arc, and further east there was the accretionary prism (Narooma terrane) (Cas et al, 1980; Powell, 1983a, 1983b).

During the Early Ordovician, the volcanic arc was split east of Parkes and an interarc basin formed (Initial Cowra Trough). During the later part of Ordovician the Mariana-type subduction changed into Chilean-type. I speculate that the change in the mode of subduction was caused by the arrival from the east of a hypothetical intraoceanic, probably late Proterozoic, volcanic island arc (oceanic plateau), which was subsequently partly subducted and tectonically underplated the accretionary prism (Narooma terrane) and the eastern part of the Monaro Forearc Basin up to the S-I line of White et al (1976). This underplated igneous material could have later become the source rock for the large Devonian I-type granite batholiths (Beams et al, in preparation).

After arrival (docking and accretion occurred later) of the hypothetical oceanic plateau, subduction was relocated (flipped), and eastward subduction started to the west of the arc. This resulted in closure of the Wagga Marginal Basin. The volcanic arc collided and underthrust the marginal basin complexes, and these in turn collided with the Victorian microcontinent and overthrust it. The Gilmore and Kiewa Sutures came into existence. The deformation and thrust pile-up of the Wagga Marginal Basin resulted in high-T/low-P metamorphism, and, in areas of ultrametamorphism, in the formation of anatectic S-type granites (Pogson, 1982; Scheibner, 1982). Subsequent granite plutonism lasted throughout the Silurian, for about 30 Ma (Fagan, 1979). Many plutons were emplaced in environments affected by stresses caused by new plate interactions and limited terrane dispersal, subsequent to the Benambran collision.

During the Benambran collision (orogeny), the hypothetical oceanic plateau on the east was accreted and tectonically underplated.

In the area west of the Kiewa Suture the rocks of the Tabberabbera Subzone of VandenBerg (1978) and the Wellington Greenstone Belt (cf. VandenBerg, in Cooper and Grindley, 1982), representing cover of the Victorian microcontinent, were thrust westward. They form a structural zone that could be best described as the Tabberabbera Foreland Fold and Thrust Belt. It developed in the front of the rising Benambran orogen. Sedimentation in the actual foreland basin (Melbourne Trough), protected by the rigid basement, was continuous from the Ordovician to the Silurian. The collisional stresses, however, were transmitted westward, and the east-verging allochthonous Stawell–Bendigo Foreland Fold and Thrust Belt of the Kanmantoo Fold Belt developed (Cox et al, 1983). In the rest of the Kanmantoo Fold Belt, this collision caused a weak metamorphic event (Milnes et al, 1977). Because only small plates were involved in the Benambran collision, it did not reach the magnitudes observed during major plate collisions. The Wagga–Omeo metamorphic belt remained mostly emergent during the Silurian.

The described plate interaction that resulted in accretion of several terranes cannot be recognized in the Tasmanian segment of the Tasmanides. This again supports the existence of the Gambier–Tamar Fracture Zone of Harrington et al (1973) and Harrington (1983), which separated the Tasmanian segment. Tasmania was hardly affected by the Benambran orogeny. Shallow-water sedimentation existed in the western part of Tasmania, while the flyschlike Mathinna Beds (Ordovician to Early Devonian) were deposited east of the Tamar Zone (Williams, 1978; Corbett, in Cooper and Grindley, 1982), indicating a separate terrane (cf. below).

Early Silurian orogenic deformation and subsequent granite emplacement occurred in the "Western Belt" of the Tuhua orogen in New Zealand (cf. Cooper and Grindley, 1982), but the plate interaction seems to have been similar to that in the Tasmanian segment. The tectonic model for the Ordovician of Victoria and New South Wales described above seems to be applicable to the area south of the Darling River Lineament and north of the Gambier-Beaconsfield Fracture Zone.

Silurian to Carboniferous Tectonic Development

To understand the Silurian to Carboniferous tectonic development of the region that subsequently became the Lachlan Fold Belt, data from adjacent orogenic regions have to be considered.

East of the Lachlan Fold Belt, in the New England Fold Belt, the rock record indicates that, from west to east, a well-developed volcanic arc–forearc basin–accretionary prism was established from the Late Silurian until the Carboniferous (Day et al, 1978; Leitch, 1974, 1975; Oversby, 1971; Scheibner, 1972b, 1974). The New England Fold Belt comprises several terranes (Fig. 3), and the critical area of contact between the Lachlan and New England Fold Belts is mostly concealed by the Sydney–Bowen Basin. As a result, the precise relationship between these orogenic regions is not clear.

Recently, Powell (1983b) suggested that a simple and constant plate geometry, which puts southeastern Australia in regional dextral shear, can explain the transition from the Late Ordovician to the Late Silurian paleogeography. However, major paleogeographic changes point to a radical rearrangement of plate motions in eastern Australia during that time. These changes point to stepping out of the subduction and development of a new subduction zone outboard (Scheibner, 1972b, 1974, 1976). In general, the Lachlan region would have been in a backarc position, behind the mentioned frontal volcanic arc (unnamed in New South Wales, cf. Crook, 1960; called the Calliope Island Arc in Queensland, Day et al, 1978). Backarc regions of Mariana-type subduction zones are characterized by tensional stresses causing basin formation, rifting, breakup, and ultimately sea-floor spreading. From the terrane concept point of view, terrane dispersal occurs in the backarc areas.

Subsequent to the Benambran (Early Silurian) terrane accretion, the region of the Lachlan Fold Belt underwent terrane dispersal. During Middle and Late Silurian times, several near-meridional extensional volcanic rifts and troughs formed (Fig. 6). At least one of these troughs, the Tumut Trough, was partly floored by ophiolites (Ashley et al, 1979). Judging from the thickness of sediments in the Hill End and Cowra Troughs, extensive thinning of the crust is indicated. Locally, new crust may have been created by diking or diffused sea-floor spreading, but there is no direct evidence for it. At this initiation these troughs had the character of volcanic rifts or volcano-tectonic depressions containing some Kuroko-type massive sulfide deposits (Scheibner and Markham, 1976).

L. Wyborn (1977) suggested that the Tumut and Buchan Troughs formed by limited left-lateral strike-slip displacement along the Long Plain–Indi Fault Zones, while Powell (1983b) suggested an opposing sense of transtension. More structural work is needed to enable full understanding of the kinematic development of these features.

The formation of extensional features indicates either operation of B-subduction of Mariana type (Uyeda, 1981) or transtension, or probably both. Terrane dispersal (possibly limited; however, there is no quantitative data support) during this period was quite effective, as some crustal blocks became detached, enabling differences in tectonic history to develop. For example, the Tumut Trough was closed before the Early Devonian, whereas the Hill End Trough continued to receive sediments up to Middle Devonian. During closure of the Tumut Trough, the southern part of the Gilmore Suture, which was formed during the Benambran accretion, took its present form.

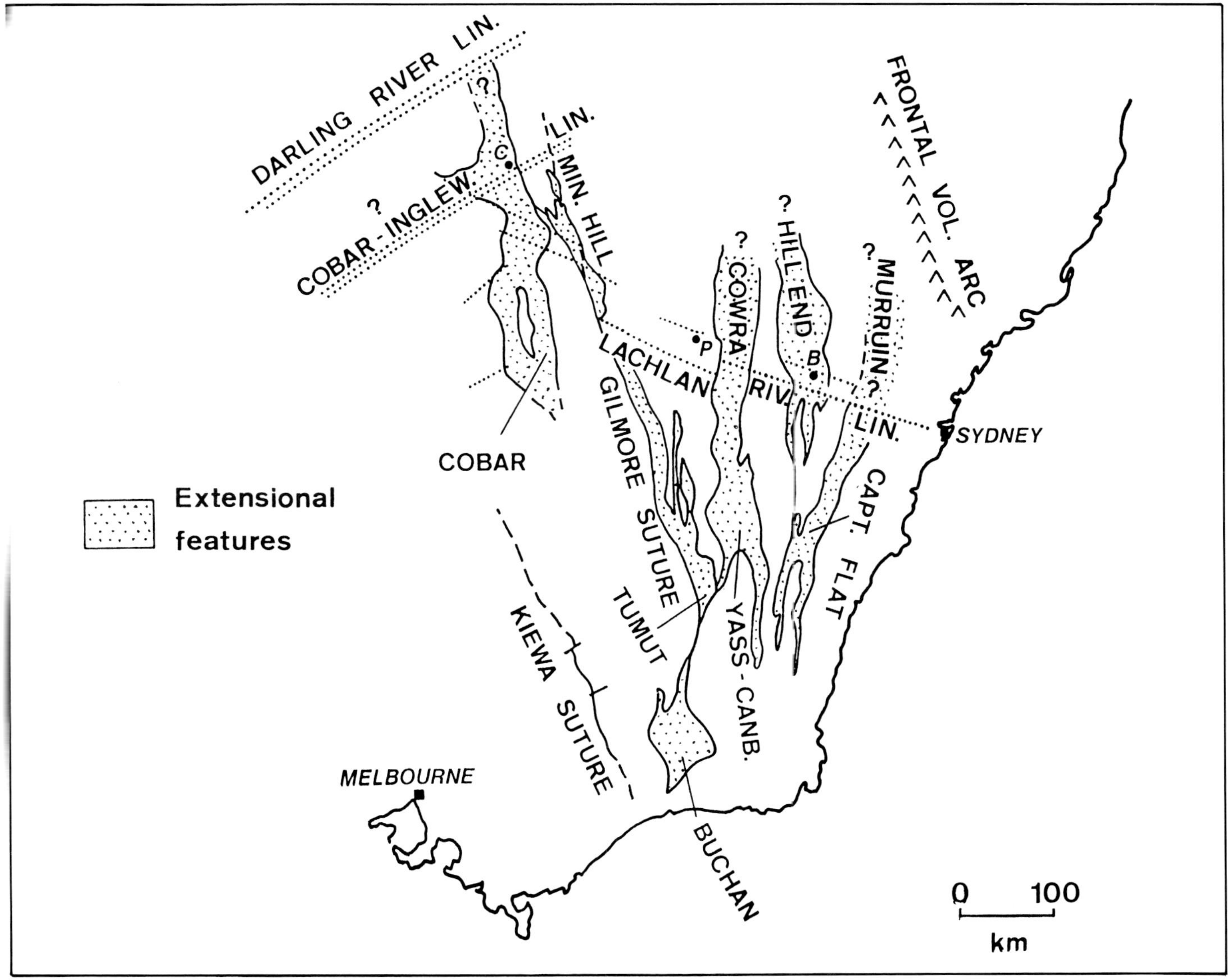

Figure 6—Silurian–Devonian extensional features in the Lachlan Fold Belt.

Some authors (D. Wyborn, 1977) tried to explain crustal fusion leading to Silurian–Early Devonian granite magmatism as a consequence of rifting and high heat flow facilitated by emplacement of basic intrusions, like the Micalong Swamp Basic Igneous Complex. If this is correct, crustal fusion would be expected to happen in areas of highest heat flow—in the rifts and troughs. However, the bulk of the orogenic granites were emplaced into the highs. The troughs first had to be inverted during orogenic deformation, and their crustal thickness had to increase before extensive granite magma generation could occur. However, it is probable that the general extensional environment during the Silurian facilitated the emplacement of granite melts that formed during the Benambran collisional event. In the Wagga–Omeo terrane, the major Silurian granites have en echelon arrangement, in agreement with the left-lateral transtension suggested for the formation of the neighboring Tumut Trough (L. Wyborn, 1977).

During the latest Silurian and Early Devonian, new extensional features, the Cobar and Mineral Hill Troughs, were created, probably also as a result of transtension. Initial volcanic rifting was developed in parts of these troughs. In the Mt. Hope area, the earliest igneous activity had A-type character and was succeeded by S-type rocks (Barron et al, 1982). Sedimentation continued into the late Early to Middle Devonian. After this time the generally tensional regime changed to a compressional one in the Lachlan region, which, at that stage, besides central New South Wales included the attached part of central Victoria, Tasmania, and also New Zealand. We can speculate that the mode of subduction changed into Chilean-type, probably due to the arrival and subsequent accretion of a hypothetical terrane, possibly an oceanic plateau, at the Australian plate margin. This feature has not been identified with certainty, but Solomon and Griffiths (1972) suggested that the Lord Howe Block was a Precambrian crustal block that collided with the Lachlan orogen. The collision-accretion causing the Tabberabberan–Tuhua orogenic event seems to have progressed from south to north, with deformational effects dying out northward and westward (Powell and Edgecombe, 1978; Powell and

Fergusson, 1979; Powell et al, 1980). The distribution of the effects of this orogenic event, which progressed from the foreland eastward, suggests that the convergent movement of the Australian plate played an active role in the collisional deformation. This collision resulted in the progressive emergence of the southern part of the Lachlan orogen, giving rise to widespread molassic deposits that accumulated in late orogenic transitional tectonic basins. The onset of molassic sedimentation was diachronous, starting in the late Early to Middle Devonian in the west (Glen, 1982) and progressing to the Late Devonian in the east (Packham, 1969). During formation of some of the transitional tectonic troughs, bimodal volcanism (McIlveen, 1975; Fergusson et al, 1979) was associated with rifting and limited terrane dispersal. In some areas, especially in Victoria, cauldron subsidence structures were typical (cf. VandenBerg, 1978). Felsic volcanics were associated with I- and A-type plutons (Collins et al, 1982).

The Carboniferous terminal paroxysm (called the Kanimblan orogeny in eastern Australia) affected large parts not only of the Australian active plate margin but also intraplate areas. The Amadeus Aulacogene was deformed (cf. Plumb, 1979), and major intraplate thrusting developed (Forman et al, 1967). The Lachlan Fold Belt was converted into a neocraton. The subsequent deposits are tectonically classified as cratonic cover (Geological Society of Australia, 1971).

Suspect and Tectonostratigraphic Terranes in the Tasmanian Segment of the Tasman Fold Belt System

The Tasmanian Segment was almost completely separated from the rest of the orogen by transform fault(s) (cf. Harrington et al, 1973; Harrington, 1983). For this reason the terranes differ and correlation is difficult. For example, no direct southern continuation of the Kanmantoo-Glenelg terrane, which developed from a marginal sea, can be identified. Instead, remnants of several narrow riftlike troughs (Smithton, Dundas, Dial Range, Fossey Mountains, Beaconsfield, Adamsfield) can be recognized between basement blocks (Corbett, in Cooper and Grindley, 1982). Some of these troughs contain dismembered ophiolites and Alpine-type ultramafics (Brown et al, 1980). While the presence of such rocks is usually indicative of an ensimatic setting, Brown et al (1980) interpreted them as having formed in connection with continental rifting. In a very preliminary form, four terranes are recognized in Tasmania (cf. Figs. 3 and 7).

West Tasmania Terrane

The West Tasmania terrane (including King Island and other small islands) comprises Proterozoic metamorphic basement overlain by late Proterozoic sandstone deposited in paratectonic setting. Latest Proterozoic to Cambrian sequences of the Smithton Trough, comprising dolomite, chert, rare stromatolitic horizons, a thick sequence of tholeiitic spilite, mudstone, graywacke, tuff, and volcanic breccia (cf. Corbett, in Cooper and Grindley, 1982), unconformably overlie the paratectonic sequence. All are intruded by Late Devonian to Carboniferous postkinematic granites.

Dundas-Tyennan Terrane

The Proterozoic Tyennan Nucleus (block) forms the basement for the felsic to intermediate calc-alkaline Cambrian Mt. Read Volcanics. The adjacent Dundas Trough was filled with latest Proterozoic and Cambrian sediments that can be divided into two sequences: The lower part of the fill is similar to epicontinental paratectonic rocks of the Smithton Trough, while the higher unconformably overlying sequences are turbiditic and contain abundant chert and spilitic pillow lavas. Partially serpentinized ultramafic-mafic bodies have been tectonically emplaced into the Dundas Trough complexes. A thick fossiliferous sedimentary sequence (Dundas Group correlatives) of Middle and Late Cambrian age overlies the older strata. There is some interfingering between the Mt. Read volcanics and the Dundas Trough complexes.

During the latest Cambrian, some orogenic movements caused displacements along bounding faults and resulted angular unconformities. Coarse Late Cambrian conglomerate is succeeded by shallow-marine sediments of Ordovician and later Silurian to Early Devonian age. All these rocks were deformed during the Tabberabberan orogeny and subsequently intruded by postkinematic granites. Permian and younger rocks represent cratonic cover.

Adamsfield-Jubilee Terrane

The Adamsfield-Jubilee terrane is similar to the Dundas-Tyennan terrane. It comprises a poorly outcropping belt of early Paleozoic sediments and possible dismembered ophiolites. These indicate the existence of early Paleozoic troughs extending from Beaconsfield in the north to Adamsfield in the south. All these rocks are bounded by Precambrian basement blocks (e.g., the Jubilee Block).

East Tasmania Terrane

The East Tasmania terrane is the most clearly defined tectonostratigraphic terrane in Tasmania. It is bounded by the Tamar fracture zone or contact zone on the west. The stratigraphy of this terrane is completely different from that of neighboring terranes to the west. The Mathina Beds (partly turbidite) include all the pre-Permian rocks. Poorly preserved fossils indicate an Ordovician to Devonian age. Devonian to Early Carboniferous granites are postkinematic. Permian to younger rocks represent cratonic cover (cf. Williams 1978; Corbett, in Cooper and Grindley, 1982).

Suspect Terranes in the Thomson Fold Belt

The Thomson Fold Belt or orogen (Fig. 2) forms the northwestern part of the Tasman Fold Belt System (Kirkegaard, 1974; Murray and Kirkegaard, 1978; Day et al, 1983). The western boundary follows the Diamantina River Lineament, Cork Fault, and Wetherby Structure towards the Burdekin River Fault Zone (Kirkegaard, 1974). An arcuate line marking a distinctive change in the gravity trends (Wellman, 1976) was chosen as the southern boundary (Murray and Kirkegaard, 1978). The Clark River Fault forms the northern boundary, and the Bowen Basin

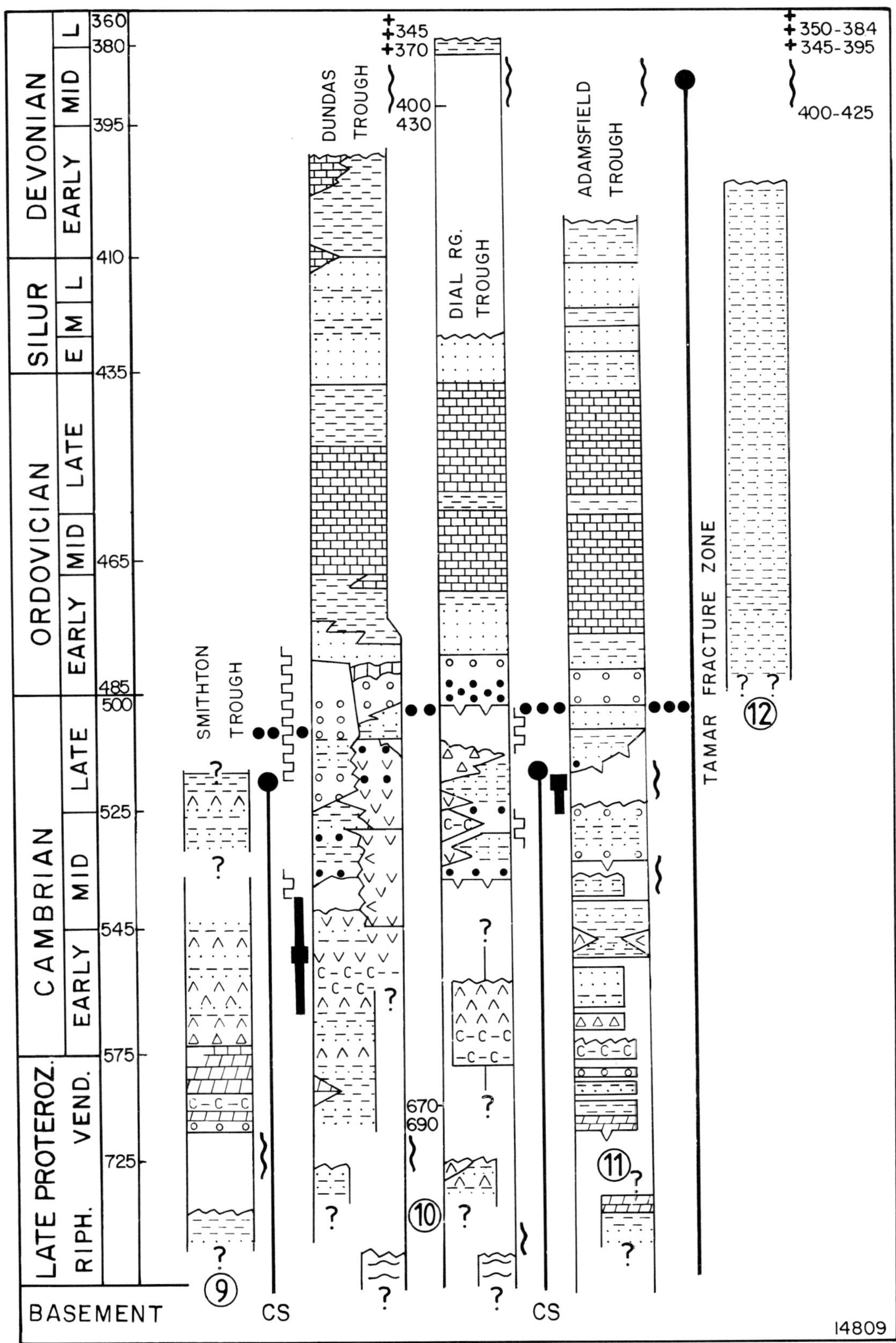

Figure 7—Flow chart of terrane accretion for Tasmania. Numbers in circles indicate terranes as shown in Figure 3. Explanation on Figure 5. Data adopted from Cooper and Grindley (1982).

covers the eastern boundary and the junction with the New England Fold Belt (Fig. 2). The Lolworth–Ravenswood Block and the Anakie Inlier are the only exposed precratonic or orogenic parts of the fold belt, the rest being concealed under the Paleozoic–Mesozoic cover. The Burdekin and Drummond Basins and the entirely subsurface Adavale Basin represent molassic transitional tectonic domains (Geological Society of Australia, 1971). The general structural trends are east-west in the Lolworth–Ravenswood Block and northeasterly elsewhere in the Thomson Fold Belt, and were imposed by a major Middle to Late Devonian deformation, while the terminal orogeny occurred during the Carboniferous. The northeasterly trend of the Thomson Fold Belt is discordant to trends in adjoining parts of the Tasmanides (Day et al, 1983).

In very preliminary form the following suspect terranes were suggested for the Thomson Fold Belt (Scheibner, in Howell et al, 1983).

Cooper Terrane

The Cooper Terrane is a composite terrane with mainly Precambrian basement and some ensimatic basement possibly formed during hypothetical early Paleozoic marginal basin development (Harrington, 1974). Above the basement are Cambrian to Ordovician metasediments and mafic volcanics. Devonian to Early Carboniferous shallow-marine to continental sediments represent molassic rocks. The terminal orogeny occurred in the Middle Carboniferous. Permian and younger rocks form the cratonic cover.

Lolworth–Ravenswood Terrane

Precambrian basement probably formed a microcontinent during early Paleozoic time. Early Paleozoic metasediments and metavolcanics (calc-alkaline) were intruded by Ordovician to Devonian granites. Devonian to Early Carboniferous shallow-marine to continental sediments represent transitional tectonic rocks. The terminal orogeny occurred in the Middle Carboniferous.

Anakie–Nebine–Wonominta Terrane

Gravity and aeromagnetic data indicate some continuity in this composite terrane, despite the fact that the Wonominta Block is structurally part of the Kanmantoo Fold Belt. This supports to some extent the concept of the "Nebine Arc" of Harrington (1974).

In the Anakie terrane, the Anakie metamorphics consist of ?late Proterozoic to Cambrian metasediments (partly turbidites), basic metavolcanics, and small lenses of serpentinized harzburgite, intruded by Ordovician granites. These rocks probably form the basement of the Nebine Ridge under Paleozoic–Mesozoic cover and connect with the Girilambone terrane in the Lachlan Fold Belt. Apparently lying unconformably on the Anakie metamorphics are unmetamorphosed Ordovician shallow-water sediments (Day et al, 1983). Devonian postkinematic granites are followed by Devonian to Early Carboniferous continental sediments. The younger rocks form cratonic cover.

Wonominta Terrane

In the Wonominta terrane, the Wonominta beds consist of late Proterozoic or older orogenic metasediments (flyschlike) and mafic metavolcanics. Above this basement there are mildly metamorphosed Early Cambrian sediments and calc-alkaline to alkaline volcanics and related intrusives (Edwards, 1979). Outboard are partly turbiditic sediments with some mafic volcanics. These rocks belong to the Kanmantoo Precratonic Province. Middle Cambrian to Ordovician shallow-marine sediments that represent molassic rocks unconformably overlie the earlier sequences. There are a few postkinematic granites of Devonian age. Devonian to Early Carboniferous continental sediments unconformably overlie the older rocks. Permian and younger rocks represent cratonic cover.

Suspect Terranes in the Hodgkinson–Broken River Fold Belt

The Hodgkinson–Broken River Fold Belt comprises the Hodgkinson Basin, the Broken River Embayment, the Clark River Basin, and the Bundock Basin, and, according to Day et al (1983), also the Gilberton Basin that developed on the adjoining Proterozoic, cratonic, Georgetown Inlier. The southerly extension of the Palmerville Fault divides the fold belt into the tectonically similar Hodgkinson and Broken River Provinces. According to Arnold and Henderson (1976), the approximately meridional-trending Gray Creek Fault divides the Broken River Province into two tectonically distinct subprovinces, the Graveyard Creek Subprovince in the west and the Camel Creek Subprovince in the east. The orogen was affected by multiple episodes of deformation and metamorphism. The terminal orogeny occurred in Early to Middle Carboniferous time.

Structural grain in the Hodgkinson Province is parallel to the Palmerville Fault and trends northeast and north, whereas in the Broken River Province it is more variable, although parallelism with the Burdekin River Fault Zone is evident. According to Arnold and Fawckner (1980), mafic-ultramafic rocks represent an inlier of Proterozoic basement within the Graveyard Creek Subprovince, rather than a Devonian intrusion or a tectonically emplaced slice of Paleozoic oceanic crust.

At present only one composite Hodgkinson–Broken River terrane has been shown on the terrane map (Fig. 3). The Graveyard Creek Subprovince could represent a separate terrane, and, as well, the Hodgkinson and Broken River Provinces could be separate entities. The quartz-rich flysch, spilitic basalt lavas, and minor chert in the Graveyard Creek Subprovince are the stratigraphically oldest rocks and were deposited in a deep oceanic area, the basement of which is represented by Proterozoic ultramafic complexes (Arnold and Rubenach, 1976). Immature quartz-intermediate flysch rocks of Ordovician age occur in the Camel Creek Subprovince. Faulted blocks of Late Ordovician limestone and andesitic volcanics occur along the western edge of this subprovince. In the Broken River Province, Silurian–Devonian marine sediments, which are similar to those in the Hodgkinson Province, lie unconformably on older strata. Sedimentation was apparently controlled by a major fault zone that formed a hinge line dividing this composite terrane from the neighboring craton to the west. Carbonate-rich shelf sediments were deposited in the western part of the terrane, while thick slope and basin flysch-type sequences accumulated to the east. A late Middle Devonian orogeny changed the pattern of sedimentation in this terrane. Flysch-type deposition continued into the Late Devonian in the Hodgkinson Province, but in the southern part of the terrane sedimentation was dominantly continental with brief marine incursions. The Late Devonian to mid-Carboniferous terminal orogeny cratonized this terrane and probably accreted it. Cratonization was completed by the emplacement of Late Carboniferous and Permian postkinematic granites. This igneous activity extended over neighboring cratonic regions to the west, and hence the plutons have stitching character. Subsequent sediments represent cover rocks.

Suspect and Tectonostratigraphic Terranes in the New England Fold Belt

The New England Fold Belt (Fig. 2) comprises many distinct fault-bounded blocks and molassic troughs (cf. Day et al, 1983).

Major Alpine-type ultramafic and ophiolite belts define suture zones (Peel and Yarrol Fault Zones), which divide the orogen into a western ensialic terrane containing shelf facies and an eastern ensimatic terrane containing oceanic facies. Along the suture zone are slices of early Paleozoic forearc and accretionary prism complexes (Leitch and Cawood, 1980), representing a separate terrane. Outboard of the deep-water complexes and separated from them by a discontinuous belt of serpentinites is the Gympie terrane (Harrington, 1983) containing shallow-water facies.

Middle Devonian orogenesis strongly affected the ensialic terrane, with the intensity of deformation decreasing southwards. Late Carboniferous to Early Permian orogenesis strongly affected the ensimatic terrane, with the intensity of metamorphism and deformation decreasing both westwards and southwards. The terminal deformation occurred during the Middle Permian and progressed northwards. Triassic folding of the eastern part of the Gympie terrane and the Esk Trough marked the end of molassic development. Subsequent rocks represent the cratonic cover. Modern reviews of the geology and stratigraphy of the New England Fold Belt have been recently published by Korsch and Harrington (1981), Leitch (1982), and Day et al, (1983).

The following terranes have been shown on the terrane map of the circum-Pacific region (Howell et al, 1983); Yarrol–Silverwood–Tamworth, Hastings, Wandilla and Texas–Woolomin, Port Macquarie, Copes Creek, Nambucca, and Gympie (Figs. 3 and 8). Independently Cawood recognized three tectonostratigraphic associations in the southern part of the orogen: the Tamworth, Tablelands, and Nambucca terranes. His Tablelands terrane is equal to the Texas–Woolomin and Port Macquarie terranes described here.

Yarrol–Silverwood–Tamworth Terrane

The Yarrol–Silverwood–Tamworth terrane comprises Late Silurian to Permian sediments and volcanics. Calc-alkaline volcanics, representing a volcanic arc, were intruded by Devonian to Carboniferous granitoids and crop out on the western edge of the terrane in Queensland. In New South Wales the volcanic arc is concealed. The junction with the Lachlan orogenic area is not clear, but the Murruin Trough, a possible marginal sea, was immediately to the west of the volcanic arc (Scheibner, 1972b). Sediments east of the arc are mostly volcaniclastic (maximum thickness 7 to 10 km [4–6 mi]) and were deposited in a possible forearc basin that had the character of a shelf or unstable shelf. During the Early Permian, volcanic rifting occurred in the Sydney–Bowen Basin region, which became the foredeep of the New England orogen; some of the bimodal volcanics preserved in the western part of this terrane belong to this rifting event.

Terminal deformation occurred during the Middle Permian. The terrane was thrust over the foredeep on the west, and renewed movement occurred on the Peel Fault System in the east. Some postkinematic Permian plutons were emplaced into this terrane. Late Permian and Triassic rocks represent molassic complexes, and subsequent rocks represent cratonic cover.

Hastings Terrane

The Hastings terrane probably represents a northward displaced and dispersed part of the Yarrol–Silverwood–Tamworth terrane, but there are some stratigraphic and facies differences.

Copes Creek Terrane

The Copes Creek terrane comprises Cambro-Ordovician sediments of volcanic arc provenance occurring as fault slivers in the Peel Fault Zone. According to Cawood (1976), rocks of the Tamworth Group (deposited in a forearc basin) lie unconformably on these Cambro-Ordovician sediments. These sediments together with the Woolomin Formation, represent pre-Late Silurian terranes.

Texas–Woolomin and Wandilla Terranes

These terranes are composite and are separated by major fault zones (Peel and Yarrol) and associated Alpine-type ultramafics from the Yarrol–Silverwood–Tamworth terrane. They comprise early Paleozoic to Carboniferous strongly deformed and metamorphosed basalts, cherts, slates, mudstones, graywackes, and melanges, which have been interpreted as having formed in subduction-related environments. The Woolomin Formation is pre-Upper Devonian (Leitch and Cawood, 1980), possibly Cambro-Ordovician, and, together with the Copes Creek terrane rocks, is related to a separate early Paleozoic plate convergence. Hopefully, more detailed mapping will enable separation of this component from the rest of the terrane, which is related to Late Silurian to Carboniferous convergence. An interesting occurrence in the Yarraman Block of limestone with warm-water Late Carboniferous conodonts (Palmieri, 1969) contrasting with generally cold-water faunas has been brought to my attention by John Roberts (personal communication). Terrane accretion occurred during the Late Carboniferous to the Early Permian, followed or accompanied by the emplacement of early S-type granites (Shaw and Flood, 1981). Subsequent marine sedimentation was limited to the margins of this

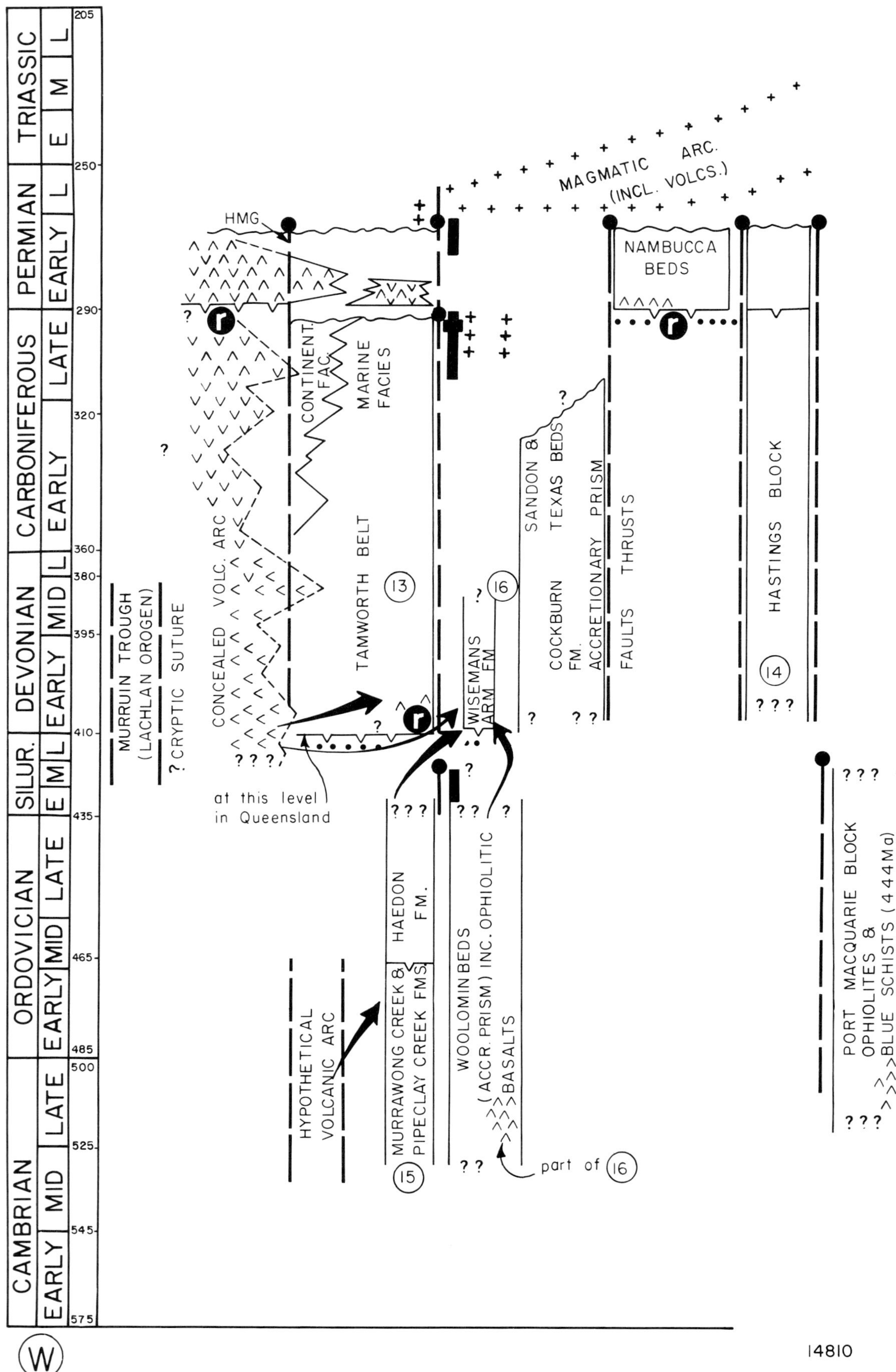

Figure 8—Schematic flow chart of terrane accretion for the southern part of the New England Fold Belt. Numbers in circles indicate terranes as shown in Figure 3. Explanation on Figure 5.

terrane. The terminal deformation was of Middle Permian age and was followed by emplacement of large I- and A-type postkinematic granitoid plutons (Shaw and Flood, 1981). During this orogeny the Wandilla terrane was displaced and separated (terrane dispersal) from the Texas–Woolomin terrane.

Port Macquarie Terrane

The Port Macquarie terrane is similar to the Texas–Woolomin terrane and especially to the early Paleozoic Woolomin Formation. A dismembered ophiolite suite with products of high-pressure metamorphism and mainly occurring in melanges characterizes this terrane. Recently, some blueschists were dated at 444 Ma by Lanphere (personal communication).

Nambucca Terrane

The Nambucca terrane comprises Late Carboniferous to Early Permian clastic rocks with minor basic lavas and acid volcanics and rare limestones. Total thickness of strata is at least 5 km (3 mi). Intense deformation and metamorphism up to low-grade greenschist facies occurred during the Middle Permian. Postkinematic intrusions are Permian to Triassic in age.

Gympie Terrane

The Gympie terrane (Day et al, 1983; Harrington, 1983) comprises Permian to Early Triassic clastic sediments, basic to intermediate volcanics, and limestone. Facies are shallow marine to continental, and the environment of deposition was a continental shelf. Greenschist-facies metamorphism locally occurred during the Triassic orogeny. Postkinematic granites and diorites are Triassic in age and are associated with some subaerial volcanics.

CONCLUSIONS

The application of the tectonostratigraphic terrane concept to the Tasman Fold Belt System is in its infancy. The situation is comparable to that of the application of plate tectonics to this region at the start of the seventies. The proposed terranes have to be vigorously tested, and good-quality paleomagnetic data, together with concentrated bioprovincial analysis, will be decisive in improving the present suggestions. Some of the suggested terranes seem to be more displaced than others. The terrane concept enables a fresh approach to the tectonic analysis and synthesis of the Tasmanides. The terrane analysis and plate tectonic models presented in this paper will hopefully evolve, provoke future lively discussions, and provide focuses for future research.

ACKNOWLEDGMENTS

Permission to publish this paper was given by the Secretary, New South Wales Department of Mineral Resources, Sydney.

I would like to thank the organizers (and mainly Dr. David G. Howell) for the invitation and the U.S. Geological Survey for financial support to attend the Circum-Pacific Terrane Conference at Stanford University in 1983. I am also indebted to Dr. Howell for the request in 1982 to compile a "Terrane Map of the SW Quadrant of the Circum-Pacific" at 1:10,000,000 scale.

Helena Basden, Richard Glen, David Howell, Denis Pogson, and Fons VandenBerg are thanked for comments and corrections of the first draft of this manuscript. I am thankful to Miss Lois Schey for deciphering my handwriting and typing the manuscript.

REFERENCES

Arnold, G. O., and J. F. Fawckner, 1980, The Broken River and Hodgkinson Provinces, *in* R. A. Henderson and P. Y. Stephenson, eds., The geology and geophysics of northeastern Australia: Brisbane, Geological Society of Australia, Queensland Division, p. 175–189.

______, and R. A. Henderson, 1976, Lower Palaeozoic history of the southwestern Broken River Province, north Queensland: Journal of the Geological Society of Australia, v. 23, p. 73–93.

______, and M. G. Rubenach, 1976, Mafic-ultramafic complexes of the Greenvale area, north Queensland. Devonian intrusions or Precambrian metamorphics?: Journal of the Geological Society of Australia, v. 23, p. 119–139.

Ashely, P. M., et al, 1979, Field and geochemical characteristics of the Coolac Ophiolite suite and its possible origin in a marginal sea: Journal of the Geological Society of Australia, v. 26, p. 45–60.

Barron, L. M., et al, 1982, The Mount Hope Group and the comagmatic granites on the Mount Allen 1:100,000 Sheet, N.S.W.: Geological Survey of N.S.W. Quarterly Notes, no. 47, p. 1–17.

Beams, S. D., 1980, Magmatic evolution of the southeast Lachlan Fold Belt, Australia: Unpublished PhD Dissertation, La Trobe University.

Brooks, C., et al, 1976, Ancient lithosphere: its role in young continental volcanism: Science, v. 193, p. 1086–1094.

Brown, A. V., et al, 1980, Geological environment petrology and tectonic significance of the Tasmanian Cambrian ophiolitic and ultramafic-mafic complexes: International Ophiolite Symposium, Cyprus, 1979, Cyprus Geological Survey Department, p. 649–659.

Cas, R. A. F., et al, 1980, Ordovician paleogeography of the Lachlan Fold Belt; a modern analogue and tectonic constraints: Journal of the Geological Society of Australia, v. 27, p. 19–31.

Cawood, P. A., 1976, Cambro-Ordovician strata, northern New South Wales: Search, v. 7, p. 378–379.

______, 1982, Structural relations in the subduction complex of the Paleozoic New England Fold Belt, eastern Australia: Journal of Geology, v. 90, p. 381–392.

Chappell, B. W., and A. J. R. White, 1974, Two contrasting granite types: Pacific Geology, v. 8, p. 173–174.

Collins, W. J., et al, 1982, Nature and origin of A-type granites with particular reference to southeastern Australia: Contributions to Mineralogy and Petrology, v. 80, p. 189–200.

Coney, P. J., et al, 1980, Cordilleran suspect terranes: Nature, v. 288, p. 329–333.

Cooper, R. A., and G. W. Grindley, eds., 1982, Proterozoic to Devonian sequences of southeastern Australia, Antarctica and New Zealand and their correlation: Geological Society of Australia Special Publication 9, 103 p.

Coulon, C., and R. S. Thorpe, 1981, Role of continental crust in petrogenesis of orogenic volcanic associations: Tectonophysics, v. 77, p. 79–93.

Cox, S. F., et al, 1983, Lower Ordovician Bendigo Trough sequence, Castlemain area, Victoria—Deformational style and implications for the tectonic evolution of the Lachlan Fold Belt: Geological Society of Australia Abstracts, v. 9, p. 41–42.

Crawford, A. J., 1983, Tectonic development of the Lachlan Foldbelt and construction of the continental crust of southeastern Australia: Geological Society of Australia Abstracts, v. 9, p. 30–32.

———, and R. R. Keays, 1978, Cambrian greenstone belts in Victoria; marginal sea-crust slices in the Lachlan Fold Belt of southeastern Australia: Earth and Planetary Science Letters, v. 41, p. 197–208.

Crook, K. A. W., 1960, Petrology of Tamworth Group, Lower and Middle Devonian Tamworth-Nundle district, New South Wales: Journal of Sedimentary Petrology, v. 30, p. 353–369.

———, 1980, Forearc evolution in the Tasman Geosyncline: the origin of the southeast Australian continental crust: Journal of the Geological Society of Australia, v. 27, p. 215–232.

Daily, B., and A. R. Milnes, 1973, Stratigraphy, structure and metamorphism of the Kanmantoo Group (Cambrian) in its type section east of Tuakallilla Beach, South Australia: Royal Society of South Australia Transactions, v. 97, p. 213–251.

Day, R. W., et al, 1978, The eastern part of the Tasman Orogenic Zone, *in* E. Scheibner, ed., The Phanerozoic structure of Australia and variations in tectonic style: Tectonophysics, v. 48, p. 327–364.

———, et al, 1983, Queensland geology. A companion volume to the 1:2,500,000 scale geological map (1975): Geological Survey of Queensland Publication 383, 194 p.

Dickinson, W. R., 1973, Widths of modern arc-trench gaps proportional to past duration of igneous activity in associated magmatic arcs: Journal of Geophysical Research, v. 78, p. 3376–3389.

Edwards, A. C., 1979, Tectonic implications of the immobile trace-element geochemistry of mafic rocks bounding the Wonominta Block: Journal of the Geological Society of Australia, v. 25, p. 459–465.

Fagan, R. K., 1979, S-type granite genesis and emplacement in N.E. Victoria and its implications: Australia Bureau of Mineral Resources Geology and Geophysics Record 1979/2, p. 29–30.

Fergusson, C. L., et al, 1979, The Upper Devonian Boyd Volcanic Complex, Eden, New South Wales: Journal of the Geological Society of Australia, v. 26, p. 87–115.

Forman, D. J., et al, 1967, Regional geology and structural of the northeastern margin of the Amadeus Basin, Northern Territory: Australia Bureau of Mineral Resources Report, 103 p.

Geological Society of Australia, 1971, Tectonic map of Australia and New Guinea, 1:5,000,000: Sydney.

Glen, R. A., 1982, The Amphitheatre Group, Cobar, New South Wales: preliminary results of new mapping and implications for ore search: Geological Survey of N.S.W. Quarterly Notes, no. 49, p. 2–14.

Harrington, H. J., 1974, The Tasman Geosyncline in Australia, *in* A. K. Denmead, et al, eds., The Tasman Geosyncline—a symposium: Geological Society of Australia, Queensland Division, p. 383–407.

———, 1983, Correlation of the Permian and Triassic Gympie Terrane of Queensland with the Brook Street and Maitai Terranes of New Zealand, *in* Permian geology of Queensland: Brisbane, Geological Society of Australia, Queensland Division, p. 431–436.

———, et al, 1973, Gambier-Beaconsfield and Gambier-Sorell Fracture Zones and the movement of plates in the Australia–Antarctic–New Zealand region: Nature (Physical Science), v. 245, 109–112.

Harrison, T. M., and L. McDougall, 1981, Excess ^{40}Ar in metamorphic rocks from Broken Hill, New South Wales: implication for ^{40}Ar/^{39}Ar age spectra and thermal history of the region: Earth and Planetary Science Letters, v. 55, p. 123–149.

Howell, D. G., and D. L. Jones, 1983, The principles of terrane analysis and some definitions: Circum-Pacific Terrane Conference, Stanford University.

———, et al, 1983, Tectonostratigraphic terrane map of the circum-Pacific region: U.S. Geological Survey Open-File Report 83-716.

Johnson, R. W., et al, 1978, Delayed partial melting of subduction-modified mantle in Papua, New Guinea: Tectonophysics, v. 46, p. 197–216.

Jones, D. L., et al, 1983, Recognition, character, and analysis of tectonostratigraphic terranes in western North America, *in* Oji Seminar Volume: Tokyo, Center for Academic Publications Japan.

Kilpatrick, D. J., and D. P. Fleming, 1980, Lower Ordovician sediments in the Wagga Trough: discovery of early Bendigonian graptolites near Eskdale, northeast Victoria: Journal of the Geological Society of Australia, v. 27, p. 69–73.

Kirkegaard, A. G., 1974, Structural elements of the northern part of the Tasman Geosyncline, *in* A. K. Denmead, et al, eds., The Tasman Geosyncline—a symposium: Brisbane, Geological Society of Australia, Queensland Division, p. 47–62.

Korsch, R. J., and H. G. Harrington, 1981, Stratigraphic and structural synthesis of the New England Orogen: Journal of the Geological Society of Australia, v. 28, p. 205–226.

Lambeck, K., 1983, Structure and evolution of the Amadeus Basin, central Australia: Geological Society of Australia Abstracts, v. 9, p. 109–110.

Leitch, E. C., 1974, The geological development of the southern part of the New England Fold Belt: Journal of the Geological Society of Australia, v. 21, p. 133–156.

______, 1975, Plate tectonic interpretation of the Palaeozoic history of the New England Fold Belt: Geological Society of America Bulletin, v. 86, p. 141-144.

______, 1982, Crustal development in New England, *in* P. G. Flood and B. Runnegar, eds., New England geology, Voisey Symposium Volume: Armidale, N.S.W., University of New England, p. 9-16.

______, and P. A. Cawood, 1980, Olistoliths and debris flow deposits of ancient consuming plate margins: An eastern Australian example: Sedimentary Geology, v. 25, p. 5-22.

McIlveen, G. R., 1975, The Eden-Comerong-Yalwal Rift Zone and the contained gold mineralisation: Geological Survey of N.S.W. Records, v. 16, p. 245-277.

Milnes, A. R., et al, 1977, Pre- to syn-tectonic emplacement of Early Palaeozoic granites in southeastern South Australia: Journal of the Geological Society of Australia, v. 24, p. 87-106.

Murray, C. G., and A. G. Kirkegaard, 1978, The Thomson Orogen of the Tasman Orogenic Zone: Tectonophysics, v. 48, p. 299-325.

Olenin, V. B., 1967, The principles of classification of oil and gas basins: Australian Oil and Gas Journal, February 1967, p. 40-46.

Oversby, B., 1971, Palaeozoic plate tectonics in the southern Tasman Geosyncline: Nature (Physical Science), v. 234, p. 45-47.

Owen, M., and D. Wyborn, 1979, Geology and geochemistry of the Tantangara and Brindabella area: Australia Bureau of Mineral Resources Bulletin 204, 52 p.

Packham, G. H., ed., 1969, The geology of New South Wales: Journal of the Geological Society of Australia, v. 16, 654 p.

______, 1973, A speculative Phanerozoic history of the south-west Pacific island areas, marginal seas: geochemistry, *in* P. J. Colemen, ed., The Western Pacific: Nedlands, University of Western Australia Press, p. 369-388.

Palmieri, V., 1969, Upper Carboniferous conodonts from limestone near Murgon, southeast Queensland: Geological Survey of Queensland Report 86.

Plumb, K. A., 1979, The tectonic evolution of Australia: Earth Science Review, v. 14, p. 205-249.

Pogson, D. J., 1982, Stratigraphy, structure, and tectonics: Nymagee-Melrose, central western New South Wales: Unpublished Thesis, N.S.W. Institute of Technology, Sydney, 193 p.

Powell, C. M., 1983a, Geology of N.S.W. South Coast: Geological Society of Australia, Specialist Group in Tectonics and Structural Geology, Field Guide 1, 118 p.

______, 1983b, Tectonic relationship between the Late Ordovician and Late Silurian palaeogeographics of southeastern Australia: Journal of the Geological Society of Australia, v. 30, p. 353-373.

______, and D. R. Edgecombe, 1978, Mid Devonian movements in the northeastern Lachlan Fold Belt: Journal of the Geological Society of Australia, v. 25, p. 165-185.

______, and C. L. Fergusson, 1979, The relationship of structures across the Lambian unconformity near Taralga, New South Wales: Journal of the Geological Society of Australia, v. 26, p. 209-219.

______, et al, 1980, Structural relationships across the Lambian unconformity in the Hervey Range Parkes Area, N.S.W.: Proceedings of the Linnean Society of N.S.W., v. 104, p. 195-210.

Preiss, W. V., et al, 1981, The Precambrian of South Australia, *in* E. R. Hunter, ed., Precambrian of the Southern Hemisphere: Amsterdam, Elsevier.

Rutland, R. W. R., 1976, Orogenic evolution of Australia: Earth Science Review, v. 12, p. 161-196.

Scheibner, E., 1972a, The Kanmantoo Pre-Cratonic Province in New South Wales: Geological Survey of N.S.W. Quarterly Notes, v. 7, p. 1-10.

______, 1972b, Actualistic models in tectonic mapping: 24th International Geological Congress, Montreal, Report 3, p. 405-422.

______, 1974, A plate tectonic model of the Palaeozoic tectonic history of New South Wales: Journal of the Geological Society of Australia, v. 20, p. 405-416.

______, 1976, Explanatory notes on the tectonic map of New South Wales: Sydney, Geological Survey of N.S.W., 283 p.

______, 1982, Some aspects of the geotectonic development of the Lachlan Fold Belt: Geological Survey of N.S.W. Unpublished Report GS1982/069. (Text of a public lecture 12th March 1982.)

______, and N. L. Markham, 1976, Tectonic setting of some strata-bound massive sulphide deposits in New South Wales, Australia, *in* K. N. Wolf, ed., Handbook of strata-bound and stratiform ore deposits, v. 6: Amsterdam, Elsevier, p. 55-77.

Shaw, S. E., and R. H. Flood, 1981, The New England Batholith, eastern Australia geochemical variations in time and space: Journal of Geophysical Research, v. 86, p. 10530-10544.

Sherwin L., 1979, Age of the Nelungaloo Volcanics near Parkes: Geological Survey of N.S.W. Quarterly Notes, v. 44, p. 2-4.

Solomon, M., and J. R. Griffiths, 1972, Tectonic evolution of the Tasman Orogenic Zone, Eastern Australia: Nature (Physical Science), v. 237, p. 3-6.

Sprigg, R. C., 1952, Sedimentation in the Adelaide Geosyncline and the formation of the continental terrace, *in* M. F. Glaessner and R. C. Sprigg, eds., Sir Douglas Mawson Anniversary Volume: University of Adelaide, p. 153-159.

Suppel, D. W., and P. R. Degeling, 1982, Comments on magmatism and metallogenesis of the Lachlan Fold Belt: Geological Society of Australia Abstracts, v. 6, p. 4-5.

Thomson, B. P., 1969, The Kanmantoo Group and early Palaeozoic tectonics, *in* L. W. Parkin, ed., Handbook of southern Australian geology: Geological Survey of South Australia, p. 97-108.

Uyeda, S., 1981, Subduction zones and back arc basins—A review: Geologische Rundschau, v. 70, p. 552-569.

VandenBerg, A. H. M., 1978, The Tasman Fold Belt System in Victoria: Tectonophysics, v. 48, p. 267-297.

Veevers, J. J., and D. Cotterill, 1978, Western margin of

Australia: evolution of a rifted arch system: Geological Society of America Bulletin, v. 89, p. 315-337.

von der Borch, C. C., 1980, Evolution of Late Proterozoic to Early Palaeozoic Adelaide Foldbelt, Australia: Comparison with post-Permian rifts and passive margins: Tectonophysics, v. 70, p. 115-134.

Wellman, P., 1976, Gravity trends and the growth of Australia, a tentative correlation: Journal of the Geological Society of Australia, v. 23, p. 11-14.

White, A. J. R., and B. W. Chappell, 1977, Ultra metamorphism and granitoid genesis: Tectonophysics, v. 43, p. 7-22.

———, et al, 1976, The Jindabyne thrust and its tectonic, physiographic and petrogenetic significance: Journal of the Geological Society of Australia, v. 23, p. 105-112.

Williams, E., 1978, Tasman Fold Belt system in Tasmania: Tectonophysics, v. 48, p. 59-205.

Williams, H., and R. D. Hatcher, Jr., 1982, Suspect terranes and accretionary history of the Appalachian orogen: Geology, v. 10, p. 530-536.

Williams, I. S., et al, 1983, Zircon and monazite U-Pb systems and the histories of I-type magmas, Berridale Batholith, Australia: Journal of Petrology, v. 24, p. 76-97.

Wyborn, D., 1977, Discussion—the Jinabyne Thrust and its tectonic, physiographic and petrogenetic significance: Journal of the Geological Society of Australia, v. 24, p. 233-236.

Wyborn, L. A. I., 1977, Aspects of the geology of the Snowy Mountains region and their implications for the tectonic evolution of the Lachlan Fold Belt: Unpublished PhD Dissertation, Australian National University.

Provisional Terrane Map of South Island, New Zealand

D. G. Bishop
New Zealand Geological Survey
Dunedin, New Zealand

J. D. Bradshaw
University of Canterbury
Christchurch, New Zealand

C. A. Landis
University of Otago
Dunedin, New Zealand

Nine tectonostratigraphic terranes are recognized in the South Island, New Zealand, and a number of accretion and amalgamation events ranging from Paleozoic to late Mesozoic are inferred. The rocks include early Paleozoic sequences with Gondwana affinities, Permian to Jurassic volcanic arc and ophiolite suites, and Carboniferous to Cretaceous accretionary complexes. Permian and Triassic faunas of probable mid- to high southern latitude characterize several terranes; other terranes and tectonic blocks, however, contain faunas with low-latitude and Tethyan affinities.

INTRODUCTION

Nine tectonostratigraphic terranes are recognized in the pre-Cretaceous rocks of South Island, New Zealand (Fig. 1). Most of these have suffered Cenozoic postaccretionary dispersion by the Alpine Fault and now occur as segments separated by 480 km (300 mi). Further work may show that some of these terranes can be grouped, or else require further subdivision; in our opinion, however, they represent the principal tectonostratigraphic units of South Island.

Selected references, providing access to the relevant literature, are given in the following brief descriptions. For a general background, the reader is referred to Suggate (1978) and the 1:1,000,000 geological map of South Island (New Zealand Geological Survey, 1972). Readers are referred to Jones et al (1982) for definitions of the terminology and an outline of the philosophy of terrane analysis.

TERRANES

Karamea Terrane

The characteristic rocks of the Karamea terrane are quartz-rich sandstones and mudstones of Late Cambrian to Late Ordovician age. In the west a thick fanlike turbidite sequence is developed, but to the east the sequence is better differentiated, and discrete quartzitic and graptolitic black mudstone units are recognized. Late Precambrian gneiss, a possible basement, is preserved locally, and the sequence was deformed and cut by granites during the Silurian. Small remnants of an unconformable shallow-marine Lower Devonian sequence occur and predate a second phase of deformation and major granite magmatism of Devono-Carboniferous age. Granitoids of Triassic–Early Cretaceous age are also widespread (Cooper, 1979; Tulloch, 1983).

Golden Bay Terrane

The western part of the Golden Bay terrane comprises a series of steeply inclined tectonic slices of diverse Cambrian–Ordovician rocks including sedimentary breccia, calc-alkaline volcanic breccia and lava, limestone, conglomerate, sandstone, and mudstone. To the east, these rocks are in tectonic contact with Ordovician to Devonian rocks, mainly quartzose sandstone and mudstone with lenticular (Ordovician) carbonate units up to 2 km (1 mi) thick. Silurian quartzites concordantly overlie Ordovician strata, but the lower Devonian marine sediments rest unconformably on the older rocks of this terrane. The similarity of stratigraphy and sedimentary facies in the Early Ordovician rocks of both the eastern and western suites suggest to us that this is one strongly deformed terrane. However, it could be argued that the western suite of tectonic slices is a separate terrane(s) (see discussion in Cooper, 1979, and Grindley, 1980).

Quartzose sediments of the Permian Parapara Group rest unconformably on deformed and metamorphosed Golden Bay rocks. Parapara rocks are themselves folded and metamorphosed, possibly contemporaneously with the emplacement of nearby Cretaceous granitoids.

Rocks correlated with the Karamea and Golden Bay terranes on the southeast side of the Alpine Fault have so far been recognized only in the far southwest of South Island (Ward, personal communication).

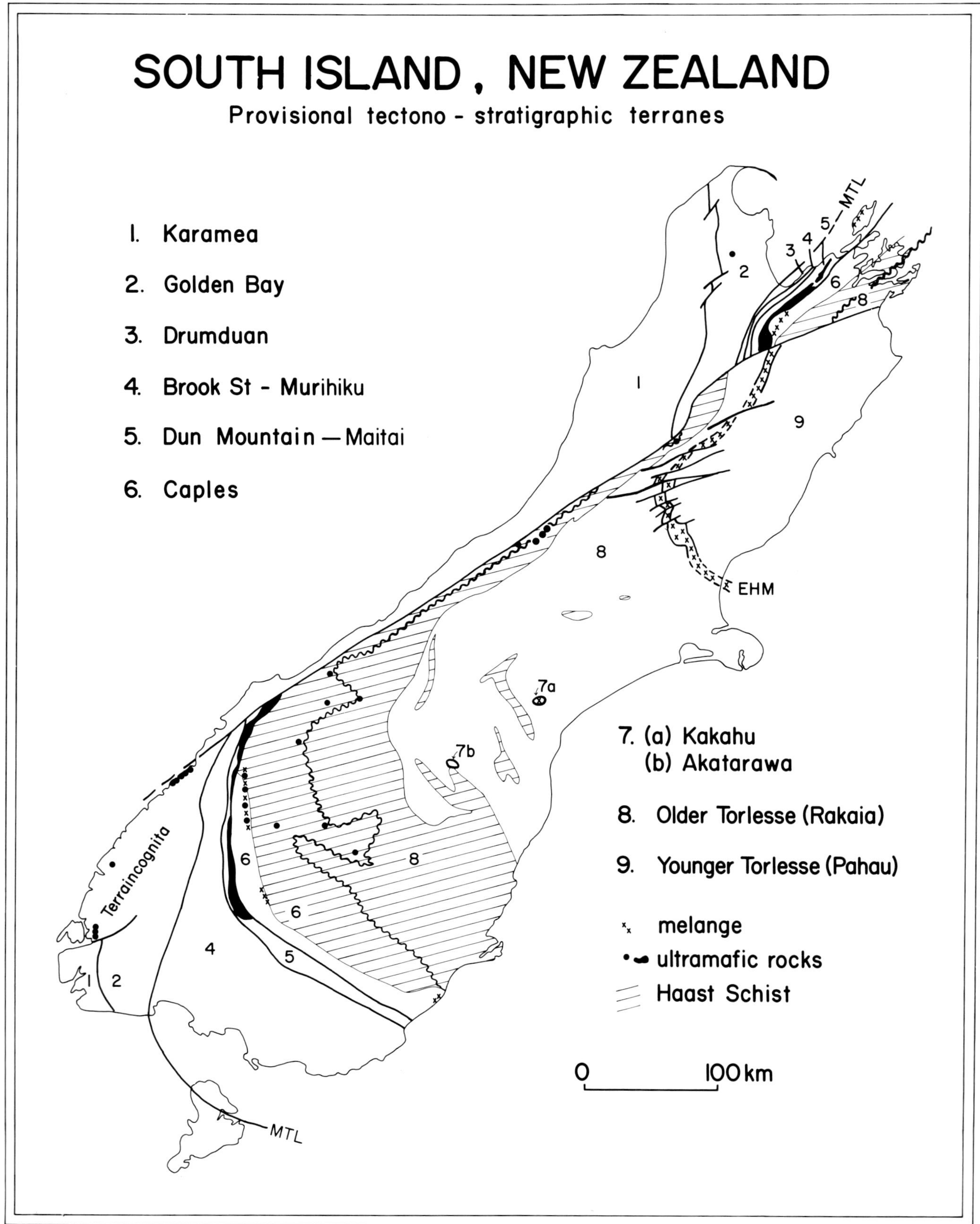

Figure 1—Provisional tectonostratigraphic terranes of South Island, New Zealand. MTL = Median Tectonic Line; EHM = Esk Head Melange.

Drumduan Terrane

The Drumduan terrane comprises a sliver of locally sheared and tectonized sedimentary rocks occurring immediately east of the Median Tectonic Line (MTL) in Nelson. Their correlatives in the south, on the other side of the Alpine Fault, have not been identified but may include enigmatic, undated fault slivers in the Te Anau–Hollyford region (Williams, 1978).

Drumduan rocks include tuffs and fine-grained, volcanogenic sediments, structurally overlain by grey mudstone, sandstone, and breccia, locally with plant fossils. The rocks are recrystallized, locally containing abundant lawsonite, and are commonly metasomatized where invaded by the Mesozoic Tasman Intrusives.

Plant fossils indicate a Late Triassic to Early Cretaceous, probably Jurassic age. The rocks structurally underlie the Permian–Triassic Brook Street–Murihiku terrane, from which they are separated by the Delaware Fault Zone (Johnston, 1981).

Brook Street–Murihiku Terrane

The Brook Street portion of the terrane represents remnants of a Permian volcano-plutonic arc and is made up of weakly metamorphosed volcanogenic sediments and basaltic-dacitic volcanics up to 16 km (10 mi) thick. Together with a suite of gabbroic to granitic intrusives, it is locally overlain in the south by thin richly fossiliferous, Late Permian limestones. In other places it is unconformably overlain by or faulted against a sequence of 10 km (6 mi) of moderately fossiliferous, abundantly tuffaceous, weakly metamorphosed Triassic and Jurassic volcaniclastic rocks, ranging from nonmarine to thick flysch sequences (Murihiku portion). Structure is simple, characteristically open synclinal folding, with no cleavage development (Boles, 1974; Houghton, 1981; Mutch, 1972).

Dun Mountain–Maitai Terrane

The Dun Mountain Ophiolite Belt, the best-developed ophiolite in New Zealand, outcrops in Nelson and West Otago and consists of serpentinized harzburgite and dunite, interlayered wehrlites, clinopyroxenites and mafic gabbros, hornblende gabbros, doleritic dike and sill complexes, and pillow and sheet lavas. These are overlain with local unconformity by the bedded volcanic breccias, volcanic sandstones, grey black laminated mudstones, and bivalvian limestones of the Upper Permian Maitai Group. The Dun Mountain rocks are variably tectonized, passing from relatively coherent sequences into tectonic melange; the overlying 4,000 m (13,000 ft) of Maitai strata, on the other hand, comprise a coherent sequence of thinly bedded, sparsely fossiliferous, fine-grained, commonly flyschlike rocks and display a remarkable continuity along the regional strike. Fossil occurrences within the Maitai strata are stratigraphically restricted and dominated by fragmental atomodesmatinid bivalve remains. Structure is simple, although regional folds may be very tight or isoclinal, with a well-developed axial plane cleavage.

The age is Upper Permian, possibly extending into the Triassic. Metamorphism is zeolite to lawsonite-albite-chlorite facies (Coombs et al, 1976; Harper and Landis, 1967; Johnston, 1981; Landis, 1980; Waterhouse, 1964).

Caples Terrane

The Caples terrane is a belt of relatively low-grade graywacke type sediments with local development of thick (meta) volcanic and pelitic units, of probably Permian to Triassic age, that outcrop between the Haast Schist and the Dun Mountain ophiolite. The rocks are variously known as the Tuapeka Group (South Otago), Caples Group (West Otago), and Pelorus Group (Nelson). The rocks differ from Torlesse graywackes (see below) in that provenance is intermediate volcanogenic and the rocks are relatively deficient in quartz and detrital micas. A typical analysis is 60.5% SiO_2, 3.2% Na_2O, and 1.8% K_2O. In many areas Caples rocks also have a coherent, mappable stratigraphy. Like the Torlesse rocks they grade through prehnite-pumpellyite, pumpellyite-actinolite, and locally lawsonite-albite-chlorite facies into the Haast Schist, and also commonly appear to young towards the higher grade rocks (greenschist facies). Structure is complex and polyphase, with synmetamorphic recumbent folding (axial plane schistosity) and subsequent folding and faulting.

In the central portion of the Caples terrane the western boundary is the Livingstone Fault, which brings basal Caples rocks—lavas and breccias, hyaloclastites, volcanogenic sandstone, metadolerite chert, and ferruginous phyllite—into contact with the Dun Mountain ophiolite. Elsewhere the western margin is marked by imbricate slabs of metasedimentary rock interleaved with ophiolite melanges. In places these melanges are faulted against the Dun Mountain ophiolite, making distinction of the two very difficult. The metasedimentary slabs are of broadly Caples character; however, positive correlation is not established, and it is possible that some of these represent fragments of an unrecognized terrane. The eastern boundary, with Torlesse sediments, lies somewhere within the schist belt (Bishop et al, 1976; Coombs et al, 1976; Turnbull, 1979). The estimated position of the suture is indicated on Figure 1.

Kakahu Terrane

A small (15 sq km [6 sq mi]) area of tectonized semischistose graywacke, argillite, and conglomerate contains conspicuous fault-bounded masses of chert, basic volcanics, and pure limestone. Conodonts from limestone samples yield Upper Carboniferous ages and are the oldest fossils east of the Median Tectonic Line (Fig. 1). Kakahu rocks have not been related convincingly, either stratigraphically or structurally, to the enclosing Torlesse rocks (8) including the adjoining Permian strata. The Kakahu terrane (Hitching, 1979), or at least the limestone-chert-volcanic portion, is therefore regarded as a suspect terrane, with possible affinities with the Permian Akatarawa terrane.

Akatarawa Terrane

A fusuline-bearing limestone lens overlying basaltic hyaloclastites outcrops within a small (5 sq km [2 sq mi]) area in the Waitaki Valley, South Canterbury. Associated limestones contain corals, crinoids, and ammonoids, and some contain nodular chert (Hornibrook and Shu, 1965). Sandstones and pebble conglomerates overlying the fossiliferous limestones are distinctly more quartzose than any other pre-Late Cretaceous sandstones east of the Median Tectonic Line (MacKinnon, 1980). The distinctive fauna is of Tethyan nature, unlike other Permian faunas in South Island, and with the unique sandstone petrography, indicates the Akatarawa terrane should be regarded as exotic. Other remnants of the Akatarawa terrane may be represented by recent discoveries of blocks of Permian chert (Jones, personal communication) and of fusulinid limestone elsewhere in the Torlesse terrane of South Canterbury.

Older Torlesse (Rakaia) Terrane

The older Torlesse (Rakaia) terrane comprises the largest part of the South Island Torlesse rocks and is dominated by nonschistose graywacke and argillite that underlie most of the Southern Alps and eastern foothills. They are predominantly of Permian and Middle and Late Triassic age. Rocks include some volcanogenic material but are dominated by quartzofeldspathic graywacke (a characteristic analysis is 69.7% SiO_2, 3.5% Na_2O, and 2.4% K_2) with minor chert, pillow lava, and limestone. The terrane is characterized by thick submarine fan-type sequences with complex lithostratigraphy, rare fossils, and complex patterns of deformation, including very steeply plunging folds and widespread overturning. Rocks in general decrease in age westwards. The metamorphic grade increases in the same direction from locally zeolite, through prehnite-pumpellyite, pumpellyite-actinolite (at which stage they grade into Haast Schist), into greenschist facies. The western boundary, a cryptic suture within the Haast Schists, separates Torlesse and Caples rocks (Fig. 1) (Andrews et al, 1976; Bishop, 1974; Campbell and Warren, 1965; MacKinnon, 1983).

Younger Torlesse (Pahau) Terrane

Younger Torlesse rocks outcrop northeast of the Esk Head Melange (Bradshaw, 1973) and resemble, in most respects, those of the older (Rakaia) Torlesse terrane. They range in age from Late Jurassic to Early Cretaceous and are predominantly alternating sequences of lithic to quartzofeldspathic sandstone and mudstone with volcanic and reworked Torlesse clasts, including schistose clasts, common in some younger rocks. Thick lenticular conglomerates composed largely of Torlesse clasts are widespread. Metamorphic grade is zeolite facies and structure is complex, including extensive broken formation and lenticular melanges.

In the northeast parts of the terrane, a brief intra-Albian hiatus separates Pahau rocks from the oldest parts of the unconformable cover. Mesozoic deformation was completed in the Pahau terrane before the intrusion of late Albian dikes.

Fluviatile to shallow-marine rocks of similar age and composition to Pahau sediments overlie deformed older Torlesse (Rakaia) rocks southwest of the Esk Head Melange and suggest general proximity of the younger and older Torlesse during the late Mesozoic (Andrews et al, 1976; MacKinnon, 1983).

AMALGAMATION AND ACCRETION

The earliest accretionary event that can be dated, albeit approximately, is the tectonic juxtapositioning of the Golden Bay terrane against the Karamea terrane, the latter probably then part of Gondwana. Both terranes were folded before the Devonian, but marked differences between the Lower Devonian sequences of each terrane suggest that amalgamation occurred later and coincides with mid-Devonian phase of metamorphism and tectonism (Tuhua orogeny). This event and the ensuing emplacement of granitoids marks the consolidation of much of the basement of New Zealand's Western province (Fig. 2).

In the Eastern Province, east of the MTL (the New Zealand Geosyncline of earlier usage), a number of collisions have been proposed or are suspected. The most widely accepted but difficult to pinpoint on the ground is the cryptic suture between the Caples and the older (Rakaia) Torlesse terrane (the latter by then including the Akatarawa and Kakahu slices). Collision probably took place in the earliest Jurassic and was accompanied by intense regional dynamo-thermal metamorphism. Amalgamation of the Caples and Dun Mountain terranes may have occurred somewhat earlier, although it is not constrained by covering strata until Albian times. The presence of lithic clasts and detrital pyroxenes characteristic of Brook Street rocks in Maitai sediments suggests a pre-Upper Permian amalgamation of Brook Street–Murihiku and Dun Mountain–Maitai terranes.

The importance of the fundamental break termed the MTL has been recognized for over 35 years (Turner, 1938); however, the timing of the events it represents is uncertain. Although not adjacent to the fault, the distinctive faunal and lithological character of the Parapara Permian as compared with Maitai rocks suggests that the Golden Bay and the Brook Street–Maitai terranes were well separated at that time. The earliest plausible time for suturing along the MTL is mid-Triassic. Three lines of evidence support this timing: (1) the influx of coarse granitic detritus probably derived from the Western Province in Middle Triassic Murihiku conglomerates; (2) the presence of faulted outliers within the Western Province of nonmarine volcaniclastic sandstone containing Triassic palymorphs and similar in character to some Murihiku sandstones; and (3) the occurrence of granitoids of Late Triassic and Jurassic age in both the Brook Street–Murihiku terrane and in the Western Province.

An upper limit is provisionally established in the southeast where an offshore basin with Albian strata overlaps the projection of several terrane boundaries including the MTL. In the southwest, however, a major metamorphic–plutonic complex of Early Cretaceous age (Kimbrough, Bradshaw, personal communication) is truncated by the MTL. Cretaceous or even Early Tertiary juxtapositioning is indicated. The oldest constraining cover beds are of Eocene age.

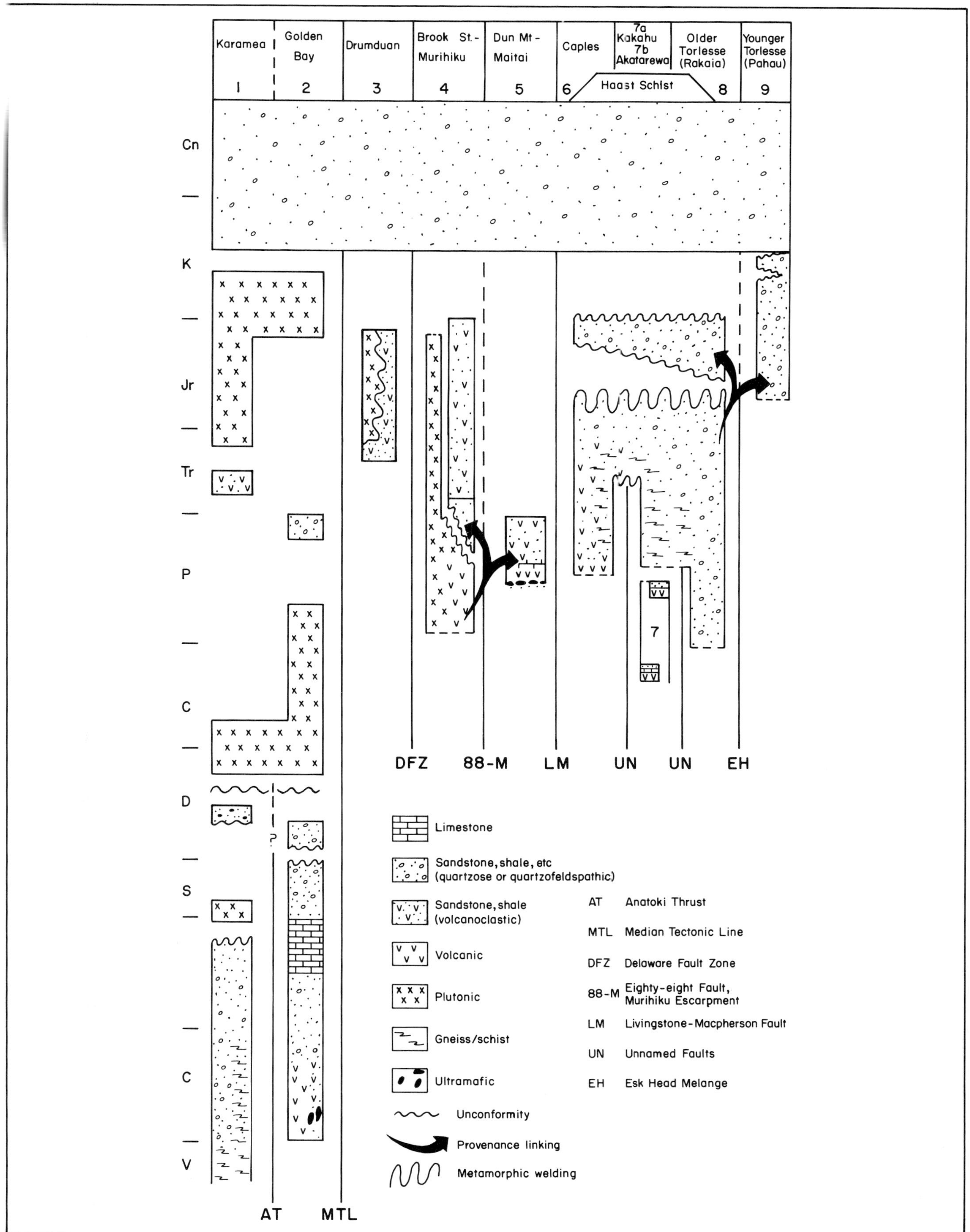

Figure 2—Time-space diagram of tectonostratigraphic terranes of South Island, showing inferred timing of amalgamation and accretion.

ORIGIN

The origin or relative original position of the South Island terranes is unknown in the absence of detailed paleontological and paleomagnetic studies on a terrane by terrane basis. Nevertheless, recognition of Tethyan faunas in the Akatarawa terrane and in isolated blocks within the younger Torlesse terrane indicates accretion of exotic material. The atomodesmatinid faunas of Maitai and Torlesse rocks are unknown elsewhere around the Pacific except for isolated localities in Australia, Alaska, and Nevada. On the other hand, late Paleozoic and Mesozoic faunas of the Torlesse and Brook Street–Murihiku terranes show affinities between terranes suggesting only limited latitudinal displacement between these belts. Middle Triassic and Middle and Late Jurassic faunas from Murihiku and Torlesse rocks also show strong affinities with faunas from the Antarctica Peninsula and New Caledonia (Grant-Mackie et al, 1976).

Available paleomagnetic data (Grindley et al, 1980) indicate that shallow intrusives of probable Jurassic age within the Murihiku portion of the Brook Street–Murihiku terrane were emplaced at latitudes of about 66°.

ACKNOWLEDGMENTS

We are grateful to D. S. Coombs, M. G. Laird, and M. R. Johnston for comments on the manuscript, which developed from a workshop held in Dunedin in 1982, following a visit by D. L. Jones and an early paper on terranes in New Zealand by D. G. Howell (Howell, 1980). Unpublished information and concepts advanced by J. Y. Bradshaw, D. L. Kimbrough, R. J. Norris, and C. M. Ward have assisted our interpretation.

REFERENCES

Andrews, P. B., et al, 1976, Lithological and paleontological content of the Carboniferous-Jurassic Canterbury Suite, South Island, New Zealand: New Zealand Journal of Geology and Geophysics, v. 19, p. 791–819.

Bishop, D. G., 1974, Stratigraphic, structural and metamorphic relationships in the Dansey Pass area, Otaga, New Zealand: New Zealand Journal of Geology and Geophysics, v. 17, p. 301–355.

______, et al, 1976, Lithostratigraphy and structure of the Caples terrane of the Humboldt Mountains, New Zealand: New Zealand Journal of Geology and Geophysics, v. 19, p. 827–848.

Boles, J. R., 1974, Structure, stratigraphy and petrology of mainly Triassic rocks, Hokonui Hills, Southland, New Zealand: New Zealand Journal of Geology and Geophysics, v. 17, p. 337–374.

Bradshaw, J. D., 1973, Allochthonous Mesozoic fossil localities in melange within the Torlesse rocks of North Canterbury: Journal of the Royal Society of New Zealand, v. 3, p. 161–167.

Campbell, J. D., and G. Warren, 1965, Fossil localities of the Torlesse Group in the South Island: Transactions of the Royal Society of New Zealand, Geology, v. 3, p. 99–137.

Coombs, D. S., et al, 1976, The Dun Mountain ophiolite belt, New Zealand: Its tectonic setting, constitution, and origin, with special reference to the southern portion: American Journal of Science, v. 276, p. 561–603.

Cooper, R. A., 1979, Lower Paleozoic rocks of New Zealand: Journal of the Royal Society of New Zealand, v. 9, p. 29–84.

Grant-Mackie, J. A., et al, 1976, Advances in correlation of Mesozoic sequences in New Zealand and New Caledonia (Abs.): 25th International Geological Congress, Sydney, v. 1, p. 268–270.

Grindley, G. W., 1980, Geological map of New Zealand, 1:63,360: Wellington, New Zealand Department of Scientific and Industrial Research, sheet S13-Cobb.

______, et al, 1980, Lower Mesozoic position of southern New Zealand determined from paleomagnetism of the Glenham Porphyry, Murihiku terrane, eastern Southland: Proceedings of the Fifth International Gondwana Symposium, Wellington, New Zealand, p. 319–326.

Harper, C. T., and C. A. Landis, 1967, K–Ar ages from regionally metamorphosed rocks, South Island, New Zealand and some tectonic implications: Earth and Planetary Science Letters, v. 2, p. 419–429.

Hitching, K. D., 1979, Torlesse geology of Kakahu, South Canterbury: New Zealand Journal of Geology and Geophysics, v. 22, p. 191–197.

Hornibrook, N. de B., and Y. K. Shu, 1965, Fusuline limestone in the Torlesse Group near Benmore Dam, Waitaki Valley, *in* J. D. Campbell and G. Warren, Fossil localities of the Torlesse Group in the South Island: Transactions of the Royal Society of New Zealand, Geology, v. 3, p. 135–137.

Houghton, B. F., 1981, Lithostratigraphy of the Takitimu Group, central Takitimu Mountains, western Southland, New Zealand: New Zealand Journal of Geology and Geophysics, v. 24, p. 333–348.

Howell, D. G., 1980, Mesozoic accretion of exotic terranes along the New Zealand segment of Gondwanaland: Geology, v. 8, p. 487–491.

Johnston, M. R., 1981, Geological map of New Zealand, 1:50,000 with notes: Wellington, New Zealand Department of Scientific and Industrial Research, part sheet 027-Dun Mountain, 1st ed.

Jones, D. L., et al, 1982, Recognition, character, and analysis of tectonostratigraphic terranes in western North America, *in* Oji Seminar Volume: Centre for Academic Publications Japan.

Landis, C. A., 1980, Little Ben Sandstone, Maitai Group (Permian): nature and extent in the Hollyford–Eglinton region, South Island, New Zealand: New Zealand Journal of Geology and Geophysics, v. 23, p. 551–567.

MacKinnon, T. C., 1980, Sedimentologic, petrographic, and tectonic aspects of Torlesse and related rocks, South Island, New Zealand: PhD Dissertation, University of Otago, 294 p.

______, 1983, Origin of the Torlesse terrane and coeval rocks, South Island, New Zealand: Geological Society of America Bulletin, v. 94, p. 967–985.

Mutch, A. R., 1972, Geology of Morley Subdivision: New Zealand Geological Survey Bulletin 76, 104 p.
New Zealand Geological Survey, 1972, Geological map of New Zealand, 1:1,000,000, South Island: Wellington, New Zealand Department of Scientific and Industrial Research, 1st ed.
Suggate, R. P., ed., 1978, The geology of New Zealand: Wellington, New Zealand Geological Survey, Government Printer, 2 v., 820 p.
Tulloch, A. J., 1983, Granitoid rocks of New Zealand—a brief review: Geological Society of America Memoir 159, p. 5–21.
Turnbull, I. M., 1979, Stratigraphy and sedimentology of the Caples terrane of the Thompson Mountains, northern Southland, New Zealand: New Zealand Journal of Geology and Geophysics, v. 22, p. 555–574.
Turner, F. J., 1938, Progressive regional metamorphism in southern New Zealand: Geological Magazine, v. 75, p. 160–174.
Waterhouse, J. B., 1964, Permian stratigraphy and faunas of New Zealand: New Zealand Geological Survey Bulletin 72, 101 p.
Williams, J. G., 1978, Eglinton Volcanics—stratigraphy, petrology, and metamorphism: New Zealand Journal of Geology and Geophysics, v. 21, p. 713–732.

The Relationship between the Tectonic Evolution of Southeast Asia and Hydrocarbon Occurrences

A. J. Barber
Chelsea College
University of London, United Kingdom

Southeast Asia consists of a cratonic core, Sundaland, built of accreted continental fragments that had stabilized toward the end of the Mesozoic. Throughout the late Mesozoic and Tertiary, additional terranes were added to this core in Sumatra, Borneo, eastern Indonesia, and the Philippines. In the Early Tertiary, the whole region was affected by a phase of expansion related to the collision of India with the southern margin of the Asian continent. Expansion resulted in the formation of subsiding sedimentary basins in which great thicknesses of sediment were deposited and in which mobile hydrocarbons were generated. In the eastern part of the region, expansion proceeded to the formation of oceanic crust to form small marginal seas. In mid-Miocene times, the northward-moving Australian continent collided with the southeastern extremity of the region in the Banda Arcs. At the same time, renewed subduction of the Pacific and Indian Ocean Plates along the eastern and western margins resulted in an overall compression of the whole of Southeast Asia. Compression generated folds and faults in the Paleogene sedimentary basins, producing structural traps for the accumulation of hydrocarbons. In eastern Indonesia and the Philippines, compression resulted in subduction of the marginal seas and the collision of many small continental blocks, derived either from the southern margin of Asia or the northwest margin of Australia, with the subduction complexes. Accumulations of hydrocarbons and seeps found in the collision zones are derived from shelf sediments associated with these continental fragments.

INTRODUCTION

Sundaland, the cratonic core of Southeast Asia, was formed by the amalgamation of a series of small continental blocks during the Permo-Trias (Stauffer, 1983; Pulunggono and Cameron, 1984). Throughout the late Mesozoic and Tertiary, subduction complexes and small continental fragments were added to this core (Fig. 1). On the western margin of Sumatra, accreted material includes an oceanic assemblage, the Woyla Group, and the Sikuleh and Natal microcontinents (Pulunggono and Cameron, 1984). On the southeastern margin the Ciletuh and Lok Ulu subduction complexes and the Meratus ophiolite have been added to the craton in Java and Kalimantan, with the North Borneo subduction complex in Sabah and Sarawak (Hamilton, 1979). The eastern part of the region, extending from China through the Philippines to Australia, consists of a complex assemblage of small ocean basins, subduction complexes, island arcs, and continental fragments, the latter derived either from the southern margin of China or the northern margin of Australia (Pigram and Panggabean, 1983). At the present time the Indo-Australian Plate is passing down a subduction zone along the western and southern margins of the region, subduction being oblique beneath Sumatra. In the southeast, collision is in progress between the Australian continent and the Banda Arc subduction zone. Along the eastern margin of the region the small ocean basins are being overridden by westward-advancing subduction complexes, while further east the Pacific Ocean and Philippine Sea Plates are being subducted obliquely northwestwards beneath the Philippines. Oblique subduction along both the eastern and western margins of Southeast Asia is generating major transcurrent movements left-lateral along the Philippine Fault System and right-lateral along the Sumatran Fault System (Fig. 1).

Sedimentary basins in the Southeast Asian region, which are potential sites for the generation and accumulation of mobile hydrocarbons, have been classified according to their relationship to the present tectonic elements (Fletcher and Soeparjadi, 1976; Nayoan et al, 1979; Fig. 1), into (1) forearc basins, (2) backarc basins, (3) intracratonic basins, and (4) continental margins. A further category of prospective regions is recognized in this account where continental margin sediments are involved in (5) collision zones (Fig. 2).

FOREARC BASINS

Forearc basins are developed as a direct result of the subduction process, where a subsiding basin forms between the rising accretionary forearc ridge and the volcanic arc. In spite of great thicknesses of sediment (more than 7 km [4.3 mi] in places), forearc basins have proved disappointing as a major source of hydrocarbons, although some gas has been discovered in the northern part of the forearc basin to the west of Sumatra. The paucity of accumulations of hydrocarbons is due primarily to low heat flow, because the forearc basin lies directly over the cold subducting lithospheric plate. Low heat flow means that temperatures are not sufficiently high to mature the carbonaceous material. Other factors are the dilution of the organic material by the abundance of volcanic detritus, derived from the erosion of the volcanic arc, and decomposition of unstable volcaniclastic components, which reduces porosity and results in low permeability in the sediments.

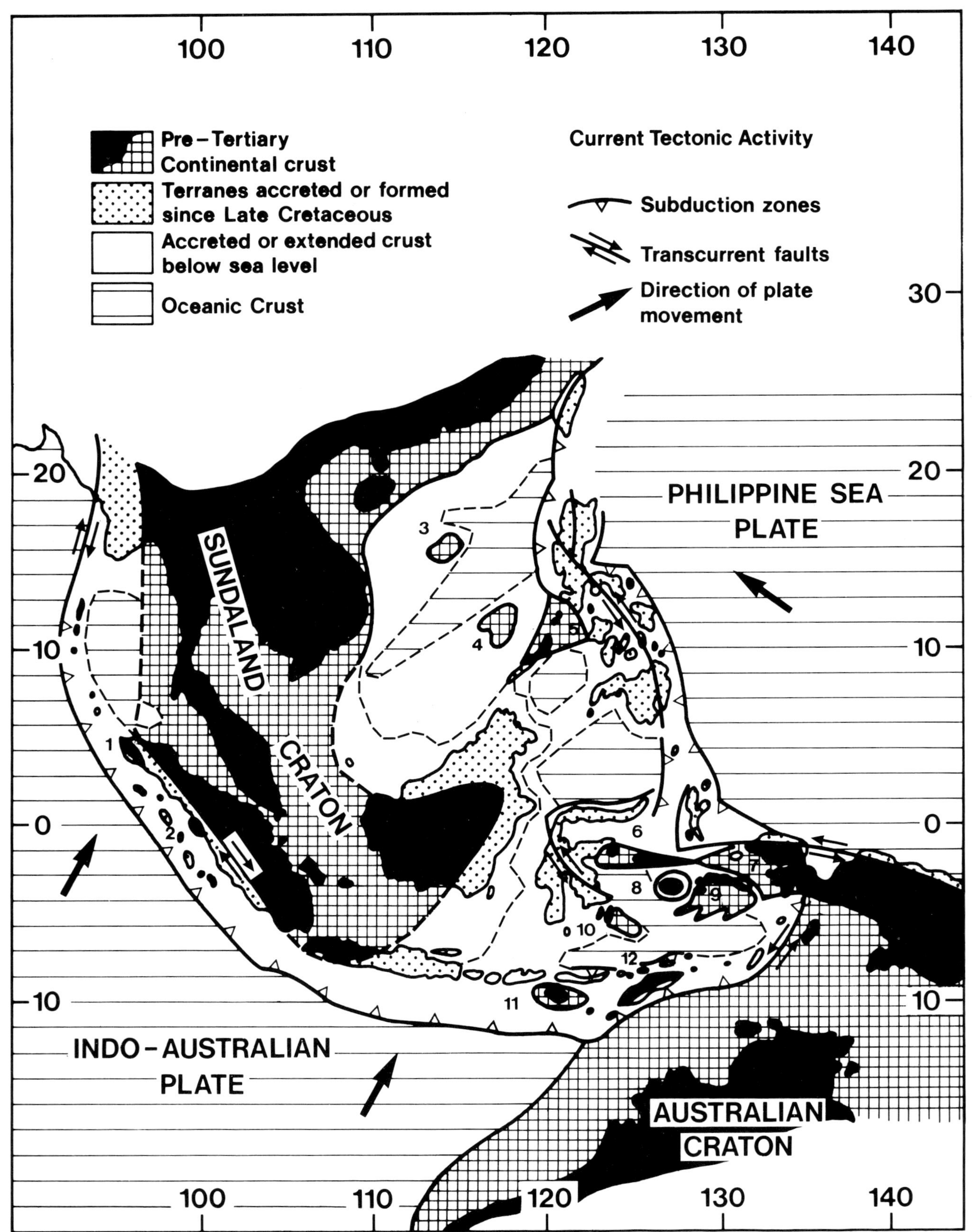

Figure 1—The accretion of Southeast Asia. Numbered microcontinental blocks: 1 = Sikuleh; 2 = Natal; 3 = Macclesfield Bank; 4 = Reed Bank; 5 = Palawan; 6 = Banggai-Sula; 7= West Irian Jaya; 8 = Buru; 9 = Seram-North Banda; 10 = Tukang Besi; 11 = Sumba; 12 = Allochthonous Timor.

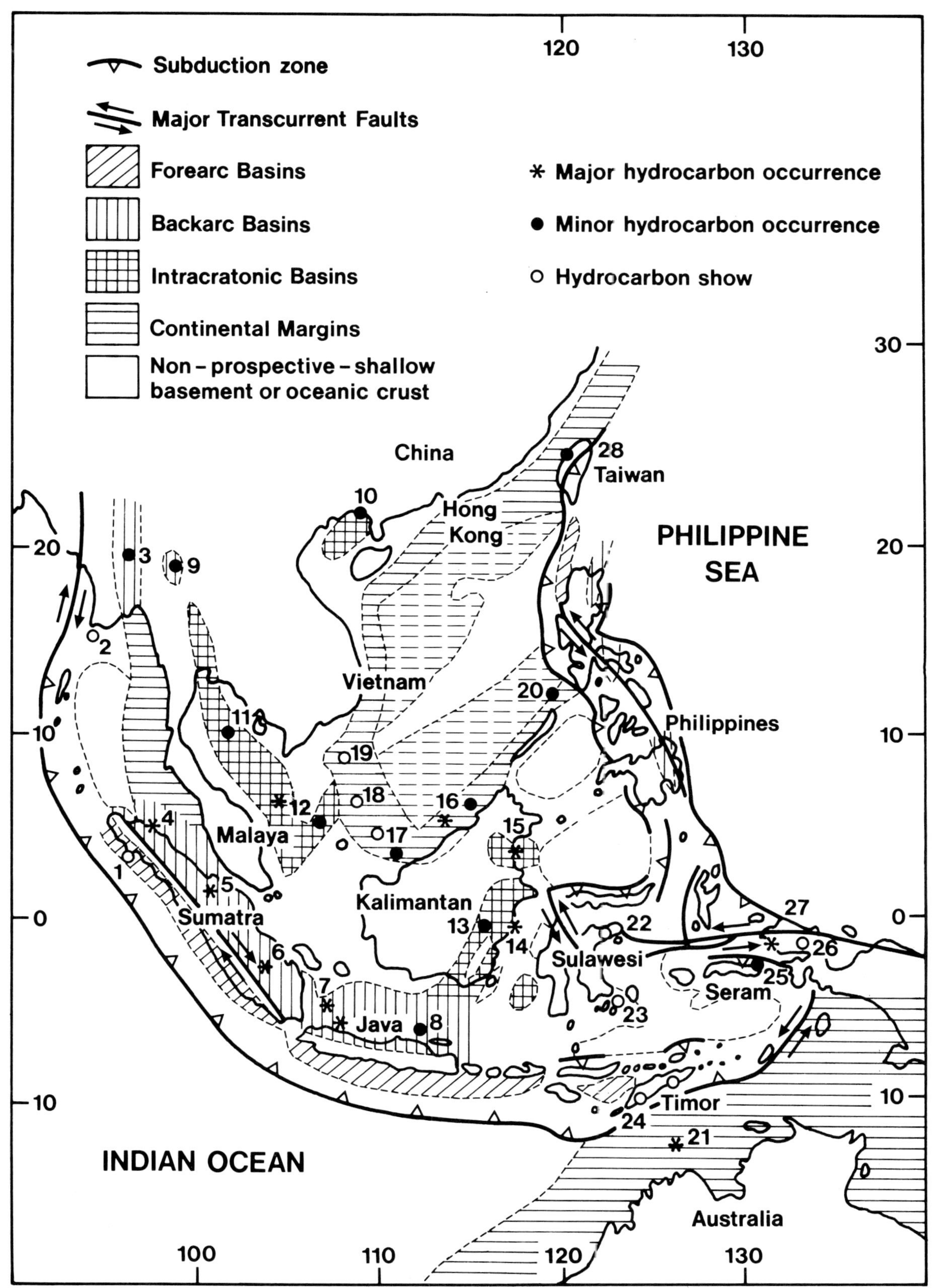

Figure 2—Tertiary sedimentary basins, continental margins and the occurrence of hydrocarbons in Southeast Asia. Numbered occurrences: 1 = West Sumatra Basin; 2 = Offshore Burma; 3 = Central Burma; 4 = North Sumatra Basin; 5 = Central Sumatra Basin; 6 = South Sumatra Basin; 7 = Sunda Strait and West Java Basins; 8 = East Java Basin; 9 = Fang Basin; 10 = Gulf of Tongking Basin; 11 = Thai Basin; 12 = Malaya and West Natuna Basins; 13 = Barito Basin; 14 = Kutai Basin; 15 = Tarakan Basin; 16 = Brunei and Sabah; 17 = North West Borneo; 18 = Saigon Basin; 19 = Mekong Basin; 20 = Palawan; 21 = Northwest Australian Shelf; 22 = East Arm Sulawesi; 23 = Buton; 24 = Timor; 25 = Bula Basin; Seram; 26 = Bituni Basin; 27 = Salawati Basin.

BACKARC BASINS

Backarc basins in eastern Sumatra and northern Java contain the most productive oil and gas fields in the region (Fig. 2). This productivity is due to the presence of prolific source rocks, suitable reservoir horizons in clastic and carbonate facies, overlain by deep-water shales that act as a seal, high heat flow during subsidence and deposition, giving early maturation, followed by a tectonic environment that generated structural traps for the accumulation of hydrocarbons.

In Sumatra the backarc basins are developed on a cratonic basement that includes Permo-Carboniferous sediments rich in carbonaceous material. These basins opened during the Paleogene as rifts with a north-south trend, resulting from extension and thinning of the continental crust in an east-west direction. Sediments were poured into the opening rifts from adjacent uplifted basement blocks. The sediments were deposited in a terrestrial environment with alluvial fan, fluviatile, lacustrine, and swamp facies, containing abundant carbonaceous material, to form the source beds. Thinning of the underlying crust, with the uprise of mantle material beneath, produced a regime with a high heat flow in the rift sediments, which led to early maturation of the organic content. Pulunggono and Cameron (1984) have pointed out recently that the major oil and gas fields in eastern Sumatra overlie an ophiolitic suture that separates two continental terranes. The suture evidently formed a zone of weakness for the opening of the rifts and a conduit for the transfer of heat from the mantle into the overlying sediments.

A marine transgression followed in the early Miocene, with the deposition of mature, siliciclastic littoral sediments, calc-arenites, and carbonate build-ups, which form the main reservoir horizons. Deposition of shale, with deepening water, formed the seal.

In the mid-Miocene, the renewal of subduction along the western margin of Sundaland resulted in the uplift of the Barisan Mountains, the construction of the volcanic arc, the commencement of transcurrent fault movements along the Sumatran Fault System, and compression of the sediments, already deposited in the backarc region, to form structural traps by folding and faulting. The folds have an en echelon arrangement with their axial traces oriented northwest-southeast, at an angle to the Sumatran Fault trend (Murphy, 1975), and are considered to have resulted from compression related to transcurrent movement along the Sumatran Fault System. Davies (1984) has suggested that early normal faults, generated during the extensional phase, have been reactivated in a transcurrent sense when the stress system changed its orientation. Continual uplift of the Barisan Mountains along the western side of Sumatra has disrupted the western margins of the backarc basins and produced a regressive sedimentary sequence, including volcaniclastic deposits derived from the volcanic arc, which has extended eastwards across the basins towards the Malacca Strait.

INTRACRATONIC BASINS

Intracratonic basins extend throughout the central part of Southeast Asia, from the Fang Basin in northern Thailand, through the Thai and Malay Basins in the Gulf of Thailand, to the Tarakan, Kutai, and Barito Basins in Kalimantan and the Makassar Strait (Fig. 2). Sedimentary sequences in these intracratonic basins record the same history of Paleogene rifting, followed by continual subsidence during the Neogene with a transgressive, followed by a regressive sedimentary sequence and a phase of compression commencing in the mid-Miocene. In Kalimantan the margins of the basins have been disrupted by the recent uplift of the Meratus Mountains.

The sequence of events in both intracratonic and backarc basins can be interpreted in terms of the subsidence model of McKenzie (1978). Both sets of basins were formed by initial isostatic subsidence, caused by extension and thinning of continental crust, with erosion of adjacent uplifted blocks to provide a thick sedimentary infill. Thinning of the crust leads to the uprise of mantle material, causing high heat flow in the crust and in the overlying sediments, raising the temperature in the lower parts of the sedimentary columns to the level at which mobile hydrocarbons are generated.

According to McKenzie's (1978) model, an active phase of extension, for the purpose of the model regarded as instantaneous, is followed by subsidence resulting from thermal relaxation as the geothermal gradient returns to normal values. Further subsidence results from sedimentary loading of the crust. Using seismic reflection profiles, controlled by borehole records, it is possible to reconstruct the history of sedimentation and subsidence for each sedimentary basin, and, from heat flow models, to establish whether temperature within the basin had ever been high enough to generate mobile hydrocarbons. This type of analysis has been used by Situmorang (1982) with respect to the South Makassar Basin, and in principle can be applied to any basin for which sufficient data are available. It is possible to distinguish hydrocarbon-prospective from nonprospective basins using this technique.

CONTINENTAL MARGINS

Continental margins represent areas in which intracratonic extension has proceeded to the point of rupture of the continental crust, with the formation of oceanic crust between separating continental fragments. During the extension phase, continental crust is broken into horsts and graben, with infilling of the graben with organic-rich terrestrial sediments. The same sequence of events described from the backarc and intracratonic basins is seen in the continental margin environment, where the potential source rocks of terrestrial facies are deposited in an area of high heat flow, followed by a transgressive sequence of marine deposits with suitable reservoir rocks, and then a prograding sequence of continental shelf sediments, extending across the subsided continental margin. This depositional sequence is seen around the margins of all the small ocean basins that characterize the Southeast Asian region and is exemplified by the South China Sea, where a complex pattern of continental breakup during the Paleogene has resulted in an extensive continental shelf environment along both its northern and southern margins.

From the common history seen in backarc, intracratonic, or continental marginal basins, it is evident that these basins all formed as a result of a phase of extension and expansion that affected the whole of the Southeast Asian region during the Paleogene. This phase of expansion has been attributed by Tapponnier et al (1982) to the collision of the Indian continent with the southern margin of Asia, which took place at about this time. As a result of the collision, segments of continental crust were displaced eastwards and extruded southeastwards along major transcurrent faults. Unrestrained southeastward movement was accompanied by lateral expansion of the moving blocks, permitting the opening up of rift systems to form sedimentary basins, which later, depending on their tectonic position, developed as backarc and intracratonic basins or evolved into small ocean basins.

In mid-Miocene times the phase of expansion ceased abruptly, when the whole region was subjected to compression, causing underthrusting of the floors of the small ocean basins and folding, faulting, and thrusting of the basin sediments in backarc and intracratonic basins and the local uplift of continental blocks.

The cause of the change from a regime of expansion to one of compression is not at all obvious. At about the same time, the Banda Arcs, at the southeastern extremity of Southeast Asia, came into contact with the northern margin of Australia. This event can have no influence on the eastern and western margins of the region. In Sumatra to the west, possible transcurrent movements were replaced by active subduction, which resulted in the formation of the forearc basins that have already been described. To the east the fragments that make up the Philippines began to advance westwards, subducting the marginal ocean basins until they came into contact with continental crust in Panay. Continued compression along the eastern margin caused the Philippine Sea Plate to commence subduction westwards beneath the Philippines.

The fundamental cause of this dramatic change from an extensional to a compressional regime must be sought in major readjustments within the earth's mantle. But, whatever the cause, it was this fortunate sequence of events that has made the western part of Southeast Asia a major oil and gas province.

COLLISION ZONES

Collision zones, where continental margin sediments are in collision with subduction complexes, constitute a potentially prospective environment that has not yet been properly explored nor properly considered in earlier surveys of the hydrocarbon potential of Southeast Asia. The Banda Sea region of eastern Indonesia contains several allochthonous continental terranes that separated from the northern margin of Australia in Jurassic times (Pigram and Panggabean, 1983).

The northern margin of Australia itself, which has been extensively surveyed for hydrocarbon occurrences and promises to be a major oil and gas province, is in collision with the Banda Arc subduction complex in the Timor Trough. This continental margin originated in the Late Jurassic by the rifting and separation of continental fragments from the northwest margin of Gondwanaland. It shows all the characteristics that have already been described for continental shelves elsewhere in the region, with the development of horsts and graben, a phase of terrestrial sedimentation overlain by transgressive and regressive continental shelf sequences. Source beds occur in the Permo-Trias and Jurassic, and reservoir sands occur in the Permian, Triassic, Jurassic, and Cretaceous sediments.

To the north, the Australian continental shelf sediments are depressed in the Timor Trough and imbricated in the accretionary wedge that forms the outer arc island of Timor and the adjacent islands of the arc.

The possibility that significant hydrocarbon accumulations might be preserved in imbricate wedges has been discussed by Thompson (1976) and Montecchi (1976). Thompson was optimistic that structural traps would form within the wedge. This possibility has yet to be seriously tested owing to the probable small size of the traps and the difficulty of locating them in an area of complex tectonics. On the other hand, Montecchi (1976) predicted that hydrocarbons would escape through the highly fractured sediments. Certainly in Timor and the outer arc islands between Savu and Kai, there are frequent mud volcanoes, indicating the escape of water, oil, and gas with entrained sediments from overpressured water-saturated continental shelf sediments beneath. Montecchi (1976) suggested that only where active wedges were rapidly buried beneath accumulating sediments would it be possible for escaping hydrocarbons to be trapped in the overlying sediments.

Seram, on the northern side of the Banda Arcs, where the Australian continental shelf is being subducted beneath the Banda Sea, may provide an example of the situation postulated by Montecchi (1976). Oil and gas have long been obtained from the Plio-Pleistocene Bula Basin (Zillman and Paten, 1976). However, carbonaceous material in the Plio-Pleistocene deposits is not sufficiently mature to be the source of the hydrocarbons. They have evidently been derived from the imbricated Australian continental margin deposits that underlie the younger sediments.

Throughout the Banda Sea region there are a number of allochthonous continental terranes (Fig. 1) that are regarded as having separated from the northern margin of Australia during Jurassic times (Pigram and Panggabean, 1983). In the mid-Miocene, several of these continental fragments collided either with the southeastern margin of Sundaland or the northern margin of Australia. All these collision zones have hydrocarbon shows.

Asphalt deposits are associated with the zone of collision between the Tukang Besi microcontinent and the southeast arm of Sulawesi in the island of Buton, while in the east arm of Sulawesi, where the Banggai–Sula microcontinent is in collision with a major ophiolite complex, oil seeps rise through the imbricated and disrupted ophiolite from the underthrust continental margin sediments beneath. Oil seeps are also associated with the Lengguru fold and thrust belt in Irian Jaya, interpreted as the site of collision between the West Irian Jaya microcontinent and the northern margin of Australia in mid-Miocene times (Pigram et al, 1982).

SUMMARY

This brief survey of the relationships between tectonics and the occurrence of hydrocarbons demonstrates that the active tectonic environment of Southeast Asia provides many targets for the explorationist. So far the most systematically investigated and productive occurrences lie in the western part of the region, where an Early Tertiary phase of extension and rifting, followed in the Late Tertiary by a phase of compression, has resulted in many favorable locations for the generation and accumulation of hydrocarbons. Less well investigated are the zones of collision in the eastern part of the region, where hydrocarbon-bearing late Paleozoic to Mesozoic continental margin sediments are involved in subduction complexes. These areas of tectonic complexity offer a challenge to the explorationist for the future.

ACKNOWLEDGMENT

R. W. Murphy has made many suggestions that resulted in the improvement of this article.

REFERENCES

Davies, P. R., 1984, Tertiary structural evolution and related hydrocarbons, North Sumatra Basin: Proceedings of the 13th Annual Convention of the Indonesian Petroleum Association, Jakarta, v. 1, p. 19–49.

Fletcher, G. L., and R. A. Soeparjadi, 1976, Indonesia's Tertiary basins—the land of plenty: Offshore Southeast Asia Conference, SEAPEX Program (Singapore). Unpublished, quoted in O. J. Bee, 1982, The Petroleum Resources of Indonesia, O.U.P., Kuala Lumpur.

Hamilton, W., 1979, Tectonics of the Indonesian region: United States Geological Survey Professional Paper 1078.

McKenzie, D., 1978, Some remarks on the development of sedimentary basins: Earth and Planetary Science Letters, v. 40, n. 1, p. 25–32.

Montecchi, P. A., 1976, Some shallow tectonic consequences of 'subduction' and their meaning to the hydrocarbon explorationist, *in* M. T. Halbouty, et al, eds., Circum-Pacific Energy and Mineral Resources: American Association of Petroleum Geologists Memoir 25, p. 189–202.

Murphy, R. W., 1975, Tertiary basins of Southeast Asia: Proceedings of the Southeast Asian Petroleum Exploration Society, v. 2, p. 1–36.

Nayoan, G. A. S., et al, 1979, Nota tentang cadangan minyak dan gas bumi Indonesia: suata pemirikiram ahli geologi: Paper presented at the 8th Annual Convention of the Indonesian Petroleum Association. Unpublished, quoted in O. J. Bee, 1982, The Petroleum Resources of Indonesia, O.U.P., Kuala Lumpur.

Pigram, C. J., and H. Panggabean, 1983, Age of the Banda Sea, eastern Indonesia: Nature, v. 301, p. 231–234.

______, et al, 1982, Late Cenozoic origin for the Bituni Basin and adjacent Lengguru fold belt, Irian Jaya: Proceedings of the 11th Annual Convention of the Indonesian Petroleum Association, Jakarta, v. 1, p. 108–126.

Pulunggono, A., and N. R. Cameron, 1984, Sumatran microplates, their characteristics and their role in the evolution of the Central and South Sumatra Basins: Proceedings of the 13th Annual Convention of the Indonesian Petroleum Association, Jakarta, v. 1, p. 121–143.

Situmorang, B., 1982, The formation of the Makassar Basin as determined from subsidence curves: Proceedings of the 11th Annual Convention of the Indonesian Petroleum Association, Jakarta, v. 1, p. 83–107.

Stauffer, P. H., 1983, Unravelling the mosaic of Palaeozoic crustal blocks in Southeast Asia: Geologische Rundschau, v. 72, n. 3, p. 1061–1080.

Tapponnier, P., et al, 1982, Propagating extrusion tectonics in Asia: new insights from simple experiments with plasticine: Geology, v. 10, p. 611–616.

Thompson, P. L., 1976, Plate tectonics in petroleum exploration and convergent continental margins, *in* M. T. Halbouty, et al, eds., Circum-Pacific Energy and Mineral Resources: American Association of Petroleum Geologists Memoir 25, p. 177–188.

Zillman, N. J., and P. J. Paten, 1976, Petroleum prospects, Bula Basin, Seram, Indonesia: Proceedings of the 4th Annual Convention of the Indonesian Petroleum Association, Jakarta, 1975, v. 2, p. 129–148.

Continental Terranes in Southeast Asia: Pieces of Which Puzzle?

Peter H. Stauffer
Palo Alto, California

Continental crust underlies most of the large pre-Tertiary core of Southeast Asia, comprising Sumatra, west Borneo, mainland Southeast Asia, and adjacent shallow sea areas. This core, joined to South China along the Song Ma-Song Da suture belt (Permo-Triassic), is composite, consisting of two to four separate terranes. The western portion (Mitchell's "Western Southeast Asia block") represents a Paleozoic continental margin subsequently rifted on its (present) west side. Its east edge is marked by suture belts apparently formed by collision of the western continent with eastward-dipping subduction zones flanking Indochina and eastern Malaya.

Late Paleozoic glacial marine deposits in western Southeast Asia indicate attachment to Gondwana, with rifting after Permian, possibly in Jurassic. Floral and paleomagnetic data are generally consistent with this and suggest land connections, at least briefly during earlier Mesozoic, from China through Southeast Asia to the main part of Gondwana, possibly as part of Pangaea. The more easterly terranes in Southeast Asia's continental core may have dispersed from a Pacifica continent, thus marking a boundary between Pacifica-derived and Gondwana-derived terranes. Geologic evidence, however, links at least eastern Malaya to western Southeast Asia, suggesting these eastern terranes were simply more outboard portions of a complex Gondwana margin.

INTRODUCTION

It has long been recognized that the geographic complexity of Southeast Asia (here taken in a broad sense to cover all the territory between India, China, and Australia and thus to include the Indonesian and Philippine Island groups) reflects an underlying geologic complexity. Since the advent of the plate tectonic paradigm and the new generally "mobilist" viewpoint, it has become clear that this complexity is in part due to separate and differing geologic histories of formerly independent constituent fragments (Stauffer, 1983).

The present highly active tectonics of the Southeast Asian region are dominated by the convergent interaction of three major lithospheric plates: the Pacific plate (with its satellite the Philippine Sea plate), moving roughly westward; the India-Australia plate, moving northward or northeastward; and the China (or Eurasia) plate, which is relatively stationary (Fig. 1). Convergent plate boundaries resulting from this interaction give rise to the presently active volcanic arcs that wrap around Southeast Asia. These plate boundaries and their precursors extending back into Late Cretaceous have built the melange terranes and volcanic island arc systems that make up most of the Philippine and eastern Indonesian island chains and have modified and overlaid the geology of some of the older parts of the region, such as Sumatra.

Inside the loop of volcanic arcs, and in part scattered among them and in the intervening small oceanic basins, are areas characterized by older (largely pre-Mesozoic) basement and continental or intermediate crustal structure. Included here are several small fragments (see Fig. 7) now lodged within the island arcs, such as the Sula Islands (inferred to have been dislocated from northwestern New Guinea; see Hamilton, 1979, pp. 156–159), the North Palawan Block (probably rifted off South China; see Holloway, 1981), and other small fragments (see Hamilton, 1979). Most of the island of New Guinea and the shallow seas between it and Australia are underlain by old continental crust, continuous with (and part of) the Australian craton.

The main area of continental crust in Southeast Asia, however, comprises almost all of mainland Southeast Asia (Indochina, Thailand, eastern Burma, and Peninsular Malaysia), most of Sumatra, western Borneo, and parts of the shallow seas adjacent to and between these land areas. This large, essentially continental region has been referred to by several names (the southern portion has been called Sundaland), and it was shown by Murphy (1975) as the "pre-Tertiary core" of Southeast Asia (Fig. 1).

The mainly continental nature of the crust underlying this old area has been inferred from its topographic level, gravity, the acidic composition of igneous rocks, and the character of Paleozoic and Mesozoic sediments. Establishing a Precambrian age for areas of "crystalline basement" has in general proved elusive (although a number of Precambrian age dates have been reported from Vietnam; see Bao et al, 1981). However, continent-derived marine shelf sediments of early Paleozoic age are now known from various parts of Indochina, Thailand, Burma, and Malaya (Peninsular Malaysia), and the inference of a widespread Precambrian continental crust seems sound.

Continental Southeast Asia adjoins the cratonic blocks of China to the northeast across a major suture belt (see below), while its boundary on the northwest is uncertain and continuity with portions of Tibet has been suggested (Mitchell, 1981; see Fig. 2). On its other margins, continental Southeast Asia adjoins oceanic crustal regions, active convergent plate boundaries, or accreted noncontinental terranes.

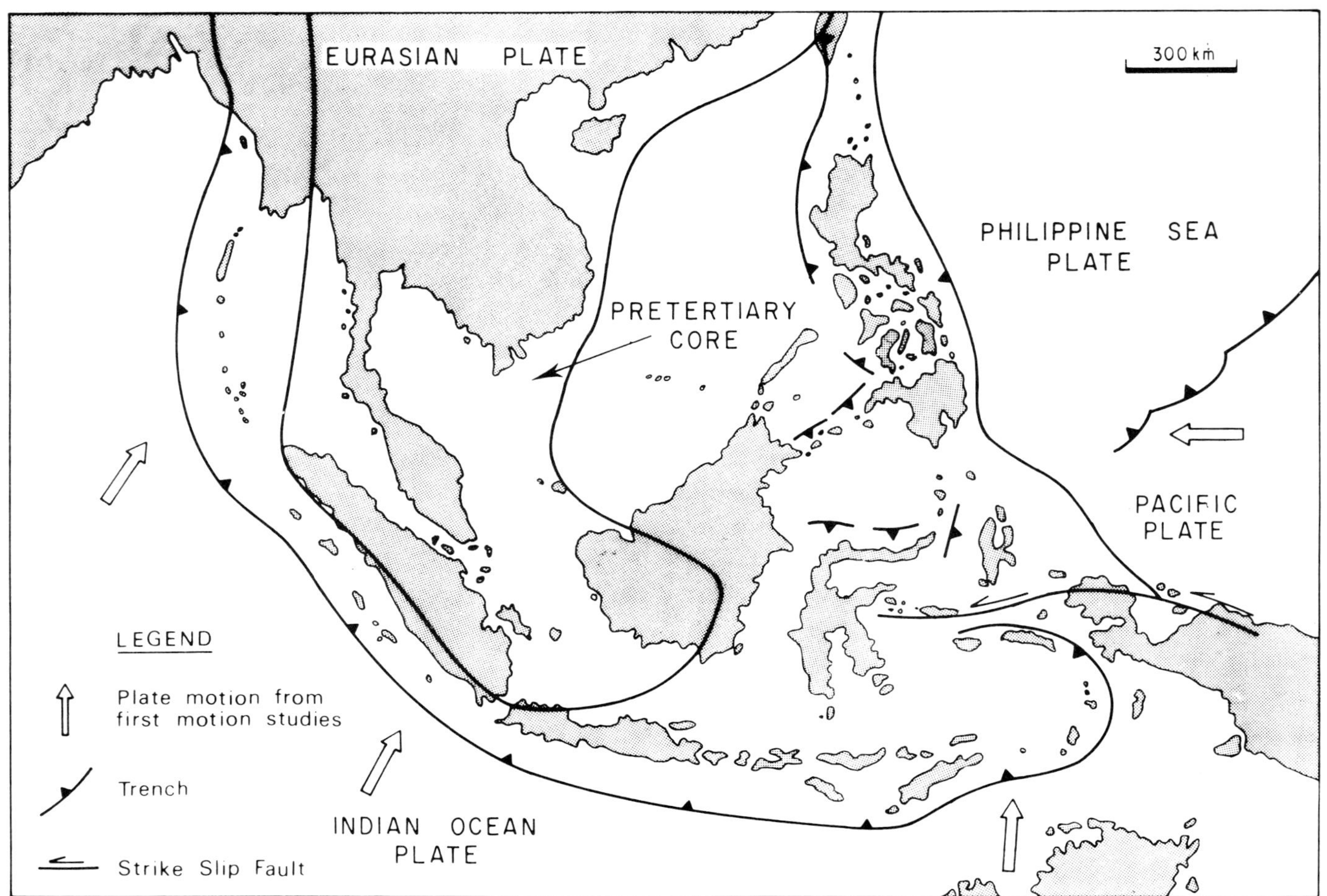

Figure 1—Southeast Asia, showing the main area of old and mainly continental crust, here called the pre-Tertiary core. Also shown are present plate boundaries and motions. (After Murphy, 1975. Copyright © 1975 Southeast Asia Petroleum Exploration Society. Used with permission.)

CONSTITUENT TERRANES IN CONTINENTAL SOUTHEAST ASIA

This old continental region in Southeast Asia is not a single craton or even a unitary fragment. Belts interpreted as sutures divide it into several terranes, which had partly independent histories prior to colliding and amalgamating. These belts are characterized by tectonic lineaments, geologic contrasts, and bodies of basic and ultrabasic rocks that have been called, with varying degrees of confidence, ophiolites (Hutchison, 1975).

Three constituent terranes (blocks) were inferred by Stauffer (1974). The West Malaya Block, comprising Sumatra, western Malaya, western Thailand, eastern Burma, and southwest Yunnan in China, and possibly extending further north, shows a fully continental character, with Cambro-Ordovician platform sediments overlying locally exposed crystalline basement, and abundant large granitic batholiths mainly of early Mesozoic age. Lower Paleozoic facies arrangements were interpreted to indicate former attachment to a continent on the (present) west side, along what is therefore a rifted continental margin, and a former continental margin at the (present) east edge of this terrane (Jones, 1968). At its east edge in Malaya, this terrane is bounded by the Bentong–Raub Line, a tectonic shear zone with discontinuous bodies of ophiolitic rocks (see Fig. 3). To the east of this zone, inferred to be continuous with the Uttaradit–Luang Prabang ophiolite belt (see Fig. 2) in Thailand and Laos (Thanasuthipitak, 1978) lies the strongly deformed central basin of Malaya (see Fig. 3) and the granitic terrane of eastern Malaya. Eastern Malaya was taken by Stauffer (1974) to be part of one terrane with western Borneo but distinct from Indochina (east of the Uttaradit–Luang Prabang belt and south of the great suture against South China, the Song Ma–Song Da or Black River belt; see Fig. 2), which formed the third terrane of Stauffer (1974). Ridd (1980) kept these three blocks, but Mitchell (1981) put Indochina together with eastern Malaya in his Indochina block, renamed the terrane west of it the Western Southeast Asia block, and extended both of them through narrow slivers into central and southern Tibet, respectively (Fig. 2).

NATURE AND SIGNIFICANCE OF SUTURES BETWEEN TERRANES

The boundary zones between these several tectonic blocks or terranes have been interpreted by most recent workers as sutures. But it has not been resolved whether

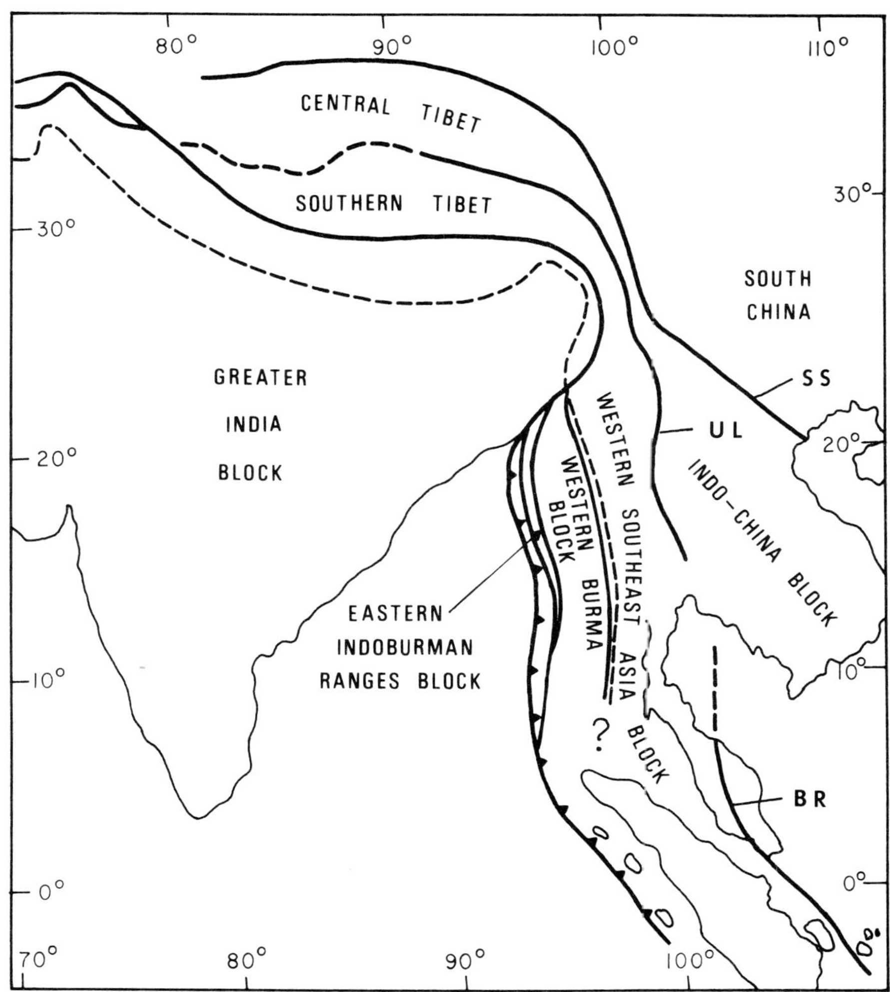

Figure 2—Component terranes and their bounding sutures in the continental part of Southeast Asia and adjacent parts of Asia, according to Mitchell (1981). Suture belts discussed in text identified by initials: Song Ma–Song Da (SS); Uttaradit–Luang Prabang (UL); Bentong–Raub (BR).

these sutures represent closed extensive oceans, collapsed small marginal basins, or merely tectonic shear zones, whose two sides have never been far separated.

The Song Ma–Song Da suture belt between Indochina and South China is a complex zone tectonically and geologically. It contains good ophiolitic suites and thick Paleozoic marine sedimentary sequences representing environments from platform to deep sea. This zone therefore appears to record the closure of an expanse of oceanic crust and compression of a well-developed continental margin (Hutchison, 1975; Salun et al, 1975). The final closure and deformation of this suture belt were mainly Triassic events (Fontaine and Workman, 1978), and by the end of Triassic South China and continental Southeast Asia were amalgamated.

Any possible suture between Indochina and eastern Malaya (with western Borneo), as inferred possibly to exist by Stauffer (1974) and Ridd (1980), is concealed beneath the South China Sea and therefore unknown, apart from its extension through Thailand and Laos, where it would coincide with the suture at the east side of Western Southeast Asia.

The nature and indeed the reality of the suture belt between Western Southeast Asia and the terrane(s) to the east of it has been controversial. The Malayan portion of this belt, the Bentong–Raub Line, was classified as a definite but dismembered ophiolite by Hutchison (1975). It shows, however, few of the features of a well-developed ophiolite (Tan and Khoo, 1981). There is a discontinuous and somewhat scattered belt of serpentinites, along with some amphibole schists of uncertain origin, and very rare peridotites. The belt is also characterized by a zone of intense tectonic shearing and by profound geologic contrasts between the areas on the two sides.

If the Bentong–Raub Line is continuous with the Uttaradit–Luang Prabang belt of Thailand and Laos (Thanasuthipitak, 1978), then the latter's well-developed ophiolites would allow a more definite interpretation of the entire belt. Such continuity is not proven, however, and if they are separate belts, their age relations are not clear, both representing late Paleozoic to Triassic events with rather imprecise dating.

Several lines of evidence do nonetheless indicate that the Bentong–Raub Line is the western edge of a broad belt

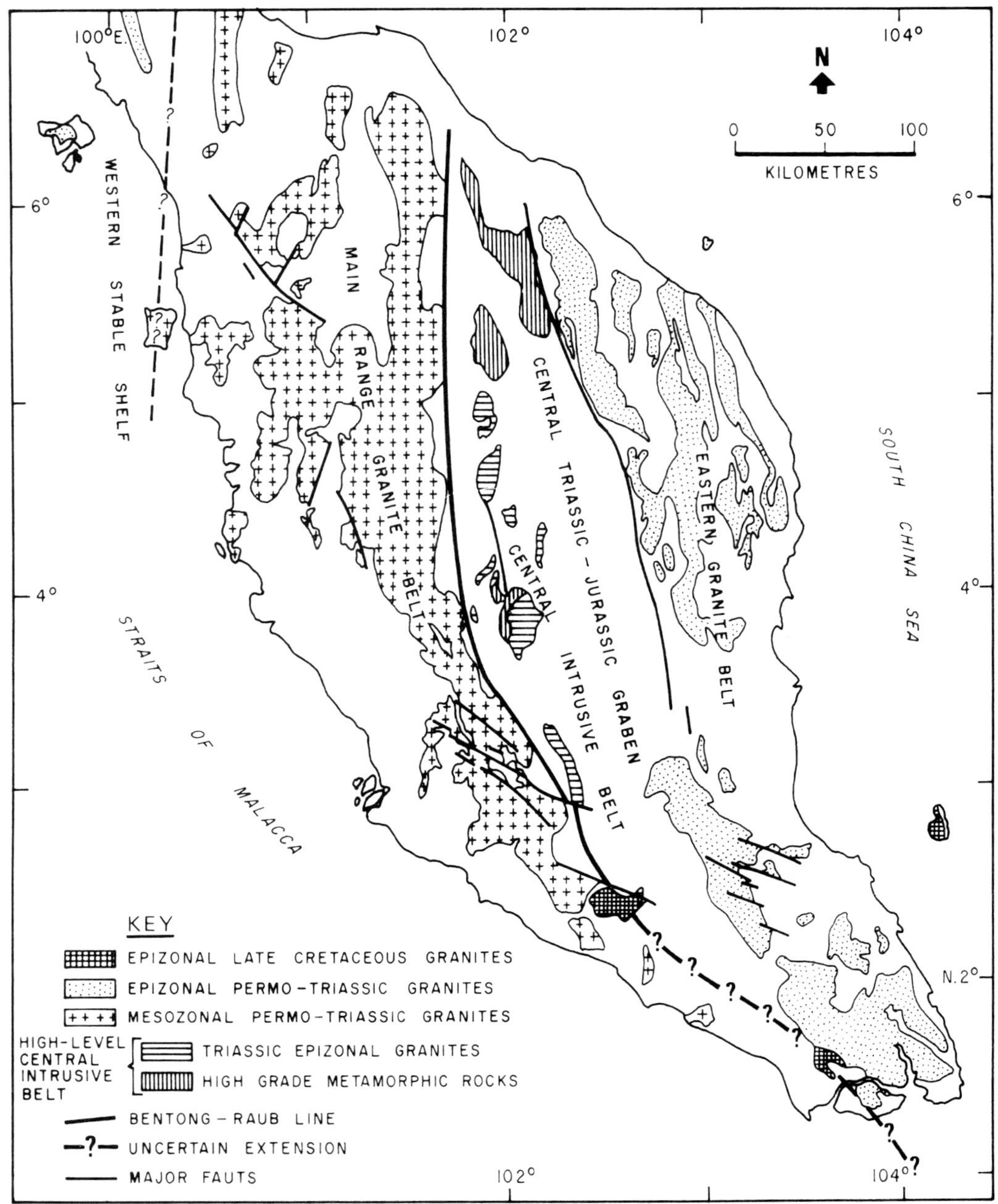

Figure 3—Granite types and tectonic belts in the Malay Peninsula, after Hutchison (1977). Note the limitation of inferred mesozonal granites to the belt just west of the Bentong-Raub Line.

that marks the site of a closed oeanic basin. First, there is the great geologic contrast across the line (Fig. 3). The western side has Paleozoic shallow-water and continent-derived sedimentary rocks and abundant granitic intrusives. Immediately east of the line are Carboniferous to Triassic sedimentary rocks, in large part of deep-marine deposition and volcanic arc derivation (e.g., Metcalfe et al, 1982).

Second, a gravity profile across the line (Ryall, 1982) shows a marked negative anomaly over the Main Range batholith, changing eastward across the line to a marked positive anomaly. The change is abrupt and of significant magnitude, and it must represent a profound geologic boundary. Depending on the densities assumed for the sediments in the central graben, the gravity values east of the Bentong-Raub Line can be interpreted to indicate a very thin or absent "granitic layer"; i.e., probably oceanic or intermediate crust (Ryall, 1982).

Third, there is the manner of occurrence of the serpentinite bodies. Among those scattered along the belt are some that have apparently been squeezed up along the faulted crests of anticlines in the Paleozoic sedimentary and metasedimentary rocks (see Krishnan, 1975). This suggests a widespread or pervasive underlying basement of already serpentinized ultrabasic or basic rocks, ready to form cold intrusions wherever fractures allow upward migration.

Finally, there is the character of the huge Main Range batholith (see Fig. 3). The petrologic and chemical

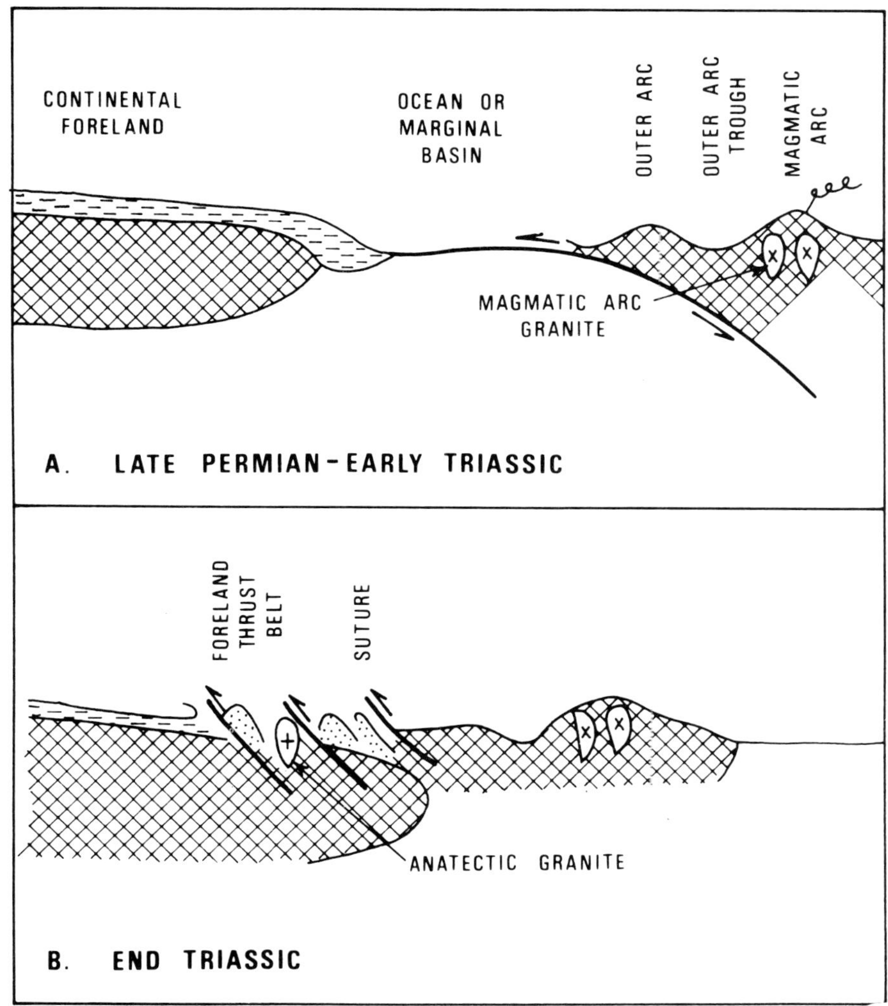

Figure 4—Schematic diagrams showing formation of the Main Range batholith in Malaya by depression of the continental edge in an eastward-dipping subduction zone, and consequent anatexis. On the left side is Western Southeast Asia with its continental margin sediments; on the right side are the subduction zone and volcanic-magmatic arc that will form central and part of eastern Malaya. (Reproduced by permission of the Geological Society of London from Mitchell, 1981, in the Journal of the Geological Society of London, v. 138.)

characteristics of this very large granitic mass (Hutchison, 1977) have been interpreted by Mitchell (1977) and Hutchison (1982) to be the result of the western continent (Western Southeast Asia) impinging against the volcanic arc of an eastward-dipping subduction zone, the collision leading to depression, heating, and anatexis of the edge of the continent. The granites of western Thailand have been analogously interpreted to result from melting of continental crust related to collision with an eastward-dipping subduction zone (Beckinsale et al, 1979). This scenario of events, as proposed for Malaya by Mitchell, is shown in Figure 4.

Even if we accept the Bentong-Raub Line and its extensions as a genuine suture, the question remains as to whether the oceanic area closed along it was only a small, previously opened sea (Helmcke, 1982), in which case the two margins might have previous geologic connections, or a large oceanic region, in which case the two sides may be totally unrelated terranes.

AFFINITIES OF THE DIFFERENT TERRANES

Each of the separate terranes making up the old continental part of Southeast Asia must have rifted off a larger continental mass at some stage in its history. Each, in other words, is a piece of a fragmented puzzle. But which puzzle?

The westernmost of these terranes, the Western Southeast Asia block of Mitchell (1981), has long been taken to represent the margin of a large continent formerly adjacent on its (present) west side (Fitch, 1952; Jones, 1968; Stauffer, 1974). That this continent was Gondwana was suggested by all these three workers and by others, but various attachment sites have been proposed: India–Antarctica (Melville, 1966), India (Burton, 1970), between India and Australia (Ridd, 1971), west Australia (Tarling, 1972), Arabia–Africa (Stauffer, 1974), and northwest Australia (Audley-Charles, 1983).

Other workers do not accept an attachment to

Gondwana (some do not accept attachment to any continent) along the western edge of Western Southeast Asia. Helmcke (1982) argues that this terrane represents the outer zone of a huge Variscan orogen, with open ocean to its west in the Paleozoic as now. Some reconstructions for Paleozoic times still ignore the evidence for the several Triassic sutures in Southeast Asia and leave the region attached to Eurasia in its present relative position (see, for example, Boucot and Gray, 1983).

What is the relevant evidence on these questions? Paleomagnetic data for the Paleozoic of Southeast Asia are still very sparse and of uncertain quality. Earlier work, reviewed in Haile (1978), had indicated a consistently low-latitude placement of Southeast Asia back through the later Paleozoic. Two more recent sets of data are at least interesting in suggesting more temperate paleolatitudes. Samples from Ordovician–Silurian limestones of the Setul Formation in northwest Malaya (Haile, 1980) yield a paleolatitude of 43°. If the field was normal, this would place Southeast Asia farther north than any other major piece of continental crust and make attachment to any continent difficult; with a reversed field, the Western Southeast Asia terrane would be rotated about 180° from its present orientation and located in a southern hemisphere position that would allow attachment to Gondwana in the general region of Iran.

Data from one site in the late Paleozoic Kaeng Krachan Formation in Thailand give an inferred paleolatitude of 24° S for central Thailand (Bunopas et al, 1978). Again, this is more consistent with attachment to Gondwana than earlier data suggesting about 15° N (McElhinny et al, 1974), but all of these data are of uncertain reliability.

Other evidence tending to support a former attachment of Western Southeast Asia to Gondwana include the late Paleozoic floras and a late Paleozoic "pebbly mudstone" belt recently reinterpreted as a glacial marine facies.

Late Carboniferous and Permian floras show a strong world-wide provinciality. Two cold-climate floras are recognized (the *Glossopteris* or Gondwana flora of the southern, and the Angara flora of the northern high paleolatitudes) and three warm-climate floras (Cathaysian, North American, and Euramerican, sharing the low-latitude belt). These floras have been mapped and discussed by Chaloner and Lacey (1973) and are shown in Figure 5. The Permian floras of Southeast Asia, reviewed in Stauffer (1974), are generally of Cathaysian type. However, it has been established that Cathaysian-type floras were growing on the more northerly parts of the Gondwana continent (Čtyroký, 1973; Stauffer, 1974; Archangelsky and Wagner, 1983), and so the floral evidence is no bar to former connection of the two areas. Indeed, such a connection is supported by the occasional occurrence of apparent elements of the Gondwana flora in Thailand (Kon'no, 1963) and Malaya (Azhar, 1977, discussed in Stauffer, 1983). A distribution map of the late Paleozoic floras on a "pre-drift" geography in which Eurasia is treated as a single block (Chaloner and Lacey, 1973) makes it clear that not only Southeast Asia, but also most of China and Korea could not have been attached to Eurasia in their present positions then, but must have been considerably farther south, closer to the northern portions of Gondwana (Fig. 5).

Perhaps a key piece of evidence on the former affinities of western Southeast Asia lies in the belt of late Paleozoic pebbly mudstones (Fig. 6) stretching from Burma down through Thailand and Malaya to Sumatra (Stauffer and Mantajit, 1981). These rocks are Early Permian near their top (Waterhouse, 1982) and probably include some Carboniferous strata. They have generally been interpreted as products of resedimentation on a submarine slope and so have assumed some importance in paleogeographic reconstructions because that slope would have to have been down to the west (Mitchell et al, 1970; Ridd, 1980; Helmcke, 1982).

Recent reexamination of these rocks, however, has yielded strong evidence of a glacial marine origin (Stauffer, 1980; Stauffer and Mantajit, 1981; Stauffer and Lee, in press; see discussion in Stauffer, 1983). The great lateral extent (>2,000 km [1,200 mi]) of this belt of tilloids, and their content of exotic cratonic megaclasts, make proximity and probable attachment of Western Southeast Asia to the Gondwana continent during late Paleozoic most likely. Note that the Permian floras of Southeast Asia, which are mainly of warm-climate character, are mostly Middle and Late Permian (i.e., younger than the tilloids). Separation from Gondwana would have to postdate the glacial marine facies (Early Permian) and has been suggested on geologic grounds to have occurred in the Jurassic (Stauffer, 1974, 1983). The glacial marine rocks do not resolve the question of the particular attachment site on the Gondwana margin, but they would be compatible with any position in the interval from Arabia–Iran around the northern edge of Greater India to northwestern Australia (see discussion in Stauffer, 1983).

The affinities and original positions of the other major Southeast Asian continental terrane(s) (Indochina, eastern Malaya, western Borneo) are much less clear. By the end of Triassic, all of these areas were amalgamated and sutured to Western Southeast Asia on one side and to South China on the other. If Western Southeast Asia did not separate from Gondwana until Jurassic, then at least South China was in land connection to the main areas of the Gondwana continent during part of the Triassic, allowing *Lystrosaurus* to migrate into Sinkiang (Simpson, 1970) and *Glossopteris* floral elements to spread to Indochina (Gothan and Weyland, 1954).

For pre-Triassic times the evidence is more ambiguous. The presence of a significant belt of tin mineralization in the eastern Malaya terrane, parallel to the major belt in Western Southeast Asia, including western Malaya (Hosking, 1973), suggests a geological relationship and perhaps an original continuity between these areas. On the other hand, the Late Permian floras of Malaya (which are from east of the Bentong–Raub Line) share close relationships to the typical northern Cathaysian floras of China (Kon'no and Asama, 1970; Kon'no et al, 1971). Perhaps major portions of China also were part of Gondwana in the Paleozoic, though they would have to rift off sometime prior to the reassembly in Triassic. If there was land connection, or at least an absence of any major

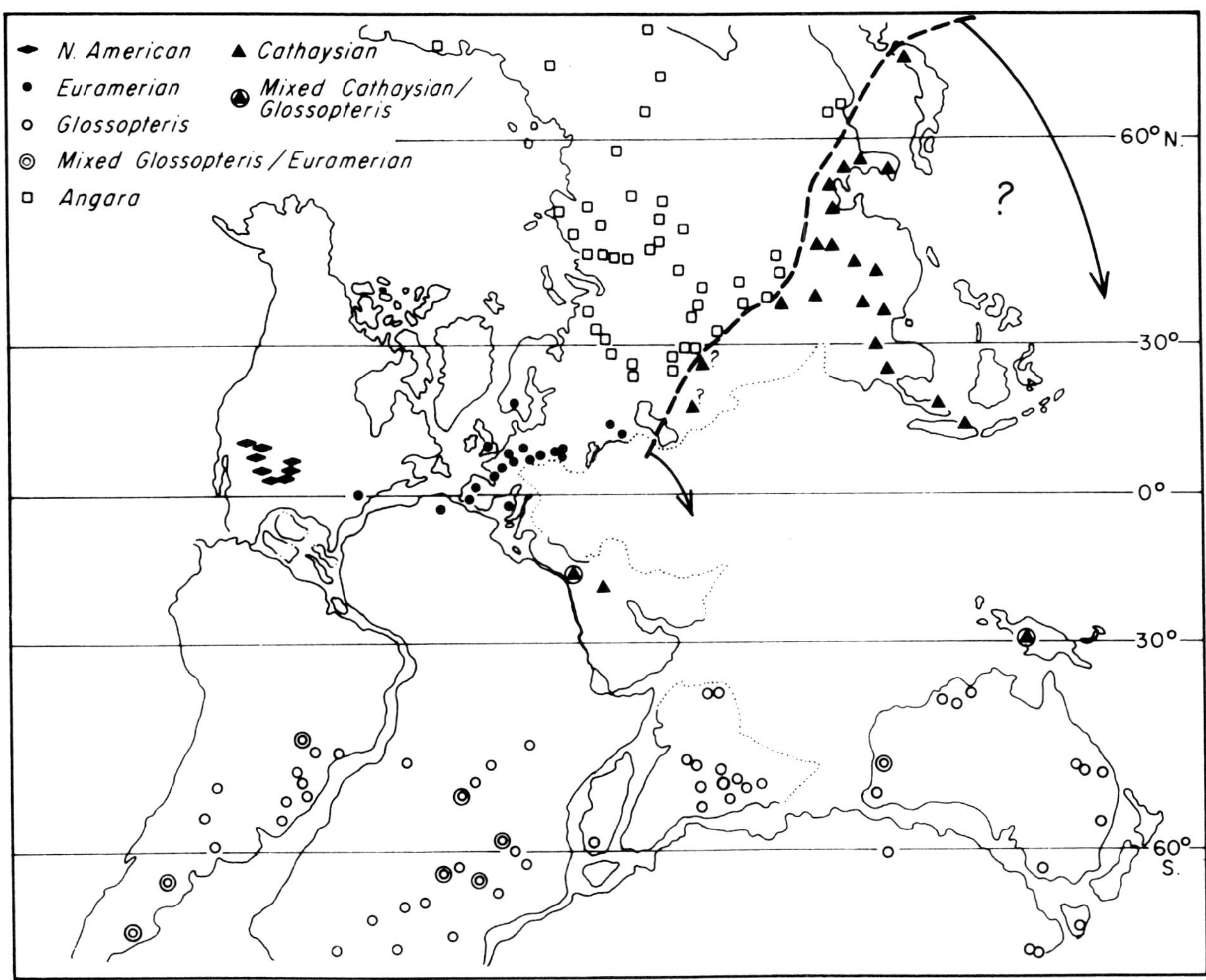

Figure 5—Distribution of some Permian and latest Carboniferous floras on a Permian reconstruction of the continents based on paleomagnetic data. After the map of Chaloner and Lacey (1973), to which some Middle and Late Permian floras from Southeast Asia and the Middle East have been added. Solid symbols are warm-climate floras. Note the anomalous distribution of the Cathaysian floras, implying incorrect placement of their areas on this reconstruction.

water gaps, between Gondwana and northern China during the Paleozoic, this would tend to support the notion of a Pangaea, though not the usual reconstructions showing a huge oceanic gulf between Southeast Asia and eastern Gondwana (e.g., Boucot and Gray, 1983).

This evidence of connection or proximity from Gondwana to North China would also suggest that the sutures in Southeast Asia may record the closure of limited oceanic areas, of marginal basin type, rather than major oceans of great width. This suggestion is in basic agreement with the paleotectonic scheme of Audley-Charles (1983), who includes all the pieces of Southeast Asia in eastern Gondwana in the Carboniferous, with Indochina and central Tibet rifting off (forming an oceanic "Tethys II") prior to the collision of these areas with Asia. Audley-Charles (1983) would also include in the Paleozoic Gondwana additional areas now to the north of the Southeast Asian terranes. Some of those areas have previously (Melville, 1966) been referred to a hypothetical continent of "Pacifica," comprising various pieces now spread around the Pacific margin, including North America, as exotic terranes. Inclusion of these areas in Gondwana would make it difficult, however, to account for the evidence of early Paleozoic volcanic arcs in western Southeast Asia (Stauffer, 1974) in such an "inboard" position, unless previous episodes of opening and closing of oceanic areas (for which there is no evidence) are invoked.

In the classical Pacifica continent of Melville (1966), which was proposed for the Jurassic and on mainly floral evidence, China and the rest of East Asia were included, as were Indochina, central Asia, and virtually all the then-known areas of Cathaysian Permo-Triassic floras. If such a Pacifica continent existed and has fragmented and dispersed to meet the dispersing fragments of Gondwana in Southeast Asia, then this region should contain a boundary between the pieces of these two puzzles (Fig. 7). Certainly in the Cenozoic the picture is one of collision and incipient collision between Gondwana fragments (India, Australia) and the composite mass of Eurasia (Fig. 1). But some Gondwana fragments (specifically the Western Southeast Asia terrane) experienced collision with parts of Asia and rifting off the Gondwana much earlier and probably in that chronological order. Therefore, the composition of a hypothetical Pacifica may have been different at different

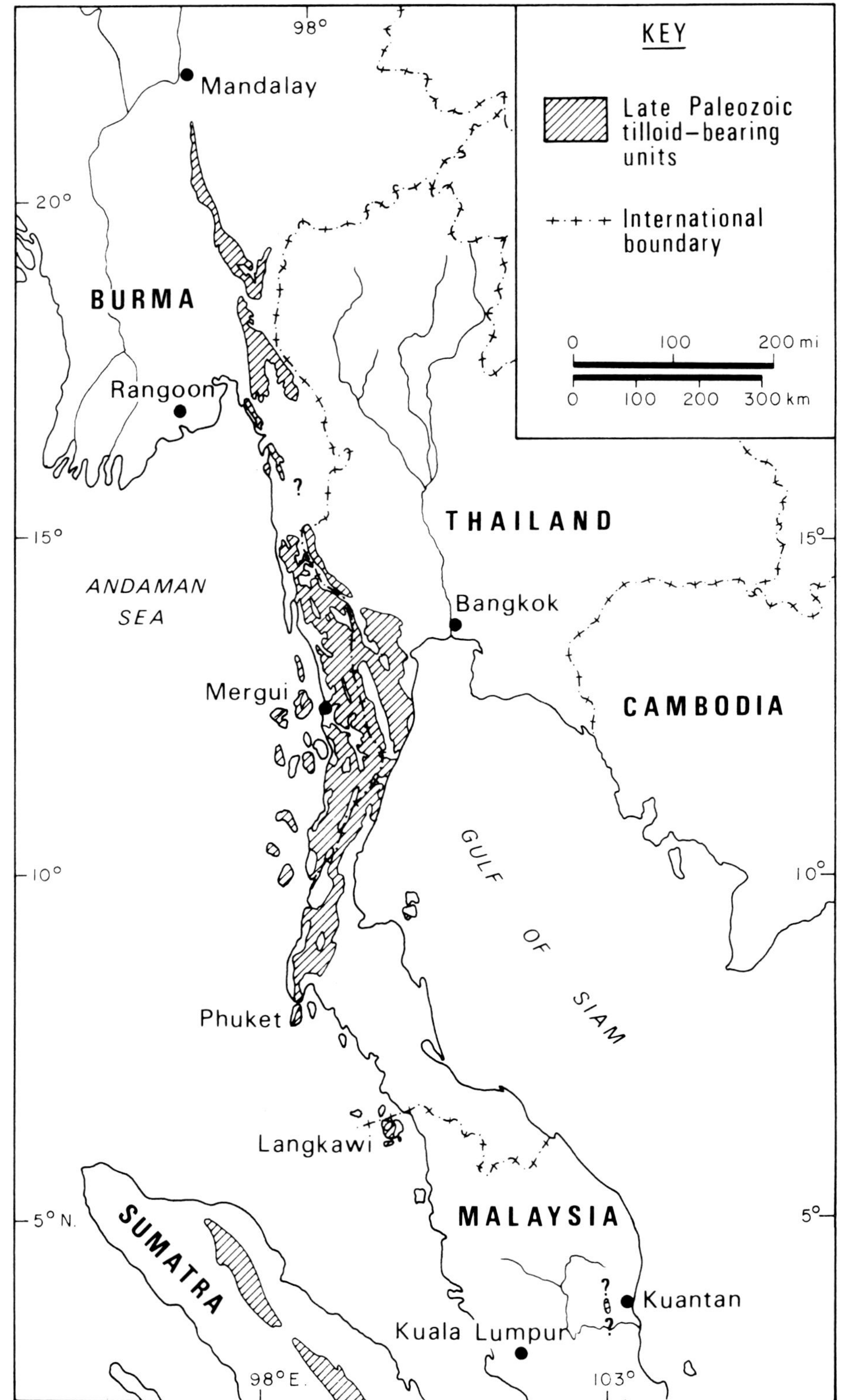

Figure 6—Distribution of late Paleozoic stratigraphic units containing pebbly mudstones in Western Southeast Asia. After Stauffer and Mantajit (1981). The pebbly mudstone facies (tilloids of inferred glacial marine origin) forms only a part of the units whose extent is shown here.

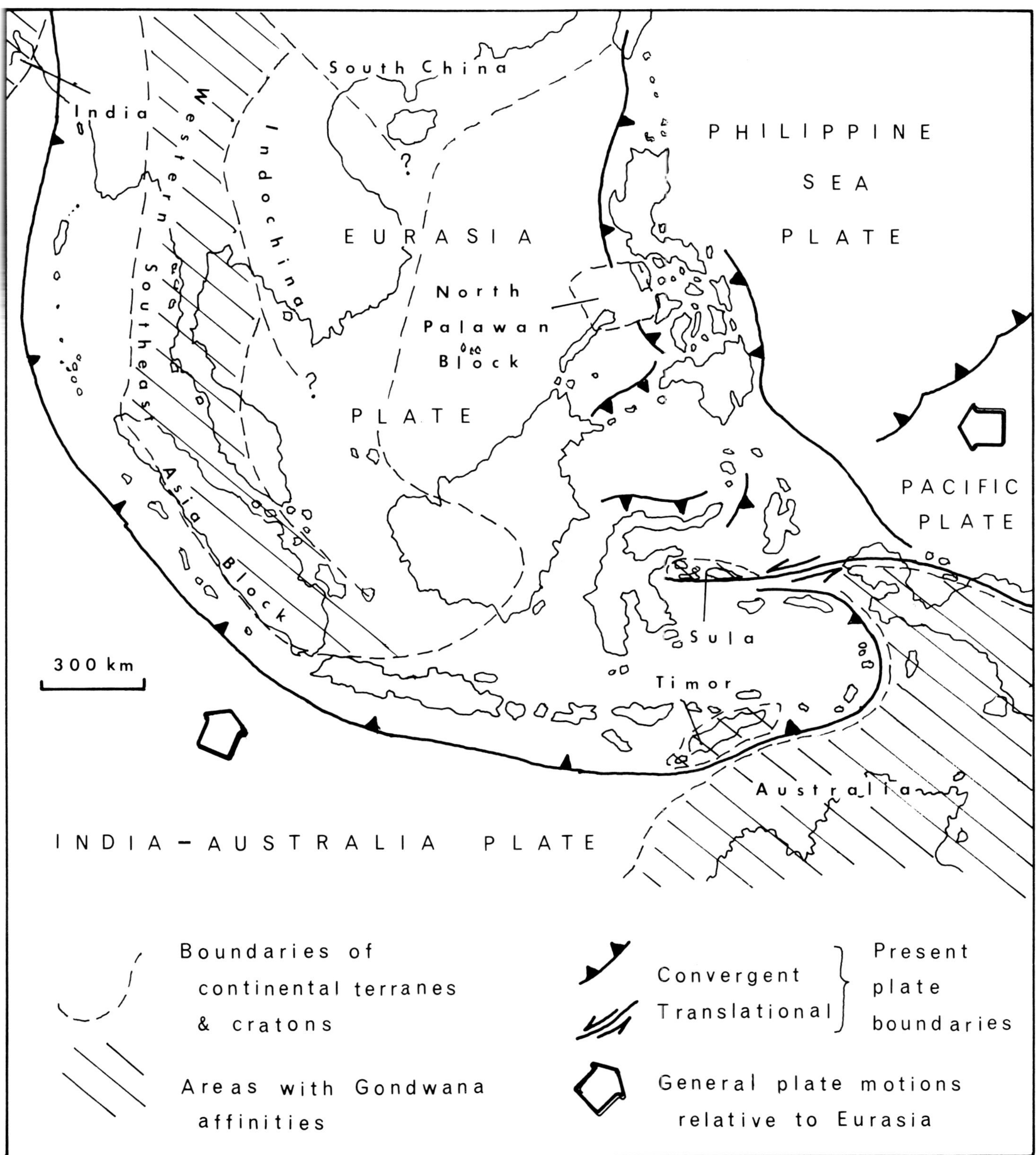

Figure 7—Major continental terranes in Southeast Asia, showing those derived from the breakup of Gondwana. Present plate boundaries and motions are also shown.

times, and at least briefly it may have been united to Gondwana in a possible Pangaea. These data and interpretations imply a scheme of paleotectonic evolution, similar to that of Parker and Gealey (in press), involving the sequential existence of a Paleozoic Tethys, the trace of whose former site may lie in the Triassic suture zones of Southeast Asia, and a Mesozoic Tethys opened between southeast Asia and the nuclear areas of eastern Gondwana (India, Antarctica, Australia) after the Triassic.

ACKNOWLEDGMENTS

The work and ideas embodied in this paper were largely developed at the University of Malaya in Kuala Lumpur, and I am grateful to my former colleagues there for many illuminating discussions. I would also like to thank G. A. Thompson, at whose suggestion I was invited to participate in the 1983 Terranes Conference at Stanford. Most of the figures were originally drafted by Ching Yu Hay and photographed by Jaafar Haji Abdullah.

REFERENCES

Archangelsky, S., and R. H. Wagner, *Glossopteris anatolica* sp. nov. from uppermost Permian strata in south-east Turkey: Bulletin of the British Museum of Natural History (Geology), v. 37, p. 81-91.

Audley-Charles, M. G., 1983, Reconstruction of eastern Gondwanaland: Naure, v. 306, p. 48-50.

Azhar Haji Hussin, 1977, Geology of the Bukit Jaya area, Pahang: B.S. Thesis, Department of Geology, University of Malaya, Kuala Lumpur, 63 p.

Bao, N. X., et al, 1981, The main stages of tectonomagmatic activities in Vietnam (Abs.), *in* Abstracts of Papers, Fourth Regional Conference on the Geology, Mineral and Energy Resources of Southeast Asia: Manila, November 1981, p. 1.

Beckinsale, R. D., et al, 1979, Geochronology and geochemistry of granite magmatism in Thailand in relation to a plate tectonic model: Journal of the Geological Society of London, v. 136, p. 529-540.

Boucot, A. J., and J. Gray, 1983, A Paleozoic Pangaea: Science, v. 222, p. 571-581.

Bunopas, S., et al, 1978, Preliminary paleomagnetic results from Thailand sedimentary rocks, *in* P. Nutalaya, ed., Proceedings of the Third Regional Conference on the Geology and Mineral Resources of Southeast Asia: Bangkok, November 1978, p. 25-32.

Burton, C. K., 1970, The paleotectonic status of the Malay Peninsula: Palaeogeography, Palaeoclimatology, Palaeoecology, v. 7, p. 51-60.

Chaloner, W. G., and W. S. Lacey, 1973. The distribution of Late Palaeozoic floras: Palaeontological Association Special Papers in Palaeontology, 12, p. 271-289.

Čtyroký, P., 1973, Permian flora from the Ga'ara region (Western Iraq): Neues Jahrbuch für Geologie und Paläontologie Monatshefte, p. 383-388.

Fitch, F. H., 1952, The geology and mineral resources of the neighbourhood of Kuantan, Pahang: Geological Survey Department, Federation of Malaya Memoir 6, 143 p.

Fontaine, H., and D. R. Workman, 1978, Review of the geology and mineral resources of Kampuchea, Laos and Vietnam, *in* P. Nutalaya, ed., Proceedings of the Third Regional Conference on the Geology and Mineral Resources of Southeast Asia: Bangkok, November 1978, p. 539-603.

Gothan, W., and H. Weyland, 1954, Lehrbuch der Palaeobotanik: Berlin, Akademie-Verlag.

Haile, N. S., 1978, Progress report on palaeomagnetic research in Southeast Asia, *in* Proceedings of the Regional Conference on the Geology and Mineral Resources of Southeast Asia: Jakarta 1975, p. 33-36.

______, 1980, Palaeomagnetic evidence from the Ordovician and Silurian of northwest Peninsular Malaysia: Earth and Planetary Science Letters, v. 48, p. 233-236.

Hamilton, W., 1979, Tectonics of the Indonesian region: U.S. Geological Survey Professional Paper 1078, 345 p.

Helmcke, D., 1982, On the Variscan evolution of central mainland Southeast Asia: Earth Evolution Sciences, v. 2, p. 309-319.

Holloway, N. H., 1981, The North Palawan Block, Philippines: its relation to the Asian Mainland and its role in the evolution of the South China Sea: Geological Society of Malaysia Bulletin, v. 14, p. 19-58.

Hosking, K. F. G., 1973, Primary mineral deposits, *in* D. J. Gobbett and C. S. Hutchison, eds., Geology of the Malay Peninsula (West Malaysia and Singapore): New York, Wiley-Interscience, p. 335-390.

Hutchison, C. S., 1975, Ophiolite in Southeast Asia: Geological Society of American Bulletin, v. 86, p. 797-806.

______, 1977, Granite emplacement and tectonic subdivision of Peninsular Malaysia: Geological Society of Malaysia Bulletin, v. 9, p. 187-207.

______, 1982, Southeast Asia, *in* A. E. M. Nairn and F. G. Stehli, eds., The ocean basins and margins, v. 6, The Indian Ocean: New York, Plenum Press, p. 451-512.

Jones, C. R., 1968, Lower Paleozoic rocks of Malay Peninsula: Bulletin of the American Association of Petroleum Geologists, v. 52, p. 1259-1278.

Kon'no, E., 1963, Some Permian plants from Thailand: Japanese Journal of Geology and Geography, v. 34, p. 139-159.

______, and K. Asama, 1970, Some Permian plants from the Jengka Pass, Pahang, West Malaysia: Geology and Paleontology of Southeast Asia, v. 8, p. 97-132.

______, et al, 1971, The Late Permian Linggiu flora from the Gunong Blumut area, Johore, Malaysia: Geology and Paleontology of Southeast Asia, v. 9, p. 1-85.

Krishnan, D., 1975, Geology of the Bentong area, Pahang, West Malaysia: B.S. Thesis, Department of Geology, University of Malaya, Kuala Lumpur, 83 p.

McElhinny, M. W., et al, 1974, Palaeomagnetic evidence shows Malay Peninsula was not a part of Gondwanaland: Nature, v. 252, p. 641-645.

Melville, R., 1966, Continental drift, Mesozoic continents and the migration of the angiosperms: Nature, v. 211, p. 116-120.

Metcalfe, I., et al, 1982, Stratigraphy and sedimentology of Middle Triassic rocks exposed near Lanchang, Pahang, Peninsular Malaysia: Geological Society of Malaysia Bulletin, v. 15, p. 19–30.

Mitchell, A. H. G., 1977, Tectonic setting for emplacement of Southeast Asian tin granites: Geological Society of Malaysia Bulletin, v. 9, p. 123–140.

______, 1981, Phanerozoic plate boundaries in mainland SE Asia, the Himalayas and Tibet: Journal of the Geological Society of London, v. 138, p. 109–122.

______, et al, 1970, The Phuket Group, Peninsular Thailand: a Paleozoic ?geosynclinal deposit: Geological Magazine, v. 107, p. 411–428.

Murphy, R. W., 1975, Tertiary basins of Southeast Asia: Southeast Asia Petroleum Exploration Society Proceedings, v. 2, p. 1-36.

Parker, E. S., and W. K. Gealey, in press, Plate tectonic evolution of the Western Pacific-Indian Ocean region, *in* Proceedings of the EAPI/ASCOPE/CCOP/IOC Workshop on the Geology and Hydrocarbon Potential of the South China Sea and Possibilities of Joint Development: Honolulu, August 1983.

Ridd, M. F., 1971, Southeast Asia as a part of Gondwanaland: Nature, v. 234, p. 531-533.

______, 1980, Possible Palaeozoic drift of SE Asia and Triassic collision with China: Journal of the Geological Society of London, v. 137, p. 635–640.

Ryall, P. J. C., 1982, Some thoughts on the crustal structure of Peninsular Malaysia—results of a gravity traverse: Geological Society of Malaysia Bulletin, v. 15, p. 9–18.

Salun, S. A., et al, 1975, Experiment in characterization of synclinal suture zones, as in certain structures of East Asia (translated from Russian): International Geology Review, v. 17, p. 1266–1274.

Simpson, G. G., 1970, Drift theory: Antarctica and Central Asia: Science, v. 170, p. 678.

Stauffer, P. H., 1974, Malaya and Southeast Asia in the pattern of continental drift: Geological Society of Malaysia Bulletin, v. 7, p. 89–138.

______, 1980, The Singa Formation: is it a glacial deposit? (Abs.): Warta Geologi, v. 6, p. 33–34.

______, 1983, Unraveling the mosaic of Paleozoic crustal blocks in Southeast Asia: Geologische Rundschau, v. 72, p. 1061–1080.

______, and C. P. Lee, in press,, Late Paleozoic glacial marine facies in Southeast Asia and its implications, *in* Proceedings of the Fifth Regional Conference on the Geology, Mineral and Energy Resources of Southeast Asia: Kuala Lumpur, April 1984.

______, and N. Mantajit, 1981, Late Palaeozoic tilloids of Malaya, Thailand and Burma, *in* M. J. Hambrey and W. B. Harland, eds., Earth's pre-Pleistocene glacial record: Cambridge, Cambridge University Press, p. 331–337.

Tan, B. K., and T. T. Khoo, 1981, Serpentinites in Peninsular Malaysia and their tectonic implications (Abs.), *in* Abstracts of Papers, Fourth Regional Conference on the Geology, Mineral and Energy Resources of Southeast Asia: Manila, November 1981, p. 21.

Tarling, D. H., 1972, Another Gondwanaland: Nature, v. 238, p. 92–93.

Thanasuthipitak, T., 1978, Geology of Uttaradit area and its implications on tectonic history of Thailand, *in* Proceedings of the Third Regional Conference on the Geology and Mineral Resources of Southeast Asia: Bangkok, November 1978, p. 187–197.

Waterhouse, J. B., 1982, An early Permian cool-water fauna from pebbly mudstones in south Thailand: Geological Magazine, v. 119, p. 337–354.

Tectonostratigraphic Terranes, Pacific Southeast Quadrant

The Accretion of Gorgona Island, Colombia: Multichannel Seismic Evidence

Susan McGeary*
Zvi Ben-Avraham**
Stanford University
Stanford, California

Gorgona Island is one of several terranes of oceanic rock found in western Colombia and Ecuador. This island on the outer shelf of southern Colombia is composed almost entirely of mafic and ultramafic rocks and is best known for its komatiitic lava flows of Late Cretaceous age. The tectonic history of this unusual oceanic block is not known. A multichannel seismic reflection profile, collected by Texpet (a subsidiary of Texaco, Inc.) 70 km (44 mi) south of Gorgona Island, reveals some details about the continental margin in this area. A basement high on the outer shelf is interpreted to be a continuation of the high-velocity Gorgona basement. The Gorgona block is separated from the folded deposits of the forearc basin by a landward-dipping thrust fault. This fault forms the current Gorgona terrane boundary and may be the original accretion structure of the Gorgona block. It is unaffected by the overlying folding and has probably been reactivated.

Gorgona Island has been thought to be an extension of a large terrane in northern Colombia, the Serrania de Baudo. However, a comparison of gravity, multichannel seismic reflection, and structural data from the two regions suggests that the two terranes are separated by a major structural break. The stratigraphy of the two regions suggests that Gorgona may have accreted before the Early Miocene, whereas the Serrania de Baudo was probably accreted during the Miocene-Pliocene as part of the collision of the Panama arc with Colombia.

INTRODUCTION

The Andes of South America extend continuously from Colombia to Patagonia and are one of the major orogenic belts in the world. Yet considerable heterogeneity exists within this mountain chain (Gansser, 1973). In particular, the northern Andes of Colombia and Ecuador differ significantly from the Andes of Peru and Chile. One of the most unusual characteristics of the northern Andes is the presence of terranes of mafic and ultramafic igneous rock within both the coastal cordilleras and the western Andean cordillera of Colombia and Ecuador (Fig. 1). Gravity and seismic refraction data show high-density, high-velocity material underlying both the Western Cordillera and the coastal ranges, suggesting that all of Colombia and Ecuador west of the Central Cordillera is underlain by mafic oceanic material of Late Cretaceous-Paleogene age (Case et al, 1971; Mooney et al, 1979). Geologic data suggest that the accretion of the oceanic terranes has occurred since the Late Cretaceous (Barrero, 1979).

One of the most unusual of these terranes is exposed on Gorgona and Gorganilla Islands on the outermost continental shelf of southern Colombia (Fig. 2). These islands contain an "ophiolitic" sequence of wehrlite and dunite, poikilitic gabbro, diabase, massive and pillow basalt, and tuff breccia (Gansser, 1950; Echeverria, 1980). What makes this exposure unusual is the presence of high-magnesium, spinifex-textured komatiitic lavas within the Late Cretaceous basalt sequence (Gansser et al, 1979; Echeverria, 1980; Espinosa et al, 1981). The origin of these young komatiitic rocks in an oceanic setting is not yet understood. The location of Gorgona Island at the outermost shelf in the forearc of a convergent margin and the oceanic lithologies and geochemistry suggest that it was probably accreted to the continental margin as part of the subduction process. Yet the terrane-bounding structures are obscured by water, and the time of accretion is not known. It is also not known what relationship, if any, exists between this small block and other larger terranes in Colombia. Gorgona Island has been postulated (Gansser, 1950) to be the southern extension of the Serrania de Baudo (Fig. 1), the coastal range of northwestern Colombia (Case et al, 1971), and to be part of a coastal ophiolitic belt extending continuously from Panama to Ecuador.

Unfortunately, the dense tropical rainforest and the annual rainfall of ~760 cm (300 in.) in western Colombia has made it difficult to investigate the nature of the stratigraphy, structural boundaries, and history of accretion for the terranes in Colombia. Only a few areas in the Western Cordillera (Barrero, 1979; Nelson, 1957) and the Serrania de Baudo (Interoceanic Canal Study Commission, 1968, 1969) have been mapped in detail. Much of the available information has been necessarily geophysical in

*Present address: Bullard Laboratory, University of Cambridge, Cambridge United Kingdom.
**On leave from the Department of Geophysics and Planetary Sciences, Tel Aviv University, Tel Aviv, Israel.

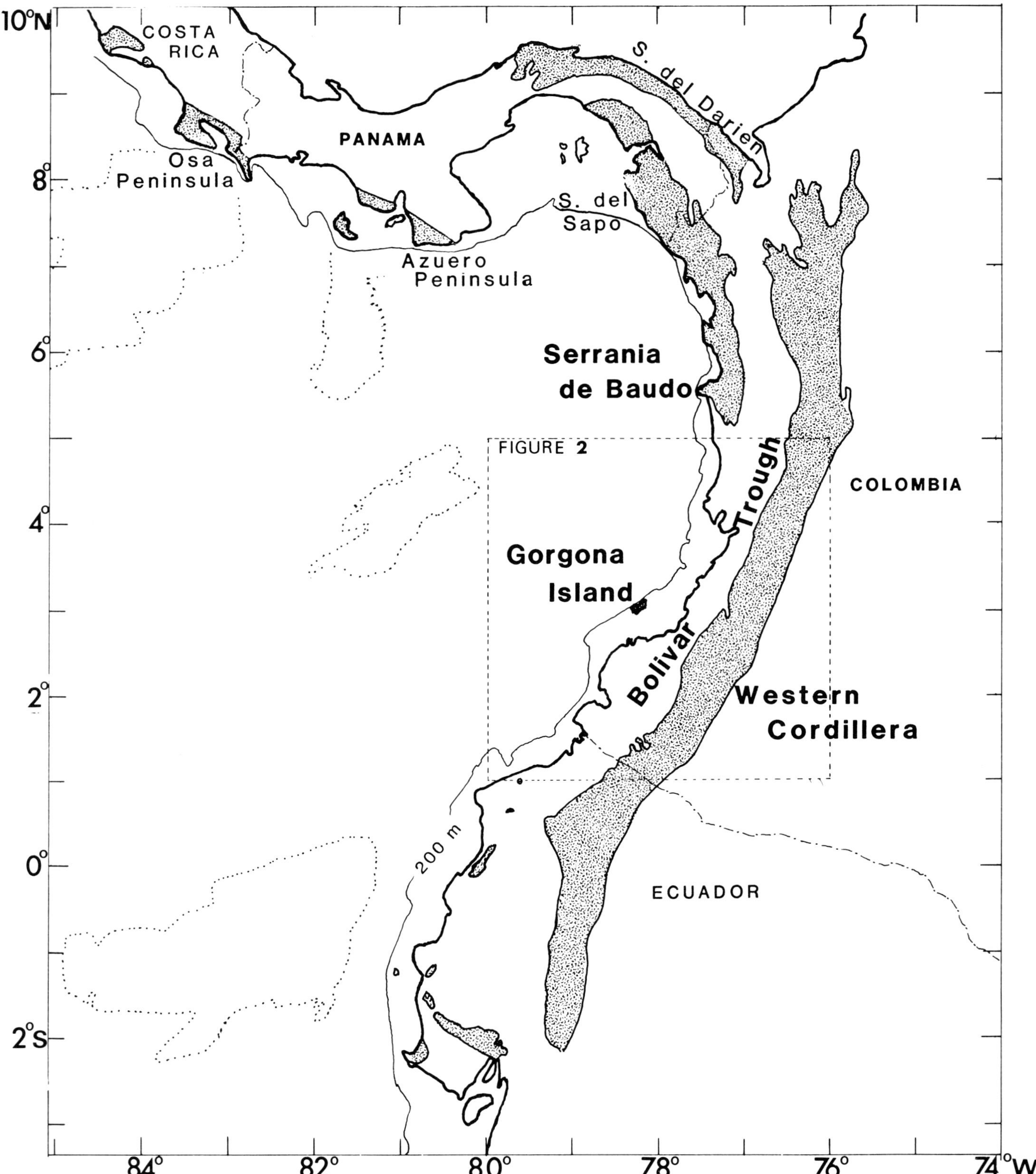

Figure 1—Northwestern South America and southern Central America region, showing the location of oceanic terranes. The terranes are shaded and include exposures of mafic igneous rocks on Gorgona Island, the coast of Costa Rica (probably part of the Nicoya Complex), the Azuero Peninsula and Serrania del Darien of Panama, the Western Cordillera of Colombia and Ecuador, the Serrania de Baudo/Serrania del Sapo, and the coast of Ecuador. Location of Figure 2 is indicated by the dotted box.

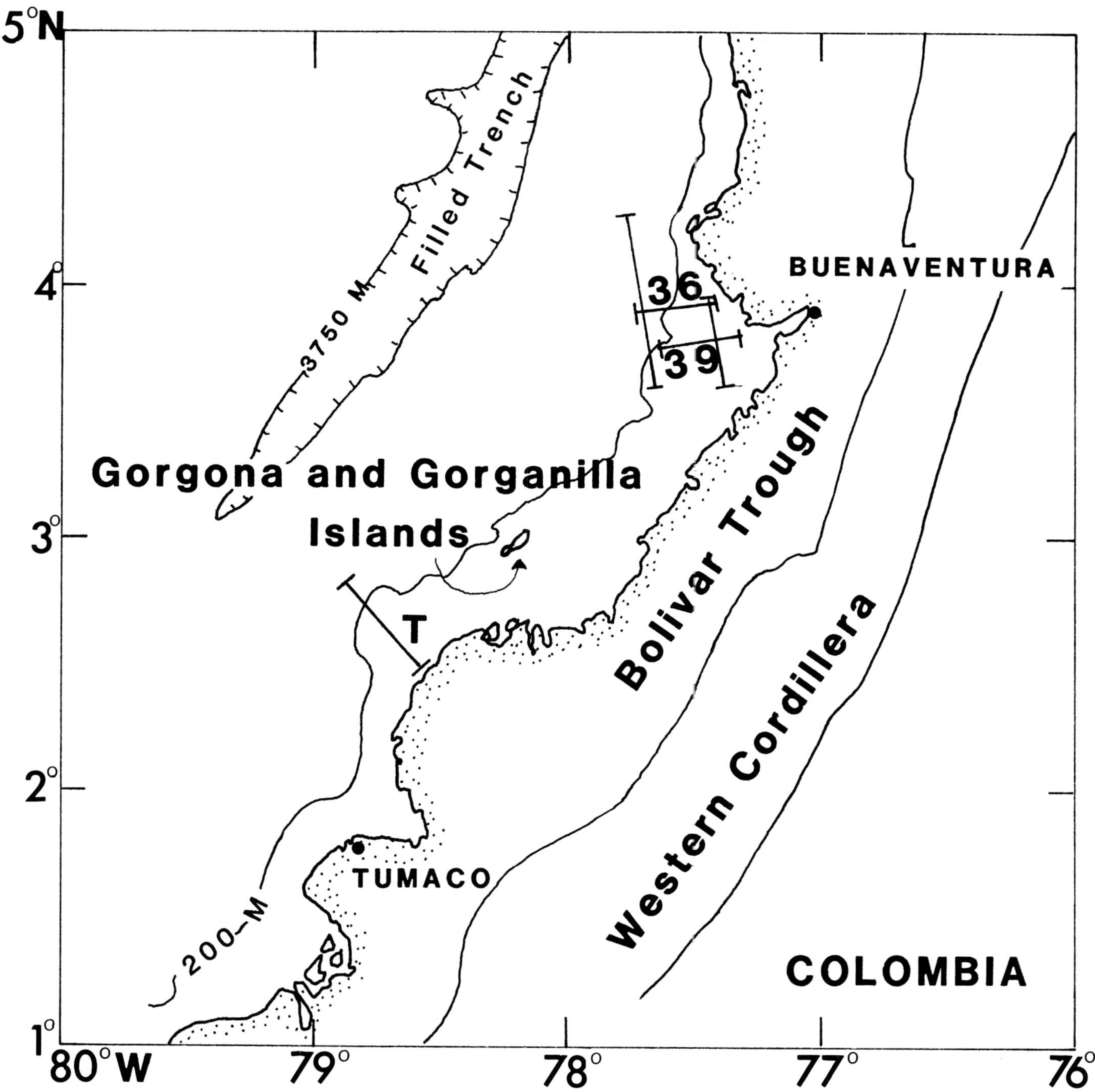

Figure 2—Location of multichannel seismic reflection Profile T (this paper) and Lines 36 and 39 (Flüh, 1982). Also shows location of Gorgona and Gorgonilla Islands, the Bolivar trough, Tumaco, and Buenaventura. The 3,750 m (12,303 ft) contour outlines the deepest part of the sediment-filled Colombia Trench.

nature, including gravity (Case et al, 1971, 1973; Case, 1974; Meissner et al, 1976), refraction (Meyer et al, 1976; Meissner et al, 1976; Mooney et al, 1979), and radar imagery data (Barlow, 1981).

In this paper we use a multichannel seismic reflection profile collected by Texpet (a subsidiary of Texaco, Inc.) on the continental shelf of Colombia, south of Gorgona Island, to describe the structure of the continental shelf near Gorgona Island. This information is used in conjunction with other geophysical and geologic data to discuss three major problems: (1) In what ways might Gorgona Island have been accreted? (2) Is Gorgona Island an extension of the Serrania de Baudo or a separate terrane? (3) When did Gorgona Island become part of the continental margin of Colombia?

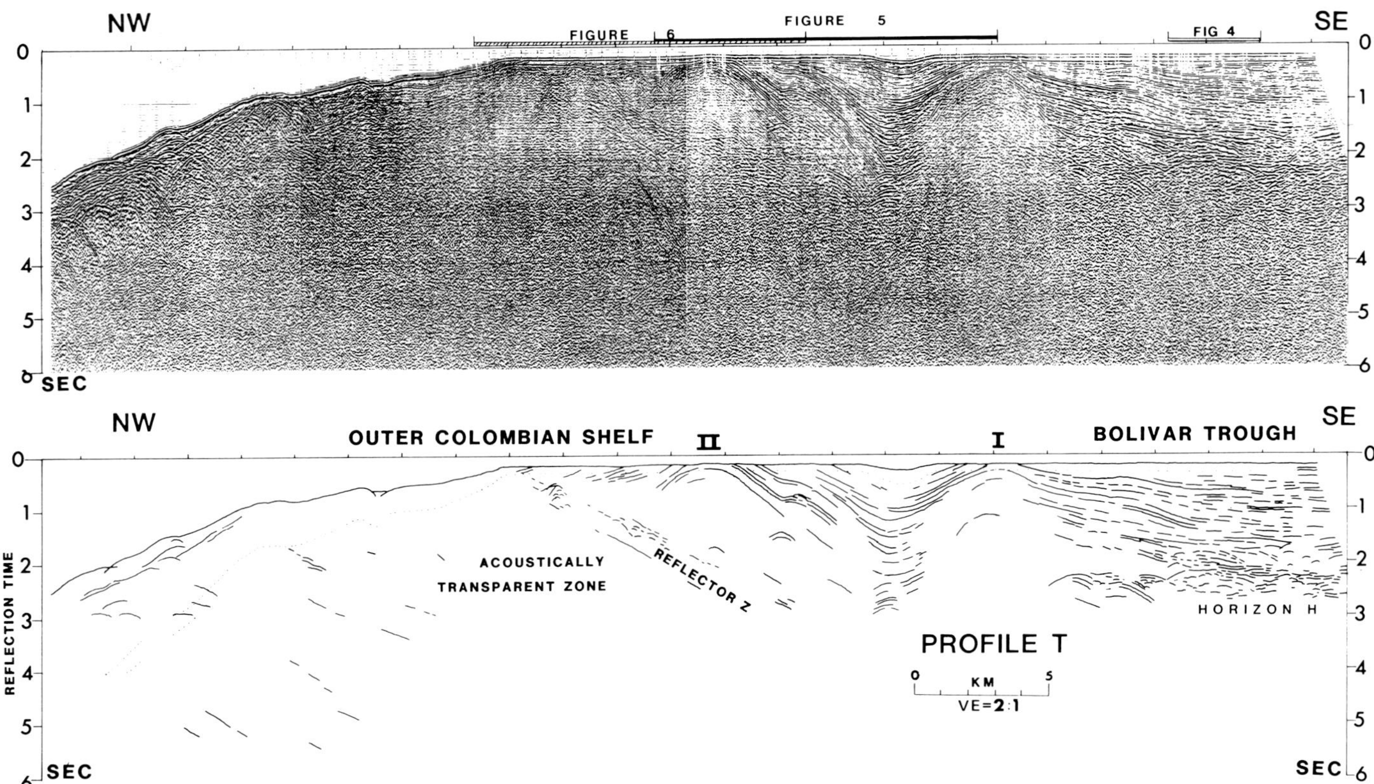

Figure 3—Texpet, multichannel Profile T and interpretive line drawing. Notice the locations of Reflector Z, Horizon H, and anticlines I and II. Horizon H is the base of the sedimentary reflectors within this part of the Bolivar Trough. The deposits are folded into two anticlines, I and II. Reflector Z separates the folded deposits of the basin from an acoustically transparent zone on the outer shelf and is interpreted to be a thrust fault. The dotted line shows the sea-floor multiple. The locations of the sections in Figures 4 through 6 are shown above the profile. The vertical exaggeration of 2 to 1 is obtained by assuming a velocity of 2.0 km/sec (6,561 ft/sec) to convert to depth.

STRUCTURE OF THE CONTINENTAL SHELF NEAR GORGONA ISLAND

A multichannel seismic reflection line (Profile T) was recorded across the continental shelf of southern Colombia near Tumaco in 1971 by Texpet. The location of the line, about 70 km (44 mi) south of Gorgona Island, is shown on Figure 2. The profile is 48 km (30 mi) in length and extends over the continental shelf and part of the continental slope. The 24-fold multichannel data were stacked and processed by Texpet, and sections were further processed, including frequency domain migration (Stolt, 1978), using the programs and facilities of the Stanford Exploration Project. Two-way arrival times were recorded to 6 seconds representing up to 12 km (7.5 mi) of sampling in depth.

The stacked seismic section clearly outlines the general features of the continental shelf (Fig. 3). The shelf is divided into two areas of distinctly different acoustic character by a zone of landward-dipping reflectors that we call Reflector Z. Seaward of Reflector Z, the outermost shelf is acoustically transparent with few visible coherent reflectors. The closest exposures of outer shelf material are the mafic and ultramafic rocks of Gorgona Island, 70 km (44 mi) to the north. A small, positive gravity anomaly parallels the outer continental shelf in this area, crossing both Gorgona Island and the outer shelf of Profile T (Fig. 8). Both the transparent acoustic character on the seismic line and the positive gravity anomaly on the outer shelf suggest that the high-density mafic basement of Gorgona Island may continue as far south as the location of Profile T. Farther down the slope, beneath the transparent zone, there are several reflectors that dip landward and are nearly parallel to Reflector Z.

Landward of Reflector Z, the seismic line shows the folded sediments of the Tumaco section of the Bolivar trough (Fig. 3). This forearc basin is part of a continuous series of basins, termed the Bolivar geosyncline by Nygren (1950), that extends from Panama to Ecuador and includes the drainages of the San Juan and Atrato rivers in Colombia and the Daule river in Ecuador. The sediment-filled basins can be as deep as 10 km (6 mi) as shown by gravity data (Case et al, 1971) and refraction data (Mooney et al, 1979). Several names have been used for this forearc basin: we follow the usage of Case et al (1973). The stratigraphy of the Tumaco section is discussed by Bueno and Govea (1976), with some control provided by three wells. The closest well to Profile T reached a depth of 3,995 m (13,107 ft) without reaching Eocene strata.

Profile T crosses the offshore western margin of the Bolivar trough, west of the onshore basin axis, and shows a thick section (~3 km [2 mi]) of sedimentary reflectors. The upper sedimentary reflectors within the southeastern side of the section are generally horizontal, yet most reflectors

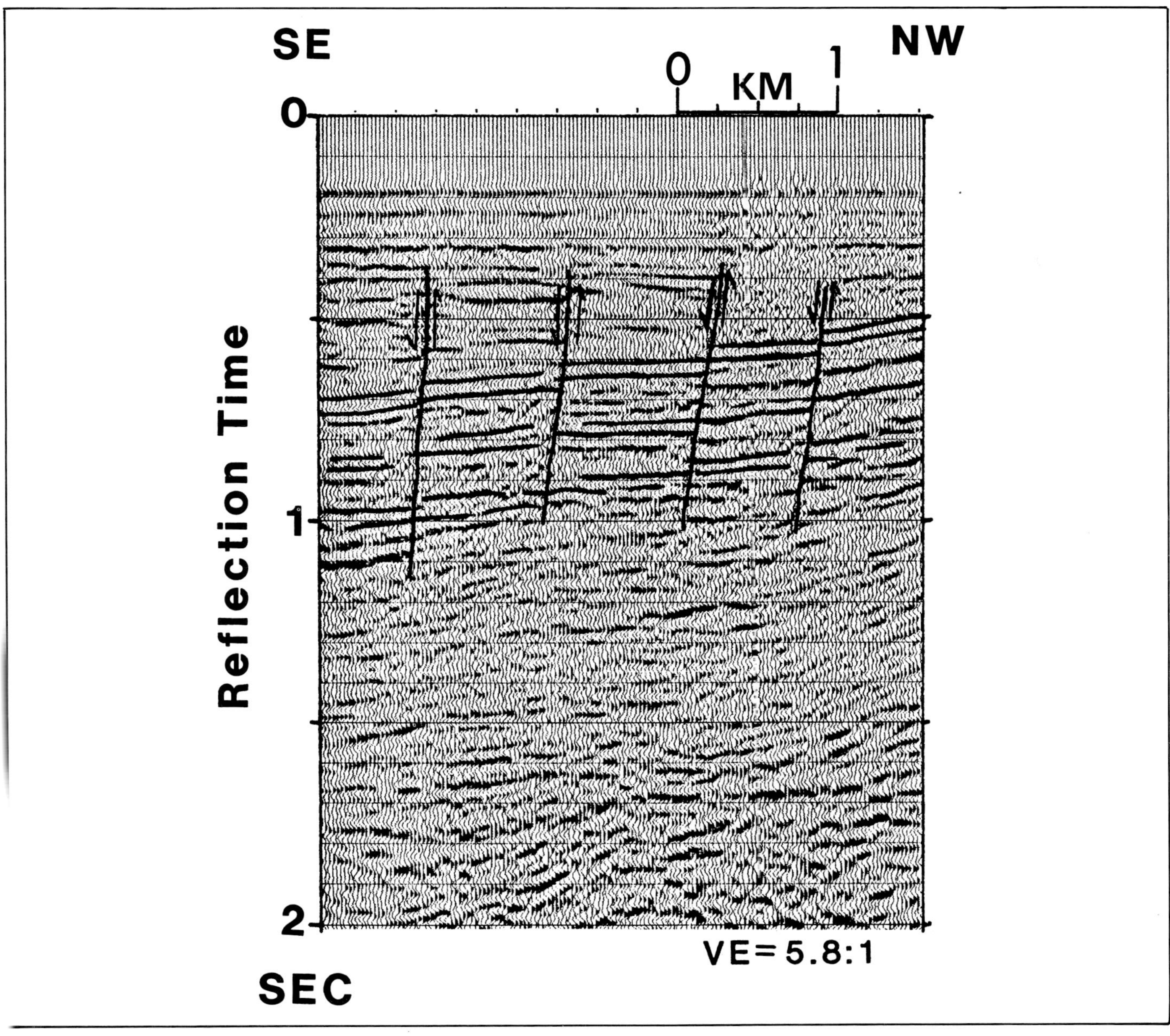

Figure 4—Constant velocity (Stolt, 1978) migration (1.7 km/sec [5,577 ft/sec]) of southeastern part of Profile T. The figure is reversed with respect to Figure 3. Notice the normal faults, down to the southeast interpreted to be within the sediment section. The vertical exaggeration of 5.9 to 1 assumes an average velocity of 1.7 km/sec.

are undulatory and fade in and out. This may partially be a result of rapid facies changes (Bueno and Govea, 1976). A F-K (frequency domain) migration (Stolt, 1978) of a small section of the basin at a constant velocity of 1.7 km/sec (5,600 ft/sec), however, shows a series of nearly perpendicular normal faults, down to the southeast, breaking the reflector continuity (Fig. 4). Normal faults are also found in the exposed northern part of the Bolivar trough (Bueno and Govea, 1976). The acoustic basement of the sedimentary section in the southeastern part of the profile is defined by a set of diffractions (Horizon H; Fig. 3) at 2.3 seconds that line up in a relatively horizontal zone.

The flat-lying deposits of the inner shelf are folded seaward into two anticlines (I and II) on the central shelf (Fig. 3 and 5). Reflectors within the folded structures are largely obscured both by numerous diffractions and by multiples of the sea floor and dipping sedimentary beds. No evidence from diffractions or truncations indicates faults within this segment of the seismic section, and individual reflectors can be traced continuously across both anticlines and the intervening depression (Fig. 5). Thrust faults may exist within the core of the anticlines but cannot be seen in the seismic data. Such compressional folds are common along the exposed western border of the Bolivar trough to the north (Barlow, 1981; Bueno and Govea, 1976). These structures usually trend north to northeast. Although the folding in the region of the Serrania de Baudo was dated by Barlow (1981) to be Pleistocene, folding has been recently active in the area of the seismic line as evidenced by truncation of the youngest shelf

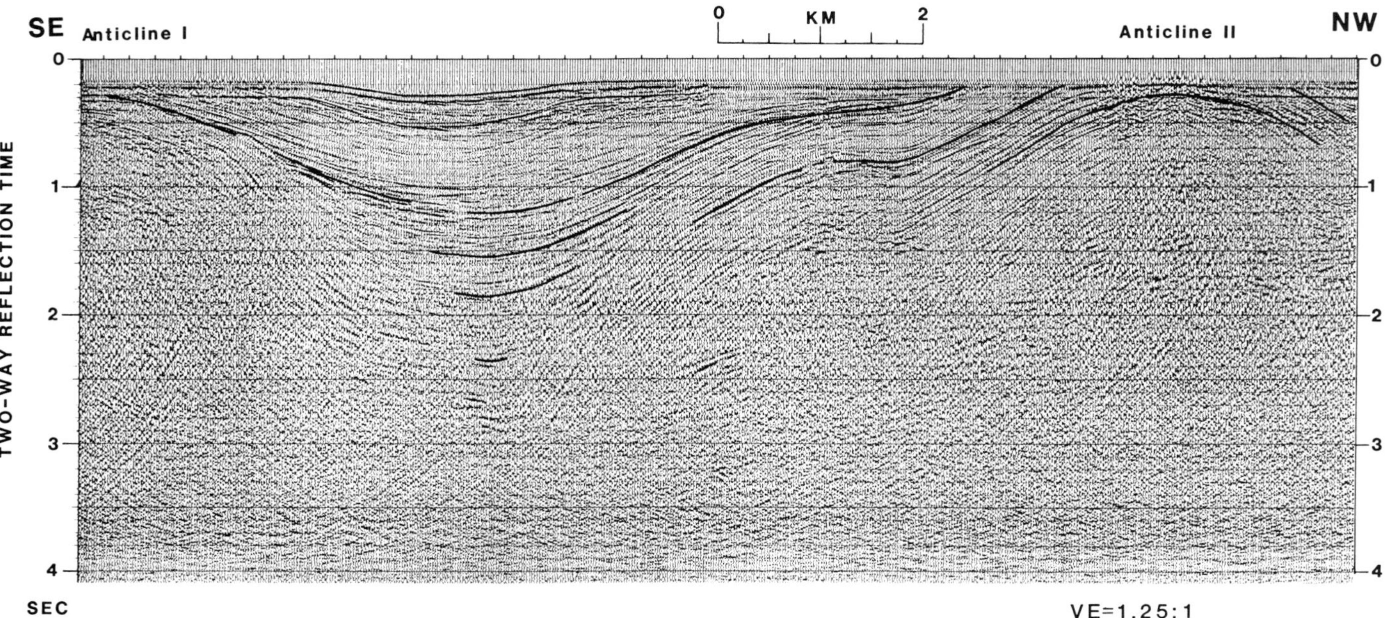

Figure 5—Stolt constant velocity migration (1.5 km/sec [4,921 ft/sec]) showing continuous reflectors across anticlines I and II. The thickening of sediment packages within the basin suggests that the folding has been active during sedimentation on the shelf. The vertical exaggeration of 1.25 to 1 assumes an average velocity of 2.5 km/sec (6,562 ft/sec).

sediments and bathymetric relief of 100 m (328 ft) between the two anticlines and the small intervening syncline. Packages of reflectors within the larger basin to the east that thin onto the anticlines suggest that the folding has been active during sedimentation on the shelf (Fig. 5). Reflectors on the seaward side of anticline II are highly disrupted and dip gently seaward, where they seem to be truncated by Reflector Z (Fig. 6).

The most unusual feature of the seismic profile, Reflector Z, is actually a zone of relatively continuous reflectors that dip about 20° landward (Fig. 6). Although these reflections are best determined at about 1.5 seconds depth, they can be tentatively traced from the surface to a depth of about 2.5 seconds. We interpret Reflector Z to be a zone of thrusting between the outer shelf "Gorgona" basement and the forarc basin sediments. This thrust zone therefore separates the outer shelf Gorgona basement from the sediments and could be considered to be a terrane boundary. If the Gorgona block is a sliver of oceanic material accreted at the trench and later uplifted, this fault zone could be the original structure of the accretion. The fact that the reflectors are not affected by the folding within the overlying sediments suggests that the fault zone has been reactivated. In fact, movement on this fault may be the cause of the folding and may explain the close juxtaposition of normal faults within the undeformed basin and folds near the basin margin.

An interpretive cross section of the structure of the margin is shown in Figure 7. The strata of the Tumaco section of the Bolivar trough (Fig. 2) were folded into two anticlines. These strata were truncated near the shelf edge by a fault zone (Reflector Z) that thrust the outer shelf basement beneath the sediments. The slope is interpreted to be underlain by an accretionary package with landward-dipping sediments and faults. As material was accreted at the trench at the base of the slope, the overlying packages were uplifted and possibly rotated. This, in turn, may have resulted in folding of the deposits of the basin and thrusting of them on top of the outer shelf. The acoustically transparent basement on the shelf edge is interpreted to correspond to the basement exposures of nearby Gorgona Island as discussed above. The uplift of Gorgona Island on the outermost shelf is shown by its geomorphology to be a recent event (Echeverria, 1980).

COMPARISON OF GORGONA SHELF WITH THAT OF SERRANIA DE BAUDO

The assumption that Gorgona Island is connected to both the Serrania de Baudo and the exposures in Ecuador as part of an ophiolitic "Coastal Cordillera" rests largely on the similarities in ages and lithologies of the mafic igneous rocks that characterize all three features. Comparison of geophysical and structural data for the two regions, however, suggests that there is a major tectonic break between Gorgona Island and the Serrania de Baudo.

Gravity data for the region have been collected by Case et al (1971, 1973), Case (1974), Meissner et al (1976), and have been compiled in a gravity anomaly map (Fig. 8) by Meissner et al (1976). The dense rocks of the Serrania de Baudo are reflected by a large positive anomaly. The Bouguer anomaly is as high as +130 mgals despite elevations within the coastal range. This anomaly can be followed south to 5° N. If the basement of the Serrania de Baudo were continuous from Panama to Ecuador, one might expect the associated gravity anomaly high to also be continuous, in a manner similar to the extension of the Nicoya Peninsula anomaly 1,000 km (620 mi) northward along the outer shelf of the Guatemalan margin (Couch and Woodcock, 1979). South of 3° N, Gorgona Island and the

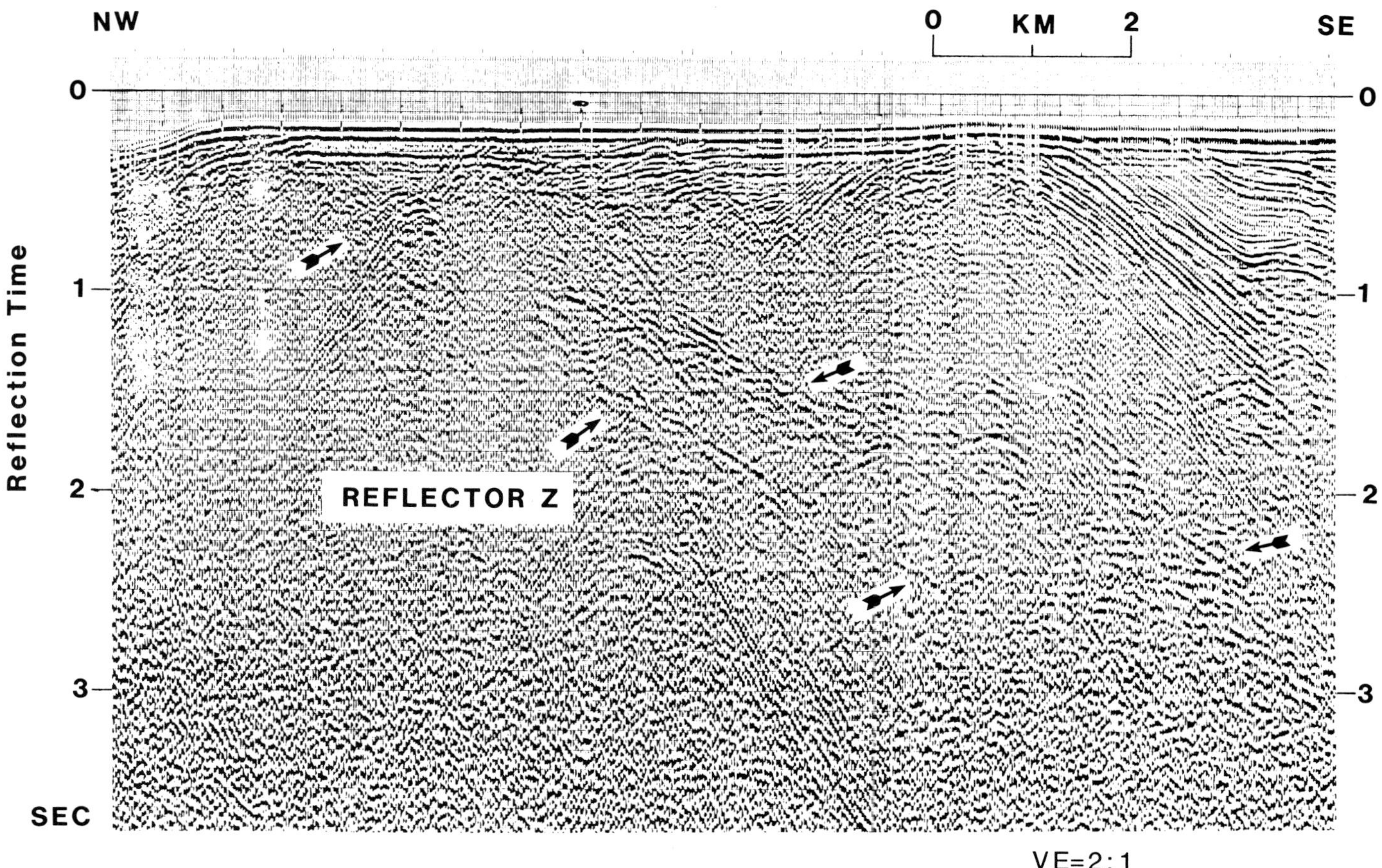

Figure 6—Close-up view of Reflector Z. Reflector Z is composed of two sets of parallel reflectors that continue to at least 2.5 seconds. The reflector is not affected by the folding of the overlying sedimentary deposits and is interpreted to be a thrust zone.

outer shelf where Profile T is located do exhibit a small positive anomaly. However, between 5° N and 3° N, the north-south or northeast-southwest trends of the gravity anomalies are disrupted and truncated by east-west-trending anomalies.

This disruption of the north-south gravity anomaly trends is also reflected in the structural data. The structural grain changes at 5° N to a complex series of northeast-trending faults and folds (Bueno and Govea, 1976). This structurally complex zone ends abruptly just north of Gorgona Island at what Bueno and Govea term a "wrench fault."

Two multichannel profiles (Lines 36 and 39; locations shown on Fig. 2) collected by Ecopetrol and interpreted by Flüh (1982) are located on the continental shelf south of the Serrania de Baudo near Buenaventura and are shown on Figure 9. They can be compared with Profile T, which shows the structure of the continental shelf south of Gorgona Island. All three profiles show the shelf and upper slope of the Colombian margin, but major differences between the two regions can be seen from comparison of the seismic data. The outer structural high that is prominent in Profile T and exposed on Gorgona Island is absent from Profiles 36 and 39. Only a small structural uplift has been described by Flüh beneath the outer shelf of Profile 36. In contrast to the folded and truncated sediments of the Gorgona region shown in Profile T, the thick sedimentary sequence of the Buenaventura area continues uninterrupted all the way to the shelf break and upper slope. It is interpreted to be part of a delta structure by Flüh. The differences between the seismic data from the two areas could be mainly caused by the changes in sedimentation patterns near a major drainage system. However, extensional growth faulting, down to the east, has created most of the structure seen in Profiles 36 and 39 on the northern part of the shelf as compared to the thrust fault and folding of the Gorgona area. The structure of the shelf in each region seems to reflect different histories.

The geophysical and structural information, therefore, may show that the Serrania de Baudo and Gorgona Island are not currently part of a continuous feature. The two terranes could have formed either as two completely separate features or as part of the same feature that later became disrupted during or after accretion in a way similar to that proposed for Wrangellia (Jones et al, 1977).

DISCUSSION

All of the oceanic terranes in northern South America and Central America have been grouped together as part of the Basic Igneous Complex (BIC) of Goossens and Rose (1973) on the basis of similar ages of formation (Cretaceous to Early Tertiary) and similar tholeiitic basaltic geochemistries. The BIC includes the Western Cordillera and Serrania de Baudo of Colombia, the coastal ranges of Ecuador, the Azuero Peninsula and basement of eastern

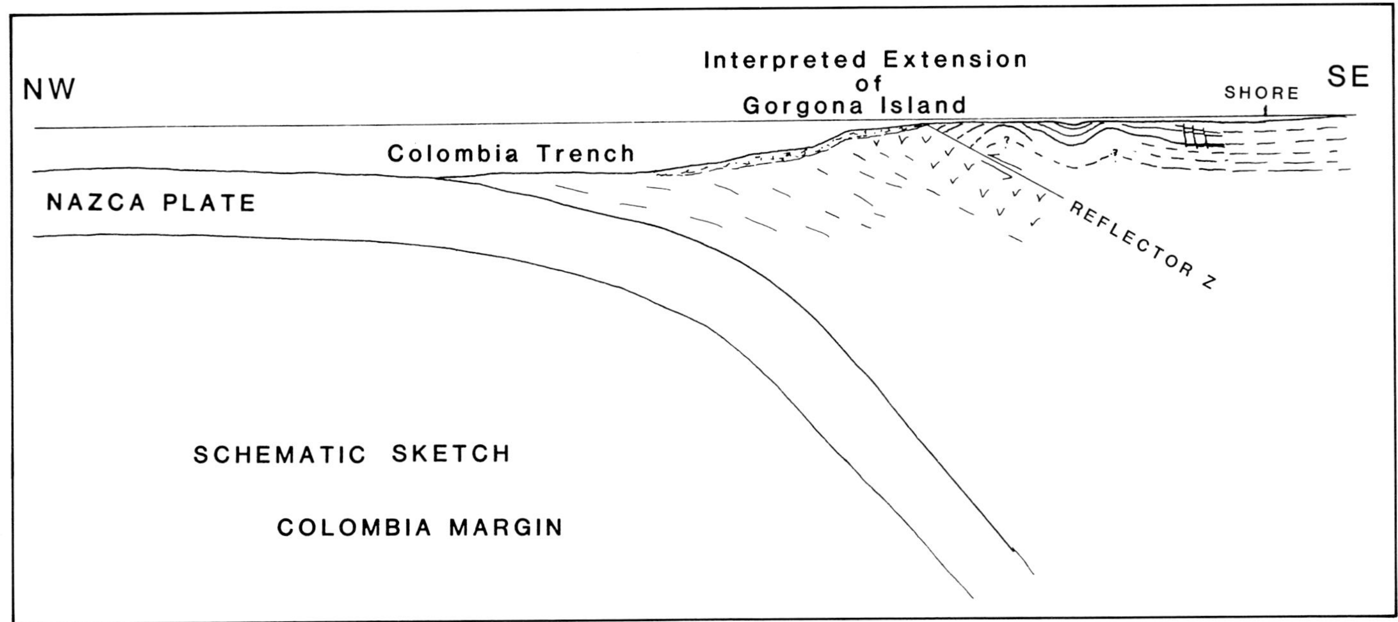

Figure 7—Interpretive cross section of the structure of the Colombian margin based on bathymetry and the multichannel seismic reflection profile T. The sketch is vertically exaggerated 2 to 1.

Panama (Fig. 1), and the Nicoya Complex of Costa Rica. This conceptual grouping has proved particularly useful in various geochemical studies of the Pacific margin of Central America and northwestern South America (Goossens et al, 1977; Pichler et al, 1974). It should not be assumed, however, that all of these features had a common origin or that they form a structurally continuous terrane. Each of the terranes has a distinct tectonic history.

Of the terranes in Colombia, the Western Cordillera has the oldest rocks and was probably the first accreted to the continental margin. It has been variously interpreted to represent imbricated oceanic crust with or without a buoyant aseismic ridge (Mooney, 1980), remnant oceanic spreading ridges (Barlow, 1981), a hot spot trace (Flüh, 1982), or an immature island arc (Barrero, 1979). However, the stratigraphic relationships and the geochemistry of the mafic igneous rock favor a primitive island arc origin (Barrero, 1979). This volcanic arc, the Western Cordillera, was isoclinally folded, uplifted, and faulted, probably as a result of collision and accretion, during the Late Cretaceous and earliest Paleocene, the Calima orogeny of Barrero (1979). The terranes west of the Western Cordillera were either accreted at the same time or during a later collisional event.

The Serrania de Baudo has also been described as an immature island arc (Barlow, 1981). The ages of the rocks in the Serrania de Baudo overlap those of the Western Cordillera. However, the youngest igneous rocks in the Serrania de Baudo intrude shallow-water limestones of middle Eocene age and Paleocene and Eocene shales (Case et al, 1971) and are younger than the mafic rocks in the Western Cordillera. In fact, they are younger than the proposed accretion of the Western Cordillera (Barrero, 1979). Because the Serrania de Baudo is structurally continuous with the Serrania del Sapo of eastern Panama, which is in turn part of the oceanic basement of the Panama arc (Case, 1974), it is possible that the Serrania de Baudo is itself part of the Panama arc and was accreted during collision with the Colombia continental margin (Pindell and Dewey, 1982). Several stratigraphic studies have shown a regional hiatus for eastern Panama and the Serrania de Baudo corresponding to uplift from bathyal to neritic levels during the late Miocene and the Pliocene (Bandy, 1970; Bandy and Casey, 1973; Duque-Caro, 1971). This fits in well with the proposed collision of Panama with Colombia at that time (Pindell and Dewey, 1982).

If Gorgona Island formed as part of the Serrania de Baudo, it would also have been accreted during the Miocene or Pliocene. The stratigraphy of the deposits on Gorgona Island, however, suggest that it was already part of the margin by Miocene time. The stratigraphy of Gorgona and Gorganilla Islands was described by Gansser (1950) and Echeverria (1980). The oldest strata exposed are upper Eocene calcareous sandstones and sandy limestones overlain by radiolarian-rich tuffaceous shales. The contact of this unit with the mafic igneous rocks is strongly disturbed and often marked by layers of chalcedony. Lower Miocene deposits are exposed nearby and probably transgress on the upper Eocene rocks (Gansser, 1950), with the unconformable contact covered by water. The base of the lower Miocene sequence is marked by the presence of fine to medium conglomerates that contain mostly well-rounded pebbles of chert and mafic igneous rocks. If the mafic Gorgona basement formed the source area for these pebbles, then Gorgona Island was already uplifted and being eroded by at least earliest Miocene time. This uplift precedes both the collision of the Panama arc and the main pulse of the Andean orogeny (Campbell, 1974) in late

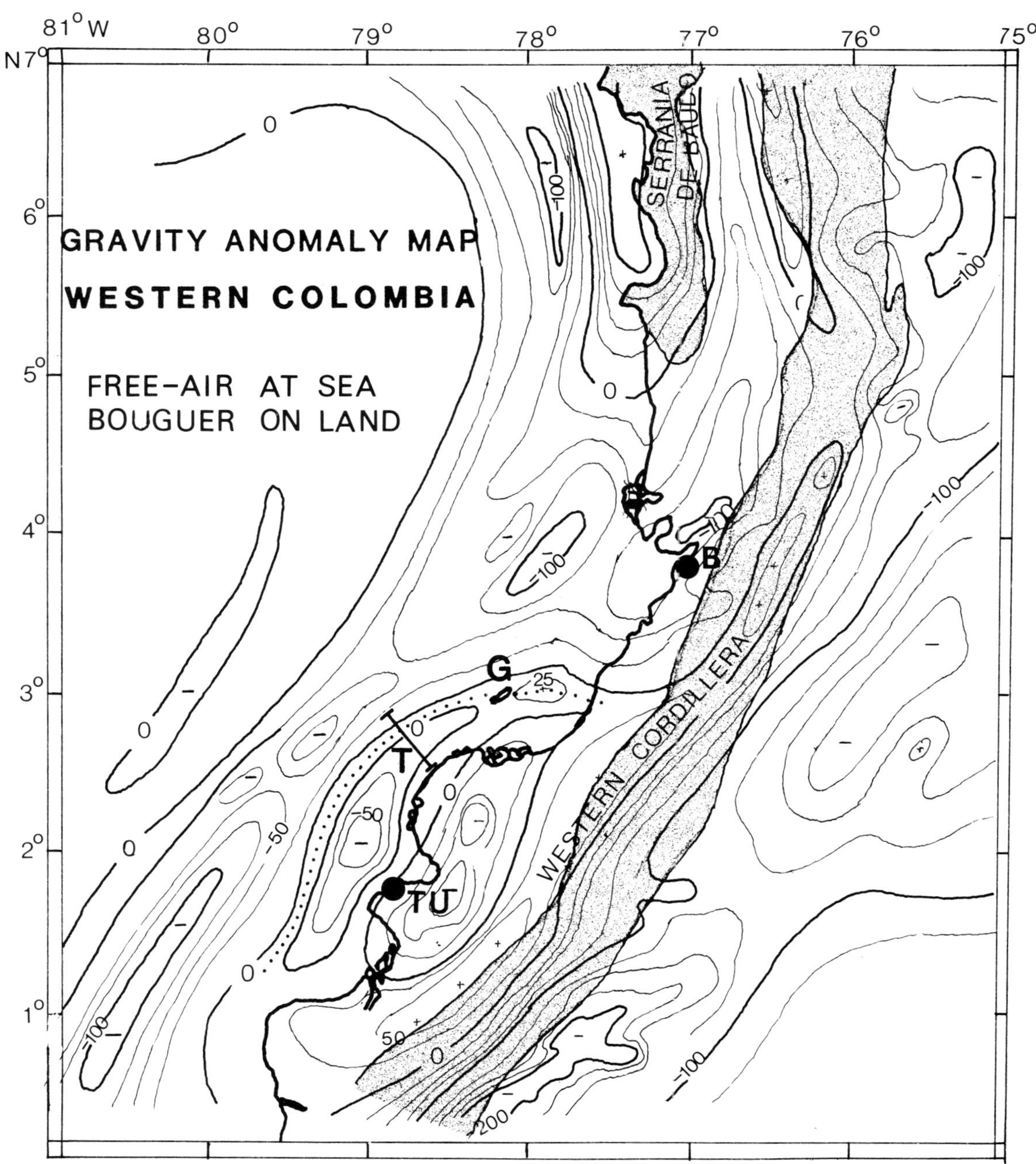

Figure 8—Gravity anomaly map of Panama and Colombia compiled by Meissner et al, 1976. (Tectonophysics, v. 35, p. 115–136, copyright © 1976 by Elsevier Science Publishers. Used with permission.) Notice the small gravity high over the outer continental shelf near Gorgona Island and crossed by Profile T. The axis of the anomaly is marked by the dotted line. Notice the large positive highs over the Serrania de Baudo and the Western Cordillera. The gravity anomaly trends are disrupted between Gorgona Island (G) and Buenaventura (B).

Miocene to Pliocene time. The deep-water sediments of late Eocene age were probably deposited before the uplift of the Gorgona basement and possibly before its accretion to the convergent margin. It is therefore unlikely that Gorgona Island formed with the Serrania de Baudo as part of the Panama arc.

The comparison presented in this paper of the geophysics, structure, and stratigraphy of the two regions therefore suggests that the Gorgona terrane is currently structurally separated from the Serrania de Baudo terrane, was accreted at an earlier time, and was probably formed in a way distinct from the proposed Panama–Serrania de Baudo arc. Where and how the komatiite sequence formed remains an unanswered question.

Other known komatiites found in South Africa and Canada formed during the Archean. They are thought to have formed in a backarc setting within crust that differs significantly from crust today, with higher heat flow during the Archean (Weaver and Tarney, 1979). Detailed analysis of rare earth element compositions show that the Gorgona komatiite may have formed from mantle already depleted from the production of midocean ridge basalt (Echeverria

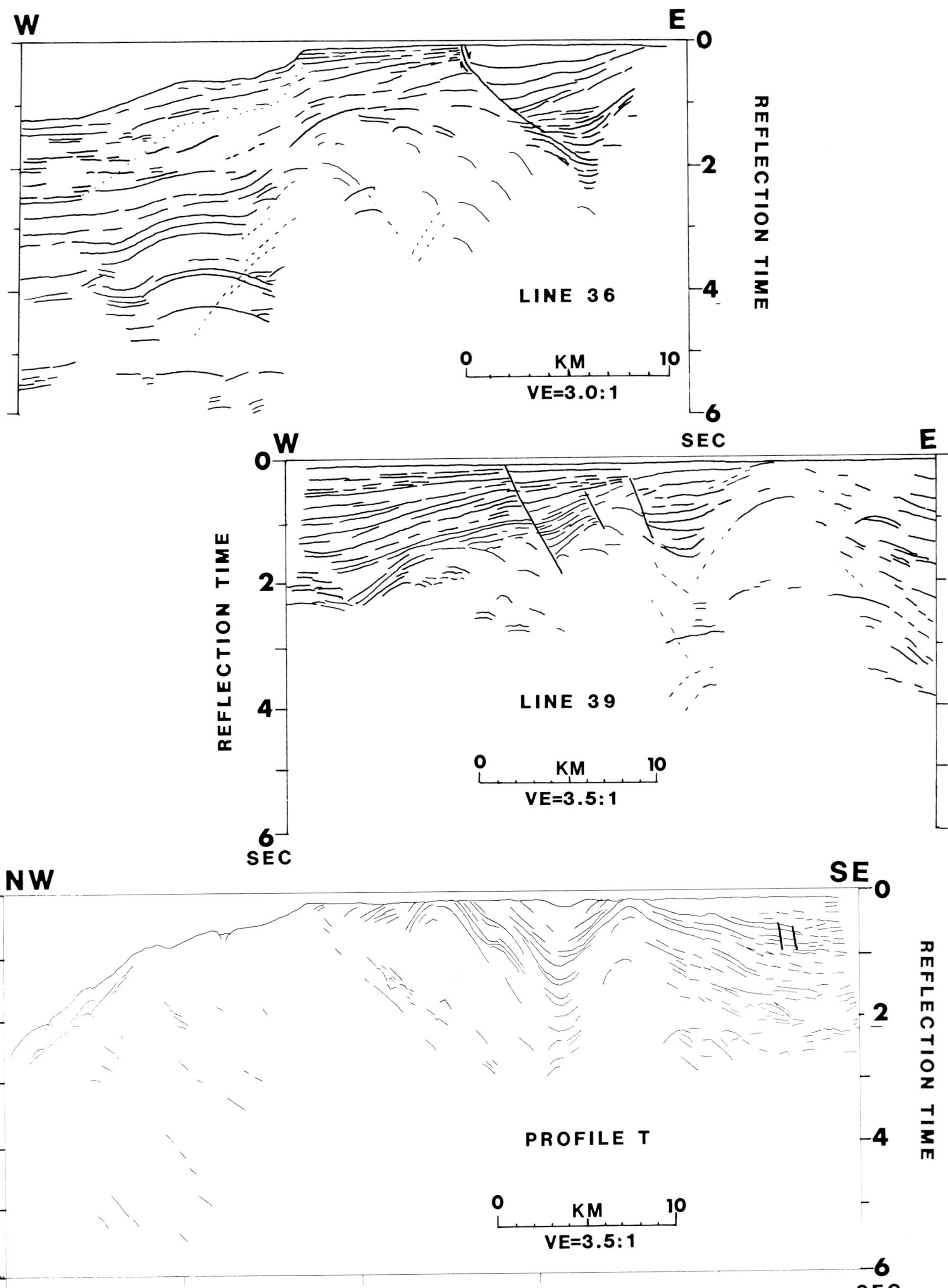

Figure 9—Line drawings of multichannel seismic reflection Lines 36 and 39 from the continental shelf near Buenaventura (from Flüh, 1982) compared to Profile T.

and Hofmann, in press). This suggests some kind of off-ridge volcanism. The tholeiitic geochemistry with low titanium content is not characteristic of a hot spot trace but could perhaps result from a leaky transform. Another possibility is that the Gorgona rocks formed in a forearc or backarc setting with respect to the volcanic arc that forms the basement of the Western Cordillera. (The polarity of the arc is not known.) This speculation implies that Gorgona Island and the basement of the Bolivar trough were accreted to the continental margin as part of the Western Cordillera during the Late Cretaceous or Early Tertiary. Either model is possible given available information for western Colombia.

The nature of the thrust zone that marks the Gorgona terrane boundary in Profile T is somewhat ambiguous. It is recently active and may be a structure entirely unrelated to the accretion of the Gorgona terrane. However, if the fault zone is the original structure along which the Gorgona block was accreted, the landward dip of the fault shows that Gorgona was thrust beneath the slope at the trench rather than obducted as many ophiolites seem to be. Later packages would presumably be thrust in turn beneath the Gorgona block, as possibly shown by deeper reflectors on Profile T. The Gorgona block seems to have been uplifted without major rotation on the reactivated thrust zones.

CONCLUSIONS

The presence of oceanic terranes along the continental margins of northern South America and Central America is well established. Although the basic stratigraphy of each terrane is known, the history of formation, transport, and accretion of each terrane is still not determined. One of the most intriguing of the terranes is Gorgona Island, known for the exposures of unique Late Cretaceous or Early Tertiary komatiitic lavas. A multichannel seismic profile near Gorgona shows the outer shelf to be underthrusting the sedimentary strata of the forearc basin, which is in turn folded. A thrust zone forms the current Gorgona terrane boundary and may be the reactivated structure along which the terrane was accreted.

A comparison of the geophysical, structural, and stratigraphic data for Gorgona Island with that of the Serrania de Baudo, the northern coastal range, suggests that Gorgona is probably a separate terrane rather than a continuation of the Serrania de Baudo as has been previously proposed. Much more data, especially along the boundaries of the Bolivar trough with the Western Cordillera, the Serrania de Baudo, and Gorgona Island is needed in order to resolve some of the problems in reconstructing the Cenozoic tectonics of western Colombia.

ACKNOWLEDGMENTS

We express our appreciation to Texaco, Inc., for generously providing the multichannel seismic reflection profile T and to Barry Katz at Texaco, Inc., for his cooperation and especially his time. We thank John Claerbout of the Stanford Exploration Project (SEP) for the use of valuable SEP equipment and programs. Discussions and manuscript reviews by James E. Case, James N. Kellogg, and Walter Mooney constructively aided the preparation of this manuscript. David Okaya was quite helpful with technical advice, and Norm Sleep contributed his unique perspective and ideas during discussion. This research was funded by National Science Foundation Grant EAR 80-25879.

REFERENCES

Bandy, O. L., 1970, Upper Cretaceous-Cenozoic paleobathymetric cycles, eastern Panama and northern Colombia: Gulf Coast Association of Geological Societies Transactions, v. 20, p. 181–193.

———, and R. E. Casey, 1973, Reflector horizons and paleobathymetric cycles, eastern Panama: Geological Society of America Bulletin, v. 84, p. 3081–3086.

Barlow, C. A., 1981, Radar geology and tectonic implications of the Choco Basin, Colombia, South America: MS Thesis, University of Arkansas, 96 p.

Barrero L., Dario., 1979, Geology of the central Western Cordillera, west of Buga and Roldanillo, Colombia: Instituto Nacional de Investigaciones Geológicos-Mineras, Publicaciones Geológicas Especiales del Ingeominas 4, p. 1–75.

Bueno, S. R., and R. C. Govea, 1976, Potential for exploration and development of hydrocarbons in Atrato Valley and Pacific coastal and shelf basins of Colombia, *in* M. T. Halbouty, et al, eds., Circum-Pacific energy and mineral resources: American Association of Petroleum Geologists Memoir 25, p. 318–327.

Campbell, C. J., 1974, Colombian Andes, *in* A. M. Spencer, ed., Mesozoic-Cenozoic orogenic belts: Data for orogenic studies: Geological Society of London Special Publication 4, p. 705–724.

Case, J. E., 1974, Oceanic crust forms basement of eastern Panama: Geological Society of America Bulletin, v. 85, p. 645–652.

———, et al, 1971, Tectonic investigations in western Colombia and eastern Panama: Geological Society of America Bulletin, v. 82, p. 2685–2711.

———, et al, 1973, Trans-Andean geophysical profile, southern Colombia: Geological Society of America Bulletin, v. 84, p. 2895–2904.

Couch, R., and S. Woodcock, 1981, Gravity and structure of the continental margins of southwestern Mexico and northwestern Guatemala: Journal of Geophysical Research, v. 86, p. 1829–1840.

Duque-Caro, H., 1971, Relaciones entre la bioestratigrafia y la cronoestratigrafia en al llamado geosinclinal de Bolívar: Instituto Nacional de Investigaciones Geológicos-Mineras, Boletín Geológico, v. 19, p. 25–68.

Echeverria, L. M., 1980, Tertiary or Mesozoic komatiites from Gorgona Island, Colombia: Field relations and geochemistry: Contributions to Mineralogy and Petrology, v. 73, p. 253–266.

———, and A. W. Hofmann, in press, The relationship of komatiites and basalts on Gorgona Island, Colombia and a comparison with Reykjanes Ridge picrites: Geology.

Espinosa, A., M. DeLaloye, and J. J. Wagner, 1981, Radiometric ages of the Gorgona Island (Colombia) komatiitic ophiolite, Ophiolites and Actuatism Meeting, Florence, Italy.

Flüh, E. R., 1982, Geodynamische Entwicklung der nördlichen Anden: PhD Dissertation, Christian-Albrechts-Universität, Kiel, 164 p.

Gansser, A., 1950, Geological and petrographical notes on Gorgona Island in relation to northwestern South America: Schweizerische Mineralogische und Petrographische Mitteilungen, v. 30, p. 219–236.

______, 1973, Facts and theories on the Andes: Journal of the Geological Society of London, v. 129, p. 93–131.

______, et al, 1979, Paleogene komatiites from Gorgona Island: Nature, v. 278, p. 545–546.

Goossens, P. J., and W. I. Rose, Jr., 1973, Chemical composition and age determination of tholeiitic rocks in the Basic Igneous Complex, Ecuador: Geological Society of America Bulletin, v. 84, p. 1043–1052.

______, et al, 1977, Geochemistry of tholeiites of the Basic Igneous Complex of northwestern South America: Geological Society of America Bulletin, v. 88, p. 1711–1720.

Interoceanic Canal Study Commission (ICSC), 1968, Geology, Final Report, Route 25, v. 1: Canal Zone, Panama, Field Director, Office of Interoceanic Canal Studies, 208 p.

______, 1969, Geology, Final Report, Route 25, v. 2: Canal Zone, Panama, Field Director, Office of Interoceanic Canal Studies, 52 p.

Jones, D. L., et al, 1977, Wrangellia—A displaced terrane in northwestern North America: Canadian Journal of Earth Sciences, v. 14, p. 2565–2577.

Meissner, R. O., et al, 1976, Dynamics of the active plate boundary in southwest Colombia according to recent geophysical measurements: Tectonophysics, v. 35, p. 115–136.

Meyer, R. P., et al, 1976, Project Narino III: Refraction observations across a leading edge: Malpelo Island to the Colombia Cordillera Occidental, *in* The geophysics of the Pacific Ocean Basin and its margin: American Geophysical Union Geophysical Monograph 19, p. 105–132.

Mooney, W. D., 1980, An East Pacific-Caribbean ridge during the Jurassic and Cretaceous and the evolution of western Colombia, *in* R. H. Pilger, ed., The origin of the Gulf of Mexico and the early opening of the Central North Atlantic Ocean: Symposium Proceedings, Louisiana State University, p. 55–73.

______, et al, 1979, Seismic refraction studies of the Western Cordillera, Colombia: Bulletin of the Seismological Society of America, v. 69, p. 1745–1761.

Nelson, H. W., 1957, Contribution to the geology of the central and western Cordillera of Colombia in the sector between Ibaque and Cali: Leidse Geologische Mededelingen, v. 22, p. 1–76.

Nygren, W. E., 1950, Bolivar geosyncline of northwestern South America: Bulletin of the American Association of Petroleum Geologists, v. 34, p. 1998–2006.

Pichler, V. H., et al, 1974, Basicher Magmatismus und Krustenbau im südlichen Mittelamerika, Kolumbien und Ecuador: Neues Jahrbuch für Geologie und Paläontologie Monatshefte, p. 102–128.

Pindell, J., and J. F. Dewey, 1982, Permo-Triassic reconstruction of western Pangea and the evolution of the Gulf of Mexico/Caribbean region: Tectonics, v. 1, p. 179–211.

Stolt, R. H., 1978, Migration by Fourier transform: Geophysics, v. 43, p. 23–48.

Weaver, B. L., and J. Tarney, 1979, Thermal aspects of komatiite generation and greenstone belt models: Nature, v. 279, p. 689–692.

The Pacific Margin of Antarctica: Terranes within Terranes within Terranes

Ian W. D. Dalziel
Anne M. Grunow
Lamont-Doherty Geological Observatory of Columbia University
Palisades, New York

The terrane concept is applicable to the Pacific margin of Antarctica in three respects. First, West (or Lesser) Antarctica presently appears to consist of at least four discrete or semidiscrete microcontinents that have moved relative to each other and relative to the East (Greater) Antarctic craton during and/or since Gondwanaland breakup. These are the Antarctic Peninsula, Ellsworth Mountains–Whitmore Mountains, Thurston Island–Eights Coast, and Marie Byrd Land. All seem to have been originally part of the Pacific margin of the supercontinent. Second, rocks along the Pacific margin of the Antarctic Peninsula form a subduction complex accreted to the continent during the late Paleozoic(?), Mesozoic, and possibly the Cenozoic. The manner of accretion, gradual offscraping and/or as exotic blocks, is unclear. Third, Cenozoic spreading has resulted in the separation of the South Shetland Islands (partially) and South Orkney Islands (completely) from the Antarctic Peninsula. The former is therefore parautochthonous and could in the future be reunited with the continent; the latter has already become an exotic terrane. Thus, like other parts of the Pacific margin, the subduction complexes of the South Shetland and South Orkney Islands are, according to the tenets set out in this volume, terranes within terranes within terranes.

The available paleomagnetic data suggest that while the Antarctic Peninsula and Ellsworth Mountains–Whitmore Mountains microcontinents have not moved far with respect to the East Antarctic craton, significant rotations may have occurred. A program is being undertaken to obtain more data bearing on the origin and evolution of the West Antarctic microcontinents.

INTRODUCTION

The continental margin of Antarctica is the least well-known segment of the circum-Pacific mobile belt. Nonetheless, enough is known about it to make some meaningful points with regard to present and future terrane analysis, of the type advocated by Howell and Jones (1984). It will be apparent that we consider the terrane concept to be applicable to the Pacific margin of Antarctica and to be of general tectonic importance. Nonetheless, we do have reservations regarding what we see as currently prevalent, overzealous applications of the concept and presumptions concerning its orogenic significance.

To date, terrane analysts have merely divided the entire Antarctic continent along classical lines into the East (Greater) Antarctic craton and the West (Lesser) Antarctic "fold belt" (Howell et al, 1984). This was the distinction made by the geologists of Captain Robert Falcon Scott's expedition in the early "heroic" days of Antarctic exploration. The distinction is, nevertheless, an important one, and the problem of the relationship of the East Antarctic craton and the West Antarctic segment of the circum-Pacific mobile belt is not only one of the longest standing problems in Antarctic geology, but also one with immense global significance. It bears on global plate interactions, paleocirculation in the southern oceans, paleoclimate, the causes of glaciation, and paleobiogeography (see discussion by Dalziel and Elliot, 1982).

In recent years, it has come to be appreciated that West Antarctica probably consists of at least four discrete or semidiscrete microcontinents (Dalziel and Elliot, 1982; see Fig. 1). The boundaries between these entities, concealed by deep ice-filled troughs (Fig. 2), were recognized by geophysicists undertaking cross-ice traverses during the International Geophysical Year (1957–1958). Most of the troughs are probably fault controlled (Jankowski and Drewry, 1981; Doake et al, 1983). Analysis of the geology of West Antarctica leading to attempts to reconstruct the Pacific margin of Gondwanaland with all its important implications (Fig. 3) therefore constitutes terrane analysis in the sense of Howell et al (1984).

More detailed terrane analysis has a role to play in the attempt to understand the evolution of subduction complexes along the Pacific margin of the Antarctic Peninsula and adjacent islands (Fig. 2). These complexes were partly accreted to the continent prior to the Late Jurassic to Early Cretaceous Gondwanaland breakup. The accretionary zone was probably adjacent to, if not actually continuous with, the contemporaneous forearc accretionary prism of southern South America (see Dalziel and Forsythe, this volume). Other parts of the Antarctic Peninsula margin subduction complex may be no older than mid-Mesozoic and as young as Cenozoic (Tanner et al, 1982). The rocks of the subduction complexes consist of pelagic and hemipelagic sedimentary strata, ocean floor and/or magmatic arc volcanics, and at least one tectonic slice of oceanic mantle (dunite-serpentinite). As in the case of the South American late Paleozoic to early Mesozoic forearc prism, these complexes consist of a myriad of terranes. Metachert layers 0.5 cm (0.2 in.) or less in thickness are important because they were probably derived from the now-vanished floor of Panthalassa as

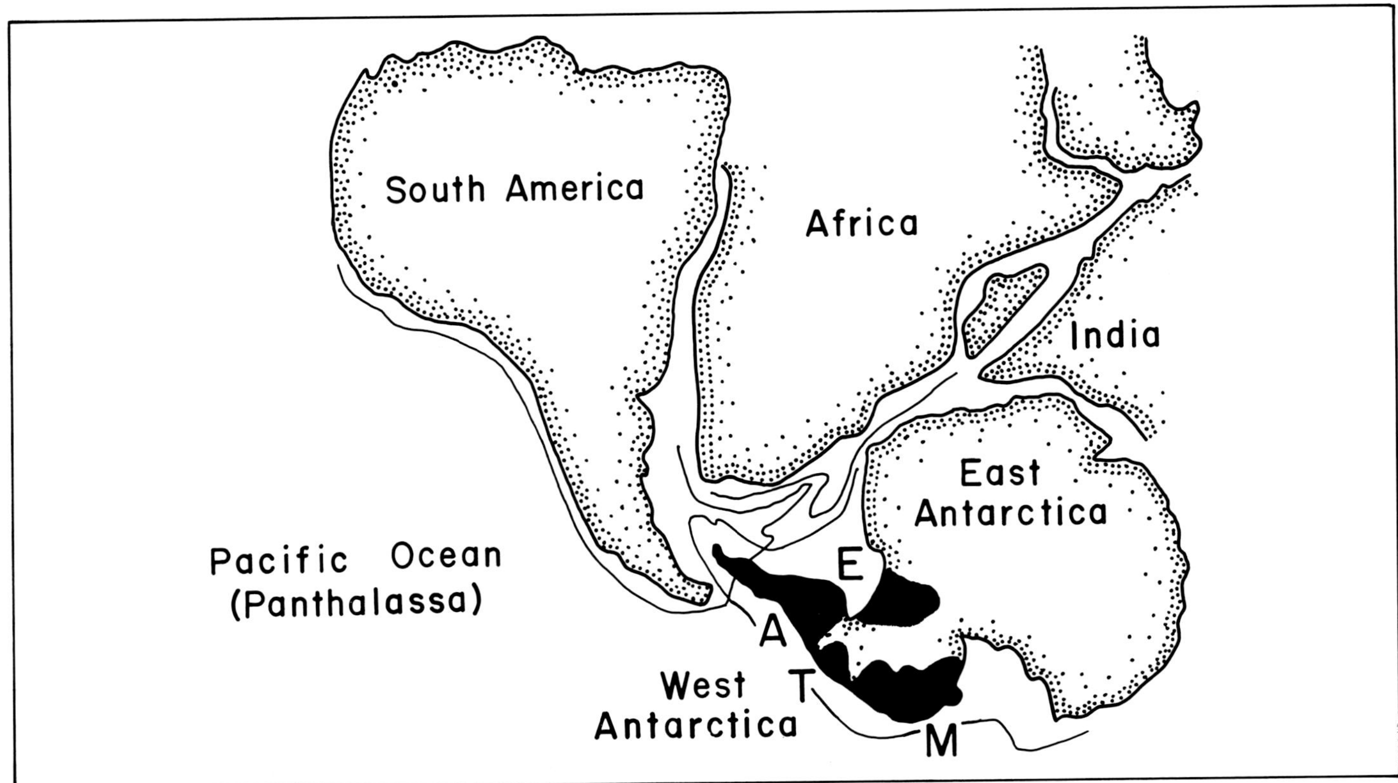

Figure 1—Gondwanaland reconstruction of Norton and Sclater (1979). West Antarctic microcontinents. (From Dalziel and Elliot, 1982. Tectonics, v. 1, p. 3–19, copyright © 1982 by the American Geophysical Union.) A = Antarctic Peninsula; E = Ellsworth Mountains–Whitmore Mountains; M = Marie Byrd Land; T = Thurston Island–Eights Coast.

material underplated onto the accretionary wedge along the Gondwanaland margin (Dalziel, 1984a). Should each one, however, be classified as a terrane because it is perhaps bounded by ductile shears?

Finally, the Antarctic segment of the Pacific margin is important with regard to concepts analyzed in the present volume because microcontinents constituting terranes have been, and indeed are being, generated there. The South Shetland Islands (SSI, Fig. 2) and South Orkney Islands (SOI, Fig. 2) are now separated from the Antarctic Peninsula by the Bransfield Strait and the Powell Basin, respectively (Fig. 4). The former is a currently active backarc spreading center, the latter a mid-late Cenozoic spreading center (Barker and Dalziel, 1983). The continental blocks or microcontinents of the South Shetland Islands and South Orkney Islands clearly, therefore, constitute terranes that are respectively parautochthonous and exotic with respect to their parent continent.

In this article, we present a brief discussion of these three aspects of the geology of the Antarctic margin of the Pacific that are obviously significant with regard to the topic of this volume. We review the relevant paleomagnetic data and outline a program we have currently under way to further elucidate the tectonic history of the Pacific margin of the Antarctic continent.

WEST ANTARCTIC MICROCONTINENTS

The problem concerning the relationship of West Antarctica to East Antarctica is illustrated in Figure 1. Modern reconstructions of Gondwanaland based on marine geophysical data still lead to a seemingly unacceptable overlap between the Antarctic Peninsula and the South American continent. The solution seems to lie in relative motion between the four major areas of exposed continental rocks in West Antarctica, namely the Antarctic Peninsula, the Ellsworth Mountains–Whitmore Mountains, Thurston Island–Eights Coast, and Marie Byrd Land (see also Fig. 1). This problem has been discussed as fully as the available data allowed in a recent paper by Dalziel and Elliot (1982), and we will touch here on only a few points of significance with regard to the study of circum-Pacific terranes.

Clearly the rocks of the four areas fall into the category of separate composite terranes. They are separated by subice troughs with depths of over 1,500 m (5,000 ft) below sea level (Fig. 2). The Antarctic Peninsula is separated from the East Antarctic craton by the Mesozoic oceanic basin of the Weddell Sea. While there appears to be continuous continental crust beneath the Ross Sea that separates Marie Byrd Land and the craton, there could have been major crustal extension there (Grindley and Oliver, 1983). There is evidence of large-scale rifting between the Ellsworth Mountains–Whitmore Mountains block and the three other West Antarctic terranes (Jankowski and Drewry, 1981; Doake et al, 1983).

Analysis of the geology strongly indicates that the Antarctic Peninsula was part of the Pacific margin of Gondwanaland before breakup, and that the Ellsworth Mountains–Whitmore Mountains block was once situated along the margin of the East Antarctic craton between the

Figure 2—Subglacial topography of West Antarctica. BI = Berkner Island; EL = Ellsworth Land; EM = Ellsworth Mountains; HN = Haag Nunataks; PIB = Pine Island Bay; SOI = South Orkney Islands; SSI = South Shetland Islands; W = Whitmore Mountains. (From Dalziel and Elliot, 1982. Tectonics, v. 1, p. 3–19, copyright © 1982 by the American Geophysical Union.)

Pensacola Mountains and the Cape fold belt of southern Africa (Schopf, 1969; Watts and Bramall, 1981; Dalziel and Elliot, 1982; Fig. 3). The original locations of the Thurston Island–Eights Coast block and Marie Byrd Land are less certain, although there is no real geologic reason to doubt that they too were originally part of the Pacific margin of Gondwanaland (Dalziel and Elliot, 1982). Some ambiguous paleomagnetic results, however, do suggest that Marie Byrd Land might be exotic with respect to West Antarctica. This will be discussed later.

A program to investigate the relationships between the Antarctic Peninsula, the Thurston Island–Eights Coast block, and the Ellsworth Mountains–Whitmore Mountains block is currently being undertaken jointly by the authors and Drs. Robert Pankhurst and Bryan Storey of the British Antarctic Survey (Dalziel and Pankhurst, 1984). The work will be discussed below.

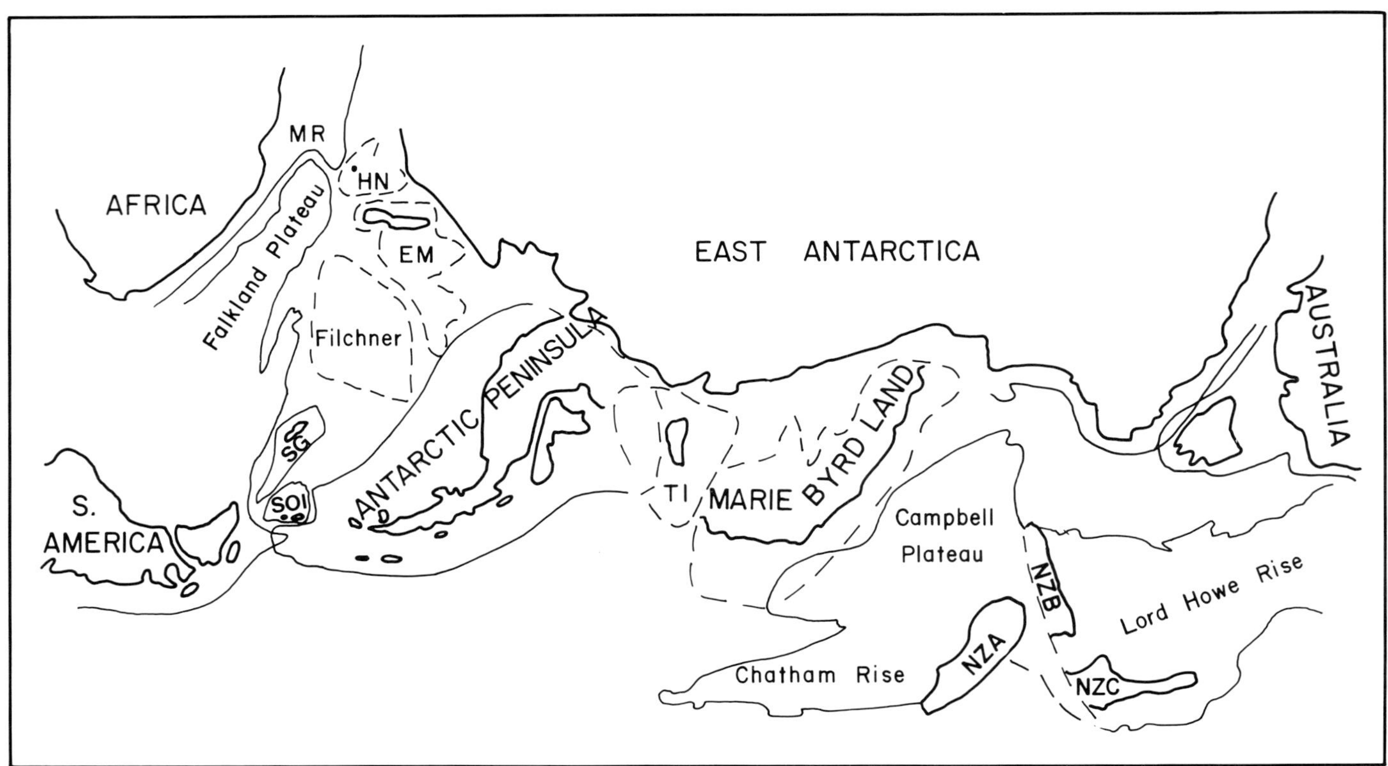

Figure 3—A speculative reconstruction of part of the Pacific margin of Gondwanaland. (From Dalziel and Elliot, 1982. Tectonics, v. 1, p. 3–19, copyright © 1982 by the American Geophysical Union.) EM = Ellsworth Mountains; HN = Haag Nunataks; MR = Mozambique Ridge; NZA, NZB, NZC = parts of New Zealand; SG = South Georgia; SOI = South Orkney Islands; TI = Thurston Island block.

SOUTHERN SCOTIA RIDGE SUBDUCTION COMPLEXES

The Antarctic Peninsula appears to have been a Pacific margin magmatic arc since the earliest Mesozoic or possibly late Paleozoic (Thomson et al, 1983). Metamorphic complexes interpreted as subduction complexes are located along the southern Scotia Ridge in the South Shetland and South Orkney Islands (Figs. 2, 4; Dalziel, 1982, 1984a). On the South Orkney Islands, one metamorphic complex is unconformably overlain by Upper Jurassic or Lower Cretaceous strata (see Dalziel et al, 1981). Otherwise the field relations of these complexes are unknown, and in that respect they constitute tectono-stratigraphic terranes of a suspect nature. In fact, they are largely composed of rocks of oceanic affinities including metachert, calc-silicate, ocean floor or island arc metavolcanics, and a dunite-serpentinite complex. Together with their location oceanward of Mesozoic–Cenozoic magmatic arc rocks along the Antarctic Peninsula and their structural style and metamorphic grade, this leads to their being widely interpreted as representing offscraped material from Pacific ocean-floor lithosphere subducted beneath the Antarctic Peninsula segment of the Gondwanaland continent (Barker et al, 1976; de Wit, 1977; Smellie, 1981; Dalziel, 1982). Graywacke-shale sequences of late Paleozoic (?) to early Mesozoic (Late Triassic) age in the South Orkney and South Shetland Islands and in the northern Antarctic Peninsula are locally in tectonic contact with the subduction complexes and are unconformably overlain by Middle to Upper Jurassic volcanic and sedimentary strata (Dalziel, 1982). These are interpreted to be the infilling of trench, trench-slope, or forearc basin(s) (Smellie, 1981; Dalziel, 1982, 1984a; Hyden and Tanner, 1981).

The subduction complexes are characterized by tectonites that have undergone polyphase ductile deformation with pervasive shearing. Melanges are found only very locally (Dalziel 1984a; Storey and Meneilly, 1983). Thus, all the discrete lithologic layers of the tectonites are bounded by ductile shears, and the distinction of separate terranes on this scale seems virtually meaningless.

There is a significant distinction within the metamorphic complexes as a whole between the relatively low-temperature blueschist and greenschist facies rocks of Smith Island, northern Elephant Island, and Clarence Island in the South Shetland Islands (Terrane A of Tanner et al, 1982; see Fig. 5) and relatively high-temperature albite-epidite amphibolite facies rocks of the South Orkney Islands and the remainder of the Elephant Island group in the South Shetland Islands (Terrane B, Figs. 5, 6). The former are not likely to be older than mid-Mesozoic and may be as young as Cenozoic; the latter are unconformably overlain by Upper Jurassic or Lower Cretaceous strata in the South Orkney Islands and could be as old as upper Paleozoic (Dalziel, 1982, 1984a; Tanner et al, 1982).

Smith Island is an isolated physiographic unit (60 × 7 × 1.5 km [37 × 4.3 × .93 mi]). Northern Elephant Island and Clarence Island are separated from the rest of the

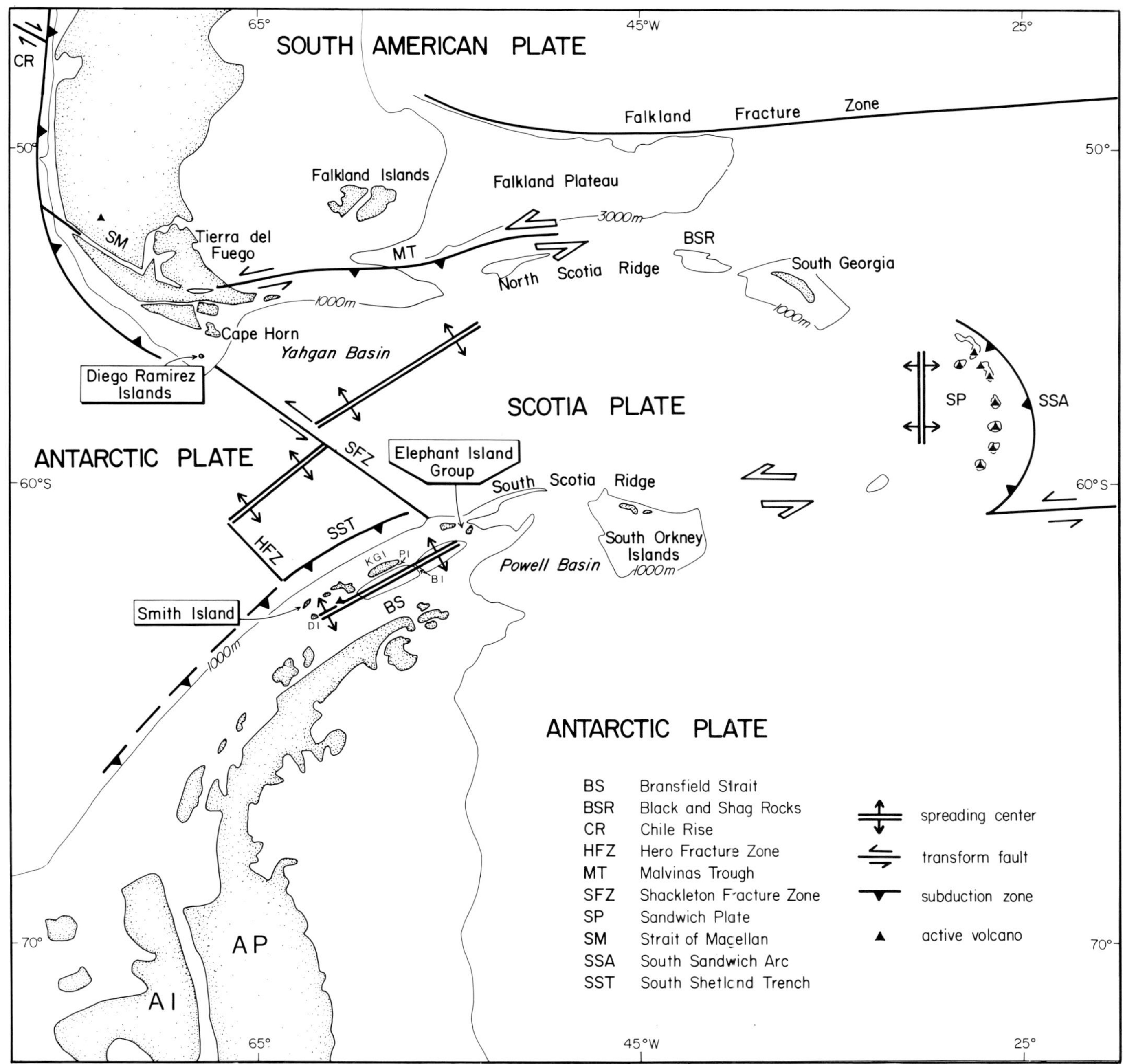

Figure 4—Tectonic setting of the southern Scotia Ridge. AI = Alexander Island; AP = Antarctic Peninsula; BI = Bridgeman Island; DI = Deception Island; KGI = King George Island; PI = Penguin Island.

Elephant Island group by a tectonic break (Dalziel, 1982; see Fig. 5). Hence, this is the boundary between the two terranes. It should be noted that both Smith Island and the Elephant Island group occur at the termination of fracture zones (Fig. 4). This may have a bearing on the substantial uplift that has occurred in both places (Dalziel, 1984a, 1984b).

SOUTH SHETLAND AND SOUTH ORKNEY ISLAND MICROCONTINENTS

The Antarctic segment of the Pacific margin has generated a new microcontinental terrane during the mid-late Cenozoic and, indeed, is currently generating a second one. The South Orkney Islands on the southern Scotia Ridge are situated on a submerged platform of continental shelf depth that is separated from the Antarctic Peninsula by the Powell Basin (Fig. 4). The South Orkney Islands are mainly composed of rocks indistinguishable, as has long been known (Ferguson, 1921; Wordie, 1921), from the high T (Terrane B) metamorphic complex of the southern part of the Elephant Island group of the South Shetland Islands (Fig. 5). The seismic velocity structure of the South Orkney Islands platform is essentially continental while that of the Powell Basin is oceanic (Harrington et al, 1972). While marine magnetic anomalies have yet to be mapped

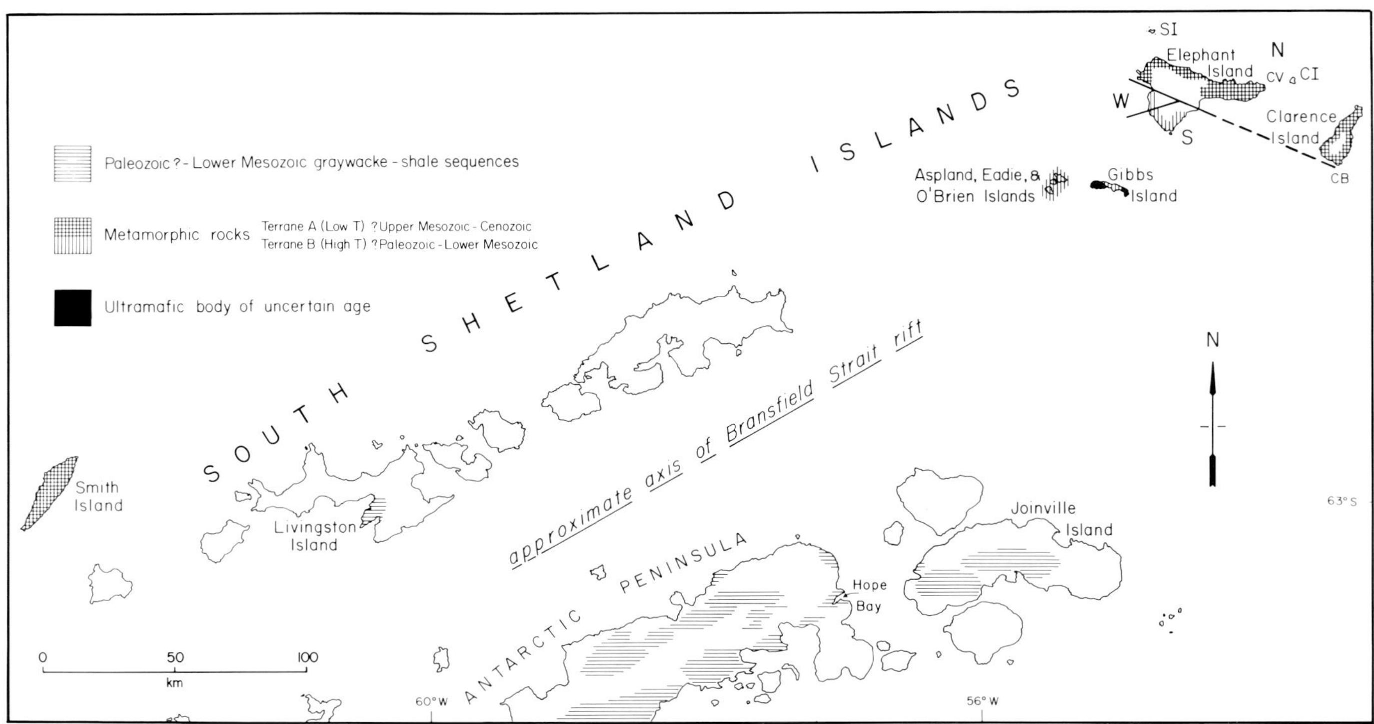

Figure 5—Geologic map of the South Shetland Islands with Bransfield Strait spreading removed (from Dalziel, 1984a). Coordinates are for a fixed Antarctic Peninsula.

over the Powell Basin (it is usually ice covered), its depth indicates that it is probably a mid-late Cenozoic spreading center (Barker and Dalziel, 1983). Hence, the South Orkney Islands appear to be the emergent part of a microcontinental terrane that has recently separated from the Antarctic Peninsula.

The South Shetland Islands are situated on a segment of the southern Scotia Ridge that is presently separating from the Antarctic Peninsula by rifting along the Bransfield Strait (Fig. 4; see Barker and Dalziel, 1983, for review). The active nature of the rifting is revealed by seismicity and especially by the active Deception Island volcano (Fig. 4), one of a line of sea mounts along the floor of Bransfield Strait. Other recently active volcanoes include Penguin Island (on the northwest side of the rift zone) and Bridgeman Island. The Bransfield Strait appears to be a backarc or marginal basin related to the mid-late Cenozoic calc-alkaline magmatic arc that existed along the line of the South Shetland Islands between Smith Island and the Elephant Island group, especially on King George Island (Fig. 4). In fact, Bransfield Strait is merely the best developed of a series of rifts along the Pacific side of the Antarctic Peninsula, the next best developed of which lies beneath King George VI Sound that separates Alexander Island from the mainland of the Antarctic Peninsula (Fig. 4). Hence, the South Shetland Islands are the emergent part of a new microcontinental (or extinct island arc) terrane that is currently coming into being.

From the point of view of interpreting the past geologic record of the circum-Pacific and other mobile belts, it is of interest to speculate on the future of the South Orkney Islands and South Shetland Islands microcontinental terranes. Clearly, the South Shetland Islands are still only parautochthonous with respect to their parent Antarctic Peninsula segment of the Antarctic continent. Other segments of the Pacific margin of the Antarctic Peninsula such as Alexander Island have barely started to separate. The rifting in the Bransfield Strait is calculated to have taken place over the past 1.5 m.y. in response to a virtual cessation of spreading on the northeast-trending Drake Passage spreading center, and hence of subduction along the South Shetland Islands trench (Barker and Dalziel, 1983; Figs. 4, 5). Reinitiation of spreading of this center, or some other cause, could in the future result in an increase in the compressive stress on the upper (Antarctic) plate at the inactive South Shetland Islands subduction zone. This could result in the closure of the Bransfield Strait rift with intense compression and orogenic uplift of its sedimentary infill and deformation of its margins.

This is exactly the type of origin postulated for the Paleozoic rift system of the eastern Cordillera of Peru and Bolivia and its southerly continuation, the so-called eugeosynclinal trough of northern Argentina and Chile (see Dalziel and Forsythe, this volume). An analogy with Early Cretaceous "rocas verdes" marginal basin of Chile and South Georgia (Dalziel, 1981; Storey et al, 1977) is even more striking because of the superior documentation of the field relations in this younger feature. Indeed, Dalziel (in press) has emphasized that the "rocas verdes" basin south of 50°S latitude is only in the southern extremity of a composite backarc or intra-arc basin that existed in the Early Cretaceous from the Huancabamba deflection at 5°S to Cape Horn and along the northern Scotia Ridge (Dalziel and Forsythe, this volume, Fig. 9). Closure of this basin resulted in the initiation of orogenic uplift in the Andes.

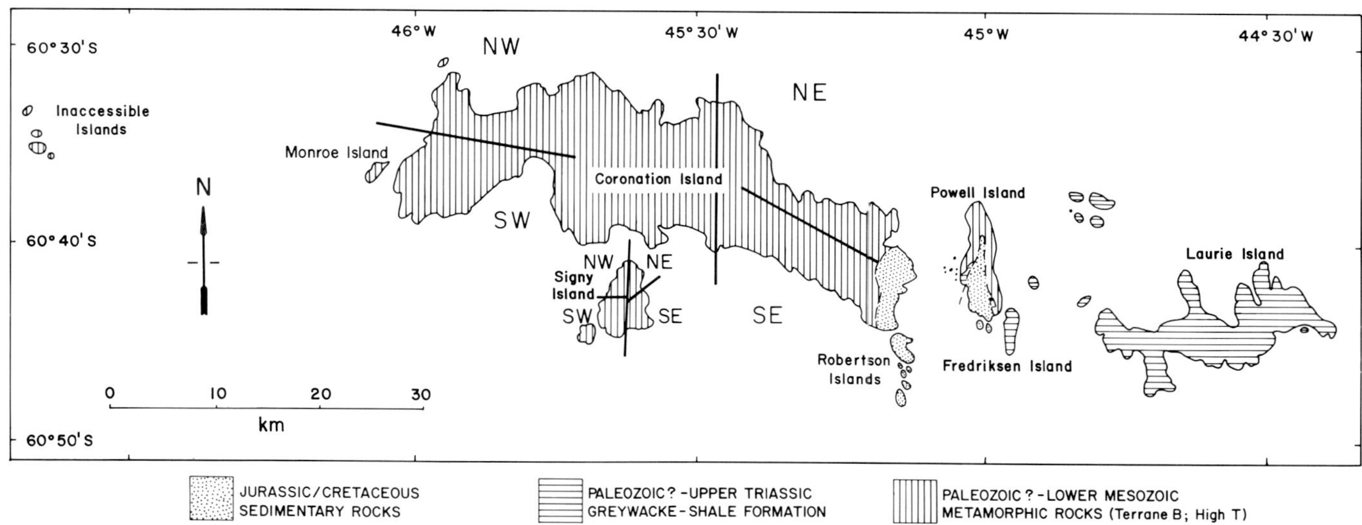

Figure 6—Geologic map of the South Orkney Islands (from Dalziel, 1984a).

Hence, the incipient South Shetland Islands microcontinent is an excellent example of a terrane that could eventually be identifiable along the Pacific margin of the Antarctic Peninsula as parautochthonous after the closing of Bransfield Strait. Its docking would undoubtedly result in deformation and orogenic uplift along the line of the former rift, but the deformation would be a function of the former existence of the present-day rift, a zone of weakness, not a function of original wide separation of a docked exotic terrane and the continental margin.

Should the plate tectonic regime along the Pacific margin of the South Shetland Islands change other than by the reinitiation of essentially downdip subduction, then the South Shetland Islands crustal fragment might be moved laterally from its present position to one more analogous to that of the South Orkney Islands microcontinent. This latter terrane, now exotic with respect to its parent continent, does not seem likely to become reunited with the Antarctic Peninsula in any simple way. Rather, it seems to be a piece of continental "flotsam or jetsam" caught up in the complex transform boundary between two major plates, the Antarctic and South American plates. It is not easy to predict where or when it will eventually dock against a major continent, but most likely if this is to happen in the future, it will be recognizable as an allochthonous terrane, if not because of its lithology and geologic history, then because of the fauna and flora of its rocks and/or their magnetic record; for example that of the mafic dikes cutting the metamorphic complex and its Mesozoic cover sequence.

Finally, it is of interest in the context of this volume to consider the history that might be deduced for these two crustal fragments in a future resting place. The rocks of the South Shetland Islands might correctly be recognized as a mid-late Cenozoic magmatic arc terrane founded on an older subduction complex (southern and western Elephant Island Group) with younger subduction-related terranes (Smith Island, northern Elephant Island, and Clarence Island) incorporated. The South Orkney Islands terrane would be diagnosed as a forearc terrane unconformably overlain by Upper Jurassic to Lower Cretaceous strata. The fact that both microcontinents had themselves been derived from a single magmatic arc (the Antarctic Peninsula), one of several terranes forming a major continent, could very likely go unrecognized.

PALEOMAGNETIC DATA

Solutions to many of the problems concerning Antarctic terranes should clearly be sought with the help of systematic and detailed paleomagnetic studies. The difficulty of doing paleomagnetic work in the Antarctic, however, can be readily imagined even by those unfamiliar with working in such extreme conditions. The immense difficulties of access to critical areas, along with frigid temperatures and high winds, crevasse hazards, and time limitations, often wreak havoc with a previously planned systematic sampling program. Many areas have only been visited by geologists once or twice, and usually without a paleomagnetist. Indeed, some potentially critical areas of outcrop have yet to be visited. Few areas have detailed mapping and geochronologic control.

The more recent paleomagnetic studies, however, have been undertaken on rocks of known age and field relations. Most recently, new data on rocks from West Antarctica have been published, with reviews of the older literature, by Longshaw and Griffiths (1983) and by Grindley and Oliver (1983).

Antarctic Peninsula

The Antarctic Peninsula has had semicontinuous igneous activity since the Triassic (Thomson et al, 1983). The possibility of resetting of the magnetization because of later igneous activity therefore constitutes an added problem. Most of the paleomagnetic work has been on intrusives and volcanics, and the majority of studies undertaken to date have determined Antarctic Peninsula paleopoles for the Cretaceous and Tertiary. These are not significantly different from those obtained from Tertiary rocks on the East Antarctic craton (Turnbull, 1959; Delisle, 1983). This implies that there has been little relative movement between the Antarctic Peninsula and the East

Antarctic craton over the past 100 m.y. (Kellogg and Reynolds, 1978; Valencio et al, 1979; Watts, 1982; Watts et al, 1984).

Most recently Longshaw and Griffiths (1983) collected three or more samples per site from granitic rocks approximately 178 m.y. old, acid to basic dikes 174 m.y. old, and lavas approximately 160 m.y. old, all Jurassic. They used alternating field (AF) and stepwise thermal demagnetization to obtain a stable component. They did not apply a dip correction to the measurements because a test assuming the dikes were intruded vertically did not improve the clustering of directions. By averaging the results from all of the sites, Longshaw and Griffiths determined a mean paleopole at 238° E, 48° S, while the mean Jurassic paleopole for the now widely accepted Gondwanaland reconstruction of Norton and Sclater (1979) is 219° E, 57° S. This suggests limited motion of the Antarctic Peninsula relative to the East Antarctic craton during or since the Jurassic. For example, one possibility is a clockwise rotation of 15° about a pole located at 0° E, 65° S (Longshaw and Griffiths, 1983, p. 952).

Ellsworth Mountains–Whitmore Mountains

The Ellsworth Mountains consist of an uppermost Precambrian(?) or lower Paleozoic to Permian sedimentary succession. The rocks were affected by a syn- or post-Permian deformational event (the Gondwanide orogeny). They are similar in lithologic character and age to rocks found in the Transantarctic Mountains (Schopf, 1969). The Antarctic Peninsula, on the other hand, contains no known Upper Precambrian through Permian sediments.

The only published study on the paleomagnetism of rocks from the Ellsworth Mountains–Whitmore Mountains block is that by Watts and Bramall (1981). They studied Upper Cambrian hematitic argillites from the Ellsworth Mountains, collecting cores from five sites located at various positions around a syncline. The samples were demagnetized by stepwise thermal heating and AF methods which showed that the magnetization was prefolding in origin. The remnance of the samples resulted in a shallow inclination, suggesting deposition in equatorial regions. However, the cones of 95% confidence are rather high (α_{95} between 11 and 26) and Fisher's precision parameter is low (k between 6 and 25). Watts and Bramall assumed that the isolated magnetic components from the argillites are of reversed polarity since Late Cambrian data from North America are mostly of reversed polarity. Based on this assumption, they obtained a paleomagnetic south pole at 296° E, 4° N, which is 70° away from the Gondwanaland polar wander path for the early Paleozoic. Clockwise rotation of the Ellsworth Mountains by 82° would not only align the paleopole with those of "stable" Gondwanaland, but would also align the structural trend with that of the equivalant rocks in the Transantarctic Mountains along the margin of East Antarctica. The remarkable similarity of the stratigraphic sections in these two localities argues strongly for an original location for the Ellsworth Mountains close to that margin, such as that speculatively shown on Figure 3.

Marie Byrd Land

Only two paleomagnetic studies have been published on rocks from Marie Byrd Land. The first by Scharnberger and Scharon (1972) gave ambiguous results. There were 4 sites out of 27 in granitic plutons and 5 sites out of 5 in basalt and diabase dikes that were suitable for paleomagnetic study. They collected three samples per site and cleaned them by AF demagnetization and then determined site means by giving unit weight to each specimen. Two of the granitic samples yielded ages of 116 ± 10 m.y. and 118 ± 6 m.y. (Rb–Sr whole rock and biotite, respectively), and they suggested a virtual geomagnetic pole for the Early Cretaceous at 116° E, 36° S. This would be indicative of an exotic origin for Marie Byrd Land with respect to East Antarctica. The results appear unreliable to us, however, partly because of the low number of samples per site and partly because only two out of four granitic rocks and none of the five mafic rocks were dated.

A more recent study by Grindley and Oliver (1983) on 26 sites in Upper Cretaceous rhyolites and mafic dikes has better age and geologic constraints on the samples. They cleaned three to eight samples per site by AF methods and calculated a paleomagnetic pole position for the Late Cretaceous at 241° E, 66° S based on a good grouping of paleomagnetic directions. These findings indicate that there could have been extension of 200 to 500 km (125–310 mi) between the East Antarctic margin and Marie Byrd Land since the Early Cretaceous and that 10 to 45° of rotation of Marie Byrd Land in that time interval is also possible. A 10° tilt correction could, they note, obviate the need for any extension or rotation.

Thurston Island–Eights Coast

Only one paleomagnetic study has been done on rocks from the Thurston Island–Eights Coast area. Scharnberger and Scharon (1982) sampled "granitic and dioritic" plutons, basaltic volcanics, and mafic dikes over a wide area. They had little age control, cleaned the samples only in 100 Oe fields, and combined results from similar lithologies even though the sampling localities were far apart. A pole similar to that of the present day was obtained, but given the lack of age control and limited cleaning this is of uncertain significance.

DISCUSSION AND CONCLUSIONS

It is clear that the terrane concept has significant applicability to several aspects of the geology of the Antarctic segment of the circum-Pacific mobile belt. Recognition that West (Lesser) Antarctica consists of several microcontinental composite terranes has significance far beyond the confines of Antarctic geology. There are paleoenvironmental and paleobiogeographic as well as global plate tectonic implications (Dalziel and Elliot, 1982). In addition, the accretionary forearc terrane of the Gondwanaland margin is part of one of the world's biggest fossil subduction complexes extending from central Chile to New Zealand. It is potentially an extremely valuable laboratory for the study of subduction-related processes such as underplating. The present-day South

Orkney Islands and South Shetland Islands microplates provide valuable examples of the birth of terranes, and it is interesting to predict their future and to speculate on the interpretations that would be made if they were eventually to be found incorporated within an orogenic belt.

Major steps forward in the understanding of the West Antarctic terranes will come from paleontologic studies of all suitable rocks, geophysical surveys of the ice- and sea-covered areas intervening between the nunataks and islands, and, most immediately, paleomagnetic work. Despite the problems inherent in the latter in the Antarctic environment, an attempt to shed significant light on the relationship between the four major West Antarctic microcontinental terranes is under way. Together with our colleagues Drs. Robert Pankhurst and Bryan Storey of the British Antarctic Survey, we are involved in a study of potentially useful rocks in the Ellsworth Mountains–Whitmore Mountains, Thurston Island–Eights Coast, and Antarctic Peninsula terranes together with some from the Transantarctic Mountains. One field season has already been undertaken in the Transantarctic Mountains–Ellsworth Mountains–Whitmore Mountains area (Fig. 2). A second will take place during the 1984–1985 austral field season in the Thurston Island–Eights Coast region. This work is part of a joint project that will provide detailed control on the field relations and geochronology of the paleomagnetic samples (Dalziel and Pankhurst, 1984). We hope that preliminary results will be available during 1985.

ACKNOWLEDGMENTS

Our work in Antarctica, including the preparation of this paper, is supported by the Division of Polar Programs, National Science Foundation, under Grant No. DPP 82 13798 to Ian W. D. Dalziel. Anne M. Grunow is supported by an ARCO graduate fellowship from the Department of Geological Sciences of Columbia University.

REFERENCES

Barker, P. F., and I. W. D. Dalziel, 1983, Progress in geodynamics of the Scotia Arc region, *in* R. S. J. Cabre, ed., Geodynamics of the eastern Pacific Region, Caribbean and Scotia Arcs: American Geophysical Union Geodynamic Series, v. 9, p. 137–170.

______, et al, 1976, The evolution of the southwestern Atlantic Ocean Basin: Leg 36 data, *in* Initial Reports of the Deep Sea Drilling Project, v. 36: Washington, DC, U.S. Government Printing Office.

Dalziel, I. W. D., 1981, Back-arc extension in the southern Andes: a review and critical reappraisal: Philosophical Transactions of the Royal Society of London, v. A300, p. 319–335.

______, 1982, Pre-Jurassic history of the Scotia Arc region, *in* C. Craddock, ed., Antarctic geoscience: Madison, University of Wisconsin Press, p. 111–126.

______, 1984a, Tectonic evolution of a forearc terrane, southern Scotia Ridge, Antarctica: Geological Society of America Special Paper 200, 32 p.

______, 1984b, A geologic transect through the southernmost Andes: report of R/V *Hero* Cruise 83/4: Antarctic Journal of the United States, v. 18, p. 8–12.

______, in press, Collision and cordilleran orogenesis: An Andean perspective: Geological Society of London, Special Publication on Collision Tectonics.

______, and D. H. Elliot, 1982, West Antarctica: problem child of Gondwanaland: Tectonics, v. 1, p. 3–19.

______, and R. J. Pankhurst, 1984, Tectonics of West Antarctica and its relation to East Antarctica: joint USARP-British Antarctic Survey geology/ geophysics project: Antarctic Journal of the United States, v. 19, n. 5.

______, et al, 1981, Triassic microfossils from the South Orkney Islands, Scotia Ridge and their geological significance: Geological Magazine, v. 118, p. 15–25.

Delisle, G., 1983, Results of paleomagnetic investigations in Northern Victoria Land, Antarctica, *in* R. L. Oliver, et al, eds., Antarctic earth science: Canberra, Australian Academy of Science, p. 146–149.

de Wit, M. J., 1977, The evolution of the Scotia Arc as a key to the reconstruction of southwestern Gondwanaland: Tectonophysics, v. 37, p. 58–81.

Doake, C. S. M., et al, 1983, Sub-glacial morphology between Ellsworth Land and Antarctic Peninsula: New data and tectonic significance, *in* R. L. Oliver, et al, eds., Antarctic earth science: Canberra, Australian Academy of Science, p. 270–273.

Ferguson, D., 1921, Geological observations in the South Shetlands, the Palmer Archipelago, and Graham Land, Antarctica: Proceedings of the Royal Society of Edinburgh, v. 53/1, p. 29–36.

Grindley, G. W., and P. J. Oliver, 1983, Paleomagnetism of Cretaceous volcanic rocks from Marie Byrd Land, *in* R. L. Oliver, et al, eds., Antarctic earth science: Canberra, Australian Academy of Science, p. 573–578.

Harrington, P. K., et al, 1972, Crustal structure of the South Orkney Islands area from seismic refraction and magnetic measurements, *in* R. J. Adie, ed., Antarctic geology and geophysics: Oslo, Universitetsforlaget, p. 27–32.

Howell, D. G., and D. L. Jones, 1984, Tectonostratigraphic terrane analysis and some terrane vernacular, *in* D. G. Howell, et al, eds., Proceedings of the Circum-Pacific Terrane Conference: Stanford University Publications, Geological Sciences, v. 18, p. 6–9.

______, et al, 1984, Preliminary tectonostratigraphic terrane map of the circum-Pacific region, *in* D. G. Howell, et al, eds., Proceedings of the Circum-Pacific Terrane Conference: Stanford University Publications, Geological Sciences, v. 18, p. 227–242.

Hyden, G. M., and P. W. G. Tanner, 1981, Late Paleozoic-early Mesozoic forearc basin sedimentary rocks at the Pacific margin in Western Antarctica: Geologische Rundschau, v. 70, p. 529–541.

Jankowski, E. J., and D. J. Drewry, 1981, The structure of West Antarctica from geophysical studies: Nature, v. 291, p. 17–21.

Kellogg, K. S., and R. L. Reynolds, 1978, Paleomagnetic results from the Lassiter Coast, Antarctica and a test for oroclinal bending of the Antarctic Peninsula: Journal of Geophysical Research, v. 83, p. 2293–2299.

Longshaw, S. K., and D. H. Griffiths, 1983, A paleomagnetic study of Jurassic rocks from the Antarctic Peninsula and its implications: Quarterly Journal of the Geological Society of London, v. 140, p. 945–954.

Natural Environment Research Council, 1982, Annual report of the British Antarctic Survey: Cambridge, England.

Norton, I. O., and J. G. Sclater, 1979, A model for the evolution of the Indian Ocean and the breakup of Gondwanaland: Journal of Geophysical Research, v. 84, p. 6803–6830.

Scharnberger, C. H., and L. Scharon, 1972, Paleomagnetism and plate tectonics of Antarctica, *in* R. J. Adie, ed., Antarctic geology and geophysics, Oslo, Universitetsforlaget, p. 843–847.

———, 1982, Paleomagnetism of rocks from Graham Land and western Ellsworth Land, Antarctica, *in* C. Craddock, ed., Antarctic geoscience: Madison, University of Wisconsin Press, p. 371–375.

Schopf, J. M., 1969, Ellsworth Mountains: position in West Antarctica due to sea floor spreading: Science, v. 164, p. 63–66.

Smellie, J. L., 1981, A complete arc-trench system recognized in Gondwana sequence of the Antarctic Peninsula region: Geological Magazine, v. 118, p. 139–159.

Storey, B. C., and A. W. Meneilly, 1983, Melange within subduction-accretion complex rocks of Fredriksen Island, South Shetland Islands: Geological Magazine, v. 120, p. 555–566.

———, et al, 1977, The occurrence of Mesozoic oceanic floor and ancient continental crust on South Georgia: Geological Magazine, v. 114, p. 203–208.

Tanner, P. W. G., et al, 1982, Radiometric evidence for the age of the subduction complex in the South Orkney and South Shetland Islands, West Antarctica: Quarterly Journal of the Geological Society of London, v. 139, p. 683–690.

Thomson, M. R. A., et al, 1983, The Antarctic Peninsula—a Late Mesozoic-Cenozoic Arc, *in* R. L. Oliver, et al, eds., Antarctic earch science: Canberra, Australian Academy of Science, p. 289–294.

Turnbull, G., 1959, Some paleomagnetic measurements in Antarctica: Arctic, v. 12, p. 151–157.

Valencio, O. A., et al, 1979, Paleomagnetism and K–Ar age of Mesozoic and Cenozoic igneous rocks from Antarctica: Earth and Planetary Science Letters, v. 45, p. 61–68.

Watts, D. R., 1982, Potassium-argon and paleomagnetic results from King George Island, South Shetland Islands and the Antarctic Peninsula, *in* C. Craddock, ed., Antarctic geoscience: Madison, University of Wisconsin Press, p. 255–261.

———, and A. M. Bramall, 1981, Paleomagnetic evidence for a displaced terrain in western Antarctica: Nature, v. 293, p. 638–641.

———, et al, 1984, Cretaceous and early Tertiary paleomagnetic results from the Antarctic Peninsula: Tectonics, v. 3, p. 333–346.

Wordie, J. M., 1921, Shackleton Antarctic Expedition 1914–17: geological observations in the Weddell Sea area: Transactions of the Royal Society of Edinburgh, v. 53/1, p. 17–27.

Andean Evolution and the Terrane Concept

Ian W. D. Dalziel
Lamont-Doherty Geological Observatory of Columbia University
Palisades, New York

Randall D. Forsythe
Rutgers University
New Brunswick, New Jersey

From the point of view of terrane analysis, three regimes can be recognized in the tectonic history of the Pacific margin of South America south of Ecuador. Prior to the Mesozoic breakup of Gondwanaland, terranes were accreted as material including sea mounts and possibly microcontinents scraped from the downgoing plate along the continental margin subduction zone. Accretion of allochthonous terranes is not known to have occurred during the late Mesozoic and Cenozoic evolution of the present Andean Cordillera. The neotectonic regime involves movement of fault blocks and the separation of at least one microcontinental terrane (the South Georgia platform). Accretion is limited to the development of a comparatively narrow forearc wedge.

INTRODUCTION

South America has posed a puzzle for terrane analysts (Howell et al, 1984). First, there is a dearth of suspect terranes, let alone of bona fide allochthonous terranes compared with North America and the rest of the circum-Pacific mobile belt. Second, the Andes form an imposing cordillera 10,000 km (6,000 mi) long and over 6 km (20,000 ft) high, seemingly formed for the most part by tectonic and igneous processes related to subduction of oceanic lithosphere. The very existence of this noncollisional mechanism of orogenesis, so-called Andean-type, has repeatedly been questioned by some devotees of the terrane concept in recent years (e.g., Nur and Ben-Avraham, 1977, 1982, 1983). Some rocks along the Pacific margin of the South American continent are obvious candidates for allochthonous terranes whose collisional docking could have triggered tectonism leading to at least local Andean orogenesis. Yet south of Ecuador (i.e., south of the Golfo de Guayaquil and the Huancabamba deflection at approximately 5° S latitude, Fig. 1), all of the supposedly "allochthonous terranes" of the Pacific margin (e.g., Nur and Ben-Avraham, 1977, 1982, 1983) were certainly part of the continent prior to the Mesozoic to Cenozoic development of the Andean Cordillera. Terrane accretion therefore has not played a critical role in Andean orogenesis.

In this review, we take a brief look at Andean evolution in the light of the terrane concept, and vice versa. We distinguish three periods in the history of the Pacific margin of South America: (1) the period prior to the initiation of Gondwanaland fragmentation and Andean orogenesis in the early Mesozoic; (2) the time of development of the Andean Cordillera in the Mesozoic and Cenozoic; and (3) the neotectonic regime. We concentrate our discussion on the Andes south of the Huancabamba deflection, partly because we are most familiar with the geology and literature of Chile (especially), Argentina, and Peru. It is clear, however, that a different tectonic situation exists in the northern Andes of Ecuador, Colombia, and Venezuela where rocks of oceanic or island arc affinities are located west of the Romeral suture along the immediate Pacific margin (Stibane, 1980, 1981; Pitcher, 1983; see Fig. 1).

Allochthonous terranes were accreted to the craton of the South American segment of Gondwanaland in the latest Precambrian, Paleozoic, and early Mesozoic prior to breakup of the supercontinent. The accretion occurred both in the form of docking of microcontinents and sea mounts and in the form of offscraping of pelagic sediments and slivers of oceanic lithosphere along a Pacific margin subduction zone. There is no evidence, in our view, that development of the present Andean Cordillera in Triassic through Cenozoic time resulted from terrane accretion. Rather, the orogenesis seems to have been triggered by the closing of a composite backarc basin in the mid-Cretaceous. The neotectonic regime involves major fracture development, relative movement of fault blocks, and terrane dispersal involving the separation of at least one microcontinent from the South American continental lithosphere. Accretion was limited to development of a narrow forearc wedge composed mainly of offscraped oceanic sediment.

GONDWANALAND MARGIN DEVELOPMENT

The continuity of the Paleozoic to early Mesozoic orogenic belts (the Gondwanides) along the ancestral Pacific (Panthalassic) margin of the Gondwanaland supercontinent was first clearly expounded by du Toit (1937). He drew analogies between parts of the Gondwanides and various tectonic settings of the present Pacific Ocean margin, acknowledging the possibility that outer platforms or "island festoons" (du Toit, 1937, p. 58) could have been incorporated into the Gondwanide belt. Du Toit's "island festoons" are analogous to "suspect terranes" in today's literature. For the next 30 years or more, however, the geology of the South American

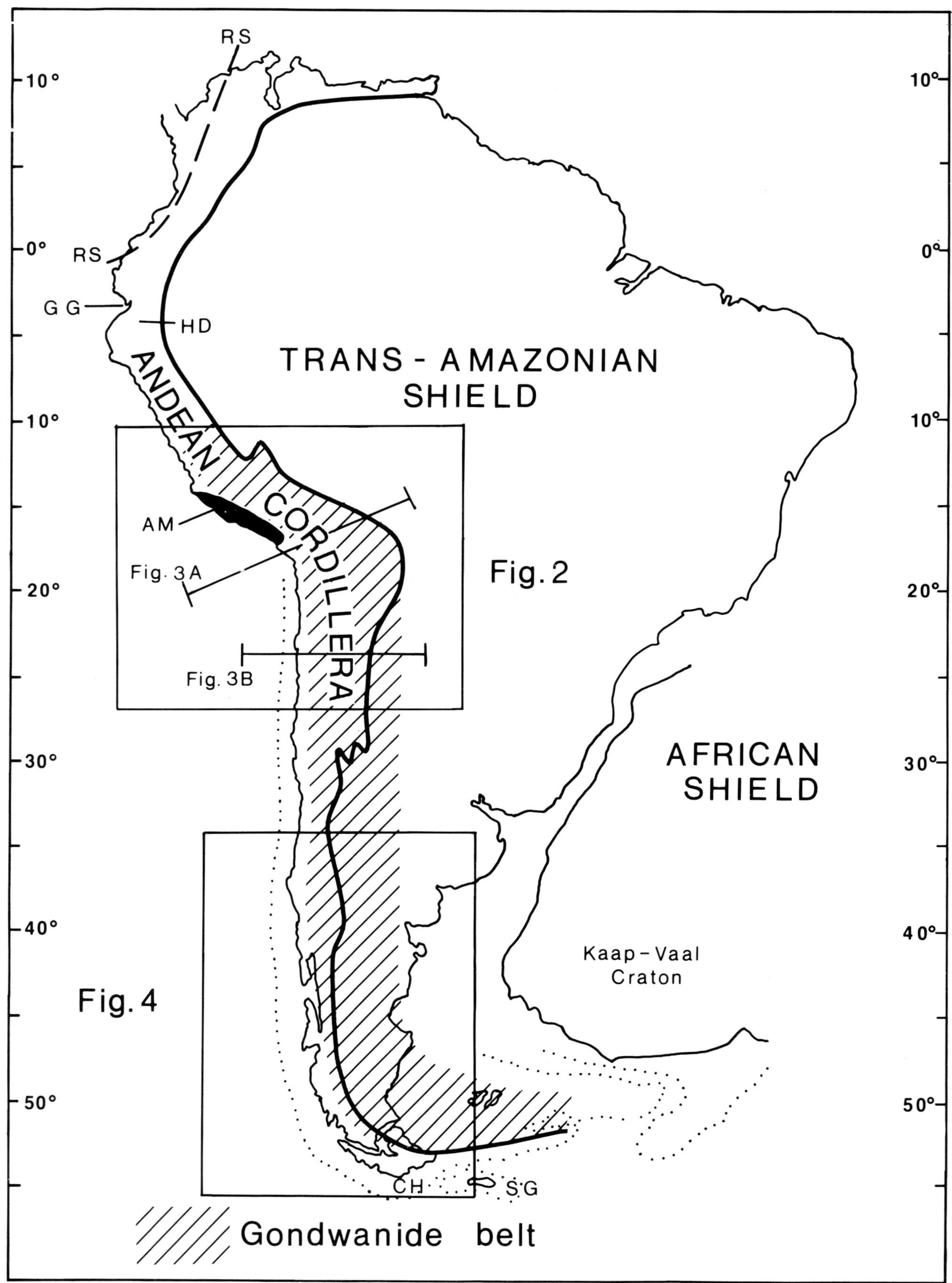

Figure 1—Location map of South America. AM = Arequipa massif; CH = Cape Horn; GG = Golfo de Guayaquil; HD = Huancabamba deflection; RS = Romeral suture.

orogenic regions was interpreted according to "stabilistic" concepts. These concepts formed the basis of paleogeographic reconstructions such as those of Weeks (1947) and Harrington (1962). Even as late as the 1970s, the presence of marine rocks in the Paleozoic metamorphic belt along the coast of Chile, thought by others to be part of a subduction complex (Aguirre et al, 1972), was taken to indicate the extent of the Paleozoic South American epicontinental sea.

In the past 15 years, however, large-scale programs of regional mapping and radiometric dating have been undertaken and the results interpreted according to more actualistic concepts of circum-Pacific tectonics. Several working hypotheses that to varying degrees invoke plate tectonic and tectonostratigraphic terrane analysis concepts have emerged. We shall briefly review three such hypotheses that pertain to three geographic segments of the South American Gondwanides: (1) The Paleozoic geology of the Eastern Cordillera of Peru and Bolivia has been interpreted in terms of an ensialic orogen. (2) A continental margin like that of present-day Japan has been proposed for the northern portions of Chile and Argentina in the early to mid-Paleozoic. (3) A margin with a wide forearc accretionary prism incorporating numerous potentially far-traveled exotic terranes has been proposed for the late Paleozoic to early Mesozoic in southernmost South America. The compatibility of these hypotheses and their implications will be discussed later.

Peru and Bolivia: An Ensialic Orogen

The early to middle Paleozoic record of sedimentation and deformation preserved in the Eastern Cordillera (Cordillera Oriental) of Peru and Bolivia has led to a two-stage model of an ensialic orogen (Dalmayrac et al, 1980; Carlier et al, 1982). The first stage is represented by deposition of Cambrian(?), Ordovician, and Silurian marine strata in a basin located between the Precambrian Arequipa massif to the west and the Brazilian portion of the Transamazonian shield to the east (Fig. 2). The strata within this basin thin towards and onlap the Precambrian massifs (Dalmayrac et al, 1980). Sedimentation is thought to have been initiated by a Cambrian phase of rifting within the Precambrian basement, although the lowest cover strata are unfossiliferous. Subsidence, sedimentation, and volcanism in the basin continued into the Devonian (Dalmayrac et al, 1980).

The second stage envisaged by the above-mentioned authors is one of basin inversion. This is recorded by deformation and uplift of the lower to middle Paleozoic basin sequence and marked by an angular unconformity. Carboniferous and Permian marine and subaerial sedimentary and volcanic strata that have only been tilted and block-faulted rest on the tightly folded and cleaved older rocks (Dalmayrac et al, 1980).

The two-stage evolution of the Eastern Cordillera of Peru and Bolivia is hence viewed by the above-cited authors as the creation and destruction of an ensialic basin (see Fig. 3a). Subsidence is thought to have been driven by crustal rifting and attenuation initiated sometime in the Cambrian. The middle to late Paleozoic deformation is not viewed as the product of collision tectonics beyond the shortening of previously distended continental crust.

The location of the Eastern Cordillera between the Arequipa massif and the Brazilian shield makes it an obvious place to search for a cryptic suture. As noted above, however, accretion of the Arequipa massif to the Brazilian segment of the Gondwanaland shield does not seem to have occurred during the Paleozoic. The lower Paleozoic strata thin towards the massif, and the rocks of the latter fall into the general age pattern of the rocks in the Transamazonian shield (Shackleton et al, 1979). Within the Eastern Cordillera of Peru there are small gabbro, norite, anorthosite and serpentinized dunite, harzburgite, and pyroxenite bodies (Aumaitre et al, 1977). On the basis of regional relationships, these are believed to be Upper Precambrian to Cambrian (Carlier et al, 1982).

The most likely hypothesis at present is that these mafic and ultramafic slivers represent a Late Precambrian suture of the Arequipa massif to the Brazilian shield. This would be consistent with the only available paleomagnetic data obtained from specimens of the sedimentary cover of the Arequipa massif. Although limited, the data suggest no latitudinal motion relative to cratonic South America at least since the Devonian (Heki et al, 1983; Knight et al, 1984). The mafic and ultramafic rocks should be studied in an attempt to test this hypothesis by geochronology.

Northern Chile and Argentina: A Japan-Type Margin

Cambrian and Ordovician sedimentation in northern Chile and Argentina took place in three tectonic environments (Fig. 2). In the east, sedimentation occurred in an epicratonic seaway across the Argentina–Paraguay border, reaching its maximum extent probably during the Llanvirnian (approximately 470 Ma) (Harrington, 1962; Suarez-Soruco, 1976; Turner, 1972). To the west in the Precordillera and Puna, Cambrian to Ordovician strata accumulated on a west-facing slope or continental shelf. These strata are correlative with those on the eastern margin of the ensialic basin in Peru. They do not, however, show any reversal in thickness or facies to the west. Here there is no apparent western margin to the marine basin. As in Peru, the sequence is initiated in the Cambrian with deposition of conglomerates and sandstones (Meson Group; Turner, 1960) over a marked angular unconformity on Precambrian metamorphic rocks, or over a nonconformity developed on upper Precambrian granites. It is argued, as for Peru, that a rift-related phase of subsidence commenced sometime in the Cambrian. Unlike the Peruvian zone, however, the rift event here appears to have led to the development of a more mature ocean-continent margin. Shallow water, neritic shelf facies, and distal or slope deposits all indicate a west-facing paleoslope. In further support of the developed nature of this continent/ocean transition, Borello (1972) has shown that the calcareous units (dolomites and limestones) of the Early Ordovician shelf sequence in central-western Argentina are bordered to the west by a "eugeosynclinal" belt of deposition (from about 29° S to 34° S). Here graywacke and shale units are tectonically(?) mixed with ophiolitic material (see Fig. 2). At these latitudes, therefore, the shelf was bordered to the west by a basin floored at least in part by oceanic crust.

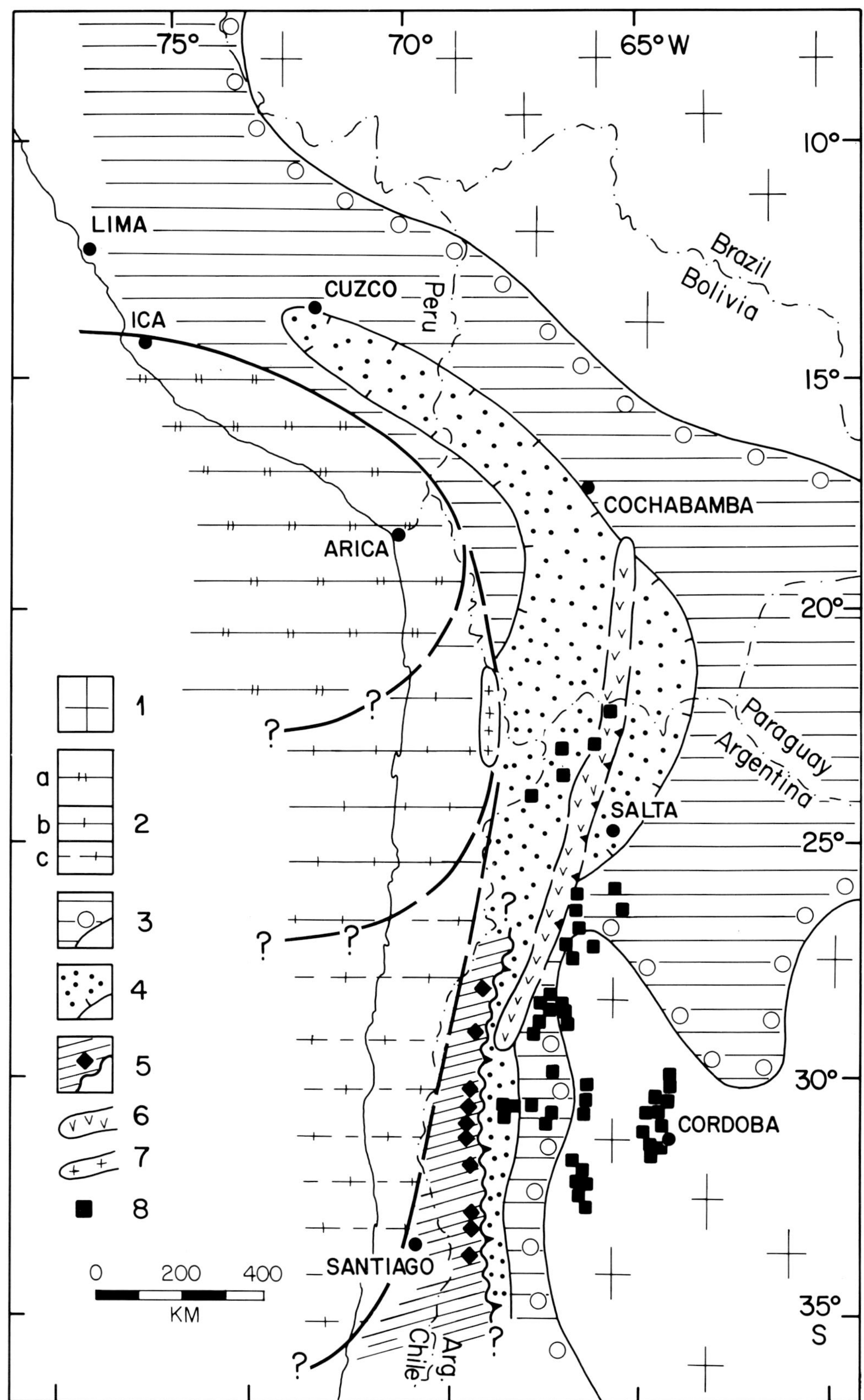

Figure 2—Sketch map of Paleozoic geology in Peru, Bolivia, northern Chile, northern Argentina, and western Paraguay. For location see Figure 1. 1 = Precambrian rocks of the Transamazonian shield; 2a, b, and c = three hypothetical southern extensions of the Precambrian Arequipa massif; 3 = approximate extent of Middle Orodovician shelf sedimentation; 4 = axial zone of Ordovician shelf sedimentation; 5 = "eugeosynclinal" belt of flysch, chert, and mafic igneous rocks; 6 = "Faja eruptive de la Puna" (a belt of Silurian volcanics); 7 = zone of Ordovician granitoids; 8 = distributed localities of Ordovician and Silurian intrusions.

a.(?) An ensialic orogen for the Eastern Cordillera of Peru and Bolivia

L. Devonian - E. Carboniferous uplift and deformation

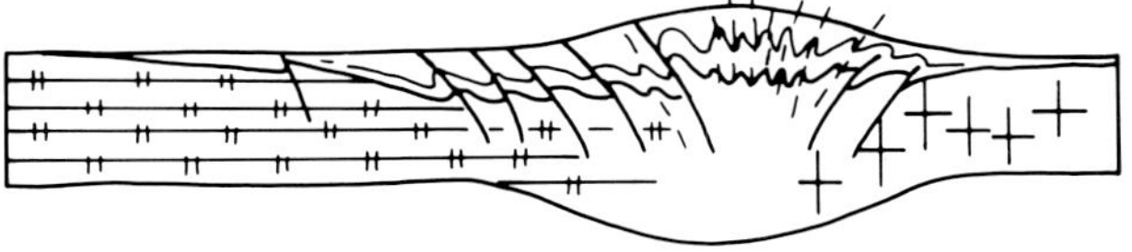

(?) Cambrian and Silurian rifting and subsidence

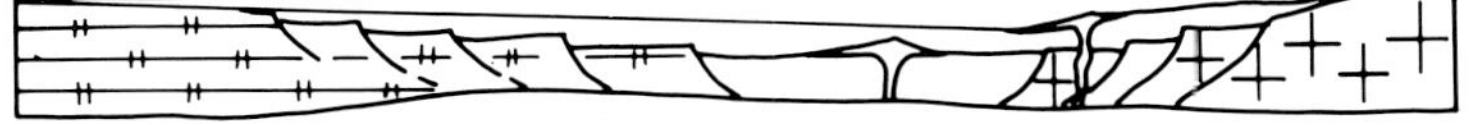

(?) L. Precambrian collision of Arequipa Massif

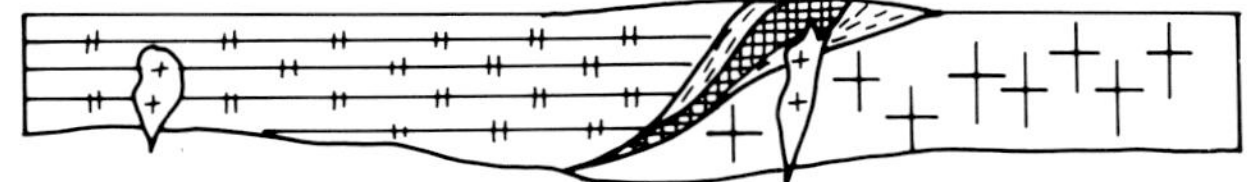

b. Japan type margin - Antler type orogen for Northern Chile and Argentina

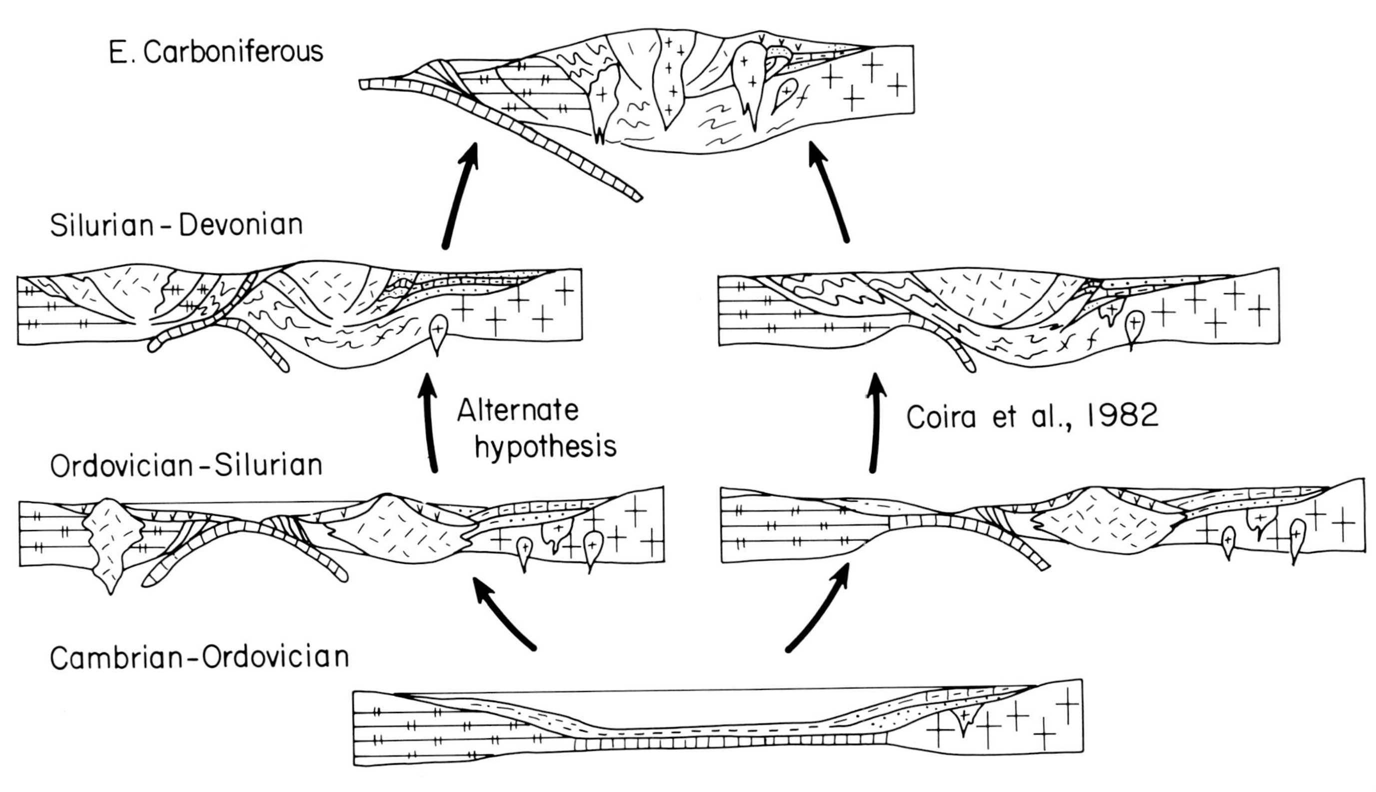

Figure 3—(**a**) Cartoons showing a hypothesis for the development of the Panthalassic margin of Peru and Bolivia. (After Dalmayrac et al, 1980, Travaux et Documents de L'ORSTROM 122.) (**b**) Cartoons showing alternative hypotheses for the development of the Panthalassic margin of northern Chile and northern Argentina (see text for discussion).

The passive margin or shelf stage of sedimentation appears, however, rather short lived. Prior to the Late Ordovician, or Early Silurian, an intense phase of calc-alkaline magmatism had started within the previous zone of shelf deposition. This magmatism is well displayed, for example, in the "Faja Eruptiva de la Puna" (Ramos and Ramos, 1979) that can be traced from the Precordillera of northwest Argentina into the Eastern Cordillera of Bolivia (see Fig. 2). While this distinctive belt of calc-alkaline rocks is only found in tectonic contact with sedimentary strata deposited along its eastern margin, plutons of similar age and composition are widely distributed throughout and to the east of the shelf zone (see Fig. 2). The Silurian to Devonian strata that continued to accumulate to the east of the "Faja Eruptiva" reflect the transformation of the lower Paleozoic west-facing shelf into a foreland basin isolated between the Silurian arc and the Paraguayan shield. Thus, in contrast to the early to middle Paleozoic evolution of the Eastern Cordillera of Peru, the rifted margin in the northwest Argentina zone was transformed during the Late Ordovician to Early Silurian into an active margin with a west-facing subduction zone. This system had forearc units represented by the eugeosynclinal assemblages, magmatic arc units represented by the "Faja Eruptiva de la Puna" and other distributed plutons, and backarc or foreland components represented by the Silurian to Devonian strata of northwest Argentina.

Much like the Peruvian segment of the Gondwanides, a culminating phase of deformation affected the northwest regions of Argentina during the Late Devonian or Early Carboniferous (Chanic phase of Coira et al, 1982). Following this, the regions to the east of the "Faja Eruptiva de la Puna" were primarily the site of Carboniferous to Permian continental deposition. To the west, in northern Chile, there are isolated blocks of pre-Devonian metamorphic (gneissic) basement. One of these has been dated in the Arica area as Precambrian (Pacci et al, 1980). Therefore, one hypothesis is that the culminating phase of deformation in the Late Devonian or Early Carboniferous represents the closure of a marginal basin between a southern extension of the Arequipa massif and the shelf sequence of northwest Argentina (Fig. 3). Recently, however, Ordovician plutons and pre-Devonian volcanics have been discovered in the Atacama desert and Puna of northern Chile (latitude 24° S; Mpodozis et al, 1983). These lie to the northwest of the distal Ordovician slope deposits and suggest that a double arc system may have been developed at these latitudes. Thus, an alternate hypothesis is that closure of the oceanic basin occurred along at least two subduction zones, perhaps, as suggested in Figure 3, with opposing polarities.

Because of the sparsity of pre-Devonian exposures in Chile, the nature of terranes and terrane movements responsible for the culminating phase of deformation in the Gondwanide fold belt of northern Chile and Argentina is difficult to determine. At present it appears that this Early Carboniferous phase represents the final docking phases of an outer pre-Devonian basement terrane, which itself could have been a parautochthonous southern extension of the Arequipa massif. Late Devonian to Early Carboniferous plutons intrude many of the older terranes (Herve et al, 1980; Coira et al, 1982). These younger plutons are undeformed, yet they cut penetratively deformed lower and middle Paleozoic strata, suggesting that the lower Paleozoic terranes had docked by the Late Devonian.

Southern South America: A Late Paleozoic to Early Mesozoic Andean-Type Margin

It is difficult to extrapolate the early and middle Paleozoic tectonic environment of northern Chile and Argentina, let alone of Bolivia and Peru, into southern South America where there are very few exposures of pre-Carboniferous rock. There are, however, sufficient occurrences of upper Paleozoic and lower Mesozoic rocks to delineate tectonic provinces in the south. For Late Devonian to Early Triassic times, three such provinces can be distinguished. These appear to represent, from west to east, the forearc, magmatic arc, and backarc environments of an Andean-type convergent margin along the Panthalassic margin of the continent (Forsythe, 1982; see Fig. 4).

The autochthonous nature of the backarc deposits and of the magmatic arc rocks seems clear. The backarc sedimentary sequences rest with striking angular unconformity on Precambrian metamorphic basement, for example in the Falkland (Malvinas) Islands (Greenway, 1972) and the Falkland Plateau (Barker et al, 1976). The rocks contain detritus derived from both cratonic and magmatic arc sources. The arc rocks form a magmatic province comparable to the Mesozoic–Cenozoic Andean suite (Forsythe, 1982). The upper Paleozoic to lower Mesozoic plutons and volcanics intrude and overlie lower Paleozoic rocks independent of age and trend. No ophiolitic material of similar or younger age has been documented from the continental side of this magmatic belt.

The forearc terrane of this late Paleozoic margin, on the other hand, contains abundant ophiolitic material. Strictly speaking, this forearc is composed of hundreds of small accreted terranes. We subdivide the forearc into zones representing in situ forearc basins and those composed of accreted forearc basement. This twofold division is shown on Figure 4. The perched basin elements are located between the accretionary belt and the arc. They contain detritus derived from the underlying basement as well as the arc, contain no oceanic material, and are generally less deformed than material in the underlying accretionary prism.

The accretionary or subduction complex is structurally and stratigraphically complex. Viewed in its entirety, it forms a mass similar in size to the Franciscan of California and like the Franciscan is likely to record at least 100 m.y. of subduction-related accretion, deformation, and metamorphism. The northern half of the complex (29° S to 42° S) is a thin coastal belt of metamorphic rocks that has yielded Late Devonian to Early Carboniferous radiometric ages (Munizaga et al, 1973). It represents the HP/LT side of the Chilean coastal paired metamorphic belt (Gonzalez-Bonarino, 1971). Oceanic material, such as pillow basalts, chert sequences, and ultramafic units are still locally recognizable. Perched basin assemblages in the southern portions of this northern metamorphic zone have Late Carboniferous to Early Permian fossils (Minato and Tanai,

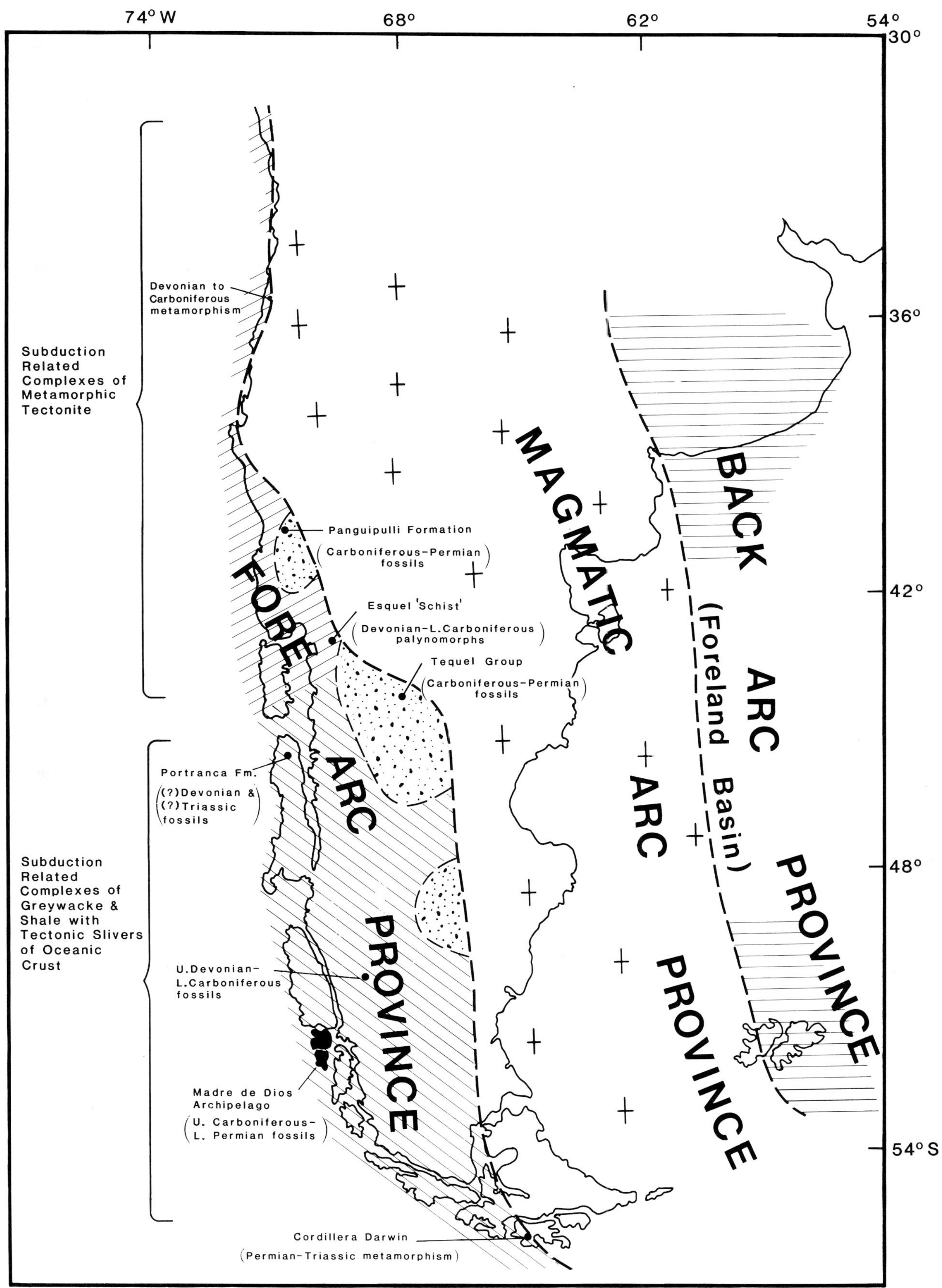

Figure 4—Late Paleozoic to early Mesozoic tectonostratigraphic zones in Patagonia. (After Forsythe, 1982, in the Quarterly Journal of the Geological Society of London, v. 139, used with the permission of the Geological Society of London; and after Forsythe and Mpodozis, 1983, used with the permission of the Servicio Nacional de Geologia y Mineria, Chile).

1977). This indicates pre-Late Carboniferous accretion for this portion of the subduction complex. South of the 42° S latitude the grade of metamorphism in the subduction complex is generally lower, with zeolite facies to middle greenschist facies assemblages dominating. Locally even here, however, one finds blueschist and amphibolite facies rocks with a penetrative tectonite fabric (Forsythe et al, 1981).

One of the areas of this southern zone that has been studied in some detail is the Madre de Dios Archipelago (Fig. 4). Here the collage of accretionary rocks has been subdivided into three mappable units: the Denaro Complex, Tarlton Limestone, and Duque de York Complex (see Fig. 5). The most prevalent unit is the Duque de York Complex that consists of conglomerate, graywacke, and shale. It forms the matrix of the collage into which the other two units have been tectonically mixed to form a "macromelange." The Tarlton Limestone is a massively bedded, fusulinid-bearing limestone of Late Carboniferous to Early Permian age (Douglass and Nestell, 1976). The Denaro Complex is a complicated but distinctive association of basalt, hydrothermal chert, pelagic chert, hemipelagic siliceous argillite, and detrital calcareous beds. Stratigraphic and geochemical arguments outlined by Mpodozis and Forsythe (1983) suggest that the Tarlton Limestone and the Denaro Complex were formed in an intraoceanic setting. The Denaro Complex appears to represent the basalt/sediment interface of the Panthalassic Ocean, and the Tarlton Limestone represents a carbonate platform built on a basaltic foundation. Paleontologic work on new radiolarians found in the Madre de Dios Archipelago (Ling et al, in preparation) suggests that the Tarlton Limestone is as old as the Denaro Complex. Thus, stratigraphic, geochemical, and paleontologic evidence now suggest that both are likely to have formed as an on-ridge carbonate/volcanic edifice located within the Panthalassic Ocean. This offers a better explanation for the occurrence of remarkably similar fusulinid and radiolarian assemblages in accretionary complexes in Japan, South America, and western Canada (Ozawa and Kanmera, 1984; Ling et al, in preparation). It suggests that a shallow or emergent intraoceanic rift system may be the parent to many accreted terranes with mixtures of carbonate and volcanic rocks.

The oceanic blocks of the Madre de Dios archipelago are not the only exotic terranes to have been accreted to the late Paleozoic to early Mesozoic Gondwanaland margin forearc. Elsewhere the metamorphic grade is higher and degree of structural transposition greater, but greenstone and metachert layers are still recognizable and also likely to be exotic material. While outcrops in the northern half of the forearc are limited, even there aeromagnetic data suggest the presence of terranes. Small, elongate bodies of highly magnetic rocks are sporadically distributed within the metamorphic terrane. Where outcrops permit, these can locally be correlated with ultramafic and mafic lenses. Again this argues for the incorporation of numerous ophiolitic fragments within the accretionary prism. Judging from geologic mapping of the Madre de Dios Archipelago and the distribution of magnetic bodies in the coast ranges south of Concepcion (Fig. 4), perhaps 20 to 30% of the forearc may consist of exotic, oceanic material. This material was likely accreted in a semicontinuous but episodic fashion from the Late Devonian to the Triassic. Individual blocks of accreted oceanic material range in size from a few hundred cubic meters to tens of cubic kilometers and are likely to number in the hundreds. Attempting to define the nature, extent, and docking history of each would present a real challenge given the paucity of fossils, limited rock exposure, and degree of deformation and metamorphism.

Development of the Panthalassic Margin: Summary

The working hypotheses developed by different groups for Peru–Bolivia, northern Chile–northern Argentina, and southern South America for the Paleozoic and early Mesozoic provide a reasonable, if incomplete, picture of the development of the Panthalassic margin of Gondwanaland. This picture is not unlike that which emerges for the Mesozoic and Cenozoic Andean margin, except for the extent of the accretionary forearc, suggesting variation in tectonic style with space and time. From the perspective of the terrane analyst, the Paleozoic and lower Mesozoic rocks of South America seem to involve increasing numbers of terranes, with increasing complexity and with potential for greater displacements southward from the equator toward Cape Horn. This is also the direction in which accretion appears to have occurred at later times. It occurred in the early Paleozoic in the north and the latest Paleozoic or early Mesozoic in the south.

In Peru, despite recent speculations that the Arequipa massif was accreted during the Mesozoic and Cenozoic Andean orogenesis (Nur and Ben-Avraham, 1977, 1982, 1983), evidence supports the Paleozoic development of an ensialic basin between the Precambrian rocks of the Arequipa massif and the Transamazonian shield. The Arequipa massif may be a terrane accreted to the shield during late Precambrian times. In northern Chile and Argentina there is evidence that an early Paleozoic magmatic arc was accreted during the mid- to late Paleozoic.

Upper Paleozoic posttectonic calc-alkaline igneous rocks "stitch" and overstep all the deformed lower to middle Paleozoic rocks of the central Andes. The late Paleozoic to early Mesozoic magmatic arc continues into southern South America (Fig. 6) where it is bordered to the west by a forearc accretionary wedge that can be traced from 29° S practically to Cape Horn (56° S). Judging by selected areas, the individual allochthonous blocks, or terranes, within this forearc are likely to number in the hundreds.

ANDEAN OROGENESIS

The Andes were built during the Mesozoic and Cenozoic on an eroded surface of uplifted older rocks that have been described in the foregoing section. The Mesozoic and Cenozoic strata are referred to by many workers in South America as belonging to the "Andean cycle" (see for example Aubouin et al, 1973; Coira et al, 1982). A major distinction compared with the earlier history of the Pacific margin appears to be that little or no material was tectonically accreted to the Andean margin during this period of time. The following account summarizes an article currently in press (Dalziel, in press).

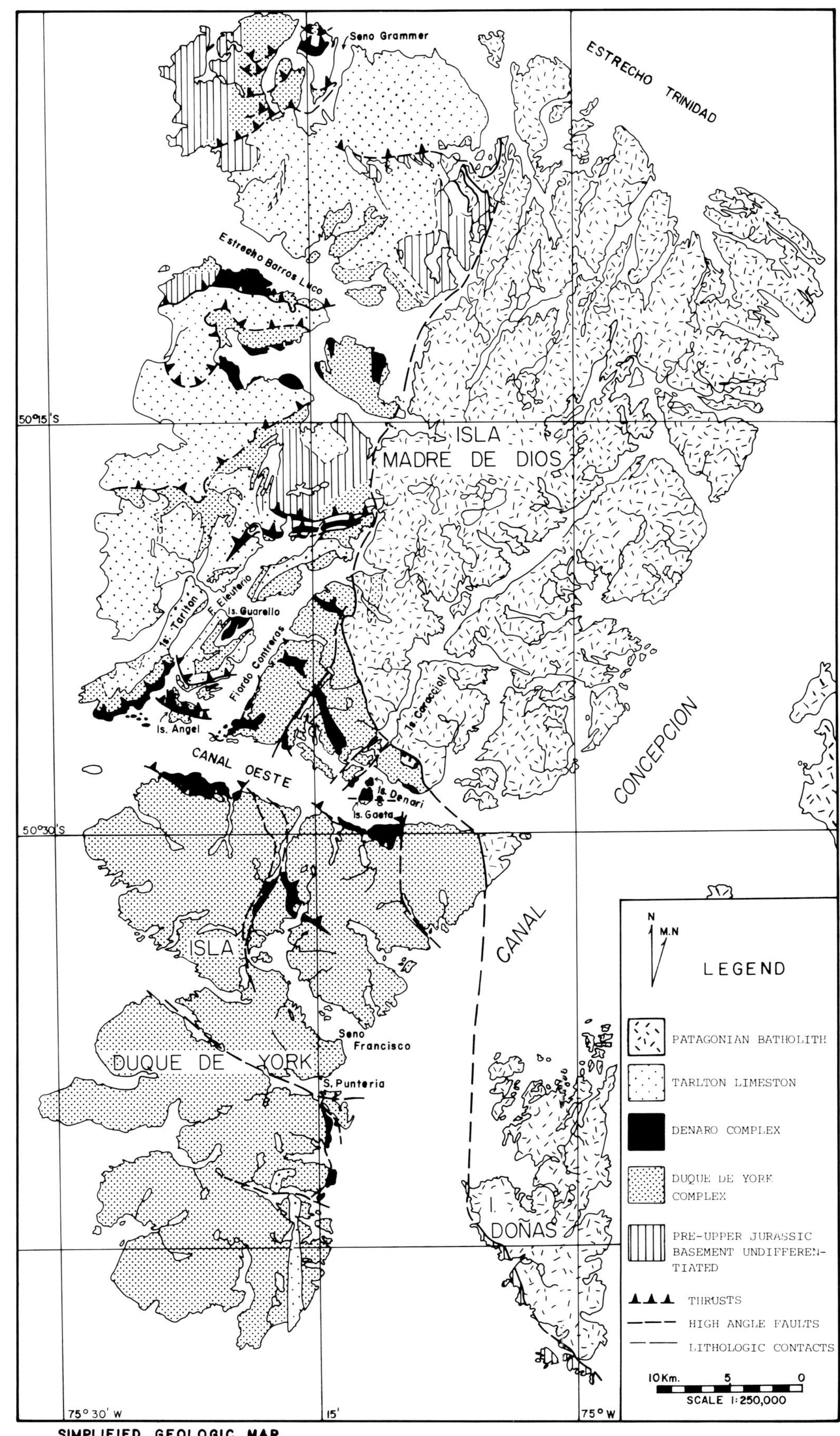

Figure 5—Geologic map of the Madre de Dios Archipelago (after Mpodozis and Forsythe, 1983). For location see Figure 4.

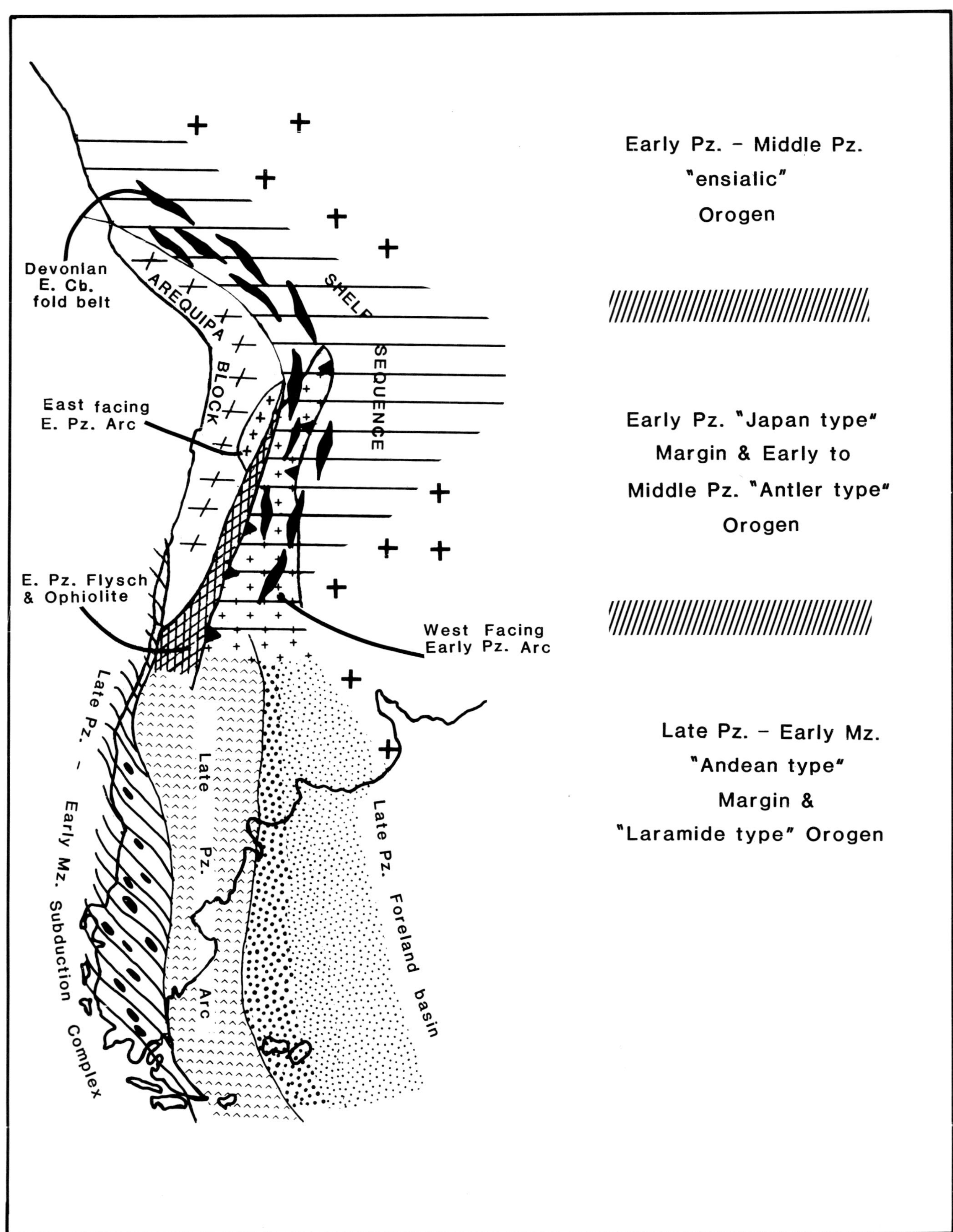

Figure 6—Summary of tectonic models for Paleozoic to early Mesozoic Gondwanide development of the Panthalassic margin of South America (for discussion and references see text).

The evolution of the present Andean Cordillera from a low-lying continental margin started in the Middle to Late Triassic times in northern Chile and Peru and Middle to Late Jurassic in the south (45° S–56° S). Mesozoic sedimentation began along the length of the chain with subaerial to shallow marine deposits intercalated with dominantly silicic and andesitic volcanic and volcaniclastic rocks. Silicic volcanism also occurred well to the east of the present Andes, and it is not easy to define precisely the time of inception of subduction-related volcanism along the Pacific margin. The age and composition of plutons and extrusives, however, indicate that a magmatic arc was established along much of the Pacific margin by the Late Triassic or Early Jurassic. Jurassic and Early Cretaceous subsidence occurred within or behind the volcanic arc along almost the entire length of the Pacific margin (Dalziel, in press). The amount of extension and subsidence varied, however, with major individual segments of the backarc trough being separated by basement ridges trending dominantly northwest-southeast (Fig. 7). Deep-marine turbidites ("flysch"), chert, and ophiolitic bodies are present only north of 5° S ("Transversale de Huancabamba") and south of 40° S ("Transversale de Bariloche"). The island arc on the Pacific side of the basin in the south has been classified as a terrane (Howell et al, 1984), and it has been suggested by some authors (e.g., Nur and Ben-Avraham, 1981) that it is allochthonous to South America. The latter interpretation conflicts, however, with the following facts:

1. the batholith on the Pacific margin is continuous with that north of the ophiolitic rocks and deep-marine turbidites;
2. the basement intruded by the batholith is the same forearc terrane as that exposed east of the ophiolites; and
3. that basement is unconformably overlain by Upper Jurassic silicic volcanics lithologically identical to, and coeval with, those of the Tobifera Formation that overlies the basement east of the Cordillera (Forsythe and Allen, 1980).

Uplift and deformation of deposits in the Pacific margin composite backarc trough took place along the entire length of the Cordillera in the mid-Cretaceous. This deformation phase has been termed the "sub-Hercynian" (Aubouin et al, 1973; Zeil, 1979). The event clearly represents the first major uplift of the Andean Cordillera involving tectonic compression. Uplift is reported all along the chain at this time with folding and cleavage formation of varying intensity (Fig. 8; Dalziel, in press). The event appears to have been remarkably synchronous given the enormous length of the backarc basin (at least 7,500 km [4,500 mi]; Fig. 7). This makes it difficult to envisage a mechanism that involved collision of an allochthonous terrane or terranes.

Since the mid-Cretaceous, the locus of Andean magmatism migrated eastward (Zeil, 1979). The major batholiths are intruded to varying degrees within the Jurassic–Early Cretaceous backarc trough (Cobbing et al, 1981; Dalziel, 1981; Levi and Aguirre, 1981; Dalziel, in press), while Cenozoic magmatic centers lie on its continental side (Dalziel, 1981; Drake et al, 1982). Foredeeps developed inboard of the rising Cordillera, and the deformation front migrated eastward (see, for example, Winslow, 1982; Wilson, 1983). Deformation on the eastern side of the Cordillera in places involved basement, for example in the Pampean Ranges of northwest Argentina; elsewhere thin-skinned tectonics seem to have prevailed (Jordan et al, 1983). Most of the vergence is continentward (but see Jordan et al, 1983) and presumably reflects what Bally (1975) refers to as "A-subduction," that is to say, the underthrusting of continental basement towards the Cordillera.

NEOTECTONIC REGIME

The dominant morphologic features of the Pacific coast of southern South America were developed during the Cenozoic. Uplift of the main Cordillera, the development of the coastal ranges and longitudinal valleys, as well as the subsidence and uplift of localized regions along the Pacific coast, were influenced strongly by crustal movements operating primarily in the Middle to Late Tertiary (Paskoff, 1970; Bruggen, 1950).

Among these morphologic elements, the most striking structural features are the longitudinal fault systems. These faults run roughly parallel to the coast and are generally found bounding segments of coast ranges or central valleys. The better known are: the Atacama fault system (Allen et al, 1971), the Linquine–Ofqui fault system (Herve, 1976), and the Magellan fault system (Winslow, 1982). They are shown on Figure 9. These high-angle faults isolate coastal blocks from the main portions of the South American plate. While strike-slip movements have been suggested for each, geologic units can be correlated from one side to the other. Thus, while these faults separate rock packages that have distinctive Tertiary and Quaternary histories, they do not appear at first hand to mark suture boundaries of allochthonous terranes. Instead, it appears that the faults are zones of weakness that allow the leading edge of the South American plate to adjust locally to the heterogeneous stress field created by varying kinematic and rheologic factors along the zone of convergence with the Pacific Ocean. In a sense they can be thought of as "tectonic bumpers" that absorb and smooth out the stress field in the zone of convergence. It may only be a rare "accident" that will rip off a "bumper" and leave it as a separate lithospheric plate. Such dispersal processes may be due to major periods during which the margin has a significant component of oblique convergence (Fitch, 1972) or where the geometry of the margin is such that a coastal block has a lateral margin unconstrained by continental lithosphere (e.g., westward extrusion of Indonesia and China blocks, Tapponier et al, 1982). Combinations of both of these conditions are met along the South American margins of the Caribbean and Scotia Seas. Recent discussions of the neotectonics of Venezuela, for example those by Munro and Smith (1984) and Vierbuchen (1984), elucidate aspects of the disruption and eastward dispersal of the coastal mountains.

It is also clear that the island of South Georgia was once part of the Pacific margin of South America (Fig. 1). Geologically, it is virtually identical to the islands between Cape Horn and the Beagle Channel (Dalziel et al, 1975;

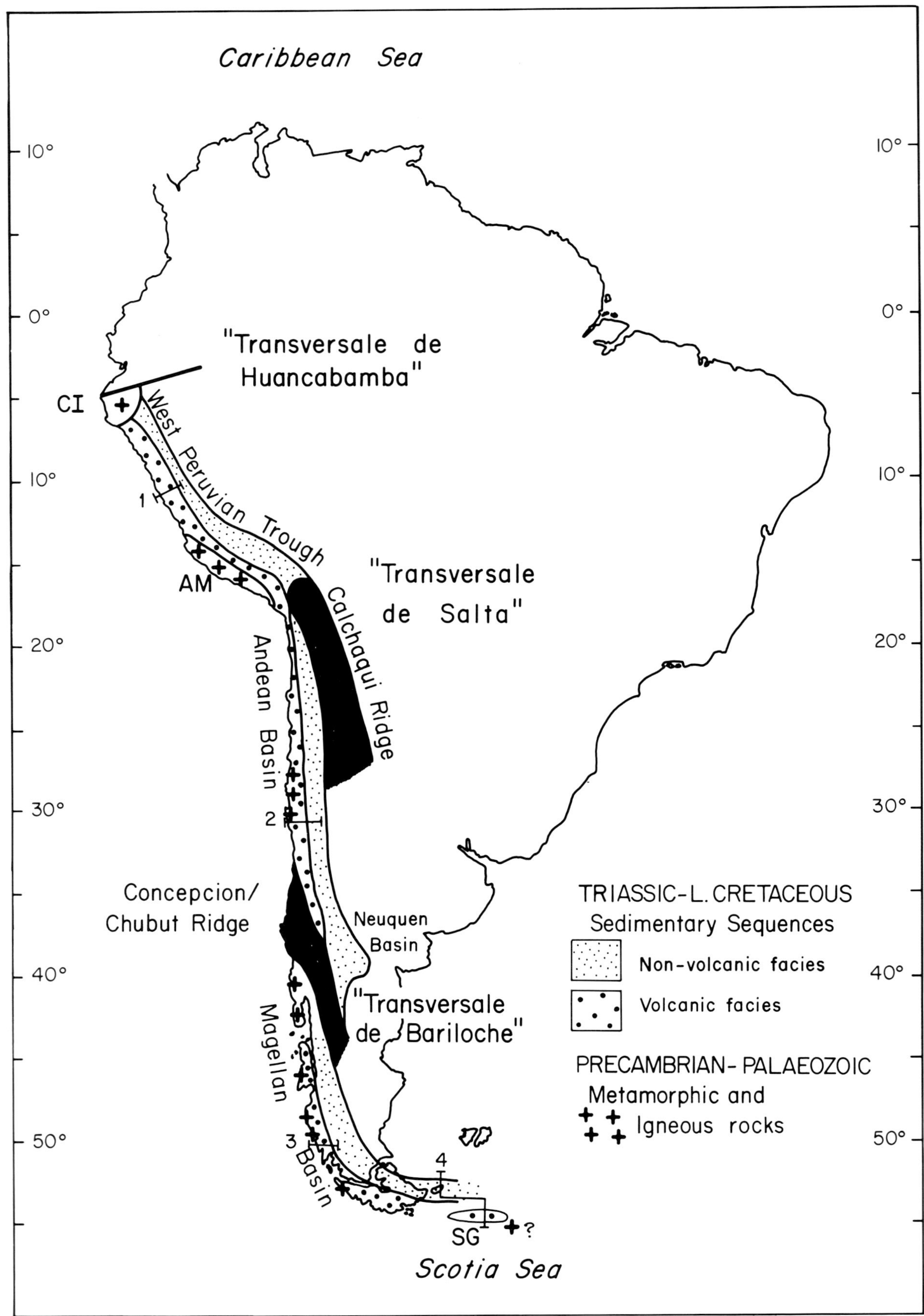

Figure 7—Subdivisions of the Andean Cordillera. (After Dalziel, in press, in the Geological Society of London Special Publication on Collision Tectonics.)

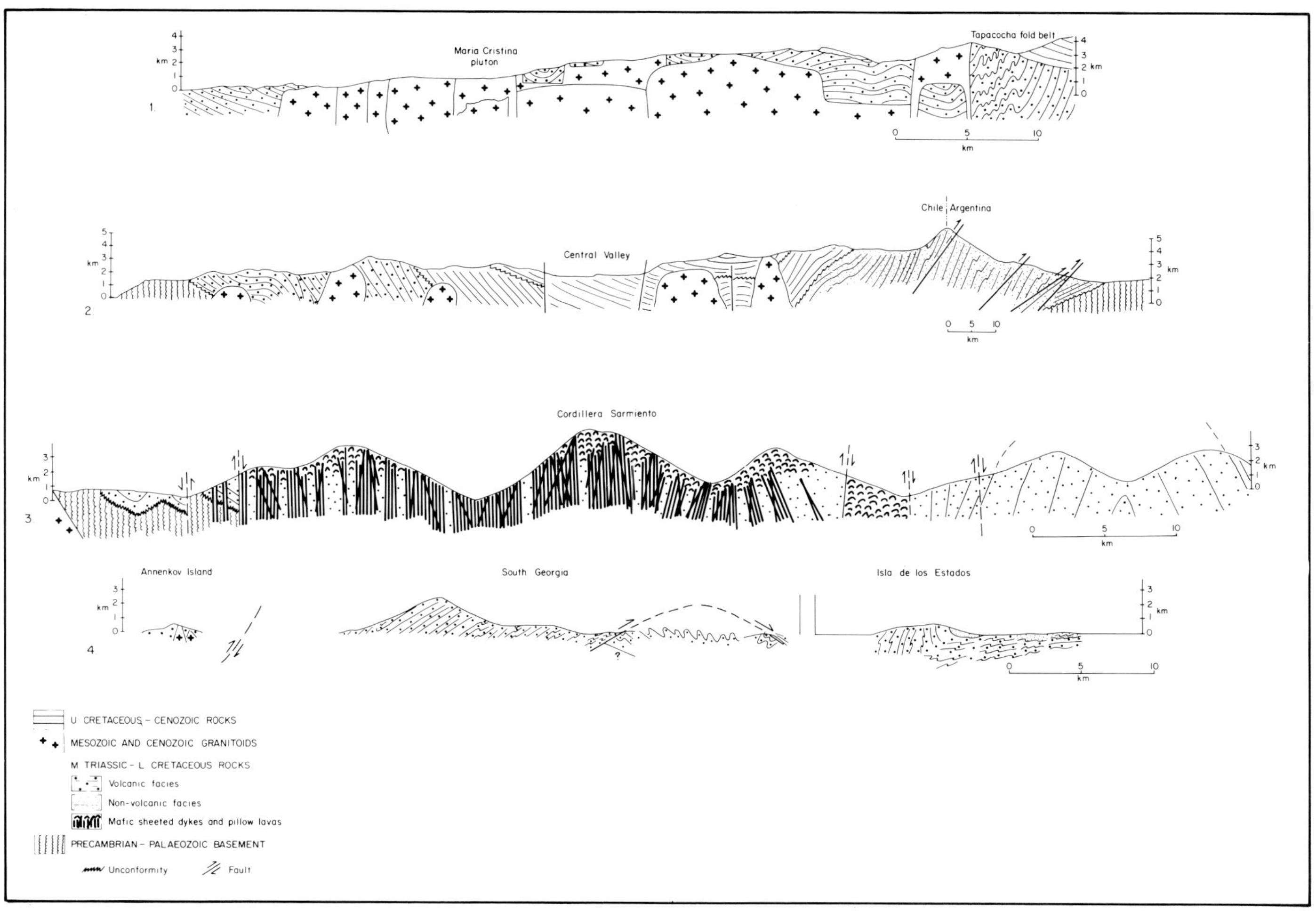

Figure 8—Geologic cross sections of the Andean Cordillera. (After Dalziel, in press, in the Geological Society of London Special Publication on Collision Tectonics.) For locations, see Figure 9.

Tanner, 1982). The exact kinematics of the plate motions that separated the South Georgia microcontinent from South America are still unclear (Barker and Dalziel, 1983). Nonetheless, South Georgia is now an allochthonous terrane along the northern transform margin of the otherwise oceanic Scotia plate (see Dalziel and Grunow, this volume, Fig. 4). The microcontinent will presumably one day collide with one of the continents bordering the present-day South Atlantic Ocean, Weddell Sea, or Indian Ocean and become an accreted terrane with the geologic characteristics of a marginal basin and magmatic arc (Tanner, 1982).

Finally, although the area of the South American continent does not appear to have increased significantly in size as a result of forearc accretion during Jurassic and younger times, this process has not been totally inactive. Recent studies appear to confirm the often-proposed truncation of the South American continent along much of the Peru–Chile trench, but also reveal local development of a narrow accretionary complex (von Huene et al, in press).

DISCUSSION

The role of terranes in the Mesozoic–Cenozoic evolution of the Andean Cordillera is relatively insignificant and has been greatly exaggerated in some of the recent literature. Growth of the South American continent, south of Ecuador, during the past 200 m.y. has been limited to: (1) local forearc accretion of material scraped off the downgoing Pacific Ocean lithosphere; (2) incorporation within the continent of the products of calc-alkaline magmatism; and (3) basin inversion incorporating the products of backarc basin igneous activity and sedimentary basin-fill. There has also been "shuffling" of fault blocks along the continental margin. At least one major microcontinental terrane, the South Georgia platform on the northern Scotia Ridge, has been dispersed.

There has been major accretion to the South American segment of Gondwanaland, but mainly before breakup of the supercontinent. A wide forearc accretionary prism was built up between an Andean-type magmatic arc and the Pacific Ocean floor during the late Paleozoic to early Mesozoic incorporating sea mounts and slivers of oceanic lithosphere as well as offscraped pelagic and trench sediments. Earlier events may have included the late Precambrian accretion of the Precambrian rocks forming the Arequipa massif of southern Peru, and the closure of a Japan-type marginal basin in north Chile and Argentina during the lower to middle Paleozoic.

Thus, major changes have taken place in one segment of the Pacific Ocean margin over a time scale of 10 to 200 m.y. These changes are comparable to those that can be

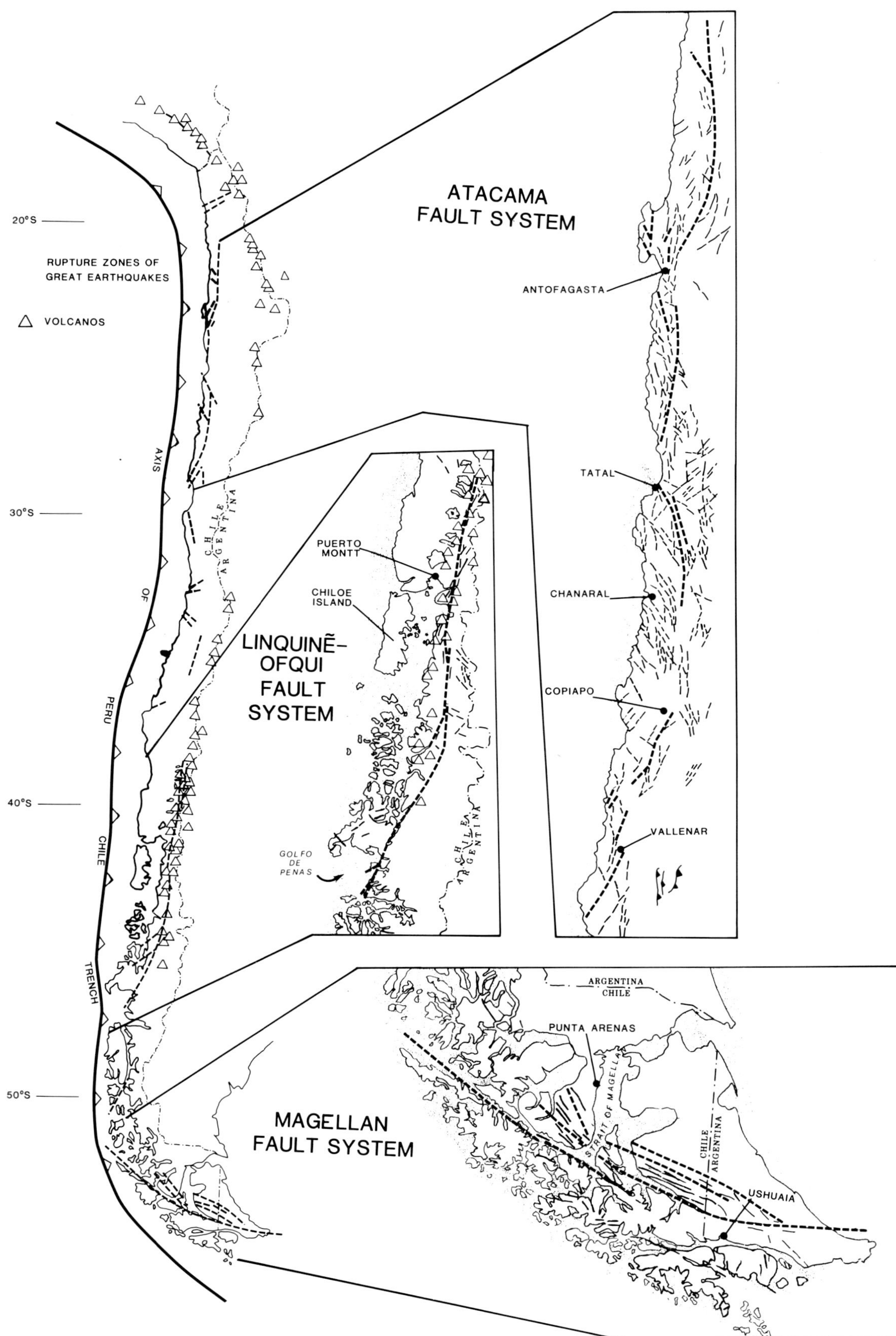

Figure 9—Major fault systems of the central, southern, and Patagonian Andes (after Herve, 1976; Winslow, 1982, and the 1:1,000,000 scale geologic map of Chile, Ser. Nac. Geol. Min., 1980).

observed around the margins of the Pacific Ocean at the present time, for example between the Andean margin and the Aleutian margin. The latter with its broad accretionary prism may be an analog for the late Paleozoic–early Mesozoic margin of southern South America.

If collision was not a major factor in Andean orogenesis, what is the mechanism? Clearly calc-alkaline magmatism played a major role. Yet in both Peru and Patagonia the granitic rocks of the batholith are largely undeformed and cross-cut deformed, often highly deformed, country rock. The relative horizontal shortening and vertical thickening of low-density continental rocks needed to form a major mountain chain must have yet another driving mechanism.

It seems to us that the process is critically dependent on two factors: the strength of the continental lithosphere and the stress field applied to it. The strength of the lithosphere can be reduced by both mechanical and thermal processes, for example by extensional faulting and by the emplacement of hot magma. The stress field applied to the continental lithosphere can be increased in several ways, for example by increase in the convergence rate at the subducting plate boundary or by increased buoyancy of the downgoing plate.

Andean tectonic uplift appears to have been initiated with a compressional event in the mid-Cretaceous that is classically known as the "sub-Hercynian" event. This followed a Jurassic through Early Cretaceous extensional phase along the margin that resulted in the formation of a composite backarc basin that extended at least from northern Peru (5° S latitude) to Cape Horn (56°) and out along the northern Scotia Ridge. It has been suggested that weakening of the Pacific edge of the South American continent by extension and basin formation, followed by mid-Cretaceous increase in convergence rate, resulted in deformation of the continental lithosphere involving horizontal shortening and vertical thickening (Dalziel, in press). Other more local compressional phases following extension and basin formation, such as occurred in the Cenozoic in Peru (Bussell, 1983), may have continued the orogenic process.

Thus, rather than orogenic uplift resulting from terrane accretions, it may in this case have resulted from compression in the continental lithosphere leading to collapse of a previously formed basin and the reuniting of a parautochthonous terrane. The degree of allochthoneity of the terrane need not be a significant factor here since momentum will be the same (insignificantly small) for both far-traveled and parautochthonous crustal blocks.

Finally, in considering the causes of circum-Pacific orogenesis through time, it is instructive to consider the present situation with regard to the allochthoneity of terranes. The southwestern Pacific margin is festooned with island arcs, microcontinents, and oceanic plates. Amalgamation of those with the southeast Asian continental margin will result in a mountain belt consisting of a collage of allochthonous terranes analogous to that of Mesozoic North America (Silver and Smith, 1983).

The northwestern Pacific, although also festooned with island arcs, would if "telescoped" form a single mountain range with no significant allochthonous terranes but several major parautochthonous blocks. Two of these would be the present Japanese Islands and Kamchatka Peninsula. Increase in compression along the Pacific margin of the South American plate at the present time would perhaps result in the shortening and thickening of sedimentary sequences in such basins as the longitudinal valley of central Chile, an uplifted forearc basin, and the foredeeps east of the Cordillera. It could perhaps result in further truncation of the continental margin at the subduction zone and in the incorporation of slivers of aseismic ridges such as the Nazca Ridge into the narrow forearc accretionary wedge. No major parautochthonous, let alone allochthonous, terranes would collide with South America. Nor are any likely to dock there for a long time because the southeastern quadrant of the Pacific is remarkably free of island arcs, oceanic plateaus, microcontinents, and aseismic ridges.

Hence, we conclude that collision of allochthonous terranes with a continental margin may be sufficient cause for at least local orogenesis. It is not, however, a necessary condition and did not play a significant role in Andean orogenesis.

ACKNOWLEDGMENTS

Our work in the Andes has been supported by NSF Grants DPP 79 20629, DPP 82 13798, and INT 79 20213 (IWDD), and EAR 82-06100 and INT 83-00521 (RDF). We thank David Howell for the opportunity to present our views and for his editorial help, particularly regarding the ways of the terrane analyst.

REFERENCES

Aguirre, L., et al, 1972, Distribution of metamorphic facies in Chile: Krystalinikum, v. 9, p. 7–9.

Allen, C. R., et al, 1971, Comparison of features of active faulting in northern Chile, California, and Japan, *in* Symposium papers on recent crustal movements and associated seismicity: Royal Society of New Zealand, Wellington, New Zealand.

Aubouin, J., et al, 1973, Esquisse paleogeographique et structurale des Andes Meridionales: Revue de Géologie Dynamique, v. 15, p. 11–72.

Aumaitre, R., et al, 1977, Donnees lithologiques et structurales relatives a un bloc precambrien sureleve de la Cordillere andine orientale (Perou central). Les corps de roches ultrabasique qui y sont presents: Bulletin de Société Géologique de France, v. 19, n. 5.

Bally, A. W., 1975, A geodynamic scenario for hydrocarbon occurrences: Proceedings of the World Petroleum Congress, v. 9, p. 33–44.

Barker, P. F., and I. W. D. Dalziel, 1983, Progress in geodynamics in the Scotia arc region, *in* R. Cabre, ed., Geodynamics of the eastern Pacific Region, Caribbean and Scotia Arcs: American Geophysical Union Geodynamics Series, v. 9, p. 137–170.

______, et al, 1976, The evolution of the southwestern Atlantic Ocean Basin: Leg 36 data, *in* Initial Reports of the Deep Sea Drilling Project, v. 36: Washington, DC, U.S. Government Printing Office.

Borello, A. V., 1972, The Precordillera as a type of geosyncline in Argentina: Ottawa, 24th International Geologic Congress, Montreal, v. 3, p. 293–299.

Bruggen, Y., 1950, Fundamentos de la geologia de Chile, Santiago (Chile), 374 p.

Bussell, M. A., 1983, Timing of tectonic and magmatic events in the Central Andes of Peru: Quarterly Journal of the Geological Society of London, v. 140, p. 279–286.

Carlier, G., et al, 1982, Present knowledge of the magmatic evolution of the eastern Cordillera of Peru: Earth Science Reviews, v. 18, p. 253–283.

Cobbing, E. J., et al, 1981, The geology of the western Cordillera of northern Peru: Institute of Geological Sciences Overseas Memoir 5, 143 p.

Coira, B., et al, 1982, Tectonic and magmatic evolution of the Andes of northern Argentina and Chile: Earth Science Reviews, v. 18, p. 303–332.

Dalmayrac, B., et al, 1980, Caracteres generaux de l'evolution geologique des Andes peruviennes: Travaux et Documents de L'ORSTROM 122, 501 p.

Dalziel, I. W. D., 1981, Back-arc extension in the southern Andes, a review and critical reappraisal: Philosophical Transactions of the Royal Society of London, v. A300, p. 319–335.

______, in press, Collision and cordilleran orogenesis, an Andean perspective: Geological Society of London Special Publication on Collision Tectonics.

______, et al, 1975, Tectonic relations of South Georgia Island to the southernmost Andes: Geological Society of America Bulletin, v. 86, p. 1034–1040.

Douglass, R. C., and M. K. Nestell, 1976, Late Paleozoic foraminifera from southern Chile. U.S. Geological Survey Professional Paper 858, 47 p.

Drake, R., et al, 1982, Geochronology of Mesozoic-Cenozoic magmatism in central Chile (lat. 31°–36° S): Earth Science Reviews, v. 18, p. 353–364.

du Toit, A. L., 1937, Our wandering continents: Edinburgh, Oliver and Boyd, 366 p.

Fitch, T. Y., 1972, Plate convergence, transcurrent faults, and internal deformation adjacent to southeast Asia and the western Pacific: Journal of Geophysical Research, v. 77, p. 4432–4460.

Forsythe, R. D., 1982, The late Paleozoic to early Mesozoic evolution of southern South America: a plate tectonic interpretation: Quarterly Journal of the Geological Society of London, v. 139, p. 671–682.

______, and R. B. Allen, 1980, The basement rocks of Peninsula Staines, Region XII, Province of Ultima Esperanza, Chile: Revista Geologica de Chile, v. 10, p. 3–15.

______, and C. Mpodozis, 1983, Geología del basamento Pre-Jurasico superior en el Archipielago Madre de Dios, Magallanes, Chile: Santiago, Chile, Servicio Nacional de Geologia y Mineria Boletin 39.

______, et al, 1981, Geologic studies in the outer Chilean fijords, R/V HERO Cruise 79-5: Antarctic Journal of the United States, v. 15, p. 109–111.

Gonzalez-Bonarino, F., 1971, Metamorphism of the crystalline basement of central Chile: Journal of Petrology, v. 12, p. 149–175.

Greenway, M. E., 1972, The geology of the Falkland Islands: British Antarctic Survey Science Report 76, 42 p.

Harrington, H. J., 1962, Paleogeographic development of South America: Bulletin of the American Association of Petroleum Geologists, v. 46, p. 1773–1814.

Heki, K., et al, 1983, Rotation of the Peruvian Block from paleomagnetic studies of the central Andes: Nature, v. 305, p. 514–516.

Herve, F., 1976, Estudio geologico del la falla Linguine-Reloncavi en el area de Linguine: antecedentes de un movimiento 1976 transcurrente: Santiago, Chile, Acta I Congreso Geologico Chileno, v. 1, p. B39–B56.

______, et al, 1980, The Late Paleozoic in Chile: stratigraphy, structure and possible tectonic framework: Revista da Academia Brasileira de Ciências, v. 53, p. 362–373.

Howell, D. G., and D. L. Jones, 1984, Tectonostratigraphic terrane analysis and some terrane vernacular, *in* D. G. Howell, et al, eds., Proceedings of the Circum-Pacific Terrane Conference: Stanford University Publications, Geological Sciences, v. 18, p. 6–9.

______, et al, 1984, Preliminary tectonostratigraphic terrane map of the circum-Pacific region, *in* D. G. Howell, et al, eds., Proceedings of the Circum-Pacific Terrane Conference: Stanford University Publications, Geological Sciences, v. 18, p. 227–242.

Jordan, T. E., et al, 1983, Andean tectonics related to geometry of subducted Nazea Plate: Geological Society of America Bulletin, v. 94, p. 341–361.

Knight, R. J., et al, 1984, Paleomagnetic study of the Arequipa massif, *in* D. G. Howell, et al, eds., Proceedings of the Circum-Pacific Terrane Conference: Stanford University Publications, Geological Sciences, v. 18, p. 134–136.

Levi, B., and L. Aguirre, 1981, Ensialic spreading-subsidence in the Mesozoic and Palaeogene Andes of central Chile: Quarterly Journal of the Geological Society of London, v. 138, p. 75–81.

Megard, F., 1978, Étude geologique des Andes de Perou central: contribution à l'étude des Andes: 1. Memoir ORSTROM, n. 86, 310 p.

Minato, M., and T. Tanai, 1977, Carboniferous-Permian plant remains found at the border of Lake Panguipulli, Valdivia, *in* T. Ishikawa and L. Aguirre, eds., Comparative studies on the geology of the circum-Pacific orogenic belt in Japan and Chile. Tokyo, Japan Society for the Promotion of Sciences, p. 69–80.

Mpodozis, C., and R. D. Forsythe, 1983, Stratigraphy and geochemistry of accreted fragments of the ancestral Pacific Ocean floor in southern South America: Palaeogeography, Palaeoclimatology, Palaeoecology, v. 41, p. 103–124.

______, et al, 1983, Los granitoids de Cerros de Lila, manifestaciones de un episodio intrusivo y termal del Paleozoico inferior en los Andes del norte de Chile. Revista Geologica de Chile, n. 18, p. 3–14.

Munizaga, F., et al, 1973, Rb-Sr ages of rocks from the Chilean metamorphic basement: Earth and Planetary Science Letters, v. 18, p. 87–91.

Munro, S. E., and F. D. Smith, Jr., 1984, The Urica fault

zone, northeastern Venezuela, *in* W. E. Bonini, et al, The Caribbean-South American plate boundary and regional tectonics: Geological Society of America Memoir 162, 421 p.

Nur, A., and Z. Ben-Avraham, 1977, Lost Pacifica continent: Nature, v. 270, p. 41–43.

______ and ______, 1981, Volcanic gaps and the consumption of aseismic ridges in South America: Geological Society of America Memoir 154, p. 729–740.

______ and ______, 1982, Oceanic plateaus, the fragmentation of continents and mountain building: Journal of Geophysical Research, v. 87, p. 3644-3661.

______ and ______, 1983, Displaced terranes and mountain building, *in* K. J. Hsu, ed., Mountain building processes: London, Academic Press, p. 73–84.

Ozawa, T., and K. Kanmera, 1984, Tectonic terranes of Late Paleozoic rocks and their accretionary history in the circum-Pacific region viewed from Fusulinacean paleobiogeography, *in* D. G. Howell, et al, eds., Proceedings of the Circum-Pacific Terrane Conference: Stanford University Publications, Geological Sciences, v. 18, p. 158–160.

Pacci, O., et al, 1980, Acerca de la edad Rb-Sr Precambrica de rocas de la formation esquistos de Belen, Dept. de Parinacota, Chile: Revista Geologica de Chile, n. 11, p. 43–50.

Paskoff, R., 1970, Recherches geomorphologiques dans le Chili semi-aride: Paris, Biscaye Frères, 420 p.

Pitcher, W. S., 1983, Granite type and tectonic environment, *in* K. J. Hsu, ed., Mountain building processes: London, Academic Press, p. 19–40.

Ramos, E. D., and V. A. Ramos, 1979, Los ciclos magmaticos de la Republica Argentina: Acta 4th Congreso Geologico Argentina, v. 1, p. 771–786.

Shackleton, R. M., et al, 1979, Structure, metamorphism and geochronology of the Arequipa massif of coastal Peru: Quarterly Journal of the Geological Society of London, v. 136, p. 195–214.

Silver, E. A., and R. B. Smith, 1983, Comparison of terrane accretion in modern southeast Asia and the Mesozoic of the North American Cordillera: Geology, v. 11, p. 198–202.

Stibane, F. R., 1980, Tectonic de los Andes Septentrionales, *in* Nuevos resultados de la investigacion geoscientifica alemanaen Latinoamericana: Bonn, Doursche Forchungs Gemeinschafts, p. 91–97.

______, 1981, K/Ar-Alter von Tonaliten der Cordillera Occidental Kolumbiens und ihre tektonische: Deutung. Zentralbl. Geol. Palaont., v. 1, p. 252–259.

Suarez-Soruco, R., 1976, El sistema Ordovicico en Bolivia: Rev. Tech., VPFB 2, p. 111–223.

Tanner, P. W. G., 1982, Geologic evolution of South Georgia, *in* C. Craddock, ed., Antarctic geoscience: Madison, University of Wisconsin Press, p. 167–176.

Tapponnier, P., et al, 1982, Propagating extrusion tectonics in Asia: New insights from simple experiments with plasticine: Geology, v. 10, p. 611–616.

Turner, J. C. M., 1960, Estratigrafia de la Sierra de Santa Victoria: Boletin Academia Nacional de Ciencias, v. 41, p. 163–196 (Cordoba, Argentina).

______, 1972, Puna, *in* A. F. Leanza, ed., Geologia regional Argentina: Academia Nacionale Ciencias, p. 91–116.

Vierbuchen, R. C., 1984, The geology of the El Pilar fault zone and adjacent areas in northeastern Venezuela, *in* W. E. Bonini, et al, The Caribbean-South American plate boundary and regional tectonics: Geological Society of America Memoir 162, 421 p.

von Huene, R., et al, in press, Structure of the frontal part of the Andean convergent margin: Journal of Geological Research.

Weeks, L. G., 1947, Paleogeography of South America: Bulletin of the American Association of Petroleum Geologists, v. 31, p. 1194–1241.

Wilson, T. J., 1983, Stratigraphic and structural evolution of the Ultima Esperanza foreland fold thrust belt, Patagonian Andes, southern Chile: Unpublished PhD Dissertation, Columbia University, 360 p.

Winslow, M. A., 1982, The structural evolution of the Magallanes basin and neotectonics in the southernmost Andes, *in* C. Craddock, ed., Antarctic geoscience: Madison, University of Wisconsin Press, p. 143–154.

Zeil, W., 1979, The Andes: Berlin, Gebruder Borntraeger, 260 p.

INDEX

A reference is indexed according to its important, or "key," words.

Three columns are to the left of a keyword entry. The first column, a letter entry, represents the Circum-Pacific book series from which the reference originated. In this case, ES stands for Earth Science series. The following number is the series number. In this case 1 represents a reference for Circum-Pacific Earth Science Series 1. The third column lists the page number of this volume on which the reference can be found.